2021 UNIFORM MECHANICAL CODE®

Illustrated Training Manual™

IAPMO®

First Printing, August 2020

Published by the International Association of Plumbing and Mechanical Officials
4755 E. Philadelphia Street • Ontario, CA 91761-2816 – USA
Main Phone: (909) 472-4100 • Main Fax: (909) 472-4150

UNIFORM MECHANICAL CODE

PROCEDURES TO BE FOLLOWED TO PLACE GAS EQUIPMENT IN OPERATION
APPENDIX B

INSTALLATION & TESTING OF OIL (LIQUID) FUEL-FIRED EQUIPMENT
APPENDIX C

REFERENCED STANDARDS
CHAPTER 17

GEOTHERMAL ENERGY SYSTEMS AND AMBIENT TEMPERATURE LOOPS
CHAPTER 17

FUEL GAS PIPING
CHAPTER 13

PROCESS PIPING
CHAPTER 14

INSTALLATION OF SPECIFIC APPLIANCES
CHAPTER 9

BOILERS AND PRESSURE VESSELS
CHAPTER 10

EXHAUST SYSTEMS
CHAPTER 5

DUCT SYSTEMS
CHAPTER 6

ADMINISTRATION
CHAPTER 1

DEFINITIONS
CHAPTER 2

REFERENCED STANDARDS
CHAPTER 18

SUSTAINABLE PRACTICES
APPENDIX E

RESI DENTIAL PLANS EXAMINER REVIEW FORM FOR HVAC SYSTEM DESIGN
APPENDIX A

SOLAR ENERGY SYSTEMS
CHAPTER 15

STATIONARY POWER PLANTS
CHAPTER 16

REFRIGERATION
CHAPTER 11

HYDRONICS
CHAPTER 12

COMBUSTION AIR
CHAPTER 7

CHIMNEYS AND VENTS
CHAPTER 8

GENERAL REGULATIONS
CHAPTER 3

VENTILATION AIR
CHAPTER 4

2024 UMC ~ A

FOREWORD

IAPMO is proud to present this 2021 edition of the Uniform Mechanical Code (UMC) Illustrated Training Manual (ITM). As in the previous edition, the ITM contains the entire 2021 UMC text as developed and published through the ANSI consensus process. The 2021 edition follows the double column text similar to the Code, and each commentary narrative is tagged with an icon (tin snips) underneath the applicable code section serving as a pointer to direct the reader toward important information.

Included in this edition are Learning Links. Learning Links are URLs to be typed into your Internet browser that will link you with a video presentation to better illustrate the commentary. The Learning Links in the ITM eBook will be hyperlinks that your cursor can hover over and click on to automatically open your browser to the video presentation. A list of Learning Links follow the Table of Contents.

The commentary narrative is a living document that continues to grow and mature. This 2021 edition displays qualities that have progressed beyond its previous edition. Every narrative has been re-evaluated for its content both technical and grammatical. The rewriting has been extensive throughout the ITM and the reader will be pleasantly surprised to find fresh narrative in old places accompanied with new illustrations.

The development of the ITM is the responsibility of the Publication Development Committee and IAPMO staff. The committee has painstakingly reviewed the narrative to ensure technical accuracy.

This publication has many uses. Its value lies in its use as a reference document or a training document. However, care should be taken by users to differentiate between the code text and the narrative and remember that the narrative is not code, has not been legally adopted, and should not be applied as such.

The Publication Development Committee and IAPMO staff welcomes your comments and suggestions for this and future ITMs.

The 2018, 2019 and 2020 Publication Development Committees shared in the development of and final approval of the commentary narrative of this edition. Since that time, changes in the committee membership may have occurred.

2018 IAPMO PUBLICATION DEVELOPMENT COMMITTEE

Martin Cooper, *Chairman*
City of Foster City

Keith Bonenfant, State of California
Kenneth Borski, City of Houston/Planning & Dev.
Shane Peters, City of Santa Monica
Bruce Pfeiffer, Plumbing Inspector - Retired
Jeremy Stettler, Davis School District, Utah
Phillip White, Plumbers Local 68
Daniel Cole, IAPMO Staff Liaison
Doug Kirk, IAPMO Staff

2019 IAPMO PUBLICATION DEVELOPMENT COMMITTEE

Martin Cooper, *Chairman*
City of Foster City

Kenneth Borski, City of Houston/Planning & Dev.
Michael Durfee, County of Salt Lake
Shane Peters, City of Santa Monica
Shannon Ramey, Sr. Forensic Engineer, Jensen Hughes
Matthew Schmidt, City of Lincoln
Jeremy Stettler, Davis School District, Utah
Phillip White, Plumbers Local 68
Daniel Cole, IAPMO Staff Liaison
Doug Kirk, IAPMO Staff

2020 IAPMO PUBLICATION DEVELOPMENT COMMITTEE

Martin Cooper, *Chairman*
City of Foster City

Kenneth Borski, City of Houston/Planning & Dev.
Jacob James Cooper, Apprentice, Power Plumbing
Michael Durfee, Chief Building Official-Retired, County of Salt Lake
Joshua Nyman, Plumbing Inspector, State of Idaho
Shannon Ramey, Sr. Forensic Engineer, Jensen Hughes
Matthew Schmidt, Plumbing Inspector, City of Lincoln
Jeremy Stettler, Davis School District, Utah
Phillip White, Plumbers Local 68
Daniel Cole, IAPMO Staff Liaison
Doug Kirk, IAPMO Staff

TABLE OF CONTENTS

LEARNING LINKS

These learning links to websites are not affiliated with IAPMO. IAPMO is not responsible for information housed on those sites and IAPMO does not endorse any linked websites or products. Use of information from third-party websites is at your risk. These links have been provided to supplement your knowledge on the subject matter.

Section 224.0
Video: Vented Wall Furnace/Wall Heater: goo.gl/oMcXDJ

Section 228.0
Video: Zeotropic: https://bit.ly/2wIFZIO

Section 302.1.1
Marking: http://bit.ly/396xvse

Section 317.3
Open Trenches http://bit.ly/2GRseZS
OSHA Trenching and Excavation Safety

Links to IAPMO websites

To submit a request for an Informal Clarification of a Plumbing or Mechanical Code: http://bit.ly/2Vezq8w

To access the IAPMO R&T Product Certification Directory: http://bit.ly/39qjcyX

To become an IAPMO Member: http://bit.ly/2Sj2xYp

The link below requires IAPMO membership

IAPMO Answers & Analysis database: http://bit.ly/2vkOtov

CHAPTER 1
ADMINISTRATION

101.0 General.

101.1 Title. This document shall be known as the "Uniform Mechanical Code," may be cited as such, and will be referred to herein as "this code."

In addition to title and scope, Chapter 1 covers general subjects such as the purpose of the code, applications to existing mechanical systems, performance provisions relating to alternate materials and methods of construction, modifications and tests. The provisions in Chapter 1 are of such a general nature that they apply to the entire Uniform Mechanical Code (UMC).

Local jurisdictions may adopt different administrative codes so that model codes can be adopted by reference rather than by transcription. Transcription requires that the full text of an ordinance often be printed two or three times.

101.2 Scope. The provisions of this code shall apply to the erection, installation, alteration, repair, relocation, replacement, addition to, use, or maintenance of mechanical systems within this jurisdiction.

The intent of this code is to regulate the design and installation of systems and equipment that directly and indirectly have an effect on the building environment. This includes virtually all equipment or systems that use any form of energy. One area of enforcement that is often overlooked and neglected is industrial processing equipment. Even though the code does not directly address the many varieties of such equipment, it is the intent of the code and responsibility of the Authority Having Jurisdiction (AHJ) to regulate the installation and maintenance of this equipment.

101.3 Purpose. This code is an ordinance providing minimum requirements and standards for the protection of the public health, safety, and welfare.

The requirements of the UMC are regarded as minimum provisions to safeguard life, health, property and public welfare. The specifications for a particular project may call for a more stringent standard or for a higher quality than the minimum code provisions provide. Requirements for quality above the minimum provisions specified in the code are a matter of contractual agreement between the owner and contractor; these requirements are not imposed by the AHJ, but are enforceable when included on approved plans.

101.4 Unconstitutional. Where a section, subsection, sentence, clause, or phrase of this code is, for a reason, held to be unconstitutional, such decision shall not affect the validity of the remaining portions of this code. The legislative body hereby declares that it would have passed this code, and each section, subsection, sentence, clause, or phrase thereof, irrespective of the fact that one or more sections, subsections, sentences, clauses, and phrases are declared unconstitutional.

101.5 Validity. Where a provision of this code, or the application thereof to a person or circumstance, is held invalid, the remainder of the code, or the application of such provision to other persons or circumstances, shall not be affected thereby.

102.0 Applicability.

102.1 Conflicts Between Codes. Where the requirements within the jurisdiction of this mechanical code conflict with the requirements of the plumbing code, the plumbing code shall prevail. In instances where this code, applicable standards, or the manufacturer's installation instructions conflict, the more stringent provisions shall prevail. Where there is a conflict between a general requirement and a specific requirement, the specific requirement shall prevail.

Should the user of this code and another code used in that jurisdiction (e.g. building, fire or plumbing code) feel there is a conflict between the two codes, it is recommended that he or she contact the AHJ. The AHJ will then determine if there is a conflict and how best to resolve it; however, if there is a conflict between this code and the Uniform Plumbing Code (UPC) as published by IAPMO, then the UPC should be followed. It is still advisable to contact the AHJ for guidance or contact IAPMO for an interpretation of the sections in conflict.

102.2 Existing Installations. Mechanical systems lawfully in existence at the time of the adoption of this code shall be permitted to have their use, maintenance, or repair continued where the use, maintenance, or repair is in accordance with the original design and location and no hazard to life, health, or property has been created by such mechanical system.

The concern for mechanical systems lawfully in existence is that their use, maintenance and repair has remained in accordance with the original design and location, and that hazards to life, health and property have not been created. If it is discovered that a system has been relocated, further investigation into the code in effect at the time of the relocation will be necessary to establish the true nature of the system's nonconforming entitlement.

102.3 Maintenance. Mechanical systems, materials, and appurtenances, both existing and new, of a premise under the Authority Having Jurisdiction shall be maintained in operating condition. Devices or safeguards required by this code shall be maintained in accordance with the code edition under which installed.

The owner or the owner's designated agent shall be responsible for maintenance of mechanical systems. To determine compliance with this subsection, the Authority Having Jurisdiction shall be permitted to cause a mechanical system to be reinspected.

This section has the effect of charging the owner of the building with the responsibility for seeing that all buildings, both existing and new, are maintained properly. This section does give the AHJ the authority to make a reinspection of any structure if there is reason to believe that the building has been improperly maintained. Vigorous enforcement of this section will have the effect of reducing existing deficient or unsafe conditions.

An owner is responsible for maintenance of a building's systems in a safe and hazard-free condition. Often the owner is difficult to locate or identify. In cases of a life-safety hazard, it may become necessary for the AHJ to order a utility disconnected; this action may also have the secondary effect of identifying the owner or the owner's representative, but it must not be used indiscriminately. The jurisdiction's legal counsel must be consulted for procedural advice when issuing a notice and order to repair a mechanical system.

102.3.1 Commercial HVAC Systems. Commercial HVAC systems both existing and new, and parts thereof shall be inspected and maintained in operating condition in accordance with ASHRAE/ACCA 180. The owner or the owner's designated agent shall be responsible for maintenance of mechanical systems and equipment. To determine compliance with this subsection, the Authority Having Jurisdiction shall be permitted to cause a HVAC system to be reinspected.

102.3.2 Residential HVAC Systems. Residential HVAC systems both existing and new, and parts thereof shall be inspected in accordance with ACCA 4 QM. The owner or the owner's designated agent shall be responsible for maintenance of mechanical systems and equipment. To determine compliance with this subsection, the Authority Having Jurisdiction shall be permitted to cause a HVAC system to be reinspected.

102.4 Additions, Alterations, Renovations, or Repairs. Additions, alterations, renovations, or repairs shall conform to that required for a new system without requiring the existing mechanical system to be in accordance with the requirements of this code. Additions, alterations, renovations, or repairs shall not cause an existing system to become unsafe, insanitary or overloaded.

Additions, alterations, renovations, or repairs to existing mechanical system installations shall comply with the provisions for new construction, unless such deviations are found to be necessary and are first approved by the Authority Having Jurisdiction.

The general rule is that existing mechanical systems may be retained in service and new additions are required to conform to the current code. That is, application of the UMC is not retroactive; it does not require that existing systems be upgraded each time a new edition of the code is adopted. With the approval of the AHJ, minor repairs, additions or alterations may be made to existing mechanical systems in accordance with the code in effect at the time the original system was installed. Alterations, additions or repairs must not cause an existing system to become unsafe or overloaded, or to create unhealthy conditions for the occupants.

Checking for unsafe conditions includes inspection for excessive pressure, temperature and improper flue use. Spillage of flue gas could indicate a need for increased vent or flue diameter, the abnormal chilling of flue gases or a need for a requirement for power venting. Overloading may indicate the need for larger diameter piping, larger motors, larger ducts or electrical wiring.

A mechanical system that has been performing safely and satisfactorily may, with the approval of the AHJ, be repaired or altered, or have minor additions made in accordance with the laws in effect at the time the original installation was made.

102.5 Health and Safety. Where compliance with the provisions of this code fails to eliminate or alleviate a nuisance, or other dangerous or insanitary condition that involves health or safety hazards, the owner or the owner's agent shall install such additional mechanical system facilities or shall make such repairs or alterations as ordered by the Authority Having Jurisdiction.

102.6 Changes in Building Occupancy. Mechanical systems that are a part of a building or structure undergoing a change in use or occupancy, as defined in the building code, shall be in accordance with the requirements of this code that are applicable to the new use or occupancy.

A change in building occupancy refers to a building being used for a different purpose than originally intended or designed. A single-family residence converted into a restaurant, or a store changed into a doctor's office are examples of a change in building occupancy or use. Whenever this happens, the mechanical systems must be suitable for the new use as defined in the building code adopted by the jurisdiction and the current mechanical code.

102.7 Moved Structures. Parts of the mechanical system of a building and part thereof that is moved from one foundation to another, or from one location to another, shall be in accordance with the provisions of this code for new installations and completely tested as prescribed elsewhere in this section for new work, except that walls or floors need not be removed during such test where equivalent means of inspection acceptable to the Authority Having Jurisdiction are provided.

This section takes into consideration that the movement of a building may often involve changes in use and occupancy. There is also concern for damage resulting from lifting and transportation, so testing of all systems is essential. The intent of the code states that moving a building essentially creates a new structure that is required to comply with all current mechanical code requirements.

102.8 Appendices. The provisions in the appendices are intended to supplement the requirements of this code and shall not be considered part of this code unless formally adopted as such.

The appendices are reference material, and although highly valuable, is not legally a part of the code as it stands. If a local jurisdiction chooses to make all or part of these appendices part of the code, it may do so through its adoption process. It should also be noted that there may even be certain chapters of this code that are not adopted by some jurisdictions and that those chapters will also not be a part of the legal code document.

103.0 Duties and Powers of the Authority Having Jurisdiction.

103.1 General. The Authority Having Jurisdiction shall be the Authority duly appointed to enforce this code. For such purposes, the Authority Having Jurisdiction shall have the powers of a law enforcement officer. The Authority Having Jurisdiction shall have the power to render interpretations of this code and to adopt and enforce rules and regulations supplemental to this code as deemed necessary in order to clarify the application of the provisions of this code. Such interpretations, rules, and regulations shall comply with the intent and purpose of this code.

In accordance with the prescribed procedures and with the approval of the appointing authority, the Authority Having Jurisdiction shall be permitted to appoint such number of technical officers, inspectors, and other employees as shall be authorized from time to time. The Authority Having Jurisdiction shall be permitted to deputize such inspectors or employees as necessary to carry out the functions of the code enforcement agency.

The Authority Having Jurisdiction shall be permitted to request the assistance and cooperation of other officials of this jurisdiction so far as required in the discharge of the duties required by this code or other pertinent law or ordinance.

This section authorizes and directs the AHJ to enforce the provisions of the UMC. Enforcement of the code can vary considerably among jurisdictions. Local policies are dictated by such variables as climactic conditions; flooding; freezing; snowfall; water quality or hardness; soil conditions; wind direction and intensity; type of structures; age of structures; surface topography and altitude. Unusual conditions may prevail within certain areas or regions of a jurisdiction. Special applications of the code must be stressed in these regions to ensure satisfactory installation and operation of mechanical systems. In addition, the enforcement of codes can vary due to policies established by the AHJ.

The AHJ in most instances is a building and safety department. A governing body, a city council, commission or, for state agencies, the state legislature will appoint a code official to administer the department. This individual will then hire or appoint assistants, inspectors, plans check personnel and other staff as needed to enforce the various codes adopted by the governing body. They are also the officials who will interpret and enforce the code requirements on behalf of the code official designated by the AHJ.

103.2 Liability. The Authority Having Jurisdiction charged with the enforcement of this code, acting in good faith and without malice in the discharge of the Authority Having Jurisdiction's duties, shall not thereby be rendered personally liable for damage that accrues to persons or property as a result of an act or by reason of an act or omission in the discharge of such duties. A suit brought against the Authority Having Jurisdiction or employee because of such act or omission performed in the enforcement of provisions of this code shall be defended by legal counsel provided by this jurisdiction until final termination of such proceedings.

It is the intent of the UMC that the AHJ shall not become personally liable for any damage that may occur to persons or property as a result of the AHJ's acts as long as its representative(s) acts in good faith and without malice or fraud. However, there seems to be an increasing trend in the courts to find civil officers personally liable for careless acts. This section requires that the jurisdiction defend the AHJ if a suit is brought against its representative(s). Furthermore, the code requires that any judgment resulting from a suit be assumed by the jurisdiction.

103.3 Applications and Permits. The Authority Having Jurisdiction shall be permitted to require the submission of plans, specifications, drawings, and such other information in accordance with the Authority Having Jurisdiction, prior to the commencement of, and at a time during the progress of, work regulated by this code.

The issuance of a permit upon construction documents shall not prevent the Authority Having Jurisdiction from thereafter requiring the correction of errors in said construction documents or from preventing construction operations being carried on thereunder where in violation of this code or of other pertinent ordinance or from revoking a certificate of approval where issued in error.

103.3.1 Licensing. Provision for licensing shall be determined by the Authority Having Jurisdiction.

103.4 Right of Entry. Where it is necessary to make an inspection to enforce the provisions of this code, or where the Authority Having Jurisdiction has reasonable cause to believe that there exists in a building or upon a premises a condition or violation of this code that makes the building or premises unsafe, insanitary, dangerous, or hazardous, the Authority Having Jurisdiction shall be permitted to enter the building or premises at reasonable times to inspect or to perform the duties imposed upon the Authority Having Jurisdiction by this code, provided that where such building or premises is occupied, the Authority Having Jurisdiction shall present credentials to the occupant and request entry. Where such building or premises is unoccupied, the Authority Having Jurisdiction shall first make a reasonable effort to locate the owner or other person having charge or control of the building or premises and request entry. Where entry is refused, the Authority Having Jurisdiction has recourse to every remedy provided by law to secure entry.

Where the Authority Having Jurisdiction shall have first obtained an inspection warrant or other remedy provided by law to secure entry, no owner, occupant, or person having charge, care or control of a building or premises shall fail or neglect, after a request is made as herein provided, to promptly permit entry herein by the Authority Having Jurisdiction for the purpose of inspection and examination pursuant to this code.

Protecting the health and safety of a community is serious business. For this reason, the code has given the AHJ the right to enter a building, for the purpose of inspection, when there is reason to believe that within the building or on the surrounding premises a condition exists that may be unsafe, dangerous, or hazardous to the public's health, safety and welfare.

This subsection is compatible with United States Supreme Court decisions regarding inspection personnel seeking entry to buildings for the purpose of making inspections. Under current case law, an inspection may not be made of property, whether it be a private residence or a business establishment, without first having secured permission from the owner or person in charge of the premises. If entry is refused by the person having control of the property, the AHJ must obtain an inspection warrant from a court having jurisdiction in order to secure entry. The important feature of the law regarding right of entry is that entry must be made only by permission of the person having control of the property, or lacking this permission, entry may be gained only through the use of an inspection warrant.

If entry is again refused after an inspection warrant has been obtained, the jurisdiction now has recourse through the courts to remedy this situation. One avenue is to obtain a civil injunction in which the court directs the person having control of the property to allow an inspection. Alternatively, the jurisdiction can initiate proceedings in criminal court for punishment of the person having control of the property. It cannot be repeated too strongly that criminal court proceedings should never be initiated against an owner or other person having control of the property if an inspection warrant has not been obtained. Also, because the consequences of not following proper procedures can be so devastating to a jurisdiction if a suit is brought against it, the jurisdiction's legal officer should always be consulted in these matters.

104.0 Permits.

104.1 Permits Required. It shall be unlawful for a person, firm, or corporation to make an installation, alteration, repair, replacement, or remodel a mechanical system regulated by this code except as permitted in Section 104.2, or to cause the same to be done without first obtaining a separate mechanical permit for each separate building or structure.

The intent of this section is that the requirement to obtain a permit applies to persons, firms or corporations doing the work or causing the work to be done. More than one permit may be issued when different contractors are working on different systems within the same structure. There must be no duplication of permits for the same portion of a mechanical system. The person holding the permit is responsible for the work. See **Figure 104.1** for an example of a permit.

104.2 Exempt Work. A permit shall not be required for the following:

(1) A portable heating appliance, portable ventilating equipment, a portable cooling unit, or a portable evaporative cooler.

Usually, this type of equipment is of low capacity and output, and the appliances are packaged units manufactured to standards not included in Chapter 17. Many portable appliances are listed; however, operating instructions and safe clearances are often indicated on the manufacturer's identification plate or label. The owner or occupant is provided with information on the safe use of portable appliances. The user of the equipment is responsible for ensuring the equipment is maintained and operated as specified by the manufacturer. Due to the portable nature of the equipment, inspection by the jurisdiction would be ineffective because conditions within the premises could change frequently.

(2) A closed system of steam, hot, or chilled water piping within heating or cooling equipment regulated by this code.

A closed system is a piping system where the same hot, warm or cooled water is contained within and is circulated whenever the equipment is operated.

(3) Replacement of a component part that does not alter its original approval and is in accordance with other applicable requirements of this code.

The original appliance or equipment must have been approved and installed in compliance with the code. The replacement parts must maintain the same operating safety features, temperature and pressure range, and the same fuel, voltage and all other features as the original parts. Any changes that alter the original specifications or approved listing may void the equipment listing and trigger the need for inspection.

(4) Refrigerating equipment that is part of the equipment for which a permit has been issued pursuant to the requirements of this code.

A permit for a refrigeration compressor system includes the component parts of that system. Replacement of parts to maintain original operating specifications would not require a permit.

(5) A unit refrigerating system.

A portable unit refrigeration system is defined as one that is not attached to ductwork, does not exceed 3 horsepower and has been factory assembled and tested before its installation. When repairs are made, only factory-authorized replacement parts must be used.

Exemption from the permit requirements of this code shall not be deemed to grant authorization for work to be done in violation of the provisions of the code or other laws or ordinances of this jurisdiction.

104.3 Application for Permit. To obtain a permit, the applicant shall first file an application therefore in writing on a form furnished by the Authority Having Jurisdiction for that purpose. Such application shall:

In this section, the UMC requires that a permit be applied for and describes the information required not only on the permit application itself, but information to be filed with the permit application. The AHJ is permitted to waive the requirement for the filing of plans and other data (see Section 104.3.1), provided it is assured that the work for which the permit is applied is of such a nature that plans or other data are not necessary in order to obtain compliance with the code.

(1) Identify and describe the work to be covered by the permit for which application is made.

Mechanical Permit Application

Jurisdiction name:

Address:

Phone: - - Fax: - -

Email: Web:

DEPARTMENT USE ONLY
Permit no.:
Date:

CATEGORY OF CONSTRUCTION		
☐ Residential	☐ Government	☐ Commercial
JOB SITE INFORMATION AND LOCATION		
Job site address:		
City:	State:	ZIP:
Subdivision:	Lot no.:	
DESCRIPTION OF WORK		
PROPERTY OWNER INSTALLATION		
Name:		
Address:		
City:	State:	ZIP:
Phone: - -	Fax: - -	
Email:		
This installation is being made on property owned by me or a member of my immediate family, and is exempt from licensing requirements under ORS 701.010. Signature:		
CONTRACTOR INSTALLATION		
Business name:		
Address:		
City:	State:	ZIP:
Phone: - -	Fax: - -	
Email:		
CCB license no.:		
Print name:		
Signature:		

CREDIT CARD INFORMATION	
☐ Visa ☐ MasterCard ☐ Discover	Phone: - -
Credit card number	/ Expiration
Name of cardholder as shown on credit card	
Cardholder signature	$ Amount

FEE SCHEDULE			
Residential	**Qty.**	**Cost ea.**	**Total cost**
Furnace/burner including ducts and vents			
Up to 100k BTU/hr.		$	$
Over 100k BTU/hr.		$	$
Heaters/stoves/vents			
Unit heater		$	$
Wood/pellet/gas stove/flue		$	$
Repair/alter/add to heating appliance/ refrigeration unit or cooling system/ absorption system		$	$
Evaporated cooler		$	$
Vent fan with one duct/appliance vent		$	$
Hood with exhaust and duct		$	$
Floor furnace including vent		$	$
Gas piping			
One to four outlets		$	$
Additional outlets (each)		$	$
Air-handling units, including ducts			
Up to 10,000 CFM		$	$
Over 10,000 CFM		$	$
Compressor/absorption system/heat pump			
Up to 3 hp/100k BTU		$	$
Up to 15 hp/500k BTU		$	$
Up to 30 hp/1,000 BTU		$	$
Up to 50 hp/1,750 BTU		$	$
Over 50 hp/1,750 BTU		$	$
Incinerators			
Domestic incinerator		$	$
Commercial			
Enter total valuation of mechanical system and installation costs $ _______			
Enter fee based on valuation of mechanical system, etc.			$
Miscellaneous fees	**Items**	**Cost ea.**	**Total cost**
Reinspection		$	$
Specially requested inspections (per hr.)		$	$
Regulated equipment (unclassed)		$	$

APPLICANT USE	
(A) Enter subtotal of above fees (or enter set minimum fee of $ _______)	$
(B) Investigative fee	$
(C) Enter 12% surcharge (.12 x [A+B])	$
(D) Seismic fee, 1% (.01 x [A])	$
(E) Plan review (% of [A])	$
TOTAL fees and surcharges (A+B+C+D+E):	$

FIGURE 104.1
SAMPLE PERMIT

Identification and description of the work that will be done must be provided on the application. Detailed project specifications, drawings or directions are frequently not prepared for small installations and renovations. For this reason, the information supplied on the permit application is the only guidance available for inspection authorities to determine the type and location of the equipment installed.

(2) Describe the land upon which the proposed work is to be done by legal description, street address, or similar description that will readily identify and definitely locate the proposed building or work.

The description must readily identify and definitely locate where the work will be done. If the only description is the street address, the result can be very confusing. A complete description is essential in new residential subdivisions. An applicant, a contractor or an inspector may be easily confused by multiple sites in a new subdivision. References must be made to the legal land description, including lot, block and subdivision numbers. When applicable, the description should also include the unit designator.

(3) Indicate the use or occupancy for which the proposed work is intended.

The use of occupancy for which the proposed installation is intended must be included on the application. This information enables a plan checker to determine if the application complies with zoning regulations. Some jurisdictions have zones where certain uses are restricted or prohibited due to fire, chemical or other environmental hazards. These may include the storage of certain paints, thinners, flammable or combustible liquids, chemicals, compressed gases, lumber or furniture and cabinet manufacturing where shavings and sawdust would constitute a hazard. The requirements for approval may vary, depending on the size of the mechanical system, occupancy of the building and maximum number of persons who may occupy the building.

(4) Be accompanied by construction documents and other data in accordance with Section 104.3.1.

When required, a permit application must be accompanied by plans, diagrams, computations or other data. This requirement is amplified in Section 104.3.1.

(5) Be signed by the permittee or the permittee's authorized agent. The Authority Having Jurisdiction shall be permitted to require evidence to indicate such authority.

The application must be signed by the permit applicant or an authorized agent. The signature on the permit application may be considered as acceptance of responsibility for permit and code compliance.

(6) Give such other data and information in accordance with the Authority Having Jurisdiction.

The applicant is required to provide other data and information that is required and requested by the AHJ. This might include product or material procedures data or information to assess demands on the municipal or public services such as electrical energy, fuel gas, water, sewer or other services, in addition to information on grade, location, etc.

104.3.1 Construction Documents. Construction documents, engineering calculations, diagrams, and other data shall be submitted in two or more sets with each application for a permit. The construction documents, computations, and specifications shall be prepared by, and the mechanical system designed by, a registered design professional. Construction documents shall be drawn to scale with clarity to identify that the intended work to be performed is in accordance with the code.

Exception: The Authority Having Jurisdiction shall be permitted to waive the submission of construction documents, calculations, or other data where the Authority Having Jurisdiction finds that the nature of the work applied for is such that reviewing of construction documents is not necessary to obtain compliance with the code.

Most jurisdictions require a minimum of two sets of construction documents — one set retained by the building department until the completion of the job and the second set maintained on the job site by the permit holder. Notations, tests, inspections and approvals are documented on the plans, inspection records or both. The remarks are dated and signed as inspections or alterations progress. This procedure is especially important on extensive projects where a part of the system is completed, installed and concealed or buried before work on other sections is even commenced. A complete set of plans and specifications should be constantly marked up to indicate the inspection progress on the project. A great deal of time and effort is required for accurate records of inspection. Time spent in this documentation may prove to be extremely valuable at a later date.

The exception in this section is usually applied where simple systems are installed or altered and the inspection staff is sufficiently knowledgeable about the system to ensure code compliance.

104.3.2 Plan Review Fees. Where a plan or other data is required to be submitted in accordance with Section 104.3.1, a plan review fee shall be paid at the time of submitting construction documents for review.

The plan review fees for mechanical system work shall be determined and adopted by this jurisdiction.

The plan review fees specified in this subsection are separate fees from the permit fees specified in Section 104.5.

Where plans are incomplete or changed so as to require additional review, a fee shall be charged at the rate shown in Table 104.5.

The plan review fee may be a separate fee or a part of the building or mechanical permit fee. This fee pays for the time it takes for the review of a project's plans and specifications to ensure compliance with the code. Ultimately, a thorough plan review is an aid to both the AHJ and the contractor interested in complying with the code.

104.3.3 Time Limitation of Application. Applications for which no permit is issued within 180 days following the date of application shall expire by limitation, plans and other data submitted for review thereafter, shall be returned to the applicant or destroyed by the Authority Having Jurisdiction. The

Authority Having Jurisdiction shall be permitted to extend the time for action by the applicant for a period not to exceed 180 days upon request by the applicant showing that circumstances beyond the control of the applicant have prevented action from being taken. No application shall be extended more than once. In order to renew action on an application after expiration, the applicant shall resubmit plans and pay a new plan review fee.

In the event that a permit application cannot be approved within 180 days because the application or pertinent plans and data are incomplete or contain uncorrected code violations, or for any other reason caused by the applicant, the application expires and is either returned or the plans destroyed. The applicant may receive a one-time extension of 180 days if he or she can prove that the extension is needed for reasons beyond his or her control.

104.4 Permit Issuance. The application, construction documents, and other data filed by an applicant for a permit shall be reviewed by the Authority Having Jurisdiction. Such plans shall be permitted to be reviewed by other departments of this jurisdiction to verify compliance with applicable laws under their jurisdiction. Where the Authority Having Jurisdiction finds that the work described in an application for permit and the plans, specifications, and other data filed therewith are in accordance with the requirements of the code and other pertinent laws and ordinances and that the fees specified in Section 104.5 have been paid, the Authority Having Jurisdiction shall issue a permit therefore to the applicant.

104.4.1 Approved Plans or Construction Documents. Where the Authority Having Jurisdiction issues the permit where plans are required, the Authority Having Jurisdiction shall endorse in writing or stamp the construction documents "APPROVED." Such approved construction documents shall not be changed, modified, or altered without authorization from the Authority Having Jurisdiction, and the work shall be done in accordance with approved plans.

The Authority Having Jurisdiction shall be permitted to issue a permit for the construction of a part of a mechanical system before the entire construction documents for the whole system have been submitted or approved, provided adequate information and detailed statements have been filed in accordance with pertinent requirements of this code. The holder of such permit shall be permitted to proceed at the holder's risk without assurance that the permit for the entire building, structure, or mechanical system will be granted.

The permit issuance process is intended to provide records for the code enforcement agency to ensure orderly controls of the inspection process. Thus, the permit application is intended to describe in detail the work to be done, while the plans and other data filed with the application are intended to graphically depict the work to be done. The AHJ is directed to review the permit application and the plans and specifications filed with the permit. This review of plans and specifications is not a discretionary procedure, but rather one that is mandated by the code. The AHJ is not at liberty to check only a portion of the plans. The drawings as well as the "specifications and other data filed" must be checked in order for the AHJ to comply with the code. In fact, if the submitted information is not reviewed by the AHJ, the AHJ is in effect presuming that the specifications are in compliance with the code, and violating the code him/herself.

The code also charges the AHJ with the issuance of a permit if it has been determined that the information filed with the application shows compliance with the code and other laws and ordinances applicable to the building at its location in the jurisdiction. The AHJ may not withhold the issuance of a permit if these conditions are met. Thus, the AHJ would be in violation of the code if it withholds the issuance of a permit because of failure of the applicant to comply with the code in an unrelated area.

This section also provides information for issuing partial permits. This provision is normally used on large projects to reduce the delay in beginning construction while waiting for the entire system to be designed or on a speculative building where future tenant requirements are unknown at the time of partial permit issuance. This provision should be reserved for occasions when waiting for complete drawings and specifications would be a hardship on the applicant. Usually, the final drawings for the mechanical systems cannot be completed until the architectural drawings are finalized. Delay caused by the approval of architectural and structural drawings before the commencement of electrical and mechanical trades may cause delays in meeting completion target dates. The code enforcement agency cannot compensate for a designer's failure to provide an adequate time schedule for a project; however, it can compensate when a delay is caused by unusual workloads internally (i.e., in the code enforcement agency).

104.4.2 Validity of Permit. The issuance of a permit or approval of construction documents shall not be construed to be a permit for, or an approval of, a violation of the provisions of this code or other ordinance of the jurisdiction. No permit presuming to give authority to violate or cancel the provisions of this code shall be valid.

The issuance of a permit based upon plans, specifications, or other data shall not prevent the Authority Having Jurisdiction from thereafter requiring the correction of errors in said plans, specifications, and other data or from preventing building operations being carried on thereunder where in violation of this code or of other ordinances of this jurisdiction.

While it may be poor public relations to suspend or revoke a permit or to require corrections of the plans after they have been approved, it is clearly the intent of the code that the approval of plans or the issuance of a permit does not eliminate the need to comply with the code or other pertinent laws or ordinances.

It is often difficult for a plans examiner to check every detail of every set of plans and specifications, including the location, elevation and details of public services. The designer of a mechanical system cannot place the responsibility for compliance with the codes and regulations on the building department.

104.4.3 Expiration. A permit issued by the Authority Having Jurisdiction under the provisions of this code shall expire by limitation and become null and void where the work authorized by such permit is not commenced within 180 days from the date of such permit, or where the work authorized by such permit is suspended or abandoned at a time after the work is commenced for a period of 180 days. Before such work is recommenced, a new permit shall first be obtained to do so, and the fee, therefore, shall be one-half the amount required for a new permit for such work, provided no changes have been made or will be made in the original construction documents for such work, and provided further that such suspension or abandonment has not exceeded 1 year.

The UMC anticipates that once a permit has been issued, construction will soon follow and proceed expeditiously until its completion. However, this ideal procedure is not always the case; therefore, the code makes provisions for those cases where work is not started or where the work in progress has been suspended for a period of time. It is assumed by the code that the code enforcement agency will have expended some effort and conducted follow-up inspections of the work, etc.; therefore, at least half of the permit fee must be obtained in order to compensate the agency for the work. See the discussion of fee refunds under Section 104.5.3.

104.4.4 Extension. A permittee holding an unexpired permit shall be permitted to apply for an extension of the time within which work shall be permitted to commence under that permit where the permittee is unable to commence work within the time required by this section. The Authority Having Jurisdiction shall be permitted to extend the time for action by the permittee for a period not exceeding 180 days upon written request by the permittee showing that circumstances beyond the control of the permittee have prevented action from being taken. No permit shall be extended more than once. In order to renew action on a permit after expiration, the permittee shall pay a new full permit fee.

104.4.5 Suspension or Revocation. The Authority Having Jurisdiction shall be permitted to, in writing, suspend or revoke a permit issued under the provisions of this code where the permit is issued in error or on the basis of incorrect information supplied or in violation of other ordinance or regulation of the jurisdiction.

As in Section 104.4.2, issuance of a permit based on plans, specifications and other data does not prevent the AHJ from subsequently requiring corrections of errors to plans and specifications. Even after a permit has been issued and errors are noticed, the AHJ is required to ensure code compliance.

If the permit holder refuses to make corrections, a stop order may be issued, as provided for in Section 106.4.

104.4.6 Retention of Plans. One set of approved construction documents and computations shall be retained by the Authority Having Jurisdiction until final approval of the work is covered therein.

One set of approved construction documents, computations, and manufacturer's installation instructions shall be returned to the applicant, and said set shall be kept on the site of the building or work at times during which the work authorized thereby is in progress.

The inspector should request a copy of the approved plans when making an inspection. This practice will assist in determining that the installation is in compliance with code requirements and the permit. The permit holder's approved plans should be marked and signed as the inspection progresses. It may be helpful to mark the system on the approved plans with colored pencils whenever a certain part of the system has been tested, inspected and approved.

104.5 Fees. Fees shall be assessed in accordance with the provisions of this section and as set forth in the fee schedule, Table 104.5. The fees are to be determined and adopted by this jurisdiction.

Permit fees are the means by which the code enforcement services provided by the AHJ are funded. Without these fees there would be no plan reviews or inspections and compliance to the codes would become the responsibility of the owner or contractor. Permit fees are therefore necessary and may vary greatly among jurisdictions.

The UMC anticipates that many jurisdictions will establish their own fee schedules and, therefore, recognizes that the fees to be charged for permits and for plan review will be established by the jurisdiction through related ordinances.

104.5.1 Work Commencing Before Permit Issuance. Where work for which a permit is required by this code has been commenced without first obtaining said permit, a special investigation shall be made before a permit is issued for such work.

104.5.2 Investigation Fees. An investigation fee, in addition to the permit fee, shall be collected whether or not a permit is then or subsequently issued. The investigation fee shall be equal to the amount of the permit fee that is required by this code if a permit were to be issued. The payment of such investigation fee shall not exempt a person from compliance with other provisions of this code, nor from a penalty prescribed by law.

When work requiring a permit is started without a permit, the code directs the AHJ to launch an investigation of the work already done. The intent of the investigation is to determine to what extent the work completed complies with the code and to describe with as much detail as possible the work that has been completed.

Because it is anticipated that the investigation may require considerable time and effort, the code specifies that a fee should be paid by the person doing the work equal in amount to the permit fee that would be required. The investigation fee is to be paid in addition to the regular permit and plan check fees. Moreover, the investigation fee is to be paid whether or not a permit is later issued.

104.5.3 Fee Refunds. The Authority Having Jurisdiction shall be permitted to authorize the refunding of a fee as follows:

(1) The amount paid hereunder that was erroneously paid or collected.

(2) Refunding of not more than a percentage, as determined by this jurisdiction where no work has been done under a permit issued in accordance with this code.

The Authority Having Jurisdiction shall not authorize refunding of a fee paid except upon written application filed by the original permittee not to exceed 180 days after the date of fee payment.

This section authorizes the AHJ to refund a portion of the permit fee, the plan check fee or both for good cause. One instance would be where the permit fee is collected in error. Another reason for authorizing the refund of the fees paid would be because circumstances beyond the control of the applicant cause delays and the eventual expiration of either the permit or the plan review.

105.0 Inspections and Testing.

105.1 General. Mechanical systems for which a permit is required by this code shall be inspected by the Authority Having Jurisdiction.

No mechanical system or portion thereof shall be covered, concealed, or put into use until inspected and approved as prescribed in this code. Neither the Authority Having Jurisdiction nor the jurisdiction shall be liable for expense entailed in the removal or replacement of material required to permit inspection. Mechanical systems regulated by this code shall not be connected to the energy fuel supply lines until authorized by the Authority Having Jurisdiction.

The inspection function of the building department is one of the most important activities of a governmental unit. A department can have the best plan checking operation possible, but if the field inspection does not require construction to be in compliance with the code and the approved plans, it is wasted effort.

Inspections during construction are referred to as "rough inspections," with the "final inspection" being performed when a system is complete and ready to be used. Even though the code states that "mechanical systems shall not be connected to the energy fuel-supply lines until authorized by the AHJ," the intent is to allow connection for purposes of testing before final inspection (see Section 105.4).

The important consideration is that a mechanical system cannot be put in normal and continuous operation and the premises cannot be occupied or open to the public without final inspection. An occupancy permit or business license should not be issued before completion of final inspections and approvals from appropriate regulatory agencies.

105.2 Required Inspections. New mechanical system work and such portions of existing systems as affected by new work, or changes, shall be inspected by the Authority Having Jurisdiction to ensure compliance with the requirements of this code and to ensure that the installation and construction of the mechanical system are in accordance with approved plans. The Authority Having Jurisdiction shall make the following inspections and other such inspections as necessary. The permittee or the permittee's authorized agent shall be responsible for the scheduling of such inspections as follows:

(1) Underground inspection shall be made after trenches or ditches are excavated and bedded, piping installed, and before backfill is put in place.

(2) Rough-in inspection shall be made prior to the installation of wall or ceiling membranes.

(3) Final inspection shall be made upon completion of the installation.

All mechanical work done under a permit must be inspected by the AHJ to make sure that it complies with this code and approved plans prior to the work being covered. If there is an addition to an existing system, the portions of the existing system that are affected by the addition shall also be inspected and approved.

105.2.1 Uncovering. Where a mechanical system, or part thereof, which is installed, altered, or repaired, is covered or concealed before being inspected, tested, and approved as prescribed in this code, it shall be uncovered for inspection after notice to uncover the work has been issued to the responsible person by the Authority Having Jurisdiction. The requirements of this section shall not be considered to prohibit the operation of mechanical systems installed to replace existing equipment serving an occupied portion of the building in the event a request for inspection of such equipment has been filed with the Authority Having Jurisdiction not more than 72 hours after such replacement work is completed, and before a portion of such mechanical system is concealed by a permanent portion of the building.

Mechanical systems may be operated when installations or repairs are made to existing equipment or fixtures in an occupied portion of a building, provided a request for inspection is made within 72 hours of the completion of repairs. However, no work may be permanently concealed until inspection is completed and approval is granted. This provision allows emergency maintenance functions to be performed and the system to be operated for normal service. If this provision was not included, the result could be damage to structures and systems and food spoilage and disruption of various industrial or commercial services.

105.2.2 Other Inspections. In addition to the inspections required by this code, the Authority Having Jurisdiction shall be permitted to require other inspections to ascertain compliance with the provisions of this code and other laws that are enforced by the Authority Having Jurisdiction.

The AHJ has discretionary power to conduct other than called for inspections of mechanical systems if it suspects code or other regulation violations. If the AHJ suspects a system or part thereof is defective or there are additional laws or ordinances that must be met, it may require that system to undergo additional inspection and/or testing to prove its integrity.

105.2.3 Inspection Requests. It shall be the duty of the person doing the work authorized by a permit to notify the Authority Having Jurisdiction that such work is ready for inspection. The Authority Having Jurisdiction shall be permitted to require that a request for inspection be filed not less than 1 working day before such inspection is desired. Such request shall be permitted to be made in writing or by telephone, at the option of the Authority Having Jurisdiction.

It shall be the duty of the person requesting inspections

in accordance with this code to provide access to and means for inspection of such work.

Request for inspection must be made by the person doing the work authorized by the permit. That person knows the progress that has been made. Inspection requests are usually required to be made at least 24 hours in advance and commonly made via phone, fax, or digitally depending on the policies of the various AHJs.

The permittee should test the system and make sure that it will hold the required pressure before calling for the inspection. If he or she fails to do this, leaks may prevent the test from passing and possibly resulting in another inspection and a reinspection fee.

The person requesting the inspection is responsible for providing access to the work and a means whereby it may be inspected. The permittee or the person requesting an inspection is responsible for providing all the equipment needed for the test. For example, a lift or scaffolding may be required for the inspection of piping installed in a high location. Ladders must be adequate to comply with federal or state safety requirements and must be of sufficient length so the inspector can safely climb up and reach the equipment or system being inspected. A ladder that reaches 3 feet below the roof so the inspector can only see the top of a flat roof is not considered adequate access to equipment on that roof. Provisions for access include lighting, walking space and drainage of water from the adjacent areas or pathways to equipment. An inspection door providing access for a visual inspection of a mechanical system is required.

It is desirable that the person requesting inspection be present to discuss the installation with the inspector. However, whether the permit holder is present or not, all corrections should be clearly listed in writing and a copy should be retained by the department.

105.2.4 Advance Notice. It shall be the duty of the person doing the work authorized by the permit to notify the Authority Having Jurisdiction, orally or in writing that said work is ready for inspection. Such notification shall be given not less than 24 hours before the work is to be inspected.

105.2.5 Responsibility. It shall be the duty of the holder of a permit to make sure that the work will stand the test prescribed before giving the notification.

The equipment, material, and labor necessary for inspection or tests shall be furnished by the person to whom the permit is issued or by whom inspection is requested.

105.2.6 Reinspections. A reinspection fee shall be permitted to be assessed for each inspection or reinspection where such portion of work for which inspection is called is not complete or where required corrections have not been made.

This provision shall not be interpreted as requiring reinspection fees the first time a job is rejected for failure to be in accordance with the requirements of this code, but as controlling the practice of calling for inspections before the job is ready for inspection or reinspection.

Reinspection fees shall be permitted to be assessed where the approved plans are not readily available to the inspector, for failure to provide access on the date for which the inspection is requested, or for deviating from plans requiring the approval of the Authority Having Jurisdiction.

To obtain reinspection, the applicant shall file an application therefore in writing upon a form furnished for that purpose and pay the reinspection fee in accordance with Table 104.5.

In instances where reinspection fees have been assessed, no additional inspection of the work will be performed until the required fees have been paid.

If reinspection is necessary because work has not progressed to the point where it is ready for inspection, the code authorizes the AHJ to charge a reinspection fee. The reinspection fee is not mandated and normally would not be charged unless there is a pattern of inspection calls and repeated disapprovals for portions of the work for which inspection was requested, failure to provide access, failure of the testing procedure, deviation from the approved plans or using unapproved materials.

105.3 Testing of Systems. Mechanical systems shall be tested and approved in accordance with this code or the Authority Having Jurisdiction. Tests shall be conducted in the presence of the Authority Having Jurisdiction or the Authority Having Jurisdiction's duly appointed representative.

No test or inspection shall be required where a mechanical system, or part thereof, is set up for exhibition purposes and has no connection with water or an energy fuel supply. In cases where it would be impractical to provide the required water or air tests, or for minor installations and repairs, the Authority Having Jurisdiction shall be permitted to make such inspection as deemed advisable in order to be assured that the work has been performed in accordance with the intent of this code. Joints and connections in the mechanical system shall be airtight, gastight, or watertight for the pressures required by the test.

This section refers to the tests required to be performed on mechanical systems. Mechanical systems may be composed of, but not limited to, air moving systems, HVAC, process piping, refrigeration, hydronics, steam and hot-water boilers and piping, and fuel-gas piping, or other ancillary components necessary for the proper function.

If a test is required for a particular system, then the installation must be tested and withstand the test before that system is approved. When a system test is required, the inspector must be present during the test to observe it in progress and verify that the system has indeed withstood the test.

105.3.1 Defective Systems. In buildings or premises condemned by the Authority Having Jurisdiction because of an insanitary condition of the mechanical system, or part thereof, the alterations in such system shall be in accordance with the requirements of this code.

The AHJ has discretionary power to conduct other than ordinary inspections of mechanical systems if it suspects code or other regulation violations. If the AHJ suspects a system or part thereof is defective, it may require that system to undergo a pressure test to prove its integrity. When a condemned system is restored, the work must comply in all ways with the provisions of the code.

105.3.2 Retesting. Where the Authority Having Jurisdiction finds that the work will not pass the test, necessary corrections shall be made, and the work shall be resubmitted for test or inspection.

105.3.3 Approval. Where prescribed tests and inspections indicate that the work is in accordance with this code, a certificate of approval shall be issued by the Authority Having Jurisdiction to the permittee on demand.

105.4 Connection to Service Utilities. No person shall make connections from a source of energy or fuel to a mechanical system or equipment regulated by this code and for which a permit is required until approved by the Authority Having Jurisdiction. The Authority Having Jurisdiction shall be permitted to authorize temporary connection of the mechanical system equipment to the source of energy or fuel for the purpose of testing the equipment.

When an energy source is connected without authorization from the building department or appliances or equipment are operated in violation of an order by the AHJ, the result is usually a notification by phone and confirmation in writing to the owner and to the utility stating that the energy source is going to be disconnected at the property line and the meter removed. The enforcement tool is available to inspection agencies to allow for the control of unsafe installations.

Before connection of a mechanical system to an energy or fuel source can be legally made, it first must have passed inspection and been approved by the AHJ. However, temporary connections during the progress of the work may be allowed to facilitate testing or to provide power to the job site, but only after approval of the AHJ and after all precautions are made for safety.

106.0 Violations and Penalties.

The installation, repair or maintenance of mechanical systems that do not comply with the various sections of the code or a violation of requirements found in other documents referenced by this code constitutes a violation of the code. Since the violation could cause property damage, sickness or even death, no one is allowed to violate the code or knowingly permit it to be violated by others. This code, once adopted, becomes law and, just like any other law enacted by a governing body, enforcement and penalty provisions for violations are included in the code.

106.1 General. It shall be unlawful for a person, firm, or corporation to erect, construct, enlarge, alter, repair, move, improve, remove, convert, demolish, equip, use, or maintain a mechanical system or permit the same to be done in violation of this code.

106.2 Notices of Correction or Violation. Notices of correction or violation shall be written by the Authority Having Jurisdiction and shall be permitted to be posted at the site of the work or mailed or delivered to the permittee or their authorized representative.

Refusal, failure, or neglect to comply with such notice or order within 10 days of receipt thereof, shall be considered a violation of this code and shall be subject to the penalties set forth by the governing laws of the jurisdiction.

106.3 Penalties. A person, firm, or corporation violating a provision of this code shall be deemed guilty of a misdemeanor, and upon conviction thereof, shall be punishable by a fine, imprisonment, or both set forth by the governing laws of the jurisdiction. Each separate day or a portion thereof, during which a violation of this code occurs or continues, shall be deemed to constitute a separate offense.

106.4 Stop Orders. Where work is being done contrary to the provisions of this code, the Authority Having Jurisdiction shall be permitted to order the work stopped by notice in writing served on persons engaged in the doing or causing such work to be done, and such persons shall forthwith stop work until authorized by the Authority Having Jurisdiction to proceed with the work.

The most common reasons for the issuance of a stop order are because a permit for the work has not been issued or for noncompliance with code requirements.

One reason for failure to obtain a permit for the installation is that the person performing the work thought that another person would secure the permit. Sometimes permits are denied due to building, zoning or licensing restrictions. Another reason is that the person performing the work may not possess the required trade, business or contractor's license.

An AHJ will seldom place a stop order for permitted work that is found defective during the first inspection. However, if the defective condition is not corrected, a stop order may be issued during a subsequent inspection, provided reasonable time has elapsed for correction of the problem noted during the first inspection.

106.5 Authority to Disconnect Utilities in Emergencies. The Authority Having Jurisdiction shall have the authority to disconnect a mechanical system to a building, structure, or equipment regulated by this code in case of emergency where necessary to eliminate an immediate hazard to life or property.

This authority is generally used only when a building is on fire or has been severely damaged due to fire or explosion or has suffered structural damage. Action must be taken when the AHJ has an imminent concern for public safety. An AHJ is required, when possible, to give written notification to the serving utility, as well as to the owner and occupant, of a decision to disconnect utilities. When utilities are to be disconnected, the AHJ must immediately notify the owner or occupant and the utility supplier that such services have been disconnected at a certain time and date. All concerned must be cautious to avoid damage to structure, contents or equipment (including freezing or flooding in certain areas) due to the loss of energy supply. The decision to interrupt utilities is critical even when buildings are burning because shutting off fuel or electrical supplies will generally interrupt lighting, controls, fans, pumps and emergency water supplies; such equipment and services are often vital to firefighting.

106.6 Authority to Condemn. Where the Authority Having Jurisdiction ascertains that a mechanical system or portion thereof, regulated by this code, has become hazardous to life, health, or property, or has become insanitary, the Authority Having Jurisdiction shall order in writing that such mechanical system either be removed or placed in a safe or sanitary condition. The order shall fix a reasonable time limit for compliance. No person shall use or maintain a defective mechanical system after receiving such notice.

Where such mechanical system is to be disconnected, written notice shall be given. In cases of immediate danger to life or property, such disconnection shall be permitted to be made immediately without such notice.

When equipment or a portion thereof regulated by the code has become hazardous to life, health or property, a written order to have it removed or restored to a safe and sanitary working condition is required. The written order should always contain a time limit for compliance (24, 48 or 72 hours, for example). If a time limit is not stipulated, there is a tendency to operate the system or equipment even with the written notice attached to it. Defective equipment must always be checked to ensure that it has either been repaired or removed by the expiration date of the notice. An accident resulting from the operation of defective equipment can be more than a personal tragedy; it is a failure to protect the public health and safety and is often the beginning of lengthy litigation. The building department must be certain that unsafe equipment is not operated. The person who operates equipment may not be aware of the consequences of a violation or not anticipate the damage that may result from operation of defective equipment. Some defects are obvious (leaking piping, slipping belts, defective safety controls, etc.). Many mechanical failures or suspected defects, such as spillage of flue gases at draft hoods, cracked heat exchangers, defective aqua stats, defective thermostats or limit switches or low-water or high-temperature and pressure indicators, remain undetected by unskilled persons but nevertheless constitute a major hazard.

The AHJ should determine that disconnected equipment is left in a safe condition. If fuel-gas piping is disconnected from an appliance, the piping must be securely capped or plugged.

Proper electrical lockout and tagging policies shall be observed. Pumps, piping, boilers, heaters and cooling systems must be drained to avoid corrosion or freezing. The ends of all piping systems must be plugged or capped whenever equipment is removed or disconnected. Closing shutoff valves is not adequate as valves may be opened through ignorance or vandalism. Provisions must be made to avoid the entry of foreign materials, including insects, into piping and equipment.

When the AHJ is aware that equipment on a premises is hazardous to life, health or property, it can be ordered removed or restored to a safe condition. The code provisions for condemnation of equipment may be applied to specific components of a mechanical system.

107.0 Board of Appeals.

107.1 General. In order to hear and decide appeals of orders, decisions, or determinations made by the Authority Having Jurisdiction relative to the application and interpretations of this code, there shall be and is hereby created a Board of Appeals consisting of members who are qualified by experience and training to pass upon matters pertaining to mechanical system design, construction, and maintenance and the public health aspects of mechanical systems and who are not employees of the jurisdiction. The Authority Having Jurisdiction shall be an ex-officio member and shall act as secretary to said board but shall have no vote upon a matter before the board. The Board of Appeals shall be appointed by the governing body and shall hold office at its pleasure. The board shall adopt rules of procedure for conducting its business and shall render decisions and findings in writing to the appellant with a duplicate copy to the Authority Having Jurisdiction.

The UMC intends that the Board of Appeals has very limited authority to hear and decide appeals of orders and decisions of the AHJ relative to the application and interpretations of the code. Moreover, the code now specifically limits the authority of the board relative to the administrative provisions of the code and does not permit waivers of the code requirements. Any broader authority to be granted to the Board of Appeals must be granted in the adoption ordinance by a modification of this section.

107.2 Limitations of Authority. The Board of Appeals shall have no authority relative to interpretation of the administrative provisions of this code, nor shall the board be empowered to waive requirements of this code.

TABLE 104.5
MECHANICAL PERMIT FEES

Permit Issuance

1. For the issuance of each permit .... ______[1]
2. For issuing each supplemental permit for which the original permit has not expired or been canceled or finalized .... ______[1]

Unit Fee Schedule

1. Furnaces:
   - For the installation or relocation of each forced-air or gravity-type furnace or burner, including ducts and vents attached to such appliance, not exceeding 100 000 British thermal units per hour (Btu/h) .... ______[1]
   - For the installation or relocation of each forced-air or gravity-type furnace or burner, including ducts and vents attached to such appliance, exceeding 100 000 Btu/h .... ______[1]
   - For the installation or relocation of each floor furnace, including vent .... ______[1]
   - For the installation or relocation of each suspended heater, recessed wall heater, or floor-mounted unit heater .... ______[1]
2. Appliance Vents:
   - For the installation, relocation, or replacement of each appliance vent installed and not included in an appliance permit .... ______[1]
3. Repairs or Additions:
   - For the repair of, alteration of, or addition to each heating appliance, refrigeration unit, cooling unit, absorption unit, or each heating, cooling, absorption, or evaporative cooling, system including installation of controls regulated by this code .... ______[1]
4. Boilers, Compressors, and Absorption Systems:
   - For the installation or relocation of each boiler or compressor, not exceeding 3 horsepower (hp), or each absorption system not exceeding 100 000 Btu/h .... ______[1]
   - For the installation or relocation of each boiler or compressor exceeding 3 hp, not exceeding 15 hp, or each absorption system exceeding 100 000 Btu/h and including 500 000 Btu/h .... ______[1]
   - For the installation or relocation of each boiler or compressor exceeding 15 hp, not exceeding 30 hp, or each absorption system exceeding 500 000 Btu/h, not exceeding 1 000 000 Btu/h .... ______[1]
   - For the installation or relocation of each boiler or compressor exceeding 30 hp, not exceeding 50 hp, or for each absorption system exceeding 1 000 000 Btu/h, not exceeding 1 750 000 Btu/h .... ______[1]
   - For the installation or relocation of each boiler or compressor exceeding 50 hp, or each absorption system exceeding 1 750 000 Btu/h .... ______[1]
5. Air Handlers:
   - For each air-handling unit not exceeding 10 000 cubic feet per minute (CFM), including ducts attached thereto .... ______[1, 2]
6. Evaporative Coolers:
   - For each air-handling unit exceeding 10 000 CFM .... ______[1]
   - For each evaporative cooler other than portable type .... ______[1]
7. Ventilation and Exhaust:
   - For each ventilation fan connected to a single duct .... ______[1]
   - For each ventilation system that is not a portion of a heating or air-conditioning system authorized by a permit .... ______[1]
   - For the installation of each hood that is served by mechanical exhaust, including the ducts for such hood .... ______[1]
8. Incinerators:
   - For the installation or relocation of each domestic-type incinerator .... ______[1]
   - For the installation or relocation of each commercial or industrial-type incinerator .... ______[1]

TABLE 104.5 (continued)
MECHANICAL PERMIT FEES

9. Miscellaneous:
   For each appliance or piece of equipment regulated by this code, but not classed in other appliance categories, or for which no other fee is listed in this table ______[1]
10. Fuel Gas Piping:
   Where Chapter 13 or Appendix B is applicable (See Section 101.2), permit fees for fuel-gas piping shall be as follows:
   For each gas piping system of one to five outlets ______[1]
   For each additional gas piping system, per outlet ______[1]
11. Process Piping:
   For each hazardous process piping system (HPP) of one to four outlets ______[1]
   For each HPP piping system of five or more outlets, per outlet ______[1]
   For each nonhazardous process piping system (NPP) of one to four outlets ______[1]
   For each NPP piping system of five or more outlets, per outlet ______[1]

Other Inspections and Fees

1. Inspections outside of normal business hours, per hour (minimum charge – 2 hours) ______[1]
2. Reinspection fees assessed under provisions of Section 105.2.6, per inspection ______[1]
3. Inspections for which no fee is specifically indicated, per hour (minimum charge – ½ hour) ______[1]
4. Additional plan review required by changes, additions, or revisions to plans or to plans for which an initial review has been completed, per hour (minimum charge – ½ hour) ______[1]

For SI units: 1000 British thermal units per hour = 0.293 kW, 1 horsepower = 0.746 kW, 1 cubic foot per minute = 0.00047 m^3/s

Notes:

[1] Jurisdiction will indicate their fees here.

[2] This fee shall not apply to an air-handling unit that is a portion of a factory-assembled appliance, cooling unit, evaporative cooler, or absorption unit for which a permit is required elsewhere in this code.

CHAPTER 2
DEFINITIONS

201.0 General.

201.1 Applicability. For the purpose of this code, the following terms have the meanings indicated in this chapter.

No attempt is made to define ordinary words, which are used in accordance with their established dictionary meanings, except where a word has been used loosely, and it is necessary to define its meaning as used in this code to avoid misunderstanding.

This chapter provides definitions of terms that are applicable specifically to the code and may not have an appropriate dictionary meaning. Most of these definitions are very specific and have a meaning that must be carefully considered when interpreting the intent of the code.

202.0 Definition of Terms.

202.1 General. The definitions of terms are arranged alphabetically according to the first word of the term.

203.0 – A –

Absorption Unit. An absorption refrigeration system that has been factory-assembled and tested prior to its installation.

Accepted Engineering Practice. That which conforms to technical or scientific-based principles, test, or standards that are accepted by the engineering profession.

Access Panel. A closure device used to cover an opening into a duct, an enclosure, equipment, or an appurtenance. [NFPA 96:3.3.1]

Accessible. Where applied to a device, appliance, or equipment, "accessible" means having access thereto, but which first may require the removal of an access panel, door, or similar obstruction.

Accessible, Readily. Having a direct access without the necessity of removing a panel, door, or similar obstruction.

Air, Class 1. Air with low contaminant concentration, low sensory-irritation intensity, and inoffensive odor. [ASHRAE 62.1:5.16.1]

Air, Class 2. Air with moderate contaminant concentration, mild sensory-irritation intensity, or mildly offensive odors. Class 2 air also includes air that is not necessarily harmful or objectionable, but that is inappropriate for transfer or recirculation to spaces used for different purposes. [ASHRAE 62.1:5.16.1]

Air, Class 3. Air with significant contaminant concentration, significant sensory-irritation intensity, or offensive odor. [ASHRAE 62.1:5.16.1]

Air, Class 4. Air with highly objectionable fumes or gases or with potentially dangerous particles, bioaerosols, or gases, at concentrations high enough to be considered harmful. [ASHRAE 62.1:5.16.1]

Air, Combustion. See Combustion Air.

Air, Conditioned. Air that has been treated to achieve a desired level of temperature, humidity, or cleanliness.

Air, Dilution. Air that enters a draft hood or draft regulator and mixes with the flue gases. [NFPA 54:3.3.2.2]

Air, Exhaust. Air being removed from any space or piece of equipment and conveyed directly to the atmosphere by means of openings or ducts.

Air, Makeup. Air that is provided to replace air being exhausted.

Air, Outside. Air from outside the building intentionally conveyed by openings or ducts to rooms or to conditioning equipment.

Air, Return. Air from the conditioned area that is returned to the conditioning equipment for reconditioning.

Air, Supply. Air being conveyed to a conditioned area through ducts or plenums from a heat exchanger of a heating, cooling, absorption, or evaporative cooling system.

Air Dispersion Systems. Materials intended for use in air handling systems in exposed locations operating under positive pressure.

Air-Handling Unit. A blower or fan used for the purpose of distributing supply air to a room, space, or area.

See **Figure 203.0**.

FIGURE 203.0
AIR HANDLING UNIT

Air Intakes. An opening in a building's envelope whose purpose is to allow outside air to be drawn into the structure to replace inside air that is removed by exhaust systems or to improve the quality of the inside air by providing a source of air having a lower concentration of odors, suspended particles, or heating content. [NFPA 96:3.3.2]

Air-Moving System. A system designed to provide heating, cooling, or ventilation in which one or more air-handling units are used to supply air to a common space or are drawing air from a common plenum or space.

Air Pollution Control Devices. Equipment and devices used for the purpose of cleaning air passing through them or by them in such a manner as to reduce or remove the impurities contained therein. [NFPA 96:3.3.3]

Anodeless Riser. An assembly of steel-cased plastic pipe used to make the transition between plastic piping installed underground and metallic piping installed aboveground. [NFPA 54:3.3.4]

Appliance. A device that utilizes an energy source to produce light, heat, power, refrigeration, air conditioning, or compressed fuel gas. This definition also shall include a vented decorative appliance.

Appliance, Fan-Assisted Combustion. An appliance equipped with an integral mechanical means to either draw or force products of combustion through the combustion chamber or heat exchanger. [NFPA 54:3.3.5.4]

Appliance, Low-Heat. A fuel-burning appliance that produces a continuous flue gas temperature, at the point of entrance to the flue, of not more than 1000°F (538°C).

Appliance, Medium-Heat. A fuel-burning appliance that produces a continuous flue gas temperature, at the point of entrance to the flue, of more than 1000°F (538°C) and less than 2000°F (1093°C).

Appliance Categorized Vent Diameter/Area. The minimum vent diameter/area permissible for Category I appliances to maintain a nonpositive vent static pressure when tested in accordance with nationally recognized standards. [NFPA 54:3.3.6]

Appliance Flue Outlet. The opening or openings in a cooking device where vapors, combustion gases, or both leave the cooking device. [NFPA 96:3.3.4] There might or might not be ductwork attached to this opening.

This ducting may need to meet the requirements of an appliance vent if ducted to the exterior.

Appliance Fuel Connector. An assembly of listed semi-rigid or flexible tubing and fittings to carry fuel between a fuel-piping outlet and a fuel-burning appliance.

Approved. Acceptable to the Authority Having Jurisdiction.

Approved Testing Agency. An organization primarily established for purposes of testing to approved standards and approved by the Authority Having Jurisdiction.

Appurtenance. An accessory or a subordinate part that enables the primary device to perform or improves its intended function. [NFPA 96:3.3.5]

Assembly Building. A building or a portion of a building used for the gathering together of 50 or more persons for such purposes as deliberation, education, instruction, worship, entertainment, amusement, drinking, dining, or awaiting transportation.

Authority Having Jurisdiction. The organization, office, or individual responsible for enforcing the requirements of a code or standard, or for approving equipment, materials, installations, or procedures. The Authority Having Jurisdiction shall be a federal, state, local, or other regional department or an individual such as a plumbing official, mechanical official, labor department official, health department official, building official, or others having statutory authority. In the absence of statutory authority, the Authority Having Jurisdiction may be some other responsible party. This definition shall include the Authority Having Jurisdiction's duly authorized representative.

Automatic. That which provides a function without the necessity of human intervention.

Automatic Boiler. A boiler equipped with certain controls and limit devices.

Azeotrope. A refrigerant blend containing two or more refrigerants whose equilibrium vapor and liquid phase compositions are the same at a given pressure. At this pressure, the slope of the temperature vs. composition curve equals zero, which mathematically is expressed as $(dt/dx)_p = 0$, which, in turn, implies the occurrence of a maximum, minimum, or saddle point temperature. Azeotropic blends exhibit some segregation of components at other conditions. The extent of the segregation depends on the particular azeotrope and the application. [ASHRAE 34:3]

204.0 – B –

Baffle Plate. An object placed in or near an appliance to change the direction or retard the flow of air, air-fuel mixtures, or flue gases.

Boiler. A closed vessel used for heating water or liquid, or for generating steam or vapor by direct application of heat from combustible fuels or electricity.

Boiler, High-Pressure. A boiler for generating steam at gauge pressures in excess of 15 psi (103 kPa), or for heating water to a temperature in excess of 250°F (121°C) or at a gauge pressure in excess of 160 psi (gauge pressure of 1103 kPa). [NFPA 211:3.3.14.2]

See **Figure 204.0**.

FIGURE 204.0
HIGH PRESSURE BOILER

Boiler Room. A room where boilers are installed.

Bonding Conductor or Jumper. A reliable conductor to ensure the required electrical conductivity between metal parts required to be electrically connected. [NFPA 70:100(I)]

Breathing Zone. The region within an occupiable space between planes 3 inches and 72 inches (76 mm and 1829 mm) above the floor and exceeds 2 feet (610 mm) from the walls or fixed air-conditioning equipment. [ASHRAE 62.1:3]

Breathing Zone Outdoor Airflow. The outdoor airflow required in the breathing zone of the occupiable space or spaces in a ventilation zone. [ASHRAE 62.1:6.2.2.1]

Breeching. A metal connector for medium- and high-heat appliances.

Broiler. A general term including broilers, salamanders, barbecues, and other devices cooking primarily by radiated heat, excepting toasters. [NFPA 54:3.3.15]

BTU/H. The listed maximum capacity of any appliance, absorption unit, or burner expressed in British thermal units input per hour, unless otherwise noted.

Building Code. The building code that is adopted by this jurisdiction.

Building Official. See Authority Having Jurisdiction.

Burner, Automatic Boiler. A device to convey fuel into the combustion chamber in proximity to its combustion air supply so as to permit a stable, controlled heat release compatible with the burner design and that is equipped with an ignition system to reliably ignite the entire heat release surface of the burner assembly.

205.0 – C –

CAS Number. The Chemical Abstract System registry number.

Central Heating Plant or Heating Plant. Environmental heating equipment installed in a manner to supply heat by means of ducts or pipes to areas other than the room or space in which the equipment is located.

Certified. A formally stated recognition and approval of an acceptable level of competency, acceptable to the Authority Having Jurisdiction. [NFPA 96:3.3.10]

Certified Person. A person trained and certified by the equipment manufacturer, or by a recognized organization through a formal certification program for the system to be serviced or cleaned; that is acceptable to the Authority Having Jurisdiction.

Chimney. One or more passageways, vertical or nearly so, for conveying flue or vent gases to the outdoors. [NFPA 54:3.3.18]

Chimney, Factory-Built. A chimney composed of listed factory-built components assembled in accordance with the manufacturer's installation instructions to form the completed chimney. [NFPA 54:3.3.18.2]

See Figure 205.0a.

Chimney, Masonry. A field-constructed chimney of solid masonry units, bricks, stones, listed masonry chimney units, or reinforced Portland cement concrete, lined with suitable chimney flue liners. [NFPA 54:3.3.18.3]

FIGURE 205.0A
FACTORY-BUILT CHIMNEY

Chimney, Metal. A field-constructed chimney of metal. [NFPA 54:3.3.18.4]

Chimney Classifications:

Chimney, High-Heat Appliance-Type. A factory-built, masonry, or metal chimney suitable for removing the products of combustion from fuel-burning high-heat appliances producing combustion gases in excess of 2000°F (1093°C), measured at the appliance flue outlet.

Chimney, Low-Heat Appliance-Type. A factory-built, masonry, or metal chimney suitable for removing the products of combustion from fuel-burning low-heat appliances producing combustion gases not in excess of 1000°F (538°C) under normal operating conditions, but capable of producing combustion gases of 1400°F (760°C) during intermittent forced firing for periods up to one hour. All temperatures are measured at the appliance flue outlet.

Chimney, Medium-Heat Appliance-Type. A factory-built, masonry, or metal chimney suitable for removing the products of combustion from fuel-burning medium-heat appliances producing combustion gases, not in excess of 2000°F (1093°C), measured at the appliance flue outlet.

Chimney, Residential Appliance-Type. A factory-built or masonry chimney suitable for removing products of combustion from residential-type appliances producing combustion gases not in excess of 1000°F (538°C), measured at the appliance flue outlet. Factory-built Type HT chimneys have high-temperature thermal shock resistance.

Chimney Connector. The pipe that connects a fuel-burning appliance to a chimney. [NFPA 211:3.3.48.1]

Circulators (Circulating Pump). A device that circulates liquids within a closed circuit for an intended purpose.

Classified. See Listed (Third Party Certified).

Clean(ing). For kitchen exhaust systems and cooking equipment, the act of removing grease, oil deposits, and other residue. [NFPA 96:3.3.12]

Clearly Identified. Capable of being recognized by a person of normal vision without causing uncertainty and indecisiveness about the location or operating process of the identified item. [NFPA 96:3.3.13]

Closed Combustible Construction. Combustible building construction, including walls, structural framing, roofs, roof ceilings, floors, and floor-ceiling assemblies, continuously enclosing a grease duct on four sides where one or more sides require protection in accordance with Section 507.4.

Closed Combustion Solid-Fuel-Burning Appliance. A heat-producing appliance that employs a combustion chamber that has no openings other than the flue collar, fuel-charging door, and adjustable openings provided to control the amount of combustion air that enters the combustion chamber.

Closet. See Confined Space.

Clothes Dryer. An appliance used to dry wet laundry by means of heat. [NFPA 54:3.3.19]

Clothes Dryer, Type 1. Primarily used in family living environment. May or may not be coin-operated for public use. [NFPA 54:3.3.19.1]

Clothes Dryer, Type 2. Used in business with direct intercourse of the function with the public. May or may not be operated by public or hired attendant. May or may not be coin-operated. [NFPA 54:3.3.19.2]

Coastal High Hazard Areas. An area within the flood hazard area that is subject to high-velocity wave action, and shown on a Flood Insurance Rate Map or other flood hazard map as Zone V, VO, VE, or V1-30.

Code. A standard that is an extensive compilation of provisions covering broad subject matter or that is suitable for adoption into law independently of other codes and standards.

Combustible Material. A material that, in the form in which it is used and under the conditions anticipated, will ignite and burn; a material that does not meet the definition of noncombustible. [NFPA 54:3.3.64.1]

Combustion Air. The total amount of air provided to the space that contains fuel-burning equipment. Includes air for fuel combustion, draft hood dilution, and ventilation of the equipment enclosure.

Combustion Chamber. The portion of an appliance within which combustion occurs. [NFPA 54:3.3.21]

Commercial Food Heat-Processing Equipment. Equipment used in a food establishment for heat-processing food or utensils and that produces grease vapors, steam, fumes, smoke, or odors that are required to be removed through a local exhaust ventilation system.

Compensating Hood. A hood for commercial food heat-processing equipment that has an outside-air supply with air delivered below or within the hood. Where makeup air is diffused directly into the exhaust within the hood cavity, it becomes a short-circuit hood.

Listed compensating hoods commonly are equipped with a fire damper to be installed at the penetration of the replacement air into the hood. This protects the building from increased risk of fire during an incident of a grease fire in a hood.

Compressor, Positive Displacement. A compressor in which increase in pressure is attained by changing the internal volume of the compression chamber.

Compressor, Refrigerant. A machine, with or without accessories, for compressing a refrigerant vapor.

Concealed Spaces. That portion(s) of a building behind walls, over suspended ceilings, in pipe chases, attics, and elsewhere whose size might normally range from 1¾ inch (44 mm) stud spaces to 8 foot (2438 mm) interstitial truss spaces and that might contain combustible materials such as building structural members, thermal, electrical insulation, or both, and ducting. Such spaces have sometimes been used as HVAC plenum chambers.

Condensate. The liquid phase produced by condensation of a particular gas or vapor.

Condenser. The part of the system designed to liquefy refrigerant vapor by removal of heat.

Condensing Appliance. An appliance that condenses part of the water vapor generated by the burning of hydrogen in fuels.

Condensing Unit. A mechanical refrigeration system, consisting of one or more power-driven compressors, condensers, liquid receivers where provided, and the regularly furnished accessories that have been factory assembled and tested prior to its installation.

Conditioned Space. An area, room, or space normally occupied and being heated or cooled for human comfort by any equipment.

Confined Space. A room or space having a volume less than 50 cubic feet per 1000 British thermal units per hour (Btu/h) (4.83 m^3/kW) of the aggregate input rating of all fuel-burning appliances installed in that space.

Construction Documents. Plans, specifications, written, graphic, and pictorial documents prepared or assembled for describing the design, location, and physical characteristics of the elements of a project necessary for obtaining a permit.

Continuous Enclosure. A recognized architectural or mechanical component of a building having a fire resistance rating as required for the structure and whose purpose is to enclose the vapor removal duct for its full length to its termination point outside the structure without any portion of the enclosure having a fire resistance rating less than the required value. [NFPA 96:3.3.22.1]

Continuous Pilot. A pilot that burns without turndown throughout the entire period that the boiler is in service, whether or not the main burner is firing.

See Figure 205.0b.

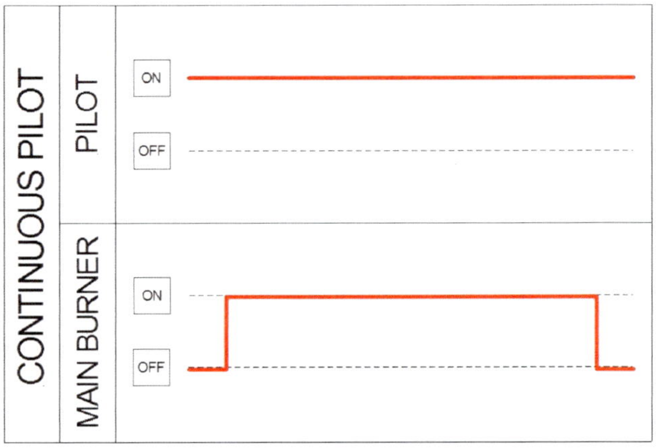

FIGURE 205.0B
CONTINUOUS PILOT

Continuous Weld. A metal-joining method that produces a product without visible interruption or variation in quality. [NFPA 96:3.3.15] For the purpose of the definition, it specifically includes the exhaust compartment of hoods and welded joints of exhaust ducts, yet specifically does not include filter support frames or appendages inside hoods.

Conversion Burner, Gas. A unit consisting of a burner

and its controls utilizing gaseous fuel for installation in an appliance originally utilizing another fuel. [NFPA 54:3.3.17.2]

Cooling. Air cooling to provide a room or space temperature of 68°F (20°C) or above.

Cooling System. All of the equipment, ducts and components, including associated refrigeration, intended or installed for the purpose of cooling air by mechanical means and discharging such air into any room or conditioned space. This definition shall not include an evaporative cooler.

Cooling Unit. A self-contained refrigeration system that has been factory assembled tested, and installed with or without conditioned air and ducts, without connecting any refrigerant-containing parts. This definition shall not include a portable cooling unit or an absorption unit.

Copper Alloy. A homogenous mixture of two or more metals in which copper is the primary component, such as brass and bronze.

Crawl Space. In a building, an area accessible by crawling, having a clearance less than human height, for access to plumbing or wiring, storage, etc.

CSST. An acronym for corrugated stainless steel tubing.

206.0 – D –

Damper. A valve or plate for controlling draft or the flow of gases, including air. [NFPA 211:3.3.52]

Ceiling Radiation Damper. A listed device installed in a ceiling membrane of a fire-resistance-rated floor-ceiling or roof-ceiling assembly to automatically limit the radiative heat transfer through an air inlet/outlet opening. [NFPA 5000:3.3.140.1]

Combination Fire/Smoke Damper. An automatic-closing metal assembly consisting of one or more louvers, blades, slats, or vanes that closes upon detection of heat or smoke as to restrict the passage of flame and smoke.

Fire Damper. An automatic-closing metal assembly consisting of one or more louvers, blades, slats, or vanes that closes upon detection of heat so as to restrict the passage of flame.

See **Figure 206.0a**.

Smoke Damper. A damper arranged to seal off airflow automatically through a part of an air duct system so as to restrict the passage of smoke.

See **Figure 206.0b**.

Volume Damper. A device that, when installed, will restrict, retard, or direct the flow of air in any duct, or the products of combustion in any heat-producing equipment, its vent connector, vent, or chimney.

See **Figure 206.0c**.

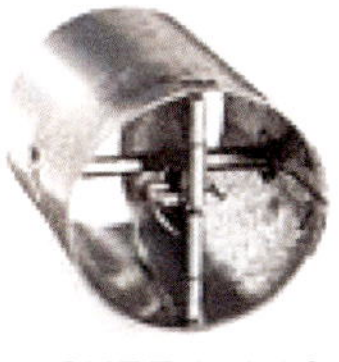

FIGURE 206.0A
FIRE DAMPER

FIGURE 206.0B
SMOKE DAMPER

FIGURE 206.0C
VOLUME DAMPER

Design Flood Elevation. The elevation of the "design flood," including wave height, relative to the datum specified on the community's legally designated flood hazard map. In areas designated as Zone AO, the design flood elevation is the elevation of the highest existing grade of the building's perimeter plus the depth number (in feet) specified on the flood hazard map. In areas designated as Zone AO where a depth number is not specified on the map, the depth number is taken as being equal to 2 feet (610 mm).

Detection Devices. Electrical, pneumatic, thermal, mechanical, or optical sensing instruments, or subcomponents of such instruments, whose purpose is to cause an automatic action upon the occurrence of some preselected event. [NFPA 96:3.3.17] In the context of this document, the event in question could be excessive temperature or flame, and the action could be the operation of a fire-extinguishing system.

Dips. Depression or cup like places in horizontal duct runs in which liquids could accumulate.

Direct Gas-Fired Nonrecirculating Industrial Air Heater. A nonrecirculating industrial air heater in which all the products of combustion generated by the appliance are released into the outdoor airstream being heated. [NFPA 54:3.3.56.1]

Direct Gas-Fired Recirculating Industrial Air Heater. An air recirculating heater in which all of the products of combustion generated by the appliance are released into the airstream being heated. [NFPA 54:3.3.56.2]

Direct Vent Appliances. Appliances that are constructed and installed so that all air for combustion is derived directly from the outdoors and all flue gases are discharged to the outdoors. [NFPA 54:3.3.5.3]

Discharge. The final portion of a duct or pipe where the product being conveyed is emptied or released from confinement; the termination point of the pipe or duct. [NFPA 96:3.3.18]

Discrete Products in Plenums. Individual, distinct products which are non-continuous such as pipe hangers, duct registers, duct fittings, and duct straps.

District Heating Plant. A power boiler plant designed to distribute hot water or steam to users located off the premises.

Draft Hood. A nonadjustable device built into an appliance, or made a part of the vent connector from an appliance, that is designed to:

(1) Provide for the ready escape of the flue gases from the appliance in the event of no draft, backdraft, or stoppage beyond the draft hood.

(2) Prevent a backdraft from entering the appliance.

(3) Neutralize the effect of stack action of the chimney or gas vent upon the operation of the appliance. [NFPA 54:3.3.31]

See **Figure 206.0d**.

Duct. A tube or conduit for transmission of air, fumes, vapors, or dust. This definition shall not include:

(1) A vent, vent connector, or chimney connector.

(2) A tube or conduit wherein the pressure of the air exceeds 1 psi (7 kPa).

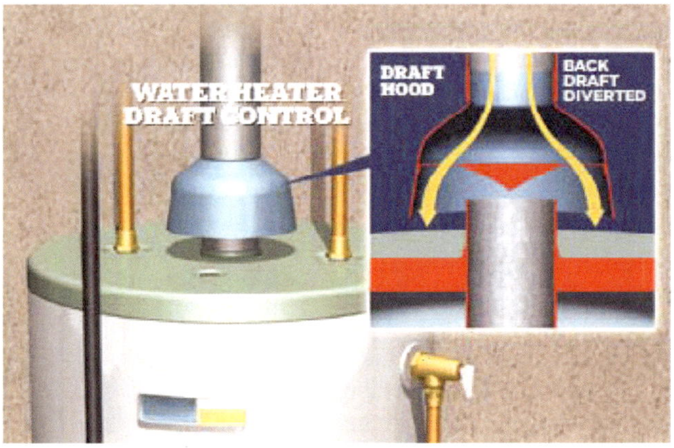

FIGURE 206.0D
DRAFT HOOD

(3) The air passages of listed self-contained systems.

Duct Furnace. A furnace normally installed in distribution ducts of air-conditioning systems to supply warm air for heating. This definition applies only to an appliance that, for air circulation, depends on a blower not furnished as part of the furnace. [NFPA 54:3.3.45.3]

See **Figure 206.0e**.

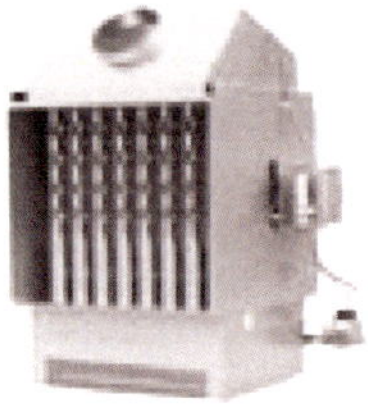

FIGURE 206.0E
DUCT FURNACE

Duct System. A continuous passageway for the transmission of air and vapors that, in addition to the containment components themselves, might include duct fittings, dampers, plenums, other items, and air-handling equipment. [NFPA 96:3.3.20]

Ductless Mini-Split System. A heating and cooling equipment that includes one or multiple indoor evaporator, air handler, or both units, an outdoor condensing unit that is connected by refrigerant piping, and electrical wiring. A ductless mini-split system is capable of cooling or heating one or more rooms without the use of traditional ductwork.

Dwelling. A building or portion thereof that contains not more than two dwelling units.

Dwelling Unit. A building or portion thereof that contains living facilities, including provisions for sleeping, eating, cooking, and sanitation, as required by this code, for not more than one family.

207.0 – E –

Easily Accessible. See Accessible, Readily.

Effective Ground-Fault Current Path. An intentionally constructed, low impedance electrically conductive path designed and intended to carry current under ground-fault conditions from the point of a ground fault on a wiring system to the electrical supply source and that facilitates the operation of the overcurrent protective device or ground-fault detectors. [NFPA 54:3.3.34]

Electric Duct Heaters. A heater located in the airstream of a forced-air system where the air-moving unit is not provided as an integral part of the equipment.

Electric Heating Appliance. A device that produces heat energy to create a warm environment by the application of electric power to resistance elements, refrigerant compressors, or dissimilar material junctions.

Electrical Code. The National Electrical Code promulgated by the National Fire Protection Association, as adopted by this jurisdiction.

Emergency Alarm System. A system intended to provide notification and warning of abnormal conditions and summon appropriate aid.

Emergency Control Station. An approved location on the premises where signals from emergency equipment are received.

Environmental Air Duct. Ducting used for conveying air at temperatures not exceeding 250°F (121°C) to or from occupied areas of any occupancy through other than heating or air-conditioning systems, such as ventilation for human usage, domestic kitchen range exhaust, bathroom exhaust ducts, and domestic-type clothes dryer exhaust ducts.

Equipment. A general term including materials, fittings, devices, appliances, and apparatus used as part of or in connection with installations regulated by this code.

Evaporative Cooler. A device used for reducing the sensible heat of air for cooling by the process of evaporation of water into an airstream.

See **Figure 207.0**.

FIGURE 207.0
EVAPORATIVE COOLER

Evaporative Cooling System. Equipment intended or installed for the purpose of environmental cooling by an evaporative cooler from which the conditioned air is distributed through ducts or plenums to the conditioned area.

Evaporator. Part of a refrigeration system in which liquid refrigerant is vaporized to produce refrigeration.

Excess Flow Valve (EFV). A valve designed to activate when the fuel gas passing through it exceeds a prescribed flow rate. [NFPA 54:3.3.99.3]

208.0 – F –

Fabrication Area (Fab Area). An area within a Group H Occupancy semiconductor fabrication facility and related research and development areas in that there are processes involving hazardous production materials. Such areas are allowed to include ancillary rooms or areas such as dressing rooms and offices that are directly related to the fab area processes.

Factory-Built Grease Duct Enclosures. A listed factory-built grease duct system evaluated as an enclosure system for reduced clearances to combustibles and as an alternative to a duct with its fire-rated enclosure. [NFPA 96:3.3.22.2.1]

Field-Applied Grease Duct Enclosures. A listed system evaluated for reduced clearances to combustibles and as an alternative to a duct with its fire-rated enclosure. [NFPA 96:3.3.22.2.2]

Fire Code. The fire code adopted by this jurisdiction.

Fire Partition. An interior wall or partition of a building that separates two areas and serves to restrict the spread of fire but does not qualify as a fire wall.

Fire Resistance Rating. The time, in minutes or hours, that materials or assemblies have withstood a fire exposure as established in accordance with ASTM E119 or UL 263.

Fire-Resistive Construction. Construction in accordance with the requirements of the building code for the time period specified.

Fire Wall. A wall separating buildings or subdividing a building to prevent the spread of the fire and having a fire resistance rating and structural stability. [NFPA 96:3.3.26]

Fireplace Stove. A chimney-connected, solid-fuel-burning stove (appliance) having part of its fire chamber open to the room.

See **Figure 208.0a**.

FIGURE 208.0A
FIREPLACE STOVE

Flammable Vapor or Fumes. The concentration of flammable constituents in air that exceeds 25 percent of its Lower Flammability Limit (LFL).

Flood Hazard Area. The greater of the following two areas:

(1) The area within a floodplain subject to a 1 percent or greater chance of flooding in any given year.

(2) The area designated as a flood hazard area on a community's flood hazard map, or otherwise legally designated.

Floor Furnace. A completely self-contained unit furnace suspended from the floor of the space being heated, taking air for combustion from outside this space. [NFPA 54:3.3.45.5] With means for observing flames and lighting the appliance from such space.

See **Figure 208.0b**.

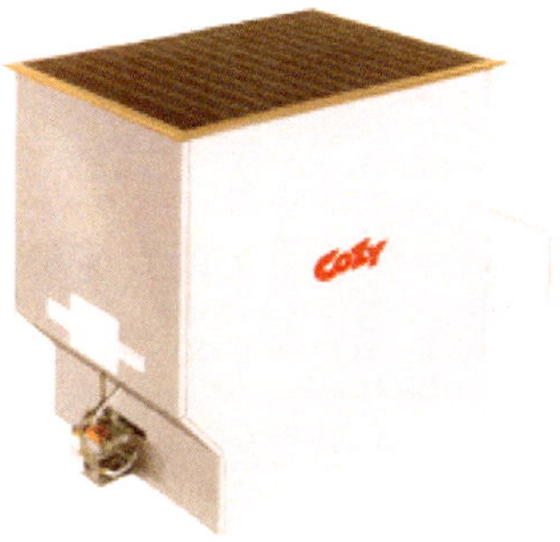

FIGURE 208.0B
FLOOR FURNACE

Forced-Air Furnace. A furnace equipped with a fan or blower that provides the primary means for circulation of air. [NFPA 54:3.3.45.6]

Downflow-Type Furnace. A forced-air-type furnace designed with airflow essentially in a vertical path, discharging air at or near the bottom of the furnace. [NFPA 211:3.3.79.2]

Enclosed Furnace. A specific heating or heating and ventilating furnace incorporating an integral total enclosure and using only outside air for combustion.

Horizontal-Type Furnace. A forced-air-type furnace designed with airflow through the furnace, essentially in a horizontal path. [NFPA 211:3.3.79.3]

Upflow-Type Furnace. A forced-air-type furnace designed with airflow essentially in a vertical path, discharging air at or near the top of the furnace. [NFPA 211:3.3.79.4]

Fractionation. A change in composition of a blend by preferential evaporation of the more volatile component or condensation of the less volatile component.

Fuel Gas. Natural, manufactured, liquefied petroleum, or a mixture of these.

Fume Incinerators. Devices utilizing intense heat or fire to break down, oxidize, or both vapors and odors contained in gases or air being exhausted into the atmosphere. [NFPA 96:3.3.27]

Furnace, Central. A self-contained appliance for heating air by transfer of heat of combustion through metal to the air and designed to supply heated air through ducts to spaces remote from or adjacent to the appliance location. [NFPA 54:3.3.45.1]

See **Figure 208.0c**.

Fusible Link. A form of fixed-temperature heat-detecting device sometimes employed to restrain the operation of an electrical or mechanical control until its designed temperature is reached. [NFPA 96:3.3.28] Such devices are to be replaced following each operation.

Fusible Plug. A device arranged to relieve pressure by operation of a fusible member at a predetermined temperature.

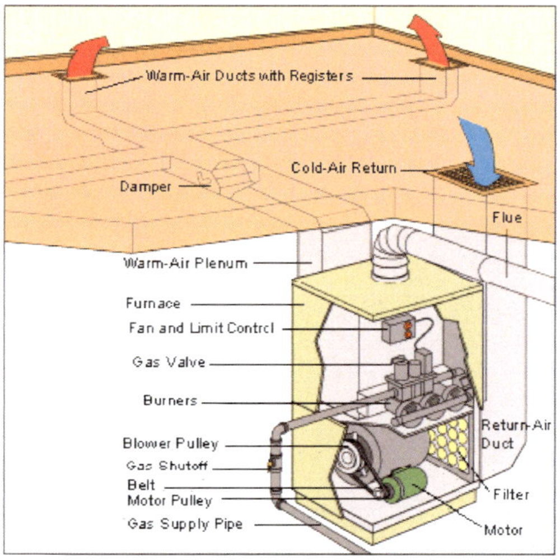

FIGURE 208.0C
CENTRAL FURNACE

209.0 – G –

Galvanized Steel. A steel that has been coated with a thin layer of zinc for corrosion protection.

Gas Convenience Outlet. A permanently mounted, hand-operated device providing a means for connecting and disconnecting an appliance or an appliance connector to the gas supply piping. [NFPA 54:3.3.48]

Gas Piping. An installation of pipe, valves, or fittings that are used to convey fuel gas, installed on any premises or in a building, but shall not include:

(1) A portion of the service piping.

(2) An approved piping connection 6 feet (1829 mm) or less in length between an existing gas outlet and a gas appliance in the same room with the outlet.

Gas Piping System. An arrangement of gas piping or regulators after the point of delivery and each arrangement of gas piping serving a building, structure, or premises, whether individually metered or not.

Generator. A device equipped with a means of heating used in an absorption system to drive refrigerant out of solution.

Gravity Heating System. A heating system consisting of a gravity-type warm air furnace, together with all air ducts or pipes and accessory apparatus installed in connection therewith.

Gravity-Type Floor Furnace. A floor furnace depending primarily on circulation of air by gravity. This classification also includes floor furnaces equipped with booster-type fans that do not materially restrict free circulation of air by gravity flow when such fans are not in operation. [NFPA 211:3.3.79.12.2]

Grease. Rendered animal fat, vegetable shortening, and other such oily matter used for the purposes of and resulting from cooking, preparing foods, or both. [NFPA 96:3.3.29] Grease might be liberated and entrained with exhaust air or might be visible as a liquid or solid.

Grease Ducts. A containment system for the transportation of air and grease vapors that is designed and installed to reduce the possibility of the accumulation of combustible condensation and the occurrence of damage if a fire occurs within the system. [NFPA 96:3.3.20.2]

Grease Filter. A removable component of the grease removal system designed to capture grease and direct it to a safe collection point. [NFPA 96:3.3.24.1]

Grease Filter, Mesh-Type. A filter construction consisting of a net made from intersecting strands with a space between each strand. [NFPA 96:3.3.24.2]

Grease Removal Devices. A system of components designed and intended to process vapors, gases, or air as it is drawn through such devices by collecting the airborne grease particles and concentrating them for further action at some future time, leaving the exiting air with a lower amount of combustible matter.

Greasetight. Constructed and performing in such a manner as not to permit the passage of grease under normal cooking conditions. [NFPA 96:3.3.31]

Ground-Source Heat Pump. A term that is applied to a variety of systems that use the ground, groundwater, or surface water as a heat source and sink. The general terms include ground-coupled (GCHP), groundwater (GWHP), and surface-water (SWHP) heat pumps. Many parallel terms exist [e.g., geothermal heat pumps (GHP), geo-exchange, and ground-source (GS) systems] and are used to meet a variety of marketing or institutional needs.

Grounding Electrode. A conducting object through which a direct connection to earth is established. [NFPA 70:100(I)]

210.0 – H –

Hazardous Location. An area or space where combustible dust, ignitable fibers, flammable liquids, volatile liquids, gases, vapors, or mixtures are or may be present in the air in quantities sufficient to produce explosive or ignitable mixtures.

Hazardous Process Piping (HPP). A process material piping or tubing conveying a liquid or gas that has a degree-of-hazard rating in health, flammability, or reactivity of Class 3 or 4, as ranked by the fire code.

Heat Pump. A refrigeration system that extracts heat from one substance and transfers it to another portion of the same substance or to a second substance at a higher temperature for a beneficial purpose.

Heat (Energy) Recovery Ventilator. A device intended to remove air from buildings, replace it with outside air, and in the process transfer heat from the warmer to the colder airstreams.

Heating Degree Day. A unit, based upon temperature difference and time, used in estimating fuel consumption and specifying nominal annual heating load of a building. For any one day when the mean temperature is less than 65°F (18°C), there exist as many degree days as there is Fahrenheit degrees difference in temperature between mean temperature for the day and 65°F (18°C).

Heating Equipment. Includes warm air furnaces, warm air heaters, combustion products vents, heating air-distribution ducts and fans, and all steam and hot water piping, together with all control devices and accessories installed as part of, or in connection with, any environmental heating system or appliance regulated by this code.

Heating System. A warm air heating plant consisting of a heat exchanger enclosed in a casing, from which the heated air is distributed through ducts to various rooms and areas. A heating system includes the outside air, return air, and supply air system, and all accessory apparatus and equipment installed in connection therewith.

High Limit Control Device. An operating device installed and serving as an integral component of a deep-fat fryer that provides secondary limitation of the grease temperature by automatically disconnecting the thermal energy input when the temperature limit is exceeded. [NFPA 96:3.3.32]

High Purity Piping. A form of process piping but is usually specified for critical clean applications in the semiconductor, pharmaceutical, biotechnology, chemical, fiber optics, food, and dairy industries.

Highside. The parts of a refrigeration system subjected to approximately condenser pressure.

Hood. An air-intake device connected to a mechanical exhaust system for collecting and removing grease-laden vapors, fumes, smoke, steam, heat, or odors from commercial food heat-processing equipment.

Fixed Baffle. A listed unitary exhaust hood design where the grease removal device is a nonremovable assembly that contains an integral fire-activated water-wash fire-extinguishing system listed for this purpose. [NFPA 96-2014:3.3.33.1]

Type I. A kitchen hood for collecting and removing grease and smoke.

Type II. A general kitchen hood for collecting and removing steam, vapor, heat, or odors.

Hot-Water-Heating Boiler. A boiler having a volume exceeding 120 gallons (454 L), a heat input exceeding 200 000 Btu/h (58.6 kW), or an operating temperature exceeding 210°F (99°C) that provides hot water to be used externally to itself.

HPM Storage Room. A room used for the storage or dispensing of hazardous production material (HPM) and that is classified as a Group H, Division 1, Division 2, Division 3, Division 4, or Division 5 Occupancy.

Hydronics. Of or relating to a heating or cooling system that transfers energy by circulating a fluid through a system of pipes or tubing.

Hydronic System. Relating to, or being a system of, heating or cooling that involves the transfer of heat by a circulating fluid (such as water or vapor).

211.0 – I –

IDLH (Immediately Dangerous to Life and Health). A concentration of airborne contaminant's, normally expressed in parts per million (ppm) or milligrams per cubic meter (mg/m^3), that represents the maximum level from which one is capable of escaping within 30 minutes without escape-impairing symptoms or irreversible health effects. This level is established by the National Institute of Occupational Safety and Health (NIOSH).

Incinerator. An appliance or combustion chamber for the reduction, by burning, of rubbish, garbage, and other wastes. [NFPA 211:3.3.91]

Industrial Heating Equipment. Includes appliances, devices, or equipment used, or intended to be used, in an industrial, manufacturing, or commercial occupancy for applying heat to any material being processed, but shall not include water heaters, boilers, or portable equipment used by artisans in pursuit of a trade.

Insanitary Location. An area, space, or room where the air is unfit or undesirable for circulation to occupiable parts of a building.

Interconnected. Mutually assembled to another component in such a manner that the operation of one directly affects the other or that the contents of one specific duct system are allowed to encounter or contact the products being moved by another duct system. [NFPA 96:3.3.34]

Interlock. A device that senses a limit or off-limit condition or improper sequence of events and shuts down the offending or related piece of equipment or prevents proceeding in an improper sequence in order to prevent a hazardous condition from developing.

Intermittent Pilot. A pilot that burns during light-off and while the main burner is firing, and that is shut off with the main burner.

See Figure 211.0a.

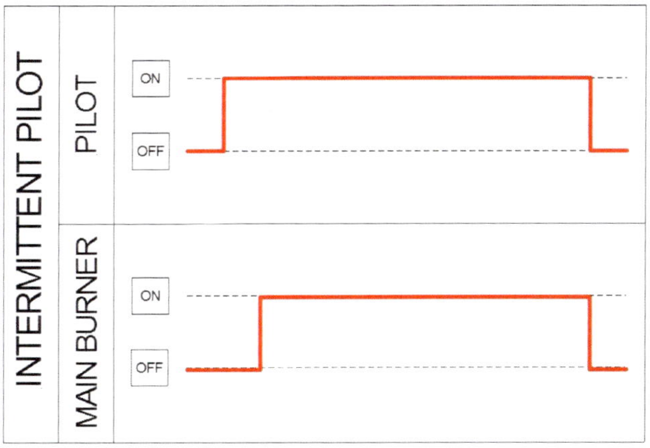

FIGURE 211.0A
INTERMITTENT PILOT

Interrupted Pilot. A pilot that burns during light-off and that is shut off during normal operation of the main burner.

See Figure 211.0b.

212.0 – J –

Joint, Brazed. A joint obtained by joining of metal parts with alloys that melt at temperatures exceeding 840°F (449°C) but less than the melting temperature of the parts being joined.

Joint, Compression. A multipiece joint with cup-shaped threaded nuts that, when tightened, compress tapered sleeves so that they form a tight joint on the periphery of the tubing they connect.

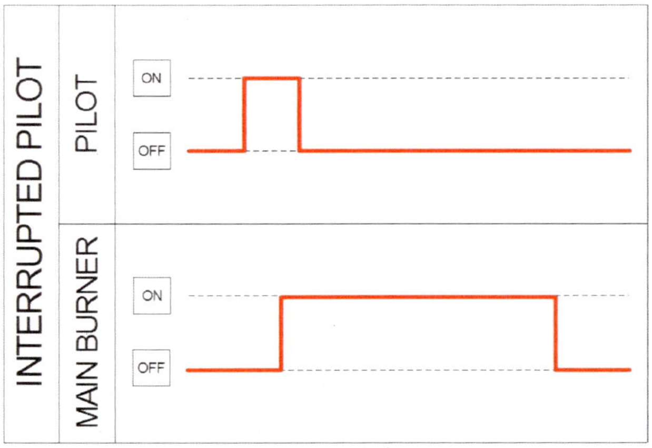

FIGURE 211.0B
INTERRUPTED PILOT

Joint, Flanged. One made by bolting together a pair of flanged ends.

Joint, Flared. A metal-to-metal compression joint in which a conical spread is made on the end of a tube that is compressed by a flare nut against a mating flare.

Joint, Mechanical. General form for gastight or liquid-tight joints obtained by the joining of parts through a positive holding mechanical construction.

Joint, Press-Connect. A permanent mechanical joint consisting of an elastomeric seal or an elastomeric seal and corrosion-resistant grip ring. The joint is made with a pressing tool and jaw or ring approved by the fitting manufacturer.

Joint, Soldered. A joint obtained by the joining of metal parts with metallic mixtures or alloys that melt at a temperature up to and including 840°F (449°C).

Joint, Welded. A gastight joint obtained by the joining of metal parts in the plastic molten state.

213.0 – K –

No definitions.

214.0 – L –

Labeled. Equipment or materials bearing a label of a listing agency (accredited conformity assessment body). See Listed (Third Party Certified).

LEL (Lower Explosive Limit). See LFL.

LFL (Lower Flammable Limit or Lower Limit of Flammability). The minimum concentration of a substance that propagates a flame through a homogeneous mixture of the substance and air under the specified test conditions. The LFL is sometimes referred to as LEL (Lower Explosive Limit). For the purposes of this definition, LFL and LEL are identical.

Limited-Combustible Material. Refers to a building construction material that does not comply with the definition of noncombustible material that, in the form in which it is used, has a potential heat value not exceeding 3500 British thermal units per pound-force (Btu/lb) (8141 kJ/kg), where tested in accordance with NFPA 259, and includes either of the following:

(1) Materials having a structural base of noncombustible material, with a surfacing not exceeding a thickness of ⅛ of an inch (3.2 mm), that has a flame-spread index not greater than 50.

(2) Materials, in the form and thickness used, having neither a flame-spread index greater than 25 nor evidence of continued progressive combustion, and of such composition that surfaces that would be exposed by cutting through the material on any plane would have neither a flame-spread index greater than 25 nor evidence of continued progressive combustion, where tested in accordance with ASTM E84.

Line Contact Installation. An installation in which a furnace is installed so that building joists, studs, or framing are contacted by the furnace jacket upon the lines formed by the intersection of the jacket sides with the top surface.

Lineset. A set of two refrigerant pipes that extends from the condenser to the evaporator (cooling coil) in direct systems, consisting of a suction line and a liquid line.

Liquefied Petroleum Gas (LP-Gas). Means and includes a material composed predominantly of any of the following hydrocarbons or mixtures of them: propane, propylene, butanes (normal butane or isobutane), and butylenes. When reference is made to liquefied petroleum gas in this code, it shall refer to liquefied petroleum gases in either the liquid or gaseous state.

Liquefied Petroleum Gas (LP-Gas) Facilities. Liquefied petroleum gas (LP-Gas) facilities include tanks, containers, container valves, regulating equipment, meters, appurtenances, or any combination thereof for the storage and supply of liquefied petroleum gas for a building, structure, or premises.

Liquid-Tight. Constructed and performing in such a manner as not to permit the passage of liquid at any temperature. [NFPA 96:3.3.35]

Listed (Third Party Certified). Equipment or materials included in a list published by a listing agency (accredited conformity assessment body) that maintains periodic inspection of current production of listed equipment or materials and whose listing states either that the equipment or material complies with approved standards or has been tested and found suitable for use in a specified manner.

Listing Agency. An agency accredited by an independent and authoritative conformity assessment body to operate a material and product listing and labeling (certification) system and that are accepted by the Authority Having Jurisdiction, which is in the business of listing or labeling. The system includes initial and ongoing product testing, a periodic inspection on current production of listed (certified) products, and that makes available a published report of such listing in which specific information is included that the material or product is in accordance with applicable standards and found safe for use in a specific manner.

Low-Pressure Hot-Water-Heating Boiler. A boiler furnishing hot water at pressures not exceeding 160 psi (1103 kPa) and at temperatures not exceeding 250°F (121°C).

Low-Pressure Steam-Heating Boiler. A boiler furnishing steam at pressures not exceeding 15 psi (103 kPa).

Low-Probability Pump. A pump that (a) is permanently sealed to prevent atmospheric release of the pumped fluid, (b) incorporates a static seal to prevent atmospheric release of the pumped fluid, or (c) incorporates not less than two sequential dynamic shaft seals and automatically shuts down upon failure of any seal to prevent atmospheric release of the pumped fluid. [ASHRAE 15:3]

Lowside. Refers to the parts of a refrigeration system subjected to approximate evaporator pressure.

215.0 – M –

Machinery. The refrigeration equipment forming a part of the refrigeration system, including, but not limited to, the following: compressors, condensers, liquid receivers, evaporators, and connecting piping.

Mechanical Ventilation. Ventilation provided by mechanically powered equipment, such as motor-driven fans and blowers, but not by devices such as wind-driven turbine ventilators and mechanically operated windows. [ASHRAE 62.1:3]

Miniature Boiler. A power boiler having an internal shell diameter of 16 inches (406 mm) or less, a gross volume of 5 cubic feet (0.14 m^3) or less, a heating surface of 20 square feet (1.86 m^2) or less (not applicable to electric boilers), and not exceeding 100 psi (689 kPa).

216.0 – N –

Natural Ventilation. Ventilation provided by thermal, wind, or diffusion effects through doors, windows, or other intentional openings in the building. [ASHRAE 62.1:3]

Noncombustible Material. As applied to building construction material, means a material that in the form in which it is used is either one of the following:

(1) A material that, in the form in which it is used and under the conditions anticipated, will not ignite, burn, support combustion, or release flammable vapors when subjected to fire or heat. Materials that are reported as passing ASTM E136 are considered noncombustible material.

(2) Material having a structural base of noncombustible material as defined in item 1 above, with a surfacing material not over ⅛ of an inch (3.2 mm) thick that has a flame-spread index not higher than 50.

Noncombustible does not apply to surface finish materials. Material required to be noncombustible for reduced clearances to flues, heating appliances, or other sources of high temperature shall refer to material in accordance with item 1 above. No material shall be classed as noncombustible that is subject to increase in combustibility or flame-spread index beyond the limits herein established, through the effects of age, moisture, or other atmospheric condition.

Nonhazardous Process Piping (NPP). Production material piping or tubing conveying a liquid or gas that is not classified as hazardous production material piping.

217.0 – O –

Occupancy. The purpose for which a building or part thereof is used or intended to be used.

Occupancy, Nontransient. Occupancy of a dwelling unit or sleeping unit for more than 30 days. [ASHRAE 62.1:3]

Occupancy Classification. Classifications are defined in the building code.

Occupational Exposure Limit (OEL). The time-weighted average (TWA) concentration for a normal 8-hour workday and a 40-hour workweek to which nearly all workers can be repeatedly exposed without adverse effect, based on the OSHA PEL, ACGIH TLV-TWA, TERA OARS-WEEL, or consistent value. [ASHRAE 34:3]

Occupiable Space. An enclosed space intended for human activities, excluding spaces that are intended to be occupied occasionally and for short periods of time, such as storage rooms, equipment rooms, and emergency exitways. [ASHRAE 62.1:3]

Open Combustible Construction. Combustible building construction, including wall, structural framing, roof, roof ceiling, floor, and floor-ceiling assemblies, adjacent to a grease duct on three or fewer sides where one or more sides require protection in accordance with Section 507.4.

218.0 – P –

Package Boiler. A class of boiler defined herein and shall be a boiler equipped and shipped complete with fuel-burning equipment, automatic controls and accessories, and mechanical draft equipment.

PE. Polyethylene.

PE-AL-PE. Polyethylene-aluminum-polyethylene.

PE-RT. Polyethylene of raised temperature.

PEL (Permissible Exposure Limit). The time-weighted average concentration [set by the U.S. Occupational Safety and Health Administration (OSHA)] for a normal 8-hour workday and a 40-hour workweek to which nearly all workers can be repeatedly exposed without adverse effect. Chemical manufacturers publish similar recommendations [e.g., acceptable exposure level (AEL), industrial exposure limit (IEL), or occupational exposure limit (OEL), depending on the company], generally for substances for which PEL has not been established. [ASHRAE 34:3] The maximum permitted time-weighted average exposures to be utilized are those published in 29 CFR 1910.1000.

PEX. Cross-linked polyethylene.

PEX-AL-PEX. Cross-linked polyethylene-aluminum-cross-linked polyethylene.

Pilot. A burner smaller than the main burner that is ignited by a spark or other independent and stable ignition source, and that provides ignition energy required to immediately light off the main burner.

Piping. The pipe or tube mains for interconnecting the various parts of a system. Piping includes pipe, tube, flanges, bolting, gaskets, valves, fittings the pressure-containing parts of other components such as expansion joints, strainers, and devices that serve such purposes as mixing, separating, snubbing, distributing, metering, or controlling flow, pipe-supporting fixtures and structural attachments.

Pitched. To be fixed or set at a desired angle or inclination. [NFPA 96:3.3.39]

Plenum. An air compartment or chamber including uninhabited crawl space areas above a ceiling or below a floor, including air spaces below raised floors of computer/data processing centers or attic spaces, to one or more ducts are connected and that forms part of either the supply-air, return-air, or exhaust-air system, other than the occupiable space being conditioned.

Plumbing Code. The Uniform Plumbing Code promulgated by the International Association of Plumbing and Mechanical Officials, as adopted by this jurisdiction.

Portable Cooling Unit. A self-contained refrigerating system, not over 3 horsepower (hp) (2.2 kW) rating that has been factory assembled and tested, installed without supply-air ducts and without connecting any refrigerant-containing parts. This definition shall not include an absorption unit.

Portable Evaporative Cooler. An evaporative cooler that discharges the conditioned air directly into the conditioned area without the use of ducts and can be readily transported from place to place without dismantling any portion thereof.

Portable Heating Appliance. A heating appliance designed for environmental heating that may have a self-contained fuel supply and is not secured or attached to a building by any means other than by a factory-installed power supply cord.

Portable Ventilating Equipment. Ventilating equipment that can be readily transported from place to place without dismantling a portion thereof and that is not connected to a duct.

Power Boiler. A boiler in which steam is generated at pressures exceeding 15 psi (103 kPa).

Power Boiler Plant. One or more power steam boilers or power hot water boilers and connecting piping and vessels within the same premises.

Power Hot Water Boiler (High Temperature Water Boiler). A boiler used for heating water or liquid to a pressure exceeding 160 psi (1103 kPa) or to a temperature exceeding 250°F (121°C).

PP. Polypropylene.

Pressure, Design. The maximum working pressure for which a specific part of a refrigeration system is designed.

Pressure, Field Test. A test performed in the field to prove system tightness.

Pressure-Imposing Element. A device or portion of the equipment used for the purpose of increasing the pressure of the refrigerant vapor.

Pressure-Limiting Device. A pressure-responsive mechanism designed to automatically stop the operation of the pressure-imposing element at a predetermined pressure.

Pressure-Relief Device. A pressure-actuated valve or rupture member or fusible plug designed to automatically relieve excessive pressure.

Pressure Test. The minimum gauge pressure to which a specific system component is subjected under test condition.

Pressure Vessel (Unfired). A closed container, having a nominal internal diameter exceeding 6 inches (152 mm) and a volume exceeding 1½ cubic feet (0.04 m^3), for liquids, gases, vapors subjected to pressures exceeding 15 psi (103 kPa), or steam under a pressure.

Pressure Vessel, Refrigerant. A refrigerant-containing receptacle that is a portion of a refrigeration system, but shall not include evaporators, headers, or piping of certain limited size and capacity.

Process Piping. Piping or tubing that conveys liquid or gas, which is used directly in research, laboratory, or production processes.

Product-Conveying Duct. Ducting used for conveying solid particulates, such as refuse, dust, fumes, and smoke; liquid particulate matter, such as spray residue, mists, and fogs; vapors, such as vapors from flammable or corrosive liquids; noxious and toxic gases; and air at temperatures exceeding 250°F (121°C).

Environmental air ducts are typically installed in structures intended for human occupancy. Product-conveying duct systems are usually installed in industrial and commercial occupancies.

Purge. The acceptable method of scavenging the combustion chamber, boiler passes, and breeching to remove combustible gases.

PVC. Polyvinyl chloride.

219.0 – Q –

Qualified. A competent and capable person or company that has met the requirements and training for a given field acceptable to the Authority Having Jurisdiction.

Quick-Disconnect Device, Fuel Gas. A hand-operated device that provides a means for connecting and disconnecting an appliance or an appliance connector to a gas supply and that is equipped with an automatic means to shut off the gas supply when the device is disconnected. [NFPA 54:3.3.28.3]

220.0 – R –

Radiant Room Heater. A room heater designed to transfer heat primarily by direct radiation. [NFPA 211:3.3.88.2.2]

Receiver, Liquid. A vessel permanently connected to a refrigeration system by inlet and outlet pipes for storage of liquid.

Recirculating Systems. Systems for control of smoke or grease-laden vapors from commercial cooking equipment that do not exhaust to the outside. [NFPA 96:3.3.41]

Reclaimed Refrigerants. Refrigerants reprocessed to the same specifications as new refrigerants by any means, including distillation. Such refrigerants have been chemically analyzed to verify that those specifications have been met. [ASHRAE 15:3]

Recovered Refrigerants. Refrigerants removed from a system in any condition without necessarily testing or processing them. [ASHRAE 15:3]

Recycled Refrigerants. Refrigerants for which contaminants have been reduced by oil separation, removal of noncondensible gases, and single or multiple passes through filter driers or other devices that reduce moisture, acidity, and particulate matter. [ASHRAE 15:3]

Refrigerant Designation. The unique identifying alphanumeric value assigned to an individual refrigerant.

Refrigerant Safety Classifications. Made up of a letter (A or B), that indicates the toxicity class, followed by a number (1, 2, or 3), that indicates the flammability class. Refrigerant blends are similarly classified, based on the compositions at their worst cases of fractionation, as separately determined for toxicity and flammability. In some cases, the worst case of fractionation is the original formulation.

Flammability Classification. Refrigerants shall be classified for flammability in accordance with one of the following:

Class 1. Refrigerants that do not show flame propagation where tested in air at 14.7 pound-force per square inch absolute (psia) (101 kPa) and 140°F (60°C).

Class 2. Refrigerants having a lower flammability limit (LFL) of more than 0.00625 pound per cubic foot (lb/ft^3) (0.10012 kg/m^3) at 140°F (60°C), 14.7 psia (101 kPa), and a heat of combustion of less than 8169 British thermal units per pound (Btu/lb) (1.8988 E+07 J/kg).

Class 3. Refrigerants that are highly flammable having a LFL of not more than 0.00625 lb/ft^3 (0.10012 kg/m^3) at 140°F (60°C) and 14.7 psia (101 kPa) or a heat of combustion not less than 8169 Btu/lb (1.8988 E+07 J/kg).

Toxicity Classification. Refrigerants shall be classified for the toxicity in accordance with one of the following:

Class A. Refrigerants have an occupational exposure limit (OEL) of not less than 400 parts per million (ppm).

Class B. Refrigerants have an OEL of less than 400 ppm.

Refrigeration Machinery Room. A room designed to house compressors and refrigerant pressure vessels.

Refrigeration Room or Space. A room or space in which an evaporator or brine coil is located for the purpose of reducing or controlling the temperature within the room or space to less than 68°F (20°C).

Refrigeration System, Absorption. A heat-operated closed refrigeration cycle in which a secondary fluid, the absorbent, absorbs a primary fluid, the refrigerant that has been vaporized in the evaporator.

Refrigeration System, Direct. A system in which the evaporator or condenser of the refrigerating system is in direct contact with the air or other substances to be cooled or heated. [ASHRAE 15:5.1.1]

Refrigeration System, Indirect. A system in which a secondary coolant cooled or heated by the refrigerating system is circulated to the air or other substance to be cooled or heated. Indirect systems are distinguished by the method of application given below. [ASHRAE 15:5.1.2]

Indirect Open Spray System. A system in which a secondary coolant is in direct contact with the air or other substance to be cooled or heated. [ASHRAE 15:5.1.2.1]

Double Indirect Open Spray System. A system in which the secondary substance for an indirect open spray system is heated or cooled by the secondary coolant circulated from a second enclosure. [ASHRAE 15:5.1.2.2]

Indirect Closed System. A system in which a secondary coolant passes through a closed circuit in the air or other substance to be cooled or heated. [ASHRAE 15:5.1.2.3]

Refrigeration System, Mechanical. A combination of interconnected refrigerant-containing parts constituting one closed refrigerant circuit in which a refrigerant is circulated for the purpose of extracting heat and in which a compressor is used for compressing the refrigerant vapor.

Refrigeration System, Self-Contained. A complete factory-assembled and tested system that is shipped in one or more sections and has no refrigerant-containing parts that are joined in the field by other than companion or block valves.

Registered Design Professional. An individual who is registered or licensed by the laws of the state to perform such design work in the jurisdiction.

Relief Valve, Vacuum. A device which automatically opens or closes for relieving a vacuum with the system, depending on whether the vacuum is above or below a predetermined value.

Removable. Capable of being transferred to another location with a limited application of effort and tools. [NFPA 96:3.3.42]

Replacement Air. See Air, Makeup.

Residential Building. A building or portion thereof designed or used for human habitation.

Riser Heat Pipe. A duct that extends at an angle of 45 degrees (0.79 rad) from the horizontal. This definition shall not include any boot connection.

Room Heater. A freestanding, nonrecessed, environmental heating appliance installed in the space being heated and not connected to ducts.

Room Heater, Unvented. An unvented, self-contained, freestanding, nonrecessed, fuel gas-burning appliance for furnishing warm air by gravity or fan circulation to the space in which installed, directly from the heater without duct connection. [NFPA 54:3.3.56.6]

Rupture Member. A pressure-relief device that operates by the rupture of a diaphragm within the device on a rise to a predetermined pressure.

221.0 – S –

Seam, Welded. See Joint, Welded.

Secondary Filtration. Fume incinerators, thermal recovery units, air pollution control devices or other filtration media installed in ducts or hoods located in the path of travel of exhaust products after the initial filtration.

Self-Contained. Having all essential working parts, except energy and control connections, so contained in a case or framework that they do not depend on appliances or fastenings outside of the machine.

Service Corridor. A fully enclosed passage used for transporting hazardous production materials and purposes other than required exiting.

Service Piping. The piping and equipment between the street gas main and the gas piping system inlet that is installed by, and is under the control and maintenance of, the serving gas supplier.

Shaft. An interior space enclosed by walls or construction extending through one or more stories or basements that connect openings in successive floors, or floors and roof, to accommodate elevators, dumbwaiters, mechanical equipment, or similar devices to transmit light or ventilation air.

Shaft Enclosure. The walls or construction forming the boundaries of a shaft.

Shall. Indicates a mandatory requirement.

Should. Indicates a recommendation or that which is advised but not required.

Single Hazard Area. Where two or more hazards can be simultaneously involved in fire by reason of their proximity, as determined by the Authority Having Jurisdiction. [NFPA 96:3.3.44]

Smoke Detector. An approved device that senses visible or invisible particles of combustion.

Solid Cooking Fuel. A solid, organic, consumable fuel such as briquettes, mesquite, hardwood, or charcoal. [NFPA 96:3.3.45]

Solid-Fuel Cooking Equipment. Cooking equipment that utilizes solid fuel. [NFPA 96:3.3.23.2] This equipment includes ovens, tandoori charcoal pots, grills, broilers, rotisseries, barbecue pits, or other type of cooking equipment that derives all or part of its heat source from the burning of solid cooking fuel.

Solvent. A substance (usually liquid) capable of dissolving or dispersing another substance; a chemical compound designed and used to convert solidified grease into a liquid or semiliquid state in order to facilitate a cleaning operation. [NFPA 96:3.3.46]

Spark Arrester. A device or method that minimizes the passage of airborne sparks and embers into a plenum, duct, and flue. [NFPA 96:3.3.48]

Standard. A document, the main text of which contains only mandatory provisions using the word "shall" to indicate requirements and that is in a form generally suitable for mandatory reference by another standard or code or for adoption into law. Nonmandatory provisions shall be located in an appendix, footnote, or fine-print note and are not to be considered a part of the requirements of a standard.

Stationary Fuel Cell Power Plant. A self-contained package or factory-matched packages that constitute an automatically operated assembly of integrated systems for generating useful electrical energy and recoverable energy that is permanently connected and fixed in place.

Steam-Heating Boiler. A boiler operated at pressures not exceeding 15 psi (103 kPa) for steam.

Strength, Ultimate. The highest stress level that the component can tolerate without rupture.

System Outdoor Airflow. The rate of outdoor airflow required at the ventilation system outdoor air intake.

222.0 – T –

Termination, Duct. The final or intended end portion of a duct system that is designed and functions to fulfill the obligations of the system in a satisfactory manner. [NFPA 96:3.3.19]

Thermal Recovery Unit. A device or series of devices whose purpose is to reclaim only the heat content of air, vapors, gases, or fluids that are being expelled through the exhaust system and to transfer the thermal energy so reclaimed to a location whereby a useful purpose can be served. [NFPA 96:3.3.49]

Trained. A person who has become proficient in performing a skill reliably and safely through instruction and practice/field experience acceptable to the Authority Having Jurisdiction. [NFPA 96:3.3.50]

Transition Gas Riser. A listed or approved section or sections of pipe and fittings used to convey fuel gas and installed in a gas piping system for the purpose of providing a transition from belowground to aboveground.

Trap. A cuplike or U-shaped configuration located on the inside of a duct system component where liquids can accumulate. [NFPA 96:3.3.51]

Type B Gas Vent. A factory-made gas vent listed by a nationally recognized testing agency for venting listed or approved appliances equipped to burn only gas.

See Figure 222.0a.

Type B-W Gas Vent. A factory-made gas vent listed by a nationally recognized testing agency for venting listed or approved gas-fired vented wall furnaces.

See Figure 222.0b.

FIGURE 222.0A
TYPE B GAS VENT

FIGURE 222.0B
TYPE B-W GAS VENT

Type L Gas Vent. A venting system consisting of listed vent piping and fittings for use with oil-burning appliances listed for use with Type L or with listed gas appliances.

223.0 – U –

Unit Heater. A heating appliance designed for nonresidential space heating and equipped with an integral means for circulation of air.

Unusually Tight Construction. Construction where:

(1) Walls and ceilings exposed to the outdoors have a continuous water vapor retarder with a rating of 1 perm or less with openings gasketed or sealed.

(2) Weatherstripping is on openable windows and doors.

(3) Caulking or sealants are applied to areas such as joints around window and door frames, between sole plates and floors, between wall-ceiling joints, between wall panels, and at penetrations for plumbing, electrical, and gas lines and at other openings.

Use (Material). The placing in action or making available for service by opening or connecting a container utilized for confinement of material, whether a solid, liquid, or gas.

224.0 – V –

Vacuum. A pressure less than that exerted by the atmosphere.

Valve, Pressure-Relief. A pressure-actuated valve held closed by a spring or other means and designed to automatically relieve pressure in excess of its setting.

See Figure 224.0a.

FIGURE 224.0A
PRESSURE RELIEF VALVE

Valve, Stop. A device in a piping system to shut off the flow of the fluid.

Valve, Three-Way-Type Stop. A manually operated valve with one inlet that alternately can stop flow to either of two outlets.

Valves, Companion or Block. Pairs of mating stop valves valving off sections of refrigeration systems and arranged so that these sections may be joined before opening these valves or separated after closing them.

Vent, Gas. A passageway composed of listed factory-built components assembled in accordance with the manufacturer's installation instructions for conveying vent gases from appliances or their vent connectors to the outdoors. [NFPA 54:3.3.53]

Vent Connector, Gas. That portion of a gas-venting system that connects a listed gas appliance to a gas vent and is installed within the space or area in which the appliance is located.

See Figure 224.0b.

Vent Offset. An arrangement of two or more fittings and pipe installed for the purpose of locating a vertical section of vent pipe in a different but parallel plane with respect to an adjacent section of vertical vent pipe. [NFPA 54:3.3.102]

Vented Appliance Categories.

Category I. An appliance that operates with a nonpositive vent static pressure and with a vent gas temperature that avoids excessive condensate production in the vent. [NFPA 54:3.3.5.11.1]

Category II. An appliance that operates with a nonpositive vent static pressure and with a vent gas temperature that can cause excessive condensate production in the vent. [NFPA 54:3.3.5.11.2]

Category III. An appliance that operates with a positive vent static pressure and with a vent gas temperature that avoids excessive condensate production in the vent. [NFPA 54:3.3.5.11.3]

Category IV. An appliance that operates with a positive vent static pressure and with a vent gas temperature that can cause excessive condensate production in the vent. [NFPA 54:3.3.5.11.4]

Vented Decorative Appliance. A vented appliance whose only function is providing an aesthetic effect of flames.

Vented Wall Furnace. A self-contained, vented, fuel gas-burning appliance complete with grilles or equivalent, designed for incorporation in or permanent attachment to the structure of a building and furnishing heated air, circulated by gravity or by a fan, directly into the space to be heated through openings in the casing. [NFPA 54:3.3.45.7]

See Learning Link goo.gl/oMcXDJ.

Ventilating Ceiling. A suspended ceiling containing many small apertures through which air, at low pressure, is forced downward from an overhead plenum dimensioned by the concealed space between the suspended ceiling and the floor or roof above.

Ventilation System. All of that equipment intended or installed for the purpose of supplying air to or removing air from, any room or space by mechanical means, other than equipment that is a portion of an environmental heating, cooling, absorption, or evaporative cooling system.

Venting Collar. The outlet opening of an appliance provided for connection of the vent system.

Venting System. The vent or chimney and its connectors, assembled to form a continuous open passageway from an

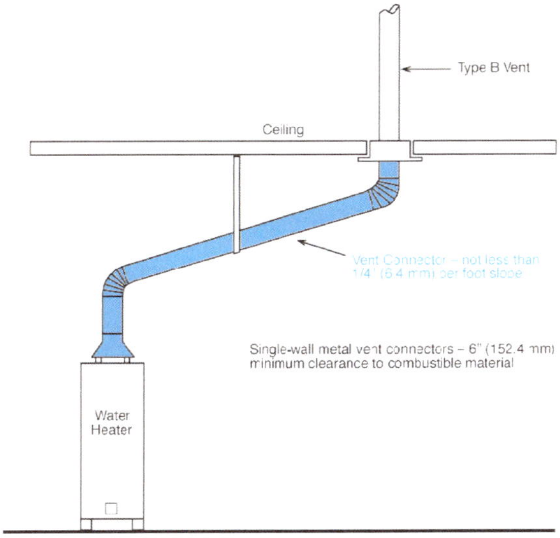

FIGURE 224.0B
GAS VENT CONNECTOR

appliance to the outdoors for the purpose of removing products of combustion. This definition also shall include a venting assembly that is an integral part of an appliance.

Venting System, Gravity-Type. A system that depends entirely on the heat from the fuel being used to provide the energy required to vent an appliance.

Venting System, Power-Type. A system that depends on a mechanical device to provide a positive draft within the venting system.

Volume, Internal Gross. The volume as determined from internal dimensions of the container, with no allowance for the volume of the internal parts.

225.0 – W –

Wall Heater. See Vented Wall Furnace.

Warm Air Furnace. An environmental heating appliance designed or arranged to discharge heated air through any duct or ducts. This definition shall not include a unit heater.

Water Heater or Hot-Water-Heating Boiler. An appliance designed primarily to supply hot water for domestic or commercial purposes and equipped with automatic controls limiting water temperature to a maximum of 210°F (99°C).

226.0 – X –

No definitions.

227.0 – Y –

No definitions.

228.0 – Z –

Zeotropic. Blends comprising multiple components of different volatilities that, when used in refrigeration cycles, change volumetric composition and saturation temperatures as they evaporate or condense at constant pressure. [ASHRAE 34:3]

See **Learning Link bit.ly/2MMJkMe**.

CHAPTER 3

GENERAL REGULATIONS

301.0 General.

301.1 Applicability. This chapter covers general requirements for heating, ventilating, air-conditioning, refrigeration, miscellaneous heat-producing, and energy-utilizing equipment or appliances. Such equipment or appliances shall comply with the requirements of this code.

301.2 Approval. Equipment or appliance shall be approved by the Authority Having Jurisdiction for safe use or comply with applicable nationally recognized standards as evidenced by the listing and label of an approved agency. A list of accepted standards is included in Chapter 17. Defective materials or parts shall be replaced in such a manner as not to invalidate an approval.

To be approved by the Authority Having Jurisdiction (AHJ) for installation in a jurisdiction, each piece of equipment or appliance must bear a label to ensure that it has been manufactured and tested in accordance with nationally recognized standards. The label provides several important pieces of information, including where the appliance could be installed (outdoors or indoors; in a large room or in an alcove, etc.); however, this should not prohibit the installation of unlisted appliances. It is important to note that some commercial and industrial equipment may not be produced in sufficient quantities to justify standardized design and acceptance testing. The AHJ should have the flexibility to accept equipment lacking standardized design if it finds that the unlisted equipment is equivalent in quality, strength, durability and safety to those prescribed by the code or to similarly listed equipment. Provisions in Section 302.2, Alternate Materials and Methods of Construction Equivalency, of the code allow the approval of alternates, which includes the approval of unlisted appliances. In addition, this chapter includes information to guide the AHJ in the approval process, such as Section 303.3 that addresses unlisted appliances and Table 303.10.1 that covers unlisted appliances' required clearances from combustibles. Lacking national or recognized standards, the AHJ is to specify the testing procedures to be used. Furthermore, and to the extent applicable, the AHJ is to apply accepted engineering principles and test protocols developed by recognized authorities and technical and scientific organizations. In addition, it is advised that an approved testing laboratory be utilized to assist the AHJ in setting and specifying the conditions of approval and also to test the equipment after its installation to ensure its proper and safe operation.

301.3 Design of Equipment. The installer or contractor shall furnish satisfactory evidence that the appliance is constructed in accordance with the requirements of this code. The permanently attached label of an approved agency shall be permitted to be accepted as such evidence.

This section requires the installer to ensure that either the appliance is listed (with a label indicating that it has been constructed and tested in accordance with nationally recognized standards) or to provide all the necessary information the AHJ would require to allow its installation. The label typically contains valuable information, such as clearances from combustibles, appliance rating and where the appliance could be installed (outdoors, indoors, in an alcove, etc.). **Figure 301.3** is an image of an appliance label listing that information. Although the label is for an old furnace, it is shown here to illustrate how the important information is usually displayed, such as the Btu/h rating of the furnace, the required clearances from combustibles, that the furnace could be installed in an alcove (closet) and that it could be installed on combustible floors. In addition, the manufacturer's installation instructions, which are attached to the equipment, could also include some of the conditions of the listing that should be followed.

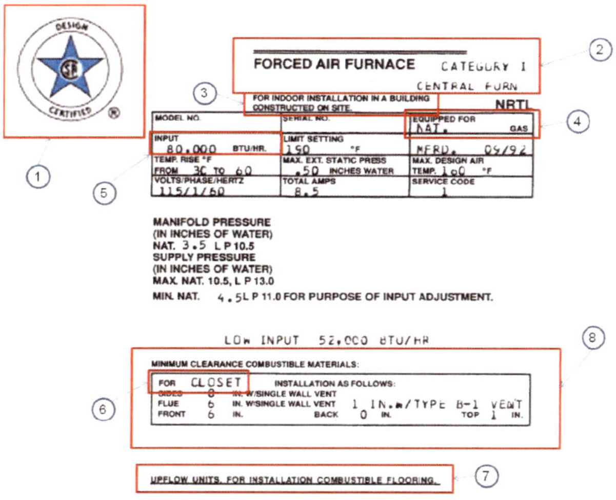

FIGURE 301.3
APPLIANCE LABEL

301.4 Electrical Connections. For equipment regulated by this code:

(1) Equipment requiring electrical connections of more than 50 volts shall have a positive means of disconnect adjacent to and in sight from the equipment served.

   Exception: Other power disconnect means shall be acceptable where in accordance with NFPA 70.

(2) A 120 volt receptacle shall be located within 25 feet (7620 mm) of the equipment for service and maintenance purposes. The receptacle outlet shall be on the supply side of the disconnect switch. The receptacle need not be located on the same level as the equipment.

(3) Electrical wiring, controls, and connections to equipment and appliances regulated by this code shall be in accordance with NFPA 70.

This section indicates that the positive means of disconnect are only required for equipment of more than 50 volts. Other methods of disconnecting equipment over 50 volts shall be allowed as long as they are in accordance NFPA 70. In addition, it requires that the receptacle needed for service and maintenance purposes is located within 25 feet of the equipment and on the supply side of the disconnect switch; it does not need to be located on the same level as the equipment. Furthermore, it requires low-voltage of 50 volts or less within a structure to be to be installed in manner to prevent physical damage.

301.5 Oil-Burning Appliances. The tank, piping, and valves for appliances burning oil shall be installed in accordance with the requirements of NFPA 31.

Currently, the most common types of appliances are those that are natural gas fired. Very special attention should be paid when encountering oil-burning appliances as they typically burn hotter than natural gas-fired appliances, requiring a different type of vent ("L" instead of "B") and have different requirements for clearances from combustibles. NFPA 31 is referenced in this section as the standard to be followed for the installation of the tank, piping and valves that characteristically accompanies such installations. NFPA 31 must be made available to the AHJ when an oil-fired appliance is proposed so its provisions could be enforced.

301.6 Personnel Protection. A metal guard shall be provided around exposed flywheels, fans, pulleys, belts, and moving machinery that are portions of a heating, ventilating, or refrigerating system.

This section provides for the safety of personnel operating machinery. It requires guards around revolving or moving parts of mechanical equipment. These are typically provided with the equipment and are usually required before an appliance or equipment is listed or approved.

302.0 Materials – Standards and Alternates.

302.1 Minimum Standards. Listed pipe, pipe fittings, appliances, appurtenances, equipment, materials, and devices used in a mechanical system shall be listed (third-party certified) by a listing agency (accredited conformity assessment body) as complying with the approved applicable recognized standards referenced in this code, and shall be free from defects. Unless otherwise provided for in this code, materials, appurtenances, or devices used or entering into the construction of mechanical systems, or parts thereof, shall be submitted to the Authority Having Jurisdiction for approval prior to being installed.

All materials used in the mechanical system must have the approval of the AHJ. This is accomplished by two methods. First, the architect or engineer will specify materials that meet or conform to an applicable standard for the use intended, which should be referenced in the code. The second method is for materials that are not made to standards referenced in the code. The architect or engineer must then submit these materials to the AHJ for approval. Sections 302.2 and 302.3 explain how these materials achieve approval from the AHJ.

Every pipe, fitting, appliance, etc., should be a good, quality product; however, this is not always the case. The material in question could be made to the proper standard and used for its intended use and still violate the requirement "... and shall be free from defects." This violation most often occurs because the item is damaged. To comply with this section of the code the materials must be in good order and installed per recognized standards and other applicable sections in the code.

302.1.1 Marking. Each length of pipe and each pipe fitting, material, and device used in a mechanical system shall have cast, stamped, or indelibly marked on it any markings required by the applicable referenced standards and listing agency, and the manufacturer's mark or name, which shall readily identify the manufacturer to the end user of the product. Where required by the approved standard that applies, the product shall be marked with the weight and the quality of the product. Materials and devices used or entering into the construction of mechanical systems, or parts thereof, shall be marked and identified in a manner satisfactory to the Authority Having Jurisdiction. Such marking shall be done by the manufacturer. Field markings shall not be acceptable.

Exception: Markings shall not be required on nipples created from cutting and threading of approved pipe.

In order for the contractor, installer, or inspector to verify if a mechanical product meets the requirements of the code, the item must be marked or labeled (see **Figure 302.1.1a**). The code does not specifically state the marking requirements, except for the manufacturer's identification. The referenced standard states the minimum marking requirements. The identification requirements typically include any of the following: the name of the manufacturer or trademark, type or model number, maximum rated pressure and temperature, serial number, nominal size, and standard designation.

FIGURE 302.1.1A
PIPE MARKING

Another type of marking on some piping products is the IAPMO product markings. The Research and Testing (R&T) Division of IAPMO tests and evaluates products to determine whether they comply with select standards and applicable codes. Each tested product is then assessed as meeting or not meeting the standards. Compliant products may qualify for the application of one of seven possible IAPMO marking labels.

By its shape or markings, an IAPMO label may denote that a product is either "listed" or "classified." Each of these labels carries considerably different implications. A product that bears the "listed" label has demonstrated compliance with the specific standard or standards to which it was tested. Listed products require no special approval or acceptance by local authorities when installed within the limits of their listing.

Additionally, a listed product is defined as compliant with code stipulations, engineering concepts, or other fundamental principles found in the Uniform Plumbing Code (UPC) or Uniform Mechanical Code (UMC). Products bearing the IAPMO labels may be safely used in the plumbing or mechanical system.

Conversely, a "classified" label represents compliance with all standards to which the product was tested but does not ensure compliance with UPC or UMC code stipulations or requirements. Classified products are published in a separate Product Directory, the objective being to avoid user confusion regarding the propriety of their application. UMC Section 103.0 allows the use of these products, but only with specific approval from the local AHJ. Products bearing the label (see **Figure 302.1.1b**) must have approval from the AHJ before being used in the plumbing or mechanical system.

The use of specialized labeling is intended to provide the consumer with guidance regarding the limits of product application. Compliance with an evaluation standard does not suggest that a product is suitable for any specific application; nor does it suggest that a product is code compliant or approved for use in a particular application. IAPMO R&T tests to standards specified by the person who submits the product for evaluation. Because all test standards are not necessarily related to code requirements or standard engineering practices, test compliance is essentially unrelated to the appropriate utilization of a particular product.

See **Learning link: What the IAPMO R&T Marks Means: http://bit.ly/396xvse**.

FIGURE 302.1.1B
IAPMO PRODUCT LABELS

302.1.2 Standards. Standards listed or referred to in this chapter or other chapters cover materials that will conform to the requirements of this code, where used in accordance with the limitations imposed in this or other chapters thereof and their listing. Where a standard covers materials of various grades, weights, quality, or configurations, the portion of the listed standard that is applicable shall be used. Design and materials for special conditions or materials not provided for herein shall be permitted to be used by special permission of the Authority Having Jurisdiction after the Authority Having Jurisdiction has been satisfied as to their adequacy. A list of mechanical standards that appear in specific sections of this code is referenced in Table 1701.1. Standards referenced in Table 1701.1 shall be applied as indicated in the applicable referenced section. A list of additional approved standards, publications, practices and guides that are not referenced in specific sections of this code appear in Table 1701.2.

Standards referenced in Table 1701.1 shall be applied as indicated in the applicable reference section. A list of additional standards, publications, and guides that are not referenced in specific sections of this code appear in Table 1701.2. The standards from Table 1701.2 shall be permitted only after they have been approved by the AHJ.

A standard contains scoping requirements, technical definitions, specifications, test methods, and guidelines for the manufacture of devices and products. A standard is considered a basis for an approved model. Their purpose is to ensure that products are made to be safe and efficient for the consumer.

Codes and standards work together to protect public health and safety and are legal requirements that are enforceable by local jurisdictions. Codes tell the user what to do, when to do it and under what circumstances to do it. Standards within in the code inform the user how to verify products that are approved, having been tested to performance requirements. The code will in turn determine the extent the standard can be practically applied. A standard may have performance specifications for materials and installation requirements. However, the standard is only applicable to the extent suggested in the text of the code. The code text takes precedence when the requirements of the standard conflict with the requirements of the code.

302.1.3 Existing Buildings. In existing buildings or premises in which mechanical installations are to be altered, repaired, or renovated, the Authority Having Jurisdiction has discretionary powers to permit deviation from the provisions of this code, provided that such proposal to deviate is first submitted for proper determination in order that health and safety requirements, as they pertain to mechanical systems, shall be observed.

302.2 Alternate Materials and Methods of Construction Equivalency. Nothing in this code is intended to prevent the use of systems, methods, or devices of equivalent or superior quality, strength, fire resistance, effectiveness, durability, and safety over those prescribed by this code. Technical documentation shall be submitted to the Authority Having Jurisdiction to demonstrate equivalency. The Authority Having Jurisdiction shall have the authority to approve or disapprove the system, method, or device for the intended purpose.

However, the exercise of this discretionary approval by the Authority Having Jurisdiction shall have no effect beyond the jurisdictional boundaries of said Authority Having Jurisdiction. An alternate material or method of construction so approved shall not be considered as in accordance with the requirements, intent, or both of this code for a purpose other than that granted by the Authority Having Jurisdiction where the submitted data does not prove equivalency.

New materials and methods of construction are constantly being developed by the mechanical industry. It is not the intent of the UMC to prohibit the use of these alternate materials and methods before they are specified in the code as long as the AHJ approves their use. It should also be noted that the approval of the AHJ in one jurisdiction does not give approval for the use of these alternates in other jurisdictions. Each jurisdiction will have to evaluate the alternate material or method for its own use.

Research reports cover many materials that will prove to be of assistance to personnel administering the UMC, particularly in fuel-gas piping and hydronic systems. A few large jurisdictions operate their own testing facilities, but, in general, medium-sized and small jurisdictions will not enjoy this advantage. Listings by agencies that have been approved by the AHJ in connection with the reference standards listed in Chapter 17 will, in most instances, provide the criteria needed to determine the acceptability of mechanical equipment. In discharging responsibility for approving equipment, it is important to recognize that the standards used by listing agencies may not always coincide with the requirements of the UMC. Thus, it is necessary to review reports of testing to be sure that the standard employed is consistent with code requirements.

302.2.1 Testing. The Authority Having Jurisdiction shall have authority to require tests, as proof of equivalency.

302.2.1.1 Tests. Tests shall be made in accordance with approved testing standards, by an approved testing agency at the expense of the applicant. In the absence of such standards, the Authority Having Jurisdiction shall have the authority to specify the test procedure.

This section provides the AHJ with discretionary authority to require tests to substantiate proof of compliance with code requirements and the intent of the code.

It is quite possible to have a good product and a substandard or inappropriate application. Inspectors can often prevent misapplication of products or equipment by discussing with an applicant the scope and the extent of the installation before any work is started. Evaluation of products by an independent agency is not normally something that can be done efficiently at a job site. The tests frequently require laboratory conditions and equipment that cannot be duplicated in the field.

The following is a list of generally accepted testing organizations and the product directories they provide:

American Gas Association Laboratories (AGA)

Directory of Certified Appliances and Accessories

Underwriters Laboratories Inc. (UL)

Gas and Oil Equipment Directory

Fire Protection Equipment Directory

Electrical Appliance and Utilization Equipment Directory

Building Materials Directory

Fire Resistance Directory

Hazardous Location Electrical Equipment Directory

Underwriters Laboratories of Canada (Issues product directories covering many of the same equipment categories as UL in the United States does.)

In addition to the above testing organizations, IAPMO provides certification for a wide variety of testing laboratories, which are usually recognized as "approved" agencies as defined in the UMC. It is important to recognize that many of these smaller testing agencies may be competent in more restricted fields of testing than are the larger nationally recognized agencies.

302.2.1.2 Request by the Authority Having Jurisdiction. The Authority Having Jurisdiction shall have the authority to require tests to be made or repeated where there is reason to believe that a material or device no longer is in accordance with the requirements on which its approval was based.

In an instance where a permitted job has existing materials or installations that no longer perform the function due to age or a history of failure, the AHJ may require additional testing or proof of compliance to a current standard.

302.3 Alternative Engineered Design. An alternative engineered design shall comply with the intent of the provisions of this code and shall provide an equivalent level of quality, strength, effectiveness, fire resistance, durability, and safety. Material, equipment, or components shall be designed and installed in accordance with the manufacturer's installation instructions.

The UMC allows the professional engineering community to design mechanical systems that may not strictly conform to the code. This may be due to new innovative methods or computer engineering that has not as yet been recognized by the code. These new systems must provide the equivalent protection of health and safety that is required from a standard mechanical system.

The professional engineer shall provide all the documentation and data to prove to the AHJ that the system meets the intent of the code. If the system is approved by the AHJ, then it must be accompanied by an inspection and testing plan designed by the engineer. The installation must meet the inspection and testing requirements of the engineer's plan and the pertinent requirements of the code.

302.3.1 Permit Application. The registered design professional shall indicate on the design documents that the mechanical system, or parts thereof, is an alternative engineered design so that it is noted on the construction permit application. The permit and permanent permit records shall indicate that an alternative engineered design was part of the approved installation.

302.3.2 Technical Data. The registered design professional shall submit sufficient technical data to substantiate the proposed alternative engineered design and to prove that the performance meets the intent of this code.

302.3.3 Design Documents. The registered design professional shall provide two complete sets of signed and sealed design documents for the alternative engineered design for submittal to the Authority Having Jurisdiction. The design documents shall include floor plans of the work. Where appropriate, the design documents shall indicate location, sizing, and loading of appurtenances, equipment, appliances, and devices.

302.3.4 Design Approval. An approval of an alternative engineered design shall be at the discretion of the Authority Having Jurisdiction. The exercise of this discretionary approval by the Authority Having Jurisdiction shall have no effect beyond the jurisdictional boundaries of said Authority Having Jurisdiction. An alternative engineered design so approved shall not be considered as in accordance with the requirements, intent, or both of this code for a purpose other than that granted by the Authority Having Jurisdiction.

302.3.5 Design Review. The Authority Having Jurisdiction shall have the authority to require testing of the alternative engineered design in accordance with Section 302.2.1, including the authority to require an independent review of the design documents by a registered design professional selected by the Authority Having Jurisdiction and at the expense of the applicant.

302.3.6 Inspection and Testing. The alternative engineered design shall be tested and inspected in accordance with the submitted testing and inspection plan and the requirements of this code.

303.0 Installation.

303.1 Listed Appliances. The installation of equipment and appliances regulated by this code shall be in accordance with the conditions of the listing, the manufacturer's installation instructions and this code. The manufacturer's installation and operating instructions shall be attached to the appliance. Clearances of listed equipment and appliances from combustible materials shall be as specified in the listing or on the rating plate.

The requirement that the installation of appliances is to conform to the conditions of the listing, manufactures instruction and this code prevents the use of a product that has a listing for a particular application from being used in one for which it has not been tested. Also, the section places the responsibility of providing the manufacturer's installation and operating instructions on the installer, requiring the instructions to be attached to the appliance. This is to assist the inspector in ensuring that the appliance is installed in accordance with the manufacturer's instructions and provides the owner with needed information to operate and maintain the appliance.

303.2 Closet or Alcove Installations. Central heating furnaces and boilers installed in closets or alcoves shall be listed for such installation. Central heating furnaces not listed for closet or alcove installation shall be installed in a room or space having a volume not less than 12 times the total volume of the furnace. Central heating boilers not listed for closet or alcove installation shall be installed in a room or space having a volume 16 times the volume of the boiler. Where the ceiling height of the room or space exceeds 8 feet (2438 mm), the volume shall be calculated on the basis of an 8 foot (2438 mm) height.

The installation clearances shall be in accordance with the appliance listing, shall not be reduced, and shall be installed in accordance with Section 904.1.

Listing of certain appliances specifies whether they could be installed in a small compartment alcove or must be installed in a large room (in comparison to the physical volume of the appliance). There are two reasons for requiring the installation to be in a large room: (1) to provide adequate space for maintenance and servicing of the equipment; and (2) to prevent excessive heat build-up in a confined space. Otherwise, the appliance must be tested and listed for installation in a closet or an alcove to ensure that these concerns have been dealt with.

To determine whether an appliance could be installed in a small room, the label must be examined to determine whether the appliance is listed for alcove or closet installation. If not, the volume of the room or compartment must be calculated (using a maximum ceiling height of 8 feet). The calculated volume must be equal to or be greater than 12 times the exterior physical volume (length x width x height) of the furnace (or 16 times if the appliance is a boiler) to determine if the room is "large in comparison to the size of appliance" so that the appliance could be installed there.

Please note that the 8-foot ceiling height limitation does not apply when determining whether a room has sufficient combustion air volume (Chapter 7) to allow the use of the interior for combustion, an issue that has been commonly misunderstood and incorrectly applied.

The section also includes important provisions that require appliances listed for alcove installations to adhere to the listed clearances, regardless of whether the enclosure is constructed of combustible or noncombustible material. Table 303.10.1 may not be used to reduce the required installation clearances in this case.

303.3 Unlisted Appliances. Except as otherwise permitted in this code, unlisted equipment and appliances shall be approved by the Authority Having Jurisdiction prior to being installed. Unlisted equipment and appliances shall be installed in accordance with the manufacturer's installation instructions and with clearances from combustible materials in accordance with Section 303.10 or Section 303.10.1.

Section 303.3 requires that all equipment that is not listed in the code be approved by the authority having jurisdiction. The manufacturer's installation instructions are to determine the standard clearances for unlisted appliances as referenced in Section 303.10. Reduced clearances for unlisted appliances are to comply with Table 303.10.1 if permitted by the manufacturer according to Section 303.10.1. If not permitted by the manufacturer, unlisted appliances may not use Table 303.10.1 to reduce the standard clearance.

Table 303.10.1 designates approved methods of protecting combustible assemblies, thus allowing the reduction of the required clearances specified in the manufacturer's installation instructions for unlisted appliances. Although Please note that reduction of clearances from combustibles (or noncombustible material) for listed appliances installed in alcoves or closets are prohibited in Section 303.2.

Note 2 requires that when using the assemblies in the table, the reduced clearance is to be measured from the appliance to the combustible material. In addition, with very few exceptions, listed assemblies require a 1-inch air gap between the protection and the combustible material. This is an important provision since it allows air to flow freely in the air gap, cooling it down.

The eight types of protection listed in the table make use of all three methods of heat transfer: conduction, convection and radiation. Installing insulating material (such as millboard) retards the rate of heat transfer conducted from the hot outer surface of the appliance to the combustibles. Installing steel sheets on the combustibles or the insulation reflects and lessens the radiated heat from the surface of the appliance. The air gap between the insulation and the combustibles required in most methods is to allow for the air to flow upward (convection) to escape near the top, and forcing cooler air to replace the escaping air, thus cooling the protective assembly. To be effective, the air gap must be ventilated at the top and bottom, or at least at the sides and bottom to allow the air to carry the heat out.

Please note that **Figure 303.10.1(1)** shows the distance the protection must extend beyond the appliance (or vent).

303.4 Anchorage of Appliances. Appliances designed to be fixed in position shall be securely fastened in place in accordance with the manufacturer's installation instructions. Supports for appliances shall be designed and constructed to sustain vertical and horizontal loads within the stress limitations specified in the building code.

Usually, the method of securing an appliance in place, if required, is provided in the manufacturer's installation instructions as one of the conditions of listing. Generally speaking, equipment regulated by the code should be securely fastened in place except for residential washers, dryers, kitchen ranges and equipment defined as Portable by Section 218.0. In Seismic Zones C, D, E and F (see the building code for definitions and location of these active seismic zones), it is suggested that a competent engineer evaluate the methods of securing the equipment in the event the manufacturer's instructions do not include guidance.

A major reason for requiring equipment to be secured in place is to prevent damage to electrical, gas or refrigerant piping connections due to vibration or shifting of equipment during a seismic event. **Figure 303.4** illustrates various methods of anchoring appliances.

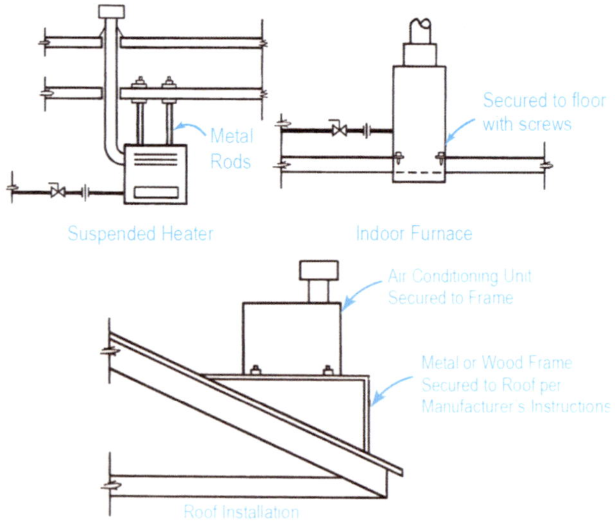

FIGURE 303.4
ANCHORING OF APPLIANCES

303.5 Movement. Movement of appliances with casters shall be limited by a restraining device installed in accordance with the connector and appliance manufacturer's installation instructions.

To protect against damaging the special gas connector that is connected to appliances on casters, this section requires a restraining device to prevent the movement of the equipment beyond the connector's length (see **Figure 303.5**).

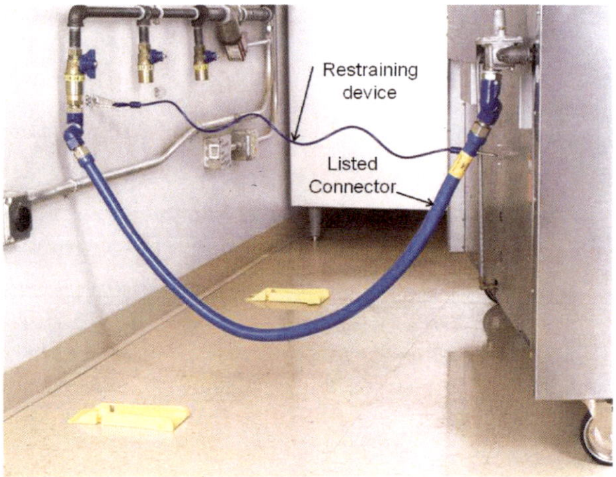

FIGURE 303.5
RESTRAINING DEVICE FOR APPLIANCES WITH CASTERS

303.6 Identification of Equipment. Where more than one heating, cooling, ventilating, or refrigerating system is installed on the roof of a building or within a building, it shall be permanently identified as to the area or space served by the equipment.

This requirement allows for inspection, servicing, repair or replacement of the correct equipment in the event of an emergency or scheduled maintenance, without interrupting neighboring spaces' equipment. When appliances are not marked, service personnel typically cycle the equipment on and off to determine which equipment to service, unless the equipment or appliance is easily identifiable. **Figure 303.6** illustrates examples of identification when more than one appliance is installed on the roof of a building or when appliances are grouped within a building.

303.7 Liquefied Petroleum Gas Facilities. Containers, container valves regulating equipment, and appurtenances for the storage and supply of liquefied petroleum gas shall be installed in accordance with NFPA 58.

NFPA 58 should be reviewed when inspecting LP-Gas installations. Chapter 13 of this code indicates that the regulations covering piping systems for LP-Gas apply to piping downstream of the final pressure regulator. Therefore, NFPA 58 should be available to the inspecting jurisdiction to enforce provisions pertaining to storage tanks, regulating valves and all other appurtenances usually installed upstream of the final pressure regulator. In addition, the fire code of the jurisdiction should be referenced for additional requirements, especially when attempting to determine the approved location of LP-Gas tank(s) and appurtenances with respect to structures or buildings on the property.

FIGURE 303.6
IDENTIFICATION OF EQUIPMENT AS TO SPACE, AREA OR UNIT SERVED

303.8 Appliances on Roofs. Appliances on roofs shall be designed or enclosed so as to withstand climatic conditions in the area in which they are installed. Where enclosures are provided, each enclosure shall permit easy entry and movement, shall be of reasonable height, and shall have at least a 30 inch (762 mm) clearance between the entire service access panel(s) of the appliance and the wall of the enclosure. [NFPA 54:9.4.1.1]

The provision of this section states that if an appliance is not listed for exterior locations and installed on a roof, it should be enclosed in an approved enclosure protecting it from the elements. The enclosure should be of dimensions to allow service of the equipment, especially at the service side (access panel) where the clearance must be at least 30 inches.

303.8.1 Load Capacity. Roofs on which appliances are to be installed shall be capable of supporting the additional load or shall be reinforced to support the additional load. [NFPA 54:9.4.1.2]

Typically, most roofs are capable of supporting small appliances (such as bathroom exhaust fans) without the need for reinforcement. However, when it is obvious that the appliance is either large or of substantial weight (such as rooftop heating and/or cooling units), the plans reviewer or inspector should require the designer to review the roof structure to ensure its adequacy for the additional imposed load. In addition, when the appliance could be seen from the public street, most jurisdictions' zoning regulations have special requirements to conceal the appliance. There, care should be exercised not to allow the installation of structures intended to conceal the appliance from interfering with the required clearances from combustibles, access to the appliance or vent termination requirements of the code.

303.8.2 Fasteners. All access locks, screws, and bolts shall be of corrosion-resistant material. [NFPA 54:9.4.1.3]

Screws and bolts that tend to corrode may not be used as fasteners for accessing equipment. They are required to be corrosion-resistant, especially when exposed to the outdoor elements or subject to a corrosive environment.

303.8.3 Installation of Appliances on Roofs. Appliances shall be installed in accordance with the manufacturer's installation instructions. [NFPA 54:9.4.2.1]

303.8.4 Edge of Roof Clearance. Appliances shall be installed on a well-drained surface of the roof. At least 6 feet (1829 mm) of clearance shall be available between any part of the appliance and the edge of a roof or similar hazard, or rigidly fixed rails, guards, parapets, or other building structures at least 42 inches (1067 mm) in height shall be provided on the exposed side. [NFPA 54:9.4.2.2]

Flat roofs are usually designed to slope to roof drains or scuppers; however, sometimes due to construction problems or other factors, certain areas of the roof are rendered relatively flat, resulting in pooling of rainwater. Care should be exercised not to allow appliance to be installed in these areas. Repeated or prolonged wetting of appliance, even those listed for exterior locations, could cause excessive corrosion, resulting in the malfunctioning of the appliance or in a much shorter appliance life.

When rooftop appliances are installed close to the roof edge, they should be placed at least 6 feet away from the edge to ensure the safety of maintenance personnel. Please note that certain building codes require 10 feet instead. If such code is adopted in the jurisdiction, the more restrictive provision (10 feet from the edge) should apply. If the appliance is closer than the required distance, railings at least 42 inches high should be constructed between the appliance and the edge of the roof (see **Figure 303.8.4**).

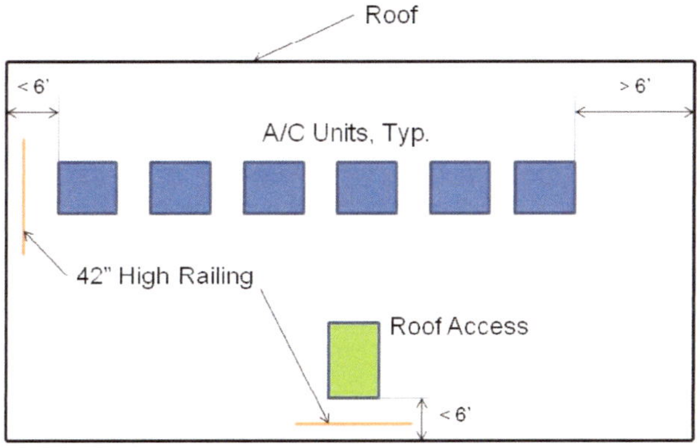

FIGURE 303.8.4
MECHANICAL EQUIPMENT AND ROOF ACCESS CLEARANCES FROM EDGE OF ROOF

303.8.4.1 Guards and Rails. Guards or rails shall be required where the following exist:

(1) The clearance between the appliance and a roof edge or open end of an equipment platform is less than 6 feet (1829 mm).

(2) The open end of the equipment platform is located more than 30 inches (762 mm) above the roof, floor, or grade below.

Where guards or rails are installed, they shall be constructed so as to prevent the passage of a 21 inch (533 mm) diameter ball, resist the imposed loading conditions, and shall extend not less than 30 inches (762 mm) beyond each side of the equipment or appliance.

Exception: Guards shall not be required where a permanent fall arrest anchorage connector system in accordance with ASSE Z359.1 is installed.

Although not specifically mentioned in this section, the railing should not obstruct the service access panel of the appliance and maintenance personnel should have at least 30 inches of working space at that side as required by Section 304.1.

303.8.5 Electrical Power. All appliances requiring an external source of electrical power for its operation shall be provided with the following:

(1) A readily accessible electrical disconnecting means within sight of the appliance that completely de-energizes the appliance.

(2) A 120-V ac grounding-type receptacle outlet on the roof adjacent to the appliance on the supply side of the disconnect switch. [NFPA 54:9.4.2.3]

The code indicates that appliance requiring electrical connections must have positive means of disconnect adjacent to or within sight of the appliance served. In addition, a 120-volt receptacle shall be located adjacent to the appliance for service and maintenance purposes. The receptacle should be wired in such a way that when the electrical power is disconnected from the appliance, the receptacle should remain energized to allow the use of diagnostic appliance and power tools. (Please note that Section 301.4 has additional requirements and should be referenced in conjunction with this section.)

303.8.6 Platform or Walkway. Where water stands on the roof at the appliance or in the passageways to the appliance, or where the roof is of a design having a water seal, a suitable platform, walkway, or both shall be provided above the water line. Such platform(s) or walkway(s) shall be located adjacent to the appliance and control panels so that the appliance can be safely serviced where water stands on the roof. [NFPA 54:9.4.2.4]

This section compliments Section 303.8.4, which requires the installation of rooftop appliances in a well-drained area. It provides for cases where the passage to either the appliance or the service area next to the appliance is subject to pooling of rainwater. In these situations, suitable raised walkways or service platforms must be provided to allow for safe maintenance of appliances and to prevent the likelihood of electric shock to service personnel standing on wet surfaces while working on energized appliances.

303.9 Avoiding Strain on Gas Piping. Appliances shall be supported and connected to the piping so as not to exert undue strain on the connections. [NFPA 54:9.1.17]

If the connections exert undue strain on the appliance it can threaten the integrity of the appliance. The forces exerted by the strain on the piping from the appliance or from the appliance piping can be eliminated by properly supporting the piping.

303.10 Clearances. Appliances and their vent connectors shall be installed with clearances from combustible material so their operation does not create a hazard to persons or property. Minimum clearances between combustible walls and the back and sides of various conventional types of appliances and their vent connectors are specified in Chapter 8 and Chapter 9. [NFPA 54:9.2.2] Where not provided in this code, listed and unlisted equipment or appliances shall be installed to maintain the required clearances for servicing and to combustible construction in accordance with the listing and the manufacturer's installation instructions.

Section 303.10 references Chapters 8 and 9 of the UMC, to set the required minimum clearances allowed for appliances and their vents. Table 303.10.1 and Table 904.2.2 shows the minimum amount of clearance from an appliance to combustibles so the operation does not create a hazard to persons or property.

303.10.1 Clearance Reduction. Reduced clearances to combustible construction for listed equipment and appliances shall comply with the listing and Table 303.10.1. Where permitted by the manufacturer, and not provided in this code, reduced clearances to combustible construction for unlisted equipment and appliances shall comply with Table 303.10.1.

303.10.1.1 Type I Hood Exhaust System. Reduced clearances for Type I exhaust systems shall be in accordance with Section 507.4.2 through Section 507.4.2.3. Clearances from the duct or the exhaust fan to the interior surface of enclosures of combustible construction shall be in accordance with Section 510.7.3 and clearances shall not be reduced.

303.10.1.2 Product Conveying Ducts. Reduced clearances to combustibles construction for product conveying ducts shall be in accordance with Section 506.10.3 through Section 506.11.6.3.

303.10.1.3 Solid-Fuel Burning Appliances. For solid-fuel burning appliances, the clearance shall not be less than 12 inches (305 mm) to combustible walls and not less than 18 inches (457 mm) to combustible ceilings. The clearance, after reduction, shall be permitted to be less than 12 inches (305 mm) to combustible walls and less than 18 inches (457 mm) to combustible ceilings. Solid-fuel burning appliances listed for lesser clearances shall be permitted to be installed in accordance with the manufacturer's instructions and their listing.

303.11 Installation in Commercial Garages. Appliances installed in enclosed, basement, and underground parking structures shall be installed in accordance with NFPA 88A. [NFPA 54:9.1.11.1]

303.11.1 Repair Garages. Appliances installed in repair garages shall be installed in accordance with NFPA 30A. [NFPA 54:9.1.11.2]

303.12 Installation in Aircraft Hangars. Heaters in aircraft hangars shall be installed in accordance with NFPA 409. [NFPA 54:9.1.12]

304.0 Accessibility for Service.

304.1 General. All appliances shall be located with respect to building construction and other equipment so as to permit access to the appliance. Sufficient clearance shall be maintained to permit cleaning of heating surfaces; the replacement of filters, blowers, motors, burners, controls, and vent connections; the lubrication of moving parts where necessary; the adjustment and cleaning of burners and pilots; and the proper functioning of explosion vents, if provided. For attic installation, the passageway and servicing area adjacent to the appliance shall be floored. [NFPA 54:9.2.1]

Unless otherwise specified, not less than 30 inches (762 mm) in depth, width, and height of working space shall be provided.

Exception: A platform shall not be required for unit heaters or room heaters.

This section requires that all appliances be accessible. It has been common practice that if a room is sufficiently large to allow disassembling of the appliance, the access opening or door should be large enough to remove the largest piece.

The section also specifies the different types of maintenance that could be performed to further clarify the extent of the required accessibility and requires a minimum 30 inches in depth, width and height of working space.

It should be observed that the code no longer repeats the access provisions in other sections when covering the various types of regulated appliances. The most notably omitted is the requirement for closets containing furnaces to have a 24-inch wide door with 30-inch deep clear area when the door is open. However, inspection must ensure that the 30-inch requirement is enforced as required by this section (when the door is open) and that the door is sufficiently wide to allow the removal and replacement of the furnace. **Figures 304.1a and 304.1b** illustrate access requirements for some appliances regulated by the code.

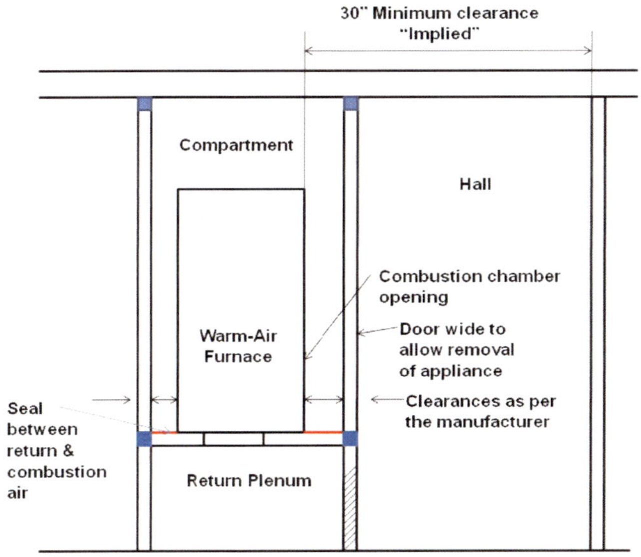

FIGURE 304.1A
ACCESS AND CLEARANCE REQUIREMENTS

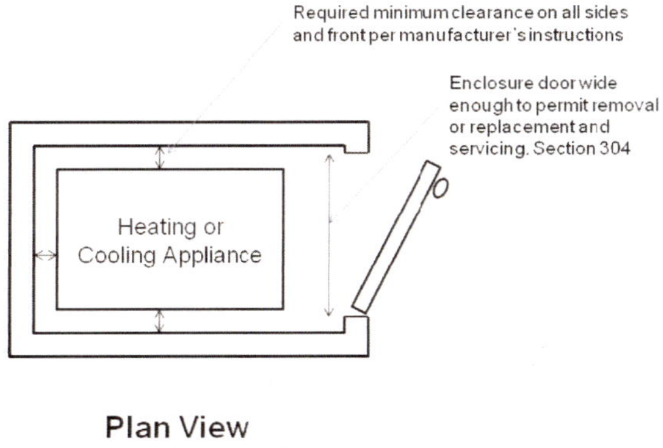

FIGURE 304.1B
ACCESS TO APPLIANCES FOR INSPECTION, SERVICE, OR REPAIR

304.2 Sloped Roof. Where equipment or appliances that require service are installed on a roof having a slope of 4 units vertical in 12 units horizontal (33 percent slope) or more, a level platform of not less than 30 inches by 30 inches (762 mm by 762 mm) shall be provided at the service side of the equipment or appliance.

To convert slope to an angle in degrees, take the arctangent of the slope. For example, to convert 4:12 slope to an angle, take the arctangent of 4/12. The slope is 18.43 degrees. See **Figure 304.2**.

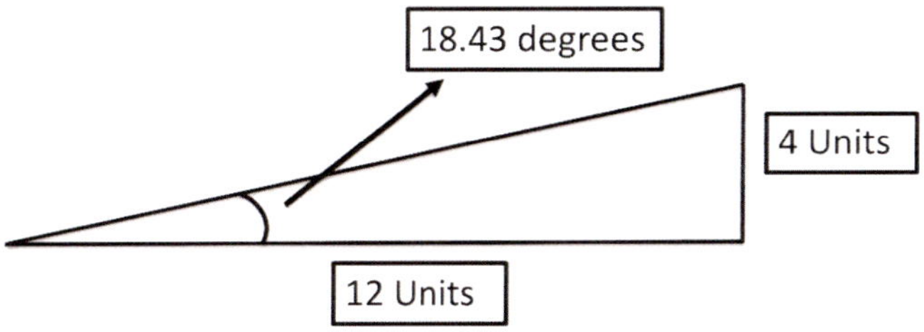

FIGURE 304.2
ANGLE OF SLOPED ROOF

304.3 Access to Appliances on Roofs. Appliances located on roofs or other elevated locations shall be accessible. [NFPA 54:9.4.3.1]

Appliances on roofs must be accessible, which means that portable ladders could be used for access (unless the building is over 15 feet in height, then the provision of Section 304.3.1 would apply).

304.3.1 Access. Buildings of more than 15 feet (4572 mm) in height shall have an inside means of access to the roof unless other means acceptable to the Authority Having Jurisdiction are used. [NFPA 54:9.4.3.2]

This section requires interior means of access to roofs of buildings exceeding 15 feet in height. However, it specifically allows approval of alternates (such as an exterior ladder), provided the AHJ approves it. The 15-foot height could be determined by measuring the distance from the finished floor elevation to the underside of the roof inside the building. However, if the alternate is used (an exterior ladder as was allowed in previous versions of the code), the distance should be measured from finished grade to the top of the roof or parapet. The specifications for exterior ladders listed in Section 304.3.1.2 could be used as a guide by the AHJ in approving the alternate.

304.3.1.1 Access Type. The inside means of access shall be a permanent or foldaway inside stairway or ladder, terminating in an enclosure, scuttle, or trapdoor. Such scuttles or trapdoors shall be at least 22 inches by 24 inches (559 mm by 610 mm) in size, shall open easily and safely under all conditions, especially snow, and shall be constructed so as to permit access from the roof side unless deliberately locked on the inside.

At least 6 feet (1829 mm) of clearance shall be available between the access opening and the edge of the roof or similar hazard, or rigidly fixed rails or guards a minimum of 42 inches (1067 mm) in height shall be provided on the exposed side. Where parapets or other building structures are utilized in lieu of guards or rails, they shall be a minimum of 42 inches (1067 mm) in height. [NFPA 54:9.4.3.3]

The section requires a permanent ladder or a foldaway stairway or ladder as an interior access to the roof (see **Figure 304.3.1.1a**). Please note that the scuttle or trap door dimensions (see **Figure 304.3.1.1b**) are different from those for attic access. It is required to be 22 inches by 24 inches while attic access is required to be 22 inches by 30 inches (see Section 304.4).

FIGURE 304.3.1.1A
INTERIOR FOLDAWAY LADDER

304.3.1.2 Permanent Ladders. Permanent ladders required by Section 304.3.1.1 shall be constructed in accordance with the following:

(1) Side railings shall extend not less than 30 inches (762 mm) above the roof or parapet wall.

(2) Landings shall not exceed 18 feet (5486 mm) apart measured from the finished grade.

(3) Width shall be not less than 14 inches (356 mm) on center.

FIGURE 304.3.1.1B
ROOF ACCESS

(4) Rungs spacing shall not exceed 12 inches (305 mm) on center, and each rung shall be capable of supporting a 300 pound (136.1 kg) load.

(5) Toe space shall be not less than 6 inches (152 mm).

This section provides specific minimum requirements for providing an access to mechanical equipment located on the roof.

304.3.2 Permanent Lighting. Permanent lighting shall be provided at the roof access. The switch for such lighting shall be located inside the building near the access means leading to the roof. [NFPA 54:9.4.3.4]

Lighting should be provided at the interior roof access opening, with a switch located inside the building near the access opening (see **Figure 304.3.1.1a**). Please note that, unlike other codes, this code does not require a light near the equipment on the roof, but rather, just means to illuminate the path through the roof access opening.

304.4 Appliances in Attics and Under-Floor Spaces. An attic or under-floor space in which an appliance is installed shall be accessible through an opening and passageway not less than the largest component of the appliance, and not less than 22 inches by 30 inches (559 mm by 762 mm).

304.4.1 Length of Passageway. Where the height of the passageway is less than 6 feet (1829 mm), the distance from the passageway access to the appliance shall not exceed 20 feet (6096 mm) measured along the centerline of the passageway. [NFPA 54:9.5.1.1]

304.4.2 Width of Passageway. The passageway shall be unobstructed and shall have solid flooring not less than 24 inches (610 mm) wide from the entrance opening to the appliance. [NFPA 54:9.5.1.2]

304.4.3 Work Platform. A level working platform not less than 30 inches by 30 inches (762 mm by 762 mm) shall be provided in front of the service side of the appliance. [NFPA 54:9.5.2]

Exception: A working platform need not be provided where the furnace is capable of being serviced from the required access opening. The furnace service side shall not exceed 12 inches (305 mm) from the access opening.

304.4.4 Lighting and Convenience Outlet. A permanent 120 V receptacle outlet and a luminaire shall be installed near the appliance. The switch controlling the luminaire shall be located at the entrance to the passageway. {NFPA 54:9.5.3}

305.0 Location.

305.1 Installation in Residential Garages. Appliances in residential garages and in adjacent spaces that open to the garage and are not part of the living space of a dwelling unit shall be installed so that all burners and burner-ignition devices are located not less than 18 inches (457 mm) above the floor unless listed as flammable vapor ignition resistant. [NFPA 54:9.1.10.1]

Garage areas are especially vulnerable to flammable vapors from potential fuel leaks from vehicles and leaks from stored fuels, paints and solvents. The 18-inch height requirement in this section applies to appliances that could provide a source of ignition, such as those having standing pilots, electronic ignitions, relays, switches, starting switches found inside some types of motors and similar devices. The 18 inches is to be measured from the floor level to the source of ignition (see **Figure 305.1**). Please note that there are some newer appliances that are listed for installation at garage floor level (Flammable Vapor Ignition Resistant), as they are designed to shield the ignition source within.

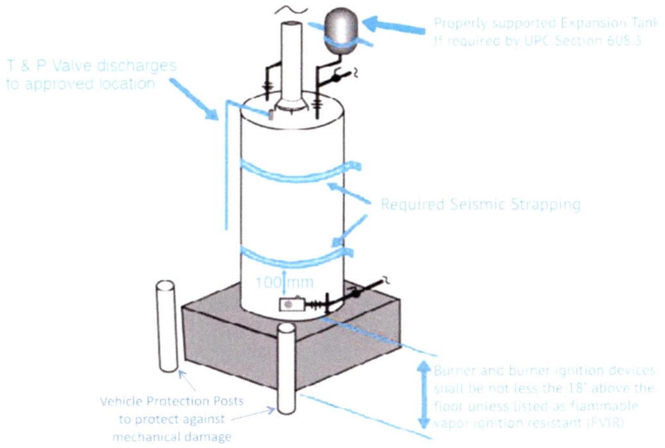

FIGURE 305.1
PROTECTION OF APPLIANCES IN GARAGES

305.1.1 Physical Damage. Appliances installed in garages, warehouses, or other areas subject to mechanical damage shall be guarded against such damage by being installed behind protective barriers or by being elevated or located out of the normal path of vehicles.

Physical damage to appliances may be avoided by elevating the appliances above or out of the path of moving equipment or vehicles. Protection against physical damage may also be provided by erecting concrete or pipe barriers. Wheel stops are not a favorable mean of protection, but could be used with the AHJ's approval. If an appliance were to be struck by a moving vehicle, the appliance may be damaged or displaced, which could cause the release of fuel from a ruptured fuel line.

305.1.2 Access from the Outside. Where appliances are installed in a separate, enclosed space having access only from outside of the garage, such appliances shall be permitted to be installed at floor level, providing the required combustion air is taken from the exterior of the garage. [NFPA 54:9.1.10.3]

The code allows for the installation of an appliance at floor level, if it is in a separate compartment and has access and combustion air from outside the garage (see **Figure 305.1.2**).

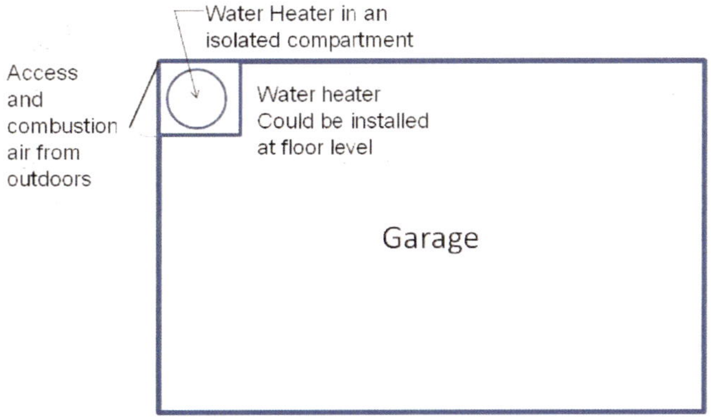

FIGURE 305.1.2
EXCEPTION TO ELEVATING OF APPLIANCE ABOVE GARAGE FLOOR

305.1.3 Cellulose Nitrate Plastic Storage. Heating equipment located in rooms where cellulose nitrate plastic is stored or processed shall be in accordance with the fire code.

The fire code of the jurisdiction should be consulted when heating appliances are installed in rooms or areas where cellulose nitrate plastic is stored or processed.

305.2 Pit Location. Where excavation is necessary to install an appliance, the depth shall extend not less than 6 inches (152 mm) below and 12 inches (305 mm) on all sides of the appliance, except on the service side, which shall have 30 inches (762 mm). Where the depth of the excavation for either the appliance or passageway exceeds 12 inches (305 mm), walls shall be lined with concrete or masonry 4 inches (102 mm) above the adjoining ground level.

305.3 Flood Hazard Areas. For buildings located in flood hazard areas, heating, ventilating, air-conditioning, refrigeration, miscellaneous heat-producing, and energy-utilizing equipment and appliances shall be elevated at or above the elevation in accordance with the building code for utilities and attendant equipment or the elevation of the lowest floor, whichever is higher.

Exception: Equipment and appliances shall be permitted to be located below the elevation in accordance with the building code for utilities and attendant equipment or the elevation of the lowest floor, whichever is higher, provided that the systems are designed and installed to prevent water from entering or accumulating within their components and the systems are constructed to resist hydrostatic and hydrodynamic loads and stresses, including the effects of buoyancy, during the occurrence of flooding to such elevation.

This section addresses protection requirements for installations in areas subject to flood damage. Equipment installed in buildings located in flood hazard areas must be elevated above the design flood elevation for that specific location. The design flood elevation is the elevation reached by floodwaters during an intense storm. Designers usually use an elevation that is called the 100-year floodplain. A 100-year floodplain is calculated to be the level of floodwater expected to be equaled or exceeded every 100 years on average. The 100-year flood is more accurately referred to as the 1 percent flood, since it is a flood that has a 1 percent chance of being equaled or exceeded in any single year.

If it is impractical to raise the equipment above that elevation, the equipment must be installed (or designed) in such a way as to prevent floodwaters from entering or pooling inside the equipment. Prolonged exposure to water damages critical components of the equipment, such as the controls. In addition, equipment must be anchored so it would resist the buoyancy effect of the floodwaters.

The building code of the jurisdiction should be consulted for additional requirements to ASCE 24, *Flood Resistant Design and Construction*, which specifies elevations of buildings, including mechanical equipment and systems, as a function of flood zone and structure category. ASCE 24 specifies the performance under flood loads that needs to be provided if equipment is located below such elevations.

305.3.1 Coastal High Hazard Areas. Mechanical systems in buildings located in coastal high hazard areas shall be in accordance with the requirements of Section 305.3, and mechanical systems, pipes, and appurtenances shall not be mounted on or penetrate through walls that are intended to breakaway under flood loads in accordance with the building code.

Coastal high hazard areas tend to impose severe loads on foundations or supporting walls of buildings not typically designed to withstand such loads. Breakaway walls are designed to fail under flood loads to avoid transferring these loads to the foundation or the supporting structure of the building. Since these walls are designed to fail at certain times, the code prohibits mechanical equipment and appurtenances, such as piping, from being mounted or attached to these walls, as this might not allow the walls to break away as designed.

305.3.2 Air Exhaust and Intake Openings. Outside air exhaust openings and air intake openings shall be located at or above the elevation required by the building code for utilities and attendant equipment or the elevation of the lowest floor, whichever is higher.

Air exhaust and intake openings are required to be located above the design flood elevation to prevent floodwaters from entering the building or flooding the mechanical equipment attached to these openings.

305.4 Elevator Shaft. Unless required for the functionality and safety of the elevator system, mechanical systems shall not be located in an elevator shaft.

305.5 Drainage Pan. Where a water heater is located in an attic, in or on an attic ceiling assembly, floor-ceiling assembly, or floor-subfloor assembly where damage results from a leaking water heater, a watertight pan of corrosion-resistant materials shall be installed beneath the water heater with not less than ¾ of an inch (20 mm) diameter drain to an approved location. Such pan shall be not less than 1½ inches (38 mm) in depth.

306.0 Automatic Control Devices.

306.1 General. Heating appliances shall be equipped with a listed device or devices that will shut off the fuel supply to the main burner or burners in the event of pilot or ignition failure. Liquefied petroleum gas-air-burning heating appliances shall be equipped with a listed automatic device or devices that will shut off the flow of gas to the pilot in the event of ignition failure.

Exception: The listed shutoff devices shall not be required on range or cooking tops, log lighters, lights, or other open-burner manually operated appliances, or listed appliances not requiring such devices and specific industrial appliances as approved by the Authority Having Jurisdiction.

Heating appliances whose manual fuel controls are not readily accessible from the main portion of the building being heated shall be equipped with remote controls.

Forced-air and gravity-type warm air furnaces shall be equipped with a listed air outlet temperature limit control that cannot be set for temperatures exceeding 250°F (121°C). Such controls shall be located in the bonnet or plenum, within 2 feet (610 mm) of the discharge side of the heating element of gravity furnaces or in accordance with the conditions of listing.

Electric duct heaters shall be equipped with an approved automatic reset air outlet temperature limit control that will limit the outlet air temperature to not exceed 200°F (93°C). The electric elements of the heater shall be equipped with fusible links or a manual reset temperature limit control that will prevent outlet air temperature in excess of 250°F (121°C).

There are three basic types of control devices required in this section:

1. Devices that cut off fuel supply to the main burners of the appliance in the event of a pilot (or electronic ignition) failure. This is a very common safety feature found in most, if not all, listed automatic appliances to protect against the danger of large quantities of gas released without the availability of an ignition source. In appliances with pilot lights, such as water heaters, the pilot light heats a thermocouple, which holds a solenoid valve in an open position, thus, allowing fuel to flow to the burners when the appliance thermostat calls for heating. If the pilot light is off, the thermocouple cools down and the solenoid valve is held in the closed position, allowing no fuel to flow. **Figure 306.1a** is a cross-sectional view of this type of device.

The safety device in electronic ignition-type appliances shuts down the fuel supply after a certain number of electronic ignition attempts that do not result in successful ignition.

The code reasonably exempts certain manually operated appliances, such as ranges, cooking tops, log lighters and certain industrial appliances, from this safety feature.

2. Devices that will also shut off the gas supply to the pilot light when it fails. Although the pilot light itself is not

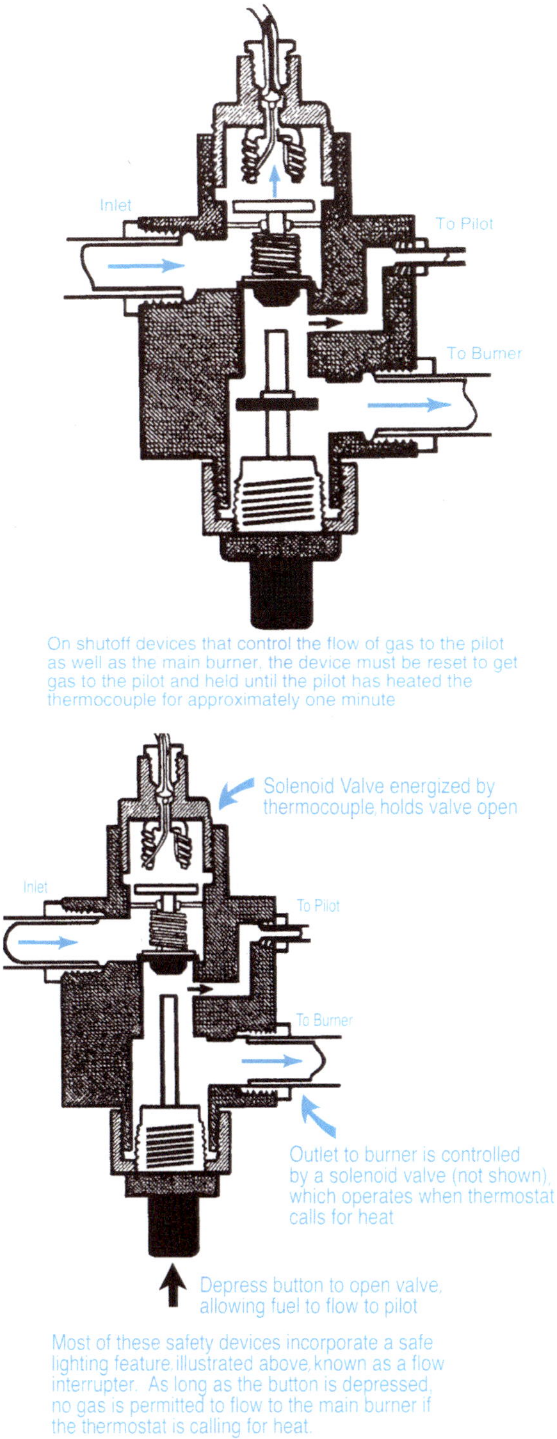

FIGURE 306.1A
AUTOMATIC CONTROL DEVICES

required to be protected by such safety device, most listed automatic appliances are equipped with such a device. However, the code specifically requires this safety feature for LP-Gas-fueled appliances. This additional safety feature is necessary since LP-Gas is heavier than air and would pool and accumulate rather than dissipate when there is even a small leak due to pilot light failure.

3. Devices that shut off fuel or electrical power at a predetermined temperature setting. These controls have several variations and usually depend on thermal expansion of metal to make or break electrical contact, resulting in shutting down a fuel valve, disconnecting electric power or both. **Figure 306.1b** shows examples of temperature-limit control devices. The helix-type device is a metal coil that expands and contracts longitudinally. The bimetal-type device consists of two dissimilar metals that are bonded together and have different thermal expansion rates that cause the bending of the strip, thus making or breaking electrical contact. These devices are required to ensure that heating appliances do not exceed predetermined safe-operating temperatures. **Figure 306.1c** shows the required location of the air outlet temperature-limit control device in a forced-air furnace.

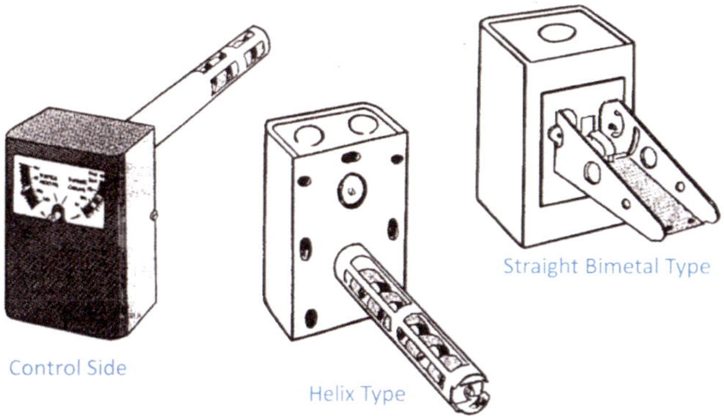

FIGURE 306.1B
TEMPERATURE LIMIT CONTROLS

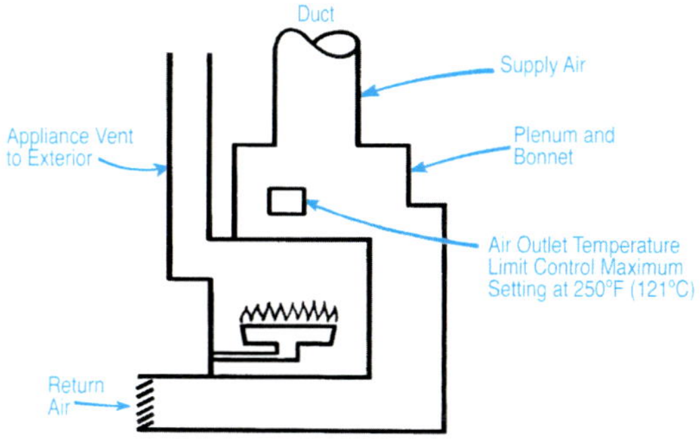

FIGURE 306.1C
AIR OUTLET LIMIT CONTROL
(Courtesy of Southern California Gas Company)

Electric duct heaters require two devices for protection against high outlet air temperatures. One device disconnects power to the unit at outlet temperatures of 200°F (93°C). The second device is a backup for the first and is required to melt a fusible link and disrupt the power supply to the heating elements when the outlet temperature reaches 250°F (121°C). When the fusible link requires replacement, it will be evident to maintenance personnel that something serious has happened and that close examination of the heater is necessary. **Figure 306.1d** illustrates a duct heater with the temperature-limiting devices.

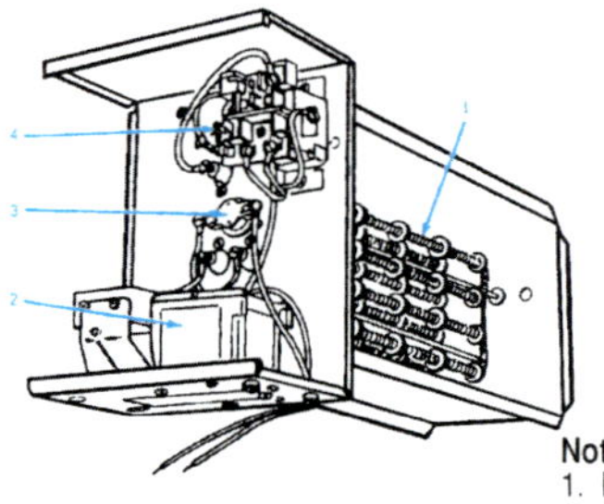

Notes:
1. Heating Elements
2. Control Panel
3. Limit Control (High Temperature)
4. Manual Reset and Secondary Contactor directly connected to Heating Elements

FIGURE 306.1D
ELECTRIC DUCT HEATER TEMPERATURE CONTROL

307.0 Labeling.

307.1 Fuel-Burning Appliances. Fuel-burning heating appliances shall bear a permanent and legible factory-applied nameplate on which shall appear:

(1) The name or trademark of the manufacturer.

(2) The approved fuel input rating of the appliance, expressed in Btu/h (kW).

(3) The model number or equivalent.

(4) The serial number.

(5) Instructions for the lighting, operation, and shutdown of the appliance.

(6) The type of fuel approved for use with the appliance.

(7) The symbol of an approved agency certifying compliance of the equipment with recognized standards.

(8) Required clearances from combustible surfaces on which or adjacent to which it is permitted to be mounted.

307.2 Electric Heating Appliances. Electric heating appliances shall bear a permanent and legible factory-applied nameplate on which shall appear:

(1) The name or trademark of the manufacturer.

(2) The model number or equivalent.

(3) The serial number.

(4) The electrical rating in volts, amperes (or watts), and, for other than single phase, the number of phases.

(5) The output rating in Btu/h (kW).

(6) The electrical rating in volts, amperes, or watts of each field-replaceable electrical component.

(7) The symbol of an approved agency certifying compliance of equipment with recognized standards.

(8) Required clearances from combustible surfaces on which or adjacent to which it is permitted to be mounted.

An appliance shall be accompanied by clear and complete installation instructions, including required clearances from combustibles other than mounting or adjacent surfaces, and temperature rating of field-installed wiring connections exceeding 140°F (60°C).

307.3 Heat Pump and Electric Cooling Appliances. Heat pumps and electric cooling appliances shall bear a permanent and legible factory-applied nameplate on which shall appear:

(1) The name or trademark of the manufacturer.

(2) The model number or equivalent.

(3) The serial number.

(4) The amount and type of refrigerant.

(5) The factory test pressures or pressures applied.

(6) The electrical rating in volts, amperes, and, for other than single phase, the number of phases.

(7) The output rating in Btu/h (kW).

(8) The electrical rating in volts, amperes, or watts of each field replaceable electrical component.

(9) The symbol of an approved agency certifying compliance of the equipment with recognized standards.

(10) Required clearances from combustible surfaces on which or adjacent to which it is permitted to be mounted.

This section requires a permanent nameplate for appliances and equipment. This means that the nameplate should be of metal or other material of durable nature, and be water resistant. The section describes specific requirements for the information that must be provided on the nameplate of fuel-burning appliances, electric heating appliances, heat pump appliances and electric cooling appliances. It is not the intent of this section to prevent the AHJ from approving unlisted appliances as provided for in Section 302.2, as might be assumed from the requirement that the label shall contain the symbol of an approved agency certifying compliance of equipment with recognized standards. However, when applicable, the label could indicate that the equipment has been approved by the AHJ in compliance with the requirements of Section 302.2.

An appliance shall be accompanied by clear and complete installation instructions, including required clearances from combustible other than mounting or adjacent surfaces, and temperature rating of field-installed wiring connections exceeding 140°F (60°C).

307.4 Absorption Units. Absorption units shall bear a permanent and legible factory-applied nameplate on which shall appear:

(1) The name or trademark of the manufacturer.

(2) The model number or equivalent.

(3) The serial number.

(4) The amount and type of refrigerant.

(5) Hourly rating in Btu/h (kW).

(6) The type of fuel approved for use with the unit.

(7) Cooling capacity Btu/h (kW).

(8) Required clearances from combustible surfaces on which or adjacent to which it is permitted to be mounted.

(9) The symbol of an approved agency certifying compliance of the equipment with recognized standards.

Unlike vapor compression systems that utilize compressors to operate, absorption equipment requires a source of heat to operate. Therefore, the hourly rating and the type of fuel should be in the nameplate so that the user can have the critical information necessary to safely operate the system.

308.0 Improper Location.

308.1 General. Piping or equipment shall not be so located as to interfere with the normal use thereof or with the normal operation and use of windows, doors, or other required facilities.

You might think that a regulation such as this would not be necessary; however, there are installers who try to take shortcuts, save money or who just do not know any better that install ducts, appliances and piping improperly so that the use of doors and windows are restricted. This commonly occurs during renovations or conversions of older buildings. See **Figure 308.1** for a drawing of a prohibited installation in which the piping blocks the door, hatch and stairs.

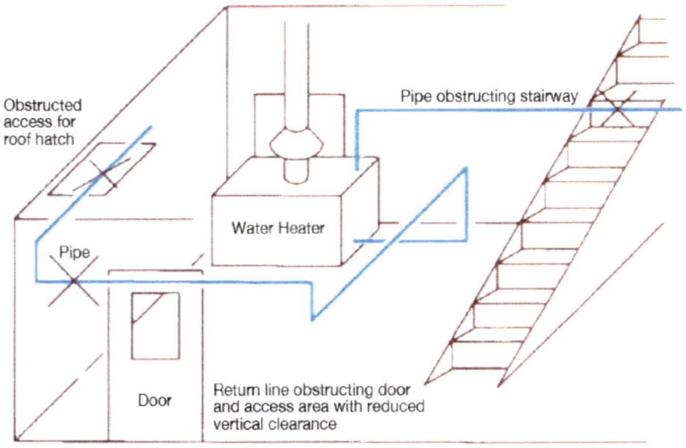

FIGURE 308.1
IMPROPER LOCATION OF PIPING

309.0 Workmanship.

309.1 Engineering Practices. Design, construction, and workmanship shall comply with accepted engineering practices and shall be of such character as to secure the results sought to be obtained by this code.

Workmanship is an important issue that concerns quality trade practices. The goal here is neat and professional-appearing mechanical systems with a minimum amount of labor that will function well during its expected life cycle. Good workmanship can mean different things to different people. It is difficult to describe specifically when an installation shows good workmanship. A journeyman or inspector who has worked in the trade for several years on various types of projects and in different geographic areas will be better able to judge what constitutes good workmanship than one with little or no experience.

Evaluation of workmanship must never be allowed to become subjective. It is not a matter of whether the work is good looking – it is a matter of code compliance. As it turns out, work that is fully compliant with all of the code's provisions will exhibit good workmanship in both looks and function. Good workmanship results in the safe and satisfactory operation of a plumbing system for an expected duration of time, with maximum economy and minimum maintenance. Good workmanship is much more than a job with a neat appearance.

309.2 Concealing Imperfections. It shall be unlawful to conceal cracks, holes, or other imperfections in materials by welding, brazing, or soldering, by using therein or thereon paint, wax, tar, solvent cement, other leak-sealing or repair agent.

Defective pipe, fixtures and fittings must be replaced. These imperfections can be the result of damage during transit, defects in manufacture, sand holes, damage due to freezing, corrosion of the system, or damage during installation. A temporary fix is just that, temporary.

309.3 Installation Practices. Mechanical systems shall be installed in a manner that is in accordance with this code, applicable standards, and the manufacturer's installation instructions.

310.0 Condensate Wastes and Control.

310.1 Condensate Disposal. Condensate from air washers, air-cooling coils, condensing appliances, and the overflow from evaporative coolers and similar water-supplied equipment or similar air-conditioning equipment shall be collected and discharged to an approved plumbing fixture or disposal area. Where discharged into the drainage system, equipment shall drain by means of an indirect waste pipe. The waste pipe shall have a slope of not less than $\frac{1}{8}$ inch per foot (10.4 mm/m) or 1 percent slope and shall be of approved corrosion-resistant material not smaller than the outlet size in accordance with Section 310.3 or Section 310.4 for air-cooling coils or condensing appliances, respectively. Condensate or wastewater shall not drain over a public way.

Condensate from evaporators, cooling coils and fuel-burning condensing appliances (such as Category IV heaters) is to be collected and discharged to a plumbing fixture or to an approved disposal area. When discharging into the drainage system, equipment shall drain through an indirect waste piping via either an air gap or an air break (see the UPC). The indirect waste pipe shall have a slope of not less than 1/8 inch per foot, or 1 percent. This area is most often overlooked by installers and needs to receive special attention, as appropriate supports are necessary to ensure that piping maintains the needed slope for positive drainage.

The piping is required to be of approved corrosion-resistant material not smaller than the outlet size as required in either Section 310.3 or 310.4.

Acceptable points of discharge of condensate piping include:

1. The tailpiece of a lavatory (must be in a space or dwelling served by the appliance).
2. The drain and overflow of a bathtub (must be in a space or dwelling served by the appliance; the drain and overflow should be provided with an access panel, even if there is no slip joint—see the UPC).
3. A floor sink.
4. A gravel pit (should be restricted to small residential units and the soil should be the type that can absorb the condensate. This method should not be used with clay or expansive soil).

See **Figures 310.1a and 310.1b**.

Disposal of the condensate onto the roof should not be

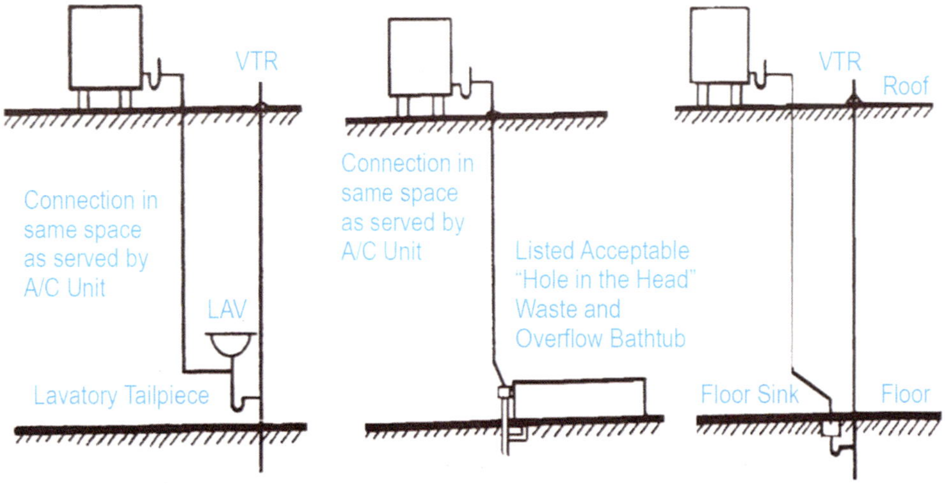

FIGURE 310.1A
CONDENSATE DRAIN TERMINATIONS

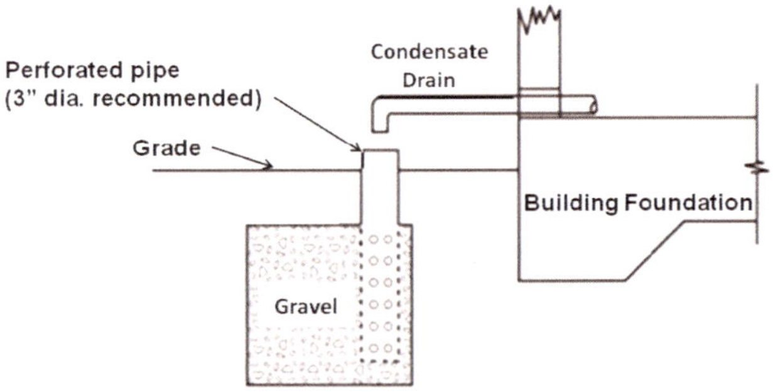

Notes:
1. Size condensate drain in accordance with Table 310.3.
2. Alternate termination point (in lieu of terminating in an approved plumbing fixture).
3. Acceptable for small residential units.
4. Must have soil with suitable absorption. Excludes clay or expansive soil.

FIGURE 310.1B
DRY WELL FOR CONDENSATE DRAIN

accepted. However, numerous jurisdictions would accept condensate to be piped to roof drains, if the roof drains are connected to a storm drainage piping system, which will ensure that the condensate will not create a nuisance.

Please note that certain areas of the country may have overloaded sewage treatment plants and may prevent the discharge of condensate drains to the sanitary drainage system. In this case, the condensate drain should be routed to the rain collection system (storm drains) or other locations as approved by the AHJ.

In any case, condensate piping should not drain over a public way or walkway, whether directly or indirectly, to prevent nuisance and possible slipping accidents.

310.1.1 Condensate Pumps. Where approved by the Authority Having Jurisdiction, condensate pumps shall be installed in accordance with the manufacturer's installation instructions. Pump discharge shall rise vertically to a point where it is possible to connect to a gravity condensate drain and discharged to an approved disposal point. Each condensing unit shall be provided with a separate sump and interlocked with the equipment to prevent the equipment from operating during a failure. Separate pumps shall be permitted to connect to a single gravity indirect waste where equipped with check valves and approved by the Authority Having Jurisdiction.

310.2 Condensate Control. Where any equipment or appliance is installed in a space where damage is capable of resulting from condensate overflow, a drain line shall be provided and shall be drained in accordance with Section 310.1. An additional protection method for condensate overflow shall be provided in accordance with one of the following:

(1) A water level detecting device that will shut off the equipment or appliance in the event the primary drain is blocked.

(2) An additional watertight pan of corrosion-resistant material, with a separate drain line, installed beneath the cooling coil, unit, or the appliance to catch the overflow condensate due to a clogged primary condensate drain.

(3) An additional drain line at a level that is higher than the primary drain line connection of the drain pan.

(4) An additional watertight pan of corrosion-resistant material with a water level detection device installed beneath the cooling coil, unit, or the appliance to catch the overflow condensate due to a clogged primary condensate drain and to shut off the equipment.

The additional pan or the additional drain line connection shall be provided with a drain pipe of not less than ¾ of an inch (20 mm) nominal pipe size, discharging at a point that is readily observed.

When appliances or equipment that produce condensate are located in a space where damage may result from a blocked primary condensate drain, an overflow or secondary drain must be provided in accordance with Section 310.1. In addition, a secondary means of protection for condensate overflow must be provided in accordance to one of the four methods provided in this section.

There are cooling coil manufacturers that provide coils

with sensors that would shut down the operation of the unit when they sense water resulting from a blocked primary condensate drain; therefore, these coils meet the first provision for additional protection. There are also manufacturers that provide devices that would be mounted on the cooling coil piping outlets to perform that same function.

310.2.1 Protection of Appurtenances. Where insulation or appurtenances are installed where damage is capable of resulting from a condensate drain pan overfill, such installations shall occur above the rim of the drain pan with supports. Where the supports are in contact with the condensate waste, the supports shall be of approved corrosion-resistant material.

310.3 Condensate Waste Pipe Material and Sizing. Condensate waste pipes from air-cooling coils shall be sized in accordance with the equipment capacity as specified in Table 310.3. The material of the piping shall comply with the pressure and temperature rating of the appliance or equipment, and shall be approved for use with the liquid being discharged.

TABLE 310.3
MINIMUM CONDENSATE PIPE SIZE

EQUIPMENT CAPACITY IN TONS OF REFRIGERATION	MINIMUM CONDENSATE PIPE DIAMETER (inches)
Up to 20	¾
21 – 40	1
41 – 90	1¼
91 – 125	1½
126 – 250	2

For SI units: 1 ton of refrigeration = 3.52 kW, 1 inch = 25 mm

The size of condensate waste pipes is for one unit or a combination of units, or as recommended by the manufacturer. The capacity of waste pipes assumes a ⅛ inch per foot (10.4 mm/m) or 1 percent slope, with the pipe running three-quarters full at the following pipe conditions:

Outside Air – 20%		Room Air – 80%	
DB	WB	DB	WB
90°F	73°F	75°F	62.5°F

For SI units: °C = (°F-32)/1.8

Condensate drain sizing for other slopes or other conditions shall be approved by the Authority Having Jurisdiction.

The sizing table in this section could be used in most areas, except for those areas with excessively high humidity weather patterns. In determining the size of the pipe, convert the Btu/hr capacity of the cooling equipment into tons of refrigeration. 12,000 Btu/h of cooling capacity = 1 ton of refrigeration.

310.3.1 Cleanouts. Condensate drain lines shall be configured or provided with a cleanout to permit the clearing of blockages and for maintenance without requiring the drain line to be cut.

310.4 Appliance Condensate Drains. Condensate drain lines from individual condensing appliances shall be sized as required by the manufacturer's instructions. Condensate drain lines serving more than one appliance shall be approved by the Authority Having Jurisdiction prior to installation.

Condensing furnaces (e.g., Category IV) are very efficient. They extract exceedingly more heat from the burners and the produced flue gasses. Flue gasses contain a large quantity of water vapor (steam), and extracting more heat cools down the flue gasses, producing sizable quantities of condensation. Therefore, condensing furnaces should have means of disposal of the condensation similar to cooling coils. Usually, the manufacturer of the furnace provides detailed instructions on how to dispose of the condensation, including the minimum size of condensate piping (based on the predicted amounts of condensation).

Since the condensation tends to be corrosive in nature, the installation instructions should be followed closely as to the type of condensate piping material to be used and also what the suitability is of drainage piping material receiving the corrosive discharge.

310.5 Point of Discharge. Air-conditioning condensate waste pipes shall connect indirectly, except where permitted in Section 310.6, to the drainage system through an air gap or air break to trapped and vented receptors, dry wells, leach pits, or the tailpiece of plumbing fixtures. A condensate drain shall be trapped in accordance with the appliance manufacturer's instructions or as approved.

Since many air moving appliances have different negative pressure drops across the cooling coil it is important to use the manufacture's recommendation on trap sizing. A properly functioning and designed condensate trap provides for discharge of water from the cooling coil drain pan, while the water seal (the water level maintained in the trap) prevents the flow of ambient air into or out of the air handler. Several problems result from improperly trapped systems, some of which can severely impact indoor air quality. Where a trap is improperly installed or designed, an incoming air stream may be introduced through the drain and the air flowing through the coil can possibly spray condensate into the fan intake, which can propel the moisture into other parts of the system; which can then be carried through the ducts and into the conditioned space possibly causing bacterial growth and transmission. Furthermore, improperly trapped systems can include the trap outlet is too short or the trap outlet is too tall, or other potential incorrect sizing. The manufacturer of the appliance typically provides trapping requirements that should be followed when sizing the condensate drainage trap.

310.6 Condensate Waste From Air-Conditioning Coils. Where the condensate waste from air-conditioning coils discharges by direct connection to a lavatory tailpiece or to an approved accessible inlet on a bathtub overflow, the connection shall be located in the area controlled by the same person controlling the air-conditioned space.

Damage to the structure or personal property may result from condensate discharge into a defective or inoperable plumbing system. This section helps to ensure that the end user of the mechanical equipment is the responsible party for controlling the condensate discharge.

310.7 Plastic Fittings. Female plastic screwed fittings shall be used with plastic male fittings and plastic male threads.

The most common thread on pipes used in plumbing systems is tapered (NPT) and designed to seal by an interference fit. Experience has shown that using a threaded female plastic fitting with a metal male fitting often leads to failure of the joint by splitting if it is tightened beyond the tensile strength of the plastic.

311.0 Heating or Cooling Air System.

311.1 Source. A heating or cooling air system shall be provided with return air, outside air, or both. A heating or cooling air system regulated by this code and designed to replace required ventilation shall be arranged to discharge into a conditioned space not less than the amount of outside air specified in Chapter 4.

311.2 Air Filters. Air filters shall be installed in a heating, cooling, or makeup air system. Media-type air filters shall comply with UL 900. Electrostatic and high efficiency particulate filters shall comply with Section 936.0.

This section requires filters to be installed in heating and cooling or air-moving equipment. Filters not only clean the air and safeguard the interior environment from contamination, but also protect the moving parts of equipment (such as blowers) from damage due to large particles that could otherwise enter the air system (see **Figure 311.2**).

FIGURE 311.2
AIR-CONDITIONING FILTERS

Exceptions:

(1) Systems serving single guest rooms or dwelling units shall not require a listed filter.

(2) Air filters used in listed appliances and in accordance with the manufacturer's instructions.

311.3 Prohibited Source. Outside or return air for a heating or cooling air system shall not be taken from the following locations:

(1) Less than 10 feet (3048 mm) in distance from an appliance vent outlet, a vent opening of a plumbing drainage system, or the discharge outlet of an exhaust fan, unless the outlet is 3 feet (914 mm) above the outside-air inlet.

(2) Less than 10 feet (3048 mm) above the surface of an abutting public way, sidewalk, street, alley, or driveway.

(3) A hazardous or insanitary location, or a refrigeration machinery room as defined in this code.

(4) An area, the volume of which is less than 25 percent of the entire volume served by such system, unless there is a permanent opening to an area the volume of which is equal to 25 percent of the entire volume served.

Exception: Such openings where used for a heating or cooling air system in a dwelling unit shall be permitted to be reduced to not less than 50 percent of the required area, provided the balance of the required return air is taken from a room or hall having not less than three doors leading to other rooms served by the furnace.

(5) A closet, bathroom, toilet room, or kitchen.

(6) Rooms or spaces containing a fuel-burning appliance therein. Where such room or space serves as source of return-air.

Exceptions:

(1) This shall not apply to fireplaces, fireplace appliances, residential cooking appliances, direct vent appliances, enclosed furnaces, and domestic-type clothes dryers installed within the room or space.

(2) This shall not apply to a gravity-type or listed vented wall heating or cooling air system.

(3) This shall not apply to a blower-type heating or cooling air system installed in accordance with the following requirements:

(a) Where the return air is taken from a room or space having a volume exceeding 1 cubic foot (0.03 m^3) for each 10 Btu/h (0.003 kW) fuel input rating of fuel-burning appliances therein.

(b) Not less than 75 percent of the supply air is discharged back into the same room or space.

(c) Return-air inlets shall not be located within 10 feet (3048 mm) from an appliance firebox or draft diverter in the same enclosed room or confined space.

This section prohibits outdoor air and return air from being taken from locations that are potential sources of contamination, odor, flammable vapors or toxic substances and also from locations that would negatively affect the operation of the furnace itself or other fuel-burning appliances.

Item (1) is the commonly held provision that either the outdoor air should be taken at least 10 feet away from an appliance vent outlet, an exhaust outlet or a plumbing vent termination, or the exhaust outlet should be at least 3 feet above the air intake. This will ensure that the outdoor air intake will not pick up flue gasses, sewer gasses or harmful exhaust. When the vent outlet is 3 feet above the air intake, the harmful gasses exhausted tend to travel upward and, therefore, will not be drawn into the building through the outdoor air intake.

Similarly, item (2) ensures that exhaust fumes from vehicles traveling in public streets or alleys will not enter the building through the outdoor air intake.

It is also the intent of this section to avoid arrangements that could cause an air pressure imbalance, which could cause fuel-fired appliances to spill combustion products into the occupied space. Pressure imbalances can be avoided by making sure that the amount of supply air discharged into a room or space is approximately equal to the amount of return air taken from that room or space.

The intent of item (4) is to prevent the system from being starved for return air by the placement of the main or only return air intake in an area not meeting the minimum volume requirements (25%). Previously deleted code language allowed for the return to be placed in an area that does not meet the minimum volume requirement provided there are permanent openings to other areas in the building with an aggregate volume meeting the minimum required volume. The permanent openings were required to meet the minimum return air area (also deleted from earlier editions of the code) of 2 square inches per 1,000 Btu/h of the input rating of the furnace. The exception would allow the reduction of that area of the permanent openings, in dwelling units, to 50 percent of the required 2 square inches per 1,000 Btu/h of the input rating of the furnace, provided the opening is to a room meeting the conditions outlined in the exception, see **Figure 311.3a** for an example.

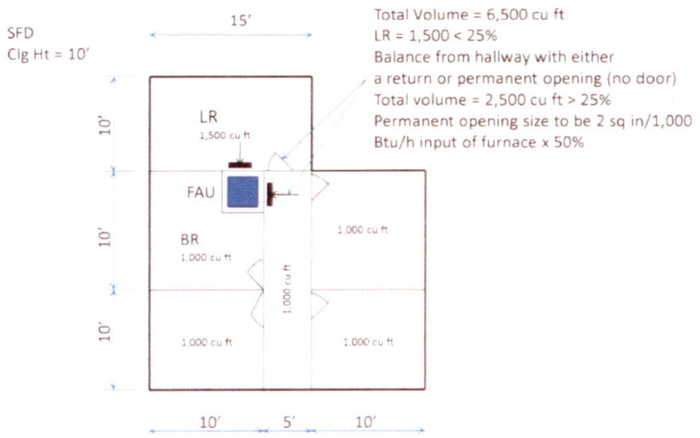

FIGURE 311.3A
RETURN AIR FROM OTHER SPACES

Item (6) addresses a similar issue. A blower-type furnace is typically required to be enclosed in a compartment to isolate the return air from its fire box, combustion air source, and flue vent draft diverter. Item (6) Exception (3) would allow such an installation with very specific conditions:

(a) The return air is taken from a room with a volume exceeding 1 cu. ft. per 10 Btu/h input rating of the furnace.

(b) Minimum 75 percent of supply air is discharged back into this room.

(c) Return air to be at least 10 feet from the fire box and the flue vent draft diverter

See **Figure 311.3b** for an example.

311.4 Return-Air Limitations. Return air from one dwelling unit shall not discharge into another dwelling unit through the heating or cooling air system.

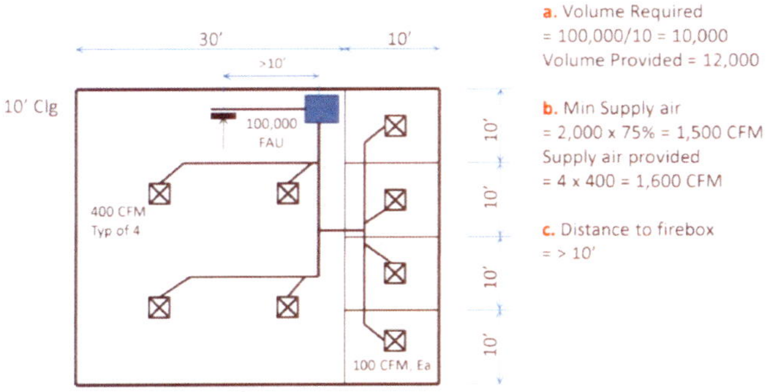

FIGURE 311.3B
BLOWER-TYPE FURNACE RETURN AIR EXCEPTION

312.0 Plumbing Connections.

312.1 General. Water supply, sanitary drainage, and backflow protection shall be in accordance with the plumbing code.

The section references the UPC for specific and detailed requirements of backflow protection on the water supply to mechanical equipment.

313.0 Hangers and Supports.

This section provides support requirements for any piping or tubing used in the HVAC system. These requirements cover hydronic systems and refrigerant systems as well as fuel gas systems. Another consideration for designing a system of pipe hangers, supports and riser clamps is to determine whether the pipe is scheduled to be insulated. Whenever insulation is applied, the type and thickness of the insulating material must be considered. The size of the hangers must be increased if the pipe is insulated, and the hangers are to be placed externally over the insulation. There must also be sufficient space for installers to apply and repair the insulation (see **Figure 313.0**).

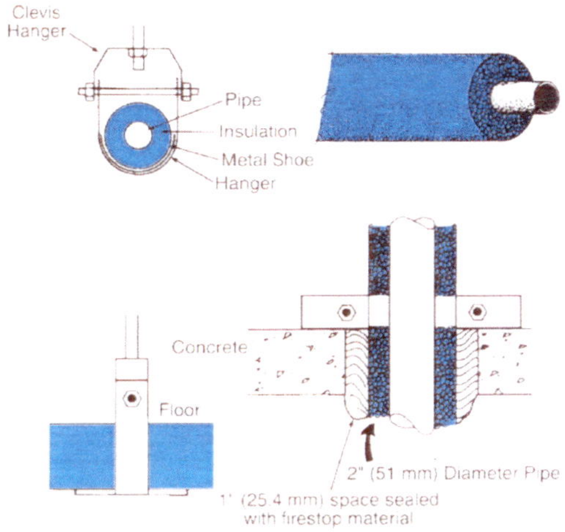

FIGURE 313.0
PIPE SUPPORTS WITH INSULATION

313.1 General. Piping, tubing, appliances, and appurtenances shall be supported in accordance with this code, the manufacturer's installation instructions, and in accordance with the Authority Having Jurisdiction.

313.2 Material. Hangers and anchors shall be of sufficient strength to support the weight of the pipe or tubing and its contents. Piping or tubing shall be isolated from incompatible materials.

313.3 Suspended Piping. Suspended piping or tubing shall be supported at intervals not to exceed those shown in Table 313.3.

313.4 Alignment. Piping or tubing shall be supported in such a manner as to maintain its alignment and prevent sagging.

313.5 Underground Installation. Piping or tubing in the ground shall be laid on a firm bed for its entire length; where other support is otherwise provided, it shall be approved in accordance with Section 302.0.

313.6 Hanger Rod Sizes. Hanger rod sizes shall be not smaller than those shown in Table 313.6.

TABLE 313.6
HANGER ROD SIZES

PIPE AND TUBE SIZE (inches)	ROD SIZES (inches)
½ - 4	⅜
5 - 8	½
10 - 12	⅝

For SI units: 1 inch = 25.4 mm

313.7 Gas Piping. Gas piping shall be supported by metal straps or hooks at intervals not to exceed those shown in Table 1310.3.5.1.

314.0 Balancing.

314.1 General. Heating, ventilating, and air-conditioning systems (including hydronic systems) shall be balanced in accordance with one of the following methods:

(1) AABC National Standards for Total System Balance

(2) ACCA Manual B

(3) ASHRAE 111

(4) NEBB Procedural Standards for Testing Adjusting Balancing of Environmental Systems

(5) SMACNA HVAC Systems Testing, Adjusting, and Balancing

This section provides requirements for mechanical equipment balancing in accordance with recognized standards. These standards provide basic information for testing, adjusting, balancing (TAB) – techniques for measurement of temperature, pressure, velocity, and quantities in various systems. In addition, these standards establish the following:

- Minimum system configuration requirements to ensure that the system can be field tested and balanced.
- Minimum instrumentation required for field measurements.
- Procedures for obtaining field measurements in HVAC testing and balancing and equipment testing; and formats for recording and reporting results.

The field data collected and reported under this standard are intended for use by the Authority Having Jurisdiction, building designers, operators, and code users, and by manufacturers and installers of HVAC systems.

The Associated Air Balance Council (AABC) - 2002 National Standard for Total System Balance includes the following:

- Fan Systems: Supply / Return / Relief / Exhaust
- Constant and Variable Air Systems
- Dual Duct Systems and High Pressure Systems
- Multi-Zone Systems
- Laboratories and Positively Pressurized Areas
- Hydronic Systems
- Sound Testing
- Vibration Testing
- Duct Leakage Testing
- Cooling Tower Testing
- Smoke Control Testing
- Commissioning
- Temperature Control System

ASHRAE Standard 111 applies to building heating, ventilating, and air-conditioning (HVAC) systems of the air-moving and hydronic types and their associated heat transfer, distribution, refrigeration, electrical power, and control subsystems. This standard includes:

- Methods for determining thermodynamic, hydraulic, hydronic, mechanical, and electrical conditions.
- Methods for determining room air-change rates, room pressurization, and cross contamination of spaces.
- Procedures for measuring and adjusting outdoor ventilation rates to meet specified requirements; and
- Methods for validating collected data while considering system effects.

SMACNA HVAC Systems Testing, Adjusting and Balancing covers variable frequency drives (VFD), direct digital control (DDC) systems, and lab hood exhaust balancing. The standard provides system testing, adjusting, and balancing fundamentals with reference tables and charts in the Appendix. This handbook will provide user with the information necessary to balance most heating, ventilation, and air conditioning (HVAC) systems.

315.0 Louvers in Hurricane Prone Regions.

315.1 General. Louvers located in areas within hurricane-prone regions that are within 1 mile (2 km) of the coastal mean high water line where the basic wind speed is 110 miles per hour (mi/h) (49.2 m/s) or more; or portions of hurricane-prone regions where the basic wind speed is 120 mi/h (53.6 m/s) or more; or Hawaii, as described in ASCE 7 shall be tested in accordance with Section 315.1.1 and Section 315.1.2.

315.1.1 Testing. Louvers that protect air intake or exhaust openings shall be tested in accordance with AMCA 550 for resistance to wind-driven rain.

315.1.2 Impact Resistance Test. Upon request by the Authority Having Jurisdiction, louvers protecting intake and exhaust ventilation ducts that are not fixed in the open position and located within 30 feet (9144 mm) of the grade shall be tested for impact resistance in accordance with AMCA 540.

This section and the referenced standards in Chapter 17, defines laboratory test methods and minimum performance ratings for water rejection capabilities of louvers intended to be used in high velocity wind conditions. This helps to ensure facilities (essential and nonessential) will remain in operation during a high velocity wind condition and where water infiltration must be kept to manageable amounts.

316.0 Protection of Piping, Tubing, Materials, and Structures.

316.1 General. Piping or tubing passing under or through walls shall be protected from breakage. Piping passing through or under cinders or other corrosive materials shall be protected from external corrosion in an approved manner. Approved provisions shall be made for expansion of hot water piping. Voids around piping or tubing passing through concrete floors on the ground shall be sealed.

Care should always be taken to ensure that mechanical systems passing through or under a wall will be protected from the weight or the expansion and contraction of the wall, which may cause damage to the pipe.

Corrosion of hydronic piping may occur when the pipe is placed in soil containing cinders or other corrosive materials. Cinders from coal fires often contain sulfur. The sulfur dissolves in groundwater to form sulfuric acid, which is corrosive to many piping materials.

There are several methods to provide for the expansion of hot water piping systems such as expansion loops, expansion joints, swing joints, and expansion coils. Consult the water supply and hydronic heating and cooling manuals for methods of piping for expansion and contraction.

Voids around piping passing through concrete floors on the ground must be sealed in a manner that is durable and waterproof to protect the interior of the building. This will also prevent the entry of vermin and insects into the room or building.

316.2 Installation. Piping or tubing shall be installed so that the piping, tubing, or connections will not be subject to undue strains or stresses, and provisions shall be made for expansion, contraction, and structural settlement. No piping or tubing, unless designed and listed for such use, shall be directly embedded in concrete or masonry. No structural member shall be seriously weakened or impaired by cutting, notching, or otherwise as defined in the building code.

The installation of the mechanical system may require piping through walls and floors, in the ground and under footings, and in many other conditions that could exert pressure or strain on the piping. Care must be taken so that the piping is installed to move freely, preventing structural settlement or expansion and contraction of piping from detrimentally affecting the system. Often, failure of the piping system is caused by the pipe being tightly secured in hangers, straps, or the structure itself. Expansion and contraction of either the pipe or the structure causes the pipe to crack or to slip out of its fittings.

Pipe passing perpendicularly through or parallel within a concrete or masonry wall or floor should always pass through a sleeve or box per Section 316.1. The intent is that piping shall not be located within or pass through concrete walls or slabs where a slight shifting or settlement can result in a major pipe failure. A pipe chase must be provided in the concrete wall when piping is to be recessed. Regardless of the method used to pass pipe through these conditions there must be provisions to allow movement of the pipe. This includes, but is not limited to, slabs, walls or footings. In no case shall the building impose a load on the pipe that could result in breakage or leaks in the piping system. **Figure 316.2** shows some recommended and prohibited piping installations.

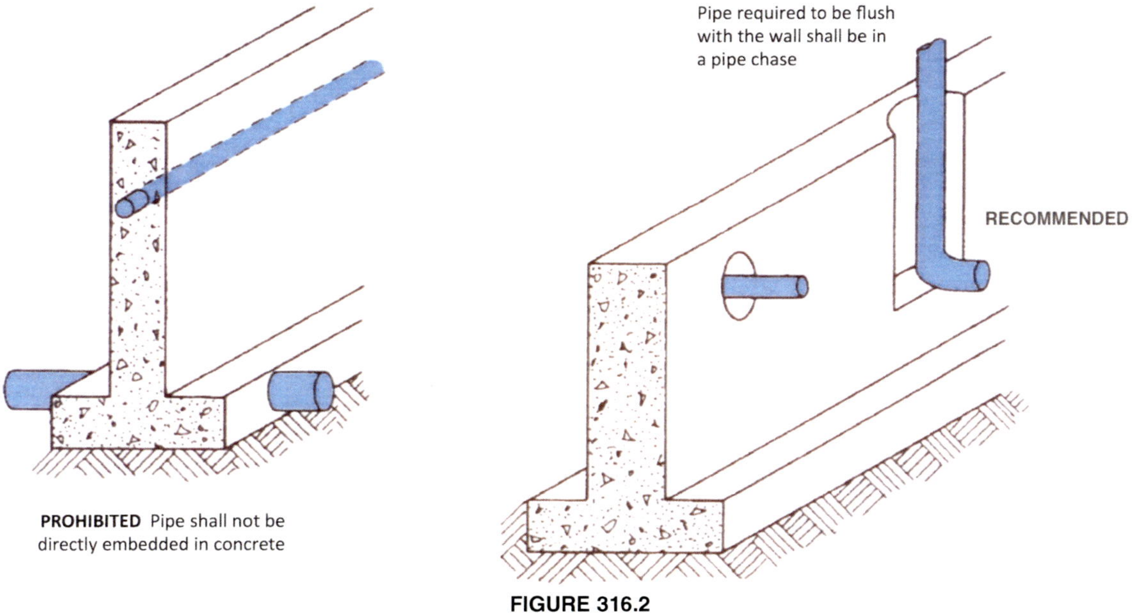

FIGURE 316.2
PROPER PLACEMENT OF PIPING WITHIN WALLS

When pipe passes or is placed within walls, the wall or structure should not be weakened by notching or cutting. The building code has provisions for how much material can be removed from the wall or stud before it is weakened to the point of failure.

316.3 Corrosion, Erosion, and Mechanical Damage. Piping or tubing subject to corrosion, erosion, or mechanical damage shall be protected in an approved manner.

Common methods of protecting piping from corrosion above and below ground include PVC sleeving or wrapping, painting, asphalt coating, or enclosing the pipe within approved material. Approved machine (factory) wrapped piping may be required.

Protection from mechanical damage usually requires the installation of sleeves, steel barricades and guards, or placing the pipe in recesses, pipe chases or gutters. Suspended ceilings or raceways may be required to protect overhead piping.

316.4 Protectively Coated Pipe. Protectively coated pipe or tubing shall be inspected and tested, and a visible void, damage, or imperfection to the pipe coating shall be repaired in an approved manner.

316.5 Fire-Resistant Construction. Piping, tubing, and duct system penetrations of fire-resistance-rated walls, partitions, floors, floor/ceiling assemblies, roof/ceiling assemblies, or shaft enclosures shall be protected in accordance with the requirements of the building code.

Piping and duct membrane penetrations and through penetrations of fire-resistance-rated walls must return the wall to its original fire-rated integrity by use of listed fire stopping systems. Not doing so will weaken the building's ability to resist the spread of fire within the building.

316.6 Steel Nail Plates. Plastic piping or tubing, copper or copper alloy piping or tubing, and ducts penetrating framing members to within 1 inch (25.4 mm) of the exposed framing shall be protected by steel nail plates not less than No. 18 gauge (0.0478 inches) (1.2141 mm) in thickness. The steel nail plate shall extend along the framing member not less than 1½ inches (38 mm) beyond the outside diameter of the pipe or tubing.

Exception: See Section 1310.4.3.

The protection of ducts or piping in the interior of walls is of utmost importance. The penetration of plastic and copper pipe by nails, screws or staples is very common. The methods provided here will ensure the pipe is not punctured either during construction or later by the owner of the building.

Section 1310.4.3, Tubing in Partitions, discusses the requirements for fuel gas piping protection. These requirements will be more stringent because of the volatility of the contents of the pipe. The pipe manufacturer's installation requirements should also be taken into consideration. These requirements may be even more stringent and, if so, should therefore be followed instead (see **Figure 316.6**).

316.7 Sleeves. Sleeves shall be provided to protect piping through concrete and masonry walls and concrete floors.

FIGURE 316.6
USE OF NAIL PLATES

Exception: Sleeves shall not be required where openings are drilled or bored.

316.7.1 Building Loads. Piping or tubing through concrete or masonry walls shall not be subject to a load from building construction.

316.7.2 Exterior Walls. In exterior walls, annular space between sleeves and pipes or tubing shall be sealed and made watertight, as approved by the Authority Having Jurisdiction. A penetration through fire-resistive construction shall be in accordance with Section 316.5.

316.8 Firewalls. A pipe sleeve through a firewall shall have the space around the pipe or tubing completely sealed with an approved fire-resistive material in accordance with other codes.

These provisions for piping sleeves are almost identical to the requirements in the previous sections concerning piping protection. They are repeated here so that there will be no doubt that the sleeves and their wall penetrations will be protected as any pipe penetration should.

316.9 Structural Members. A structural member weakened or impaired by cutting, notching, or otherwise shall be reinforced, repaired, or replaced so as to be left in a safe structural condition in accordance with the requirements of the building code.

In Section 316.2 the code prohibits the weakening of a structural member. By the drilling, boring or notching, the weakening of wall supports is unavoidable. Because of this fact, this section adds the requirement to return the member to its original integrity. The building code will state the methods used to accomplish this.

316.10 Rodentproofing. Mechanical system shall be constructed in such a manner as to restrict rodents or vermin from entering a building by following the ductwork from the outside into the building.

316.11 Metal Collars. In or on buildings where openings have been made in walls, floors, or ceilings for the passage of ductwork or pipes, such openings shall be closed and protected by the installation of approved metal collars securely fastened to the adjoining structure.

The infestation of rats, mice or other vermin can lead to serious health hazards. These sections of the code give the users guidance in protecting the penetrations in floors or walls from these pests.

317.0 Trenching, Excavation, and Backfill.

317.1 Trenches. Trenches deeper than the footings of a building or structure, and paralleling the same, shall be located not less than 45 degrees (0.79 rad) from the bottom exterior edge of the footing, or as approved in accordance with Section 302.0.

317.2 Tunneling and Driving. Tunneling and driving shall be permitted to be done in yards, courts, or driveways of a building site. Where sufficient depth is available to permit, tunnels shall be permitted to be used between open-cut trenches. Tunnels shall have a clear height of 2 feet (610 mm) above the pipe and shall be limited in length to one-half the depth of the trench, with a maximum length of 8 feet (2438 mm). Where pipes are driven, the drive pipe shall be not less than one size larger than the pipe to be laid.

It may be necessary at times to place piping under existing walkways and driveways. This is especially true in residential and repair installations. Tunneling or pipe driving may be required to complete the installation. In that case, the above requirements should be followed.

317.3 Open Trenches. Excavations required to be made for the installation of a mechanical system or part thereof, within the walls of a building, shall be open trench work and shall be kept open until it has been inspected, tested, and accepted.

Tunneling and driving are not permitted within the building. Instead trenches shall be used to install piping within the walls of a building. The trenches must be kept open and the pipe joints visible for inspection. Be sure to follow Occupational Safety and Health Administration (OSHA) safety requirements for trench work.

See **Learning Link http://bit.ly/2GRseZS**.

317.4 Excavations. Excavations shall be completely backfilled as soon after inspection as practicable. Precaution shall be taken to ensure compactness of backfill around piping without damage to such piping. Trenches shall be backfilled in thin layers to 12 inches (305 mm) above the top of the piping with clean earth, which shall not contain stones, boulders, cinderfill, frozen earth, construction debris, or other materials that will damage or break the piping or cause corrosive action. Mechanical devices such as bulldozers, graders, etc., shall be permitted to then be used to complete backfill to grade. Fill shall be properly compacted. Precautions shall be taken to ensure permanent stability for pipe laid in filled or made ground.

TABLE 303.10.1
REDUCTION OF CLEARANCES WITH SPECIFIED FORMS OF PROTECTION[1, 2, 3, 4, 5, 6, 7, 8, 9, 10, 11]
[NFPA 54: TABLE 10.2.3]

TYPE OF PROTECTION APPLIED TO AND COVERING ALL SURFACES OF COMBUSTIBLE MATERIAL WITHIN THE DISTANCE SPECIFIED AS THE REQUIRED CLEARANCE WITH NO PROTECTION	WHERE THE REQUIRED CLEARANCE WITH NO PROTECTION FROM APPLIANCE, VENT CONNECTOR, OR SINGLE-WALL METAL PIPE IS:									
	36 (INCHES)		18 (INCHES)		12 (INCHES)		9 (INCHES)		6 (INCHES)	
	ALLOWABLE CLEARANCES WITH SPECIFIED PROTECTION (INCHES)									
	USE COLUMN 1 FOR CLEARANCES ABOVE APPLIANCE OR HORIZONTAL CONNECTOR. USE COLUMN 2 FOR CLEARANCES FROM APPLIANCES, VERTICAL CONNECTOR, AND SINGLE-WALL METAL PIPE.									
	ABOVE (COLUMN 1)	SIDES AND REAR (COLUMN 2)	ABOVE (COLUMN 1)	SIDES AND REAR (COLUMN 2)	ABOVE (COLUMN 1)	SIDES AND REAR (COLUMN 2)	ABOVE (COLUMN 1)	SIDES AND REAR (COLUMN 2)	ABOVE (COLUMN 1)	SIDES AND REAR (COLUMN 2)
(1) 3½ inch thick masonry wall without ventilated air space	—	24	—	12	—	9	—	6	—	5
(2) ½ of an inch insulation board over 1 inch glass fiber or mineral wool batts	24	18	12	9	9	6	6	5	4	3
(3) 0.024 inch (nominal 24 gauge) sheet metal over 1 inch glass fiber or mineral wool batts reinforced with wire on rear face with ventilated air space	18	12	9	6	6	4	5	3	3	3
(4) 3½ inch thick masonry wall with ventilated air space	—	12	—	6	—	6	—	6	—	6
(5) 0.024 inch (nominal 24 gauge) sheet metal with ventilated air space	18	12	9	6	6	4	5	3	3	2
(6) ½ of an inch thick insulation board with ventilated air space	18	12	9	6	6	4	5	3	3	3
(7) 0.024 inch (nominal 24 gauge) sheet metal with ventilated air space over 0.024 inch (nominal 24 gauge) sheet metal with ventilated air space	18	12	9	6	6	4	5	3	3	3
(8) 1 inch glass fiber or mineral wool batts sandwiched between two sheets 0.024 inch (nominal 24 gauge) sheet metal with ventilated air space	18	12	9	6	6	4	5	3	3	3

For SI units: 1 inch = 25.4 mm, °C = (°F-32)/1.8

Notes:

[1] Reduction of clearances from combustible materials shall not interfere with combustion air, draft hood clearance and relief, and accessibility of servicing.

[2] All clearances shall be measured from the outer surface of the combustible material to the nearest point on the surface of the appliance, disregarding any intervening protection applied to the combustible material.

[3] Spacers and ties shall be of noncombustible material. No spacer or tie shall be used directly opposite the appliance or connector.

[4] Where all clearance reduction systems use a ventilated air space, adequate provision for air circulation shall be provided as described. [See Figure 303.10.1(2) and Figure 303.10.1(3)]

[5] At least 1 inch (25.4 mm) shall be between clearance reduction systems and combustible walls and ceilings for reduction systems using a ventilated air space.

[6] Where a wall protector is mounted on a single flat wall away from corners, it shall have a minimum 1 inch (25.4 mm) air gap. To provide adequate air circulation, the bottom and top edges, or only the side and top edges, or all edges shall be left open.

[7] Mineral wool batts (blanket or board) shall have a minimum density of 8 pounds per cubic foot (lb/ft^3) (128 kg/m^3) and a minimum melting point of 1500°F (816°C).

[8] Insulation material used as part of a clearance reduction system shall have a thermal conductivity of 1 British thermal unit inch per hour square foot degree Fahrenheit [$Btu \cdot in/(h \cdot ft^2 \cdot °F)$] [0.1W/(m•K)] or less.

[9] At least 1 inch (25.4 mm) shall be between the appliance and the protector. In no case shall the clearance between the appliance and the combustible surface be reduced below that allowed in Table 303.10.1.

[10] All clearances and thicknesses are minimum; larger clearances and thicknesses are acceptable.

[11] Listed single-wall connectors shall be installed in accordance with the manufacturer's installation instructions.

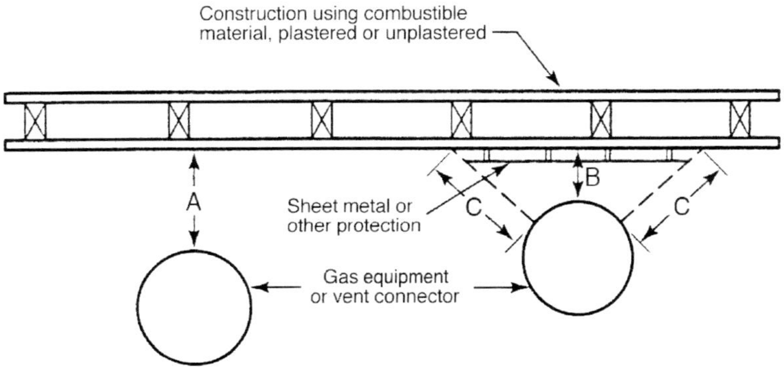

Notes:

1 A – Equals the clearance with no protection specified in Table 802.7.3.3 and Table 904.2.2 and in the sections applying to various types of equipment.

2 B – Equals the reduced clearance permitted in accordance with Table 303.10.1.

3 The protection applied to the construction using combustible material shall extend far enough in each direction to make C equal to A.

FIGURE 303.10.1(1)[1, 2, 3]
EXTENT OF PROTECTION NECESSARY T25O REDUCE CLEARANCES FROM GAS APPLIANCES OR VENT CONNECTORS
[NFPA 54: FIGURE 10.3.2.3(a)]

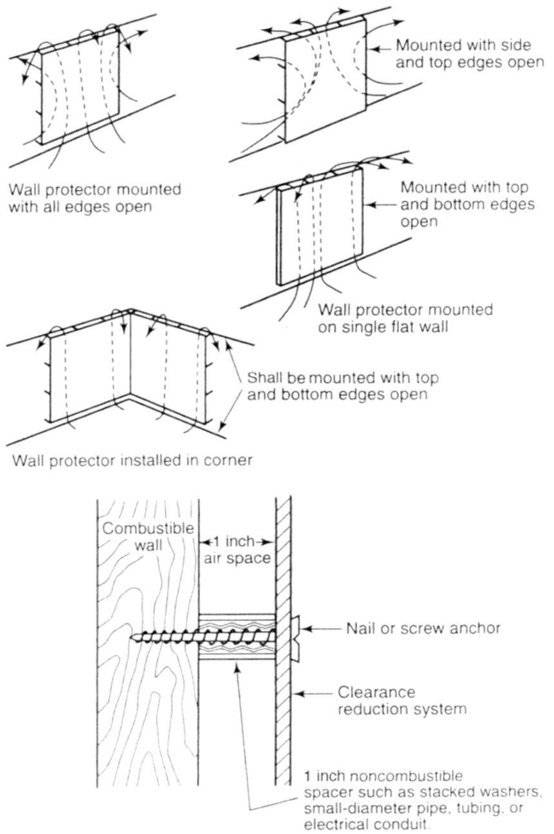

For SI units: 1 inch = 25.4 mm

Note: Masonry walls can be attached to combustible walls using wall ties. Spacers shall not be used directly behind appliance or connector.

FIGURE 303.10.1(2)
WALL PROTECTOR CLEARANCE REDUCTION SYSTEM
[NFPA 54: FIGURE 10.3.2.3(b)]

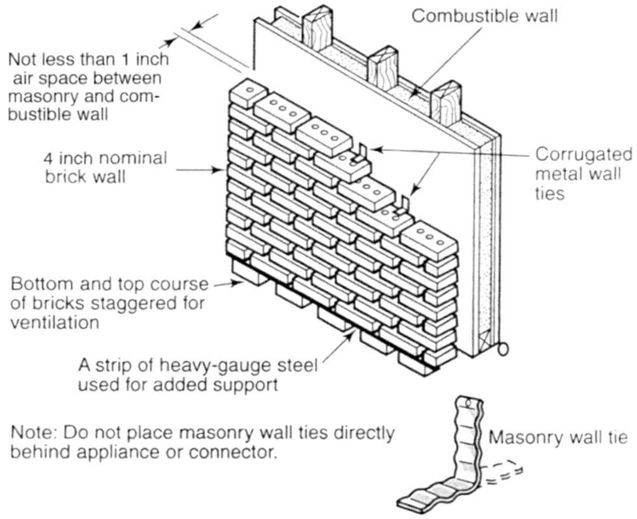

For SI units: 1 inch = 25.4 mm

FIGURE 303.10.1(3)
MASONRY CLEARANCE REDUCTION SYSTEM
[NFPA 54: FIGURE 10.3.2.3(c)]

TABLE 313.3
HANGERS AND SUPPORTS

MATERIALS	TYPES OF JOINTS	HORIZONTAL	VERTICAL
Cast	Lead and Oakum	5 feet, except 10 feet where 10 foot lengths are installed[1, 2, 3]	Base and each floor, not to exceed 15 feet
	Compression Gasket	Every other joint, unless over 4 feet then support each joint[1, 2, 3]	Base and each floor, not to exceed 15 feet
Cast-Iron Hubless	Shielded Coupling	Every other joint, unless over 4 feet then support each joint[1, 2, 3, 4]	Base and each floor, not to exceed 15 feet
Copper & Copper Alloys	Soldered, Brazed, Threaded, or Mechanical	1½ inches and smaller, 6 feet; 2 inches and larger, 10 feet	Each floor, not to exceed 10 feet[5]
Steel Pipe for Water DWV	Threaded or Welded	¾ inch and smaller, 10 feet; 1 inch and larger, 12 feet	Every other floor, not to exceed 25 feet[5]
Steel Pipe for Gas	Threaded or Welded	½ inch, 6 feet; ¾ inch and 1 inch, 8 feet; 1¼ inches and larger, 10 feet	½ inch, 6 feet; ¾ inch and 1 inch, 8 feet; 1¼ inches every floor level
Schedule 40 PVC and ABS	Solvent Cemented	All sizes, 4 feet; allow for expansion every 30 feet[3]	Base and each floor; provide mid-story guides; provide for expansion every 30 feet
CPVC	Solvent Cemented	1 inch and smaller, 3 feet; 1¼ inches and larger, 4 feet	Base and each floor; provide mid-story guides
CPVC-AL-CPVC	Solvent Cemented	½ inch, 5 feet; ¾ inch, 65 inches; 1 inch, 6 feet	Base and each floor; provide mid-story guides
Lead	Wiped or Burned	Continuous Support	Not to exceed 4 feet
Steel	Mechanical	In accordance with standards acceptable to the Authority Having Jurisdiction	
PEX	Cold Expansion, Insert and Compression	1 inch and smaller, 32 inches; 1¼ inches and larger, 4 feet	Base and each floor; provide mid-story guides
PEX-AL-PEX	Metal insert and metal compression	½ inch, ¾ inch, 1 inch } All sizes 98 inches	Base and each floor; provide mid-story guides
PE-AL-PE	Metal insert and metal compression	½ inch, ¾ inch, 1 inch } All sizes 98 inches	Base and each floor; provide mid-story guides
PE-RT	Insert and Compression	1 inch and smaller, 32 inches; 1¼ inches and larger, 4 feet	Base and each floor; provide mid-story guides
Polypropylene (PP)	Fusion weld (socket, butt, saddle, electrofusion), threaded (metal threads only), or mechanical	1 inch and smaller, 32 inches; 1¼ inches and larger, 4 feet	Base and each floor; provide mid-story guides

For SI unit: 1 inch = 25.4 mm, 1 foot = 304.8 mm

Notes:

1 Support adjacent to joint, not to exceed 18 inches (457 mm).

2 Brace not to exceed 40 feet (12 192 mm) intervals to prevent horizontal movement.

3 Support at each horizontal branch connection.

4 Hangers shall not be placed on the coupling.

5 Vertical water lines shall be permitted to be supported in accordance with recognized engineering principles with regard to expansion and contraction, where first approved by the Authority Having Jurisdiction.

CHAPTER 4
VENTILATION AIR

401.0 General.

401.1 Applicability. This chapter contains requirements for ventilation air supply, exhaust, and makeup air requirements for occupiable spaces within a building. Spaces within buildings, except those within a dwelling unit in residential occupancies where occupants are nontransient, shall comply with Section 402.0 through Section 404.0. Requirements for ventilation air rate for dwelling units in residential occupancies, where the occupants are nontransient, shall be in accordance with Section 405.0.

This chapter focuses mainly on the provisions of ventilation air. Although the title of the chapter is Ventilation Air (which is the predominant topic), the chapter also includes minimum rates of exhaust for certain uses, such as toilet rooms, both private and public, and for parking garages.

Ventilation of the building interior with outdoor air is for the comfort and health of the occupants. The main purpose is to replenish the oxygen consumed and to exhaust the moisture and carbon dioxide produced during the breathing process. In addition, ventilation removes contaminants produced inside the building, both unintentionally by the emission of materials used within the building, such as carpeting, surface finishes and furniture, or deliberately by manufacturing or industrial processes performed within the building. When there are substantial amounts of contaminants produced by a manufacturing process, the building code of the jurisdiction, the fire code of the jurisdiction, and Chapter 5 of this code should be used to determine the required rates and method of exhausting the contaminants to the outdoors, and the rates of corresponding makeup air supply or outdoor air requirements.

As stated above, the majority of the provisions in this chapter are dedicated to ventilation air supply for human health and comfort. The chapter discusses two approved means of providing outdoor air to a building: natural ventilation and mechanical ventilation. Natural ventilation can be provided through operable windows and/or skylights, provided they have minimum area requirements, proximity to the ventilated spaces and are readily accessible to the occupants to operate. When natural ventilation is not provided, the code requires the use of mechanical ventilation with minimum flow rates that are dependent on the type and number of expected occupants and the floor area of the zones in the building.

Outdoor mechanical ventilation (for the occupancies covered in the table) is required to continually operate when the building is occupied and is not allowed to cycle on and off along with the heating or cooling units serving the building.

Mechanical ventilation shall be provided to exhaust moisture and odors from each dwelling. This is in addition to, not in lieu of, any natural ventilation openings that are provided.

402.0 Ventilation Air.

Indoor air pollution is a health threat common to buildings, especially relatively modern ones. Numerous studies by international health and environmental scientists have documented the widespread nature of the problem. The World Health Organization estimated that in 1989, 30 percent of all new or refurbished buildings have indoor air problems.

Today, the term Sick Building Syndrome has become synonymous with tight building construction. This term has been historically applied to buildings when more than 20 percent of the employees experience symptoms such as headaches, fatigue or eye, nose or throat irritation that cease once the employee leaves the building.

In 1987, the United States National Institute for Occupational Safety and Health (NIOSH) studied 446 buildings following staff complaints of upper respiratory illness and poor air quality. The study concluded that more than half the problems were due to inadequate ventilation.

The American Society of Heating, Refrigerating and Air-Conditioning Engineers (ASHRAE) published a position statement and paper on Indoor Air Quality (IAQ) in February 1989. In it, ASHRAE asserted the importance of IAQ and related energy conservation and public health issues. Since then, the standard ASHRAE 62.2, Ventilation for Acceptable Indoor Air Quality, has been updated several times. A major portion of the latest standard is now included in this chapter.

IAPMO includes easy-to-use tables and formulas in this Chapter to calculate ventilation rates for common building uses. The tables and accompanying provisions enable the user to determine the appropriate ventilation (outside) air required to avoid poor IAQ.

In certain areas of the country, such as in the state of California, the outdoor air requirement of this chapter is superseded by other regulations, such as energy conservation regulations, which provide similar but somewhat different methods of calculating the outdoor air requirements.

402.1 Occupiable Spaces. Occupiable spaces listed in Table 402.1 shall be designed to have ventilation (outdoor) air for occupants in accordance with this chapter.

Mechanical ventilation shall be provided to the types of occupied spaces listed in Table 402.1 at the corresponding specified rates. The most common categories listed are: Correctional Facilities, Educational Facilities, Food Services, Hotels, Offices, Public Assemblies, Retail and Sports. For those nonresidential occupancies not found in the table, the Authority Having Jurisdiction (AHJ) should select and use the rates for a similar occupancy or request an analysis by the designer to determine the most appropriate table category that can be used.

402.1.1 Construction Documents. The outdoor air ventilation rate and air distribution assumptions made in the design of the ventilation system shall be clearly identified on the construction documents.

Since the calculations for minimum mechanical ventilation requirements are somewhat involved and would include certain assumptions of the occupant load and use, the information must be shown on the design plans to allow for review and verification. In addition, the calculated outdoor air rates should be shown to allow field verification by the installer.

402.1.2 Ventilation in Health Care Facilities. Mechanical ventilation for health care facilities shall be designed and installed in accordance with this code and ASHRAE 170.

Ventilation in healthcare facilities is more complex than requiring a certain amount of ventilation in a space. ASHRAE has published a new standard that sets requirements for ventilation in healthcare facilities (ANSI/ASHRAE Standard 170) which has more comprehensive requirements for ventilation, including design parameter requirements. The Facility Guidelines Institute has incorporated ASHRAE Standard 170 into the ventilation design requirements at healthcare facilities.

402.2 Natural Ventilation. Natural ventilation systems shall be designed in accordance with this section and shall include mechanical ventilation systems designed in accordance with Section 403.0, Section 404.0, or both.

Exceptions:

(1) An engineered natural ventilation system where approved by the Authority Having Jurisdiction need not comply with Section 402.2.

(2) The mechanical ventilation systems shall not be required where:

(a) natural ventilation openings that comply with the requirements of Section 402.2 and are permanently open or have controls that prevent the openings from being closed during periods of expected occupancy or

(b) the zone is not served by heating or cooling equipment. [ASHRAE 62.1:6.4]

Natural ventilation is allowed in lieu of mechanical ventilation. Natural ventilation is dependent on wind direction and speed with respect to the location of the openings into the building. Natural ventilation is not automatic, so it is dependent on the occupant to manually operate the openings. As a result, natural ventilation is not as reliable as mechanical ventilation in ventilating spaces, especially those in critical need of ventilation to expel harmful emissions. Conversely, natural ventilation does not rely on mechanical means (that is subject to failure) to provide the needed ventilation.

Exception (1) allows utilizing an engineered natural ventilation system in lieu of the above prescriptive method.

402.2.1 Floor Area to be Ventilated. Spaces, or portions of spaces, to be naturally ventilated shall be located within a distance based on the ceiling height, as determined in accordance with Section 402.2.1.1, Section 402.2.1.2, or Section 402.2.1.3, from operable wall openings in accordance with the requirements of Section 402.2.2. For spaces with ceilings that are not parallel to the floor, the ceiling height shall be determined in accordance with Section 402.2.1.4. [ASHRAE 62.1:6.4.1]

402.2.1.1 Single Side Opening. For spaces with operable openings on one side of the space, the distance from the operable openings shall be not more than $2H$, where H is the ceiling height. [ASHRAE 62.1:6.4.1.1]

402.2.1.2 Double Side Opening. For spaces with operable openings on two opposite sides of the space, the distance from the operable openings shall be not more than $5H$, where H is the ceiling height. [ASHRAE 62.1:6.4.1.2]

402.2.1.3 Corner Openings. For spaces with operable openings on two adjacent sides of a space, the distance from the operable openings shall be not more than $5H$ along a line drawn between the two openings that are farthest apart. Floor area outside that line shall comply with Section 402.2.1.1. [ASHRAE 62.1:6.4.1.3]

402.2.1.4 Ceiling Height. The ceiling height, H, to be used in Section 402.2.1.1 through Section 402.2.1.3 shall be the minimum ceiling height in the space.

Exception: For ceilings that are increasing in height as distance from the openings is increased, the ceiling height shall be determined as the average height of the ceiling within 20 feet (6096 mm) from the operable openings. [ASHRAE 62.1:6.4.1.4]

402.2.2 Location and Size of Openings. Spaces or portions of spaces to be naturally ventilated shall be permanently open to operable wall openings directly to the outdoors. The openable area shall be not less than 4 percent of the net occupiable floor area. Where openings are covered with louvers or otherwise obstructed, openable area shall be based on the net free unobstructed area through the opening. Where interior rooms, or portions of rooms, without direct openings to the outdoors are ventilated through adjoining rooms, the opening between rooms shall be permanently unobstructed and have a free area of not less than 8 percent of the area of the interior room or less than 25 square feet (2.3 m^2). [ASHRAE 62.1:6.4.2]

To provide natural ventilation to a space, the prescriptive requirement of the code is for the space to have windows and/or skylights with an openable area of at least 4 percent of the floor area of the subject space or room. The calculated floor area should not include any inaccessible spaces, such as shafts and chases, and should not include storage areas and equipment rooms not normally occupied (see Chapter 2 for the definition of Occupiable Space).

The ventilated space must also be within 25 feet of the opening. (Please note that the minimum area requirement is included in most building codes in the country; however, the maximum distance requirement from the opening is not.) The required minimum must be free area, excluding any obstructions, such as louvers or grilles.

The code also allows ventilating an interior adjoining room as long as the aggregate area of the openings to the outdoors meets the minimum area requirement for both spaces. Furthermore, there needs to be a permanent opening(s) between the two adjoining spaces that has an aggregate area of at least 8 percent of the floor area of the interior room, with a minimum of 25 square feet.

Example 1 (See **Figure 402.2.2a**):

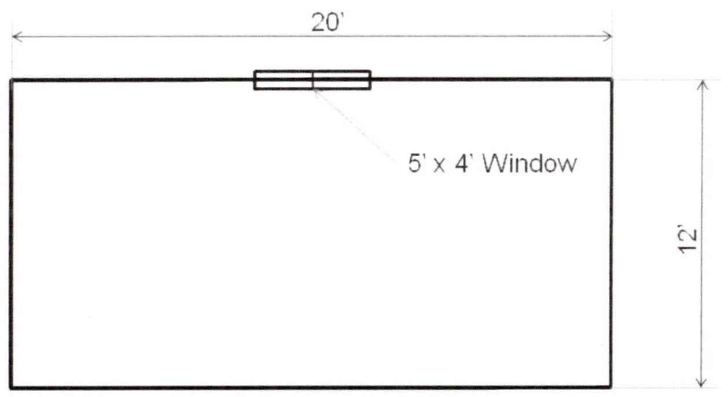

FIGURE 402.2.2A
EXAMPLE

Consider a room 20 feet long and 12 feet wide with one window that is 5 feet wide and 4 feet high. The openable area of the window is approximately 50 percent of its gross area. Is the window area sufficient to meet the natural ventilation requirements of the code?

Room area = 20 x 12 = 240 square feet

Required natural ventilation area = 240 x .04 = 9.6 square feet

Window openable area = 5 x 4 x 0.5 = 10 square feet

The window area is sufficient to provide natural ventilation for the room.

Example 2 (see **Figure 402.2.2b**):

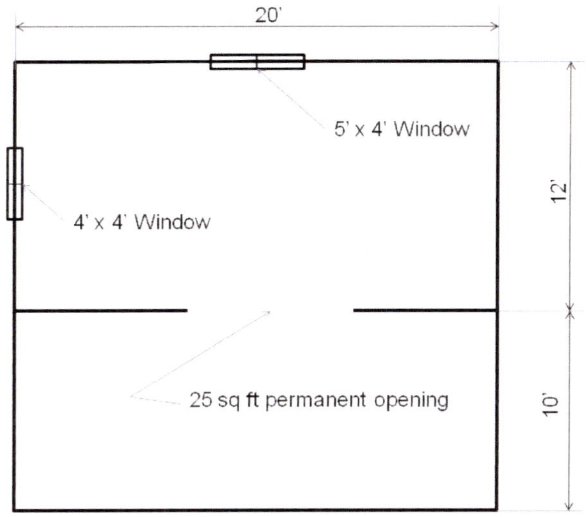

FIGURE 402.2.2B
EXAMPLE

An interior room is adjacent to the room in the previous example and has a floor area of 200 square feet. What is the minimum window size that should be added and what is the minimum area of the required permanent opening between the two adjoining rooms?

Adjoining interior room area = 200 square feet

Total area of the two rooms = 200 + 240 = 440 square feet

Minimum total opening area = 440 x 0.04 = 17.6 square feet

Window already provided has a net opening = 10 square feet

Required minimum openable area = 17.6 – 10 = 7.6 square feet

Assuming another window with 50 percent openable area:

Minimum required window size = 7.6/0.5 = 15.2 square feet

Use one 4 foot by 4 foot openable window.

If a skylight is to be provided, the minimum openable area of the skylight must be at least 7.6 square feet.

In addition, there must be a device to open the skylight within reach of the occupant to meet the definition of readily accessible.

The minimum permanent opening between the two rooms must be at least = 200 x .08 = 16 square feet,

Use 25 square feet minimum required by the code.

402.2.3 Control and Accessibility. The means to open required operable openings shall be readily accessible to building occupants where the space is occupied. Controls shall be designed to coordinate operation of the natural and mechanical ventilation systems. [ASHRAE 62.1:6.4.3]

The code states that the means to open the operable openings must be readily accessible to the building occupants whenever the space is occupied. Readily accessible is defined as having the openings operable without the need to use portable ladders or tools that may not be available to the occupants. In the case of a window, this would not be an issue of concern; however, in the case of an openable skylight, there must be a device within reach of the occupants to open and close the skylight to meet this requirement.

402.3 Mechanical Ventilation. Where natural ventilation is not permitted by this section or the building code, mechanical ventilation systems shall be designed, constructed, and installed to provide a method of supply air and exhaust air. Mechanical ventilation systems shall include controls, manual or automatic, that enable the fan system to operate wherever the spaces served are occupied. The system shall be designed to maintain minimum outdoor airflow as required by Section 403.0 under any load conditions.

When natural ventilation is not permitted or not feasible, such as in buildings with fixed windows, mechanical ventilation is to be provided. All occupied spaces in the building (as listed in Table 402.1) are to be provided with continuous ventilation when the occupants are present. For the types of occupancy where the occupants are present only during certain times of day (such as in office buildings), ventilation is to be continuous during working hours. The heating and cooling systems would typically cycle on and off in response to the heating and cooling needs of the building; however, the system controls should be set to force the ventilation to operate continuously while the occupants are present to supply the building with the minimum outdoor ventilation requirements. This should not prevent the entire system (i.e. heating, cooling and ventilation) from completely shutting down during periods of nonuse (off hours), as required by most energy conservation regulations in the country.

402.4 Outdoor Air Intake Protection. Required outdoor-air intakes shall be covered with a screen having not less than ¼ of an inch (6.4 mm) openings, and shall have not more than ½ of an inch (12.7 mm) openings.

402.4.1 Weather Protections. Outdoor air intakes that are part of the mechanical ventilation system shall be designed to manage rain entrainment, to prevent rain intrusion, and manage water from snow in accordance with ASHRAE 62.1.

Reference to ASHRAE Standard 62.1 is for protection of an outside air intake from rain and snow. This Standard states the following:

5.5.2 Rain Entrainment. Outdoor air intakes that are part of the mechanical ventilation system shall be designed to manage rain entrainment in accordance with any one of the following:

(a) Limit water penetration through the intake to 0.07 oz/ft²·h (21.5 g/m²·h) of inlet area when tested using the rain test apparatus described in Section 58 of UL 1995.

(b) Select louvers that limit water penetration to a maximum of 0.01 oz/ft² (3 g/m²) of louver free area at the maximum intake velocity. This water penetration rate shall be determined for a minimum 15-minute test duration when subjected to a water flow rate of 0.25 gal/min (16 mL/s) as described under the Water Penetration Test in AMCA 500-L13 or equivalent. Manage the water that penetrates the louver by providing a drainage area and/or moisture removal devices.

(c) Select louvers that restrict wind-driven rain penetration to less than 2.36 oz/ft²·h (721 g/m²·h) when subjected to a simulated rainfall of 3 in. (75 mm) per hour and a 29 mph (13 m/s) wind velocity at the design outdoor air intake rate with the air velocity calculated based on the louver face area.

Note: This performance corresponds to Class A (99% effectiveness) when rated according to AMCA 511 and tested per AMCA 500-L.

(d) Use rain hoods sized for no more than 500 fpm (2.5 m/s) face velocity with a downward-facing intake such that all intake air passes upward through a horizontal plane that intersects the solid surfaces of the hood before entering the system.

(e) Manage the water that penetrates the intake opening by providing a drainage area and/or moisture removal devices.

5.5.3 Rain Intrusion. Air-handling and distribution equipment mounted outdoors shall be designed to prevent rain intrusion into the airstream when tested at design airflow and with no airflow, using the rain test apparatus described in Section 58 of UL 1995.12

5.5.4 Snow Entrainment. Where climate dictates, outdoor air intakes that are part of the mechanical ventilation system shall be designed to manage water from snow, which is blown or drawn into the system, as follows:

(a) Suitable access doors to permit cleaning of wetted surfaces shall be provided.

(b) Outdoor air ductwork or plenums shall pitch to drains designed in accordance with the requirements of Section 5.10.

403.0 Ventilation Rates.

403.1 General. The design outdoor air intake flow rate for a ventilation system shall be determined in accordance with Section 403.2 through Section 403.9.4.

403.2 Zone Calculations. Ventilation zone parameters shall be determined in accordance with Section 403.2.1 through Section 403.2.3 for each ventilation zone served by the ventilation system. [ASHRAE 62.1:6.2.2]

The code requires that each unique zone in the building with certain outdoor air needs be calculated separately as outlined below. Simple buildings with one type of occupancy and similar outdoor air needs throughout could be calculated as one zone; however, when the building has several zones with varying outdoor air needs but with a single air-handling unit serving all zones, the calculations tend to become more complicated. It is usually impractical and costly to design a system that can supply each of the zones with its exact outdoor air needs all of the time. In addition, adding all the outdoor air needs for all the zones and setting the unit to provide that quantity will likely be inadequate, as all the zones will receive the same ratio of outdoor air to the total circulating air. Although the excess air in one zone would be transferred to the critical zones that have more needs than the average, this process is inherently inefficient and could create areas with insufficient outdoor air supply. On the other hand, setting the outdoor air ratio to the highest ratio required by the most critical zone will introduce excessive amounts of outdoor air, resulting in wasteful energy that will be expended in heating and cooling the excess air.

The method outlined below accounts for the inherent deficiency in transferring the outdoor air among zones and calculates the adequate rates that will ensure fulfilling the needs of all the critical zones while not wasting additional energy in heating or cooling excessive outdoor air.

403.2.1 Breathing Zone Outdoor Airflow. The outdoor airflow required in the breathing zone (V_{bz}) of the occupiable space or spaces in a ventilation zone shall be not less than the value determined in accordance with Equation 403.2.1.

$$V_{bz} = R_p \cdot P_z + R_a \cdot A_z \qquad \text{(Equation 403.2.1)}$$

Where:

A_z = zone floor area, the net occupiable floor area of the ventilation zone, square feet (m²).

P_z = zone population, the number of people in the ventilation zone during typical usage.

R_p = outdoor airflow rate required per person as determined from Table 402.1.

R_a = outdoor airflow rate required per unit area as determined from Table 402.1. [ASHRAE 62.1:6.2.2.1]

This section calculates the outdoor airflow rate required in each breathing zone of the occupied space.

Breathing Zone is defined in Chapter 2 as the region within an occupied space between three planes; 3 inches to 6 feet above the floor and 2 feet from the walls. This definition is to assist the mechanical system designers in locating the area where the occupants of the zone will be present in order to direct the airflow to these areas. There is no intention to restrict the location of the air supply registers within this area. The definition's intent is merely to require the airflow to be delivered where it is needed.

Occupied space is defined as an enclosed space intended for human activities, but not including those spaces intended primarily for other purposes (e.g., storage rooms and equipment rooms) that are only occupied occasionally and for short periods of time.

The equation to be used is (Equation 403.2.1):

Zone outdoor airflow =

(Rate/person* x No. of occupants) + (Rate/sq ft* x Zone area)

* Rates from Table 402.1

The number of occupants in the zone (P_z) is either the number of fixed seats in the zone or the number as per Table 402.1. If the designer provides a different number than those, the provided number should exceed those in the table. Otherwise, reasonable and rational justification for using a lesser number of occupants should be submitted to the AHJ.

403.2.2 Zone Air Distribution Effectiveness. The zone air distribution effectiveness (E_z) shall be not greater than the default value determined in accordance with Table 403.2.2. [ASHRAE 62.1:6.2.2.2]

This factor (E_z) accounts for the location of the air supply and return registers in the zone. The effectiveness of air delivery is accounted for in Table 403.2.2.

403.2.3 Zone Outdoor Airflow. The zone outdoor airflow (V_{oz}) provided to the ventilation zone by the supply air distribution system shall be determined in accordance with Equation 403.2.3. [ASHRAE 62.1:6.2.2.3]

$$V_{oz} = V_{bz}/E_z \qquad \text{(Equation 403.2.3)}$$

The corrected zone outdoor air flow rate is found using the result of the equation that calculates the zone outdoor air flow rate, in conjunction with the zone air distribution effectiveness factor from Table 403.2.2.

403.3 Single-Zone Systems. For ventilation systems where one or more air handlers supply a mixture of outdoor air and recirculated air to only one ventilation zone, the outdoor air intake flow (V_{ot}) shall be determined in accordance with Equation 403.3. [ASHRAE 62.1:6.2.3]

$$V_{ot} = V_{oz} \qquad \text{(Equation 403.3)}$$

In the case where the air handler is supplying outdoor air to one zone (or a number of zones with similar outdoor air needs that could be considered as one zone), the required outdoor air rates will be calculated according to Section 403.2.3.

Example of Sizing Ventilation (Outside) Air for a Single-Zone System: A building has a dedicated duct system providing ventilation (outdoor) air to the various floors. A floor is being remodeled into a call center. The floor area is 10,000 square feet and is served by five air handlers (similar zones). The duct delivering ventilation air will be ducted to each of the air handlers' return air plenum, and then the conditioned air is ducted to ceiling diffusers.

How much ventilation air should be delivered to each air handler?

Partial Table 402.1
MINIMUM VENTILATION RATES IN BREATHING ZONE

Occupancy Category	People Outdoor Air Rate (cfm/person)	Area Outdoor Air Rate (cfm/ft²)	Default Occupant Density (people/1,000ft²)
Telephone/ Data Entry	5	0.06	60

RP is the outdoor air rate required per person. Table 402.1 has a listing for telephone/data entry. This matches the example's call center designation.

Rp = 5 cubic feet per minute (cfm) per person.

Pz is the zone population. The designer is using the table's value because a seating chart is not available to use for the actual number. The table designates 60 persons per 1,000 square feet.

P_z = 60 x 10,000/1,000 = 600 occupants

R_a is the outdoor airflow rate.

R_a = 0.06 cfm per sq ft.

Az is the net floor area = 10,000 square feet.

V_{bz} = (5 cfm x 600) + (.06 cfm x 10,000)

V_{bz} = 3,000 cfm + 600 cfm

V_{bz} = 3,600 cfm

Table 403.2.2 indicates that the effectiveness of air delivery (E_z) is 1.

$V_{oz} = V_{bz}/E_z$

V_{oz} = 3,600 cfm divided by 1

V_{oz} = 3,600 cfm

There are five similar zones:

3,600 / 5 = 720 cfm per air handler.

It should be noted that this is a significant amount of outdoor air being introduced. Some method of relieving or exhausting air from the zones should be provided.

403.4 One Hundred Percent Outdoor Air Systems. For ventilation systems where one or more air handlers supply only outdoor air to one or more ventilation zones, the outdoor air intake flow (V_{ot}) shall be determined in accordance with Equation 403.4. [ASHRAE 62.1:6.2.4]

$$V_{ot} = \Sigma_{all\ zones} V_{oz} \qquad \text{(Equation 403.4)}$$

This is a case when a building requires 100-percent outdoor air and there is no recirculation. The air handler will supply only outdoor air and there would be provisions to exhaust 100 percent of the air supplied. In this case, the aggregate outdoor air rate for all the zones will be the required flow rate at the air handler.

403.5 Multiple-Zone Recirculating Systems. For ventilation systems where one or more air handlers supply a mixture of outdoor air and recirculated air to more than one ventilation zone, the outdoor air intake flow (V_{ot}) shall be determined in accordance with Section 403.5.1 through Section 403.5.2. [ASHRAE 62.1:6.2.5]

403.5.1 Uncorrected Outdoor Air Intake. The uncorrected outdoor air intake (V_{ou}) flow shall be determined in accordance with Equation 403.5.1. [ASHRAE 62.1:6.2.5.1]

(Equation 403.5.1)

$$V_{ou} = D \sum_{all\ zones} (R_p \cdot P_z) + \sum_{all\ zones} (R_a \cdot A_z)$$

This equation calculates the total outdoor airflow rate for the entire system before it is corrected for the system ventilation efficiency (which accounts for the inefficiency in transferring air among the zones).

The term D in the equation, which accounts for occupant diversity, is used when the occupants fluctuate among the zones. For example, if the occupants gather at a certain area in the building at a certain time of day then return to other areas for the remainder of the day, the system design must consider the location of the occupants and provide the needed rate of outdoor air to the gathering area, as well as to the remainder of the building when the occupants return. This will create diversity in the number of occupants as the total system occupants will be less than the aggregate number of occupants when the zones are totaled. In this case, the D factor will account for that diversity and allow proper sizing of the system. In all other cases, D will be equal to 1 and would not have any impact on the calculations.

The equation below sums outdoor air rate requirements for each zone to derive the total aggregate rate of outdoor air to all the zones.

$$V_{ou} = D \sum_{all\ zones} R_p \cdot P_z + \sum_{all\ zones} R_a \cdot A_z$$

Where $D = P_s / \sum_{all\ zones} P_z$

Where P_s is the system's total occupants.

403.5.1.1 Occupant Diversity. The occupant diversity ratio (D) shall be determined in accordance with Equation 403.5.1.1 to account for variations in population within the ventilation zones served by the system.

$$D = P_s / \sum_{all\ zones} P_z \qquad \text{(Equation 403.5.1.1)}$$

Where the system population (P_s) is the total population in the area served by the system.

Exception: Alternative methods to account for occupant diversity shall be permitted, provided that the resulting (V_{ou}) value is not less than that determined in accordance with Equation 403.5.1. [ASHRAE 62.1:6.2.5.1.1]

403.5.1.2 System Ventilation Efficiency. The system ventilation efficiency (E_v) shall be determined in accordance with Section 403.5.1.3 for the simplified procedure or Section 404.0 for the alternate procedure. These procedures also establish zone minimum primary airflow rates for VAV systems. [ASHRAE 62.1:6.2.5.2]

403.5.1.3 Simplified Procedure for System Ventilation Efficiency. System ventilation efficiency (E_v) shall be determined in accordance with Equation 403.5.1.3(1) or Equation 403.5.1.3(2). [ASHRAE 62.1:6.2.5.3 - 6.2.5.3.1]

$$E_v = 0.88 \cdot D + 0.22 \text{ for } D < 0.60 \qquad \text{[Equation 403.5.1.3(1)]}$$

$$E_v = 0.75 \text{ for } D \geq 0.60 \qquad \text{[Equation 403.5.1.3(2)]}$$

403.5.1.4 Zone Minimum Primary Airflow. For each zone, the minimum primary airflow ($V_{pz\text{-}min}$) shall be determined in accordance with Equation 403.5.1.4. [ASHRAE 62.1:6.2.5.3.2]

$$V_{pz\text{-}min} = V_{oz} \cdot 1.5 \qquad \text{(Equation 403.5.1.4)}$$

403.5.2 Outdoor Air Intake. The design outdoor air intake flow (V_{ot}) shall be determined in accordance with Equation 403.5.2. [ASHRAE 62.1:6.2.5.4]

$$V_{ot} = V_{ou} / E_v \qquad \text{(Equation 403.5.2)}$$

The uncorrected outdoor air intake is then adjusted for the system ventilation efficiency to determine the total outdoor air rate needed at the air handler to satisfy all the outdoor air needs of all the zones.

403.6 Design for Varying Operating Conditions. Ventilation systems shall be designed to be capable of providing not less than the minimum ventilation rates required in the breathing zone where the zones served by the system are occupied, including all full and part-load conditions. The minimum outdoor air intake flow shall be permitted to be less than the design value at part-load conditions. [ASHRAE 62.1:6.2.6.1]

This section addresses situations when the building has varying loads. Typically, these buildings will be designed with mechanical systems that will vary the heating and cooling depending on the load [usually, using a variable air volume (VAV) system]. These mechanical systems meet the varying loads by modulating the quantities of air delivered to a zone through a VAV box volume damper. The damper will be fully opened when the load is at its maximum design condition and will close down to a minimum setting when the load is at its lowest level. The minimum setting of the VAV box damper will also affect the amount of outdoor air flow to the area. Since zone population could be constant while the load varies (for example, at different times of the day), this section mandates that the designer consider these conditions and ensure that the minimum outdoor air rates be delivered, even at part-load conditions.

403.6.1 Short-Term Conditions. Where it is known that peak occupancy will be of short duration, ventilation will be varied or interrupted for a short period of time, or both, the design shall be permitted to be based on the average conditions over a time period (T) determined by Equation 403.6.1.

$$T = 3v/V_{bz} \qquad \text{(Equation 403.6.1)}$$

Where:

T = averaging time period, minutes.

v = the volume of the ventilation zone where averaging is being applied, cubic foot (m³).

V_{bz} = the breathing zone outdoor airflow calculated in

accordance with Equation 403.2.1 and design value of the zone population (P_z), cubic foot per minute (CFM) (m^3/min).

Acceptable design adjustments based on this optional provision including the following:

(1) Zones with fluctuating occupancy: The zone population (P_z) shall be permitted to be averaged over time (T).

(2) Zones with intermittent interruption of supply air: The average outdoor airflow supplied to the breathing zone over time (T) shall be not less than the breathing zone outdoor airflow (V_{bz}) calculated using Equation 403.2.1.

(3) Systems with intermittent closure of the outdoor air intake: The average outdoor air intake over time (T) shall be not less than the minimum outdoor air intake (V_{ot}) calculated using Equation 403.3, Equation 403.4, or Equation 403.5.2. [ASHRAE 62.1:6.2.6.2]

This section allows for the averaging of the occupant load when the use of a zone peaks for a short period of time. The equation would be used to determine the time period that could be used to average the occupant load in the zone. This provision is not often used by design engineers as zone occupancy fluctuation is not usually well known during the design phase and could substantially deviate from the original assumptions, resulting in poor air quality in these zones. A more often used method is that specified in Section 403.6.

403.7 Exhaust Ventilation. Exhaust airflow shall be provided in accordance with the requirements in Table 403.7. Exhaust makeup air shall be permitted to be a combination of outdoor air, recirculated air, and transfer air.

Table 403.7 lists several occupancy categories and the corresponding required exhaust rates. The exhaust rates are in the form of cfm/square foot, with few exceptions. The exhaust is to be replaced directly by outdoor air or recirculating air (that includes outdoor air), or indirectly through transfer air from other spaces.

One example of indirect replacement of exhaust is when transfer air from an adjacent space is provided to a toilet room to replace the exhausted air. This is a common practice intended to create a slightly negative pressure within the toilet room so that odors are kept within. However, supplying makeup air through an adjacent corridor that happens to be of fire-resistive construction must be avoided as this is prohibited by Section 602.6 of this code.

403.7.1 Parking Garages. Exhaust rate for parking garages shall be in accordance with Table 403.7. Exhaust rate shall not be required for enclosed parking garages having a floor area of 1000 square feet (92.9 m^2) or less and used for the storage of 5 or less vehicles.

Enclosed parking garages shall be defined in accordance with the AHJ.

403.7.2 Enclosed Parking Garages. Mechanical ventilation systems for enclosed parking garages shall operate continuously.

Exceptions:

(1) Mechanical ventilation systems shall be permitted to operate intermittently where the system is designed to operate automatically upon detection of vehicle operation or the presence of occupants by approved automatic detection devices.

(2) Approved automatic carbon monoxide sensing devices, and nitrogen dioxide detectors shall be permitted to modulate the ventilation system to not exceed a maximum average of 50 parts per million of carbon monoxide, or 1 part per million nitrogen dioxide during an eight-hour period with a concentration of not more than 200 parts per million for carbon monoxide, or 5 parts per million nitrogen dioxide, for a period not exceeding 15 minutes. Automatic sensing devices installed in modulated parking garage ventilation systems shall be approved in accordance with Section 301.2.

Most newer diesel vehicle exhaust cannot be detected by carbon monoxide detectors. Nitrogen dioxide detectors have been developed which will detect diesel emissions which will solve the concern about diesel vehicle emissions in enclosed parking garages.

403.8 Dynamic Reset. The system shall be permitted to be designed to reset the outdoor air intake flow (V_{ot}), the space or ventilation zone airflow (V_{oz}) as operating conditions change, or both. [ASHRAE 62.1:6.2.7]

If the building has varying operating conditions, the code allows for modulating the outdoor airflow to decrease the rate when the building is either unoccupied or has a much lower occupancy. Typically, designers use carbon dioxide monitors to modulate the air to a zone depending on the percentage of carbon dioxide in the room air, which is indicative of either an increase or decrease in the number of occupants in the zone.

The previous three sections are principally intended to allow the designer flexibility in designing a system that would fit the operating condition of a building when it deviates from the standard. When this occurs, the AHJ should scrutinize the design and consider requiring justification that is reasonable and logical before it would allow for such deviation.

403.9 Air Classification and Recirculation. Air shall be classified, and its recirculation shall be limited in accordance with Section 403.9.1 through Section 403.9.4. [ASHRAE 62.1:5.16] Recirculated air shall not be taken from prohibited locations in accordance with Section 311.3.

403.9.1 Class 1 Air. Recirculation or transfer of Class 1 air to other spaces shall be permitted. [ASHRAE 62.1:5.16.3.1]

403.9.2 Class 2 Air. Recirculation of Class 2 air within the space of origin shall be permitted. Recirculation or transfer of Class 2 air to other Class 2 or Class 3 spaces shall be permitted, provided the other spaces are used for the same or similar purpose or task and involve the same or similar pollutant sources as the Class 2 space. Transfer of Class 2 air to toilet rooms shall be permitted. Recirculation or transfer of Class 2 air to Class 4 spaces shall be permitted. Class 2 air shall not be recirculated or transferred to Class 1 spaces. Where using an energy recover device, recirculation from leakage, carryover, or transfer from the exhaust side of the energy recovery device shall be permitted and the recirculated Class 2 air shall not exceed 10 percent of the outdoor air intake flow. [ASHRAE 62.1:5.16.3.2]

403.9.3 Class 3 Air. Recirculation of Class 3 air within the space of origin shall be permitted. Class 3 air shall not be recirculated or transferred to other spaces. Where using an energy recover device, recirculation from leakage, carryover, or transfer from the exhaust side of the energy recovery device shall be permitted and the recirculated Class 3 air shall not exceed 5 percent of the outdoor air intake flow. [ASHRAE 62.1:5.16.3.3]

403.9.4 Class 4 Air. Class 4 air shall not be recirculated or transferred to any other space or be recirculated within the space of origin. [ASHRAE 62.1:5.16.3.4]

404.0 Alternative Procedure for Multiple-Zone Systems Ventilation Efficiency.

404.1 General. This section presents an alternative procedure for calculating the system ventilation efficiency (E_v) for multiple zone recirculating systems that shall be used when Section 403.5.1.3 is not used. The system ventilation efficiency shall equal the lowest zone ventilation efficiency among the ventilation zones served by the air handler in accordance with Equation 404.1. [ASHRAE 62.1:A1.3]

$E_v = \text{minimum } (E_{vz})$ (Equation 404.1)

This section is extracted from ASHRAE 62.1 Appendix A and provides an option to calculate system ventilation efficiency (E_v). ASHRAE section 6.2.5.3.2 also requires that the system be designed for the largest number of people to occupy all ventilation zones during use.

404.2 Average Outdoor Air Fraction. The average outdoor air fraction (X_s) for the ventilation system shall be determined in accordance with Equation 404.2.

$X_s = V_{ou}/V_{ps}$ (Equation 404.2)

The uncorrected outdoor air intake (V_{ou}) shall be determined in accordance with Section 403.5.1, and the system primary airflow (V_{ps}) shall be determined at the condition analyzed. [ASHRAE 62.1:A1.1]

404.3 Zone Ventilation Efficiency. The zone ventilation efficiency (E_{vz}) shall be determined in accordance with Section 404.3.1 or Section 404.3.2. [ASHRAE 62.1:A1.2]

404.3.1 Single Supply Systems. For single supply systems, where the air supplied to a ventilation zone is a mixture of outdoor air and system-level recirculated air, zone ventilation efficiency (E_{vz}) shall be determined in accordance with Equation 404.3.1. Examples of single supply systems include constant-volume reheat, single-duct VAV, single-fan dual-duct, and multizone systems.

$E_{vz} = 1+X_s-Z_{pz}$ (Equation 404.3.1)

The average outdoor air fraction for the system (X_s) shall be determined in accordance with Equation 404.2 and the primary outdoor air fraction for the zone (Z_{pz}) shall be determined in accordance with Equation 404.3.1. [ASHRAE 62.1:A1.2.1]

$Z_{pz} = V_{oz}/V_{pz}$ (Equation 404.3.1)

For VAV systems, V_{pz} is the lowest zone primary airflow value expected at the design condition analyzed.

404.3.2 Secondary-Recirculation Systems. For secondary-recirculation systems where the supply air or a portion thereof to each ventilation zone is recirculated air (air that has not been directly mixed with outdoor air) from other zones, zone ventilation efficiency (E_{vz}) shall be determined in accordance with Equation 404.3.2(1). Examples of secondary-recirculation systems include dual-fan dual-duct and fan-powered mixing-box systems, and systems that include transfer fans for conference rooms.

$E_{vz} = (F_a+X_s \bullet F_b-Z_{pz} \bullet E_p \bullet F_c)/F_a$ [Equation 404.3.2(1)]

The system air fractions F_a, F_b, and F_c shall be determined in accordance with Equation 404.3.2(2), Equation 404.3.2(3), and Equation 404.3.2(4). The zone primary air fraction (E_p) shall be determined in accordance with Equation 404.3.2(5). For single-zone and single-supply systems E_p shall equal to 1.0. The zone secondary recirculation fraction (E_r) shall be determined by the designer based on system configuration. The zone air distribution effectiveness (E_z) shall be determined in accordance with Section 403.2.2. [ASHRAE 62.1:A1.2.2]

$F_a = E_p+(1-E_p) \bullet E_r$ [Equation 404.3.2(2)]

$F_b = E_p$ [Equation 404.3.2(3)]

$F_c = 1-(1-E_z) \bullet (1-E_r) \bullet (1-E_p)$ [Equation 404.3.2(4)]

$E_p = V_{pz}/V_{dz}$ [Equation 404.3.2(5)]

Where:

E_p = Primary air fraction: The fraction of primary air in the discharge air to the ventilation zone.

E_r = Secondary recirculation fraction: In systems with secondary recirculation of return air, the fraction of secondary recirculated air to the zone that is representative of average system return air rather than air directly recirculated from the zone.

E_v = System ventilation efficiency: the efficiency with which the system distributes air from the outdoor air intake to the breathing zone in the ventilation-critical zone, which requires the largest fraction of outdoor air in the primary airstream.

E_{vz} = Zone ventilation efficiency: The efficiency with which the system distributes air from the outdoor air intake to the breathing zone in any particular ventilation zone.

E_z = Zone air distribution effectiveness: A measure of the effectiveness of supply air distribution to the breathing zone. E_z is determined in accordance with Section 403.2.2.

F_a = Supply air fraction: The fraction of supply air to the ventilation zone from sources or air outside the zone.

F_b = Mixed air fraction: The fraction of supply air to the ventilation zone from fully mixed primary air.

F_c = Outdoor air fraction: The fraction of outdoor air to the ventilation zone from sources of air outside the zone.

V_{dz} = Zone discharge airflow: The expected discharge (supply) airflow to the zone that includes primary airflow and secondary recirculated airflow, CFM (m^3/min).

V_{pz} = Zone primary airflow: The zone primary airflow to the ventilation zone, including outdoor air and recirculated air.

X_s = Average outdoor air fraction: At the primary air handler, the fraction of outdoor air intake flow in the system primary airflow.

Z_{pz} = Primary outdoor air fraction: The outdoor air fraction required in the primary air supplied to the ventilation zone prior to the introduction of secondary recirculation air. [ASHRAE 62.1:A3]

Secondary recirculation systems supply air wholly or partially to individual ventilation zones without receiving air directly from the outside. It is recirculated air from other zones, which may have already mixed with outside air.

405.0 Indoor Air Quality for Residential Occupancies.

405.1 General. Rooms or occupied spaces within residential occupancies, where the occupants are nontransient, shall be designed to have mechanical ventilation and exhaust air in accordance with Section 405.2 through Section 405.5.

405.1.1 Natural Ventilation. Where approved by the Authority Having Jurisdiction, natural ventilation shall be permitted for Climate Zone 1, Climate Zone 2, or for thermally conditioned buildings for less than 876 hours per year.

405.2 Ventilation Air Rate. The required mechanical ventilation outdoor air rate (Q_{tot}) shall be as calculated in accordance with Equation 405.2.

Exception: For existing buildings and where permitted by the Authority Having Jurisdiction, the total mechanical ventilation (Q_{tot}) is not required where Q_{tot} is calculated to be less than 15 ft^3/min (0.007 m^3/s).

$$Q_{tot} = 0.03A_{floor} + 7.5\ (N_{br} + 1) \qquad \text{(Equation 405.2)}$$

Where:

Q_{tot} = Total required ventilation outdoor air rate, CFM

A_{floor} = Floor area, ft^2

N_{br} = Number of bedrooms more than 1

For SI Units: 1 cubic foot per minute = 0.00047 m^3/s

Example of sizing where mechanical ventilation is not required:

Given an existing building with a floor area of 400 square feet with one bedroom.

Using equation $Q_{tot} = 0.03A_{floor} + 7.5\ (N_{br} + 1)$

Where:

Q_{tot} = Total required ventilation outdoor air rate, cfm

A_{floor} = Floor area, ft^2

N_{br} = Number of bedrooms more than 1

Then $Q_{tot} = 0.03(400) + 7.5\ (0 + 1) =$

$Q_{tot} = 12 + 7.5 = 19.5$

Therefore since the calculated Q_{tot} is greater than 15 then mechanical ventilation would be required.

405.2.1 Reduced Ventilation Air Rate. Where permitted by the Authority Having Jurisdiction, the mechanical ventilation air rate required in Section 405.2 shall be permitted to be reduced where an infiltration rate is determined in accordance with ASTM E779.

405.3 Bathroom Exhaust. A mechanical exhaust directly to the outdoors shall be provided in each room containing a bathtub, shower, or tub/shower combination. The fan shall run intermittently (on demand) or continuously. A readily accessible manual control designed to be operated as needed or an automatic control shall be provided for intermittent operations.

405.3.1 Exhaust Rate. The exhaust rate shall be not less than 50 ft^3/min (0.02 m^3/s) for intermittent operation and 20 ft^3/min (0.009 m^3/s) for continuous operation.

405.4 Kitchen Exhaust. A mechanical exhaust directly to the outdoors shall be provided in each kitchen. The fan shall run intermittently (on demand) or continuously. A readily accessible manual control designed to be operated as needed or an automatic control shall be provided for intermittent operations.

405.4.1 Exhaust Rate. For intermittent-controlled operations, the exhaust rate shall be not less than 100 ft^3/min (0.047 m^3/s) for range hoods or 300 ft^3/min (0.141 m^3/s) for mechanical exhaust fans including downdraft appliances. For continuous operated ventilation, the exhaust rate shall be not less than 5 air changes per hour based on kitchen volume for enclosed kitchens.

405.5 Ventilation Openings. Occupiable spaces shall be provided with a readily accessible ventilation opening openable to the outdoors. The opening shall be not less than 5 square feet (0.464 m^2) or 4 percent of the occupied floor area. The openable area shall be based on free, unobstructed area through the opening.

TABLE 402.1
MINIMUM VENTILATION RATES IN BREATHING ZONE[1, 2]
[ASHRAE 62.1: TABLE 6.2.2.1]

OCCUPANCY CATEGORY[4]	PEOPLE OUTDOOR Air Rate R_p (CFM/person)	AREA OUTDOOR Air Rate R_a (CFM/ft^2)	DEFAULT OCCUPANT DENSITY[3] (people/1000 ft^2)	AIR CLASS
CORRECTIONAL FACILITIES				
Booking/waiting	7.5	0.06	50	2
Cell	5	0.12	25	2
Day room	5	0.06	30	1
Guard stations	5	0.06	15	1
EDUCATIONAL FACILITIES				
Art classroom	10	0.18	20	2
Classrooms (ages 5-8)	10	0.12	25	1
Classrooms (age 9 plus)	10	0.12	35	1
Computer lab	10	0.12	25	1
Daycare (through age 4)	10	0.18	25	2
Daycare sickroom	10	0.18	25	3
Lecture classroom[h]	7.5	0.06	65	1
Lecture hall (fixed seats)[h]	7.5	0.06	150	1
Media center[a]	10	0.12	25	1
Multi-use assembly[h]	7.5	0.06	100	1
Music/theater/dance[h]	10	0.06	35	1
Science laboratories	10	0.18	25	2
University/college laboratories	10	0.18	25	2
Wood/metal shop	10	0.18	20	2
FOOD AND BEVERAGE SERVICE				
Bars, cocktail lounges	7.5	0.18	100	2
Cafeteria/fast food dining	7.5	0.18	100	2
Kitchen (cooking)	7.5	0.12	20	2
Restaurant dining rooms	7.5	0.18	70	2
GENERAL				
Break rooms[h]	5	0.06	25	1
Coffee stations[h]	5	0.06	20	1
Conference/meeting[h]	5	0.06	50	1
Corridors[h]	–	0.06	–	1
Occupiable storage rooms for liquids or gels[b]	5	0.12	2	2
HOTELS, MOTELS, RESORTS, DORMITORIES				
Barracks sleeping areas[h]	5	0.06	20	1
Bedroom/living room[h]	5	0.06	10	1
Laundry rooms, central	5	0.12	10	2
Laundry rooms within dwelling units	5	0.12	10	1
Lobbies/pre-function[h]	7.5	0.06	30	1
Multipurpose assembly[h]	5	0.06	120	1
OFFICE BUILDINGS				
Break Rooms	5	0.12	50	1
Main entry lobbies[h]	5	0.06	10	1
Occupiable storage rooms for dry materials	5	0.06	2	1
Office space[h]	5	0.06	5	1
Reception areas[h]	5	0.06	30	1
Telephone/data entry[h]	5	0.06	60	1
MISCELLANEOUS SPACES				
Bank or bank lobbies[h]	7.5	0.06	15	1
Bank vaults/safe deposit[h]	5	0.06	5	2
Computer (not printing)[h]	5	0.06	4	1
Freezer and refrigerated spaces (<50°F)[c]	10	–	–	2
General manufacturing (excludes heavy industrial and processes using chemicals)	10	0.18	7	3
Pharmacy (prep. area)	5	0.18	10	2
Photo studios	5	0.12	10	1

TABLE 402.1 (continued)
MINIMUM VENTILATION RATES IN BREATHING ZONE[1, 2]
[ASHRAE 62.1: TABLE 6.2.2.1]

OCCUPANCY CATEGORY[4]	PEOPLE OUTDOOR Air Rate R_p (CFM/person)	AREA OUTDOOR Air Rate R_a (CFM/ft^2)	DEFAULT OCCUPANT DENSITY[3] (people/1000 ft^2)	AIR CLASS
Shipping/receiving[b]	10	0.12	2	2
Sorting, packing, light assembly	7.5	0.12	7	2
Telephone closets	–	–	–	1
Transportation waiting[h]	7.5	0.06	100	1
Warehouses[b]	10	0.06	–	2
PUBLIC ASSEMBLY SPACES				
Auditorium seating area[h]	5	0.06	150	1
Courtrooms[h]	5	0.06	70	1
Legislative chambers[h]	5	0.06	50	1
Libraries	5	0.12	10	1
Lobbies[h]	5	0.06	150	1
Museums (children's)	7.5	0.12	40	1
Museums/galleries[h]	7.5	0.06	40	1
Places of religious worship[h]	5	0.06	120	1
RESIDENTIAL				
Common corridors[h]	–	0.06	–	1
Dwelling unit[f, g, h]	5	0.06	See footnote[f]	1
RETAIL				
Sales (except as below)	7.5	0.12	15	2
Barber shop[h]	7.5	0.06	25	2
Beauty and nail salons	20	0.12	25	2
Coin-operated laundries	7.5	0.12	20	2
Mall common areas[h]	7.5	0.06	40	1
Pet shops (animal areas)	7.5	0.18	10	2
Supermarket[h]	7.5	0.06	8	1
SPORTS AND ENTERTAINMENT				
Bowling alley (seating)	10	0.12	40	1
Disco/dance floors[h]	20	0.06	100	2
Gambling casinos	7.5	0.18	120	1
Game arcades	7.5	0.18	20	1
Gym, sports arena (play area)[e]	20	0.18	7	2
Health club/aerobics room	20	0.06	40	2
Health club/weight rooms	20	0.06	10	2
Spectator areas[h]	7.5	0.06	150	1
Stages, studios[d, h]	10	0.06	70	1
Swimming (pool & deck)[c]	–	0.48	–	2

For SI units: 1 cubic foot per minute = 0.0283 m^3/min, 1 square foot = 0.0929 m^2

Notes:

1 This table applies to no-smoking areas. Rates for smoking-permitted spaces shall be determined using other methods.

2 Volumetric airflow rates are based on dry air density of 0.075 pounds of dry air per cubic foot (lbda/ft^3) (1.201 kgda/m^3) at a barometric pressure of 1 atm (101 kPa) and an air temperature of 70°F (21°C). Rates shall be permitted to be adjusted for actual density.

3 The default occupant density shall be used where actual occupant density is not known.

4 Where the occupancy category for a proposed space or zone is not listed, the requirements for the listed occupancy category that is most similar in terms of occupant density, activities, and building construction shall be used.

ITEM-SPECIFIC NOTES FOR TABLE 402.1

a For high school and college libraries, the values shown for "Public Assembly Spaces – Libraries" shall be used.

b Rate may not be sufficient where stored materials include those having potentially harmful emissions.

c Rate does not allow for humidity control. "Deck area" refers to the area surrounding the pool that is capable of being wetted during pool use or when the pool is occupied. Deck area that is not expected to be wetted shall be designated as an occupancy category.

d Rate does not include special exhaust for stage effects such as dry ice vapors and smoke.

e Where combustion equipment is intended to be used on the playing surface or in the space, additional dilution ventilation, source control, or both shall be provided.

f Default occupancy for dwelling units shall be two persons for studio and one-bedroom units, with one additional person for each additional bedroom.

g Air from one residential dwelling shall not be recirculated or transferred to other spaces outside of that dwelling.

h Ventilation air for this occupancy category shall be permitted to be reduced to zero where the space is in occupied-standby mode.

TABLE 403.2.2
ZONE AIR DISTRIBUTION EFFECTIVENESS[1, 2, 3, 4, 5]
[ASHRAE 62.1: TABLE 6.2.2.2]

AIR DISTRIBUTION CONFIGURATION	E_z
Ceiling supply of cool air.	1.0
Ceiling supply of warm air and floor return.	1.0
Ceiling supply of warm air 15°F or more above space temperature and ceiling return.	0.8
Ceiling supply of warm air less than 15°F above space temperature and ceiling return provided that the 150 feet per minute (fpm) supply air jet reaches to within 4.5 feet of floor level.	1.0[6]
Floor supply of cool air and ceiling return provided that the vertical throw is more than 50 fpm at a height of 4.5 feet or more above the floor.	1.0
Floor supply of cool air and ceiling return, provided low-velocity displacement ventilation achieves unidirectional flow and thermal stratification, or underfloor air distribution systems where the vertical throw is 50 fpm or less at a height of 4.5 feet above the floor.	1.2
Floor supply of warm air and floor return.	1.0
Floor supply of warm air and ceiling return.	0.7
Makeup supply drawn in on the opposite side of the room from the exhaust, return, or both.	0.8
Makeup supply drawn in near to the exhaust, return, or both locations.	0.5

For SI units: °C = (°F-32)/1.8, 1 foot per minute = 0.005 m/s, 1 foot = 304.8 mm

Notes:

1 "Cool air" is air cooler than space temperature.

2 "Warm air" is air warmer than space temperature.

3 "Ceiling supply" includes any point above the breathing zone.

4 "Floor supply" includes any point below the breathing zone.

5 As an alternative to using the above values, E_Z shall be permitted to be regarded as equal to air change effectiveness determined in accordance with ASHRAE 129 for air distribution configurations except unidirectional flow.

6 For lower velocity supply air, $E_Z = 0.8$

TABLE 403.7
MINIMUM EXHAUST RATES
[ASHRAE 62.1: TABLE 6.5]

OCCUPANCY CATEGORY[8]	EXHAUST RATE (CFM/unit)	EXHAUST RATE (CFM/ft^2)	AIR CLASS
Arenas[2]	–	0.50	1
Art classrooms	–	0.70	2
Auto repair rooms[1]	–	1.50	2
Barber shops	–	0.50	2
Beauty and nail salons	–	0.60	2
Cells with toilet	–	1.00	2
Copy, printing rooms	–	0.50	2
Darkrooms	–	1.00	2
Educational science laboratories	–	1.00	2
Janitor closets, trash rooms, recycling	–	1.00	3
Kitchens – commercial	–	0.70	2
Kitchenettes	–	0.30	2
Locker rooms for athletic, industrial and health care facilities	–	0.50	2
Other locker rooms	–	0.25	2
Shower rooms[7,10]	20/50	–	2
Paint spray booths	–	–	4
Parking garages[3]	–	0.75	2
Pet shops (animal areas)	–	0.90	2
Refrigerating machinery rooms[6]	–	–	3
Residential – kitchens[7]	50/100	–	2
Soiled laundry storage rooms	–	1.00	3
Storage rooms, chemical	–	1.50	4
Toilets – private[5, 9]	25/50	–	2
Toilets – public[4, 9]	50/70	–	2
Woodwork shop/classrooms	–	0.50	2

For SI units: 1 cubic foot per minute = 0.0283 m^3/min, 1 square foot = 0.0929 m^2

Notes:

1 Stands where engines are run shall have exhaust systems that directly connect to the engine exhaust and prevent escape of fumes.

2 Where combustion equipment is intended to be used on the playing surface, additional dilution ventilation, source control, or both shall be provided.

3 Exhaust shall not be required where two or more sides comprise walls that are at least 50 percent open to the outside.

4 Rate is per water closet, urinal, or both. Provide the higher rate where periods of heavy use are expected to occur. The lower rate shall be permitted to be used otherwise.

5 Rate is for a toilet room intended to be occupied by one person at a time. For continuous system operation during normal hours of use, the lower rate shall be permitted to be used. Otherwise the higher rate shall be used.

6 For refrigeration machinery rooms, the exhaust rate shall comply with Chapter 11.

7 For continuous system operation, the lower rates shall be permitted. Otherwise the higher rate shall be used.

8 For unlisted occupancies for a proposed space not listed in the table, the requirements for the listed occupancy that is most similar in terms of occupant density and occupancy type shall be used.

9 Exhaust air that has been cleaned in accordance with the criteria of Class 1 shall be permitted to be recirculated.

10 Rate is per shower head.

CHAPTER 5
EXHAUST SYSTEMS

501.0 General.

501.1 Applicability. This chapter includes requirements for environmental air ducts, product-conveying systems, and commercial hoods and kitchen ventilation. Part I addresses environmental air ducts and product conveying systems. Part II addresses commercial hoods and kitchen ventilation.

Many industrial processes produce air contaminants in the form of dusts, fumes, smokes, mists, vapors and gases. Contaminants should be controlled at the source so they are not dispersed through the workplace or allowed to increase to flammable or toxic concentrations; however, zero concentration of contaminants is not economically feasible. Absolute control of contaminants cannot be maintained. Industrial hygiene science is based on the fact that most air contaminants become toxic only if their concentration exceeds a maximum allowable limit for a specified period.

Flammable gases and vapors can also be products of industrial processes. A flammable liquid's vapor pressure and volatility or rate of evaporation determines its ability to form an explosive mixture. These properties can be expressed by the flash point, which is the temperature to which combustible liquid must be heated to produce a flash when a small flame is passed across the surface of the liquid. In practice, the air-vapor or air-gas mixture must be in the explosive range before it can be ignited. If the concentration is limited to a certain percentage of the lower explosive limit of the material, the resulting factor of safety allows latitude for imperfections in air distribution and variations in temperature or mixture and guards against unpredictable or unrecognized sources of ignition.

Combustible dust is another product of some industrial or food-processing operations. Many organic and some mineral dusts can produce dust explosions. Often, a primary explosion results from a small amount of dust in suspension that has been exposed to an ignition source; the pressure and vibration created can dislodge larger accumulations from dust on horizontal surfaces, creating a larger secondary explosion. Explosive dusts are potential hazards whenever uncontrolled dust escapes, dispersing in the atmosphere or settling on horizontal surfaces such as beams and ledges.

Exhaust systems should be designed to dispose of these contaminants, flammable vapors and dusts. Exhaust systems have four basic components:

1. The hood, or entry point of the system;
2. The duct system, which transfers air;
3. The air-cleaning device, which removes contaminants from the air; and
4. The air-moving device, which provides motive power for overcoming system resistance.

This chapter covers the requirements for the installation of product-conveying duct systems. The provisions are meant to cover the transfer and disposal of not only solid particles, but also the transfer and disposal of contaminants, flammable vapors, fumes, dusts, smokes, spray, mists, fogs, etc.

In addition, a section is included to regulate environmental air ventilation systems that are not part of heating or cooling systems.

Please note that provisions for commercial hood and kitchen ventilation systems are covered separately by Chapter 5, Part II beginning in Section 507.0.

502.0 Termination.

502.1 Exhaust Opening Protection. Exhaust openings terminating to the outdoors shall be covered with a corrosion-resistant screen having not less than ¼ of an inch (6.4 mm) openings, and shall have not more than ½ of an inch (12.7 mm) openings.

Exception: Clothes dryers.

502.2 Termination of Exhaust Ducts. Exhaust ducts shall terminate in accordance with Section 502.2.1 through Section 502.2.3.

502.2.1 Environmental Air Ducts. Environmental air duct exhaust shall terminate not less than 3 feet (914 mm) from a property line, 10 feet (3048 mm) from a forced air inlet, 10 feet (3048 mm) above a public walkway, and 3 feet (914 mm) from openings into the building. The discharge of environmental exhaust ducts shall not be directed onto a public walkway.

For the same reasons that they are not allowed to terminate in attics and crawl spaces, the 3-foot clearance to openings should include attic and crawl space vents.

502.2.2 Product Conveying Ducts. Ducts conveying explosive or flammable vapors, fumes, or dusts shall terminate not less than 30 feet (9144 mm) from a property line, 10 feet (3048 mm) from openings into the building, 6 feet (1829 mm) from exterior walls or roofs, 30 feet (9144 mm) from combustible walls or openings into the building that are in the direction of the exhaust discharge, and 10 feet (3048 mm) above adjoining grade.

Other product-conveying outlets shall terminate not less than 10 feet (3048 mm) from a property line, 3 feet (914 mm) from exterior walls or roofs, 10 feet (3048 mm) from openings into the building, and 10 feet (3048 mm) above adjoining grade.

Garage ventilation systems in parking garages should terminate in accordance with the provisions of this section since ducts routinely convey exhaust products containing carbon monoxide, smoke, soot, water vapor and particles of rubber from tires. Because garage ventilation systems may also be required to dissipate fumes from vehicular fuel spills, it is recommended that they should be regarded as ducts conveying flammable vapors.

502.2.3 Commercial Kitchen Ducts. Commercial kitchens exhaust ducts shall terminate in accordance with Section 510.9 or Section 519.5.

Part I – Environmental Air Ducts and Product-Conveying Systems.

503.0 Motors, Fans, and Filters.

503.1 General. Motors and fans shall be sized to provide the required air movement. Motors in areas that contain flammable vapors or dusts shall be of a type approved for such environments. A manually operated remote control installed at an approved location shall be provided to shut off fans or blowers in flammable vapor or dust systems. Equipment used in operations that generate explosive or flammable vapors, fumes, or dusts shall be interlocked with the ventilation system so that the equipment cannot be operated unless the ventilation fans are in operation. Motors for fans used to convey flammable vapors or dusts shall be located outside the duct or shall be protected with approved shields and dust-proofing. Where belts are used, they shall not enter the duct unless the belt and pulley within the duct are enclosed. Motors and fans shall be accessible for servicing and maintenance.

Motors and fans must be sized to provide the air movement needed for efficient operation of the system. Among the factors to be considered are required volume, required velocities, duct and fitting pressure losses and fan characteristics. The code requires that motors in areas containing flammable vapors or dusts be of types approved for such an environment.

A manually operated remote control switch installed in an approved location (one that is well marked and accessible) must be provided to shut down the fans and blowers in flammable vapor or flammable dust-conveying systems.

Any electrical equipment (saws, grinders, buffing wheels, etc.) used in industrial operations that generate explosive or flammable vapors, fumes or dusts must be electrically interlocked with the product-conveying duct system (makeup and exhaust air). This interlock prevents operation of contaminant-producing power equipment when the ventilation system is not operating. In this way, a dangerous build-up of vapor or dust may be avoided since they can be ignited by a spark or heat from power equipment. Motors or fans used to convey flammable dust or vapors must be located outside the duct or be protected with an approved shield and dust-proofing method.

An acceptable design can be a fan within the duct, with the motor located outside and connected via a belt (see **Figure 503.1**). Such motors and fans must be accessible for servicing and maintenance. This is essential for reliable operation and long system life.

503.2 Fans. Parts of fans in contact with explosive or flammable vapors, fumes, or dusts shall be of nonferrous or nonsparking materials, or their casing shall be lined or constructed of such material. Where the size and hardness of materials passing through a fan are capable of producing a spark, both the fan, and the casing shall be of nonsparking materials. Where fans are required to be spark-resistant, their bearings shall not be within the airstream, and parts of the fan shall be grounded. Fans in systems handling materials that

FIGURE 503.1
PRODUCT-CONVEYING DUCT

are likely to clog the blades, and fans in buffing or woodworking exhaust systems, shall be of the radial-blade or tube-axial type.

Equipment used to exhaust explosive or flammable vapors, fumes, or dusts shall bear an identification plate stating the ventilation rate for which the system was designed.

Fans located in systems conveying corrosives shall be of materials that are resistant to the corrosive or shall be coated with corrosion-resistant materials.

Either all parts of fans in contact with explosive or flammable vapors, fumes or dusts must be made of nonferrous or non-sparking materials or their casing must be constructed or lined with such materials. However, when the size and hardness of any material passing through the fan(s) are capable of producing a spark, both the fan and the casing must be made from non-sparking materials. When spark-resistant fans are required, their bearings must be outside the air stream carrying explosive or flammable vapors. In addition, all fan parts must be electrically grounded. These safety precautions limit the likelihood of sparking that can ignite the flammable vapors or dusts. Such special-purpose fans should be listed for their intended use by an approved agency.

For the following applications, fans must be of the radial-blade or tube-axial type (see **Figure 503.2**).

1. To handle materials likely to clog the blades (metals, wet or heavy dusts); and
2. For buffing or woodworking systems.

Equipment used to exhaust explosive or flammable vapors, fumes or dusts must bear an identification plate showing the ventilation in cubic feet per minute (cfm) for which the system was designed.

Fans located in a product-conveying duct system containing corrosive vapors must be built from, or coated thoroughly with, corrosion-resistant materials. The manufacturer of the fan and the listing agency should verify that the chosen fan is acceptable for the particular corrosive vapor to be vented. Each corrosive vapor has its own properties; therefore, no single fan material can possess universal application for all types of vapors.

Not all fans can operate effectively under high resistance. Duct velocities required in Table 505.9 dictate the maximum size of a duct system needed to maintain that velocity. Long duct systems increase the total static resistance that the fan needs to overcome. Simply spinning a fan faster may not achieve the required result of more airflow. Designing a fan and motor capable of moving product-laden air within a duct system is critical to the overall operation of the product-conveying and removal system.

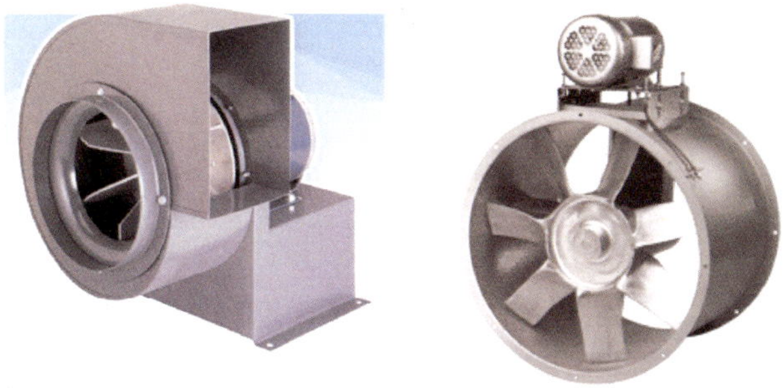

FIGURE 503.2
RADIAL FAN AND TUBE AXIAL FAN

504.0 Environmental Air Ducts.

504.1 General. Where not specified in this chapter, exhaust ducts shall be constructed and installed in accordance with Chapter 6 and shall be airtight as approved by the Authority Having Jurisdiction. Environmental air ducts that have an alternate function as a part of an approved smoke-control system do not require design as Class 1 product-conveying ducts.

Exceptions:

(1) Ductless range hoods where installed in accordance with the manufacturer's installation instructions.

(2) Condensing clothes dryers where installed in accordance with the manufacturer's installation instructions.

Environmental air ducts not elsewhere regulated in the code must comply with Chapter 6 and this section. Exhaust ventilation ducts must terminate outside the building in a location complying with the building and mechanical codes and should be equipped with backdraft dampers. Ducts used as part of an approved smoke-control system do not require to be designed as Class 1 product-conveying ducts.

Ducts used for domestic kitchen range ventilation and domestic clothes dryers must be of metal and must have a smooth interior surface, except that domestic dryer exhaust can be ducted by a flexible duct connector not more than 6 feet in length, as long as no part of that connector is concealed within construction. For length limitations and additional information, refer to Section 504.0.

Bathroom and laundry exhaust ducts may be flexible, nonmetallic, corrugated or gypsum wall board subject to the temperature and moisture limitations of Section 504.6 and Section 602.4.2.

Exhaust ducts for commercial dryers, laundries, dry cleaning and other commercial establishments must be installed in accordance with their listing. Verify requirements of combustion or makeup air with the installation instructions.

No exhaust ducts are permitted to extend into or through ducts or plenums. A failure in the exhaust duct would introduce the environmental contaminant back into the building's conditioning system by simply blowing into the duct.

Environmental air duct exhaust shall terminate a minimum of 3 feet from the property line and 3 feet from openings into the building (refer to Section 502.2.1).

504.1.1 Backdraft Protection. Exhaust ducts shall terminate outside the building and shall be equipped with backdraft dampers or with motorized dampers that automatically shut where the systems or spaces served are not in use.

Exception: Where the exhaust duct does not discharge into a common exhaust plenum and one of the following:

(1) The exhaust fan runs continuously.

(2) The exhaust duct serves space(s) that are not mechanically heated or cooled.

(3) The space served is maintained at positive pressure.

504.2 Independent Exhaust Systems. Single or combined mechanical exhaust systems shall be independent of other exhaust systems.

The majority of exhaust systems are required to exhaust separately because of the risk of dispersing harmful exhaust to adjoining areas within a building. For example, exhaust from a kitchen range hood must be separate from bathroom exhaust and must operate independently of each other. Subject to the approval of the AHJ, certain systems that exhaust air may be combined with other similar systems depending on the exhaust type.

504.3 Domestic Range. Ducts used for domestic kitchen range or cooktop ventilation shall be of metal and shall have smooth interior surfaces.

Exception: Ducts for domestic kitchen downdraft grill-range ventilation installed under a concrete slab floor shall be permitted to be of approved Schedule 40 PVC provided:

(1) The under-floor trench in which the duct is installed shall be completely backfilled with sand or gravel.

(2) Not more than 1 inch (25.4 mm) of 6 inch diameter (152 mm) PVC coupling shall be permitted to protrude above the concrete floor surface.

(3) PVC pipe joints shall be solvent cemented to provide an air and greasetight duct.

(4) The duct shall terminate above grade outside the building and shall be equipped with a backdraft damper.

504.4 Clothes Dryers. A clothes dryer exhaust duct shall not be connected to a vent connector, gas vent, chimney, and shall not terminate into a crawl space, attic, or other concealed space. Exhaust ducts shall not be assembled with screws or other fastening means that extend into the duct and that are capable of catching lint, and that reduce the efficiency of the exhaust system. Exhaust ducts shall be constructed of rigid metallic material. Transition ducts used to connect the dryer to the exhaust duct shall be listed and labeled in accordance with UL 2158A, or installed in accordance with the clothes dryer manufacturer's installation instructions. Clothes dryer exhaust ducts shall terminate to the outside of the building in accordance with

Section 502.2.1 and shall be equipped with a backdraft damper. Screens shall not be installed at the duct termination. Devices, such as fire or smoke dampers that will obstruct the flow of the exhaust shall not be used. Where joining of ducts, the male end shall be inserted in the direction of airflow.

Domestic clothes dryer exhaust ducts must terminate outside of the building and be equipped with a backdraft damper. The point of termination through the wall or roof shall be sealed to prevent moisture, dust, insects and rodents from entering the building. Screens are not allowed at the duct termination. It should be noted that the exhaust duct should not be terminated in an attic, even if it is well ventilated, because the moisture vapor may condense on the roof sheathing, rafters or insulation, particularly in cold climates, and bypassed lint may cause fire hazard in the attic or clog other required attic ventilation screens. Exhaust ducts for clothes dryers must not be connected with metal screws or fastening devices that may extend inside the duct. This is to prevent the accumulation of lint, which may create a fire hazard. See **Figure 504.4**.

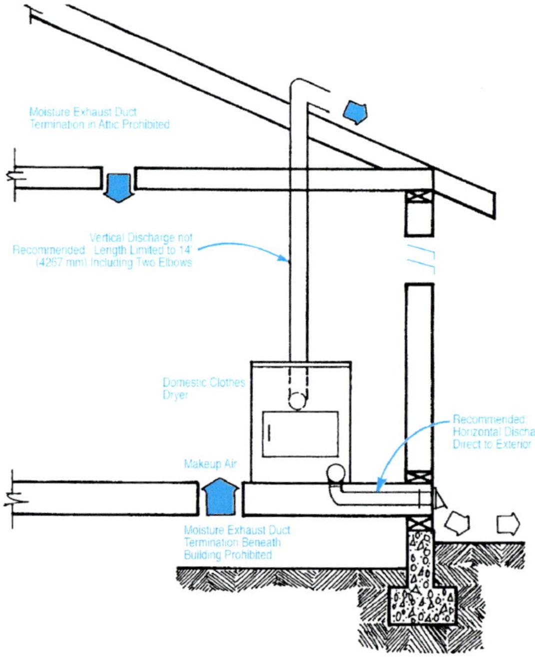

FIGURE 504.4
DOMESTIC CLOTHES DRYER AND MOISTURE EXHAUST DUCT

504.4.1 Provisions for Makeup Air. Makeup air shall be provided in accordance with the following:

(1) Makeup air shall be provided for Type 1 clothes dryers in accordance with the manufacturer's installation instructions. [NFPA 54:10.4.3.1] Where a closet is designed for the installation of a clothes dryer, an opening of not less than 100 square inches (0.065 m^2) for makeup air shall be provided in the door or by other approved means.

(2) Provision for makeup air shall be provided for Type 2 clothes dryers, with a minimum free area of 1 square inch (0.0006 m^2) for each 1000 British thermal units per hour (Btu/h) (0.293 kW) total input rating of the dryer(s) installed. [NFPA 54:10.4.3.2]

Designated dryer spaces shall be provided with a 4-inch moisture duct system that is capable of being connected to an owner-provided dryer at a future point in time. This will limit the possibility of noncomplying duct systems being installed by well-meaning but uninformed installers that may violate the provisions of this code and the installation instructions of the appliance.

The operation of Type 1 clothes dryer exhausting to the outside will cause makeup air to eventually come from the outside. Normal building infiltration may not be enough to meet the demand of a dryer, especially if the dryer is located in a closet and operating behind a closed door. The building code of the jurisdiction defines the minimum size of a room. Rooms smaller than the minimum room dimensions may be considered a closet. These smaller spaces require 100 square inches of makeup air. However, per Section 403.7, Exhaust Ventilation, "Exhaust makeup air shall be permitted to be any combination of outdoor air, recirculated air and transfer air."

It may seem like a waste of energy to allow nonconditioned air to be introduced into the building. Needs of an appliance to operate in a safe manner and removal of environmental air contaminants should override the desire to keep the house "leakproof" for energy use concerns (see **Figure 504.4.2**).

Dryer vent temperatures commonly operate at a temperature exceeding the normal operating temperature of a downflow domestic range vent. See Section 603.11 if ducting must be installed below slabs. Note that ground temperatures will greatly increase condensation within the dryer (moisture) exhaust duct.

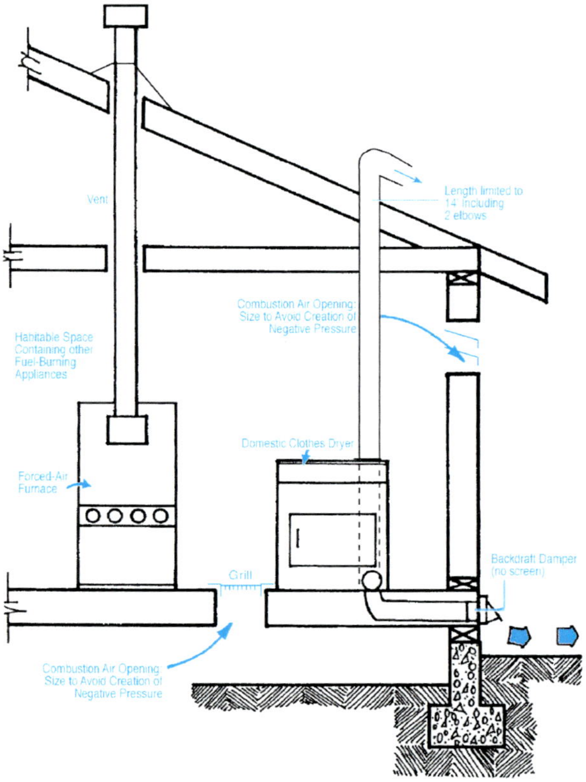

FIGURE 504.4.2
CLOTHES DRYER AND FORCED-AIR FURNACE

504.4.2 Domestic Clothes Dryers. Where a compartment or space for a Type 1 clothes dryer is provided, not less than a 4 inch diameter (102 mm) exhaust duct of approved material shall be installed in accordance with Section 504.0.

Type 1 clothes dryer exhaust ducts shall be of rigid metal and shall have smooth interior surfaces. The diameter shall be not less than 4 inches nominal (100 mm), and the thickness shall be not less than 0.016 of an inch (0.406 mm).

504.4.2.1 Length Limitation. Unless otherwise permitted or required by the dryer manufacturer's instructions and approved by the Authority Having Jurisdiction, domestic dryer moisture exhaust ducts shall not exceed a total combined horizontal and vertical length of 14 feet (4267 mm), including two 90 degree (1.57 rad) elbows. A length of 2 feet (610 mm) shall be deducted for each 90 degree (1.57 rad) elbow in excess of two.

Exception: Where an exhaust duct power ventilator, in accordance with Section 504.4.2.3, is used, the maximum length of the dryer exhaust duct shall be permitted to be in accordance with the dryer exhaust duct power ventilator manufacturer's installation instructions.

Although this subsection allows longer lengths if permitted by the listing of the clothes dryer, it is prudent for the Authority Having Jurisdiction (AHJ) to use discretion since the dryers are not usually installed at the time of final inspection and the homeowner/occupant may switch the appliance later.

As noted in the exception, the use of exhaust duct power ventilators to increase the distance to discharge is permitted when that equipment is in accordance with Section 504.4.2.3. The use of longer and/or larger off-the-shelf ducting material to extend a dryer vent beyond the 14-foot limit is a code violation. Negative effects caused by increased condensation within the ductwork and a drop in velocity originally intended to keep bypassed lint suspended in the airflow to its point of termination may increase system failure due to clogged ducting.

504.4.2.2 Transition Ducts. Listed clothes dryer transition ducts not more than 6 feet (1829 mm) in length shall be permitted to be used to connect the Type 1 dryer to the exhaust ducts. Transition ducts and flexible clothes dryer transition ducts shall not be concealed within construction, and shall be installed in accordance with the manufacturer's installation instructions.

504.4.2.3 Exhaust Duct Power Ventilators. Dryer exhaust duct power ventilators for single residential clothes dryers shall be listed and labeled in accordance with UL 705 and installed in accordance with the manufacturer's installation instructions.

Each manufacturer has a specific listing to UL 705. The manufacturer's installation instructions are required as an integral part of the approval and installation of these systems. The appliance and exhaust ventilator may or may not be interlocked electrically with each other.

504.4.3 Commercial Clothes Dryers. Commercial dryer exhaust ducts shall be installed in accordance with their listings. The installation of commercial clothes dryer exhaust ducts shall comply with the appliance manufacturer's installation instructions.

Installations of commercial clothes dryers must comply with the manufacturers' installation instructions, which often specify size openings for replacement air.

When clothes dryers are installed in a room with other fuel-burning appliances, combustion air openings must be sized to prevent the creation of negative pressure within the compartment (see Section 701.3).

504.4.3.1 Exhaust Ducts for Type 2 Clothes Dryers. Exhaust ducts for Type 2 clothes dryers shall comply with the following:

(1) Exhaust ducts for Type 2 clothes dryers shall comply with Section 504.4. [NFPA 54:10.4.5.1]
(2) Exhaust ducts for Type 2 clothes dryers shall be constructed of sheet metal or other noncombustible material. Such ducts shall be equivalent in strength and corrosion resistance to ducts made of galvanized sheet steel not less than 0.0195 of an inch (0.4953 mm) thick. [NFPA 54:10.4.5.2]
(3) Type 2 clothes dryers shall be equipped or installed with lint-controlling means. [NFPA 54:10.4.5.3]
(4) Exhaust ducts for unlisted Type 2 clothes dryers shall be installed with a minimum clearance of 6 inches (152 mm) from adjacent combustible material. Where exhaust ducts for Type 2 clothes dryers are installed with reduced clearances, the adjacent combustible material shall be protected in accordance with Table 303.10.1. [NFPA 54:10.4.5.4]
(5) Where ducts pass through walls, floors, or partitions, the space around the duct shall be sealed with noncombustible material. [NFPA 54:10.4.5.5]
(6) Multiple installations of Type 2 clothes dryers shall be made in a manner to prevent adverse operation due to back pressures that might be created in the exhaust systems. [NFPA 54:10.4.5.6] The exhaust fan shall operate continuously or shall be interlocked to exhaust air where a clothes dryer is in operation.

Exhaust fans shall operate continuously or shall be interlocked with the clothes dryer to prevent the clothes dryer from starting in case of a fan failure. If a fan is not operating, an accumulation of lint can occur in the exhaust duct and put the occupants at risk due to the potential fire hazard.

504.4.4 Common Exhaust. Where permitted by the clothes dryer manufacturer's installation instructions, multiple clothes dryers shall be permitted to be installed with a common exhaust. The common exhaust duct shall be constructed of rigid metal and shall be installed in a fire-resistant rated enclosure in accordance with the building code. The duct material shall be of rigid metal with a thickness of not less than 0.020 of an inch (0.508 mm) (24 gauge). The duct enclosure shall be provided with a cleanout opening at the base of not less than 12 inches by 12 inches (305 mm by 305 mm). The exhaust fan shall be located downstream of branch connections and operated continuously and shall be monitored by an approved means.

504.4.5 Duct Supports. Ducts shall be supported in accordance with SMACNA HVAC Duct Construction Standards – Metal and Flexible.

504.5 Heat (Energy) Recovery Ventilators. Heat (energy) recovery ventilators shall be installed in accordance with their listings and comply with the appliance manufacturer's installation instructions. Non-ducted heat recovery ventilators shall comply with UL 1815. Ducted heat recovery ventilators shall comply with UL 1812. Heat (energy) recovery ventilator ducts shall comply with Chapter 6.

Installation of these devices must conform to the manufacturer's installation instructions. Typically, these would include mounting instructions, location of related registers and grilles and installation guidelines that include clearance for service. Ducts associated with this equipment must be in accordance with Chapter 6, Duct Systems (see **Figure 504.5**).

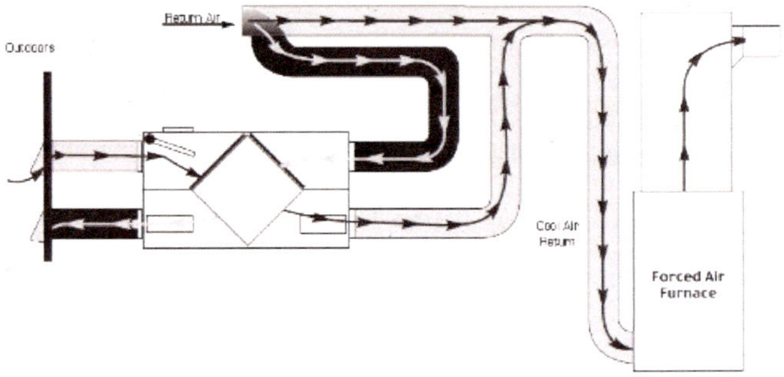

FIGURE 504.5
HEAT RECOVERY VENTILATOR

504.6 Gypsum Wallboard Ducts. Bathroom and laundry room exhaust ducts shall be permitted to be of gypsum wallboard subject to the limitations of Section 602.4.2.

505.0 Product-Conveying Systems.

505.1 General. A mechanical ventilation or exhaust system shall be installed to control, capture, and remove emissions generated from product use or handling where required in accordance with the building code or fire code and where such emissions result in a hazard to life or property. The design of the system shall be such that the emissions are confined to the area in which they are generated by air currents, hoods, or enclosures and shall be exhausted by a duct system to a safe location or treated by removing contaminants. Ducts conveying explosives or flammable vapors, fumes, or dusts shall extend directly to the exterior of the building without entering other spaces and shall not extend into or through ducts and plenums.

Exception: Ducts conveying vapor or fumes having flammable constituents less than 25 percent of their Lower Flammability Limit (LFL) shall be permitted to pass through other spaces.

When product use or handling generates emissions that must be removed, as required by the building or fire codes, or pose a hazard to life and property, a mechanical ventilation or exhaust system must be installed to control, capture and remove these emissions (see **Figures 505.1a, 505.1b, 505.1c**). Ducts that convey explosives or flammable vapors, fumes or dusts must extend from the generating source and area of first containment directly to the building exterior without entering or passing through other spaces. In addition, exhaust ducts may not extend into or through ducts or plenums.

The exception to this requirement states that ducts conveying vapor or fumes at less than 25 percent of their lower flammability limit (LFL) may pass through other spaces. Section 505.1 Exception requires the design to not exceed 25 percent LFL for vapors and fumes. Dusts may be of higher concentration.

Separate and distinct systems must be provided for incompatible materials. Mixing incompatible materials could cause explosion, chemical reaction or precipitation of solids or liquids in the duct system or could lessen system life. Contaminated air must not be circulated to occupied areas unless the designer can show that contaminants have been removed. Obviously, recirculation of air contaminated with explosive or flammable vapors, fumes or dusts; flammable or toxic gases; or radioactive material must not be allowed, regardless of the level of removal that can be achieved.

Users should spend time evaluating the type of product being exhausted (vapors or particulates), flammability and normal operating temperature. This discovery/determination is sometimes difficult and may involve a consultation with fire district personnel, the system designer, other available standards, equipment listings or a visit to other operational systems. Application of the following codes revolves around these three factors.

505.1.1 Mechanical Ventilation. A mechanical ventilation system shall be interlocked to operate with the equipment used to produce vapors, fumes, or dusts that are flammable or hazardous.

505.2 Incompatible Materials. Incompatible materials shall not be conveyed in the same system. [NFPA 91:4.2.2]

505.3 Flammability Limit. Unless the circumstances stipulated in Section 505.3.1, Section 505.3.2, or Section 505.3.3 exist, in systems conveying flammable vapors, gases, or mists, the concentration shall not exceed 25 percent of the lower flammability limit (LFL). [NFPA 91:4.2.3]

505.3.1 Higher Concentrations. Higher concentrations shall be permitted if the exhaust system is designed and protected in accordance with NFPA 69, using one or more of the following techniques:

(1) Combustible concentration reduction

(2) Oxidant concentration reduction

FIGURE 505.1A
VEHICLE EXHAUST COLLECTION SYSTEM

CABINET SHOP

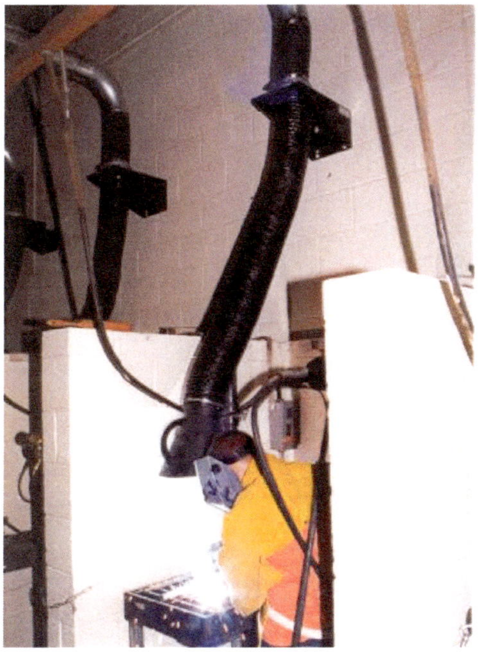

WELDING

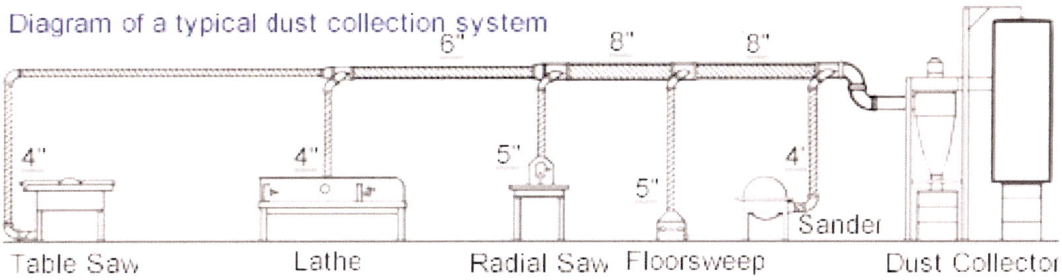

FIGURE 505.1B
EXHAUST COLLECTION SYSTEM

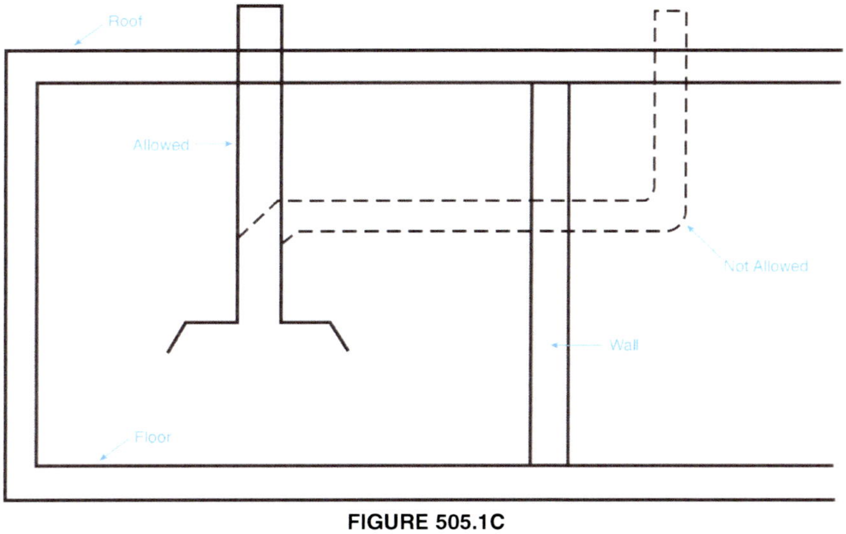

FIGURE 505.1C
VENTILATION DUCT DIRECTLY TO EXTERIOR

(3) Deflagration suppression

(4) Deflagration pressure containment [NFPA 91:4.2.3.1]

Contaminated air shall not be recirculated to occupied areas unless contaminants have been removed. Air contaminated with explosive or flammable vapors, fumes, or dusts; flammable or toxic gases; or radioactive material shall not be recirculated.

505.3.2 Ovens and Furnaces. Higher concentrations shall be permitted for ovens and furnaces designed and protected in accordance with NFPA 86. [NFPA 91:4.2.3.2]

505.3.3 Deflagration. Higher concentrations shall be permitted where deflagration venting is provided in accordance with NFPA 68. [NFPA 91:4.2.3.3]

505.4 Air-Moving Devices. Air-moving devices shall be sized to establish the velocity required to capture, control, and convey materials through the exhaust system. [NFPA 91:4.2.5]

505.5 Generating Flames, Sparks, or Hot Materials. Operations generating flames, sparks, or hot material such as from grinding wheels and welding shall not be manifolded into any exhaust system that air conveys flammable or combustible materials. [NFPA 91:4.2.6]

505.6 Fire Dampers. Fire dampers shall be permitted to be installed in exhaust systems in accordance with the following:

(1) Where ducts pass through fire barriers

(2) Where a collection system installed on the end of the system is protected with an automatic extinguishing system

(3) Where the duct system is protected with an automatic extinguishing system

(4) Where ducts have been listed with interrupters

(5) Where necessary to facilitate the control of smoke pursuant to the applicable NFPA standards [NFPA 91:4.2.9]

505.6.1 Prohibited. Fire dampers shall not be installed if the material being exhausted is toxic and if a risk evaluation indicates that the toxic hazard is greater than the fire hazard. [NFPA 91:4.2.10]

Although related, fire dampers and smoke dampers are not the same. For commentary on smoke dampers see Section 606.1.

505.7 Fire Detection and Alarm Systems. Unless the circumstances stipulated in Section 505.7.1 or Section 505.7.2 exist, fire detection and alarm systems shall not be interlocked to shut down air-moving devices. [NFPA 91:4.2.14]

505.7.1 Automatic Extinguishing System. Where shutdown is necessary for the effective operation of an automatic extinguishing system, it shall be permitted to interlock fire detection and alarm systems to shut down air-moving devices. [NFPA 91:4.2.14.1]

505.7.2 Shut Down Permitted. Where a documented risk analysis acceptable to the Authority Having Jurisdiction shows that the risk of damage from fire and the products of combustion would be higher with air-moving devices operating, it shall be permitted to interlock fire detection and alarm systems to shut down air-moving devices. [NFPA 91:4.2.14.2]

505.8 Product-Conveying Ducts Classification. Product-conveying ducts shall be classified according to their use, as follows:

Class 1 - Ducts conveying nonabrasives, such as smoke, spray, mists, fogs, noncorrosive fumes and gases, light fine dusts, or powders.

Class 2 - Ducts conveying moderately abrasive particulate in light concentrations, such as sawdust and grain dust, and buffing and polishing dust.

Class 3 - Ducts conveying Class 2 materials in high concentrations and highly abrasive materials in low concentrations, such as manganese, steel chips, and coke.

Class 4 - Ducts conveying highly abrasive material in high concentrations.

Class 5 - Ducts conveying corrosives, such as acid vapors.

This section identifies five classes of product-conveying ducts according to their use. Each higher class and resulting duct thickness requirements found in Tables 506.2(1) and 506.2(2) reflect a resistance to the abrasive nature of the product being conveyed.

505.9 Minimum Velocities and Circulation. The velocity and circulation of air in work areas shall be such that contaminant's are captured by an airstream at the area where the emissions are generated and conveyed into a product-conveying duct system. Mixtures within work areas where contaminants are generated shall be diluted to be accordance with Section 505.3 with air that does not contain other contaminants. The velocity of air within the duct shall be not less than set forth in Table 505.9.

Systems conveying particulate matter shall be designed by employing the constant velocity method. Systems conveying explosive or radioactive materials shall be pre-balanced through duct sizing. Other systems shall be permitted to be designed with balancing devices such as dampers. Dampers provided to balance airflow shall be provided with securely fixed minimum-position blocking devices to prevent restricting flow below the required volume or velocity.

In conveying solid particulates, vapors or noxious, corrosive, flammable or explosive gases, adequate velocity must be maintained throughout the system. Otherwise, many contaminants will tend to precipitate, condense and separate from the air stream; if this occurs, the deposit will block the duct system. Mixtures within the work area must be diluted below 25 percent of their LFL with air that does not contain other contaminants. Minimum velocities are set forth in Table 505.9 for various common industrial materials.

The entire work area ventilation system must be designed and constructed so that contaminants are captured by an air stream at the area of generation (typically a single workstation) and conveyed directly into the product-conveying duct system.

Systems for particulate matter are to be designed using the Constant Velocity Method. Common duct sizes limit the ability to maintain constant velocity. Certain systems allow

TABLE 505.9
RANGE OF MINIMUM DUCT DESIGN VELOCITIES[1]
[NFPA 91: TABLE A.4.2.5]

NATURE OF CONTAMINANTS	EXAMPLES	FEET PER MINUTE DESIGN VELOCITY (feet per minute)
Vapors, gases, smoke	All vapors, gases, and smoke	1000 - 2000[2]
Fumes	Welding	2000 - 2500
Very fine light dust	Cotton lint, wood flour, litho powder	2500 - 3000
Dry dusts and powders	Fine rubber dust, molding powder dust, jute lint, cotton dust, shavings (light), soap dust, leather shavings	3000 - 4000
Average industrial dusts	Grinding dust, buffing lint (dry), wool jute dust (shaker waste), coffee beans, shoe dust, granite dust, silica flour, general material handling, brick cutting, clay dust, foundry (general), limestone dust, packaging and weighing asbestos dust in textile industries	3500 - 4000
Heavy dusts	Sawdust (heavy and wet), metal turnings, foundry tumbling barrels and shake-out, sandblast dust, wood blocks, hog waste, brass turning, cast-iron boring dust, lead dust	4000 - 4500
Heavy or moist dusts	Lead dust with chips, moist cement dust, asbestos chunks from transite pipe cutting machines, buffing lint (sticky), quick-lime dust	4500 and up

For SI units: 1 foot per minute = 0.005 m/s

[1] Systems that are handling combustible particulate solids shall be accordance with NFPA 654.

[2] Any desired velocity (economic optimum velocity usually within this range).

openings intended to maintain minimum velocities in larger ducts without causing excessive velocities in smaller attached ducts (see **Figure 505.9**). Systems conveying explosive or radioactive materials are to be pre-balanced through duct sizing. Other systems may be designed with balancing devices such as dampers, provided that minimum position blocking devices are installed to prevent restricting flow below the minimum required volume or velocity.

While the duct system can be designed and installed with predictable results, humidity and temperatures within the industrial plant can be difficult to control or predict. Large openings, such as windows, doors or vents, can drastically alter the environment. Similarly, subsequent remodeling of the structure or the heating, ventilation and air-conditioning (HVAC) system may affect the product-conveying system and are difficult to predict. Care should be taken when reviewing proposed remodeling or alterations to these installations so that the existing exhaust systems are not adversely impacted.

A common question posed by those who design product-conveying duct systems is whether a small cyclone-type dust collector or self-contained or portable mechanical dust-collection system can be used to collect flammable dusts in bags or containers at each piece of woodworking equipment, in lieu of the central system required by the building code of the jurisdiction. The bag acts as a filter. The smallest of the particles pass through the bag and settle on horizontal surfaces, thereby increasing the risk of ignition and combustion. Since storage of wood residues should be in exterior noncombustible containers or in other approved locations as specified in the building code of the jurisdiction, a central product-conveying duct system that will transport hazardous dusts to the outside of the building is required.

Note that under Section 506.6 ducts conveying explosive dust must have explosion vents located outside the building.

FIGURE 505.9
OPENING USED TO MAINTAIN A CONSTANT VELOCITY

505.10 Makeup Air. Makeup air shall be provided to replenish air exhausted by the ventilation system. Makeup air intakes shall be located so as to avoid recirculation of contaminated air within enclosures.

In order to maintain a balanced-pressure environment, makeup air must be provided to replenish the air exhausted by the ventilation system. The makeup air intakes or outside air sources must be located to avoid recirculation of contaminated air within enclosures.

505.11 Hoods and Enclosures. Hoods and enclosures shall be used where contaminants originate in a concentrated area. The design of the hood or enclosure shall be such that air currents created by the exhaust systems will capture the contaminants and transport them directly to the exhaust duct. The volume of air shall be sufficient to dilute explosive or flammable vapors, fumes, or dusts in accordance with Section 505.9. Hoods of steel shall have a base metal thickness not less than 0.027 of an inch (0.686 mm) (No. 22 gauge) for Class 1 and Class 5 metal duct systems; 0.033 of an inch (0.838 mm) (No. 20 gauge) for hoods serving a Class 2 duct system; 0.044 of an inch (1.118 mm) (No. 18 gauge) for hoods serving a Class 3 duct system; and 0.068 of an inch (1.727 mm) (No. 14 gauge) for hoods serving a Class 4 duct system.

Approved nonmetallic hoods and duct systems shall be permitted to be used for Class 5 corrosive systems where the corrosive mixture is nonflammable. Metal hoods used with Class 5 duct systems shall be protected with an approved corrosion-resistant material. Edges of hoods shall be rounded. The minimum clearance between hoods and combustible construction shall be the clearance required by the duct system.

When air contaminants are produced in a concentrated area, hoods and enclosures must be used to capture these contaminants (see **Figure 505.11**). A concentrated source area is typical of many industrial processes that generate product contaminants. The hood or enclosure design must be such that the air currents created by the exhaust system will capture the contaminants and transport them directly to the exhaust duct. This is usually achieved by creating a partial vacuum within the hood or a slight positive pressure around the hood. The volume of air must be great enough to dilute explosive or flammable vapors, fumes or dusts to below 25 percent of the LFL or lower explosive limit (LEL) of the contaminant. The LFL is unique to each material to be exhausted and is also a function of temperature, humidity and other factors. Research in mines, refineries, chemical and other industrial plants, and farm environments of all types, has led to the current body of scientific knowledge and field experience in what constitutes a "safe lower limit." Much of this knowledge has been acquired the hard way, through fires, explosions, catastrophic injury and loss of life.

Hoods must be made of steel and must have a base metal thickness as follows:

Class 1 and Class 5 Duct Systems — 22 gauge [0.027 inch]

Class 2 Duct Systems — 20 gauge [0.033 inch]

Class 3 Duct Systems — 18 gauge [0.044 inch)]

Class 4 Duct Systems — 14 gauge [0.068 inch]

Class 5 corrosive systems may use approved nonmetallic hoods only when the corrosive mixture is nonflammable. Metal hoods used in such installations must be protected with corrosion-resistant material.

Hood edges must be rounded. The clearance of hood to combustible construction must not be less than that required for the exhaust duct.

506.0 Product-Conveying Ducts.

506.1 Materials. Materials used in product-conveying duct systems shall be suitable for the intended use and shall be of metal.

Exceptions:

(1) Asbestos-cement, concrete, clay, or ceramic materials shall be permitted to be used where it is shown that these materials will be equivalent to metal ducts installed in accordance with this chapter.

(2) Ducts serving a Class 5 system shall be permitted to be constructed of approved nonmetallic material where the corrosive characteristics of the material being conveyed make a metal system unsuitable and where the mixture being conveyed is nonflammable.

Approved nonmetallic material shall be either a listed product having a flame-spread index not exceeding 25 and a smoke-developed rating of 50 or less on both inside and outside surfaces without evidence of continued progressive combustion, or shall have a flame-spread index not exceeding 25 and shall be installed with an automatic fire-sprinkler protection system inside the duct.

(3) Ducts used in central vacuum cleaning systems within a dwelling unit shall be constructed of materials in accordance with ASTM F2158 or the applicable standards referenced in Chapter 17. Penetrations of fire walls or floor-ceiling or roof-ceiling assemblies shall be in accordance with the building code.

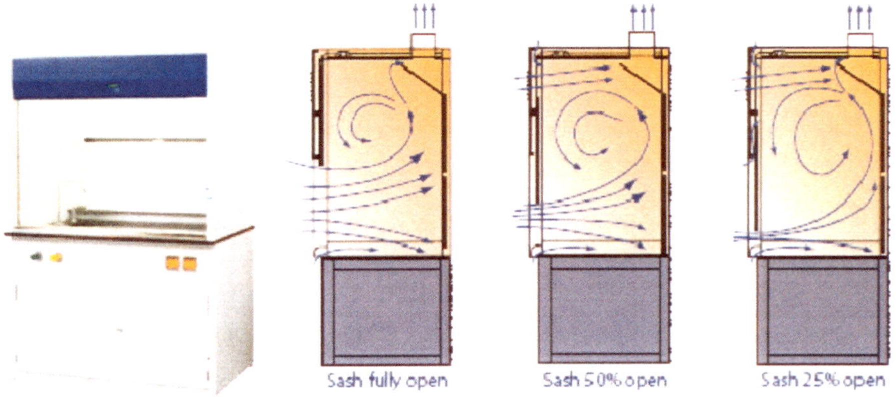

FIGURE 505.11
HOODS AND ENCLOSURES

Copper or ferrous pipes or conduits extending from within the separation between a garage and dwelling unit to the central vacuuming unit shall be permitted to be used.

Aluminum ducts shall not be used in systems conveying flammable vapors, fumes, or explosive dusts, nor in Class 2, 3, or 4 systems. Galvanized steel and aluminum ducts shall not be used where the temperature of the material being conveyed exceeds 400°F (204°C).

Metal ducts used in Class 5 systems that are not resistant to the corrosiveness of the product shall be protected with an approved corrosion-resistant material.

The exceptions to this are as follows:

(1) Asbestos cement, concrete, clay or ceramic materials may be used if it can be shown that their performance will be equivalent to that of metal ducts. These materials are most commonly used for plumbing, sewer and water system supply piping. They may sometimes provide appropriate and long-lasting service as product-conveying ducts, especially if burial is necessary. The manufacturer must be consulted for instructions on suitability of such materials as ducts for the specific product to be conveyed. Plastic sewer piping used underground may not be suitable for systems containing flammable products.

(2) Ducts serving Class 5 systems conveying corrosives such as acid vapors may be constructed of approved materials, but only when the material to be conveyed is determined to be nonflammable and a metal system is not suitable. Approved nonmetallic material must either be a listed product having a flame/smoke rating of 25/50 or less, on both inside and outside surfaces, without evidence of continued progressive combustion, or must have a flame spread index of 25 or less and installed with an automatic fire sprinkler system inside the duct. Ducts serving Class 5 systems, conveying corrosives such as acid vapors, may be constructed of approved nonmetallic materials, but only when the material to be conveyed is determined to be nonflammable and a metal system is not suitable.

(3) In dwelling units, ducts used in central vacuum cleaning systems may be of PVC pipe or tubing. The materials must meet the new standard developed for vacuum cleaning systems.

Penetration of fire walls or floor/ceiling or roof/ceiling assemblies must comply with the building code of the jurisdiction. Copper tube or galvanized or black steel pipes or conduits extending from within the separation between the garage and the dwelling unit to the central vacuum cleaning unit may be used. Consult the building code of the jurisdiction to determine if the wall is required to be constructed as a fire wall or fire barrier.

Aluminum ducts must not be used where flammable or explosive dusts, vapors or fumes are to be conveyed or in Class 2, 3 or 4 systems. Installation of aluminum ducts is limited to Class 1 systems conveying nonabrasive and noncorrosive contaminants. Aluminum ducts as well as galvanized steel ducts must not be used where the temperature of the material to be conveyed exceeds 400°F. These metals (aluminum and zinc) have been found to constitute a hazard in high-temperature operations.

Metal ducts used in Class 5 systems that are not resistant to the corrosiveness of the conveyed product must be protected with a corrosion-resistant material. The ducts should be coated or lined with a nonpermeable, corrosion-resistant material as recommended by the manufacturer for the type of product being conveyed.

506.2 Construction. Ducts used for conveying products shall be airtight construction as approved by the Authority Having Jurisdiction, and shall not have openings other than those required for operation and maintenance of the system. Ducts constructed of steel shall comply with Table 506.2(1) or Table 506.2(2).

Exceptions:

(1) Class 1 product-conveying ducts that operate at less than 4 inches water column (0.9 kPa) negative pressure and convey noncorrosive, nonflammable and nonexplosive materials at temperatures not exceeding 250°F (121°C) shall be permitted to be constructed in accordance with SMACNA HVAC Duct Construction Standards–Metal and Flexible.

(2) Ducts used in central vacuuming systems within a dwelling unit shall be constructed of materials in accordance with ASTM F2158 or the applicable standards referenced in Chapter 17. Penetrations of fire-resistive walls or floor-ceiling or roof-ceiling assemblies shall be in accordance with the building code. Copper or ferrous pipes or conduit extending from within the separation between a garage and dwelling unit to the central vacuum unit shall be permitted to be used.

The use of rectangular ducts conveying particulates shall be subject to approval of the Authority Having Jurisdiction. The design of rectangular ducts shall consider the adhesiveness and buildup of products being conveyed within the duct.

Aluminum construction shall be permitted to be used in Class 1 duct systems. The thickness of aluminum ducts shall be not less than two Brown and Sharpe gauges thicker than the gauges required for steel ducts set forth in Table 506.2(1) and Table 506.2(2).

Product-conveying ducts must be of substantially air-tight construction. They must not have any openings except as required for system operation and maintenance. Steel ducts must comply with Table 506.2(1) or 506.2(2) for round and rectangular ducts, respectively. These tables specify required thickness, negative internal pressure and reinforcement spacing. No standard is referenced for the connection and reinforcement. See Sheet Metal and Air Conditioning National Association (SMACNA) round duct construction standards.

The exceptions to this requirement are as follows:

1. Class 1 product-conveying ducts that operate at less than 4 inches water column negative pressure and convey noncorrosive, nonflammable and non-explosive materials at temperatures not exceeding 250°F may be

TABLE 506.2(1)
MINIMUM SHEET METAL THICKNESS FOR ROUND DUCTS FOR PRODUCT-CONVEYING SYSTEM DUCTS

NEGATIVE PRESSURE (inches water column)	REINF. SPACING (inches)	CLASS 1 (inches)								
		Up to 7	8 to 11	12 to 15	16 to 19	20 to 23	24 to 35	36 to 47	48 to 59	60
To 7	0	0.021 (24 ga.)	0.021 (24 ga.)	0.033 (20 ga.)	0.044 (18 ga.)	0.055 (16 ga.)	0.068 (14 ga.)	0.127 (10 ga.)	—	—
	96	0.021 (24 ga.)	0.021 (24 ga.)	0.021 (24 ga.)	0.027 (22 ga.)	0.033 (20 ga.)	0.044 (18 ga.)	0.055 (16 ga.)	0.068 (14 ga.)	0.068 (14 ga.)
	48	0.021 (24 ga.)	0.021 (24 ga.)	0.021 (24 ga.)	0.021 (24 ga.)	0.021 (24 ga.)	0.027 (22 ga.)	0.033 (20 ga.)	0.044 (18 ga.)	0.055 (16 ga.)
	24	0.021 (24 ga.)	0.021 (24 ga.)	0.021 (24 ga.)	0.021 (24 ga.)	0.021 (24 ga.)	0.021 (24 ga.)	0.027 (22 ga.)	0.033 (20 ga.)	0.044 (18 ga.)
8 to 11	0	0.021 (24 ga.)	0.027 (22 ga.)	0.044 (18 ga.)	0.055 (16 ga.)	0.068 (14 ga.)	0.097 (12 ga.)	—	—	—
	96	0.021 (24 ga.)	0.027 (22 ga.)	0.027 (22 ga.)	0.044 (18 ga.)	0.044 (18 ga.)	0.044 (18 ga.)	0.068 (14 ga.)	0.097 (12 ga.)	0.097 (12 ga.)
	48	0.021 (24 ga.)	0.021 (24 ga.)	0.021 (24 ga.)	0.027 (22 ga.)	0.033 (20 ga.)	0.033 (20 ga.)	0.055 (16 ga.)	0.068 (14 ga.)	0.068 (14 ga.)
	24	0.021 (24 ga.)	0.021 (24 ga.)	0.021 (24 ga.)	0.021 (24 ga.)	0.027 (22 ga.)	0.027 (22 ga.)	0.044 (18 ga.)	0.055 (16 ga.)	0.055 (16 ga.)
12 to 15	0	0.021 (24 ga.)	0.033 (20 ga.)	0.055 (16 ga.)	0.068 (14 ga.)	0.097 (12 ga.)	0.097 (12 ga.)	—	—	—
	96	0.021 (24 ga.)	0.027 (22 ga.)	0.044 (18 ga.)	0.044 (18 ga.)	0.055 (16 ga.)	0.055 (16 ga.)	0.097 (12 ga.)	0.112 (11 ga.)	0.112 (11 ga.)
	48	0.021 (24 ga.)	0.027 (22 ga.)	0.027 (22 ga.)	0.033 (20 ga.)	0.044 (18 ga.)	0.044 (18 ga.)	0.068 (14 ga.)	0.068 (14 ga.)	0.097 (12 ga.)
	24	0.021 (24 ga.)	0.021 (24 ga.)	0.021 (24 ga.)	0.027 (22 ga.)	0.027 (22 ga.)	0.027 (22 ga.)	0.055 (16 ga.)	0.055 (16 ga.)	0.055 (16 ga.)
16 to 20	0	0.021 (24 ga.)	0.044 (18 ga.)	0.068 (14 ga.)	0.097 (12 ga.)	0.112 (11 ga.)	—	—	—	—
	96	0.021 (24 ga.)	0.033 (20 ga.)	0.055 (16 ga.)	0.055 (16 ga.)	0.068 (14 ga.)	0.068 (14 ga.)	0.112 (11 ga.)	0.112 (11 ga.)	0.112 (11 ga.)
	48	0.021 (24 ga.)	0.027 (22 ga.)	0.033 (20 ga.)	0.044 (18 ga.)	0.055 (16 ga.)	0.055 (16 ga.)	0.068 (14 ga.)	0.097 (12 ga.)	0.112 (11 ga.)
	24	0.021 (24 ga.)	0.021 (24 ga.)	0.027 (22 ga.)	0.033 (20 ga.)	0.044 (18 ga.)	0.044 (18 ga.)	0.055 (16 ga.)	0.068 (14 ga.)	0.097 (12 ga.)
		CLASS 2 (inches)								
To 7	0	0.027 (22 ga.)	0.027 (22 ga.)	0.033 (20 ga.)	0.044 (18 ga.)	0.055 (16 ga.)	0.068 (14 ga.)	0.127 (10 ga.)	—	—
	96	0.027 (22 ga.)	0.027 (22 ga.)	0.033 (20 ga.)	0.033 (20 ga.)	0.044 (18 ga.)	0.044 (18 ga.)	0.055 (16 ga.)	0.068 (14 ga.)	0.068 (14 ga.)
	48	0.027 (22 ga.)	0.027 (22 ga.)	0.033 (20 ga.)	0.033 (20 ga.)	0.044 (18 ga.)	0.044 (18 ga.)	0.055 (16 ga.)	0.055 (16 ga.)	0.055 (16 ga.)
	24	0.027 (22 ga.)	0.027 (22 ga.)	0.033 (20 ga.)	0.033 (20 ga.)	0.044 (18 ga.)	0.044 (18 ga.)	0.055 (16 ga.)	0.055 (16 ga.)	0.055 (16 ga.)
8 to 11	0	0.027 (22 ga.)	0.027 (22 ga.)	0.044 (18 ga.)	0.055 (16 ga.)	0.068 (14 ga.)	0.097 (12 ga.)	—	—	—
	96	0.027 (22 ga.)	0.027 (22 ga.)	0.033 (20 ga.)	0.044 (18 ga.)	0.044 (18 ga.)	0.044 (18 ga.)	0.068 (14 ga.)	0.097 (12 ga.)	0.097 (12 ga.)
	48	0.027 (22 ga.)	0.027 (22 ga.)	0.033 (20 ga.)	0.033 (20 ga.)	0.044 (18 ga.)	0.044 (18 ga.)	0.055 (16 ga.)	0.068 (14 ga.)	0.068 (14 ga.)
	24	0.027 (22 ga.)	0.027 (22 ga.)	0.033 (20 ga.)	0.033 (20 ga.)	0.044 (18 ga.)	0.044 (18 ga.)	0.055 (16 ga.)	0.055 (16 ga.)	0.055 (16 ga.)
12 to 15	0	0.027 (22 ga.)	0.033 (20 ga.)	0.055 (16 ga.)	0.068 (14 ga.)	0.097 (12 ga.)	0.097 (12 ga.)	—	—	—
	96	0.027 (22 ga.)	0.033 (20 ga.)	0.044 (18 ga.)	0.044 (18 ga.)	0.055 (16 ga.)	0.055 (16 ga.)	0.097 (12 ga.)	0.112 (11 ga.)	0.112 (11 ga.)
	48	0.027 (22 ga.)	0.027 (22 ga.)	0.033 (20 ga.)	0.033 (20 ga.)	0.044 (18 ga.)	0.044 (18 ga.)	0.068 (14 ga.)	0.068 (14 ga.)	0.097 (12 ga.)
	24	0.027 (22 ga.)	0.027 (22 ga.)	0.033 (20 ga.)	0.033 (20 ga.)	0.044 (18 ga.)	0.044 (18 ga.)	0.055 (16 ga.)	0.055 (16 ga.)	0.055 (16 ga.)
16 to 20	0	0.027 (22 ga.)	0.044 (18 ga.)	0.068 (14 ga.)	0.097 (12 ga.)	0.112 (11 ga.)	—	—	—	—
	96	0.027 (22 ga.)	0.033 (20 ga.)	0.055 (16 ga.)	0.055 (16 ga.)	0.068 (14 ga.)	0.068 (14 ga.)	0.112 (11 ga.)	0.112 (11 ga.)	0.112 (11 ga.)
	48	0.027 (22 ga.)	0.033 (20 ga.)	0.033 (20 ga.)	0.044 (18 ga.)	0.055 (16 ga.)	0.055 (16 ga.)	0.068 (14 ga.)	0.068 (14 ga.)	0.097 (12 ga.)
	24	0.027 (22 ga.)	0.033 (20 ga.)	0.033 (20 ga.)	0.033 (20 ga.)	0.044 (18 ga.)	0.044 (18 ga.)	0.055 (16 ga.)	0.055 (16 ga.)	0.055 (16 ga.)
		CLASS 3 (inches)								
To 7	0	0.033 (20 ga.)	0.033 (20 ga.)	0.044 (18 ga.)	0.044 (18 ga.)	0.055 (16 ga.)	0.068 (14 ga.)	0.127 (10 ga.)	—	—
	96	0.033 (20 ga.)	0.033 (20 ga.)	0.044 (18 ga.)	0.044 (18 ga.)	0.055 (16 ga.)	0.055 (16 ga.)	0.068 (14 ga.)	0.068 (14 ga.)	0.068 (14 ga.)
	48	0.033 (20 ga.)	0.033 (20 ga.)	0.044 (18 ga.)	0.044 (18 ga.)	0.055 (16 ga.)	0.055 (16 ga.)	0.068 (14 ga.)	0.068 (14 ga.)	0.068 (14 ga.)
	24	0.033 (20 ga.)	0.033 (20 ga.)	0.044 (18 ga.)	0.044 (18 ga.)	0.055 (16 ga.)	0.055 (16 ga.)	0.068 (14 ga.)	0.068 (14 ga.)	0.068 (14 ga.)
8 to 11	0	0.033 (20 ga.)	0.033 (20 ga.)	0.044 (18 ga.)	0.055 (16 ga.)	0.068 (14 ga.)	0.097 (12 ga.)	—	—	—
	96	0.033 (20 ga.)	0.033 (20 ga.)	0.044 (18 ga.)	0.044 (18 ga.)	0.055 (16 ga.)	0.055 (16 ga.)	0.068 (14 ga.)	0.097 (12 ga.)	0.097 (12 ga.)
	48	0.033 (20 ga.)	0.033 (20 ga.)	0.044 (18 ga.)	0.044 (18 ga.)	0.055 (16 ga.)	0.055 (16 ga.)	0.068 (14 ga.)	0.068 (14 ga.)	0.068 (14 ga.)
	24	0.033 (20 ga.)	0.033 (20 ga.)	0.044 (18 ga.)	0.044 (18 ga.)	0.055 (16 ga.)	0.055 (16 ga.)	0.068 (14 ga.)	0.068 (14 ga.)	0.068 (14 ga.)

TABLE 506.2(1) (continued)
MINIMUM SHEET METAL THICKNESS FOR ROUND DUCTS FOR PRODUCT-CONVEYING SYSTEM DUCTS

NEGATIVE PRESSURE (inches water column)	REINF. SPACING (inches)	CLASS 3 (continued) (inches)								
		Up to 7	8 to 11	12 to 15	16 to 19	20 to 23	24 to 35	36 to 47	48 to 59	60
12 to 15	0	0.133 (20 ga.)	0.133 (20 ga.)	0.155 (16 ga.)	0.168 (14 ga.)	0.197 (12 ga.)	0.197 (12 ga.)	—	—	—
	96	0.133 (20 ga.)	0.133 (20 ga.)	0.144 (18 ga.)	0.144 (18 ga.)	0.155 (16 ga.)	0.155 (16 ga.)	0.197 (12 ga.)	0.112 (11 ga.)	0.112 (11 ga.)
	48	0.133 (20 ga.)	0.133 (20 ga.)	0.144 (18 ga.)	0.144 (18 ga.)	0.155 (16 ga.)	0.155 (16 ga.)	0.168 (14 ga.)	0.168 (14 ga.)	0.168 (14 ga.)
	24	0.133 (20 ga.)	0.133 (20 ga.)	0.144 (18 ga.)	0.144 (18 ga.)	0.155 (16 ga.)	0.155 (16 ga.)	0.168 (14 ga.)	0.168 (14 ga.)	0.168 (14 ga.)
16 to 20	0	0.133 (20 ga.)	0.144 (18 ga.)	0.168 (14 ga.)	0.197 (12 ga.)	0.112 (11 ga.)	—	—	—	—
	96	0.133 (20 ga.)	0.133 (20 ga.)	0.155 (16 ga.)	0.155 (16 ga.)	0.168 (14 ga.)	0.168 (14 ga.)	0.112 (11 ga.)	0.112 (11 ga.)	0.112 (11 ga.)
	48	0.133 (20 ga.)	0.133 (20 ga.)	0.144 (18 ga.)	0.144 (18 ga.)	0.155 (16 ga.)	0.155 (16 ga.)	0.168 (14 ga.)	0.168 (14 ga.)	0.197 (12 ga.)
	24	0.133 (20 ga.)	0.133 (20 ga.)	0.144 (18 ga.)	0.144 (18 ga.)	0.155 (16 ga.)	0.155 (16 ga.)	0.168 (14 ga.)	0.168 (14 ga.)	0.168 (14 ga.)
		CLASS 4 (inches)								
To 7	0	0.155 (16 ga.)	0.155 (16 ga.)	0.155 (16 ga.)	0.155 (16 ga.)	0.155 (16 ga.)	0.168 (14 ga.)	0.168 (14 ga.)	—	—
	96	0.155 (16 ga.)	0.155 (16 ga.)	0.155 (16 ga.)	0.155 (16 ga.)	0.168 (14 ga.)	0.168 (14 ga.)	0.197 (12 ga.)	0.197 (12 ga.)	0.197 (12 ga.)
	48	0.155 (16 ga.)	0.155 (16 ga.)	0.155 (16 ga.)	0.155 (16 ga.)	0.168 (14 ga.)	0.168 (14 ga.)	0.197 (12 ga.)	0.197 (12 ga.)	0.197 (12 ga.)
	24	0.155 (16 ga.)	0.155 (16 ga.)	0.155 (16 ga.)	0.155 (16 ga.)	0.168 (14 ga.)	0.168 (14 ga.)	0.197 (12 ga.)	0.197 (12 ga.)	0.197 (12 ga.)
8 to 11	0	0.155 (16 ga.)	0.155 (16 ga.)	0.155 (16 ga.)	0.155 (16 ga.)	0.168 (14 ga.)	0.197 (12 ga.)	—	—	—
	96	0.155 (16 ga.)	0.155 (16 ga.)	0.155 (16 ga.)	0.155 (16 ga.)	0.168 (14 ga.)	0.168 (14 ga.)	0.197 (12 ga.)	0.197 (12 ga.)	0.197 (12 ga.)
	48	0.155 (16 ga.)	0.155 (16 ga.)	0.155 (16 ga.)	0.155 (16 ga.)	0.168 (14 ga.)	0.168 (14 ga.)	0.197 (12 ga.)	0.197 (12 ga.)	0.197 (12 ga.)
	24	0.155 (16 ga.)	0.155 (16 ga.)	0.155 (16 ga.)	0.155 (16 ga.)	0.168 (14 ga.)	0.168 (14 ga.)	0.197 (12 ga.)	0.197 (12 ga.)	0.197 (12 ga.)
12 to 15	0	0.155 (16 ga.)	0.155 (16 ga.)	0.155 (16 ga.)	0.168 (14 ga.)	0.197 (12 ga.)	0.197 (12 ga.)	—	—	—
	96	0.155 (16 ga.)	0.155 (16 ga.)	0.155 (16 ga.)	0.155 (16 ga.)	0.168 (14 ga.)	0.168 (14 ga.)	0.197 (12 ga.)	0.112 (11 ga.)	0.112 (11 ga.)
	48	0.155 (16 ga.)	0.155 (16 ga.)	0.155 (16 ga.)	0.155 (16 ga.)	0.168 (14 ga.)	0.168 (14 ga.)	0.197 (12 ga.)	0.197 (12 ga.)	0.097 (12 ga.)
	24	0.055 (16 ga.)	0.055 (16 ga.)	0.055 (16 ga.)	0.055 (16 ga.)	0.068 (14 ga.)	0.068 (14 ga.)	0.097 (12 ga.)	0.097 (12 ga.)	0.097 (12 ga.)
16 to 20	0	0.055 (16 ga.)	0.055 (16 ga.)	0.068 (14 ga.)	0.097 (12 ga.)	0.112 (11 ga.)	—	—	—	—
	96	0.055 (16 ga.)	0.055 (16 ga.)	0.055 (16 ga.)	0.055 (16 ga.)	0.068 (14 ga.)	0.068 (14 ga.)	0.112 (11 ga.)	0.112 (11 ga.)	0.112 (11 ga.)
	48	0.055 (16 ga.)	0.055 (16 ga.)	0.055 (16 ga.)	0.055 (16 ga.)	0.068 (14 ga.)	0.068 (14 ga.)	0.097 (12 ga.)	0.097 (12 ga.)	0.097 (12 ga.)
	24	0.055 (16 ga.)	0.055 (16 ga.)	0.055 (16 ga.)	0.055 (16 ga.)	0.068 (14 ga.)	0.068 (14 ga.)	0.097 (12 ga.)	0.097 (12 ga.)	0.097 (12 ga.)

For SI units: 1 inch = 25.4 mm, 1 inch water column = 0.249 kPa

constructed in accordance with SMACNA HVAC Duct Construction Standards Metal and Flexible.

SMACNA requires connections that are 2- to 4-inch slip joints for systems operating in excess of 2 inch negative water column pressure. No additional reinforcement is required until the duct is in excess of 37 inches in diameter. This reinforcement is noted as an angle iron flange that is rolled to the diameter of the duct.

2. Ducts used in central vacuuming systems within a dwelling unit may be of PVC pipe or tubing. Penetrations of fire-resistive walls or floor/ceiling or roof/ceilings assemblies must comply with the building code of the jurisdiction. However, some building codes no longer require a fire-resistive wall between the garage and the dwelling. Copper tubing or galvanized steel or black steel pipes or conduit extending from within the separation between a garage and dwelling unit to the central vacuum unit may be used. A short metal sleeve encasing the vacuum pipe as it penetrates the membrane of the fire-resistive construction may not comply with building code of the jurisdiction requirements.

Use of rectangular ducts is subject to the approval of the AHJ. The deposit or collection of the conveyed product in the corners of the rectangular ducts is a concern. The designer should consider the adhesiveness and likelihood of buildup of the conveyed product within the duct system.

Aluminum ducts are limited for service as Class 1 duct (nonabrasive) only. In addition to this limitation, the thickness of an aluminum duct must be at least two B & S gauges thicker than the gauges required for steel ducts, as set forth in Tables 506.2(1) and 506.2(2).

506.3 Penetrations. Exhaust ducts shall not pass through fire walls. {NFPA 91:4.2.11}

TABLE 506.2(2)
MINIMUM SHEET METAL THICKNESS FOR RECTANGULAR DUCTS

NEGATIVE PRESSURE (inches water column)	REINF. SPACING (inches)	LONGEST SIDE OF DUCT									
		CLASS 1 (inches)					CLASS 2 (inches)				
		Up to 12	13 to 24	25 to 36	37 to 48	49 to 60	Up to 12	13 to 24	25 to 36	37 to 48	49 to 60
To 7	48	0.021 (24 ga.)	0.033 (20 ga.)	0.055 (16 ga.)	0.068 (14 ga.)	—	0.033 (20 ga.)	0.044 (18 ga.)	0.055 (16 ga.)	0.068 (14 ga.)	—
To 7	24	0.021 (24 ga.)	0.027 (22 ga.)	0.033 (20 ga.)	0.033 (20 ga.)	0.033 (20 ga.)	0.033 (20 ga.)	0.044 (18 ga.)	0.055 (16 ga.)	0.055 (16 ga.)	0.055 (16 ga.)
To 7	12	0.021 (24 ga.)	0.021 (24 ga.)	0.021 (24 ga.)	0.021 (24 ga.)	0.021 (24 ga.)	0.033 (20 ga.)	0.044 (18 ga.)	0.055 (16 ga.)	0.055 (16 ga.)	0.055 (16 ga.)
8 to 11	48	0.027 (22 ga.)	0.068 (14 ga.)	0.097 (12 ga.)	0.097 (12 ga.)	—	0.033 (20 ga.)	0.068 (14 ga.)	0.097 (12 ga.)	0.097 (12 ga.)	—
8 to 11	24	0.027 (22 ga.)	0.055 (16 ga.)	0.055 (16 ga.)	0.068 (14 ga.)	0.068 (14 ga.)	0.033 (20 ga.)	0.055 (16 ga.)	0.055 (16 ga.)	0.068 (14 ga.)	0.068 (14 ga.)
8 to 11	12	0.021 (24 ga.)	0.027 (22 ga.)	0.027 (22 ga.)	0.027 (22 ga.)	0.027 (22 ga.)	0.033 (20 ga.)	0.044 (18 ga.)	0.055 (16 ga.)	0.055 (16 ga.)	0.055 (16 ga.)
12 to 15	48	0.044 (18 ga.)	0.097 (12 ga.)	—	—	—	0.044 (18 ga.)	0.044 (18 ga.)	—	—	—
12 to 15	24	0.044 (18 ga.)	0.055 (16 ga.)	0.097 (12 ga.)	0.097 (12 ga.)	0.097 (12 ga.)	0.044 (18 ga.)	0.044 (18 ga.)	0.097 (12 ga.)	0.097 (12 ga.)	0.097 (12 ga.)
12 to 15	12	0.027 (22 ga.)	0.044 (18 ga.)	0.044 (18 ga.)	0.044 (18 ga.)	0.044 (18 ga.)	0.033 (20 ga.)	0.044 (18 ga.)	0.055 (16 ga.)	0.055 (16 ga.)	0.055 (16 ga.)
16 to 20	48	0.068 (14 ga.)	0.112 (11 ga.)	—	—	—	0.068 (14 ga.)	0.112 (11 ga.)	—	—	—
16 to 20	24	0.068 (14 ga.)	0.068 (14 ga.)	0.112 (11 ga.)	0.112 (11 ga.)	0.112 (11 ga.)	0.068 (14 ga.)	0.068 (14 ga.)	0.112 (11 ga.)	0.112 (11 ga.)	0.112 (11 ga.)
16 to 20	12	0.033 (20 ga.)	0.068 (14 ga.)	0.068 (14 ga.)	0.068 (14 ga.)	0.068 (14 ga.)	0.033 (20 ga.)	0.044 (18 ga.)	0.068 (14 ga.)	0.068 (14 ga.)	0.068 (14 ga.)
		CLASS 3 (inches)					CLASS 4 (inches)				
To 7	48	0.044 (18 ga.)	0.055 (16 ga.)	0.068 (14 ga.)	0.068 (14 ga.)	—	0.055 (16 ga.)	0.068 (14 ga.)	0.097 (12 ga.)	0.097 (12 ga.)	—
To 7	24	0.044 (18 ga.)	0.055 (16 ga.)	0.068 (14 ga.)	0.068 (14 ga.)	0.068 (14 ga.)	0.055 (16 ga.)	0.068 (14 ga.)	0.097 (12 ga.)	0.097 (12 ga.)	0.097 (12 ga.)
To 7	12	0.044 (18 ga.)	0.055 (16 ga.)	0.068 (14 ga.)	0.068 (14 ga.)	0.068 (14 ga.)	0.055 (16 ga.)	0.068 (14 ga.)	0.097 (12 ga.)	0.097 (12 ga.)	0.097 (12 ga.)
8 to 11	48	0.044 (18 ga.)	0.068 (14 ga.)	0.097 (12 ga.)	0.097 (12 ga.)	—	0.055 (16 ga.)	0.068 (14 ga.)	0.097 (12 ga.)	0.097 (12 ga.)	—
8 to 11	24	0.044 (18 ga.)	0.055 (16 ga.)	0.068 (14 ga.)	0.068 (14 ga.)	0.068 (14 ga.)	0.055 (16 ga.)	0.068 (14 ga.)	0.097 (12 ga.)	0.097 (12 ga.)	0.097 (12 ga.)
8 to 11	12	0.044 (18 ga.)	0.055 (16 ga.)	0.068 (14 ga.)	0.068 (14 ga.)	0.068 (14 ga.)	0.055 (16 ga.)	0.068 (14 ga.)	0.097 (12 ga.)	0.097 (12 ga.)	0.097 (12 ga.)
12 to 15	48	0.044 (18 ga.)	0.097 (12 ga.)	—	—	—	0.055 (16 ga.)	0.097 (12 ga.)	—	—	—
12 to 15	24	0.044 (18 ga.)	0.055 (16 ga.)	0.097 (12 ga.)	0.097 (12 ga.)	0.097 (12 ga.)	0.055 (16 ga.)	0.068 (14 ga.)	0.097 (12 ga.)	0.097 (12 ga.)	0.097 (12 ga.)
12 to 15	12	0.044 (18 ga.)	0.055 (16 ga.)	0.068 (14 ga.)	0.068 (14 ga.)	0.068 (14 ga.)	0.055 (16 ga.)	0.068 (14 ga.)	0.097 (12 ga.)	0.097 (12 ga.)	0.097 (12 ga.)
16 to 20	48	0.068 (14 ga.)	0.112 (11 ga.)	—	—	—	0.055 (16 ga.)	0.112 (11 ga.)	—	—	—
16 to 20	24	0.068 (14 ga.)	0.068 (14 ga.)	0.112 (11 ga.)	0.112 (11 ga.)	0.112 (11 ga.)	0.055 (16 ga.)	0.068 (14 ga.)	0.112 (11 ga.)	0.112 (11 ga.)	0.112 (11 ga.)
16 to 20	12	0.044 (18 ga.)	0.068 (14 ga.)	0.068 (14 ga.)	0.068 (14 ga.)	0.068 (14 ga.)	0.055 (16 ga.)	0.068 (14 ga.)	0.097 (12 ga.)	0.097 (12 ga.)	0.097 (12 ga.)

For SI units: 1 inch = 25.4 mm, 1 inch water column = 0.249 kPa

506.3.1 Fire Barriers. Exhaust ducts passing through a fire barrier having a fire resistance rating of 2 hours or greater shall meet one of the following specifications:

(1) Wrapped or encased with listed or approved materials having a fire resistance rating equal to the fire barrier for 10 feet (3048 mm) of the duct on each side of the fire barrier including duct supports within this span.

(2) Constructed of materials and supports having a minimum fire resistance rating equal to the fire barrier.

(3) Enclosed with a shaft that is constructed of material having a fire resistance rating equal to the fire barrier for 10 feet (3048 mm) of the duct on each side of the fire barrier with no inlets to the duct within this distance, and the duct entry into and exit from the shaft is protected in accordance with Section 506.3.2. [NFPA 91:4.2.12]

506.3.2 Protection. Exhaust ducts passing through fire barriers of any fire resistance rating shall be protected by sealing the space around the duct with listed or approved fire stopping having a fire resistance rating equal to the fire resistance rating of the fire barrier. [NFPA 91:4.2.13]

The local jurisdiction will define a fire barrier and firewall when adopting their codes for enforcement. Generally, a fire wall has a fire resistance rating and is able to sustain itself when other structural elements have failed, whereas a fire barrier is simply a wall assembly with a fire-resistant rating.

506.4 Condensate. Joints in duct construction shall be liquidtight when the conveying system contains condensable vapors or liquids in suspension. [NFPA 91:4.3.6.1]

506.4.1 Drainage. Provisions shall be made for drainage of condensate at low points in the duct. [NFPA 91:4.3.6.2]

506.5 Fittings. Fittings in Class 2, 3, and 4 systems shall be not less than two gauges thicker than the thickness required for straight runs. Flexible metallic duct shall be permitted to be used for connecting ductwork to vibrating equipment. Duct systems subject to wide temperature fluctuations shall be provided with expansion joints.

Branches shall connect to main ducts at the large end of transitions at an angle not exceeding 45 degrees (0.79 rad).

Except for ducts used to convey noncorrosive vapors with no particulate, accessible cleanouts shall be provided at 10 foot (3048 mm) intervals and at changes in direction. Access openings shall also be provided for access to sprinklers and other equipment within the duct that require servicing.

In Class 2, 3 and 4 duct systems, the fittings must be not less than two gauges thicker than the thickness required for straight runs. The increase in turbulence experienced at fittings could cause a greater friction within the fitting.

Expansion joints are required when wide temperature variations are encountered in a product-conveying duct system, such as one carrying heated gases.

Duct branches must connect to main ducts (or plenums) at the large end of the transitions at an angle not exceeding 45 degrees. The purpose is to avoid excessive turbulence and possible deposits of the material being conveyed on duct interior surfaces (see **Figure 506.3**).

Accessible cleanouts are required at 10-foot intervals for product-conveying ducts, except for those conveying noncorrosive vapors with no particulates. Access openings are also required to fire sprinklers and to other equipment requiring servicing within the duct. The size of the opening should be adequate in size to facilitate equipment used for cleaning.

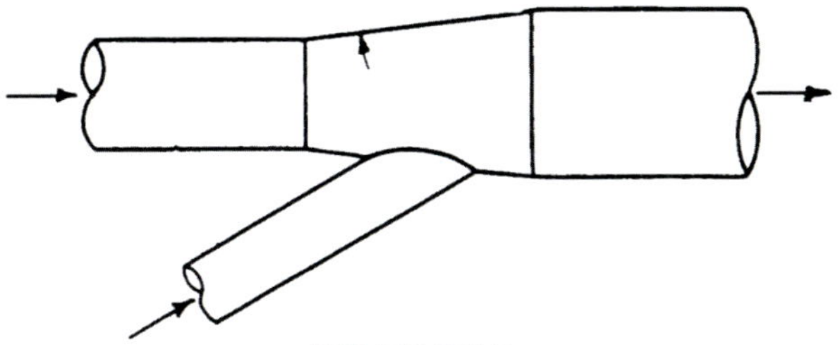

FIGURE 506.3
DUCT BRANCH

506.6 Explosion Venting. Ducts conveying explosive dusts shall have explosion vents, openings protected by anti-flashback swing valves, or rupture diaphragms. Openings to relieve explosive forces shall be located outside the building. Where relief devices cannot provide sufficient pressure relief, ductwork shall be designed to withstand an internal pressure of not less than 100 pounds-force per square inch (psi) (689 kPa).

Where a room or building contains a dust explosion hazard that is external to protected equipment, as defined in NFPA 654, such areas shall be provided with deflagration venting to a safe outside location.

Duct systems that carry explosive dusts require mechanical devices on the duct system to act as a relief in the event of an explosion. These devices include explosion vents, openings protected by anti-flashback swing valves and rupture diaphragms. Such openings must always be located outside the building. When relief devices cannot provide sufficient pressure relief, the product-conveying duct system must be designed to withstand a positive internal pressure of not less than 100 pounds per square inch (psi).

Provisions for excessive vacuum should also be considered, although this is not required by the code. If fans are operating, and the hoods, dampers, or intake ducts suddenly become blocked or covered, considerable negative pressure can result, collapsing all or part of the duct system. Weighted pivot doors and spring-loaded disks similar to poppet check valves are solutions recommended by SMACNA (see **Figure 506.6**).

FIGURE 506.6
EXPLOSION VENTS

506.7 Supports. Supports shall be of noncombustible materials, and the spacing shall not exceed 12 feet (3658 mm) for 8 inch (203 mm) ducts and 20 feet (6096 mm) for larger ducts.

Support spacing cannot exceed 12 feet for 8-inch ducts or 20 feet for larger ducts, unless justified by the system's design. In the design of supports, the designer must assume that at least 50 percent of the duct is full of the particulate to be conveyed. Supports for product-conveying systems demand the same careful consideration given to the selection and design of the ducting. Ducting exposed to the outside needs to be evaluated for wind, seismic and snow and internal product loading. Supporting systems for ducting inside buildings need to be designed to resist seismic and internal loading. In the case of ducting routed across or under roofs, the designer is cautioned to consider the structure's capacity to carry these additional loads. In some cases, the contractor is required to provide a duct support system independent of the building structure. In such cases, the considerations mentioned above have a design impact on the duct system and its supporting structure only.

506.8 Fire Protection. Sprinklers or other fire-protection devices shall be installed within ducts having a cross-sectional dimension exceeding 10 inches (254 mm) where the duct conveys flammable vapors or fumes. Sprinklers shall be installed at 12 foot (3658 mm) intervals in horizontal ducts and at changes in direction. In vertical runs, sprinklers shall be installed at the top and at alternate floor levels.

When ducts carrying flammable vapors or fumes have a cross-sectional dimension in excess of 10 inches, sprinklers or other fire sprinkler devices must be installed. In horizontal ducts, sprinklers must be installed at 12-foot intervals and at changes in direction. In vertical runs, sprinklers are required at the top and at alternate floor levels.

506.8.1 Loads. Duct supports shall be designed to carry the weight of the duct half filled with material. Where sprinkler protection is provided or cleaning of the duct will be performed, the hanger's design shall include the weight of any expected liquid accumulation. Duct supports shall be designed to prevent placing loads on connected equipment. [NFPA 91:4.6.1 – 4.6.3]

The designer must verify the weight of the material being conveyed and design the supports and anchors accordingly. This additional weight will increase the load on the supporting structure. If the supporting structure is a roof, the designer shall consider the gravity loading of the roof. As an example, a 20-foot run of 12-inch round duct half full of wood chips can weigh 234 pounds, plus the weight of the duct. The weight of water in the same 12-inch duct would weigh 468 pounds.

506.8.2 Corrosion. Hangers and supports exposed to corrosive atmospheres shall be resistant to the corrosive atmospheres. [NFPA 91:4.6.4]

506.8.3 Vibration and Stress. To avoid vibration and stress on the duct, hangers and supports shall be securely fastened to the building or structure. [NFPA 91:4.6.5]

Selection of the hangers involves a significant portion of the labor. An inadequate hanging system can be disastrous. In any hanging system, the failure of one support transfers loading to adjacent hangers. The result can be a cascading failure resulting in potential injury and property damage. There are many hanger alternatives. Besides the load-carrying capacity, the choice must take into account the particulars of the building and the recommendations of the hanger manufacturer.

506.8.4 Expansion and Contraction. Hangers and supports shall be designed to allow for expansion and contraction. [NFPA 91:4.6.6]

Duct support of systems operating at higher temperatures need to account for the expansion of the duct.

506.9 Protection from Physical Damage. Ducts installed in locations where they are subject to physical damage shall be protected by guards.

Ducts installed where it may be damaged from vehicles or material handling are examples of locations where they are subject to physical damage. This protection may be in the form of bollards, fences or curb.

506.10 Duct Clearances. Unless the conditions stipulated in Section 506.10.1 or Section 506.10.2 exist, all duct systems and system components shall have a clearance of at least 6 inches (152 mm) from stored combustible materials, and not less than ½ of an inch (13 mm) clearance from combustible construction. [NFPA 91:4.7.1]

506.10.1 Protection Provided. Where stored combustible material or combustible construction is protected from ductwork by the use of materials or products listed for protection purposes, clearance shall be maintained in accordance with those listings. [NFPA 91:4.7.1.1]

506.10.2 Systems Conveying Combustible Materials. Unless the conditions stipulated in Section 506.10.3 exist, all duct systems and system components handling combustible materials shall have a clearance of not less than 18 inches (457 mm) from stored combustible materials or combustible construction. [NFPA 91:4.7.2]

All ductwork and system components handling combustible material and operating at less than 140°F shall have a clearance of not less than 18 inches from combustible construction or any combustible material. This required clearance can be reduced only when the combustible material is protected in accordance with Table 303.10.1. Note that the prescriptive material is applied to the combustible construction and not wrapped around the product-conveying duct.

Systems conveying noncombustible product at less than 140°F do not require clearance from combustible material or construction.

506.10.3 Reduced Clearance Permitted. When the ductwork system is operating at 140°F (60°C) or below and is equipped with an approved automatic extinguishing system designed for the specific hazard, the clearance shall be permitted to be reduced to 6 inches (152 mm) from combustible materials and ½ of an inch (12.7 mm) from combustible construction. [NFPA 91:4.7.2.1]

506.10.4 Clearance Increases. All duct systems and system components operating at temperatures above 140°F (60°C) shall have clearances from stored combustible materials or combustible construction not less than those listed in Table 506.10.4. [NFPA 91:4.7.3]

TABLE 506.10.4
BASIC MINIMUM CLEARANCES TO UNPROTECTED SURFACES
[NFPA 91: TABLE 4.7.3]

DUCT GAS TEMPERATURE	LARGEST DUCT DIMENSION (inches)	CLEARANCE (inches)
140°F – 600°F incl.	8	8
	>8	12
>600°F – 900°F incl.	8	18
	>8	24
>900°F	All ducts lined with refractory	24

For SI units: 1 inch = 25.4 mm, °C=(°F-32)/1.8

506.10.4.1 Temperatures Over 900°F. Ducts handling materials at temperatures in excess of 900°F (482°C) shall be lined with refractory material or the equivalent. [NFPA 91:4.7.3.1]

506.10.4.2 Clearance Reduction. When stored combustible materials or combustible construction are protected from ductwork in accordance with Section 506.11, the clearance established in Table 506.10.4 shall be permitted to be reduced in accordance with Table 506.11, but not to less than specified in Section 506.10. [NFPA 91:4.7.3.2]

506.11 Clearance Reduction Methods. It shall be permitted to protect stored combustible material or combustible construction from ductwork in accordance with Table 506.11 and Section 506.11.1 through Section 506.11.6. In no case shall the clearance between the duct and the combustible surface be reduced below that allowed in Table 506.11. [NFPA 91:4.7.4, 4.7.4.1]

506.11.1 Spacers and Ties. Spacers and ties for protection materials shall be of noncombustible material and shall not be installed on the duct side of the protection system. [NFPA 91:4.7.4.2]

506.11.2 Wool Batts Insulation. Mineral wool batts (blanket or board) shall have a minimum density of 8 pounds per cubic feet (lb/ft^3) (128 kg/m^3) and have a minimum melting point of 1500°F (816°C). [NFPA 91:4.7.4.3]

506.11.3 Insulation Board. Insulation board used as a part of a clearance-reduction system shall meet the following criteria:

(1) Have a thermal conductivity of 1 British thermal unit inch per hour square foot degree Fahrenheit [$Btu•in/(h•ft^2•°F)$] [0.14 W/(m•K)] or less.

(2) Be formed of noncombustible material. [NFPA 91:4.7.4.4]

506.11.4 Duct and Thermal Shield. With all clearance reduction systems, at least 1 inch (25.4 mm) clear space shall be provided between the duct and the thermal shield. [NFPA 91:4.7.4.5]

506.11.5 Thermal Shield and Combustible Surface. When using clearance reduction systems that include an air gap, at least 1 inch (25.4 mm) clear space shall be provided between the thermal shield and the combustible surface. [NFPA 91:4.7.4.6]

506.11.6 Reduced Clearance with Air Gaps. When using clearance reduction systems that include an air gap between the combustible surface and the selected means of protection, air circulation shall be provided by one of the methods in Section 506.11.6.1 through Section 506.11.6.3. [NFPA 91:4.7.4.7]

506.11.6.1 Air Circulation. Air circulation shall be permitted to be provided by leaving all edges of the protecting system open with at least a 1 inch (25.4 mm) air gap. [NFPA 91:4.7.4.7.1]

506.11.6.2 Single Flat Wall. If the means of protection is mounted on a single flat wall away from corners, air circulation shall be permitted to be provided by one of the following:

(1) Leaving only the top and bottom edges open to circulation by maintaining the 1 inch (25.4 mm) air gap.

(2) Leaving the top and both side edges open to circulation by maintaining the 1 inch (25.4 mm) air gap. [NFPA 91:4.7.4.7.2]

TABLE 506.11
REDUCTION OF DUCT CLEARANCE WITH SPECIFIED FORMS OF PROTECTION
[NFPA 91: TABLE 4.7.4]

FORM OF PROTECTION*	MAXIMUM ALLOWABLE REDUCTION IN CLEARANCE (percent)	
	AS WALL PROTECTOR OR VERTICAL SURFACE	AS CEILING PROTECTOR OR HORIZONTAL SURFACE
3½ inch thick masonry wall without ventilated air space	33	None
½ inch thick noncombustible insulation board over 1 inch glass fiber or mineral wool batts without ventilated air space	50	33
0.024 inch (24 gauge) sheet metal over 1 inch glass fiber or mineral wool batts reinforced with wire, or equivalent on rear face with at least a 1 inch air gap	66	66
3½ inch (90 mm) thick masonry wall with at least a 1 inch air gap	66	None
0.024 inch (24 gauge) sheet metal with at least a 1 inch air gap	66	50
½ inch thick noncombustible insulation board with at least a 1 inch air gap	66	50
0.024 inch (24 gauge) sheet metal with ventilated air space over at least 0.024 inch (24 gauge) sheet metal with at least a 1 inch air gap	66	50
1 inch glass fiber or mineral wool batts sandwiched between two sheets of 0.024 inch (24 guage) sheet metal with at least a 1 inch air gap	66	50

For SI units: 1 inch = 25.4 mm

* Clearance reduction applied to and covering all combustible surfaces within the distance specified as required clearance with no protection in Table 506.10.4.

506.11.6.3 Thermal Shielding. Thermal shielding that covers two walls in a corner shall be permitted to be open at the top and bottom edges with at least 1 inch (25.4 mm) air gap. [NFPA 91:4.7.4.7.3]

Part II - Commercial Hoods and Kitchen Ventilation.

507.0 General Requirements.

507.1 Type I Hood Exhaust System. Exhaust systems serving Type I hoods shall comply with Section 507.0 through Section 518.0.

507.2 Exhaust System. Cooking equipment used in processes producing smoke or grease-laden vapors shall be equipped with an exhaust system that complies with all the equipment and performance requirements of this chapter. [NFPA 96:4.1.1] All such equipment and its performance shall be maintained in accordance with the requirements of this chapter during all periods of operation of the cooking equipment. [NFPA 96:4.1.2] The following equipment shall be kept in working condition:

(1) Cooking equipment

(2) Hoods

(3) Ducts (if applicable)

(4) Fans

(5) Fire-extinguishing equipment

(6) Special effluent or energy control equipment [NFPA 96:4.1.3]

Maintenance and repairs shall be performed on all components at intervals necessary to maintain good working conditions. [NFPA 96:4.1.3.1]

507.2.1 Airflow. All airflows shall be maintained. [NFPA 96:4.1.4]

507.2.2 Responsibility. The responsibility for inspection, testing, maintenance, and cleanliness of the ventilation control and fire protection of the commercial cooking operations, including cooking appliances, shall ultimately be that of the owner of the system, provided that this responsibility has not been transferred in written form to a management company, tenant, or other party. [NFPA 96:4.1.5]

507.2.3 Solid-fuel Cooking Equipment. All solid-fuel cooking equipment shall comply with the requirements of Section 517.0. [NFPA 96:4.1.6]

507.2.4 Multitenant Applications. Multitenant applications shall require the concerted cooperation of design, installation, operation, and maintenance responsibilities by tenants and by the building owner. [NFPA 96:4.1.7]

507.2.5 Interior Surfaces. All interior surfaces of the exhaust system shall be accessible for cleaning and inspection purposes. [NFPA 96:4.1.8]

507.2.6 Used in Other Applications. Cooking equipment used in fixed, mobile, or temporary concessions, such as trucks, buses, trailers, pavilions, tents, or any form of roofed enclosure, shall comply with this chapter. [NFPA 96:4.1.9]

507.3 Listed Devices. Penetrations shall be sealed with listed devices in accordance with the requirements of Section 507.3.1.

507.3.1 Penetration. Devices that require penetration of a Type I hood or grease duct, such as pipe and conduit penetration fittings and fasteners, shall be listed in accordance with UL 710 or UL 1978. Seams, joints, and penetrations of the hood enclosure shall comply with Section 508.3.2. Seams, joints, and penetrations of the ductwork shall comply with Section 510.5.3.

507.4 Clearance. Where enclosures are not required, hoods, grease removal devices, exhaust fans, and ducts shall have a clearance of at least 18 inches (457 mm) to combustible material, 3 inches (76 mm) to limited-combustible material, and 0 inches (0 mm) to noncombustible material. [NFPA 96:4.2.1]

These clearances may be reduced when they are in compliance with Section 507.4.2. See Section 510.7 where enclosures are required or provided.

See **Figure 507.4** to define material when determining clearance distances.

507.4.1 Listed. Where a hood, duct, or grease removal device is listed for clearances less than those required in Section 507.4, the listing requirements shall be permitted. [NFPA 96:4.2.2]

507.4.2 Clearance Reduction. Where a clearance reduction system consisting of 0.013 of an inch (0.33 mm) (28 gauge) sheet metal spaced out 1 inch (25.4 mm) on noncombustible spacers is provided, there shall be a minimum of 9 inches (229 mm) clearance to combustible material. [NFPA 96:4.2.3.1]

The manufacturer's gauge number shall supersede. Metal placed directly on combustible material provides no reduction in clearance at all. Integral to the reduction in clearance is the required air space. Care should be taken to avoid fastener/spacers directly behind the duct or hood.

507.4.2.1 Mineral Wool Batts or Ceramic Fiber Blanket. Where a clearance reduction system consisting of 0.027 of an inch (0.686 mm) (22 gauge) sheet metal on 1 inch (25.4 mm) mineral wool batts or ceramic fiber blanket reinforced with wire mesh or equivalent spaced 1 inch (25.4 mm) on noncombustible spacers is provided, there shall be a minimum of 3 inches (76 mm) clearance to combustible material. [NFPA 96:4.2.3.2]

507.4.2.2 Field-Applied Grease Duct Enclosure. Where a clearance reduction system consisting of a listed and labeled field-applied grease duct enclosure material, system, product, or method of construction specifically evaluated for such purpose in accordance with ASTM E2336, the required clearance shall be in accordance with the listing. [NFPA 96:4.2.3.3]

507.4.2.3 Zero Clearance. Zero clearance to limited-combustible materials shall be permitted where protected by one of the following:

(1) Metal lath and plaster.

(2) Ceramic tile.

(3) Quarry tile.

Table A.3.3.37 Types of Construction Assemblies Containing Noncombustible, Limited-Combustible, and Combustible Materials

Type of Assembly	Classifications for Determining Hood and Grease Duct Clearance*		
	Non-combustible	Limited-Combustible	Combustible
Wall assemblies			
Brick, clay tile, or concrete masonry products	X		
Plaster, ceramic, or quarry tile on brick, clay tile, or concrete masonry products	X		
Plaster on metal lath on metal studs	X		
Gypsum board on metal studs		X	
Solid gypsum board†		X	
Plaster on wood or metal lath on wood studs			X
Gypsum board on wood studs			X
Plywood or other wood sheathing on wood or metal studs			X
Floor-ceiling or roof-ceiling assemblies			
Plaster applied directly to underside of concrete slab	X		
Suspended membrane ceiling			
With noncombustible mineral wool acoustical material	X		
With combustible fibrous tile			X
Gypsum board on steel joists beneath concrete slab		X	
Gypsum board on wood joists			X

Notes:
(1) Wall assembly descriptions assume same facing material on both sides of studs.
(2) Categories are not changed by use of fire retardant–treated wood products.
(3) Categories are not changed by use of Type X gypsum board.
(4) See definitions in 3.3.37 of *combustible material*, *limited-combustible material*, and *noncombustible material*.
*See clearance requirements in Section 4.2.
†Solid gypsum walls and partitions, 50.8 mm (2 in.) or 57.15 mm (2¼ in.) thickness, are described in the Gypsum Association publication *Fire Resistance Design Manual*.

FIGURE 507.4
COMBUSTIBLE CLASSIFICATIONS
[NFPA 96:TABLE A.3.3.37]

(4) Other noncombustible materials or assembly of noncombustible materials that are listed for the purpose of reducing clearance.

(5) Other materials and products that are listed for the purpose of reducing clearance. [NFPA 96:4.2.3.4]

The zero-inch clearance reduction above is to limited-combustibles only. Other general provisions for reduced clearances shown in Table 303.10.1 are not applicable where specific reduction measures are shown above.

507.4.3 Clearance Integrity. In the event of damage, the material or product shall be repaired and restored to meet its intended listing or clearance requirements and shall be acceptable to the Authority Having Jurisdiction. [NFPA 96:4.2.4.1]

507.4.3.1 Fire. In the event of a fire within a kitchen exhaust system, the duct and its enclosure (rated shaft, factory-built grease duct enclosure, or field-applied grease duct enclosure) shall be inspected by qualified personnel to determine whether the duct and protection method are structurally sound, capable of maintaining their fire protection function, and in compliance with this chapter for continued operation. [NFPA 96:4.2.4.2]

Fire within the exhaust system can reduce the ability of the exhaust system to safely handle another fire. As such, a qualified person is required to inspect the system to ensure it is safe to put back in service.

507.4.3.2 Required Protection. Protection shall be provided on the wall from the bottom of the hood to the floor, or to the top of the noncombustible material extending to the floor, to the same level as required in Section 507.4. [NFPA 96:4.2.4.3]

Any material that was used to reduce clearance from the non-enclosed hood or duct should be extended from the floor to 18 inches or 3 inches above the hood or duct as appropriate.

507.4.3.3 Protection Methods. The protection methods for ducts to reduce clearance shall be applied to the combustible or limited-combustible construction, not to the duct itself. [NFPA 96:4.2.4.4]

This section speaks only to the reduced clearance from grease ducts; however, as Section 507.4 references both hoods and ducts, one could reasonably infer that the method of application of clearance-reducing materials applies to construction around the hood as well.

507.4.4 Factory Built. Factory-built grease duct enclosures shall be protected with a through-penetration firestop system classified in accordance with ASTM E814 or UL 1479 having an "F" and a "T" rating equal to the fire resistance rating of the assembly being penetrated from the point at which the duct penetrates a ceiling, wall, or floor to the outlet terminal. [NFPA 96:4.3.3]

Listed factory-built and field-applied grease duct enclosures still need to maintain the integrity of the fire resistiveness of the building. Where penetrations of the building assemblies are allowed, the "F" (flame transmission) and "T" (temperature transmission) ratings of these listed grease duct enclosures shall not reduce the time-resistance rating of the wall or floor/ceiling assembly required by the building code of the jurisdiction. The prescriptive method of penetration shown in the listed installation instructions should be followed exactly. These instructions should be left on the job site for inspection purposes.

507.4.4.1 Listing. The factory-built grease duct protection system shall be listed in accordance with UL 2221. [NFPA 96:4.3.3.1]

507.4.4.2 Single Wall. Listed single wall factory-built grease ducts shall be permitted to be enclosed with field-applied grease duct enclosure material where the material and the assembly of duct and material are listed for that application and installed in accordance with the grease duct manufacturer's listing and their installation instructions. [NFPA 96:4.3.3.2]

507.4.4.3 Installation. The factory-built grease duct protection system shall be installed in accordance with the manufacturer's instructions and the listing requirements. [NFPA 96:4.3.3.3]

507.4.5 Field Applied. Field-applied grease duct enclosures shall be protected with a through penetration firestop system classified in accordance with ASTM E814 or UL 1479 having an "F" and a "T" rating equal to the fire resistance rating of the assembly being penetrated. The surface of the field fabricated grease duct shall be continuously covered on all sides from the point at which the duct enclosure penetrates a ceiling, wall, or floor to the outlet terminal. The field-applied grease duct shall be listed in accordance with ASTM E2336 and installed in accordance with the manufacturer's instructions and the listing requirements. [NFPA 96:4.3.1 – 4.3.1.2]

This exception does not allow the installation of damaged field-applied or factory-built grease duct enclosure systems. Follow the installation instructions for the protection of these products from weather and normal commercial kitchen operation hazards.

507.4.6 Both Field-Applied and Factory Built. Field-applied grease duct enclosures and factory-built grease duct enclosures shall demonstrate that they provide mechanical and structural integrity, resiliency, and stability where subjected to expected building environmental conditions, duct movement under general operating conditions, and duct movement due to fire conditions. [NFPA 96:4.3.4]

Expect grease ducts installed in multiple-story buildings to be enclosed within building construction. The above clearance options are necessary for proper installation guidance to maintain the clearances required when the product is enclosed.

507.4.6.1 Physical Damage. Measures shall be taken to prevent physical damage to a material or product used for the purpose of reducing clearances.

Exception: Where the duct is protected with a field-applied grease duct enclosure or factory-built grease duct enclosure.

507.4.6.2 Specification. The specifications of material, gauge, and construction of the duct used in the testing and listing of field-applied grease duct enclosures and factory-built grease duct enclosures shall be included as minimum requirements in their listing and installation documentation. [NFPA 96:4.3.5]

507.4.6.3 Clearance Options. The following clearance options for which field-applied grease duct enclosures and factory-built grease duct enclosures have been successfully evaluated shall be clearly identified in their listing and installation documentation and on their labels:

(1) Open combustible construction clearance at manufacturer's requested dimensions.

(2) Closed combustible construction clearance at manufacturer's requested dimensions, with or without specified ventilation.

(3) Rated shaft clearance at manufacturer's requested dimensions, with or without specified ventilation. [NFPA 96:4.3.6]

507.4.7 Building and Structural Contact. A duct shall be permitted to contact noncombustible floors, interior walls, and other noncombustible structures or supports, but it shall not be in contact for more than 50 percent of its surface area for each linear foot of contact length. [NFPA 96:4.4.1]

In essence, a rectangular hood or duct could be in contact with noncombustible material for the entire length, but only on two sides. Ducts that penetrate noncombustible walls or ceiling/floor assemblies in contact on the entire duct perimeter should be limited to 6 inches in length. Thicker walls or floor assemblies should maintain a minimum 3-inch clearance for the protection of the duct and building. Consult the building code of the jurisdiction or follow a listed means of penetration protection.

507.4.7.1 Corrosion Protection. Where duct contact must exceed the requirements of Section 507.4.7, the duct shall be protected from corrosion. [NFPA 96:4.4.2]

Ducts in contact with noncombustible material tend to develop condensation during periods of nonuse. They also will have different rates of heat transmission during periods of use that may also hasten corrosion and failure of the grease duct.

507.4.7.2 Zero Clearance. Where the duct is listed for zero clearance to combustibles or is otherwise protected with a material or product listed for the purpose of reducing clearance to zero, the duct shall be permitted to exceed the contact limits of Section 507.4.7 without additional corrosion protection. [NFPA 96:4.4.3]

See also Section 510.7.3 where the grease duct clearance from an enclosure constructed of limited combustible material shall not be less than 6 inches.

507.4.8 Clearance Between Duct and Interior Surfaces. Clearances between the duct and interior surfaces of enclosures shall meet the requirements of Section 507.4 through Section 507.4.3.3. [NFPA 96:4.5]

507.5 Drawings. A drawing(s) of the exhaust system installation along with copies of operating instructions for subassemblies and components used in the exhaust system, including electrical schematics, shall be on the premises. [NFPA 96:4.6]

Coordination of requirements from the fire and health authorities cannot be achieved without plans that are specific and detailed. The better the plan, the greater overall compliance can be expected.

507.6 Notification of Change. If required by the Authority Having Jurisdiction, notification in writing shall be given of any alteration, replacement, or relocation of any exhaust or extinguishing system or part thereof or cooking equipment. [NFPA 96:4.7]

Satisfaction shall be provided to the Authority Having Jurisdiction that the complete exhaust system as addressed in this chapter is installed and operable in accordance with the approved design and the manufacturer's installation instructions.

508.0 Type I Hoods.

508.1 Where Required. Type I hoods shall be installed at or above commercial-type deep-fat fryers, broilers, grills, hot-top ranges, ovens, barbecues, rotisseries, and similar equipment that emits comparable amounts of smoke or grease in a food-processing establishment. For the purpose of this section, a food-processing establishment shall include a building or portion thereof used for the processing of food, but shall not include a dwelling unit.

Exceptions:

(1) Cooking appliance that is in accordance with UL 710B for reduced emissions where the grease discharge does not exceed 2.9 E-09 ounces per cubic inch (oz/in^3) (5.0 E-06 kg/m^3) where operated with a total airflow of 500 cubic feet per minute (CFM) (0.236 m^3/s).

(2) Recirculating systems listed in accordance with UL 710B and installed in accordance with Section 516.0.

A "food-processing establishment" includes a building or portion thereof used for the processing of food except for a dwelling unit or its kitchen.

In general, a Type I hood should be required over grease- or smoke-producing commercial-type cooking equipment, such as deep fat fryers, broilers, fry grills, hot-top ranges, ovens, barbeques, rotisseries, etc., installed in commercial businesses.

When a residential-type cooking range is installed in locations other than commercial businesses, a Type II hood may be allowed with prior approval from the AHJ. However, when commercial-type cooking equipment is installed in these types of occupancies, the issue is not quite so clear cut, since these occupancies are not usually considered "businesses engaged in processing food." Intensity of use and types of cooking appliances should be a consideration at the time of plan review, and a determination of hood types should be made at that time.

508.2 Listed Type I Hood Assemblies. Listed hood assemblies shall be installed in accordance with the terms of their listing and the manufacturer's instructions. Listed hood assemblies shall be tested in accordance with UL 710 or equivalent. [NFPA 96:5.4.1, 5.4.2]

508.2.1 Listed Ultraviolet Hoods. Listed ultraviolet hoods shall be installed and maintained in accordance with the terms of their listing and the manufacturer's instructions. Duct systems connected to ultraviolet hoods shall comply with Section 510.0. Ultraviolet hoods shall be tested and listed in accordance with UL 710 and UL 710C. [NFPA 96:5.5 – 5.5.2]

508.2.2 Listed Ventilated Ceiling Technology. Listed ventilated ceiling technology shall be installed and maintained in accordance with the terms of its listing and the manufacturer's instructions. [NFPA 96:5.6]

508.2.3 Construction of Listed Exhaust Hoods. Listed exhaust hoods with or without exhaust dampers shall be permitted to be constructed of materials required by the listing. [NFPA 96:5.1.6]

508.2.4 Assembly of Listed Exhaust Hoods. Listed exhaust hoods with or without exhaust dampers shall be permitted to be assembled in accordance with the listing requirements. [NFPA 96:5.1.7]

508.3 Construction of Type I Hoods. The hood or that portion of a primary collection means designed for collecting cooking vapors and residues shall be constructed of and be supported by steel not less than 0.048 of an inch (1.219 mm) (No. 18 MSG), in thickness, stainless steel not less than 0.036 of an inch (0.914 mm) (No. 20 MSG) in thickness, or other approved material of equivalent strength and fire and corrosion resistance. [NFPA 96:5.1.1]

Exception: Listed exhaust hoods.

Exhaust dampers in a listed hood shall meet the terms of the listing and be provided with the hood. These dampers may be a requirement of a manifold (common duct) system as per Section 511.4 and may be located in the hood as per Section 510.3.1.

508.3.1 Grease Vapor. Wall-mounted exhaust hood assemblies shall be tight fitting against the back wall so as to not permit passage of grease vapor behind the hood, or between the back wall and the hood assembly. [NFPA 96:5.1.13]

Type I hoods require a 3-inch separation from limited combustible material. This air space is commonly built into the back of a hood. Hoods lacking this air space need to be flashed to the wall on the sides and bottom (see **Figure 508.3.1**).

FIGURE 508.3.1
EXHAUST HOOD MOUNTING

508.3.2 Seams, Joints, and Penetrations. All seams, joints, and penetrations of the hood enclosure that direct and capture grease-laden vapors and exhaust gases shall have a liquid-tight continuous external weld to the hood's lower outermost perimeter. [NFPA 96:5.1.2]

Exceptions:

(1) Seams, joints, and penetrations of the hood shall be permitted to be internally welded, provided that the weld is formed smooth or ground smooth, so as to not trap grease, and is cleanable. [NFPA 96:5.1.3]

(2) Penetrations shall be permitted to be sealed by devices that are listed for such use and whose presence does not detract from the hood's or duct's structural integrity. [NFPA 96:5.1.5]

Grease hoods are intended to capture grease without leaks into areas that may otherwise be inaccessible to cleaning (see **Figure 508.3.2**). Grease that may leak will cause a greater hazard in the event of a fire.

FIGURE 508.3.2
EXHAUST SYSTEM SEAMS

508.3.2.1 Sealed. Internal hood joints, seams, filter support frames, and appurtenances attached inside the hood shall be sealed or otherwise made greasetight. [NFPA 96:5.1.4]

508.3.3 Eyebrow-Type Hoods. Eyebrow-type hoods over gas or electric ovens shall be permitted to have a duct constructed as required in Section 510.0 from the oven flue(s) connected to the hood canopy upstream of the exhaust plenum, as shown in **Figure 508.3.3**. [NFPA 96:5.1.8.1]

508.3.3.1 Duct Connection. The duct connecting the oven flue(s) to the hood canopy shall be connected with a continuous weld or have a duct-to-duct connection. [See Figure 511.1.2(2) through **Figure 511.1.2(4)**] [NFPA 96:5.1.8.2]

Selection of the gauge of material to use as the oven flue shall be consistent with the means of attachment to the hood and oven.

508.3.4 Insulation. Insulation materials other than electrical insulation shall have a flame spread index of 25 or less, when tested in accordance with ASTM E84 or UL 723. Adhesives or cements used in the installation of insulating materials shall comply with the requirements of this section when tested with the specific insulating material. [NFPA 96:5.1.9, 5.1.10]

Exposed building insulation materials within 18 inches of the hood or duct are subject to higher temperatures

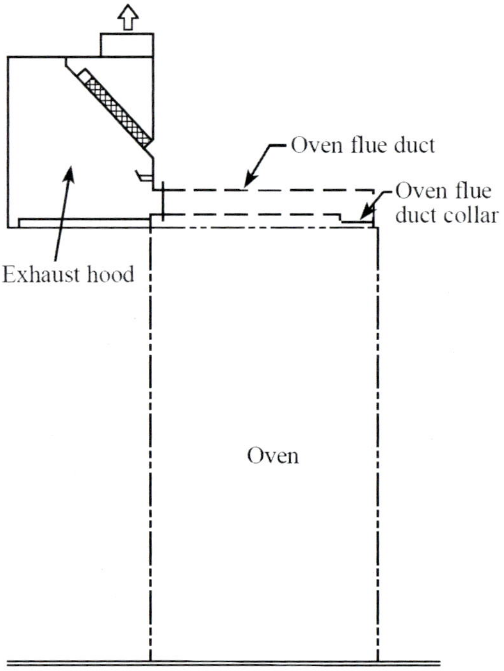

FIGURE 508.3.3
TYPICAL SECTION OF EYEBROW-TYPE HOOD
[NFPA 96: FIGURE 5.1.8.1]

in the event of a grease hood fire. Insulation with a flame spread rating of 25 or less should not adversely spread a fire. Insulation materials in contact with a grease duct shall be listed for that purpose. Grease hoods are a lighter gauge than required for grease ducts. Insulation materials used as a field-applied grease duct enclosure may not necessarily be listed for applications in contact with a hood.

508.3.5 Exhaust Hood Assemblies with Integrated Supply-Air Plenums. The construction and size of exhaust hood assemblies with integrated supply air plenums shall comply with the requirements of Section 508.2.3 through Section 508.3.4 and Section 508.5. [NFPA 96:5.3.1]

See Chapter 2 for the definition of "Compensating Hood."

Unless listed otherwise, the same principles of minimum airflow and materials apply to both compensating hoods and exhaust-only hoods (see **Figure 508.3.5**).

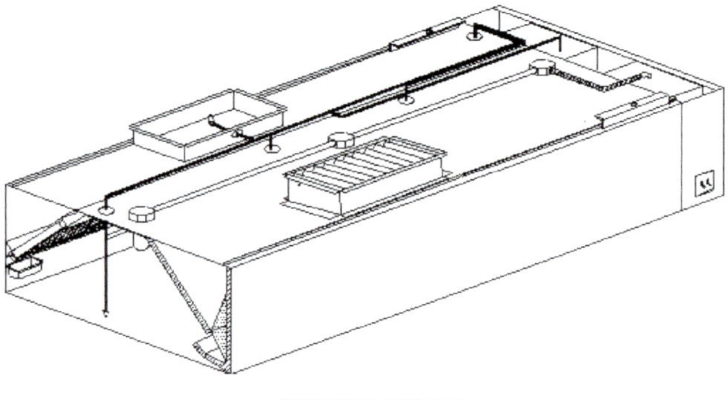

FIGURE 508.3.5
INTEGRATED SUPPLY AIR HOOD

508.3.5.1 Outer Shell. The construction of the outer shell or the inner exhaust shell shall comply with Section 508.2.3 through Section 508.3.4. [NFPA 96:5.3.2]

508.3.5.2 Inner Shell. Where the outer shell is welded, the inner shell shall be of greasetight construction. [NFPA 96:5.3.3]

508.3.5.3 Fire Dampers. A fire-actuated damper shall be installed in the supply air plenum at each point where a supply air duct inlet or a supply air outlet penetrates the continuously welded shell of the assembly. [NFPA 96:5.3.4.1]

Compensating hoods commonly have fire dampers that are provided from the manufacturer. The installation at the replenishing air inlet is meant to protect the non-grease ducting and the building in the event of a hood fire.

508.3.5.3.1 Listing. The fire damper shall be listed for such use or be part of a listed exhaust hood with or without exhaust damper. [NFPA 96:5.3.4.2]

508.3.5.3.2 Actuating Temperature. The actuation device shall have a maximum temperature rating of 286°F (141°C). [NFPA 96:5.3.4.3]

508.3.5.3.3 Exemption. Supply air plenums that discharge air from the face rather than from the bottom or into the exhaust hood and that are isolated from the exhaust hood by the continuously welded shell extending to the lower outermost perimeter of the entire hood assembly shall not require a fire-actuated damper. [NFPA 96:5.3.4.4]

508.4 Supports. Hoods shall be secured in place by noncombustible supports. The supports shall be capable of supporting the expected weight of the hood and plus 800 pounds (362.9 kg).

508.5 Hood Size. Hoods shall be sized in accordance with the airflow capacity in accordance with Section 508.5.1.1 and installed to provide for the removal of heat, and capture and removal of grease-laden vapors in accordance with Section 511.2.2.

The sizing criteria of this section also include the capture and removal of heat and steam for Type II hoods.

508.5.1 Canopy Size and Location. For canopy type commercial cooking hoods, the inside edge thereof shall overhang or extend a horizontal distance of not less than 6 inches (152 mm) beyond the edge of the cooking surface on open sides, and the vertical distance between the lip of the hood and the cooking surface shall not exceed 4 feet (1219 mm).

Use the edge of the cooking surface where appliances have an exposed cooking surface. Take into account the clearance required to the grease removal devices internal to the hood when determining the height of the hood. The lower lip of a canopy hood is not regulated as to its distance above the floor by this code.

Exception: Listed exhaust hoods are to be installed in accordance with the terms of their listings and the manufacturer's installation instructions.

Listed hoods matched with specific equipment may lead to the exclusion of the 6-inch overhang or varying heights. However, the requirement to capture and remove cooking and grease laden vapors still exists. Normally, listed hoods just specify the maximum temperature range of the cooking equipment they protect. The overhang and vertical distance more typically fall within the same range as required for unlisted hoods.

508.5.1.1 Capacity of Hoods. Canopy-type commercial cooking hoods shall exhaust through the hood with a quantity of air not less than determined by the application in accordance with Section 508.5.1.2 through Section 508.5.1.5. The exhaust quantity shall be the net exhaust from the hood determined in accordance with Equation 508.5.1.1. The duty level for the hood shall be the duty level of the appliance that has the highest (heaviest) duty level of appliances installed underneath the hood.

Exception: Listed exhaust hoods installed in accordance with the manufacturer's installation instructions.

$$E_{NET} = E_{HOOD} - MA_{ID} \qquad \text{(Equation 508.5.1.1)}$$

Where:

E_{NET} = net hood exhaust

E_{HOOD} = total hood exhaust

MA_{ID} = makeup air, internal discharge

Commercial kitchen hood exhaust airflow rates are mainly determined by the effluent and thermal plume generated by the cooking process. Other factors that could affect hood performance include the introduction of makeup air, appliance positioning under the hood, and the style and size of hood. An increased depth of a hood (front-to-back) has a very beneficial effect on hood performance. If the appliance line is positioned against the wall of a 4 foot deep or 5 foot deep wall-mounted canopy hood, research has found that the capture and containment rate of the 5 foot deep hood will be approximately 40% less than the 4 foot deep canopy hood. The extra foot of hood front overhang allows the thermal plume to be exhausted more efficiently by allowing it to recirculate and not escape from the front of the hood. In result, the threshold of capture and containment is less than that for larger hoods with more front overhang. The current formulas calculate exhaust airflow rate based on cubic foot per minute per linear foot of hood basis and do not penalize larger depths of hoods. It encourages greater hood front overhang that enhances hood performance.

Equation 508.5.1.1 and Table 508.5.1.2 through Table 508.5.1.5 are to be used for unlisted hoods to determine exhaust airflow rates. For hoods listed to UL 710, the listed exhaust airflow rates will be used as the minimum exhaust airflow rates to determine the capacity of the hoods.

For unlisted hoods, the first step to determine the airflow rate is to identify the style and length of the hood. This can be found in the mechanical drawings, equipment schedule, or kitchen drawings. If the drawings are not available, sum the lengths of the individual appliances underneath the hood to determine the length of the hood. Include also the hood side overhang on each end of the hood and any spaces in between the appliances. Second, choose the appropriate Duty level of the appliances. Where there are multiple appliances under a single hood, the appliance with the highest duty level shall determine the appropriate table. For example, if there are three appliances under a single hood comprising of an electric convection oven (light duty), an electric flat griddle (medium duty) and an electric wok range (heavy duty), Table 508.5.1.3 Heavy-Duty Cooking Appliances would be selected as the highest duty level and the heavy-duty level, on a cfm per linear foot basis, would be applied to the entire length of the hood.

If the total length of the hood is ten feet, Table 508.5.1.3 requires a wall mounted canopy to have a total capacity of 4000 cfm (400 cfm per linear foot multiplied by 10 feet). If a double island canopy is used, then the front of the hood measured from end-to-end on both sides will determine the total capacity. In this example, the total length would be twenty feet and the total capacity would then be 8000 cfm (400 cfm per linear foot multiplied by 20 feet).

Where there is a short circuit, i.e. internal makeup air which is typically untempered makeup air introduced directly into the hood cavity, the makeup air rate can be subtracted from the total hood exhaust rate to determine the net hood exhaust rate. The net hood exhaust rate will then determine the capacity of the hood. Where a Type I or Type II hood has a short circuit, ASHRAE 154 recommends that the makeup airflow shouldn't exceed 10% of the exhaust airflow. In the example above, the wall mounted canopy required 4000 cfm. If the short circuit air flow is 400 cfm (best practice does not exceed 10% of 4000), then 400 can be deducted from 4000 to derive the net hood exhaust of 3600 cfm.

508.5.1.2 Extra-Heavy-Duty Cooking Appliances. The minimum net airflow for hoods used for solid fuel cooking appliances such as charcoal, briquette, and mesquite to provide the heat source for cooking shall be in accordance with Table 508.5.1.2.

508.5.1.3 Heavy-Duty Cooking Appliances. The minimum net airflow for hoods used for cooking appliances such as gas under-fired broilers, gas chain (conveyor) broilers, electric and gas wok ranges, and electric and gas over-fired (upright) broilers shall be in accordance with Table 508.5.1.3.

508.5.1.4 Medium-Duty Cooking Appliances. The minimum net airflow for hoods used for cooking appliances such as electric and gas hot-top ranges, gas open-burner ranges (with or without oven), electric and gas flat griddles, electric and gas double-sided griddles, electric and gas fryers (including open deep fat fryers, donut fryers, kettle fryers, and pressure fryers), and electric and gas conveyor pizza ovens shall be in accordance with Table 508.5.1.4.

508.5.1.5 Light-Duty Cooking Appliances. The minimum net airflow for hoods used for cooking appliances such as gas and electric ovens (including standard, bake, roasting, revolving, retherm, convection, combination convection/steamer, rotisserie, countertop conveyorized baking/finishing, deck, and pastry), discrete element ranges (with or without oven), electric and gas steam-jacketed kettles less than 20 gallons (76 L), electric and gas pasta cookers, electric and gas compartment steamers (both pressure and atmospheric), electric and gas cheese melters, electric and gas tilting skillets (braising pans) electric and gas rotisseries, and electric and gas salamanders shall be in accordance with Table 508.5.1.5.

508.5.2 Noncanopy-Type Hoods. Noncanopy-type commercial cooking hoods shall be installed and sized in accordance with the manufacturer's installation instructions, and Section 508.5.2.1 and Section 508.5.2.2.

Exception: Listed hood assemblies designed and installed specifically for the intended use.

TABLE 508.5.1.2
EXTRA-HEAVY-DUTY COOKING APPLIANCE AIRFLOW

TYPE OF HOOD	AIRFLOW (cubic foot per minute per linear foot of hood)
Backshelf/pass-over	Not permitted
Double island canopy (per side)	550
Eyebrow	Not permitted
Single island canopy	700
Wall-mounted canopy	550

For SI units: 1 cubic foot per minute = 0.00047 m^3/s, 1 foot = 304.8 mm

TABLE 508.5.1.3
HEAVY-DUTY COOKING APPLIANCE AIRFLOW

TYPE OF HOOD	AIRFLOW (cubic foot per minute per linear foot of hood)
Backshelf/pass-over	400
Double island canopy (per side)	400
Eyebrow	Not permitted
Single island canopy	600
Wall-mounted canopy	400

For SI units: 1 cubic foot per minute = 0.00047 m^3/s, 1 foot = 304.8 mm

TABLE 508.5.1.4
MEDIUM-DUTY COOKING APPLIANCE AIRFLOW

TYPE OF HOOD	AIRFLOW (cubic foot per minute per linear foot of hood)
Backshelf/pass-over	300
Double island canopy (per side)	300
Eyebrow	250
Single island canopy	500
Wall-mounted canopy	300

For SI units: 1 cubic foot per minute = 0.00047 m^3/s, 1 foot = 304.8 mm

TABLE 508.5.1.5
LIGHT-DUTY COOKING APPLIANCE AIRFLOW

TYPE OF HOOD	AIRFLOW (cubic foot per minute per linear foot of hood)
Backshelf/pass-over	250
Double island canopy (per side)	250
Eyebrow	250
Single island canopy	400
Wall-mounted canopy	200

For SI units: 1 cubic foot per minute = 0.00047 m^3/s, 1 foot = 304.8 mm

Listed hoods, installed and used in accordance to their listing, normally operate at a lesser capacity of air (see **Figure 508.5.2**). While an intended result may be energy savings, care should be taken to stay within the terms of the listing for the cooking temperature of the equipment below the hood.

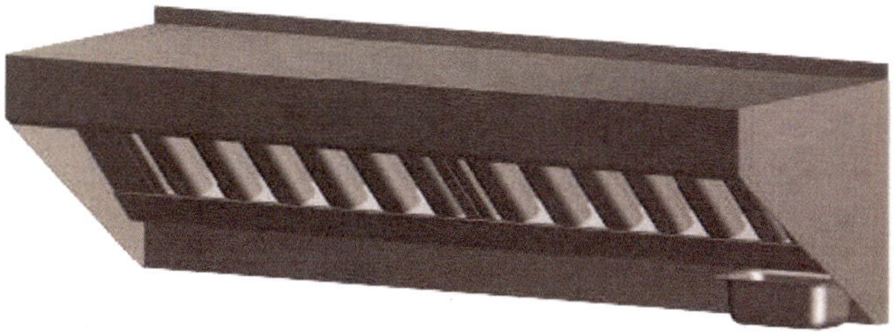

FIGURE 508.5.2
LISTED HOOD

508.5.2.1 Installation. Noncanopy-type commercial cooking hoods shall be installed with the edge of the hood set back not more than 1 foot (305 mm) from the edge of the cooking surface, and the vertical distance between the lip of the hood and the cooking surface shall not exceed 3 feet (914 mm).

508.5.2.2 Capacity. In addition to other requirements for hoods specified in this section, the volume of air exhausting through a noncanopy-type hood to the duct system shall be not less than 300 cubic feet per minute per lineal foot [$(ft^3/min)/ft$)] [0.464 $(m^3/s)/m$] of cooking equipment. Listed noncanopy exhaust hoods and filters shall be sized and installed in accordance with the terms of their listing and the manufacturer's installation instructions.

508.5.3 Labeling. Type I hoods shall bear a label indicating the exhaust flow rate in cubic feet per minute per lineal foot [$(m^3/s)/m$].

Hoods are designed in accordance with the appliance in which they are installed and the addition of a label will ensure, if the existing appliance is replaced, that the hood is of sufficient capacity to handle the exhaust of the replacement appliance. This would apply to unlisted and listed hoods.

508.6 Solid-Fuel Hood Assemblies. Where solid-fuel cooking equipment is to be used, the solid-fuel hood assembly shall be in accordance with Section 517.0.

508.7 Exhaust Outlets. An exhaust outlet within an unlisted hood shall be located so as to optimize the capture of particulate matter. Each outlet shall serve not more than a 12 foot (3658 mm) section of an unlisted hood.

This section requires an outlet for every 12 linear feet of hood on unlisted hoods. A 13-foot unlisted hood would require two outlets.

509.0 Grease Removal Devices in Hoods.

509.1 Grease Removal Devices. Listed grease filters or other listed grease removal devices intended for use with commercial cooking operations shall be provided. Listed grease filters and grease removal devices that are removable but not an integral component of a specific listed exhaust hood shall be listed in accordance with UL 1046 and shall be designated on the filter. [NFPA 96:6.1.1, 6.1.2]

Even if the hood is not listed, the filters are required to be listed (see **Figure 509.1**).

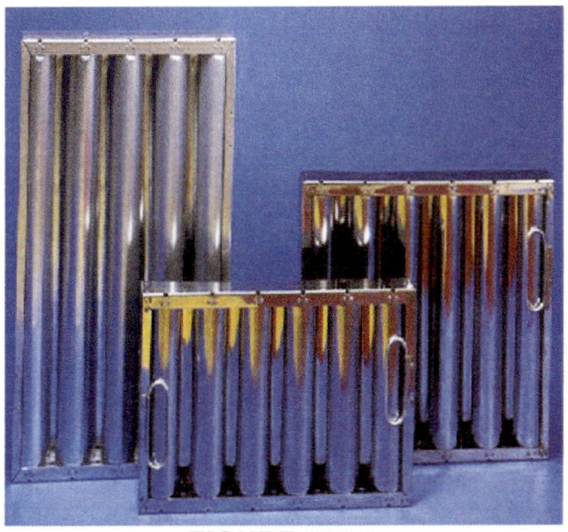

FIGURE 509.1
EXHAUST SYSTEM GREASE FILTERS

509.1.1 Grease Filters, Mesh-Type. Mesh filters shall not be used unless evaluated as an integral part of a listed exhaust hood or listed in conjunction with a primary filter in accordance with UL 1046. [NFPA 96:6.1.3]

See **Figure 509.1.1**.

FIGURE 509.1.1
HEAVY DUTY ALUMINUM MESH

509.2 Installation. The distance between the grease removal device and the cooking surface shall be as great as possible but not less than 18 inches (457 mm). [NFPA 96:6.2.1.1]

509.2.1 Vertical Distance. Where grease removal devices are used in conjunction with solid fuel or solid fuel-type broilers, including gas or electrically heated charbroilers, a minimum vertical distance of 4 feet (1219 mm) shall be maintained between the lower edge of the grease removal device and the cooking surface. [NFPA 96:6.2.1.2]

The term "grease-burning charbroiler" is used to describe a broiler in which drippings from the product being cooked generally fall onto the fuel bed. The equipment is designed so the grease and the drippings will flame and smoke adding flavor to the product being cooked. The result may cause flame to flare above the cooking surface, adding to the hazard that the grease on the filters may ignite.

Exceptions:

(1) For cooking equipment without exposed flame and where flue gases bypass grease removal devices, the minimum vertical distance shall be permitted to be reduced to not less than 6 inches (152 mm). [NFPA 96:6.2.1.3]

(2) Where a grease removal device is listed for separation distances less than those required in Section 509.2 and Section 509.2.1, the listing requirements shall be permitted. [NFPA 96:6.2.1.4]

(3) Grease removal devices supplied as part of listed hood assemblies shall be installed in accordance with the terms of the listing and the manufacturer's instructions. [NFPA 96:6.2.1.5]

Listed grease removal devices used in an unlisted hood may be installed according to their listing.

509.2.2 Grease Removal Device Protection. Where the distance between the grease removal device and the appliance flue outlet (heat source) is less than 18 inches (457 mm), grease removal devices shall be protected from combustion gas outlets and from direct flame impingement occurring during normal operation of cooking appliances producing high flue gas temperatures. [NFPA 96:6.2.2.1]

509.2.2.1 Installation. This protection shall be permitted to be accomplished by the installation of a steel or stainless steel baffle plate between the heat source and the grease removal device. [NFPA 96:6.2.2.2]

509.2.2.2 Size and Location. The baffle plate shall be sized and located so that flames or combustion gases travel a distance not less than 18 inches (457 mm) from the heat source to the grease removal device. [NFPA 96:6.2.2.3]

509.2.2.3 Clearance. The baffle shall be located not less than 6 inches (152 mm) from the grease removal device. [NFPA 96:6.2.2.4]

509.2.3 Grease Filters. Grease filters shall be listed and constructed of steel or other non-combustible material, and shall be of rigid construction that will not distort or crush under normal operation, handling, cleaning, or replacement.

509.2.3.1 Arrangement. Grease filters shall be arranged so that all exhaust air passes through the grease filters. [NFPA 96:6.2.3.4]

509.2.3.2 Accessibility. Grease filters shall be easily accessible for removal. [NFPA 96:6.2.3.5]

509.2.3.3 Angled Installation. Grease filters shall be installed at an angle not less than 45 degrees (0.79 rad) from the horizontal. [NFPA 96:6.2.3.6]

An angle of 45 degrees will encourage the collected grease to overcome the upward airflow and drain by gravity to the drip tray below.

509.2.4 Grease Drip Trays. Grease filters shall be equipped with a grease drip tray beneath their lower edges. [NFPA 96:6.2.4.1]

509.2.4.1 Size and Pitch. Grease drip trays shall be kept to the minimum size needed to collect grease. Grease drip trays shall be pitched to drain into an enclosed metal container having a capacity not exceeding 1 gallon (4 L). [NFPA 96:6.2.4.2, 6.2.4.3]

See **Figure 509.2.4.1**.

FIGURE 509.2.4.1
GREASE FILTER DRIP TRAY

509.2.5 Grease Filter Orientation. Grease filters that require a specific orientation to drain grease shall be clearly so designated on the face of the filter as to be visible with the filter installed, or the hood or filter shall be constructed so that filters cannot be installed in the wrong orientation. [NFPA 96:6.2.5]

The orientation of the grease filter is crucial to allow weep holes located in the filter frame to function properly and to allow grease to drain through.

509.3 Solid-Fuel Grease Removal Devices. Where solid-fuel cooking equipment is provided with grease removal devices, these devices shall be in accordance with Section 517.0.

510.0 Exhaust Duct Systems.

510.1 General. Ducts shall not pass through fire walls. [NFPA 96:7.1.1]

Firewalls and fire partitions are designed to keep the hazard found in one portion of a building from adversely affecting the adjoining tenant or portion of the building. Because of the distinct hazard related to grease ducts, these fire-resistive features of the building shall not be breached even if protected by grease duct enclosures.

510.1.1 Fire Hazards. All ducts shall lead directly to the exterior of the building, so as not to unduly increase any fire hazard. [NFPA 96:7.1.2]

510.1.2 Interconnection. Duct systems shall not be interconnected with any other building ventilation or exhaust system. [NFPA 96:7.1.3]

Duct systems include the grease duct, the grease

duct enclosure and replenishing air. Other building ventilation or exhaust systems may include but are not limited to environmental air systems, product conveying systems, appliance vents and HVAC systems.

510.1.3 Duct Installation. All ducts shall be installed with a minimum 2 percent slope on horizontal runs up to 75 feet (22 860 mm) and a minimum 8 percent slope on horizontal runs greater than 75 feet (22 860 mm). Factory-built grease ducts shall be permitted to be installed at a lesser slope in accordance with the listing and the manufacturer's instructions. All horizontal ducts shall be provided with access in accordance with Section 510.3.3.

Drains shall be provided at low points in horizontal ducts. Where provided, drains shall be continuously welded to the exhaust duct in accordance with the terms of the listing and the manufacturer's installation manual.

All ducts shall be installed without forming dips or traps. In manifold (common duct) systems, the lowest end of the main duct shall be connected flush on the bottom with the branch duct. [NFPA 96:7.1.4 – 7.1.4.5]

The approved grease reservoir needs to be located under the hood where the collected grease (should it become ignited in the event of a fire) will not affect other portions of the building. The operational temperature within the grease duct and the prescriptive slope will encourage the majority of grease that may bypass the grease removal devices to drain back to the hood. See also Section 511.1.2 for in-line fan systems.

510.1.4 Accessibility. Openings required for accessibility shall comply with Section 510.3 through Section 510.3.2. [NFPA 96:7.1.5]

510.1.5 Sign. A sign stating the following shall be placed on all access panels:

ACCESS PANEL – DO NOT OBSTRUCT [NFPA 96:7.1.6]

510.1.6 Bracing and Supports. Duct bracing and supports shall be of noncombustible material, securely attached to the structure and designed to carry gravity and lateral loads within the stress limitations of the building code. Bolts, screws, rivets, and other mechanical fasteners shall not penetrate duct walls.

The design of the hood may not include the ability to support ducting connected to it. Support the ducting independent of the hood itself. Refer to SMACNA Industrial Duct Construction Standards or the installation instruction for listed factory-built grease duct enclosures. The instructions for field-installed grease duct enclosures include information regarding the protection of the grease duct support.

510.1.7 Type I Exhaust Duct Systems. Listed grease ducts shall be installed in accordance with the terms of their listing and the manufacturer's instructions. [NFPA 96:7.1.7]

510.2 Clearance. Clearance between ducts and combustible materials shall be provided in accordance with the requirements of Section 507.4 through Section 507.4.3.3. [NFPA 96:7.2]

Clearances prescribed in Section 507.4 are for ducts that are not enclosed.

510.3 Openings. Openings shall be provided at the sides or at the top of the duct, whichever is more accessible, and at changes of direction. Openings shall be protected by approved access constructed and installed in accordance with the requirements of Section 510.3.7. [NFPA 96:7.3.1, 7.3.2]

Exception: Openings shall not be required in portions of the duct that are accessible from the duct entry or discharge. [NFPA 96:7.3.3]

Grease duct systems cannot have openings except those that are necessary for the proper maintenance of the system. A portion of a duct system that is inaccessible from the duct entry at the hood or from the discharge at the fan must be provided with cleanout openings. The entire interior of the duct system must be accessible for cleaning. Cleanout openings must be equipped with tight-fitting doors constructed of steel at least as thick as that required for the duct system. The doors must be equipped with a substantial latching method that is strong enough to hold the doors tightly closed and designed so the doors can be opened without the use of tools. These doors must also be able to resist leakage due to cleaning agents and residual grease in the duct system. A greater amount of grease will naturally collect at the bottom of a horizontal duct, making it more susceptible to leakage. Openings in the bottom of the duct make cleaning of the duct system more problematic, as the cleaning agents will drain out of the access opening that is being worked through. See **Figure 510.3**.

FIGURE 510.3
EXAMPLE OF OPENING IN A DUCT

510.3.1 Access Panel. For hoods with dampers in the exhaust or supply collar, an access panel for cleaning and inspection shall be provided in the duct or the hood within 18 inches (457 mm) of the damper. [NFPA 96:7.3.4]

Exception: Dampers that are accessible from under the hood.

Disassembly of the damper to clean the grease duct system is prohibited. Care must be exercised to maintain the original damper adjustments during the cleaning process.

510.3.2 Access for Cleaning and Inspection. Exhaust fans with ductwork connected to both sides shall have access for cleaning and inspection within 3 feet (914 mm) of each side of the fan. Wall-mounted exhaust fans shall have access for cleaning and inspection within 3 feet (914 mm) of the exhaust fan. [NFPA 96:7.3.7, 7.3.8]

The disassembly of the fan to access the duct for cleaning and maintenance is prohibited. Both roof and wall-mounted fans are required to swing away for internal access and cleaning of the connected grease ducting. In-line exhaust fans and utility set exhaust fans have ducting on both sides of the fan that require access for cleaning.

510.3.3 Horizontal Ducts. On horizontal ducts, at least one 20 inch by 20 inch (508 mm by 508 mm) opening shall be provided for personnel entry. [NFPA 96:7.4.1.1]

510.3.3.1 Cleaning. Where an opening of the size specified in Section 510.3.3 is not possible, openings large enough to permit thorough cleaning shall be provided at 12 feet (3658 mm) intervals. [NFPA 96:7.4.1.2]

Openings "large enough" shall include the size necessary to allow for tools, agents and equipment necessary to permit thorough cleaning of the grease duct.

510.3.3.2 Safe Access and Work Platform. If not easily accessible from a 10 foot (3048 mm) stepladder, openings on horizontal grease duct systems shall be provided with safe access and a work platform. [NFPA 96:7.4.1.3]

510.3.3.3 Support. Support systems for horizontal grease duct systems 24 inches (610 mm) and larger in any cross-sectional dimension shall be designed for the weight of the ductwork plus 800 pounds (362.9 kg) at any point in the duct systems. [NFPA 96:7.4.1.4]

510.3.4 Vertical Ducts. On vertical ductwork where personnel entry is possible, access shall be provided at the top of the vertical riser to accommodate descent. [NFPA 96:7.4.2.1]

510.3.4.1 Access. Where personnel entry is not possible, adequate access for cleaning shall be provided on each floor. [NFPA 96:7.4.2.2]

510.3.4.2 Safe Access and Work Platform. If not easily accessible from the floor or a 10 foot (3048 mm) stepladder, openings on vertical grease ducts shall be provided with safe access and a work platform. [NFPA 96:7.4.2.3]

510.3.5 Nonlisted Ductwork. On nonlisted ductwork, the edge of the opening shall be not less than 1½ inches (38 mm) from all outside edges of the duct or welded seams. [NFPA 96:7.4.1.5]

Stresses exerted on the corners, edges and welded seams of cold rolled steel ducting may cause premature failure especially in the event of a fire. Openings that are less than 1½ inches from edges may tend to warp and pull away from the sealant used at the opening.

510.3.6 Access Panels. Access panels shall be of the same material and thickness as the duct. Access panels shall have a gasket or sealant that is rated for 1500°F (816°C) and shall be greasetight. Fasteners, such as bolts, weld studs, latches, or wing nuts, used to secure the access panels shall be carbon steel or stainless steel and shall not penetrate duct walls. [NFPA 96:7.4.3.1 – 7.4.3.3]

Sealants or gaskets shall remain grease tight or be replaced after the access panel has been used.

510.3.6.1 Listed Grease Ducts. Listed grease duct access door assemblies (access panels) shall be installed in accordance with the terms of the listing and the manufacturer's instructions. [NFPA 96:7.4.3.4]

510.3.6.2 Within an Enclosure. Where openings are located in ducts within an enclosure, the access panel including its components shall be of the same fire rating as the enclosure.

510.3.7 Fire Protection System Devices. Openings for installation, servicing, and inspection of listed fire protection system devices and for duct cleaning shall be provided in ducts and enclosures and shall be in accordance with the requirements of Section 510.3 through Section 510.3.2 and Section 510.7.7. Enclosure openings required to reach access panels in the ductwork shall be large enough for removal of the access panel through the enclosure opening. [NFPA 96:7.4.4.1 – 7.4.4.2]

In addition to proper sizing, enclosure openings shall have the same fire-resistive rating as required for the enclosure.

510.4 Listed Grease Ducts. Listed grease ducts shall be installed in accordance with the terms of the listing and the manufacturer's instructions. [NFPA 96:7.1.7]

Not all listed grease ducts are intended to be "factory built grease duct enclosures." However, these listed grease ducts shall have the manufacturers' instructions at the job site for installation and inspection purposes.

510.4.1 Factory-Built Grease Ducts. Factory-built grease ducts in accordance with UL 1978 shall be permitted to incorporate non-welded joints in accordance with their listings.

510.5 Other Grease Ducts. Other grease ducts shall comply with the requirements of Section 510.5.1 through Section 510.5.5. [NFPA 96:7.5]

510.5.1 Materials. Ducts shall be constructed of and supported by carbon steel not less than 0.060 of an inch (1.524 mm) (No. 16 MSG) in thickness or stainless steel not less than 0.048 of an inch (1.219 mm) (No. 18 MSG) in thickness. [NFPA 96:7.5.1.1]

Any section of the duct between a Type I hood and the ceiling above is not considered part of the hood and must comply with the thickness requirements for ducts.

510.5.2 Factory-Built Grease Ducts. Factory-built grease ducts listed in accordance with UL 1978 shall be permitted to use materials in accordance with their listing. [NFPA 96:7.5.1.2]

510.5.3 Installation. All seams, joints, penetrations, and duct-to-hood collar connections shall have a liquid-tight continuous external weld. [NFPA 96:7.5.2.1]

Because visual verification is important, many jurisdictions require the installer to provide a light and extension cord to internally illuminate welded seams for liquid-tight joint inspection.

Exceptions:

(1) Factory-built grease ducts listed in accordance with UL 1978 shall be permitted to incorporate nonwelded joint construction in accordance with their listings. [NFPA 96:7.5.2.1.1]

(2) Duct-to-hood collar connections as shown in **Figure 510.5.3** shall not require a liquidtight continuous external weld. [NFPA 96:7.5.2.2]

(3) Penetrations shall be permitted to be sealed by other

listed devices that are tested to be greasetight and are evaluated under the same conditions of fire severity as the hood or enclosure of listed grease extractors and whose presence does not detract from the hood's or duct's structural integrity. [NFPA 96:7.5.2.3]

(4) Internal welding shall be permitted, provided the joint is formed or ground smooth and is readily accessible for inspection. [NFPA 96:7.5.2.4]

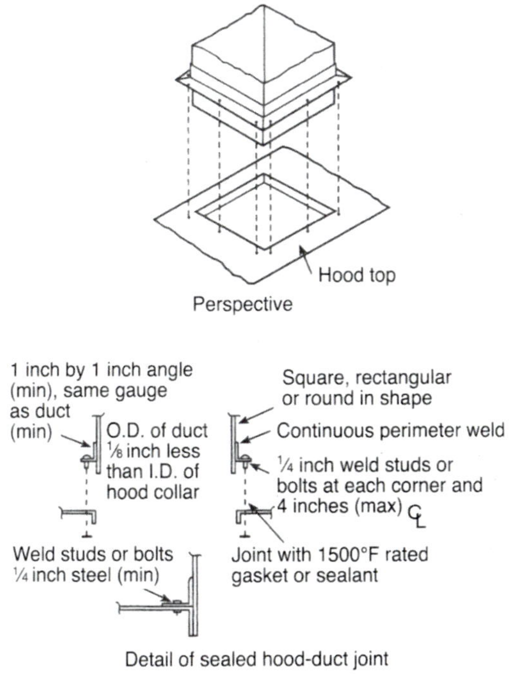

For SI units: 1 inch = 25.4 mm, °C = (°F-32)/1.8

FIGURE 510.5.3
PERMITTED DUCT-TO HOOD COLLAR CONNECTION
[NFPA 96: FIGURE 7.5.2.2]

Figure 510.5.3 details a portion of the hood that needs to be turned down and a 1-inch by 1-inch angle spaced above the bottom of the duct. This overlapping material in the hood will protect the sealant or gasket used at the duct connection from exposure to grease, cleaning agents and fire.

510.5.3.1 Duct Leakage Test. Prior to the use of or concealment of any portion of a grease duct system, a leakage test shall be performed to determine that all welded joints and seams are liquid tight. [NFPA 96:7.5.2.1.2]

510.5.3.2 Welded Duct Connection. Acceptable duct-to-duct connection shall be as follows:

(1) Telescoping joint, as shown in **Figure 510.5.3.2(1)**.

(2) Bell-type joint, as shown in **Figure 510.5.3.2(2)**.

(3) Flange with edge weld, as shown in **Figure 510.5.3.2(3)**.

(4) Flange with filled weld, as shown in **Figure 510.5.3.2(4)**. [NFPA 96:7.5.5.1]

Overlapping ducts minimize leakage due to failed weld. In most instances, welds in the shop are done under more ideal conditions where the welder can turn the duct for

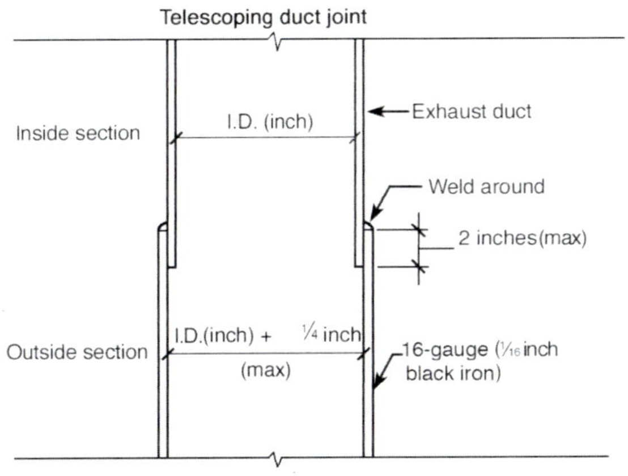

For SI units: 1 inch = 25.4 mm

Notes:

1. Duct size decreases (going upward) with each telescope.
2. Smaller (inside) duct is always above or uphill (on sloped duct), to be self draining into larger (outside) duct.

FIGURE 510.5.3.2(1)
TELESCOPING -TYPE DUCT CONNECTION
[NFPA 96: FIGURE 7.5.5.1(a)]

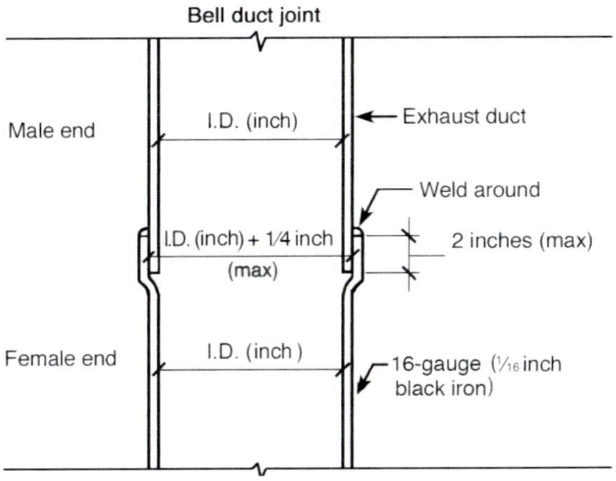

For SI units: 1 inch = 25.4 mm

Notes:

1. Duct size stays the same throughout the duct system.
2. Smaller (inside) male duct end is always above or uphill (on sloped duct), to be self-draining into larger (outside) female duct end.

FIGURE 510.5.3.2(2)
BELL-TYPE DUCT CONNECTION
[NFPA 96: FIGURE 7.5.5.1(b)]

access for proper weld penetration. Clearances on the job site with the duct in place make a proper weld more difficult to perform. Limiting the overlap to 2 inches maximum decreases the possibility that excessive gaps in the duct could develop during the welding process. Although not included in the text here, NFPA 96: 7.5.5.1 (3) and (4) call a flange weld an acceptable duct-to-duct connection [see **Figures 510.5.3.2(1) through 510.5.3.2(4)**]. Horizontal ducting (with flange welds) is particularly susceptible to pocketing grease and leaking where the welds are not completely liquid tight.

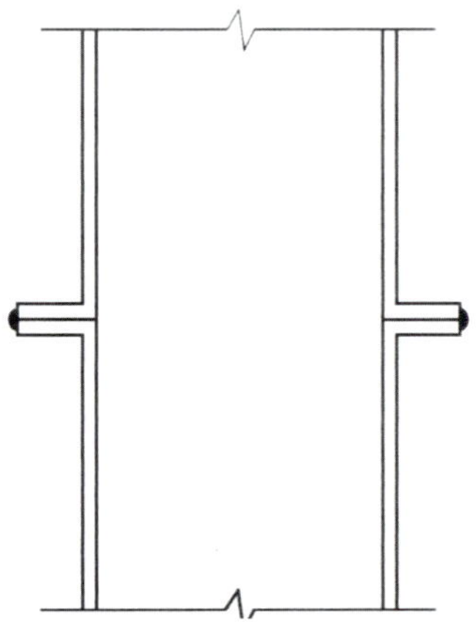

FIGURE 510.5.3.2(3)
FLANGE WITH EDGE WELD DUCT CONNECTION
[NFPA 96: FIGURE 7.5.5.1(c)]

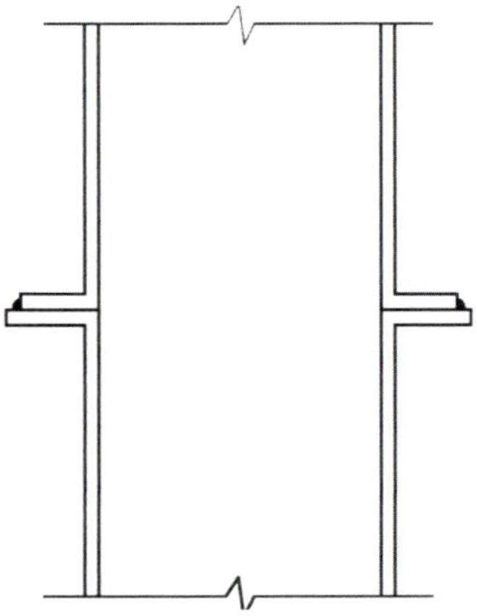

FIGURE 510.5.3.2(4)
FLANGE WITH FILLED WELD DUCT CONNECTION
[NFPA 96: FIGURE 7.5.5.1(d)]

510.5.4 Butt Welded Connections. Butt welded connections shall not be permitted. [NFPA 96:7.5.5.2]

510.5.5 Telescoping and Bell-Type Connections. For telescoping and bell-type connections, the inside duct section shall always be uphill of the outside duct section. [NFPA 96:7.5.5.3] The overlap shall not exceed 2 inches (51 mm) as shown in **Figure 510.5.3.2(1)**.

510.5.6 Duct Leakage Test. Prior to the use of or concealment of a grease duct system, a leakage test shall be performed to determine that welded joints and seams are liquid tight. The leakage test shall consist of a light test, water pressure test, or an approved equivalent test. The permit holder shall be responsible for providing the necessary equipment and for performing the test. Such test shall be conducted in accordance with ASHRAE 154.

To be in accordance with ASHRAE 154, a light, air, or water test shall be performed on grease ducts to ensure liquid tight seal.

A light test shall consist of passing a minimum 100W lamp through the entire section of ductwork. The lamp shall be open and light shall be emitted equally in all directions perpendicular to the interior duct walls. No light shall be visible from the exterior.

When performing the air test the duct shall be pressurized with 1-inch WC for 20 minutes and hold the initial set pressure.

When performing a water test, a pressure washer operating at a minimum 1500psi shall be used to simulate duct cleaning. No water shall be visible on the exterior surface during the test.

510.6 Exterior Installations. The exterior portion of the ductwork shall be vertical wherever possible and shall be installed and supported on the exterior of a building. Bolts, screws, rivets, and other mechanical fasteners shall not penetrate duct walls. Clearance of ducts shall comply with Section 507.4 through Section 507.4.3.3. [NFPA 96:7.6.1 – 7.6.3]

Ducts shall not pass through fire walls where openings are prohibited (see Section 510.1). However, where ducting is allowed to be installed exterior to the building, vertical portions are less likely to retain any grease residue that could become fuel for a fire.

510.6.1 Weather Protection. All ducts shall be protected on the exterior by paint or other suitable weather-protective coating. Ducts constructed of stainless steel shall not be required to have additional paint or weather-protective coatings. Ductwork subject to corrosion shall have minimal contact with the building surface. [NFPA 96:7.6.4 – 7.6.6]

Welding of galvanized ducting depletes the zinc coating, leaving the welded edge exposed to corrosion. Cold rolled steel ducting without coatings shall be coated at all points exposed to the outside. Grease ducting is subject to expansion and vibration; sections in contact with building surfaces tend to wear off necessary coatings.

510.7 Interior Installations. In all buildings more than one story in height and in one-story buildings where the roof-ceiling assembly is required to have a fire resistance rating, the ducts shall be enclosed in a continuous enclosure extending from the lowest fire-rated ceiling or floor above the hood, through any concealed spaces, to or through the roof, to maintain the integrity of the fire separations required by the applicable building code provisions. The enclosure shall be sealed around the duct at the point of penetration of the first fire-rated barrier after the hood, to maintain the fire resistance rating of the enclosure. The enclosure shall be vented to the exterior of the building through weather-protected openings. [NFPA 96:7.7.1.2 – 7.7.1.4]

Exception: The continuous enclosure provisions shall not be required where a field-applied grease duct enclosure or a factory-built grease duct enclosure (see Section 507.4.4 through Section 507.4.6) is protected with a listed duct-through-penetration protection system equivalent to the fire resistance rating of the assembly being penetrated and where the materials are installed in accordance with the conditions of the listings and the manufacturer's instructions and are acceptable to the Authority Having Jurisdiction. [NFPA 96:7.7.1.5]

Notably absent from this section is ducting that penetrates a nonrated ceiling or a nonrated ceiling/roof assembly. Single-story buildings that have a hood or grease duct penetrating a nonrated ceiling or a ceiling/roof assembly shall maintain clearances required as per Section 507.4.

510.7.1 Less than Four Stories. Buildings less than four stories in height shall have an enclosure with a fire resistance rating of not less than 1 hour. [NFPA 96:7.7.2.1.1]

510.7.2 Four Stories or More. Buildings four stories or more in height shall have an enclosure with a fire resistance rating of not less than 2 hours. [NFPA 96:7.7.2.1.2]

510.7.3 Clearance. Clearance from the duct or the exhaust fan to the interior surface of enclosures of combustible construction shall be not less than 18 inches (457 mm). Clearance from the duct to the interior surface of enclosures of noncombustible or limited-combustible construction shall be not less than 6 inches (152 mm). Provisions for reducing clearances as described in Section 507.4 through Section 507.4.3.3 shall not be applicable to enclosures. [NFPA 96:7.7.2.2.1 – 7.7.2.2.3]

Exception: Clearance from the outer surfaces of field-applied grease duct enclosures and factory-built grease duct enclosures to the interior surfaces of construction installed around them shall be permitted to be reduced where the field-applied grease duct enclosure materials and factory-built grease duct enclosures are installed in accordance with the conditions of the listing and the manufacturer's instructions and are acceptable to the Authority Having Jurisdiction. [NFPA 96:7.7.2.2.4]

Operational temperature and fire-event temperature cannot be easily dissipated to the surrounding environment when the grease duct is enclosed. The only relief is through the vent to the exterior required at the termination of the grease duct enclosure; therefore, the provisions to reduce clearances (the space around the duct) are not applicable.

510.7.4 Mechanical and Structural Integrity. Field-applied grease duct enclosures and factory-built grease duct enclosures shall provide mechanical and structural integrity, resiliency, and stability when subjected to expected building environmental conditions, duct movement under general operating conditions, and duct movement as a result of interior and exterior fire conditions. [NFPA 96:7.7.2.2.5]

510.7.5 Materials. For field-applied grease duct enclosures and factory-built grease duct enclosures, the materials and products shall be provided in accordance with Section 510.7.5.1 and Section 510.7.5.2.

510.7.5.1 Protection from Physical Damage. Measures shall be taken to prevent physical damage to any covering or enclosure material. Any damage to the covering or enclosure shall be repaired, and the covering or enclosure shall be restored to meet its intended listing and fire resistance rating and to be acceptable to the Authority Having Jurisdiction. [NFPA 96:7.7.3.1, 7.7.3.2]

510.7.5.2 Inspection. In the event of a fire within a kitchen exhaust system, the duct, the enclosure, and the covering directly applied to the duct shall be inspected by qualified personnel to determine whether the duct, the enclosure, and the covering directly applied to the duct are structurally sound, capable of maintaining their fire protection functions, suitable for continued operation, and acceptable to the Authority Having Jurisdiction. [NFPA 96:7.7.3.3]

Some destructive analysis may be required to complete a thorough inspection and determination of the field-applied or factory-built grease duct enclosure. If the grease hood and duct system are not suitable for continued use, components need to be replaced as per the requirements of this code.

510.7.6 Listed. For listed grease ducts, see Section 510.4.

510.7.7 Fire Doors. Where openings in the enclosure walls are provided, they shall be protected by listed fire doors of proper rating. Fire doors shall be installed in accordance with NFPA 80. Openings on other listed materials or products shall be clearly identified and labeled according to the terms of the listing and the manufacturer's instructions and shall be acceptable to the Authority Having Jurisdiction. [NFPA 96:7.7.4.1 – 7.7.4.3] The fire door shall be readily accessible, aligned, and of sufficient size to allow access to the rated access panels on the ductwork. [NFPA 96:7.7.4.4]

Floor/ceiling assemblies in a multiple-story building need to provide a firestop system. Grease ducts shall not penetrate a fire wall or fire barrier as per Section 510.1. This is the type of construction normally used to create a fire zone in buildings. Grease ducts penetrating the floor/ceiling shall not increase the hazard of the spread of fire by connecting to another hood on a different floor or have another grease duct in the same enclosure.

510.7.8 Ducts with Enclosure(s). Each duct system shall constitute an individual system serving only exhaust hoods in one fire zone on one floor. Multiple ducts shall not be permitted in a single enclosure unless acceptable to the Authority Having Jurisdiction. [NFPA 96:7.7.5.1 – 7.7.5.2]

510.8 Underground Installations. Grease ducts installed underground shall be approved for underground installation. The material of the grease duct shall be corrosion-resistant and shall comply with Section 510.5.1.

510.8.1 Grease Receptacle. The grease duct shall be sloped to drain the grease back to an approved grease collection device. A grease collection device shall be located at the base of the vertical riser.

510.8.2 Cleanouts. For horizontal installations, cleanouts for cleaning and maintenance shall be provided on the top portion of the grease duct in accordance with Section 510.3 and shall be labeled at the interior portion of the duct.

510.9 Termination of Type I Hood Exhaust System. The exhaust system shall terminate as follows:

(1) Outside the building with a fan or duct.

(2) Through the roof or to the roof from outside, as in Section 510.9.1, or through a wall, as in Section 510.9.2. [NFPA 96:7.8.1]

510.9.1 Rooftop Terminations. Rooftop terminations shall be arranged with or provided with the following:

(1) A minimum of 10 feet (3048 mm) of horizontal clearance from the outlet to adjacent buildings, property lines, and air intakes.

Extension of grease outlets (after the grease fan) may adversely affect the operation of the fan and conflict with the terms of the fan listing. Designers should verify that an extension meets the listing of the fan.

(2) A minimum of 5 feet (1524 mm) of horizontal clearance from the outlet (fan housing) to any combustible structure.

(3) A vertical separation of 3 feet (914 mm) above any air intakes within 10 feet (3048 mm) of the exhaust outlet.

(4) The ability to drain grease out of any traps or low points formed in the fan or duct near the termination of the system into a collection container that is noncombustible, closed, rainproof, and structurally sound for the service to which it is applied and that will not sustain combustion.

Regular service of this grease collection container is often neglected as it is out of site and out of mind. It is the owner's responsibility to keep this container from overflowing onto the roof, which would negate its effectiveness.

(5) A grease collection device that is applied to exhaust systems that does not inhibit the performance of any fan.

(6) Listed grease collection systems that meet the requirements of Section 510.9.1(4) and Section 510.9.1(5).

(7) A listed grease duct complying with Section 507.4.7 or ductwork complying with Section 507.4.8.

(8) A hinged upblast fan supplied with flexible weatherproof electrical cable and service hold-open retainer to permit inspection and cleaning that is listed for commercial cooking equipment with the following conditions:

(a) Where the fan attaches to the ductwork, the ductwork is a minimum of 18 inches (457 mm) away from any roof surface, as shown in **Figure 510.9.1**.

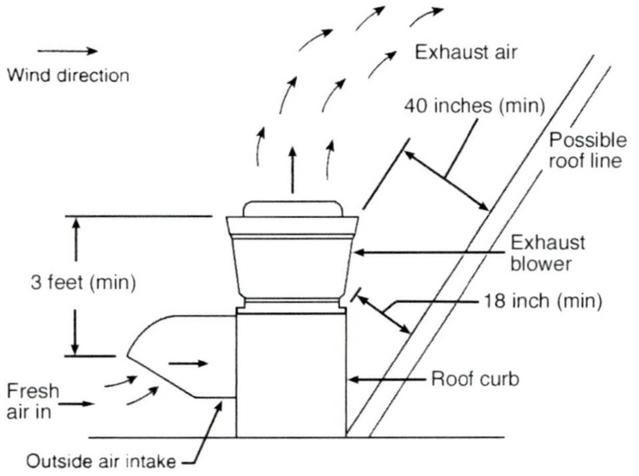

For SI units: 1 inch = 25.4 mm, 1 foot = 304.8 mm

FIGURE 510.9.1
UPBLAST FAN CLEARANCES
[NFPA 96: FIGURE 7.8.2.1]

This one section contains seven unique requirements: hinged fan for duct cleaning; upblast to limit grease on the roof; flexible electrical cable to allow for uninterrupted service after opening; hold-open retainer for safety during cleaning; fans listed for use; fans with 18-inch clearance above the roof and discharge not less than 40 inches above the roof.

(b) The fan discharges a minimum of 40 inches (1016 mm) away from any roof surface, as shown in **Figure 510.9.1**.

(9) Other approved fan, provided it meets all of the following criteria:

(a) The fan meets the requirements of Section 510.9.1(3) and Section 511.1.3.

(b) Its discharge or its extended duct discharge meets the requirements of Section 510.9.1(2). (See Section 511.1.3)

(c) Exhaust fan discharge is directed up and away from the roof surface. [NFPA 96:7.8.2.1]

An example of "other approved fan" includes a utility fan set when the fan and motor are located beside the grease duct riser. If the fan housing is not able to rotate to exhaust the air up and away from the roof surface, an extension or elbow may be added. Unlisted extensions should be consistent with materials allowed for grease ducts as per Section 510.5.1.

510.9.1.1 Listed Vibration Isolation Connectors. Listed vibration isolation connectors shall be permitted to be used on exterior roof locations where required for proper equipment vibration isolation.

510.9.1.2 Inspection and Cleaning. Fans shall be provided with safe access and a work surface for inspection and cleaning. [NFPA 96:7.8.2.2]

The AHJ shall approve the access and the minimum size of the work surface.

510.9.2 Wall Terminations. Wall terminations shall be arranged with or provided with the following properties:

(1) The termination shall be through a noncombustible wall with a minimum of 10 feet (3048 mm) of clearance from the outlet to adjacent buildings, property lines, grade level, combustible construction, electrical equipment or lines, and with the closest point of any air intake or operable door or window at or below the plane of the exhaust termination. The closest point of any air intake or operable door or window above the plane of the exhaust termination shall be a minimum of 10 feet (3048 mm) in distance, plus 3 inches (76 mm) for each 1 degree (0.017 rad) from horizontal, the angle of degree being measured from the center of the exhaust termination to the center of the air intake or operable door or window, as indicated in **Figure 510.9.2**.

Wall terminations shall be through a noncombustible wall with a minimum of 10 feet of clearance from the outlet to adjacent buildings, property lines, grade level, combustible construction, electrical equipment or lines, and the closest point of any air intake or operable door or window at or below the plane of the exhaust termination (see **Figure 510.9.2**).

Exception: A wall termination in a secured area shall be permitted to be at a lower height above grade if acceptable to the Authority Having Jurisdiction.

Combustible walls that have a noncombustible wall surface are not considered to be noncombustible (see Section 216.0). Wind and natural draft conditions will cause the grease duct discharge to eventually extend upward. Products of combustion, heat, steam and minor grease residue may

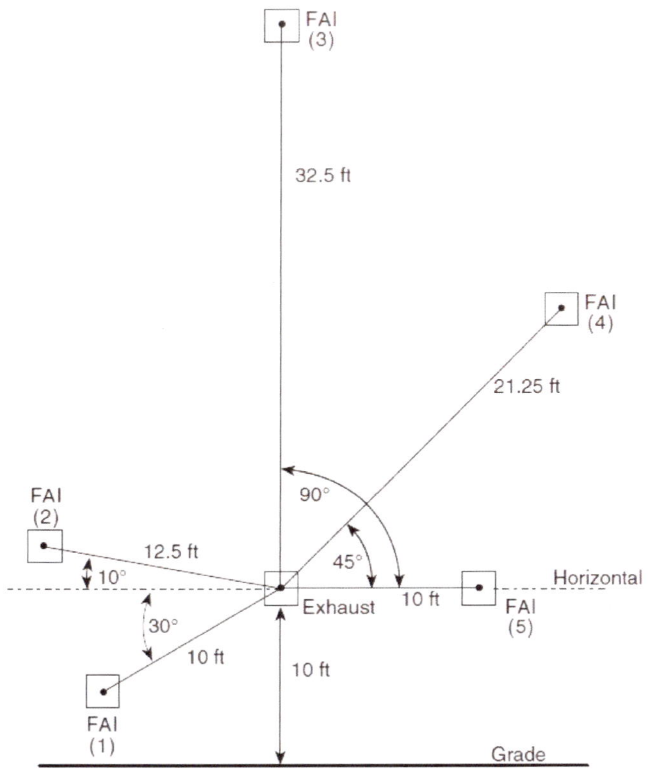

For SI units: 1 inch = 25.4 mm, 1 foot = 304.8 mm

Notes:

1. Fresh air intake (FAI) applies to an air intake, including an operable door or window.
2. Example:
   (a) FAIs 1 and 5 are on the same plane of exhaust or lower: 10 feet (3048 mm) min. between closest edges.
   (b) FAIs 2, 3, and 4 are above plane of exhaust fan:
   10 feet + 0.25 feet (3048 mm + 76 mm) per degree between closest edges.

FIGURE 510.9.2
EXHAUST TERMINATION DISTANCE FROM FRESH AIR INTAKE (FAI) OR OPERABLE DOOR OR WINDOW
[NFPA 96: FIGURE 7.8.3]

be blown back to the wall surface. Openings beside and above the grease duct outlet would be subject to a greater change of recirculating the exhausted air back into the building. Where the exhaust opening is more than 10 feet below an overhanging portion of a building, any opening above will need to maintain the prescribed vertical increased distance above from the exhaust. The intent is to maintain the vertical distance required regardless of the distance of the overhang.

(2) The exhaust flow shall be directed perpendicularly outward from the wall face or upward.

(3) All the ductwork shall be pitched to drain the grease back into the hood(s) or with a drain provided to bring the grease back into a container within the building or into a remote grease trap.

(4) A listed grease duct shall comply with Section 510.3.3 through Section 510.3.7; other ducts shall comply with Section 510.5.

(5) An approved fan shall meet the requirements of Section 510.9.2(3), and Section 511.1.1 or Section 511.1.3. [NFPA 96:7.8.3]

Grease ducts must be able to be cleaned. Openings through the discharge may provide the best access for cleaning. The exhaust fan needs to be hinged in order to swing away to access the duct. A noncombustible wall will not need to maintain an 18-inch clearance of the fan housing, but the listed fan will keep the point of discharge sufficiently clear of the wall surface. If the duct has traversed through a grease duct enclosure, a means to ventilate the enclosure to the outside must be provided as per Section 510.7.

510.10 Solid-Fuel Duct Systems. Where solid-fuel cooking equipment is to be vented, the duct system shall be in accordance with Section 517.0.

511.0 Air Movement.

511.1 Exhaust Fans for Commercial Cooking Operations. Exhaust fans shall be installed in accordance with Section 511.1.1 through Section 511.1.6. Exhaust fans shall comply with UL 762 and be installed in accordance with the manufacturer's installation instructions.

UL 762 includes a comprehensive set of construction and performance requirements that are used to evaluate and list power ventilators for restaurant exhaust appliances.

Fans used in a Type I hood application must include provisions for handling grease and access for cleaning. With the inclusion of this new standard roof, wall and inline ventilators could be approved by the AHJ. Review and strict adherence to the terms of the listing for power ventilators is necessary to ensure the ambient temperature and clearance to combustibles are adequate and do not create an unsafe condition. Roof ventilators would need to terminate a minimum of 40" above the roof line.

511.1.1 Upblast Fans. Upblast fans with motors surrounded by the airstream shall be hinged and supplied with flexible weatherproof electrical cable, and service hold-open retainers. Installation shall conform to the requirements of Section 510.9. Upblast fans shall have a drain directed to a readily accessible and visible grease receptacle not to exceed 1 gallon (4 L). [NFPA 96:8.1.2.1 – 8.1.2.3]

Regular service of the fan is required to remove any grease residue. All of the requirements in this section ensure that the duct system will be accessible through the point of termination without disassembling the fan or electrical components. The hold-open devices allow for a measure of safety to prohibit the closing of the fan while a service person is cleaning the duct below. Working surfaces may need to be provided for the safe access of the exhaust system.

511.1.2 In-Line Exhaust Fans. In-line fans shall be of the type with the motor located outside the airstream and with belts and pulleys protected from the airstream by a greasetight housing. In-line fans shall be connected to the exhaust duct by flanges securely bolted as shown in **Figure 511.1.2(1) through Figure 511.1.2(4)**, or by a system specifically listed for such use. Flexible connectors shall not be used. [NFPA 96:8.1.3.1 – 8.1.3.3]

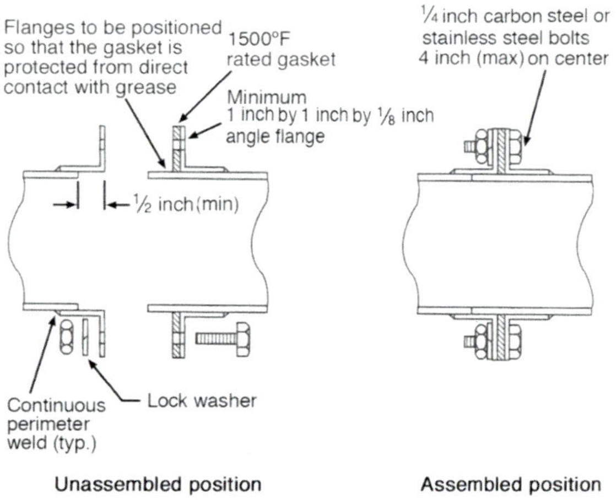

For SI units: 1 inch = 25.4 mm, °C = (°F-32)/1.8

FIGURE 511.1.2(1)
TYPICAL SECTION OF DUCT-TO-FAN CONNECTION-BUTT JOINT METHOD
[NFPA 96: FIGURE 8.1.3.2(a)]

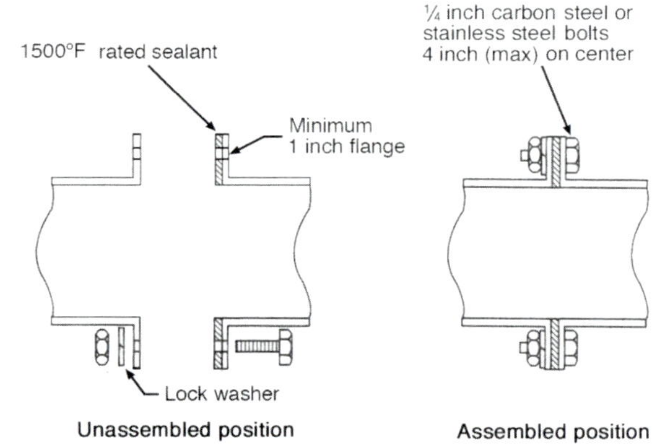

For SI units: 1 inch = 25.4 mm, °C = (°F-32)/1.8

FIGURE 511.1.2(3)
TYPICAL SECTION OF DUCT-TO-FAN CONNECTION–SEALANT METHOD
[NFPA 96: FIGURE 8.1.3.2(c)]

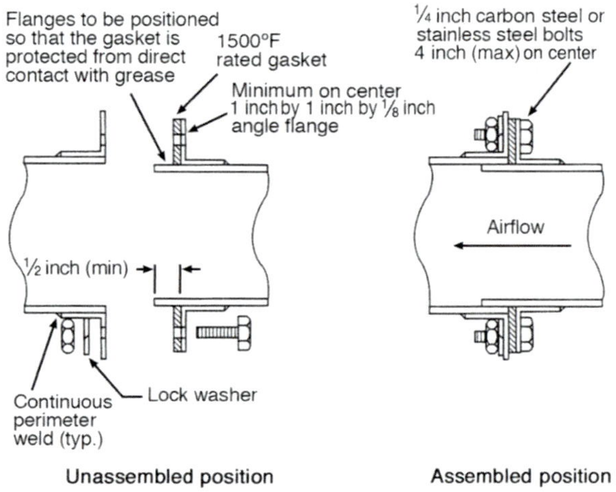

For SI units: 1 inch = 25.4 mm, °C = (°F-32)/1.8

FIGURE 511.1.2(2)
TYPICAL SECTION OF DUCT-TO-FAN CONNECTION-OVERLAPPING METHOD
[NFPA 96: FIGURE 8.1.3.2(b)]

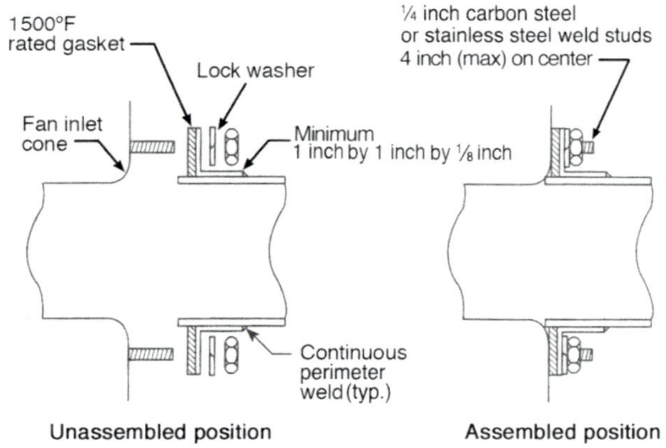

For SI units: 1 inch = 25.4 mm, °C = (°F-32)/1.8

FIGURE 511.1.2(4)
TYPICAL SECTION OF DUCT-TO-FAN CONNECTION-DIRECT TO FAN INLET CONE METHOD
[NFPA 96: FIGURE 8.1.3.2(d)]

511.1.2.1 Accessibility. If the design or positioning of the fan allows grease to be trapped, a drain directed to a readily accessible and visible grease receptacle not exceeding 1 gallon (4 L), shall be provided. In-line exhaust fans shall be located in easily accessible areas of adequate size to allow for service or removal. Where the duct system connected to the fan is in an enclosure, the space or room in which the exhaust fan is located shall have the same fire resistance rating as the enclosure. [NFPA 96:8.1.3.4 – 8.1.3.6]

Due to the flammability of the collected grease, accessibility and visibility is required to keep the receptacle from overflowing. The receptacle is limited to 1 gallon for the same fuel-loading reason.

These sections also allow the designer to provide a space in the building for the fan. In this case, the room or enclosure is required to be large enough to allow for service and removal of the fan and the enclosure shall meet the same rating as required by Section 510.7. Grease duct and fans are required to be inspected for grease build-up at an annual or more frequent interval as per Section 514.3. Access to the fan enclosed in the building or located external to the building should be provided with no less than required for access to other HVAC equipment according to Chapter 9 of this code.

511.1.3 Utility Set Exhaust Fans. Utility set exhaust fans shall be installed in accordance with Section 511.1.3.1 through Section 511.1.3.3.

Examples of in-line fans meeting the description of Sections 511.1.2 and 511.1.3 include utility-set and turbo-axial types that are listed for grease duct systems, although these fan systems may transmit vibrations into the duct system or the building. Flexible duct connectors shall not be used. While the in-line fan is required to be securely bolted to the ducting, the designer may employ vibration isolation techniques depending on how the fan is connected to the building structure (see **Figures 511.1.3a and 511.1.3b**).

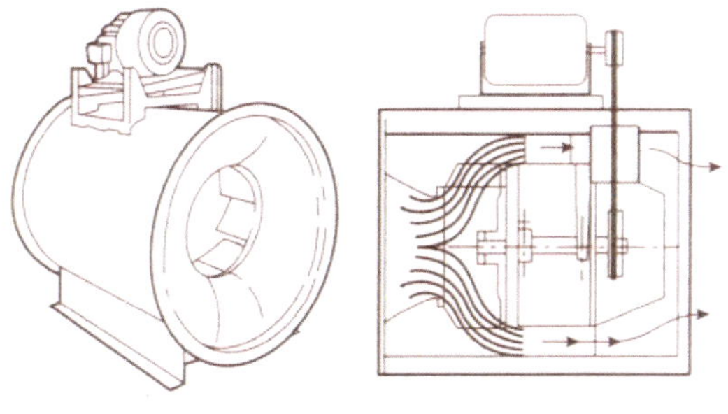

FIGURE A.8.1.2 In-Line Fan.

FIGURE 511.1.3A
IN-LINE FAN

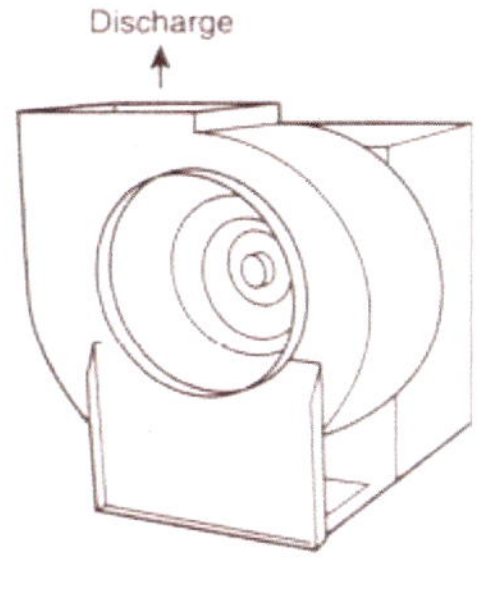

FIGURE A.8.1.3 Typical Utility Set Fan.

FIGURE 511.1.3B
TYPICAL UTILITY SET FAN

511.1.3.1 At the Rooftop. Fans installed at the rooftop termination point shall be in accordance with the following:

(1) Section 510.9.1 and Section 510.9.1.2.

(2) Vibration isolation connectors shall be permitted.

(3) A drain shall be directed to a readily accessible and visible grease receptacle not to exceed 1 gallon (4 L).

511.1.3.2 Within the Building. Fans installed within the building shall be in accordance with the following:

(1) Located in an accessible area of adequate size to allow for service or removal. [NFPA 96:8.1.4.2]

(2) Flexible connectors shall not be used. [NFPA 96:8.1.4.5]

(3) Exhaust fans shall have a drain directed to a readily accessible and visible grease receptacle not to exceed 1 gallon (4 L). [NFPA 96:8.1.4.6]

511.1.3.3 Duct Systems. Where the duct system connected to the fan is in an enclosure, the space or room in which the exhaust fan is located shall have the same fire resistance rating as the enclosure. The fan shall be connected to the exhaust duct by flanges securely bolted as shown in Figure 511.1.2(1) through Figure 511.1.2(4) or by a system specifically listed for such use. [NFPA 96:8.1.4.3, 8.1.4.4]

511.1.4 Construction. Exhaust fan housings shall be constructed of carbon steel not less than 0.060 inch (1.524 mm) (No. 16 MSG) in thickness, of stainless steel not less than 0.048 of an inch (1.219 mm) (No. 18 MSG) in thickness, or, if listed, in accordance with the terms of the listing. [NFPA 96:8.1.5]

Fan housings are required to be the same gauge as required for the grease ducts that they serve. Other listed fans and component housings shall be specifically listed for grease duct use.

511.1.5 Openings. Openings for cleaning, servicing, and inspection shall conform to the requirements of Section 510.3.2. Clearances shall conform to the requirements of Section 507.4 through 507.4.3.3, or Section 510.7.3 and Section 510.7.4 if installed within an enclosure. [NFPA 96:8.1.6.1, 8.1.6.2]

Although clearance may be reduced, adequate access to remove the fan may not be reduced. Grease residue will be more likely to collect near the fan. When the fan is installed in-line, the access for cleaning the ducts shall not exceed 3 feet from the fan housing as per Section 510.3.2.

511.1.6 Standard. All wiring and electrical equipment shall comply with NFPA 70. [NFPA 96:8.1.7]

511.2 Airflow. The air velocity through a duct shall be not less than 500 feet per minute (ft/min) (2.54 m/s) and not exceed 2500 ft/min (12.7 m/s).

A grease duct serving a Type I hood must be designed, installed and adjusted so that it provides an air velocity within the duct system capable of keeping the grease laden particle suspended until the point of termination.

The exception states that transition duct sections shall be permitted to be connected to hoods and exhaust fans that do not meet the velocity, provided that they do not exceed 3 feet in length and do not contain traps for grease.

511.2.1 Exceptions. Transition duct sections that do not exceed 3 feet (914 mm) in length and do not contain grease traps shall be permitted to be connected to hoods and exhaust fans that do not meet this velocity. [NFPA 96:8.2.1.2]

If the fan is not hinged to provide access to the duct for cleaning, providing an access within 3 feet of the fan is consistent with allowing a transition that does not exceed 3 feet. Transitions that are necessary to meet site conditions may increase both velocity and resistance to air moving through the system. Routine maintenance coupled with greater accessibility to these transitions will minimize hazards associated with them as long as they are not so restrictive as to affect the exhaust air volume.

511.2.2 Exhaust-Air Volumes. Exhaust air volumes for hoods shall be of sufficient level to provide for capture and removal of grease-laden cooking vapors. Test data, performance tests acceptable to the Authority Having Jurisdiction, or both shall be displayed, provided on request, or both. [NFPA 96:8.2.2.1, 8.2.2.2] Lower exhaust air volumes shall be permitted during no-load and partial load cooking conditions, provided they are sufficient to capture and remove flue gases and cooking effluent from cooking equipment.

When demand control grease duct exhaust is approved, replacement air shall be automatically controlled by variable-speed fans, dampers, or equivalent controls to ensure the proper air balance of the building. Prior to approval, the AHJ would review the design plans for a facility with a commercial kitchen ventilation system, which would typically include a table or diagram indicating the design outdoor air balance over the full range of anticipated airflow.

The total replacement airflow rate would need to equal the total exhaust airflow rate plus the net exfiltration. It could be permissible to supply replacement air to the kitchen space by using transfer air from areas other than the kitchen.

Continuous exhaust air volumes for hoods shall be maintained at a sufficient level to provide for the capture and removal of grease-laden cooking vapors. Lower exhaust-air volumes shall be permitted during no-load and partial load cooking conditions provided they are sufficient to capture and remove flue gases and cooking effluent from cooking equipment. Designers and operators should consider the effect of the multiple pilot lights and grease vapors emitted from cooking equipment during long periods of non-use or occasional use and adjust ventilation timing or equipment types to compensate.

511.2.2.1 Performance Test. A performance test shall be conducted upon completion and before final approval of the installation of a ventilation system serving commercial cooking appliances. The test shall verify the rate of exhaust airflow in accordance with Section 508.5.1.2 through Section 508.5.1.5. The permit holder shall furnish the necessary test equipment and devices required to perform the tests. [ASHRAE 154:4.7.1]

511.2.2.2 Capture and Containment Test. The permit holder shall verify the capture and containment performance of Type I hoods. A field test shall be conducted with all appliances under the hood at operating temperatures, all the hoods operating at design airflows, and with all sources of replacement air operating at design airflows for the restaurant. Capture and containment shall be verified visually by observing smoke or steam produced by actual cooking operation or by simulating cooking using devices such as smoke candles or smoke puffers. Smoke bombs shall not be used. [ASHRAE 154:4.7.2]

511.2.3 Exhaust Fan Operation. A hood exhaust fan(s) shall continue to operate after the extinguishing system has been activated unless fan shutdown is required by a listed component of the ventilation system or by the design of the extinguishing system. The hood exhaust fan shall start upon activation of the extinguishing system if the exhaust fan and all cooking equipment served by the fan have been shut down, unless fan shutdown is required by a listed component of the ventilation system or by the listing of the extinguishing system. The exhaust fan shall be provided with a means so that the fan is activated when any heat-producing cooking appliance under the hood is turned on. [NFPA 96:8.2.3.1 – 8.2.3.3]

Ventilation of the building is employed by fire department personnel to reduce the combustible gases in the building. Use of the grease hood and duct to automatically restart after activation of the hood's extinguishing system is not the intent of this code.

511.3 Makeup Air. The makeup air quantity shall prevent negative pressures in the commercial cooking area(s) from exceeding 0.02 inch water column (0.005 kPa). Where the fire-extinguishing system activates, makeup air supplied internally to a hood shall be shut off.

For compensating hoods, where a Type I or Type II hood has an internal discharge of makeup air, the makeup air flow shall not exceed 10 percent of the exhaust airflow, the exhaust airflow shall be the net exhaust from the hood in accordance with Section 508.5.1.2 through Section 508.5.1.5. The total hood exhaust shall be determined in accordance with Equation 511.3.

$$E_{NET} = E_{HOOD} - MA_{ID} \quad \text{(Equation 511.3)}$$

Where:

E_{NET} = net hood exhaust

E_{HOOD} = total hood exhaust

MA_{ID} = makeup air, internal discharge

Makeup air quantity shall be adequate to prevent negative pressures in the commercial cooking area(s) from exceeding 0.02-inch water column. Makeup air diffusers must be located to prevent short circuiting of air furnished to the exhaust system. Windows and doors shall not be used for the purpose of providing makeup air.

In general, makeup air requirements in this section do not necessitate the installation of supply fans. It is possible to design a makeup air system that relies solely on the negative pressure in the kitchen space to produce airflow of outside air; it is also possible to design a system that mechanically forces the makeup air into the kitchen. The intent of this section is to indicate desired performance, not to specify how it is to be accomplished. Either a duct or an electrically dampered opening that is not a window or door could be employed for this purpose.

Makeup air location in the kitchen may adversely affect the ability of the hood to capture the heat and grease from equipment under the hood. Registers having air directed at the hood may encourage roll out of vapors, while registers having air directed away may encourage a venturi effect and siphon vapors out of the hood. Many manufacturers of listed hoods have recommendations for the optimal makeup air locations.

To reduce the possibility of providing oxygen to an ongoing grease fire, makeup air that is provided internally to a hood shall be interlocked to shut off with the discharge of the hoods fire-suppression system (see **Figure 511.3**).

511.3.1 Air Balance. Design plans for a facility with a commercial kitchen ventilation system shall include a schedule or diagram indicating the design outdoor air balance. The design outdoor air balance shall indicate the exhaust and replacement air for the facility and the net exfiltration where applicable. The total replacement airflow rate shall equal the total exhaust airflow rate and the net exfiltration.

511.4 Common Duct (Manifold) Systems. Master kitchen exhaust ducts that serve multiple tenants shall include provision to bleed air from outdoors or from adjacent spaces into the master exhaust duct where required to maintain the necessary minimum air velocity in the master exhaust duct. [NFPA 96:8.4.1]

Master kitchen exhaust ducts that serve multiple tenants shall include provisions to bleed air from outdoors or from adjacent spaces into the master exhaust duct where required to maintain the necessary minimum velocity in the

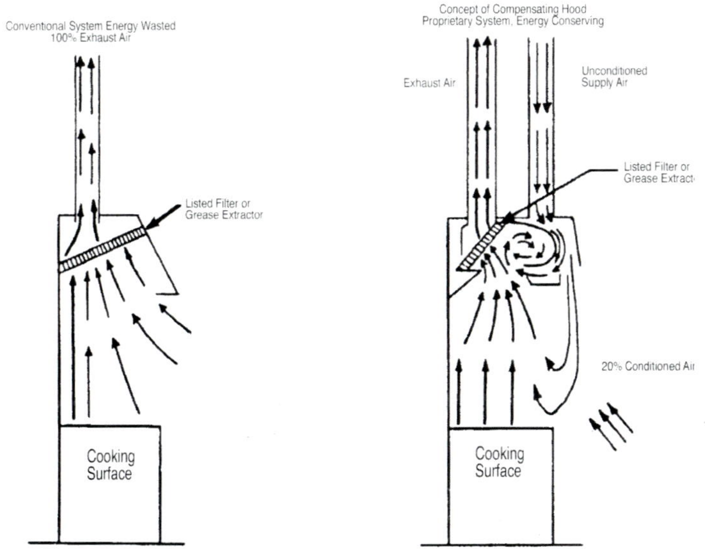

FIGURE 511.3
HOOD REPLACEMENT AIR

master exhaust duct (see **Figure 511.4**). This bleed air duct shall be connected to the top or side of the master exhaust duct. The bleed air duct shall have a fire damper at least 12 inches from the exhaust duct connection. The bleed air duct shall have the same construction and clearance requirements as the main exhaust duct from the connection to the exhaust duct of at least 12 inches on both sides of the fire damper. Each bleed air duct shall have a volume damper installed to provide a means of adjusting the bleed air quantity. The volume damper shall be installed between the fire damper and the bleed air source. The bleed air ducts shall not be used as a means of exhausting grease-laden vapors and shall be so labeled. Unused tenant exhaust connections to the master exhaust duct that are not used as bleed air connections shall be disconnected and sealed at the main duct.

511.4.1 Connections. Bleed air ducts shall connect to the top or side of the master exhaust duct. [NFPA 96:8.4.2]

511.4.2 Fire Damper. The bleed-air duct shall have a fire damper at least 12 inches (305 mm) from the exhaust duct connection. [NFPA 96:8.4.3]

511.4.3 Construction and Clearance. The bleed-air duct shall have the same construction and clearance requirements as the main exhaust duct from the connection to the exhaust duct to at least 12 inches (305 mm) on both sides of the fire damper. [NFPA 96:8.4.4]

511.4.4 Adjustment. Each bleed air duct shall have a means of adjusting (e.g., by using volume dampers) the bleed air quantity. [NFPA 96:8.4.5]

511.4.5 Adjustment Location. Means to adjust the bleed air quantity shall be installed between the fire damper and the source of bleed air. [NFPA 96:8.4.6]

511.4.6 Bleed Air Duct. A bleed air duct shall not be used for the exhaust of grease-laden vapors and shall be so labeled. [NFPA 96:8.4.7]

511.4.7 Disconnect. Unused tenant exhaust connections to the master exhaust duct that are not used as bleed air connections shall be disconnected and sealed at the main duct. [NFPA 96:8.4.8]

Designers shall consider the time of operation of the hood systems that are connected to a master kitchen (manifold) exhaust serving multiple tenants. While it may be possible to "lower exhaust-air volumes permitted during no-load cooking conditions, provided they are sufficient to capture and remove flue gases and residual vapors from cooking equipment" as per Section 511.2.2, the additional control features required to maintain maximum velocities in branches may become prohibitive.

511.5 Solid-Fuel Air Movement Requirements. Where solid-fuel cooking equipment is used, exhaust and replacement air also shall be in accordance with Section 517.0.

512.0 Auxiliary Equipment.

512.1 Dampers. Dampers shall not be installed in exhaust ducts or exhaust duct systems. [NFPA 96:9.1.1]

512.1.1 Use. Where specifically listed for such use or where required as part of a listed device or system, dampers in exhaust ducts or exhaust duct systems shall be permitted. [NFPA 96:9.1.2]

As an example, control dampers that may be necessary for master kitchen exhaust duct systems shall be listed for use in grease ducts. Dampers exposed to airflow containing grease residue give extra surface area for the grease to attach. Installation that complies with the terms of the instructions will ensure easy cleaning.

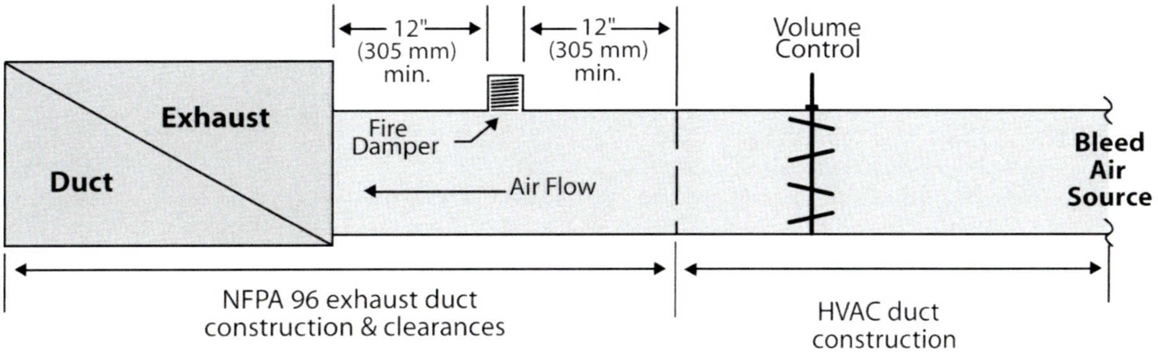

FIGURE 511.4
BLEED AIR SYSTEM

512.2 Electrical Equipment. Wiring systems of any type shall not be installed in ducts. [NFPA 96:9.2.1]

Wiring systems, motors, lights and other electrical devices shall not be installed in the exhaust ducts or hoods or located in the path of travel of exhaust products unless they are specifically approved for such use. Any excess heat or source of ignition from the operation of an electrical equipment or wiring shall be prohibited unless listed. The listing process evaluates the hazard associated with the operation of the device and ensures operation in a safe manner. The excessive temperature caused by a grease duct during a fire will cause greater risk to short circuiting the electrical components and wiring systems.

512.2.1 Device Installation in Ducts. Motors, lights, and other electrical devices shall be permitted to be installed in ducts or hoods or to be located in the path of travel of exhaust products only where specifically listed for such use. [NFPA 96:9.2.2]

512.2.2 Lighting Units. Lighting units on hoods shall not be located in concealed spaces except as permitted by Section 512.2.3 and Section 512.2.4. [NFPA 96:9.2.3.2]

512.2.3 Concealed Spaces. Lighting units shall be permitted in concealed spaces where such units are part of a listed exhaust hood. [NFPA 96:9.2.3.3]

512.2.4 Listed Lighting Units. Listed lighting units specifically listed for such use and installed in accordance with the terms of the listing shall be permitted to be installed in concealed spaces. [NFPA 96:9.2.3.4]

Lighting units in, on or in concealed spaces of the hood are a normal and necessary part of hood equipment. Listed hoods or unlisted hoods shall use listed lighting units.

512.2.5 Standard. Electrical equipment shall be installed in accordance with NFPA 70, with due regard to the effects of heat, vapor, and grease on the equipment.

512.3 Other Equipment. Fume incinerators, thermal recovery units, air pollution control devices, or other devices shall be permitted to be installed in ducts or hoods or to be located in the path of travel of exhaust products where specifically listed for such use. [NFPA 96:9.3.1]

512.3.1 Access Required. Equipment shall have space provided to all access panels or doors for the safe removal and servicing of control devices, such as filters, electrostatic precipitator cells, and odor control media beds, and for cleaning of the equipment housing. [NFPA 96:9.3.1.3]

512.3.2 Downgrading. Downgrading other parts of the exhaust system due to the installation of approved devices, whether listed or not, shall not be allowed. [NFPA 96:9.3.2]

512.3.3 Fire-Extinguishing System. Any equipment installed in the path of exhaust products that provides secondary filtration or air pollution control shall be provided with an approved automatic fire-extinguishing system, installed in accordance with the fire-extinguishing system manufacturer's instructions. [NFPA 96:9.3.3]

512.3.3.1 Protection. The fire-extinguishing system required by Section 512.3.3 shall provide protection for the component sections of the equipment, and ductwork downstream of the equipment. [NFPA 96:9.3.3.1]

512.3.3.2 Filter Media. Filter media used in secondary filtration or air pollution control units and not complying with Section 509.2.3 shall have fire protection that is adequate for the filter media being used in accordance with the fire-extinguishing system manufacturer's instructions. [NFPA 96:9.3.3.2]

512.3.4 Source of Ignition. If the equipment provides a source of ignition, it shall be provided with detection to operate the fire-extinguishing system protecting the equipment. [NFPA 96:9.3.4]

512.3.5 Air Recirculation. Where a cooking exhaust system employs an air pollution control device that recirculates air into the building, the requirements of Section 516.0 shall apply. [NFPA 96:9.3.5]

512.3.6 Carbon Monoxide Detector Required. If the heat source is non-electric and open flames are used, a carbon monoxide detector shall be installed in both the kitchen and dining areas. [NFPA 96:9.3.7]

512.4 Solid-Fuel Auxiliary Equipment. Where solid fuel cooking comprises a part of a cooking operation, additional provisions, and equipment as described in Section 517.0 shall be used where required.

513.0 Fire-Extinguishing Equipment.

513.1 General. Fire-extinguishing equipment for the protection of grease removal devices, hood exhaust plenums, and exhaust duct systems shall be provided. [NFPA 96:10.1.1]

Fire-extinguishing equipment shall be provided for the protection of grease-removal devices, hood exhaust plenums and exhaust duct systems.

513.1.1 Devices in Exhaust Ducts. Fume incinerators, thermal recovery units, air pollution control devices, or other devices installed in the exhaust duct, shall be protected by an automatic fire-extinguishing system. [NFPA 96:10.1.3]

513.1.2 Protection. Cooking equipment that produces grease-laden vapors and that might be a source of ignition of grease in the hood, grease removal device, or duct shall be protected by fire-extinguishing equipment. [NFPA 96:10.1.2]

The fire-extinguishing system shall be used to protect cooking equipment that produces grease-laden vapors (such as but not limited to deep-fat fryers, ranges, griddles, broilers, woks, tilting skillets and braising pans).

513.2 Types of Equipment. Fire-extinguishing equipment shall include both automatic fire-extinguishing systems as primary protection and portable fire extinguishers as secondary backup. [NFPA 96:10.2.1]

Many jurisdictions have separate inspections and inspectors for fire and mechanical code provisions. The types of fire-suppression equipment and systems found in this code are not meant to conflict with recognized fire codes.

513.2.1 Identification. A placard shall be conspicuously placed near each Class K extinguisher that states that the fire protection system shall be activated prior to using the fire extinguisher. [NFPA 96:10.2.2]

The placard required here is not meant to be part of the portable extinguisher. The placard shall be provided near the location where the portable extinguisher is mounted.

513.2.2 Standard. Automatic fire-extinguishing systems shall comply with UL 300 or other equivalent standards and shall be installed in accordance with the requirements of the listing. In existing dry or wet chemical systems not in compliance with UL 300, the fire-extinguishing system shall be made to comply with this section when any of the following occurs:

(1) The cooking medium is changed from animal oils and fats to vegetable oils.

(2) The positioning of the cooking equipment is changed.

(3) Cooking equipment is replaced.

(4) The equipment is no longer supported by the manufacturer. [NFPA 96:10.2.3, 10.2.3.1]

Exception: Automatic fire-extinguishing equipment provided as part of listed recirculating systems shall comply with UL 710B. [NFPA 96:10.2.5]

The type of automatic fire-extinguishing system used for protection of the cooking equipment shall comply with UL 300, standard for Fire Testing of Fire Extinguishing Systems for Protection of Commercial Cooking Equipment, or other equivalent standards and shall be installed in accordance with its listing. The standard contains a series of tests addressing the extinguishing agent, fire detectors, piping limitations, nozzle coverage, installation, and maintenance.

513.2.3 Installation. Automatic fire-extinguishing systems shall be installed in accordance with the terms of their listing, the manufacturer's installation instructions, and the following standards where applicable:

(1) NFPA 12

(2) NFPA 13

(3) NFPA 17

(4) NFPA 17A

513.2.4 Modification of Existing Hood Systems. Any abandoned pipe or conduit from a previous installation shall be removed from within the hood, plenum, and exhaust duct. [NFPA 96:10.2.7.1]

513.2.4.1 Sealing. Penetrations and holes resulting from the removal of conduit or piping shall be sealed with listed or equivalent liquid-tight sealing devices. [NFPA 96:10.2.7.2]

513.2.4.2 Obstructions. The addition of obstructions to spray patterns from the cooking appliance nozzle(s) such as baffle plates, shelves, or any modification shall not be permitted. [NFPA 96:10.2.7.3]

513.2.4.3 System Re-evaluation. Changes or modifications to the hazard after installation of the fire-extinguishing systems shall result in re-evaluation of the system design by a properly trained, qualified, and certified person(s). [NFPA 96:10.2.7.4]

513.2.5 Baffle Hoods with Water Wash. Areas requiring protection in accordance with Section 513.1 shall be permitted to be protected by a water-wash system that is listed as a fire-extinguishing system in compliance with UL 300 or other equivalent standards and installed in accordance with the requirements of its listing. [NFPA 96:10.2.8.1]

Water and grease do not mix; however, a "listed fixed baffle hood containing a constant or fire-actuated water wash system that is listed and in compliance with UL 300 or other equivalent standards" provides equivalent protection from grease-related fires.

513.2.5.1 Listed for the Purpose. Each such area not provided with a listed water-wash fire-extinguishing system shall be provided with a fire-extinguishing system listed for the purpose. [NFPA 96:10.2.8.2]

513.2.5.2 Domestic Water Supply. The water supply for water-wash fire-extinguishing systems shall be permitted to be supplied from the domestic water supply when the minimum water pressure and flow are provided in accordance with the terms of the listing. [NFPA 96:10.2.8.3]

See Section 513.2.6 for the options of supervision required.

513.2.5.3 Control Valve. The water supply for water-wash fire-extinguishing systems shall be controlled by a listed indicating valve. [NFPA 96:10.2.8.4]

513.2.5.4 Activation. Where a separate fire-extinguishing system is used for protection of cooking equipment only, a water-wash fire-extinguishing system listed for protection of the grease removal device(s), hood exhaust plenum(s), exhaust duct(s), or combination thereof shall be provided with instructions and appropriate means for electrical interface for simultaneous activation. [NFPA 96:10.2.8.5]

513.2.5.5 Water-Wash System. A water-wash system approved to be used for protection of the grease removal device(s), hood exhaust plenum(s), exhaust duct(s), or combination thereof shall include instructions and appropriate electrical interface for simultaneous activation of the water-wash system from an automatic fire-extinguishing system, where the automatic fire-extinguishing system is used for cooking equipment protection only. [NFPA 96:10.2.8.6]

513.2.5.6 Exception. Where the automatic fire-extinguishing system in accordance with NFPA 17A provides protection for the hood and duct in a fixed baffle hood containing a water-wash system, the water-wash system shall be made inoperable or delayed for a minimum of 60 seconds upon operation of the automatic fire-extinguishing system. [NFPA 96:10.2.8.7]

513.2.5.7 Water Supply. The water required for listed automatic fire-extinguishing systems shall be permitted to be supplied from the domestic water supply where the minimum water pressure and flow are provided in accordance with the terms of the listing. The water supply shall be controlled by a supervised water supply control valve. Where the water supply is from a dedicated fire protection water supply in a building with one or more fire sprinkler systems, separate indicating control valves and drains shall be provided and arranged so that the hood system and sprinkler systems can be controlled individually. [NFPA 96:10.2.9.1, 10.2.9.2]

513.2.6 Water Valve Supervision. Valves controlling the water supply to listed water-wash fire-extinguishing systems, automatic fire-extinguishing systems, or both shall be listed indicating type of valve and shall be supervised open by one of the following methods:

(1) Central station, proprietary, or remote station alarm service.

(2) Local alarm service that will cause the sounding of an audible signal at a constantly attended point.

(3) Locking valves open.

(4) Sealing of valves and approved weekly recorded inspection. [NFPA 96:10.2.10]

513.3 Simultaneous Operation. Fixed pipe extinguishing systems in a single hazard area shall be arranged for simultaneous automatic operation upon actuation of any one of the systems. [NFPA 96:10.3.1]

513.3.1 Hoods. Hoods installed end to end, back to back, or both, or sharing a common ductwork, not exceeding 75 feet (22 860 mm) in distance from the farthest hood, and having a grease producing appliance(s) located under one or more of the hoods, shall be considered a single hazard area requiring simultaneous automatic fire protection in all hoods and ducts. [NFPA 96:10.3.1.1]

Extinguishing systems for hoods in a single hazard area must operate simultaneously upon detection in any of the hoods or their ductwork up to 75 feet from the hood. For protection of duct further than 75 feet from the hood(s) see Common Ductwork.

513.3.1.1 Common Ductwork. In hoods that are installed end to end, back to back, or both, and that share a common ductwork, the ductwork beyond 75 feet (22 860 mm) from the farthest hood shall be protected by an independent fire-extinguishing system with its own detection system or by a fire-extinguishing system that activates simultaneously with the fire-extinguishing system(s) protecting the hoods. [NFPA 96:10.3.1.1.1]

Ductwork more than 75 feet from a hood or multiple hoods in a single hazard zone must have an independent extinguishing system that does not automatically operate with the hood(s) and/or upstream duct protection system.

513.3.2 Independent Systems. Hoods installed end to end, back to back, or both that do not share a common exhaust duct and are separated by a wall(s) or other means to ensure that grease-laden vapors exhausted under one hood cannot propagate to the other hoods, the hoods' fire-extinguishing system(s) shall be independent and shall not be required to simultaneously discharge. [NFPA 96:10.3.1.2]

Separate independent fire systems for multiple hoods on a single duct system are not required to simultaneously discharge.

513.3.3 Exempt Equipment. Fume incinerators, thermal recovery units, air pollution control devices, or other devices installed in the exhaust duct shall not be required to comply with Section 513.3.1. [NFPA 96:10.3.1.3]

As an example, pollution control unit or other device installed in the duct shall have it's own independent fire extinguishing system so if the fire reached the unit it would then discharge and extinguish the fire, but it is not required that the PCU extinguishing system discharge simultaneously with the hood extinguishing system.

513.3.4 Automatic Sprinkler System. Simultaneous operation shall not be required where the one fixed pipe extinguishing system is an automatic sprinkler system. Where an automatic sprinkler system is used in conjunction with a water-based fire-extinguishing system served by the same water supply, hydraulic calculations shall consider both systems operating simultaneously. [NFPA 96:10.3.2, 10.3.2.1]

513.3.5 Dry or Wet Chemical Systems. Simultaneous operation shall be required where a dry or wet chemical system is used to protect common exhaust ductwork by one of the methods specified in NFPA 17 or NFPA 17A. [NFPA 96:10.3.3]

513.4 Fuel and Electric Power Shutoff. Upon activation of any fire-extinguishing system for a cooking operation, all sources of fuel and electrical power that produce heat to all equipment requiring protection by that system shall automatically shut off. [NFPA 96:10.4.1]

Exception: Solid-fuel cooking operations.

The fuel gases and electricity providing heat to appliances that process food are required to be shut off in the event of a fire and activation of any fire-extinguishing system. If the building or the cooking area is equipped with a fire-suppression system other than the one under the Type I hood, an automatic shutoff valve for the gas line or a shunt for the electrical power is required to stop the flow of gas and or electricity. These gas valves shall be tested in accordance with the manufacturers' instructions. Most valves can be locked open and tested with the remainder of the gas line. See Section 1314.0, Pressure Testing and Inspection, of this code.

513.4.1 Steam. Steam supplied from an external source shall not be required to automatically shut off. [NFPA 96:10.4.2]

513.4.2 Protection Not Required. Any gas appliance not requiring protection but located under ventilating equipment where protected appliances are located, shall be automatically shut off upon activation of the extinguishing system. [NFPA 96:10.4.3]

Any appliances under the hood may inadvertently cause a reignition of a fire after the extinguishing activation due to a gas leak that may develop during the heat of a fire. Once the extinguishing system has been depleted, no further protection shall be automatically provided.

513.4.3 Manual Reset. Shutoff devices shall require manual reset. [NFPA 96:10.4.4]

Contractors must use only manual reset-type devices to minimize personal safety and property loss. A complete evaluation by properly trained and qualified person(s) or a company and approval of the AHJ is required prior to the reestablishment of gas or electrical power after the activation of the extinguishing system.

513.5 Manual Activation. All systems shall have both automatic and manual methods of actuation. At least one manual actuation device shall be located in a means of egress or at a location acceptable to the Authority Having Jurisdiction.

The manual actuation device shall clearly identify the hazard protected. An automatic sprinkler system shall not require a method of manual actuation. [NFPA 96:10.5.1 – 10.5.2]

513.6 System Annunciation. Upon activation of an automatic fire-extinguishing system, an audible alarm or visual indicator shall be provided to show that the system has activated. [NFPA 96:10.6.1]

513.6.1 Signaling. Where a fire alarm signaling system is serving the occupancy where the extinguishing system is located, the activation of the automatic fire-extinguishing system shall activate the fire alarm signaling system. [NFPA 96:10.6.2]

513.7 Special Design and Application. Hoods containing automatic fire-extinguishing systems are protected areas; therefore, these hoods shall not be considered obstructions to overhead sprinkler systems and shall not require floor additional sprinkler coverage underneath. [NFPA 96:10.7.1]

513.8 Review and Certification. Where required, complete drawings of the system installation, including the hood(s), exhaust duct(s), and appliances, along with the interface of the fire-extinguishing system detectors, piping, nozzles, fuel and electric power shutoff devices, agent storage container(s), and manual actuation device(s), shall be submitted to the Authority Having Jurisdiction. [NFPA 96:10.8.1]

513.9 Installation Requirements. Installation of systems shall be performed only by persons properly trained and qualified to install the specific system being provided. The installer shall provide certification to the Authority Having Jurisdiction that the installation is in agreement with the terms of the listing and the manufacturer's instructions and/or approved design. [NFPA 96:10.8.2.1, 10.8.2.2]

513.10 Portable Fire Extinguishers. Portable fire extinguishers shall be selected and installed in kitchen cooking areas in accordance with NFPA 10 and shall be specifically listed for such use. Class K fire extinguishers shall be provided for cooking appliance hazards that involve combustible cooking media (vegetable oils and animal oils and fats). [NFPA 96:10.9.1, 10.9.2]

513.10.1 Other Fire Extinguishers. Portable fire extinguishers shall be provided for other hazards in kitchen areas and shall be selected and installed in accordance with NFPA 10. [NFPA 96:10.9.3]

513.10.2 Carbon Dioxide-Type. Carbon dioxide-type extinguishers shall not be permitted. [NFPA 96:10.9.4]

513.11 Maintenance. Portable fire extinguishers shall be maintained in accordance with NFPA 10. [NFPA 96:10.9.5]

513.11.1 Permitted Use. Portable fire extinguishers listed specifically for use in the kitchen cooking areas shall also be permitted.

513.12 Solid-Fuel Fire-Extinguishing Equipment. Where solid-fuel cooking equipment is served by fire extinguishing equipment, the provisions of Section 517.0 shall apply.

514.0 Procedures for the Use, Inspection, Testing, and Maintenance of Equipment.

514.1 Operating Procedures. Exhaust systems shall be operated whenever cooking equipment is turned on. [NFPA 96:11.1.1]

514.1.1 Filters. Filter-equipped exhaust systems shall not be operated with filters removed. [NFPA 96:11.1.2]

514.1.2 Openings. Openings provided for replacing air exhausted through ventilating equipment shall not be restricted by covers, dampers, or other means that would reduce the operating efficiency of the exhaust system. [NFPA 96:11.1.3]

514.1.3 Posting of Instructions. Instructions for manually operating the fire-extinguishing system shall be posted conspicuously in the kitchen and shall be reviewed with employees by the management. [NFPA 96:11.1.4.3]

514.1.4 Listing and Manufacturer's Instructions. Listed exhaust hoods shall be operated in accordance with the terms of their listings and the manufacturer's instructions. [NFPA 96:11.1.5]

514.1.5 Nonoperational. Cooking equipment shall not be operated while its fire-extinguishing system or exhaust system is nonoperational or impaired. [NFPA 96:11.1.6]

514.1.6 Secondary Control Equipment. Secondary filtration and pollution control equipment shall be operated in accordance with the terms of its listing and the manufacturer's recommendations. [NFPA 96:11.1.7]

514.1.7 Inspection Frequency. Inspection and maintenance of "other equipment" as allowed in Section 512.3 shall be conducted by properly trained and qualified persons at a frequency determined by the manufacturer's instructions or the equipment listing. [NFPA 96:11.1.8]

514.2 Inspection, Testing, and Maintenance. Maintenance of the fire-extinguishing systems and listed exhaust hoods containing a constant or fire-activated water system that is listed to extinguish a fire in the grease removal devices, hood exhaust plenums, and exhaust ducts shall be made by properly trained, qualified, and certified person(s) acceptable to the Authority Having Jurisdiction at least every 6 months. [NFPA 96:11.2.1]

514.2.1 Requirements. All actuation and control components, including remote manual pull stations, mechanical and electrical devices, detectors, and actuators, shall be tested for proper operation during the inspection in accordance with the manufacturer's procedures. The specific inspection and maintenance requirements of the extinguishing system standards as well as the applicable installation and maintenance manuals for the listed system and service bulletins shall be followed. [NFPA 96:11.2.2, 11.2.3]

514.2.2 Fusible Links and Sprinklers. Fusible links of the metal alloy type and automatic sprinklers of the metal alloy type shall be replaced at least semiannually. [NFPA 96:11.2.4]

514.2.3 Inspection Tag. The year of manufacture and the date of installation of the fusible links shall be marked on the system inspection tag. The tag shall be signed or initialed by the installer.

Detection devices that are bulb-type automatic sprinklers and fusible links other than the metal alloy type shall be examined and cleaned or replaced annually. [NFPA 96:11.2.5, 11.2.5.1, 11.2.6]

514.2.4 Temperature-Sensing Elements. Fixed temperature-sensing elements other than the fusible metal alloy type shall be permitted to remain continuously in service, provided they are inspected and cleaned, or replaced if necessary in accordance with the manufacturer's instructions, every 12 months or more frequently to ensure proper operation of the system. [NFPA 96:11.2.7]

514.2.5 Certification. Where required, certificates of inspection and maintenance shall be forwarded to the Authority Having Jurisdiction. [NFPA 96:11.2.8]

514.3 Inspection for Grease Buildup. The entire exhaust system shall be inspected for grease buildup by a properly trained, qualified, and certified person(s) acceptable to the Authority Having Jurisdiction and in accordance with Table 514.3. [NFPA 96:11.4]

TABLE 514.3
SCHEDULE OF INSPECTION FOR GREASE BUILDUP
[NFPA 96: TABLE 11.4]

TYPE OR VOLUME OF COOKING	INSPECTION
Systems serving solid fuel cooking operations.	Monthly
Systems serving high-volume cooking operations.[1]	Quarterly
Systems serving moderate-volume cooking operations.	Semiannually
Systems serving low-volume cooking operations.[2]	Annually

Notes:

[1] High-volume cooking operations include 24-hour cooking, charbroiling, and wok cooking.

[2] Low-volume cooking operations include churches, day camps, seasonal businesses, and senior centers.

514.4 Cleaning of Exhaust Systems. If, upon inspection, the exhaust system is found to be contaminated with deposits from grease-laden vapors, the contaminated portions of the exhaust system shall be cleaned by a properly trained, qualified, and certified person(s) acceptable to the Authority Having Jurisdiction. [NFPA 96:11.6.1]

514.4.1 Measurement System. A measurement system of deposition shall be established to trigger a need to clean when the exhaust system is inspected at the frequencies in Table 514.3. [NFPA 96:11.6.1.1]

514.4.1.1 Combustible Contaminants. Hoods, grease removal devices, fans, ducts, and other appurtenances shall be cleaned to remove combustible contaminants to a minimum of 0.002 of an inch (50 μm). [NFPA 96:11.6.1.1.1]

514.4.1.2 Gauge Comb. A grease depth gauge comb as shown in **Figure 514.4.1.2** shall be placed upon the surface to measure grease depth. [NFPA 96:11.6.1.1.2]

514.4.1.3 Cleaning Method. Where a measured depth of 0.078 of an inch (2000 μm) is observed, the surfaces shall be cleaned in accordance with Section 514.4. [NFPA 96:11.6.1.1.3]

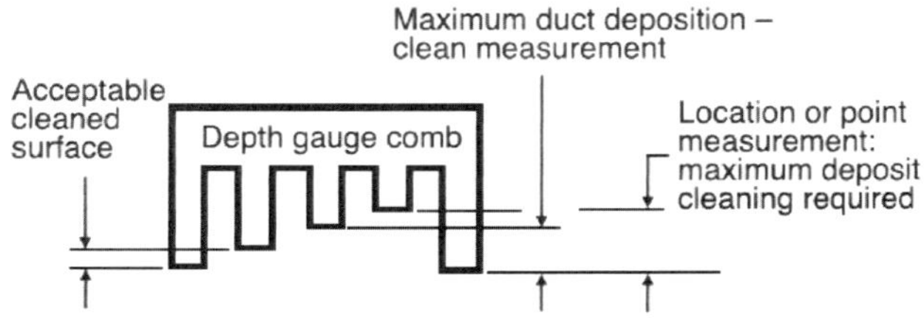

FIGURE 514.4.1.2
DEPTH GAUGE COMB
[NFPA 96: FIGURE 11.6.1.1.2]

514.4.1.4 Combustible Contaminants. Where a measured depth of 0.125 of an inch (3175 μm) is observed in a fan housing, the surfaces shall be cleaned in accordance with Section 514.4. [NFPA 96:11.6.1.1.4]

514.4.2 Removal of Contaminants. Hoods, grease removal devices, fans, ducts, and other appurtenances shall be cleaned to remove combustible contaminants prior to surfaces becoming heavily contaminated with grease or oily sludge. [NFPA 96:11.6.2]

514.4.3 Electrical Switches. At the start of the cleaning process, electrical switches that could be activated accidentally shall be locked out. [NFPA 96:11.6.3]

514.4.4 Fire Suppression System. Components of the fire suppression system shall not be rendered inoperable during the cleaning process. [NFPA 96:11.6.4]

514.4.5 Inoperable. Fire-extinguishing systems shall be permitted to be rendered inoperable during the cleaning process where serviced by properly trained and qualified persons. [NFPA 96:11.6.5]

514.4.6 Solvents/Cleaning Aids. Flammable solvents or other flammable cleaning aids shall not be used. [NFPA 96:11.6.6]

514.4.7 Cleaning Chemicals. Cleaning chemicals shall not be applied on fusible links or other detection devices of the automatic extinguishing system. [NFPA 96:11.6.7]

514.4.8 Coating. After the exhaust system is cleaned, it shall not be coated with powder or other substance. [NFPA 96:11.6.8]

514.4.9 Access Panels and Cover Plates. When cleaning procedures are completed, all access panels (doors) and cover plates shall be restored to their normal operational condition. [NFPA 96:11.6.9]

514.4.10 Date of Inspection. When an access panel is removed, a service company label or tag preprinted with the name of the company and giving the date of inspection or cleaning shall be affixed near the affected access panels. [NFPA 96:11.6.10]

514.4.11 Airflow. Dampers and diffusers shall be positioned for proper airflow. [NFPA 96:11.6.11]

514.4.12 Operable State. When cleaning procedures are completed, all electrical switches and system components shall be returned to an operable state. [NFPA 96:11.6.12]

514.4.13 Certification of Service. When an exhaust system is inspected or cleaned, a certificate showing the name of the servicing company, the name of the person performing the work, and the date of inspection or cleaning shall be maintained on the premises. [NFPA 96:11.6.13]

514.4.14 Report Provided. After cleaning or inspection is completed, the exhaust cleaning company and the person performing the work at the location shall provide the owner of the system with a written report that also specifies areas that were inaccessible or not cleaned. [NFPA 96:11.6.14]

514.4.15 Unclean Area. Where required, certificates of inspection and cleaning and reports of areas not cleaned shall be submitted to the Authority Having Jurisdiction. [NFPA 96:11.6.15]

514.4.16 Metal Containers. Metal containers used to collect grease drippings shall be inspected or emptied at least weekly. [NFPA 96:11.6.16]

514.5 Cooking Equipment Maintenance. Inspection and servicing of the cooking equipment shall be made at least annually by properly trained and qualified persons. [NFPA 96:11.7.1]

514.5.1 Cleaning. Cooking equipment that collects grease below the surface, behind the equipment, or in cooking equipment flue gas exhaust, such as griddles or charbroilers, shall be inspected and, if found with grease accumulation, cleaned by a properly trained, qualified, and certified person(s) acceptable to the Authority Having Jurisdiction. [NFPA 96:11.7.2]

515.0 Minimum Safety Requirements for Cooking Equipment.

515.1 Cooking Equipment. Cooking equipment shall be approved based on one of the following criteria:

(1) Listings by a testing laboratory.

(2) Test data acceptable to the Authority Having Jurisdiction. [NFPA 96:12.1.1]

515.1.1 Installation. All listed appliances shall be installed in accordance with the terms of their listings and the manufacturer's instructions. Solid fuel used for flavoring within a gas-operated appliance shall be in a solid fuel holder (smoker box) that is listed with the equipment. [NFPA 96:12.1.2.1, 12.1.2.1.1]

Listed cooking equipment provides clearances from combustible material found on the affixed label. These distances must not be reduced by use of clearance reduction methods found in this code. Some gas-fired equipment use large pots for cooking where heat is directed around the perimeter of the pot or receptacle. These types of appliances may require more than a 6-inch clearance common to other types of cooking equipment.

515.1.1.1 Re-evaluation. Cooking appliances requiring protection shall not be moved, modified, or rearranged without prior re-evaluation of the fire-extinguishing system by the system installer or servicing agent, unless otherwise allowed by the design of the fire-extinguishing system. A solid fuel holder shall not be added to an existing appliance until the fire-extinguishing system has been evaluated by the fire-extinguishing system service provider. [NFPA 96:12.1.2.2, 12.1.2.2.1]

515.1.1.2 Prior Location. The fire-extinguishing system shall not require re-evaluation where the cooking appliances are moved for the purposes of maintenance and cleaning, provided the appliances are returned to approved design location prior to cooking operations, and any disconnected fire-extinguishing system nozzles attached to the appliances are reconnected in accordance with the manufacturer's listed design manual. [NFPA 96:12.1.2.3]

515.1.1.3 Minimum Space. All deep-fat fryers shall be installed with at least a 16 inch (406 mm) space between the fryer and surface flames from adjacent cooking equipment. [NFPA 96:12.1.2.4]

Fat dripping from recently removed baskets with fried foods is flammable. If the cook turns attention to one side, the basket may turn as well. If the basket is suspended over open flame, the fat drippings will ignite and the contents of the basket may also ignite. A 16-inch separation or an 8-inch-high baffle will tend to stop or remove the hazard that may result from one moment of inattention.

515.1.1.4 Space Not Required. Where a steel or tempered glass baffle plate is installed at a minimum 8 inches (203 mm) in height between the fryer and surface flames of the adjacent appliance, the requirement for a 16 inch (406 mm) space shall not apply. [NFPA 96:12.1.2.5]

515.1.1.5 Minimum Height. If the fryer and the surface flames are at different horizontal planes, the minimum height of 8 inches (203 mm) shall be measured from the higher of the two. [NFPA 96:12.1.2.5.1]

515.2 Operating Controls. Deep-fat fryers shall be equipped with a separate high-limit control in addition to the adjustable operating control (thermostat) to shut off fuel or energy when the fat temperature reaches 475°F (246°C) at 1 inch (25.4 mm) below the surface. [NFPA 96:12.2]

516.0 Recirculating Systems.

516.1 General Requirements. Recirculating systems containing or for use with appliances used in processes producing smoke or grease-laden vapors shall be equipped with components complying with the following:

(1) The clearance requirements of Section 507.4 through Section 507.4.3.3.

(2) A hood complying with the requirements of Section 508.0.

(3) Grease removal devices complying with Section 509.0.

(4) The air movement requirements of Section 511.2.1 and Section 511.2.2.

(5) Auxiliary equipment (such as particulate and odor removal devices) complying with Section 512.0.

(6) Fire-extinguishing equipment complying with the requirements of Section 513.0.

Exception: Fire-extinguishing equipment in accordance with Section 513.1 and Section 513.5.

(7) The use and maintenance requirements of Section 514.0.

(8) The minimum safety requirements of Section 515.0.

(9) All the requirements of Section 516.0. [NFPA 96:13.1]

516.2 Design Restrictions. All recirculating systems shall comply with the requirements of Section 516.2.1 through Section 516.2.9. [NFPA 96:13.2]

516.2.1 Gas/Electrically Fueled Cooking Appliances. Only gas-fueled or electrically fueled cooking appliances shall be used. Listed gas-fueled equipment designed for use with specific recirculating systems shall have the flue outlets connected in the intended manner. Gas-fueled appliances shall have a minimum 18 inches (457 mm) clearance from the flue outlet to the filter inlet in accordance with Section 509.2.2 through Section 509.2.2.3 and shall meet the installation requirements of this code, NFPA 54 or NFPA 58. {NFPA 96:13.2.1 – 13.2.3}

516.2.2 Recirculation. Recirculating systems shall be listed with a testing laboratory in accordance with UL 710B or equivalent. [NFPA 96:13.2.4]

516.2.3 Protection. Cooking appliances that require protection and that are under a recirculating hood shall be protected by either the integral fire protection system in accordance with UL 710B or a system in accordance with Section 513.0. [NFPA 96:13.2.4.2]

516.2.4 Maximum Limits. A recirculating system shall not use cooking equipment that exceeds that recirculating system's labeled maximum limits for that type of equipment, stated in maximum energy input, maximum cooking temperature, and maximum square area of cooking surface or cubic volume of cooking cavity. [NFPA 96:13.2.6]

516.2.5 Label. The listing label shall show the type(s) of cooking equipment tested and the maximum limits specified in Section 516.2.4. [NFPA 96:13.2.7]

Although listed equipment together with a listed recirculating hood meets the intent of the code, more than one such system in a single establishment will add heat and moisture loading beyond the capability of an air-conditioning system to remove. The designer should provide information to the AHJ proving that excessive moisture will not build up within the building. Much the same as unvented dishwashing machines, moisture can collect on surfaces to the extent of causing dripping onto foods or onto locations where bacteria growth may be a result.

516.2.6 Fire Damper. A fire-actuated damper shall be installed at the exhaust outlet of the system. [NFPA 96:13.2.8] The actuation device for the fire damper shall have a maximum temperature rating of 375°F (191°C). [NFPA 96:13.2.10]

516.2.7 Installation of Electrical Wiring. No electrical wiring shall be installed in the interior sections of the hood plenum that is capable of becoming exposed to grease.

Exception: As permitted by NFPA 70.

516.2.8 Power Supply. The power supply of any electrostatic precipitator (ESP) shall be of the "cold spark," ferroresonant type in which the voltage falls off as the current draw of a short increases. [NFPA 96:13.2.11]

516.2.9 Listing Evaluation. Listing evaluation shall include the following:

(1) Capture and containment of vapors at published and labeled airflows.

(2) Grease discharge at the exhaust outlet of the system not to exceed an average of 2.9 E-09 (oz/in^3) (5.0 E-06 kg/m^3) of exhausted air sampled from that equipment at maximum amount of product that is capable of being processed over a continuous 8 hour test with the system operating at its minimum listed airflow.

(3) Listing and labeling of clearance to combustibles from all sides, top, and bottom.

(4) Electrical connection in the field in accordance with NFPA 70.

(5) Interlocks on all removable components that lie in the path of airflow within the unit to ensure that they are in place during operation of the cooking appliance. {NFPA 96:13.2.12}

516.3 Interlocks. The recirculating system shall be provided with interlocks of critical components and operations as indicated in Section 516.3.1 through Section 516.3.3.1 such that, if any of these interlocks are interrupted, the cooking appliance will not be able to operate. [NFPA 96:13.3.1]

516.3.1 Airflow Sections. All closure panels encompassing airflow sections shall have interlocks to ensure that the panels are in place and fully sealed. [NFPA 96:13.3.2]

516.3.2 Filter Component. Each filter component (grease and odor) shall have an interlock to prove the component is in place. [NFPA 96:13.3.3]

516.3.3 ESP Interlocks. Each ESP shall have a sensor to prove its performance is as designed, with no interruption of the power to exceed 2 minutes. [NFPA 96:13.3.4.1]

516.3.3.1 Manual Reset. The sensor shall be a manual reset device or circuit. [NFPA 96:13.3.4.2]

516.3.4 Airflow Switch or Transducer. An airflow switch or transducer shall be provided after the last filter component to ensure that a minimum airflow is maintained. The airflow switch or transducer shall open the interlock circuit when the airflow falls 25 percent below the system's normal operating flow or 10 percent below its listed minimum rating, whichever is lower. The airflow switch or transducer shall be a manual reset device or circuit. [NFPA 96:13.3.5.1 – 13.3.5.3]

516.4 Location and Application Restrictions. The location of recirculating systems shall be approved by the Authority Having Jurisdiction. Items to be reviewed in the fire risk assessment shall include, but not be limited to, life safety, combustibility of surroundings, proximity to air vents, and total fuel load. [NFPA 96:13.4.1 – 13.4.2]

516.5 Additional Fire Safety Requirements. In addition to the appliance nozzle(s), a recirculating system shall be listed with the appropriate fire protection for grease filters, grease filtration, odor filtration units, and ductwork, where applicable. [NFPA 96:13.5.1]

516.5.1 Installation Downstream. In addition to any other fire-extinguishing system activation device, there shall be a fire-extinguishing system activation device installed downstream of any ESP. [NFPA 96:13.5.2]

516.5.2 Locations. The requirements of Section 513.6 shall also apply to recirculating system locations. [NFPA 96:13.5.3]

516.6 Use and Maintenance. Automatic or manual covers on cooking appliances, especially fryers, shall not interfere with the application of the fire suppression system. [NFPA 96:13.6.1]

516.6.1 Manufacturer's Instructions. All filters shall be cleaned or replaced in accordance with the manufacturer's instructions. [NFPA 96:13.6.2]

516.6.2 Cleaning Schedule. All ESPs shall be cleaned a minimum of once per week and according to the manufacturer's cleaning instructions. [NFPA 96:13.6.3]

516.6.3 Hood Plenum and Blower Section Cleaning Schedule. The entire hood plenum and the blower section shall be cleaned a minimum of once every 3 months. [NFPA 96:13.6.4]

516.6.4 Inspection of Safety Interlocks. Inspection and testing of the total operation and all safety interlocks in accordance with the manufacturer's instructions shall be performed by qualified service personnel a minimum of once every 6 months or more frequently if required. [NFPA 96:13.6.5]

516.6.5 Inspection. Fire-extinguishing equipment shall be inspected in accordance with Section 514.2. [NFPA 96:13.6.6]

516.6.6 Maintenance Log. A signed and dated log of maintenance as performed in accordance with Section 516.6.3 and Section 516.6.4 shall be available on the premises for use by the Authority Having Jurisdiction. [NFPA 96:13.6.7]

517.0 Solid-Fuel Cooking Operations.

Similar to grease-burning charbroilers, solid-fuel cooking equipment will derive all or part of its heat source from the burning of solid cooking fuel. Gas or electric fuel equipment that use smoke from wood or chips to add flavoring but not cook the food need to follow venting provisions found in Section 508.0.

517.1 Venting Application. Venting requirements of solid-fuel cooking operations shall be determined in accordance with Section 517.1.1 through Section 517.1.6. [NFPA 96:14.1]

517.1.1 Natural Draft. Where solid-fuel cooking equipment is required by the manufacturer to have a natural draft, the vent shall comply with Section 517.4. [NFPA 96:14.1.1]

517.1.2 System Compliance. Where the solid-fuel cooking equipment has a self-contained top, is the appliance to be vented in an isolated space (except for a single water heater with its own separate vent), has a separate makeup air system, and is provided with supply and return air (not supplied or returned from other spaces), the system shall comply with Section 517.4 and Section 517.6. [NFPA 96:14.1.2]

517.1.3 Makeup Air System. Where the solid-fuel cooking equipment is located in a space with other vented equipment, all vented equipment shall have an exhaust system interlocked with a makeup air system for the space per Section 517.6. [NFPA 96:14.1.3]

Solid-fuel equipment depending on natural draft (warm air rises) to exhaust products of combustion may easily become subject to backdraft when other ventilation systems are installed in the same room (see **Figure 517.1.3**).

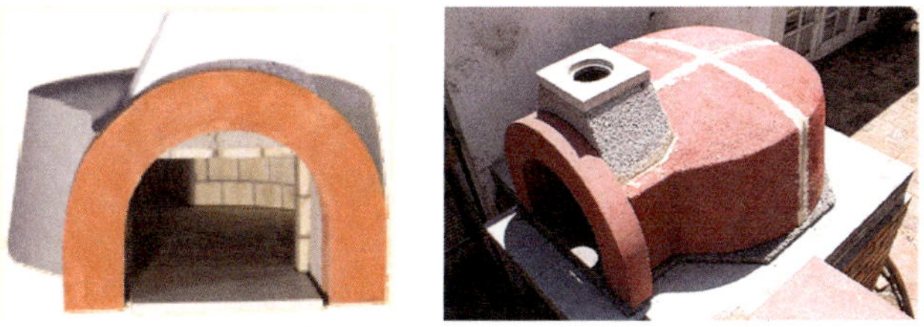

FIGURE 517.1.3
SOLID-FUEL COOKING EQUIPMENT

517.1.4 Natural Draft Ventilation Systems. Natural draft ventilation systems and power-exhausted ventilation systems shall comply with Section 517.3, Section 517.4, and Section 517.6. [NFPA 96:14.1.4]

517.1.5 Opening Requirements. Where a solid-fuel cooking appliance allows effluent to escape from the appliance opening, this opening shall be covered by a hood and an exhaust system that meets the requirements of Section 517.3, Section 517.4, and Section 517.6. [NFPA 96:14.1.5]

517.1.6 Spark Arresters. Solid-fuel cooking operations shall have spark arresters to minimize the passage of airborne sparks and embers into plenums and ducts. Where the solid-fuel cooking operation is not located under a hood, a spark arrester shall be provided to minimize the passage of sparks and embers into flues and chimneys. [NFPA 96:14.1.6, 14.1.7]

517.2 Location of Appliances. Every appliance shall be located with respect to building construction and other equipment so as to permit access to the appliance. [NFPA 96:14.2.1]

517.2.1 Prohibited Location. Solid-fuel cooking appliances shall not be installed in confined spaces. [NFPA 96:14.2.2]

Exception: Solid-fuel cooking appliances listed for installation in confined spaces such as alcoves shall be installed in accordance with the terms of the listing and the manufacturer's instructions. [NFPA 96:14.2.3]

517.2.2 Flammable Vapors. Solid-fuel cooking appliances shall not be installed in any location where gasoline or any other flammable vapors or gases are present. [NFPA 96:14.2.4]

517.3 Hoods for Solid-Fuel Cooking. Hoods shall be sized and located in a manner capable of capturing and containing all the effluent discharging from the appliances. The hood and its exhaust system shall comply with the requirements of Section 508.0 through Section 513.0. [NFPA 96:14.3.1, 14.3.2]

517.3.1 Separation. Except as permitted in Section 517.3.1.1, exhaust systems serving solid-fuel cooking equipment, including gas or electrically operated equipment, shall be separate from all other exhaust systems. [NFPA 96:14.3.3]

Exception: Cooking equipment not requiring automatic fire-extinguishing equipment (per Section 513.0) shall be permitted to be installed under a common hood with solid-fuel cooking equipment that is served by a duct system separate from all other exhaust systems. [NFPA 96:14.3.5]

For the purposes of this section, any provision for makeup or replenishment air shall also be separate from the replenishment air requirements for other exhaust systems (see Sections 517.1.3 and 517.6.2).

517.3.1.1 Equipment with Solid Fuel for Flavoring. Gas-operated equipment utilizing solid fuel for flavoring that meets all the following conditions shall not be required to have a separate exhaust system:

(1) The solid fuel holder (smoker box) shall be listed with the gas-operated equipment.

(2) The solid fuel holder shall be located underneath the gas burners.

(3) Spark arresters conforming with Section 517.1.6 shall be provided.

(4) The maximum quantity of solid fuel consumed shall not exceed 1 pound (0.45 kg) per hour per 100 000 Btu/h (29 kW) of gas burner capacity.

(5) The gas-operated equipment shall be protected by a fire suppression system listed for the equipment, including the solid fuel holder.

(6) Gas-operated equipment with integral solid fuel holder(s) intended for flavoring, such as radiant charbroiler(s), shall comply simultaneously with the requirements of UL 300 that address the gas radiant charbroiler(s) and mesquite wood charbroiler(s).

(7) A fire suppression system nozzle(s) shall be installed to protect the solid fuel holder.

(8) The fire suppression system shall be designed and installed to protect the entire cooking operation.

(9) Each solid fuel holder shall be limited to a size of 150 cubic inches (2.5 L), with no dimension to exceed 20 inches (508 mm).

(10) A maximum of one solid fuel holder for each 100 000 Btu/h (29 kW), or portion thereof, of burner capacity shall be permitted.

(11) Solid fuel shall be immersed in water for a continuous period of at least 24 hours immediately prior to being placed in the cooking equipment.

(12) The inspection frequency shall be the same as for solid fuel cooking operations in Table 514.3. [NFPA 96:14.3.4]

517.4 Exhaust Systems for Solid-Fuel Cooking. Where a hood is not required, in buildings where the duct system is three stories or less in height, a duct complying with Section 510.0 shall be provided. [NFPA 96:14.4]

517.4.1 Hood. If a hood is used in buildings where the duct system is three stories or less in height, the duct system shall comply with Section 510.0. [NFPA 96:14.4.1]

517.4.2 Building Exceeding Four Stories. A listed or approved grease duct system that is four stories in height or greater shall be provided for solid-fuel cooking exhaust systems. [NFPA 96:14.4.2]

517.4.3 Prohibited. Wall terminations of solid-fuel exhaust systems shall be prohibited. [NFPA 96:14.4.4]

517.5 Grease Removal Devices for Solid-Fuel Cooking. Grease removal devices shall be constructed of steel or stainless steel or be approved for solid-fuel cooking. [NFPA 96:14.5.1]

517.5.1 Spark Arrester Devices. If airborne sparks and embers can be generated by the solid fuel cooking operation, spark arrester devices shall be used prior to using the grease removal device, to minimize the entrance of these sparks and embers into the grease removal device and into the hood and the duct system. [NFPA 96:14.5.2]

517.5.2 Filters. Filters shall be a minimum of 4 feet (1219 mm) above the appliance cooking surface. [NFPA 96:14.5.3]

517.6 Air Movement for Solid-Fuel Cooking. Exhaust system requirements shall comply with Section 511.0 for hooded operation or shall be installed in accordance with the manufacturer's recommendations for unhooded applications. [NFPA 96:14.6.1]

517.6.1 Replacement Air. A replacement or makeup air system shall be provided to ensure a positive supply of replacement air at all times during cooking operations. [NFPA 96:14.6.2]

517.6.2 Operation. Makeup air systems serving solid-fuel cooking operations shall be interlocked with the exhaust air system and powered, if necessary, to prevent the space from attaining a negative pressure while the solid-fuel appliance is in operation. [NFPA 96:14.6.3]

517.7 Fire-Extinguishing Equipment for Solid-Fuel Cooking. Solid-fuel cooking appliances that produce grease-laden vapors shall be protected by listed fire-extinguishing equipment.

Exception: Where acceptable to the Authority Having Jurisdiction, solid-fuel cooking appliances constructed of solid masonry or reinforced Portland or refractory cement concrete and vented in accordance with NFPA 211 shall not require fixed automatic fire-extinguishing equipment. [NFPA 96:14.7.1, 14.7.2]

517.7.1 Grease Removal Devices, Hoods, and Duct Systems. Listed fire-extinguishing equipment shall be provided for the protection of grease removal devices, hoods, and duct systems. [NFPA 96:14.7.3]

Exception: Where acceptable to the Authority Having Jurisdiction, solid-fuel cooking appliances constructed of solid masonry or reinforced Portland or refractory cement concrete and vented in accordance with NFPA 211 shall not require automatic fire-extinguishing equipment for the protection of grease removal devices, hoods, and duct systems. [NFPA 96:14.7.4]

517.7.2 Listed Fire-Extinguishing Equipment. Listed fire-extinguishing equipment for solid-fuel-burning cooking appliances, where required, shall comply with Section 513.0 and shall use water-based agents. [NFPA 96:14.7.5]

517.7.3 Rating and Design. Fire-extinguishing equipment

shall be rated and designed to extinguish solid-fuel cooking fires. The fire-extinguishing equipment shall be of sufficient size to totally extinguish fire in the entire hazard area and prevent reignition of the fuel. [NFPA 96:14.7.6, 14.7.7]

517.7.4 Listing/Class. All solid fuel appliances (whether under a hood or not) with fireboxes of 5 cubic feet (0.14 m^3) volume or less shall have at least a listed 2-A rated water-spray fire extinguisher or a 1.6 gallon (6.1 L) wet chemical fire extinguisher listed for Class K fires in accordance with NFPA 10 with a maximum travel distance of 20 feet (6096 mm) to the appliance. [NFPA 96:14.7.8]

517.7.5 Fixed-Water Pipe System. Solid fuel appliances with fireboxes exceeding 5 cubic feet (0.14 m^3) shall be provided with a fixed-water pipe system with a hose in the kitchen capable of reaching the firebox. The hose shall be equipped with an adjustable nozzle capable of producing a fine to medium spray or mist. The nozzle shall be of the type that cannot produce a straight stream. The system shall have a minimum operating pressure of 40 psi (276 kPa) and shall provide a minimum of 5 gallons per minute (gpm) (0.3 L/s). [NFPA 96:14.7.9.1 – 14.7.9.2]

517.7.6 Fuel Storage. All fuel storage areas shall be provided with a sprinkler system meeting the requirements of NFPA 13 except as permitted in accordance with the following:

(1) Where acceptable to the Authority Having Jurisdiction, fuel storage areas shall be permitted to be protected with a fixed water pipe system with a hose capable of reaching all parts of the area.

(2) In lieu of the sprinkler system outlined in Section 517.7.6, a listed 2-A rated water spray fire extinguisher or a 1.6 gallon (6.1 L) wet chemical fire extinguisher listed for Class K fires with a maximum travel distance of 20 feet (6096 mm) to the solid fuel piles shall be permitted to be used for a solid fuel pile, provided that the fuel pile does not exceed 5 cubic feet (0.14 m^3) volume. [NFPA 96:14.9.2.8 – 14.9.2.8.2]

517.7.7 Auxiliary Fuel. In addition to the requirements of Section 517.7.4 through Section 517.8.3, where any solid-fuel cooking appliance is also provided with auxiliary electric, gas, oil, or other fuel for ignition or supplemental heat and the appliance is also served by any portion of a fire-extinguishing system complying with Section 513.0, such auxiliary fuel shall be shut off on actuation of the fire-extinguishing system. [NFPA 96:14.7.11]

517.8 Other Safety Requirements. Metal-fabricated solid-fuel cooking appliances shall be listed for the application where produced in practical quantities or shall be approved by the Authority Having Jurisdiction. Where listed, metal-fabricated solid fuel cooking appliances shall be installed in accordance with the terms of their listings and with the applicable requirements of this chapter. [NFPA 96:14.9.4.1, 14.9.4.2]

517.8.1 Site-Built Solid Fuel Cooling Appliances. Site-built solid-fuel cooking appliances shall be submitted for approval to the Authority Having Jurisdiction before being considered for installation. All units submitted to the Authority Having Jurisdiction shall be installed, operated, and maintained in accordance with the approved terms of the manufacturer's instructions and any additional requirements set forth by the Authority Having Jurisdiction. [NFPA 96:14.9.4.3.1 – 14.9.4.3.2]

517.8.2 Additional Devices. Except for the spark arresters required in Section 517.1.6, there shall be no additional devices of any type in any portion of the appliance, flue pipe, and chimney of a natural draft solid-fuel operation. [NFPA 96:14.9.4.4]

517.8.3 Prohibited. No solid fuel cooking device of any type shall be permitted for deep fat frying involving more than 1 quart (qt) (1 L) of liquid shortening, nor shall any solid fuel cooking device be permitted within 3 feet (914 mm) of any deep fat frying unit. [NFPA 96:14.9.4.5]

518.0 Downdraft Appliances.

Section 518.0 addresses such appliances as hoodless griddle type cooking appliances. Typically, the portion of food served is small and prepared in front of the customers directly at their table. These types of tables have built-in downdraft exhaust systems with two fans, one to push the smoke and one to draw air across the table. The cooking vapors are then captured and conveyed to a grease duct that is attached to the bottom of the table.

518.1 General. Downdraft appliance ventilation system containing or for use with appliances used in processes producing smoke or grease-laden vapors shall be equipped with components complying with the following:

(1) The clearance requirements of Section 507.4 through Section 507.4.3.3.

(2) The primary collection means designed for collecting cooking vapors and residues complying with the requirements of Section 508.0.

(3) Grease removal devices complying with Section 509.0.

(4) Special-purpose filters as listed in accordance with UL 1046.

(5) Exhaust ducts complying with Section 510.0.

(6) The air movement requirements of Section 511.2.1 and Section 511.2.2.

(7) Auxiliary equipment (such as particulate and odor removal devices) complying with Section 512.0.

(8) Fire-extinguishing equipment complying with the requirements of Section 513.0, and as specified in Section 518.3.

(9) The use and maintenance requirements of Section 514.0.

(10) The minimum safety requirements of Section 515.0. [NFPA 96:15.1.1]

518.2 Ventilation System. The downdraft appliance ventilation system shall be capable of capturing and containing all the effluent discharge from the appliance(s) it is serving. [NFPA 96:15.1.2]

518.3 Fire-Extinguishing Equipment. For fire-extinguishing equipment on downdraft appliance ventilation systems, the following shall apply:

(1) Cooking surface, duct, and plenum protection shall be provided.

(2) At least one fusible link or heat detector shall be installed within each exhaust duct opening in accordance with the manufacturer's listing.

(3) A fusible link or heat detector shall be provided for each protected cooking appliance located in the plenum area of that appliance or in accordance with the extinguishing system manufacturer's listing.

(4) A manual activation device shall be provided as part of each appliance at a height acceptable to the Authority Having Jurisdiction.

(5) Portable fire extinguishers shall be provided in accordance with Section 513.10 through Section 513.11. [NFPA 96:15.2]

518.3.1 Integral Fire-Extinguishing System. A listed downdraft appliance ventilation system employing an integral fire-extinguishing system including detection systems that has been evaluated for grease and smoke capture, fire extinguishing, and detection shall be considered as complying with Section 518.3. [NFPA 96:15.2.1]

518.3.2 Interlocks. The downdraft appliance ventilation system shall be provided with interlocks such that the cooking fuel supply will not be activated unless the exhaust and supply air systems have been activated. [NFPA 96:15.2.2]

518.4 Airflow Switch or Transducer. An airflow switch or transducer shall be provided after the last filter component to ensure that a minimum airflow is maintained. [NFPA 96:15.3.1]

518.4.1 Interlocks. The airflow switch or transducer shall open the interlock circuit when the airflow falls 25 percent below the system's normal operating flow or less than 10 percent its listed minimum rating, whichever is lower. [NFPA 96:15.3.2]

518.4.2 Manual Reset. The airflow switch or transducer shall be a manual reset device or circuit. [NFPA 96:15.3.3]

518.5 Surface Materials. Any surface located directly above the cooking appliance shall be of noncombustible or limited-combustible materials. [NFPA 96:15.4]

519.0 Type II Hood Exhaust System Requirements.

519.1 Where Required. Type II hoods shall be installed above equipment and dishwashers that generate steam, heat, or products of combustion, and where grease or smoke is not present.

Exceptions:

(1) Dishwashing machines connected to a Type II duct system and exhausted directly to the outdoors.

(2) Dishwashing machines with a self-contained condensing system listed in accordance with UL 921 and installed in a space where the HVAC system has been engineered to accommodate the latent and sensible heat load emitted from such appliances as approved by the Authority Having Jurisdiction. Such equipment shall be provided with an interlocking device to prevent opening of the appliance prior to completion of its cycle.

Type II hoods are required over other cooking equipment producing steam, heat, vapors or odors, such as bakery ovens, steam tables, coffee urns, dishwashers, etc.

Unless meeting Exceptions 1 or 2, this section also requires that Type II hoods be installed over commercial dishwashing machines, regardless of rinsing temperature. The amount of water vapor released by a tray of drying dishes is exactly the same whether they have been processed in a high- or low-temperature dishwasher. The total amount of moisture released into the air will be the same for the same wetted area of the dishes. When the dishes are hot, the water vapor is not so readily seen, but it is nonetheless being released into the air.

519.2 Construction of Type II Hoods. Type II hoods constructed of steel shall be not less than 0.024 of an inch (0.61 mm) (No. 24 gauge). Hoods constructed of copper shall be of copper sheets weighing not less than 0.17 ounces per square inch (oz/in^2) (7.47 kg/m^2). Joints and seams shall be substantially tight. Solder shall not be used except for sealing a joint or seam.

519.3 Dishwashing Appliances. The net airflow for Type II hoods used for dishwashing equipment shall be not less than 200 cubic feet per minute (0.094 m^3/s) per linear foot (m) of hood length.

519.4 Type II Exhaust Duct Systems. Ducts and plenums serving Type II hoods shall be constructed of rigid metallic materials in accordance with Chapter 6. Duct bracing and supports shall comply with Chapter 6. Ducts subject to positive pressure shall be adequately sealed.

Ducting utilized with moisture-related processes such as a dishwasher shall have overlap of traverse joints configured to allow moisture to drain back to the hood.

519.5 Termination of Type II Hood Exhaust System. The exhaust system shall terminate as follows:

(1) Rooftop terminations shall terminate not less than 10 feet (3048 mm) from a property line, and the exhaust flow shall be directed away from the roof surface of the roof, not less than 40 inches (1016 mm).

(2) Horizontal terminations shall terminate not less than 10 feet (3048 mm) from adjacent buildings, property lines, operable openings, and from grade level.

(3) The discharge outlet shall not be directed onto a public walkway.

519.6 Makeup Air. Makeup air shall be provided in accordance with Section 511.3.

CHAPTER 6

DUCT SYSTEMS

601.0 General.

601.1 Applicability. Ducts and plenums that are portions of a heating, cooling, ventilation, or exhaust system shall comply with the requirements of this chapter and Chapter 5.

Chapter 6, along with Chapter 5, regulates the installation of duct systems (as defined in Section 206.0) that are used to convey indoor environmental air that has been heated or cooled for purposes of occupant comfort, or for air exhausted from a building.

601.2 Sizing Requirements. Duct systems shall be sized in accordance with applicable standards in Chapter 17 or by other approved methods.

Exception: Residential duct systems shall be sized in accordance with ACCA Manual D, ACCA Manual Zr, as applicable, or by other approved methods.

602.0 Material.

602.1 General. Materials used for duct systems shall comply with Section 602.2 through Section 602.6 as applicable.

Concealed building spaces or independent construction within buildings shall be permitted to be used as ducts or plenums. Gypsum board shall not be used for positive pressure ducts.

Exception: In healthcare facilities, concealed spaces shall not be permitted to be used as ducts or plenums.

SMACNA reflects the current industry practices in duct construction. The Sheet Metal and Air Conditioning Contractors National Association (SMACNA) has developed the Duct Construction Standards for supply air, return air, and outside air for heating, cooling, or evaporative cooling systems. Shop drawings should be submitted when using the standard HVAC duct construction because of the wide variety of options available.

Except in healthcare facilities, concealed building spaces may be used as negative pressure ducts or plenums, which include spaces within fire-rated assemblies, provided the penetrations into these spaces are protected by approved membrane-penetration firestops or fire dampers as required by the building code. The gypsum product must be listed as mold or mildew resistant. **Figure 602.1** illustrates a panned joist space that could be used for return air in a dwelling.

The required flame spread and smoke-developed rating stipulated in Section 602.2 would preclude this construction in uses other than dwellings (see Exception 1).

602.2 Combustibles Within Ducts or Plenums. Materials exposed within ducts or plenums shall be noncombustible or shall have a flame spread index not to exceed 25 and a smoke-developed index not to exceed 50, where tested as a composite product in accordance with ASTM E84 or UL 723. Plastic piping installed in plenums shall be tested in accordance with all requirements of ASTM E84 or UL 723. Mounting methods, supports and sample sizes of materials for testing that are not specified in ASTM E84 or UL 723 shall be prohibited.

Exceptions:

(1) Return-air and outside-air ducts, plenums, or concealed spaces that serve a dwelling unit.

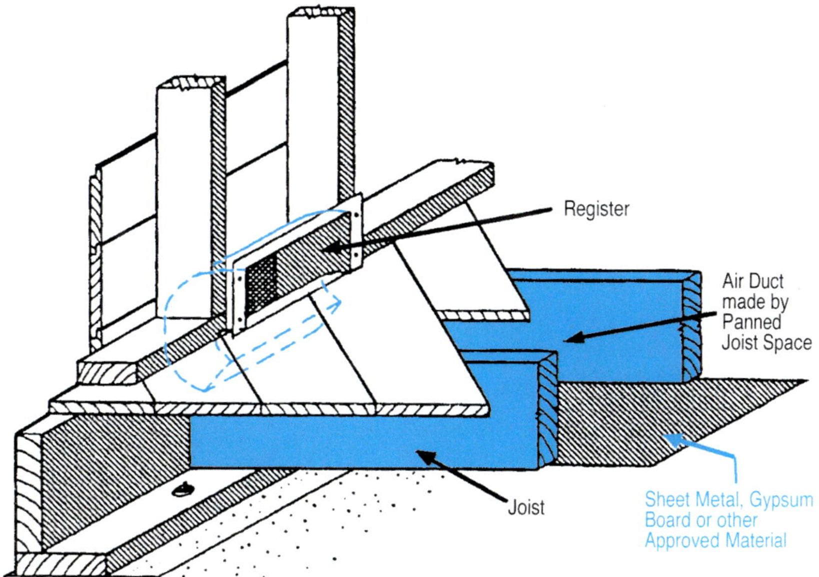

FIGURE 602.1
DETAIL OF PANNED JOIST SPACE (RESIDENTIAL)

(2) Air filters in accordance with the requirements of Section 311.2.

(3) Water evaporation media in an evaporative cooler.

(4) Charcoal filters where protected with an approved fire suppression system.

(5) Products listed and labeled for installation within plenums in accordance with Section 602.2.1 through Section 602.2.3.

(6) Smoke detectors.

(7) Duct insulation, coverings, and linings and other supplementary materials installed in accordance with Section 605.0.

(8) Materials in a hazardous fabrication area including the areas above and below the fabrication area sharing a common air recirculation path with the fabrication area.

This section requires that materials exposed to the airstream in ducts and plenums exhibit specified surface-burning characteristics, except in certain locations and products as indicated. The purpose of the requirement is to limit propagation of flames through the duct system or plenum and to limit the quantity of smoke resulting from involvement of insulating and acoustic materials. The method of testing to determine surface-burning characteristics of materials is commonly referred to as the "Steiner Tunnel Test," as described in ASTM E 84 and (UL 723. The Underwriters Laboratories UL) label indicates only that the material would be suitable based on its flame spread index and smoke developed value. The listing cautions that thermal conductivity, vapor resistance and other properties of the material have not been investigated. **Figure 602.2** shows a typical label that is required to be attached to tested materials that comply with the flame spread index and smoke-developed rating required in this section.

If ASTM E84 or UL 723 standard test methods are modified and used as a basis to determine flame and smoke ratings, then the tests don't comply with the standard and the materials may not comply with the required flame and smoke ratings. An inspector or installer seeing ASTM E84 on a pipe would likely assume that it meets the requirement of the code. Simply adding the ASTM E 84 or UL 723 marking to the wall of the pipe does not necessarily mean that the product was tested in full compliance with the standard in the manner that the code intends. Especially for PVC and CPVC pipe, the inspector or installer should examine the material listing and see if the testing agency has modified the testing method and rendered the pipe material noncompliant.

602.2.1 Electrical. Electrical wiring in plenums shall comply with NFPA 70. Electrical wires and cables and optical fiber cables shall be listed and labeled for use in plenums and shall have a flame spread distance not exceeding 5 feet (1524 mm), an average optical density not exceeding 0.15, and a peak optical density not exceeding 0.5, where tested in accordance with NFPA 262.

602.2.2 Fire Sprinkler Piping. Nonmetallic fire sprinkler piping in plenums shall be listed and labeled for use in plenums and shall have a flame spread distance not exceeding 5 feet (1524 mm), an average optical density not exceeding 0.15 and, a peak optical density not exceeding 0.5, where tested in accordance with UL 1887.

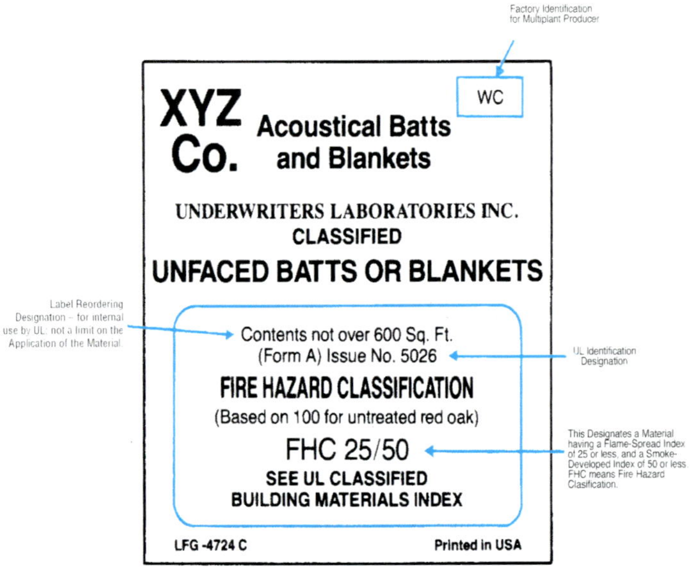

FIGURE 602.2
TYPICAL LABEL FOR COMBUSTABLE MATERIALS WITHIN DUCTS AND PLENUMS

602.2.3 Pneumatic Tubing. Nonmetallic pneumatic tubing in plenums shall be listed and labeled for use in plenums and shall have a flame spread distance not exceeding 5 feet (1524 mm), an average optical density not exceeding 0.15, and a peak optical density not exceeding 0.5, where tested in accordance with UL 1820.

602.2.4 Discrete Products in Plenums. Discrete plumbing, mechanical, and electrical products that are located in a plenum and have exposed combustible material shall be listed and labeled in accordance with UL 2043.

602.3 Metallic. Ducts, plenums, or fittings of metal shall comply with SMACNA HVAC Duct Construction Standards – Metal and Flexible. Flexible metallic ducts shall comply with UL 181.

602.4 Nonmetallic Ducts. Nonmetallic ducts shall comply with Section 602.4.1, Section 602.4.2, Section 602.4.3 or Section 602.4.4.

602.4.1 Phenolic. Phenolic duct, plenum, or fitting material shall comply with UL 181. Ducts, plenums, or fittings of phenolic shall be constructed in accordance with SMACNA Phenolic Duct Construction Standards or the conditions of its listing.

Phenolic duct is a Class 1 Air Duct listed to UL 181 with SMACNA leakage class of 3 and a thermal conductivity and R-value tested to ASTM C 518. Flexible isolation connectors on a phenolic system cannot be any longer than ten inches.

602.4.2 Gypsum. Where gypsum products are exposed in ducts or plenums, the air temperature shall be restricted to a range from 50°F (10°C) to 125°F (52°C), and moisture content shall be controlled so that the material is not adversely affected. All gypsum products shall have a mold or mildew resistant surface. For the purpose of this section, gypsum products shall not be exposed in supply ducts.

602.4.3 Air Dispersion Systems. Air dispersion systems shall be listed and labeled in accordance with UL 2518.

602.4.4 Other Materials. Flexible and rigid ducts, plenums, or fittings for use in heating, ventilation, and air conditioning systems of other nonmetallic materials listed and labeled to UL 181 shall be permitted.

Exception: Plastic ducts shall comply with Section 603.5.

602.5 Vibration Isolators. Vibration isolation connectors installed between mechanical equipment and metal ducts (or casings) shall be made of an approved material and shall not exceed 10 inches (254 mm) in length.

Vibration isolators (flexible connections) minimize noise and vibration from mechanical equipment. Without them, excessive operating noise and premature weakening of joints and supports could result. Vibration isolators shall be made of approved materials. Neoprene and rubber are the common materials for this purpose, and good installation practices would provide adequate supports on both sides of these isolators. **Figure 602.5** illustrates typical isolator details and points of installation within a system.

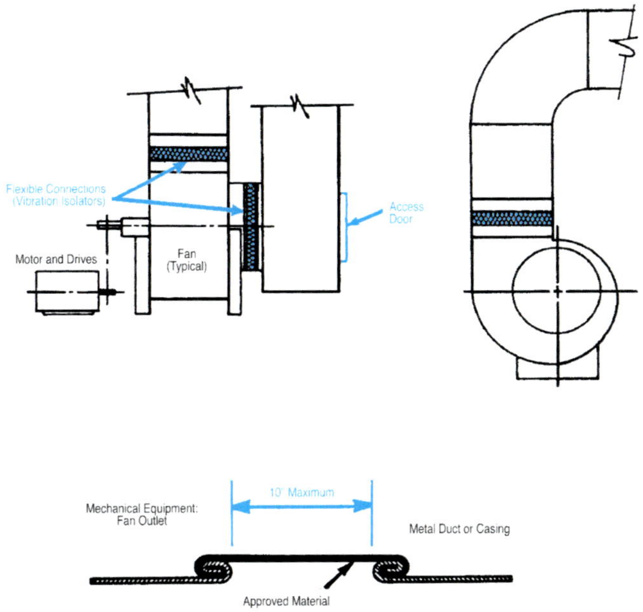

FIGURE 602.5
VIBRATION ISOLATOR DETAILS

602.6 Corridors. Corridors shall not be used to convey air to or from rooms where the corridor is required to be of fire-resistive construction in accordance with the building code except where permitted by the building code.

Corridors may not be used as a part of an air distribution system when required by the building code of the jurisdiction to be of fire-resistive construction. This requirement is to assist in maintaining a smoke-free environment in the exit passageway in the case of fire.

603.0 Installation of Ducts.

603.1 General. Air ducts shall be installed in accordance with this chapter and the installation instructions.

The installation of duct, including flexible or factory-made duct, is regulated by this section. Approved factory-made air ducts may be installed in any occupancy, but shall not be used as vertical risers in air duct systems serving more than two stories. Factory-made air ducts shall not penetrate fire resistant rated construction. Note that, except for plastic duct, a 4-inch minimum separation from earth is necessary unless otherwise protected.

Factory-made air ducts must be approved by the jurisdiction for the intended use and must comply with the requirements of the referenced standard for factory-made air ducts in UL 181. The broad range of products covered under this standard are often of proprietary design. They typically come in flexible or rigid lengths, or in sheets or boards for field assembly into a rigid duct. Their composition is limited to metal or mineral materials. Nonmetallic or organic materials may only be used as "binders, adhesives, sealants, or finishes."

The manufacturer must mark every portion of a factory-made air duct system to show compliance with the standard. The contractor must install the system in accordance with the terms of its listing and the installation instructions shipped with the duct material. The SMACNA-HVAC Duct Construction Standard, which includes directions for the installation of factory-made air ducts, should be consulted.

One of the most critical areas of inspection of factory-made air ducts is to determine that proper methods and materials have been used for joining and splicing. The manufacturer's instructions should include a complete and detailed description of the methods to be used for joining and splicing the ducts. The materials used for this purpose must comply with the requirements of the referenced standard for closure systems found in UL 181A and UL 1818.

Not all duct tapes or sheet metals are approved materials for joining and splicing ducts. The materials and methods must conform to their listing, the manufacturer's installation instructions and approved standards.

Figure 603.1 illustrates two examples of UL labels for factory-made air ducts.

603.1.1 Pressure Classification. The pressure classification of ducts shall be not less than the design operating pressure of the air distribution in which the duct is utilized.

603.1.2 Air Temperature. The temperature of the air to be conveyed in a duct shall not exceed 250°F (121°C).

603.1.3 Protection. Air ducts, other than plastic ducts, shall be installed with not less than 4 inches (102 mm) of separation from earth, except where installed as a liner inside of concrete, tile, or metal pipe and shall be protected from physical damage.

603.1.4 Vertical Risers. Ducts listed and labeled to UL 181 shall not be used for vertical risers in air-duct systems serving more than two stories.

603.1.5 Penetrations. Ducts listed and labeled to UL 181 shall not penetrate a fire-resistance-rated assembly or construction.

FIGURE 603.1
TYPICAL LABELS FOR FACTORY-MADE AIR DUCTS

603.2 Under Floor or Crawl Space. Air ducts installed under a floor in a crawl space shall be installed in accordance with the following:

(1) Shall not prevent access to an area of the crawl space.

(2) Where it is required to move under ducts for access to areas of the crawl space, a vertical clearance of not less than 18 inches (457 mm) shall be provided.

603.3 Metal Ducts. Ducts shall be supported at each change of direction and in accordance with SMACNA HVAC Duct Construction Standards – Metal and Flexible. Riser ducts shall be held in place by means of metal straps or angles and channels to secure the riser to the structure.

Ducts shall be installed in a building with clearances that will retain the full thickness of fireproofing on structural members.

This section addresses the support of metal ducts. Vertical and horizontal ducts are covered as well as horizontal ducts with trapeze supports in both rectangular and round shapes. Horizontal sheet metal ducts are required to be braced and guyed to prevent lateral or horizontal swing, which could affect their ability to conduct air efficiently.

603.3.1 Earthquake Loads. Ducts located in structures that are installed in areas classified as seismic design category C, D, E, or F shall be in accordance with the building code.

603.4 Flexible Air Ducts. Flexible air ducts shall comply with UL 181, and shall be installed in accordance with the manufacturer's installation instructions and SMACNA HVAC Duct Construction Standards – Metal and Flexible.

Flexible air duct installations shall comply with the following:

(1) Ducts shall be installed using the minimum required length to make the connection.

(2) Horizontal duct runs shall be supported at not more than 4 feet (1219 mm) intervals.

(3) Vertical risers shall be supported at not more than 6 feet (1829 mm) intervals.

(4) Sag between support hangers shall not exceed ½ inch (12.7 mm) per foot (305 mm) of support spacing.

(5) Supports shall be rigid and shall be not less than 1½ inches (38 mm) wide at point of contact with the duct surface.

(6) Duct bends shall be not less than one duct diameter bend radius.

(7) Screws shall not penetrate the inner liner of non-metallic flexible ducts unless permitted in accordance with the manufacturer's installation instructions.

(8) Fittings for attaching non-metallic ducts shall be beaded and have a collar length of not less than 2 inches (51 mm) for attaching the duct. Metal worm-gear clamps shall be used.

Exception: Where attaching metallic ducts using screws in accordance with the manufacturer's installation instructions.

(9) Duct inner liner shall be installed at not less than 1 inch (25.4 mm) on the collar and past the bead prior to the application of the tape and mechanical fastener. Where mastic is used instead of tape, the mastic shall be applied in accordance the mastic manufacturer's instructions.

(10) Duct outer vapor barriers shall be secured using two wraps of approved tape. A mechanical fastener shall be permitted to be used in place of, or in combination with, the tape.

(11) Flexible air ducts shall not penetrate a fire-resistance-rated assembly or construction.

(12) The temperature of the air to be conveyed in a flexible air duct shall not exceed 250°F (121°C).

(13) Flexible Air ducts shall be sealed in accordance with Section 603.9.

The user of the UMC is to be aware that SMACNA is an installation standard and UL 181 is a material standard. Both standards would need to be referenced to ensure compliance with the UMC. Many HVAC installations utilize flexible air ducts and connectors, and the UMC provides specific requirements for flexible air ducts and connectors that can be used by designers for factory-made air ducts. Flexible air ducts are considered factory-made air ducts. Unlike flexible air ducts and connectors, not all factory-made air ducts are required to comply with UL 181. UL 181 addresses requirements which include preformed lengths of flexible or rigid ducts, materials in the form of boards for field fabrication of lengths of rigid ducts, and preformed flexible air connectors. The user is to refer to the manufacturer's installation instructions and SMACNA for the installation of flexible air ducts since both sources will provide installation instructions for flexible air ducts and connectors. The temperature limitation is the industry standard for warm air. The material in flexible air ducts and air connectors are only tested to resist a temperature up to 250°F (121°C).

603.4.1 Length Limitation. Flexible air ducts shall be not more than 5 feet (1524 mm) in length and shall not be used in lieu of rigid elbows or fittings. Flexible air ducts shall be permitted to be used as an elbow at a terminal device.

Exception: Residential occupancies.

603.4.2 Flexible Air Connectors. Flexible air connectors shall not be permitted.

603.5 Plastic Ducts. Plastic air ducts and fittings shall be permitted where installed underground and listed for such use.

603.6 Protection of Ducts. Ducts installed in locations where they are exposed to mechanical damage by vehicles or from other causes shall be protected by approved barriers.

603.7 Support of Ducts. Ducts shall be supported in accordance with the manufacturer's installation instructions; and Section 603.7.1, Section 603.7.2 or Section 603.7.3.

This section provides requirements for support of flexible air ducts (see **Figure 603.7**). The manufactures installation instructions shall be provided at the time of inspection. For all other duct systems the SMACNA HVAC Duct Construction Standards - Metal and Flexible would be the guide as a referenced standard.

603.7.1 Metal Ducts. Ducts shall be supported at each change of direction and in accordance with SMACNA HVAC Duct Construction Standards – Metal and Flexible. Riser ducts shall be held in place by means of metal straps or angles and channels to secure the riser to the structure.

603.7.1.1 Rectangular Ducts. Supports for rectangular ducts shall be installed on two opposite sides of each duct and shall be riveted, bolted, or metal screwed to each side of the duct at intervals specified.

603.7.1.2 Horizontal Round Ducts. Horizontal round ducts not more than 40 inches (1016 mm) in diameter where suspended from above shall be supported with one hanger per interval and in accordance with Section 603.7.1.3 through Section 603.7.1.5.

603.7.1.3 Tight-Fitting Around the Perimeter. Ducts shall be equipped with tight-fitting circular bands extending around the entire perimeter of the duct at each specified support interval.

603.7.1.4 Size of Circular Bands. Circular bands shall be not less than 1 inch (25.4 mm) wide nor less than equivalent to the gauge of the duct material it supports.

Exception: Ducts not more than 10 inches (254 mm) in diameter shall be permitted to be supported by No. 18 gauge galvanized steel wire.

603.7.1.5 Connection. Each circular band shall be provided with means of connecting to the suspending support.

603.7.2 Flexible Air Ducts. Flexible air ducts shall be supported in accordance with Section 603.4.

603.7.3 Other Ducts. Other approved ducts shall be supported in accordance with the manufacturer's installation instructions.

603.8 Protection Against Flood Damage. In flood hazard areas, ducts shall be located above the elevation required by the building code for utilities and attendant equipment or the elevation of the lowest floor, whichever is higher, or shall be designed and constructed to prevent water from entering or accumulating within the ducts during floods up to such elevation. Where the ducts are located below that elevation, the ducts shall be capable of resisting hydrostatic and hydrodynamic loads and stresses, including the effects of buoyancy, during the occurrence of flooding to such elevation.

This section provides consistency with the regulations of the National Flood Insurance Program (NFIP) and the flood-resistant provisions of ASCE 24. In particular, the NFIP regulations require that service systems and equipment be elevated above the design flood elevation, but also provides an option to elevating. The alternative is set forth in regulation at 44 CFR §60.3(a)(3)(iii) and (iv), and requires that buildings and structures be constructed with methods and practices that minimize flood damages and, specifically, that electrical, heating, ventilation, plumbing and air-conditioning equipment and other service facilities be designed and/or located so as to prevent water from entering or accumulating within the components during conditions of flooding.

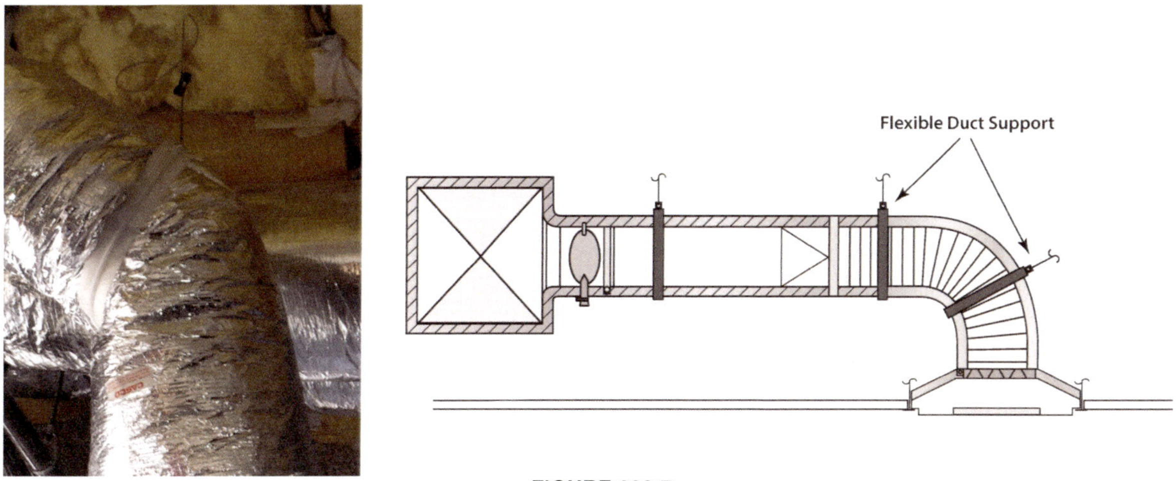

FIGURE 603.7
FLEXIBLE DUCT SUPPORT

603.9 Joints and Seams of Ducts. Joints and seams for duct systems shall comply with SMACNA HVAC Duct Construction Standards – Metal and Flexible. Joints of duct systems shall be made substantially airtight by means of tapes, mastics, gasketing, or other means. Crimp joints for round ducts shall have a contact lap of not less than 1½ inches (38 mm) and shall be mechanically fastened by means of not less than three sheet-metal screws equally spaced around the joint, or an equivalent fastening method.

Air tightness is critical in duct systems to maintain overall energy and operating efficiency; therefore, the installer must make all joints "substantially airtight" with tapes, mastics, gasketing or other means. Note that allowable air loss is not defined, so this becomes a judgment call regarding sound practice.

Figure 603.9a illustrates acceptable methods for connecting residential round ducts.

Figure 603.9b illustrates several other types of joints and seams.

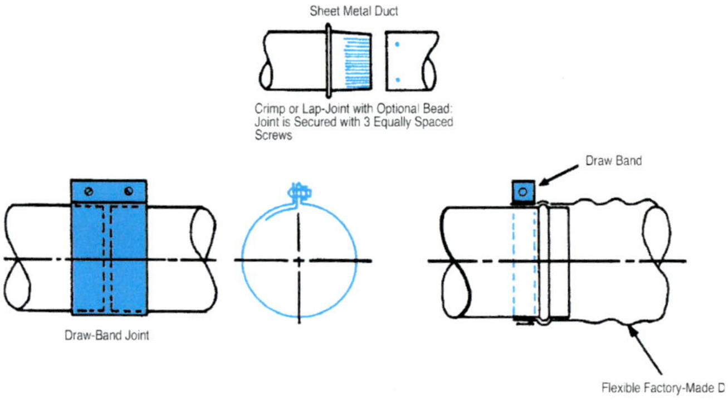

FIGURE 603.9A
RESIDENTIAL ROUND DUCT CONNECTIONS

603.9.1 Closure Systems. Joints and seams and reinforcements for air ducts and plenums listed and labeled to UL 181, shall be in accordance with the manufacturer's installation instructions. Closure systems for sealing air ducts and plenums shall be listed and labeled in accordance with UL 181A or UL 181B, and marked in accordance with Table 603.9.1.

TABLE 603.9.1
CLOSURE MARKINGS

TYPE OF DUCTWORK	STANDARD	TYPE OF CLOSURE SYSTEM	MARKING
Rigid Metallic or Rigid Fiberglass	UL 181A	Pressure Sensitive Tape	181A-P
Rigid Metallic or Rigid Fiberglass	UL 181A	Mastic Tape	181A-M
Rigid Metallic or Rigid Fiberglass	UL 181A	Heat Sensitive Tape	181A-H
Flexible Air Ducts	UL 181B	Pressure Sensitive Tape*	181B-FX*
Flexible Air Ducts	UL 181B	Mastic*	181B-M*

* Mechanical fasteners shall be used in conjunction with a listed pressure sensitive tape or mastic in accordance with UL 181. Nonmetallic mechanical fasteners shall be listed and labeled in accordance with UL 181B and labeled "181B-C."

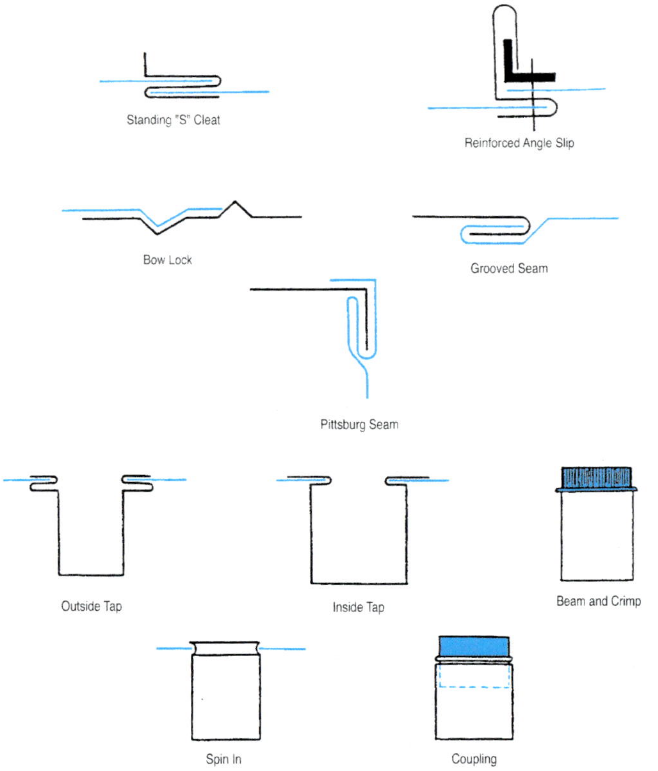

FIGURE 603.9B
JOINTS AND CONNECTIONS

603.9.2 Duct Leakage Tests. Ductwork shall be leak-tested in accordance with the SMACNA HVAC Air Duct Leakage Test Manual. Representative sections totaling not less than 10 percent of the total installed duct area shall be tested. Where the tested 10 percent fail to comply with the requirements of this section, then 40 percent of the total installed duct area shall be tested. Where the tested 40 percent fail to comply with the requirements of this section, then 100 percent of the total installed duct area shall be tested. Sections shall be selected by the building owner or designated representative of the building owner. Positive pressure leakage testing shall be permitted for negative pressure ductwork. The permitted duct leakage shall be not more than the following:

$$L_{max} = C_L P^{0.65} \quad \text{(Equation 603.9.2)}$$

Where:

L_{max} = maximum permitted leakage, $(ft^3/min)/100$ square feet $[0.0001\ (m^3/s)/m^2]$ duct surface area.

C_L = six, duct leakage class, $(ft^3/min)/100$ square feet $[0.0001\ (m^3/s)/m^2]$ duct surface area at 1 inch water column (0.2 kPa).

P = test pressure, which shall be equal to the design duct pressure class rating, inch water column (kPa).

Exception: Transfer air duct operating at less than 1 inch of water column (0.25 kPa).

The SMACNA HVAC Air Duct Leakage Test manual includes procedures for testing air leakage at ducts along with other equipment and accessories such as variable air volume boxes, fire dampers, control/ volume dampers, access doors, etc.

603.10 Cross Contamination. Exhaust ducts that convey Class 4 air shall be negatively pressurized relative to ducts, plenums, or occupiable spaces through which the ducts pass. Exhaust ducts under positive pressure that convey Class 2 or Class 3 air shall not extend into or pass through ducts, plenums, or occupiable spaces other than the space from which the exhaust air is drawn.

Exception: Exhaust ducts conveying Class 2 air and exhaust ducts conveying air from residential kitchen hoods that are sealed in accordance with Seal Class A of the SMACNA HVAC Air Duct Leakage Test Manual. {ASHRAE 62.1:5.2.1, 5.2.2}

603.11 Underground Installation. Ducts installed underground shall be approved for the installation and shall have a slope of not less than ⅛ inch per foot (10.4 mm/m) back to the main riser. Ducts, plenums, and fittings shall be permitted to be constructed of concrete, clay, or ceramics where installed in the ground or in a concrete slab, provided the joints are sealed and duct is secured in accordance with SMACNA HVAC Duct Construction Standards – Metal and Flexible. Metal ducts where installed in or under a concrete slab shall be encased in not less than 2 inches (51 mm) of concrete, secured in accordance with SMACNA HVAC Duct Construction Standards – Metal and Flexible.

Ducts are required to "be made substantially airtight" in Section 603.9; however, this requirement is primarily intended to prevent moisture from the ground or moisture and concrete from entering the duct system if embedded in concrete. **Figure 603.11** illustrates a typical detail for embedded ducts. Underground ducts must be sloped to drain to an accessible point in the event that water enters the duct through the openings. The slope requirement is consistent with SMACNA HVAC Duct Construction Standards.

603.12 Air Dispersion Systems. Where installed, air dispersion systems shall be completely in exposed locations in duct systems under positive pressure, and not pass through or penetrate fire-resistant-rated construction. Air dispersion systems shall be listed and labeled in accordance with UL 2518.

603.13 Clearances. Duct clearances shall be in accordance with Section 603.13.1 through Section 603.13.6.

603.13.1 General. Duct clearances to combustible construction shall be in accordance with the listing and the manufacturer's installation instructions.

603.13.2 Supply Air Ducts to Listed Furnaces. Supply air ducts connecting to listed central heating furnaces shall have the same minimum clearance to combustibles as required for the furnace supply plenum for a distance of not less than 3 feet (914 mm) from the supply plenum. Clearance shall not be required beyond the 3 feet (914 mm) distance. [NFPA 54:10.3.2.7]

603.13.3 Supply Air Ducts to Unlisted Furnaces. Supply air ducts connecting to unlisted central heating furnaces equipped with temperature limit controls with a maximum setting of 250°F (121°C) shall have a minimum clearance to combustibles of 6 inches (152 mm) for a distance of not less than 6 feet (1829 mm) from the furnace supply plenum. Clearance shall not be required beyond the 6 feet (1829 mm) distance. [NFPA 54:10.3.2.8]

603.13.4 Furnace Plenums and Air Ducts. A furnace plenum supplied as a part of the air-conditioning appliance shall be installed in accordance with the manufacturer's instructions. Where a furnace plenum is not supplied with the appliance, any fabrication and installation instructions provided by the manufacturer shall be followed. The method of connecting supply and return ducts shall facilitate proper circulation of air. Where the air conditioner is installed within an enclosure, the installation shall comply with Section 604.4. [NFPA 54:10.2.5]

603.13.5 Duct Furnaces. Duct furnaces shall be installed in accordance with Section 905.0.

603.13.6 Fuel-Burning Kilns. A hood and duct serving a fuel-burning kiln shall have a clearance from combustible construction of not less than 18 inches (457 mm). This clear-

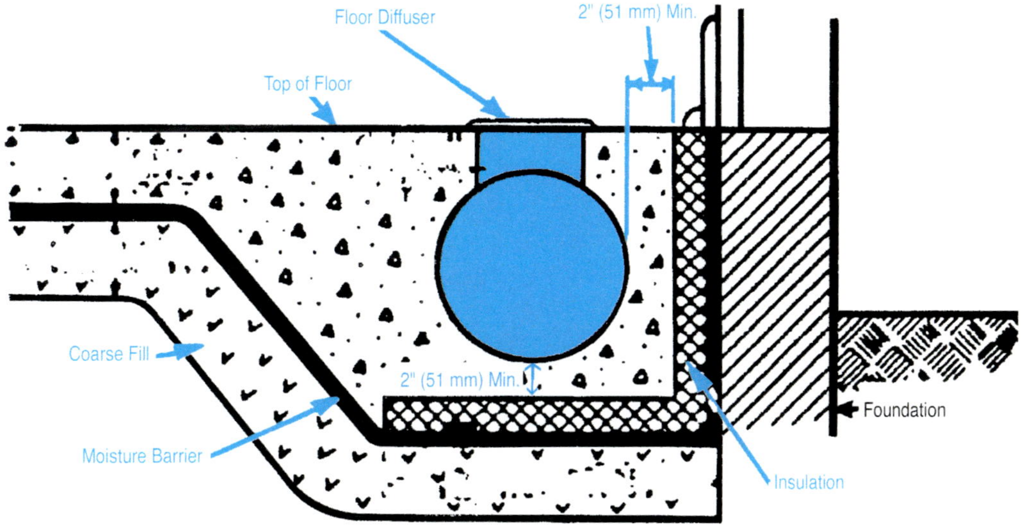

FIGURE 603.11
DETAIL OF EMBEDDED DUCT

ance shall be permitted to be reduced in accordance with Table 303.10.1.

604.0 Furnace Plenums and Ducts Used in Fuel-Gas Appliances.

604.1 Furnace Plenums and Air Ducts. Furnace plenums and air ducts shall be installed in accordance with this Chapter, and NFPA 90A or NFPA 90B. [NFPA 54:10.3.7.1]

604.2 Supplied as a Part of Furnace. A furnace plenum supplied as a part of a furnace shall be installed in accordance with the manufacturer's instructions. [NFPA 54:10.3.7.2]

604.3 Not Supplied with the Furnace. Where a furnace plenum is not supplied with the furnace, any fabrication and installation instructions provided by the manufacturer shall be followed. The method of connecting supply and return ducts shall facilitate proper circulation of air. [NFPA 54:10.3.7.3]

604.4 Return Air. Where a furnace is installed so supply ducts carry air circulated by the furnace to areas outside the space containing the furnace, the return air shall also be handled by a duct(s) sealed to the furnace casing and terminating outside the space containing the furnace. [NFPA 54:10.3.7.4]

605.0 Insulation of Ducts.

605.1 General. Air ducts conveying air at temperatures exceeding 140°F (60°C) shall be insulated to maintain an insulation surface temperature of not more than 140°F (60°C). Insulation material on the exterior of ducts shall be legibly printed with the name of the manufacturer, the thermal resistance (R) value at installed thickness, flame-spread index, and smoke developed index of the composite material. Internal duct liners and insulation shall be installed in accordance with SMACNA HVAC Duct Construction Standards – Metal and Flexible.

Exceptions:

(1) Factory-installed plenums, casings, or ductwork furnished as a part of HVAC equipment tested and rated in accordance with approved energy efficiency standards.

(2) Ducts or plenums located in conditioned spaces where heat gain or heat loss will not increase energy use.

(3) For runouts less than 10 feet (3048 mm) in length to air terminals or air outlets, the rated R-value of insulation need not exceed R-3.5.

(4) Backs of air outlets and outlet plenums exposed to unconditioned or indirectly conditioned spaces with face areas exceeding 5 square feet (0.5 m^2) need not exceed R-2; those 5 square feet (0.5 m^2) or smaller need not be insulated.

(5) Ducts and plenums used exclusively for evaporative cooling systems.

Section 605.0 describes the specific required installation and performance characteristics of duct-insulating materials. SMACNA HVAC Duct Construction Standards – Metal and Flexible provide insulation requirements by duct location and heating zone (obtainable from the local United States Weather Service office) for both cooling-only and heating-only modes. Three insulation types are described: mineral fiber blankets, mineral fiber blanket duct liners and mineral fiber boards. Vapor barrier and weatherproof barrier requirements are also provided.

This section requires that duct-insulating materials meet the following test requirements:

1. Insulating materials installed inside ducts "shall have a mold-, humidity- and erosion-resistant surface that meets the requirements of the referenced standard UL 181."
2. Insulating materials installed on the exterior of ducts (inside buildings) are required to have specific flame spread and smoke-developed ratings. In addition, "faced" insulation "shall be legibly printed with the name of the manufacturer, the thermal resistance (R) value at installed thickness and the flame-spread and smoke-developed ratings of the composite material."

In most cases, these provisions will require that duct-insulating materials and the methods of installation be listed by an approved agency, and this is the only practical way that an inspector will have to ensure that test requirements are met.

605.1.1 Within Ducts or Plenums. Materials installed within ducts and plenums for insulating, sound deadening, or other purposes shall have a mold, humidity, and erosion-resistant surface where tested in accordance with UL 181. Duct liners in systems operating with air velocities exceeding 2000 feet per minute (10.16 m/s) shall be fastened with both adhesive and mechanical fasteners, and exposed edges shall have approved treatment to withstand the operating velocity. Where the internal insulation is capable of being in contact with condensates or other liquids, the material shall be water-resistant. Pipe and duct insulation shall not be used to reduce the maximum flame and smoke requirements in Section 602.2 unless listed for application in plenums and tested in accordance with ASTM E84 or UL 723 as a composite assembly of the duct or pipe and its associated insulation, coatings and adhesives.

If ASTM E84 or UL 723 standard test methods are modified and used as a basis to determine flame and smoke ratings, then the tests don't comply with the standard and the materials may not comply with the required flame and smoke ratings. An inspector or installer seeing ASTM E84 on a pipe would likely assume that it meets the requirement of the code. Simply adding the ASTM E 84 or UL 723 marking to the wall of the pipe does not necessarily mean that the product was tested in full compliance with the standard in the manner that the code intends. Especially for PVC and CPVC pipe, the inspector or installer should examine the material listing and see if the testing agency has modified the testing method and rendered the pipe material noncompliant.

605.1.2 Duct Coverings and Linings. Insulation applied to the surface of ducts, including duct coverings, linings, tapes, and adhesives, located in buildings shall have a flame-spread index not to exceed 25 and a smoke-developed index not to exceed 50, where tested in accordance with ASTM E84 or UL 723. The specimen preparation and mounting procedures of ASTM E2231 shall be used. Air duct coverings and linings shall not flame, glow, smolder, or smoke where tested in accordance with ASTM C411 at the temperature to which they are exposed in service. In no case shall the test temperature be less than 250°F (121°C). Coverings shall not penetrate a fire-resistance-rated assembly.

606.0 Smoke Dampers, Fire Dampers, and Ceiling Dampers.

606.1 Smoke Dampers. Smoke dampers shall comply with UL 555S, and shall be installed in accordance with the manufacturer's installation instructions where required by the building code.

Air conveyance systems (ducts) represent a significant source of transmitting smoke and heat throughout a building in a fire situation. Smoke dampers are intended for installation in ducts and air transfer openings that are designed to resist the passage of air and smoke. The devices are installed to operate automatically, be controlled by a smoke detection system and, where required, be allowed to be positioned from a remote command station. Accordingly, smoke dampers are required to reduce — if not eliminate — this risk in buildings where timely evacuation is not always feasible.

Smoke dampers may be required where ducts penetrate though smoke barriers, or at other locations within an engineered smoke control system. A smoke barrier is a continuous membrane, either vertical or horizontal, such as a wall, floor or ceiling assembly, which is designed and constructed to restrict the movement of smoke. Smoke dampers can be used in heating, ventilating, and air-conditioning (HVAC) systems where the fans are shut down in the event of fire, and can also be used in smoke control systems designed to operate during a fire incident. Smoke dampers are designed to operate against air velocity and fan pressure.

To meet the requirements of the UMC and the building code of the jurisdiction, these smoke dampers must meet specific standards of performance, including fire tests. UL 555S tests the damper under simulated fire conditions of elevated temperatures, [250 to 350°F (121°C to 177°C)] and under simulated airflow rates and pressures.

606.2 Fire Dampers. Fire dampers shall comply with UL 555, and shall be installed in accordance with the manufacturer's installation instructions where required by the building code. Fire dampers shall have been tested for closure under airflow conditions and shall be labeled for both maximum airflow permitted and direction of flow. Where more than one damper is installed at a point in a single air path, the entire airflow shall be assumed to be passing through the smallest damper area.

Ductwork shall be connected to damper sleeves or assemblies in accordance with the fire damper manufacturer's installation instructions.

Fire dampers are used to prevent transmission of flame where air ducts penetrate fire barriers. A fire barrier is a fire-resistant-rated vertical or horizontal assembly of materials designed to restrict the spread of fire. Fire dampers can also be employed in air transfer openings in walls and partitions. Building codes specify where fire dampers are required.

Fire dampers are available in two types, static fire dampers and dynamic fire dampers. Fire dampers for use in static systems, as their name implies, are used in duct systems or penetrations where there is no or negligible airflow when the damper closes. Fire dampers for use in dynamic systems are required at locations in which fan pressure will be on during a fire incident, and are expected to be able to operate (close) against the air velocity and pressure produced by the system fan.

Fire dampers carry an hourly fire-resistance rating of, usually, 1-1/2 or 3 hours. Fire dampers are also provided with an airflow rating that indicates the maximum velocity and static pressure that the damper is designed for. Refer to the UL 555 standard for a more detailed explanation of the limitations of the ratings.

606.3 Ceiling Radiation Dampers. Ceiling radiation dampers shall comply with UL 555C, and shall be installed in accordance with the manufacturer's installation instructions in the fire-resistive ceiling membrane of floor-ceiling and roof-ceiling assemblies where required by the building code. Fire dampers not meeting the temperature limitation of ceiling radiation dampers shall not be used as a substitute.

Ceiling radiation dampers are designed to function as heat barriers in duct openings that penetrate fire-resistive membrane (suspended) ceilings. A fire-rated ceiling protects the structure above from excessive heat and the potential for collapse. Properly installed, it provides a barrier to heat between the fire area and the floor above. Without openings (for lights, grilles, diffusers, etc.), no problems are encountered; however, when openings are introduced to convey environmental air, ceiling dampers are required to maintain the fire-resistive integrity.

606.4 Multiple Arrangements. Where size requires the use of multiple dampers, each damper shall be listed for use in multiple arrangements and installed in accordance with the manufacturer's installation instructions.

606.5 Access and Identification. Fire and smoke dampers shall be provided with an approved means of access large enough to allow inspection and maintenance of the damper and its operating parts. The access shall not affect the integrity of the fire-resistance-rated assembly. The access openings shall not reduce the fire-resistance rating of the assembly.

Access shall not require the use of tools. Access doors in ducts shall be tight fitting and approved for the required duct construction. Access points shall be permanently identified on the exterior by a label with letters not less than ½ of an inch (12.7 mm) in height reading as one of the following:

(1) Smoke Damper

(2) Fire Damper

(3) Fire/Smoke Damper

All dampers are mechanical devices and can only be expected to operate as designed when maintained in working order. To ensure proper inspection and maintenance, access is required to the damper location. This section requires not only access, but access that does not require a key or special tool (see **Figure 606.5**).

606.6 Freedom from Interference. Dampers shall be installed in a manner to ensure positive closing or opening as required by function. Interior liners or insulation shall be held

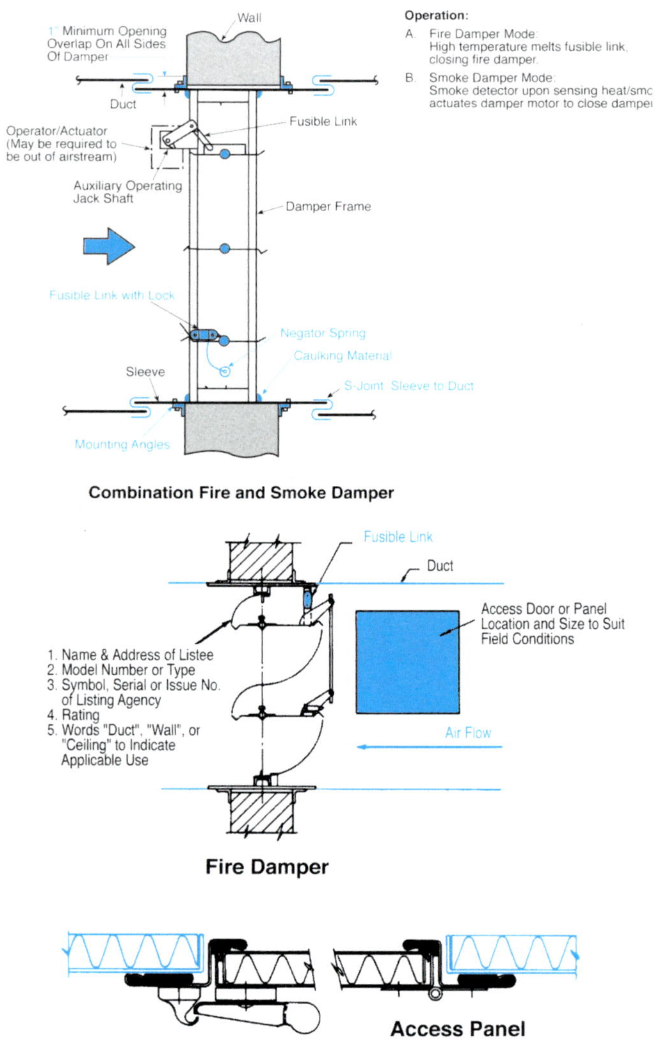

FIGURE 606.5
TYPICAL FIRE DAMPER DETAILS

back from portions of a damper, its sleeve, or adjoining duct that would interfere with the damper's proper operation. Exterior materials shall be installed so as to not interfere with the operation or maintenance of external operating devices needed for the function of the damper.

606.7 Temperature Classification of Operating Elements. Fusible links, thermal sensors, and pneumatic or electric operators shall have a temperature rating or classification as in accordance with the building code.

607.0 Ventilating Ceilings.

607.1 General. Perforated ceilings shall be permitted to be used for air supply within the limitations of this section. Exit corridors, where required to be of fire-resistive construction by the building code, shall not have ventilating ceilings.

607.2 Requirements. Ventilating ceilings shall comply with the following:

(1) Suspended ventilating ceiling material shall have a Class 1 flame spread classification on both sides, determined in accordance with the building code. Suspended ventilating ceiling supports shall be of noncombustible materials.

(2) Luminaires recessed into ventilating ceilings shall be of a type approved for that purpose.

This section regulates the installation of ventilating ceilings, as defined in Section 224.0. In addition to the requirements of this section, the space above a ventilating ceiling must comply with the definition of a "Plenum," as defined in Section 218.0, and, therefore, must also comply with the following applicable requirements of Sections 603.10 and 602.2:

- "Exhaust ducts and venting systems under positive pressure shall not extend through ducts or plenums."
- "Materials exposed within ducts or plenums shall be noncombustible or shall have a flame-spread index not to exceed 25 and a smoke-developed index not to exceed 50 where tested as a composite product in accordance with ASTM E84 or UL 723."

Figure 607.2 illustrates the requirements for installation of a ventilating ceiling.

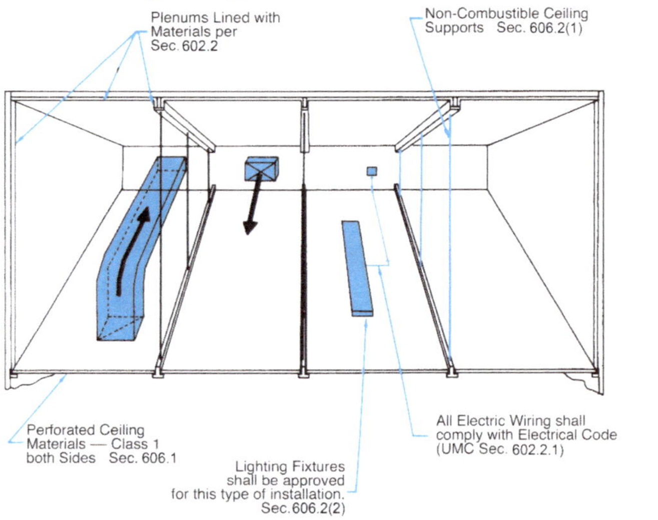

FIGURE 607.2
VENTILATING CEILING REQUIREMENTS

608.0 Use of Under-Floor Space as Supply Plenum for Dwelling Units.

608.1 General. An under-floor space shall be permitted to be used as a supply plenum.

608.2 Dwelling Units. The use of under-floor space shall be limited to dwelling units not more than two stories in height. Except for the floor immediately above the under-floor plenum, supply ducts shall be provided extending from the plenum to registers on other floor levels.

Exception: In flood hazard areas, under-floor spaces shall not be used as supply plenums unless the flood opening requirements in the building code are met.

This section regulates the use of under-floor space as an HVAC supply plenum in single-family residences. The requirements for supplying heated air to under-floor space limits air temperature that comes into contact with exposed wood construction and, at the same time, limits the volume of under-floor air to be heated each time the circulating fan is turned on. **Figure 608.2** illustrates the requirements of this section.

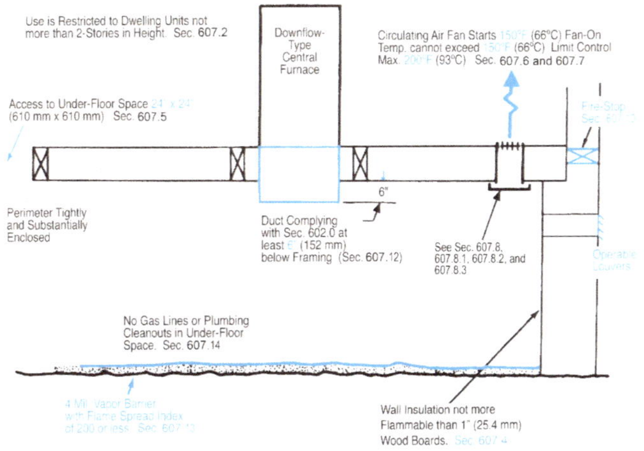

FIGURE 608.2
UNDER-FLOOR SPACE AS SUPPLY PLENUM

608.3 Enclosed. Such spaces shall be cleaned of all loose combustible scrap material and shall be tightly enclosed.

608.4 Flammable Materials. The enclosing material of the under-floor space, including the sidewall insulation, shall be not more flammable than 1 inch (25.4 mm) (nominal) wood boards (flame-spread index of 200). Installation of foam plastics is regulated by the building code.

608.5 Access. Access shall be through an opening in the floor and shall be not less than 24 inches by 24 inches (610 mm by 610 mm).

608.6 Automatic Control. A furnace supplying warm air to under-floor space shall be equipped with an automatic control that will start the air-circulating fan where the air in the furnace bonnet reaches a temperature not exceeding 150°F (66°C). Such control shall be one that cannot be set to exceed 150°F (66°C).

608.7 Temperature Limit. A furnace supplying warm air to such space shall be equipped with an approved temperature limit control that will limit outlet air temperature to 200°F (93°C).

608.8 Noncombustible Receptacle. A noncombustible receptacle shall be placed below each floor opening into the air chamber, and such receptacle shall comply with Section 608.8.1 through Section 608.8.3.

608.8.1 Location. The receptacle shall be securely suspended from the floor members and shall be not more than 18 inches (457 mm) below the floor opening.

608.8.2 Area. The area of the receptacle shall extend 3 inches (76 mm) beyond the opening on all sides.

608.8.3 Perimeter. The perimeter of the receptacle shall have a vertical lip not less than 1 inch (25.4 mm) high at the open sides where it is at the level of the bottom of the joists, or 3 inches (76 mm) high where the receptacle is suspended.

608.9 Floor Registers. Floor registers shall be designed for easy removal in order to give access for cleaning the receptacles.

608.10 Exterior Wall and Interior Stud Partitions. Exterior walls and interior stud partitions shall be fire blocked at the floor.

608.11 Wall Register. Each wall register shall be connected to the air chamber by a register box or boot.

608.12 Distance from Combustible. A duct complying with Section 602.0 shall extend from the furnace supply outlet not less than 6 inches (152 mm) below combustible framing.

608.13 Vapor Barrier. The entire ground surface of the under-floor space shall be covered with a vapor barrier having a thickness not less than 4 mils (0.1 mm) and a flame-spread index of not more than 200.

608.14 Prohibited. Fuel gas lines and plumbing waste cleanouts shall not be located within the space.

609.0 Automatic Shutoffs.

609.1 Air-Moving Systems and Smoke Detectors. Air-moving systems supplying air in excess of 2000 cubic feet per minute (ft^3/min) (0.9439 m^3/s) to enclosed spaces within buildings shall be equipped with an automatic shutoff. Automatic shutoff shall be accomplished by interrupting the power source of the air-moving equipment upon detection of smoke in the main supply-air duct served by such equipment. Duct smoke detectors shall comply with UL 268A and shall be installed in accordance with the manufacturer's installation instructions. Such devices shall be compatible with the operating velocities, pressures, temperatures, and humidities of the system. Where fire-detection or alarm systems are provided for the building, the smoke detectors shall be supervised by such systems in an approved manner.

Exceptions:

(1) Where the space supplied by the air-moving equipment is served by a total coverage smoke-detection system in accordance with the fire code, interconnection to such system shall be permitted to be used to accomplish the required shutoff.

(2) Automatic shutoff is not required where occupied rooms served by the air-handling equipment have direct exit to the exterior, and the travel distance does not exceed 100 feet (30 480 mm).

(3) Automatic shutoff is not required for Group R, Division 3 and Group U Occupancies.

(4) Automatic shutoff is not required for approved smoke-control systems or where analysis demonstrates shutoff would create a greater hazard, such as shall be permitted to be encountered in air-moving equipment supplying specialized portions of Group H Occupancies. Such equipment shall be required to have smoke detection with remote indication and manual shutoff capability at an approved location.

(5) Smoke detectors that are factory installed in listed air-moving equipment shall be permitted to be used in lieu of smoke detectors installed in the main supply-air duct served by such equipment.

This section specifies when and where smoke detectors are to be installed. The occupancies requiring smoke detectors in HVAC systems as specified in this section are defined and described in the building code of the jurisdiction and Section 217.0 (see the definition of "Occupancy").

Automatic shutoff must be accomplished by shutting off the power source of the air-moving system upon detection of smoke in the main supply duct served by such equipment. Smoke detectors must be labeled by an approved agency for air duct installation and installed per the manufacturer's installation instructions. The detectors must be capable of operating at velocities, pressures, temperatures and moisture conditions of the system. **Figure 609.1** illustrates the proper location of smoke detectors in the supply duct.

Exceptions to these requirements include systems in which all rooms with direct access to exterior and travel distance is less than 100 feet. An exception permits the use of factory-installed smoke detectors placed in listed air-moving equipment in lieu of a detector in the main supply duct.

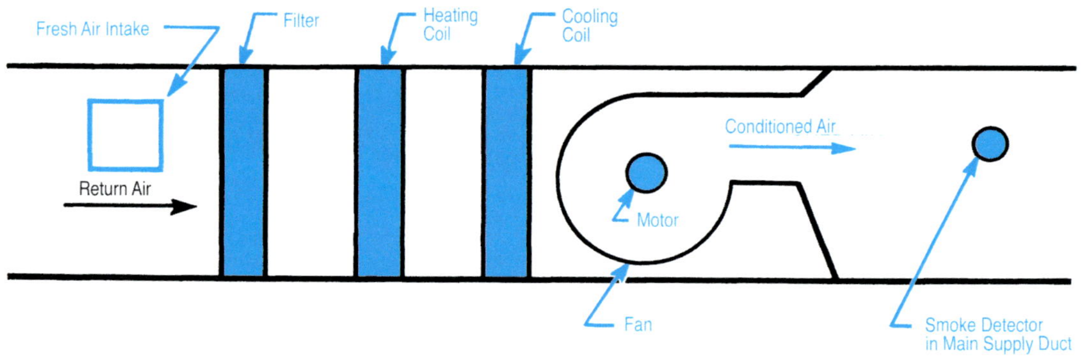

FIGURE 609.1
PROPER LOCATION OF SMOKE DETECTOR IN SUPPLY DUCT

CHAPTER 7

COMBUSTION AIR

701.0 General.

This chapter includes the requirements for providing combustion air to fuel-burning appliances. Specifically, the requirements apply to natural draft and Category I-type appliances. Those are typically natural gas- and liquefied petroleum-gas- (LP-gas) fired appliances. The chapter directs the reader to the manufacturer's instructions for other types of appliances, such as Category II, III and IV, and oil-fired appliances to determine their combustion air needs (see also Section 301.5, Oil-Burning Appliances).

"Combustion air" is a generic term that includes air for combustion, air for dilution of flue gases (combustion products at the drafthood) and ventilation of the space that encloses the appliance(s).

Fuel Consumption: Assuming standard air (air at 59°F and 14.7 psi, or atmospheric pressure) and ideal conditions, complete combustion of fuel gases requires 2 cubic feet of oxygen for each cubic foot of fuel gas. Air is roughly 20-percent oxygen, 79-percent nitrogen and 1-percent trace gases; consequently, it takes 10 cubic feet of air (containing 2 cubic feet of oxygen) to completely burn 1 cubic foot of fuel gas. Deviation from these ideal conditions will affect the air density, which will in turn affect the required amount of air for complete combustion.

Dilution and Ventilation: Combustion of fuel gas produces hot waste products comprised mainly of nitrogen, carbon dioxide and water vapor. The function of the drafthood on natural-draft gas appliances, besides preventing backdraft, is two-fold: to infuse air that will cool down the flue gases entering the vent; and to blend dry air with the wet products of combustion, reducing the moisture content per unit volume of flue gases, thus limiting condensation within the vent or connector. (The amount of water vapor created by burning 100 cubic feet of natural gas is in excess of 1 gallon of liquid, if all the vapor is allowed to cool down and condense.) The air necessary for drafthood dilution, and for circulation within the compartment to cool it down and replenish the oxygen necessary for combustion, creates the need for an additional 15 cubic feet of air for each cubic foot of gas. As a result, to provide sufficient air for basic fuel combustion, drafthood dilution and equipment room ventilation, at least 25 cubic feet of air is needed for each cubic foot of fuel gas (see **Figure 701.0a**).

Two-Openings and One-Opening Overview: The burning of fuel inside an appliance heats the air, causing it to rise within the appliance, along with the products of combustion, and exit through the flue outlet. As the hot flue gases exits the flue outlet and enters the vent through the drafthood, a "venturi" effect is created, drawing cool, almost stagnant air from the compartment to mix with the faster moving hot flue gases (see **Figure 701.0b**).

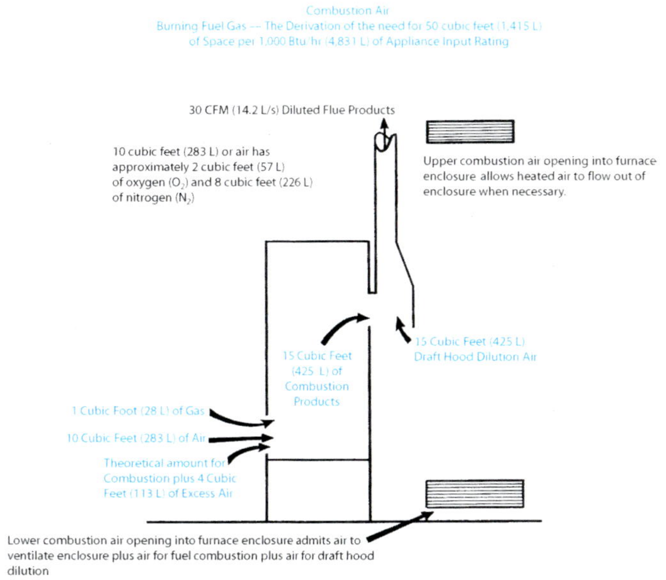

FIGURE 701.0A
COMBUSTION AIR

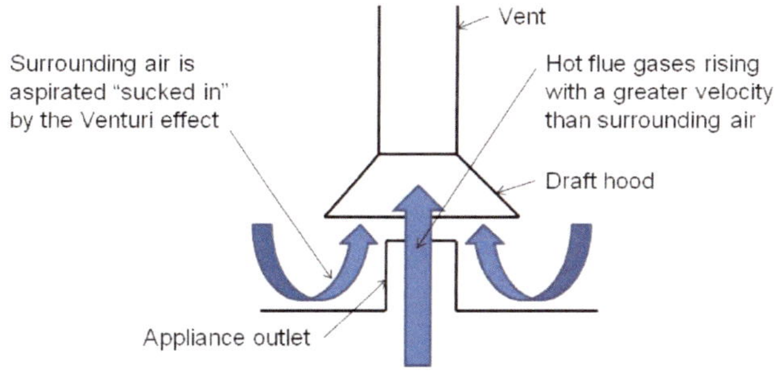

FIGURE 701.0B
THE "VENTURI" EFFECT

In addition, during the operation of the appliance, the heat generated within the compartment heats the air, making it buoyant (less dense), which allows it to rise and exit through the top combustion air opening or duct. This creates a "thermo-siphon" action, which draws cool air from the bottom or lower combustion air opening or duct to replace the air that exits. It has been found that this air replenishment process works best with two openings or ducts, providing uniform and adequate circulation of air within the compartment. This will in turn provide sufficient oxygen for complete combustion of fuel, air to dilute the flue gases entering the vent and air to cool down the compartment in the process. Placing the upper combustion air opening within 12 inches of the ceiling of the compartment ensures that the hottest air is exhausted. Placing the lower combustion air opening within 12 inches of the floor of the compartment ensures the greatest temperature differential and the maximum "stack" effect that can be achieved, thus generating the greatest air circulation possible and avoiding the formation of stagnant pockets of air.

In summary, proper functioning of fuel-fired equipment in an enclosure is dependent upon the availability of sufficient air within the enclosure to adequately serve all the aforementioned needs. Combustion air deficiencies will create problems ranging from inefficient performance of equipment to possible health hazards, such as the production and discharge of carbon monoxide (the consequence of incomplete combustion), carbonization of flue pipes (soot) and development of condensation within the vent (resulting in premature vent failure). The byproducts of incomplete combustion are both poisonous (carbon monoxide) and corrosive (hydrocarbons), and could be easily avoided by placing combustion air openings of adequate sizes in proper locations.

In cold climates, freezing water pipes in mechanical rooms can be a problem and attempts have been made to reduce openings to the outdoors to the absolute minimum. A study done by Battelle (GRI Report Number 93/0316, Analysis of Combustion Air Openings to the Outdoors) states that one upper combustion air opening sized at 1 square inch per 3,000 Btu/h will provide adequate combustion, dilution and ventilation air to the space or room containing an appliance(s). This opening or duct must communicate directly with the outdoors or to spaces that communicate with the outdoors. The appliance must have clearances of at least 1 inch on the sides and back and 6 inches in the front for this method to work properly. There is also the requirement that the combustion air opening be no smaller than the sum of the areas of all vent connectors in the space or room.

As the room warms up, the pressure in the room increases, and the warmer air flows out through that upper opening. After a short period, the room will develop a slightly negative pressure and cooler outside air will be drawn into the enclosure through the same opening until the pressure in the room becomes positive again. This cycle keeps repeating in a type of puffing action. This method has proven to be adequate, but less efficient than the two-opening method.

701.1 Applicability. Air for combustion, ventilation, and dilution of flue gases for appliances installed in buildings shall be obtained by application of one of the methods covered in Section 701.4 through Section 701.9.3. Where the requirements of Section 701.4 are not met, outdoor air shall be introduced in accordance with methods covered in Section 701.6 through Section 701.9.3.

Exceptions:

(1) This provision shall not apply to direct vent appliances.

(2) Type 1 clothes dryers that are provided with makeup air in accordance with Section 504.4.1. [NFPA 54:9.3.1.1]

The methods for obtaining combustion air outlined in this chapter are only intended for gas utilization equipment that is the natural-draft type and for Category I-vented appliances. A Category I appliance is defined in Chapter 2 as: an appliance that operates with nonpositive vent static pressure and with a vent gas temperature that avoids excessive condensate production in the vent. To be classified as a Category I appliance, the pressure at the vent outlet of the appliance must not be positive. This will allow the use of a "B" vent (that is not airtight) and the venting process will be dependent on the buoyancy of hot air to conduct the flue gases to the outside. However, this category also includes "fan-assisted" type appliances that utilize a small fan to create a uniform stream of combustion air across the burners (thus more efficient) (see **Figure 701.1a**).

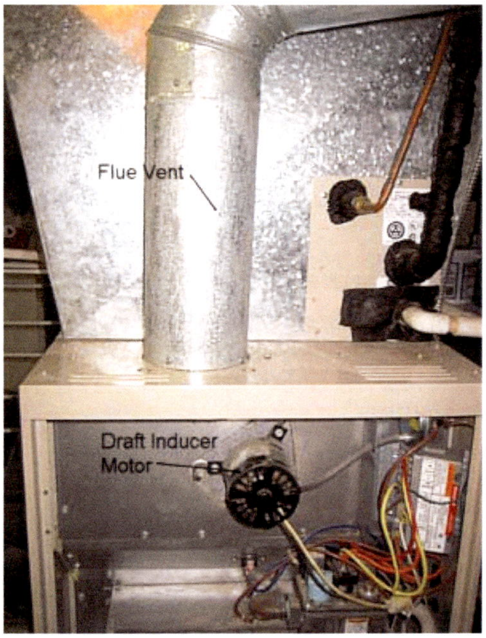

FIGURE 701.1A
FAN-ASSISTED-TYPE FURNACES

However, the fan pressure is just enough to overcome the friction losses in the combustion chamber and the attached conduit leading to the outlet of the appliance. The flue gases will be delivered to the outlet at an atmospheric or nonpositive pressure.

The provisions of this chapter should not apply to other types of appliances, such as Categories II, III and IV or to oil-fired equipment. These types should follow the requirements of the manufacturer. In addition, direct-vent appliances are exempt from the requirements of this chapter. The process of drawing combustion air and venting for these types of appliances is typically done through the use of two concentric pipes connecting the appliance to the outside air, the outer pipe drawing combustion air and the inner pipe exhausting flue gases to the outdoors. The combustion chamber is isolated and sealed from the indoor air, allowing the combustion process to be independent and to have no effect on the interior of the building (see **Figure 701.1b**).

Clothes dryers are also exempt from the chapter requirement as they utilize fans to draw combustion air into the dryer and force the products of combustion, along with lint and moisture, to the outdoors through a moisture exhaust duct. Provisions for the dryer exhaust and make-up air are found in Section 504.4 Clothes Dryers.

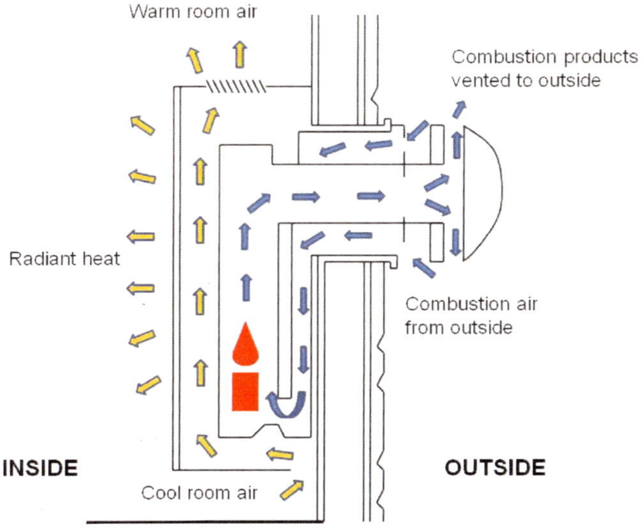

FIGURE 701.1B
DIRECT VENT APPLIANCE

701.1.1 Other Types of Appliances. Appliances of other than natural draft design, appliances not designated as Category I vented appliances, and appliances equipped with power burners shall be provided with combustion, ventilation, and dilution air in accordance with the appliance manufacturer's instructions. [NFPA 54:9.3.1.2]

This section repeats the requirements of the previous section, stating that the manufacturer's instructions must be followed for combustion air requirements for vented appliances other than natural draft and Category I.

701.2 Pressure Difference. Where used, a draft hood or a barometric draft regulator shall be installed in the same room or enclosure as the appliance served so as to prevent any difference in pressure between the hood or regulator and the combustion air supply. [NFPA 54:9.3.1.4]

Drafthoods that are used to draw cool dry air to mix with hot moist flue gases also function as devices that allow the escape of flue gases from the appliance in the event of a blocked vent or backdraft. They also regulate the stack effect of the vent and ensure uniform and adequate draw of combustion air across the burners.

Barometric dampers have a similar function. They regulate the air flow through the vent by modulating the vent draft with varying amounts of room air.

Both these devices are critical for the flow of air across the flames of the burners. They are required to be in the same room or enclosure of the appliance served to prevent differences in pressure between the hood or regulator and the combustion air supply. The difference in pressure could negatively impact the combustion and venting process of the appliance.

701.3 Makeup Air. Where exhaust fans, clothes dryers, and kitchen ventilation systems interfere with the operation of appliances, makeup air shall be provided. [NFPA 54:9.3.1.5]

As can be seen from the narrative above, the combustion process for natural-draft-type and Category I appliances is very delicate and any interference from blowers or fans of other appliances in the area, such as exhaust fans, kitchen exhaust or clothes dryers, could adversely affect the combustion and venting process. Fans could create negative pressure in the appliance room or enclosure, resulting in insufficient flow of air across the burners. It could even be so severe as to draw flue gases away from their intended path to the vent. Therefore, if appliances with fans are present, the simplest way to avoid problems is to balance the pressure in the room or compartment by providing make-up air in sufficient quantities to neutralize the exhaust fan effect.

701.4 Indoor Combustion Air. The required volume of indoor air shall be determined in accordance with the method in Section 701.4.1 or Section 701.4.2 except that where the air infiltration rate is known to be less than 0.40 ACH (air change per hour), the method in Section 701.4.2 shall be used. The total required volume shall be the sum of the required volume calculated for appliances located within the space. Rooms communicating directly with the space in which the appliances are installed through openings not furnished with doors, and through combustion air openings sized and located in accordance with Section 701.5, are considered a part of the required volume. [NFPA 54:9.3.2]

Combustion air for appliances could be obtained from the indoors, provided that sufficient infiltration from the outdoors is available in order to continuously replenish the indoor air. The rate of infiltration is dependent on the number and size of gaps and cracks around openings in the building (e.g., windows and doors, and around piping and duct penetrations through the building's exterior walls, raised floors, ceiling/roof and ceiling/attic assemblies). Infiltration also takes place through walls and ceilings (even through gypsum wallboard) unless these assemblies are protected by an impervious membrane intended to prevent such infiltration.

To determine whether there is sufficient infiltration to the interior to accommodate sustained combustion of an appliance, the code has historically used an empirical formula that calculates a certain minimum interior volume (50 cubic feet) per 1,000 Btu/h of appliance rating. This empirical formula was based on the assumption that 1 cubic feet of gas (that gives approximately 1,000 Btu) requires 10 cubic feet of air to achieve complete combustion. To account for the additional air required for dilution and ventilation, the 10 cubic feet of air was increased by 15 cubic feet to a total of 25 cubic feet. If infiltration to the appliance room is at one air change per hour (ACH), 25 cubic feet of interior volume would be sufficient to continue the combustion process indefinitely. However, assuming a worst-case scenario of 0.5 ACH infiltration rate, the 25 would be doubled to 50.

The formula has been satisfactory and proved its accuracy until a few years ago when energy conservation regulations were adopted throughout the country and construction practices changed. As these regulations limit the amount of infiltrated air into the building (that would waste additional energy in heating or cooling the infiltrated air), it was necessary to change the code to account for severe cases of limited infiltration. The term "unusually tight construction" was inserted in the code and defined as:

UNUSUALLY TIGHT CONSTRUCTION is construction where:

1) Walls and ceilings exposed to the outside atmosphere have a continuous water vapor retarder with a rating of one perm or less with any openings gasketed or sealed;
2) Weatherstripping is on openable windows and doors; and
3) Caulking or sealants are applied to areas such as joints around window and door frames, between sole plates and floors, between wall-ceiling joints, between wall panels, and at penetrations for plumbing, electrical, and gas lines and at other openings.

Items 2 and 3 of the definition are standard requirements of virtually all energy conservation codes adopted in the country. However, Item 1 is not, but is usually provided where severe weather conditions exist or when tighter infiltration rates are desired. For wood construction-type buildings, this is usually achieved by either installing a vapor retarder on the inside of all exterior assemblies (using a plastic membrane rated at one perm or less) or installing "batt" insulation that has a vapor retarder backing with that rating and overlapping the joints of the backing onto the studs and joists to create a "continuous" membrane.

In the past, the code required that if a building meets the definition of "Unusually Tight Construction," the interior should not be used as a source of combustion air, unless provisions are made to allow free communication with the outdoors through permanent opening(s) to the outside. However, more recently, studies and tests proved that for "unusually tight" buildings, it is still possible to use the interior as a source of combustion air without any openings to the exterior, albeit with an increased factor (more than 50). To determine that factor, calculations for the infiltration rate are required and the amount of ACH is more precisely determined. If the infiltration rate is found to be less than 0.4 ACH, the building is determined to be "tight" and the formulas in Section 701.4.2 must be used to calculate the factor that will determine the minimum interior volume required for combustion air.

Studies have also shown that the old empirical formula may have been too conservative for ordinary buildings, requiring a larger volume than needed. Therefore, the code now specifies that if the infiltration rates are known for a building (through calculations), the formulas in Section 701.4.2 could be used to decrease the volume requirements from 50 to as low as 25 for fan-assisted combustion-type appliances and to 35 for other types of Category I appliances. These reduced volumes are achieved when the ventilation rates are at their highest values.

However, the section still allows the use of the "Standard Method" outlined in Section 701.4.1 (factor of 50 cubic feet per 1,000 Btu/h of appliance input) if the infiltration rates are not known, provided that the building is not ruled as "tight." If it is suspected that a building is "tight," the infiltration rate must then be calculated to either prove that the rate is more than 0.4 ACH, or if found to be 0.4 or less, to mandate the use of the equations in Section 701.4.2 instead of using the Standard Method (50).

When the use of the Standard Method is desired for a building, it may not be practical to require calculations of the infiltration rates on every occasion to eliminate the possibility that the building is "tight." Therefore, it is suggested to apply the definition of "Unusually Tight Construction" to determine whether calculations should be submitted. (Although that term can no longer be found in Chapter 7, the definition of the term can still be found in Chapter 2.) If the building meets the definition as a "tight" building, and the interior of the building is to be used as a source of combustion air, engineering calculations must be submitted to compute ACH. Based on these calculations, the minimum volume would be determined using the appropriate equation.

The section goes on to reference Section 701.5, which allows the use of adjacent interior spaces through openings not furnished with doors, such as "cased" openings, or through the use of combustion air openings in accordance with Section 701.5.

701.4.1 Standard Method. The minimum required volume shall be 50 cubic feet per 1000 British thermal units per hour (Btu/h) (4.83 m^3/kW). [NFPA 54:9.3.2.1]

Example using the Standard Method:

An 80,000 Btu/h furnace and a 50,000 Btu/h water heater with drafthoods are installed in a room 35 feet by 35 feet by 8 feet. (The building has been ruled to be of ordinary construction; not "tight.")

The Standard Method, which requires 50 cubic feet per 1,000 Btu/hr of all appliances in the space, is to be used to determine the minimum volume. First, add the capacities of the furnace and water heater.

80,000 Btu/h + 50,000 Btu/h = 130,000 Btu/h

130,000 x 50/1,000 = 6,500 cubic feet

Available room volume = 35 x 35 x 8 = 9,800 cubic feet

Combustion air may be obtained from indoors.

701.4.2 Known Air Infiltration Rate Method. Where the air infiltration rate of a structure is known, the minimum required volume shall be determined as follows [NFPA 54:9.3.2.2]:

(1) For appliances other than fan-assisted, calculate using the following Equation 701.4.2(1). [NFPA 54:9.3.2.2(1)]

[Equation 701.4.2(1)]

$$\text{Required Volume}_{other} \geq \frac{21 \text{ ft}^3}{ACH}\left(\frac{I_{other}}{1000 \text{ Btu/h}}\right)$$

(2) For fan-assisted appliances, calculate using the following Equation 701.4.2(2). [NFPA 54: 9.3.2.2(2)]

[Equation 701.4.2(2)]

$$\text{Required Volume}_{fan} \geq \frac{15 \text{ ft}^3}{ACH}\left(\frac{I_{fan}}{1000 \text{ Btu/h}}\right)$$

Where:

I_{other} = All appliances other than fan-assisted input (Btu/h)

I_{fan} = Fan-assisted appliance input (Btu/h)

ACH = Air change per hour (percent of volume of space exchanged per hour, expressed as a decimal)

For SI units: 1 cubic foot = 0.0283 m^3, 1000 British thermal units per hour = 0.293 kW

(3) For purposes of these calculations, an infiltration rate greater than 0.60 ACH shall not be used in the equations in Section 701.4.2(1) and Section 701.4.2(2). [NFPA 54:9.3.2.2(3)]

This section prohibits the use of the air change per hour (ACH) factor exceeding 0.60 (as the calculations would provide results less than the allowed minimums).

If an installation includes both types of appliances (natural draft and fan assisted), a separate calculation is performed for each type to determine the volume. Adding the results will determine the total required volume.

Example using the Known Infiltration Rate Method:

An engineer submitted calculations showing that four proposed buildings have the following infiltration rates:

Building 1: 0.35 ACH

Building 2: 0.48 ACH

Building 3: 0.60 ACH

Building 4: 0.75 ACH

In each of the buildings, drafthood appliances with aggregate input ratings of 150,000 Btu/h are to be installed. What is the minimum required volume of the appliance room in each building?

Since appliances are the natural-draft type (other), 701.4.2 (1) is to be used.

Building 1: Since ACH is less than 0.4, this building is of unusually tight construction and this method must be used:

$$\frac{21}{0.35} \times \frac{150{,}000}{1{,}000} = 60 \times 150 = 9{,}000 \text{ cubic feet}$$

(Note that the factor is 60 > 50)

Building 2:

$$\frac{21}{0.48} \times \frac{150{,}000}{1{,}000} = 43.75 \times 150 = 6{,}562.5 \text{ cubic feet}$$

Building 3:

$$\frac{21}{0.6} \times \frac{150{,}000}{1{,}000} = 35.00 \times 150 = 5{,}250 \text{ cubic feet}$$

Building 4*:

$$\frac{21}{0.6} \times \frac{150{,}000}{1{,}000} = 35 \times 150 = 5{,}250 \text{ cubic feet}$$

* Section 701.4.2 (3) does not allow the use of an infiltration rate greater than 0.6.

Example using the Known Infiltration Rate Method with different types of appliances:

An 80,000 Btu/h fan-assisted furnace and a 50,000 Btu/h water heater with a drafthood are installed in a room 35 feet by 35 feet by 8 feet.

Engineering calculations were submitted showing that the ACH is 0.65 for the room. (However, this method does not allow the use of ACH larger than 0.6; therefore, 0.6 ACH will be used in the equation.)

Since a drafthood appliance and a fan-assisted appliance are proposed, the required volume must be calculated separately.

Furnace, using Equation 701.4.2 (2):

Required volume for furnace:

$$80{,}000 \text{ Btu/h: } \frac{15 \text{ ft}^3}{0.60} \times \frac{80{,}000 \text{ Btu/h}}{1{,}000 \text{ Btu/h}} = 2{,}000 \text{ cubic feet}$$

Water heater, using Equation 701.4.2 (1):

Required volume for water heater:

$$50{,}000 \text{ Btu/h: } \frac{21 \text{ ft}^3}{0.60} \times \frac{50{,}000 \text{ Btu/h}}{1{,}000 \text{ Btu/h}} = 1{,}750 \text{ cubic feet}$$

Total volume required:

$2{,}000 \text{ ft}^3 + 1{,}750 \text{ ft}^3 = 3{,}750$ cubic feet

Available volume of the room:

35 x 35 x 8 = 9,800 cubic feet

Interior volume is sufficient and no outdoor combustion air is required.

701.5 Indoor Opening Size and Location. Openings used to connect indoor spaces shall be sized and located in accordance with the following:

(1) Combining spaces on the same story. Each opening shall have a minimum free area of 1 square inch per 1000 Btu/h (0.002 m^2/kW) of the total input rating of all appliances in the space, but not less than 100 square inches (0.065 m^2). One permanent opening shall commence within 12 inches (305 mm) of the top of the enclosure and one permanent opening shall commence within 12 inches (305 mm) of the bottom of the enclosure (see **Figure 701.5**). The minimum dimension of air openings shall not be less than 3 inches (76 mm).

(2) Combining spaces in different stories. The volumes of spaces in different stories shall be considered as communicating spaces where such spaces are connected by one or more permanent openings in doors or floors having a total minimum free area of 2 square inches per 1000 Btu/h (0.004 m^2/kw) of total input rating of all appliances. [NFPA 54:9.3.2.3]

When the volume of the room containing the appliance(s) is smaller than the required minimum interior volume needed for combustion air, the volume of an adjacent space may be combined with the appliance room volume, if combustion air openings of sufficient sizes are constructed in the walls between the room and adjacent space. Other levels or stories of the building could also be combined with the room volume if openings of sufficient sizes are constructed. Please note that combustion air openings should not breach walls (or ceilings/floor assemblies) that are required to be of fire-resistive construction since fire dampers are prohibited in combustion air openings or ducts (see Section 701.12).

When combining spaces on the same story, two openings must be provided; each combustion air opening shall have a net free area of 1 square inch per 1,000 Btu/h of the total input rating of all gas utilization equipment. Each opening shall not be smaller than 100 square inches. One opening shall commence within 12 inches of the ceiling of the room where the appliances are located, and one opening shall commence within 12 inches of the floor of that room. Note that the opening needs to only "commence" within the

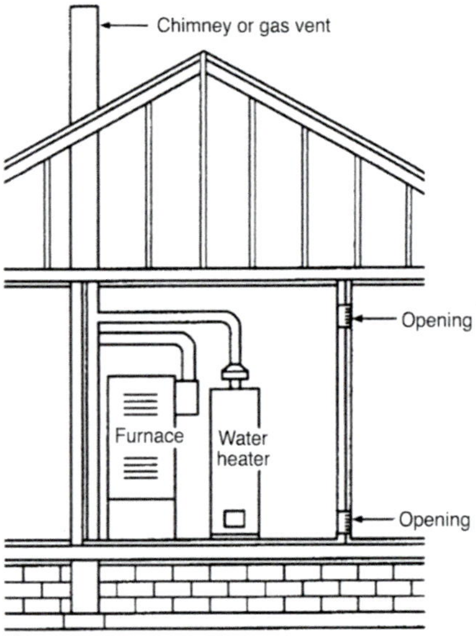

FIGURE 701.5
ALL COMBUSTION AIR FROM ADJACENT INDOOR SPACES THROUGH INDOOR COMBUSTION AIR OPENINGS [NFPA 54: FIGURE A.9.3.2.3(1)]

top or bottom 12 inches and does not need to be entirely within the 12 inches.

Please note that a fully louvered door between the two spaces would not be sufficient to meet the requirement as the top of a typical 6-foot 8-inch door in a room with an 8-foot ceiling height does not commence within 12 inches of the ceiling. Even if the net free opening in the louvered door exceeds the total required combustion air openings, an additional opening, commencing within 12 inches of the ceiling, must also be provided to meet the code requirement.

When combining spaces on different stories, openings shall have a minimum free area sized at 2 square inches per 1,000 Btu/h of all gas utilization equipment in the space. There is no requirement here as to where the openings should be or where they should "commence." Horizontal or vertical openings are acceptable as long as they meet the total required free area. The openings should have a minimum dimension of 3 inches.

Example: See **Figure 701.5a**.

A 60,000 Btu/h furnace is installed in a single-car garage 25 feet long and 12 foot wide with an 8-foot ceiling height. The garage is not of tight construction and no infiltration rates were submitted. Please determine whether the interior volume is sufficient for combustion air.

Volume required = 60,000 x 50/1,000 = 3,000 cubic feet

Volume available = 25 x 12 x 8 = 2,400 cubic feet

Interior volume is insufficient for combustion air.

Either request the ventilation rate to be calculated, require that air be obtained from the outdoors or obtain air from an interior adjacent room.

An adjacent workroom (with no fire-rated walls to the garage) is available with the following dimensions: 10 feet long, 12 feet wide and 8-foot ceiling height. There is no permanent "cased" opening between the two spaces.

Volume of adjacent work room = 12 x 10 x 8 = 960 cubic feet

Total combined volume = 2,400 + 960 = 3,300 cubic feet

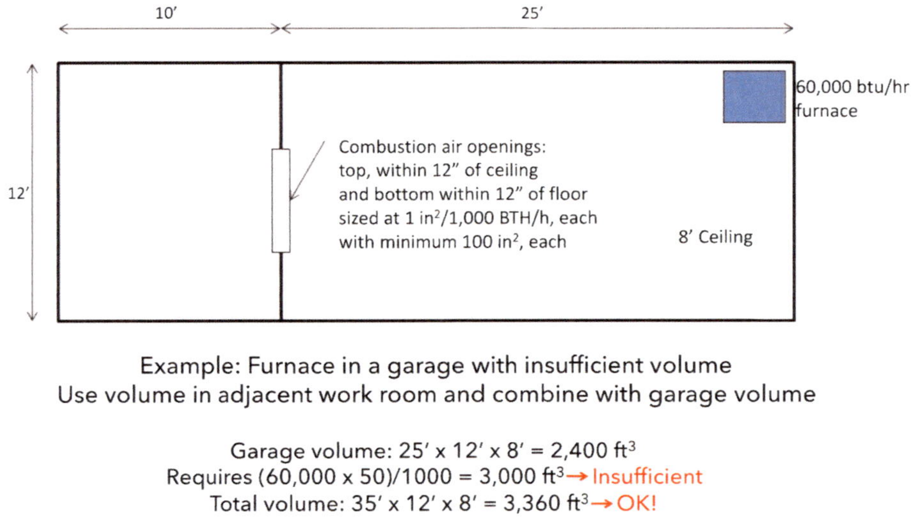

FIGURE 701.5A
TWO OPENINGS TO INDOORS EACH SIZED AT 1 SQ IN/1,OOO BTU/H MINIMUM 100 SQ IN, EACH

To combine the two spaces (that have sufficient volume: 3,300 available > 3,000 required), there needs to be two permanent openings in the adjoining wall: one commencing within 12 inches of the ceiling of the garage (containing the furnace) and one commencing within 12 inches of the garage floor, each sized at 1 square inch/1,000 Btu/h of furnace input rating, with a minimum of 100 square inch each of free area (deduct any obstructions, such as louvers or grilles).

Each opening = 60,000 x 1/1,000 = 60 sq in (less than minimum)

Two-100 square inch combustion air openings are needed in the wall between the two spaces.

701.6 Outdoor Combustion Air. Outdoor combustion air shall be provided through opening(s) to the outdoors in accordance with the methods in Section 701.6.1 or Section 701.6.2. The minimum dimension of air openings shall not be less than 3 inches (76 mm). [NFPA 54:9.3.3]

Combustion air can be obtained from outdoors if one of the following methods is used:

1. Two openings directly communicating with the outdoors or two openings to areas communicating with the outdoors, such as ventilated attics or ventilated crawl spaces. Each opening shall be sized at 1 square inch/4,000 Btu/h of appliance input rating [see **Figure 701.6.1(1)**].
2. Two vertical ducts to the outdoors or to areas communicating with the outdoors. Each duct shall be sized at 1 square inch/4,000 Btu/h [see **Figure 701.6.1 (2)**].
3. Two horizontal ducts to the outdoors or to areas communicating with the outdoors. Each duct shall be sized at 1 square inch/2,000 Btu/h [see **Figure 701.6.1(3)**].
4. Combination of the above, such as openings, ducts or vertical and horizontal ducts. Each opening or duct shall be sized based on the above. For example, if the upper duct is horizontal, it must be sized at 1 square inch/2,000 Btu/h, while if the lower is an opening to a ventilated crawl space, it could be sized as 1 square inch/4,000 Btu/h.
5. One "upper" opening or duct to the outdoors or communicating with the outdoors. The opening shall be sized at 1 square inch/3,000 Btu/h (see **Figure 701.6.2**). There is no minimum size required for the duct or opening to the outdoors (as required for indoor openings) except that the minimum dimension of an opening (or duct) shall not be less than 3 inches. Also, note that the one opening method has additional requirements regulating the minimum size of the opening and restrictions on the minimum size of the compartment where the appliance is installed. See Section 701.6.2.

701.6.1 Two Permanent Openings Method. Two permanent openings, one commencing within 12 inches (305 mm) of the top of the enclosure and one commencing within 12 inches (305 mm) of the bottom of the enclosure, shall be provided. The openings shall communicate directly, or by ducts, with the outdoors or spaces that freely communicate with the outdoors as follows:

(1) Where directly communicating with the outdoors or where communicating to the outdoors through vertical ducts, each opening shall have a minimum free area of 1 square inch per 4000 Btu/h (0.0005 m^2/kW) of total input rating of all appliances in the enclosure. [See **Figure 701.6.1(1) and Figure 701.6.1(2)**]

(2) Where communicating with the outdoors through horizontal ducts, each opening shall have a minimum free area of not less than 1 square inch per 2000 Btu/h (0.001 m^2/kW) of total input rating of all appliances in the enclosure. [See **Figure 701.6.1(3)**] [NFPA 54:9.3.3.1]

As mentioned previously, the two permanent openings method requires two openings or ducts communicating directly with the outdoors or routed to spaces that communicate directly with the outdoors, such as ventilated attics or crawl spaces. The two openings or ducts shall have one commencing within 12 inches of the ceiling of the compartment or room where the appliances are located and the other commencing within 12 inches of the floor of the compartment or room.

Direct openings or vertical ducts are to be sized at 1 square inch/4,000 Btu/h (free opening) of the total appliances input rating. The smaller sizing factor for direct openings to the outdoors is due to the almost negligible friction losses encountered as the air moves directly across the openings to or from the outdoors. Also, this smaller factor for vertical ducts is due to the natural convection process of warm air rising in the upper vertical duct and colder air descending in the lower vertical duct.

Horizontal ducts are to be sized at 1 square inch/2,000 Btu/h (or twice as large as vertical ducts). The increase in size is to offset the increased friction losses encountered in the horizontal ducts. The losses should be kept to a minimum since the only driving force for air to move across these horizontal ducts is the temperature differential inside the compartment between the warm air in the upper portion of the compartment and the slightly cooler air at the bottom. It should be noted that if a vertical duct runs in the horizontal direction at any point, the size of the entire duct should be the same as required for a horizontal duct, or twice as required for a vertical duct.

Unlike openings to the indoors, there are no minimum area requirements for the combustion air openings or ducts. However, the minimum dimension of the opening in any direction must not be less than 3 inches (the smallest opening could either be a net area of 3 inches by 3 inches rectangular opening or a circular opening with a "net" diameter of 3 inches).

If ducts are routed to attics or crawl spaces, these spaces must be adequately ventilated. Attics should have ventilation openings to the outside as required by the building code of the jurisdiction. However, the net clear area must be at least equal to or exceed the total aggregate area of the combustion air openings or ducts connecting to the attic. In addition, if any of the attic ventilation openings are subject to closure due to snow or other weather-related factors, the attic would not be an acceptable source of combustion air. The same logic applies to a crawl space if it is used as a source of combustion air. It should be noted that previous versions of this code required the combustion air ducts to the attic to rise at least 6 inches above the joist or insulation, whichever was higher, to avoid the possibility of insulation or other construction material blocking the combustion air ducts. Although this provision is no longer in the code, it might be worthwhile to amend local codes to include this important requirement.

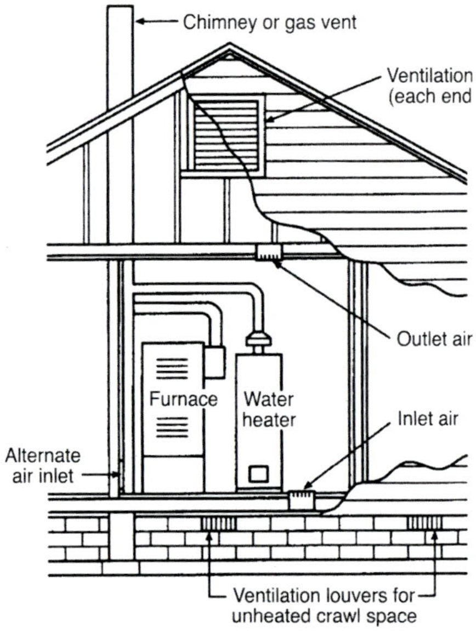

FIGURE 701.6.1(1)
ALL COMBUSTION AIR FROM OUTDOORS – INLET AIR FROM VENTILATED CRAWL SPACE AND OUTLET AIR TO VENTILATED ATTIC
[NFPA 54: FIGURE A.9.3.3.1(1)(a)]

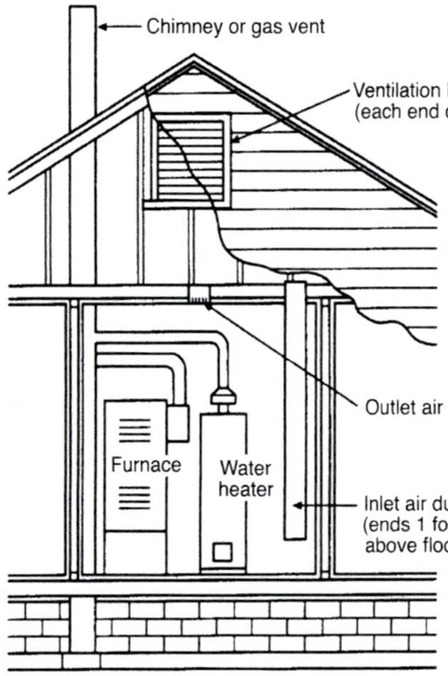

FIGURE 701.6.1(2)
ALL COMBUSTION AIR FROM OUTDOORS THROUGH VENTILATED ATTIC
[NFPA 54: FIGURE A.9.3.3.1(1)(b)]

Example:

An appliance with an input rating of 80,000 Btu/h is installed in a small compartment. Indoor combustion air is not available. Calculate the minimum free area for:

1. Two direct openings to the outdoors.
2. Two vertical ducts to the outdoors.
3. Two horizontal ducts to the outdoors.
4. Upper horizontal duct to the outdoors and lower opening to an adequately ventilated crawl space.
5. Upper vertical duct that has a 12-inch horizontal section and a lower vertical duct with no offset.

1. 80,000 x 1/4,000 = 20 sq in, each
2. 80,000 x 1/4,000 = 20 sq in, each
3. 80,000 x 1/2,000 = 40 sq in, each
4. Upper = 80,000 x 1/2,000 = 40 sq in
   Lower = 80,000 x 1/4,000 = 20 sq in
5. Upper (must use horizontal equation) = 80,000 x 1/2,000 = 40 sq in
   Lower = 80,000 x 1/4,000 = 20 sq in

Note that there is no minimum area requirement; however, the minimum dimension of the combustion air opening (or duct) must not be less than 3 inches.

701.6.2 One Permanent Opening Method. One permanent opening, commencing within 12 inches (305 mm) of the top of the enclosure, shall be provided. The appliance shall have clearances of at least 1 inch (25.4 mm) from the sides and back and 6 inches (152 mm) from the front of the appliance. The opening shall directly communicate with the outdoors or shall communicate through a vertical or horizontal duct to the outdoors or spaces that freely communicate

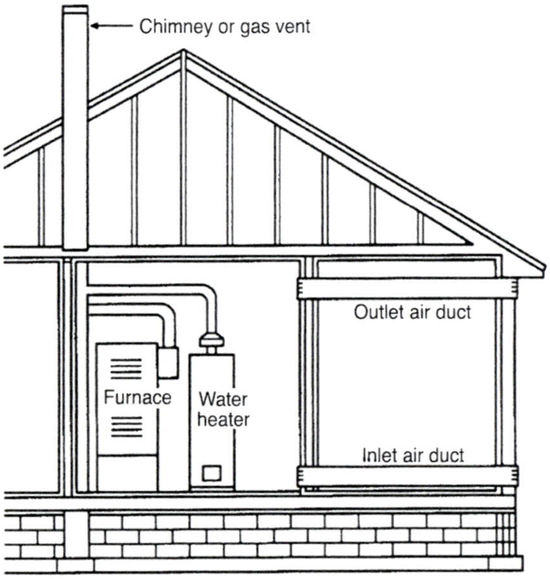

FIGURE 701.6.1(3)
ALL COMBUSTION AIR FROM OUTDOORS THROUGH HORIZONTAL DUCTS
[NFPA 54: FIGURE A.9.3.3.1(2)]

with the outdoors (see **Figure 701.6.2**) and shall have a minimum free area of the following:

(1) One square inch per 3000 Btu/h (0.0007 m^2/kW) of the total input rating of all appliances located in the enclosure.

(2) Not less than the sum of the areas of all vent connectors in the space. [NFPA 54:9.3.3.2]

This method allows one upper opening or duct commencing within 12 inches of the ceiling of the compartment and sized at 1 square inch/3,000 Btu/h (free area) of the total input rating of all appliances located in the compartment or room. The size of the opening or duct shall not be less than the total areas of the vent connectors in the space. In addition, the compartment containing the appliances must be of a size so that the equipment would have at least 1-inch clearance on the back and sides of the appliance(s) and 6-inch clearance in front (see **Figure 701.6.2**). Please note that the minimum 3-inch dimension of the combustion air opening is also applicable here.

Example:

A 140,000 Btu/h fan-assisted furnace and two 50,000 Btu/h water heaters are installed in a room 35 feet by 35 feet by 8 feet; all air will be from the outdoors. Determine the minimum diameter of an upper opening to supply combustion air.

Total appliance rating:

140,000 Btu/h + 50,000 Btu/h + 50,000 Btu/h = 240,000 Btu/h

240,000 Btu/h ÷ 3,000 = 80 sq in of combustion air opening is required.

To determine the minimum diameter of the opening, begin with using the following equation for finding the diameter if the area of a circle is known:

$D = \frac{\sqrt{4A}}{\pi}$ Where A= area

A = 80 square inches

$D = \frac{\sqrt{4 \times 80}}{\pi}$

D = 10.09 inches

A 10-inch round opening would not be acceptable. At least an 11-inch round opening would be required. A rectangular opening or duct with clear dimensions of 9 inches by 9 inches would also be acceptable. The area of all vent connectors combined must not exceed 80 square inches.

701.7 Combination Indoor and Outdoor Combustion Air. The use of a combination of indoor and outdoor combustion air shall be in accordance with Section 701.7.1 through Section 701.7.3. [NFPA 54:9.3.4] (see Appendix G for example calculations)

701.7.1 Indoor Openings. Where used, openings connecting the interior spaces shall comply with Section 701.5. [NFPA 54:9.3.4(1)]

701.7.2 Outdoor Opening(s) Location. Outdoor opening(s) shall be located in accordance with Section 701.6. [NFPA 54: 9.3.4(2)]

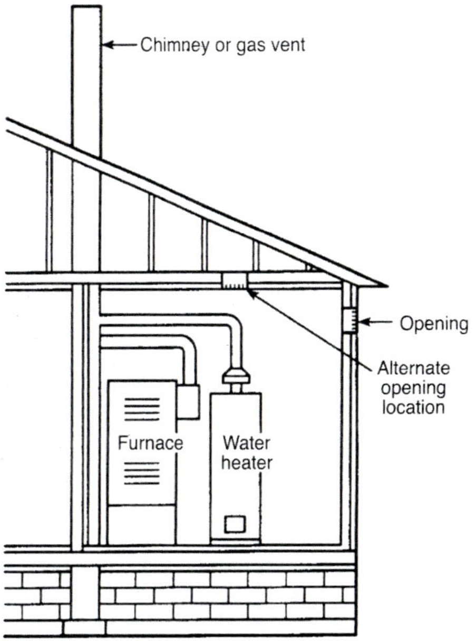

FIGURE 701.6.2
ALL COMBUSTION AIR FROM OUTDOORS THROUGH SINGLE COMBUSTION AIR OPENING
[NFPA 54: FIGURE A.9.3.3.2]

701.7.3 Outdoor Opening(s) Size. The outdoor opening(s) size shall be calculated in accordance with the following:

(1) The ratio of the interior spaces shall be the available volume of all communicating spaces divided by the required volume.

(2) The outdoor size reduction factor shall be 1 minus the ratio of interior spaces.

(3) The minimum size of outdoor opening(s) shall be the full size of outdoor opening(s) calculated in accordance with Section 701.6, multiplied by the reduction factor. The minimum dimension of air openings shall not be less than 3 inches (76 mm). [NFPA 54:9.3.4(3)]

In cold climates, it is not desirable to have combustion air openings to the outdoors; therefore, every attempt is made to utilize the interior of the building as a source of combustion air. The volume of the room is calculated and the Standard Method is used to determine if this volume is adequate for the appliance input rating (factor of 50). If the volume of the room is insufficient, an adjacent room (if available) is used to combine its volume with the appliance room volume to achieve the required minimum. Another possible approach is to calculate the infiltration rate. Once the infiltration rates are known, the Known Infiltration Method could then be used to calculate the minimum required volume. This will usually result in a lesser volume requirement. Instead of the Standard Method factor of 50, the Known Infiltration Method, depending on ACH calculated, could result in factors as low as 35 for gravity-type appliances and as low as 25 for fan-assisted-type appliances.

If these approaches do not yield the desired result, this

section allows the option of using the interior volume (that is not adequate), coupled with openings to the outdoors that are just large enough to account for the "deficiency" of the interior volume. For example, if the interior volume is only 70 percent of the minimum required volume, as calculated by Section 701.4 (using either method: Standard or Known Infiltration), the full outdoor combustion air opening size, as calculated in accordance with Section 701.4 (using one of the methods: Two Openings or One Opening), could be reduced by 70 percent or be sized for the 30-percent deficiency of the volume requirement.

The steps to be followed are:

1. Calculate the available volume (room).
2. Calculate the required volume:

   Appliance rating in Btu/h times factor (50 if Standard Method, or calculated factor if Known Infiltration), divided by 1,000.
3. Divide the available volume by the required volume (For example, the result is 0.7).
4. Subtract the result from the number 1 (1 - 0.7 = 0.3). This will be the "deficiency" of the space volume.
5. Determine the "full" required outdoor combustion air opening size, using either the Two-openings or the One-opening Method (For example, if two openings, 40 square inches, each).
6. Multiply the required opening size (40) by the deficiency (0.3) to determine the minimum required opening size to the outdoors (40 x 0.3 = 12). Therefore, this example results in requiring two openings (top and bottom) sized at 12 square inches, each (instead of 40 square inches, each). Please note that the minimum dimension of combustion air opening should not be less than 3 inches.

Example:

A 140,000 Btu/hr fan-assisted furnace and two 50,000 Btu/hr water heaters are installed in a room 35 feet by 35 feet by 8 feet.

Using the Standard Method, the required volume for combustion air would be 50 cubic feet per 1,000 Btu/hr.

Available room volume:

35 x 35 x 8 = 9,800 cubic feet

Required room volume:

140,000 + 50,000 + 50,000 = 240,000 Btu/hr

240,000 x 50 ÷ 1,000 = 12,000 cubic feet

The volume of room is not sufficient to provide the required combustion air.

The ratio of available volume to required volume:

9,800 ÷ 12,000 = 0.8167

1.0 - 0.8167 = 0.1833 (This is the deficiency and the reduction factor to be applied.)

Assuming the One-opening Method is used. Minimum area of the opening:

240,000 x 1 ÷ 3,000 = 80 square inches

Applying the deficiency or the reduction factor to the required size:

0.1833 x 80 in^2 = 14.67 square inches

A single upper opening or duct directly to the outside sized at 14.67 square inches minimum clear area is needed.

Another method (that might be simpler) is suggested:

1. Calculate the available room volume:

   35 x 35 x 8 = 9,800 cu ft
2. Calculate the maximum input rating of the appliances that could be installed in this room using the factor 50 (or any other factor if the Known Infiltration Rate Equations are used):

   9,800 x (1,000 ÷ 50) = 196,000 Btu/h [note that the reciprocal of the factor (50/1,000) is used here].
3. Calculate the total appliances input rating:

   140,000 + 50,000 + 50,000 = 240,000 Btu/hr
4. Calculate the remaining capacity in Btu/h "needing" combustion air from outdoors:

   240,000 – 196,000 = 44,000 Btu/h
5. Using the One-opening Method, determine required combustion air opening or duct size to outdoors:

   44,000 x 1 ÷ 3,000 = 14.67 square inches.

701.8 Engineered Installations. Engineered combustion air installations shall provide an adequate supply of combustion, ventilation, and dilution air and shall be approved by the Authority Having Jurisdiction. [NFPA 54:9.3.5]

When first approved by the Authority Having Jurisdiction (AHJ), combustion, ventilation and dilution air supply may be designed in accordance with recognized engineering principles. Although this method is not used often, it is typically considered in very large or complicated and non-conventional type systems (such as in large commercial and industrial applications). Although the combustion air provisions in the code would be adequate for most gravity and Category I appliances, some systems may require a special design to yield an economic but still safe installation. It is recommended that in such installations, the engineered system be evaluated against the manufacturer's instructions and any national standard used by the engineer or that is relevant to the installation.

701.9 Mechanical Combustion Air Supply. Where all combustion air is provided by a mechanical air supply system, the combustion air shall be supplied from outdoors at the minimum rate of not less than 0.35 cubic feet per minute per 1000 Btu/h [0.034 (m^3/min)/kW] for all appliances located within the space. [NFPA 54:9.3.6]

This section allows for the use of mechanical blowers or fans to supply combustion air in lieu of the "gravity" type combustion air openings and ducts outlined in the previous sections. The fan or blower controls must be interlocked with each of the appliances in the room so that appliances will not operate if the fan is not providing the combustion air. The manufacturer's data sheets for the appliances in the room must be reviewed in the event they have larger rate requirements. If not, the minimum rate required by this section is 0.35 cubic feet per minute (cfm)/1,000 Btu/h of appliance input rating.

Example:

A mechanical room has three furnaces rated at 200,000 Btu/h, each and four "tankless" water heaters rated at 150,000 Btu/h, each. Please calculate the required mechanical ventilation rate needed for combustion air.

Total input rating = (200,000 x 3) + (150,000 x 4)

= 1,200,000 Btu/h

Cubic feet per minute for combustion air

= 1,200,000 x 0.35/1,000

= 420 cfm minimum

701.9.1 Exhaust Fans. Where exhaust fans are installed, additional air shall be provided to replace the exhausted air. [NFPA 54:9.3.6.1]

This section is similar to Section 701.3 which requires make-up air to replace exhausted air. The requirement is still valid even when mechanical methods are used to supply combustion air to the appliance(s).

701.9.2 Interlock. Each of the appliances served shall be interlocked to the mechanical air supply system to prevent main burner operation where the mechanical air supply system is not in operation. [NFPA 54:9.3.6.2]

As mentioned previously, to ensure that combustion air will be available when the appliance burners are on, the fan or blower electrical control must be interlocked with the burners to ensure that the main burners will not operate unless the combustion air fan is operational.

701.9.3 Specified Combustion Air. Where combustion air is provided by the building's mechanical ventilation system, the system shall provide the specified combustion air rate in addition to the required ventilation air. [NFPA 54:9.3.6.3]

The building mechanical system is typically designed to supply a certain amount of outside air for ventilation of the building and for the fresh air needs of the occupants. Most design procedures calculate these rates as cfm/occupant or cfm/square feet, plus any additional quantities to replace exhausted air. If the building mechanical system is to also supply combustion air to appliances and equipment, these quantities of combustion air needs must be added to the design ventilation quantities. Needless to say, the burners of all the appliances needing combustion air must be interlocked with the building mechanical systems to prevent the operation of the appliances when the building mechanical system is down or not operating.

701.10 Louvers, Grilles, and Screens. The required size of openings for combustion, ventilation, and dilution air shall be based on the net free area of each opening. Where the free area through a design of louver, grille, or screen is known, it shall be used in calculating the size opening required to provide the free area specified. Where the louver and grille design and free area are not known, it shall be assumed that wood louvers have 25 percent free area and metal louvers and grilles have 75 percent free area. Nonmotorized louvers and grilles shall be fixed in the open position. [NFPA 54:9.3.7.1]

Usually, louvers or grilles are installed on combustion air openings. This is to protect the openings against entry of rain, birds or rodents. Excessive amounts of water (rain) could have an adverse effect on the appliance or equipment and birds nesting in the airway could obstruct the combustion air source and could make the air unsanitary. In these cases, the solid portion of the louver or grille must be considered when determining the net free area of the opening. If the free area of the grille is known (some prefabricated grilles are provided with labels indicating the free area), the label is to be used to determine the net free area. If the free area is not known, the overall "gross area" should be measured and a factor of 25 percent should be applied when the grille has wood louvers, and a factor of 75 percent should be applied when the grille has metal louvers to account for the blocking effect of the louvers and to determine the free area. If the opening is covered with a wire mesh, the openings should not be smaller than 1/4 inch, as smaller size mesh is subject to obstruction by the accumulation of dirt or lint.

Example:

A grille has an overall dimension of 30 inches by 25 inches.

If the grille has wood louvers, what is the free area?

Wood grille: 30 x 25 x 0.25 = 187.5 square inches

If the grille has metal louvers, what is the free area?

Metal grille: 30 x 25 x 0.75 = 562.5 square inches

Example:

If 100 square inches of net free opening is required and the grille has metal louvers (75 percent free area), what size grille would be required?

100 ÷ 0.75 = 133.3 square inches

Grille size: 10 x 14 = 140 square inches (exceeds minimum)

Manual and fire dampers are prohibited on combustion air openings (see Section 701.12); however, motorized dampers are allowed, provided there is assurance that they will open, and remain open, when the main burners of the appliance are operational. This section requires that the motorized damper be interlocked so the burners will not operate if the damper is closed, and that they will shut down if the damper closes during the equipment operation. As mechanical devices installed on combustion air openings or ducts are subject to failure, this requirement will provide the necessary assurance that the appliance will not operate unless the combustion air is made available.

701.10.1 Minimum Screen Mesh Size. Screens shall not be smaller than ¼ of an inch (6.4 mm) mesh. [NFPA 54:9.3.7.2]

701.10.2 Motorized Louvers. Motorized louvers shall be interlocked with the appliance so they are proven in the full open position prior to main burner ignition and during main burner operation. Means shall be provided to prevent the main burner from igniting should the louver fail to open during burner startup and to shut down the main burner if the louvers close during burner operation. [NFPA 54:9.3.7.3]

701.11 Combustion Air Ducts. Combustion air ducts shall comply with the following [NFPA 54:9.3.8]:

(1) Ducts shall be constructed of galvanized steel or a material having equivalent corrosion resistance, strength, and rigidity.

Exception: Within dwellings units, unobstructed stud and joist spaces shall not be prohibited from conveying combustion air, provided that not more than one fireblock is removed. [NFPA 54:9.3.8.1]

Combustion air ducts must be of metal or equivalent corrosion-resistant material so they will not be subject to corrosion, which would develop holes and cracks in the ducts, thus affecting the airflow within. The duct material should also be equivalent to metal in strength and rigidity so ducts will not be crushed, bent or subject to physical damage, potentially affecting the air flow within. Flexible air ducts should not be acceptable as they are easily deformed or crushed. The exception allows the use of an unobstructed stud or joist cavity to convey combustion air in dwelling units. It is imperative that care be exercised in inspecting these installations to ensure that no obstruction exists in the cavity and that fire blocks are installed between the stud cavity and any adjacent return or circulating air for forced air units (see **Figure 701.11a**). As fire blocking is important in preventing the spread of fire inside stud or joist cavities, the code allows only one fire block to be removed in any one cavity to allow its use as a combustion air duct.

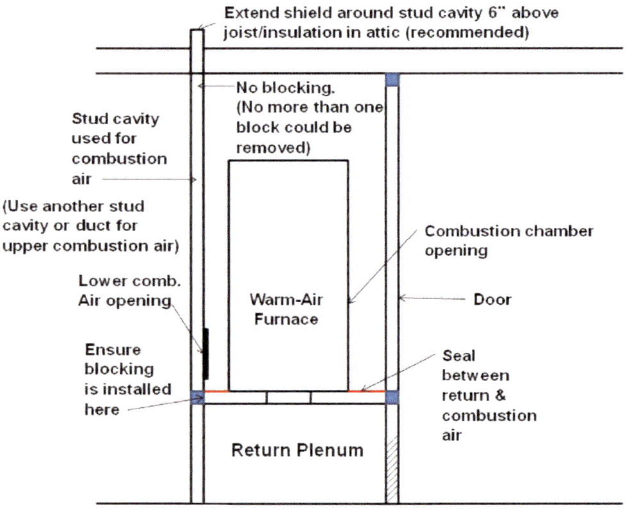

FIGURE 701.11A
BLOCKING

(2) Ducts shall terminate in an unobstructed space, allowing free movement of combustion air to the appliances. [NFPA 54:9.3.8.2]

In order to ensure the free flow of air throughout the appliance enclosure, the combustion air duct terminations inside the compartment should be kept away from any obstructions, including the floor of the appliance enclosure itself. The minimum dimension to the floor of the compartment should be at least the minimum width of the combustion air duct (typically 3 inches). In addition, if the duct terminates in the back or one of the sides of the compartment, there should be a clear pathway around the appliance to the front where the fire box is located (see **Figure 701.11b**). This pathway is typically 3 inches, and should be maintained even if the manufacturer's installation instructions allow lesser clearances from the appliance to the sides and back of the compartment.

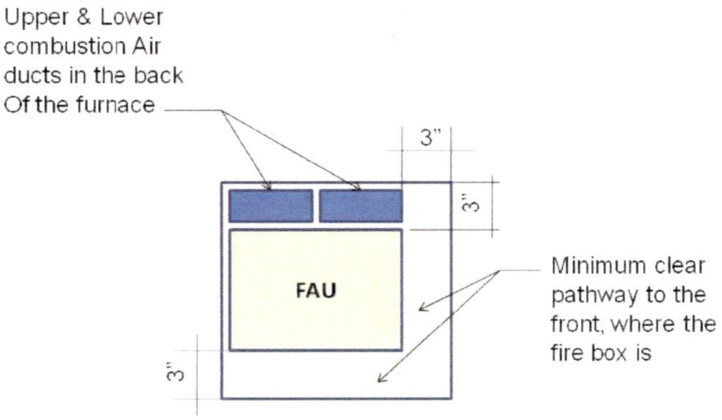

FIGURE 701.11B
COMBUSTION AIR PATHWAY

(3) Ducts shall serve a single space. [NFPA 54:9.3.8.3]

Ducts shall serve a single space. Due to the fact that separate compartments could have different convection and thermo-siphon rates of flow, providing one duct to serve both compartments could lead to disruption of this delicate process and result in an uneven or insufficient rate of combustion airflow for either or both compartments.

(4) Ducts shall not serve both upper and lower combustion air openings where both such openings are used. The separation between ducts serving upper and lower combustion air openings shall be maintained to the source of combustion air. [NFPA 54:9.3.8.4]

The separation of the two ducts ensures a free flow of air from inside the enclosure to the outside. The upper opening will allow warm air to flow out, and the lower opening will draw cooler air in. Allowing one duct to serve both openings would result in the disruption of the convection and thermo-siphon process within the duct and, in turn, within the compartment, with the possible result of uneven or insufficient combustion airflow. Note that if a stud cavity is used as a combustion air duct, one stud cavity should be used as the upper duct and a separate stud cavity should be used as the lower combustion air duct.

(5) Ducts shall not be screened where terminating in an attic space. [NFPA 54:9.3.8.5]

Ducts shall not be screened when terminating in an attic. This is to prevent the possibility of obstruction of the ducts when insulation is blown in the attic, a popular method of insulating existing buildings that have insufficient or no insulation. If no screens are installed, the blown-in insulation would simply fall in the compartment and alert the observer instead of settling on top of the duct screen and blocking the flow of combustion air.

(6) Combustion air intake openings located on the exterior of the building shall have the lowest side of the combustion air intake openings located at least 12 inches (305 mm) vertically from the adjoining finished ground level. [NFPA 54:9.3.8.8]

The lowest side of the combustion air opening to the outside must be at least 12 inches above grade. This is to ensure that the combustion air opening will not be blocked with snow or dirt. If local conditions indicate that snow accumulation could exceed 12 inches, the height of the combustion air opening should be raised higher than the expected snow level.

(7) Horizontal upper combustion air ducts shall not slope downward toward the source of combustion air. [NFPA 54:9.3.8.6]

A horizontal upper combustion air duct shall not slope downward toward the source of combustion air. The duct should be either horizontal or slope upward from the appliance compartment until it reaches the source of combustion air. This will not allow the trapping of warm air rising within the duct to the outside of the compartment and will help maintain continuous circulation of combustion air.

(8) The remaining space surrounding a chimney liner, gas vent, special gas vent, or plastic piping installed within a masonry, metal, or factory built chimney shall not be used to supply combustion air.

Exception: Direct vent appliances designed for installation in a solid fuel-burning fireplace where installed in accordance with the manufacturer's installation instructions. [NFPA 54:9.3.8.7]

The space around a chimney liner installed within a masonry chimney shall not be used to supply combustion air. Since the chimney could be carrying hot flue gases from a fireplace or similar equipment, the heat radiating to the space could easily disrupt the natural convection process of the combustion air within or prevent the thermo-siphon action of the supply combustion air from flowing to the appliance compartment. There is also the possibility of a leak of flue gases into the liner area that could later be drawn into the appliance compartment. This code provision does not apply to appliances that have listed venting systems, and the installation instructions show that it could be installed in this manner.

701.12 Dampers Prohibited. Combustion air ducts or plenums shall not be installed so as to require openings in or penetrations through construction where fire dampers are required. Manually operated dampers shall not be installed in combustion air openings. With prior approval, power-actuated movable louvers admitting combustion air shall be permitted to be used and, where installed, shall be electrically interlocked with the main burner fuel-supply valve so as to prevent fuel delivery unless the louvers are in the fully open position.

Manual or fire dampers are prohibited from being installed on combustion air openings or ducts. The installation of combustion air ducts should avoid penetrating any fire-rated assembly, unless a rated shaft could be constructed around the combustion air duct to avoid the need for fire dampers. Motorized dampers are allowed provided there is assurance that they will open, and remain open, when the main burners of the appliance are operational. This section requires that the motorized damper be interlocked so the burners will not operate if the damper is closed and that they will shut down in the event the damper closes during the equipment operation. As mechanical devices installed on combustion air openings or ducts are subject to failure, this requirement will provide the necessary assurance that the appliance will not operate unless combustion air is made available.

702.0 Extra Device or Attachment.

702.1 General. No device or attachment shall be installed on any appliance that could in any way impair the combustion of gas. [NFPA 54:9.1.15]

CHAPTER 8

CHIMNEYS AND VENTS

801.0 General.

801.1 Applicability. The requirements of this chapter shall govern the venting of fuel-burning appliances.

801.2 Venting of Gas Appliances. Low-heat and medium-heat gas appliances shall be vented in accordance with this chapter. Other gas appliances shall be vented in accordance with NFPA 211 or other applicable standards.

801.3 Appliances Fueled by Other Fuels. Appliances fueled by fuels other than gas shall be vented in accordance with NFPA 211 and the appliance manufacturer's instructions.

802.0 Venting of Appliances.

802.1 Listing. Type B and Type B-W gas vents shall comply with UL 441, Type L gas vents shall comply with UL 641.

802.1.1 Installation. Listed vents shall be installed in accordance with this chapter and the manufacturer's installation instructions. [NFPA 54:12.2.1]

802.1.2 Prohibited Discharge. Appliance vents shall not discharge into a space enclosed by screens having openings less than ¼ of an inch (6.4 mm) mesh.

802.2 Connection to Venting Systems. Except as permitted in Section 802.2.1 through Section 802.2.7, all appliances shall be connected to venting systems. [NFPA 54:12.3.1]

802.2.1 Appliances Not Required to be Vented. The following appliances shall not be required to be vented:

(1) Listed ranges.

(2) Built-in domestic cooking units listed and marked for optional venting.

(3) Listed hot plates.

(4) Listed Type 1 clothes dryers exhausted in accordance with Section 504.4.

This section identifies that clothes dryers are exhausted, rather than vented. This means that the products of combustion and the moisture removed from the clothing are combined and removed together. This exhaust is powered with a fan to ensure proper operation. Clothes dryer exhausts and appliance vents must never be combined. Clothes dryer exhausts are under positive pressure and Category I vents are under negative pressure; therefore, the two are incompatible.

(5) A single listed booster-type (automatic instantaneous) water heater, when designed and used solely for the sanitizing rinse requirements of a dishwashing machine, provided that the appliance is installed with the draft hood in place and unaltered, if a draft hood is required, in a commercial kitchen having a mechanical exhaust system. [Where installed in this manner, the draft hood outlet shall not be less than 36 inches (914 mm) vertically and 6 inches (152 mm) horizontally from any surface other than the appliance.]

(6) Listed refrigerators.

(7) Counter appliances.

(8) Room heaters listed for unvented use.

(9) Direct gas-fired makeup-air heaters.

(10) Other appliances listed for unvented use and not provided with flue collars.

(11) Specialized appliances of limited input such as laboratory burners or gas lights. [NFPA 54:12.3.2]

Some of the appliances listed in Sections 802.2.1(5) through 802.2.1(11) must be vented if the total input of all the appliances in this class exceeds 20 Btu/hr per ft^3 (207 W per m^3) of room volume. This ratio is the same as is provided by the Standard Method of calculation of the room volume needed to provide adequate air for combustion and ventilation. See Section 701.1 for more on combustion air.

Listed ranges, domestic cooking appliances with provision for optional venting, listed hot plates and laundry stoves and listed clothes dryers can be installed in rooms with a smaller volume than is required by Section 701.4.1. These appliances have one of the following characteristics:

- Limited input (e.g., hot plates and laundry stoves)
- Short operating period (e.g., cooking appliances and dryers)
- Removal of products of combustion (e.g., dryers)
- Locations that have additional air supply (e.g., kitchens)

802.2.2 Maximum Input Rating. Where any or all of the appliances in Section 802.2.1(5) through Section 802.2.1(11) are installed so the aggregate input rating exceeds 20 Btu/h/ft^3 (207 W/m^3) of room or space in which it is installed, one or more shall be provided with venting systems or other approved means for conveying the vent gases to the outdoors so that the aggregate input rating of the remaining unvented appliances does not exceed 20 Btu/h/ft^3 (207 W/m^3). [NFPA 54:12.3.2.1]

802.2.3 Adjacent Room or Space. Where the calculation includes the volume of an adjacent room or space, the room or space in which the appliances are installed shall be directly connected to the adjacent room or space by a doorway, archway, or other opening of comparable size that cannot be closed. [NFPA 54:12.3.2.2]

802.2.4 Ventilating Hoods. The use of ventilating hoods and exhaust systems to vent appliances shall be limited to industrial appliances and appliances installed in commercial applications. [NFPA 54:12.3.3]

A common application of the provision in this Section is in restaurant kitchens, where, for sanitation reasons, very hot water is required for dishwashing. The gas supply to the booster is interlocked with the hood, usually with a normally closed solenoid valve in the gas line that is opened only when the hood is running. This interlock is needed to meet the requirements of Sections 802.3.3 and 802.3.4.

The method of venting appliances in a building using ventilation hoods is used frequently in restaurant kitchens to vent booster water heaters used to provide the very hot water required for sanitation. Information on the design and installation of ventilating hoods in industrial plants can be obtained from NFPA 91, Standard for Exhaust Systems for Air Conveying of Vapors, Gases, Mists, and Noncombustible Particulate Solids.

802.2.5 Well-Ventilated Spaces. The flue gases from industrial-type appliances shall not be required to be vented to the outdoors where such gases are discharged into a large and well-ventilated industrial space. [NFPA 54:12.3.4]

This provision is an exception to the requirement in Section 802.2 that all gas utilization equipment be connected to a venting system. It recognizes common, safe practice in metal-treating and other industries. Some of the equipment normally installed under the requirements of this Section includes heat-treating furnaces, radiant tube burners and pot heaters in foundries.

Building ventilation must be designed to dilute flue gases to a safe level by means of natural draft or mechanical ventilation. In many of these installations, ventilation for fume removal or process reasons exceeds the ventilation level needed for safe removal of the products of combustion.

802.2.6 Direct Vent Appliances. Listed direct vent appliances shall be installed in accordance with the manufacturer's installation instructions and Section 802.8.2. [NFPA 54:12.3.5]

802.2.7 Appliances with Integral Vents. Appliances incorporating integral venting means shall be installed in accordance with the manufacturer's installation instructions and Section 802.8 and Section 802.8.1. [NFPA 54:12.3.6]

An integral vent is a vent that is supplied by the manufacturer and is part of the gas appliance.

802.2.8 Incinerators, Commercial–Industrial. Commercial industrial-type incinerators shall be vented in accordance with NFPA 82. [NFPA 54:12.3.7]

802.3 Minimum Safe Performance. Venting systems shall be designed and constructed to convey all flue and vent gases to the outdoors. [NFPA 54:12.1]

Although gas is a clean-burning fuel, the products of combustion must not be allowed to accumulate inside of a building. (See **Figure 802.3** for the products of combustion formed under ideal conditions.) Therefore, venting of most gas utilization equipment is required (see Section 802.2.1 for a list of appliances not requiring venting). A properly installed and maintained venting system will perform the following functions to provide for proper appliance function and the safety of building occupants:

1. Convey all of the combustion products to the outside atmosphere.
2. Prevent damage to the gas equipment, from water vapor condensation in the flue gases.
3. Prevent overheating of walls, building structure and other combustible materials that are installed with required clearance to the appliance and venting system.
4. Provide fast priming of natural draft-venting systems to minimize spillage of combustion products into the building.

The venting system's job is to channel combustion products out of the building. To do so, the venting system must keep the combustion products as warm as possible in the vent. Heat retention maximizes the draft produced.

High-efficiency condensing appliances have a seasonal efficiency of 90 percent or higher, which reduces vent gas temperatures to a point where the water vapor produced as a product of combustion condenses to liquid water in the appliance or in the vent. These condensing appliances carry a vented appliance category of Category IV. This type of appliance produces much cooler vent gases, resulting in water condensing in the vent. Venting must be accomplished with a fan because the vent gases are not hot enough to operate the natural draft vent. Water will condense in the vent and will dissolve some of the gases produced during combustion, which are slightly acidic. The vent materials used with these appliances must be able to resist the acidic condensate.

The advantage of high-efficiency appliances is that they reduce the amount of gas consumed with no loss in output. A mid-efficiency appliance uses one-third less gas than a conventional appliance, and a condensing appliance uses only one-half of the gas of a conventional appliance. The savings in fuel are offset by higher first cost and the higher maintenance requirements of high-efficiency appliances.

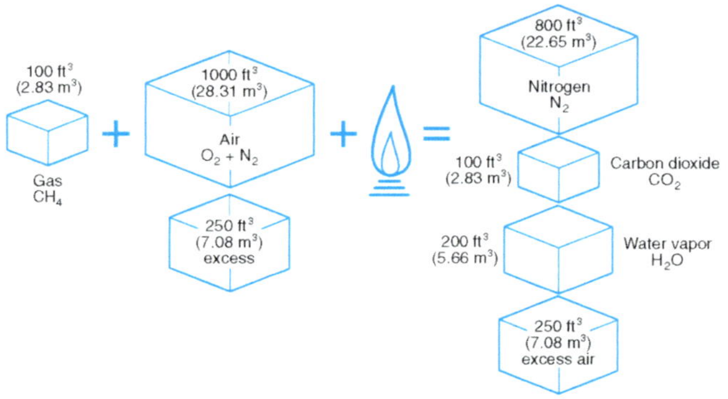

FIGURE 802.3
COMBUSTION OF NATURAL GAS

802.3.1 Appliance Draft Requirements. A venting system shall satisfy the draft requirements of the appliance in accordance with the manufacturer's instructions. [NFPA 54:12.4.1]

This section provides an overall performance requirement for a natural draft vent system. The vent system must remove all the products of combustion from gas appliances in the building. The principle on which natural draft vents operate is simple. Combustion gases rise in the chimney or vent only because they are hotter, and, therefore, lighter than the surrounding air. The hotter the gases and the higher the vent, the more swiftly and powerfully the gases will rise. Conversely, the cooler the gases and the shorter the vent, the more sluggish the gas movement will be. The flue gases in the vent must remain hot enough over the length of the vent to provide a strong draft. If cooled enough, the upward motion of the gases stops altogether and combustion gases can spill into the building through the relief opening of the appliance draft hood.

When the combustion products and dilution air rise in the vent, this volume must be replaced by air from outside the building. This replacement air can be supplied through normal air infiltration (through small openings in the building walls), through outdoor openings purposely installed in outside walls or by a mechanical air system. Thus, to ensure proper vent operation, the buildings air tightness and other devices exhausting air from the building must be taken into account.

Blockage of a natural draft vent may cause flue gases to spill into the building through the relief opening of the draft hood. Spillage can also be caused if the pressure in the vicinity of the appliance is much lower than the pressure outside the building, referred to as "depressurization." Depressurization may be caused by mechanical exhausts, fireplaces, wind—anything that removes air from the building. Consequently, periodically verifying the performance of appliance venting systems is wise. Verification can be done readily with a natural draft system, described as follows:

1. Operate the appliance for at least 5 minutes to allow the flue gases to heat the vent.
2. Move a lighted match across the entire width of the draft hood relief opening (see **Figure 802.3.1**).

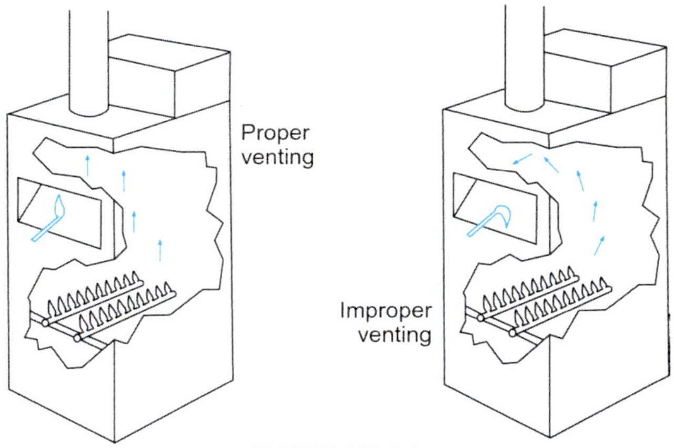

FIGURE 802.3.1
MATCH TEST TO DETERMINE PROPER AND IMPROPER VENTING

3. If the match flame is blown downward or extinguished, have the venting system or chimney checked by a qualified agency.
4. If the appliance is not equipped with a draft hood, the match test might not be able to be performed. In this case, consult the appliance instruction manual.

Good air circulation in adequate amounts is also vital for good venting and effective appliance operation. Signs of improper operation include a yellow or wavering flame; discoloration around access doors; a pungent odor in the building; excessive humidity, condensation or mold; corrosion of the vent material or soot near the burner or vent area. If any of these conditions are observed, a qualified agency should be contacted to have the installation inspected.

Under extremely adverse conditions, carbon monoxide can be produced as a result of either improper venting of combustion products, insufficient fresh air to support the proper burning of gas or improperly adjusted appliances.

802.3.2 Appliance Venting Requirements. Appliances required to be vented shall be connected to a venting system designed and installed in accordance with the provisions of Section 802.4 through Section 802.15.1. [NFPA 54:12.4.2]

802.3.3 Mechanical Draft Systems. Mechanical draft systems shall be listed in accordance with UL 378 and installed in accordance with both the appliance and the mechanical draft system manufacturer's installation instructions. [NFPA 54:12.4.3.1]

802.3.3.1 Venting. Appliances requiring venting shall be permitted to be vented by means of mechanical draft systems of either forced or induced draft design. [NFPA 54:12.4.3.2]

Mechanical draft systems use the mechanical force from a fan or blower to vent the combustion products. Natural buoyancy is not needed or significant. There are two types of mechanical draft systems: forced draft and induced draft. Forced draft exists wherever the static pressure in the vent or chimney is greater than atmospheric pressure. Typically, any section of vent downstream of the blower will be under a positive gauge pressure. Therefore, care must be taken to ensure that the vent pipe used is appropriate for positive-pressure combustion gases. Otherwise, leaks of the combustion products may occur. A mechanical draft water heater is shown in **Figure 802.3.3.1**.

An induced draft system has the exhaust fan (induced draft fan) mounted outside of the building. The fan creates a negative pressure in the insulated exhaust pipe "pulling" the hot gas from the appliance.

FIGURE 802.3.3.1
MECHANICALLY VENTED WATER HEATER
(Courtesy of A. O. Smith Water Products)

802.3.3.2 Leakage. Forced draft systems and all portions of induced draft systems under positive pressure during operation shall be designed and installed so as to prevent leakage of flue or vent gases into a building. [NFPA 54:12.4.3.3]

802.3.3.3 Vent Connectors. Vent connectors serving appliances vented by natural draft shall not be connected into any portion of mechanical draft systems operating under positive pressure. [NFPA 54:12.4.3.4]

The key term in Sections 802.3.3.2 and 802.3.3.3 is "positive pressure." Vents serving fan-assisted appliances operate under negative pressure when vented into a chimney, even though the appliances use a mechanical fan. Fan-assisted appliances can be common-vented with other fan-assisted appliances, draft hood-equipped appliances and all other Category I-vented appliances.

Sections 802.8.1 and 802.8.5 specify the location of vent terminals in relation to air inlets, doors and windows.

A minimum elevation of 7 feet above grade is required so that combustion gases do not create a nuisance for pedestrians. Potential nuisances include the high temperature of the vent and the possibility of condensation of water in the vent gases that can form a pool of water, ice or snow on the pavement. The definition of a "public walkway" is not specified in the code. The dictionary definition and common sense should be used.

802.3.3.4 Operation. Where a mechanical draft system is employed, provision shall be made to prevent the flow of gas to the main burners when the draft system is not performing so as to satisfy the operating requirements of the appliance for safe performance. [NFPA 54:12.4.3.5]

802.3.3.5 Exit Terminals. The exit terminals of mechanical draft systems shall be not less than 7 feet (2134 mm) above finished ground level where located adjacent to public walkways and shall be located as specified in Section 802.8 and Section 802.8.1. [NFPA 54:12.4.3.6]

802.3.4 Ventilating Hoods and Exhaust Systems. Where automatically operated appliances, other than commercial cooking appliances, are vented through a ventilating hood or exhaust system equipped with a damper or with a power means of exhaust, provisions shall be made to allow the flow of gas to the main burners only when the damper is open to a position to properly vent the appliance and when the power means of exhaust is in operation. [NFPA 54:12.4.4.1]

This is a safe, alternative method for venting of gas utilization equipment in commercial applications. For example, in restaurant kitchens a "booster" water heater is used to provide very hot water for sanitation. In this application, the booster water heater is operated only while the restaurant kitchen is in operation, and the range hood is used to vent both the range and the water heater.

Range hoods are covered under this code, which requires an interlock that operates when the airflow falls below a predetermined value. This circuit can be interlocked with the main gas valve of the water heater (or other gas appliance vented via the range hood) so that gas will not flow to the main burners unless a minimum airflow is achieved in the vent system.

802.3.5 Circulating Air Ducts, Above-Ceiling Air-Handling Spaces, and Furnace Plenums. Venting systems shall not extend into or pass through any fabricated air duct or furnace plenum. [NFPA 54:12.4.5.1]

802.3.6 Above-Ceiling or Nonducted Air Handling System. Where a venting system passes through an above-ceiling air space or other nonducted portion of an air-handling system, it shall conform to one of the following requirements:

(1) The venting system shall be a listed special gas vent, other system serving a Category III or Category IV appliance, or other positive pressure vent, with joints sealed in accordance with the appliance or vent manufacturer's instructions.

(2) The vent system shall be installed such that no fittings or joints between sections are installed in the above-ceiling space.

(3) The venting system shall be installed in a conduit or enclosure with joints between the interior of the enclosure and the ceiling space sealed. [NFPA 54:12.4.5.2]

802.4 Type of Venting System to be Used. The type of venting system to be used shall be in accordance with Table 802.4. [NFPA 54:12.5.1]

To use Table 802.4, locate in the left-hand column the type of gas utilization equipment to be vented and read across the row to the second and third columns to find the type(s) of venting system(s) and location of requirements. For example, listed Category I equipment can be vented using a Type B gas vent as required in Section 802.6, chimney as required in Section 802.5, single-wall metal pipe as required in Section 802.7, chimney lining system that is listed for gas venting as required in 802.5.3 or a special gas vent listed for these appliances as required in 802.4.3.

Table 802.4 refers to Category I through Category IV appliances. The categories are based on vent temperature and pressure. A specific temperature is not provided because it is not the same for all appliances. The ANSI Z21 standards can be referenced for the manufacture and testing of appliances for this information.

Incinerators as listed in Table 802.4 referenced in NFPA 82. NFPA 82 Section 5.3.1.6 allows masonry chimneys to vent the incinerator; and NFPA 82 Section 5.3.1.7 allows factory-built chimneys to vent the incinerator.

The criteria in the ANSI Z21 standards for appliance categorization are based on a flue loss of 17 percent of total energy. The 17 percent flue loss is the same flue loss built into the vent sizing tables for fan-assisted appliances. In this way, the standards ensure that the appliance will work properly with the vent system.

In the following **Commentary Table**, Appliance Vent Categories, the term "nonpositive vent pressure" means that the pressure in the vent will be lower than the surrounding atmosphere if the vent system meets the requirements of Sections 802.0 and 803.0. The incorporation of a fan into the appliance does not always mean that the vent pressure is positive. If unsure, check the appliance nameplate or manufacturer's instructions for the venting category or check the vent pressure with a manometer or other pressure gauge when the appliance is operating.

802.4.1 Plastic Piping. Where plastic piping is used to vent an appliance, the appliance shall be listed for use with such venting materials and the appliance manufacturer's installation instructions shall identify the specific plastic piping material. The plastic pipe venting materials shall be labeled in accordance with the product standards specified by the appliance manufacturer or shall be listed and labeled in accordance with UL 1738. [NFPA 54:12.5.2]

Before the introduction of high-efficiency gas utilization equipment (90+ percent efficiency), plastic piping was prohibited as a vent material. High-efficiency (Category IV) appliances reduce vent temperatures, resulting in condensate formation. As accumulation of condensate became a source of corrosion of metal vents, plastic piping became the preferred material. Plastic vent materials are required to be used for listed gas utilization equipment only when specified in the manufacturer's instructions.

TABLE 802.4
TYPE OF VENTING SYSTEM TO BE USED
[NFPA 54: TABLE 12.5.1]

APPLIANCES	TYPE OF VENTING SYSTEM	LOCATION OF REQUIREMENTS
Listed Category I appliances Listed appliances equipped with draft hood Appliances listed for use with Type B gas vent	Type B gas vent	Section 802.6
	Chimney	Section 802.5
	Single-wall metal pipe	Section 802.7
	Listed chimney lining system for gas venting	Section 802.5.3
	Special gas vent listed for these appliances	Section 802.4.3
Listed vented wall furnaces	Type B-W gas vent	Section 802.6, Section 907.0
Category II, Category III, and Category IV appliances	As specified or furnished by manufacturers of listed appliances	Section 802.4.1, Section 802.4.3
Incinerators	In accordance with NFPA 82	
Appliances that can be converted to use of solid fuel Unlisted combination gas- and oil-burning appliances Combination gas- and solid-fuel-burning appliances Appliances listed for use with chimneys only Unlisted appliances	Chimney	Section 802.5
Listed combination gas- and oil-burning appliances	Type L vent	Section 802.6
	Chimney	Section 802.5
Decorative appliances in vented fireplace	Chimney	Section 911.2
Gas-fired toilets	Single-wall metal pipe	Section 802.7, Section 929.3
Direct vent appliances	—	Section 802.2.6
Appliances with integral vents	—	Section 802.2.7

COMMENTARY TABLE
Appliance Vent Categories

Appliance Category[1]	Vent Pressure	Temperature Above/Below Defined in Relevant ANSI Standard	Comment
I	Nonpositive[2]	Above	Natural draft venting
II	Nonpositive[2]	Below	Materials must be corrosion-resistant. Condensate must be drained.
III	Positive[3]	Above	Vent must be gastight.
IV	Positive[3]	Below	Vent must be liquid-tight and gastight. Condensate must be drained.

1 The newer models of appliances will be identified as Category I, II, III, or IV on the nameplate on the appliance and will be stated as such in the manufacturer's installation instructions.

2 The term nonpositive vent pressure means that even if fans or blowers are used in the appliance or vent systems, venting is accomplished by natural draft. (The vent pressure is lower than the atmospheric pressure.)

3 The term positive vent pressure means that fans, blowers, or other means are used to propel vent gases through the vent at above atmospheric pressure.

802.4.2 Plastic Vent Joints. Plastic pipe and fittings used to vent appliances shall be installed in accordance with the appliance manufacturer's installation instructions. Plastic pipe venting materials listed and labeled in accordance with UL 1738 shall be installed in accordance with the vent manufacturer's installation instructions. Where primer is required, it shall be of a contrasting color. [NFPA 54:12.5.3]

802.4.3 Special Gas Vents. Special gas vents shall be listed and labeled in accordance with UL 1738 and installed in accordance with the special gas vent manufacturer's installation instructions. [NFPA 54:12.5.4]

All special gas vents are listed vent materials. Special gas vents are listed in accordance with UL 1738, Standard for Venting Systems for Gas-Burning Appliances, Categories II, III and IV. See **Figure 802.4.3a**. Installation instructions for special gas vents include limitations on operating temperature, categories of appliance to be used with each vent, clearance to combustible materials, types of fittings and joint sealant to be used and vent termination requirements.

FIGURE 802.4.3a
VENT LISTING LABEL

Special attention should be given to the following areas:

1. Proper support for the special vent to prevent sagging and to allow for expansion, contraction and condensate drainage. See **Figure 802.4.3b**.
2. Proper cutting and cleaning of joints and fittings, and the use of recommended joint sealants (substitutes are not usually permitted).
3. Construction of a condensate trap (see appliance manufacturer's instruction for special requirements).
4. Wall penetrations (the pipe should not be secured at a thimble, as the pipe must be allowed to move to accommodate expansion and contraction).
5. Insulation [the vent pipe or the fittings of the inside of a wall thimble must not be insulated when polymeric (nonmetallic) vent materials are used].

802.5 Masonry, Metal, and Factory-Built Chimneys. Chimneys shall be installed in accordance with Section 802.5.1 through Section 802.5.3.

FIGURE 802.4.3B
VENT SUPPORTS

802.5.1 Factory-Built Chimneys. Factory-built chimneys shall be listed in accordance with UL 103, UL 959, or UL 2561. Factory-built chimneys shall be installed in accordance with the manufacturer's installation instructions. Factory-built chimneys used to vent appliances that operate at positive vent pressure shall be listed for such application. [NFPA 54:12.6.1.1]

Factory-built chimneys shown in **Figures 802.5.1a through 802.5.1c**, consist entirely of factory-built parts, such as

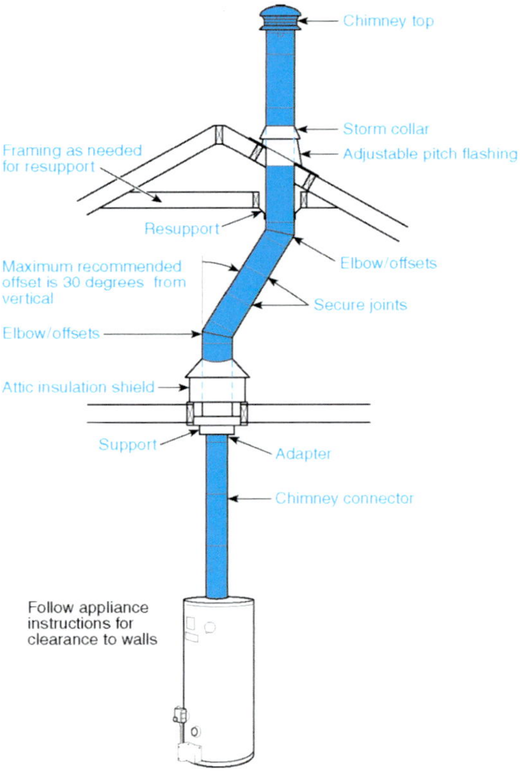

FIGURE 802.5.1A
TYPICAL INSTALLATION OF RESIDENTIAL FACTORY-BUILT CHIMNEY

chimney sections, supports, thimbles, flashings, caps and other parts required to complete a particular installation. The parts are designed to be assembled with other parts of the same model without requiring field alteration or construction. All of the parts for a particular model as described in the chimney manufacturer's instructions are to be used. **Figure 802.5.1d** shows the hardware for a positive-pressure factory-built chimney.

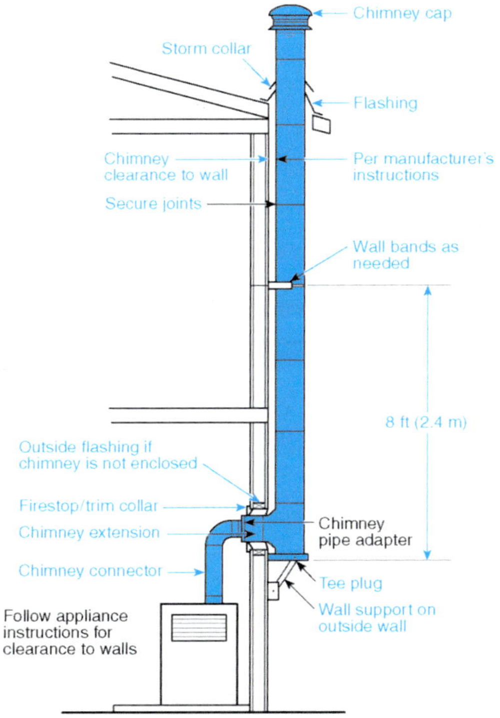

FIGURE 802.5.1B
ALTERNATE INSTALLATION OF RESIDENTIAL FACTORY-BUILT CHIMNEY

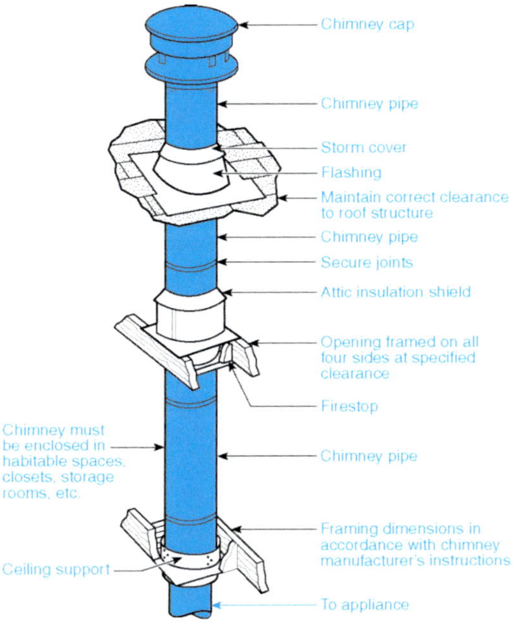

FIGURE 802.5.1C
ALTERNATE INSTALLATION OF RESIDENTIAL FACTORY-BUILT CHIMNEY

FIGURE 802.5.1D
HARDWARE FOR A POSITIVE-PRESSURE FACTORY-BUILT CHIMNEY

The following three types of factory-built chimneys are available:

1. Residential-type and building heating appliance chimneys that are intended for venting flue gases at a temperature not exceeding 1,000°F (538°C) under continuous operating conditions.
2. 1,400°F (760°C) chimneys for venting flue gases at a temperature not exceeding 1,400°F (760°C) under continuous operating conditions. These chimneys must comply with UL 2561, Standard for 1400 Degree Fahrenheit Factory-Built Chimneys.
3. Medium-heat appliance chimneys for venting flue gases at a temperature not exceeding 1,800°F (982°C). These chimneys must comply with UL 959, Standard for Medium Heat Appliance Factory-Built Chimneys.

The installation of chimneys serving appliances with vent temperatures over 1,000°F (538°C) is not covered in this code. See NFPA 211, Standard for Chimneys, Fireplaces, Vents, and Solid Fuel-Burning Appliances, which also describes the various types of heating appliances mentioned here.

Factory-built chimneys must be installed in accordance with the terms of their listing and the manufacturer's instructions, which are furnished with each chimney assembly and have been verified by the listing agency. The product standard for chimneys, UL 103, Standard for Factory-Built Chimneys for Residential Type and Building Heating Appliances, includes separate tests to verify the proper operation of factory-built chimneys that are designed for positive combustion-product pressure. Under this listing, wood burning appliances would be considered Low-heat Appliances, as defined in Chapter 2. The factory-built chimney would need to be installed vertically with no more than a 30 degree maximum offset.

Category II, III and IV appliances are not suitable for venting by the conventional method of natural draft venting. These appliances must not be connected to chimneys. These appliances should be vented in accordance with the appliance manufacturer's instructions, which cover the type, size and length of material to be used; how the venting system is to be installed; and the minimum clearance to combustible material. These instructions have been verified by the listing agency.

802.5.1.1 Decorative Shrouds. Decorative shrouds addressed in Section 802.5.4.3 shall be listed or labeled in accordance with UL 103 for factory-built residential chimneys, UL 127 for factory-built fireplaces, or UL 1482 for solid-fuel room heaters.

A decorative shroud is an aesthetic-pleasing ornamental piece to cover the chimney termination cap or rain pan. These shrouds have been a cause of roof and chase fires by allowing the creation of temperatures in excess of those permitted in the UL standards referenced in this code section. Therefore, the decorative shroud must be UL listed in compliance with the intended use, whether a factory-built residential chimney, a factory built fireplace, or a solid-fuel room heater. It is imperative that these shrouds are installed according to the manufacturer's installation instructions so as to prevent a fire hazard.

802.5.2 Metal Chimneys. Metal chimneys shall be built and installed in accordance with NFPA 211. [NFPA 54:12.6.1.2]

802.5.3 Masonry Chimneys. Masonry chimneys shall be built and installed in accordance with NFPA 211 and lined with one of the following:

(1) Approved clay flue lining.

(2) A chimney lining system listed and labeled in accordance with UL 1777.

(3) Other approved material that resists corrosion, erosion, softening, or cracking from vent gases at temperatures up to 1800°F (982°C).

Exception: Masonry chimney flues lined with a chimney lining system specifically listed for use with listed appliances with draft hoods, Category I appliances, and other appliances listed for use with Type B vents shall be permitted. The liner shall be installed in accordance with the liner manufacturer's installation instructions. A permanent identifying label shall be attached at the point where the connection is to be made to the liner. The label shall read: "This chimney liner is for appliances that burn gas only. Do not connect to solid-or liquid-fuel-burning appliances or incinerators." [NFPA 54:12.6.1.3]

Masonry chimneys can be used for all appliances permitted in Section 802.4 that can be vented using chimneys.

The exception permits installation of liners in masonry chimneys that may not be covered in NFPA 211. The labeling requirement notifies future installers of the vent liners' limitations. Notification is important if a change in appliance fuels is considered.

The installation of gas vents in existing masonry chimneys, as shown in **Figure 802.5.3**, requires specialized techniques and hardware that are different from those used in the roof applications. The special considerations include inspection and cleaning of the chimney, hardware for supporting the pipe, sealing of the top of the chimney and termination. Note the requirement to permanently mark the chimney, identifying the intended use of the vent.

802.5.4 Termination. A chimney for residential-type or low-heat appliances shall extend at least 3 feet (914 mm) above the highest point where it passes through a roof of a building

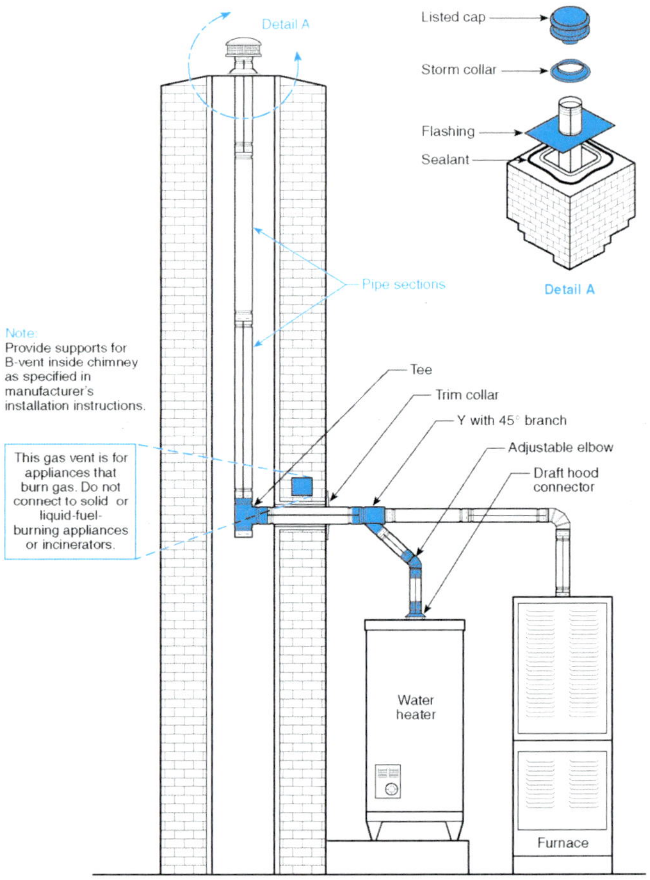

FIGURE 802.5.3
TYPICAL INSTALLATION USING TYPE B VENT TO LINE CHIMNEY

and at least 2 feet (610 mm) higher than a portion of any building within a horizontal distance of 10 feet (3048 mm). [NFPA 54:12.6.2.1] (See **Figure 802.5.4**)

802.5.4.1 Medium-Heat Gas Appliances. A chimney for medium-heat appliances shall extend at least 10 feet (3048 mm) higher than any portion of any building within 25 feet (7620 mm). [NFPA 54:12.6.2.2]

802.5.4.2 Chimney Height. A chimney shall extend at least 5 feet (1524 mm) above the highest connected appliance draft hood outlet or flue collar. [NFPA 54:12.6.2.3]

802.5.4.3 Decorative Shrouds. Decorative shrouds shall not be installed at the termination of factory-built chimneys except where such shrouds are listed and labeled for use with the specific factory-built chimney system and are installed in accordance with the manufacturer's installation instructions. [NFPA 54:12.6.2.4]

This requirement prohibits the installation of unlisted shrouds around factory-built chimneys, thereby affecting the flow of flue gases and trapping hot flue gases in close proximity to the combustible roof. Shrouds are used to hide chimneys. Shrouds can interfere with vent operation if they extend above the top of the vent.

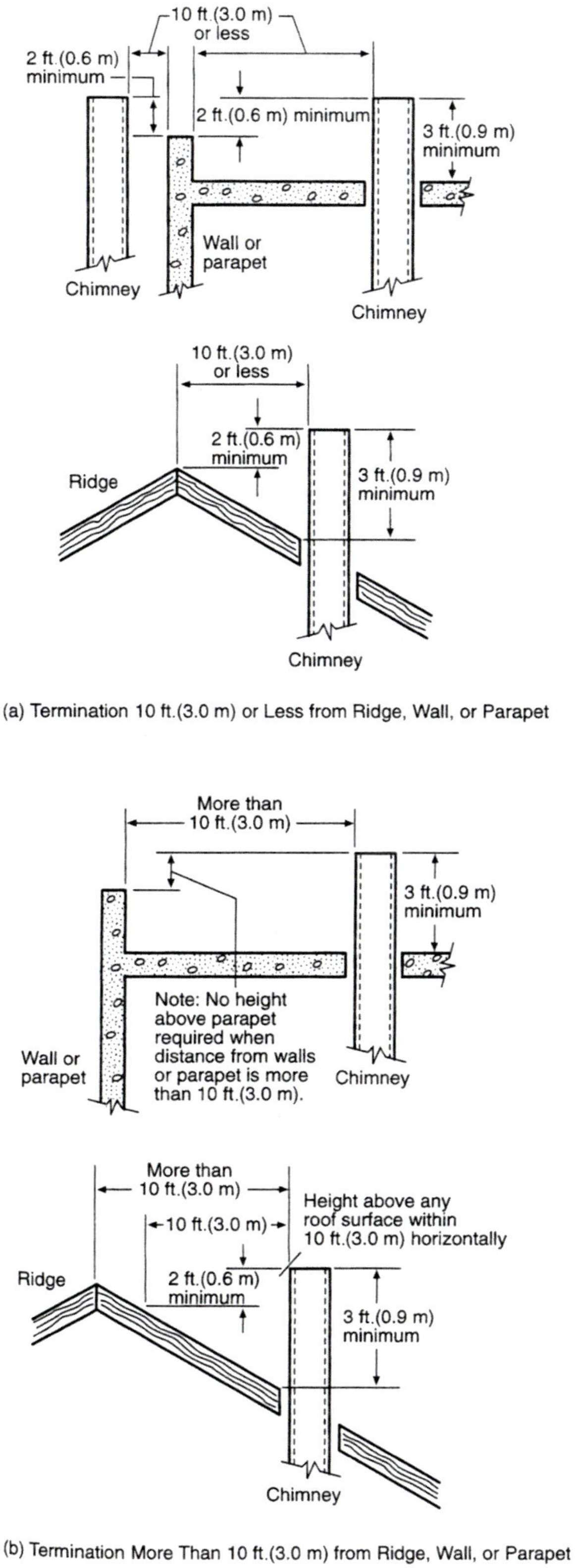

FIGURE 802.5.4
TYPICAL TERMINATION LOCATIONS FOR CHIMNEYS AND SINGLE-WALL METAL PIPES SERVING RESIDENTIAL-TYPE AND LOW-HEAT APPLIANCE [NFPA 54: FIGURE A.12.6.2.1]

802.5.5 Size of Chimneys. The effective area of a chimney venting system serving listed appliances with draft hoods, Category I appliances, and other appliances listed for use with Type B vents shall be in accordance with one of the following methods:

(1) Those listed in Section 803.0.

(2) For sizing an individual chimney venting system for a single appliance with a draft hood, the effective areas of the vent connector and chimney flue shall be not less than the area of the appliance flue collar or draft hood outlet or greater than seven times the draft hood outlet area.

(3) For sizing a chimney venting system connected to two appliances with draft hoods, the effective area of the chimney flue shall be not less than the area of the larger draft hood outlet plus 50 percent of the area of the smaller draft hood outlet or greater than seven times the smaller draft hood outlet area.

(4) Chimney venting systems using mechanical draft shall be sized in accordance with approved engineering methods.

(5) Other approved engineering methods. [NFPA 54:12.6.3.1]

Information on the sizing of chimneys is also provided in Section 803.0 in the form of tables and additional requirements. These venting requirements were developed by a research project funded by the Gas Research Institute and conducted by Battelle.

The requirements in Section 803.0 recognize that there is a significant difference between the operation of interior and exterior chimneys. An exterior chimney is one in which one or more sides are exposed to the outdoors below the roof line. An interior chimney is not exposed to the outdoors below the building's roof line. (Chimneys that pass through unheated attics or garages are not considered to be outdoors.) An interior chimney is largely isolated from weather changes. An exterior chimney is affected by the outside ambient temperature; therefore, the requirements for exterior chimneys are keyed to the lowest temperature expected in different parts of the country.

The requirements for sizing chimneys were developed using a flue loss of 17 percent for the heating appliance. Most real furnaces and boilers operate at a higher flue loss; therefore, the code requirements are conservative. Several manufacturers offer customized vent kits, which include venting requirements for masonry chimneys that are keyed to the exact performance of their appliances. In this case, their recommendations should be used.

Figure 802.5.5 shows an example of a custom vent kit. The figure illustrates a furnace with a field-installed modification utilizing customized venting instructions for venting using a masonry chimney. Note the draft hood to the right of the induced draft blower, which has been added. The kit includes the hardware and venting instructions.

The sizing methods in Sections 802.5.5(2) and 802.5.5(3) permit a simple, alternative, chimney-sizing method that limits the vent to a maximum size of 7 times the smallest draft hood outlet area and a minimum size based on the draft hood outlet area. These methods limit the maximum size of the vent connector and gas vent to minimize condensation in the gas vent caused by insufficient flow of hot vent gases to heat an excessively large gas vent. Excessive condensation can cause premature failure of a chimney.

Note that Section 802.5.5(3) limits the use of this alternative to chimney venting systems serving only two appliances. Sample calculations using the venting tables in Section 803.0 and the alternative method showed that the alternative method results in vent sizes too small when four or more appliances are involved, and for some tall vents. The alternate always provides acceptable vent sizes for two appliances. Venting systems serving more than two appliances must use the vent sizing tables in Section 803.0 or other engineering methods.

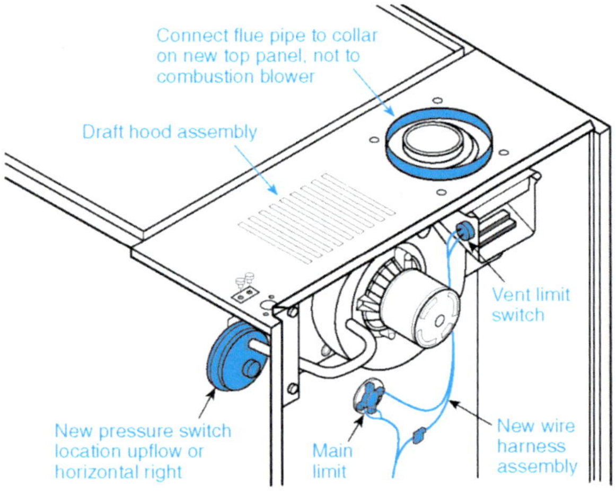

FIGURE 802.5.5
CUSTOMIZED VENT KIT

802.5.6 Inspection of Chimneys or Vents. This inspection shall be made after chimneys, vents, or parts thereof, authorized by the permit, have been installed and before such vent or part thereof has been covered or concealed.

802.5.7 Inspection of Chimneys. Before replacing an existing appliance or connecting a vent connector to a chimney, the chimney passageway shall be examined to ascertain that it is clear and free of obstructions and shall be cleaned if previously used for venting solid- or liquid-fuel-burning appliances or fireplaces. [NFPA 54:12.6.4.1]

Inspections are required prior to the connection of a new appliance to a chimney that previously served appliances using fuels other than gas fuels. It is important that this inspection be done because a buildup of soot or creosote that may have been caused by oil- or solid-fuel-burning appliances can be softened by the products of gas combustion. These deposits, if softened, can break away and block the chimney.

Some common chimney troubles and ways to detect and remedy them are shown in **Figure 802.5.7**.

802.5.7.1 Standard. Chimneys shall be lined in accordance with NFPA 211.

Exception: Existing chimneys shall be permitted to have their use continued when an appliance is replaced by an appliance of similar type, input rating, and efficiency, where the chimney complies with Section 802.5.7 through Section 802.5.7.3 and the sizing of the chimney is in accordance with Section 802.5.5. [NFPA 54:12.6.4.2]

802.5.7.2 Cleanouts. Cleanouts shall be examined and where they do not remain tightly closed when not in use, they shall be repaired or replaced. [NFPA 54:12.6.4.3]

802.5.7.3 Existing Chimney. When inspection reveals that an existing chimney is not safe for the intended application, it shall be repaired, rebuilt, lined, relined, or replaced with a vent or chimney to conform to NFPA 211 and shall be suitable for the appliances to be attached. [NFPA 54:12.6.4.4]

802.5.8 Chimney Serving Appliances Burning Other Fuels. An appliance shall not be connected to a chimney flue serving a separate appliance designed to burn solid fuel. [NFPA 54:12.6.5.1]

802.5.8.1 Gas and Liquid Fuel-Burning Appliances. Where one chimney serves gas appliances and liquid fuel-burning appliances, the appliances connected through separate openings or connected through a single opening where joined by a suitable fitting located as close as practical to the chimney. Where two or more openings are provided into one chimney flue, they shall be at different levels. Where the gas appliance is automatically controlled, it shall be equipped with a safety shutoff device. [NFPA 54:12.6.5.2]

802.5.8.2 Gas and Solid Fuel-Burning Appliances. A listed combination gas- and solid fuel-burning appliance connected to a single chimney flue shall be equipped with a manual reset device to shut off gas to the main burner in the event of sustained backdraft or flue gas spillage. The chimney flue shall be sized to properly vent the appliance. [NFPA 54:12.6.5.3]

802.5.8.3 Combination Gas- and Oil-Burning Appliances. A single chimney flue serving a listed combination gas- and oil-burning appliance shall be sized in accordance with the appliance manufacturer's instructions. [NFPA 54:12.6.5.4]

802.5.9 Support of Chimneys. All portions of chimneys shall be supported for the design and weight of the materials employed. Listed factory-built chimneys shall be supported and spaced in accordance with the manufacturer's installation instructions. [NFPA 54:12.6.6]

802.5.10 Cleanouts. Where a chimney that formerly carried flue products from liquid or solid fuel-burning appliances is used with an appliance using fuel gas, an accessible cleanout shall be provided. The cleanout shall have a tight-fitting cover and be installed so its upper edge is at least 6 inches (152 mm) below the lower edge of the lowest chimney inlet opening. [NFPA 54:12.6.7]

When a gas-heating unit or conversion burner is connected to a chimney that previously served for venting another fuel, after a period of operation on natural gas, carbon deposits or mortar materials can become dislodged from the chimney wall and drop to the base of the chimney. The accumulation of this debris, if great enough, can result in blockage of the venting system, sometimes with fatal results.

A 6-inch drop leg and inspection/maintenance cleanout

is required. The cleanout provides a sump to collect a considerable amount of debris before a potential accident occurs. The cleanout also enables homeowners and service personnel to clean debris easily and to inspect conditions.

802.5.11 Space Surrounding Lining or Vent. The remaining space surrounding a chimney liner, gas vent, special gas vent, or plastic piping installed within a masonry chimney shall not be used to vent another appliance.

Exception: The insertion of another liner or vent within the chimney as provided in this code and the liner or vent manufacturer's instructions. [NFPA 54:12.6.8.1]

The installation of nonmetallic gas vents in an abandoned chimney or a chase with a gas vent in it is a practical solution to an existing unlined, oversized or failed chimney. When installing vents in a chimney space, all of the vent material manufacturer's installation requirements must be followed. If required by the vent manufacturer, penetrating the masonry chimney to install supports might be necessary.

When using an abandoned chimney as a chase for multiple vents, care should be taken to ensure that the vent materials are compatible and that vent failure does not result inadvertently. Category I or III gas appliance vent gas temperatures can exceed 300°F (150°C). Contact between the high-temperature Category I or III vent material and the PVC vent pipe can cause the PVC pipe to degrade or soften to the extent of failure. With the chimney opening sealed around the vent pipes, the products of combustion, which can escape from the failed PVC pipe, will be trapped in the chimney. These gases could eventually leak back into the living space or corrode the other nearby vents.

802.5.11.1 Combustion Air. The remaining space surrounding a chimney liner, gas vent, special gas vent, or plastic piping installed within a masonry, metal or factory-built chimney flue shall not be used to supply combustion air.

Exception: Direct vent appliances designed for installation in a solid-fuel-burning fireplace where installed in accordance with the manufacturer's installation instructions. [NFPA 54:12.6.8.2]

802.6 Gas Vents. The installation of gas vents shall meet the following requirements:

(1) Gas vents shall be installed in accordance with the manufacturer's installation instructions.

(2) A Type B-W gas vent shall have a listed capacity not less

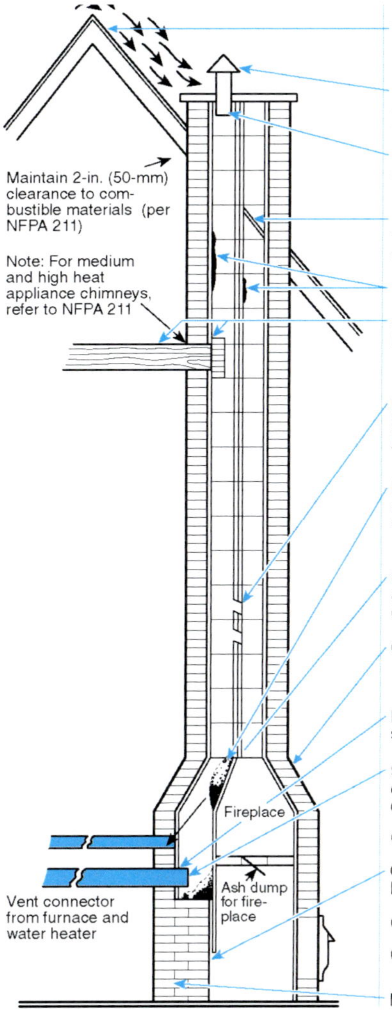

Top of chimney lower than surrounding objects	Observation	Extend chimney
Undersized chimney cap or ventilator	Observation	Remove or replace
Restricted opening	Measurement	Make opening as large as inside of chimney
Piece of broken tile wedged in chimney	By lowering a light or weight down chimney	Break tile with a rod or weight on string or wire
Creosote buildup	Observation	Clean chimney
Joist protruding into chimney	By lowering a light or weight down chimney	Change support for joist so that chimney will be clear
Leakage between loose joined tiles	Start smoke fire with top of chimney closed; look for leakage from chimney	Rebuild chimney with a course of brick between flue tiles
Debris accumulated in offset	By lowering a light or weight down chimney	Breakout with rod or weight may be necessary to open chimney
Solid fuel-burning appliance connection	Observation	Remove and seal opening
Offset	By lowering a light or weight down chimney	Change to straight or to long offset
Loosely fitted smoke pipe	Smoke test	Clean leaks with cement
Smoke pipe extends into chimney	By lowering a light or weight down chimney	Make end flush with inside of chimney
Chimney too short	Measurement	Extend
Opening between flues	Smoke test	Close opening permanently
Chimney too small	Measurement	Rebuild
Chimney too large	Measurement	Rebuild or insert liner
No cleanout	Observation	Add cleanout

FIGURE 802.5.7
COMMON CHIMNEY TROUBLES

than that of the listed vented wall furnace to which it is connected.

(3) Gas vents installed within masonry chimneys shall be installed in accordance with the manufacturer's installation instructions. Gas vents installed within masonry chimneys shall be identified with a permanent label installed at the point where the vent enters the chimney. The label shall contain the following language: "This gas vent is for appliances that burn gas. Do not connect to solid or liquid fuel–burning appliances or incinerators."

(4) Screws, rivets, and other fasteners shall not penetrate the inner wall of double-wall gas vents, except at the transition from the appliance draft hood outlet, flue collar, or single-wall metal connector to a double-wall vent. [NFPA 54:12.7.2]

Section 802.6 applies to gas vents of all types. Not included are chimneys, which are covered in Section 802.5, and single-wall metal pipe, which is covered in Section 802.7.

Type B and B-W gas vents that currently are listed are of double-wall construction: basically, a pipe within a pipe, with airspace between the two walls. This construction reduces the heat loss from the vent gases in much the same way that an insulated coffee mug works.

Usually, the inner wall is made of aluminum to resist corrosion, and the outer wall is made of galvanized steel for strength. The parts of the vent—pipe sections, supports, spacers, caps and flashings—are furnished for field erection to form a continuous passageway from the gas appliance to the terminus of the roof, including the cap or roof assembly. These parts are manufactured by several companies that furnish the vent pipe and other related parts needed to erect a complete vent. The piping and parts produced by the different manufacturers are not interchangeable without the use of special, listed adapters.

Type B gas vents can be round or oval (see **Figure 802.6a**). Type B-W gas vents are always oval because they are designed to be installed within a stud space. Type L vents are similar in construction to Type B gas vents, except that the inner pipe of a Type L vent is stainless steel. Type L vents are intended for venting flue gases at higher temperatures than are normally produced by gas appliances, such as for oil furnaces listed for use with Type L vents.

Typical Type B and B-W gas vent installations are diagrammed in **Figures 802.6b and 802.6c**, respectively.

FIGURE 802.6A
CONSTRUCTION OF DOUBLE-WALL TYPE B GAS VENT

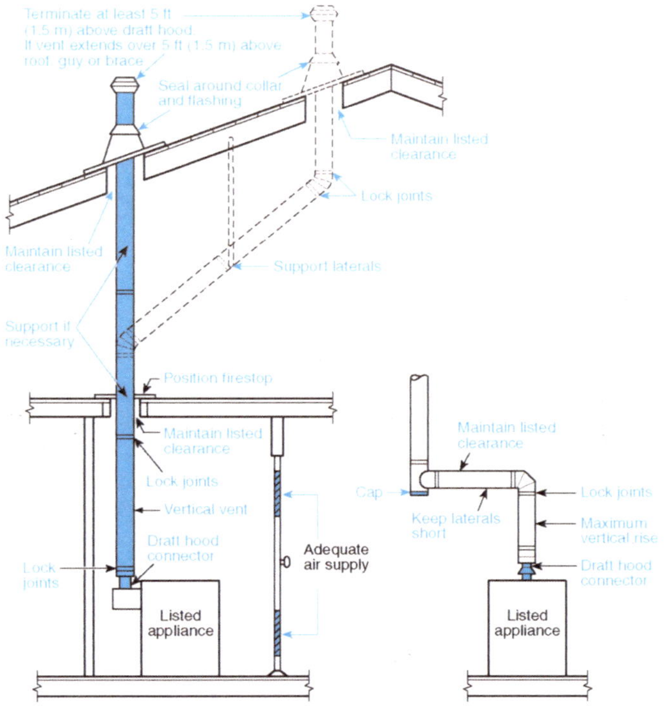

FIGURE 802.6B
TYPICAL TYPE B GAS VENT: SINGLE APPLIANCE INSTALLATION

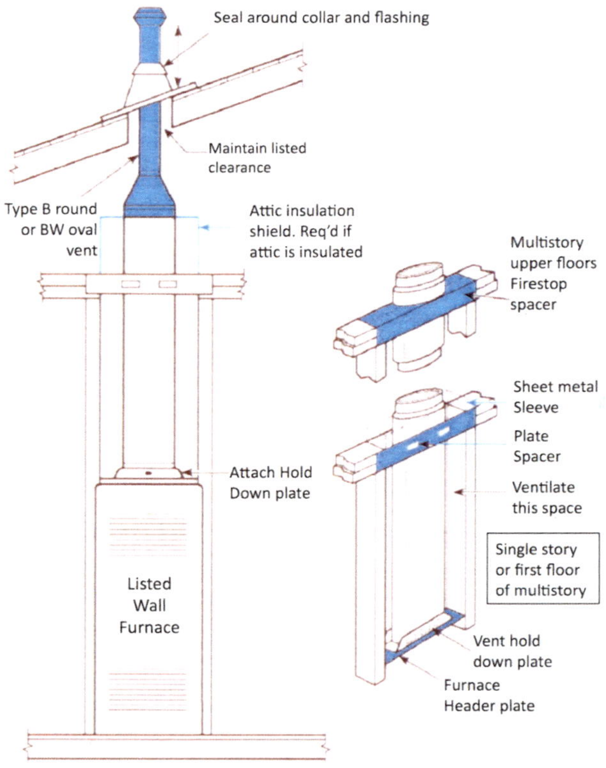

FIGURE 802.6C
TYPICAL TYPE B-W GAS VENT INSTALLATION

802.6.1 Gas Vent Termination. The termination of gas vents shall comply with the following requirements:

(1) A gas vent shall terminate in accordance with one of the following:

(a) Gas vents that are 12 inches (300 mm) or less in size and located not less than 8 feet (2438 mm) from a vertical wall or similar obstruction shall terminate above the roof in accordance with **Figure 802.6.1** and Table 802.6.1.

(b) Gas vents that are over 12 inches (300 mm) in size or are located less than 8 feet (2438 mm) from a vertical wall or similar obstruction shall terminate not less than 2 feet (610 mm) above the highest point where they pass through the roof and not less than 2 feet (610 mm) above a portion of a building within 10 feet (3048 mm) horizontally.

(c) Industrial appliances as provided in Section 802.2.5.

(d) Direct vent systems as provided in Section 802.2.6.

(e) Appliances with integral vents as provided in Section 802.2.7.

(f) Mechanical draft systems as provided in Section 802.3.3 through Section 802.3.3.5.

(g) Ventilating hoods and exhaust systems as provided in Section 802.3.4.

(2) A Type B or a Type L gas vent shall terminate at least 5 feet (1524 mm) in vertical height above the highest connected appliance draft hood or flue collar.

(3) A Type B-W gas vent shall terminate at least 12 feet (3658 mm) in vertical height above the bottom of the wall furnace.

(4) A gas vent extending through an exterior wall shall not terminate adjacent to the wall or below eaves or parapets, except as provided in Section 802.2.6 and Section 802.3.3 through Section 802.3.3.5.

(5) Decorative shrouds shall not be installed at the termination of gas vents except where such shrouds are listed for use with the specific gas venting system and are installed in accordance with the manufacturer's installation instructions.

(6) All gas vents shall extend through the roof flashing, roof jack, or roof thimble and terminate with a listed cap or listed roof assembly.

(7) A gas vent shall terminate at least 3 feet (914 mm) above a forced air inlet located within 10 feet (3048 mm). [NFPA 54:12.7.3]

Sections 802.6.1(c) through 802.6.1(g) are alternatives to the normal vent termination described in Section 802.6.1(a) & (b) and shown in **Figure 802.6.1**. Section 802.6.1(c), which references Section 802.2.5, indicates that vent termination may not apply to spaces in which industrial gas equipment is installed if they are well-ventilated.

Safety can be achieved in high-rise, multilevel appliance installations by separating the appliance from any habitable or occupied space. The separation resolves the question of safety that arises when inter-communication of vents exists between various floors of a building. Separation ensures that no vent gases will enter the occupied space from the appliance room if the common vent becomes obstructed at any level or if the outlet is blocked. When such stoppage occurs, all vent gases from appliances operating below the obstruction will exit through the upper draft hood relief opening rather than through the vent outlet. Large quantities of vent gases will be dumped into the space containing those appliances immediately below the obstruction, while at the same time the appliances located at lower levels will appear to be operating normally.

One practical plan to separate or isolate appliances, if installation can be made adjacent to an outside wall, is shown by **Figure 802.6.1a** and **Figure 802.6.3.1**.

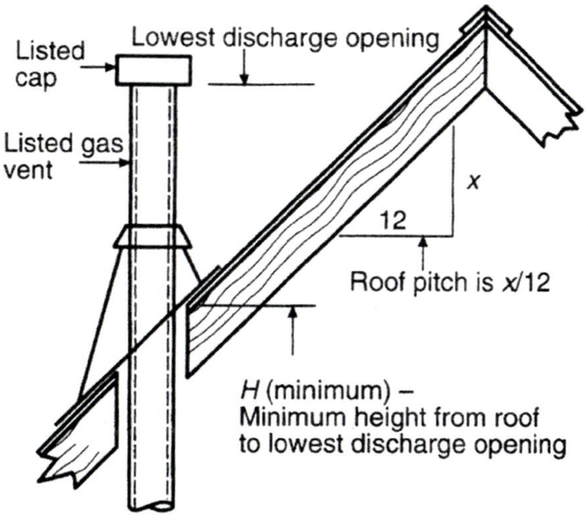

FIGURE 802.6.1
TERMINATION LOCATIONS FOR GAS VENTS WITH LISTED CAPS 12 INCHES OR LESS IN SIZE NOT LESS THAN 8 FEET FROM A VERTICAL WALL [NFPA 54: FIGURE 12.7.3]

TABLE 802.6.1
ROOF PITCH HEIGHT
[NFPA 54: TABLE 12.7.3]

ROOF SLOPE	H (minimum) (feet)
Flat to $^{6}/_{12}$	1.0
Over $^{6}/_{12}$ to $^{7}/_{12}$	1.25
Over $^{7}/_{12}$ to $^{8}/_{12}$	1.5
Over $^{8}/_{12}$ to $^{9}/_{12}$	2.0
Over $^{9}/_{12}$ to $^{10}/_{12}$	2.5
Over $^{10}/_{12}$ to $^{11}/_{12}$	3.25
Over $^{11}/_{12}$ to $^{12}/_{12}$	4.0
Over $^{12}/_{12}$ to $^{14}/_{12}$	5.0
Over $^{14}/_{12}$ to $^{16}/_{12}$	6.0
Over $^{16}/_{12}$ to $^{18}/_{12}$	7.0
Over $^{18}/_{12}$ to $^{20}/_{12}$	7.5
Over $^{20}/_{12}$ to $^{21}/_{12}$	8.0

For SI units: 1 inch = 25.4 mm, 1 foot = 304.8 mm

Access to the appliance room is through a door that opens onto an outdoor balcony. A panel separates the appliance room from the inside of the building. If the appliance is a central furnace, the cold air return and outlet ducts are attached to the furnace.

No requirements are given in the code for installation of Category II, III or IV appliances using a common vent in multiple-story buildings. A Category II, III or IV appliance must be installed in accordance with the terms of its listing and the manufacturer's instructions. Category I appliance common vents are not usually suitable for venting Category II, III or IV appliances.

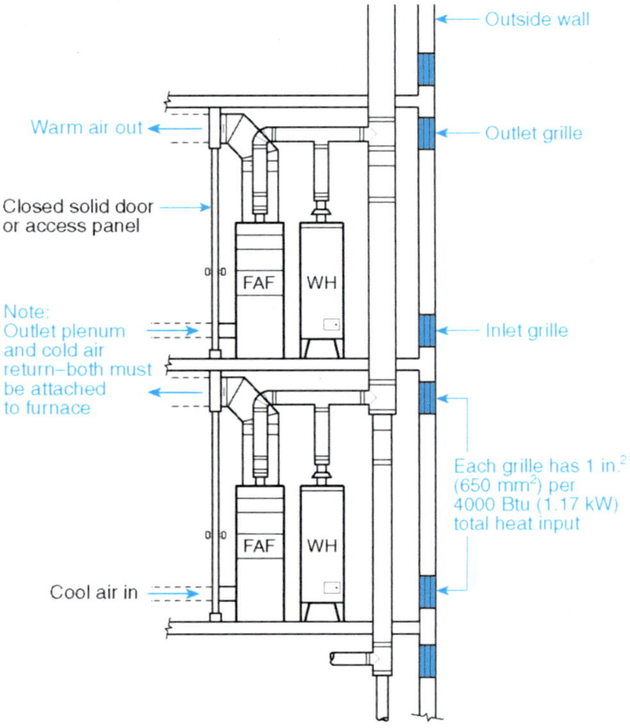

FIGURE 802.6.1A
OUTSIDE WALL AIR SUPPLY IN MULTISTORY BUILDING

802.6.1.1 Insulation Shield. Where a vent passes through an insulated assembly, an approved metal shield shall be installed between the vent and insulation. The shield shall extend not less than 2 inches (51 mm) above the insulation and be secured to the structure in accordance with the manufacturer's installation instructions.

802.6.2 Size of Gas Vents. Venting systems shall be sized and constructed in accordance with Section 802.6.2.1 through Section 802.6.2.3 and the appliance manufacturer's instructions. [NFPA 54:12.7.4]

802.6.2.1 Category I Appliances. The sizing of natural draft venting systems serving one or more listed appliances equipped with a draft hood or appliances listed for use with a Type B gas vent, installed in a single story of a building, shall be in accordance with one of the following:

(1) The provisions of Section 803.0.

(2) Vents serving fan-assisted combustion system appliances, or combinations of fan-assisted combustion system and draft hood-equipped appliances, shall be sized in accordance with Section 803.0 or other approved engineering methods.

(3) For sizing an individual gas vent for a single, draft hood-equipped appliance, the effective area of the vent connector and the gas vent shall be not less than the area of the appliance draft hood outlet or greater than seven times the draft hood outlet area.

(4) For sizing a gas vent connected to two appliances with draft hoods, the effective area of the vent shall be not less than the area of the larger draft hood outlet plus 50 percent of the area of the smaller draft hood outlet or greater than seven times the smaller draft hood outlet area.

(5) Other approved engineering practices. [NFPA 54:12.7.4.1]

The tables in Section 803.0 provide a safe method for sizing vents for all Category I and draft hood-equipped appliances. The alternate methods permitted in this section offer simple techniques for installers to follow for commonly encountered vent arrangements. These methods provide a maximum size of seven times the smallest draft hood outlet area and a minimum size based on the draft hood outlet area. This provision limits the maximum size of the vent connector and gas vent to minimize gas vent condensation caused by insufficient flow of hot vent gases to heat an excessively large gas vent. Excessive condensation can cause premature failure of a gas vent.

Note that these alternative methods cannot be used for fan-assisted combustion appliances. The tables in Section 803.0 (or other engineering methods) must be used for all installations of fan-assisted combustion appliances. Vents for Category II, III, and IV appliances must be in accordance with the equipment manufacturer's instructions. These appliances are listed by definition, and the manufacturer's instructions reflect the test results determined by the listing laboratory.

Note that Section 802.6.2.1(4) allows the use of a simpler method for sizing a venting system than the tables in Section 803.0. This simpler method is applicable only up to two appliances. Venting systems serving more than two appliances must use the vent sizing tables in Section 803.0 or other engineering methods.

802.6.2.2 Vent Offsets. Type B and Type L vents sized in accordance with Section 802.6.2.1(3) or Section 802.6.2.1(4) shall extend in a generally vertical direction with offsets not exceeding 45 degrees except that a vent system having not more than one 60 degree offset shall be permitted. Any angle greater than 45 degrees from the vertical is considered horizontal. The total horizontal distance of a vent plus the horizontal vent connector serving draft hood-equipped appliances shall not be greater than 75 percent of the vertical height of the vent. [NFPA 54:12.7.4.2]

802.6.2.3 Category II, Category III, and Category IV Appliances. The sizing of gas vents for Category II, Category III, and Category IV appliances shall be in accordance with the appliance manufacturer's instructions. The sizing of plastic pipe specified by the appliance manufacturer as a venting material for Category II, III, and IV appliances shall be in accordance with the appliance manufacturers' instructions. [NFPA 54:12.7.4.3]

802.6.2.4 Sizing. Chimney venting systems using mechanical draft shall be sized in accordance with approved engineering methods. [NFPA 54:12.7.4.4]

802.6.3 Gas Vents Serving Appliances on More than One Floor. A common vent shall be permitted in multistory installations to vent Category I appliances located on more than one floor level, provided the venting system is designed and installed in accordance with approved engineering methods.

For the purpose of this section, crawl spaces, basements, and attics shall be considered as floor levels. [NFPA 54:12.7.5.1]

802.6.3.1 Occupiable Space. All appliances connected to the common vent shall be located in rooms separated from occupiable space. Each of these rooms shall have provisions for an adequate supply of combustion, ventilation, and dilution air that is not supplied from occupiable space. [NFPA 54:12.7.5.2] (See **Figure 802.6.3.1**)

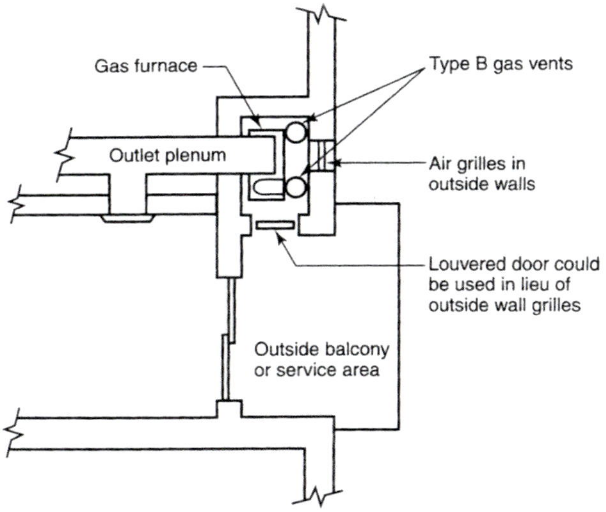

FIGURE 802.6.3.1
PLAN VIEW OF PRACTICAL SEPARATION METHOD FOR MULTISTORY GAS VENTING
[NFPA 54: FIGURE A.12.7.5.2]

Occupiable is an enclosed space intended for human activities excluding those spaces intended primarily for other purposes such as storage rooms and equipment rooms that are only occupied occasionally and for short periods of time. Occupiable spaces are more inclusive than habitable spaces and therefore increase the scope and occupancy types for the provisions of this section.

Appliances are to be separated from the occupiable space (shown in **Figure 802.6.3.1**). Further information on sizing vents that serve appliances on more than one floor of a building can be found in Sections 803.2.12 through 803.2.15.

802.6.3.2 Multistory Venting System. The size of the connectors and common segments of multistory venting systems for appliances listed for use with a Type B double-wall gas vent shall be in accordance with Table 803.2(1), provided all of the following apply:

(1) The available total height (*H*) for each segment of a multistory venting system is the vertical distance between the level of the highest draft hood outlet or flue collar on that floor and the centerline of the next highest interconnection tee.
(2) The size of the connector for a segment is determined from the appliance's gas input rate and available connector rise and shall not be smaller than the draft hood outlet or flue collar size.
(3) The size of the common vertical vent segment, and of the interconnection tee at the base of that segment, is based on the total appliance's gas input rate entering that segment and its available total height. [NFPA 54:12.7.5.3]

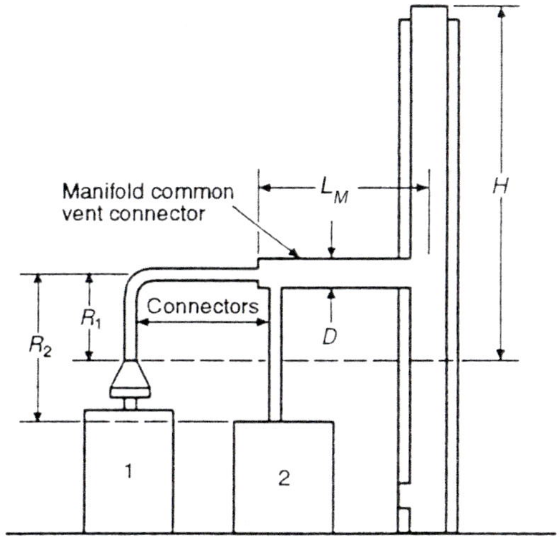

FIGURE 802.6.3.2
USE OF MANIFOLDED COMMON VENT CONNECTOR
[NFPA 54: FIGURE F.1(k)]

802.6.4 Support of Gas Vents. Gas vents shall be supported and spaced in accordance with the manufacturer's installation instructions. [NFPA 54:12.7.6]

802.6.5 Marking. In those localities where solid and liquid fuels are used extensively, gas vents shall be permanently identified by a label attached to the wall or ceiling at a point where the vent connector enters the gas vent. The label shall read: "This gas vent is for appliances that burn gas. Do not connect to solid or liquid fuel-burning appliances or incinerators." The Authority Having Jurisdiction shall determine whether its area constitutes such a locality. [NFPA 54:12.7.7]

802.7 Single-Wall Metal Pipe. Single-wall metal pipe shall be constructed of galvanized sheet steel not less than 0.0304 of an inch (0.7722 mm) thick or of other approved, noncombustible, corrosion-resistant material. [NFPA 54:12.8.1]

The use of single-wall metal pipe (e.g., stovepipe) as a vent for gas utilization equipment is limited by the code, especially in dwellings, for several reasons, including the following:

1. The surface of the pipe is usually hot enough to burn a person who accidentally contacts the pipe.
2. The high surface temperature can cause overheating of adjacent combustible material.
3. The heat loss from a long length of pipe in a cool area can result in reduced draft and the condensation of water vapor in the flue gases.

The use of single-wall metal pipe as a vent, where permitted in Section 802.7, is limited to Category I appliances. Note that some building codes do not permit single-wall metal chimneys and unlisted metal chimneys inside one- and two- family dwellings.

Single-wall metal pipe is a venting material that has limitations due to its inherently higher heat loss than chimneys and gas vents. **Figure 802.7** shows the limitations of single-wall gas vents when compared with the characteristics of double-wall vents.

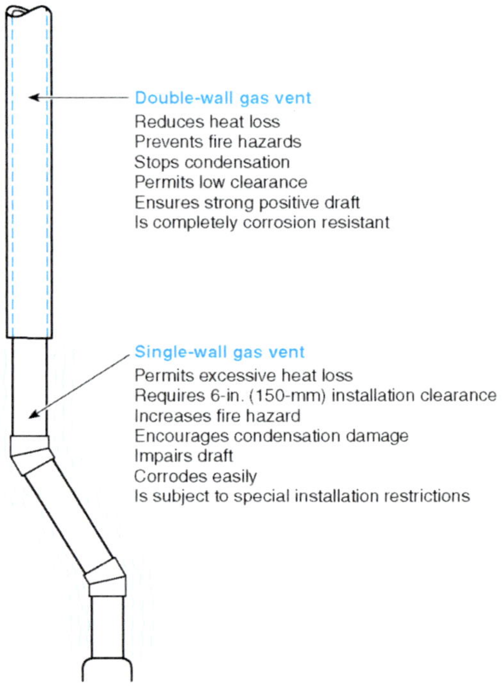

FIGURE 802.7
COMPARISON OF DOUBLE-WALL AND SINGLE-WALL GAS VENT CHARACTERISTICS

802.7.1 Cold Climate. Uninsulated single-wall metal pipe shall not be used outdoors for venting appliances in regions where the 99 percent winter design temperature is below 32°F (0°C). [NFPA 54:12.8.2]

The outdoor use of single-wall metal pipe in cold climates is restricted by Section 802.7.1 to prevent premature corrosion of the vent. In cold climates, the low ambient temperature can cool vent gases to the point at which water vapor condenses in the vent and causes corrosion. The code does not attempt to define a "cold climate" because the particular installation and local conditions often determine the acceptability of using single-wall metal pipe. One simple guideline is checking to see how often the replacement of single-wall metal pipe is required in the area in question.

Table 803.2(7) and Table 803.2(9) quantify the effects of cold ambient temperatures on masonry chimneys. They cannot be used for single-wall metal pipe.

802.7.2 Termination. The termination of single-wall metal pipe shall meet the following requirements:

(1) Single-wall metal pipe shall terminate at least 5 feet (1524 mm) in vertical height above the highest connected appliance draft hood outlet or flue collar.

(2) Single-wall metal pipe shall extend at least 2 feet (610 mm) above the highest point where it passes through a roof of a building and at least 2 feet (610 mm) higher than any portion of a building within a horizontal distance of 10 feet (3048 mm). (See **Figure 802.5.4**)

(3) An approved cap or roof assembly shall be attached to the terminus of a single-wall metal pipe. [NFPA 54:12.8.3]

Sections 802.7.1 through 802.7.2 provide the limitations under which single-wall metal pipe can be installed. The limitations are based on the fact that single-wall metal pipe can operate at a high temperature that easily can ignite combustible material placed in contact with it. When this type of pipe passes through a combustible roof, protection of the roof is required and a thimble is specified.

802.7.3 Installation with Appliances Permitted by Section 802.4. Single-wall metal pipe shall not be used as a vent in dwellings and residential occupancies. [NFPA 54:12.8.4.1]

802.7.3.1 Limitations. Single-wall metal pipe shall be used only for runs directly from the space in which the appliance is located through the roof or exterior wall to the outer air. A pipe passing through a roof shall extend without interruption through the roof flashing, roof jacket, or roof thimble. [NFPA 54:12.8.4.2]

802.7.3.2 Attic or Concealed Space. Single-wall metal pipe shall not originate in any unoccupied attic or concealed space and shall not pass through any attic, inside wall, concealed space, or floor. [NFPA 54:12.8.4.3]

802.7.3.3 Clearances. Minimum clearances from single-wall metal pipe to combustible material shall be in accordance with Table 802.7.3.3. Reduced clearances from single-wall metal pipe to combustible material shall be as specified for vent connectors. [NFPA 54:12.8.4.4]

802.7.3.4 Combustible Exterior Wall. Single-wall metal pipe shall not pass through a combustible exterior wall unless guarded at the point of passage by a ventilated metal thimble not smaller than the following:

(1) For listed appliances with draft hoods and appliances listed for use with Type B gas vents, the thimble shall be a minimum of 4 inches (102 mm) larger in diameter than the metal pipe. Where there is a run of not less than 6 feet (1829 mm) of metal pipe in the opening between the draft hood outlet and the thimble, the thimble shall be a minimum of 2 inches (51 mm) larger in diameter than the metal pipe.

(2) For unlisted appliances having draft hoods, the thimble

TABLE 802.7.3.3
CLEARANCE FOR CONNECTORS*
[NFPA 54: TABLE 12.8.4.4]

MINIMUM DISTANCE FROM COMBUSTIBLE MATERIAL (inches)				
APPLIANCE	LISTED TYPE B GAS VENT MATERIAL	LISTED TYPE L VENT MATERIAL	SINGLE-WALL METAL PIPE	FACTORY-BUILT CHIMNEY SECTIONS
Listed appliance with draft hoods and appliance listed for use with Type B gas vents	As listed	As listed	6	As listed
Residential boilers and furnaces with listed gas conversion burner and with draft hood	6	6	9	As listed
Residential appliances listed for use with Type L vents	Not permitted	As listed	9	As listed
Listed gas-fired toilets	Not permitted	As listed	As listed	As listed
Unlisted residential appliances with draft hood	Not permitted	6	9	As listed
Residential and low-heat appliance other than those above	Not permitted	9	18	As listed
Medium-heat appliance	Not permitted	Not permitted	36	As listed

For SI units: 1 inch = 25.4 mm

* These clearances shall apply unless the installation instructions of a listed appliance or connector specify different clearances, in which case the listed clearances shall apply.

shall be a minimum of 6 inches (152 mm) larger in diameter than the metal pipe.

(3) For residential and low-heat appliances, the thimble shall be a minimum of 12 inches (305 mm) larger in diameter than the metal pipe.

Exception: In lieu of thimble protection, all combustible material in the wall shall be removed a sufficient distance from the metal pipe to provide the specified clearance from such metal pipe to combustible material. Any material used to close up such opening shall be noncombustible. [NFPA 54:12.8.4.6]

802.7.3.5 Roof Thimble. Where a single-wall metal pipe passes through a roof constructed of combustible material, a noncombustible, nonventilating thimble shall be used at the point of passage. The thimble shall extend at least 18 inches (457 mm) above and 6 inches (152 mm) below the roof with the annular space open at the bottom and closed only at the top. The thimble shall be sized in accordance with Section 802.7.3.4. [NFPA 54:12.8.4.5]

802.7.4 Size of Single-Wall Metal Pipe. Single-wall metal piping shall comply with the following requirements:

(1) A venting system of a single-wall metal pipe shall be sized in accordance with one of the following methods and the appliance manufacturer's instructions:

  (a) For a draft hood-equipped appliance, in accordance with Section 803.0.

  (b) For a venting system for a single appliance with a draft hood, the areas of the connector and the pipe each shall not be less than the area of the appliance flue collar or draft hood outlet, whichever is smaller. The vent area shall not be greater than seven times the draft hood outlet area.

  (c) Other approved engineering methods.

(2) Where a single-wall metal pipe is used and has a shape other than round, it shall have an equivalent effective area equal to the effective area of the round pipe for which it is substituted and the minimum internal dimension of the pipe shall be 2 inches (51 mm).

(3) The vent cap or a roof assembly shall have a venting capacity not less than that of the pipe to which it is attached. [NFPA 54:12.8.5]

802.7.5 Support of Single-Wall Metal Pipe. All portions of single-wall metal pipe shall be supported for the design and weight of the material employed. [NFPA 54:12.8.6]

802.7.6 Marking. Single-wall metal pipe shall comply with the marking provisions of Section 802.6.5. [NFPA 54:12.8.7]

802.8 Through-the-Wall Vent Termination. A mechanical draft venting system shall terminate at least 3 feet (914 mm) above any forced air inlet located within 10 feet (3048 mm). (See **Figure 802.8**)

Exceptions:

(1) This provision shall not apply to the combustion air intake of a direct vent appliance.

The intent of Section 802.8.1 is to prevent gases from being drawn back into a building. This requirement recognizes that vent gases are lighter than air. Exception 1 recognizes that direct-vent appliance inlets do not communicate with the air in a building.

(2) This provision shall not apply to the separation of the integral outdoor air inlet and flue gas discharge of listed outdoor appliances. [NFPA 54:12.9.1]

Exception 2 prevents confusion in the installation of outdoor gas equipment. Some authorities have misinterpreted the code to prohibit such equipment or to require it to be modified in the field, which is not the intent of this section. An example of this type of equipment is a packaged rooftop air conditioner, which incorporates a gas vent and a circulation air inlet used for the building air supply.

802.8.1 Mechanical Draft Venting System. A mechanical draft venting system of other than direct vent type shall

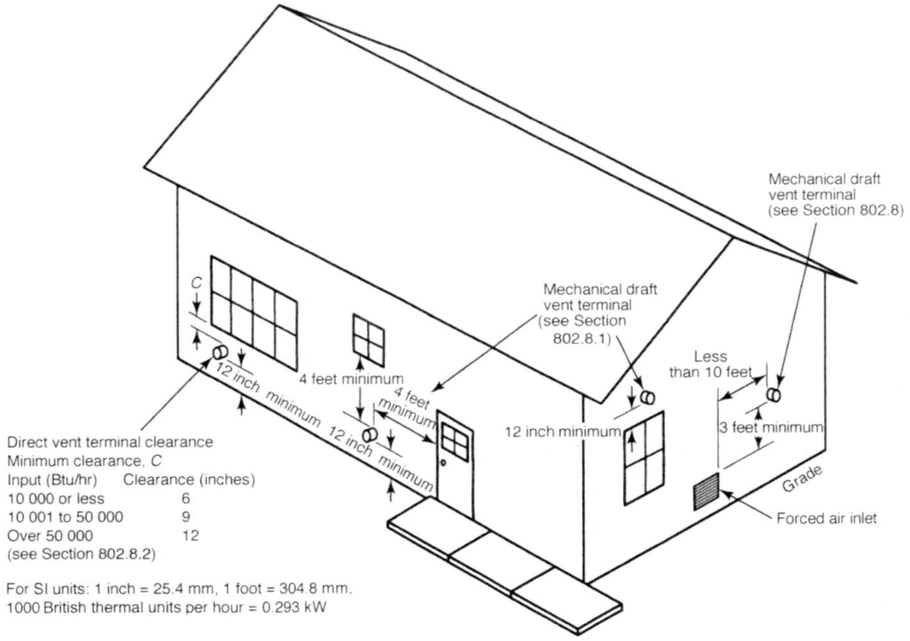

FIGURE 802.8
EXIT TERMINALS OF MECHANICAL DRAFT AND DIRECT-VENT VENTING SYSTEMS

terminate not less than 4 feet (1219 mm) below, 4 feet (1219 mm) horizontally from, or 1 foot (305 mm) above any door, operable window, or gravity air inlet into any building. The bottom of the vent terminal shall be located at least 12 inches (305 mm) above finished ground level. [NFPA 54:12.9.2]

802.8.2 Direct Vent Appliance. The clearances for through-the-wall direct vent terminals shall be in accordance with Table 802.8.2. The bottom of the vent terminal and the air intake shall be located not less than 12 inches (305 mm) above finished ground level. [NFPA 54:12.9.3]

802.8.3 Category I through Category IV and Noncategorized Appliances. Through-the-wall vents for Category II and Category IV appliances and noncategorized condensing appliances shall not terminate over public walkways or over an area where condensate or vapor could create a nuisance or hazard or could be detrimental to the operation of regulators, relief valves, or other equipment. Where local experience indicates that condensate is a problem with Category I and Category III appliances, this provision shall also apply.

Drains for condensate shall be installed in accordance with the appliance and the vent manufacturer's installation instructions. [NFPA 54:12.9.4]

The installer is responsible to locate vent terminations for Category II and IV appliances away from walkways and gas equipment for the protection of pedestrians and equipment, including gas meters. These appliances are high-efficiency with low vent temperatures.

TABLE 802.8.2
THROUGH-THE-WALL DIRECT VENT TERMINATION CLEARANCES
[NFPA 54: TABLE 12.9.3]

DIRECT VENT APPLIANCE INPUT RATING	THROUGH THE WALL VENT TERMINAL CLEARANCE FROM ANY AIR OPENING INTO A BUILDING (inches)
10 000 Btu/h and less	6
Greater than 10 000 Btu/h and not exceeding 50 000 Btu/h	9
Greater than 50 000 Btu/h and not exceeding 150 000 Btu/h	12
> 150 000 Btu/h	In accordance with the appliance manufacturer's instructions and in no case less than the clearances specified in Section 802.8.1.

For SI Units: 1 inch = 25.4 mm, 1000 British thermal units per hour = 0.293 kW

802.8.4 Annular Spaces. Where vents, including those for direct vent appliances or combustion air intake pipes, penetrate outside walls of buildings, the annular spaces around such penetrations shall be permanently sealed using approved materials to prevent entry of combustion products into the building. [NFPA 54:12.9.5]

802.8.5 Vent Terminals. Vent systems for Category IV appliances that terminate through an outside wall of a building and discharge flue gases perpendicular to the adjacent wall shall be located not less than 10 feet (3048 mm) horizontally from an operable opening in an adjacent building.

Exception: This shall not apply to vent terminals that are 2 feet (610 mm) or more above or 25 feet (7620 mm) or more below operable openings. [NFPA 54:12.9.6]

802.9 Condensation Drain. Provision shall be made to collect and dispose of condensate from venting systems serving Category II and Category IV appliances and noncategorized condensing appliances in accordance with Section 802.8.3. [NFPA 54:12.10.1]

802.9.1 Local Experience. Where local experience indicates that condensation is a problem, provision shall be made to drain off and dispose of condensate from venting systems serving Category I and Category III appliances in accordance with Section 802.8.3. [NFPA 54:12.10.2]

Typically, Category I and III vent systems do not have excessive condensation because of high gas temperatures. However, where there is condensation that could cause a problem, then it must be disposed of in accordance with Section 802.9.

802.10 Vent Connectors for Category I Appliances. A vent connector shall be used to connect an appliance to a gas vent, chimney, or single-wall metal pipe, except where the gas vent, chimney, or single-wall metal pipe is directly connected to the appliance. [NFPA 54:12.11.1]

Vent connectors are relatively short runs of single-wall metal pipe, Type B vent or other vent materials that are used to connect an appliance to a chimney or vent. When two or more appliances are connected to a chimney or vent, at least one connector is used. When one appliance is connected to a vent with no change in size or venting material, there is no vent connector.

The vent connector has an important effect on the overall operation of the venting system. Gas Research Institute (GRI)-sponsored research has proven that a vent primes better and has a stronger draft if there is a connector rise directly over the appliance. Retaining as much heat as possible in the flue gases as they pass through the connector is also very important. For more information, refer to the GRI report by S. M. Ricci, et al., "Vent Oversizing and the 'Seven-Times' Rule: Analysis and Recommendations," GRI-98/0285.

802.10.1 Materials. A vent connector shall be made of noncombustible, corrosion-resistant material capable of withstanding the vent gas temperature produced by the appliance and of sufficient thickness to withstand physical damage. [NFPA 54:12.11.2.1]

802.10.1.1 Unconditioned Area. Where the vent connector used for an appliance having a draft hood or a Category I appliance is located in or passes through an unconditioned area, attic, or crawl space, that portion of the vent connector shall be listed Type B, Type L, or listed vent material having equivalent insulation qualities.

Exception: Single-wall metal pipe located within the exterior walls of the building and located in an unconditioned area other than an attic or a crawl space having a local 99 percent winter design temperature of 5°F (-15°C) or higher. [NFPA 54:12.11.2.2]

Attics and unconditioned areas of buildings, such as garages and some basements, are commonly used for storage. Consequently, Type B and L vent material are specified for vent connectors of appliances installed in attics because fires occur as a result of the ignition of combustible material that has been carelessly placed on or near single-wall metal vent connectors. Type B or equivalent materials are also needed to make sure that excessive condensation does not occur inside the vent connector.

802.10.1.2 Residential-Type Appliances. Vent connectors for residential-type appliances shall comply with the following:

(1) Vent connectors for listed appliances having draft hoods, appliances having draft hoods and equipped with listed conversion burners, and Category I appliances that are not installed in attics, crawl spaces, or other unconditioned areas shall be one of the following:
   (a) Type B or Type L vent material.
   (b) Galvanized sheet steel not less than 0.018 of an inch (0.457 mm) thick.
   (c) Aluminum (1100 or 3003 alloy or equivalent) sheet not less than 0.027 of an inch (0.686 mm) thick.
   (d) Stainless steel sheet not less than 0.012 of an inch (0.305 mm) thick.
   (e) Smooth interior wall metal pipe having resistance to heat and corrosion equal to or greater than that of Section 802.10.1.2(1)(b), Section 802.10.1.2(1)(c), or Section 802.10.1.2(1)(d).
   (f) A listed vent connector.
(2) Vent connectors shall not be covered with insulation.

Exception: Listed insulated vent connectors shall be installed in accordance with the manufacturer's installation instructions. [NFPA 54:12.11.2.3]

802.10.1.3 Nonresidential Low-Heat Appliances. A vent connector for a nonresidential low-heat appliance shall be a factory-built chimney section or steel pipe having resistance to heat and corrosion equivalent to that for the appropriate galvanized pipe as specified in Table 802.10.1.3. Factory-built chimney sections shall be joined together in accordance with the chimney manufacturer's instructions. [NFPA 54:12.11.2.4]

TABLE 802.10.1.3
MINIMUM THICKNESS FOR GALVANIZED STEEL VENT CONNECTORS FOR LOW-HEAT APPLIANCES
[NFPA 54: TABLE 12.11.2.4]

DIAMETER OF CONNECTOR (inches)	MINIMUM THICKNESS (inches)
Less than 6	0.019
6 to less than 10	0.023
10 to 12 inclusive	0.029
14 to 16 inclusive	0.034
Over 16	0.056

For SI units: 1 inch = 25.4 mm, 1 square inch = 0.000645 m^2

802.10.1.4 Medium-Heat Appliances. Vent connectors for medium-heat appliances shall be constructed of factory-built, medium-heat chimney sections or steel of a thickness not less than that specified in Table 802.10.1.4 and shall comply with the following:

(1) A steel vent connector for an appliance with a vent gas temperature in excess of 1000°F (538°C) measured at the

entrance to the connector shall be lined with medium-duty fire brick or the equivalent.

(2) The lining shall be at least 2½ inches (64 mm) thick for a vent connector having a diameter or greatest cross-sectional dimension of 18 inches (457 mm) or less.

(3) The lining shall be at least 4½ inches (114 mm) thick laid on the 4½ inches (114 mm) bed for a vent connector having a diameter or greatest cross-sectional dimension greater than 18 inches (457 mm).

(4) Factory-built chimney sections, if employed, shall be joined together in accordance with the chimney manufacturer's instructions. [NFPA 54:12.11.2.5]

TABLE 802.10.1.4
MINIMUM THICKNESS FOR STEEL VENT CONNECTORS FOR MEDIUM-HEAT APPLIANCES
[NFPA 54: TABLE 12.11.2.5]

VENT CONNECTOR SIZE		
DIAMETER (inches)	AREA (square inches)	MINIMUM THICKNESS (inches)
Up to 14	Up to 154	0.053
Over 14 to 16	154 to 201	0.067
Over 16 to 18	201 to 254	0.093
Over 18	Larger than 254	0.123

For SI units: 1 inch = 25.4 mm, 1 square inch = 0.000645 m^2

802.10.2 Size of Vent Connector. A vent connector for an appliance with a single draft hood or for a Category I fan-assisted combustion system appliance shall be sized and installed in accordance with Section 803.0 or other approved engineering methods. [NFPA 54:12.11.3.1]

When sizing connectors serving fan-assisted Category I appliances, the user will note that the selection of single-wall metal pipe connectors is limited. This limit is not included in Section 802.0 but becomes evident when the tables in Section 803.0 are used to size the vent and connector. The limited selection is due to the lower vent-operating temperature of these appliances, which reduces the allowable heat loss in the connector compared to nonfan-assisted combustion appliances. Single-wall connectors are not permitted when Tables 803.2(6) through 803.2(9) are being used to size connectors to exterior masonry chimneys.

802.10.2.1 Manifold. For a single appliance having more than one draft hood outlet or flue collar, the manifold shall be constructed according to the instructions of the appliance manufacturer. Where there are no instructions, the manifold shall be designed and constructed in accordance with approved engineering practices. As an alternative method, the effective area of the manifold shall equal the combined area of the flue collars or draft hood outlets, and the vent connectors shall have a minimum 1 foot (305 mm) rise. [NFPA 54:12.11.3.2]

In the absence of specific instructions, it is important to provide an adequate rise above the draft hoods. This rise is needed to ensure proper flue priming on startup. The minimum 1-foot connector rise minimizes spillage.

802.10.2.2 Size. Where two or more appliances are connected to a common vent or chimney, each vent connector shall be sized in accordance with Section 803.0 or other approved engineering methods. [NFPA 54:12.11.3.3]

As an alternative method applicable only where all of the appliances are draft hood-equipped, each vent connector shall have an effective area not less than the area of the draft hood outlet of the appliance to which it is connected. [NFPA 54:12.11.3.4]

802.10.2.3 Height. Where two or more appliances are vented through a common vent connector or vent manifold, the common vent connector or vent manifold shall be located at the highest level consistent with available headroom and clearance to combustible material and sized in accordance with Section 803.0 or other approved engineering methods. [NFPA 54:12.11.3.5]

As an alternative method applicable only where there are two draft hood-equipped appliances, the effective area of the common vent connector or vent manifold and all junction fittings shall be not less than the area of the larger vent connector plus 50 percent of the area of the smaller flue collar outlet. [NFPA 54:12.11.3.6]

802.10.2.4 Size Increase. Where the size of a vent connector is increased to overcome installation limitations and obtain connector capacity equal to the appliance input, the size increase shall be made at the appliance draft hood outlet. [NFPA 54:12.11.3.7]

802.10.3 Two or More Appliances Connected to a Single Vent. Where two or more openings are provided into one chimney flue or vent, either of the following shall apply:

(1) The openings shall be at different levels.

(2) The connectors shall be attached to the vertical portion of the chimney or vent at an angle of 45 degrees or less relative to the vertical. [NFPA 54:12.11.4.1]

802.10.3.1 Height of Connector. Where two or more vent connectors enter a common vent, chimney flue, or single-wall metal pipe, the smaller connector shall enter at the highest level consistent with the available headroom or clearance to combustible material. [NFPA 54:12.11.4.2]

802.10.3.2 Pressure. Vent connectors serving Category I appliances shall not be connected to any portion of a mechanical draft system operating under positive static pressure, such as those serving Category III or Category IV appliances. [NFPA 54:12.11.4.3]

802.10.4 Clearance. Minimum clearances from vent connectors to combustible material shall be in accordance with Table 802.7.3.3.

Exception: The clearance between a vent connector and combustible material shall be permitted to be reduced where the combustible material is protected as specified for vent connectors in Table 303.10.1. [NFPA 54:12.11.5]

802.10.5 Joints. Joints between sections of connector piping and connections to flue collars or draft hood outlets shall be fastened in accordance with one of the following methods:

(1) Sheet metal screws.

(2) Vent connectors of listed vent material assembled and connected to flue collars or draft hood outlets in accordance with the manufacturer's instructions.

(3) Other approved means. [NFPA 54:12.11.6]

An appliance must be connected to the vent system by three methods of connection. The most common are screws to hold the vent material to the appliance. Vent connectors can also be used. Other methods are allowed but must be approved.

802.10.6 Slope. A vent connector shall be installed without any dips or sags and shall slope upward toward the vent or chimney at least ¼ inch per foot (20.8 mm/m).

Exception: Vent connectors attached to a mechanical draft system installed in accordance with appliance and the draft system manufacturers' instructions. [NFPA 54:12.11.7]

802.10.7 Length of Vent Connector. The length of vent connectors shall comply with Section 802.10.7.1 or Section 802.10.7.2.

802.10.7.1 Single Wall Connector. The maximum horizontal length of a single-wall connector shall be 75 percent of the height of the chimney or vent, except for engineered systems. [NFPA 54:12.11.8.1]

802.10.7.2 Type B Double Wall Connector. The maximum horizontal length of a Type B double-wall connector shall be 100 percent of the height of the chimney or vent, except for engineered systems. The maximum length of an individual connector for a chimney or vent system serving multiple appliances, from the appliance outlet to the junction with the common vent or another connector, shall be 100 percent of the height of the chimney or vent. [NFPA 54:12.11.8.2]

802.10.8 Support. A vent connector shall be supported for the design and weight of the material employed to maintain clearances and prevent physical damage and separation of joints. [NFPA 54:12.11.9]

802.10.9 Chimney Connection. Where entering a flue in a masonry or metal chimney, the vent connector shall be installed above the extreme bottom to avoid stoppage. Where a thimble or slip joint is used to facilitate removal of the connector, the connector shall be firmly attached to or inserted into the thimble or slip joint to prevent the connector from falling out. Means shall be employed to prevent the connector from entering so far as to restrict the space between its end and the opposite wall of the chimney flue. [NFPA 54:12.11.10]

802.10.10 Inspection. The entire length of a vent connector shall be readily accessible for inspection, cleaning, and replacement. [NFPA 54:12.11.11]

802.10.11 Fireplaces. A vent connector shall not be connected to a chimney flue serving a fireplace unless the fireplace flue opening is permanently sealed. [NFPA 54:12.11.12]

802.10.12 Passage through Ceilings, Floors, or Walls. A vent connector shall not pass through a ceiling, floor, or fire-resistance-rated wall. A single-wall metal pipe connector shall not pass through an interior wall.

Exception: Vent connectors made of listed Type B or Type L vent material and serving listed appliances with draft hoods and other appliances listed for use with Type B gas vents that pass through walls or partitions constructed of combustible material shall be installed with not less than the listed clearance to combustible material.

In penetrating an interior wall, floor or ceiling, a single-wall metal connector (e.g., stove pipe) would be entering a space or room of the building other than that in which the appliance is located. Such space could be a storeroom or other part of the building that is not normally occupied. In such cases, at least some portion of the connector would be out of sight. Therefore, any damage to the connector, such as separation of joints, perforation by corrosion with consequent leakage of flue gases into the building or placement of combustible material near or on the connector, would escape early detection, creating a potentially hazardous situation.

802.10.12.1 Medium-Heat Appliances. Vent connectors for medium-heat appliances shall not pass through walls or partitions constructed of combustible material. [NFPA 54:12.11.13.2]

Note that the term "combustible," is defined in Section 205.0 as material subject to an increase in combustibility or flame-spread rating beyond the limits established in the definition of Limited-Combustible Material.

802.11 Vent Connectors for Category II, Category III, and Category IV Appliances. The vent connectors for Category II, Category III, and Category IV appliances shall be in accordance with Section 802.4 through Section 802.4.3. [NFPA 54:12.12]

802.12 Draft Hoods and Draft Controls. Vented appliances shall be installed with draft hoods.

Exception: Dual oven-type combination ranges; incinerators; direct vent appliances; fan-assisted combustion system appliances; appliances requiring chimney draft for operation; single-firebox boilers equipped with conversion burners with inputs greater than 400 000 Btu/h (117 kW); appliances equipped with blast, power, or pressure burners that are not listed for use with draft hoods; and appliances designed for forced venting. [NFPA 54:12.13.1]

Draft hoods on vent systems perform the following three functions:

1. The negative pressure in the vent system created by the hot exhaust gases draws in dilution air at the draft hood opening. This dilution air is taken from the room in which the draft hood is located. This room air is much cooler than the exhaust gases, thereby lowering the net stack temperature and reducing fire hazards. Dilution air is also much drier than the exhaust gases, thereby raising the dewpoint and reducing any condensation.

2. The draft hood acts as a break between the vent system and the appliance and eliminates stack action. Appliance manufacturers design their equipment to operate with a specific range of airflow through the appliance. If there

were no separation between the appliance and the vent system, excessive drafts created by tall chimneys would affect the combustion process and flame stability, possibly even pilot outage. Excessive drafts would also lower efficiency by moving the products of combustion through the heat exchanger before optimal heat transfer. Wind effects can also create temporary downdrafts.

3. Finally, a draft hood provides a relief opening in the event of a downdraft. Vent systems may temporarily experience poor venting at startup (before the vent heats up) or during windy conditions. Under these conditions, some of the products of combustion may "spill out" at the draft hood. The principal products of combustion from a properly burning appliance are carbon dioxide and water vapor and should cause no immediate harm. Once draft is established (or reestablished when the wind subsides), all of the combustion products are vented safely up the vent. During a sustained downdraft, such as in a blockage, all of the combustion products may spill into the living space and may eventually displace the oxygen in the room, potentially leading to incomplete combustion and the formation of carbon monoxide. New central heating appliances equipped with draft hoods must have safety switches, such as the spill switches, which will shut off the burner in the event of a sustained downdraft.

Draft hoods are an integral part of the equipment design and should never be altered. The height of the draft hood above the flue collar will affect combustion and the airflow through the appliance. If a draft hood were removed entirely, even a temporary downdraft would immediately affect the combustion process, potentially creating carbon monoxide.

Barometric draft regulators perform the same functions as a draft hood, but are generally used in connections with power burners and conversion burners. Where power burners are used, the gas input, combustion air, flame pattern and draft all must be carefully set to match the equipment they serve.

Barometric draft regulators are usually adjustable so that the amount of draft can be set for maximum efficiency and safe burner operation. Barometric draft regulators, when used with gas appliances, are double-acting so that they will act as a relief opening in the event of a downdraft. Safety shutoff devices are required on all conversion burners installed after 1990.

802.12.1 Installation. A draft hood supplied with or forming a part of a listed vented appliance shall be installed without alteration, exactly as furnished and specified by the appliance manufacturer. [NFPA 54:12.13.2]

If a draft hood is not supplied by the appliance manufacturer where one is required, a draft hood shall be installed, be of a listed or approved type, and, in the absence of other instructions, be of the same size as the appliance flue collar. Where a draft hood is required with a conversion burner, it shall be of a listed or approved type. [NFPA 54:12.13.2.1]

Where a draft hood of special design is needed or preferable, the installation shall be approved and in accordance with the recommendations of the appliance manufacturer. [NFPA 54:12.13.2.2]

802.12.2 Draft Control Devices. Where a draft control device is part of the appliance or is supplied by the appliance manufacturer, it shall be installed in accordance with the manufacturer's instructions. In the absence of manufacturer's instructions, the device shall be attached to the flue collar of the appliance or as near to the appliance as practical. [NFPA 54:12.13.3]

802.12.3 Additional Devices. Appliances requiring controlled chimney draft shall be permitted to be equipped with listed double-acting barometric draft regulators installed and adjusted in accordance with the manufacturer's instructions. [NFPA 54:12.13.4]

802.12.4 Location. Draft hoods and barometric draft regulators shall be installed in the same room or enclosure as the appliance in such a manner as to prevent any difference in pressure between the hood or regulator and the combustion air supply. [NFPA 54:12.13.5]

802.12.5 Positioning. Draft hoods and draft regulators shall be installed in the position for which they were designed with reference to the horizontal and vertical planes and shall be located so that the relief opening is not obstructed by any part of the appliance or adjacent construction. The appliance and its draft hood shall be located so that the relief opening is accessible for checking vent operation. [NFPA 54:12.13.6]

802.12.6 Clearance. A draft hood shall be located so that its relief opening is not less than 6 inches (152 mm) from any surface except that of the appliance it serves and the venting system to which the draft hood is connected. Where a greater or lesser clearance is indicated on the appliance label, the clearance shall not be less than that specified on the label. Such clearances shall not be reduced. [NFPA 54:12.13.7]

802.13 Manually Operated Dampers. A manually operated damper shall not be placed in any appliance vent connector. Fixed baffles shall not be classified as manually operated dampers. [NFPA 54:12.14]

802.14 Obstructions. Devices that retard the flow of vent gases shall not be installed in a vent connector, chimney, or vent. The following shall not be considered as obstructions:

(1) Draft regulators and safety controls specifically listed for installation in venting systems and installed in accordance with the manufacturer's installation instructions.

(2) Approved draft regulators and safety controls designed and installed in accordance with approved engineering methods.

(3) Listed heat reclaimers and automatically operated vent dampers installed in accordance with the manufacturer's installation instructions.

(4) Vent dampers serving listed appliances installed in accordance with Section 803.1 or Section 803.2 or other approved engineering methods.

(5) Approved economizers, heat reclaimers, and recuperators installed in venting systems of appliances not required to be equipped with draft hoods, provided the appliance manufacturer's instructions cover the installation of such a device in the venting system and performance in accordance with Section 802.3 and Section 802.3.1 is obtained. [NFPA 54:12.16]

802.15 Automatically Operated Vent Dampers. An automatically operated vent damper shall be of a listed type. [NFPA 54:12.15]

Preferably, automatically operated vent dampers, such as the one shown in **Figure 802.15**, should be included as part of the listed gas utilization equipment. Manually operated dampers are not permitted in vent connectors, and the requirement clarifies that fixed baffles, which are sometimes used to balance system draft on startup, are allowed.

FIGURE 802.15
VENT DAMPER

802.15.1 Listing. Automatically operated vent dampers for oil-fired appliances shall comply with UL 17. The automatic damper control shall comply with UL 378.

803.0 Sizing of Category I Venting Systems.

The tables in Section 803.0 provide sizes for Category I appliances serving natural gas appliances, fan-assisted draft appliances and combinations of both in one venting system. The tables incorporate the Fan Min and Fan Max columns to provide minimum and maximum capacities for each connector and vent serving fan-assisted combustion appliances. The other columns in the tables contain the following:

- The Nat Max columns provide maximum capacities for each connector and vent serving natural draft appliances.
- The Fan + Fan column provides maximum capacities for each connector and vent serving multiple fan-assisted appliances.
- The Fan + Nat columns provide maximum capacities for each connector and vent serving combinations of natural draft and fan-assisted appliances.
- The Nat + Nat columns provide maximum capacities for each connector and vent serving multiple natural draft appliances.

Category I appliances are defined as appliances that operate with a nonpositive vent static pressure and with a vent gas temperature that avoids excessive condensate in the vent.

Venting systems serving Category I appliances are conventional venting systems, in which the heat of the flue gases is the force that operates the vent. Section 803.0 includes the following:

- Six tables are listed in **Commentary Table 1** for single-appliance installations.

COMMENTARY TABLE 1

Table No.	Vent	Connector
803.1.2 (1)	Type B	None
803.1.2 (2)	Type B	Single wall
803.1.2 (3)	Masonry chimney	Type B
803.1.2 (4)	Masonry chimney	Single wall
803.1.2 (5)	Single-wall metal pipe	None
803.1.2 (6)	Exterior Masonry	Type B

- Additional requirements for single-appliance vents are given. These qualifications must be considered when using single-appliance vent Tables 803.1.2(1) through 803.1.2(6).
- Nine vent tables are listed in **Commentary Table 2** and **Commentary Table 3** for multiple-appliance venting installations. The five tables listed in **Commentary Table 2** are used for interior chimney or vent installations.

COMMENTARY TABLE 2

Table No.	Vent	Connector
803.2 (1)	Type B	Double wall
803.2 (2)	TypeB	Single wall
803.2 (3)	Masonry chimney	Type B
803.2 (4)	Masonry chimney	Single wall
803.2 (5)	Single-wall metal pipe	None

- Four tables covering exterior masonry chimneys are listed in **Commentary Table 3**. The tables are organized by geographical regions of the United States based on lowest anticipated winter design temperature.

COMMENTARY TABLE 3

Table No.	Vent	Appliance
803.2 (6)	Exterior masonry chimney	NAT + NAT
803.2 (7)	Exterior masonry chimney	NAT + NAT
803.2 (8)	Exterior masonry chimney	FAN + NAT
803.2 (9)	Exterior masonry chimney	FAN + NAT

- Additional requirements for multiple-appliance vents are provided. These qualifications must be considered when using multiple-appliance vent Tables 803.2(1) through 803.2(9).

Tables 803.2(6) through 803.2(9) of the code provide sizing information for exterior masonry chimneys serving heating appliances. The tables provide sizing options based on local 99-percent winter design temperature in six temperature ranges.

803.1 Single Appliance Vent Table 803.1.2(1) through Table 803.1.2(6). Venting Table 803.1.2(1) through Table 803.1.2(6) shall not be used where obstructions are installed in the venting system. The installation of vents serving listed appliances with vent dampers shall be in accordance with the appliance manufacturer's instructions or in accordance with the following:

(1) The maximum capacity of the vent system shall be determined using the "NAT Max" column.

(2) The minimum capacity shall be determined as though the appliance were a fan-assisted appliance, using the "FAN Min" column to determine the minimum capacity of the vent system. Where the corresponding "FAN Min" is "NA," the vent configuration shall not be permitted and an alternative venting configuration shall be utilized. [NFPA 54:13.1.1]

The tables were developed with the assumption that there are no restrictions in the path of the vent gas flow, as indicated by the symbol shown in **Figure 803.1** An additional assumption is that there is no deliberate attempt to remove heat. Therefore, devices such as heat economizers may not be used in conjunction with the tables. Vent dampers installed as part of a listed appliance are to be installed in accordance with the manufacturer's instructions.

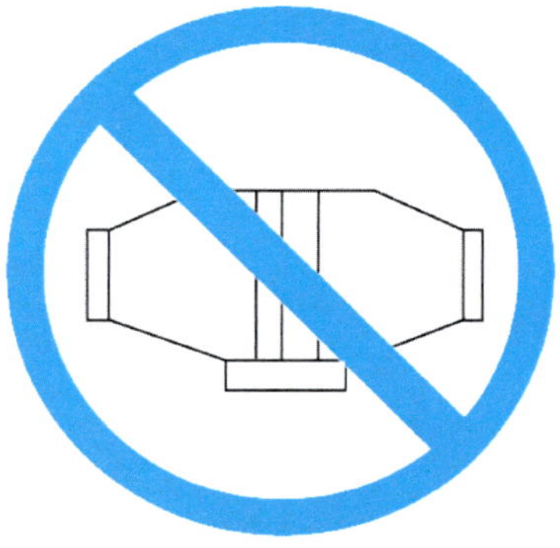

FIGURE 803.1
NO OBSTRUCTIONS

803.1.1 Vent Downsizing. Where the vent size determined from the tables is smaller than the appliance draft hood outlet or flue collar, the use of the smaller size shall be permitted, provided that the installation complies with all of the following requirements:

(1) The total vent height (*H*) is at least 10 feet (3048 mm).

(2) Vents for appliance draft hood outlets or flue collars 12 inches (305mm) in diameter or smaller are not reduced more than one table size.

(3) Vents for appliance draft hood outlets or flue collars larger than 12 inches (305 mm) in diameter are not reduced more than two table sizes.

(4) The maximum capacity listed in the tables for a fan-assisted appliance is reduced by 10 percent (0.90 x maximum table capacity).

(5) The draft hood outlet is greater than 4 inches (102 mm) in diameter. A 3 inch (76 mm) diameter vent shall not be connected to a 4 inch (102 mm) diameter draft hood outlet. This provision shall not apply to fan-assisted appliances. [NFPA 54:13.1.2]

There is an economic incentive to using smaller vents wherever possible. If the vent is smaller than the draft hood or flue collar, venting problems can occur. The restrictions recognize and avoid these venting problems. Limits are placed on "downsizing" (see **Figure 803.1.1**). In particular, note that a 4-inch draft hood outlet may not be reduced 3 inches.

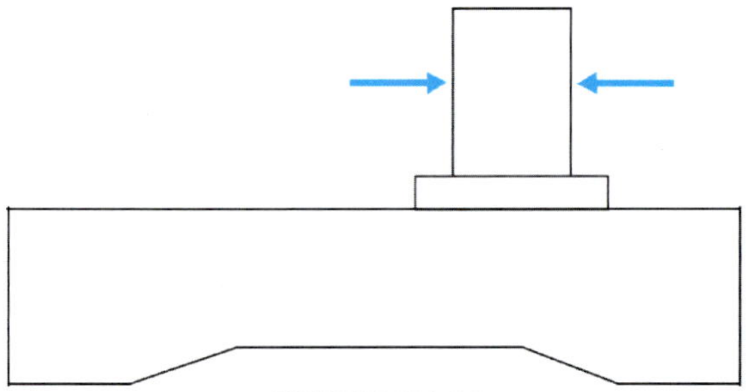

FIGURE 803.1.1
DOWNSIZING

803.1.2 Elbows. Single-appliance venting configurations with zero (0) lateral lengths in Table 803.1.2(1), Table 803.1.2(2), and Table 803.1.2(5) shall not have elbows in the venting system. Single-appliance venting with lateral lengths include two 90 degree elbows. For each additional elbow up to and including 45 degrees, the maximum capacity listed in the venting tables shall be reduced by 5 percent. For each additional elbow greater than 45 degrees up to and including 90 degrees, the maximum capacity listed in the venting tables shall be reduced by 10 percent. Where multiple offsets occur in a vent, the total lateral length of all offsets combined shall not exceed that specified in Table 803.1.2(1) through Table 803.1.2(5). [NFPA 54:13.1.3]

The sizing tables were designed with an assumption that up to two 90-degree turns were part of the venting system, except for the zero lateral length case addressed in Section 803.1.3.The zero lateral case was assumed to extend straight up from the appliance outlet to the vent termination. Adding additional elbows to the system is possible only by reducing the maximum capacity listed in the venting tables (See **Figure 803.1.2**).

The following **Commentary Table** shows the capacity reduction for elbows of less than 90 degrees.

803.1.3 Zero Lateral. Zero (0) lateral (*L*) shall apply only to a straight vertical vent attached to a top outlet draft hood or flue collar. [NFPA 54:13.1.4]

Where zero lateral is used, the use of elbows in the venting system is not permitted (see **Figure 803.1.3**). If elbows are needed, the table rows for 2-feet lateral length

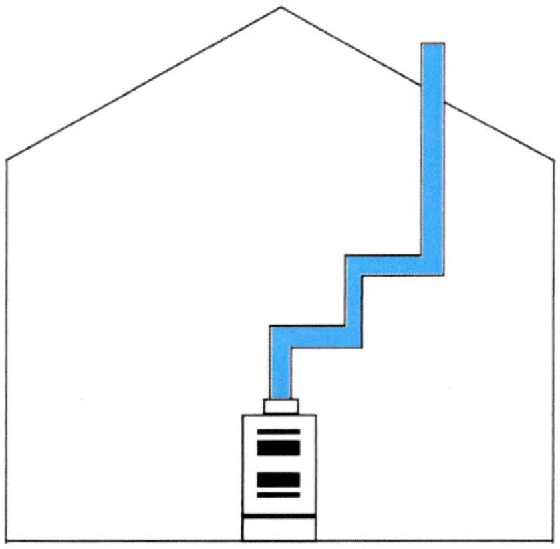

FIGURE 803.1.2
ELBOWS

COMMENTARY TABLE

Capacity Reduction for Elbows of Less Than 90 Degrees

Elbow	Vent Table Capacity Reduction, Per Elbow (%)
0–45	5
>45–90	10

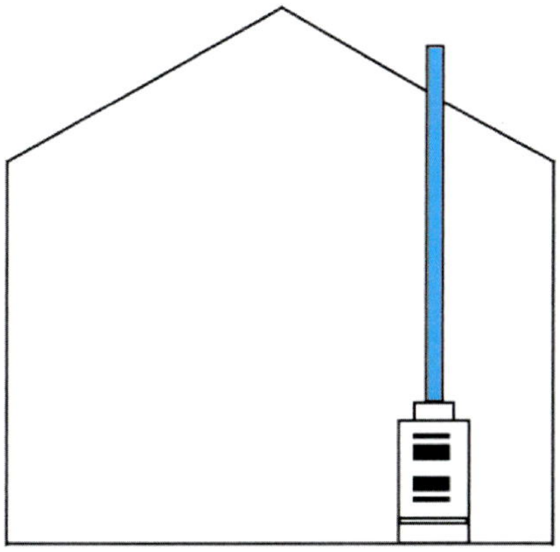

FIGURE 803.1.3
NO ELBOWS

must be used. Elbows in a vent system with zero offset are not common but may be needed to route the vent to avoid a building obstruction, such as a beam.

803.1.4 High-Altitude Installations. Sea level input ratings shall be used when determining maximum capacity for high-altitude installation. Actual input (derated for altitude) shall be used for determining minimum capacity for high-altitude installation. [NFPA 54:13.1.5]

Using the sea level input rating for the maximum capacity is a conservative measure because less draft is produced at high altitudes. The derating process will also make condensation more likely. The reduced input rate should be used for minimum capacity.

803.1.5 Multiple Input Ratings. For appliances with more than one input rate, the minimum vent capacity (FAN Min) determined from Table 803.1.2(1) through Table 803.2(9) shall be less than the lowest appliance input rating, and the maximum vent capacity (FAN Max/NAT Max) determined from the tables shall be greater than the highest appliance rating input. [NFPA 54:13.1.6]

803.1.6 Corrugated Chimney Liner Reduction. Listed corrugated metallic chimney liner systems in masonry chimneys shall be sized by using Table 803.1.2(1) or Table 803.1.2(2) for Type B vents, with the maximum capacity reduced by 20 percent (0.80 x maximum capacity) and the minimum capacity as shown in Table 803.1.2(1) or Table 803.1.2(2).

Corrugated metallic liner systems installed with bends or offsets shall have their maximum capacity further reduced in accordance with Section 803.1.2. The 20 percent reduction for corrugated metallic chimney liner systems includes an allowance for one long radius 90 degree (1.57 rad) turn at the bottom of the liner. [NFPA 54:13.1.7]

Because properly installed corrugated chimney liners have a heat loss similar to Type B vent, they are sized using Table 803.1.2(1) or 803.1.2(2). However, such liners' corrugations and their tendency to spiral in the chimney require a 20 percent maximum capacity reduction.

Many liners begin at the breaching and then bend up vertically. This 90-degree elbow at the beginning of the liner is included in the table capacity.

803.1.7 Connection to Chimney Liners. Connections between chimney liners and listed double-wall connectors shall be made with listed adapters designed for such purpose. [NFPA 54:13.1.8]

Guidance is provided for using exterior chimneys (see **Figure 803.1.7**) for heating appliances. These restrictions are needed because of the high heat loss in chimneys exposed to the outdoors below the roofline in cold climates. These restrictions do not apply to chimneys and vents not exposed to the outdoors below the roofline because these chimneys and vents are considered to be interior chimneys.

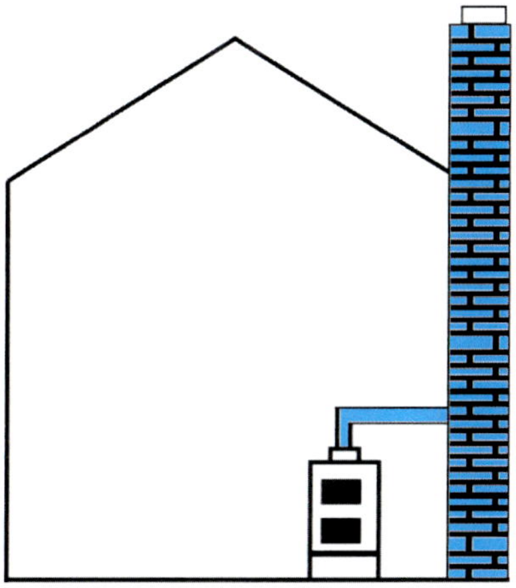

FIGURE 803.1.7
EXTERIOR CHIMNEYS

The following items are required when sizing exterior masonry chimneys:

- To size chimneys and vents serving only water heaters, the outdoor design temperature must be greater than 5°F (15°C). In colder climates, either the chimney must be lined with a metallic liner or another vent must be used.
- To size chimneys and vents serving space heating appliances, use Table 803.2(7) for the minimum capacity and use Table 803.2(8) for the combined vent capacity of all appliances providing heat to a building.

Note that only B-vent connectors are allowed to minimize heat loss through vent connectors.

Water heaters are treated differently from heating appliances because their total operating hours are not sufficient to keep a chimney warm in cold periods, resulting in condensation of water in the chimney vent.

803.1.8 Vertical Vent Upsizing Using 7 x Rule. Where the vertical vent has a larger diameter than the vent connector, the vertical vent diameter shall be used to determine the minimum vent capacity, and the connector diameter shall be used to determine the maximum vent capacity. The flow area of the vertical vent shall not exceed seven times the flow area of the listed appliance categorized vent area, flue collar area, or draft hood outlet area unless designed in accordance with approved engineering methods. [NFPA 54:13.1.9]

In a vent system with a larger vent than the vent connector, the draft is limited by the small diameter of the vent connector (see **Figure 803.1.8**). Condensation will begin in the larger diameter. The maximum and minimum capacities must be determined accordingly. Practical limits exist as to how large the vertical vent may be relative to its source of vent gas flow; therefore, the flow area of the vent may not be more than seven times the flow area of the outlet of the appliance or draft hood.

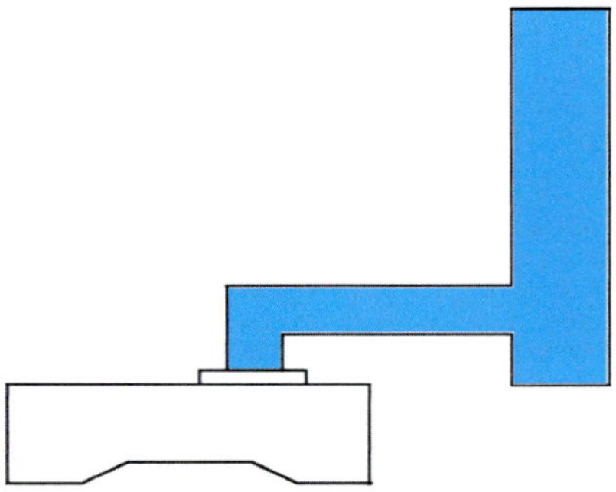

FIGURE 803.1.8
UPSIZING VERTICAL VENT

803.1.9 Draft Hood Conversion Accessories. Draft hood conversion accessories for use with masonry chimneys venting listed Category I fan-assisted appliances shall be listed and installed in accordance with the listed accessory manufacturer's installation instructions. [NFPA 54:13.1.10]

803.1.10 Chimney and Vent Locations. Table 803.1.2(1) through Table 803.1.2(5) shall be used only for chimneys and vents not exposed to the outdoors below the roof line. A Type B vent or listed chimney lining system passing through an unused masonry chimney flue shall not be considered to be exposed to the outdoors. Where vents extend outdoors above the roof more than 5 feet (1524 mm) higher than required by Table 802.6.1, and where vents terminate in accordance with Section 802.6.1(1)(b), the outdoor portion of the vent shall be enclosed as required by this paragraph for vents not considered to be exposed to the outdoors, or such venting system shall be engineered. A Type B vent passing through an unventilated enclosure or chase insulated to a value of not less than R8 shall not be considered to be exposed to the outdoors. Table 803.1.2(3) in combination with Table 803.1.2(6) shall be used for clay tile-lined exterior masonry chimneys, provided all of the following requirements are met:

(1) The vent connector is Type B double wall.

(2) The vent connector length is limited to 18 in./in. (18 mm/mm) of vent connector diameter.

(3) The appliance is draft hood equipped.

(4) The input rating is less than the maximum capacity given in Table 803.1.2(3).

(5) For a water heater, the outdoor design temperature shall not be less than 5°F (-15°C).

(6) For a space-heating appliance, the input rating is greater than the minimum capacity given by Table 803.1.2(6). [NFPA 54:13.1.11]

803.1.11 Residential and Low-Heat Appliances. Flue lining system for residential and low heat appliance shall be in accordance with Section 803.1.11.1 and Section 803.1.11.2.

803.1.11.1 Clay Flue Lining. Clay flue lining shall be manufactured in accordance with ASTM C315 or other approved standard.

803.1.11.2 Chimney Lining. Chimney lining shall be listed in accordance with UL 1777.

803.1.12 Corrugated Vent Connector Size. Corrugated vent connectors shall not be smaller than the listed appliance categorized vent diameter, flue collar diameter, or draft hood outlet diameter. [NFPA 54:13.1.12]

803.1.13 Upsizing. Vent connectors shall not be upsized more than two sizes greater than the listed appliance categorized vent diameter, flue collar diameter, or draft hood outlet diameter. [NFPA 54:13.1.13]

A sudden, large expansion of the vent connector diameter creates a pressure drop that may limit the draft and encourage condensation. Therefore, a limit is placed on upsizing.

803.1.14 Single Run of Vent. In a single run of vent or vent connector, more than one diameter and type shall be permitted to be used, provided that all the sizes and types are permitted by the tables. [NFPA 54:13.1.14]

The installer is required to check the minimum and maximum capacities for the vent section for each type, as if the entire vent section were made of that type (see **Figure 803.1.14**). For example, suppose a vent connector was half single-wall and half Type B vent. The installer must show that a vent connector of that length would be allowed if it was all single-wall and also if it was all Type B vent.

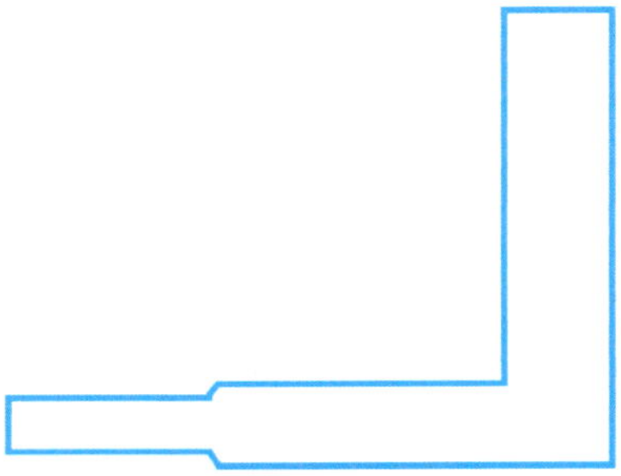

FIGURE 803.1.14
MULTIPLE VERTICAL VENT SIZES

803.1.15 Interpolation. Interpolation shall be permitted in calculating capacities for vent dimensions that fall between table entries. [NFPA 54:13.1.15]

If the installation dimensions fall between two table entries for which are defined values, the designer may calculate the "in between" value. This is called "interpolation." An example follows:

Table 803.1.2(1) provides the capacity of a 3-inch diameter, 15-foot high vent with zero lateral serving a natural draft appliance of 58,000 Btu/hr, and of 3-inch diameter, 20-foot high vent with zero lateral of 61,000 Btu/hr. The capacity of a 3-inch diameter, 18-foot high vent with zero lateral can be interpolated by taking 3⁄5 of the difference between 58,000 and 61,000 and adding it to the capacity of the 15-foot high vent. In this case, the difference is 3,000 Btu/hr, 3⁄5 of 3,000 is 1,800 and the capacity of the 18-foot vent is 58,800 Btu/hr.

803.1.16 Extrapolation. Extrapolation beyond the table entries shall not be permitted. [NFPA 54:13.1.16]

803.1.17 Engineering Methods. For vent heights lower than 6 feet (1829 mm) and higher than shown in Table 803.1.2(1) through Table 803.2(9), engineering methods shall be used to calculate vent capacities. [NFPA 54:13.1.17]

803.1.18 Height Entries. Where the actual height of a vent falls between entries in the height column of the applicable table in Table 803.1.2(1) through Table 803.1.2(6), either of the following shall be used:

(1) Interpolation.

(2) The lower appliance input rating shown in the table entries for FAN MAX and NAT MAX column values, and the higher appliance input rating for the FAN MIN column values. [NFPA 54:13.1.18]

803.2 Multiple Appliance Vent Table 803.2(1) through Table 803.2(9). Venting Table 803.2(1) through Table 803.2(9) shall not be used where obstructions are installed in the venting system. The installation of vents serving listed appliances with vent dampers shall be in accordance with the appliance manufacturer's instructions, or in accordance with the following:

(1) The maximum capacity of the vent connector shall be determined using the NAT Max column.

(2) The maximum capacity of the vertical vent or chimney shall be determined using the FAN + NAT column when the second appliance is a fan-assisted appliance, or the NAT + NAT column when the second appliance is equipped with a draft hood.

(3) The minimum capacity shall be determined as if the appliance were a fan-assisted appliance, as follows:

   (a) The minimum capacity of the vent connector shall be determined using the FAN Min column.

   (b) The FAN + FAN column shall be used when the second appliance is a fan-assisted appliance, and the FAN + NAT column shall be used when the second appliance is equipped with a draft hood, to determine whether the vertical vent or chimney configuration is not permitted (NA). Where the vent configuration is NA, the vent configuration shall not be permitted and an alternative venting configuration shall be utilized. [NFPA 54:13.2.1]

Tables 803.2(1) through 803.2(9) were developed with the assumption that there are no restrictions in the path of the vent gas flow. An additional assumption is that there is no deliberate attempt to remove heat. Therefore, devices such as heat economizers may not be used in conjunction with the tables. Vent dampers installed as part of a listed appliance are to be installed in accordance with the manufacturer's instructions or the general requirements of Section 803.2.

Section 803.2 provides guidance for using the venting tables with draft hood appliances equipped with a vent damper. An example of such an appliance is a boiler, which uses a vent damper to obtain higher efficiencies as follows:

- The maximum capacity of the vent connector is found in the Nat Max column of the Vent Connector Capacity sections of Tables 803.2(1) through 803.2(4), as stated in Section 803.2.
- The maximum capacity of the vent is found using the Fan Max column of the Common Vent Capacity sections of Tables 803.2 (1) through 803.2(4), as stated in Section 803.2.

The following treats the vent as one serving a natural draft appliance when the appliance is operating:

- The minimum capacity of the vent connector is found using the Fan Min column of the Vent Connector Capacity sections of Tables 803.2(1) through 803.2(4), as stated in Section 803.2.

The minimum capacity of the vent depends on the venting category of the second appliance (the first appliance is the boiler with the vent damper or other obstruction):

- If the second appliance is a fan-assisted appliance, the Fan + Fan column of the Common Vent Capacity portion of Tables 803.2(1) through 803.2(4) is used. If the second appliance is equipped with a draft hood, the Fan + Nat column of the Common Vent Capacity of Tables 803.2(1) through 803.2(4) is used.

This treats the vent as one serving a fan-assisted (increased efficiency) appliance when the appliance is not operating. The reason for this unusual combination is that the appliance operates as a natural draft appliance, but when the appliance is not operating and the vent damper is closed, it is similar to a fan-assisted (energy saving) appliance in its propensity to condense water.

803.2.1 Vent Connector Maximum Length. The maximum vent connector horizontal length shall be 18 inches per inch (18 mm/mm) of connector diameter as shown in Table 803.2.1, or as permitted by Section 803.2.2. [NFPA 54:13.2.2]

TABLE 803.2.1
VENT CONNECTOR MAXIMUM LENGTH
[NFPA 54: TABLE 13.2.2]

CONNECTOR DIAMETER (inches)	MAXIMUM CONNECTOR HORIZONTAL LENGTH (feet)
3	4½
4	6
5	7½
6	9
7	10½
8	12
9	13½
10	15
12	18
14	21
16	24
18	27
20	30
22	33
24	36

For SI units: 1 inch = 25.4 mm, 1 foot = 304.8 mm

803.2.2 Vent Connector Exceeding Maximum Length. The vent connector shall be routed to the vent utilizing the shortest possible route. Connectors with longer horizontal lengths than those listed in Table 803.2.1 are permitted under the following conditions:

(1) The maximum capacity (FAN Max or NAT Max) of the vent connector shall be reduced 10 percent for each additional multiple of the length listed in Table 803.2.1. For example, the maximum length listed for a 4 inch (100 mm) connector is 6 feet (1829 mm). With a connector length greater than 6 feet (1829 mm) but not exceeding 12 feet (3658 mm), the maximum capacity must be reduced by 10 percent (0.90 x maximum vent connector capacity). With a connector length greater than 12 feet (3658 mm) but not exceeding 18 feet (5486 mm), the maximum capacity must be reduced by 20 percent (0.80 x maximum vent capacity).

(2) For a connector serving a fan-assisted appliance, the minimum capacity (FAN Min) of the connector shall be determined by referring to the corresponding single appliance table. For Type B double-wall connectors, Table 803.1.2(1) shall be used. For single-wall connectors, Table 803.1.2(2) shall be used. The height (*H*) and lateral (*L*) shall be measured according to the procedures for a single-appliance vent, as if the other appliances were not present. [NFPA 54:13.2.3]

For brevity, the common venting tables (the lower portion of Tables 803.2(7) through 803.2(10)) are designed with the assumption that the vent connector is no more than 18 inches long for each inch of diameter. Sections 803.2.1 and 803.2.3, summarized as follows, provide guidance on how to handle connectors that are longer than this:

- The maximum capacity for vent connectors that exceed 18 inches per inch of diameter is reduced by 10 percent for each length of 18 inches per inch of vent diameter.
- The minimum capacity for connectors serving fan-assisted appliances is determined differently for single-wall and Type B double-wall connectors, as described in Section 803.2.3.

803.2.3 Vent Connector Manifolds. Where the vent connectors are combined prior to entering the vertical portion of the common vent to form a common vent manifold, the size of the common vent manifold and the common vent shall be determined by applying a 10 percent reduction (0.90 x maximum common vent capacity) to the common vent capacity part of the common vent tables. The length of the common vent manifold (*LM*) shall not exceed 18 inches per inch (18 mm/mm) of common vent diameter (*D*). [NFPA 54:13.2.4] (See **Figure 802.6.3.2**)

803.2.4 Vent Offsets. Where the common vertical vent is offset, the maximum capacity of the common vent shall be reduced in accordance with Section 803.2.5, and the horizontal length of the common vent offset shall not exceed 18 inches per inch (18 mm/mm) of common vent diameter (*D*). Where multiple offsets occur in a common vent, the total horizontal length of all offsets combined shall not exceed 18 inches per inch (18 mm/mm) of the common vent diameter. [NFPA 54:13.2.5]

The "common vent" is the vertical vent or chimney only. A combination of vent connections prior to entering a vertical chimney or vent is considered to be a vent manifold (see **Figure 803.2.4a**), and guidance for sizing such is located here. Note that a chimney liner system that comes out through the breaching, as many do, should be considered a manifold. The capacity of the vent manifold is reduced by 10 percent from the table value. Also note that the length of a vent manifold is limited to 18 inches per inch of vent manifold diameter.

Vent offsets, addressed in this section and illustrated in **Figure 803.2.4b**, typically occur high in the venting system and tend to reduce the draft produced. This reduction in draft is offset by the requirement for a 10 percent capacity reduction.

803.2.5 Elbows in Vents. For each elbow up to and including 45 degrees (0.79 rad) in the common vent, the maximum common vent capacity listed in the venting tables shall be reduced by 5 percent. For each elbow greater than 45 degrees (0.79 rad) up to and including 90 degrees (1.57 rad), the maximum common vent capacity listed in the venting tables shall be reduced by 10 percent. [NFPA 54:13.2.6]

See commentary for Section 803.1.2.

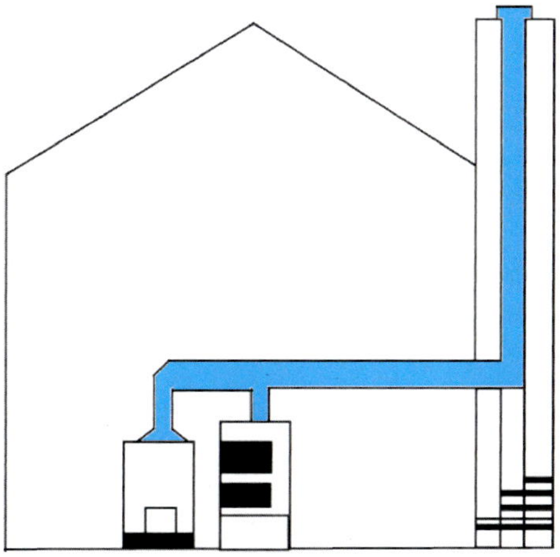

FIGURE 803.2.4A
VENT MANIFOLDS

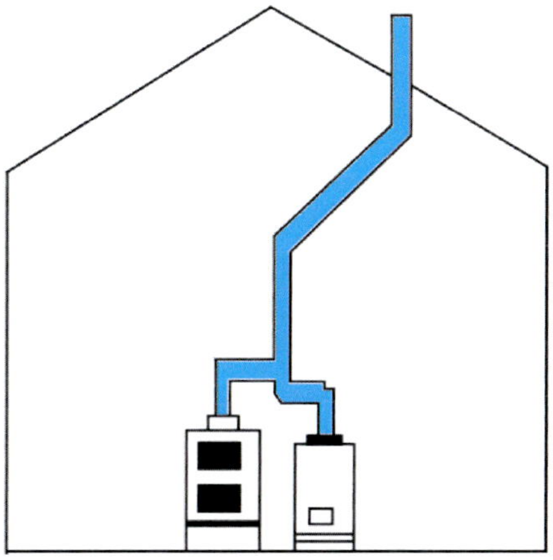

FIGURE 803.2.4B
VENT OFFSETS

803.2.6 Elbows in Connectors. The vent connector capacities listed in the common vent sizing tables include allowance for two 90 degree elbows. For each additional elbow up to and including 45 degrees, the maximum vent connector capacity listed in the venting tables shall be reduced by 5 percent. For each elbow greater than 45 degrees up to and including 90 degrees, the maximum vent connector capacity listed in the venting tables shall be reduced by 10 percent. [NFPA 54:13.2.7]

803.2.7 Common Vent Minimum Size. The cross-sectional area of the common vent shall be equal to or greater than the cross-sectional area of the largest connector. [NFPA 54:13.2.8]

Refer to Section and **Figure 803.1.2** for commentary.

803.2.8 Tee and Wye Fittings. Tee and wye fittings connected to a common gas vent shall be considered as part of the common gas vent and constructed of materials consistent with that of the common gas vent. [NFPA 54:13.2.9]

803.2.9 Size of Fittings. At the point where tee or wye fittings connect to a common gas vent, the opening size of the fitting shall be equal to the size of the common vent. Such fittings shall not be prohibited from having reduced size openings at the point of connection of appliance gas vent connectors. [NFPA 54:13.2.10]

803.2.10 High-Altitude Installations. Sea level input ratings shall be used when determining maximum capacity for high-altitude installation. Actual input (derated for altitude) shall be used for determining minimum capacity for high-altitude installation. [NFPA 54:13.2.11]

Using the sea level input rating required by this section for the maximum capacity is a conservative measure because less draft is produced at high altitudes. The derating process will also make condensation more likely. The reduced input rate should be used for the minimum capacity.

803.2.11 Connector Rise. The connector rise (*R*) for each appliance connector shall be measured from the draft hood outlet or flue collar to the centerline where the vent gas streams come together. [NFPA 54:13.2.12]

803.2.12 Vent Height. For multiple appliances all located on one floor, available total height (*H*) shall be measured from the highest draft hood outlet or flue collar up to the level of the outlet of the common vent. [NFPA 54:13.2.13]

803.2.13 Multistory Installations. For multistory installations, available total height (*H*) for each segment of the system shall be the vertical distance between the highest draft hood outlet or flue collar entering that segment and the centerline of the next higher interconnection tee. [NFPA 54:13.2.14]

803.2.14 Size of Vents for Multistory Installations. The size of the lowest connector and of the vertical vent leading to the lowest interconnection of a multistory system shall be in accordance with Table 803.1.2(1) or Table 803.1.2(2) for available total height (*H*) up to the lowest interconnection. [NFPA 54:13.2.15]

Following is an example of the use of tables in Section 803.0 to size a multistory vent. Assume a four-story apartment building that has a listed fan-assisted combustion furnace installed on each floor. Each furnace has a 4-inch flue collar and an input of 80,000 Btu/hr. All the furnaces are installed outside the conditioned space (i.e. isolated combustion) and are to be vented into the common vent, which is located 5 feet from the furnaces. For the purpose of calculation, the overall system is divided into smaller, simple vent systems for each level.

The structure is such that the vent connector rise is to be 2 feet and the available total height for each level is 10 feet, except for the top floor, which is 6 feet. The vent connector for the first floor or lowest furnace to the common vent is considered to be an individual vent terminating at the first tee or interconnection, and this vent is sized in accordance with Table 803.1.2(1). Every other vent connector is sized in accordance with the Vent Connector Capacity section of Table 803.2(1). Each section of the common vent is sized in accordance with the Common Vent Capacity section of Table 803.2(1) to accommodate the accumulated total input of all appliances discharging into it, but sections are never smaller than the largest section below them.

Procedure for Sizing Connectors

First-Floor Connector (Furnace 1). The input is 80,000 Btu/hr and the total height is 10 feet. This section of the vent is treated as a single-appliance vent with 5 feet of vent connector. Using Table 803.1.2(1), read across the line for a 10-foot height (H) and lateral (L) of 5 feet. Look under the FAN columns for a range that includes 80,000. The 4-inch column has a range of 32,000 Btu/hr to 113,000 Btu/hr, and the 5-inch column has a range of 41,000 Btu/hr to 187,000 Btu/hr. Both sizes are acceptable, and a 4-inch size is selected because it is smaller and less expensive.

Table 803.2.1 limits a 4-inch vent connector to a maximum length of 6-feet (1.5 feet length for each inch of diameter), which is not exceeded in this example.

Second- and Third-Floor Connectors (Furnaces 2 and 3). The input is 80,000 Btu/hr, the connector rise is 2 feet, and the vent height is 10 feet. Using the Vent Connector Capacity section of Table 803.2(1), read across the 10-foot height, 2-foot rise row in the FAN columns. It shows that a 4-inch connector has a range of 36,000 Btu/hr to 86,000 Btu/hr, a 5-inch connector has a range of 51,000 Btu/hr to 136,000 Btu/hr and a 6-inch connector has a range of 67,000 Btu/hr to 206,000 Btu/hr. A connector with a diameter of 4 inches, 5 inches or 6 inches will work, and a 4-inch connector is selected.

Fourth-Floor Connector (Furnace 4). The input is 80,000 Btu/hr, the connector rise is 2 feet and the vent height is 6 feet. Using the Vent Connector Capacity section of Table 803.2(1), read across the 6-foot height, 2-foot rise row. It shows that a 5-inch connector has a range of 48,000 Btu/hr to 121,000 Btu/hr and a 6-inch connector has a range of 60,000 Btu/hr to 183,000 Btu/hr. Either a 5-inch or a 6-inch connector will work, and a 5-inch connector is selected because it is smaller and less expensive.

Table 803.2.1 limits a 5-inch vent connector to a maximum length of 7½ feet (1.5 feet length for each inch of diameter), which is not exceeded by the connector size selected.

Procedure for Sizing Common Vents

Common Vent for Furnaces 1 and 2. The input is 160,000 Btu/hr and the vent height is 10 feet. Using the Common Vent Capacity section of Table 803.2(1), FAN + FAN columns only, read across the 10-foot line. It shows that a 5-inch vent has a capacity of 169,000 Btu/hr, which exceeds the 160,000 Btu/hr input. A 5-inch vent is used.

Common Vent for Third-Floor Furnace (Furnace 3). The input is 240,000 Btu/hr and the height is 10 feet. Using the Common Vent Capacity section of Table 803.2(1) read across the 10-foot line in the FAN + FAN columns. It shows that a 6-inch common vent has a capacity of 243,000 Btu/hr. A 6-inch vent is used.

Common Vent for Fourth-Floor Furnace (Furnace 4). The input is 320,000 Btu/hr and the height is 6 feet. Using the FAN + FAN columns in the Common Vent Capacity section of Table 803.2(1), read across the 6-foot line. It shows that an 8-inch vent has a capacity of 404,000 Btu/hr. An 8-inch vent is selected.

The foregoing sizing procedures are summarized in the following two **Commentary Tables**, *Common Vent Serving Appliances on Four Floors*, and *Common Vent Serving Appliances on Three Floors; Independent Vent Serving Appliances on the Fourth Floor*.

Check for Excessive Vent Area.

The vent connector and common vent have been sized using Table 803.2(1) and Section 803.2. Refer to Section 803.2.16.

In the following example, the smallest vent connector diameter is 4 inches and the largest common vent diameter is 8 inches.

For a 4-inch diameter connector:

Area = $\pi r^2 = 3.14(2)^2 = 12.56$ in.^{2} where:

$\pi = 3.14$

r = 2 (½ the 4-inch diameter)

For an 8-inch diameter common vent:

Area = $\pi r^2 = 3.14(4)^2 = 50.25$ in.^{2} where:

$\pi = 3.14$

r = 4 (½ the 8-inch diameter)

Section 803.2.16 limits the common vent to seven times the smallest connector area. The ratio of the area of the smallest connector to the common vent area in this example must be checked to be sure it is less than 7:

Ratio = area (largest common vent section)/area (smallest connector)

Ratio = 50.25/1 2.56 = 4

Because 4 is less than 7, the sizing of the common vent is acceptable.

803.2.15 Vent Type Multistory Installation. Where used in multistory systems, vertical common vents shall be Type B double wall and shall be installed with a listed vent cap. [NFPA 54:13.2.16]

803.2.16 Offsets in Multistory Installations. Offsets in multistory common vent systems shall be limited to a single offset in each system, and systems with an offset shall comply with all of the following:

(1) The offset angle shall not exceed 45 degrees (0.79 rad) from vertical.

(2) The horizontal length of the offset shall not exceed 18 inches per inch (18 mm/mm) of common vent diameter of the segment in which the offset is located.

(3) For the segment of the common vertical vent containing the offset, the common vent capacity listed in the common venting tables shall be reduced by 20 percent (0.80 x maximum common vent capacity).

(4) A multistory common vent shall not be reduced in size above the offset. [NFPA 54:13.2.17]

803.2.17 Vertical Vent Size Limitation. Where two or more appliances are connected to a vertical vent or chimney, the flow area of the largest section of vertical vent or chimney shall not exceed seven times the smallest listed appliance

COMMENTARY TABLE
Common Vent Serving Appliances on Four Floors

Furnace	Total Input to Common Vent (Btu/hr)	Available Total Height (ft)	Vent Connector Size (in.)	Common Vent Size (in.)
1	80,000	10	4	4
2	160,000	10	4	5
3	240,000	10	4	6
4	320,000	6	5	8

COMMENTARY TABLE
Common Vent Serving Appliances on Three Floors; Independent Vent Serving Appliance on the Fourth Floor

Furnace	Total Input to Common Vent (Btu/hr)	Available Total Height (ft)	Vent Connector Size (in.)	Common Vent Size (in.)
1	80,000	10	4	4
2	160,000	10	4	5
3	240,000	10	4	6
4	320,000	6	4	None

categorized vent areas, flue collar area, or draft hood outlet area unless designed in accordance with approved engineering methods. [NFPA 54:13.2.18]

803.2.18 Multiple Input Ratings. For appliances with more than one input rate, the minimum vent connector capacity (FAN Min) determined from the tables shall be less than the lowest appliance input rating, and the maximum vent connector capacity (FAN Max or NAT Max) determined from the tables shall be greater than the highest appliance input rating. [NFPA 54:13.2.19]

803.2.19 Corrugated Metallic Chimney Liner Reduction. Listed corrugated metallic chimney liner systems in masonry chimneys shall be sized by using Table 803.2(1) or Table 803.2(2) for Type B vents, with the maximum capacity reduced by 20 percent (0.80 x maximum capacity) and the minimum capacity as shown in Table 803.2(1) or Table 803.2(2). Corrugated metallic liner systems installed with bends or offsets shall have their maximum capacity further reduced in accordance with Section 803.2.4 and Section 803.2.5. The 20 percent reduction for corrugated metallic chimney liner systems includes an allowance for one long radius 90 degree (1.57 rad) turn at the bottom of the liner. [NFPA 54:13.2.20]

803.2.20 Chimneys and Vents. Table 803.2(1) through Table 803.2(5) shall be used only for chimneys and vents not exposed to the outdoors below the roof line. A Type B vent or listed chimney lining system passing through an unused masonry chimney flue shall not be considered to be exposed to the outdoors. A Type B vent passing through an unventilated enclosure or chase insulated to a value of not less than R8 shall not be considered to be exposed to the outdoors. Where vents extend outdoors above the roof more than 5 feet (1524 mm) higher than required by Table 802.6.1, and where vents terminate in accordance with Section 802.6.1(1)(b), the outdoor portion of the vent shall be enclosed as required by this section for vents not considered to be exposed to the outdoors, or such venting system shall be engineered. Table 803.2(6) through Table 803.2(9) shall be used for clay tile lined exterior masonry chimneys, provided all the following conditions are met:

(1) The vent connector is Type B double wall.

(2) At least one appliance is draft hood equipped.

(3) The combined appliance input rating is less than the maximum capacity given by Table 803.2(6) (for NAT+NAT) or Table 803.2(8) (for FAN+NAT).

(4) The input rating of each space-heating appliance is greater than the minimum input rating given by Table 803.2(7) (for NAT+NAT) or Table 803.2(9) (for FAN+NAT).

(5) The vent connector sizing is in accordance with Table 803.2(3). [NFPA 54:13.2.22]

803.2.21 Vent Connector Sizing. Vent connectors shall not be increased more than two sizes greater than the listed appliance categorized vent diameter, flue collar diameter, or draft hood outlet diameter. Vent connectors for draft hood-equipped appliances shall not be smaller than the draft hood outlet diameter. Where a vent connector size(s) determined from the tables for a fan-assisted appliance(s) is smaller than the flue collar diameter, the use of the smaller size(s) shall be permitted, provided that the installation complies with all of the following conditions:

(1) Vent connectors for fan-assisted appliance flue collars 12 inches (300 mm) in diameter or smaller are not

reduced by more than one table size [e.g., 12 inches to 10 inches (300 mm to 250 mm) is a one-size reduction] and those larger than 12 inches (300 mm) in diameter are not reduced more than two table sizes [e.g., 24 inches to 20 inches (600 mm to 500 mm) is a two-size reduction].

(2) The fan-assisted appliance(s) is common vented with a draft hood-equipped appliance(s).

(3) The vent connector has a smooth interior wall. [NFPA 54:13.2.24]

A sudden, large expansion of the vent connector diameter creates a pressure drop that may limit the draft and encourage condensation. Therefore, this section places a limit on upsizing, limiting the vent connector diameter to two sizes larger than one of the following that is applicable:

- Appliance categorized vent connector diameter
- Flue collar diameter
- Draft hood outlet diameter

Note that an appliance includes either a draft hood or flue collar for connection to the vent or vent connector. The term "Appliance categorized vent diameter/area" is defined as follows:

The minimum vent area/diameter permissible for Category I appliances to maintain a nonpositive vent static pressure when tested in accordance with nationally recognized standards.

Therefore, where connecting an appliance to a vent connector, use the applicable tables in Section 803.0 to find the minimum size vent connector.

Note that only smooth wall vent connectors may be downsized. It would also be advisable to check with the manufacturer to make sure it allows a downsizing.

803.2.22 Combination of Pipe Types and Sizes. All combinations of pipe sizes, single-wall metal pipe, and double-wall metal pipe shall be allowed within any connector run(s) or within the common vent, provided ALL of the appropriate tables permit ALL of the desired sizes and types of pipe, as if they were used for the entire length of the subject connector or vent. Where single-wall and Type B double-wall metal pipes are used for vent connectors within the same venting system, the common vent shall be sized using Table 803.2(2) or Table 803.2(4) as appropriate. [NFPA 54:13.2.25]

803.2.23 Multiple Connector and Vent Sizes. Where Table 803.1.2(1) through Table 803.2(9) permits more than one diameter of pipe to be used for a connector or vent, all the permitted sizes shall be permitted to be used. [NFPA 54:13.2.26]

803.2.24 Interpolation. Interpolation shall be permitted in calculating capacities for vent dimensions that fall between table entries. [NFPA 54:13.2.27]

803.2.25 Extrapolation. Extrapolation beyond the table entries shall not be permitted. [NFPA 54:13.2.28]

Extrapolation is estimating the values outside of the tables. For example, Table 803.2(1) provides common vent capacities for vents up to 100 feet. The installer is not allowed to estimate the capacity of a vent 110 feet high.

803.2.26 Engineering Methods. For vent heights lower than 6 feet (1829 mm) and higher than shown in the tables, engineering methods shall be used to calculate vent capacities. [NFPA 54:13.2.29]

This reinforces the requirement that vent heights outside the parameters of the tables must be calculated and that the tables cannot be used.

803.2.27 Height Entries. Where the actual height of a vent falls between entries in the height column of the applicable table in Table 803.2(1) through Table 803.2(9), either of the following shall be used:

(1) Interpolation.

(2) The lower appliance input rating shown in the table entries, for FAN MAX and NAT MAX column values; and the higher appliance input rating for the FAN MIN column values. [NFPA 54:13.2.30]

99% Winter Design Temperatures for the Contiguous United States

This map is a necessarily generalized guide to temperatures in the contiguous United States. Temperatures shown for areas such as mountainous regions and large urban centers are not necessarily accurate. The climate data used to develop this map are from the ASHRAE Handbook – Fundamentals (Climate Conditions for the United States).

For 99% winter design temperatures in Alaska, consult the ASHRAE Handbook — Fundamentals.
99% winter design temperatures for Hawaii are greater than 37°F
For SI units: °C = (°F-32)/1.8

FIGURE 803.1.2(6)
RANGE OF WINTER DESIGN TEMPERATURES USED IN ANALYZING EXTERIOR MASONRY CHIMNEYS IN THE UNITED STATES
[NFPA 54: FIGURE F.2.4]

TABLE 803.1.2(1)
TYPE B DOUBLE-WALL GAS VENT [NFPA 54: TABLE 13.1(a)]*

							NUMBER OF APPLIANCES:	SINGLE								
							APPLIANCE TYPE:	CATEGORY I								
							APPLIANCE VENT CONNECTION:	CONNECTED DIRECTLY TO VENT								
		VENT DIAMETER – *D* (inch)														
		3			4			5			6			7		
		APPLIANCE INPUT RATING IN THOUSANDS OF BTU PER HOUR														
HEIGHT *H* (feet)	LATERAL *L* (feet)	FAN		NAT	FAN		NAT	FAN		NAT	FAN		NAT	FAN		NAT
		Min	Max	Max	Min	Max	Max	Min	Max	Max	Min	Max	Max	Min	Max	Max
6	0	0	78	46	0	152	86	0	251	141	0	375	205	0	524	285
	2	13	51	36	18	97	67	27	157	105	32	232	157	44	321	217
	4	21	49	34	30	94	64	39	153	103	50	227	153	66	316	211
	6	25	46	32	36	91	61	47	149	100	59	223	149	78	310	205
8	0	0	84	50	0	165	94	0	276	155	0	415	235	0	583	320
	2	12	57	40	16	109	75	25	178	120	28	263	180	42	365	247
	5	23	53	38	32	103	71	42	171	115	53	255	173	70	356	237
	8	28	49	35	39	98	66	51	164	109	64	247	165	84	347	227
10	0	0	88	53	0	175	100	0	295	166	0	447	255	0	631	345
	2	12	61	42	17	118	81	23	194	129	26	289	195	40	402	273
	5	23	57	40	32	113	77	41	187	124	52	280	188	68	392	263
	10	30	51	36	41	104	70	54	176	115	67	267	175	88	376	245
15	0	0	94	58	0	191	112	0	327	187	0	502	285	0	716	390
	2	11	69	48	15	136	93	20	226	150	22	339	225	38	475	316
	5	22	65	45	30	130	87	39	219	142	49	330	217	64	463	300
	10	29	59	41	40	121	82	51	206	135	64	315	208	84	445	288
	15	35	53	37	48	112	76	61	195	128	76	301	198	98	429	275
20	0	0	97	61	0	202	119	0	349	202	0	540	307	0	776	430
	2	10	75	51	14	149	100	18	250	166	20	377	249	33	531	346
	5	21	71	48	29	143	96	38	242	160	47	367	241	62	519	337
	10	28	64	44	38	133	89	50	229	150	62	351	228	81	499	321
	15	34	58	40	46	124	84	59	217	142	73	337	217	94	481	308
	20	48	52	35	55	116	78	69	206	134	84	322	206	107	464	295
30	0	0	100	64	0	213	128	0	374	220	0	587	336	0	853	475
	2	9	81	56	13	166	112	14	283	185	18	432	280	27	613	394
	5	21	77	54	28	160	108	36	275	176	45	421	273	58	600	385
	10	27	70	50	37	150	102	48	262	171	59	405	261	77	580	371
	15	33	64	NA	44	141	96	57	249	163	70	389	249	90	560	357
	20	56	58	NA	53	132	90	66	237	154	80	374	237	102	542	343
	30	NA	NA	NA	73	113	NA	88	214	NA	104	346	219	131	507	321
50	0	0	101	67	0	216	134	0	397	232	0	633	363	0	932	518
	2	8	86	61	11	183	122	14	320	206	15	497	314	22	715	445
	5	20	82	NA	27	177	119	35	312	200	43	487	308	55	702	438
	10	26	76	NA	35	168	114	45	299	190	56	471	298	73	681	426
	15	59	70	NA	42	158	NA	54	287	180	66	455	288	85	662	413
	20	NA	NA	NA	50	149	NA	63	275	169	76	440	278	97	642	401
	30	NA	NA	NA	69	131	NA	84	250	NA	99	410	259	123	605	376
100	0	NA	NA	NA	0	218	NA	0	407	NA	0	665	400	0	997	560
	2	NA	NA	NA	10	194	NA	12	354	NA	13	566	375	18	831	510
	5	NA	NA	NA	26	189	NA	33	347	NA	40	557	369	52	820	504
	10	NA	NA	NA	33	182	NA	43	335	NA	53	542	361	68	801	493
	15	NA	NA	NA	40	174	NA	50	321	NA	62	528	353	80	782	482
	20	NA	NA	NA	47	166	NA	59	311	NA	71	513	344	90	763	471
	30	NA	NA	NA	NA	NA	NA	78	290	NA	92	483	NA	115	726	449
	50	NA	NA	NA	NA	NA	NA	NA	NA	NA	147	428	NA	180	651	405

For SI units: 1 inch = 25.4 mm, 1 foot = 304.8 mm, 1000 British thermal units per hour = 0.293 kW, 1 square inch = 0.000645 m^2

* NA: Not applicable.

TABLE 803.1.2(1)
TYPE B DOUBLE-WALL GAS VENT [NFPA 54: TABLE 13.1(a)] (continued)

		NUMBER OF APPLIANCES: SINGLE														
		APPLIANCE TYPE: CATEGORY I														
		APPLIANCE VENT CONNECTION: CONNECTED DIRECTLY TO VENT														
		VENT DIAMETER – *D* (inch)														
		8			9			10			12			14		
		APPLIANCE INPUT RATING IN THOUSANDS OF BTU PER HOUR														
HEIGHT *H* (feet)	LATERAL *L* (feet)	FAN		NAT	FAN		NAT	FAN		NAT	FAN		NAT	FAN		NAT
		Min	Max	Max	Min	Max	Max	Min	Max	Max	Min	Max	Max	Min	Max	Max
6	0	0	698	370	0	897	470	0	1121	570	0	1645	850	0	2267	1170
	2	53	425	285	63	543	370	75	675	455	103	982	650	138	1346	890
	4	79	419	279	93	536	362	110	668	445	147	975	640	191	1338	880
	6	93	413	273	110	530	354	128	661	435	171	967	630	219	1330	870
8	0	0	780	415	0	1006	537	0	1261	660	0	1858	970	0	2571	1320
	2	50	483	322	60	619	418	71	770	515	98	1124	745	130	1543	1020
	5	83	473	313	99	607	407	115	758	503	154	1110	733	199	1528	1010
	8	99	463	303	117	596	396	137	746	490	180	1097	720	231	1514	1000
10	0	0	847	450	0	1096	585	0	1377	720	0	2036	1060	0	2825	1450
	2	48	533	355	57	684	457	68	852	560	93	1244	850	124	1713	1130
	5	81	522	346	95	671	446	112	839	547	149	1229	829	192	1696	1105
	10	104	504	330	122	651	427	142	817	525	187	1204	795	238	1669	1080
15	0	0	970	525	0	1263	682	0	1596	840	0	2380	1240	0	3323	1720
	2	45	633	414	53	815	544	63	1019	675	86	1495	985	114	2062	1350
	5	76	620	403	90	800	529	105	1003	660	140	1476	967	182	2041	1327
	10	99	600	386	116	777	507	135	977	635	177	1446	936	227	2009	1289
	15	115	580	373	134	755	491	155	953	610	202	1418	905	257	1976	1250
20	0	0	1057	575	0	1384	752	0	1756	930	0	2637	1350	0	3701	1900
	2	41	711	470	50	917	612	59	1150	755	81	1694	1100	107	2343	1520
	5	73	697	460	86	902	599	101	1133	738	135	1674	1079	174	2320	1498
	10	95	675	443	112	877	576	130	1105	710	172	1641	1045	220	2282	1460
	15	111	654	427	129	853	557	150	1078	688	195	1609	1018	248	2245	1425
	20	125	634	410	145	830	537	167	1052	665	217	1578	990	273	2210	1390
30	0	0	1173	650	0	1548	855	0	1977	1060	0	3004	1550	0	4252	2170
	2	33	826	535	42	1072	700	54	1351	865	74	2004	1310	98	2786	1800
	5	69	811	524	82	1055	688	96	1332	851	127	1981	1289	164	2759	1775
	10	91	788	507	107	1028	668	125	1301	829	164	1944	1254	209	2716	1733
	15	105	765	490	124	1002	648	143	1272	807	187	1908	1220	237	2674	1692
	20	119	743	473	139	977	628	160	1243	784	207	1873	1185	260	2633	1650
	30	149	702	444	171	929	594	195	1189	745	246	1807	1130	305	2555	1585
50	0	0	1297	708	0	1730	952	0	2231	1195	0	3441	1825	0	4934	2550
	2	26	975	615	33	1276	813	41	1620	1010	66	2431	1513	86	3409	2125
	5	65	960	605	77	1259	798	90	1600	996	118	2406	1495	151	3380	2102
	10	86	935	589	101	1230	773	118	1567	972	154	2366	1466	196	3332	2064
	15	100	911	572	117	1203	747	136	1536	948	177	2327	1437	222	3285	2026
	20	113	888	556	131	1176	722	151	1505	924	195	2288	1408	244	3239	1987
	30	141	844	522	161	1125	670	183	1446	876	232	2214	1349	287	3150	1910
100	0	0	1411	770	0	1908	1040	0	2491	1310	0	3925	2050	0	5729	2950
	2	21	1155	700	25	1536	935	30	1975	1170	44	3027	1820	72	4313	2550
	5	60	1141	692	71	1519	926	82	1955	1159	107	3002	1803	136	4282	2531
	10	80	1118	679	94	1492	910	108	1923	1142	142	2961	1775	180	4231	2500
	15	93	1095	666	109	1465	895	126	1892	1124	163	2920	1747	206	4182	2469
	20	105	1073	653	122	1438	880	141	1861	1107	181	2880	1719	226	4133	2438
	30	131	1029	627	149	1387	849	170	1802	1071	215	2803	1663	265	4037	2375
	50	197	944	575	217	1288	787	241	1688	1000	292	2657	1550	350	3856	2250

For SI units: 1 inch = 25.4 mm, 1 foot = 304.8 mm, 1000 British thermal units per hour = 0.293 kW, 1 square inch = 0.000645 m^2

TABLE 803.1.2(1)
TYPE B DOUBLE-WALL GAS VENT [NFPA 54: TABLE 13.1(a)] (continued)

		NUMBER OF APPLIANCES: SINGLE														
		APPLIANCE TYPE: CATEGORY I														
		APPLIANCE VENT CONNECTION: CONNECTED DIRECTLY TO VENT														
		VENT DIAMETER – *D* (inch)														
		16			18			20			22			24		
		APPLIANCE INPUT RATING IN THOUSANDS OF BTU PER HOUR														
HEIGHT *H* (feet)	LATERAL *L* (feet)	FAN Min	FAN Max	NAT Max	FAN Min	FAN Max	NAT Max	FAN Min	FAN Max	NAT Max	FAN Min	FAN Max	NAT Max	FAN Min	FAN Max	NAT Max
6	0	0	2983	1530	0	3802	1960	0	4721	2430	0	5737	2950	0	6853	3520
	2	178	1769	1170	225	2250	1480	296	2782	1850	360	3377	2220	426	4030	2670
	4	242	1761	1160	300	2242	1475	390	2774	1835	469	3370	2215	555	4023	2660
	6	276	1753	1150	341	2235	1470	437	2767	1820	523	3363	2210	618	4017	2650
8	0	0	3399	1740	0	4333	2220	0	5387	2750	0	6555	3360	0	7838	4010
	2	168	2030	1340	212	2584	1700	278	3196	2110	336	3882	2560	401	4634	3050
	5	251	2013	1330	311	2563	1685	398	3180	2090	476	3863	2545	562	4612	3040
	8	289	2000	1320	354	2552	1670	450	3163	2070	537	3850	2530	630	4602	3030
10	0	0	3742	1925	0	4782	2450	0	5955	3050	0	7254	3710	0	8682	4450
	2	161	2256	1480	202	2868	1890	264	3556	2340	319	4322	2840	378	5153	3390
	5	243	2238	1461	300	2849	1871	382	3536	2318	458	4301	2818	540	5132	3371
	10	298	2209	1430	364	2818	1840	459	3504	2280	546	4268	2780	641	5099	3340
15	0	0	4423	2270	0	5678	2900	0	7099	3620	0	8665	4410	0	10 393	5300
	2	147	2719	1770	186	3467	2260	239	4304	2800	290	5232	3410	346	6251	4080
	5	229	2696	1748	283	3442	2235	355	4278	2777	426	5204	3385	501	6222	4057
	10	283	2659	1712	346	3402	2193	432	4234	2739	510	5159	3343	599	6175	4019
	15	318	2623	1675	385	3363	2150	479	4192	2700	564	5115	3300	665	6129	3980
20	0	0	4948	2520	0	6376	3250	0	7988	4060	0	9785	4980	0	11 753	6000
	2	139	3097	2000	175	3955	2570	220	4916	3200	269	5983	3910	321	7154	4700
	5	219	3071	1978	270	3926	2544	337	4885	3174	403	5950	3880	475	7119	4662
	10	273	3029	1940	334	3880	2500	413	4835	3130	489	5896	3830	573	7063	4600
	15	306	2988	1910	372	3835	2465	459	4786	3090	541	5844	3795	631	7007	4575
	20	335	2948	1880	404	3791	2430	495	4737	3050	585	5792	3760	689	6953	4550
30	0	0	5725	2920	0	7420	3770	0	9341	4750	0	11 483	5850	0	13 848	7060
	2	127	3696	2380	159	4734	3050	199	5900	3810	241	7194	4650	285	8617	5600
	5	206	3666	2350	252	4701	3020	312	5863	3783	373	7155	4622	439	8574	5552
	10	259	3617	2300	316	4647	2970	386	5803	3739	456	7090	4574	535	8505	5471
	15	292	3570	2250	354	4594	2920	431	5744	3695	507	7026	4527	590	8437	5391
	20	319	3523	2200	384	4542	2870	467	5686	3650	548	6964	4480	639	8370	5310
	30	369	3433	2130	440	4442	2785	540	5574	3565	635	6842	4375	739	8239	5225
50	0	0	6711	3440	0	8774	4460	0	11 129	5635	0	13 767	6940	0	16 694	8430
	2	113	4554	2840	141	5864	3670	171	7339	4630	209	8980	5695	251	10 788	6860
	5	191	4520	2813	234	5826	3639	283	7295	4597	336	8933	5654	394	10 737	6818
	10	243	4464	2767	295	5763	3585	355	7224	4542	419	8855	5585	491	10 652	6749
	15	274	4409	2721	330	5701	3534	396	7155	4511	465	8779	5546	542	10 570	6710
	20	300	4356	2675	361	5641	3481	433	7086	4479	506	8704	5506	586	10 488	6670
	30	347	4253	2631	412	5523	3431	494	6953	4421	577	8557	5444	672	10 328	6603
100	0	0	7914	4050	0	10 485	5300	0	13 454	6700	0	16 817	8600	0	20 578	10 300
	2	95	5834	3500	120	7591	4600	138	9577	5800	169	11 803	7200	204	14 264	8800
	5	172	5797	3475	208	7548	4566	245	9528	5769	293	11 748	7162	341	14 204	8756
	10	223	5737	3434	268	7478	4509	318	9447	5717	374	11 658	7100	436	14 105	8683
	15	252	5678	3392	304	7409	4451	358	9367	5665	418	11 569	7037	487	14 007	8610
	20	277	5619	3351	330	7341	4394	387	9289	5613	452	11 482	6975	523	13 910	8537
	30	319	5505	3267	378	7209	4279	446	9136	5509	514	11 310	6850	592	13 720	8391
	50	415	5289	3100	486	6956	4050	572	8841	5300	659	10 979	6600	752	13 354	8100

For SI units: 1 inch = 25.4 mm, 1 foot = 304.8 mm, 1000 British thermal units per hour = 0.293 kW, 1 square inch = 0.000645 m^2

TABLE 803.1.2(2)
TYPE B DOUBLE-WALL GAS VENT [NFPA 54: TABLE 13.1(b)]*

							NUMBER OF APPLIANCES:		SINGLE							
							APPLIANCE TYPE:		CATEGORY I							
							APPLIANCE VENT CONNECTION:		SINGLE-WALL METAL CONNECTOR							
		VENT DIAMETER – *D* (inch)														
		3			4			5			6			7		
		APPLIANCE INPUT RATING IN THOUSANDS OF BTU PER HOUR														
HEIGHT *H* (feet)	LATERAL *L* (feet)	FAN		NAT	FAN		NAT	FAN		NAT	FAN		NAT	FAN		NAT
		Min	Max	Max	Min	Max	Max	Min	Max	Max	Min	Max	Max	Min	Max	Max
6	0	38	77	45	59	151	85	85	249	140	126	373	204	165	522	284
	2	39	51	36	60	96	66	85	156	104	123	231	156	159	320	213
	4	NA	NA	33	74	92	63	102	152	102	146	225	152	187	313	208
	6	NA	NA	31	83	89	60	114	147	99	163	220	148	207	307	203
8	0	37	83	50	58	164	93	83	273	154	123	412	234	161	580	319
	2	39	56	39	59	108	75	83	176	119	121	261	179	155	363	246
	5	NA	NA	37	77	102	69	107	168	114	151	252	171	193	352	235
	8	NA	NA	33	90	95	64	122	161	107	175	243	163	223	342	225
10	0	37	87	53	57	174	99	82	293	165	120	444	254	158	628	344
	2	39	61	41	59	117	80	82	193	128	119	287	194	153	400	272
	5	52	56	39	76	111	76	105	185	122	148	277	186	190	388	261
	10	NA	NA	34	97	100	68	132	171	112	188	261	171	237	369	241
15	0	36	93	57	56	190	111	80	325	186	116	499	283	153	713	388
	2	38	69	47	57	136	93	80	225	149	115	337	224	148	473	314
	5	51	63	44	75	128	86	102	216	140	144	326	217	182	459	298
	10	NA	NA	39	95	116	79	128	201	131	182	308	203	228	438	284
	15	NA	NA	NA	NA	NA	72	158	186	124	220	290	192	272	418	269
20	0	35	96	60	54	200	118	78	346	201	114	537	306	149	772	428
	2	37	74	50	56	148	99	78	248	165	113	375	248	144	528	344
	5	50	68	47	73	140	94	100	239	158	141	363	239	178	514	334
	10	NA	NA	41	93	129	86	125	223	146	177	344	224	222	491	316
	15	NA	NA	NA	NA	NA	80	155	208	136	216	325	210	264	469	301
	20	NA	NA	NA	NA	NA	NA	186	192	126	254	306	196	309	448	285
30	0	34	99	63	53	211	127	76	372	219	110	584	334	144	849	472
	2	37	80	56	55	164	111	76	281	183	109	429	279	139	610	392
	5	49	74	52	72	157	106	98	271	173	136	417	271	171	595	382
	10	NA	NA	NA	91	144	98	122	255	168	171	397	257	213	570	367
	15	NA	NA	NA	115	131	NA	151	239	157	208	377	242	255	547	349
	20	NA	NA	NA	NA	NA	NA	181	223	NA	246	357	228	298	524	333
	30	NA	NA	NA	NA	NA	NA	NA	NA	NA	NA	NA	NA	389	477	305
50	0	33	99	66	51	213	133	73	394	230	105	629	361	138	928	515
	2	36	84	61	53	181	121	73	318	205	104	495	312	133	712	443
	5	48	80	NA	70	174	117	94	308	198	131	482	305	164	696	435
	10	NA	NA	NA	89	160	NA	118	292	186	162	461	292	203	671	420
	15	NA	NA	NA	112	148	NA	145	275	174	199	441	280	244	646	405
	20	NA	NA	NA	NA	NA	NA	176	257	NA	236	420	267	285	622	389
	30	NA	NA	NA	NA	NA	NA	NA	NA	NA	315	376	NA	373	573	NA
100	0	NA	NA	NA	49	214	NA	69	403	NA	100	659	395	131	991	555
	2	NA	NA	NA	51	192	NA	70	351	NA	98	563	373	125	828	508
	5	NA	NA	NA	67	186	NA	90	342	NA	125	551	366	156	813	501
	10	NA	NA	NA	85	175	NA	113	324	NA	153	532	354	191	789	486
	15	NA	NA	NA	132	162	NA	138	310	NA	188	511	343	230	764	473
	20	NA	NA	NA	NA	NA	NA	168	295	NA	224	487	NA	270	739	458
	30	NA	NA	NA	NA	NA	NA	231	264	NA	301	448	NA	355	685	NA
	50	NA	NA	NA	NA	NA	NA	NA	NA	NA	NA	NA	NA	540	584	NA

For SI units: 1 inch = 25.4 mm, 1 foot = 304.8 mm, 1000 British thermal units per hour = 0.293 kW, 1 square inch = 0.000645 m^2

* NA: Not applicable.

TABLE 803.1.2(2)
TYPE B DOUBLE-WALL GAS VENT [NFPA 54: TABLE 13.1(b)] (continued)*

				NUMBER OF APPLIANCES:			SINGLE						
				APPLIANCE TYPE:			CATEGORY I						
				APPLIANCE VENT CONNECTION:			SINGLE-WALL METAL CONNECTOR						
		VENT DIAMETER – *D* (inch)											
		8			9			10			12		
		APPLIANCE INPUT RATING IN THOUSANDS OF BTU PER HOUR											
HEIGHT *H* (feet)	LATERAL *L* (feet)	FAN		NAT	FAN		NAT	FAN		NAT	FAN		NAT
		Min	Max	Max	Min	Max	Max	Min	Max	Max	Min	Max	Max
6	0	211	695	369	267	894	469	371	1118	569	537	1639	849
	2	201	423	284	251	541	368	347	673	453	498	979	648
	4	237	416	277	295	533	360	409	664	443	584	971	638
	6	263	409	271	327	526	352	449	656	433	638	962	627
8	0	206	777	414	258	1002	536	360	1257	658	521	1852	967
	2	197	482	321	246	617	417	339	768	513	486	1120	743
	5	245	470	311	305	604	404	418	754	500	598	1104	730
	8	280	458	300	344	591	392	470	740	486	665	1089	715
10	0	202	844	449	253	1093	584	351	1373	718	507	2031	1057
	2	193	531	354	242	681	456	332	849	559	475	1242	848
	5	241	518	344	299	667	443	409	834	544	584	1224	825
	10	296	497	325	363	643	423	492	808	520	688	1194	788
15	0	195	966	523	244	1259	681	336	1591	838	488	2374	1237
	2	187	631	413	232	812	543	319	1015	673	457	1491	983
	5	231	616	400	287	795	526	392	997	657	562	1469	963
	10	284	592	381	349	768	501	470	966	628	664	1433	928
	15	334	568	367	404	742	484	540	937	601	750	1399	894
20	0	190	1053	573	238	1379	750	326	1751	927	473	2631	1346
	2	182	708	468	227	914	611	309	1146	754	443	1689	1098
	5	224	692	457	279	896	596	381	1126	734	547	1665	1074
	10	277	666	437	339	866	570	457	1092	702	646	1626	1037
	15	325	640	419	393	838	549	526	1060	677	730	1587	1005
	20	374	616	400	448	810	526	592	1028	651	808	1550	973
30	0	184	1168	647	229	1542	852	312	1971	1056	454	2996	1545
	2	175	823	533	219	1069	698	296	1346	863	424	1999	1308
	5	215	806	521	269	1049	684	366	1324	846	524	1971	1283
	10	265	777	501	327	1017	662	440	1287	821	620	1927	1243
	15	312	750	481	379	985	638	507	1251	794	702	1884	1205
	20	360	723	461	433	955	615	570	1216	768	780	1841	1166
	30	461	670	426	541	895	574	704	1147	720	937	1759	1101
50	0	176	1292	704	220	1724	948	295	2223	1189	428	3432	1818
	2	168	971	613	209	1273	811	280	1615	1007	401	2426	1509
	5	204	953	602	257	1252	795	347	1591	991	496	2396	1490
	10	253	923	583	313	1217	765	418	1551	963	589	2347	1455
	15	299	894	562	363	1183	736	481	1512	934	668	2299	1421
	20	345	866	543	415	1150	708	544	1473	906	741	2251	1387
	30	442	809	502	521	1086	649	674	1399	848	892	2159	1318
100	0	166	1404	765	207	1900	1033	273	2479	1300	395	3912	2042
	2	158	1152	698	196	1532	933	259	1970	1168	371	3021	1817
	5	194	1134	688	240	1511	921	322	1945	1153	460	2990	1796
	10	238	1104	672	293	1477	902	389	1905	1133	547	2938	1763
	15	281	1075	656	342	1443	884	447	1865	1110	618	2888	1730
	20	325	1046	639	391	1410	864	507	1825	1087	690	2838	1696
	30	418	988	NA	491	1343	824	631	1747	1041	834	2739	1627
	50	617	866	NA	711	1205	NA	895	1591	NA	1138	2547	1489

For SI units: 1 inch = 25.4 mm, 1 foot = 304.8 mm, 1000 British thermal units per hour = 0.293 kW, 1 square inch = 0.000645 m^2

* NA: Not applicable.

TABLE 803.1.2(3)
MASONRY CHIMNEY [NFPA 54: TABLE 13.1(c)]*

						NUMBER OF APPLIANCES:				SINGLE						
						APPLIANCE TYPE:				CATEGORY I						
						APPLIANCE VENT CONNECTION:				TYPE B DOUBLE-WALL CONNECTOR						
		TYPE B DOUBLE-WALL CONNECTOR DIAMETER – *D* (inch) TO BE USED WITH CHIMNEY AREAS WITHIN THE SIZE LIMITS AT BOTTOM														
		3			4			5			6			7		
		APPLIANCE INPUT RATING IN THOUSANDS OF BTU PER HOUR														
HEIGHT *H* (feet)	LATERAL *L* (feet)	FAN		NAT	FAN		NAT	FAN		NAT	FAN		NAT	FAN		NAT
		Min	Max	Max	Min	Max	Max	Min	Max	Max	Min	Max	Max	Min	Max	Max
6	2	NA	NA	28	NA	NA	52	NA	NA	86	NA	NA	130	NA	NA	180
	5	NA	NA	25	NA	NA	49	NA	NA	82	NA	NA	117	NA	NA	165
8	2	NA	NA	29	NA	NA	55	NA	NA	93	NA	NA	145	NA	NA	198
	5	NA	NA	26	NA	NA	52	NA	NA	88	NA	NA	134	NA	NA	183
	8	NA	NA	24	NA	NA	48	NA	NA	83	NA	NA	127	NA	NA	175
10	2	NA	NA	31	NA	NA	61	NA	NA	103	NA	NA	162	NA	NA	221
	5	NA	NA	28	NA	NA	57	NA	NA	96	NA	NA	148	NA	NA	204
	10	NA	NA	25	NA	NA	50	NA	NA	87	NA	NA	139	NA	NA	191
15	2	NA	NA	35	NA	NA	67	NA	NA	114	NA	NA	179	53	475	250
	5	NA	NA	35	NA	NA	62	NA	NA	107	NA	NA	164	NA	NA	231
	10	NA	NA	28	NA	NA	55	NA	NA	97	NA	NA	153	NA	NA	216
	15	NA	NA	NA	NA	NA	48	NA	NA	89	NA	NA	141	NA	NA	201
20	2	NA	NA	38	NA	NA	74	NA	NA	124	NA	NA	201	51	522	274
	5	NA	NA	36	NA	NA	68	NA	NA	116	NA	NA	184	80	503	254
	10	NA	NA	NA	NA	NA	60	NA	NA	107	NA	NA	172	NA	NA	237
	15	NA	NA	NA	NA	NA	NA	NA	NA	97	NA	NA	159	NA	NA	220
	20	NA	NA	NA	NA	NA	NA	NA	NA	83	NA	NA	148	NA	NA	206
30	2	NA	NA	41	NA	NA	82	NA	NA	137	NA	NA	216	47	581	303
	5	NA	NA	NA	NA	NA	76	NA	NA	128	NA	NA	198	75	561	281
	10	NA	NA	NA	NA	NA	67	NA	NA	115	NA	NA	184	NA	NA	263
	15	NA	NA	NA	NA	NA	NA	NA	NA	107	NA	NA	171	NA	NA	243
	20	NA	NA	NA	NA	NA	NA	NA	NA	91	NA	NA	159	NA	NA	227
	30	NA	NA	NA	NA	NA	NA	NA	NA	NA	NA	NA	NA	NA	NA	188
50	2	NA	NA	NA	NA	NA	92	NA	NA	161	NA	NA	251	NA	NA	351
	5	NA	NA	NA	NA	NA	NA	NA	NA	151	NA	NA	230	NA	NA	323
	10	NA	NA	NA	NA	NA	NA	NA	NA	138	NA	NA	215	NA	NA	304
	15	NA	NA	NA	NA	NA	NA	NA	NA	127	NA	NA	199	NA	NA	282
	20	NA	NA	NA	NA	NA	NA	NA	NA	NA	NA	NA	185	NA	NA	264
	30	NA	NA	NA	NA	NA	NA	NA	NA	NA	NA	NA	NA	NA	NA	NA
Minimum internal area of chimney (square inches)		12			19			28			38			50		
Maximum internal area of chimney (square inches)		Seven times the listed appliance categorized vent area, flue collar area, or draft hood outlet areas.														

For SI units: 1 inch = 25.4 mm, 1 foot = 304.8 mm, 1000 British thermal units per hour = 0.293 kW, 1 square inch = 0.000645 m^2

* NA: Not applicable.

TABLE 803.1.2(3)
MASONRY CHIMNEY [NFPA 54: TABLE 13.1(c)] (continued)*

				NUMBER OF APPLIANCES:				SINGLE					
				APPLIANCE TYPE:				CATEGORY I					
				APPLIANCE VENT CONNECTION:				TYPE B DOUBLE-WALL CONNECTOR					
		TYPE B DOUBLE-WALL CONNECTOR DIAMETER – *D* (inch) TO BE USED WITH CHIMNEY AREAS WITHIN THE SIZE LIMITS AT BOTTOM											
		8			9			10			12		
		APPLIANCE INPUT RATING IN THOUSANDS OF BTU PER HOUR											
HEIGHT *H* (feet)	LATERAL *L* (feet)	FAN		NAT	FAN		NAT	FAN		NAT	FAN		NAT
		Min	Max	Max	Min	Max	Max	Min	Max	Max	Min	Max	Max
6	2	NA	NA	247	NA	NA	320	NA	NA	401	NA	NA	581
	5	NA	NA	231	NA	NA	298	NA	NA	376	NA	NA	561
8	2	NA	NA	266	84	590	350	100	728	446	139	1024	651
	5	NA	NA	247	NA	NA	328	149	711	423	201	1007	640
	8	NA	NA	239	NA	NA	318	173	695	410	231	990	623
10	2	68	519	298	82	655	388	98	810	491	136	1144	724
	5	NA	NA	277	124	638	365	146	791	466	196	1124	712
	10	NA	NA	263	155	610	347	182	762	444	240	1093	668
15	2	64	613	336	77	779	441	92	968	562	127	1376	841
	5	99	594	313	118	759	416	139	946	533	186	1352	828
	10	126	565	296	148	727	394	173	912	567	229	1315	777
	15	NA	NA	281	171	698	375	198	880	485	259	1280	742
20	2	61	678	375	73	867	491	87	1083	627	121	1548	953
	5	95	658	350	113	845	463	133	1059	597	179	1523	933
	10	122	627	332	143	811	440	167	1022	566	221	1482	879
	15	NA	NA	314	165	780	418	191	987	541	251	1443	840
	20	NA	NA	296	186	750	397	214	955	513	277	1406	807
30	2	57	762	421	68	985	558	81	1240	717	111	1793	1112
	5	90	741	393	106	962	526	125	1216	683	169	1766	1094
	10	115	709	373	135	927	500	158	1176	648	210	1721	1025
	15	NA	NA	353	156	893	476	181	1139	621	239	1679	981
	20	NA	NA	332	176	860	450	203	1103	592	264	1638	940
	30	NA	NA	288	NA	NA	416	249	1035	555	318	1560	877
50	2	51	840	477	61	1106	633	72	1413	812	99	2080	1243
	5	83	819	445	98	1083	596	116	1387	774	155	2052	1225
	10	NA	NA	424	126	1047	567	147	1347	733	195	2006	1147
	15	NA	NA	400	146	1010	539	170	1307	702	222	1961	1099
	20	NA	NA	376	165	977	511	190	1269	669	246	1916	1050
	30	NA	NA	327	NA	NA	468	233	1196	623	295	1832	984
Minimum internal area of chimney (square inches)		63			78			95			132		
Maximum internal area of chimney (square inches)		Seven times the listed appliance categorized vent area, flue collar area, or draft hood outlet areas.											

For SI units: 1 inch = 25.4 mm, 1 foot = 304.8 mm, 1000 British thermal units per hour = 0.293 kW, 1 square inch = 0.000645 m^2

* NA: Not applicable.

TABLE 803.1.2(4) MASONRY CHIMNEY [NFPA 54: TABLE 13.1(d)]*

								NUMBER OF APPLIANCES:	SINGLE							
								APPLIANCE TYPE:	CATEGORY I							
								APPLIANCE VENT CONNECTION:	SINGLE-WALL METAL CONNECTOR							
		SINGLE-WALL METAL CONNECTOR DIAMETER – *D* (inch) TO BE USED WITH CHIMNEY AREAS WITHIN THE SIZE LIMITS AT BOTTOM														
		3			4			5			6			7		
		APPLIANCE INPUT RATING IN THOUSANDS OF BTU PER HOUR														
HEIGHT *H* (feet)	LATERAL *L* (feet)	FAN		NAT	FAN		NAT	FAN		NAT	FAN		NAT	FAN		NAT
		Min	Max	Max	Min	Max	Max	Min	Max	Max	Min	Max	Max	Min	Max	Max
6	2	NA	NA	28	NA	NA	52	NA	NA	86	NA	NA	130	NA	NA	180
	5	NA	NA	25	NA	NA	48	NA	NA	81	NA	NA	116	NA	NA	164
8	2	NA	NA	29	NA	NA	55	NA	NA	93	NA	NA	145	NA	NA	197
	5	NA	NA	26	NA	NA	51	NA	NA	87	NA	NA	133	NA	NA	182
	8	NA	NA	23	NA	NA	47	NA	NA	82	NA	NA	126	NA	NA	174
10	2	NA	NA	31	NA	NA	61	NA	NA	102	NA	NA	161	NA	NA	220
	5	NA	NA	28	NA	NA	56	NA	NA	95	NA	NA	147	NA	NA	203
	10	NA	NA	24	NA	NA	49	NA	NA	86	NA	NA	137	NA	NA	189
15	2	NA	NA	35	NA	NA	67	NA	NA	113	NA	NA	178	166	473	249
	5	NA	NA	32	NA	NA	61	NA	NA	106	NA	NA	163	NA	NA	230
	10	NA	NA	27	NA	NA	54	NA	NA	96	NA	NA	151	NA	NA	214
	15	NA	NA	NA	NA	NA	46	NA	NA	87	NA	NA	138	NA	NA	198
20	2	NA	NA	38	NA	NA	73	NA	NA	123	NA	NA	200	163	520	273
	5	NA	NA	35	NA	NA	67	NA	NA	115	NA	NA	183	NA	NA	252
	10	NA	NA	NA	NA	NA	59	NA	NA	105	NA	NA	170	NA	NA	235
	15	NA	NA	NA	NA	NA	NA	NA	NA	95	NA	NA	156	NA	NA	217
	20	NA	NA	NA	NA	NA	NA	NA	NA	80	NA	NA	144	NA	NA	202
30	2	NA	NA	41	NA	NA	81	NA	NA	136	NA	NA	215	158	578	302
	5	NA	NA	NA	NA	NA	75	NA	NA	127	NA	NA	196	NA	NA	279
	10	NA	NA	NA	NA	NA	66	NA	NA	113	NA	NA	182	NA	NA	260
	15	NA	NA	NA	NA	NA	NA	NA	NA	105	NA	NA	168	NA	NA	240
	20	NA	NA	NA	NA	NA	NA	NA	NA	88	NA	NA	155	NA	NA	223
	30	NA	NA	NA	NA	NA	NA	NA	NA	NA	NA	NA	NA	NA	NA	182
50	2	NA	NA	NA	NA	NA	91	NA	NA	160	NA	NA	250	NA	NA	350
	5	NA	NA	NA	NA	NA	NA	NA	NA	149	NA	NA	228	NA	NA	321
	10	NA	NA	NA	NA	NA	NA	NA	NA	136	NA	NA	212	NA	NA	301
	15	NA	NA	NA	NA	NA	NA	NA	NA	124	NA	NA	195	NA	NA	278
	20	NA	NA	NA	NA	NA	NA	NA	NA	NA	NA	NA	180	NA	NA	258
	30	NA	NA	NA	NA	NA	NA	NA	NA	NA	NA	NA	NA	NA	NA	NA
Minimum internal area of chimney (square inches)		12			19			28			38			50		
Maximum internal area of chimney (square inches)		Seven times the listed appliance categorized vent area, flue collar area, or draft hood outlet areas.														

For SI units: 1 inch = 25.4 mm, 1 foot = 304.8 mm, 1000 British thermal units per hour = 0.293 kW, 1 square inch = 0.000645 m^2

* NA: Not applicable.

TABLE 803.1.2(4)
MASONRY CHIMNEY [NFPA 54: TABLE 13.1(d)] (continued)*

		NUMBER OF APPLIANCES:							SINGLE				
		APPLIANCE TYPE:							CATEGORY I				
		APPLIANCE VENT CONNECTION:							SINGLE-WALL METAL CONNECTOR				
		SINGLE-WALL METAL CONNECTOR DIAMETER – *D* (inch) TO BE USED WITH CHIMNEY AREAS WITHIN THE SIZE LIMITS AT BOTTOM											
		8			9			10			12		
		APPLIANCE INPUT RATING IN THOUSANDS OF BTU PER HOUR											
HEIGHT *H* (feet)	LATERAL *L* (feet)	FAN		NAT	FAN		NAT	FAN		NAT	FAN		NAT
		Min	Max	Max	Min	Max	Max	Min	Max	Max	Min	Max	Max
6	2	NA	NA	247	NA	NA	319	NA	NA	400	NA	NA	580
	5	NA	NA	230	NA	NA	297	NA	NA	375	NA	NA	560
8	2	NA	NA	265	NA	NA	349	382	725	445	549	1021	650
	5	NA	NA	246	NA	NA	327	NA	NA	422	673	1003	638
	8	NA	NA	237	NA	NA	317	NA	NA	408	747	985	621
10	2	216	518	297	271	654	387	373	808	490	536	1142	722
	5	NA	NA	276	334	635	364	459	789	465	657	1121	710
	10	NA	NA	261	NA	NA	345	547	758	441	771	1088	665
15	2	211	611	335	264	776	440	362	965	560	520	1373	840
	5	261	591	312	325	755	414	444	942	531	637	1348	825
	10	NA	NA	294	392	722	392	531	907	504	749	1309	774
	15	NA	NA	278	452	692	372	606	873	481	841	1272	738
20	2	206	675	374	258	864	490	252	1079	625	508	1544	950
	5	255	655	348	317	842	461	433	1055	594	623	1518	930
	10	312	622	330	382	806	437	517	1016	562	733	1475	875
	15	NA	NA	311	442	773	414	591	979	539	823	1434	835
	20	NA	NA	292	NA	NA	392	663	944	510	911	1394	800
30	2	200	759	420	249	982	556	340	1237	715	489	1789	1110
	5	245	737	391	306	958	524	417	1210	680	600	1760	1090
	10	300	703	370	370	920	496	500	1168	644	708	1713	1020
	15	NA	NA	349	428	884	471	572	1128	615	798	1668	975
	20	NA	NA	327	NA	NA	445	643	1089	585	883	1624	932
	30	NA	NA	281	NA	NA	408	NA	NA	544	1055	1539	865
50	2	191	837	475	238	1103	631	323	1408	810	463	2076	1240
	5	NA	NA	442	293	1078	593	398	1381	770	571	2044	1220
	10	NA	NA	420	355	1038	562	447	1337	728	674	1994	1140
	15	NA	NA	395	NA	NA	533	546	1294	695	761	1945	1090
	20	NA	NA	370	NA	NA	504	616	1251	660	844	1898	1040
	30	NA	NA	318	NA	NA	458	NA	NA	610	1009	1805	970
Minimum internal area of chimney (square inches)		63			78			95			132		
Maximum internal area of chimney (square inches)		Seven times the listed appliance categorized vent area, flue collar area, or draft hood outlet areas.											

For SI units: 1 inch = 25.4 mm, 1 foot = 304.8 mm, 1000 British thermal units per hour = 0.293 kW, 1 square inch = 0.000645 m^2

* NA: Not applicable.

TABLE 803.1.2(5)
SINGLE-WALL METAL PIPE OR TYPE B ASBESTOS-CEMENT VENT [NFPA 54: TABLE 13.1(e)]*

		NUMBER OF APPLIANCES:			SINGLE				
		APPLIANCE TYPE:			DRAFT HOOD-EQUIPPED				
		APPLIANCE VENT CONNECTION:			CONNECTED DIRECTLY TO PIPE OR VENT				
		DIAMETER – *D* (inch) TO BE USED WITH CHIMNEY AREAS WITHIN THE SIZE LIMITS AT BOTTOM							
		3	4	5	6	7	8	10	12
HEIGHT *H* (feet)	LATERAL *L* (feet)	APPLIANCE INPUT RATING IN THOUSANDS OF BTU PER HOUR							
		MAXIMUM APPLIANCE INPUT RATING IN THOUSANDS OF BTU PER HOUR							
6	0	39	70	116	170	232	312	500	750
	2	31	55	94	141	194	260	415	620
	5	28	51	88	128	177	242	390	600
8	0	42	76	126	185	252	340	542	815
	2	32	61	102	154	210	284	451	680
	5	29	56	95	141	194	264	430	648
	10	24	49	86	131	180	250	406	625
10	0	45	84	138	202	279	372	606	912
	2	35	67	111	168	233	311	505	760
	5	32	61	104	153	215	289	480	724
	10	27	54	94	143	200	274	455	700
	15	NA	46	84	130	186	258	432	666
15	0	49	91	151	223	312	420	684	1040
	2	39	72	122	186	260	350	570	865
	5	35	67	110	170	240	325	540	825
	10	30	58	103	158	223	308	514	795
	15	NA	50	93	144	207	291	488	760
	20	NA	NA	82	132	195	273	466	726
20	0	53	101	163	252	342	470	770	1190
	2	42	80	136	210	286	392	641	990
	5	38	74	123	192	264	364	610	945
	10	32	65	115	178	246	345	571	910
	15	NA	55	104	163	228	326	550	870
	20	NA	NA	91	149	214	306	525	832
30	0	56	108	183	276	384	529	878	1370
	2	44	84	148	230	320	441	730	1140
	5	NA	78	137	210	296	410	694	1080
	10	NA	68	125	196	274	388	656	1050
	15	NA	NA	113	177	258	366	625	1000
	20	NA	NA	99	163	240	344	596	960
	30	NA	NA	NA	NA	192	295	540	890
50	0	NA	120	210	310	443	590	980	1550
	2	NA	95	171	260	370	492	820	1290
	5	NA	NA	159	234	342	474	780	1230
	10	NA	NA	146	221	318	456	730	1190
	15	NA	NA	NA	200	292	407	705	1130
	20	NA	NA	NA	185	276	384	670	1080
	30	NA	NA	NA	NA	222	330	605	1010

For SI units: 1 inch = 25.4 mm, 1 foot = 304.8 mm, 1000 British thermal units per hour = 0.293 kW, 1 square inch = 0.000645 m^2

* NA: Not applicable.

TABLE 803.1.2(6)
EXTERIOR MASONRY CHIMNEY [NFPA 54: TABLE 13.1(f)][1,2]

				NUMBER OF APPLIANCES:	SINGLE			
				APPLIANCE TYPE:	NAT			
				APPLIANCE VENT CONNECTION:	TYPE B DOUBLE-WALL CONNECTOR			
MINIMUM ALLOWABLE INPUT RATING OF SPACE-HEATING APPLIANCE IN THOUSANDS OF BTU PER HOUR								
VENT HEIGHT *H* (feet)	INTERNAL AREA OF CHIMNEY (square inches)							
	12	19	28	38	50	63	78	113
	Local 99% winter design temperature: 37°F or greater							
6	0	0	0	0	0	0	0	0
8	0	0	0	0	0	0	0	0
10	0	0	0	0	0	0	0	0
15	NA	0	0	0	0	0	0	0
20	NA	NA	123	190	249	184	0	0
30	NA	NA	NA	NA	NA	393	334	0
50	NA	NA	NA	NA	NA	NA	NA	579
	Local 99% winter design temperature: 27°F to 36°F							
6	0	0	68	116	156	180	212	266
8	0	0	82	127	167	187	214	263
10	0	51	97	141	183	201	225	265
15	NA	NA	NA	NA	233	253	274	305
20	NA	NA	NA	NA	NA	307	330	362
30	NA	NA	NA	NA	NA	419	445	485
50	NA	NA	NA	NA	NA	NA	NA	763
	Local 99% winter design temperature: 17°F to 26°F							
6	NA	NA	NA	NA	NA	215	259	349
8	NA	NA	NA	NA	197	226	264	352
10	NA	NA	NA	NA	214	245	278	358
15	NA	NA	NA	NA	NA	296	331	398
20	NA	NA	NA	NA	NA	352	387	457
30	NA	NA	NA	NA	NA	NA	507	581
50	NA	NA	NA	NA	NA	NA	NA	NA
	Local 99% winter design temperature: 5°F to 16°F							
6	NA	NA	NA	NA	NA	NA	NA	416
8	NA	NA	NA	NA	NA	NA	312	423
10	NA	NA	NA	NA	NA	289	331	430
15	NA	NA	NA	NA	NA	NA	393	485
20	NA	NA	NA	NA	NA	NA	450	547
30	NA	NA	NA	NA	NA	NA	NA	682
50	NA	NA	NA	NA	NA	NA	NA	972
	Local 99% winter design temperature: -10°F to 4°F							
6	NA	NA	NA	NA	NA	NA	NA	484
8	NA	NA	NA	NA	NA	NA	NA	494
10	NA	NA	NA	NA	NA	NA	NA	513
15	NA	NA	NA	NA	NA	NA	NA	586
20	NA	NA	NA	NA	NA	NA	NA	650
30	NA	NA	NA	NA	NA	NA	NA	805
50	NA	NA	NA	NA	NA	NA	NA	1003
Local 99% winter design temperature: -11°F or lower Not recommended for any vent configurations								

For SI units: 1 inch = 25.4 mm, 1 foot = 304.8 mm, 1000 British thermal units per hour = 0.293 kW, 1 square inch = 0.000645 m^2, °C = (°F-32)/1.8

Notes:

[1] See Figure 803.1.2(6) for a map showing local 99 percent winter design temperatures in the United States.

[2] NA: Not applicable.

TABLE 803.2(1)
TYPE B DOUBLE-WALL VENT [NFPA 54: TABLE 13.2(a)]*

		NUMBER OF APPLIANCES:	TWO OR MORE													
		APPLIANCE TYPE:	CATEGORY I													
		APPLIANCE VENT CONNECTION:	TYPE B DOUBLE-WALL CONNECTOR													
		VENT CONNECTOR CAPACITY														
		TYPE B DOUBLE-WALL VENT AND CONNECTOR DIAMETER – *D* (inch)														
		3			4			5			6			7		
		APPLIANCE INPUT RATING LIMITS IN THOUSANDS OF BTU PER HOUR														
VENT HEIGHT *H* (feet)	CONNECTOR RISE *R* (feet)	FAN		NAT	FAN		NAT	FAN		NAT	FAN		NAT	FAN		NAT
		Min	Max	Max	Min	Max	Max	Min	Max	Max	Min	Max	Max	Min	Max	Max
6	1	22	37	26	35	66	46	46	106	72	58	164	104	77	225	142
	2	23	41	31	37	75	55	48	121	86	60	183	124	79	253	168
	3	24	44	35	38	81	62	49	132	96	62	199	139	82	275	189
8	1	22	40	27	35	72	48	49	114	76	64	176	109	84	243	148
	2	23	44	32	36	80	57	51	128	90	66	195	129	86	269	175
	3	24	47	36	37	87	64	53	139	101	67	210	145	88	290	198
10	1	22	43	28	34	78	50	49	123	78	65	189	113	89	257	154
	2	23	47	33	36	86	59	51	136	93	67	206	134	91	282	182
	3	24	50	37	37	92	67	52	146	104	69	220	150	94	303	205
15	1	21	50	30	33	89	53	47	142	83	64	220	120	88	298	163
	2	22	53	35	35	96	63	49	153	99	66	235	142	91	320	193
	3	24	55	40	36	102	71	51	163	111	68	248	160	93	339	218
20	1	21	54	31	33	99	56	46	157	87	62	246	125	86	334	171
	2	22	57	37	34	105	66	48	167	104	64	259	149	89	354	202
	3	23	60	42	35	110	74	50	176	116	66	271	168	91	371	228
30	1	20	62	33	31	113	59	45	181	93	60	288	134	83	391	182
	2	21	64	39	33	118	70	47	190	110	62	299	158	85	408	215
	3	22	66	44	34	123	79	48	198	124	64	309	178	88	423	242
50	1	19	71	36	30	133	64	43	216	101	57	349	145	78	477	197
	2	21	73	43	32	137	76	45	223	119	59	358	172	81	490	234
	3	22	75	48	33	141	86	46	229	134	61	366	194	83	502	263
100	1	18	82	37	28	158	66	40	262	104	53	442	150	73	611	204
	2	19	83	44	30	161	79	42	267	123	55	447	178	75	619	242
	3	20	84	50	31	163	89	44	272	138	57	452	200	78	627	272

	COMMON VENT CAPACITY											
	TYPE B DOUBLE-WALL COMMON VENT DIAMETER – *D* (inch)											
	4			5			6			7		
	COMBINED APPLIANCE INPUT RATING IN THOUSANDS OF BTU PER HOUR											
VENT HEIGHT *H* (feet)	FAN +FAN	FAN +NAT	NAT +NAT	FAN +FAN	FAN +NAT	NAT +NAT	FAN +FAN	FAN +NAT	NAT +NAT	FAN +FAN	FAN +NAT	NAT +NAT
6	92	81	65	140	116	103	204	161	147	309	248	200
8	101	90	73	155	129	114	224	178	163	339	275	223
10	110	97	79	169	141	124	243	194	178	367	299	242
15	125	112	91	195	164	144	283	228	206	427	352	280
20	136	123	102	215	183	160	314	255	229	475	394	310
30	152	138	118	244	210	185	361	297	266	547	459	360
50	167	153	134	279	244	214	421	353	310	641	547	423
100	175	163	NA	311	277	NA	489	421	NA	751	658	479

For SI units: 1 inch = 25.4 mm, 1 foot = 304.8 mm, 1000 British thermal units per hour = 0.293 kW, 1 square inch = 0.000645 m^2

* NA: Not applicable.

TABLE 803.2(1)
TYPE B DOUBLE-WALL VENT [NFPA 54: TABLE 13.2(a)] (continued)

		NUMBER OF APPLIANCES: TWO OR MORE								
		APPLIANCE TYPE: CATEGORY I								
		APPLIANCE VENT CONNECTION: TYPE B DOUBLE-WALL CONNECTOR								
		VENT CONNECTOR CAPACITY								
		TYPE B DOUBLE-WALL VENT AND CONNECTOR DIAMETER – *D* (inch)								
		8			9			10		
		APPLIANCE INPUT RATING LIMITS IN THOUSANDS OF BTU PER HOUR								
VENT HEIGHT *H* (feet)	CONNECTOR RISE *R* (feet)	FAN		NAT	FAN		NAT	FAN		NAT
		Min	Max	Max	Min	Max	Max	Min	Max	Max
6	1	92	296	185	109	376	237	128	466	289
	2	95	333	220	112	424	282	131	526	345
	3	97	363	248	114	463	317	134	575	386
8	1	100	320	194	118	408	248	138	507	303
	2	103	356	230	121	454	294	141	564	358
	3	105	384	258	123	492	330	143	612	402
10	1	106	341	200	125	436	257	146	542	314
	2	109	374	238	128	479	305	149	596	372
	3	111	402	268	131	515	342	152	642	417
15	1	110	389	214	134	493	273	162	609	333
	2	112	419	253	137	532	323	165	658	394
	3	115	445	286	140	565	365	167	700	444
20	1	107	436	224	131	552	285	158	681	347
	2	110	463	265	134	587	339	161	725	414
	3	113	486	300	137	618	383	164	764	466
30	1	103	512	238	125	649	305	151	802	372
	2	105	535	282	129	679	360	155	840	439
	3	108	555	317	132	706	405	158	874	494
50	1	97	627	257	120	797	330	144	984	403
	2	100	645	306	123	820	392	148	1014	478
	3	103	661	343	126	842	441	151	1043	538
100	1	91	810	266	112	1038	341	135	1285	417
	2	94	822	316	115	1054	405	139	1306	494
	3	97	834	355	118	1069	455	142	1327	555

	COMMON VENT CAPACITY								
	TYPE B DOUBLE-WALL COMMON VENT DIAMETER – *D* (inch)								
	8			9			10		
	COMBINED APPLIANCE INPUT RATING IN THOUSANDS OF BTU PER HOUR								
VENT HEIGHT *H* (feet)	FAN +FAN	FAN +NAT	NAT +NAT	FAN +FAN	FAN +NAT	NAT +NAT	FAN +FAN	FAN +NAT	NAT +NAT
6	404	314	260	547	434	335	672	520	410
8	444	348	290	602	480	378	740	577	465
10	477	377	315	649	522	405	800	627	495
15	556	444	365	753	612	465	924	733	565
20	621	499	405	842	688	523	1035	826	640
30	720	585	470	979	808	605	1209	975	740
50	854	706	550	1164	977	705	1451	1188	860
100	1025	873	625	1408	1215	800	1784	1502	975

For SI units: 1 inch = 25.4 mm, 1 foot = 304.8 mm, 1000 British thermal units per hour = 0.293 kW, 1 square inch = 0.000645 m^2

TABLE 803.2(1)
TYPE B DOUBLE-WALL VENT [NFPA 54: TABLE 13.2(a)] (continued)*

		NUMBER OF APPLIANCES: TWO OR MORE											
		APPLIANCE TYPE: CATEGORY I											
		APPLIANCE VENT CONNECTION: TYPE B DOUBLE-WALL CONNECTOR											
		VENT CONNECTOR CAPACITY											
		TYPE B DOUBLE-WALL VENT AND CONNECTOR DIAMETER – *D* (inch)											
		12			14			16			18		
		APPLIANCE INPUT RATING LIMITS IN THOUSANDS OF BTU PER HOUR											
VENT HEIGHT	CONNECTOR RISE	FAN		NAT	FAN		NAT	FAN		NAT	FAN		NAT
H (feet)	*R* (feet)	Min	Max	Max	Min	Max	Max	Min	Max	Max	Min	Max	Max
6	2	174	764	496	223	1046	653	281	1371	853	346	1772	1080
	4	180	897	616	230	1231	827	287	1617	1081	352	2069	1370
	6	NA	NA	NA	NA	NA	NA	NA	NA	NA	NA	NA	NA
8	2	186	822	516	238	1126	696	298	1478	910	365	1920	1150
	4	192	952	644	244	1307	884	305	1719	1150	372	2211	1460
	6	198	1050	772	252	1445	1072	313	1902	1390	380	2434	1770
10	2	196	870	536	249	1195	730	311	1570	955	379	2049	1205
	4	201	997	664	256	1371	924	318	1804	1205	387	2332	1535
	6	207	1095	792	263	1509	1118	325	1989	1455	395	2556	1865
15	2	214	967	568	272	1334	790	336	1760	1030	408	2317	1305
	4	221	1085	712	279	1499	1006	344	1978	1320	416	2579	1665
	6	228	1181	856	286	1632	1222	351	2157	1610	424	2796	2025
20	2	223	1051	596	291	1443	840	357	1911	1095	430	2533	1385
	4	230	1162	748	298	1597	1064	365	2116	1395	438	2778	1765
	6	237	1253	900	307	1726	1288	373	2287	1695	450	2984	2145
30	2	216	1217	632	286	1664	910	367	2183	1190	461	2891	1540
	4	223	1316	792	294	1802	1160	376	2366	1510	474	3110	1920
	6	231	1400	952	303	1920	1410	384	2524	1830	485	3299	2340
50	2	206	1479	689	273	2023	1007	350	2659	1315	435	3548	1665
	4	213	1561	860	281	2139	1291	359	2814	1685	447	3730	2135
	6	221	1631	1031	290	2242	1575	369	2951	2055	461	3893	2605
100	2	192	1923	712	254	2644	1050	326	3490	1370	402	4707	1740
	4	200	1984	888	263	2731	1346	336	3606	1760	414	4842	2220
	6	208	2035	1064	272	2811	1642	346	3714	2150	426	4968	2700

	COMMON VENT CAPACITY											
	TYPE B DOUBLE-WALL COMMON VENT DIAMETER – *D* (inch)											
	12			14			16			18		
	COMBINED APPLIANCE INPUT RATING IN THOUSANDS OF BTU PER HOUR											
VENT HEIGHT *H* (feet)	FAN +FAN	FAN +NAT	NAT +NAT	FAN +FAN	FAN +NAT	NAT +NAT	FAN +FAN	FAN +NAT	NAT +NAT	FAN +FAN	FAN +NAT	NAT +NAT
6	900	696	588	1284	990	815	1735	1336	1065	2253	1732	1345
8	994	773	652	1423	1103	912	1927	1491	1190	2507	1936	1510
10	1076	841	712	1542	1200	995	2093	1625	1300	2727	2113	1645
15	1247	986	825	1794	1410	1158	2440	1910	1510	3184	2484	1910
20	1405	1116	916	2006	1588	1290	2722	2147	1690	3561	2798	2140
30	1658	1327	1025	2373	1892	1525	3220	2558	1990	4197	3326	2520
50	2024	1640	1280	2911	2347	1863	3964	3183	2430	5184	4149	3075
100	2569	2131	1670	3732	3076	2450	5125	4202	3200	6749	5509	4050

For SI units: 1 inch = 25.4 mm, 1 foot = 304.8 mm, 1000 British thermal units per hour = 0.293 kW, 1 square inch = 0.000645 m^2

* NA: Not applicable.

TABLE 803.2(1)
TYPE B DOUBLE-WALL VENT [NFPA 54: TABLE 13.2(a)] (continued)*

NUMBER OF APPLIANCES: TWO OR MORE
APPLIANCE TYPE: CATEGORY I
APPLIANCE VENT CONNECTION: TYPE B DOUBLE-WALL CONNECTOR

		VENT CONNECTOR CAPACITY								
		TYPE B DOUBLE-WALL VENT AND CONNECTOR DIAMETER – *D* (inch)								
		20			22			24		
		APPLIANCE INPUT RATING LIMITS IN THOUSANDS OF BTU PER HOUR								
VENT HEIGHT *H* (feet)	CONNECTOR RISE *R* (feet)	FAN		NAT	FAN		NAT	FAN		NAT
		Min	Max	Max	Min	Max	Max	Min	Max	Max
6	2	NA	NA	NA	NA	NA	NA	NA	NA	NA
	4	NA	NA	NA	NA	NA	NA	NA	NA	NA
	6	NA	NA	NA	NA	NA	NA	NA	NA	NA
8	2	NA	NA	NA	NA	NA	NA	NA	NA	NA
	4	471	2737	1800	560	3319	2180	662	3957	2590
	6	478	3018	2180	568	3665	2640	669	4373	3130
10	2	NA	NA	NA	NA	NA	NA	NA	NA	NA
	4	486	2887	1890	581	3502	2280	686	4175	2710
	6	494	3169	2290	589	3849	2760	694	4593	3270
15	2	NA	NA	NA	NA	NA	NA	NA	NA	NA
	4	523	3197	2060	624	3881	2490	734	4631	2960
	6	533	3470	2510	634	4216	3030	743	5035	3600
20	2	NA	NA	NA	NA	NA	NA	NA	NA	NA
	4	554	3447	2180	661	4190	2630	772	5005	3130
	6	567	3708	2650	671	4511	3190	785	5392	3790
30	2	NA	NA	NA	NA	NA	NA	NA	NA	NA
	4	619	3840	2365	728	4861	2860	847	5606	3410
	6	632	4080	2875	741	4976	3480	860	5961	4150
50	2	NA	NA	NA	NA	NA	NA	NA	NA	NA
	4	580	4601	2633	709	5569	3185	851	6633	3790
	6	594	4808	3208	724	5826	3885	867	6943	4620
100	2	NA	NA	NA	NA	NA	NA	NA	NA	NA
	4	523	5982	2750	639	7254	3330	769	8650	3950
	6	539	6143	3350	654	7453	4070	786	8892	4810

	COMMON VENT CAPACITY								
	TYPE B DOUBLE-WALL COMMON VENT DIAMETER – *D* (inch)								
	20			22			24		
	COMBINED APPLIANCE INPUT RATING IN THOUSANDS OF BTU PER HOUR								
VENT HEIGHT *H* (feet)	FAN +FAN	FAN +NAT	NAT +NAT	FAN +FAN	FAN +NAT	NAT +NAT	FAN +FAN	FAN +NAT	NAT +NAT
6	2838	2180	1660	3488	2677	1970	4206	3226	2390
8	3162	2439	1860	3890	2998	2200	4695	3616	2680
10	3444	2665	2030	4241	3278	2400	5123	3957	2920
15	4026	3133	2360	4971	3862	2790	6016	4670	3400
20	4548	3552	2640	5573	4352	3120	6749	5261	3800
30	5303	4193	3110	6539	5157	3680	7940	6247	4480
50	6567	5240	3800	8116	6458	4500	9837	7813	5475
100	8597	6986	5000	10 681	8648	5920	13 004	10 499	7200

For SI units: 1 inch = 25.4 mm, 1 foot = 304.8 mm, 1000 British thermal units per hour = 0.293 kW, 1 square inch = 0.000645 m^2

* NA: Not applicable.

TABLE 803.2(2)
TYPE B DOUBLE-WALL VENT [NFPA 54: TABLE 13.2(b)]*

					NUMBER OF APPLIANCES:						TWO OR MORE					
					APPLIANCE TYPE:						CATEGORY I					
					APPLIANCE VENT CONNECTION:						SINGLE-WALL METAL CONNECTOR					
		VENT CONNECTOR CAPACITY														
		SINGLE-WALL METAL VENT CONNECTOR DIAMETER – *D* (inch)														
		3			4			5			6			7		
		APPLIANCE INPUT RATING LIMITS IN THOUSANDS OF BTU PER HOUR														
VENT HEIGHT *H* (feet)	CONNECTOR RISE *R* (feet)	FAN Min	FAN Max	NAT Max	FAN Min	FAN Max	NAT Max	FAN Min	FAN Max	NAT Max	FAN Min	FAN Max	NAT Max	FAN Min	FAN Max	NAT Max
6	1	NA	NA	26	NA	NA	46	NA	NA	71	NA	NA	102	207	223	140
	2	NA	NA	31	NA	NA	55	NA	NA	85	168	182	123	215	251	167
	3	NA	NA	34	NA	NA	62	121	131	95	175	198	138	222	273	188
8	1	NA	NA	27	NA	NA	48	NA	NA	75	NA	NA	106	226	240	145
	2	NA	NA	32	NA	NA	57	125	126	89	184	193	127	234	266	173
	3	NA	NA	35	NA	NA	64	130	138	100	191	208	144	241	287	197
10	1	NA	NA	28	NA	NA	50	119	121	77	182	186	110	240	253	150
	2	NA	NA	33	84	85	59	124	134	91	189	203	132	248	278	183
	3	NA	NA	36	89	91	67	129	144	102	197	217	148	257	299	203
15	1	NA	NA	29	79	87	52	116	138	81	177	214	116	238	291	158
	2	NA	NA	34	83	94	62	121	150	97	185	230	138	246	314	189
	3	NA	NA	39	87	100	70	127	160	109	193	243	157	255	333	215
20	1	49	56	30	78	97	54	115	152	84	175	238	120	233	325	165
	2	52	59	36	82	103	64	120	163	101	182	252	144	243	346	197
	3	55	62	40	87	107	72	125	172	113	190	264	164	252	363	223
30	1	47	60	31	77	110	57	112	175	89	169	278	129	226	380	175
	2	51	62	37	81	115	67	117	185	106	177	290	152	236	397	208
	3	54	64	42	85	119	76	122	193	120	185	300	172	244	412	235
50	1	46	69	34	75	128	60	109	207	96	162	336	137	217	460	188
	2	49	71	40	79	132	72	114	215	113	170	345	164	226	473	223
	3	52	72	45	83	136	82	119	221	123	178	353	186	235	486	252
100	1	45	79	34	71	150	61	104	249	98	153	424	140	205	585	192
	2	48	80	41	75	153	73	110	255	115	160	428	167	212	593	228
	3	51	81	46	79	157	85	114	260	129	168	433	190	222	603	256

	COMMON VENT CAPACITY											
	TYPE B DOUBLE-WALL COMMON VENT DIAMETER – *D* (inch)											
	4			5			6			7		
	COMBINED APPLIANCE INPUT RATING IN THOUSANDS OF BTU PER HOUR											
VENT HEIGHT *H* (feet)	FAN +FAN	FAN +NAT	NAT +NAT	FAN +FAN	FAN +NAT	NAT +NAT	FAN +FAN	FAN +NAT	NAT +NAT	FAN +FAN	FAN +NAT	NAT +NAT
6	NA	78	64	NA	113	99	200	158	144	304	244	196
8	NA	87	71	NA	126	111	218	173	159	331	269	218
10	NA	94	76	163	137	120	237	189	174	357	292	236
15	121	108	88	189	159	140	275	221	200	416	343	274
20	131	118	98	208	177	156	305	247	223	463	383	302
30	145	132	113	236	202	180	350	286	257	533	446	349
50	159	145	128	268	233	208	406	337	296	622	529	410
100	166	153	NA	297	263	NA	469	398	NA	726	633	464

For SI units: 1 inch = 25.4 mm, 1 foot = 304.8 mm, 1000 British thermal units per hour = 0.293 kW, 1 square inch = 0.000645 m^2

* NA: Not applicable.

TABLE 803.2(2)
TYPE B DOUBLE-WALL VENT [NFPA 54: TABLE 13.2(b)] (continued)

NUMBER OF APPLIANCES: TWO OR MORE
APPLIANCE TYPE: CATEGORY I
APPLIANCE VENT CONNECTION: SINGLE-WALL METAL CONNECTOR

VENT CONNECTOR CAPACITY — SINGLE-WALL METAL VENT CONNECTOR DIAMETER – *D* (inch) — APPLIANCE INPUT RATING LIMITS IN THOUSANDS OF BTU PER HOUR

VENT HEIGHT *H* (feet)	CONNECTOR RISE *R* (feet)	8 FAN Min	8 FAN Max	8 NAT Max	9 FAN Min	9 FAN Max	9 NAT Max	10 FAN Min	10 FAN Max	10 NAT Max
6	1	262	293	183	325	373	234	447	463	286
	2	271	331	219	334	422	281	458	524	344
	3	279	361	247	344	462	316	468	574	385
8	1	285	316	191	352	403	244	481	502	299
	2	293	353	228	360	450	292	492	560	355
	3	302	381	256	370	489	328	501	609	400
10	1	302	335	196	372	429	252	506	534	308
	2	311	369	235	381	473	302	517	589	368
	3	320	398	265	391	511	339	528	637	413
15	1	312	380	208	397	482	266	556	596	324
	2	321	411	248	407	522	317	568	646	387
	3	331	438	281	418	557	360	579	690	437
20	1	306	425	217	390	538	276	546	664	336
	2	317	453	259	400	574	331	558	709	403
	3	326	476	294	412	607	375	570	750	457
30	1	296	497	230	378	630	294	528	779	358
	2	307	521	274	389	662	349	541	819	425
	3	316	542	309	400	690	394	555	855	482
50	1	284	604	245	364	768	314	507	951	384
	2	294	623	293	376	793	375	520	983	458
	3	304	640	331	387	816	423	535	1013	518
100	1	269	774	249	345	993	321	476	1236	393
	2	279	788	299	358	1011	383	490	1259	469
	3	289	801	339	368	1027	431	506	1280	527

COMMON VENT CAPACITY — TYPE B DOUBLE-WALL COMMON VENT DIAMETER – *D* (inch) — COMBINED APPLIANCE INPUT RATING IN THOUSANDS OF BTU PER HOUR

VENT HEIGHT *H* (feet)	8 FAN +FAN	8 FAN +NAT	8 NAT +NAT	9 FAN +FAN	9 FAN +NAT	9 NAT +NAT	10 FAN +FAN	10 FAN +NAT	10 NAT +NAT
6	398	310	257	541	429	332	665	515	407
8	436	342	285	592	473	373	730	569	460
10	467	369	309	638	512	398	787	617	487
15	544	434	357	738	599	456	905	718	553
20	606	487	395	824	673	512	1013	808	626
30	703	570	459	958	790	593	1183	952	723
50	833	686	535	1139	954	689	1418	1157	838
100	999	846	606	1378	1185	780	1741	1459	948

For SI units: 1 inch = 25.4 mm, 1 foot = 304.8 mm, 1000 British thermal units per hour = 0.293 kW, 1 square inch = 0.000645 m^2

TABLE 803.2(3)
MASONRY CHIMNEY [NFPA 54: TABLE 13.2(c)]*

		NUMBER OF APPLIANCES:							TWO OR MORE							
		APPLIANCE TYPE:							CATEGORY I							
		APPLIANCE VENT CONNECTION:							TYPE B DOUBLE-WALL CONNECTOR							
		VENT CONNECTOR CAPACITY														
		TYPE B DOUBLE-WALL VENT CONNECTOR DIAMETER – *D* (inch)														
		3			4			5			6			7		
		APPLIANCE INPUT RATING LIMITS IN THOUSANDS OF BTU PER HOUR														
VENT HEIGHT *H* (feet)	CONNECTOR RISE *R* (feet)	FAN		NAT	FAN		NAT	FAN		NAT	FAN		NAT	FAN		NAT
		Min	Max	Max	Min	Max	Max	Min	Max	Max	Min	Max	Max	Min	Max	Max
6	1	24	33	21	39	62	40	52	106	67	65	194	101	87	274	141
	2	26	43	28	41	79	52	53	133	85	67	230	124	89	324	173
	3	27	49	34	42	92	61	55	155	97	69	262	143	91	369	203
8	1	24	39	22	39	72	41	55	117	69	71	213	105	94	304	148
	2	26	47	29	40	87	53	57	140	86	73	246	127	97	350	179
	3	27	52	34	42	97	62	59	159	98	75	269	145	99	383	206
10	1	24	42	22	38	80	42	55	130	71	74	232	108	101	324	153
	2	26	50	29	40	93	54	57	153	87	76	261	129	103	366	184
	3	27	55	35	41	105	63	58	170	100	78	284	148	106	397	209
15	1	24	48	23	38	93	44	54	154	74	72	277	114	100	384	164
	2	25	55	31	39	105	55	56	174	89	74	299	134	103	419	192
	3	26	59	35	41	115	64	57	189	102	76	319	153	105	448	215
20	1	24	52	24	37	102	46	53	172	77	71	313	119	98	437	173
	2	25	58	31	39	114	56	55	190	91	73	335	138	101	467	199
	3	26	63	35	40	123	65	57	204	104	75	353	157	104	493	222
30	1	24	54	25	37	111	48	52	192	82	69	357	127	96	504	187
	2	25	60	32	38	122	58	54	208	95	72	376	145	99	531	209
	3	26	64	36	40	131	66	56	221	107	74	392	163	101	554	233
50	1	23	51	25	36	116	51	51	209	89	67	405	143	92	582	213
	2	24	59	32	37	127	61	53	225	102	70	421	161	95	604	235
	3	26	64	36	39	135	69	55	237	115	72	435	180	98	624	260
100	1	23	46	24	35	108	50	49	208	92	65	428	155	88	640	237
	2	24	53	31	37	120	60	51	224	105	67	444	174	92	660	260
	3	25	59	35	38	130	68	53	237	118	69	458	193	94	679	285

	COMMON VENT CAPACITY														
	MINIMUM INTERNAL AREA OF MASONRY CHIMNEY FLUE (square inches)														
	12			19			28			38			50		
	COMBINED APPLIANCE INPUT RATING IN THOUSANDS OF BTU PER HOUR														
VENT HEIGHT *H* (feet)	FAN +FAN	FAN +NAT	NAT +NAT	FAN +FAN	FAN +NAT	NAT +NAT	FAN +FAN	FAN +NAT	NAT +NAT	FAN +FAN	FAN +NAT	NAT +NAT	FAN +FAN	FAN +NAT	NAT +NAT
6	NA	74	25	NA	119	46	NA	178	71	NA	257	103	NA	351	143
8	NA	80	28	NA	130	53	NA	193	82	NA	279	119	NA	384	163
10	NA	84	31	NA	138	56	NA	207	90	NA	299	131	NA	409	177
15	NA	NA	36	NA	152	67	NA	233	106	NA	334	152	523	467	212
20	NA	NA	41	NA	NA	75	NA	250	122	NA	368	172	565	508	243
30	NA	NA	NA	NA	NA	NA	NA	270	137	NA	404	198	615	564	278
50	NA	NA	NA	NA	NA	NA	NA	NA	NA	NA	NA	NA	NA	620	328
100	NA	NA	NA	NA	NA	NA	NA	NA	NA	NA	NA	NA	NA	NA	348

For SI units: 1 inch = 25.4 mm, 1 foot = 304.8 mm, 1000 British thermal units per hour = 0.293 kW, 1 square inch = 0.000645 m^2

* NA: Not applicable.

TABLE 803.2(3)
MASONRY CHIMNEY [NFPA 54: TABLE 13.2(c)] (continued)*

		NUMBER OF APPLIANCES: TWO OR MORE APPLIANCE TYPE: CATEGORY I APPLIANCE VENT CONNECTION: TYPE B DOUBLE-WALL CONNECTOR								
		VENT CONNECTOR CAPACITY								
		TYPE B DOUBLE-WALL VENT CONNECTOR DIAMETER – *D* (inch)								
		8			9			10		
		APPLIANCE INPUT RATING LIMITS IN THOUSANDS OF BTU PER HOUR								
VENT HEIGHT *H* (feet)	**CONNECTOR RISE *R* (feet)**	**FAN**		**NAT**	**FAN**		**NAT**	**FAN**		**NAT**
		Min	**Max**	**Max**	**Min**	**Max**	**Max**	**Min**	**Max**	**Max**
6	1	104	370	201	124	479	253	145	599	319
	2	107	436	232	127	562	300	148	694	378
	3	109	491	270	129	633	349	151	795	439
8	1	113	414	210	134	539	267	156	682	335
	2	116	473	240	137	615	311	160	776	394
	3	119	517	276	139	672	358	163	848	452
10	1	120	444	216	142	582	277	165	739	348
	2	123	498	247	145	652	321	168	825	407
	3	126	540	281	147	705	366	171	893	463
15	1	125	511	229	153	658	297	184	824	375
	2	128	558	260	156	718	339	187	900	432
	3	131	597	292	159	760	382	190	960	486
20	1	123	584	239	150	752	312	180	943	397
	2	126	625	270	153	805	354	184	1011	452
	3	129	661	301	156	851	396	187	1067	505
30	1	119	680	255	145	883	337	175	1115	432
	2	122	715	287	149	928	378	179	1171	484
	3	125	746	317	152	968	418	182	1220	535
50	1	115	798	294	140	1049	392	168	1334	506
	2	118	827	326	143	1085	433	172	1379	558
	3	121	854	357	147	1118	474	176	1421	611
100	1	109	907	334	134	1222	454	161	1589	596
	2	113	933	368	138	1253	497	165	1626	651
	3	116	956	399	141	1282	540	169	1661	705

	COMMON VENT CAPACITY								
	MINIMUM INTERNAL AREA OF MASONRY CHIMNEY FLUE (square inches)								
	63			78			113		
	COMBINED APPLIANCE INPUT RATING IN THOUSANDS OF BTU PER HOUR								
VENT HEIGHT *H* (feet)	**FAN +FAN**	**FAN +NAT**	**NAT +NAT**	**FAN +FAN**	**FAN +NAT**	**NAT +NAT**	**FAN +FAN**	**FAN +NAT**	**NAT +NAT**
6	NA	458	188	NA	582	246	1041	853	NA
8	NA	501	218	724	636	278	1144	937	408
10	606	538	236	776	686	302	1226	1010	454
15	682	611	283	874	781	365	1374	1156	546
20	742	668	325	955	858	419	1513	1286	648
30	816	747	381	1062	969	496	1702	1473	749
50	879	831	461	1165	1089	606	1905	1692	922
100	NA	NA	499	NA	NA	669	2053	1921	1058

For SI units: 1 inch = 25.4 mm, 1 foot = 304.8 mm, 1000 British thermal units per hour = 0.293 kW, 1 square inch = 0.000645 m^2

* NA: Not applicable.

TABLE 803.2(4)
MASONRY CHIMNEY [NFPA 54: TABLE 13.2(d)]*

								NUMBER OF APPLIANCES:	TWO OR MORE							
								APPLIANCE TYPE:	CATEGORY I							
								APPLIANCE VENT CONNECTION:	SINGLE-WALL METAL CONNECTOR							
		VENT CONNECTOR CAPACITY														
		SINGLE-WALL METAL VENT CONNECTOR DIAMETER – *D* (inch)														
		3			4			5			6			7		
		APPLIANCE INPUT RATING LIMITS IN THOUSANDS OF BTU PER HOUR														
VENT HEIGHT *H* (feet)	CONNECTOR RISE *R* (feet)	FAN		NAT	FAN		NAT	FAN		NAT	FAN		NAT	FAN		NAT
		Min	Max	Max	Min	Max	Max	Min	Max	Max	Min	Max	Max	Min	Max	Max
6	1	NA	NA	21	NA	NA	39	NA	NA	66	179	191	100	231	271	140
	2	NA	NA	28	NA	NA	52	NA	NA	84	186	227	123	239	321	172
	3	NA	NA	34	NA	NA	61	134	153	97	193	258	142	247	365	202
8	1	NA	NA	21	NA	NA	40	NA	NA	68	195	208	103	250	298	146
	2	NA	NA	28	NA	NA	52	137	139	85	202	240	125	258	343	177
	3	NA	NA	34	NA	NA	62	143	156	98	210	264	145	266	376	205
10	1	NA	NA	22	NA	NA	41	130	151	70	202	225	106	267	316	151
	2	NA	NA	29	NA	NA	53	136	150	86	210	255	128	276	358	181
	3	NA	NA	34	97	102	62	143	166	99	217	277	147	284	389	207
15	1	NA	NA	23	NA	NA	43	129	151	73	199	271	112	268	376	161
	2	NA	NA	30	92	103	54	135	170	88	207	295	132	277	411	189
	3	NA	NA	34	96	112	63	141	185	101	215	315	151	286	439	213
20	1	NA	NA	23	87	99	45	128	167	76	197	303	117	265	425	169
	2	NA	NA	30	91	111	55	134	185	90	205	325	136	274	455	195
	3	NA	NA	35	96	119	64	140	199	103	213	343	154	282	481	219
30	1	NA	NA	24	86	108	47	126	187	80	193	347	124	259	492	183
	2	NA	NA	31	91	119	57	132	203	93	201	366	142	269	518	205
	3	NA	NA	35	95	127	65	138	216	105	209	381	160	277	540	229
50	1	NA	NA	24	85	113	50	124	204	87	188	392	139	252	567	208
	2	NA	NA	31	89	123	60	130	218	100	196	408	158	262	588	230
	3	NA	NA	35	94	131	68	136	231	112	205	422	176	271	607	255
100	1	NA	NA	23	84	104	49	122	200	89	182	410	151	243	617	232
	2	NA	NA	30	88	115	59	127	215	102	190	425	169	253	636	254
	3	NA	NA	34	93	124	67	133	228	115	199	438	188	262	654	279

	COMMON VENT CAPACITY														
	MINIMUM INTERNAL AREA OF MASONRY CHIMNEY FLUE (square inches)														
	12			19			28			38			50		
	COMBINED APPLIANCE INPUT RATING IN THOUSANDS OF BTU PER HOUR														
VENT HEIGHT *H* (feet)	FAN +FAN	FAN +NAT	NAT +NAT	FAN +FAN	FAN +NAT	NAT +NAT	FAN +FAN	FAN +NAT	NAT +NAT	FAN +FAN	FAN +NAT	NAT +NAT	FAN +FAN	FAN +NAT	NAT +NAT
6	NA	NA	25	NA	118	45	NA	176	71	NA	255	102	NA	348	142
8	NA	NA	28	NA	128	52	NA	190	81	NA	276	118	NA	380	162
10	NA	NA	31	NA	136NA	56	NA	205	89	NA	295	129	NA	405	175
15	NA	NA	36	NA	NA	66	NA	230	105	NA	335	150	NA	400	210
20	NA	NA	NA	NA	NA	74	NA	247	120	NA	362	170	NA	503	240
30	NA	NA	NA	NA	NA	NA	NA	NA	135	NA	398	195	NA	558	275
50	NA	NA	NA	NA	NA	NA	NA	NA	NA	NA	NA	NA	NA	612	325
100	NA	NA	NA	NA		NA	NA	NA	NA	NA	NA	NA	NA	NA	NA

For SI units: 1 inch = 25.4 mm, 1 foot = 304.8 mm, 1000 British thermal units per hour = 0.293 kW, 1 square inch = 0.000645 m^2

* NA: Not applicable.

TABLE 803.2(4)
MASONRY CHIMNEY [NFPA 54: TABLE 13.2(d)] (continued)*

		NUMBER OF APPLIANCES: TWO OR MORE								
		APPLIANCE TYPE: CATEGORY I								
		APPLIANCE VENT CONNECTION: SINGLE-WALL METAL CONNECTOR								
		VENT CONNECTOR CAPACITY								
		SINGLE-WALL METAL VENT CONNECTOR DIAMETER – *D* (inch)								
		8			9			10		
		APPLIANCE INPUT RATING LIMITS IN THOUSANDS OF BTU PER HOUR								
VENT HEIGHT	CONNECTOR RISE	FAN		NAT	FAN		NAT	FAN		NAT
H (feet)	*R* (feet)	Min	Max	Max	Min	Max	Max	Min	Max	Max
6	1	292	366	200	362	474	252	499	594	316
	2	301	432	231	373	557	299	509	696	376
	3	309	491	269	381	634	348	519	793	437
8	1	313	407	207	387	530	263	529	672	331
	2	323	465	238	397	607	309	540	766	391
	3	332	509	274	407	663	356	551	838	450
10	1	333	434	213	410	571	273	558	727	343
	2	343	489	244	420	640	317	569	813	403
	3	352	530	279	430	694	363	580	880	459
15	1	349	502	225	445	646	291	623	808	366
	2	359	548	256	456	706	334	634	884	424
	3	368	586	289	466	755	378	646	945	479
20	1	345	569	235	439	734	306	614	921	387
	2	355	610	266	450	787	348	627	986	443
	3	365	644	298	461	831	391	639	1042	496
30	1	338	665	250	430	864	330	600	1089	421
	2	348	699	282	442	908	372	613	1145	473
	3	358	729	312	452	946	412	626	1193	524
50	1	328	778	287	417	1022	383	582	1302	492
	2	339	806	320	429	1058	425	596	1346	545
	3	349	831	351	440	1090	466	610	1386	597
100	1	315	875	328	402	1181	444	560	1537	580
	2	326	899	361	415	1210	488	575	1570	634
	3	337	921	392	427	1238	529	589	1604	687

	COMMON VENT CAPACITY								
	MINIMUM INTERNAL AREA OF MASONRY CHIMNEY FLUE (square inches)								
	63			78			113		
	COMBINED APPLIANCE INPUT RATING IN THOUSANDS OF BTU PER HOUR								
VENT HEIGHT *H* (feet)	FAN +FAN	FAN +NAT	NAT +NAT	FAN +FAN	FAN +NAT	NAT +NAT	FAN +FAN	FAN +NAT	NAT +NAT
6	NA	455	187	NA	579	245	NA	846	NA
8	NA	497	217	NA	633	277	1136	928	405
10	NA	532	234	771	680	300	1216	1000	450
15	677	602	280	866	772	360	1359	1139	540
20	765	661	321	947	849	415	1495	1264	640
30	808	739	377	1052	957	490	1682	1447	740
50	NA	821	456	1152	1076	600	1879	1672	910
100	NA	NA	494	NA	NA	663	2006	1885	1046

For SI units: 1 inch = 25.4 mm, 1 foot = 304.8 mm, 1000 British thermal units per hour = 0.293 kW, 1 square inch = 0.000645 m^2

* NA: Not applicable.

TABLE 803.2(5)
SINGLE-WALL METAL PIPE OR TYPE B ASBESTOS-CEMENT VENT [NFPA 54: TABLE 13.2(e)]*

				NUMBER OF APPLIANCES:	TWO OR MORE		
				APPLIANCE TYPE:	DRAFT HOOD-EQUIPMENT		
				APPLIANCE VENT CONNECTION:	DIRECT TO PIPE OR VENT		
		VENT CONNECTOR CAPACITY					
		VENT CONNECTOR DIAMETER – *D* (inch)					
TOTAL VENT HEIGHT *H* (feet)	CONNECTOR RISE *R* (feet)	3	4	5	6	7	8
		APPLIANCE INPUT RATING IN THOUSANDS OF BTU PER HOUR					
6-8	1	21	40	68	102	146	205
	2	28	53	86	124	178	235
	3	34	61	98	147	204	275
15	1	23	44	77	117	179	240
	2	30	56	92	134	194	265
	3	35	64	102	155	216	298
30 and up	1	25	49	84	129	190	270
	2	31	58	97	145	211	295
	3	36	68	107	164	232	321

	COMMON VENT CAPACITY						
	COMMON VENT DIAMETER – *D* (inch)						
TOTAL VENT HEIGHT *H* (feet)	4	5	6	7	8	10	12
	COMBINED APPLIANCE INPUT RATING IN THOUSANDS OF BTU PER HOUR						
6	48	78	111	155	205	320	NA
8	55	89	128	175	234	365	505
10	59	95	136	190	250	395	560
15	71	115	168	228	305	480	690
20	80	129	186	260	340	550	790
30	NA	147	215	300	400	650	940
50	NA	NA	NA	360	490	810	1190

For SI units: 1 inch = 25.4 mm, 1 foot = 304.8 mm, 1000 British thermal units per hour = 0.293 kW, 1 square inch = 0.000645 m^2
* NA: Not applicable.

TABLE 803.2(6)
EXTERIOR MASONRY CHIMNEY [NFPA 54: TABLE 13.2(f)]*

				NUMBER OF APPLIANCES:	TWO OR MORE			
				APPLIANCE TYPE:	NAT + NAT			
				APPLIANCE VENT CONNECTION:	TYPE B DOUBLE-WALL CONNECTOR			
COMBINED APPLIANCE MAXIMUM INPUT RATING IN THOUSANDS OF BTU PER HOUR								
VENT HEIGHT *H* (feet)	INTERNAL AREA OF CHIMNEY (square inches)							
	12	19	28	38	50	63	78	113
6	25	46	71	103	143	188	246	NA
8	28	53	82	119	163	218	278	408
10	31	56	90	131	177	236	302	454
15	NA	67	106	152	212	283	365	546
20	NA	NA	NA	NA	NA	325	419	648
30	NA	NA	NA	NA	NA	NA	496	749
50	NA	NA	NA	NA	NA	NA	NA	922
100	NA	NA	NA	NA	NA	NA	NA	NA

For SI units: 1 inch = 25.4 mm, 1 foot = 304.8 mm, 1000 British thermal units per hour = 0.293 kW, 1 square inch = 0.000645 m^2
* NA: Not applicable.

TABLE 803.2(7)
EXTERIOR MASONRY CHIMNEY [NFPA 54: TABLE 13.2(g)][1, 2]

				NUMBER OF APPLIANCES:	TWO OR MORE			
				APPLIANCE TYPE:	NAT + NAT			
				APPLIANCE VENT CONNECTION:	TYPE B DOUBLE-WALL CONNECTOR			
MINIMUM ALLOWABLE INPUT RATING OF SPACE-HEATING APPLIANCE IN THOUSANDS OF BTU PER HOUR								
VENT HEIGHT *H* (feet)	INTERNAL AREA OF CHIMNEY (square inches)							
	12	19	28	38	50	63	78	113
Local 99% winter design temperature: 37°F or greater								
6	0	0	0	0	0	0	0	NA
8	0	0	0	0	0	0	0	0
10	0	0	0	0	0	0	0	0
15	NA	0	0	0	0	0	0	0
20	NA	NA	NA	NA	NA	184	0	0
30	NA	NA	NA	NA	NA	393	334	0
50	NA	NA	NA	NA	NA	NA	NA	579
100	NA	NA	NA	NA	NA	NA	NA	NA
Local 99% winter design temperature: 27°F to 36°F								
6	0	0	68	NA	NA	180	212	NA
8	0	0	82	NA	NA	187	214	263
10	0	51	NA	NA	NA	201	225	265
15	NA	NA	NA	NA	NA	253	274	305
20	NA	NA	NA	NA	NA	307	330	362
30	NA	NA	NA	NA	NA	NA	445	485
50	NA	NA	NA	NA	NA	NA	NA	763
100	NA	NA	NA	NA	NA	NA	NA	NA
Local 99% winter design temperature: 17°F to 26°F								
6	NA	NA	NA	NA	NA	NA	NA	NA
8	NA	NA	NA	NA	NA	NA	264	352
10	NA	NA	NA	NA	NA	NA	278	358
15	NA	NA	NA	NA	NA	NA	331	398
20	NA	NA	NA	NA	NA	NA	387	457
30	NA	NA	NA	NA	NA	NA	NA	581
50	NA	NA	NA	NA	NA	NA	NA	862
100	NA	NA	NA	NA	NA	NA	NA	NA
Local 99% winter design temperature: 5°F to 16°F								
6	NA	NA	NA	NA	NA	NA	NA	NA
8	NA	NA	NA	NA	NA	NA	NA	NA
10	NA	NA	NA	NA	NA	NA	NA	430
15	NA	NA	NA	NA	NA	NA	NA	485
20	NA	NA	NA	NA	NA	NA	NA	547
30	NA	NA	NA	NA	NA	NA	NA	682
50	NA	NA	NA	NA	NA	NA	NA	NA
100	NA	NA	NA	NA	NA	NA	NA	NA
Local 99% winter design temperature: 4°F or lower Not recommended for any vent configurations								

For SI units: 1 inch = 25.4 mm, 1 foot = 304.8 mm, 1000 British thermal units per hour = 0.293 kW, 1 square inch = 0.000645 m^2, °C = (°F-32)/1.8

Notes:

1 See Figure 803.1.2(6) for a map showing local 99 percent winter design temperatures in the United States.

2 NA: Not applicable.

TABLE 803.2(8)
EXTERIOR MASONRY CHIMNEY [NFPA 54: TABLE 13.2(h)]*

				NUMBER OF APPLIANCES:	TWO OR MORE			
				APPLIANCE TYPE:	FAN + NAT			
				APPLIANCE VENT CONNECTION:	TYPE B DOUBLE-WALL CONNECTOR			
COMBINED APPLIANCE MAXIMUM INPUT RATING IN THOUSANDS OF BTU PER HOUR								
VENT HEIGHT *H* (feet)	INTERNAL AREA OF CHIMNEY (square inches)							
	12	19	28	38	50	63	78	113
6	74	119	178	257	351	458	582	853
8	80	130	193	279	384	501	636	937
10	84	138	207	299	409	538	686	1010
15	NA	152	233	334	467	611	781	1156
20	NA	NA	250	368	508	668	858	1286
30	NA	NA	NA	404	564	747	969	1473
50	NA	NA	NA	NA	NA	831	1089	1692
100	NA	NA	NA	NA	NA	NA	NA	1921

For SI units: 1 inch = 25.4 mm, 1 foot = 304.8 mm, 1000 British thermal units per hour = 0.293 kW, 1 square inch = 0.000645 m^2

* NA: Not applicable.

TABLE 803.2(9)
EXTERIOR MASONRY CHIMNEY [NFPA 54: TABLE 13.2(i)][1, 2]

				NUMBER OF APPLIANCES:		TWO OR MORE		
				APPLIANCE TYPE:		FAN + NAT		
				APPLIANCE VENT CONNECTION:		TYPE B DOUBLE-WALL CONNECTOR		
MINIMUM ALLOWABLE INPUT RATING OF SPACE-HEATING APPLIANCE IN THOUSANDS OF BTU PER HOUR								
VENT HEIGHT *H* (feet)	INTERNAL AREA OF CHIMNEY (square inches)							
	12	19	28	38	50	63	78	113
Local 99% winter design temperature: 37°F or greater								
6	0	0	0	0	0	0	0	0
8	0	0	0	0	0	0	0	0
10	0	0	0	0	0	0	0	0
15	NA	0	0	0	0	0	0	0
20	NA	NA	123	190	249	184	0	0
30	NA	NA	NA	334	398	393	334	0
50	NA	NA	NA	NA	NA	714	707	579
100	NA	NA	NA	NA	NA	NA	NA	1600
Local 99% winter design temperature: 27°F to 36°F								
6	0	0	68	116	156	180	212	266
8	0	0	82	127	167	187	214	263
10	0	51	97	141	183	201	225	265
15	NA	111	142	183	233	253	274	305
20	NA	NA	187	230	284	307	330	362
30	NA	NA	NA	330	319	419	445	485
50	NA	NA	NA	NA	NA	672	705	763
100	NA	NA	NA	NA	NA	NA	NA	1554
Local 99% winter design temperature: 17°F to 26°F								
6	0	55	99	141	182	215	259	349
8	52	74	111	154	197	226	264	352
10	NA	90	125	169	214	245	278	358
15	NA	NA	167	212	263	296	331	398
20	NA	NA	212	258	316	352	387	457
30	NA	NA	NA	362	429	470	507	581
50	NA	NA	NA	NA	NA	723	766	862
100	NA	NA	NA	NA	NA	NA	NA	1669
Local 99% winter design temperature: 5°F to 16°F								
6	NA	78	121	166	214	252	301	416
8	NA	94	135	182	230	269	312	423
10	NA	111	149	198	250	289	331	430
15	NA	NA	193	247	305	346	393	485
20	NA	NA	NA	293	360	408	450	547
30	NA	NA	NA	377	450	531	580	682
50	NA	NA	NA	NA	NA	797	853	972
100	NA	NA	NA	NA	NA	NA	NA	1833
Local 99% winter design temperature: -10°F to 4°F								
6	NA	NA	145	196	249	296	349	484
8	NA	NA	159	213	269	320	371	494
10	NA	NA	175	231	292	339	397	513
15	NA	NA	NA	283	351	404	457	586
20	NA	NA	NA	333	408	468	528	650
30	NA	NA	NA	NA	NA	603	667	805
50	NA	NA	NA	NA	NA	NA	955	1003
100	NA	NA	NA	NA	NA	NA	NA	NA
Local 99% winter design temperature: -11°F or lower Not recommended for any vent configurations								

For SI units: 1 inch = 25.4 mm, 1 foot = 304.8 mm, 1000 British thermal units per hour = 0.293 kW, 1 square inch = 0.000645 m^2, °C = (°F-32)/1.8

Notes:

[1] See Figure 803.1.2(6) for a map showing local 99 percent winter design temperatures in the United States.

[2] NA: Not applicable.

CHAPTER 9

INSTALLATION OF SPECIFIC APPLIANCES

901.0 General.

901.1 Applicability. This chapter addresses requirements for the design, construction, and installation of specific appliances. In addition to the requirements of this chapter, appliances shall comply with the general requirements of Chapter 3.

902.0 General.

902.1 Nonindustrial Appliance. This chapter is applicable primarily to nonindustrial-type appliances and installations and, unless specifically indicated, does not apply to industrial-type appliances and installations. Listed appliances shall be installed in accordance with their listing and the manufacturer's installation instructions or, as elsewhere specified in this chapter, as applicable to the appliance. Unlisted appliances shall be installed as specified in this part as applicable to the appliances.

All listed gas-fired equipment and accessories shall be installed according to the listing and the manufacturers' instructions. Code requirements for unlisted equipment installations are specified throughout this chapter.

If a gas appliance, other than a direct-vent type, is installed in a sleeping room or bathroom where combustion, ventilation, and dilution are obtained from the same space, refer to Chapter 7, Section 701.4, Indoor Combustion Air. When calculating combustion air, the minimum volume of indoor air required is 50 cubic feet per 1,000 Btu/hr.

It is important to note that many variables affect the efficient operation of a gas-fired appliance that vents hazardous byproducts of combustion to the atmosphere. Such variables include, but are not limited to, type of appliance and venting requirements, air exchanges per hour (ACH), air infiltration, and return air locations. To ensure proper ventilation, all code requirements and life/safety issues are to be considered before installing an appliance in a sleeping room or bathroom (see **Figure 902.1**).

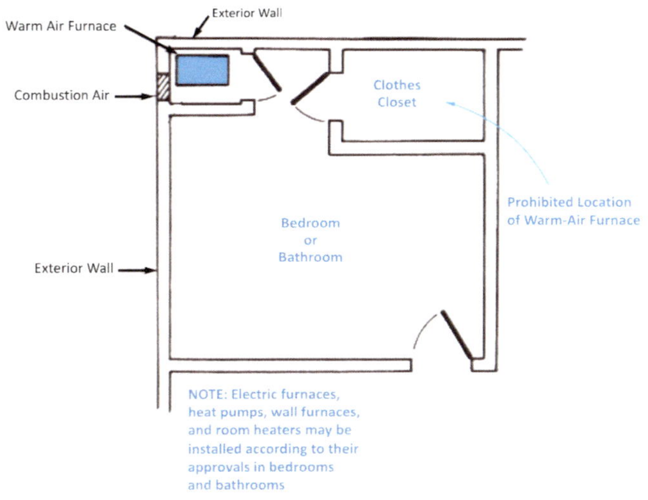

FIGURE 902.1
PROHIBITED AND ACCEPTABLE LOCATION OF WARM-AIR FURNACES

902.2 Combustion Air from Bedroom or Bathroom. Appliances shall not be installed so their combustion, ventilation, and dilution air are obtained only from a bedroom or bathroom unless the bedroom or bathroom has the required volume in accordance with Section 701.4. [NFPA 54:10.1.2]

902.3 Added or Converted Appliances. When additional or replacement appliances or equipment is installed or an appliance is converted to gas from another fuel, the location in which the appliances or equipment is to be operated shall be checked to verify the following:

(1) Air for combustion and ventilation is provided where required, in accordance with the provisions of Section 701.0. Where existing facilities are not adequate, they shall be upgraded to meet Section 701.0 specifications.

(2) The installation components and appliances meet the clearances to combustible material provisions of Section 303.10. It shall be determined that the installation and operation of the additional or replacement appliances do not render the remaining appliances unsafe for continued operation.

(3) The venting system is constructed and sized in accordance with the provisions of Section 802.0. Where the existing venting system is not adequate, it shall be upgraded to comply with Section 802.0. [NFPA 54:9.1.2]

902.4 Type of Gas(es). The appliance shall be connected to the fuel gas for which it was designed. No attempt shall be made to convert the appliance from the gas specified on the rating plate for use with a different gas without consulting the installation instructions, the serving gas supplier, or the appliance manufacturer for complete instructions. [NFPA 54:9.1.3]

902.5 Safety Shutoff Devices for Unlisted LP-Gas Appliances Used Indoors. Unlisted appliances for use with undiluted LP-Gases and installed indoors, except attended laboratory equipment, shall be equipped with safety shutoff devices of the complete shutoff type. [NFPA 54:9.1.4]

902.6 Fuel Input Rate. The fuel input rate to the appliance shall not be increased or decreased in violation of the approved rating at the altitude where it is being used.

All fuel-fired appliances are designed to operate with a maximum and minimum heat energy input capacity. The capacity can be field adjusted in accordance with the elevation due to the air density at different elevations. Alteration beyond or below the required input can result in overfiring or underfiring which can be hazardous such as overheating, vent failure, corrosion, and poor draft and combustion.

902.7 Use of Air or Oxygen Under Pressure. Where air or oxygen under pressure is used in connection with the gas supply, effective means such as a back pressure regulator and relief valve shall be provided to prevent air or oxygen from passing back into the gas piping. Where oxygen is used, installation shall be in accordance with NFPA 51. [NFPA 54:9.1.5]

902.8 Building Structural Members. Appliances and equipment shall be furnished either with load-distributing bases or with a sufficient number of supports to prevent damage to either the building structure or the appliance and the equipment. [NFPA 54:9.1.8.1]

902.8.1 Structural Capacity. At the locations selected for installation of appliances and equipment, the dynamic and static load-carrying capacities of the building structure shall be checked to determine whether they are adequate to carry the additional loads. The appliances and equipment shall be supported and shall be connected to the piping so as not to exert undue stress on the connections. [NFPA 54:9.1.8.2]

902.9 Flammable Vapors. Appliances shall not be installed in areas where the open use, handling, or dispensing of flammable liquids occurs, unless the design, operation, or installation reduces the potential of ignition of the flammable vapors. Appliances installed in compliance with Section 305.1 through Section 305.1.2, Section 303.11, or Section 303.12 shall be considered to comply with the intent of this provision. [NFPA 54:9.1.9]

902.10 Solid-Fuel Burning Appliances. Unless otherwise specified, solid-fuel burning appliances shall be installed in accordance with NFPA 211 and the manufacturer's installation instructions.

Chimneys for solid fuel appliances must be either field or factory built. Connectors for solid fuel appliances must be factory built chimney material, Type L vent or steel pipe resistant to corrosion and heat. Galvanized steel pipe is prohibited for use with solid fuel appliances. Connectors and chimneys for solid fuel appliances need to be "designed, located and installed to allow access for internal inspection and cleaning." An example of a clearance reduction system in accordance with NFPA 211 for solid fuel room heaters, fireplace stoves, room heater/fireplace stove combinations, and ranges is shown in **Figure 902.10**. Furnaces and boilers, along with nonresidential appliances, have further restrictions for clearances to combustibles.

Listed wood burning appliances shall be installed in accordance with the building code, the appliance listings and the manufacturer's instructions. Particular attention must be paid to ensure adequate combustion air, the proper anchorage of chimneys, clearance to combustibles and terminations above the building roof and adjacent structures based on the flue outlet temperatures of the combustion gases.

902.11 Combination of Appliances and Equipment. Any combination of appliances, equipment, attachments, or devices used together in any manner shall comply with the standards that apply to the individual appliance and equipment. [NFPA 54:9.1.21]

902.12 Protection of Gas Appliances from Fumes or Gases Other than Products of Combustion. Non-direct-vent appliances installed in beauty shops, barber shops, or other facilities where chemicals that generate corrosive or flammable products such as aerosol sprays are routinely used shall be located in a mechanical room separate or partitioned off from other areas with provisions for combustion and dilution air from outdoors. Direct vent appliances in such facilities shall be in accordance with the appliance manufacturer's installation instructions. [NFPA 54:9.1.6.2]

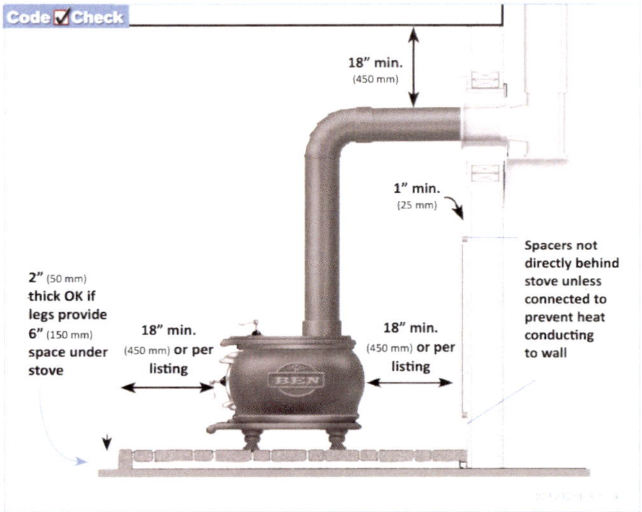

FIGURE 902.10
FIREPLACE STOVE REDUCTION CLEARANCES
(Courtesy of Code Check)

902.13 Process Air. In addition to air needed for combustion in commercial or industrial processes, process air shall be provided as required for cooling of appliances, equipment, or material; for controlling dew point, heating, drying, oxidation, dilution, safety exhaust, odor control, and air for compressors; and for comfort and proper working conditions for personnel. [NFPA 54:9.1.7]

902.14 Gas Appliance Pressure Regulators. Where the gas supply pressure is higher than that at which the appliance is designed to operate or varies beyond the design pressure limits of the appliance, a gas appliance pressure regulator shall be installed. [NFPA 54:9.1.18]

902.15 Venting of Gas Appliance Pressure Regulators. Venting of gas appliance pressure regulators shall comply with the following requirements:

(1) Appliance pressure regulators requiring access to the atmosphere for successful operation shall be equipped with vent piping leading outdoors or, if the regulator vent is an integral part of the appliance, into the combustion chamber adjacent to a continuous pilot, unless constructed or equipped with a vent limiting means to limit the escape of gas from the vent opening in the event of diaphragm failure.

(2) Vent limiting means shall be employed on listed appliance pressure regulators only.

(3) In the case of vents leading outdoors, means shall be employed to prevent water from entering this piping and also to prevent blockage of vents by insects and foreign matter.

(4) Under no circumstances shall a regulator be vented to the appliance flue or exhaust system.

(5) In the case of vents entering the combustion chamber,

the vent shall be located so the escaping gas is readily ignited by the pilot and the heat liberated thereby does not adversely affect the normal operation of the safety shutoff system. The terminus of the vent shall be securely held in a fixed position relative to the pilot. For manufactured gas, the need for a flame arrester in the vent piping shall be determined.

(6) A vent line(s) from an appliance pressure regulator and a bleed line(s) from a diaphragm-type valve shall not be connected to a common manifold terminating in a combustion chamber. Vent lines shall not terminate in positive-pressure-type combustion chambers. [NFPA 54:9.1.19]

Gas pressure regulators have a vent that allows the air above the regulator diaphragm to be displaced so the diaphragm can move. This air is vented out of the regulator. This air may also contain gas products and, if in large enough quantities or if the diaphragm fails, could cause an explosion within the building. The regulator, if required, must be vented according to the requirements of this section.

902.16 Bleed Lines for Diaphragm-Type Valves. Bleed lines shall comply with the following requirements:

(1) Diaphragm-type valves shall be equipped to convey bleed gas to the outdoors or into the combustion chamber adjacent to a continuous pilot.

(2) In the case of bleed lines leading outdoors, means shall be employed to prevent water from entering this piping and also to prevent blockage of vents by insects and foreign matter.

(3) Bleed lines shall not terminate in the appliance flue or exhaust system.

(4) In the case of bleed lines entering the combustion chamber, the bleed line shall be located so the bleed gas is readily ignited by the pilot and the heat liberated thereby does not adversely affect the normal operation of the safety shutoff system. The terminus of the bleed line shall be securely held in a fixed position relative to the pilot. For manufactured gas, the need for a flame arrester in the bleed line piping shall be determined.

(5) A bleed line(s) from a diaphragm-type valve and a vent line(s) from an appliance pressure regulator shall not be connected to a common manifold terminating in a combustion chamber. Bleed lines shall not terminate in positive-pressure-type combustion chambers. [NFPA 54:9.1.20]

903.0 Air-Conditioning Appliances.

903.1 Electric Air Conditioners. Electric air conditioning systems designed for permanent installation shall comply with UL 1995 or UL 60335-2-40.

The scope of UL 60335-2-40 includes requirements for "Household and Similar Electrical Appliances" such as: air conditioning systems equipment, package and split system air conditioners, heat pumps, heat pump water heaters and liquid chillers. Appliances not intended for normal household use but which may be a source of danger to the public, such as appliances intended to be used by laymen in shops, in light industry and on farms, are within the scope of this standard. UL 60335-2-40 does not allow units rated over 600 volts.

The appliances referenced above may consist of one or more factory-made assemblies. If provided in more than one assembly, the separate assemblies are to be used together, and the requirements are based on the use of matched assemblies. Motor-compressors incorporated with electric air-conditioning systems are designated either hermetic or semi-hermetic motor-compressors. Therefore these appliances are subject to the requirements of the National Electrical Code Section 440, Air-Conditioning and Refrigeration Equipment.

This standard does not take into account chemicals other than group A1, A2, or A3. Clause 1DV.1 of the standard prohibits the use of flammable refrigerants at this time.

903.2 Gas-Fired Air Conditioners and Heat Pumps. Gas-fired air conditioners shall comply with Section 903.2.1 through Section 903.2.6.

903.2.1 Independent Gas Piping. Gas piping serving heating appliances shall be permitted to also serve cooling appliances where heating and cooling appliances cannot be operated simultaneously. [NFPA 54:10.2.1]

903.2.2 Connection of Gas Engine-Powered Air Conditioners. To protect against the effects of normal vibration in service, gas engines shall not be rigidly connected to the gas supply piping. [NFPA 54:10.2.2]

903.2.3 Clearances for Indoor Installation. The installation of air-conditioning appliances shall comply with the following requirements:

(1) Listed air-conditioning appliances shall be installed with clearances in accordance with the terms of their listing and the manufacturer's installation instructions.

(2) Unlisted air-conditioning appliances shall be installed with clearances from combustible material of not less than 18 inches (457 mm) above the appliance and at the sides, front, and rear and in accordance with the manufacturer's installation instructions. [NFPA 54:10.2.3(2)]

(3) Listed and unlisted air-conditioning appliances shall be permitted to be installed with reduced clearances to combustible material, provided that the combustible material or appliance is protected as described in Table 303.10.1 and such reduction is allowed by the manufacturer's installation instructions. [NFPA 54:10.2.3(3)]

(4) Where the furnace plenum is adjacent to plaster on metal lath or noncombustible material attached to combustible material, the clearance shall be measured to the surface of the plaster or other noncombustible finish where the clearance specified is 2 inches (51 mm) or less. [NFPA 54:10.2.3(4)]

(5) Listed air-conditioning appliances shall have the clearance from supply ducts within 3 feet (914 mm) of the furnace plenum be not less than that specified from the furnace plenum. No clearance is necessary beyond this distance. [NFPA 54:10.2.3(5)]

903.2.4 Assembly and Installation. Air-conditioning appliances shall be installed in accordance with the manufacturer's installation instructions. Unless the appliance is listed for installation on a combustible surface, such as a floor or roof, or unless the surface is protected in an approved manner, it shall be installed on a surface of noncombustible construction with noncombustible material and surface finish and with no combustible material against the underside thereof. [NFPA 54:10.2.4]

903.2.5 Refrigeration Coils. The installation of refrigeration coils shall be in accordance with Section 904.7 and Section 904.8. [NFPA 54:10.2.6]

903.2.6 Switches in Electrical Supply Line. Means for interrupting the electrical supply to the air-conditioning appliance and to its associated cooling tower (if supplied and installed in a location remote from the air conditioner) shall be provided within sight of and not over 50 feet (15 240 mm) from the air conditioner and the cooling tower. [NFPA 54:10.2.7]

904.0 Central Heating Boilers and Furnaces.

A "heating system" is a heating plant with an enclosed heat exchanger that distributes heated air to various rooms and areas. A heating system with accessories includes the outside air, return air and supply air (see Chapter 2, Section 210.0).

Chapter 2, Section 225.0 defines a "Warm-air furnace" as an "environmental heating appliance designed to discharge heated air through ducts." Unit heaters are not designed for comfort heat and, therefore, are not included in this definition. Many of the code requirements for the installation of warm-air heating systems are contained in other chapters of this code. Chapter 4 regulates the minimum provisions for ventilation air supply to specific occupancies.

Section 305.1 Installation in Garages, provides for heating systems to be protected from potential damage in all locations. Equipment capable of igniting flammable vapors must be elevated a minimum 18 inches above finished floor in all areas exposed to such vapors.

This section and Section 304.0 Accessibility for Service, provide for "accessibility for servicing" equipment. A clear working space must be maintained to permit servicing, repair or removal of gas-fired equipment in all locations, including attics, without removing permanent construction. Unless otherwise specified, a minimum clearance of 30 inches is required for servicing.

Access requirements are sometimes confused with minimum distances for clearance to combustible materials. Understand that the equipment listing dictates minimum clearances to combustibles. Sometimes it is necessary to determine practical access to the equipment by considering what space is needed to service the equipment efficiently.

Types of Central Forced-Air Furnaces

Various forced-air furnaces derive their names from the direction of airflow through the heat exchanger and are equipped with a fan or blower to circulate airflow to the conditioned areas.

Section 208.0 refers to three types of forced-air furnaces: horizontal, upflow and downflow. For example, an appliance with a horizontal airflow through the unit is designed for installation in a horizontal position. This appliance type is usually installed in areas with low headroom where the appliance is not exposed to water inundation or other possible damage.

All listed equipment in this category shall be installed according to the manufacturers' instructions and to the specifications of the code (see **Figure 904.0**).

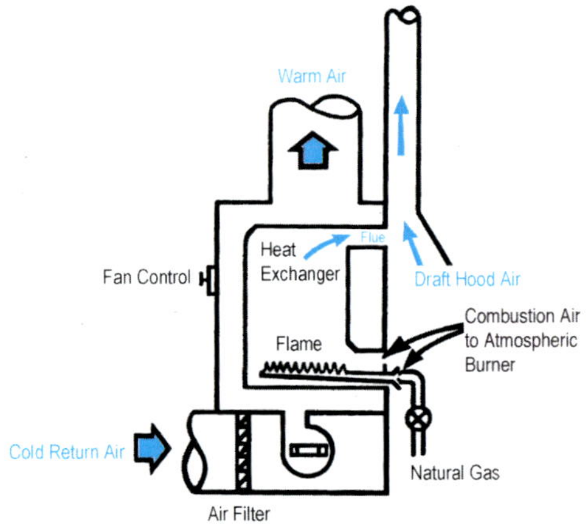

FIGURE 904.0
UPFLOW FURNACE

904.1 Location. Central heating furnace and low-pressure boiler installations in bedrooms or bathrooms shall comply with one of the following:

(1) Central heating furnaces and low-pressure boilers shall be permitted to be installed in a closet located in the bedroom or bathroom, provided the closet is equipped with a listed, gasketed door assembly, and a listed self-closing device. The self-closing door assembly shall comply with the requirements of Section 904.1.1. The door assembly shall be installed with a threshold and bottom door seal and shall comply with the requirements of Section 904.1.2. Combustion air for such installations shall be obtained from the outdoors. The closet shall be for the exclusive use of the central heating furnace or low-pressure boiler.

(2) Central heating furnaces and low-pressure boilers shall be of the direct vent type.

904.1.1 Self-Closing Doors. Self-closing doors shall swing easily and freely, and shall be equipped with a self-closing device to cause the door to close and latch each time it is opened. The closing mechanism shall not have a hold-open feature.

904.1.2 Gasketing. Gasketing on gasketed doors or frames shall be furnished in accordance with the published listings of the door, frame, or gasketing material manufacturer.

Exception: Where acceptable to the Authority Having Jurisdiction, gasketing of noncombustible or limited-combustible material shall be permitted to be applied to the frame, provided closing and latching of the door are not inhibited.

904.2 Clearance. Central heating furnaces and low-pressure boilers shall be provided with clearances in accordance with Section 904.2.1 through Section 904.2.7.

904.2.1 Listed Units. Listed central heating furnaces and low-pressure boilers shall be installed with clearances in accordance with the terms of their listings and the manufacturer's installation instructions.

904.2.2 Unlisted Units. Unlisted central heating furnaces and low-pressure boilers shall be installed with clearances from combustible material not less than those specified in Table 904.2.2. [NFPA 54:10.3.2.2]

904.2.3 Listed and Unlisted Units. Listed and unlisted central heating furnaces and low-pressure boilers shall be permitted to be installed with reduced clearances to combustible material, provided that the combustible material or appliance is protected as described in Table 303.10.1 and **Figure 303.10.1(1) through Figure 303.10.1(3)**, and such reduction is allowed by the manufacturer's installation instructions. [NFPA 54:10.3.2.3]

904.2.4 Front Clearance. Front clearance shall be sufficient for servicing the burner and the furnace or boiler. [NFPA 54:10.3.2.4]

904.2.5 Adjacent to Plaster or Noncombustible Materials. Where the furnace plenum is adjacent to plaster on metal lath or noncombustible material attached to combustible material, the clearance shall be measured to the surface of the plaster or other noncombustible finish where the clearance specified is 2 inches (51 mm) or less. [NFPA 54:10.3.2.5]

904.2.6 Interference. The clearances to these appliances shall not interfere with combustion air, draft hood clearance and relief, and accessibility for servicing. [NFPA 54:10.3.2.6]

904.2.7 Central Heating Furnaces. Central heating furnaces other than those listed in Section 603.13.2 or Section 603.13.3 shall have clearances from the supply ducts of not less than 18 inches (457 mm) from the furnace plenum for the first 3 feet (914 mm), then 6 inches (152 mm) for the next 3 feet (914 mm) and 1 inch (25.4 mm) beyond 6 feet (1829 mm). [NFPA 54:10.3.2.9]

904.3 Assembly and Installation. A central heating boiler or furnace shall be installed in accordance with the manufacturer's instructions in one of the following manners:

(1) On a floor of noncombustible construction with noncombustible flooring and surface finish and with no combustible material against the underside thereof.

(2) On fire-resistive slabs or arches having no combustible material against the underside thereof.

Exceptions:

(1) Appliances listed for installation on a combustible floor.

(2) Installation on a floor protected in an approved manner. [NFPA 54:10.3.3]

904.3.1 Under-Floor Installation. Furnaces installed in an under-floor area of the building shall be in accordance with the Section 904.3.1.1 through Section 904.3.1.3.

904.3.1.1 Supported by Ground. Where a furnace is supported by the ground, it shall be installed on a concrete slab not less than 3 inches (76 mm) above the adjoining ground level.

904.3.1.2 Supported from Above. Where a furnace is supported from above, a clearance of not less than 6 inches (152 mm) shall be provided from finished grade.

904.3.1.3 Excavation. Where excavation is necessary to install a furnace, it shall be installed in accordance with Section 303.11.

904.4 Temperature or Pressure Limiting Devices. Steam and hot water boilers, respectively, shall be provided with approved automatic limiting devices for shutting down the burner(s) to prevent boiler steam pressure or boiler water temperature from exceeding the maximum allowable working pressure or temperature. Safety limit controls shall not be used as operating controls. [NFPA 54:10.3.4]

904.5 Low-Water Cutoff. All water boilers and steam boilers shall be provided with an automatic means to shut off the fuel supply to the burner(s) if the boiler water level drops below the lowest safe water line. In lieu of the low-water cutoff, water tube or coil-type boilers that require forced circulation to prevent overheating and failure shall have an approved flow sensing device arranged to shut down the boiler when the flow rate is inadequate to protect the boiler against overheating. [NFPA 54:10.3.5]

TABLE 904.2.2
CLEARANCES TO COMBUSTIBLE MATERIAL FOR UNLISTED FURNACES AND BOILERS*
[NFPA 54: TABLE 10.3.2.2]

	MINIMUM CLEARANCE (inches)					
APPLIANCE	ABOVE AND SIDES OF FURNACE PLENUM	TOP OF BOILER	JACKET SIDES AND REAR	FRONT	DRAFT HOOD AND BAROMETRIC DRAFT REGULATOR	SINGLE-WALL VENT CONNECTOR
1. Automatically fired, forced air or gravity system, equipped with temperature limit control that cannot be set higher than 250°F.	6	–	6	18	6	18
2. Automatically fired heating boilers – steam boilers operating at not over 15 pounds-force per square inch (psi) and hot water boilers operating at 250°F or less.	6	6	6	18	18	18
3. Central heating boilers and furnaces, other than in 1 or 2.	18	18	18	18	18	18

For SI units: 1 inch = 25.4 mm, °C = (°F-32)/1.8, 1 pound-force per square inch = 6.8947 kPa
* See Section 904.1 for additional requirements for central heating boilers and furnaces.

904.6 Steam Safety and Pressure Relief Valves. Steam and hot water boilers shall be equipped, respectively, with listed or approved steam safety or pressure relief valves of appropriate discharge capacity and conforming with ASME requirements. A shutoff valve shall not be placed between the relief valve and the boiler or on discharge pipes between such valves and the atmosphere. [NFPA 54:10.3.6]

904.6.1 Discharge. Relief valves shall be piped to discharge near the floor. [NFPA 54:10.3.6.1]

904.6.2 Size. The entire discharged piping shall be at least the same size as the relief valve discharge piping. [NFPA 54:10.3.6.2]

904.6.3 End Connections. Discharge piping shall not contain threaded end connection at its termination point. [NFPA 54:10.3.6.3]

904.7 Refrigeration Coils. The installation of refrigeration coils shall comply with the following requirements:

(1) A refrigeration coil shall not be installed in conjunction with a forced air furnace where circulation of cooled air is provided by the furnace blower, unless the blower has sufficient capacity to overcome the external static pressure resistance imposed by the duct system and refrigeration coil at the air flow rate for heating or cooling, whichever is greater.

(2) Furnaces shall not be located upstream from refrigeration coils, unless the refrigeration coil is designed or equipped so as not to develop excessive temperature or pressure.

(3) Refrigeration coils shall be installed in parallel with or on the downstream side of central furnaces to avoid condensation in the heating element, unless the furnace has been specifically listed for downstream installation. With a parallel flow arrangement, the dampers or other means used to control flow of air shall be sufficiently tight to prevent any circulation of cooled air through the furnace.

(4) Means shall be provided for disposal of condensate and to prevent dripping of condensate on the heating element. [NFPA 54:10.3.8]

Before cooling coils are inserted in the air discharge ducts, it is necessary to ensure that the fan and motor have sufficient capacity to overcome the external static resistance of the cooling coils. Additional horsepower and a compatible fan design are frequently necessary when cooling coils are added to an existing heating system. This section ensures adequate airflow across the heat exchanger and cooling unit. If insufficient air moves across a heat exchanger, an overheating condition may occur. If insufficient air moves across the cooling coils, liquid refrigerant may cause damage to the compressor.

Furnaces shall not be located "upstream" from cooling units, unless the cooling unit is designed so as not to develop excessive temperature or pressure. To avoid condensate collection at the heat exchanger or element and eventual damage by oxidation, cooling coils shall be installed parallel with or on the "downstream" side of the heating element, unless permitted by design. An approved means of condensate disposal shall be provided.

Similar reasoning is applied to cooling systems used in conjunction with hot water boilers. To prevent damage, no chilled medium shall be allowed to enter the boiler. Appropriate flow controls shall be installed to prevent gravity circulation of the boiler water during the cooling cycle.

904.8 Cooling Units Used with Heating Boilers. Boilers, where used in conjunction with refrigeration systems, shall be installed so that the chilled medium is piped in parallel with the heating boiler with appropriate valves to prevent the chilled medium from entering the heating boiler. [NFPA 54:10.3.9.1]

904.8.1 Exposed to Refrigerated Air Circulation. Where hot water heating boilers are connected to heating coils located in air-handling units where they can be exposed to refrigerated air circulation, such boiler piping systems shall be equipped with flow control valves or other automatic means to prevent gravity circulation of the boiler water during the cooling cycle. [NFPA 54:10.3.9.2]

904.9 Furnace (Upright and Horizontal). Upright furnaces shall be permitted to be installed in an attic, furred, or under-floor space exceeding 5 feet (1524 mm) in height, provided the required listings and furnace and duct clearances are observed. Horizontal furnaces shall be permitted to be installed in an attic, furred, or under-floor space, provided the required listings and furnace and duct clearances are observed.

904.10 Solid-Fuel-Fired Furnaces. Factory-built solid-fuel-fired furnaces shall comply with UL 391 and shall be installed in accordance with the manufacturer's installation instructions.

904.11 Oil-Fired Central Furnaces. Oil-fired central furnaces shall comply with UL 727 and shall be installed in accordance with the manufacturer's installation instructions.

904.12 Commercial or Industrial Gas Heaters. Commercial or industrial gas-fired heaters shall comply with UL 795 and shall be installed in accordance with the manufacturer's installation instructions.

904.13 Electric Central Furnaces. Electric central heating furnaces shall comply with UL 1995 or UL 60335-2-40 and shall be installed in accordance with the manufacturer's installation instructions.

905.0 Duct Furnaces.

A duct furnace is a warm-air furnace installed in an air supply or distribution duct where the air circulation depends on an independent blower not furnished as part of the furnace (see Section 206.0). The duct furnace shall be installed on the positive pressure side of the blower (see Section 905.5). This type of furnace is required to be installed with a minimum clearance of 6 inches between the appliance and adjacent walls, ceilings, and floors constructed of combustible materials. If lesser clearance is allowed by the appliance listing, code requirements for accessibility and combustion air shall be considered (see Sections 304.0 Accessibility for Service and 701.1). **Figure 905.0** is an example of a duct furnace.

905.1 Clearances. The installation of duct furnaces shall comply with the following clearance requirements:

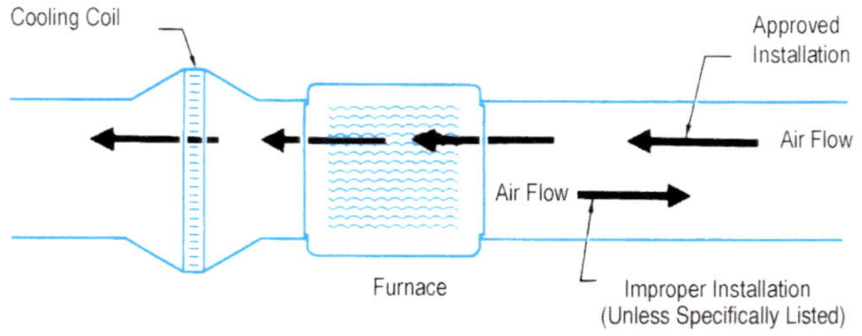

FIGURE 905.0
DUCT FURNACE

(1) Listed duct furnaces shall be installed with clearances of not less than 6 inches (152 mm) between adjacent walls, ceilings, and floors of combustible material and the furnace draft hood. Furnaces listed for installation at lesser clearances shall be installed in accordance with their listings and the manufacturer's installation instructions. In no case shall the clearance be such as to interfere with combustion air and accessibility.

(2) Unlisted duct furnaces shall be installed with clearances to combustible material in accordance with the clearances specified for unlisted furnaces and boilers in Table 904.2.2. Combustible floors under unlisted duct furnaces shall be protected in an approved manner.

905.2 Installation of Duct Furnaces. Duct furnaces shall be installed in accordance with the manufacturer's instructions. [NFPA 54:10.10.2]

905.3 Access Panels. The ducts connected to duct furnaces shall have removable access panels on both the upstream and downstream sides of the furnace. [NFPA 54:10.10.3]

905.4 Location of Draft Hoods and Controls. The controls, combustion air inlet, and draft hoods for duct furnaces shall be located outside the ducts. The draft hood shall be located in the same enclosure from which combustion air is taken. [NFPA 54:10.10.4]

905.5 Circulating Air. Where a duct furnace is installed so that supply ducts carry air circulated by the furnace to areas outside the space containing the furnace, the return air shall also be handled by a duct(s) sealed to the furnace casing and terminating outside the space containing the furnace. The duct furnace shall be installed on the positive-pressure side of the circulating air blower. [NFPA 54:10.10.5]

905.6 Duct Furnaces Used with Refrigeration Systems. A duct furnace shall not be installed in conjunction with a refrigeration coil where circulation of cooled air is provided by the blower.

Exception: Where the blower has sufficient capacity to overcome the external static resistance imposed by the duct system, furnace, and the cooling coil and the air throughput necessary for heating or cooling, whichever is greater. [NFPA 54:10.10.6.1]

A duct furnace shall not be installed in conjunction with a refrigeration coil where cooled air is circulated by a blower unless all the requirements of Section 905.6 and the manufacturer's instructions are followed.

905.6.1 In Conjunction with Cooling Appliances. Duct furnaces used in conjunction with cooling appliances shall be installed in parallel with or on the upstream side of cooling coils to avoid condensation within heating elements. With a parallel flow arrangement, the dampers or other means used to control the flow of air shall be sufficiently tight to prevent any circulation of cooled air through the unit.

Exception: Where the duct furnace has been specifically listed for downstream installation. [NFPA 54:10.10.6.2]

905.6.2 Located Upstream from Cooling Coils. Where duct furnaces are to be located upstream from cooling units, the cooling unit shall be so designed or equipped as to not develop excessive temperatures or pressures. [NFPA 54:10.10.6.3]

905.6.3 Heat Exchangers. Where a duct furnace is installed downstream of an evaporative cooler or air washer, the heat exchanger shall be constructed of corrosion-resistant materials. Stainless steel, ceramic-coated steel, and an aluminum-coated steel in which the bond between the steel and the aluminum is an iron-aluminum alloy are considered to be corrosion resistant. Air washers operating with chilled water that deliver air below the dew point of the ambient air at the duct furnace shall be considered as refrigeration systems. [NFPA 54:10.10.6.4]

905.7 Installation in Commercial Garages and Aircraft Hangars. Duct furnaces installed in garages for more than three motor vehicles or in aircraft hangars shall be of a listed type and shall be installed in accordance with Section 303.11 and Section 303.12. [NFPA 54:10.10.7]

Gas utilization equipment installed in commercial garages, aircraft hangars, parking structures and repair garages shall be installed according to the NFPA standard for such applications.

905.8 Electric Duct Heaters. Electric duct heaters installed within an air duct shall be listed and labeled in accordance with UL 1996 and designed for the maximum air temperature. The duct heater and fan shall be interlocked such that the electric duct heater operates when the fan is operating.

905.8.1 Installation. Duct heaters shall be installed in accordance with the manufacturer's installation instructions, and shall not create a hazard to persons or property. Where installed 4 feet (1219 mm) or less from a heat pump or air conditioner, the duct heater shall be listed for such installation.

905.8.2 Clearance. A working space clearance shall be maintained to permit replacement of controls and heating elements and for adjusting and cleaning of controls. The working space for energized equipment shall comply with NFPA 70.

906.0 Floor Furnaces.

906.1 Installation. The installation of floor furnaces shall comply with the following requirements:

(1) Listed floor furnaces shall be installed in accordance with their listing and the manufacturer's installation instructions.

(2) Unlisted floor furnaces shall not be installed on combustible floors.

(3) Thermostats controlling floor furnaces shall not be located in a room or space that is capable of being sepa-

rated from the room or space in which the register of the floor furnace is located.

906.2 Temperature Limit Controls. Floor furnaces shall be provided with temperature limit controls in accordance with the following requirements:

(1) Listed automatically operated floor furnaces shall be equipped with temperature limit controls. [NFPA 54:10.11.2.1]

(2) Unlisted automatically operated floor furnaces shall be equipped with a temperature limit control arranged to shut off the flow of gas to the burner in the event the temperature at the warm air outlet register exceeds 350°F (177°C) above room temperature. [NFPA 54:10.11.2.2]

906.3 Combustion and Circulating Air. Combustion and circulating air shall be provided in accordance with Section 701.0. [NFPA 54:10.11.3]

906.4 Placement. The following provisions apply to furnaces that serve one story:

(1) Floors. Floor furnaces shall not be installed in the floor of any doorway, stairway landing, aisle, or passageway of any enclosure, public or private, or in an exitway from any such room or space.

(2) Walls and Corners. The register of a floor furnace with a horizontal warm air outlet shall not be placed closer than 6 inches (152 mm) from the nearest wall. A distance of at least 18 inches (457 mm) from two adjoining sides of the floor furnace register to walls shall be provided to eliminate the necessity of occupants walking over the warm air discharge. The remaining sides shall be a minimum of 6 inches (152 mm) from a wall. Wall register models shall not be placed closer than 6 inches (152 mm) to a corner.

(3) Draperies. The furnace shall be placed so that a door, drapery, or similar object cannot be nearer than 12 inches (305 mm) to any portion of the register of the furnace. [NFPA 54:10.11.4]

906.5 Bracing. The space provided for the furnace shall be framed with doubled joists and with headers not lighter than the joists. [NFPA 54:10.11.5]

906.6 Support. Means shall be provided to support the furnace when the floor register is removed. [NFPA 54:10.11.6]

906.7 Clearance. The lowest portion of the floor furnace shall have at least a 6 inch (152 mm) clearance from the general ground level. A reduced clearance to a minimum of 2 inches (51 mm) shall be permitted, provided the lower 6 inches (152 mm) portion of the floor furnace is sealed by the manufacturer to prevent entrance of water. Where these clearances are not present, the ground below and to the sides shall be excavated to form a "basin-like" pit under the furnace so that the required clearance is provided beneath the lowest portion of the furnace. A 12 inch (305 mm) clearance shall be provided on all sides except the control side, which shall have an 18 inch (457 mm) clearance. [NFPA 54:10.11.7]

906.8 Access. The space in which any floor furnace is installed shall be accessible by an opening in the foundation not less than 24 inches by 18 inches (610 mm by 457 mm) or by a trapdoor not less than 24 inches by 24 inches (610 mm by 610 mm) in any cross-section thereof, and a passageway not less than 24 inches by 18 inches (610 mm by 457 mm) in any cross-section thereof. [NFPA 54:10.11.8]

906.9 Seepage Pan. Where the excavation exceeds 12 inches (305 mm) in depth or water seepage is likely to collect, a watertight copper pan, concrete pit, or other suitable material shall be used, unless adequate drainage is provided or the appliance is sealed by the manufacturer to meet this condition. A copper pan shall be made of not less than 16 ounces per square foot (oz/ft^2) (4.9 kg/m^2) sheet copper. The pan shall be anchored in place so as to prevent floating, and the walls shall extend at least 4 inches (102 mm) above the ground level with at least a 6 inches (152 mm) clearance on all sides, except on the control side, which shall have at least an 18 inch (457 mm) clearance. [NFPA 54:10.11.9]

906.10 Wind Protection. Floor furnaces shall be protected, where necessary, against severe wind conditions. [NFPA 54:10.11.10]

906.11 Upper Floor Installations. Listed floor furnaces shall be permitted to be installed in an upper floor, provided the furnace assembly projects below into a utility room, closet, garage, or similar nonhabitable space. In such installations, the floor furnace shall be enclosed completely (entirely separated from the nonhabitable space) with means for air intake to meet the provisions of Section 701.0, with access for servicing, minimum furnace clearances of 6 inches (152 mm) to all sides and bottom, and with the enclosure constructed of Portland cement plaster or metal lath or other noncombustible material. [NFPA 54:10.11.11]

906.12 First Floor Installation. Listed floor furnaces installed in the first or ground floors of buildings shall not be required to be enclosed unless the basements of these buildings have been converted to apartments or sleeping quarters, in which case the floor furnace shall be enclosed as specified for upper floor installations and shall project into a nonhabitable space. [NFPA 54:10.11.12]

This section ensures that floor furnaces are installed with adequate clearances to combustible materials, access for maintenance and repair and under-floor space for venting and combustion air. A "floor furnace is a completely self-contained unit suspended from the floor of the space being heated," where combustion air is provided from outside the space being heated.

The three basic types of furnaces available include; the flat-floor furnace with the supply air grille extending slightly above and parallel to the floor; the wall-register floor furnace with an L-shaped grille installed parallel to and perpendicular to the floor (a wall-register model shall not be placed closer than 6 inches to a corner); and the dual wall register, which provides heat to both sides of a wall with two L-shaped grilles centered on a single partition. Two examples are illustrated in **Figure 906.12a**.

A floor furnace shall not be installed in the floor of any doorway, stairway landing, aisle or passageway of any public or private enclosure, or in an exit from any such rooms or space (see Section 906.4). Clearance and access requirements for a typical installation, above and below the floor, are illustrated by **Figures 906.12b, 906.12c and 906.12d**.

A floor furnace is prohibited in a cement slab (see **Figure 906.12e**). As for all gas appliances, refer to the manufacturer's instructions for a safe installation.

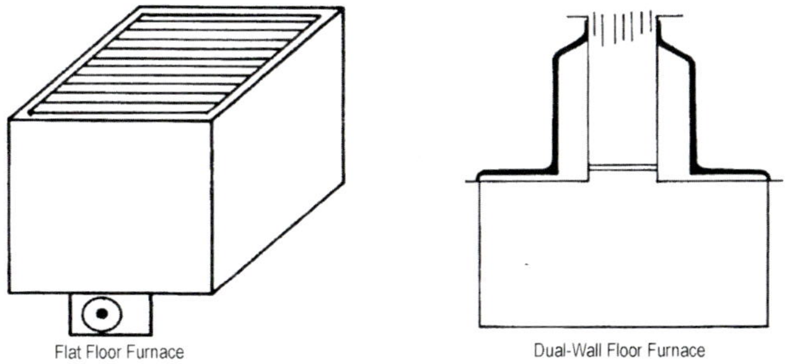

FIGURE 906.12A
EXAMPLES OF FLOOR FURNACES

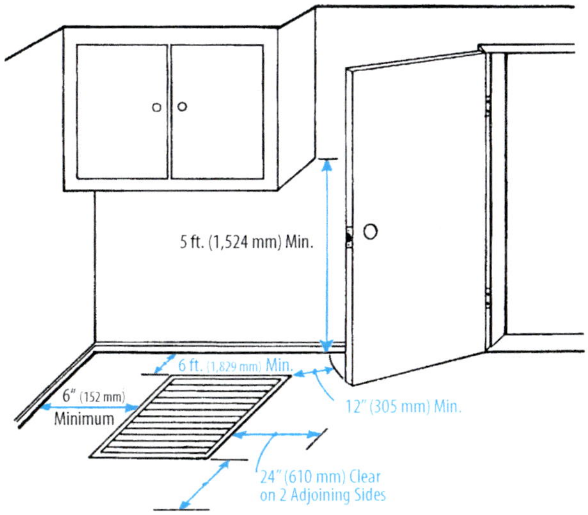

FIGURE 906.12B
FLAT FLOOR FURNACE - MINIMUM CLEARANCE

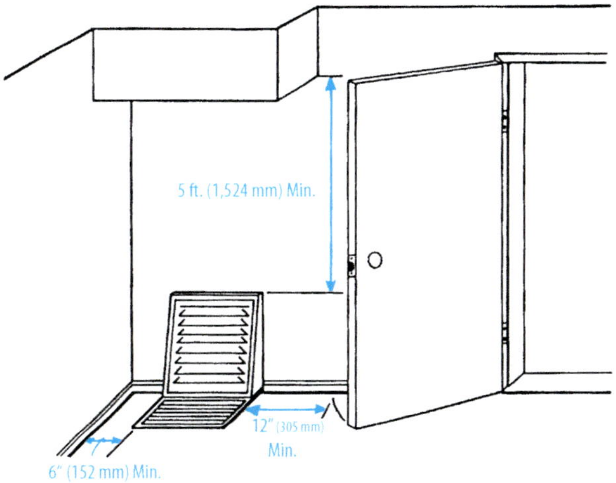

FIGURE 906.12C
WALL REGISTER FLOOR FURNACE - MINIMUM CLEARANCE

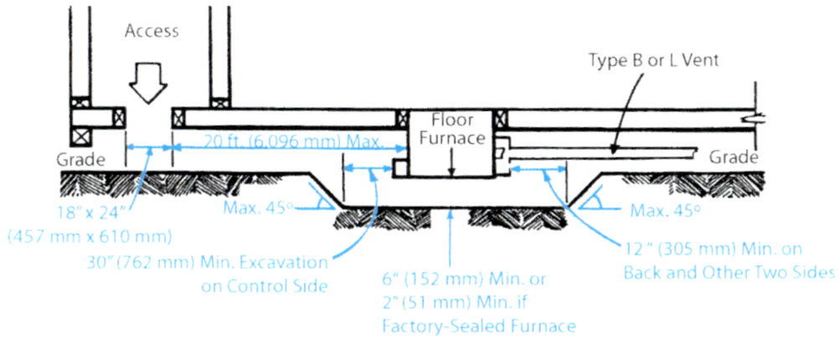

FIGURE 906.12D
FLOOR FURNACE - REQUIRED ACCESS AND CLEARANCES

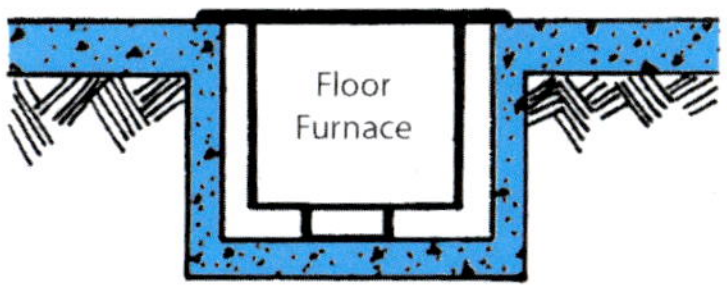

FIGURE 906.12E
FLOOR FURNACE ON SLAB - PROHIBITED

906.13 Oil-Fired Floor Furnaces. Oil-fired floor furnaces shall comply with UL 729 and installed in accordance with the manufacturer's installation instructions.

907.0 Wall Furnaces.

907.1 Installation. Listed wall furnaces shall be installed in accordance with their listings and the manufacturer's installation instructions. Wall furnaces installed in or attached to combustible material shall be listed for such installation.

907.1.1 Unlisted Wall Furnaces. Unlisted wall furnaces shall not be installed in or attached to combustible material. [NFPA 54:10.26.1.2]

907.1.2 Vented Wall Furnaces. Vented wall furnaces connected to a Type B-W gas vent system listed only for a single story shall be installed only in single-story buildings or the top story of multistory buildings. Vented wall furnaces connected to a Type B-W gas vent system listed for installation in multistory buildings shall be permitted to be installed in single-story or multistory buildings. Type B-W gas vents shall be attached directly to a solid header plate that serves as a firestop at that point and that shall be permitted to be an integral part of the vented wall furnace, as illustrated in **Figure 907.1.2**. The stud space in which the vented wall furnace is installed shall be ventilated at the first ceiling level by installation of the ceiling plate spacers furnished with the gas vent. Firestop spacers shall be installed at each subsequent ceiling or floor level penetrated by the vent. [NFPA 54:10.26.1.3]

907.1.3 Direct Vent Wall Furnaces. Direct vent wall furnaces shall be installed with the vent air intake terminal in the outdoors. The thickness of the walls on which the furnace is mounted shall be within the range of wall thickness marked on the furnace and covered in the manufacturer's installation instructions. [NFPA 54:10.26.1.4]

907.1.4 Panels, Grilles, and Access Doors. Panels, grilles, and access doors that are required to be removed for normal servicing operations shall not be attached to the build-

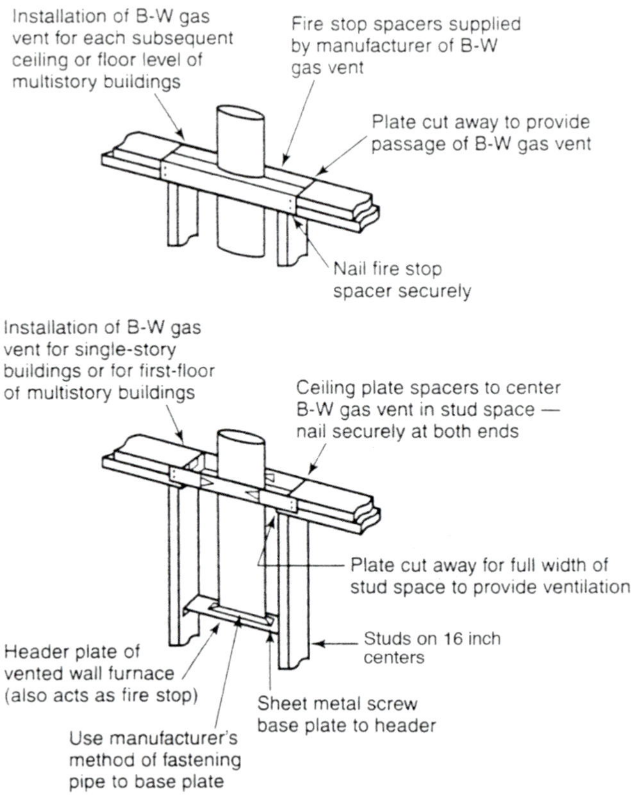

For SI units: 1 inch = 25.4 mm

FIGURE 907.1.2
INSTALLATION OF TYPE B-W GAS VENTS FOR VENTED WALL FURNACES
[NFPA 54: FIGURE 10.26.1.3]

ing. For additional information on the venting of wall furnaces, see Section 802.0. [NFPA 54:10.26.1.5]

907.2 Location. Wall furnaces shall be located so as not to cause a hazard to walls, floors, curtains, furniture, or doors. Wall furnaces installed between bathrooms and adjoining rooms shall not circulate air from bathrooms to other parts of the building. [NFPA 54:10.26.2]

907.3 Combustion and Circulating Air. Combustion and circulating air shall be provided in accordance with Section 701.0. [NFPA 54:10.26.3]

907.4 Oil-Fired Wall Furnaces. Oil-fired wall furnaces shall comply with UL 730 and installed in accordance with the manufacturer's installation instructions.

908.0 Clothes Dryers.

908.1 Electric Clothes Dryers. Commercial electric clothes dryers shall comply with UL 1240 and installed in accordance with the manufacturer's installation instructions. Residential and coin-operated electric clothes dryers shall comply with UL 2158 and installed in accordance with the manufacturer's installation instructions.

908.2 Gas-Fired Clothes Dryers. Gas-fired clothes dryers shall comply with Section 908.2.1 through Section 908.2.3.

908.2.1 Clearance. The installation of clothes dryers shall comply with the following requirements:

(1) Listed Type 1 clothes dryers shall be installed with a clearance of not less than 6 inches (152 mm) from adjacent combustible material. Clothes dryers listed for installation at reduced clearances shall be installed in accordance with their listing and the manufacturer's installation instructions. Type 1 clothes dryers installed in closets shall be listed for such installation.

(2) Listed Type 2 clothes dryers shall be installed with clearances of not less than that shown on the marking plate and in the manufacturer's instructions. Type 2 clothes dryers designed and marked, "For use only in noncombustible locations," shall not be installed elsewhere.

(3) Unlisted clothes dryers shall be installed with clearances to combustible material of not less than 18 inches (457 mm). Combustible floors under unlisted clothes dryers shall be protected in an approved manner.

908.2.2 Exhausting to the Outdoors. Type 1 and Type 2 clothes dryers shall be exhausted to the outside air in accordance with Section 504.4.

908.2.3 Multiple-Family or Public Use. All clothes dryers installed for multiple-family or public use shall be equipped with approved safety shutoff devices and shall be installed as specified for a Type 2 clothes dryer under Section 504.4.3.1. [NFPA 54:10.4.6]

A Type 1 clothes dryer is primarily manufactured for use in a single-family environment and may be coin-operated. A Type 2 clothes dryer is designed and listed for use in a business setting utilized by the public (see Section 205.0). General installation and venting requirements are specified in the code and the manufacturer's installation instructions for Type 1 and 2 dryers.

All dryers shall be exhausted to the outside atmosphere. Rigid duct is to be properly supported to prevent sagging or "traps" and the duct joints sealed in an approved manner. All dryer ducts shall be constructed of metal with a smooth interior surface (see Section 504.0, Environmental Air Ducts). No dryer duct shall be assembled with screws or other fastening methods that penetrate the duct, which can cause lint to accumulate. Lint collection in the duct causes improper dryer operation and a fire hazard. See Section 504.4.2.1, Length Limitation, for length limitations.

909.0 Conversion Burners.

909.1 General. Installation of conversion burners shall conform to CSA Z21.8. [NFPA 54:10.5]

910.0 Burner Assemblies.

910.1 Oil Burners. Oil burners shall comply with UL 296 and installed in accordance with the manufacturer's installation instructions.

910.2 Gas Burners. Commercial gas burners shall comply with UL 295 and installed in accordance with the manufacturer's installation instructions.

911.0 Decorative Appliances for Installation in Vented Fireplaces.

911.1 Prohibited Installations. Decorative appliances for installation in vented fireplaces shall not be installed in bathrooms or bedrooms unless the appliance is listed and the bedroom or bathroom has the required volume in accordance with Section 701.4. [NFPA 54:10.6.1]

A decorative appliance is "a vented appliance" where the flames provide an esthetic effect (see Section 224.0). A vented decorative appliance usually resembles a solid-fuel-burning fireplace in appearance; however, it is important to realize that it is a gas-burning appliance and is subject to all applicable installation standards. This type of appliance is usually not listed for use as a source of comfort heat.

Installation shall be prohibited in bathrooms or bedrooms unless listed for that purpose and if the room size complies with Section 701.4, Indoor Combustion Air, for required volume.

911.2 Installation. A decorative appliance for installation in a vented fireplace shall be installed only in a vented fireplace having a working chimney flue and constructed of noncombustible materials. These appliances shall not be thermostatically controlled. [NFPA 54:10.6.2]

Examples of decorative appliances are shown in **Figure 911.2**.

911.2.1 Listed Decorative Appliance. A listed decorative appliance for installation in a vented fireplace shall be installed in accordance with its listing and the manufacturer's installation instructions.

911.2.2 In Manufactured Homes. A decorative appliance for installation in a vented fireplace, where installed in a manufactured home, shall be listed for installation in manufactured homes. [NFPA 54:10.6.2.2]

911.2.3 Unlisted Decorative Appliance. An unlisted decorative appliance for installation in a vented fireplace shall be installed in a fireplace having a permanent free opening, based on appliance input rating and chimney height, equal to or greater than that specified in Table 911.2. [NFPA 54:10.6.2.3]

911.3 Fireplace Screens. A fireplace screen shall be installed with a decorative appliance for installation in a vented fireplace. [NFPA 54:10.6.3]

912.0 Gas Fireplaces, Vented.

912.1 Prohibited Installations. Vented gas fireplaces shall not be installed in bathrooms or bedrooms unless the appliance is listed and the bedroom or bathroom has the required volume in accordance with Section 701.4.

Exception: Direct vent gas fireplaces. [NFPA 54:10.7.1]

912.2 Installation. The installation of vented gas fireplaces shall comply with the following requirements:

(1) Listed vented gas fireplaces shall be installed in accordance with their listing and the manufacturer's installation instructions and where installed in or attached to combustible material shall be specifically listed for such installation.

(2) Unlisted vented gas fireplaces shall not be installed in or attached to combustible material. They shall have a clearance at the sides and rear of not less than 18 inches (457 mm). Combustible floors under unlisted vented gas fireplaces shall be protected in an approved manner. Unlisted appliances of other than the direct vent type shall be

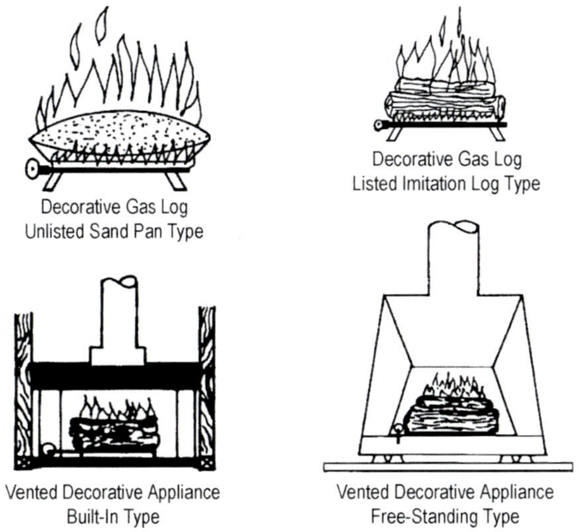

FIGURE 911.2
DECORATIVE APPLIANCES

TABLE 911.2
FREE OPENING AREA OF CHIMNEY DAMPER FOR VENTING FLUE GASES FROM UNLISTED DECORATIVE APPLIANCES FOR INSTALLATION IN VENTED FIREPLACES
[NFPA 54: TABLE 10.6.2.3]

CHIMNEY HEIGHT (feet)	MINIMUM PERMANENT FREE OPENING (square inches)*						
	8	13	20	29	39	51	64
	APPLIANCE INPUT RATING (Btu/h)						
6	7800	14 000	23 200	34 000	46 400	62 400	80 000
8	8400	15 200	25 200	37 000	50 400	68 000	86 000
10	9000	16 800	27 600	40 400	55 800	74 400	96 400
15	9800	18 200	30 200	44 600	62 400	84 000	108 800
20	10 600	20 200	32 600	50 400	68 400	94 000	122 200
30	11 200	21 600	36 600	55 200	76 800	105 800	138 600

For SI units: 1 foot = 304.8 mm, 1000 British thermal units per hour = 0.293 kW, 1 square inch = 0.000645 m^2

* The first six minimum permanent free openings [8 square inches (0.005 m^2) to 51 square inches (0.03 m^2)] correspond approximately to the cross-sectional areas of chimneys having diameters of 3 inches (76 mm) through 8 inches (203 mm), respectively. The 64 square inch (0.04 m^2) opening corresponds to the cross-sectional area of a standard 8 inch (203 mm) by 8 inch (203 mm) chimney tile.

equipped with a draft hood and shall be vented in accordance with Section 802.0. Appliances that use metal, asbestos, or ceramic material to direct radiation to the front of the appliance shall have a clearance of 36 inches (914 mm) in front and, where constructed with a double back of metal or ceramic, shall be installed with a clearance of not less than 18 inches (457 mm) at the sides and 12 inches (305 mm) at the rear.

(3) Panels, grilles, and access doors that are required to be removed for normal servicing operations shall not be attached to the building.

(4) Direct vent gas fireplaces shall be installed with the vent-air intake terminal in the outdoors and in accordance with the manufacturer's installation instructions.

912.3 Combustion and Circulating Air. Combustion and circulating air shall be provided in accordance with Section 701.0. [NFPA 54:10.7.3]

913.0 Factory-Built Fireplaces and Fireplace Stoves.

913.1 Factory-Built Fireplaces. Factory-built fireplaces shall comply with UL 127 and installed in accordance with the manufacturer's installation instructions.

913.1.1 Gasketed Fireplace Doors. A gasketed fireplace door shall not be installed on a factory-built fireplace, except where the fireplace system has been tested in accordance with UL 127.

913.2 Fireplace Stoves. Fireplace stoves shall comply with UL 737 and installed in accordance with the manufacturer's installation instructions.

913.3 Fireplace Accessories. Heat exchangers, glass doors assemblies, combustion air vents, and termination caps shall comply with UL 907 and installed in accordance with the manufacturer's installation instructions.

914.0 Non-Recirculating Direct Gas-Fired Industrial Air Heaters.

914.1 Application. Direct gas-fired industrial air heaters of the non-recirculating type shall be listed in accordance with CSA Z83.4. [NFPA 54:10.8.1]

An industrial direct-gas-fired heater is "a heater in which all the products of combustion generated by the burners are released into the airstream being heated."

For various prescriptive code requirements, such as clearances to combustibles and accessibility, pertaining to the installation of nonrecirculating and circulating direct-gas-fired industrial heaters, refer to the code, the appliance rating plate, and the manufacturers' installation instructions. **Figure 914.1** illustrates an industrial-type makeup air unit.

The purpose of the nonrecirculating industrial air heater is to offset building heat loss by heating outside air only, while the circulating type offsets heat loss by heating outdoor air and, if applicable, indoor air.

These gas appliances are permitted in industrial or commercial occupancies only and, without exception, shall not serve any area serving sleeping quarters.

914.2 Prohibited Installations. Non-recirculating direct gas-fired industrial air heaters shall not serve any area containing sleeping quarters. Non-recirculating direct gas-fired industrial air heaters shall not recirculate room air. [NFPA 54:10.8.2.1, 10.8.2.2]

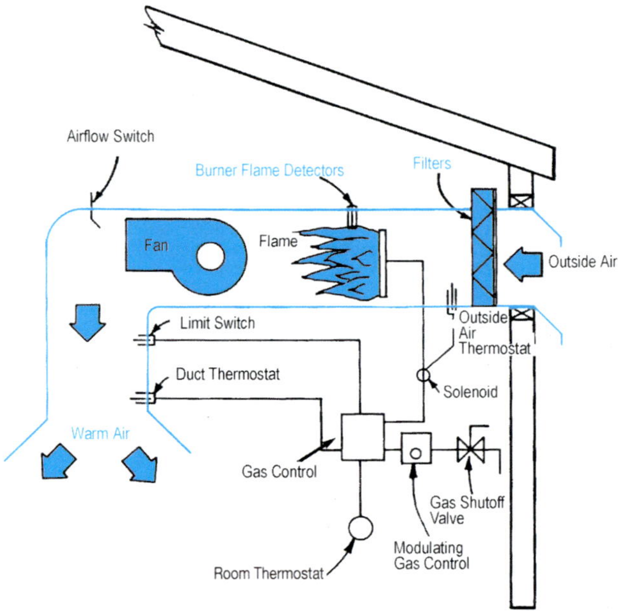

FIGURE 914.1
DIRECT-FIRED MAKEUP AIR HEATER

914.3 Installation. Non-recirculating direct gas-fired industrial air heaters shall be installed in accordance with the manufacturer's instructions. [NFPA 54:10.8.3.1]

914.3.1 Fresh Air Ventilation. Non-recirculating direct gas-fired industrial air heaters shall be permitted to provide fresh air ventilation. [NFPA 54:10.8.3.2]

914.3.2 Access Required. Non-recirculating direct gas-fired industrial air heaters shall be provided with access for removal of burners; for replacement of motors, controls, filters, and other working parts; and for adjustment and lubrication of parts requiring maintenance. [NFPA 54:10.8.3.3]

914.4 Clearance from Combustible Materials. Non-recirculating direct gas-fired industrial air heaters shall be installed with a clearance from combustible materials of not less than that shown on the rating plate and the manufacturer's instructions. [NFPA 54:10.8.4]

914.5 Air Supply. All air to the non-recirculating direct gas-fired industrial air heater shall be ducted directly from outdoors. Where outdoor air dampers or closing louvers are used, they shall be verified to be in the open position prior to main burner operation. [NFPA 54:10.8.5]

914.6 Atmospheric Vents, Gas Reliefs, or Bleeds. Non-recirculating direct gas-fired industrial air heaters with valve train components equipped with atmospheric vents, gas reliefs, or bleeds shall have their vent lines, gas reliefs, or bleeds lead to a safe point outdoors. Means shall be employed on these lines to prevent water from entering and to prevent blockage from insects and foreign matter. An atmospheric vent line shall not be required to be provided on a valve train component equipped with a listed vent limiter. [NFPA 54:10.8.6]

914.7 Relief Openings. The design of the installation shall include adequate provisions to permit the non-recirculating direct gas-fired industrial air heater to operate at its rated airflow without overpressurizing the space served by the heater by taking into account the structure's designed infiltration rate, properly designed relief openings, or an interlocked powered exhaust system, or a combination of these methods. [NFPA 54:10.8.7]

Installation design shall include adequate provisions for combustion and relief air to permit efficient operation of the equipment. Generally, an engineered system is required to be submitted for plan review to the Authority Having Jurisdiction (AHJ). Most system designers take into consideration either the structural infiltration rate or an interlocked power exhaust system to provide relief air. If adequate relief air is not provided, the pressure within the building increases, the airflow rate of the heater decreases and a potentially unsafe condition is created.

Air to direct-gas-fired industrial air heaters may be taken from the building, directly from the outdoors or a combination of both. Relief openings may be louvers or counterbalanced gravity dampers. If the dampers are motorized, they must be electrically interlocked so the equipment does not operate unless the louvers or dampers are in the open position.

The recirculation of air by an industrial-type heater may be hazardous in the presence of flammable solids, liquids and gases; explosive materials such as grain and coal dust; gunpowder and substances that are toxic when exposed to heat, like refrigerants and aerosols. Recirculation is not recommended in uninsulated buildings where outside temperatures fall below 32°F (0°C).

914.7.1 Infiltration Rate. The structure's designed infiltration rate and the size of relief opening(s) shall be determined by approved engineering methods. [NFPA 54:10.8.7.1]

914.7.2 Louver or Gravity Dampers. Louver or counterbalanced gravity damper relief openings shall be permitted. Where motorized dampers or closeable louvers are used, they shall be proved to be in their open position prior to main burner operation. [NFPA 54:10.8.7.2]

914.8 Purging. Inlet ducting, when used, shall be purged with at least four air changes prior to an ignition attempt. [NFPA 54:10.8.8]

915.0 Recirculating Direct Gas-Fired Industrial Air Heaters.

915.1 Application. Direct gas-fired industrial air heaters of the recirculating type shall be listed in accordance with CSA Z83.18. [NFPA 54:10.9.1]

915.2 Prohibited Installations. Recirculating direct gas-fired industrial air heaters shall not serve any area containing sleeping quarters. Recirculating direct gas-fired industrial air heaters shall not recirculate room air in buildings that contain flammable solids, liquids, or gases; explosive materials; or substances that can become toxic when exposed to flame or heat. [NFPA 54:10.9.2.1, 10.9.2.2]

915.3 Installation. Recirculating direct gas-fired industrial air heaters shall be installed in accordance with the manufacturer's instructions. [NFPA 54:10.9.3]

915.4 Clearance from Combustible Materials. Recirculating direct gas-fired industrial air heaters shall be installed with a clearance from combustible materials of not less than that shown on the rating plate and the manufacturer's instructions. [NFPA 54:10.9.4]

915.5 Air Supply. Ventilation air to the recirculating direct gas-fired industrial air heater shall be ducted directly from outdoors. Air to the recirculating direct gas-fired industrial air heater in excess of the minimum ventilation air specified on the heater's rating plate shall be taken from the building, ducted directly from outdoors, or a combination of both. Where outdoor air dampers or closing louvers are used, they shall be verified to be in the open position prior to main burner operation. [NFPA 54:10.9.5]

915.6 Atmospheric Vents, Gas Reliefs, or Bleeds. Recirculating direct gas-fired industrial air heaters with valve train components equipped with atmospheric vents, gas reliefs, or bleeds shall have their vent lines, gas reliefs, or bleeds lead to a safe point outdoors. Means shall be employed on these lines to prevent water from entering and to prevent blockage from insects and foreign matter. An atmospheric vent line shall not be required to be provided on a valve train component equipped with a listed vent limiter. [NFPA 54:10.9.6]

915.7 Relief Openings. The design of the installation shall include adequate provisions to permit the recirculating direct gas-fired industrial air heater to operate at its rated airflow without overpressurizing the space served by the heater by taking into account the structure's designed infiltration rate, properly designed relief openings or an interlocked powered exhaust system, or a combination of these methods. [NFPA 54:10.9.7]

915.7.1 Infiltration Rate. The structure's designed infiltration rate and the size of relief opening(s) shall be determined by approved engineering methods. [NFPA 54:10.9.7.1]

915.7.2 Louver or Gravity Dampers. Louver or counterbalanced gravity damper relief openings shall be permitted. Where motorized dampers or closeable louvers are used, they shall be proved to be in their open position prior to main burner operation. [NFPA 54:10.9.7.2]

915.8 Purging. Inlet ducting, when used, shall be purged with at least four air changes prior to an ignition attempt. [NFPA 54:10.9.8]

916.0 Room Heaters.

916.1 Electric Room Heaters. Electric room heaters shall comply with UL 2021.

These electric room heaters are not portable but are permanently fixed. UL 2021 covers fixed and location-dedicated electric room heating equipment rated 600 volts or less to be employed in ordinary locations in accordance with the National Electrical Code, ANSI/NFPA 70. It does not cover movable heaters, wall- or ceiling-hung heaters, baseboard heaters, duct heaters, central-heating furnaces, fan-coil units, panel- or cable-type radiant-heating equipment, electric boilers, or any other electric heating equipment or appliances that are covered in or as a part of separate, individual requirements.

916.2 Gas-Fired Room Heaters. Gas-fired room heaters shall comply with Section 916.2.1 through Section 916.2.4.

916.2.1 Prohibited Installations. Unless specifically permitted by the Authority Having Jurisdiction, unvented room heaters shall not be installed as primary heat sources. Unvented room heaters shall not be permitted in spaces that do not have the required volume of indoor air as defined in Section 701.4.

916.2.1.1 Unvented Room Heaters. Unvented room heaters shall not be installed in bathrooms or bedrooms.

Exceptions:

(1) Where approved by the Authority Having Jurisdiction, one listed wall-mounted unvented room heater equipped with an oxygen depletion safety shutoff system shall be permitted to be installed in a bathroom, provided that the input rating does not exceed 6000 Btu/h (1.76 kW) and combustion and ventilation air is provided as specified in Section 902.2.

(2) Where approved by the Authority Having Jurisdiction, one listed wall-mounted unvented room heater equipped with an oxygen depletion safety shutoff system shall be permitted to be installed in a bedroom, provided that the input rating does not exceed 10 000 Btu/h (3 kW) and combustion and ventilation air is provided as specified in Section 902.2. [NFPA 54:10.22.1]

(3) Portable oil fired unvented heating appliances used as supplemental heating in storage occupancies, utility occupancies, and in accordance with the fire code.

916.2.2 Installations in Institutions. Room heaters shall not be installed in the following occupancies:

(1) Residential board and care

(2) Health care [NFPA 54:10.22.3]

916.2.3 Clearance. A room heater shall be placed so as not to cause a hazard to walls, floors, curtains, furniture, doors where open, and to the free movements of persons within the room. Heaters designed and marked, "For use in noncombustible fireplace only," shall not be installed elsewhere. Listed room heaters shall be installed in accordance with their listings and the manufacturer's installation instructions. In no case shall the clearances be such as to interfere with combustion air and accessibility. Unlisted room heaters shall be installed with clearances from combustible material not less than the following:

(1) Circulating type room heaters having an outer jacket surrounding the combustion chamber, arranged with openings at top and bottom so that air circulates between the inner and outer jacket, and without openings in the outer jacket to permit direct radiation, shall have clearance at sides and rear of not less than 12 inches (305 mm).

(2) Radiating type room heaters other than those of the circulating type described in Section 916.2.3(1) shall have clearance at sides and rear of not less than 18 inches (457 mm), except that heaters that make use of metal, asbestos, or ceramic material to direct radiation to the front of the heater shall have a clearance of 36 inches (914 mm) in front and, where constructed with a double back of metal or ceramic, shall be permitted to be installed with a clearance of 18 inches (457 mm) at sides and 12 inches (305 mm) at rear. Combustible floors under unlisted room heaters shall be protected in an approved manner.

916.2.4 Wall-Type Room Heaters. Wall-type room heaters shall not be installed in or attached to walls of combustible material unless listed for such installation. [NFPA 54:10.22.5]

916.3 Solid-Fuel-Type Room Heaters. Solid-fuel type room heaters shall comply with UL 1482.

UL third party listing and labeling verifies product safety and performance. The installers and inspectors should verify the UL listing and label on the equipment. This UL listing requires room heaters to have an included blower assembly.

917.0 Unit Heaters.

917.1 Support. Suspended-type unit heaters shall be safely and adequately supported, with due consideration given to their weight and vibration characteristics. Hangers and brackets shall be of noncombustible material. [NFPA 54:10.25.1]

917.2 Clearance. Suspended-type unit heaters shall comply with the following requirements:

(1) A listed unit heater shall be installed with clearances from combustible material of not less than 18 inches (457 mm) at the sides, 12 inches (305 mm) at the bottom, and 6 inches (152 mm) above the top where the unit heater has an internal draft hood, or 1 inch (25.4 mm) above the top of the sloping side of a vertical draft hood. A unit heater listed for reduced clearances shall be installed in accordance with its listing and the manufacturer's installation instructions.

(2) Unlisted unit heaters shall be installed with clearances to combustible material of not less than 18 inches (457 mm).

(3) Clearances for servicing shall be in accordance with the manufacturer's installation instructions.

917.2.1 Floor-Mounted-Type Unit Heaters. Floor-mounted-type unit heaters shall comply with the following requirements:

(1) A listed unit heater shall be installed with clearances from combustible material at the back and one side of not less than 6 inches (152 mm). Where the flue gases are vented horizontally, the 6 inch (152 mm) clearance shall be measured from the draft hood or vent instead of the rear wall of the unit heater. A unit heater listed for reduced clearances shall be installed in accordance with its listing and the manufacturer's installation instructions.

(2) Floor-mounted-type unit heaters installed on combustible floors shall be listed for such installation.

(3) Combustible floors under unlisted floor-mounted unit heaters shall be protected in an approved manner.

(4) Clearances for servicing shall be in accordance with the manufacturer's instructions.

917.3 Combustion and Circulating Air. Combustion and circulating air shall be provided in accordance with Section 701.0. [NFPA 54:10.25.3]

917.4 Ductwork. A unit heater shall not be attached to a warm air duct system unless listed and marked for such installation. [NFPA 54:10.25.4]

917.5 Installation in Commercial Garages and Aircraft Hangars. Unit heaters installed in garages for more than three motor vehicles or in aircraft hangars shall be of a listed type and shall be installed in accordance with Section 303.11 and Section 303.12. [NFPA 54:10.25.5]

917.6 Oil-Fired Unit Heaters. Oil-fired unit heaters shall comply with UL 731 and installed in accordance with the manufacturer's installation instructions.

UL 731 includes a comprehensive set of construction and performance requirements that are used to evaluate and list oil fired unit heaters. Refer also to the requirements within Chapters 7 and 8 of this Code. The Building Code contains provisions for fire blocking at concealed spaces. NFPA 31 addresses oil storage tanks within and outside of buildings, as well as fuel oil piping systems, tank rooms, enclosures and other spaces.

918.0 Food Service Appliance, Floor-Mounted.

Food service equipment that depends on fuel gas for its operation and are installed in commercial locations such as restaurants or hotels shall comply with Sections 918.1 through 918.8.

918.1 Clearance for Listed Appliances. Listed floor-mounted food service appliances, such as ranges for hotels and restaurants, deep fat fryers, unit broilers, kettles, steam cookers, steam generators, and baking and roasting ovens, shall be installed not less than 6 inches (152 mm) from combustible material except that at least a 2 inch (51 mm) clearance shall be maintained between a draft hood and combustible material. Floor-mounted food service appliances listed for installation at lesser clearances shall be installed in accordance with its listing and the manufacturer's installation instructions. Appliances designed and marked, "For use only in noncombustible locations," shall not be installed elsewhere.

918.2 Clearance for Unlisted Appliances. Unlisted floor-mounted food service appliances shall be installed to provide a clearance to combustible material of not less than 18 inches (457 mm) from the sides and rear of the appliance and from the vent connector and not less than 48 inches (1219 mm) above cooking tops and at the front of the appliance.

Clearances for unlisted appliances installed in partially enclosed areas such as alcoves shall not be reduced. Reduced clearances for unlisted appliances installed in rooms that are not partially enclosed shall be in accordance with Table 303.10.1. [NFPA 54:10.12.2]

918.3 Mounting on Combustible Floors. Listed floor-mounted food service appliances that are listed specifically for installation on floors constructed of combustible material shall be permitted to be installed on combustible floors in accordance with its listing and the manufacturer's installation instructions.

918.3.1 Not Listed for Mounting on Combustible Floors. Floor-mounted food service appliances that are not listed for mounting on a combustible floor shall be mounted in accordance with Section 918.4 or be mounted in accordance with one of the following:

(1) Where the appliance is set on legs that provide not less than 18 inches (457 mm) open space under the base of the appliance or where it has no burners and no portion of any oven or broiler within 18 inches (457 mm) of the floor, it shall be permitted to be mounted on a combustible floor without special floor protection, provided at least one sheet metal baffle is between the burner and the floor.

(2) Where the appliance is set on legs that provide not less than 8 inches (203 mm) open space under the base of the appliance, it shall be permitted to be mounted on combustible floors, provided the floor under the appliance is protected with not less than ⅜ of an inch (9.5 mm) insulating millboard covered with sheet metal not less than 0.0195 of an inch (0.4953 mm) thick. The preceding specified floor protection shall extend not less than 6 inches (152 mm) beyond the appliance on all sides.

(3) Where the appliance is set on legs that provide not less than 4 inches (102 mm) under the base of the appliance, it shall be permitted to be mounted on combustible floors, provided the floor under the appliance is protected with hollow masonry not less than 4 inches (102 mm) in thickness covered with sheet metal not less than 0.0195 of an inch (0.4953 mm) thick. Such masonry courses shall be laid with ends unsealed and joints matched in such a way as to provide for free circulation of air through the masonry.

(4) Where the appliance does not have legs at least 4 inches (102 mm) high, it shall be permitted to be mounted on combustible floors, provided the floor under the appliance is protected by two courses of 4 inch (102 mm) hollow clay tile, or equivalent, with courses laid at right angles and with ends unsealed and joints matched in such a way as to provide for free circulation of air through such masonry courses, and covered with steel plate not less than 3⁄16 of an inch (4.8 mm) in thickness. [NFPA 54:10.12.3.2]

918.4 Installation on Noncombustible Floors. Listed floor-installed food service appliances that are designed and marked "For use only in noncombustible locations" shall be installed on floors of noncombustible construction with noncombustible flooring and surface finish and with no combustible material against the underside thereof, or on noncombustible slabs or arches having no combustible material against the underside thereof. Such construction shall in all cases extend not less than 12 inches (305 mm) beyond the appliance on all sides. [NFPA 54:10.12.4.1, 10.12.4.2]

918.5 Combustible Material Adjacent to Cooking Top. Listed and unlisted food service ranges shall be installed to provide clearance to combustible material not less than 18 inches (457 mm) horizontally for a distance of up to 2 feet (610 mm) above the surface of the cooking top where the combustible material is not completely shielded by high shelving, warming closet, or other system. Reduced combustible material clearances are permitted where protected in accordance with Table 303.10.1. [NFPA 54:10.12.5]

918.6 Use with Casters. Floor-mounted appliances with casters shall be listed for such construction and shall be installed in accordance with the manufacturer's installation instructions for limiting the movement of the appliance to prevent strain on the connection. [NFPA 54:10.12.6]

918.7 Level Installation. Floor-mounted food service appliances shall be installed level on a firm foundation. [NFPA 54:10.12.7]

918.8 Ventilation. Means shall be provided to properly ventilate the space in which a food service appliance is installed to permit proper combustion of the gas. [NFPA 54:10.12.8]

This section is applicable to permanently mounted or immovable types of cooking equipment. Equipment with casters are allowed when installed per the listing and manufacturer's instructions, provided movement is limited to where no stress can be applied to the gas connection. Ranges; deep fat fryers; unit broilers; kettles; steam cookers and generators; baking and roasting ovens and any other like appliance shall meet requirements for location, ventilation and clearances to combustibles.

919.0 Food Service Appliances, Counter Appliances.

919.1 Vertical Clearance. A vertical distance of not less than 48 inches (1219 mm) shall be provided between the top of all food service hot plates and griddles and combustible material. [NFPA 54:10.13.1]

919.2 Clearance for Listed Appliances. Listed food service counter appliances such as hot plates and griddles, food and dish warmers, and coffee brewers and urns, where installed on combustible surfaces, shall be set on their own bases or legs and shall be installed with a horizontal clearance of not less than 6 inches (152 mm) from combustible material, except that not less than a 2 inches (51 mm) clearance shall be maintained between a draft hood and combustible material. Food service counter appliances listed for installation at lesser clearances shall be installed in accordance with their listing and the manufacturer's installation instructions.

919.3 Clearance for Unlisted Appliances. Unlisted food service hot plates and griddles shall be installed with a horizontal clearance from combustible material of not less than 18 inches (457 mm). Unlisted gas food service counter appliances, including coffee brewers and urns, waffle bakers, and hot water immersion sterilizers, shall be installed with a horizontal clearance from combustible material of not less than 12 inches (305 mm). Reduced clearances for gas food service counter appliances shall be in accordance with Table 303.10.1. Unlisted food and dish warmers shall be installed with a horizontal clearance from combustible material of not less than 6 inches (152 mm). [NFPA 54:10.13.3]

919.4 Mounting of Unlisted Appliances. Unlisted food service counter appliances shall not be set on combustible material unless they have legs that provide not less than 4 inches (102 mm) of open space below the burners and the combustible surface is protected with insulating millboard at least ¼ of an inch (6.4 mm) thick covered with sheet metal not less than 0.0122 of an inch (0.3099 mm) thick, or with equivalent protection. [NFPA 54:10.13.4]

920.0 Household Cooking Appliances.

920.1 Electric Household Cooking Appliances. Electric household cooking appliances designed for permanent installations shall be installed in accordance with the manufacturer's installation instructions. Household electric ranges shall comply with UL 858.

920.2 Gas-Fired Household Cooking Appliances. Gas-fired household cooking appliances shall comply with Section 920.3 though Section 920.4.3.

920.3 Floor-Mounted Units. Floor mounted units shall be installed in accordance with Section 920.3.1 and Section 920.3.2.

920.3.1 Clearance from Combustible Material. The clearances specified as follows shall not interfere with combustion air, accessibility for operation, and servicing:

(1) Listed floor-mounted household cooking appliances, where installed on combustible floors, shall be set on their own bases or legs and shall be installed in accordance with their listing and the manufacturer's installation instructions.

(2) Listed household cooking appliances with listed gas room heater sections shall be installed so that the warm air discharge side shall have a clearance of not less than 18 inches (457 mm) from adjacent combustible material. A clearance of not less than 36 inches (914 mm) shall be provided between the top of the heater section and the bottom of cabinets.

(3) Listed household cooking appliances that include a solid or liquid fuel-burning section shall be spaced from combustible material and otherwise installed in accordance with their listing and the manufacturer's installation instructions for the supplementary fuel section of the appliance.

(4) Unlisted floor-mounted household cooking appliances shall be installed with not less than 6 inches (152 mm) clearance at the back and sides to combustible material. Combustible floors under unlisted appliances shall be protected in an approved manner.

920.3.2 Vertical Clearance Above Cooking Top. Household cooking appliances shall have a vertical clearance above the cooking top of not less than 30 inches (762 mm) to combustible material or metal cabinets. A minimum clearance of 24 inches (610 mm) is permitted where one of the following is installed:

(1) The underside of the combustible material or metal cabinet above the cooking top is protected with not less than ¼ of an inch (6.4 mm) insulating millboard covered with sheet metal not less than 0.0122 of an inch (0.3099 mm) thick.

(2) A metal ventilating hood of sheet metal not less than 0.0122 of an inch (0.3099 mm) thick is installed above the cooking top with a clearance of not less than ¼ of an inch (6.4 mm) between the hood and the underside of the combustible material or metal cabinet, and the hood is as wide as the appliance and is centered over the appliance.

(3) A listed cooking appliance or microwave oven installed over a listed cooking appliance shall be in accordance with the terms of the upper appliance's listing and the manufacturer's installation instructions. Microwave ovens shall comply with UL 923.

920.4 Built-In Units. Built-in units shall be installed in accordance with Section 920.4.1 through Section 920.4.3.

920.4.1 Installation. Listed built-in household cooking appliances shall be installed in accordance with their listing and the manufacturer's installation instructions. The installation shall not interfere with combustion air, accessibility for operation, and servicing. Unlisted built-in household cooking appliances shall not be installed in or adjacent to combustible material.

920.4.2 Vertical Clearance. Built-in top (or surface) cooking appliances shall have a vertical clearance above the cooking top of not less than 30 inches (762 mm) to combustible material or metal cabinets. A clearance of not less than 24 inches (610 mm) is permitted where one of the following is installed:

(1) The underside of the combustible material or metal cabinet above the cooking top is protected with not less than ¼ of an inch (6.4 mm) insulating millboard covered with sheet metal not less than 0.0122 of an inch (0.3099 mm) thick.

(2) A metal ventilating hood of sheet metal not less than 0.0122 of an inch (0.3099 mm) thick is installed above the cooking top with a clearance of not less than ¼ of an inch (6.4 mm) between the hood and the underside of the combustible material or metal cabinet, and the hood not less than the width of the appliance and is centered over the appliance.

(3) A listed cooking appliance or microwave oven installed over a listed cooking appliance shall be in accordance with the terms of the upper appliance listing and the manufacturer's installation instructions. Microwave ovens shall comply with UL 923.

A domestic range or cooktop requires a minimum vertical clearance to combustibles of 30 inches (see **Figure 920.4.2**). This vertical distance may be reduced to 24 inches if the under-cabinet space is protected with an approved method. Even if cabinets are constructed of steel or other noncombustible material, reduced clearances are to be considered cautiously. The contents of the cupboard and/or the packaging material could be ignited by intense heat.

920.4.3 Level Installation. Cooking appliances shall be installed so that the cooking top, broiler pan, or oven racks are level. [NFPA 54:10.14.3]

921.0 Cooking Appliances Listing.

921.1 Commercial Electric Ranges. Commercial electric ranges shall comply with UL 197 and installed in accordance with the manufacturer's installation instructions.

921.2 Commercial Wood-Fired Baking Ovens. Commercial wood-fired baking ovens (refractory type) shall comply with UL 2162 and installed in accordance with the manufacturer's installation instructions.

921.3 Oil-Burning Ranges. Oil-burning ranges shall comply with UL 896 and installed in accordance with the manufacturer's installation instructions.

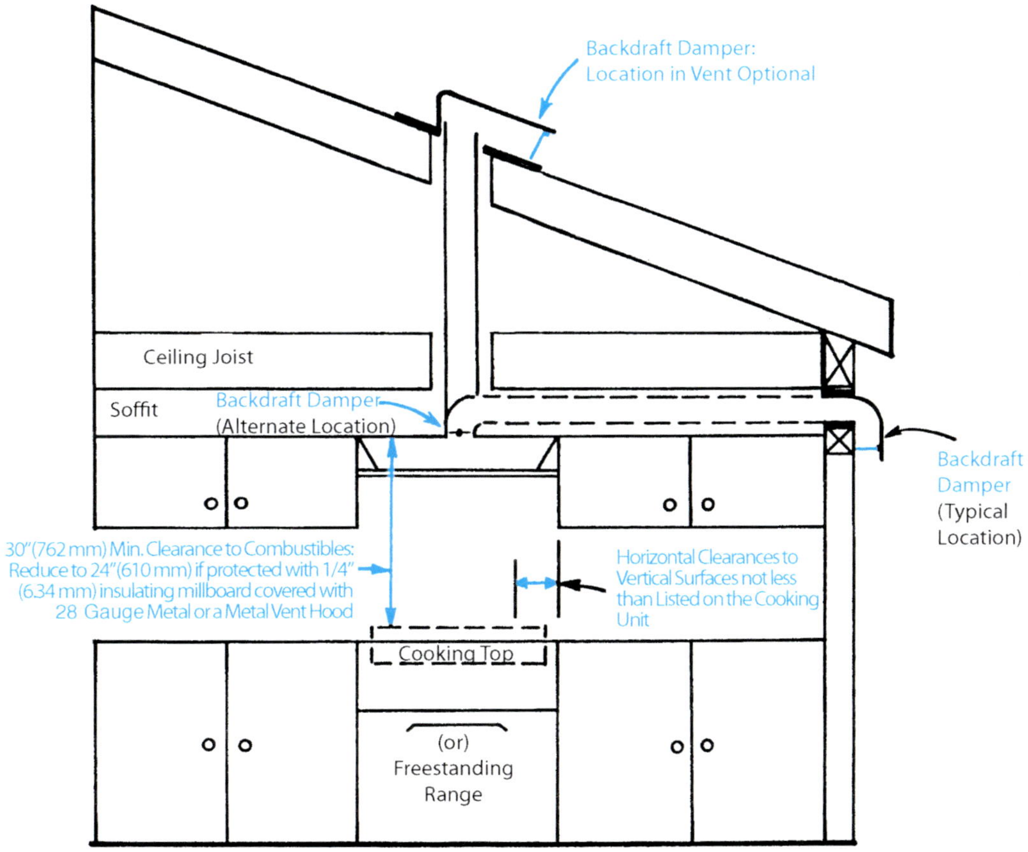

FIGURE 920.4.2
KITCHEN RANGE HOOD AND DUCT

922.0 Open-Top Broiler Units.

922.1 Listed Units. Listed open-top broiler units shall be installed in accordance with the manufacturer's installation instructions. [NFPA 54:10.18.1]

922.2 Unlisted Units. Unlisted open-top broiler units shall be installed in accordance with the manufacturer's instructions but shall not be installed in combustible material. [NFPA 54:10.18.2]

922.3 Protection Above Domestic Units. Domestic open-top broiler units shall be provided with a metal ventilating hood not less than 0.0122 of an inch (0.3099 mm) thick with a clearance of not less than ¼ of an inch (6.4 mm) between the hood and the underside of combustible material or metal cabinets. A clearance of at least 24 inches (610 mm) shall be maintained between the cooking top and the combustible material or metal cabinet, and the hood shall be at least as wide as the open-top broiler unit and centered over the unit. Listed domestic open-top broiler units incorporating an integral exhaust system and listed for use without a ventilating hood shall not be required to be provided with a ventilating hood if installed in accordance with Section 920.3.2(1). [NFPA 54:10.18.3]

922.4 Commercial Units. Commercial open-top broiler units shall be provided with ventilation in accordance with Chapter 5, Part II. {NFPA 54:10.18.4}

Listed open-top broiler units designed for indoor installation shall be installed as listed and exhausted as follows:

- The exhaust duct and fan shall have a minimum flow rate of 100 cfm per square foot. If the exhaust duct penetrates a floor or ceiling, the duct shall be enclosed in a 1-hour fire-resistive shaft and ducted to the outside. This shaft is to be separated from the duct by a minimum 1-inch air space.
- Vertical clearance of 24 inches to combustible materials above the cooking surface shall be maintained. The exhaust hood, in upward flow conditions, shall be centered above the broiler and shall be at least as wide as the unit.
- When an exhaust duct terminates at rooftop, the duct height shall be a minimum 18 inches above the roof, with an approved termination method. Exhaust duct terminated at a sidewall in a downflow application shall not be located less than 3 feet from a property line or an opening into the building (see Section 507.1 Exhaust Systems).

923.0 Outdoor Cooking Appliances.

923.1 Listed Units. Listed outdoor cooking appliances shall be installed in accordance with their listing and the manufacturer's installation instructions.

923.2 Unlisted Units. Unlisted outdoor cooking appliances shall be installed outdoors with clearances to combustible material of not less than 36 inches (914 mm) at the sides and back and not less than 48 inches (1219 mm) at the front. In no case shall the appliance be located under overhead combustible construction. [NFPA 54:10.19.2]

924.0 Illuminating Appliances.

924.1 Clearances for Listed Appliances. Listed illuminating appliances shall be installed in accordance with their listing and the manufacturer's installation instructions.

924.2 Clearances for Unlisted Appliances. Clearances for unlisted illuminating appliances shall comply with the following:

(1) Unlisted enclosed illuminating appliances installed outdoors shall be installed with clearances in any direction from combustible material of not less than 12 inches (305 mm). [NFPA 54:10.15.2.1(1)]

(2) Unlisted enclosed illuminating appliances installed indoors shall be installed with clearances in any direction from combustible material of not less than 18 inches (457 mm). [NFPA 54:10.15.2.1(2)]

924.2.1 Open-Flame Type. Clearances shall comply with the following:

(1) Unlisted open-flame illuminating appliances installed outdoors shall have clearances from combustible material not less than that specified in Table 924.2.1. The distance from ground level to the base of the burner shall be a minimum of 7 feet (2134 mm) where installed within 2 feet (610 mm) of walkways. Lesser clearances shall be permitted to be used where acceptable to the Authority Having Jurisdiction.

(2) Unlisted open-flame illuminating appliances installed outdoors shall be equipped with a limiting orifice or other limiting devices that maintain a flame height consistent with the clearance from combustible material, as given in Table 924.2.1.

(3) Appliances designed for flame heights in excess of 30 inches (762 mm) shall be permitted to be installed if acceptable to the Authority Having Jurisdiction. Such appliances shall be equipped with a safety shutoff device or automatic ignition.

(4) Unlisted open-flame illuminating appliances installed indoors shall have clearances from combustible material acceptable to the Authority Having Jurisdiction. [NFPA 54:10.15.2.2]

TABLE 924.2.1
CLEARANCES FOR UNLISTED OUTDOOR OPEN-FLAME ILLUMINATING APPLIANCES
[NFPA 54:TABLE 10.15.2.2]

FLAME HEIGHT ABOVE BURNER HEAD (inches)	MINIMUM CLEARANCE FROM COMBUSTIBLE MATERIAL (feet)*	
	HORIZONTAL	VERTICAL
12	2	6
18	3	8
24	3	10
30	4	12

For SI units: 1 inch = 25.4 mm, 1 foot = 304.8 mm

* Measured from the nearest portion of the burner head.

924.3 Mounting on Buildings. Illuminating appliances designed for wall or ceiling mounting shall be securely attached to substantial structures in such a manner that they are not dependent on the gas piping for support. [NFPA 54:10.15.3]

924.4 Mounting on Posts. Illuminating appliances designed for post mounting shall be securely and rigidly attached to a post. Posts shall be rigidly mounted. The strength and rigidity of posts greater than 3 feet (914 mm) in height shall be at least equivalent to that of a 2½ inch (64 mm) diameter post constructed of 0.064 of an inch (1.626 mm) thick steel or a 1 inch (25.4 mm) Schedule 40 steel pipe. Posts 3 feet (914 mm) or less in height shall not be smaller than a ¾ of an inch (19.1 mm) Schedule 40 steel pipe. Drain openings shall be provided near the base of posts where water collecting inside the posts is possible. [NFPA 54:10.15.4]

924.5 Appliance Pressure Regulators. Where an appliance pressure regulator is not supplied with an illuminating appliance and the service line is not equipped with a service pressure regulator, an appliance pressure regulator shall be installed in the line serving one or more illuminating appliances. [NFPA 54:10.15.5]

925.0 Incinerators and Crematories.

925.1 Field Constructed Commercial-Industrial Incinerators. Field constructed commercial-industrial incinerators shall be constructed and installed in accordance with NFPA 82.

According to this standard, an incinerator is any equipment predominantly designed for burning solid fuel and applicable to new construction and new equipment as determined by the AHJ.

925.2 Factory-Built Commercial Crematories. Factory-built commercial incinerators and crematories shall comply with UL 2790 and installed in accordance with the manufacturer's installation instructions.

925.3 Residential Incinerators. Residential incinerators shall comply with UL 791 and installed in accordance with the manufacturer's installation instructions.

926.0 Infrared Heaters.

926.1 Support. Suspended-type infrared heaters shall be fixed in position independent of gas and electric supply lines. Hangers and brackets shall be of noncombustible material. Heaters subject to vibration shall be provided with vibration-isolating hangers. [NFPA 54:10.17.1]

926.2 Clearance. The installation of infrared heaters shall comply with the following clearance requirements:

(1) Listed heaters shall be installed with clearances from combustible material in accordance with their listing and the manufacturer's installation instructions.

(2) Unlisted heaters shall be installed in accordance with clearances from combustible material acceptable to the Authority Having Jurisdiction.

(3) In locations used for the storage of combustible materials, signs shall be posted to specify the maximum permissible stacking height to maintain required clearances from the heater to the combustibles.

926.3 Combustion and Ventilation Air. Where unvented infrared heaters are used, natural or mechanical means shall be provided to supply and exhaust at least 4 $ft^3/min/1000$ Btu/h (0.38 $m^3/min/kW$) input of installed heaters. [NFPA 54:10.17.3.1]

926.3.1 Exhaust Openings. Exhaust openings for removing flue products shall be above the level of the heaters. [NFPA 54:10.17.3.2]

926.4 Installation in Commercial Garages and Aircraft Hangars. Overhead heaters installed in garages for more than three motor vehicles or in aircraft hangars shall be of a listed type and shall be installed in accordance with Section 303.11 and Section 303.12. [NFPA 54:10.17.4]

Suspended heaters of this type are required to be installed in a fixed position independent of the gas and electric supply lines. No combustible material is allowed for hangers or supports, and heaters subject to vibration shall be protected with isolating hangers.

Installations for listed and unlisted heaters shall comply with the proper clearances to combustibles and combustion and ventilation air requirements as discussed in this section.

NFPA 409 Standard on Aircraft Hangars, Section 5.12.5 also requires listed heaters for suspended or elevated heaters in aircraft hangars.

927.0 Pool Heaters.

927.1 Location. A pool heater shall be located or protected so as to minimize accidental contact of hot surfaces by persons. [NFPA 54:10.20.1]

927.2 Clearance. The installation of pool heaters shall comply with the following requirements:

(1) In no case shall the clearances be such as to interfere with combustion air, draft hood, or vent terminal clearance and relief, and accessibility for servicing.

(2) A listed pool heater shall be installed in accordance with its listing and the manufacturer's installation instructions.

(3) An unlisted pool heater shall be installed with a clearance of not less than 12 inches (305 mm) on the sides and the rear. A combustible floor under an unlisted pool heater shall be protected in an approved manner.

927.3 Temperature or Pressure-Limiting Devices. An unlisted pool heater shall be provided with overtemperature protection or overtemperature and overpressure protection by means of an approved device(s). [NFPA 54:10.20.3.1]

927.3.1 Pressure Relief Valve. Where a pool heater is provided with over-temperature protection only and is installed with any device in the discharge line of the heater that can restrict the flow of water from the heater to the pool (such as a check valve, shutoff valve, therapeutic pool valving, or flow nozzles), a pressure-relief valve shall be installed either in the heater or between the heater and the restrictive device. [NFPA 54:10.20.3.2]

927.4 Bypass Valves. Where an integral bypass system is not provided as a part of the pool heater, a bypass line and valve shall be installed between the inlet and outlet piping for use in adjusting the flow of water through the heater.[NFPA 54:10.20.4]

See **Figure 927.4**.

927.5 Venting. A pool heater listed for outdoor installation shall be installed with the venting means supplied by the man-

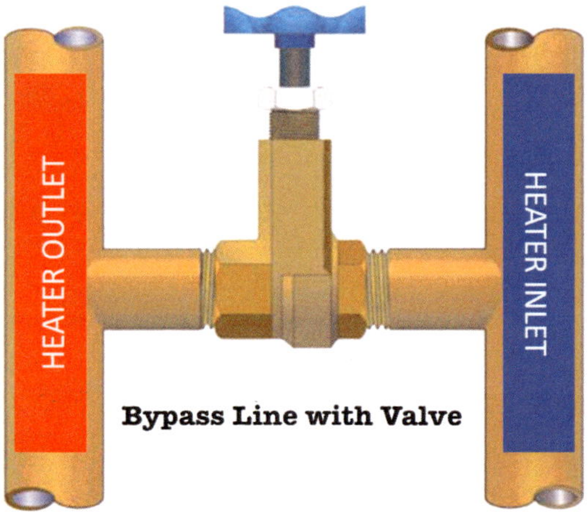

FIGURE 927.4
BYPASS LINE WITH VALVE

ufacturer and in accordance with the manufacturer's installation instructions. [NFPA 54:10.20.5]

A pool heater shall be located to prevent or minimize accidental harm to anyone in the area. Requirements for access, clearances, temperature and pressure-limiting devices shall be as described in this section. In no case shall a pool heater be located with less than the required clearance or interfere with combustion air, draft hood, vent termination and relief or accessibility for servicing.

A pool heater, installed with an over-temperature device in the outlet line of the heater that may restrict the flow to the pool, requires a pressure-relief valve to be installed upstream of the device.

A pool heater listed for outdoor installation shall be vented according to the manufacturer's instructions and Chapter 8.

928.0 Refrigerators.

928.1 Clearance. Refrigerators shall be provided with clearances for ventilation at the top and back in accordance with the manufacturer's instructions. Where such instructions are not available, at least 2 inches (51 mm) shall be provided between the back of the refrigerator and the wall at least 12 inches (305 mm) above the top. [NFPA 54:10.21.1]

928.2 Venting or Ventilating Kits Approved for Use with a Refrigerator. Where an accessory kit is used for conveying air for burner combustion or unit cooling to the refrigerator from areas outside the room in which it is located, or for conveying combustion products diluted with air containing waste heat from the refrigerator to areas outside the room in which it is located, the kit shall be installed in accordance with the refrigerator manufacturer's instructions. [NFPA 54:10.21.2]

929.0 Gas-Fired Toilets.

929.1 Clearance. A listed gas-fired toilet shall be installed in accordance with its listing and the manufacturer's installation instructions, provided that the clearance shall be such to afford ready accessibility for use, cleanout, and necessary servicing.

929.2 Installation on Combustible Floors. Listed gas-fired toilets installed on combustible floors shall be listed for such installation. [NFPA 54:10.24.2]

929.3 Vents. Vents or vent connectors that are capable of being contacted during casual use of the room in which the toilet is installed shall be protected or shielded to prevent such contact. [NFPA 54:10.24.3]

930.0 Appliances for Installation in Manufactured Housing.

930.1 General. Appliances installed in manufactured housing after the initial sale shall be listed for installation in manufactured housing, or approved, and shall be installed in accordance with the requirements of this code and the manufacturer's installation instructions. Appliances installed in the living space of manufactured housing shall be in accordance with the requirements of Section 701.0. [NFPA 54:10.29]

931.0 Small Ceramic Kilns.

931.1 General. The provisions of this section apply to kilns used for ceramics that have a maximum interior volume of 20 cubic feet (0.57 m^3) and are used for hobby or noncommercial purposes.

931.2 Installation. Kilns shall be installed in accordance with the manufacturer's installation instructions and the provisions of this code.

931.3 Fuel-Gas Controls. Fuel-gas controls shall comply with Section 306.0 and Section 902.4. Standing pilots shall not be used with gas-fired kilns.

931.4 Electrical Equipment. All electrical equipment used as part of, or in connection with, the installation of a kiln shall be in accordance with the requirements in the electrical code. Electric kilns shall be listed and labeled in accordance with UL 499.

931.5 Installations Inside Buildings. In addition to other requirements specified in this section, interior installations shall comply with the requirements of Section 603.13.6 and Section 931.5.1 through Section 931.5.4.

931.5.1 Kiln Clearances. The sides and tops of kilns shall be located not less than 18 inches (457 mm) from a noncombustible wall surface and 3 feet (914 mm) from a combustible wall surface. Kilns shall be installed on noncombustible flooring consisting of not less than 2 inches (51 mm) of solid masonry or concrete extending not less than 12 inches (305 mm) beyond the base or supporting members of the kiln.

Exception: These clearances shall be permitted to be reduced, provided the kiln is installed in accordance with its listing.

In no case shall the clearance on the gas or electrical control side of a kiln be reduced to less than 30 inches (762 mm).

931.5.2 Hoods. A canopy-type hood shall be installed directly above each kiln. The face opening area of the hood shall be equal to or greater than the top horizontal surface area of the kiln. The hood shall be constructed of not less than 0.024 of an inch (0.61 mm) (No. 24 gauge) galvanized steel or equivalent and be supported at a height of between 12 inches (305 mm) and 30 inches (762 mm) above the kiln by noncombustible supports.

Exception: Electric kilns installed with listed exhaust blowers shall be permitted to be used where marked as being suitable for the kiln and installed in accordance with the manufacturer's installation instructions.

931.5.3 Gravity Ventilation Ducts. Each hood shall be connected to a gravity ventilation duct extending in a vertical direction to outside the building. This duct shall be of the same construction as the hood and shall have a minimum cross-sectional area of not less than one-fifteenth of the face opening area of the hood. The duct shall terminate not less than 12 inches (305 mm) above a portion of a building within 4 feet (1219 mm) and terminate not less than 4 feet (1219 mm) from an openable window or other opening into the building or adjacent property line. The duct opening to the outside shall be shielded, without reduction of duct area, to prevent entrance of rain into the duct. The duct shall be supported at each section by noncombustible supports.

931.5.4 Makeup Air. Provisions shall be made for air to enter the room in which a kiln is installed at a rate not less than the air being removed through the kiln hood.

931.6 Exterior Installations. Kilns shall be installed with minimum clearances as specified in Section 931.5.1. Wherever a kiln is located under a roofed area and is partially enclosed by more than two vertical wall surfaces, a hood and gravity ventilation duct shall be installed in accordance with Section 931.5.2, Section 931.5.3, and Section 603.13.6.

932.0 Outdoor Open Flame Decorative Appliances.

932.1 General. Permanently fixed in place outdoor open flame decorative appliances shall be installed in accordance with Section 932.1.1 through Section 932.1.3. [NFPA 54:10.31]

932.1.1 Listed Units. Listed outdoor open flame decorative appliances shall be installed in accordance with the manufacturer's installation instructions. [NFPA 54:10.31.1]

932.1.2 Unlisted Units. Unlisted outdoor open flame decorative appliances shall be installed outdoors in accordance with the manufacturer's installation instructions and with clearances to combustible material of not less than 36 inches (914 mm) from the sides. In no case shall the appliance be located under overhead combustible construction. [NFPA 54:10.31.2]

932.1.3 Connection to the Piping System. The connection to the gas piping system shall be in accordance with Section 1312.1(1), Section 1312.1(2), Section 1312.1(4), or Section 1312.1(5). [NFPA 54:10.31.3]

933.0 Evaporative Cooling Systems.

933.1 General. Evaporative cooling systems, including air ducts and fire dampers that are a portion of an evaporative cooling system, shall be in accordance with Section 933.2 through Section 933.4.3. Evaporative cooling systems shall be provided with outside air as specified for cooling systems in Section 403.0.

933.2 Location. Evaporative cooling systems shall be installed so as to minimize the probability of damage from an external source.

933.3 Access, Inspection, and Repair. Evaporative coolers shall be accessible for inspection, service, and replacement without removing permanent construction.

933.4 Installation. An evaporative cooler supported by the building structure shall be installed on a level base and shall be secured directly or indirectly to the building structure, to prevent displacement of the cooler.

933.4.1 Modifications to the Supporting Structure. Modifications made to the supporting framework of buildings as a result of the installation shall be in accordance with the requirements of the building code. Openings in exterior walls shall be flashed in an approved manner in accordance with the requirements of the building code.

933.4.2 On the Ground. An evaporative cooler supported directly by the ground shall be isolated from the ground by a level concrete slab extending not less than 3 inches (76 mm) above the adjoining ground level.

933.4.3 On a Platform. An evaporative cooler supported on an aboveground platform shall be elevated not less than 6 inches (152 mm) above adjoining ground level.

934.0 Refrigeration Appliances.

934.1 Self-Contained Refrigerators and Freezers. Factory-built commercial refrigerators and freezers shall comply with UL 471 or UL 60335-2-89 and shall be installed in accordance with the manufacturer's installation instructions.

934.2 Unit Coolers. Factory-built unit coolers for use in refrigerators, freezers, refrigerated warehouses, and walk-in coolers shall comply with UL 412 or UL 60335-2-89 and shall be installed in accordance with the manufacturer's installation instructions.

934.3 Self-Contained Mechanical Refrigeration Systems. Self-contained mechanical refrigeration systems for use in walk-in coolers shall comply with UL 427 or UL 60335-2-89 and shall be installed in accordance with the manufacturer's installation instructions.

935.0 Ductless Mini-Split Systems Installation.

935.1 General. A ductless mini-split system installation shall be installed in accordance with the manufacturer's installation instructions and Section 310.2 for condensate control.

Because of the potential damage and health risk associated with uncontrolled condensation, ductless mini-split equipment must comply with the same code requirements as other condensate producing equipment. See **Figure 935.1**.

936.0 Air Filter Appliances.

936.1 Electrostatic Air Cleaners. Electrostatic air cleaners shall comply with UL 867 and installed in accordance with the manufacturer's installation instructions.

Electrostatic air cleaners are intended to remove dust and other particles from the air. UL 867 addresses the construction, protection, performance, testing, rating, and marking of electrostatic air filters. It includes a comprehensive set of construction and performance requirements that are used to evaluate and list electrostatic air cleaners.

936.2 High-Efficiency Particulate Air Filter Units. High-efficiency particulate air filter units for use in industrial and laboratory exhaust and ventilation systems shall be installed in accordance with the manufacturer's installation instructions.

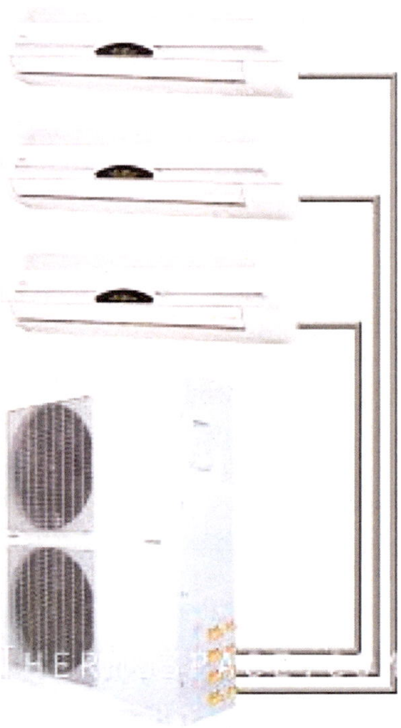

FIGURE 935.1
DUCTLESS MINI-SPLIT

937.0 Gaseous Hydrogen Systems.

937.1 General. Gaseous hydrogen systems shall be installed in accordance with NFPA 2.

938.0 Compressed Natural Gas (CNG) Vehicular Fuel Systems.

938.1 General. The installation of compressed natural gas (CNG) fueling (dispensing) systems shall conform to NFPA 52. Residential CNG fueling appliances shall be listed in accordance with CSA NGV 5.1 and installed in accordance to the appliance manufacturer's installation instructions. [NFPA 54:10.28]

CHAPTER 10

BOILERS AND PRESSURE VESSELS

1001.0 General.

1001.1 Applicability. The requirements of this chapter shall apply to the construction, installation, operation, repair, and alteration of boilers and pressure vessels. Low-pressure boilers shall comply with this chapter and Section 904.0.

Exceptions:

(1) Listed and approved potable water heaters with a nominal capacity not exceeding 120 gallons (454 L) and having a heat input not exceeding 200 000 British thermal units per hour (Btu/h) (58.6 kW) used for hot water supply at a pressure not exceeding 160 pounds-force per square inch (psi) (1103 kPa) and at temperatures not exceeding 210°F (99°C), in accordance with the plumbing code.

(2) Pressure vessels used for unheated water supply, including those containing air that serves as a cushion and is compressed by the introduction of water and tanks connected to sprinkler systems.

(3) Portable unfired pressure vessels and Interstate Commerce Commission (I.C.C.) containers.

(4) Containers for liquefied petroleum gases, bulk oxygen, and medical gas that are regulated by the fire code.

(5) Unfired pressure vessels in business, factory, hazardous, mercantile, residential, storage, and utility occupancies having a volume not exceeding 5 cubic feet (0.14 m^3) and operating at pressures not exceeding 250 psi (1724 kPa).

(6) Pressure vessels used in refrigeration systems shall comply with Chapter 11.

Part I of Chapter 11 covers refrigeration systems. Refrigeration systems, equipment and devices, including the replacement of parts, alterations and substitution of a different refrigerant, must conform to the requirements of Chapter 11 and other applicable provisions of this code. Part II covers cooling towers.

(7) Pressure tanks used in conjunction with coaxial cables, telephone cables, power cables, and other similar humidity control systems.

(8) A boiler or pressure vessel subject to regular inspection by federal inspectors or licensed by federal authorities.

1001.2 Boiler Rooms and Enclosures. Boiler rooms and enclosures shall comply with the building code.

1001.3 Air for Combustion and Ventilation. Air for combustion and ventilation shall be provided in accordance with Chapter 7.

Chapter 7 governs air for combustion and ventilation. Any fuel-burning boiler requires a proper supply of air for safe and efficient combustion, exhaust gas discharge, and boiler ventilation.

1001.4 Drainage. For heating or hot-water-supply boiler applications, the boiler room shall be equipped with a floor drain or other approved means for disposing of the accumulation of liquid wastes incident to cleaning, recharging, and routine maintenance. No steam pipe shall be directly connected to a part of a plumbing or drainage system, nor shall a water having a temperature above 140°F (60°C) be discharged under pressure directly into a part of a drainage system. Pipes from boilers shall discharge by means of indirect waste piping as determined by the Authority Having Jurisdiction or the boiler manufacturer's instructions.

The floor drain should be sized large enough and close enough to handle the temperature and pressure-relief valve at full flow.

The following should be taken into account when considering drainage size and location:

- Is there a slipping or injury hazard when the water drains from the relief pipe to the floor drain?
- What happens if the floor drain is blocked and flooding begins?
- How is the boiler condensate going to be handled? Is the piping able to handle the acid nature of the condensate? Will it be diluted or neutralized in some way?

The key phrase regarding the restriction of direct piping to the plumbing system for water above 140°F (60°C) is "under pressure." This allows for piping to a floor drain, provided it is exposed to the atmosphere prior to entry into the plumbing system.

1001.5 Mounting. Equipment shall be set or mounted on a level base capable of supporting and distributing the weight contained thereon. Boilers, tanks, and equipment shall be securely anchored to the structure. Equipment requiring vibration isolation shall be installed as designed by a registered design professional and approved by the Authority Having Jurisdiction.

Manufacturers are clear in the installation instructions about boiler foundation requirements. Here's an example:

Boiler Foundation

Never install boiler on combustible flooring or carpeting, even if a concrete or aerated foundation is used. Severe personal injury, death or substantial property damage can result.

1. A level concrete or solid brick pad is required if:
   a) There is a possibility of the floor becoming flooded.
   b) Non-level conditions exist.
2. An aerated boiler foundation is recommended if any of the following conditions exist:
   a) Electrical wiring or telephone cables buried in the concrete floor of the boiler room.
   b) Concrete floor is "green."

c) There is a history of the floor becoming flooded.

d) Water is channeled under the concrete.

This manufacturer varies the length and width dimensions based on the model of boiler. In **Figure 1001.5a**, the manufacturer felt 2-inch-thick concrete was sufficient to support the weight.

Vibration isolation devices provide deterrence to "telegraphed" boiler noises to the surrounding hard surfaces. Besides boilers, pumps and piping should be included in any vibration isolation strategy (see **Figure 1001.5b**).

BOILER FOUNDATION SIZES		
Boiler Size	**Foundation L**	**Foundation W**
5	20"	36½"
6	23"	36½"
7	26"	36½"
8	29"	36½"

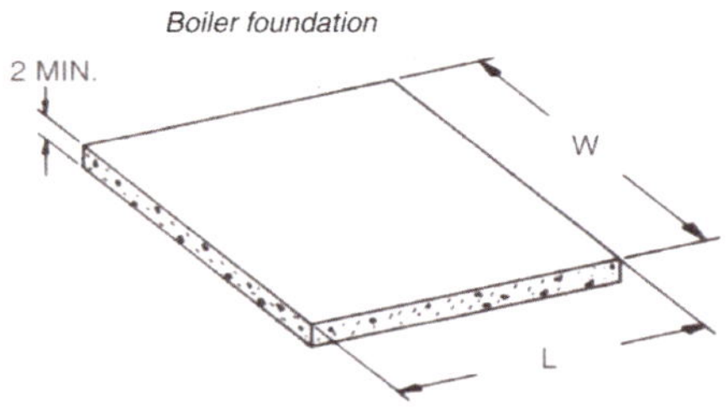

FIGURE 1001.5A
BOILER FOUNDATION SIZES

FIGURE 1001.5B
SPRING MOUNT - EXAMPLE OF VIBRATION ISOLATION STRATEGY

1001.5.1 Floors. Boilers shall be mounted on floors of noncombustible construction unless listed for mounting on combustible flooring.

1001.6 Chimneys or Vents. Boilers shall be connected to a chimney or vent, as provided for other fuel-burning equipment in Chapter 8 of this code.

Chapter 8 governs the venting of fuel-burning appliances requiring that a venting system is designed and constructed so as to develop a positive flow adequate to convey flue and vent gases to the outdoors.

1002.0 Standards.

1002.1 General. Pressure vessels shall be constructed and designed in accordance with the ASME Boiler & Pressure Vessel Code (BPVC) Section VIII. Boilers shall be constructed, designed, and installed in accordance with one of the following:

The rules of this Standard include requirements for the assembly, installation, maintenance, and operation of controls and safety devices on automatically operated boilers directly fired with gas, oil, gas-oil, or electricity, having fuel input ratings under 12,500,000 Btu/hr. See **Figure 1002.1**.

(1) ASME BPVC Section I

Section I of the BPVC provides requirements for all methods of construction of power, electric, and miniature boilers; high temperature water boilers used in stationary service; and power boilers used in locomotive, portable, and traction service. Rules pertaining to use of the V, A, M, PP, S and E Code symbol stamps are also included.

(2) ASME BPVC Section IV

Section IV of the BPVC provides requirements for design, fabrication, installation and inspection of steam heating, hot water heating, hot water supply boilers, and potable water heaters intended for low pressure service that are directly fired by oil, gas, electricity, coal or other solid or liquid fuels. It contains appendices which cover approval of new material, methods of checking safety valve and safety relief valve capacity, examples of methods of checking safety valve and safety relief valve capacity, examples of methods of calculation and computation, definitions relating to boiler design and welding, and quality control systems.

(3) NFPA 85

The following is the document scope of NFPA 85, Boiler and Combustion Systems Hazards Code:

1.1 Scope. This code shall apply to single burner boilers, multiple burner boilers, stokers, and atmospheric fluidized-bed boilers with a fuel input rating of 3.7 MWt (12.5 million Btu/hr) or greater, to pulverized fuel systems, to fired or unfired steam generators used to recover heat from combustion turbines [heat recovery steam generators (HRSGs)], and to other combustion turbine exhaust systems.

1.1.2 This code shall cover strength of the structure, operation and maintenance procedures, combustion and draft control equipment, safety interlocks, alarms, trips, and other related controls that are essential to safe equipment operation.

1.1.3 This code does not cover process heaters used in chemical and petroleum manufacture in which steam generation is incidental to the operation of a processing system.

1.1.4 Chapter 5 covers single burner boilers that fire the following fuels:

(1) Fuel gas
(2) Other gas having a calorific value and characteristics similar to natural gas
(3) Fuel oil
(4) Fuel gas and fuel oil that are fired simultaneously for fuel transfer
(5) Fuel gas and fuel oil that are fired simultaneously and continuously

FIGURE 1002.1
GAS-FIRED BOILER

1.1.5 Chapter 6 covers multiple burner boilers firing one or more of the following:

(1) Fuel gas

(2) Fuel oil

(3) Pulverized coal

(4) Simultaneous firing of more than one of the fuels

1.1.6 Chapter 7 covers atmospheric fluidized-bed boilers.

1.1.7 Chapter 8 covers HRSG systems or other combustion turbine exhaust systems.

1.1.8 Chapter 9 covers pulverized fuel systems, beginning with the raw fuel bunker, which is upstream of the pulverizer, and is the point at which primary air enters the pulverizing system, and terminating at the point where pressure can be relieved by fuel being burned or collected in a device that is built in accordance with this code. The pulverized fuel system shall include the primary air ducts, which are upstream of the pulverizer, to a point where pressure can be relieved.

1.1.9 Chapter 10 covers boilers using a stoker to fire the following fuels:

(1) Coal

(2) Wood

(3) Refuse-derived fuel (RDF)

(4) Municipal solid waste (MSW)

(5) Other solid fuels

1.1.9.1 Where solid fuel is fired simultaneously with other fuels (e.g., a solid fuel stoker fired in combination with fuel gas, fuel oil, or pulverized auxiliary fuel), additional controls and interlocks shall include those covered in Chapters 5, 6, and 9.

Exception No. 1: The purge requirements of Chapters 5 and 6 shall not be required when the stoker is firing and the boiler is on-line. In those cases, if no cooling air is being provided to the auxiliary burners, a purge of their associated air supply ducts shall be provided.

Exception No. 2: Where fuel oil or fuel gas is fired in a supervised manual system in accordance with Chapter 5, the excessive steam pressure interlock shall not be required.

1002.2 Oil-Burning Boilers. Oil-burning boilers shall comply with Section 1002.2.1 and Section 1002.2.2.

1002.2.1 Listing & Labeling. Oil-burning boilers shall be listed and labeled in accordance with UL 726.

1002.2.2 Installation. Tanks, piping, and valves for oil-burning boilers shall be installed in accordance with NFPA 31.

NFPA 31 is the standard for the Installation of Oil-Burning Equipment. The following NFPA 31 section references highlight the applicability of this standard.

- Section 1.1.1 applies to the installation of stationary oil-burning equipment and appliances, including but not limited to industrial, commercial, and residential steam, hot water, or warm air heating plants.
- Section 1.1.2 applies to all accessory equipment and control systems, whether electric, thermostatic, or mechanical, and all electrical wiring connected to oil-fired equipment.

- Section 1.1.3 applies to the installation of liquid fuel storage and supply systems connected to liquid fuel-burning appliances.
- Section 1.1.4 applies to multi-fueled appliances in which fuel oil is one of the optional fuels.

1002.3 Electric Boilers. Electric boilers shall be listed and labeled in accordance with UL 834.

UL 834, Electric Heating, Water Supply and Power Boilers covers the requirements for electric heating, water supply, and power boilers rated at 600 volts or less intended for commercial or industrial applications utilizing hot water or steam. This may also be applied to commercial, industrial, or residential use space heating applications. These boilers are intended for installation in accordance with the National Electrical Code, NFPA 70, and other applicable codes, such and the Uniform Mechanical Code. Each boiler consists of sheathed resistance-type heating elements and a vessel or a tank constructed, inspected, and stamped in accordance with the applicable sections of the ASME Boiler and Pressure Vessel Code. Each boiler is to be provided with one or more safety valves or safety relief valves conforming to ASME requirements with all necessary temperature or pressure regulating controls, including an integral limit control, wiring, and auxiliary equipment assembled as a unit. The equipment covered by the standard shall be one of the following types of water heating boilers.

1. High Pressure — A boiler furnishing:
   a. Steam at pressures in excess of 15 psi (130kPa); or
   b. Hot water at temperatures in excess of 250°F (121°C) or at pressures in excess of 160 psi (1103 kPa).
2. Low-Pressure Hot-Water and Low-Pressure Steam — A boiler furnishing:
   a. Hot water at pressures not exceeding 160 psi and at temperatures not more the 250°F; or
   b. A boiler furnishing steam at pressures not more than 15 psi.
3. Miniature — A boiler that does not exceed the following limits:
   a. 16 inches inside diameter of shell; and
   b. 5 cubic feet gross volume, exclusive of casing and insulation, and 100 psi (690 kPa) maximum allowable working pressure.

1002.4 Solid-Fuel-Fired Boilers. Solid-fuel-fired boilers shall comply with UL 2523 and shall be installed in accordance with the manufacturer's installation instructions.

1002.5 Dual Purpose Water Heater. Water heaters utilized for combined space- and water-heating applications shall be listed or labeled in accordance with the standards referenced in Table 1203.2, and shall be installed in accordance with the manufacturer's installation instructions.

1003.0 Detailed Requirements.

1003.1 Safety Requirements. The construction of boilers and pressure vessels and the installation thereof shall be in accordance with minimum requirements for safety from structural and mechanical failure and excessive pressures as established by the Authority Having Jurisdiction in accordance with nationally recognized standards.

1003.2 Controls. Required electrical, mechanical, safety, and operating controls shall carry the approval of an approved testing agency or be accepted by the Authority Having Jurisdiction. Electrical controls shall be of such design and construction as to be suitable for installation in the environment in which they are located.

Controls mentioned in Chapter 10 are the low-water cutoff (see Section 1008.0) and those listed in Table 1003.2.1 for automatic boilers. This list, while not definitive, provides a basic control array typically expected on boiler installations:

- Assured Fuel Supply Control
- Assure Air Supply Control
- Low-Fire Start-Up Control
- Pre-Purging Control
- Hot Water Systems:
  - Temperature
  - Low-Water Limit
- Steam Systems
  - Pressure
  - Low-Water Limit
- Fuel Shutoff
- Control and Limit Device System Design

Most controls are simply on/off switches. The control senses a condition and either turns on or turns off from its "normal" condition. The "normal" condition of a control is its de-energized condition, which is either normally closed (NC) or normally open (NO).

1003.2.1 Automatic Boilers. Automatic boilers shall be equipped with controls and limit devices in accordance with ASME CSD-1 or Table 1003.2.1.

The Authority Having Jurisdiction shall have the authority to approve solid-fuel-fired boilers that comply with the safety requirements for automatic gas fired boilers or oil fired boilers.

The rules of this Standard include requirements for the assembly, installation, maintenance, and operation of controls and safety devices on automatically operated boilers directly fired with gas, oil, gas-oil, or electricity, having fuel input ratings under 12,500,000 Btu/hr.

1003.3 Gauges. Steam boilers shall be provided with a pressure gauge and a water level glass. Water boilers shall be provided with a pressure gauge and a temperature gauge. Automatic boilers shall be equipped with the following gauges, as applicable:

(1) Oil temperature

(2) Oil suction pressure

(3) High and low gas pressure

(4) Stack temperature

(5) Windbox pressure

When the boiler is off, it may read 2/3 full. When it is running, it reads 1/3 full. When the water is not visible, it is probably critically close to an unsafe water level condition (see **Figure 1003.3a**).

There is no such thing as a steady water level in a gauge glass. The water level is actually a range of operation somewhere between the top 2/3 of the glass to the bottom 1/3 of the glass. It is very common to see the water in the sight glass moving up and down.

Two conditions should be avoided:

- If the water is at the top of the gauge glass while the burner is firing, steam is probably not being made.
- Boiler manufacturers' safe water level for their boilers is usually just out of sight of the bottom of the gauge glass.

Tridicators are common boiler gauges, which display both temperature and pressure on the same gauge face (see **Figure 1003.3b**). The pressure is typically reported in pounds per square inch gauge (psig).

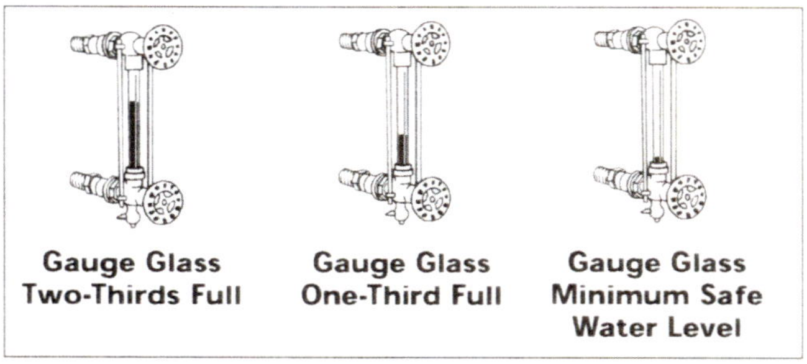

FIGURE 1003.3A
SIGHT GLASS LEVELS
(Courtesy of McDonnell & Miller)

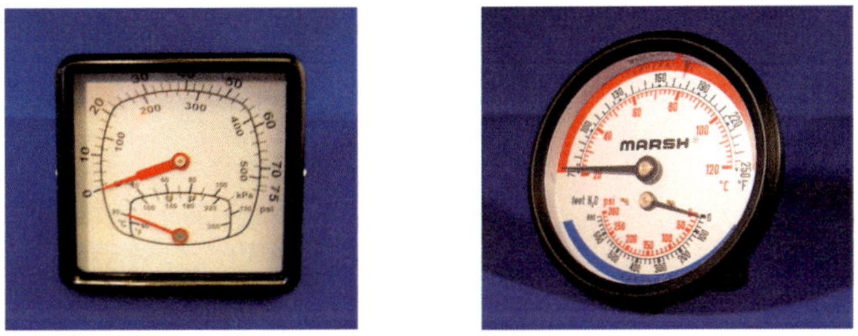

FIGURE 1003.3B
TRIDICATORS (COMBINATION GUAGES)
(Courtesy of Adron Industries)

1003.4 Stack Dampers. Stack dampers on boilers fired with oil or solid fuel shall not close off more than 80 percent of the stack area where closed, except on automatic boilers with prepurge, automatic draft control, and interlock. Operative dampers shall not be placed within a stack, flue, or vent of a gas-fired boiler, except on an automatic boiler with prepurge, automatic draft control, and interlock.

Stack dampers and vent dampers perform the same basic function – they stop chimney draft from drawing heated air from the building. Generally, stack dampers are bimetal and thermally activated while vent dampers tend to be electrical and automatically controlled (see **Figure 1003.4**).

When the vent or stack is left open when the boiler is not firing, conditioned air is pulled through the chimney due to "stack effect." Also, a closed damper substantially reduces boiler standby losses.

Electrically controlled vent dampers rotate to their open position when the boiler is going to fire. An interlock prevents burner operation unless the damper is verified open by an integral switch. If the damper does not open, the burner circuit will not be completed and the burner will not be allowed to fire. When the burner shuts down, the damper will rotate to the closed position.

Under no circumstances should a damper interlock be bypassed.

FIGURE 1003.4
AUTOMATIC VENT DAMPER
(Courtesy of Slant Fin)

1003.5 Welding. Welding on pressure vessels shall be done by certified welders in accordance with nationally recognized standards.

Welding on pressure vessels and piping shall be performed by a welder certified to the ASME Boiler and Pressure Vessel Code, Section IX, who holds a valid certificate of competence from a recognized testing laboratory.

1004.0 Expansion Tanks.

When water is heated, its density decreases and its volume increases. When a closed hydronic system is filled with cold water and then heated to a high limit, the result is about 5 percent more water than before it was heated. Since water is not compressible, the extra volume created by expansion must go someplace. There are only two options: water containment or water relief.

Water containment (generally the preferred method) gives the water someplace to go (expansion tank). Water relief allows the additional water volume to leave the system (relief valve).

The water relief solution discharges the thermally expanded water until the system reaches a pressure setting below the setting of the water heater's temperature and pressure-relief valve (30 psi, for example). The downside to the water relief strategy is that once the system water cools, the volume decreases and with it so does the system pressure. The automatic fill features are set based on system pressure and will refill the system. Not only wasteful, this creates a potentially undesirable condition where oxygen introduced into the system damages the iron parts of the boiler and components (pumps, fittings, etc.).

Water containment solutions retain additional water volume in the system by using expansion tanks. Expansion tanks work on the principle that you can't compress water, but you can compress air. As the system pressure increases and decreases due to thermal expansion, the air in the tank becomes a shock absorber or pneumatic spring. If there is no operable compression tank to absorb the additional water volume, the relief valve will probably open every time the boiler burner comes on.

Also, if the system isn't working like it should and overruns the expansion tank's ability to do its job, the relief valve is going to do its job and open to get rid of the excess pressure. Chronic opening of the relief valve is not good. Minerals in the water may clog the valve in an open or closed position.

1004.1 General. An expansion tank shall be installed in a hot-water-heating system as a means for controlling increased pressure caused by thermal expansion. Expansion tanks shall be of the closed or open type and securely fastened to the structure. Tanks shall be rated for the pressure of the system. Supports shall be capable of carrying twice the weight of the tank filled with water without placing a strain on connecting piping.

Hot-water-heating systems incorporating hot water tanks or fluid relief columns shall be installed to prevent freezing under normal operating conditions.

Providing adequate support for the expansion tank is no small matter. Besides the weight of the tank itself, water weighs 8.33 pounds per gallon. **Figure 1004.1a** is an expansion tank support table of requirements for a particular series of diaphragm expansion tanks. A 30-gallon tank weighs 75 pounds plus 250 pounds of water (if full). The total support required is 650 pounds. Of course, the diaphragm tank would never be filled completely unless the diaphragm was ruptured. **Figure 1004.1b** is an example of an expansion tank support.

To prevent freezing, propylene (food grade) glycol is mixed in with the boiler water up to a 50-50 ratio. Propylene glycol weighs slightly less than water. Ethylene glycol weighs more than water but is not recommended because of the health hazards involved.

Examples of Support Required by Tank Capacity

Capacity (Gal)	Tank Weight	Water Weight	Total Weight	Support Required
15	51	125	176	352
24	71	200	271	542
30	75	250	325	650
40	95	333	428	856
60	129	500	629	1258
80	149	666	815	1631
100	185	833	1018	2036
120	219	1000	1219	2437
135	239	1125	1364	2727
175	327	1458	1785	3570
220	392	1833	2225	4449
240	424	1999	2423	4846
304	526	2532	3058	6117
400	688	3332	4020	4020

FIGURE 1004.1A
EXPANSION TANK SUPPORT TABLE

FIGURE 1004.1B
EXPANSION TANK SUPPORT

1004.2 Open-Type Expansion Tanks. Open type expansion tanks shall be located not less than 3 feet (914 mm) above the highest point of the system. Such tanks shall be sized based on the capacity of the system. An overflow with a diameter of not less than one-half the size of the supply or not less than 1 inch (25 mm) in diameter shall be installed at the top of the tank. The overflow shall discharge through an air gap into the drainage system.

With an open expansion tank, fresh air absorbed into the water tends to corrode the system.

Open expansion tanks must be located at least three feet above the highest point of the system, sometimes on the top of buildings, where it may be exposed to freezing (see **Figure 1004.2**).

Open systems use a "feed and expansion" tank (cistern). As the name suggests, the tank is used to feed the supply and to accommodate any water expansion that is generated by the heating process. The "feed and expansion" tank is placed at the highest point in the system. An open-vented system typically uses a small-bore two pipe network, where one pipe is used to feed the system and the other allows the cooled water to return to the heat source (boiler).

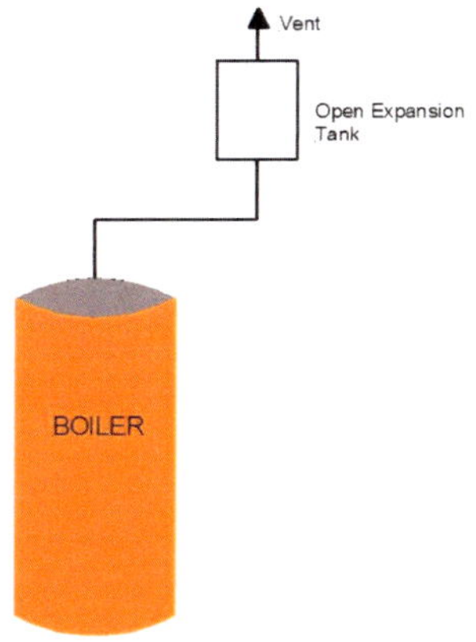

FIGURE 1004.2
OPEN EXPANSION TANK
(Courtesy of Praxis Green Inc.)

1004.3 Closed-Type Systems. Closed-type systems shall have an airtight tank or other approved air cushion that will be consistent with the volume and capacity of the system, and shall be designed for a hydrostatic test pressure of two and one-half times the allowable working pressure of the system. Expansion tanks for systems designed to operate at more than 30 pounds-force per square inch (psi) (207 kPa) shall comply with ASME BPVC Section VIII. Provisions shall be made for draining the tank without emptying the system.

There are two types of expansion tanks used on hot water boiler systems: the steel expansion tank and the diaphragm (bladder) expansion tank.

A closed-type system has an airtight tank or other suitable air cushions that are designed for a hydrostatic test pressure of two and one-half times the allowable working pressure.

Steel Tank

Although most new systems use the diaphragm-type expansion tank, many steel tanks are still in service (see **Figure 1004.3a**).

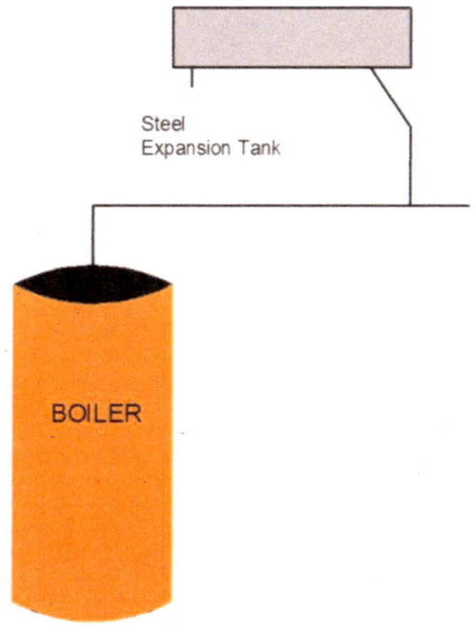

FIGURE 1004.3A
STEEL EXPANSION TANK
(Courtesy of Praxis Green Inc.)

A steel expansion tank is exactly that — an enclosed steel tank with holes for fittings. One fitting is the connection to the boiler system, while the other is typically a drain valve. There should be a gate or ball valve in the pipe between the boiler system and the expansion tank to isolate the expansion tank from the system pressure to allow for the draining of a "waterlogged" tank. Waterlogged expansion tanks have lost the air required to act as a cushion.

In a steel expansion tank the air and water touch each other, while in a bladder-type expansion tank the air and water are separated by a diaphragm (see **Figure 1004.3b**).

The sizing of the steel expansion tank is based on the size of the hydronic loop and hot water boiler system (see Section 1004.4).

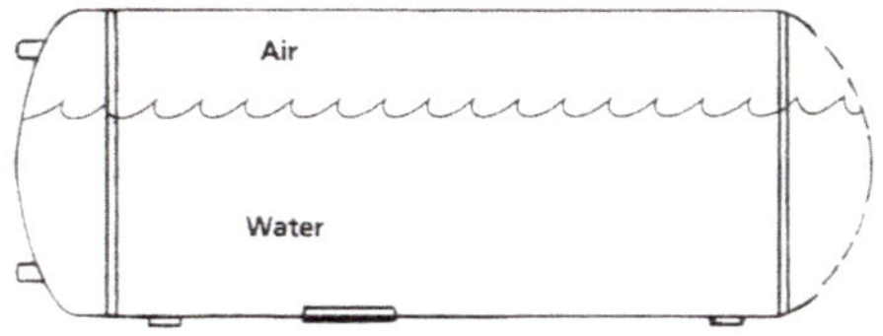

FIGURE 1004.3B
STEEL EXPANSION TANK
(Courtesy of Praxis Green Inc.)

Diaphragm Tank

See **Figures 1004.3c and 1004.3d**.

Fresh out of the shipping container, the air side of the diaphragm is fully expanded and flush against the inside of the tank. When connected to a pressurized system, the water's pressure pushes the diaphragm against the compressed air and squeezes it like a balloon.

As the system temperature and pressure increases, the diaphragm flexes against the air cushion to allow for increased water expansion. When the system water cools, the water leaves the tank and returns to the system.

Other than periodic pressure checks, diaphragm tanks should not need service. If the diaphragm develops a leak, the tank will lose its air cushion and will become waterlogged. The system will seem like there is no expansion tank, which means the relief valve is likely to open when the system heats up.

An undercharged diaphragm expansion tank will allow cold water to enter the tank when the system is cold, while an overcharged tank will not allow heated water to expand into the tank.

Diaphragm tanks separate the air from the water with a flexible rubber diaphragm. They're generally smaller because one side of the diaphragm is precharged with compressed air (see **Figure 1004.3e**).

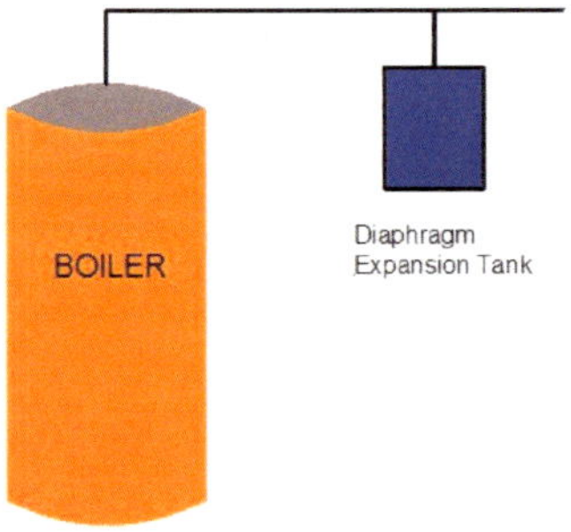

FIGURE 1004.3C
DIAPHRAGM TANK
(Courtesy of Praxis Green Inc.)

FIGURE 1004.3D
DIAPHRAGM EXPANSION TANK
(Courtesy of Watts Regulator)

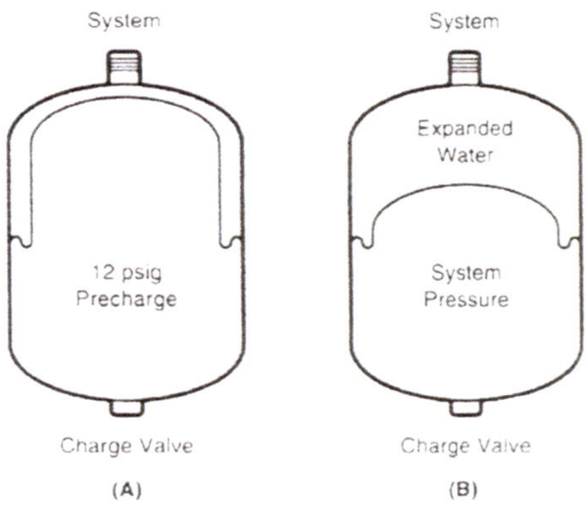

FIGURE 1004.3E
DIAPHRAGM EXPANSION TANK

1004.4 Minimum Capacity of Closed-Type Tank. The minimum capacity for a gravity-type hot water system expansion tank shall be in accordance with Table 1004.4(1). The minimum capacity for a forced-type hot water system expansion tank shall be in accordance with Table 1004.4(2), or Equation 1004.4. Equation 1004.4 shall not be used for diaphragm-type expansion tanks.

(Equation 1004.4)

$$V_t = \frac{(0.00041t - 0.0466)\ V_s}{\left(\frac{P_a}{P_f} - \frac{P_a}{P_o}\right)}$$

Where:

V_t = Minimum volume of expansion tank, gallons

V_s = Volume of system, not including expansion tank, gallons

t = Average operating temperature, °F

P_a = Atmospheric pressure, feet H_2O absolute

P_f = Fill pressure, feet H_2O absolute

P_o = Maximum operating pressure, feet H_2O absolute

For SI units: 1 gallon = 3.785 L, °C = (°F-32)/1.8, 1 foot of water = 2.99 kPa

TABLE 1004.4(1)
EXPANSION TANK CAPACITIES FOR GRAVITY HOT WATER SYSTEMS[1]

INSTALLED EQUIVALENT DIRECT RADIATION[2] (square feet)	TANK CAPACITY (gallons)
Up to 350	18
Up to 450	21
Up to 650	24
Up to 900	30
Up to 1100	35
Up to 1400	40
Up to 1600	2 to 30
Up to 1800	2 to 30
Up to 2000	2 to 35
Up to 2400	2 to 40

For SI units: 1 gallon = 3.785 L, 1 square foot = 0.0929 m^2

Notes:

[1] Based on a two-pipe system with an average operating water temperature of 170°F (77°C), using cast-iron column radiation with a heat emission rate of 150 British thermal units per square foot hour [Btu/(ft^2•h)] (473 W/m^2) equivalent direct radiation.

[2] For systems that exceed 2400 square feet (222.9 m^2) of installed equivalent direct water radiation, the required capacity of the cushion tank shall be increased on the basis of 1 gallon (4 L) tank capacity per 33 square feet (3.1 m^2) of additional equivalent direct radiation.

TABLE 1004.4(2)
EXPANSION TANK CAPACITIES FOR FORCED HOT WATER SYSTEMS[1]

SYSTEM VOLUME[2] (gallons)	TANK CAPACITY DIAPHRAGM TYPE (gallons)	TANK CAPACITY NONDIAPHRAGM TYPE (gallons)
100	9	15
200	17	30
300	25	45
400	33	60
500	42	75
1000	83	150
2000	165	300

For SI units: 1 gallon = 3.785 L

Notes:

[1] Based on an average operating water temperature of 195°F (91°C), a fill pressure of 12 psig (83 kPa), and an operating pressure of not more than 30 psig (207 kPa).

[2] Includes volume of water in boiler, radiation, and piping, not including expansion tank.

This code section is best illustrated with an example:

TABLE 1004.4(1)

SYMBOL	INITIAL INPUT	SOURCE
V_s	253 gallons	Calculation
t	160°F	Average temperature between off and high limit
P_a	14.7 psi	Standard atmospheric pressure
P_f	12 psig	Standard fill valve and expansion tank settings
P_o	30 psig	A standard small system safety "pop" rating

$$Vt = \frac{(0.00041t - 0.0466)\,Vs}{\left(\frac{Pa}{Pf} - \frac{Pa}{Po}\right)}$$

P_a, P_f and P_o are in feet H_2O absolute. What does that mean? Absolute pressure is zero referenced against a perfect vacuum, so it is equal to gauge pressure plus atmospheric pressure.

Feet of H_2O is the "weight" or pressure of a 12-inch column of water. If the user is going to use psig, he or she will have to convert those numbers into feet H_2O absolute.

To convert psi to feet H_2O absolute divide by 0.4335.

$$\frac{\text{psia}}{0.4335} = \text{Feet } H_2O \text{ Absolute}$$

P_f - cold fill pressure = 12 psig on the boiler gauge

To change that from psig to psia, add 14.7 psi which equals 26.7 psia.

$$\frac{12 + 14.7}{0.4335} = 61.6 \text{ Feet } H_2O \text{ Absolute}$$

P_o maximum operating pressure rating of the temperature and pressure valve is a good working point since the maximum pressure should not exceed the safety rating.

$$\frac{30 + 14.7}{0.4335} = 103.1 \text{ Feet } H_2O \text{ Absolute}$$

TABLE 1004.4(2)

SYMBOL	INITIAL INPUT	ADJUSTED FORMULA INPUT
V_s	253 gallons	253 gallons
t	160°F	160°F
P_a	14.7 psi	33.91
P_f	12 psig	61.6
P_o	30 psig	103.1

$$V_t = \frac{((0.0004x160) - 0.0466x253)}{\left(\frac{33.91}{61.6} - \frac{33.91}{103.1}\right)}$$

$$V_t = \frac{0.019 \times 253}{(0.55 - 0.32)}$$

$$V_t = \frac{4.807}{0.223}$$

$$V_t = 21.6$$

The expansion tank required for this system will have a capacity of 22 gallons.

"Forced hot water systems" are systems with circulators or pumps compared to gravity systems.

In Table 1004.4(2), the fill pressure of 12 psi is based on the factory setting for an automatic fill valve, which is also the typical fill pressure for a two-story residence. The maximum operating pressure of 30 psig is the typical boiler relief valve shipped with residential and small commercial boilers. For the most part, 30-pound and 50-pound relief valves are typically seen on residential and light commercial systems.

Multiple tanks may be used to accommodate the thermal expansion of higher temperature or higher pressure systems.

1005.0 Safety or Relief Valve Discharge.

1005.1 General. Pressurized vessels or boilers shall be provided with overpressure protection by means of a listed pressure relief valve installed in accordance with the manufacturer's installation instructions.

The pressure relief valve is the final means of protection in a situation where everything else has failed to limit heat production.

The opening pressure of the valve and the maximum heating capacity of the equipment the valve can support are stamped onto a permanent tag attached to the valve. During inspections, this should be checked to ensure that the capacity of the relief valve is equal to or greater than the maximum rating of the equipment (see **Figures 1005.1a and 1005.1b**).

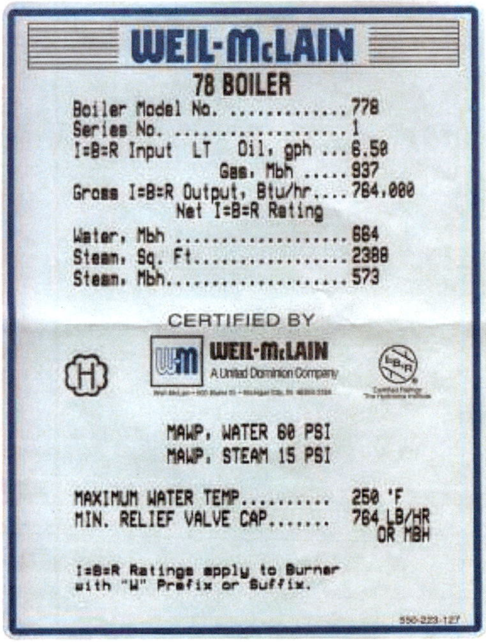

FIGURE 1005.1A
THE PLATE AFFIXED TO THE BOILER SHOWS MIN. RELIEF VALVE CAP (MINIMUM RELIEF VALVE CAPACITY) OF 764 MBH
(Courtesy of Weil-McLain)

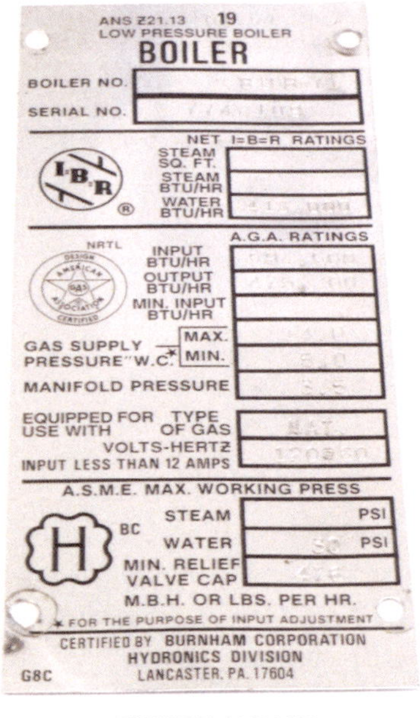

FIGURE 1005.1B
A METAL PLATE SHOWING PRESSURES
(With the serial number, this plate is unique to the boiler to which it is attached.)
(Courtesy of First Supply Corp and Praxis Green Inc.)

1005.2 Discharge Piping. The discharge piping serving a temperature relief valve, pressure relief valve, or combination of both shall have no valves, obstructions, or means of isolation and provided with the following:

(1) Equal to the size of the valve outlet and shall discharge full size to the flood level of the area receiving the discharge and pointing down.

(2) Materials shall be rated at not less than the operating temperature of the system and approved for such use.

(3) Discharge pipe shall discharge independently by gravity through an air gap into the drainage system or outside of the building with the end of the pipe not exceeding 2 feet (610 mm) and not less than 6 inches (152 mm) above the ground and pointing downwards.

(4) Discharge in such a manner that does not cause personal injury or structural damage.

(5) No part of such discharge pipe shall be trapped or subject to freezing.

(6) The terminal end of the pipe shall not be threaded.

(7) Discharge from a relief valve into a water heater pan shall be prohibited.

The waste pipe from the valve should end between two feet and six inches above the floor or drain and should be unthreaded to prevent tampering. Common field practice is to end the waste pipe not more than 6 inches above the floor to reduce splashing and scalding potential during discharge. The end of the discharge tube should be readily visible so that if the relief valve is leaking or open, it can be easily seen.

The primary purpose of a pressure-relief valve is protection of life and property by venting fluid from an over pressurized vessel. Even though it may seem obvious, a manufacturer was compelled to state in its specification sheet, "This device is designed for emergency safety release and shall not be used as an operating control."

Periodically check that the valve can open and is not corroded closed, which could make it ineffective (see **Figures 1005.2a and 1005.2b**).

1005.3 Splash Shield. Where the operating temperature exceeds 212°F (100°C), the discharge pipe shall be installed with a splash shield or centrifugal separator.

1005.4 Hazardous Discharge. Where the discharge from safety valves is capable of being hazardous, discharge of steam inside the boiler room, such discharge shall be discharged to the outside of the boiler room. Discharges from relief valves on industrial boilers shall be discharged to an approved location.

FIGURE 1005.2A
PERIODIC TESTING OF RELIEF VALVE

FIGURE 1005.2B
WATER AND SCALE BUILDUP ON FLOOR
(This indicates the relief valve has been leaking past the seat for some time. This valve should be replaced as soon as possible as it could scale up and become inoperative.)

1005.5 Vacuum Relief Valve. Hot-water heating systems that are subjected to a vacuum while in operation or during shutdown shall be protected with a vacuum relief valve. Where the piping configuration, equipment location, and valve outlets are located below the boiler elevation, the system shall be equipped with a vacuum relief valve at the highest point.

Vacuum relief valves should be installed in hot-water systems that are subjected to a vacuum. Vacuum relief valves are required to facilitate full draining of the system during shutdown and servicing.

1006.0 Shutoff Valves.

1006.1 General. An approved manual shutoff valve shall be installed upstream of all control devices on the main burner of a gas-fired boiler. The takeoff point for the gas supply to the pilot shall be upstream of the gas shutoff valve of the main burner and shall be valved separately. A union or other approved means of disconnect shall be provided immediately downstream of these shutoff valves.

See **Figure 1006.1**.

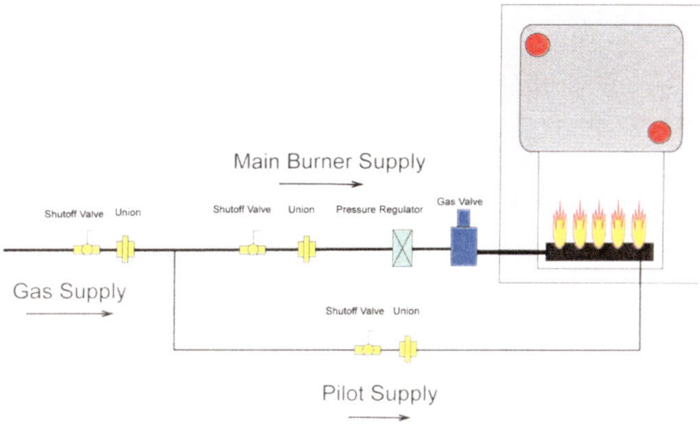

FIGURE 1006.1
GAS SHUTOFF VALVES
(Courtesy of Praxis Green Inc.)

1007.0 Gas-Pressure Regulators.

1007.1 General. An approved gas-pressure regulator shall be installed on gas-fired boilers where the gas supply pressure is exceeding that at which the main burner is designed to operate. A separate approved gas-pressure regulator shall be installed to regulate the gas pressure to the pilot or pilots.

Gas pressure regulators reduce high-pressure gas in a cylinder or line to a lower, usable level as well as maintain pressure within a gas delivery system (see **Figure 1007.1**). A pressure regulator is a normally open (NO) valve that takes a high inlet pressure and converts it to a lower, preset downstream pressure. They are not flow control devices; they are used to control delivery pressure only.

There are three basic operating components in gas pressure regulators: a loading mechanism, a sensing- and a control-element.

- Loading Element – Applies the needed force to the restricting element. This can be a weight, a spring, a piston actuator or an elastomer or metal diaphragm actuator in combination with a spring. Loading mechanisms for gas pressure regulators determine the setting of the regulator delivery pressure.
- Measuring/Sensing Element – Tells when the inlet flow is equal to the outlet flow.
- Restricting/Control Element – A globe valve, butterfly valve, poppet valve or any valve that can create a variable restriction to the flow.

There are four main types of gas pressure regulators: line, general-purpose, high-purity and special service regulators.

- Line regulators – Typically point-of-use regulator - serving low-pressure pipelines.
- General-purpose regulators – Designed for economy and longevity, and generally used in noncorrosive general plant and maintenance shop applications.
- High-purity regulators – Designed and constructed to provide diffusion, resistance and easy cleanup.
- Special service regulators – Specifically constructed for specialized applications including oxygen, acetylene and fluorine service, high-pressure, ultra high-pressure and corrosion service.

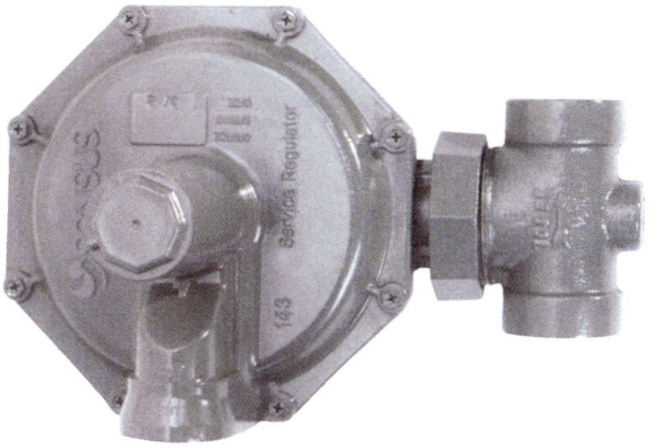

FIGURE 1007.1
GAS PRESSURE REGULATOR
(Courtesy of Sensus Metering Systems)

1008.0 Low-Water Cutoff.

1008.1 General. Hot water boilers and steam boilers shall be installed with a low-water cutoff. A coil-type boiler or a water-tube boiler that requires forced circulation to prevent overheating of the coils or tubes shall be installed with a flow-sensing device in the outlet piping in lieu of the low-water cutoff. The low-water cutoff or the flow sensing device shall be installed so as to prevent damage to the boiler and to permit testing of the fuel-supply cutoff without draining the heating system. The low-water cutoff shall shut off the combustion at a water level setpoint that is in accordance with the boiler manufacturer's instructions.

A low-water cutoff (LWCO) is a mechanical or electronic sensor that monitors the water level in a boiler or boiler system and turns off electrical power to the burner if the water level or pressure falls below a preset, safe level. The boiler needs water to move the heat away from metal surfaces.

Mechanical-Float LWCO

Original LWCO valve designs used a mechanical float that operated like a light switch using the water level as the actuator (see **Figures 1008.1a and 1008.1b**). The water level in the LWCO chamber will mimic the water level in the boiler. When the water level drops, the float valve also drops, opening the switch and turning off the burner. It will stay off until the water level raises the float to a safe level and the circuit is closed again.

Float-operated low water-cutoffs have been around since the 1920s. This type of LWCO is typically mounted in the boiler's gauge glass tappings.

Mechanical-float LWCOs can get hung up or jammed due to sludge that forms as water is lost from the boiler and mineral debris is left behind. The valve should be cleaned and flushed regularly.

Electronic LWCO

Electronic LWCO controls have generally replaced the mechanical float. The electronic LCWO uses a switch and a sensor, reducing the chances of a mechanical malfunction due to sludge or debris (see **Figures 1008.1c and 1008.1d**).

Like the mechanical-float-type LWCO, they will interrupt the electrical current to the burner if the water in the system drops below the sensor's probe. A probe uses the boiler's water to complete an electrical circuit past an insulator (the center portion of the probe) back to a ground (the threaded portion of the probe). As long as water covers the probe, the circuit is completed to the burner. When water drops off the probe for 10 seconds, the burner circuit is open, preventing the burner from firing.

Some electronic LWCO also provide an additional circuit for a low-water alarm.

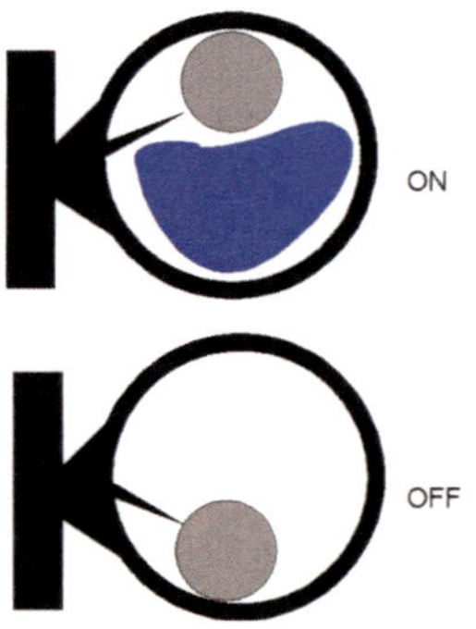

FIGURE 1008.1A
FLOAT LWCO

FIGURE 1008.1B
FLOAT TYPE LWCO
(Courtesy of McDonnell & Miller)

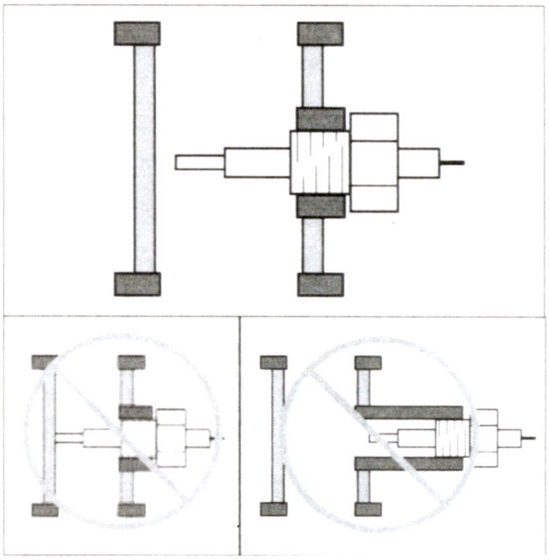

FIGURE 1008.1C
INSTALLATION OF PROBE OF ELECTRONIC LWCO
(Courtesy of McDonnell & Miller)

FIGURE 1008.1D
ELECTRONIC LWCO
(Courtesy of Taco)

1009.0 Combustion Regulators – Safety Valves.

1009.1 General. The following requirements shall be retroactive:

(1) Hot-water-heating boilers, other than manually fired, shall be equipped with two temperature combustion regulators in series. Steam-heating boilers, other than manually fired, shall be equipped with a pressure combustion regulator and a low-water cutoff. (See Section 1008.0)

Section 1003.2.1 regulates automatic boilers.

(2) Boilers and pressure vessels shall be provided with the required number, size, and capacity of safety or relief valves to ensure positive relief of overpressure in accordance with nationally recognized standards, as applicable. Valves so employed shall be constructed, sealed, and installed in accordance with nationally recognized standards, as applicable.

The tremendous violence of exploding hot water storage tanks and heaters can result in property damage and personal injury. To prevent such catastrophic accidents, all devices that heat water are required to be equipped with pressure- or temperature-relief devices, or a combination of both, called a T&P valve (see **Figure 1009.1**).

A gallon of water vaporizes into about 6,000 cubic feet of steam. Water conversion from liquid to vapor is harmless when steam vapor is gradually released from an open vessel; however, it can be catastrophically explosive when it is suddenly released from a failed boiler or water heater to the atmosphere.

Information Available From T&P Plate or Tag

Here are some important facts that a T&P plate or tag will provide:

- The date of manufacture — The four-digit dating code (e.g. 8322 was manufactured the 22nd week of 1983).
- The ASME data plate rating — Steam pressure containment or controlled release. The ASME rating reflects the maximum amount of heat input that a given pressure-relief valve is able to vent as it strives to maintain the "maximum pressure" (psi) for which it is rated.
- The AGA data plate rating — Water temperature control. Standard AGA heat-input limitations (e.g., 200,000 Btus) are based upon the assumption that there is water inflow during the overfiring condition. "AGA Temperature Steam Rating" assumes there will be no cold water coming into the water heater to replace the water flowing out through the relief valve. If there is no replacement water coming into the storage tank, the remaining stored water will heat until it eventually boils off.

Selection

When selecting a T&P relief valve:

1. Use the AGA rating.
2. The Btu/h input from the heat source must never exceed the AGA Temperature Steam Rating.

T&P valves are not operating pressure-control devices. They should not intentionally be used for relief of thermal expansion.

Installation

The probe must always be installed to sense water temperature in the hottest part of the storage tank — normally within the top 6 inches of the vessel. The probe must not be too short to extend through the hot water piping and into the top of the storage tank.

Areas that have lime and calcium in their potable water supply should watch for mineralization of the relief valve sensor probe. Minerals will insulate the probe from the actual water temperature, causing the valve not to open at the rated temperature but rather a higher temperature causing a potentially dangerous condition.

The **Inspection Checklist** is an excerpt from the 1985 Watts Regulator article entitled "*T&P Valve Re-inspection Programs That Work!*"

In conclusion, it is important to remember the following:

1) Both gas and electric boilers and water heaters require temperature- and pressure-relief valves when they are constructed with, or are connected to, a storage tank having a dimension greater than 3 inches in diameter.

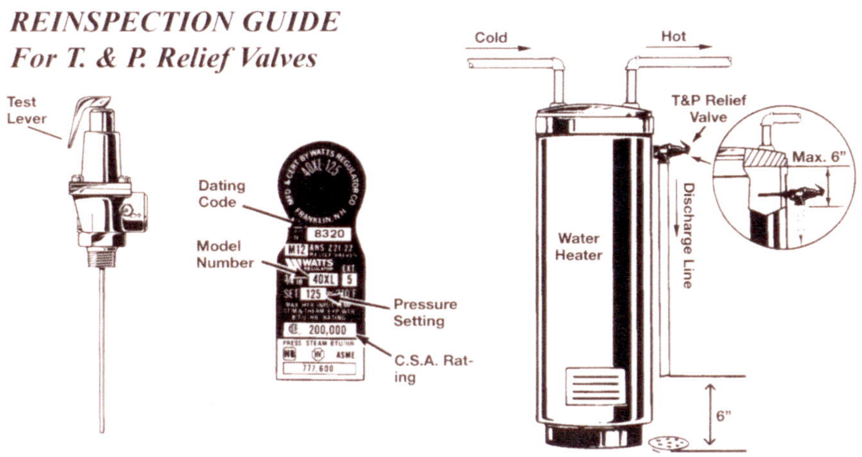

FIGURE 1009.1
TEMPERATURE AND PRESSURE-RELIEF VALVES
(Courtesy of Watts Regulator)

2) When two or more Btu ratings are referenced on the nameplate of a combination T&P valve, always use the lowest ratings shown for water temperature control purposes.
3) Electric water boilers or water heaters generally have a relatively smaller Btu input than a comparable gas-fired water heater. Nonetheless, electric heaters require installation of T&P relief valves of appropriate capacity when there is stored water to be relieved during an over-heat condition. To convert electrical energy input into equivalent Btu per hour (for purposes of sizing relief valves properly) multiply kilowatt-hours by a factor of 3.413.

To help prevent explosions, periodically inspect temperature and pressure relief valves to make sure they have been installed properly, have not been tampered with and are functioning properly.

1010.0 Clearance for Access.

1010.1 General. Where boilers are installed or replaced, clearance shall be provided to allow access for inspection, maintenance, and repair. Passageways around all sides of boilers shall have an unobstructed width of not less than 18 inches (457 mm). Clearance for repair and cleaning shall be permitted to be provided through a door or access panel into another area, provided the opening is of sufficient size.

Exception: Subject to the approval of the Authority Having Jurisdiction, boilers shall be permitted to be installed with a side clearance of less than 18 inches (457 mm), provided that the lesser clearance does not inhibit inspection, maintenance, or repair.

Boilers have a manufacturer's listing plate, which will have the minimum clearances to combustibles and access (see **Figure 1010.1**). Check the listing plate and manufacturer's installation instructions first, and then compare these clearances to the minimum clearances of the UMC.

There is an exception that allows the clearances to be reduced with prior approval by the Authority Having Jurisdiction (AHJ), provided that the lesser clearances do not inhibit inspection, maintenance and repair.

1010.2 Power Boilers. Power boilers having a steam-generating capacity in excess of 5000 pounds per hour (lb/h) (0.6299 kg/s) or having a heating surface in excess of 1000 square feet (92.9 m^2) or input in excess of 5 000 000 Btu/h (1464 kW) shall have a clearance of not less than 7 feet (2134 mm) from the top of the boiler to the ceiling.

1010.3 Steam-Heating Boilers, Hot Water Boilers, and Power Boilers. Steam-heating boilers and hot-water-heating boilers that exceed one of the following limits:

INSPECTION CHECKLIST

ITEM TO INSPECT	WHAT TO DO	PURPOSE
1. Test Lever	Manually lift lever	Following installation, the valve lever MUST be operated AT LEAST ONCE A YEAR by the water heater owner to ensure that waterways are clear. Certain naturally occurring mineral deposits may adhere to the valve, blocking waterways, rendering it inoperative. When the lever is operated, hot water will discharge if the waterways are clear. PRECAUTIONS MUST BE TAKEN TO AVOID PERSONAL INJURY FROM CONTACT WITH HOT WATER AND TO AVOID PROPERTY DAMAGE.
2. Valve Location	Observe how valve is installed	Valve thermostat must be immersed in tank water and located in top 6" of tank to accurately sense temperature.
3. Discharge Line	Observe size and direction	Discharge line must be installed to avoid water damage and scalding injury, when valve operates.
4. Nameplate	a. Observe pressure setting	Pressure-relief setting cannot exceed working pressure of tank.
	b. Observe CSA rating	CSA rating must be in excess of Btu of heater.
	c. Observe model number	To ensure that valve is temperature and pressure type rather than plain pressure relief.
	d. Observe dating code	To determine age of valve, all devices have a 4 digit serial number dating code. The first two digits are the year and the last two digits are the week of the year the valve was manufactured (i.e., 8320 is 20th week of 1983.)
5. Complete Valve	Remove valve from tank	TEMPERATURE- AND PRESSURE-RELIEF VALVES should be inspected AT LEAST ONCE EVERY THREE YEARS, and replaced if necessary, by a licensed plumbing contractor or qualified service technician to ensure that the product has not been affected by corrosive water conditions and to ensure that the valve and discharge line have not been altered or tampered with illegally. Certain naturally occurring conditions may corrode the valve or its components over time, rendering the valve inoperative. Such conditions can only be detected if the valve and its components are physically removed and inspected. Do not attempt to conduct an inspection on your own. Contact your plumbing contractor for a reinspection to assure continuing safety.
6. Inspection Log	Log inspection data.	To provide a record of inspection date and results of inspection.

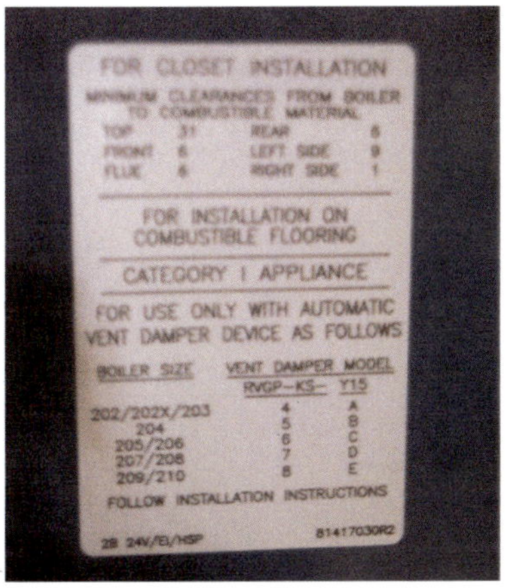

FIGURE 1010.1
SAMPLE BOILER LISTING PLATE SHOWING MINIMUM CLEARANCES FROM BOILER TO COMBUSTIBLE MATERIALS

(1) 5 000 000 Btu/h input (1464 kW)

(2) 5000 pounds steam per hour (0.6299 kg/s) capacity

(3) 1000 square foot (92.9 m^2) heating surface

Power boilers that do not exceed one of the following limits:

(1) 5 000 000 Btu/h input (1464 kW)

(2) 5000 pounds steam per hour (0.6299 kg/s) capacity

(3) 1000 square foot (92.9 m^2) heating surface

Boilers with manholes on top of the boiler, except those described in Section 1010.2 and Section 1010.4, shall have a clearance of not less than 3 feet (914 mm) from the top of the boiler to the ceiling.

1010.4 Package Boilers, Steam-Heating Boilers, and Hot-Water-Heating Boilers. Package boilers, steam-heating boilers, and hot-water-heating boilers with no manhole on top of the shell and not exceeding one of the above limits shall have a clearance of not less than 2 feet (610 mm) from the ceiling.

1011.0 Boilers, Stokers, and Steam Generators.

1011.1 General. The design, installation, and operation of single burner boilers, multiple burner boilers, stokers, and atmospheric fluidized-bed boilers with not less than a fuel input rating of 12.5 E+06 Btu/h (3.663 MW) to pulverized fuel systems, fired or unfired steam generators used to recover heat from combustion turbines and to other combustion turbine exhaust systems shall be in accordance with NFPA 85. That portion of the oil-burning system supplied on boilers and covered within the scope of NFPA 85 shall be installed in accordance with NFPA 85.

See commentary in Section 1002.1 (3) for more on NFPA 85.

1012.0 Operating Adjustments and Instructions.

1012.1 General. Hot water boiler installations, upon completion, shall have controls set, adjusted, and tested by the installing contractor. A complete control diagram of a permanent legible type, together with complete boiler operating instructions, shall be furnished by the installer for each installation.

1013.0 Inspections and Tests.

1013.1 General. An installation for which a permit is required shall not be put into service until it has been inspected and approved by the Authority Having Jurisdiction.

It shall be the duty of the owner or his authorized representative to notify the Authority Having Jurisdiction that the installation is ready for inspection and test. It also shall be the duty of the owner or his authorized representative to post in a conspicuous position on the installation a notice in substantially the following form: "Warning! This installation has not been inspected and approved by the Authority Having Jurisdiction and shall not be covered or concealed until so inspected and approved," and it shall be unlawful for anyone other than the Authority Having Jurisdiction to remove such notice. The Authority Having Jurisdiction shall require such tests as it deems necessary to determine that the installation is in accordance with the provision of this section. Such tests shall be made by the owner or his authorized representative in the presence of the Authority Having Jurisdiction.

Exception: On installations designed and supervised by a registered design professional, the Authority Having Jurisdiction shall have the authority to permit inspection and testing by such registered design professional.

Where the owner or his authorized representative requests inspection of a boiler prior to its installation, the Authority Having Jurisdiction shall make such inspection.

1013.2 Operating Permit. It shall be unlawful to operate a boiler or pressure vessel without first obtaining a valid operating permit to do so from the Authority Having Jurisdiction. Such permit shall be displayed in a conspicuous place adjacent to the boiler or vessel. The operating permit shall not be issued until the equipment has been inspected and approved by the Authority Having Jurisdiction.

Exception: The operation of steam-heating boilers, low-pressure hot-water-heating boilers, hot water supply boilers, and pressure vessels in residential occupancies of less than six dwelling units and utility occupancies.

1013.3 Maintenance Inspection. The Authority Having Jurisdiction shall inspect boilers and pressure vessels operated under a permit in accordance with ASHRAE/ACCA 180 at such intervals as deemed necessary, but not less frequently than in accordance with Section 1013.4 through Section 1013.7.

1013.4 Power and Miniature Boilers. Power boilers and miniature boilers shall be inspected externally annually. Where construction and operating conditions permit, they shall be subject to inspection internally annually.

1013.5 Steam-Heating and Water-Heating Boilers. Steam-heating boilers and hot-water-heating boilers shall be inspected externally annually. Where construction and operating conditions permit, they shall also be subject to inspection internally annually.

1013.6 Automatic Steam-Heating Boilers. Automatic steam-heating boilers shall be inspected externally biennially. Where construction and operating conditions permit, they shall be subject to inspection internally biennially.

1013.7 Unfired Pressure Vessels. Unfired pressure vessels shall be inspected externally biennially. Where subject to corrosion and construction permits, they shall be subject to inspection internally biennially.

Inspection of boilers and pressure vessels covered by insurance shall be permitted to be made by employees of the insuring company holding commissions from the National Board of Boiler and Pressure Vessel Inspectors, subject to approval of the Authority Having Jurisdiction. Approved insuring company inspectors shall make reports on prescribed forms on inspections authorized by the Authority Having Jurisdiction. The reports shall be filed in the Authority Having Jurisdiction office. Company inspectors shall notify the Authority Having Jurisdiction of suspension of insurance because of dangerous conditions, new insurance in effect, and discontinuance of insurance coverage.

It is important that boilers and pressure vessels are inspected on a routine basis. In some jurisdictions, the AHJ does not perform these inspections. In many jurisdictions, the state or insurance companies perform inspections and issue permits for boiler operation. Installers should check with the AHJ for clarification of this requirement.

1014.0 Operation and Maintenance of Boilers and Pressure Vessels.

1014.1 General. Boilers and pressure vessels shall be operated and maintained in accordance with requirements for protection of the public established by the Authority Having Jurisdiction in accordance with nationally recognized standards.

The Authority Having Jurisdiction shall notify the owner or authorized representative of defects or deficiencies and properly corrected. Where such corrections are not made, or where the operation of the boiler or pressure vessel is deemed unsafe by the Authority Having Jurisdiction, they shall have the authority to revoke the permit to operate the boiler or pressure vessel. Where the operation of a boiler or pressure vessel is deemed by the Authority Having Jurisdiction to constitute an immediate danger, the pressure on such boiler or pressure vessel shall be permitted to be relieved at the owner's cost and the boiler or pressure vessel shall not thereafter be operated without the approval of the Authority Having Jurisdiction.

TABLE 1003.2.1
CONTROLS AND LIMIT DEVICES FOR AUTOMATIC BOILERS

BOILER GROUP	FUEL	FUEL INPUT RANGE[1] (INCLUSIVE), BTU/H	TYPE OF PILOT[2]	SAFETY CONTROL TIMING (NOMINAL MAXIMUM TIME IN SECONDS)				ASSURED FUEL SUPPLY CONTROL[4]	ASSURED AIR SUPPLY CONTROL[5]	LOW FIRE START UP CONTROL[6]	PRE-PURGING CONTROL[7]	HOT WATER TEMPERATURE AND LOW WATER LIMIT CONTROLS[8]	STEAM PRESSURE AND LOW WATER LIMIT CONTROLS[9]	APPROVED FUEL SHUTOFF[10]	CONTROL AND LIMIT DEVICE SYSTEM DESIGN[11]
					TRIAL FOR MAIN BURNER FRAME										
				TRIAL FOR PILOT	DIRECT ELECTRIC IGNITION	FLAME PILOT	MAIN BURNER FLAME FAILURE[3]								
A	Gas	0 - 400 000	Any type	90	Not Required	90	90	Not Required	Required	Not Required	Not Required	Required	Required	Not Required	Required
B	Gas	400 001 - 2 500 000	Interrupted or Intermittent	15	15	15	2-4	Not Required	Required	Not Required	Not Required	Required	Required	Not Required	Required
C	Gas	2 500 001 - 5 000 000	Interrupted or Intermittent	15	15	15	2-4	Required	Required	Required	Required	Required	Required	Required	Required
D	Gas	Over 5 000 000	Interrupted	15	15	15	2-4	Required	Required	Required	Required	Required	Required	Required	Required
E	Oil	0 - 400 000	Any type	Not Required	90	90	90	Not Required	Required	Not Required	Not Required	Required	Required	Not Required	Required
F	Oil	400 001 - 1 000 000	Interrupted	Not Required	30	30	2-4	Required	Required	Not Required	Not Required	Required	Required	Not Required	Required
G	Oil	1 000 001 - 3 000 000	Interrupted	Not Required	15	15	2-4	Required	Required	Not Required	Not Required	Required	Required	Not Required	Required
H	Oil	Over 3 000 000	Interrupted	15	15	60	2-4	Required	Required	Required	Required	Required	Required	Required	Required
K	Electric	All	Not required	Not Required	Not Required	Not Required	Not Required	Not Required	Not Required	Not Required	Not Required	Required	Required	Not Required	Required
L	Gas, Oil and/or Coal	12 500 000 or more	Any	10 sec Per NFPA 85	Per NFPA 85	Per NFPA 85	Per NFPA 85	Per NFPA 85	Per NFPA 85	Per NFPA 85	Per NFPA 85	Per ASME Power Boiler Code, Section 1 and NFPA 85	Per ASME Power Boiler Code, Section 1 and NFPA 85	Per NFPA 85	Per NFPA 85
M	Heat Recovery Steam Generator	Any	None	Per NFPA 85	Per NFPA 85	Per NFPA 85	Per NFPA 85	Per NFPA 85	Per NFPA 85	Per NFPA 85	Per NFPA 85	Per ASME Boiler & Pressure Code & NFPA 85	Per ASME Boiler & Pressure Code & NFPA 85	Per NFPA 85	Per NFPA 85

For SI units: 1000 British thermal units per hour = 0.293 kW

FOOTNOTES FOR TABLE 1003.2.1 (continued)

1 Fuel input shall be determined by one of the following:
The burner input shall not exceed the input shown on the burner nameplate or as otherwise identified by the manufacturer.
The nominal boiler rating, as determined by the building official, plus 25 percent.

2 Automatic boilers shall have one flame failure device on each burner, which shall prove the presence of an ignition source at the point where it will ignite the main burner, except that boiler groups A, B, E, F, and G, which are equipped with direct electric ignition, shall monitor the main burner, and boiler groups using interrupted pilots shall monitor the main burner after the prescribed limited trial and ignition periods. Boiler group A, equipped with continuous pilot, shall accomplish 100 percent shutoff within 90 seconds upon pilot flame failure. The use of intermittent pilots in boiler group C is limited to approved burner units.

3 In boiler groups B, C, and D a 90 second main burner flame failure limit shall be permitted to be applied where continuous pilots are provided on manufacturer assembled boiler-burner units that have been approved by an approved testing agency in accordance with nationally recognized standards approved by the building official. Boiler groups F and G equipped to re-energize their ignition systems within 0.8 second after main burner flame failure will be permitted 30 seconds for group F or 15 seconds for group G to re-establish their main burner flames.

4 Boiler groups C and D shall have controls interlocked to accomplish a non-recycling fuel shutoff upon high or low gas pressure, and boiler groups F, G, and H using steam or air for fuel atomization shall have controls interlocked to accomplish a nonrecycling fuel shutoff upon low atomizing steam or air pressure. Boiler groups F, G, and H equipped with a preheated oil system shall have controls interlocked to provide fuel shutoff upon low oil temperature.

5 Automatic boilers shall have controls interlocked to shut off the fuel supply in the event of draft failure where forced or induced draft fans are used or, in the event of low combustion airflow, where a gas power burner is used. Where a single motor directly driving both the fan and the oil pump is used, a separate control is not required.

6 Boiler groups C, D, and H, where firing in excess of 400 000 Btu/h (117 kW) per combustion chamber, shall be provided with low fire start of its main burner system to permit smooth light-off. This will normally be a rate of one-third of its maximum firing rate.

7 Boiler groups C, D, and H shall not permit pilot or main burner trial for ignition operation before a purging operation of sufficient duration to permit not less than four complete air changes through the furnace, including a combustion chamber and the boiler passes. Where this is not readily determinable, five complete air changes of the furnace, including combustion chamber up to the first pass, will be considered equivalent. An atmospheric gas burner with no mechanical means of creating air movement or an oil burner that obtains two-thirds or more of the air required for combustion without mechanical means of creating air movement shall not require purge by means of four air changes, so long as its secondary air openings are not provided with means of closing. Where such burners have means of closing secondary air openings, a time delay shall be provided that puts these closures in a normally open position for four minutes before an attempt for ignition. An installation with a trapped combustion chamber shall, in every case, be provided with a mechanical means of creating air movement for purging.

8 An automatic hot-water-heating boiler, low-pressure hot-water-heating boiler, and power hot water boiler shall be equipped with two high-temperature limit controls with a manual reset on the control, with the higher setting interlocked to shut off the main fuel supply, except that manual reset on the high-temperature limit control shall not be required on an automatic package boiler not exceeding 400 000 Btu/h (117 kW) input and that has been approved by an approved testing agency. An automatic hot-water heating, power boiler, and package hot-water supply boiler shall be equipped with one low-water level limit control with a manual reset interlocked to shut off the fuel supply, so installed as to prevent damage to the boiler and to permit testing of the control without draining the heating system, except on boilers used in Group R Occupancies of less than six units and in Group U Occupancies and further, except that the low-water level limit control is not required on package hot-water supply boilers approved by a nationally recognized testing agency. However, a low- water flow limit control installed in the circulating water line shall be permitted to be used instead of the low-water level limit control for the same purpose on coil-type boilers.

9 An automatic low-pressure steam-heating boiler, small power boiler, and power steam boiler shall be equipped with two high-steam pressure limit controls interlocked to shut off the fuel supply to the main burner with manual reset on the control, with the higher setting and two low-water-level limit controls, one of which shall be provided with a manual reset device and independent of the feed water controller. Coil-type flash steam boilers shall be permitted to use two high-temperature limit controls, one of which shall be manually reset in the hot water coil section of the boiler instead of the low-water level limit control.

10 Boiler groups C, D, and H shall use an approved automatic reset safety shutoff valve for the main burner fuel shutoff, which shall be interlocked to the programming control devices required. On oil burners where the safety shutoff valve will be subjected to pressures in excess of 10 psi (69 kPa) where the burner is not firing, a second safety shutoff valve shall be provided in series with the first. Boiler groups C and D using gas in excess of 1 psi (7 kPa) pressure or having a trapped combustion chamber or employing horizontal fire tubes shall be equipped with two approved safety shutoff valves, one of which shall be an automatic reset type, one of which shall be permitted to be used as an operating control, and both of which shall be interlocked to the limit-control devices required. Boiler groups C and D using gas in excess of 1 psi (7 kPa) pressure shall be provided with a permanent and ready means for making periodic tightness checks of the main fuel safety shutoff valves.

11 Control and limit device systems shall be grounded with operating voltage not to exceed 150 volts, except that, upon approval by the building official, existing control equipment to be reused in an altered boiler control system shall be permitted to use 220 volts single phase with one side grounded, provided such voltage is used for all controls. Control and limit devices shall interrupt the ungrounded side of the circuit. A readily accessible means of manually disconnecting the control circuit shall be provided with controls so arranged that where they are de-energized, the burner shall be inoperative.

CHAPTER 11
REFRIGERATION

1101.0 General.

1101.1 Applicability. Part I governs the design, installation, and construction of refrigeration systems, equipment, refrigerant piping, pressure vessels, safety devices, replacement of parts, alterations, and substitution of different refrigerants. Part II governs the installation and construction of cooling towers.

1101.2 Equipment. Equipment for refrigerant recovery, recycling, or both shall comply with UL 1963.

UL 1963 covers refrigerant recovery and recycling equipment to be employed in accordance with the National Electrical Code, NFPA 70 and include battery operated equipment. These requirements do not cover equipment rated more than 600 volts or employing a universal motor rated more than 250 volts or intended for installation and use in a hazardous location.

Part I – Refrigeration Systems.

1102.0 Refrigeration Systems.

The basic mechanical refrigeration cycle works as follows. The refrigerant is a gas as it flows through the evaporator and compressor, and to the inlet of the condenser. The refrigerant boils in the evaporator, absorbing heat from the surrounding area, causing a cooling effect described as refrigeration. This absorbed heat plus the heat of compression are removed from the refrigerant gas at the condenser. The cooled refrigerant gas then condenses to a liquid, flows into a liquid receiver, and eventually continues to flow as a liquid to the thermostatic expansion valve.

The basic components of a mechanical refrigerating system are the compressor, condenser, liquid receiver, metering device or thermostatic expansion valve, and evaporator, plus the interconnecting piping, tubing, valves, and fittings (see **Figure 1102.0**). The refrigerant is a substance that can be readily converted from a gas to a liquid or from a liquid to a gas by changing the pressure within the system or by changing the temperature of the refrigerant.

The compressor is a pump that increases the pressure of the refrigerant. This difference in pressure within the system causes the refrigerant to flow or circulate. Evaporation and condensation of the refrigerant is determined by the temperature and pressure, which is controlled by starting or stopping the compressor. The "high side" or the high-pressure side of the refrigeration system begins at the discharge side of the compressor; includes the condenser, liquid receiver, and interconnecting piping; and ends at the thermostatic expansion valve located near the inlet to the evaporator. The refrigerant is in a liquid state from the outlet of the condenser, while it is stored in the liquid receiver, and in the piping from the receiver to the thermostatic expansion valve. The "low side" or low-pressure side of the mechanical refrigerating system begins at the outlet of the thermostatic expansion valve and

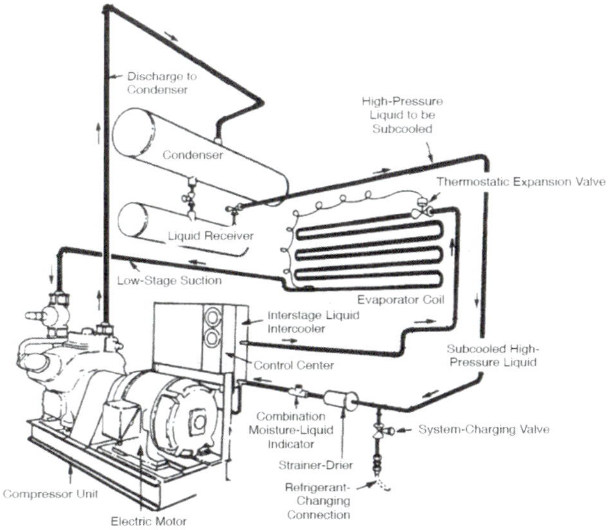

FIGURE 1102.0
REFRIGERATION COMPONENTS

includes the evaporator and connecting piping to the inlet side of the compressor. The inlet side of the compressor is one end of the low side of the refrigerating system.

Compressors are generally manufactured in two styles. In a hermetic compressor, the motor and compressor are contained in a totally enclosed airtight housing. An open compressor has the motor and compressor as separate units. The compressors are designed and manufactured as reciprocating or centrifugal units.

The condenser does the opposite of an evaporator. It removes heat from the refrigerant gas, converting the gas to a liquid. There are three basic types of condensers: air-cooled, water-cooled, and evaporative condensers. These are heat exchangers of a shell and tube or finned tube construction.

The liquid receiver is a tank located near the outlet of the condenser. It stores the liquid refrigerant to compensate for varying flow and serves as a reservoir to store refrigerant before it is pumped out of the system. A pressure-relief valve is always connected to the liquid receiver.

The thermostat expansion valve regulates the flow of liquid refrigerant to the evaporator. This flow control determines the amount of heat removed at the evaporator and separates the high-pressure and low-pressure sides of the system.

Evaporators are heat exchangers that permit the transfer or absorption of heat by the refrigerant gas from the area being cooled. Evaporators often have rows of finned tubes or coils where the refrigerant is vaporized. The liquid refrigerant boils, absorbing heat from the surrounding air flowing past the fins. This heat is then carried away in the refrigerant, and the heat is removed at the condenser.

The capacity of a refrigerating system is usually measured in tons of refrigeration. Refrigerating equipment is rated in Btu per hour of cooling capacity produced at an evaporator temperature of 5°F and a condensing temperature of 86°F. One ton of refrigeration is equal to 12,000 Btu per hour or 4.7 horsepower of refrigeration.

Absorption Refrigeration Equipment

An absorption refrigeration system operates on the principle that heat is absorbed when liquid is vaporized to a gas, and heat is released when a gas is condensed to a liquid. Liquids boil and vaporize more quickly at low pressures or in a vacuum. The refrigerating effect of an absorption system depends on the action of a refrigerant and an absorbent at various pressures and temperatures within the system, and is also dependent on the application of heat at a specific location and the affinity of refrigerant gas to be absorbed by certain chemical compounds.

There are two common types of absorption refrigeration systems. One uses water as a refrigerant with lithium bromide as an absorbent. The other type of system uses ammonia as a refrigerant, with water as an absorbent. A pump is sometimes required to operate the second system, especially for air-conditioning units.

Absorption refrigeration systems can be used for household refrigeration or air-conditioning applications. These systems can operate domestic refrigerators without a source of electrical power, which is especially advantageous in rural areas.

Storage of Refrigerants

Storage of refrigerants is regulated for safety reasons. Requirements for storage of refrigerants are in the Fire Code.

1102.1 General. Refrigeration systems using a refrigerant other than ammonia shall comply with this chapter and ASHRAE 15.

1102.2 Ammonia Refrigeration Systems. Refrigeration systems using ammonia as the refrigerant shall comply with IIAR 2, IIAR 3, IIAR 4, and IIAR 5 and shall not be required to comply with this chapter.

IIAR 2 points to the design and installation of ammonia refrigeration standards. ASHRAE 15 is the safety standard for refrigeration systems that covers occupancy classification; refrigerating system classification; refrigerant safety classification; restrictions on refrigerant use; installation restrictions; and design, construction, operating and testing requirements.

1102.3 Refrigerants. The refrigerant used shall be of a type listed in Table 1102.3 or in accordance with ASHRAE 34 where approved by the Authority Having Jurisdiction.

Exception: Lithium bromide absorption systems using water as the refrigerant.

ASHRAE 34 is the standard for the numbering; classification; safety group; refrigerant concentration limits; refrigerant designation; and application instructions for refrigerants. This standard allows for the quick understanding of the numbering and classification system. See **Figure 1102.3** for refrigerant classification.

FIGURE 1102.3
REFRIGERANT CLASSIFICATION

1103.0 Classification.

1103.1 Classification of Refrigerants. Refrigerants shall be classified in accordance with Table 1102.3 or in accordance with ASHRAE 34 where approved by the Authority Having Jurisdiction.

1103.1.1 Safety Group. Table 1102.3 classifies refrigerants by toxicity and flammability, and assigns safety groups using combinations of toxicity class and flammability class. For the purposes of this chapter, the refrigerant Groups A1, A2L, A2, A3, B1, B2L, B2, and B3 shall be considered to be individual and distinct safety groups. Each refrigerant is assigned into not more than one group.

1103.2 Classification of Refrigeration Systems. Refrigeration systems shall be classified according to the degree of probability that a leakage of refrigerant will enter an occupancy-classified area in accordance with Section 1103.2.1 and Section 1103.2.2. [ASHRAE 15:5.2]

1103.2.1 High-Probability System. Systems in which the basic design, or the location of components, is such that a leakage of refrigerant from a failed connection, seal, or component will enter the occupied space shall be classified as high-probability systems. A high-probability system shall be a direct system or an indirect open spray system in which the refrigerant is capable of producing pressure that is more than the secondary coolant. [ASHRAE 15:5.2.1]

1103.2.2 Low-Probability System. Systems in which the basic design, or the location of the components, is such that a leakage of refrigerant from a failed connection, seal, or component is not capable of entering the occupied space shall be classified as low-probability systems. A low-probability system shall be an indirect closed system, double indirect system, or an indirect open spray system. In a low-probability indirect open spray system, the secondary coolant pressure remains more than the refrigerant pressure in operating and standby conditions. [ASHRAE 15:5.2.2]

1103.3 Higher Flammability Refrigerants. Group A3 and B3 refrigerants shall not be used except where approved by the Authority Having Jurisdiction.

Exceptions:

(1) Laboratories with more than 100 square feet (9.29 m^2) of space per person.

(2) Industrial occupancies.

(3) Listed self-contained systems containing not more than 0.331 pounds (0.150 kg) of Group A3 refrigerant, provided that the equipment is installed in accordance with the listing and the manufacturer's installation instructions. [ASHRAE 15:7.5.3]

1104.0 Requirements for Refrigerant and Refrigeration System Use.

1104.1 System Selection. Refrigeration systems shall be limited in application in accordance with Table 1104.1, and the requirements of Section 1104.0.

1104.2 Refrigerant Concentration Limit. The concentration of refrigerant in a complete discharge of an independent circuit of high-probability systems shall not exceed the amounts shown in Table 1102.3, except as provided in Section 1104.3 and Section 1104.4. The volume of occupied space shall be determined in accordance with Section 1104.2.1 through Section 1104.2.3.

Exceptions:

(1) Listed equipment containing not more than 6.6 pounds (2.99 kg) of refrigerant, regardless of the refrigerant safety classification, provided the equipment is installed in accordance with the listing and with the manufacturer's installation instructions.

(2) Listed equipment for use in laboratories with more than 100 square feet (9.29 m^2) of space per person, regardless of the refrigerant safety classification, provided that the equipment is installed in accordance with the listing and the manufacturer's installation instructions. [ASHRAE 15:7.2]

1104.2.1 Volume Calculations. The volume used to convert from refrigerant concentration limits to refrigerating system quantity limits for refrigerants in Section 1104.2 shall be based on the volume of space to which refrigerant disperses in the event of a refrigerant leak. [ASHRAE 15:7.3]

1104.2.2 Nonconnecting Spaces. Where a refrigerating system or part thereof is located in one or more enclosed occupied spaces that do not connect through permanent openings or HVAC ducts, the volume of the smallest occupied space shall be used to determine the refrigerant quantity limit in the system. Where different stories and floor levels connect through an open atrium or mezzanine arrangement, the volume to be used in calculating the refrigerant quantity limit shall be determined by multiplying the floor area of the lowest space by 8.2 feet (2499 mm). [ASHRAE 15:7.3.1]

1104.2.3 Ventilated Spaces. Where a refrigerating system or a part thereof is located within an air handler, in an air distribution duct system, or in an occupied space served by a mechanical ventilation system, the entire air distribution system shall be analyzed to determine the worst-case distribution of leaked refrigerant. The worst case or the smallest volume in which the leaked refrigerant disperses shall be used to determine the refrigerant quantity limit in the system, subject to the criteria in accordance with Section 1104.2.3.1 through Section 1104.2.3.3. [ASHRAE 15:7.3.2]

1104.2.3.1 Closures. Closures in the air distribution system shall be considered. Where one or more spaces of several arranged in parallel are capable of being closed off from the source of the refrigerant leak, their volume(s) shall not be used in the calculation.

Exceptions: The following closure devices are not considered:

(1) Smoke dampers, fire dampers, and combination smoke/fire dampers that close only in an emergency not associated with a refrigerant leak.

(2) Dampers, such as variable-air-volume (VAV) boxes, that provide limited closure where airflow is not reduced below 10 percent of its maximum (with the fan running). [ASHRAE 15:7.3.2.1]

1104.2.3.2 Plenums. The space above a suspended ceiling shall not be included in calculating the refrigerating system quantity limits unless such space is part of the air supply or return system. [ASHRAE 15:7.3.2.2]

1104.2.3.3 Supply and Return Ducts. The volume of the supply and return ducts and plenums shall be included when calculating the refrigerating system quantity limits. [ASHRAE 15:7.3.2.3]

1104.3 Institutional Occupancies. The RCL value required in Section 1104.2 shall be reduced by 50 percent for the areas of institutional occupancies. The total of Group A2, B2, A3, and B3 refrigerants shall not exceed 550 pounds (249.5 kg) in the occupied areas and machinery rooms of institutional occupancies.

Exception: The total of all Group A2L refrigerants shall not be limited in machinery rooms of institutional occupancies.

1104.4 Industrial Occupancies and Refrigerated Rooms. Section 1104.2 shall not apply in industrial occupancies and refrigerated rooms where in accordance with the following:

(1) The space(s) containing the machinery is (are) separated from other occupancies by tight construction with tight-fitting doors.

(2) Access is restricted to authorized personnel.

(3) Refrigerant detectors are installed with the sensing location and alarm level as required in refrigeration machinery rooms in accordance with Section 1106.2.2.2.

(4) Open flames and surfaces exceeding 800°F (427°C) shall not be permitted where a Group A2, B2, A3, or B3 refrigerant, is used.

(5) Electrical equipment that is in accordance with Class 1, Division 2, of NFPA 70 where the quantity of a Group A2, B2, A3, or B3 refrigerant in an independent circuit is capable of exceeding 25 percent of the lower flammability limit (LFL) upon release to the space based on the volume determined in accordance with Section 1104.2.1 through Section 1104.2.3.

(6) Refrigerant containing parts in systems exceeding 100 horsepower (74.6 kW) compressor drive power, except

evaporators used for refrigeration or dehumidification, condensers used for heating, control and pressure-relief valves for either, low-probability pumps, and connecting piping, are located in a machinery room or outdoors. [ASHRAE 15:7.2.2]

1104.5 Flammable Refrigerants. The total of Group A2, B2, A3, and B3 refrigerants, other than Group A2L and B2L refrigerants shall not exceed 1100 pounds (498.9 kg) without approval by the Authority Having Jurisdiction. Institutional Occupancies shall comply with Section 1104.3.

1104.6 Applications for Human Comfort and for Nonindustrial Occupancies. In nonindustrial occupancies, Group A2, A2L, A3, B1, B2L, B2, and B3 refrigerants shall not be used in high-probability systems for human comfort.

1104.7 Refrigerant Type and Purity. Refrigerants shall be of a type specified by the equipment manufacturer. Unless otherwise specified by the equipment manufacturer, refrigerants used in new equipment shall be of purity in accordance with AHRI 700.

AHRI 700 covers the testing and sampling criteria for impurities to determine whether the recovered refrigerant can be reclaimed and used by a different owner. The testing and reporting criteria within the standard are very specific and extensive.

Under the scope of ASHRAE 15 there are four different types of refrigerants: new, recovered, recycled, and reclaimed. "Recovered" refrigerant may be used in the same system by the same owner. If contaminated, it must be reclaimed and meet the requirements of AHRI 700. "Recycled" refrigerant shall not be reused except in the same systems from which they were removed. If contaminated it must be reclaimed in accordance with AHRI 700.

1104.7.1 Recovered Refrigerants. Recovered refrigerants shall not be reused except in the system from which they were removed or as provided in Section 1104.7.2 or Section 1104.7.3. When contamination is evident by discoloration, odor, acid test results, or system history, recovered refrigerants shall be reclaimed in accordance with Section 1104.7.3 before reuse. [ASHRAE 15:7.5.1.4]

1104.7.2 Recycled Refrigerants. Recycled refrigerants shall not be reused except in systems using the same refrigerant and lubricant designation and belonging to the same owner as the systems from which they were removed. Where contamination is evident by discoloration, odor, acid test results, or system history, recycled refrigerants shall be reclaimed in accordance with Section 1104.7.3.

Exception: Drying shall not be required in order to use recycled refrigerants where water is the refrigerant, is used as an absorbent or is a deliberate additive. [ASHRAE 15:7.5.1.5]

1104.7.3 Reclaimed Refrigerants. Used refrigerants shall not be reused in a different owner's equipment unless tested and found to be in accordance with the requirements of AHRI 700. Contaminated refrigerants shall not be used unless reclaimed and is in accordance with AHRI 700. [ASHRAE 15:7.5.1.6]

1104.7.4 Mixing. Refrigerants, including refrigerant blends, with different designations as in accordance with Table 1102.3 shall not be mixed in a system.

Exception: Addition of a second refrigerant shall be permitted where specified by the equipment manufacturer to improve oil return at low temperatures. The refrigerant and amount added shall be in accordance with the manufacturer's instructions. [ASHRAE 15:7.5.1.7]

1104.8 Changing Refrigerants. A change in the type of refrigerant in a system shall not be made without notifying the Authority Having Jurisdiction, the user, and due observance of safety requirements. The refrigerant being considered shall be evaluated for suitability. [ASHRAE 15:5.3]

1105.0 General Requirements.

1105.1 Human Comfort. Cooling systems used for human comfort shall be in accordance with the return-air and outside-air provisions for furnaces in Section 604.1 and Section 904.7. Cooling equipment used for human comfort in residential buildings shall be selected in accordance with ACCA Manual S to satisfy the calculated loads determined in accordance with ACCA Manual J or other approved methods. Refrigerants used for human comfort shall be in accordance with Section 1104.6.

1105.2 Supports and Anchorage. Supports and anchorage for refrigeration equipment and piping shall be designed in accordance with the building code as Occupancy Category H (hazardous facilities). Supports shall be made of noncombustible materials.

Exceptions:

(1) Equipment containing Group A1 refrigerants shall be permitted to be supported by the same materials permitted for the building type.

(2) The use of approved vibration isolators specifically designed for the normal, wind, and seismic loads encountered, shall be permitted.

A compressor or portion of a condensing unit supported from the ground shall rest on a concrete or other approved base extending not less than 3 inches (76 mm) above the adjoining ground level.

1105.3 Access. An unobstructed readily accessible opening and passageway not less than 36 inches (914 mm) in width and 80 inches (2032 mm) in height shall be provided and maintained to the compressor, valves required by this chapter, or other portions of the system requiring routine maintenance.

Exceptions:

(1) Refrigerant evaporators, suspended overhead, shall be permitted to use portable means of access.

(2) Air filters, brine control or stop valves, fan motors or drives, and remotely de-energized electrical connections shall be permitted to be provided access to an unobstructed space not less than 30 inches (762 mm) in depth, width, and height. Where an access opening is immediately adjacent to these items and the equipment is capable of being serviced, repaired, and replaced from this opening, the dimensions shall be permitted to be reduced to 22 inches

(559 mm) by 30 inches (762 mm) provided the largest piece of equipment is removed through the opening.

(3) Cooling equipment, using Group A1 refrigerants or brine, located in an attic or furred space shall be permitted to be provided access by a minimum opening and passageway thereto of not less than 22 inches (559 mm) by 30 inches (762 mm).

(4) Cooling or refrigeration equipment, using Group A1 or B1 refrigerants or brine, located on a roof or on an exterior wall of a building, shall be permitted to be provided access as for furnaces in Section 304.3.

1105.4 Illumination and Service Receptacles. In addition to the requirements of Section 301.4, permanent luminaires shall be installed for equipment required by this code to be accessible or readily accessible. Such luminaires shall provide illumination to perform the required tasks for which access is provided. Control of the illumination source shall be provided at the access entrance.

Exceptions:

(1) Luminaires shall be permitted to be omitted where the fixed lighting of the building will provide the required illumination.

(2) Equipment located on the roof or on the exterior walls of a building.

1105.5 Ventilation of Rooms Containing Condensing Units. Where not in a refrigerant machinery room, rooms or spaces in which a refrigerant-containing portion of a condensing unit is installed shall be provided with ventilation in accordance with Section 1105.5.1 or Section 1105.5.2. Ventilation for machinery rooms shall comply with Section 1106.0.

1105.5.1 Permanent Gravity Ventilation Openings. Permanent gravity ventilation openings of not less than 2 square feet (0.2 m^2) net free area opening shall be terminated directly to the outside of the building or extend to the outside of the building by continuous ducts.

1105.5.2 Mechanical Exhaust System. A mechanical exhaust system shall be designed to provide a complete change of air not less than every 20 minutes in such room or space and shall discharge to the outside of the building.

Exceptions:

(1) A condensing unit in a room or space where the cubical content exceeds 1000 cubic feet per horsepower (ft^3/hp) (37.95 m^3/kW) of the unit.

(2) A condensing unit in a room or space that has permanent gravity ventilation having an area of 2 square feet (0.2 m^2) or more to other rooms or openings exceeding 1000 ft^3/hp (37.95 m^3/kW).

1105.6 Prohibited Locations. Refrigeration systems or portions thereof shall not be located within a required exit enclosure. Refrigeration compressors exceeding 5 horsepowers (3.7 kW) rating shall be located not less than 10 feet (3048 mm) from an exit opening in a Group A; Group B; Group E; Group F; Group I; Group R, Division 1; or Group S Occupancy, unless separated by a one-hour fire-resistive occupancy separation.

1105.7 Condensate. Condensate from air-cooling coils shall be collected and drained to an approved location. Drain pans and coils shall be arranged to allow thorough drainage and access for cleaning. Where temperatures drop below freezing, heat tracing and insulation of condensate drains shall be installed.

1105.8 Defrost. Where defrost cycles are required for portions of the system, provisions shall be made for collection and disposal of the defrost liquid in a safe and sanitary manner.

1105.9 Overflows. Where condensate or defrost liquids are generated in an attic or furred space, and structural damage will result from overflow, provisions for overflow shall be provided.

1105.10 Condensate, Defrost, and Overflow Disposal. Disposal of condensate, defrost, or overflow discharges shall comply with Section 310.0.

1105.11 Refrigerant Port Protection. Air conditioning refrigerant circuit access ports located outdoors shall be protected from unauthorized access with locking-type tamper-resistant caps or in a manner approved by the Authority Having Jurisdiction.

Exception: Refrigerant ports in secure locations protected by walls or fencing and requiring key access.

1105.12 Storage. Refrigerants and refrigerant oils not charged within the refrigeration system shall be stored in accordance with Section 1105.12.1 and the fire code. Storage of materials in a refrigeration machinery room shall comply with the fire code.

Hazardous storage and locations is governed by the Fire Code and usually falls under the jurisdiction of the Fire Marshal.

1105.12.1 Quantity. The total amount of refrigerant stored in a machinery room in all containers not provided with relief valves and piping in accordance with Section 1113.0 shall not exceed 330 pounds (149.7 kg). Refrigerant shall be stored in approved storage containers. Additional quantities of refrigerant shall be stored in an approved storage facility. [ASHRAE 15:11.5]

1106.0 Refrigeration Machinery Rooms.

1106.1 Where Required. Refrigeration systems shall be provided with a refrigeration machinery room where the conditions as outlined in Section 1106.1.1 through Section 1106.1.4 exist.

Exception: Refrigeration equipment shall be permitted to be located outdoors in accordance with ASHRAE 15.

1106.1.1 Quantity. The quantity of refrigerant in a single, independent refrigerant circuit of a system exceeds the amounts of Table 1102.3.

1106.1.2 Equipment. Direct- and indirect-fired absorption equipment is used.

Exception: Direct and indirect-fired lithium bromide absorption systems using water as the refrigerant.

1106.1.3 A1 System. An A1 system having an aggregate

combined compressor horsepower of 100 (74.6 kW) or more is used.

1106.1.4 A1 Refrigerant. The system contains other than a Group A1 refrigerant.

Exceptions:

(1) Lithium bromide absorption systems using water as the refrigerant.

(2) Systems containing less than 300 pounds (136.1 kg) of refrigerant R-123 and located in an approved exterior location.

Refrigeration machinery rooms shall house refrigerant-containing portions of the system other than the piping and evaporators permitted by Section 1104.4, discharge piping required of this chapter, and cooling towers regulated by Part II of this chapter, and their essential piping.

1106.2 Refrigeration Machinery Room, General Requirements. Where a refrigeration system is located indoors and a machinery room is required in accordance with Section 1106.1, the machinery room shall be in accordance with Section 1106.2.1 through Section 1106.2.5.2.

1106.2.1 Access. Machinery rooms shall not be prohibited from housing other mechanical equipment unless specifically prohibited elsewhere in this chapter. A machinery room shall be so dimensioned that parts are accessible with space for service, maintenance, and operations. There shall be clear head room of not less than 7.25 feet (2210 mm) below equipment situated over passageways. [ASHRAE 15:8.11.1]

1106.2.2 Openings. Each refrigeration machinery room shall have a tight-fitting door or doors opening outward, self-closing where they open into the building and adequate in number to ensure freedom for persons to escape in an emergency. With the exception of access doors and panels in air ducts and air-handling units in accordance with Section 1106.6, there shall be no openings that will permit passage of escaping refrigerant to other parts of the building. [ASHRAE 15:8.11.2]

1106.2.2.1 Detectors and Alarms. Each refrigeration machinery room shall contain one or more refrigerant detectors in accordance with Section 1106.2.2.2, located in areas where refrigerant from a leak will concentrate, that actuate an alarm and mechanical ventilation in accordance with Section 1106.2.4 at a set point not more than the corresponding Occupational Exposure Limit, OEL, in accordance with Table 1102.3, a set point determined in accordance with the OEL as defined in Chapter 2 shall be approved by the Authority Having Jurisdiction. The alarm shall annunciate visual and audible alarms inside the refrigeration machinery room and outside each entrance to the refrigeration machinery room. The alarms required in this section shall be of the manual reset type with the reset located inside the refrigeration machinery room. Alarms set at other levels, such as IDLH, and automatic reset alarms shall be permitted in addition to those required in accordance with this section. The meaning of each alarm shall be clearly marked by signage near the annunciator.

Exception: Refrigerant detectors are not required where only systems using R-718 (water) are located in the refrigeration machinery room.

1106.2.2.2 Refrigerant Detectors. Refrigerant detectors required in accordance with Section 1106.2.2.1 or Section 1107.1.7 shall meet all of the following conditions:

(1) The refrigerant detector shall perform automatic self-testing of sensors. Where a failure is detected, a trouble signal shall be activated.

(2) The refrigerant detector shall have one or more set points to activate responses in accordance with Section 1106.2.2.1 or Section 1107.1.7.

(3) The refrigerant detector as installed, including any sampling tubes, shall activate responses within a time not to exceed 30 seconds after exposure to refrigerant concentration exceeding the set point value specified in Section 1106.2.2.1 or Section 1107.1.7.

The fan will be started on Alarm 1 along with the audible and visual alarms. The exception being ammonia which is allowed to have a fan setpoint of 1000 ppm per ASHRAE 15:8.12(h).

1106.2.3 Mechanical Ventilation. Machinery rooms shall be vented to the outdoors, utilizing mechanical ventilation in accordance with Section 1106.2.4 and Section 1106.2.5.

1106.2.4 Ventilation. Mechanical ventilation referred to in Section 1106.2.3 shall be by one or more power-driven fans capable of exhausting air from the machinery room at not less than the amount shown in accordance with Section 1106.2.5.

To obtain a reduced airflow for normal ventilation, multiple fans or multispeed fans shall be used. Provision shall be made to supply makeup air to replace that being exhausted. Ducts for supply and exhaust to the machinery room shall serve no other area. The makeup air supply locations shall be positioned relative to the exhaust air locations to avoid short-circuiting. Inlets to the exhaust ducts shall be located in an area where refrigerant from a leak will concentrate, in consideration of the location of the replacement supply air paths, refrigerating machines, and the density of the refrigerant relative to air.

Inlets to exhaust ducts shall be within 1 foot (305 mm) of the lowest point of the machinery room for refrigerants that are heavier than air, and shall be within 1 foot (305 mm) of the highest point for refrigerants that are lighter than air. The discharge of the exhaust air shall be to the outdoors in such a manner as not to cause a nuisance or danger.

1106.2.5 Emergency Ventilation-Required Airflow. An emergency ventilation system shall be required to exhaust an accumulation of refrigerant due to leaks or a rupture of the system. The emergency ventilation required shall be capable of removing air from the machinery room in not less than the airflow quantity in Section 1106.2.5.1 or Section 1106.2.5.2. Where multiple refrigerants are present, then the highest airflow quantity shall apply.

1106.2.5.1 Ventilation - A1, A2, A3, B1, B2L, B2 and B3 refrigerants. The emergency ventilation for A1, A2, A3, B1, B2L, B2 and B3 refrigerants shall have the capacity to provide mechanical exhaust at a rate as determined in accordance with Equation 1106.2.5.1.

$$Q = 100 \sqrt{G} \qquad \text{(Equation 1106.2.5.1)}$$

Where:

Q = Air flow rate, cubic feet per minute.

G = Refrigerant mass in largest system, pounds.

For SI units: 1 cubic foot per minute = 0.00047 m^3/s, 1 pound = 0.453 kg

1106.2.5.2 Ventilation - Group A2L Refrigerants. The emergency ventilation for A2L refrigerants shall have the capacity to provide mechanical exhaust at a rate determined in accordance with Table 1106.2.5.2.

TABLE 1106.2.5.2
REQUIRED AIRFLOW FOR GROUP A2L REFRIGERANTS

REFRIGERANT	MINIMUM AIR FLOW* (CFM)
R-32	32 500
R-143a	28 600
R-444A	13 700
R-444B	22 400
R-445A	16 400
R-446A	50 500
R-447A	50 200
R-447B	29 600
R-451A	14 900
R-451B	14 900
R-452B	31 500
R-454A	4290
R-454B	6650
R-454C	32 800
R-455A	4770
R-457A	31 400
R-1234yf	16 500
R-1234zeE	12 600

For SI units: 1 cubic foot per minute = 0.00047 m^3/s

* The values were tabulated from the following equation:

$$Q_{A2L} \geq [(\rho \cdot v \cdot A)/(LFL \cdot 0.50)] \qquad \text{(Equation 1106.2.5.2)}$$

Where:

ρ = Refrigerant density, pounds per cubic feet (kg/m^3).

v = Refrigerant velocity equal to the refrigerant acoustic velocity (speed of sound), feet per second (m/s).

A = Cross-section flow area of refrigerant leak, square feet (m^2), A = 0.00136 ft^2 (0.000126 m^2).

LFL = Lower Flammability Limit, or $ETFL_{60}$ where no LFL exist, published value in accordance with ASHRAE 34.

Q_{A2L} = Minimum required air flow rate, conversion to other units of measures is permitted, cubic feet per second (m^3/s).

For exact ventilation rates and for refrigerants not listed, the ventilation rate shall be calculated using this equation.

1106.3 Normal Operation. A part of the refrigeration machinery room mechanical ventilation shall be in accordance with the following:

(1) Operated, where occupied, to supply not less than 0.5 CFM/ft^2 (2.54 $L/s/m^2$) of machinery room area or 20 cubic feet per minute (9.44 L/s) per person.

(2) Operable, where occupied at a volume required to not exceed the higher of a temperature rise of 18°F (10°C) above inlet air temperature or a maximum temperature of 122°F (50°C).

1106.4 Natural Ventilation. Where a refrigerating system is located outdoors more than 20 feet (6096 mm) from buildings opening and is enclosed by a penthouse, lean-to, or other open structure, natural or mechanical ventilation shall be provided. The requirements for such natural ventilation shall be in accordance with the following:

(1) The free-aperture cross section for the ventilation of a machinery room shall be not less than as determined in accordance with Equation 1106.4.

$$F = \sqrt{G} \qquad \text{(Equation 1106.4)}$$

Where:

F = The free opening area, square feet.

G = The mass of refrigerant in the largest system, any part of which is located in the machinery room, pounds.

For SI units: 1 cubic foot per minute = 0.00047 m^3/s, 1 pound = 0.453 kg

(2) The location of the gravity ventilation openings shall be based on the relative density of the refrigerant to air. [ASHRAE 15:8.11.5(a), (b)]

1106.5 Combustion Air. No open flames that use combustion air from the machinery room shall be installed where refrigerant is used. Combustion equipment shall not be installed in the same machinery room with refrigerant-containing equipment except under one of the following conditions:

(1) Combustion air shall be ducted from outside the machinery room and sealed in such a manner as to prevent refrigerant leakage from entering the combustion chamber.

(2) A refrigerant detector, that is in accordance with Section 1106.2.2.1, shall be installed to automatically shut down the combustion process in the event of refrigerant leakage.

Exception: Machinery rooms where carbon dioxide (R-744) or water (R-718) is the refrigerant.

1106.6 Airflow. There shall be no airflow to or from an occupied space through a machinery room unless the air is ducted and sealed in such a manner as to prevent a refrigerant leakage from entering the airstream. Access doors and panels in ductwork and air-handling units shall be gasketed and tight fitting. [ASHRAE 15:8.11.7]

1106.7 Ventilation Intake. Makeup air intakes to replace the exhaust air shall be provided to the refrigeration machinery room directly from outside the building. Intakes shall be located as required by other sections of the code and fitted with backdraft dampers or other approved flow-control means to prevent reverse flow. Distribution of makeup air shall be arranged to provide thorough mixing within the refrigeration machinery room to prevent short circuiting of the makeup air directly to the exhaust.

1106.8 Maximum Temperature. Ventilation or mechanical cooling systems shall be provided to maintain a temperature of not more than 104°F (40°C) in the refrigerant machinery room under design load and weather conditions.

1106.9 Refrigerant Parts in Air Duct. Joints and refrigerant-containing parts of a refrigerating system located in an air duct carrying conditioned air to and from an occupied space shall be constructed to withstand a temperature of 700°F (371°C) without leakage into the airstream. [ASHRAE 15:8.8]

1106.10 Dimensions. Refrigeration machinery rooms shall be of such dimensions that system parts are readily accessible with approved space for maintenance and operations. An unobstructed walking space not less than 36 inches (914 mm) in width and 80 inches (2032 mm) in height shall be maintained throughout, allowing free access to not less than two sides of moving machinery and approaching each stop valve. Access to refrigeration machinery rooms shall be restricted to authorized personnel and posted with a permanent sign.

1106.11 Restricted Access. Access to the refrigeration machinery room shall be restricted to authorized personnel. Doors shall be clearly marked or permanent signs shall be posted at each entrance to indicate this restriction. [ASHRAE 15:8.11.8]

1106.12 Exits. Exits shall comply with the building code for special hazards.

1107.0 Machinery Room, Special Requirements.

1107.1 General. In cases specified in the rules of Section 1106.1, a refrigeration machinery room shall comply with the special requirements in accordance with Section 1107.1.1 through Section 1107.1.10, in addition to Section 1106.2.

1107.1.1 Flame-Producing Devices. There shall be no flame-producing device or continuously operating hot surface over 800°F (427°C) permanently installed in the room.

1107.1.2 Doors. Doors communicating with the building shall be approved, self-closing, tight-fitting fire doors.

1107.1.3 Walls, Floors, and Ceilings. Walls, floor, and ceiling shall be tight and of noncombustible construction. Walls, floor, and ceiling separating the refrigeration machinery room from other occupied spaces shall be not less than one-hour fire-resistive construction.

1107.1.4 Machinery Rooms. The refrigeration machinery room shall have a door that opens directly to the outdoors or through a vestibule equipped with self-closing, tight-fitting doors.

1107.1.5 Exterior Openings. Exterior openings, where present, shall not be under a fire escape or an open stairway.

1107.1.6 Sealing. All pipes piercing the interior walls, ceiling, or floor of such rooms shall be tightly sealed to the walls, ceiling, or floor through which they pass.

1107.1.7 Group A2L and B2L Refrigerants. Where refrigerant of Groups A2L or B2L are used, the requirements of Class 1, Division 2, of NFPA 70, shall not apply to the machinery room provided that the conditions in Section 1107.1.7.1 through Section 1107.1.7.3 are met.

1107.1.7.1 Mechanical Ventilation. The mechanical ventilation system in the machinery room is run continuously in accordance with Section 1106.2.5 and failure of the mechanical ventilation system actuates an alarm, or the mechanical ventilation system in the machinery room is activated by one or more refrigerant detectors, in accordance with the requirements of Section 1106.2.2.1 and Section 1106.2.2.2.

1107.1.7.2 Refrigeration Detectors. For the refrigerant detection required in Section 1106.2.2.1, detection of refrigerant concentration that exceeds 25 percent of the LFL or the upper detection limit of the refrigerant detector, whichever is lower, shall automatically de-energize the following equipment in the machinery room:

(a) refrigerant compressors

(b) refrigerant pumps

(c) normally-closed automatic refrigerant valves

1107.1.7.3 Machinery Rooms. The machinery room shall comply with Section 1107.1.8.

1107.1.8 Group A2, A3, B2, or B3 Refrigerants. Where any refrigerant of Groups A2, A3, B2, or B3 are used, the machinery room shall comply with Class 1, Division 2, of NFPA 70.

1107.1.9 Refrigeration Systems. As part of the mechanical ventilation system in accordance with Section 1106.2.4, refrigeration systems that contain more than 110 pounds (50 kg) of any Group A2L, A2, A3, B2L, B2, or B3, refrigerant shall have not less than one exhaust air inlet located adjacent to each system not more than 9 feet (3 m) away.

1107.1.10 Remote Control. Remote control of the mechanical equipment in the refrigeration machinery room shall be provided immediately outside the machinery room door solely for the purpose of shutting down the equipment in an emergency. Ventilation fans shall be on a separate electrical circuit and have a control switch located immediately outside the machinery room door.

1108.0 Refrigeration Machinery Room Equipment and Controls.

1108.1 General. Equipment, piping, ducts, vents, or similar devices that are not essential for the refrigeration process, maintenance of the equipment, or for the illumination, ventilation, or fire protection of the room shall not be placed in or pass through a refrigeration machinery room.

1108.2 Electrical. Electrical equipment and installations shall comply with the electrical code. The refrigeration machinery room shall not be classified as a hazardous location except as provided in Section 1107.1.7 or Section 1107.1.8.

1108.3 Emergency Shut-off. A clearly identified emergency shut-off switch of the break-glass type or with an approved tamper-resistant cover shall be provided immediately adjacent to and outside of the principal refrigeration machinery room entrance. The switch shall provide off-only control of refrigerant compressors, refrigerant pumps, and normally-closed automatic refrigerant valves located in the machinery room. For other than A1 and B1 refrigerants, emergency shutoff shall be automatically activated by refrigerant Alarm 2 in accordance with Section 1106.2.2.1.

1108.4 Installation, Maintenance, and Testing. Detection and alarm systems in accordance with Section 1106.2.2.1 shall be installed, maintained, and tested in accordance with the fire code.

1108.5 Emergency Pressure Control System. Where required by the fire code, an emergency pressure control system shall be installed in accordance with applicable fire code requirements.

1109.0 Refrigeration Piping, Containers, and Valves.

1109.1 Materials. Materials used in the construction and installation of refrigerating systems shall be suitable for conveying the refrigerant used. Materials shall not be used that will deteriorate because of the refrigerant, lubricant, or their combination in presence of air or moisture to a degree that poses a safety hazard. [ASHRAE 15:9.1.1] Refrigerant piping shall be metallic.

1109.1.1 Copper and Copper Alloy Pipe. Copper and copper alloy refrigeration piping, valves, fittings, and related parts used in the construction and installation of refrigeration systems shall be approved for the intended use. Refrigeration piping shall comply with ASME B31.5.

1109.1.2 Copper Linesets. Copper linesets shall comply with ASTM B280 or ASTM B1003.

1109.1.3 Iron and Steel. Iron and steel refrigeration piping, valves, fittings, and related parts shall be approved for the intended use. Pipe exceeding 2 inches (50 mm) iron pipe size shall be electric-resistance welded or seamless pipe. Refrigeration piping shall comply with ASME B31.5.

1109.1.4 Prohibited Contact. Aluminum, zinc, magnesium, or their alloys shall not be used in contact with methyl chloride. Magnesium alloys shall not be used where in contact with halogenated refrigerants. [ASHRAE 15:9.1.2]

1109.2 Joints. Iron or steel pipe joints shall be of approved threaded, flanged, or welded types. Exposed threads shall be tinned or coated with an approved corrosion inhibitor. Copper or copper alloy pipe joints of iron pipe size shall be of approved threaded, flanged, press-connect or brazed types. Copper tubing joints and connections shall be connected by approved flared, lapped, swaged, or brazed joints, soldered joints, or mechanical joints that comply with UL 207 either individually or as part of an assembly or a system by an approved nationally recognized laboratory. Piping and tubing shall be installed so as to prevent vibration and strains at joints and connections.

1109.3 Penetration of Piping. Refrigerant piping shall not penetrate floors, ceilings, or roofs.

Exceptions:

(1) Penetrations connecting the basement and the first floor.

(2) Penetrations connecting the top floor and a machinery penthouse or roof installation.

(3) Penetrations connecting adjacent floors served by the refrigeration system.

(4) Penetrations of a direct system where the refrigerant concentration does not exceed that listed in Table 1102.3 for the smallest occupied space through which the refrigerant piping passes.

(5) In other than industrial occupancies and where the refrigerant concentration exceeds that listed in Table 1102.3 for the smallest occupied space, penetrations that connect separate pieces of equipment that are in accordance with one of the following:

  (a) Enclosed by an approved gastight, fire-resistive duct or shaft with openings to those floors served by the refrigerating system.

  (b) Located on the exterior wall of a building where vented to the outdoors or to the space served by the system and not used as an air shaft, closed court, or similar space. [ASHRAE 15:8.10.3]

1109.4 Location of Refrigeration Piping. Refrigerant piping crossing an open space that affords passageway in a building shall be not less than 7.25 feet (2210 mm) above the floor unless the piping is located against the ceiling of such space and is permitted by the Authority Having Jurisdiction. [ASHRAE 15:8.10.1]

1109.4.1 Protection from Mechanical Damage. Passages shall not be obstructed by refrigerant piping. Refrigerant piping shall not be located in an elevator, dumbwaiter, or other shaft containing a moving object, or in a shaft that has openings to living quarters, or to means of egress. Refrigerant piping shall not be installed in an enclosed public stairway, stair landing, or means of egress. [ASHRAE 15:8.10.2]

1109.5 Underground Piping. Refrigerant piping placed underground shall be protected against corrosion.

1109.5.1 Piping in Concrete Floors. Refrigerant piping installed in concrete floors shall be encased in a pipe duct. Refrigerant piping shall be isolated and supported to prevent damaging vibration, stress, or corrosion. [ASHRAE 15:8.10.4]

1109.6 Support. In addition to the requirements of Section 1105.2, piping and tubing shall be securely fastened to a permanent support within 6 feet (1829 mm) following the first bend in such tubing from the compressor and within 2 feet (610 mm) of each subsequent bend or angle. Piping and tubing shall be supported at points not more than 15 feet (4572 mm) apart.

1109.7 Pipe Enclosure. Refrigerant piping and tubing shall be installed so that it is not subject to damage from an external source. Soft annealed copper tubing shall not exceed 1⅜ inches (35 mm) nominal size. Mechanical joints, other than approved press-connect joints, shall not be made on tubing exceeding ¾ of an inch (20 mm) nominal size. Soft annealed copper tubing conveying refrigerant shall be enclosed in iron or steel piping and fittings, or in conduit, molding, or raceway that will protect the tubing against mechanical injury from an exterior source.

Exceptions:

(1) Tubing entirely within or tubing within 5 feet (1524 mm)

of a refrigerant compressor where so located that it is not subject to external injury.

(2) Copper tubing serving a dwelling unit, where such tubing contains Group A1 refrigerant and is placed in locations not subject to damage from an external source.

1109.8 Visual Inspection. Refrigerant piping and joints erected on the premises shall be exposed to view for visual inspection prior to being covered or enclosed.

Exception: Copper tubing enclosed in iron or steel piping conduit, molding, or raceway, provided there are no fittings or joints concealed therein.

1109.9 Condensation. Piping and fittings that convey brine, refrigerant, or coolants that during normal operation are capable of reaching a surface temperature below the dew point of the surrounding air and that are located in spaces or areas where condensation will cause a hazard to the building occupants or damage to the structure, electrical or other equipment shall be protected to prevent such damage.

1109.10 Identification. Piping shall be in accordance with the reference standard for identification. The type of refrigerant, function and pressure shall be indicated.

1110.0 Valves.

1110.1 More than 6.6 Pounds of Refrigerant. Systems containing more than 6.6 pounds (2.99 kg) of refrigerant shall have stop valves installed at the following locations:

(1) The suction inlet of a compressor, compressor unit, or condensing unit.

(2) The discharge of a compressor, compressor unit, or condensing unit.

(3) The outlet of a liquid receiver.

Exceptions:

(1) Systems that have a refrigerant pumpout function capable of storing the refrigerant charge, or are equipped with the provisions for pumpout of the refrigerant.

(2) Self-contained systems. [ASHRAE 15:9.12.4]

1110.2 More than 110 Pounds of Refrigerant. Systems containing more than 110 pounds (49.9 kg) of refrigerant shall have stop valves installed at the following locations:

(1) The suction inlet of a compressor, compressor unit, or condensing unit.

(2) The discharge outlet of a compressor, compressor unit, or condensing unit.

(3) The inlet of a liquid receiver, except for self-contained systems or where the receiver is an integral part of the condenser or condensing unit.

(4) The outlet of a liquid receiver.

(5) The inlets and outlets of condensers where more than one condenser is used in parallel in the systems.

Exception: Systems that have a refrigerant pumpout function capable of storing the refrigerant charge, or are equipped with the provisions for pumpout of the refrigerant or self-contained systems. [ASHRAE 15:9.12.5]

1110.3 Support. Stop valves installed in copper refrigerant lines of ¾ of an inch (20 mm) or less outside diameter shall be supported independently of the tubing or piping.

1110.4 Access. Stop valves required by Section 1110.0 shall be readily accessible from the refrigeration machinery room floor or a level platform.

1110.5 Identification. Stop valves shall be identified by tagging in accordance with the reference standard for identification. A valve chart shall be mounted under glass at an approved location near the principal entrance to a refrigeration machinery room.

1111.0 Pressure-Limiting Devices.

1111.1 Where Required. Pressure-limiting devices complying with Section 1111.2 through Section 1111.4 shall be provided for compressors on all systems operating above atmospheric pressure.

Exception: Pressure limiting devices are not required for listed factory-sealed systems containing less than 22 pounds (9.9 kg) of Group A1 refrigerant. [ASHRAE 15:9.9.1]

1111.2 Setting. Pressure limiting devices shall be set in accordance with one the following:

(1) For positive displacement compressors:

In this type of compressor, a volume of air is drawn into a chamber through an intake and then compressed by mechanically reducing the volume of the chamber, which increases its pressure.

Most notable example would be a reciprocating compressor in which air is trapped between piston and cylinder volume and then literally pressed to increase its pressure.

(a) When systems are protected by a highside pressure relief device, the compressor's pressure limiting device shall be set not more than 90 percent of the operating pressure for the highside pressure relief device.

(b) When systems are not protected by a highside pressure relief device, the compressor's pressure limiting device shall be set not more than the system's highside design pressure.

(2) For nonpositive displacement compressors:

In this type, a rotating component imparts its kinetic energy to the air then eventually converted into pressure energy.

Centrifugal compressors are non-positive displacement type. Rotating impeller imparts kinetic energy to the air which is converted to pressure energy as air passes through the diffuser.

The term 'positive displacement' emphasizes the fact that a volume is physically being displaced, whereas there is no physical displacement in centrifugal pumps.

(a) When systems are protected by a highside pressure relief device, the compressor's pressure limiting device shall be set not more than 90 percent of the operating pressure for the highside pressure relief device.

(b) When systems are protected by a lowside pressure relief device that is only subject to lowside pressure, and is provided with a permanent relief path between the systems' highside and lowside, without intervening valves, the compressor's pressure limiting device shall be set not more than the systems' highside design pressure. [ASHRAE 15:9.9.2]

1111.3 Location. Stop valves shall not be installed between the pressure imposing element and pressure limiting devices serving compressors. [ASHRAE 15:9.9.3]

1111.4 Emergency Stop. Activation of a pressure-limiting device shall stop the action of the pressure-imposing element. [ASHRAE 15:9.9.4]

1112.0 Pressure-Relief Devices.

1112.1 General. Refrigeration systems shall be protected by a pressure-relief device or other approved means to safely relieve pressure due to fire or abnormal conditions. [ASHRAE 15:9.4.1]

1112.2 Positive Displacement Compressor. A positive displacement compressor with a stop valve in the discharge connection shall be equipped with a pressure-relief device that is sized, and with a pressure setting, in accordance with the compressor manufacturer to prevent rupture of the compressor or to prevent the pressure from increasing to more than 10 percent above the maximum allowable working pressure of components located in the discharge line between the compressor and the stop valve or in accordance with Section 1113.5, whichever is larger. The pressure-relief device shall discharge into the low-pressure side of the system or in accordance with Section 1112.11.

Exception: Hermetic refrigerant motor-compressors that are listed and have a displacement not more than 50 cubic feet per minute (1.42 m^3/min).

The relief device(s) shall be sized based on compressor flow at the following conditions:

(1) For compressors in single-stage systems and high-stage compressors of other systems, the flow shall be calculated based on 50°F (10°C) saturated suction temperature at the compressor suction.

(2) For low-stage or booster compressors in compound systems, the compressors that are capable of running only where discharging to the suction of a high-stage compressor, the flow shall be calculated based on the saturated suction temperature equal to the design operating intermediate temperature.

(3) For low-stage compressors in cascade systems, the compressors that are located in the lower-temperature stage(s) of cascade systems, the flow shall be calculated based on the suction pressure being equal to the pressure setpoint of the pressure-relieving devices that protect the lowside of the stage against overpressure.

Exceptions: For Section 1112.2(1), Section 1112.2(2), and Section 1112.2(3), the discharge capacity of the relief device shall be permitted to be the minimum regulated flow rate of the compressor where the following conditions are met:

(1) The compressor is equipped with capacity regulation.

(2) Capacity regulation actuates to a flow at not less than 90 percent of the pressure-relief device setting.

(3) A pressure-limiting device is installed and set in accordance with the requirements of Section 1111.0. [ASHRAE 15:9.8]

1112.3 Liquid-Containing Portions of Systems. Liquid-containing portions of systems, including piping, that is isolated from pressure-relief devices required elsewhere, and that develops pressures exceeding their working design pressures due to temperature rise, shall be protected by the installation of pressure-relief devices.

1112.4 Evaporators. Heat exchanger coils located downstream, or upstream within 18 inches (457 mm), of a heating source and capable of being isolated shall be fitted with a pressure-relief device that discharges to another part of the system in accordance with Section 1112.5 through Section 1112.5.2 or outside any enclosed space in accordance with Section 1112.11. The pressure relief device shall be connected at the highest possible location of the heat exchanger or piping between the heat exchanger and its manual isolation valves.

Exceptions:

(1) Relief valves shall not be required on heat exchanger coils that have a design pressure more than 110 percent of refrigerant saturation pressure when exposed to the maximum heating source temperature.

(2) A relief valve shall not be required on self-contained or unit systems where the volume of the lowside of the system, which is shut off by valves, is more than the specific volume of the refrigerant at critical conditions of temperature and pressure, as determined in accordance with Equation 1112.4.

$$V_1 / [W_1 - (V_2 - V_1) / V_{gt}] \qquad \text{(Equation 1112.4)}$$

Shall be more than V_{gc}

Where:

V_1 = Lowside volume, cubic foot (m^3).

V_2 = Total volume of system, cubic foot (m^3).

W_1 = Total weight of refrigerant in system, pounds (kg).

V_{gt} = Specific volume of refrigerant vapor at 110°F (43°C), cubic feet per pound (m^3/kg).

V_{gc} = Specific volume at critical temperature and pressure, cubic feet per pound (m^3/kg). [ASHRAE 15:9.4.4]

1112.5 Hydrostatic Expansion. Pressure rise resulting from hydrostatic expansion due to temperature rise of liquid refrigerant trapped in or between closed valves shall be addressed in accordance with Section 1112.5.1 and Section 1112.5.2. [ASHRAE 15:9.4.3]

1112.5.1 Hydrostatic Expansion During Normal Operation. Where trapping of liquid with subsequent hydrostatic expansion is capable of occurring automatically during normal operation or during standby, shipping, or power failure, engineering controls shall be used that are capable of preventing the pressure from exceeding the design pressure.

Acceptable engineering controls include but are not limited to the following:

(1) Pressure relief device to relieve hydrostatic pressure to another part of the system.

(2) Reseating pressure relief valve to relieve the hydrostatic pressure to an approved treatment system. [ASHRAE 15:9.4.3.1]

1112.5.2 Hydrostatic Expansion During Maintenance. Where trapping of liquid with subsequent hydrostatic expansion is capable of occurring only during maintenance—i.e., when personnel are performing maintenance tasks—either engineering or administrative controls shall be used to relieve or prevent the hydrostatic overpressure. [ASHRAE 15:9.4.3.2]

1112.6 Actuation. Pressure-relief devices shall be direct-pressure actuated or pilot operated. Pilot-operated pressure-relief valves shall be self-actuated, and the main valve shall open automatically at the set pressure and, where an essential part of the pilot fails, shall discharge its full rated capacity. [ASHRAE 15:9.4.5]

1112.7 Stop Valves Prohibited. Stop valves shall not be located between a pressure-relief device and parts of the system protected thereby. A three-way valve, used in conjunction with the dual relief valve in accordance with Section 1113.6, shall not be considered a stop valve. [ASHRAE 15:9.4.6]

1112.8 Location. Pressure-relief devices shall be connected directly to the pressure vessel or other parts of the system protected thereby. These devices shall be connected above the liquid refrigerant level and installed so that they are accessible for inspection and repair, and so that they are not capable of being readily rendered inoperative.

Exception: Where fusible plugs are used on the highside, they shall be located above or below the liquid refrigerant level. [ASHRAE 15:9.4.8]

1112.9 Materials. The seats and discs of pressure-relief devices shall be constructed of compatible material to resist refrigerant corrosion or other chemical action caused by the refrigerant. Seats or discs of cast iron shall not be used. Seats and discs shall be limited in distortion, by pressure or other cause, to a set pressure change of not more than 5 percent in a span of five years. [ASHRAE 15:9.4.9]

1112.10 Pressure-Relief Device Settings. Pressure-relief valves shall start to function at a pressure not exceeding the design pressure of the parts of the system protected.

Exception: Relief valves that discharge into other parts of the system shall comply with Section 1112.11.3. [ASHRAE 15:9.5.1]

1112.10.1 Rupture Member Setting. Rupture members used in lieu of, or in series with, a relief valve shall have a nominal rated rupture pressure not exceeding the design pressure of the parts of the system protected. The conditions of application shall comply with ASME BPVC Section VIII. The size of rupture members installed ahead of relief valves shall not be less than the relief-valve inlet. [ASHRAE 15:9.5.2]

1112.11 Discharge from Pressure-Relief Devices. Pressure-relief systems designed for vapor shall comply with Section 1112.11.1 through Section 1112.11.4.1.

1112.11.1 Discharging Location Interior to Building. Pressure-relief devices, including fusible plugs, serving refrigeration systems shall be permitted to discharge to the interior of a building where in accordance with the following:

(1) The system contains less than 110 pounds (49.9 kg) of a Group A1 refrigerant.

(2) The system contains less than 6.6 pounds (2.99 kg) of a Group A2, B1 or B2 refrigerant.

(3) The system does not contain any quantity of a Group A3 or B3 refrigerant.

(4) The system is not required to be installed in a machinery room in accordance with Section 1106.0.

(5) The refrigerant concentration limits in Section 1104.2 are not exceeded. Refrigeration systems that do not comply with the above requirements shall comply with the requirements of Section 1112.11.2 through Section 1112.11.4. [ASHRAE 15:9.7.8.1]

1112.11.2 Discharging Location Exterior to Building. Pressure-relief devices designed to discharge external to the refrigeration system shall be arranged to discharge outside of a building and shall be in accordance with the following:

(1) The point of vent discharge shall be located not less than 15 feet (4572 mm) above the adjoining ground level.

Exception: Outdoor systems containing Group A1 refrigerant shall be permitted to discharge at any elevation where the point of discharge is located in an access-controlled area accessible to authorized personnel only.

(2) The point of vent discharge shall be located not less than 20 feet (6096 mm) from windows, building ventilation openings, pedestrian walkways, or building exits.

(3) For heavier-than-air refrigerants, the point of vent discharge shall be located not less than 20 feet (6096 mm) horizontally from below-grade walkways, entrances, pits or ramps where a release of the entire system charge into such a space would yield a concentration of refrigerant in excess of the RCL. The direct discharge of a relief vent into enclosed outdoor spaces, such as a courtyard with walls on all sides, shall not be permitted where a release of the entire system charge into such a space would yield a concentration of refrigerant in excess of the RCL. The volume for the refrigerant concentration calculation shall be determined using the gross area of the space and a height of 8.2 feet (2499 mm), regardless of the actual height of the enclosed space.

(4) The termination point of a vent discharge line shall be made in a manner that prevents discharged refrigerant from spraying directly onto personnel that are capable of being in the vicinity.

(5) The termination point of vent discharge line shall be made in a manner that prevents foreign material or debris from entering the discharge piping.

(6) Relief vent lines that terminate vertically upward and are subject to moisture entry shall be provided with a drip pocket having a length of not less than 24 inches (610 mm) and having the size of the vent discharge pipe. The drip pocket shall be installed to extend below the first change in vent pipe direction and shall be fitted with a valve or drain plug to permit removal of accumulated moisture. [ASHRAE 15:9.7.8.2]

1112.11.3 Internal Relief. Pressure-relief valves designed to discharge from a higher-pressure vessel into a lower pressure vessel internal to the system shall comply with the following:

(1) The pressure-relief valve that protects the higher-pressure vessel shall be selected to deliver capacity in accordance with Section 1113.5 without exceeding the maximum allowable working pressure of the higher-pressure vessel accounting for the change in mass flow capacity due to the elevated backpressure.

(2) The capacity of the pressure-relief valve protecting the part of the system receiving a discharge from a pressure-relief valve protecting a higher-pressure vessel shall be not less than the sum of the capacity required in Section 1113.5 plus the mass flow capacity of the pressure-relief valve discharging into that part of the system.

(3) The design pressure of the body of the relief valve used on the higher-pressure vessel shall be rated for operation at the design pressure of the higher-pressure vessel in both pressure-containing areas of the valve. [ASHRAE 15:9.7.8.3]

1112.11.4 Discharge Location, Special Requirements. Additional requirements for relief device discharge location and allowances shall apply for specific refrigerants in accordance with Section 1112.11.4.1. [ASHRAE 15:9.7.8.4]

1112.11.4.1 Water (R-718). Where water is the refrigerant, discharge to a floor drain shall be permitted where the following conditions are met:

(1) The pressure-relief device set pressure shall not exceed 15 psig (103 kPa).

(2) The floor drain shall be sized to handle the flow rate from a single broken tube in a refrigerant-containing heat exchanger.

(3) The Authority Having Jurisdiction finds it acceptable that the working fluid, corrosion inhibitor, and other additives used in this type of refrigeration system are permitted to infrequently be discharged to the sewer system, or a catch tank that is sized to handle the expected discharge shall be installed and equipped with a normally closed drain valve and an overflow line to drain. [ASHRAE 15:9.7.8.4.1]

1112.12 Discharge Piping. The piping used for pressure-relief device discharge shall be in accordance with Section 1112.12.1 through Section 1112.12.5. [ASHRAE 15:9.7.9]

1112.12.1 Piping Connection. Piping connected to the discharge side of a fusible plug or rupture member shall have provisions to prevent plugging of the pipe upon operation of a fusible plug or rupture member. [ASHRAE 15:9.7.9.1]

1112.12.2 Pipe Size. The size of the discharge pipe from the pressure-relief device or fusible plug shall be not less than the outlet size of the pressure-relief device or fusible plug. [ASHRAE 15:9.7.9.2]

1112.12.3 Maximum Length. The maximum length of the discharge piping installed on the outlet of pressure-relief devices and fusible plugs discharging to the atmosphere shall be determined in accordance with Section 1112.12.4 and Section 1112.12.5. See Table 1112.12.3 for the allowable flow capacity of various equivalent lengths of single discharge piping vents for conventional pressure-relief valves. [ASHRAE 15:9.7.9.3]

TABLE 1112.12.3
ATMOSPHERIC PRESSURE AT NOMINAL INSTALLATION ELEVATION (*Pa*)
[ASHRAE 15: TABLE 9.7.9.3.2]

ELEVATION ABOVE SEA LEVEL, FEET	POUNDS PER SQUARE INCH, ABSOLUTE (P_a)
0	14.7
500	14.4
1000	14.2
1500	13.9
2000	13.7
2500	13.4
3000	13.2
3500	12.9
4000	12.7
4500	12.5
5000	12.2
6000	11.8
7000	11.3
8000	10.9
9000	10.5
1000	10.1

For SI units: 1 foot = 304.8 mm, 1 pound-force per square inch = 6.8947 kPa

1112.12.4 Design Back Pressure. The design back pressure due to flow in the discharge piping at the outlet of pressure-relief devices and fusible plugs, discharging to atmosphere, shall be limited by the allowable equivalent length of piping determined in accordance with Equation 1112.12.4(1).

[Equation 1112.12.4(1)]

$$L = \frac{0.2146 \cdot d^5 \left(P_0^2 - P_2^2\right)}{f \cdot C_r^2} - \frac{d \cdot \ln\left(\frac{P_0}{P_2}\right)}{6 \cdot f}$$

Where:

L = Equivalent length of discharge piping, feet.

Cr = Rated capacity as stamped on the relief device in pounds per minute (lb/min), or in SCFM multiplied by 0.0764, or as calculated in Section 1112.14 for a rupture member or fusible plug, or as adjusted for reduced capacity due to piping in accordance with the manufacturer of the device, or as adjusted for reduced capacity due to piping as estimated by an approved method.

f = Moody friction factor in fully turbulent flow.

d = Inside diameter of pipe or tube, inches.

ln = Natural logarithm.

P_2 = Absolute pressure at outlet of discharge piping, psia.

P_0 = Allowed back pressure (absolute) at the outlet of pressure relief device, (psia).

For SI units: 1 foot = 304.8 mm, 1 pound-force per square inch = 6.8947 kPa, 1 pound per minute = 0.00756 kg/s

Unless the maximum allowable back pressure (P_0) is specified by the relief valve manufacturer, the following maximum allowable back pressure values shall be used for P_0, where P is the set pressure and P_a is atmospheric pressure at the nominal elevation of the installation (see Table 1112.12.3):

For conventional relief valves: 15 percent of set pressure:

$$P_0 = (0.15 \cdot P) + P_a \qquad \text{[Equation 1112.12.4(2)]}$$

For balanced relief valves: 25 percent of set pressure:

$$P_0 = (0.25 \cdot P) + P_a \qquad \text{[Equation 1112.12.4(3)]}$$

For rupture disks alone: fusible plugs, and pilot operated relief devices, 50 percent of set pressure:

$$P_0 = (0.50 \cdot P) + P_a \qquad \text{[Equation 1112.12.4(4)]}$$

For fusible plugs, P shall be the saturated absolute pressure for the stamped temperature melting point of the fusible plug or the critical pressure of the refrigerant used, whichever is smaller. [ASHRAE 15:9.7.9.3.1, 9.7.9.3.2]

1112.12.5 Simultaneous Operation. When outlets of two or more relief devices or fusible plugs, which are expected to operate simultaneously, connect to a common discharge pipe, the common pipe shall be sized large enough to prevent the back pressure at each relief device from exceeding the maximum allowable back pressure in accordance with Section 1112.12.4. [ASHRAE 15:9.7.9.3.3]

1112.13 Rating of Pressure-Relief Device. The rated discharge capacity of a pressure-relief device expressed in pounds of air per minute (kg/s), shall be determined in accordance with ASME BPVC Section VIII. Pipe and fittings between the pressure-relief valve and the parts of the system it protects shall have not less than the area of the pressure-relief valve inlet area. [ASHRAE 15:9.7.6]

1112.14 Rating of Rupture Members and Fusible Plugs. The rated discharge capacity of a rupture member or fusible plug discharging to atmosphere under critical flow conditions, in pounds of air per minute (kg/s), shall be determined in accordance with the following formulas:

$$C = 0.64 P_1 d^2 \qquad \text{[Equation 1112.14(1)]}$$

$$d = 1.25 \sqrt{C/P_1} \qquad \text{[Equation 1112.14(2)]}$$

Where:

C = Rated discharge capacity of air, pounds per minute.

d = Smallest internal diameter of the inlet pipe, retaining flanges, fusible plug, or rupture member; inches.

For rupture members:

[Equation 1112.14(3)]

P_1 = (rated pressure in psig x 1.1) + 14.7

For fusible plugs:

P_1 = Absolute saturation pressure, corresponding to the stamped temperature melting point of the fusible plug or the critical pressure of the refrigerant used, whichever is smaller, pound-force per square inch atmosphere, psia. [ASHRAE 15:9.7.7]

For SI units: 1 inch = 25.4 mm, 1 pound-force per square inch = 6.8947 kPa, 1 pound per minute = 0.00756 kg/s

1113.0 Overpressure Protection.

1113.1 General. Pressure vessels shall be provided with overpressure protection in accordance with ASME BPVC Section VIII. Pressure vessels containing liquid refrigerant that are capable of being isolated by stop valves from other parts of a refrigerating system shall be provided with overpressure protection. Pressure relief devices or fusible plugs shall be sized in accordance with Section 1113.5. [ASHRAE 15:9.7.1, 9.7.2]

1113.2 Type of Protection. Pressure vessels with an internal gross volume of 3 cubic feet (0.1 m^3) or less shall use one or more pressure relief devices or a fusible plug. Pressure vessels of more than 3 cubic feet (0.1 m^3) but less than 10 cubic feet (0.28 m^3) internal gross volume shall use one or more pressure relief devices. Fusible plugs shall not be used. [ASHRAE 15:9.7.2.1, 9.7.2.2]

1113.3 Discharging into Lowside of System. For pressure-relief valves discharging into the lowside of the system, a single relief valve (not rupture member) of the required relieving capacity shall not be used on vessels of 10 cubic feet (0.28 m^3) or more internal gross volume except under the conditions permitted in Section 1112.11.3. [ASHRAE 15:9.7.3]

1113.4 Parallel Pressure-Relief Devices. Two or more pressure-relief devices in parallel to obtain the required capacity shall be considered as one pressure-relief device. The discharge capacity shall be the sum of the capacities required for each pressure vessel being protected.

1113.5 Discharge Capacity. The minimum required discharge capacity of the pressure-relief device or fusible plug for a pressure vessel shall be determined in accordance with Equation 1113.5:

$$C = fDL \qquad \text{(Equation 1113.5)}$$

Where:

C = Minimum required discharge capacity of the relief device expressed as mass flow of air, pounds per minute (kg/s).

D = Outside diameter of vessel, feet (m).

L = Length of vessel, feet (m).

f = Factor dependent upon type of refrigerant from Table 1113.5.

Where combustible materials are used within 20 ft (6096

mm) of a pressure vessel, the value of f shall be multiply by 2.5. Equation 1113.5 is based on fire conditions, other heat sources shall be calculated separately. Where one pressure-relief device or fusible plug is used to protect more than one pressure vessel, the required capacity shall be the sum of the capacity required for every pressure vessel. [ASHRAE 15:9.7.5]

TABLE 1113.5
RELIEF DEVICES CAPACITY FACTOR*
[ASHRAE 15:TABLE 9.7.5]

REFRIGERANT	VALUE OF *f*
Where used on the lowside of a limited-charge cascade system:	
R-23, R-170, R-744, R-1150, R-508A, R-508B	1
R-13, R-13B1, R-503	2
R-14	2.5
Other applications:	
R-718	0.2
R-717	0.5
R-11, R-32, R-113, R-123, R-142b, R-152a, R-290, R-600, R-600a, R-764	1
R-12, R-22, R-114, R-124, R-134a, R-401A, R-401B, R-401C, R- 405A, R-406A, R-407C, R-407D, R-407E, R-409A, R-409B, R-411A, R-411B, R-411C, R-412A, R-414A, R-414B, R-500, R-1270	1.6
R-143a, R-402B, R-403A, R-407A, R-408A, R-413A	2
R-115, R-402A, R-403B, R-404A, R-407B, R-410A, R-410B, R- 502, R-507A, R-509A	2.5

* In accordance with Section 1102.2, ammonia refrigeration systems are not regulated by this chapter. R-717 (ammonia) is included in this table because the table is extracted from ASHRAE 15 and is not capable of being modified.

1113.6 Three-Way Valve. Pressure vessels of 10 cubic feet (0.28 m^3) or more internal gross volume shall use one or more rupture member(s) or dual pressure-relief valves where discharging to the atmosphere. Dual pressure-relief valves shall be installed with a three-way valve to allow testing or repair. Where dual relief valves are used, the valve shall comply with Section 1113.5.

Exception: A single relief valve shall be permitted on pressure vessels of 10 cubic feet (0.28 m^3) or more internal gross volume where in accordance with the following conditions:

(1) The relief valves are located on the lowside of the system.

(2) The vessel is provided with shutoff valves designed to allow pumpdown of the refrigerant charge of the pressure vessel.

(3) Other pressure vessels in the system are separately protected in accordance with Section 1113.1. [ASHRAE 15:9.7.2.3]

1114.0 Special Discharge Requirements.

1114.1 General. Systems containing other than Group A1 or B1 refrigerants shall discharge to atmosphere through an approved flaring device.

Exceptions:

(1) Where the Authority Having Jurisdiction determines upon review of a rational engineering analysis that fire, health, or environmental hazards will not result from the proposed atmospheric release.

(2) Lithium bromide absorption system using water as the refrigerant.

1114.2 Design Requirements. Flaring devices shall be designed to incinerate the entire discharge. The products of refrigerant incineration shall not pose health or environmental hazards. Incineration shall be automatic upon initiation of discharge, shall be designed to prevent blow-back, and shall not expose structures or materials to the threat of fire. Standby fuel, such as LP-Gas, and standby power shall have the capacity to operate for one and a half times the required time for complete incineration of the charge.

1114.3 Testing. Flaring systems shall be tested to demonstrate their safety and effectiveness. A report from an approved agency shall be submitted detailing the emission products from the system as installed.

1115.0 Labeling and Identification.

1115.1 General. In addition to labels required elsewhere in this chapter, a refrigeration system shall be provided with identification labels in accordance with Section 1115.2 and Section 1115.3.

1115.2 Volume and Type. A condenser, receiver, absorber, accumulator and similar equipment having an internal volume of more than 3 cubic feet (0.1 m^3) and containing refrigerant shall be equipped with a permanent label setting forth the type of refrigerant in such vessel.

1115.3 Permanent Sign. In a refrigeration machinery room and for a direct refrigerating system of more than 10 horsepower (7.5 kW), there shall be a permanent sign at an approved location giving the following information:

(1) Name of contractor installing the equipment.

(2) Name and number designation of refrigerant in system.

(3) Pounds of refrigerant in system.

1115.4 Marking of Pressure-Relief Devices. Pressure-relief valves for refrigerant containing components shall be set and sealed by the manufacturer or an assembler as defined in ASME BPVC Section VIII. Each pressure relief valve shall be marked by the manufacturer or assembler with the data required in ASME BPVC Section VIII.

Exception: Relief valves for systems with design pressures of 15 pounds-force per square inch gauge (psig) (103 kPa) or less shall be marked. [ASHRAE 15:9.6.1]

1115.4.1 Rupture Members. Rupture members for refrigerant pressure vessels shall be marked with the data required in accordance with ASME BPVC Section VIII. [ASHRAE 15:9.6.2]

1115.4.2 Fusible Plugs. Fusible plugs shall be marked with the melting temperatures in °F (°C). [ASHRAE 15:9.6.3]

1116.0 Testing of Refrigeration Equipment.

1116.1 Factory Tests. Refrigerant-containing parts of unit systems shall be tested and proved tight by the manufacturer at not less than the design pressure for which they are rated. Pressure vessels shall be tested in accordance with Section 1117.0. [ASHRAE 15:9.14.1]

1116.1.1 Testing Procedure. Tests shall be performed with dry nitrogen or another nonflammable, nonreactive, dried gas. Oxygen, air, or mixtures containing them shall not be used. The means used to build up the test pressure shall have a pressure-limiting device or a pressure-reducing device and a gage on the outlet side. The pressure-relief device shall be set above the test pressure but low enough to prevent permanent deformation of the system's components.

Exceptions:

(1) Mixtures of dry nitrogen, inert gases, nonflammable refrigerants permitted for factory tests.

(2) Mixtures of dry nitrogen, inert gases, or a combination of them with flammable refrigerants in concentrations not exceeding the lesser of a refrigerant weight fraction (mass fraction) of 5 percent or 25 percent of the LFL shall be permitted for factory tests.

(3) Compressed air without added refrigerant shall be permitted for factory tests provided the system is subsequently evacuated to less than 0.039 inch of mercury (0.132 kPa) before charging with refrigerant. The required evacuation level is atmospheric pressure for systems using R-718 (water) or R-744 (carbon dioxide) as the refrigerant. [ASHRAE 15:9.14.1.1]

1116.1.2 Applied Pressure. The test pressure applied to the highside of each factory-assembled refrigerating system shall be not less than the design pressure of the highside. The test pressure applied to the lowside of a factory assembled refrigerating system shall be not less than the design pressure of the lowside.

1116.1.3 Design Pressure of 15 psig or Less. Units with a design pressure of 15 psig (103 kPa) or less shall be tested at a pressure not less than 1.33 times the design pressure, and shall be proved leak-tight at not less than the lowside design pressure. [ASHRAE 15:9.14.3]

1116.2 Field Tests. Refrigerant-containing parts of a system that is field-erected shall be tested and proved tight after complete installation and before the operation. The high and low sides of each system shall be tested and proved tight at not less than the lower of the pressure in Table 1116.2 or the setting of the pressure-relief device.

Exceptions:

(1) Compressors, condensers, evaporators, coded pressure vessels, safety devices, pressure gauges, control mechanisms, and systems that are factory tested.

(2) Refrigeration systems containing Group R-22, not exceeding 5 tons of refrigeration capacity (18 kW), and field-piped using approved, factory-charged line sets shall be permitted to be proved tight by observing retention of pressure on a set of charging gauges and soaping connections while the system is operating.

TABLE 1116.2
FIELD LEAK TEST PRESSURES (psig)*

REFRIGERANT NUMBER	HIGHSIDE WATER COOLED	HIGHSIDE AIR COOLED	LOWSIDE
11	15	35	15
12	140	220	140
22	230	360	230
113	15	15	15
114	40	80	40
115	275	340	275
123	15	30	15
134a	150	250	150
152a	130	220	130
500	165	265	165
502	250	385	250
744*	—	—	—

For SI units: 1 pound-force per square inch gauge = 6.8947 kPa

* Special design required; test pressures typically exceed 1000 psig (6895 kPa).

1116.3 Test Gases. Tests shall be performed with dry nitrogen or other nonflammable, nonreactive, dried gas. Oxygen, air, or mixtures containing them shall not be used. The means used to build up the test pressure shall have either a pressure-limiting device or a pressure-reducing device and a gauge on the outlet side. The pressure-relief device shall be set above the test pressure but low enough to prevent permanent deformation of the system's components.

Exceptions:

(1) Mixtures of dry nitrogen, inert gases, or a combination of them with nonflammable refrigerant in concentrations of a refrigerant weight fraction (mass fraction) not exceeding 5 percent shall be permitted for tests.

(2) Mixtures of dry nitrogen, inert gases, or a combination of them with flammable refrigerants in concentrations not exceeding the lower of a refrigerant weight fraction (mass fraction) of 5 percent or 25 percent of the LFL shall be permitted for tests.

(3) Compressed air without added refrigerants shall be permitted for tests, provided the system is subsequently evacuated to less than 1000 microns (0.1333 kPa) before charging with refrigerant. The required evacuation level is atmospheric pressure for systems using R-718 (water) or R-744 (carbon dioxide) as the refrigerant.

(4) Systems erected on the premises using Group A1 refrigerant and with copper tubing not exceeding 0.62 of an inch (15.7 mm) outside diameter shall be tested by means of the refrigerant charged into the system at the saturated vapor pressure of the refrigerant at not less than 68°F (20°C). [ASHRAE 15:10.1.2]

1116.4 Declaration. A dated declaration of test shall be provided for systems containing more than 55 pounds (24.9 kg) of refrigerant. The declaration shall give the name of the refrigerant and the field test pressure applied to the highside and the lowside of the system. The declaration of test shall be signed by the installer and, where an inspector is present at the tests, the inspector shall also sign the declaration. Where requested, copies of this declaration shall be furnished to the Authority Having Jurisdiction. [ASHRAE 15:10.2]

1116.5 Brine Systems. Brine-containing portions of a system shall be tested at one and a half times the design pressure of the system using brine as the test fluid.

1117.0 Refrigerant-Containing Pressure Vessels.

1117.1 Inside Dimensions 6 Inches or Less. Pressure vessels having inside dimensions of 6 inches (152 mm) or less shall be:

(1) Listed either individually or as part of an assembly by an approved, nationally recognized testing laboratory, or

(2) Marked directly on the vessel or on a nameplate attached to the vessel with a "U" or "UM" symbol signifying compliance with ASME BPVC Section VIII, or

(3) When requested by the Authority Having Jurisdiction, the manufacturer shall provide documentation to confirm that the vessel meets the design, fabrication, and testing requirements of ASME BPVC Section VIII.

Exception: Vessels having an internal or external design pressure of 15 psig (103 kPa) or less.

Pressure vessels having inside dimensions of 6 inches (152 mm) or less shall be protected by either a pressure-relief device or a fusible plug. [ASHRAE 15:9.3.1.1]

1117.1.1 Pressure-Relief Device. Where a pressure-relief device is used to protect a pressure vessel having an inside dimension of 6 inches (152 mm) or less, the ultimate strength of the pressure vessel so protected shall withstand a pressure of not less than 3.0 times the design pressure. [ASHRAE 15:9.3.1.2]

1117.1.2 Fusible Plug. Where a fusible plug is used to protect a pressure vessel having an inside diameter of 6 inches (152 mm) or less, the ultimate strength of the pressure vessel so protected shall withstand a pressure 2.5 times the saturation pressure of the refrigerant used at the temperature stamped on the fusible plug or 2.5 times the critical pressure of the refrigerant used, whichever is less. [ASHRAE 15:9.3.1.3]

1117.2 Inside Dimensions More than 6 Inches. Pressure vessels having an inside diameter exceeding 6 inches (152 mm) and having an internal or external design pressure greater than 15 psig (103 kPa) shall be directly marked, or marked on a nameplate, with a "U" or "UM" symbol signifying compliance with the rules of ASME BPVC Section VIII. [ASHRAE 15:9.3.2]

1117.3 Pressure Vessels for 15 psig or Less. Pressure vessels having an internal or external design pressure of 15 psig (103 kPa) or less shall have an ultimate strength to withstand not less than 3.0 times the design pressure and shall be tested with a pneumatic test pressure of not less than 1.25 times the design pressure or a hydrostatic test pressure of not less than 1.5 times the design pressure. [ASHRAE 15:9.3.3]

1118.0 Maintenance and Operation.

1118.1 General. Refrigeration systems shall be operated and maintained as required by the fire code.

Part II – Cooling Towers.

Cooling towers are heat exchangers used to cool water that flows through a condenser. Heat from the circulating water is transferred to the air. Water-cooled condensers are commonly used with large air-conditioning systems and certain industrial processes. It is not common practice and is generally prohibited by regulations or cost to conduct water from a public water supply through a condenser and then waste it to the public sewer.

Cooling towers are generally constructed of a wood or metal structure with a waterproof sump at the bottom and water spray nozzles on the top. Heat in the water is transferred to the air by evaporation as the water either flows or falls over the louvers or slats. The slats or louvers are made of wood, metal, or plastic.

As the water falls from the top of the tower to the sump at the bottom, some water vaporizes. "Latent" heat is released from the water, causing a cooling effect. A water supply and float valve are usually located in the sump to replace the water lost due to wind and natural evaporation. Water is pumped from the bottom of the sump through the condenser and recirculated to the top of the cooling tower.

Air is circulated through cooling towers by natural or mechanical draft. Natural draft cooling towers, sometimes called atmospheric towers, depend on wind to circulate air through the tower. These towers should be located outdoors, on a roof, or at an elevated location away from walls or buildings that would restrict wind. These units are much larger than forced draft towers of equal heat extraction capacity. The natural draft tower is quiet and less expensive to operate and maintain, but usually more expensive to construct. It is more difficult to find a suitable location for such an installation due to size, weight, and prevailing wind direction. Mechanical draft cooling towers may have a fan located on the top to draw air through the louvers. This type is called an induced draft cooling tower. Another design uses a blower or fan to force cooling air into the bottom of the tower and is called a forced draft cooling tower. The counter flow of warm water downward and cool air upward removes heat from the water in all of these designs.

Regular maintenance of cooling towers is essential to avoid accumulation of scale due to the evaporation of water and the accumulation of ash, slime, insects, and leaves. Water treatment is often required to reduce scale deposits. Carbon dioxide and sulfur dioxide can be absorbed by the waterdrops passing through the air. These absorbed gases can form acids, contributing to corrosion problems of piping and equipment. Oxygen is also absorbed in the water, resulting in increased corrosion of the system. Many algaecides and corrosion inhibitors are toxic; backflow of these materials into the potable water system must be prevented. Cooling towers need proper drainage and overflows.

Cooling towers are commonly used. Their use is increasing with the extensive use of air conditioners.

1119.0 General.

1119.1 Applicability. Cooling towers, evaporative condensers, and fluid coolers shall be readily accessible.

Where located on roofs, such equipment having combustible exterior surfaces shall be protected with an approved automatic fire-extinguishing system.

1120.0 Support and Anchorage.

1120.1 General. Cooling towers, evaporative condensers, and fluid coolers shall be supported on noncombustible grillage designed in accordance with the building code. Seismic restraints shall be as required by the building code.

1121.0 Drainage.

1121.1 General. Drains, overflows, and blow-down provisions shall have an indirect connection to an approved disposal location. Discharge of chemical waste shall be as approved by the regulatory authority.

1122.0 Chemical Treatment Systems.

1122.1 General. Chemical treatment systems shall comply with the fire code. Where chemicals used present a contact hazard to personnel, approved emergency eye-wash and shower facilities shall be installed.

1122.2 Automated Control of Cycles of Concentration. Cooling towers, evaporative condensers, and fluid coolers shall include controls that automate system bleed based on conductivity, fraction of metered makeup volume, metered bleed volume, recirculating pump run time, or bleed time.

Digitally Controlled Automation is becoming more advanced, reducing energy costs and maintenance. It is the same concept that is being employed for digitally controlled HVAC systems and is becoming a requirement in many states for advanced energy management.

1123.0 Location.

1123.1 General. Cooling towers, evaporative condensers, and fluid coolers shall be located such that their plumes cannot enter occupied spaces. Plume discharges shall be not less than 25 feet (7620 mm) away from a ventilation inlet to a building. Location on the property shall be as required for buildings by the building code.

1124.0 Electrical.

1124.1 General. Electrical systems shall be in accordance with the electrical code. Equipment shall be provided with a vibration switch to shut off fans operating with excessive vibration. In climates commonly subject to electrical storms, lightning protection shall be provided on roof-mounted equipment.

1125.0 Refrigerants and Hazardous Fluids.

1125.1 General. Equipment containing refrigerants as a part of a closed-cycle refrigeration system shall comply with Part I of this chapter. Equipment containing other fluids that are flammable, combustible, or hazardous shall be in accordance with this code and the fire code.

1126.0 Drift Eliminators.

1126.1 General. Cooling towers, evaporative condensers, and fluid coolers shall be equipped with drift eliminators that have a drift rate of not more than 0.005 percent of the circulated water flow rate in accordance with the equipment manufacturer's instructions.

TABLE 1102.3
REFRIGERANT GROUPS, PROPERTIES, AND ALLOWABLE QUANTITIES[8]
[ASHRAE 34: TABLE 4-1, TABLE 4-2]

REFRIGERANT	CHEMICAL FORMULA	CHEMICAL NAME[1] (COMPOSITION FOR BLENDS)	SAFETY GROUP[7]	OEL[2] (ppm)	POUNDS PER 1000 CUBIC FEET OF SPACE
R-11	CCl_3F	Trichlorofluoromethane	A1	C1000	0.39
R-12	CCl_2F_2	Dichlorodifluoromethane	A1	1000	5.6
R-12B1	$CBrClF_2$	Bromochlorodifluoromethane	—	—	—
R-13	$CClF_3$	Chlorotrifluoromethane	A1	1000	—
R-13B1	$CBrF_3$	Bromotrifluoromethane	A1	1000	—
R-14	CF_4	Tetrafluoromethane (carbon tetrafluoride)	A1	1000	25
R-21	$CHCl_2F$	Dichlorofluoromethane	B1	—	—
R-22	$CHClF_2$	Chlorodifluoromethane	A1	1000	13
R-23	CHF_3	Trifluoromethane	A1	1000	7.3
R-30	CH_2Cl_2	Dichloromethane (methylene chloride)	B1	—	—
R-31	CH_2ClF	Chlorofluoromethane	—	—	—
R-32	CH_2F_2	Difluoromethane (methylene fluoride)	A2L	1000	4.8
R-40	CH_3Cl	Chloromethane (methyl chloride)	B2	—	—
R-41	CH_3F	Fluoromethane (methyl fluoride)	—	—	—
R-50	CH_4	Methane	A3	1000	—
R-113	CCl_2FCClF_2	1, 1, 2-trichloro-1, 2, 2 – trifluoroethane	A1	1000	1.2
R-114	$CClF_2CClF_2$	1, 2-dichloro-1, 1, 2, 2 tetrafluoroethane	A1	1000	8.7
R-115	$CClF_2CF_3$	Chloropentafluoroethane	A1	1000	47
R-116	CF_3CF_3	Hexafluoroethane	A1	1000	34
R-123	$CHCl_2CF_3$	2, 2-dichloro-1, 1, 1, - trifluoroethane	B1	50	3.5
R-124	$CHClFCF_3$	2-chloro-1, 1, 1, 2 - tetrafluoroethane	A1	1000	3.5
R-125	CHF_2CF_3	Pentafluoroethane	A1	1000	23
R-134a	CH_2FCF_3	1, 1, 1, 2-tetrafluoroethane	A1	1000	13
R-141b	CH_3CCl_2F	1, 1-dichloro-1-fluoroethane	—	500	0.78
R-142b	CH_3CClF_2	1-chloro-1, 1-difluoroethane	A2	1000	5.1
R-143a	CH_3CF_3	1, 1, 1-trifluoroethane	A2L	1000	4.5
R-152a	CH_3CHF_2	1, 1-difluoroethane	A2	1000	2.0
R-170	CH_3CH_3	Ethane	A3	1000	0.54
R-E170	CH_3OCH_3	Methoxymethane (Dimethyl ether)	A3	1000	1.0
R-218	CF_3CF2CF_3	Octafluoropropane	A1	1000	43
R-227ea	CF_3CHFCF_3	1, 1, 1, 2, 3, 3, 3-heptafluoropropane	A1	1000	36
R-236fa	$CF_3CH_2CF_3$	1, 1, 1, 3, 3, 3-hexafluoropropane	A1	1000	21
R-245fa	$CHF_2CH_2CF_3$	1, 1, 1, 3, 3-pentafluoropropane	B1	300	12
R-290	$CH_3CH_2CH_3$	Propane	A3	1000	0.56
R-C318	$-(CF_2)_4-$	Octafluorocyclobutane	A1	1000	41
R-400	zeotrope	R-12/114 (50.0/50.0)	A1	1000	10
R-400	zeotrope	R-12/114 (60.0/40.0)	A1	1000	11
R-401A	zeotrope	R-22/152a/124 (53.0/13.0/34.0)	A1	1000	6.6
R-401B	zeotrope	R-22/152a/124 (61.0/11.0/28.0)	A1	1000	7.2
R-401C	zeotrope	R-22/152a/124 (33.0/15.0/52.0)	A1	1000	5.2
R-402A	zeotrope	R-125/290/22 (60.0/2.0/38.0)	A1	1000	17

TABLE 1102.3 (continued)
REFRIGERANT GROUPS, PROPERTIES, AND ALLOWABLE QUANTITIES[8]
[ASHRAE 34: TABLE 4-1, TABLE 4-2]

REFRIGERANT	CHEMICAL FORMULA	CHEMICAL NAME[1] (COMPOSITION FOR BLENDS)	SAFETY GROUP[7]	OEL[2] (ppm)	POUNDS PER 1000 CUBIC FEET OF SPACE
R-402B	zeotrope	R-125/290/22 (38.0/2.0/60.0)	A1	1000	15
R-403A	zeotrope	R-290/22/218 (5.0/75.0/20.0)	A2	1000	7.6
R-403B	zeotrope	R-290/22/218 (5.0/56.0/39.0)	A1	1000	18
R-404A	zeotrope	R-125/143a/134a (44.0/52.0/4.0)	A1	1000	31
R-405A	zeotrope	R-22/152a/142b/C318 (45.0/7.0/5.5/42.5)	—	1000	16
R-406A	zeotrope	R-22/600a/142b (55.0/4.0/41.0)	A2	1000	4.7
R-407A	zeotrope	R-32/125/134a (20.0/40.0/40.0)	A1	1000	19
R-407B	zeotrope	R-32/125/134a (10.0/70.0/20.0)	A1	1000	21
R-407C	zeotrope	R-32/125/134a (23.0/25.0/52.0)	A1	1000	18
R-407D	zeotrope	R-32/125/134a (15.0/15.0/70.0)	A1	1000	16
R-407E	zeotrope	R-32/125/134a (25.0/15.0/60.0)	A1	1000	17
R-407F	zeotrope	R-32/125/134a (30.0/30.0/40.0)	A1	1000	20
R-407G	zeotrope	R-32/125/134a (2.5/2.5/95.0)	A1	1000	13
R-407H	zeotrope	R-32/125/134a (32.5/15.0/52.5)	A1	1000	19
R-407I	zeotrope	R-32/125/124a (19.5/8.5/72.0)	A1	1000	16.0
R-408A	zeotrope	R-125/143a/22 (7.0/46.0/47.0)	A1	1000	21
R-409A	zeotrope	R-22/124/142b (60.0/25.0/15.0)	A1	1000	7.1
R-409B	zeotrope	R-22/124/142b (65.0/25.0/10.0)	A1	1000	7.3
R-410A	zeotrope	R-32/125 (50.0/50.0)	A1	1000	26
R-410B	zeotrope	R-32/125 (45.0/55.0)	A1	—	27
R-411A[6]	zeotrope	R-1270/22/152a (1.5/87.5/11.0)	A2	990	2.9
R-411B[6]	zeotrope	R-1270/22/152a (3.0/94.0/3.0)	A2	980	2.8
R-412A	zeotrope	R-22/218/142b (70.0/5.0/25.0)	A2	1000	5.1
R-413A	zeotrope	R-218/134a/600a (9.0/88.0/3.0)	A2	1000	5.8
R-414A	zeotrope	R-22/124/600a/142b (51.0/28.5/4.0/16.5)	A1	1000	6.4
R-414B	zeotrope	R-22/124/600a/142b (50.0/39.0/1.5/9.5)	A1	1000	6.0
R-415A	zeotrope	R-22/152a (82.0/18.0)	A2	1000	2.9
R-415B	zeotrope	R-22/152a (25.0/75.0)	A2	1000	2.1
R-416A[6]	zeotrope	R-134a/124/600 (59.0/39.5/1.5)	A1	1000	3.9
R-417A[6]	zeotrope	R-125/134a/600 (46.6/50.0/3.4)	A1	1000	3.5
R-417B	zeotrope	R-125/134a/600 (79.0/18.3/2.7)	A1	1000	4.3
R-417C	zeotrope	R-125/134a/600 (19.5/78.8/1.7)	A1	1000	5.4
R-418A	zeotrope	R-290/22/152a (1.5/96.0/2.5)	A2	1000	4.8
R-419A	zeotrope	R-125/134a/E170 (77.0/19.0/4.0)	A2	1000	4.2
R-419B	zeotrope	R-125/134a/E170 (48.5/48.0/3.5)	A2	1000	4.6
R-420A	zeotrope	R-134a/142b (88.0/12.0)	A1	1000	12
R-421A	zeotrope	R-125/134a (58.0/42.0)	A1	1000	17
R-421B	zeotrope	R-125/134a (85.0/15.0)	A1	1000	21
R-422A	zeotrope	R-125/134a/600a (85.1/11.5/3.4)	A1	1000	18
R-422B	zeotrope	R-125/134a/600a (55.0/42.0/3.0)	A1	1000	16
R-422C	zeotrope	R-125/134a/600a (82.0/15.0/3.0)	A1	1000	18
R-422D	zeotrope	R-125/134a/600a (65.1/31.5/3.4)	A1	1000	16
R-422E	zeotrope	R-125/134a/600a (58.0/39.3/2.7)	A1	1000	16
R-423A	zeotrope	R-134a/227ea (52.5/47.5)	A1	1000	19
R-424A[6]	zeotrope	R-125/134a/600a/600/601a (50.5/47.0/0.9/1.0/0.6)	A1	970	6.2
R-425A	zeotrope	R-32/134a/227ea (18.5/69.5/12.0)	A1	1000	16

TABLE 1102.3 (continued)
REFRIGERANT GROUPS, PROPERTIES, AND ALLOWABLE QUANTITIES[8]
[ASHRAE 34: TABLE 4-1, TABLE 4-2]

REFRIGERANT	CHEMICAL FORMULA	CHEMICAL NAME[1] (COMPOSITION FOR BLENDS)	SAFETY GROUP[7]	OEL[2] (ppm)	POUNDS PER 1000 CUBIC FEET OF SPACE
R-426A[6]	zeotrope	R-125/134a/600/601a (5.1/93.0/1.3/0.6)	A1	990	5.2
R427A	zeotrope	R-32/125/143a/134a (15.0/25.0/10.0/50.0)	A1	1000	18
R428A	zeotrope	R-125/143a/290/600a (77.5/20.0/0.6/1.9)	A1	1000	23
R-429A	zeotrope	R-E170/152a/600a (60.0/10.0/30.0)	A3	1000	0.81
R-430A	zeotrope	R-152a/600a (76.0/24.0)	A3	1000	1.3
R-431A	zeotrope	R-290/152a (71.0/29.0)	A3	1000	0.69
R-432A	zeotrope	R-1270/E170 (80.0/20.0)	A3	700	0.13
R-433A	zeotrope	R-1270/290 (30.0/70.0)	A3	880	0.34
R-433B	zeotrope	R-1270/290 (5.0/95.0)	A3	950	0.51
R-433C	zeotrope	R-1270/290 (25.0/75.0)	A3	790	0.41
R-434A	zeotrope	R-125/143a/134a/600a (63.2/18.0/16.0/2.8)	A1	1000	20
R-435A	zeotrope	R-E170/152a (80.0/20.0)	A3	1000	1.1
R-436A	zeotrope	R-290/600a (56.0/44.0)	A3	1000	0.50
R-436B	zeotrope	R-290/600a (52.0/48.0)	A3	1000	0.51
R-437A	zeotrope	R-125/134a/600/601 (19.5/78.5/1.4/0.6)	A1	990	5.0
R-438A	zeotrope	R-32/125/134a/600/601a (8.5/45.0/44.2/1.7/0.6)	A1	990	4.9
R-439A	zeotrope	R-32/125/600a (50.0/47.0/3.0)	A2	990	4.7
R-440A	zeotrope	R-290/134a/152a (0.6/1.6/97.8)	A2	1000	1.9
R-441A	zeotrope	R-170/290/600a/600 (3.1/54.8/6.0/36.1)	A3	1000	0.39
R-442A	zeotrope	R-32/125/134a/152a/227ea (31.0/31.0/30.0/3.0/5.0)	A1	1000	21
R-443A	zeotrope	R-1270/290/600a (55.0/40.0/5.0)	A3	580	0.19
R-444A	zeotrope	R-32/152a/1234ze(E) (12.0/5.0/83.0)	A2L	850	5.1
R-444B	zeotrope	R-32/152a/1234ze(E) (41.5/10.0/48.5)	A2L	890	4.3
R-445A	zeotrope	R-744/134a/1234ze (E) (6.0/9.0/85.0)	A2L	930	4.2
R-446A	zeotrope	R-32/1234ze(E)/600 (68.0/29.0/3.0)	A2L	960	2.5
R-447A	zeotrope	R-32/125/1234ze(E) (68.0/3.5/28.5)	A2L	900	2.6
R-447B	zeotrope	R-32/125/1234ze(E) (68.0/8.0/24.0)	A2L	970	23
R-448A	zeotrope	R-32/125/1234yf/134a/1234ze(E) (26.0/26.0/20.0/21.0/7.0)	A1	890	24
R-449A	zeotrope	R-32 /125 /1234yf /134a (24.3/24.7/25.3/25.7)	A1	830	23
R-449B	zeotrope	R-32/125/1234yf/134a (25.2/24.3/23.2/27.3)	A1	850	23
R-449C	zeotrope	R-32/125/1234yf/134a (20.0/20.0/31.0/29.0)	A1	800	23
R-450A	zeotrope	R-134a/1234ze(E) (42.0/58.0)	A1	880	20
R-451A	zeotrope	R-1234yf/134a (89.8/10.2)	A2L	520	5.3
R-451B	zeotrope	R-1234yf/134a (88.8/11.2)	A2L	530	5.3
R-452A	zeotrope	R-32/125/1234yf (11.0/59.0/30.0)	A1	780	27
R-452B	zeotrope	R-32/125/1234yf (67.0/7.0/26.0)	A2L	870	23
R-452C	zeotrope	R-32/125/1234yf (12.5/61.0/26.5)	A1	800	27
R-453A	zeotrope	R-32/125/134a/227ea/600/601a (20.0/20.0/53.8/5.0/0.6/0.6)	A1	1000	7.8
R-454A	zeotrope	R-32/1234yf (35.0/65.0)	A2L	690	28
R-454B	zeotrope	R-32/1234yf (68.9/31.1)	A2L	850	22
R-454C	zeotrope	R-32/1234yf (21.5/78.5)	A2L	620	29

TABLE 1102.3 (continued)
REFRIGERANT GROUPS, PROPERTIES, AND ALLOWABLE QUANTITIES[8]
[ASHRAE 34: TABLE 4-1, TABLE 4-2]

REFRIGERANT	CHEMICAL FORMULA	CHEMICAL NAME[1] (COMPOSITION FOR BLENDS)	SAFETY GROUP[7]	OEL[2] (ppm)	POUNDS PER 1000 CUBIC FEET OF SPACE
R-455A	zeotrope	R-744/32/1234yf (3.0/21.5/75.5)	A2L	650	23
R-456A	zeotrope	R-32/134a/1234ze(E) (6.0/45.0/49.0)	A1	900	20
R-457A	zeotrope	R-32/1234yf/152a (18.0/70.0/12.0)	A2L	650	25
R-458A	zeotrope	R-32/125/134a/227ea/236fa (20.5/4.0/61.4/13.5/0.6)	A1	1000	18
R-459A	zeotrope	R-32/1234yf/1234ze(E) (68.0/26.0/6.0)	A2L	870	23
R-459B	zeotrope	R-32/1234yf/1234ze(E) (21.0/69.0/10.0)	A2L	640	30
R-460A	zeotrope	R-32/125/134a/1234ze(E) (12.0/52.0/14.0/22.0)	A1	650	24
R-460B	zeotrope	R-32/125/134a/1234ze(E) (28.0/25.0/20.0/27.0)	A1	950	25
R-460C	zeotrope	R-32/125/134a/1234ze(E) (2.5/2.5/46.0/49.0)	A1	900	20.0
R-461A	zeotrope	R-125/143a/134a/227ea/600a (55.0/5.0/32.0/5.0/3.0)	A1	1000	17
R-462A	zeotrope	R-32/125/143a/134a/600 (9.0/42.0/2.0/44.0/3.0)	A2	1000	3.9
R-463A	zeotrope	R-744/32/125/1234yf/134a (6.0/36.0/30.0/14.0/14.0)	A1	990	19
R-464A	zeotrope	R-32/125/1234ze(E)/227ea (27.0/27.0/40.0/6.0)	A1	930	27.0
R-465A	zeotrope	R-32/290/1234yf (21.0/7.9/71.1)	A2	660	2.5
R-500	azeotrope[3]	R-12/152a (73.8/26.2)	A1	1000	7.6
R-501	azeotrope[3]	R-22/12 (75.0/25.0)[4]	A1	1000	13
R-502	azeotrope[3]	R-22/115 (48.8/51.2)	A1	1000	21
R-503	azeotrope[3]	R-23/13 (40.1/59.9)	—	1000	—
R-504	azeotrope[3]	R-32/115 (48.2/51.8)	—	1000	28
R-505	azeotrope[3]	R-12/31 (78.0/22.0)[4]	—	—	—
R-506	azeotrope[3]	R-31/114 (55.1/44.9)	—	—	—
R-507A[5]	azeotrope[3]	R-125/143a (50.0/50.0)	A1	1000	32
R-508A[5]	azeotrope[3]	R-23/116 (39.0/61.0)	A1	1000	14
R-508B	azeotrope[3]	R-23/116 (46.0/54.0)	A1	1000	13
R-509A[5]	azeotrope[3]	R-22/218 (44.0/56.0)	A1	1000	24
R-510A	azeotrope[3]	R-E170/600a (88.0/12.0)	A3	1000	0.87
R-511A	azeotrope[3]	R-290/E170 (95.0/5.0)	A3	1000	0.59
R-512A	azeotrope[3]	R-134a/152a (5.0/95.0)	A2	1000	1.9
R-513A	azeotrope[3]	R-1234yf/134a (56.0/44.0)	A1	650	20
R-513B	azeotrope[3]	R-1234yf/134a (58.5/41.5)	A1	640	21
R-514A	azeotrope[3]	R-1336mzz(Z)/1130 (E) (74.7/25.3)	B1	320	0.86
R-515A	azeotrope[3]	R-1234ze(E)/227ea (88.0/12.0)	A1	810	19
R-516A	azeotrope[3]	R-1234yf/134a/152a (77.5/8.5/14.0)	A2L	590	7.0
R-600	$CH_3CH_2CH_2CH_3$	Butane	A3	1000	0.15
R-600a	$CH(CH_3)_2CH_3$	2-methylpropane (isobutene)	A3	1000	0.59
R-601	$CH_3CH_2CH_2CH_2CH_3$	Pentane	A3	600	0.18

TABLE 1102.3 (continued)
REFRIGERANT GROUPS, PROPERTIES, AND ALLOWABLE QUANTITIES[8]
[ASHRAE 34: TABLE 4-1, TABLE 4-2]

REFRIGERANT	CHEMICAL FORMULA	CHEMICAL NAME[1] (COMPOSITION FOR BLENDS)	SAFETY GROUP[7]	OEL[2] (ppm)	POUNDS PER 1000 CUBIC FEET OF SPACE
R-601a	$(CH_3)_2CHCH_2CH_3$	2-methylbutane (isopentane)	A3	600	0.18
R-610	$CH_3CH_2OCH_2CH_3$	Ethoxyethane (ethyl ether)	—	400	—
R-611	$HCOOCH_3$	Methyl formate	B2	100	—
R-620	—	(Reserved for future assignment)	—	—	—
R-630	CH_3NH_2	Methanamine (methyl amine)	—	—	—
R-631	$CH_3CH_2(NH_2)$	Ethanamine (ethyl amine)	—	—	—
R-702	H_2	Hydrogen	A3	—	—
R-704	He	Helium	A1	—	—
R-717	NH_3	Ammonia	B2L	25	0.014
R-718	H_2O	Water	A1	—	—
R-720	Ne	Neon	A1	—	—
R-728	N_2	Nitrogen	A1	—	—
R-732	O_2	Oxygen	—	—	—
R-740	Ar	Argon	A1	—	—
R-744	CO_2	Carbon dioxide	A1	5000	3.4
R-744A	N_2O	Nitrous oxide	—	—	—
R-764	SO_2	Sulfur dioxide	B1	—	—
R-1130(E)	CHCl=CHCl	Trans-1,2-dichloroethene	B1	200	0.25
R-1132a	$CF_2 = CH_2$	1, 1-difluoroethylene	A2	500	2.0
R-1150	$CH_2=CH_2$	Ethene (ethylene)	A3	200	—
R-1224yd(Z)	$CF_3CF=CHCl$	(Z)-1-chloro-2,3,3,3-tetrafluoropropen	A1	1000	23
R-1233zd(E)	$CF_3CH=CHCl$	Trans-1-chloro-3,3,3-trifluoro-1- propene	A1	800	5.3
R-1234yf	$CF_3CF=CH_2$	2, 3, 3, 3-tetrafluoro-1-propene	A2L	500	4.7
R-1234ze(E)	$CF_3CH=CHF$	Trans-1,3,3,3- tetrafluoro-1-propene	A2L	800	4.7
R-1270	$CH_3CH=CH_2$	Propene (propylene)	A3	500	0.11
R-1336mzz(Z)	$CF_3CHCHCF_3$	Cis-1,1,1,4,4,4-hexaflouro-2-butene	A1	500	5.4

For SI units: 1 pound = 0.453 kg, 1 cubic foot = 0.0283 m^3

Notes:

1 The preferred chemical name is followed by the popular name in parenthesis.
2 The OELs are 8-hour TWAs; a "C" designation denotes a ceiling limit.
3 Azeotropic refrigerants exhibit some segregation of components at conditions of temperature and pressure other than those at which they were formulated. The extent of segregation depends on the particular azeotrope and hardware system configuration.
4 The exact composition of this azeotrope is in question and additional experimental studies are needed.
5 R-507, R-508, and R-509 are allowed alternative designations for R-507A, R-508A, and R-509A due to a change in designations after assignment of R-500 through R-509. Corresponding changes were not made for R-500 through R-506.
6 The RCL values for these refrigerant blends are approximated in the absence of adequate data for a component comprising less than 4 percent m/m of the blend and expected to have a small influence in an acute, accidental release.
7 Refrigerant flammability classification of Class 2L shall comply with the requirements for flammability classification of Class 2.
8 In accordance with Section 1102.2, ammonia refrigeration systems are not regulated by this chapter. R-717 (ammonia) is included in this table because the table is extracted from ASHRAE 34 and is not capable of being modified.

TABLE 1104.1
PERMISSIBLE REFRIGERATION SYSTEMS[1]

OCCUPANCY GROUP[3]	HIGH-PROBABILITY SYSTEM	LOW PROBABILITY SYSTEM	MACHINERY ROOM
A-1	Group A1 only	Any	Any
A-2	Group A1 only	Any	Any
A-3	Group A1 only	Any	Any
A-4	Group A1 only	Any	Any
B	Group A1[2] only	Any	Any
E	Group A1 only	Any	Any
F-1	Group A1[2] only	Any	Any
F-2	Any[2]	Any	Any
H-1	Any	Any	Any
H-2	Any	Any	Any
H-3	Any	Any	Any
H-4	Group A1 only	Any	Any
H-5	Group A1 only	Any	Any
I-1	None	Any	Any
I-2	Group A1 only	Any	Any
I-3	None	Any	Any
I-4	Group A1 only	Any	Any
M	Group A1[2] only	Any	Any
R-1	Group A1 only	Any	Any
R-2	Group A1 only	Any	Any
R-3	Group A1 only	Any	Any
R-4	Group A1 only	Any	Any
S-1	Group A1[2] only	Any	Any
S-2	Any[2]	Any	Any
U	Any	Any	Any

Notes:

[1] See Section 1104.0.

[2] A refrigerant shall be permitted to be used within a high-probability system where the room or space is in accordance with Section 1104.4.

[3] Occupancy classifications are defined in the building code.

CHAPTER 12
HYDRONICS

1201.0 General.

1201.1 Applicability. This chapter shall apply to hydronic piping systems that are part of heating, cooling, ventilation, refrigeration, and air conditioning systems. Such piping systems include steam, hot water, chilled water, steam condensate, condenser water, and ground source heat pump systems. The regulations of this chapter shall govern the construction, location, and installation of hydronic piping systems.

Chapter 12 was revised to specifically address hydronic systems including piping systems that are part of heating, ventilation and air conditioning systems. Such piping system shall include steam, hot water, chilled water, steam condensate and ground source heat pump systems.

1201.2 Insulation. Surfaces within reach of building occupants shall not exceed 140°F (60°C). Where sleeves are installed, the insulation shall continue full size through them.

Coverings and insulation used for piping shall be of material approved for the operating temperature of the system and the installation environment. Where installed in a plenum, the insulation, jackets, and lap-seal adhesives, including pipe coverings and linings, shall have a flame-spread index not to exceed 25 and a smoke-developed index not to exceed 50 where tested in accordance with ASTM E84 or UL 723.

See **Figure 1201.2**.

The purpose of testing building materials using ASTM E84 is to minimize the smoke and flame spread during a fire, giving occupants of the building a better chance to escape, allow firefighters more time to control the fire, and to minimize building damage.

Flame spread index is a measure of the material's ability to resist flames over its surface. The standard calls for not greater than 25 on a scale of 0-100, where untreated red oak lumber has a value of 100 and noncombustible cement-asbestos board has a value of 0.

Smoke-developed index is a measure of the concentration of smoke emitted by a material as it burns. The standard calls for not greater than 50 on the same relative scale (0-100) as the flame spread index.

UL 723, Test for Surface Burning Characteristics of Building Materials (Scope): "This method of test for surface burning characteristics of building materials is applicable to any type of building material that, by its own structural quality or the manner in which it is applied, is capable of supporting itself in position or being supported in the test furnace to a thickness comparable to its intended use.

"The purpose of the test is to determine the comparative burning characteristics of the material under test by evaluating the spread of flame over its surface and the density of the smoke developed when exposed to a test fire, and thus to establish a basis on which surface burning characteristics of different materials are compared, without specific regard to all the end-use parameters that affect the surface burning characteristics."

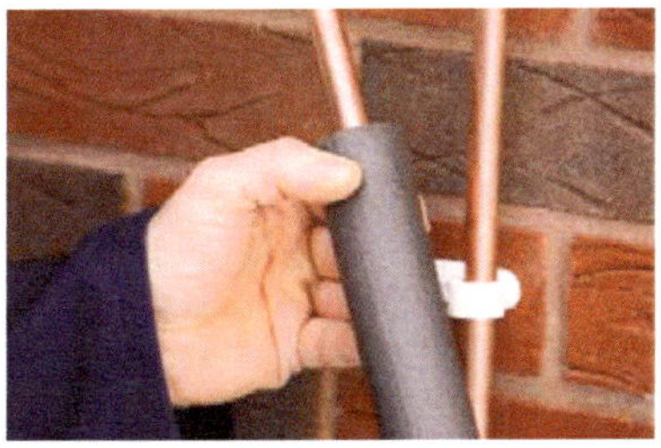

FIGURE 1201.2
FOAM TUBE INSULATION

1201.3 Water Hammer. The piping system shall be designed to prevent water hammer.

1201.4 Terminal Units. Terminal units, valves, and flow control devices shall be installed in accordance with the manufacturer's installation instructions.

1201.5 Return-Water Low-Temperature Protection. Where a minimum return-water temperature to the heat source is specified by the manufacturer, the heating system shall be designed and installed to meet or exceed the minimum return-water temperature during the normal operation of the heat source.

1202.0 Protection of Potable Water Supply.

Any fluid having the potential of imposing more than a minor or moderate hazard to the potable water supply must be separated by a double wall heat exchanger. Typically, two kinds of antifreeze are used in hydronic systems, propylene glycol or ethylene glycol which is toxic; therefore the potable water must be protected. In addition, the materials used to construct the system must be compatible with the chemicals used. There must be an acceptable method of protecting the potable water systems that interface with these systems. The method of protection is relative to the hazard imposed by the chemicals. Employing these considerations is in the best interest of public health and safety.

1202.1 Prohibited Sources. Hydronic systems or parts thereof shall be constructed in such a manner that polluted, contaminated water or substances shall not enter a portion of the potable water system either during normal use or where the system is subject to pressure that exceeds the operating pressure in the potable water system. Piping, components and devices in contact with the potable water shall be approved for such use and where an additive is used it shall not affect the performance of the system.

1202.2 Chemical Injection. Where systems include an additive, chemical injection or provisions for such injection, the potable water supply shall be protected by a reduced-pressure principle backflow prevention assembly listed or labeled in accordance with ASSE 1013. Such additive or chemical shall be compatible with system components.

1202.3 Compatibility. Fluids used in hydronic systems shall be compatible with all components that will contact the fluid. Where a heat exchanger is installed with a dual purpose water heater, such application shall comply with the requirements for a single wall heat exchanger in Section 1218.1.

1203.0 Capacity of Heat Source.

1203.1 Heat Source. The heat source shall be sized to the design load.

1203.2 Dual Purpose Water Heater. Water heaters utilized for combined space-heating and water-heating applications shall be listed or labeled in accordance with the standards referenced in Table 1203.2, and shall be installed in accordance with the manufacturer's installation instructions. The total heating capacity of a dual purpose water heater shall be based on the sum of the potable hot water requirements and the space heating design requirements corrected for hot water first-hour draw recovery.

TABLE 1203.2
WATER HEATERS

TYPE	STANDARDS
Gas-Fired, 75 000 Btu/h or less, Storage	CSA Z21.10.1
Gas-Fired, Above 75 000 Btu/h, Storage and Instantaneous	CSA Z21.10.3
Electric, Space Heating	UL 834
Solid Fuel-Fired	UL 2523

For SI units: 1000 British thermal units per hour = 0.293 kW

The installation of a dual purpose hydronic system should also give consideration to the temperature of the water with regard to health issues. In certain temperature regimes bacteria can actually reproduce quite well, thereby increasing the levels of exposure. Their ideal temperature for proliferation is between 77°F and 108°F. As water temperature increases, the bacteria's concentration decreases. At sustained temperatures above 140°F, bacteria cannot survive. Hence, industry practice has been to maintain the DHW storage tanks at 140°F and mix down using an anti-scald mixing valve to a lower, safer temperature of operation.

Combination open systems should be designed to prevent the stagnation of water on the heating coils for an entire non-heating season.

1203.3 Tankless Water Heaters. Tankless water heaters shall be rated by the manufacturer for space-heating applications, and the output performance shall be determined by the temperature rise and flow rate of water through the unit. The ratings shall be expressed by the water temperature rise at a given flow rate. Manufacturer's flow rates shall not be exceeded.

1204.0 Identification of a Potable and Nonpotable Water System.

1204.1 General. In buildings where potable water and nonpotable water systems are installed, each system shall be clearly identified in accordance with Section 1204.2 through Section 1204.6.

1204.2 Color and Information. Each system shall be identified with a colored pipe or band and coded with paint, wraps, and materials compatible with the piping.

1204.3 Potable Water. Potable water systems shall be identified with a green background with white lettering. The minimum size of letters and length of the color field shall be in accordance with Table 1204.3.

TABLE 1204.3
MINIMUM LENGTH OF COLOR FIELD AND SIZE OF LETTERS

OUTSIDE DIAMETER OF PIPE OR COVERING (inches)	MINIMUM LENGTH OF COLOR FIELD (inches)	MINIMUM SIZE OF LETTERS (inches)
½ to 1¼	8	½
1½ to 2	8	¾
2½ to 6	12	1¼
8 to 10	24	2½
over 10	32	3½

For SI units: 1 inch = 25.4 mm

Identification of the water supply system is critical to the safe functioning of the building and the protection of the occupants of that building. The first step in the protection of a safe and pure water supply is the correct labeling of the various water systems in the building. This is important not only during construction but especially after the building is occupied when it is subject to maintenance and possibly when added to or altered later on. Therefore, the requirements above must be adhered to for every installation when potable and nonpotable water systems are present.

1204.4 Nonpotable Water. Nonpotable water systems shall have a yellow background with black uppercase lettering, with the words "CAUTION: NONPOTABLE WATER, DO NOT DRINK." Each nonpotable system shall be identified to designate the liquid being conveyed, and the direction of normal flow shall be clearly shown. The minimum size of the letters and length of the color field shall comply with Table 1204.3.

1204.5 Location of Piping Identification. The background color and required information shall be indicated every 20 feet (6096 mm) but not less than once per room, and shall be visible from the floor level.

1204.6 Flow Directions. Flow directions shall be indicated on the system.

1205.0 Installation, Testing, and Inspection.

1205.1 Operating Instructions. Operating and maintenance information shall be provided to the building owner.

1205.2 Pressure Testing. System piping and components shall be tested with a pressure of not less than one and one-half times the operating pressure but not less than 100 psi (689 kPa). Piping shall be tested with water or air except that plastic pipe shall not be tested with air. Test pressures shall be held for a period of not less than 30 minutes with no perceptible drop in pressure. These tests shall be made in the presence of the Authority Having Jurisdiction.

Exceptions:

(1) For PEX, PP-R, PP-RCT, PEX-AL-PEX, PE-RT, and PE-AL-PE piping systems, testing with air shall be permitted where authorized by the manufacturer's instructions for the PEX, PP-R, PP-RCT, PEX-AL-PEX, PE-RT, and PE-AL-PE pipe and fittings products, and air testing is not prohibited by applicable codes, laws, or regulations outside this code.

(2) Copper tubing shall be permitted to be tested at not less than 80 psi (552 kPa).

System piping must be tested in order to verify there are no leaks and capable of withstanding system operating pressures before placing it into service. Materials such as PEX, PP-R, PP-RCT, PEX-AL-PEX, PE-RT, and PE-AL-PE are flexible materials. Therefore, a failure or separation of the piping may cause unrestrained piping to whip or lash about potentially causing bodily injury or property damage. Piping must be properly restrained, and all fastening and securing requirements should be installed in accordance with manufacturer's installation instructions. When allowed by the manufacturer, air pressure testing may be permitted, provided that the manufacturer's instructions are followed and all testing is performed in accordance with local code regulations.

1205.3 Flushing. Heating and cooling sources, system piping and tubing shall be flushed after installation with water or a cleaning solution. Cleaning and flushing of the heating and cooling sources shall comply with the manufacturer's instructions. The cleaning solution shall be compatible with all system components and shall be used in accordance with the manufacturer's instructions.

System flushing after installation will eliminate debris from the piping. The cleaning or flushing solution specifically designed for the system will remove fluxes and oils that are still in the system. The flushing medium and the duration of flush are to be determined by manufacturer's installation instructions and the Authority Having Jurisdiction (AHJ).

1206.0 Pressure and Safety Devices.

1206.1 General. Each closed hydronic system shall be protected against pressures exceeding design limitations with not less than one pressure relief valve. Each closed section of the system containing a heat source shall have a relief valve located so that the heat source is not capable of being isolated from a relief device. Pressure relief valves shall be installed in accordance with their listing and the manufacturer's installation instructions.

Any heated closed system is capable of developing pressures that exceed its design working pressure. Closed liquid-filled systems can develop high hydrostatic pressures with even slight temperature increases. A hydronic system is more likely to be subjected to extreme pressures that could cause system failures and the associated hazards. Pressure relief valves shall be used for the relief of thermal expansion and to prevent injury and property damage from over pressurized vessels and piping. It may be required to have multiple valves depending on system layout. Discharge piping shall discharge to an approved safe location.

1206.2 Discharge Piping. The discharge piping serving a temperature relief valve, pressure relief valve, or combination of both shall have no valves, obstructions, or means of isolation and be provided with the following:

(1) Equal to the size of the valve outlet and shall discharge full size to the flood level of the area receiving the discharge and pointing down.

(2) Materials shall be rated at not less than the operating temperature of the system and approved for such use.

(3) Discharge pipe shall discharge independently by gravity through an air gap into the drainage system or outside of the building with the end of the pipe not exceeding 2 feet (610 mm) and not less than 6 inches (152 mm) above the ground and pointing downwards.

(4) Discharge in such a manner that does not cause personal injury or structural damage.

(5) No part of such discharge pipe shall be trapped or subject to freezing.

(6) The terminal end of the pipe shall not be threaded.

(7) Discharge from a relief valve into a water heater pan shall be prohibited.

1207.0 Heating Appliances and Equipment.

1207.1 General. Heating appliances, equipment, safety and operational controls shall be listed for its intended use in a hydronic heating system and installed in accordance with the manufacturer's installation instructions.

1207.2 Boilers. Boilers and their control systems shall comply with Section 1002.0.

1207.2.1 Condensing Boilers. A condensing boiler, in which the heat exchanger and venting system are designed to operate with condensing flue gases, shall be permitted to be connected directly to the panel heating system without a protective mixing device.

Condensing boilers allow the flue gas to condense. The condensation occurs on the heating surface of the heat exchanger and venting system. Therefore, the heat exchanger and venting system must be designed to operate with condensing flue gases.

1207.2.2 Noncondensing Boilers. Where the heat exchanger and venting system are not designed to operate with condensed flue gases, the boiler shall be permitted to connect directly to the panel heating system where protected from flue gas condensation. The operating temperature of the boiler shall be more than the fluid temperature in accordance with the manufacturer's instructions.

Where a noncondensing boiler is used, corrosion will occur when the flue gases are cooled below the dew point and come in contact with a material that is not corrosion resistant. To avoid corrosion, heating systems should be designed to operate in a way that ensures a minimum return water temperature. It is important to verify that the minimum required return water temperature is in accordance with the manufacturer's instructions to avoid corrosion.

1207.3 Dual-Purpose Water Heaters. Water heaters used for combined space- and water-heating applications shall be in accordance with the standards referenced in Table 1203.2, and shall be installed in accordance with the manufacturer's installation instructions. Water used as the heat transfer fluid in the hydronic heating system shall be isolated from the potable water supply and distribution in accordance with Section 312.1, Section 1202.0, and Section 1218.0.

1207.3.1 Temperature Limitations. Where a combined space- and water-heating application requires water for space heating at temperatures exceeding 140°F (60°C), a thermostatic mixing valve in accordance with ASSE 1017 shall be installed to temper the water supplied to the potable water distribution system to a temperature of 140°F (60°C) or less.

Scalding accidents can easily occur when the potable hot water exceeds a temperature of 140°F (60°C). A temperature actuated mixing valve is required to limit the temperature of hot water to 140°F (60°C) or less when the water heater is used for both potable hot water and hot water for space heating. The plumbing code requires a maximum temperature of 120°F for bathing. Regardless of the water supply demand downstream from the valve or supply pressure fluctuations upstream from the valve, the user will be provided some protection from scalding injury because the temperature of the water supplied will not exceed 140°F. See **Figure 1207.3.1**.

Temperature/Time Burn Chart

Temp. in Deg. F	Time for 1st deg. burn	Time for 2nd Deg. burn
111 F =	270 min's.	300 min's.
113 F =	120 min's.	180 min's.
116 F =	20 min's.	45 min's.
118 F =	15 min's.	20 min's.
120 F =	8 min's.	10 min's.
124 F =	2 min's.	4.2 min's.
131 F =	17 seconds	30 seconds
140 F =	3 seconds	5 seconds
151 F =	Instant	2 Seconds

(Source: Report prepared by Dr. Moritz and Dr. Henriques at Harvard Medical School in the 1940s for adult males. Children and elderly can receive burns in less time because their skin is thinner.) 11

FIGURE 1207.3.1
BURN CHART

1207.4 Solar Heat Collector Systems. Solar water heating systems used in hydronic panel radiant heating systems shall be installed in accordance with the Uniform Solar, Hydronics and Geothermal Code (USHGC).

1208.0 Circulators and Pumps.

1208.1 General. Circulators and pumps shall be selected for their intended use based on the heat transfer fluid, intended operating temperature range and pressure. Circulators and pumps shall be installed to allow for service and maintenance. The manufacturer's installation instructions shall be followed for correct orientation and installation. Motor Operated pumps rated 600V or less shall be listed and labeled in accordance with UL 778.

1208.2 Mounting. The circulator or pump shall be installed in such a way that strain from the piping is not transferred to the circulator or pump housing. The circulator or pump shall be permitted to be directly connected to the piping, provided the piping is supported on each side of the circulator or pump. Where the installation of a circulator or pump will cause strain on the piping, the circulator or pump shall be installed on a mounting bracket or base plate. Where means for controlling vibration of a circulator or pump is required, an approved means for support and restraint shall be provided.

1208.3 Sizing. The selection and sizing of a circulator or pump shall be based on all of the following:

(1) Loop or system head pressure, feet of head (m)
(2) Capacity, gallons per minute (L/s)
(3) Maximum and minimum temperature, °F (°C)
(4) Maximum working pressure, pounds per square inch (kPa)
(5) Fluid type

Circulators, when properly sized, overcome the friction loss of the piping to provide the necessary volume of fluid flow required by the circuits they serve.

1209.0 Expansion Tanks.

See the commentary for Section 1004.0.

1209.1 General. An expansion tank shall be installed in each closed hydronic system to control system pressure due to thermal expansion and contraction. Expansion tanks shall be of the closed or open type. Expansion tanks shall be rated for the pressure of the system.

1209.2 Installation. Expansion tanks shall be accessible for maintenance and shall be installed in accordance with the manufacturer's installation instructions. Each expansion tank shall be equipped with a shutoff device that will remain open during operation of the hydronic system. Valve handles shall be locked open or removed to prevent from being inadvertently shut off. Provisions shall be made for draining the tank without emptying the system. Expansion tanks shall be securely fastened to the structure. Supports shall be capable of carrying twice the weight of the tank filled with water without placing a strain on connecting piping. Hot-water-heating systems incorporating hot water tanks or fluid relief columns shall be installed to prevent freezing under normal operating conditions.

1209.3 Open-Type Expansion Tanks. Open type expansion tanks shall be located not less than 3 feet (914 mm) above the highest point of the system. An overflow with a diameter of not less than one-half the size of the supply or not less than 1 inch (25 mm) in diameter shall be installed at the top of the tank. The overflow shall discharge through an air gap into the drainage system.

See the commentary for Section 1004.2.

1209.4 Closed-Type Tanks. Closed-type expansion tanks shall be designed for a hydrostatic test pressure of two and one-half times the allowable working pressure of the system. Expansion tanks for systems designed to operate at more than 30 pounds-force per square inch (psi) (207 kPa) shall comply with ASME BPVC Section VIII.

See the commentary for Section 1004.3. See **Figure 1209.4** for various types of closed expansion tanks.

1209.5 Sizing. Expansion tanks shall be sized to accept the full expansion volume of the fluid in the system. The minimum capacity of a closed-type expansion tank shall be sized in accordance with Section 1004.4.

1210.0 Materials.

1210.1 Piping, Tubing, and Fittings. Hydronic pipe and tubing shall comply with the applicable standards referenced in Table 1210.1 and shall be approved for use based on the intended purpose. Materials shall be rated for the operating temperature and pressure of the system and shall be compatible with the type of heat transfer fluid. Pipe fittings and valves

FIGURE 1209.4
CLOSED EXPANSION TANKS

shall be approved for the specific installation with the piping, materials to be installed and shall comply with the applicable standards referenced in Table 1210.1. Where required, exterior piping shall be protected against freezing, UV radiation, corrosion and degradation. Embedded pipe or tubing shall comply with Section 1221.2.

There are several methods to help protect exterior pipes from freezing:

1. Constant circulation.
2. Addition of "food" grade (propylene) glycol to non-potable water systems.
3. Heating of fluids.
4. Insulation of pipes.

1210.2 Expansion and Contraction. Pipe and tubing shall be so installed that it will not be subject to undue strains or stresses, and provisions shall be made for expansion, contraction, and structural settlement.

1210.3 Hangers and Supports. Pipe and tubing shall be supported in accordance with Section 313.0. Equipment that is part of the piping system shall be provided with additional support in accordance with this code and manufacturer's installation instructions.

Besides the forces of thrust and torque in a piping system, there is thermal expansion. The weight of the pipe and its contents are significant. At a little over 8 pounds a gallon, water weight adds up fast. Assuming the pipe is full, a 4-inch diameter pipe 10 feet long has over 45 pounds of water in addition to the 110 pounds for the 16-gauge galvanized pipe.

Various manufacturers have different products to allow support of piping and equipment (see **Figure 1210.3a**). Hangers and supports include: beam clamps, pipe hangers, pipe clamps, pipe rollers, pipe supports, concrete inserts, brackets, channels, struts, trays, fasteners, enclosures, etc.

Hangers and supports are susceptible to dissimilar metal corrosion.

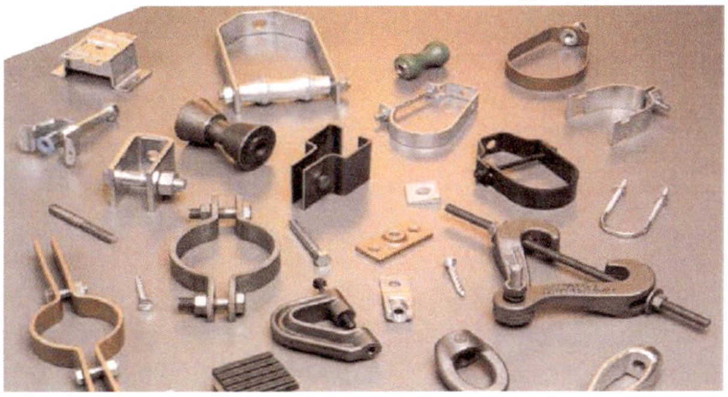

FIGURE 1210.3A
SAMPLES OF PIPE HANGERS
(Courtesy of Cooper B-Line)

Whichever method is used, "adequately supported" may include:

1. Piping should always be kept vertical (plumb) or horizontal (level) to ensure proper alignment of fittings and equipment. Pipe pitched for drainage typically is ¼-inch pitch per foot of horizontal run.
2. Providing support for components connected by piping. (e.g. circulators, low-water cutoffs, expansion tanks).
3. Distance between the supports is usually specified by the manufacturer but must always satisfy the AHJ.

In certain geographic areas, earthquakes create additional demands on the hanger and support systems. Specialized systems have been developed to survive the unique horizontal loadings (both perpendicular and parallel) that occur during earthquakes (see **Figure 1210.3b**).

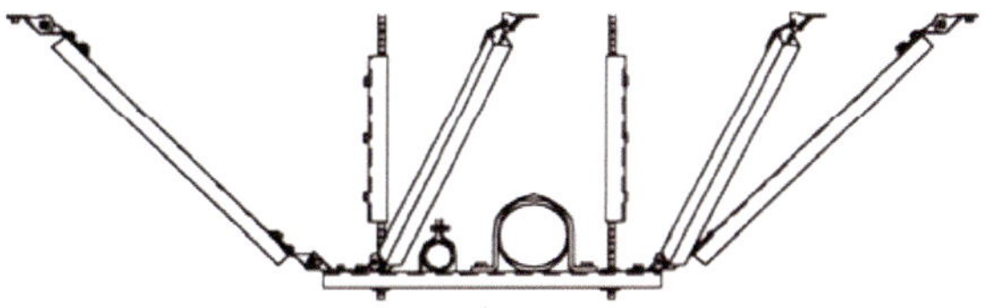

FIGURE 1210.3B
TRAPEZE TRANSVERSE AND LONGITIDINAL BRACING
(Courtesy of Cooper B-Line)

The swing joint is a joint of threaded pipes and fittings where the pipe threads allow small amounts of movement during thermal expansion and contraction (see **Figure 1210.3c**).

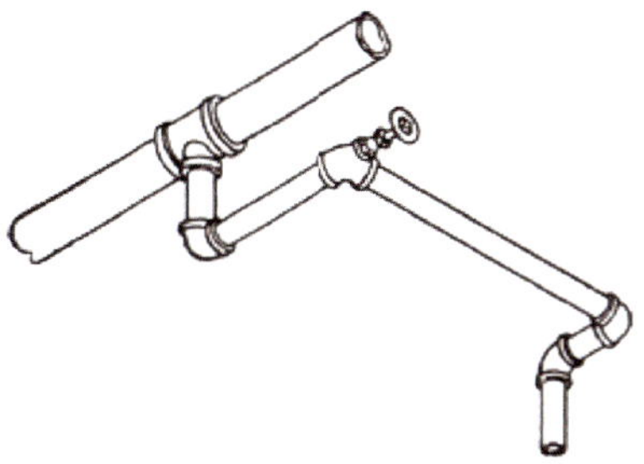

FIGURE 1210.3C
SWING JOINT
(Courtesy of Answers.com)

TABLE 1210.1
MATERIALS FOR HYDRONIC SYSTEM PIPING, TUBING, AND FITTINGS

MATERIAL	STANDARDS	
	PIPING/TUBING	FITTINGS
Copper/Copper Alloy	ASTM B42, ASTM B43, ASTM B75, ASTM B88, ASTM B135, ASTM B251[2], ASTM B302, ASTM B447	ASME B16.15, ASME B16.18, ASME B16.22, ASME B16.23, ASME B16.24, ASME B16.26, ASME B16.29, ASME B16.51, ASSE 1061, IAPMO PS 117
Ductile Iron	AWWA C115/A21.15, AWWA C151/A21.51	AWWA C110/A21.10[1], AWWA C153/A21.53
Steel	ASTM A53, ASTM A106, ASTM A254	ASME B16.5, ASME B16.9, ASME B16.11, ASTM A420
Stainless Steel	ASTM A269, ASTM A312, ASTM A554, ASTM A778	ASTM F1476, ASTM F1548, ASTM F3226, IAPMO PS 117
Gray Iron	—	ASTM A126
Malleable Iron	—	ASME B16.3
Chlorinated Polyvinyl Chloride (CPVC)	ASTM D2846, ASTM F441, ASTM F442, CSA B137.6	ASSE 1061, ASTM D2846, ASTM F437, ASTM F438, ASTM F439, ASTM F1970, CSA B137.6
Polyethylene (PE)	ASTM D1693, ASTM D2513, ASTM D2683, ASTM D2737, ASTM D3035, ASTM D3350, ASTM F714, AWWA C901, CSA B137.1, NSF 358-1	ASTM D2609, ASTM D2683, ASTM D3261, ASTM F1055, CSA B137.1, NSF 358-1
Cross-Linked Polyethylene (PEX)	ASTM F876, ASTM F3253, CSA B137.5, NSF 358-3	ASSE 1061, ASTM F877, ASTM F1055, ASTM F1807, ASTM F1960, ASTM F2080, ASTM F2098, ASTM F2159, ASTM F2735, ASTM F3253, CSA B137.5, NSF 358-3
Polypropylene (PP)	ASTM F2389, CSA B137.11, NSF 358-2	ASTM F2389, CSA B137.11, NSF 358-2
Polyvinyl Chloride (PVC)	ASTM D1785, ASTM D2241, CSA B137.3	ASTM D2464, ASTM D2466, ASTM D2467, ASTM F1970, CSA B137.2, CSA B137.3
Raised Temperature Polyethylene (PE-RT)	ASTM F2623, ASTM F2769, CSA B137.18	ASSE 1061, ASTM F1807, ASTM F2159, ASTM F2735, ASTM F2769, ASTM D3261, ASTM F1055, CSA B137.18
Cross-Linked Polyethylene/ Aluminum/Cross-Linked Polyethylene (PEX-AL-PEX)	ASTM F1281, CSA B137.10	ASTM F1281, ASTM F1974, ASTM F2434, CSA B137.10
Polyethylene/Aluminum/Polyethylene (PE-AL-PE)	ASTM F1282, CSA B137.9	ASTM F1282, ASTM F1974, CSA B137.9
Chlorinated Polyvinyl Chloride/Aluminum/ Chlorinated Polyvinyl Chloride (CPVC/AL/CPVC)	ASTM F2855	ASTM D2846

Note:
1 Ductile and gray iron.
2 Only type K, L, or M tubing allowed to be installed.

An expansion loop is a loop formed within an otherwise straight run of tubing that can absorb the movement in the pipe due to thermal expansion/contraction.

U-bends, coiled loops, and offset and returns are accepted methods of providing limited expansion of long rigid runs of pipe (see **Figure 1210.3d**).

1210.4 Oxygen Diffusion Corrosion. PEX and PE-RT tubing in closed hydronic systems shall contain an oxygen barrier.

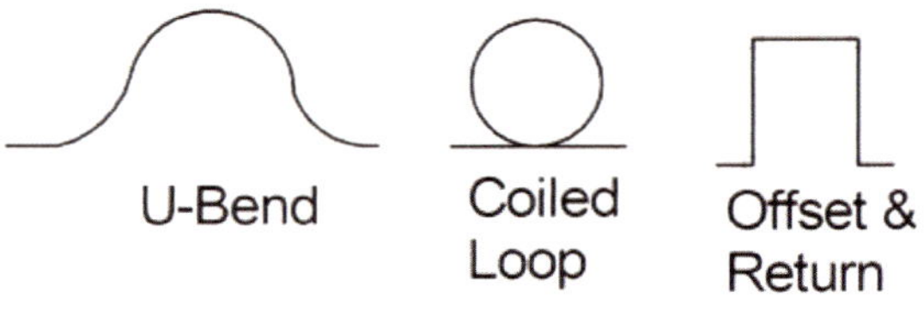

FIGURE 1210.3D
EXAMPLES OF EXPANSION LOOPS
(Courtesy of Praxis Green, Inc.)

Exception: Closed hydronic systems without ferrous components in contact with the hydronic fluid.

Tubing made from thermoplastics allows oxygen molecules to slowly pass through the tube wall and enter the water in the system. This process is called oxygen diffusion. Oxygen corrosion is a very serious corrosion problem in hydronic systems. The dissolved oxygen present in the water when the system is first filled quickly reacts with any iron or steel components. The rate of oxygen diffusion varies for different materials and higher temperatures. The solution to this problem is to create an oxygen diffusion barrier in or on the tubing. One such barrier is a thin layer of a special compound called EVOH (ethylene vinyl alcohol) that is bonded to the tubing during manufacturing. Another type of oxygen barrier is a thin layer of aluminum sandwiched between layers of PEX-AL-PEX. The use of oxygen barrier-equipped tubing does not guarantee that oxygen-related corrosion will not occur. There are several other ways for oxygen to enter a hydronic system such as improperly sized or placed expansion tank, leaky valve seals or pump gaskets, and improperly located air vents. Nonferrous materials such as copper, brass and stainless steel can be used to slow the process or eliminate the oxygen diffusion corrosion.

1211.0 Joints and Connections.

1211.1 General. Joints and connections shall be of an approved type. Joints shall be gas and watertight and designed for the pressure of the hydronic system. Changes in direction shall be made by the use of fittings or with pipe bends. Pipe bends shall have a radius of not less than six times the outside diameter of the tubing or shall be in accordance with the manufacturer's installation instructions. Joints between pipe and fittings shall be installed in accordance with the manufacturer's installation instructions.

To comply with the minimum radius required for changes in direction when bending pipe, take the outside diameter of the tubing (if it is 1-inch copper tube, then that is the pipe diameter), and multiply by 6. This gives a minimum center line turn radius (R) of 6 inches. If the pipe is making a 90 degree turn to the left, the distance from the straight piece going into the turn (A) and the straight piece coming out of the turn (B) must be not less than 6 inches. If it is making a 180 degree turn, the distance between the straight piece going into the turn (A) and the straight piece coming out of the turn (B) is 12 inches. This is twice the radius [the diameter (D)] of the semi-circle created by the bend (see **Figure 1211.1a**).

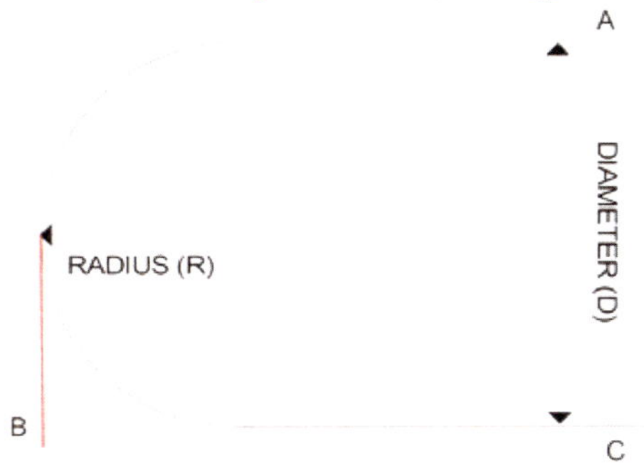

FIGURE 1211.1A
BEND RADIUS
(Courtesy of Praxis Green, Inc.)

The goal is to get a smooth bend with a relatively consistent round. Bending a tube causes the outside loop wall to get longer and thin out. The inside loop wall compresses (see **Figure 1211.1b**). The tighter the bend (for example four times the pipe diameter instead of the required six), the more distance the outside loop wall has to cover with less pipe material. The tube wall may become so thin it could collapse and become flat. Stretched far enough, the tube wall may become so thin it fractures.

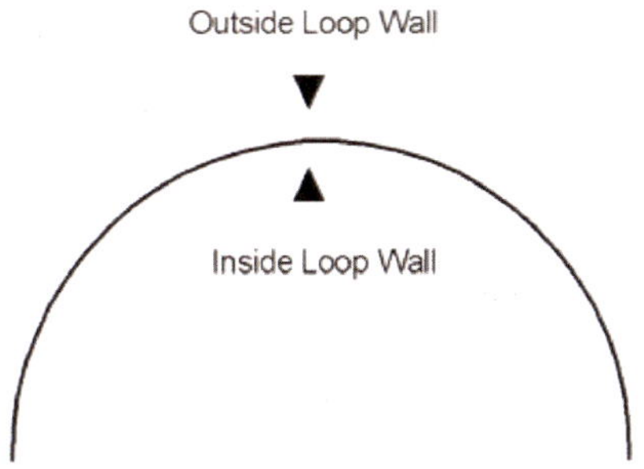

FIGURE 1211.1B
INSIDE AND OUTSIDE LOOP WALLS
(Courtesy of Praxis Green, Inc.)

The inside loop wall has the opposite problem –less distance and more material to compress, could result in wrinkling. Even with appropriate tools, tube "out of round" cannot be completely eliminated but must be minimized. Keyhole loops allow for narrower on-center applications (see **Figure 1211.1c**).

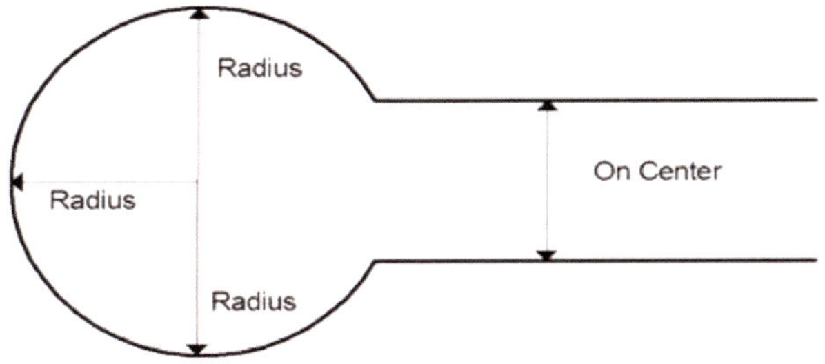

FIGURE 1211.1C
KEYHOLE LOOP
(Courtesy of Praxis Green, Inc.)

1211.2 Chlorinated Polyvinyl Chloride (CPVC) Pipe. Joints between chlorinated polyvinyl chloride (CPVC) pipe and fittings shall be installed in accordance with one of the following methods:

(1) Mechanical joints shall include flanged, grooved, and push fit fittings. Removable and non-removable push fit fittings with an elastomeric o-ring that employ quick assembly push fit connectors shall be in accordance with ASSE 1061.

(2) Solvent cement joints for CPVC pipe and fittings shall be clean from dirt and moisture. Solvent cements in accordance with ASTM F493, requiring the use of a primer shall be orange in color. The primer shall be colored and be in accordance with ASTM F656. Listed solvent cement in accordance with ASTM F493 that does not require the use of primers, yellow or red in color, shall be

permitted for pipe and fittings manufactured in accordance with ASTM D2846, ½ of an inch (15 mm) through 2 inches (50 mm) in diameter or ASTM F442, ½ of an inch (15 mm) through 3 inches (80 mm) in diameter. Apply primer where required inside the fitting and to the depth of the fitting on pipe. Apply liberal coat of cement to the outside surface of pipe to depth of fitting and inside of fitting. Place pipe inside fitting to forcefully bottom the pipe in the socket and hold together until joint is set.

(3) Threaded joints for CPVC pipe shall be made with pipe threads in accordance with ASME B1.20.1. A minimum of Schedule 80 shall be permitted to be threaded, and the pressure rating shall be reduced by 50 percent. The use of molded fittings shall not result in a 50 percent reduction in the pressure rating of the pipe provided that the molded fittings shall be fabricated so that the wall thickness of the material is maintained at the threads. Thread sealant compound that is compatible with the pipe and fitting, insoluble in water, and nontoxic shall be applied to male threads. Caution shall be used during assembly to prevent over tightening of the CPVC components once the thread sealant has been applied. Female CPVC threaded fittings shall be used with plastic male threads only.

1211.3 CPVC/AL/CPVC Plastic Pipe and Joints. Joints between chlorinated polyvinyl chloride/aluminum/ chlorinated polyvinyl chloride (CPVC/AL/CPVC) pipe and fittings shall be installed in accordance with one of the following methods:

(1) Mechanical joints shall include flanged, grooved, and push-fit fittings.

(2) Solvent cement joints for CPVC/AL/CPVC pipe and fittings shall be clean from dirt and moisture. Solvent cements in accordance with ASTM F493, requiring the use of a primer shall be orange in color. The primer shall be colored and be in accordance with ASTM F656. Listed solvent cement in accordance with ASTM F493 that does not require the use of primers, yellow in color, shall be permitted for pipe and fittings manufactured in accordance with ASTM D2846, ½ of an inch (15 mm) through 2 inches (50 mm) in diameter, ½ of an inch (15 mm) through 3 inches (80 mm) in diameter. Apply primer where required inside the fitting and to the depth of the fitting on pipe. Apply liberal coat of cement to the outside surface of pipe to depth of fitting and inside of fitting. Place pipe inside fitting to forcefully bottom the pipe in the socket and hold together until joint is set.

1211.4 Copper or Copper Alloy Pipe and Tubing. Joints between copper or copper alloy pipe or tubing and fittings shall be installed in accordance with one of the following methods:

(1) Brazed joints between copper or copper alloy pipe or tubing and fittings shall be made with brazing alloys having a liquid temperature above 1000°F (538°C). The joint surfaces to be brazed shall be cleaned bright by either manual or mechanical means. Tubing shall be cut square and reamed to full inside diameter. Brazing flux shall be applied to the joint surfaces where required by manufacturer's recommendation. Brazing filler metal in accordance with AWS A5.8 shall be applied at the point where the pipe or tubing enters the socket of the fitting.

(2) Flared joints for soft copper or copper alloy tubing shall be made with fittings that are in accordance with the applicable standards referenced in Table 1210.1. Pipe or tubing shall be cut square using an appropriate tubing cutter. The tubing shall be reamed to full inside diameter, resized to round, and expanded with a proper flaring tool.

"Flared joint" is defined as "any metal to metal compression joint in which a conical spread is made on the end of a tube that is compressed by a flare nut against a mating flare" (see **Figure 1211.3a**).

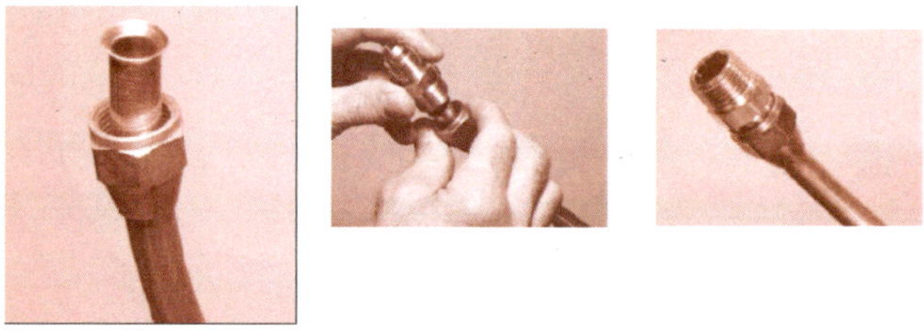

FIGURE 1211.3A
COMPLETED FLARED JOINT
(Courtesy of Copper Development Association)

(3) Mechanically formed tee fittings shall have extracted collars that shall be formed in a continuous operation consisting of drilling a pilot hole and drawing out the pipe or tube surface to form a collar having a height not less than three times the thickness of the branch tube wall. The branch pipe or tube shall be notched to conform to the inner curve of the run pipe or tube and shall have two dimple depth stops to ensure that penetration of the branch pipe or tube into the collar is of a depth for brazing and that the branch pipe or tube does not obstruct the flow in the main line pipe or tube. Dimple depth stops shall be in line with the run of the pipe or tube. The second dimple shall be ¼ of an inch (6.4 mm) above the first and shall serve as a visual point of inspection. Fittings and joints shall be made by brazing. Soldered joints shall not be permitted.

The recognized standard for mechanically formed tee fittings is ASTM F 2014-00(2006), Standard Specification for Non-Reinforced Extruded Tee Connections for Piping Applications, explains: "This specification covers the pipe materials and dimensions for producing nonreinforced extruded tee connections manufactured by mechanical forming processes (see **Figure 1211.3b**). The term "extruded tee connection" applies to butt-weld or socket-weld connections. The non-reinforced extruded pipe tee connection is an alternative to the tee fittings, nozzle, and other welded connections. The non-reinforced extruded pipe tee connection has been widely used for systems in the marine, process piping, food, pharmaceutical, and similar industries. Different materials that have acceptable forming qualities to produce extruded tee connections shall consist of copper, copper-nickel alloy, titanium, steel, and stainless steel. The extruded tee connection shall be free from burrs and cracks, which would affect the suitability for the intended service."

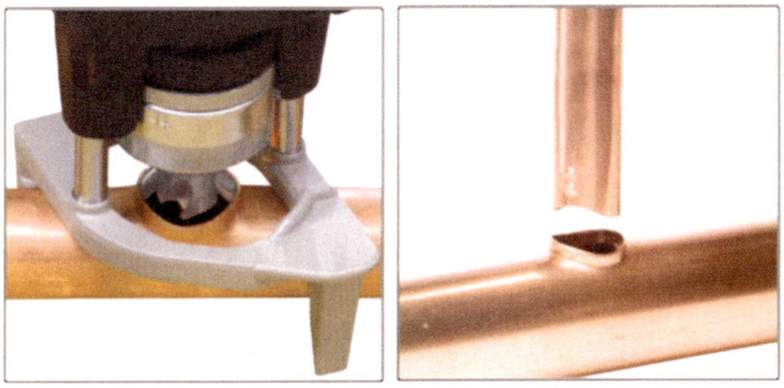

FIGURE 1211.3B
MECHANICALLY FORMED TEE FITTINGS

The definition of a "Brazed Joint" in Section 212.0 states "any joint obtained by joining of metal parts with alloys that melt at temperatures higher than 840°F but lower than the melting temperature of the parts being joined." The excluded "soft soldered joints" are defined by "Soldered Joint" which is "a joint made by the joining of metal parts with metallic mixtures of alloys that melt at a temperature up to and including 840°F."

The higher temperature joining process is allowed, but the lower temperature process is not.

(4) Pressed fittings for copper or copper alloy pipe or tubing shall have an elastomeric o-ring that forms the joint. The pipe or tubing shall be fully inserted into the fitting, and the pipe or tubing marked at the shoulder of the fitting. Pipe or tubing shall be cut square, chamfered, and reamed to full inside diameter. The fitting alignment shall be checked against the mark on the pipe or tubing to ensure the pipe or tubing is inserted into the fitting. The joint shall be pressed using the tool recommended by the manufacturer.

(5) Removable and nonremovable push fit fittings for copper or copper alloy tubing or pipe that employ quick assembly push fit connectors shall be in accordance with ASSE 1061. Push fit fittings for copper or copper alloy pipe or tubing shall have an approved elastomeric o-ring that forms the joint. Pipe or tubing shall be cut square, chamfered, and reamed to full inside diameter. The tubing shall be fully inserted into the fitting, and the tubing marked at the shoulder of the fitting. The fitting alignment shall be checked against the mark on the tubing to ensure the tubing is inserted into the fitting and gripping mechanism has engaged on the pipe.

(6) Soldered joints between copper or copper alloy pipe or tubing and fittings shall be made in accordance with ASTM B828. Pipe or tubing shall be cut square and reamed to the full inside diameter including the removal of burrs on the outside of the pipe or tubing. Surfaces to be joined shall be cleaned bright by manual or mechanical means. Flux shall be applied to pipe or tubing and fittings and shall be in accordance with ASTM B813, and shall become noncorrosive and nontoxic after soldering. Insert pipe or tubing into the base of the fitting and remove excess flux. Pipe or tubing and fitting shall be supported to ensure a uniform capillary space around the joint. Solder in accordance with ASTM B32 shall be applied to the joint surfaces until capillary action draws the molten solder into the cup. Joint surfaces shall not be disturbed until cool, and any remaining flux residue shall be cleaned.

"Soldered joint" is defined in Section 212 as "A joint obtained by the joining of metal parts with metallic mixtures or alloys that melt at a temperature up to and including 840°F (449°C)."

A soldered joint should have a clean, even, continuous bead of solder around the connection. There should be no visible scorch marks from overheating the copper tube. The connection should be pressure tested to ensure a proper sealed joint.

The standard for solder metal is ASTM B 32-2014, Standard Specification for Solder Metal. The scope of this standard covers solder metal alloys (commonly known as soft solders) used in non-electronic applications, including but not limited to, tin-lead, tin-antimony, tin-antimony-copper-silver, tin-antimony-copper-silver-nickel, tin-silver, tin-copper-silver, and lead-tin-silver, used for the purpose of joining together two or more metals at temperatures below their melting points.

(7) Threaded joints for copper or copper alloy pipe shall be made with pipe threads in accordance with ASME B1.20.1. Thread sealant tape or compound shall be applied only on male threads, and such material shall be of approved types, insoluble in water, and nontoxic.

1211.5 Cross-Linked Polyethylene (PEX) Pipe. Joints between cross-linked polyethylene (PEX) pipe and fittings shall be installed with fittings for PEX tubing that comply with the applicable standards referenced in Table 1210.1. PEX tubing labeled in accordance with ASTM F876 shall be marked with the applicable standard designation for the fittings specified for use with the tubing. Mechanical joints shall be installed in accordance with the manufacturer's installation instructions.

1211.6 Cross-Linked Polyethylene/Aluminum/Cross-Linked Polyethylene (PEX-AL-PEX) Pipe. Joints between cross-linked polyethylene/aluminum/cross-linked polyethylene (PEX-AL-PEX) pipe and fittings shall be installed in accordance with one of the following methods:

(1) Mechanical joints between PEX-AL-PEX pipe and fittings shall include mechanical and compression type fittings and insert fittings with a crimping ring. Insert fittings utilizing a crimping ring shall be in accordance with ASTM F1974 or ASTM F2434. Crimp joints for crimp insert fittings shall be joined to PEX-AL-PEX pipe by the compression of a crimp ring around the outer circumference of the pipe, forcing the pipe material into annular spaces formed by ribs on the fitting.

(2) Compression joints shall include compression insert fittings and shall be joined to PEX-AL-PEX pipe through the compression of a split ring or compression nut around the outer circumference of the pipe, forcing the pipe material into the annular space formed by the ribs on the fitting.

1211.7 Ductile Iron Pipe. Joints between ductile iron pipe and fittings shall be installed in accordance with one of the following methods:

(1) Mechanical joints for ductile iron pipe and fittings shall consist of a bell that is cast integrally with the pipe or fitting and provided with an exterior flange having bolt holes and a socket with annular recesses for the sealing gasket and the plain end of the pipe or fitting. The elastomeric gasket shall comply with AWWA C111. Lubricant recommended for the application by the pipe manufacturer shall be applied to the gasket and plain end of the pipe.

(2) Push-on joints for ductile iron pipe and fittings shall consist of a single elastomeric gasket that shall be assembled by positioning the elastomeric gasket in an annular recess in the pipe or fitting socket and forcing the plain end of the pipe or fitting into the socket. The plain end shall compress the elastomeric gasket to form a positive seal and shall be designed so that the elastomeric gasket shall be locked in place against displacement. The elastomeric gasket shall comply with AWWA C111. Lubricant recommended for the application by the pipe manufacturer shall be applied to the gasket and plain end of the pipe.

1211.8 Polyethylene (PE) Plastic Pipe/Tubing. Joints between polyethylene (PE) plastic pipe or tubing and fittings shall be installed in accordance with one of the following methods:

(1) Butt-fusion joints shall be installed in accordance with ASTM F2620 and shall be made by heating the squared ends of two pipes, pipe and fitting, or two fittings by holding ends against a heated element. The heated element shall be removed where the proper melt is obtained, and joined ends shall be placed together with applied force.

(2) Electro-fusion joints shall be heated internally by a conductor at the interface of the joint. Align and restrain fitting to pipe to prevent movement and apply electric current to the fitting. Turn off the current when the proper time has elapse to heat the joint. The joint shall fuse together and remain undisturbed until cool.

(3) Socket-fusion joints shall be installed in accordance ASTM F2620 and shall be made by simultaneously heating the outside surface of a pipe end and the inside of a fitting socket. Where the proper melt is obtained, the pipe and fitting shall be joined by inserting one into the other with applied force. The joint shall fuse together and remain undisturbed until cool.

(4) Mechanical joints between PE pipe or tubing and fittings shall include insert and mechanical compression fittings that provide a pressure seal resistance to pullout. Joints for insert fittings shall be made by cutting the pipe square, using a cutter designed for plastic piping, and removal of sharp edges. Two stainless steel clamps shall be placed over the end of the pipe. Fittings shall be checked for proper size based on the diameter of the pipe. The end of pipe shall be placed over the barbed insert fitting, making contact with the fitting shoulder. Clamps shall be positioned equal to 180 degrees (3.14 rad) apart and shall be tightened to provide a leak tight joint. Compression type couplings and fittings shall be permitted for use in joining PE piping and tubing. Stiffeners that extend beyond the clamp or nut shall be prohibited. Bends shall be not less than 30 pipe diameters, or the coil radius where bending with the coil. Bends shall not be permitted closer than 10 pipe diameters of a fitting or valve. Mechanical joints shall be designed for their intended use.

1211.9 Polyethylene/Aluminum/Polyethylene (PE-AL-PE). Joints between polyethylene/aluminum/polyethylene (PE-AL-PE) pipe and fittings shall be installed in accordance with one of the following methods:

(1) Mechanical joints for PE-AL-PE pipe or tubing and fittings shall be either of the metal insert fittings with a split ring and compression nut or metal insert fittings with copper crimp rings. Metal insert fittings shall comply with ASTM F1974. Crimp insert fittings shall be joined to the pipe by placing the copper crimp ring around the outer circumference of the pipe, forcing the pipe material into the space formed by the ribs on the fitting until the pipe contacts the shoulder of the fitting. The crimp ring shall then be positioned on the pipe so the edge of the crimp ring is 1/8 of an inch (3.2 mm) to 1/4 of an inch (6.4 mm) from the end of the pipe. The jaws of the crimping tool shall be centered over the crimp ring and tool perpendicular to the barb. The jaws shall be closed around the crimp ring and shall not be crimped more than once.

(2) Compression joints for PE-AL-PE pipe or tubing and fittings shall be joined through the compression of a split ring, by a compression nut around the circumference of the pipe. The compression nut and split ring shall be placed around the pipe. The ribbed end of the fitting shall be inserted onto the pipe until the pipe contacts the shoulder of the fitting. Position and compress the split ring by tightening the compression nut onto the insert fitting.

1211.10 Polyethylene of Raised Temperature (PE-RT). Joints between polyethylene of raised temperature (PE-RT) tubing and fittings shall be installed with fittings for PE-RT tubing that comply with the applicable standards referenced in Table 1210.1. Metal insert fittings, metal compression fittings, and plastic fittings shall be manufactured to and marked in accordance with the standards for fittings in Table 1210.1.

1211.11 Polypropylene (PP) Pipe. Joints between polypropylene pipe and fittings shall be installed in accordance with one of the following methods:

(1) Heat fusion joints for polypropylene (PP) pipe shall be installed with socket-type heat-fused polypropylene fittings, butt-fusion polypropylene fittings or pipe, or electro-fusion polypropylene fittings. Joint surfaces shall be clean and free from moisture. The joint shall be undisturbed until cool. Joints shall be made in accordance with ASTM F2389 or CSA B137.11.

(2) Mechanical and compression sleeve joints shall be installed in accordance with the manufacturer's installation instructions. Polypropylene pipe shall not be threaded. Polypropylene transition fittings for connection to other piping materials shall only be threaded by the use of copper alloy or stainless steel inserts molded in the fitting.

1211.12 Polyvinyl Chloride (PVC) Pipe. Joints between polyvinyl chloride pipe and fittings shall be installed in accordance with one of the following methods:

(1) Mechanical joints shall be designed to provide a perma-

nent seal and shall be of the mechanical or push-on joint. The mechanical joint shall include a pipe spigot that has a wall thickness to withstand without deformation or collapse; the compressive force exerted where the fitting is tightened. The push-on joint shall have a minimum wall thickness of the bell at any point between the ring and the pipe barrel. The elastomeric gasket shall comply with ASTM D3139, and be of such size and shape as to provide a compressive force against the spigot and socket after assembly to provide a positive seal.

(2) Solvent cement joints for PVC pipe and fittings shall be clean from dirt and moisture. Pipe shall be cut square and pipe shall be deburred. Where surfaces to be joined are cleaned and free of dirt, moisture, oil, and other foreign material, apply primer purple in color in accordance with ASTM F656. Primer shall be applied until the surface of the pipe and fitting is softened. Solvent cements in accordance with ASTM D2564 shall be applied to all joint surfaces. Joints shall be made while both the inside socket surface and outside surface of pipe are wet with solvent cement. Hold joint in place and undisturbed for 1 minute after assembly.

(3) Threads shall comply with ASME B1.20.1. A minimum of Schedule 80 shall be permitted to be threaded; however, the pressure rating shall be reduced by 50 percent. The use of molded fittings shall not result in a 50 percent reduction in the pressure rating of the pipe provided that the molded fittings shall be fabricated so that the wall thickness of the material is maintained at the threads. Thread sealant compound that is compatible with the pipe and fitting, insoluble in water, and nontoxic shall be applied to male threads. Caution shall be used during assembly to prevent over tightening of the PVC components once the thread sealant has been applied. Female PVC threaded fittings shall be used with plastic male threads only.

1211.13 Steel Pipe and Tubing. Joints between steel pipe or tubing and fittings shall be installed in accordance with one of the following methods:

(1) Mechanical joints shall be made with an approved and listed elastomeric gasket.

(2) Threaded joints shall be made with pipe threads that are in accordance with ASME B1.20.1. Thread sealant tape or compound shall be applied only on male threads, and such material shall be of approved types, insoluble in water, and nontoxic.

(3) Welded joints shall be made by electrical arc or oxygen/acetylene method. Joint surfaces shall be cleaned by an approved procedure. Joints shall be welded by an approved filler metal.

(4) Pressed joints shall have an elastomeric o-ring that forms the connection. The pipe or tubing shall be fully inserted into the fitting, and the pipe or tubing marked at the shoulder of the fittings. Pipe or tubing shall be cut square, chamfered, and reamed to full inside diameter. The fitting alignment shall be checked against the mark on the pipe or tubing to ensure the pipe or tubing is fully inserted into the fitting. The joint shall be pressed using the tool recommended by the manufacturer.

1211.14 Joints Between Various Materials. Joints between various materials shall be installed in accordance with the manufacturer's installation instructions and shall comply with Section 1211.14.1 and Section 1211.14.2.

1211.14.1 Copper or Copper Alloy Pipe or Tubing to Threaded Pipe Joints. Joints from copper or copper alloy pipe or tubing to threaded pipe shall be made by the use of copper alloy adapter, copper alloy nipple [minimum 6 inches (152 mm)], dielectric fitting, or dielectric union in accordance with ASSE 1079. The joint between the copper or copper alloy pipe or tubing and the fitting shall be a soldered, brazed, flared, or pressed joint and the connection between the threaded pipe and the fitting shall be made with a standard pipe size threaded joint.

1211.14.2 Plastic Pipe to Other Materials. Where connecting plastic pipe to other types of piping, approved types of adapter or transition fittings designed for the specific transition intended shall be used.

1212.0 Valves.

1212.1 General. Valves shall be rated for the operating temperature and pressure of the system. Valves shall be compatible with the type of heat transfer medium and piping material.

1212.2 Where Required. Valves shall be installed in hydronic piping systems in accordance with Section 1212.3 through Section 1212.11.

Isolation valves are required in hydronic systems to allow the isolation of any device or component that will require servicing, repair, or replacement. Valves must be located on the supply and return piping so that the component or group of components may be separated from the rest of the system. Valves are also used to take system components out of service temporarily. Draining a water hydronic system causes air to enter the system, and will require that fresh water be introduced to refill the system. The time-consuming process of purging and bleeding air from the system and the corrosion problems associated with new water make it desirable to avoid system draining whenever possible. In order to change the tank air charge pressure it is necessary to isolate the tank circuit from the main system piping. A high quality, gate type, lock-shield valve (isolation valve) should be used for this purpose.

1212.3 Heat Exchanger. Isolation valves shall be installed on the supply and return side of the heat exchanger.

1212.4 Pressure Vessels. Isolation valves shall be installed on connections to pressure vessels.

1212.5 Pressure Reducing Valves. Isolation valves shall be installed on both sides of a pressure reducing valve.

1212.6 Equipment, Components, and Appliances. Serviceable equipment, components, and appliances within the system shall have isolation valves installed upstream and downstream of such devices.

1212.7 Expansion Tank. Isolation valves shall be installed at connections to non-diaphragm-type expansion tanks.

1212.8 Flow Balancing Valves. Where flow balancing

valves are installed, such valves shall be capable of increasing or decreasing the amount of flow by means of adjustment.

1212.9 Mixing or Temperature Control Valves. Where mixing or temperature control valves are installed, such valves shall be capable of obtaining the design water temperature and design flow requirements.

1212.10 Thermosiphoning. An approved type check valve shall be installed on liquid heat transfer piping to control thermosiphoning of heated liquids.

1212.11 Air Removal Device or Air Vents. Isolation valves shall be installed where air removal devices or automatic air vents are utilized to permit cleaning, inspection, or repair without shutting the system down.

1213.0 System Controls.

System controls are used to ensure the safe operation of the heat source by preventing operation of the appliance when an unsafe condition exists. Continued appliance operation during an unsafe condition presents a life safety hazard and potential for property damage and must be avoided.

1213.1 Water Temperature Controls. A heat source or system of commonly connected heat sources shall be protected by a water-temperature-activated operating control to stop heat output of the heat source where the system water reaches a pre-set operating temperature.

1213.2 Operating Steam Controls. A steam heat source or system of commonly connected steam heat sources shall be protected by a pressure-actuated control to shut off the fuel supply where the system pressure reaches a pre-set operating pressure.

1213.2.1 Water-Level Controls. A primary water-level control shall be installed on a steam heat source to control the water level in the heat source. The control shall be installed in accordance with the manufacturer's installation instructions.

1213.3 Occupied Spaces. A temperature-sensing device shall be installed in the occupied space to regulate the operation of the hydronic system.

1213.4 Simultaneous Operation. Radiant heating and cooling systems sharing a common space temperature control shall be configured to prevent simultaneous heating and cooling.

1213.5 Temperature Reading. A temperature gauge or transmitter shall be installed for reading the fluid temperatures in the panel system supply and heat source outlet. One temperature gauge or transmitter shall be permitted where the temperature between the heat source outlet and panel system supply are the same.

1214.0 Pressure and Flow Controls.

1214.1 Balancing. A means for balancing distribution loops, heat emitting devices, and multiple boiler installations shall be provided in accordance with the manufacturer's instructions. A means for balancing and flow control shall include the piping design, pumping equipment, or balancing devices.

1214.2 Low-Water Control. Direct-fired heat sources within a closed heating system shall have a low-water fuel cut-off device, except as specified in Section 1214.3. Where a low-water control is integral with the heat source as part of the appliance's integrated control and is listed for such use, a separate low-water control shall not be required. An external cut-off device shall be installed in accordance with the heat-source manufacturer's installation instructions. No valve shall be located between the external low-water fuel cut-off and the heat-source unit. Where a pumped condensate return is installed, a second low-water cut-off shall be provided.

1214.3 Flow-Sensing Devices. A direct-fired heat source, requiring forced circulation to prevent overheating, shall have a flow-sensing device installed with the appliance, or such device shall be integral with the appliance. A low-water fuel cut-off device shall not be required.

1214.4 Automatic Makeup Fluid. Where an automatic makeup fluid supply fill device is used to maintain the fluid content of the heat-source unit, or any closed-loop in the system, the makeup supply shall be located at the expansion tank connection or other approved location.

A pressure-reducing valve shall be installed on a makeup feed line. The pressure of the feed line shall be set in accordance with the design of the system, and connections to potable water shall be in accordance with Section 1202.0 to prevent contamination due to backflow.

1214.5 Differential Pressure Regulation. Provisions shall be made to bypass zone flows in excess of design velocity in a multi-zone hydronic system where the closing of some or all of the two-way zone valves causes excess flow through the open zones or deadheading of a fixed-speed circulator or pump.

1214.5.1 Differential Pressure Bypass Valve. Where a differential pressure bypass valve is used for the purpose specified in Section 1214.5, it shall be installed and adjusted to provide bypass of the distribution system when most or all of the zones are closed.

1214.6 Air-Removal Device. Provision shall be made for the removal of air from fluid in hydronic systems. Air-removal devices shall be located in the areas of the hydronic piping system where air is likely to accumulate. Air-removal devices shall be installed to facilitate their removal for examination, repair, or replacement.

1214.7 Air-Separation Device. To assist with the removal of entrained air, an air-separation device shall be installed in hydronic systems. The device shall be located in accordance with the manufacturer's installation instructions or at the point of no mechanically-induced pressure change within the hydronic system.

1214.8 Secondary Loops. Secondary loops that are isolated from the primary heat-distribution loop by a heat exchanger are closed-loop hydronic systems and shall have an expansion tank in accordance with Section 1209.0, an air-removal device in accordance with Section 1214.6, and an air-separation device in accordance with Section 1214.7.

1215.0 Hydronic Space Heating.

1215.1 General. Based on the system design, the heat-distribution units shall be selected in accordance with the manufacturer's specifications.

1215.2 Installation. Heat-distribution units shall be installed in accordance with the manufacturer's installation instructions and this code.

1215.3 Freeze Protection. Hydronic systems and components shall be designed, installed, and protected from freezing.

1215.4 Balancing. System loops shall be installed so that the design flow rates are achieved within the system.

1215.5 Heat Transfer Fluid. The ignitable flash point of heat transfer fluid in a hydronic piping system shall be a minimum of 50°F (28°C) above the maximum system operating temperature. The heat transfer fluid shall be compatible with the makeup fluid supplied to the system.

1216.0 Steam Systems.

1216.1 Steam Traps. For other than one-pipe steam systems, each heat-distribution unit shall be supplied with a steam trap that is listed for the application.

1216.2 Sloping for Two-Pipe System. Two-pipe steam system piping and heat-distribution units shall be sloped down at not less than $^1/_8$ inch per foot (10.4 mm/m) in the direction of the steam flow.

1216.3 Sloping for One-Pipe System. One-pipe steam system piping and heat-distribution units shall be sloped down at not less than $^1/_8$ inch per foot (10.4 mm/m) towards the steam boiler, without trapping.

1216.4 Automatic Air Vents. Steam automatic air vents shall be installed to eliminate air pressure in heat-distribution units on gravity steam piping systems. Air vents shall not be used on a vacuum system.

1216.5 Condensate Flow. System piping shall be installed to allow condensate to flow to the condensate receiver or steam boiler either by gravity or pump-assisted.

1216.6 Steam-Distribution Piping. Where multi-row elements are installed in an enclosure, they shall be top fed and piped in parallel down to the steam trap. A single steam trap for each row of heating elements shall be installed. Where the size of the return header is increased by a minimum of one pipe size, a single steam trap shall be permitted to be installed for multiple rows. Where multiple steam unit heaters are installed, an individual steam trap for each unit shall be installed.

1217.0 Radiant Heating and Cooling.

1217.1 Installation. Radiant heating and cooling systems shall be installed in accordance with the system design.

1217.1.1 Manifolds. Manifolds shall be equipped with isolation valves on the supply and return lines. Manifolds shall be capable of withstanding the pressure and temperature of the system. The material of the manifold shall be compatible with the system fluid and shall be installed in accordance with the manufacturer's installation instructions.

Manifolds are mainly used to control separate zones and are required for systems that require more than one water temperature. It is crucial to maintain manifolds in proper working conditions, thus requiring isolation valves for servicing and maintenance. See Section 1212.0 for isolation valve requirements.

1217.2 Radiant Under-Floor Heating. Floor finished surface temperatures shall not exceed the following temperatures for space heating applications:

(1) 85°F (29°C) in general occupied applications.

(2) 90°F (32°C) in bathrooms, foyers, distribution areas such as hallways and indoor swimming pools.

(3) 88°F (31°C) in industrial spaces.

(4) 95°F (35°C) in radiant panel perimeter areas, i.e., up to 2.5 feet (762 mm) from outside walls.

The radiant heating system temperature shall not exceed the maximum temperature rating of the materials used in its construction.

1217.3 Radiant Cooling Systems. Radiant cooling systems shall be designed to minimize the potential for condensation. To prevent condensation on any cooled radiant surface, the supply water temperature for a radiant cooling system shall be not less than 3°F (2°C) above the anticipated space dewpoint temperature, or in accordance with the manufacturer's recommendation. Chilled water piping, valves, and fittings shall be insulated and vapor sealed to prevent surface condensation.

1217.4 Tube Placement. Hydronic radiant system tubing shall be installed in accordance with the manufacturer's installation instructions and with the tube layout and spacing in accordance with the system design. Except for distribution mains, tube spacing and the individual loop lengths shall be installed with a variance of not more than ±10 percent from the design. The maximum loop length of continuous tubing from a supply-and-return manifold shall not exceed the lengths specified by the manufacturer or, in the absence of manufacturer's specifications, the lengths specified in Table 1217.4. Actual loop lengths shall be determined by spacing, flow rate, and pressure drop requirements as specified in the system design.

For the purpose of system balancing, each individual loop shall have a tag securely affixed to the manifold to indicate the length of the loop and the room(s) and area(s) served.

TABLE 1217.4
MAXIMUM LENGTH OF CONTINUOUS TUBING FROM A SUPPLY-AND-RETURN MANIFOLD ARRANGEMENT

NOMINAL TUBE SIZE (inches)	MAXIMUM LOOP LENGTH (feet)
$^1/_4$	125
$^5/_{16}$	200
$^3/_8$	250
$^1/_2$	300
$^5/_8$	400
$^3/_4$	500
1	750

For SI units: 1 inch = 25.4 mm, 1 foot = 304.8 mm

1217.5 Poured Floor Structural Concrete Slab Systems (Thermal Mass). Where tubing is embedded in a structural concrete slab such tubes shall not be larger in outside dimension than one-third of the overall thickness of the slab and shall be spaced not less than three diameters on center except within 10 feet (3048 mm) of the distribution manifold. The top of the tubing shall be embedded in the slab not less than 2 inches (51 mm) below the surface.

1217.5.1 Slab Penetration Tube and Joint Protection. Where embedded in or installed under a concrete slab, tubing shall be protected from damage at penetrations of the slab with protective sleeving approved by the tubing manufacturer. The space between the tubing and sleeve shall be sealed with an approved sealant compatible with the tubing. The tubing at the location of an expansion joint in a concrete slab shall be encased in protective pipe sleeving that covers the tubing not less than 12 inches (305 mm) on either side of the expansion joint or the tubing shall be installed below the slab.

1217.5.2 Insulation. Where a poured concrete radiant floor system is installed in contact with the soil, insulation recommended by the manufacturer for such an application and with a minimum R-value of 5 shall be placed between the soil and the concrete; extend to the outside edges of the concrete; and be placed on all slab edges.

1217.5.3 Types of Tube Fasteners. Tubing that is embedded within concrete shall be fastened according to manufacturer's instructions. Unless prohibited by the manufacturer, tube fasteners include the following:

(1) Ties made of wire, typically fastened to anchors such as rebar or wire mesh.
(2) Plastic tube/cable ties, typically nylon, fastened to anchors such as rebar or wire mesh.
(3) Staples made of metal or plastic or combination thereof, without sharp edges that would harm tube, fastened to insulation or subfloor.
(4) Plastic rails with integrated tube holders intended for the specific type of tube.
(5) Insulation sheets with integrated knobs for holding the specific type of tube and intended for this application.
(6) Other fasteners recommended by the manufacturer.

1217.5.4 Spacing of Tube Fasteners. The maximum spacing between tube fasteners within a concrete floor shall not exceed the spacing specified by the manufacturer or, in the absence of manufacturer's specifications, 2.5 feet (762 mm).

1217.6 Joist Systems and Subfloors. Where tubing is installed below a subfloor, the tube spacing shall be in accordance with the system design and joist space limitations.

Where tubing is installed above or in the subfloor, the tube spacing shall not exceed 12 inches (305 mm) center-to-center for living areas.

Where tubing is installed in the joist cavity, the cavity shall be insulated with not less than R-12 material below the heated space.

An air space of not less than 1 inch (25.4 mm) and not more than 3 inches (76 mm) shall be maintained between the top of the insulation and the underside of the floor unless a conductive plate is installed in accordance with manufacturer's instructions.

Where tubing is installed in panels above or in the subfloor and not embedded in concrete, the floor assembly shall be insulated with not less than R-5 material below the tubing when installed over habitable space.

1217.6.1 Tubing Fasteners. Tubing that is installed within joist spaces and subfloor panel systems shall be fastened according to manufacturer's instructions. Unless prohibited by the manufacturer, tubing fasteners shall include the following:

(1) Heat transfer panel systems made of wood, aluminum or other thermally conductive materials intended for this application and the specific type of tube.
(2) Staples made of metal or plastic or combination thereof, without sharp edges that would harm tube. intended for this application and the specific type of tube fastened to subfloor.
(3) Plastic rails with integrated tube holders intended for the specific type of tube.
(4) Other fasteners recommended by the manufacturer.

1217.7 Wall and Ceiling Panels. Where piping is installed in the stud wall cavity or the ceiling joist cavity, the cavity shall be insulated with not less than R-12 material. The insulation shall be installed in such a manner as to prevent heating or cooling loss from the space intended to be controlled.

An air space of not less than 1 inch (25.4 mm) and not more than 3 inches (76 mm) shall be maintained between the insulation and the interior surface of the panel unless a conductive plate is installed.

1217.8 Radiant Heating and Cooling Panels. Radiant heating and cooling panels shall be installed in accordance with the manufacturer's installation instructions.

1217.8.1 Electric Heating Panel Systems. Clearances for electric heating panels or between outlets, junction boxes, mounting luminaries, ventilating, or other openings shall comply with NFPA 70.

1217.8.2 Radiant Wall and Ceiling Panels. Radiant panels attached to wood, steel, masonry, or concrete framing members shall be fastened by means of anchors, bolts, or approved screws of sufficient size and anchorage to support the loads applied. Panels shall be installed with corrosion-resistant fasteners. Piping systems shall be designed for thermal expansion to prevent the load being transmitted to the panel.

1218.0 Heat Exchangers.

1218.1 General. Systems utilizing heat exchangers shall protect the potable water system from being contaminated by the heat transfer medium. Systems that incorporate a single-wall heat exchanger to separate potable water from the heat-transfer fluid shall meet the following requirements:

(1) Heat transfer medium is either potable water or contains fluids recognized as safe by the Food and Drug Administration (FDA) as food grade.
(2) A tag or label shall be securely affixed to the heat source with the word, "CAUTION" and the following statements:
    (a) The heat transfer medium shall be water or other nontoxic fluid recognized as safe by the FDA.
    (b) The maximum operating pressure of the heat exchanger shall not exceed the maximum operating pressure of the potable water supply.
(3) The word "CAUTION" and the statements listed above shall have an uppercase height of not less than 0.120 of an inch (3.048 mm). The vertical spacing between lines of type shall be not less than 0.046 of an inch (1.168 mm). Lowercase letters shall be not less than compatible with the uppercase letter size specification.

Systems that do not comply with the requirements for a single-wall heat exchanger shall install a double wall heat exchanger. Double-wall heat exchangers shall separate the potable water from the heat transfer medium by providing a space between the two walls that are vented to the atmosphere.

The system must have adequate protection to ensure that the potable water supply is properly safeguarded. Systems that do not comply with the requirements for a single wall heat exchanger must be provided with a double-wall heat exchanger. The double-wall heat exchanger must have an intermediate space between the walls that is open to the atmosphere. This type of construction would allow any leakage of fluid through the walls of the heat exchanger to discharge externally from the heat exchanger where it would be observable.

1219.0 Indirect-Fired Domestic Hot-Water Storage Tanks.

1219.1 General. Domestic hot-water heat exchangers, whether internal or external to the heating appliance, shall be permitted to be used to heat water in domestic hot-water storage tanks. Tanks used to store hot water shall be listed for the intended use and constructed in accordance with nationally recognized standards. A pressure- and temperature-relief valve with a set pressure not exceeding 150 percent of the maximum operating pressure of the system, and at a temperature of 210°F (99°C), shall be installed on the storage tank.

Where the normal operating temperature of the boiler or dual-purpose water heater that provides heat input for domestic hot water exceeds 140°F (60°C), a thermostatically controlled mixing valve as specified in Section 1207.3.1 shall be installed to limit the water supplied to the potable hot water system to a temperature of 140°F (60°C) or less. The potability of the water shall be maintained throughout the system.

1220.0 Snow and Ice Melt Systems.

1220.1 Use of Chemical Additives and Corrosive Fluids. Where auxiliary systems contain chemical additives, corrosive fluids, or both not intended or designed for use in the primary system, a double wall heat exchanger shall be used in accordance with Section 1218.1. The chemical additives in the auxiliary systems shall be compatible with auxiliary system components and accepted for use by the heat exchanger manufacturer.

1220.2 Types of Tube Fasteners. Tubing that is embedded within concrete shall be fastened according to manufacturer's instructions. Unless prohibited by the manufacturer, tube fasteners include the following:

(1) Ties made of wire, typically fastened to anchors such as rebar or wire mesh.

(2) Plastic tube/cable ties, typically nylon, fastened to anchors such as rebar or wire mesh.

(3) Staples made of metal or plastic or combination thereof, without sharp edges that would harm tube, fastened to insulation or subfloor.

(4) Plastic rails with integrated tube holders intended for the specific type of tube.

(5) Insulation sheets with integrated knobs for holding the specific type of tube and intended for this application.

(6) Other fasteners recommended by the manufacturer.

1220.3 Spacing of Tube Fasteners. The maximum spacing between tube fasteners within a concrete area shall not exceed the spacing specified by the manufacturer or, in the absence of manufacturer's specifications, 2.5 feet (762 mm).

1220.4 Snow and Ice Melt Controls. An automatic thermostatically operating control device that controls the supply hydronic solution temperature to the snow and ice melt area shall be installed in the system. Snow and ice melt systems shall be protected from freezing with a mixture of propylene glycol or ethylene glycol, and water or other approved fluid. Automotive antifreeze shall not be used.

1220.4.1 Tube Placement. Snow and ice melt tubing shall be installed in accordance with the manufacturer's installation instructions and with the tube layout and spacing in accordance with the system design. Except for distribution mains, tube spacing and the individual loop lengths shall be installed with a variance of not more than ±10 percent from the design.

The maximum loop length of continuous tubing from a supply-and-return manifold arrangement shall not exceed the lengths specified by the manufacturer or, in the absence of manufacturer's specifications, the lengths specified in Table 1220.4.1. Actual loop lengths shall be determined by spacing, flow rate, and pressure drop in accordance with the system design.

TABLE 1220.4.1
LOOP LENGTHS FOR SNOW AND ICE MELT SYSTEMS[1,2]

NOMINAL TUBE SIZE (inches)	MAXIMUM ACTIVE LOOP LENGTH (feet)	TOTAL LOOP LENGTH (feet)
PE-RT and PEX Tubing		
1/2	115	140
5/8	225	250
3/4	300	325
1	450	475
Copper Tubing[3]		
1/2	–	140
3/4	–	280

For SI units: 1 inch = 25.4 mm, 1 foot = 304.8 mm

Notes:

[1] The total PE-RT and PEX loop lengths consist of two separate sections, the active loop, and the leader length. The active loop is installed within the heated slab. The leader length is the total distance to and from the manifold and heated slab, including any vertical distances.

[2] The manifolds shall be installed as close to the snow melt area as possible.

[3] In concrete use minimum Type L copper water tubing. In bituminous pavement use a Type K copper water tubing.

1220.4.2 Poured Structural Concrete Slab Systems. Where tubes are embedded in a structural concrete slab, such tubes shall not be larger in outside dimension than one-third of the overall thickness of the slab and shall be spaced not less than three diameters on center except within 10 feet (3048 mm) of the distribution manifold. The top of the tubing shall be embedded in the slab not less than 2 inches (51 mm) below the surface of the finished concrete slab.

1220.4.3 Slab Penetration Tube and Joint Protection. Where embedded in or installed under a concrete slab, tubing shall be protected from damage at penetrations of the slab with protective sleeving approved by the tubing manufacturer. The space between the tubing and sleeve shall be sealed with an approved sealant compatible with the tubing. The tubing at the location of an expansion joint in a concrete slab shall be encased in a protective pipe sleeve that covers the tubing not less than 12 inches (305 mm) on either side of the joint or the tubing shall be installed below the slab.

1220.4.4 Concrete Slab Preparation. A solid foundation shall be prepared before the tubing is installed. Compaction shall be used for slabs, sidewalks, and driveways.

1220.4.5 Insulation. Where a poured concrete snow melt system is installed in contact with the soil, insulation recommended by the manufacturer for such application and with a minimum R-value of 5 shall be placed between the concrete and the compacted grade; and be extended as close as practicable to the outside edges of the concrete.

1220.4.6 Testing and Flushing. Testing of snow and ice melt systems shall be in accordance with Section 1205.2 and flushing shall be in accordance with Section 1205.3.

1220.5 Hydronic Makeup Air Units. Hydronic makeup air units that are affected by freezing shall be protected against freezing by a hydronic solution.

1221.0 Piping Installation.

1221.1 General. Piping, fittings, and connections shall be installed in accordance with the conditions of their approval and manufacturer's installation instructions.

1221.2 Embedded Piping Materials and Joints. Piping embedded in concrete shall be steel pipe, Type L copper tubing or plastic pipe or tubing rated at not less than 80 psi at 180°F (552 kPa at 82°C). Joints of pipe or tubing that are embedded in a portion of the building, such as concrete or plaster shall be installed in accordance with the requirements of Section 1221.2.1 through Section 1221.2.3.

1221.2.1 Steel Pipe. Steel pipe shall be welded by electrical arc or oxygen/acetylene method.

1221.2.2 Copper Tubing. Copper tubing shall be joined by brazing with filler metals having a melting point not less than 1000°F (538°C).

1221.2.3 Plastics. Plastic pipe and tubing shall be installed in continuous lengths or shall be joined by heat fusion methods or other approved fittings in accordance with Table 1210.1 and the manufacturer's installation instructions.

Exception: Solvent cement joints.

1221.3 Pressure Testing. Piping to be embedded in concrete shall be pressure-tested in accordance with Section 1205.2 prior to pouring concrete. During the pour, the pipe system shall maintain the test pressure of not less than one and one-half times the hydronic system operating pressure and not less than 100 psi (689 kPa). During freezing or the possibility of freezing conditions, testing shall be done with air where permitted by the manufacturer.

1221.4 System Drainage. Hydronic piping systems shall be installed to permit the system to be drained. The system shall drain by indirect waste in accordance with Section 1001.4. Embedded piping underground or under floors is not required to be designed for draining the system.

It's important to be able to completely drain pipes to facilitate repairs, to prevent corrosion or to provide a means to completely flush system fluids.

If the piping system is unable to be completely drained when needing repair, there is the risk of spillage along with subsequent hazards. Furthermore, any residual fluid remaining in the pipe will prevent a good solder joint.

If a pipe is intended as a dry pipe, but has been pressure tested with water, it is important to be able to completely drain the pipe to help prevent corrosion from the inside.

During the initial fill, debris and other construction or manufacturing residue should be cleaned out. If the system is not completely drained, it is possible for contaminated fluid to remain behind.

1221.5 Condensate Drainage. Condensate drains from dehumidifying coils shall be constructed and sloped for condensate removal. Such drains shall be installed in accordance with Section 310.0.

1221.6 Clearance to Combustibles. Hydronic piping where the exterior temperature exceeds 250°F (121°C) shall have a clearance of not less 1 inch (25.4 mm) to combustible materials.

CHAPTER 13
FUEL GAS PIPING

1301.0 Scope of Gas Piping.

1301.1 Applicability. The regulations of this chapter shall govern the installation of fuel gas piping in or in connection with a building, structure or within the property lines of premises up to 5 pounds-force per square inch (psi) (34 kPa) for natural gas and 10 psi (69 kPa) for undiluted propane, other than service pipe. Fuel oil piping systems shall be installed in accordance with NFPA 31.

This chapter was extracted from the National Fuel Gas Code, which is the American National Standard that applies to the installation of fuel-gas piping systems and fuel-gas utilization equipment that are supplied with natural gas; manufactured gas; liquefied petroleum gas (LP-gas), in the vapor phase only; LP-gas-air mixtures; mixtures of these gases; and gas-air mixtures in the flammable range.

Figure 1301.1 gives typical properties of several fuel gases. Their compositions can vary widely, and the table should not be used for design purposes. Instead, the gas supplier should be contacted for data on the fuel gas to be used.

The storage, handling and use of liquid LP-gas are not included under the scope of NFPA 54. NFPA 58, Liquefied Petroleum Gas Code covers liquid LP-gas. LP-gas has two sources: it is a byproduct of natural gas production that is removed during the cleanup of natural gas prior to its entering the gas transmission system; and it is a product of petroleum refineries.

1302.0 Coverage of Piping System.

Fuel-gas systems covered under Chapter 13 are natural gas systems, LP-gas vapor-only systems and manufactured gas systems. The vast majority of systems installed today are natural gas systems; therefore, most of the information covered here will pertain to natural gas systems.

The system component covered by this chapter is the piping from the shutoff valve at the meter (natural gas) or at the final regulator (LP-gas) to the gas appliance. This includes all pipe, fittings, valves, regulators and connectors used in the system to deliver fuel gas to the appliance (see **Figure 1302.0**).

There are literally hundreds of different types of devices that utilize fuel gas for their operation. Some are appliances such as water heaters, furnaces and ovens. Others are not defined as appliances, such as tiki torches, decorative gas logs and man-made volcanoes (such as at the Mirage Hotel in Las Vegas, Nevada). For the purposes of this chapter, the terms "appliance" and "gas utilization device" will have the same meaning throughout the chapter.

Typical Fuel Gases % Constituents and Properties		Natural Gas	Commercial Propane	Propane HD 5	Commercial Butane	Water Gas	Producer Gas	Coke Oven Gas
Butane(s)		0.6			{95 min.			
Butylene					{combined			
Carbon dioxide 0.4					5.1		4.3	2.0
Carbon monoxide					40.2		26.7	6.2
Ethane		6.0						
Hydrogen					50.0	50.0	13.9	53.2
Methane		81.1			0.7	0.7	3.1	26.7
Nitrogen		9.2			4.0		51.5	7.0
Pentane(s)		0.2						
Propane		2.1	{95 min.	95				
Propylene			{combined					
Others, unknown, miscellaneous		0.4	5 max.	5 max.	5 max.	Some	0.5	4.9
Heat of combustion	Gross	1000	2500	2500	3200	260	165	
Btu per ft^3	Net	900	2300	2300	2950	240	155	
Specific gravity (air = 1.00)		0.65	1.52	1.45	1.95	0.70	0.86	0.40
Flammability limits	Lower	3.9	2.4	2.1	1.9	6.9	16.8	5.0
% volume in air	Upper	15.0	9.6	10.1	8.6	69.5	73.7	28.0
Combustion	Per ft^3 gas	10	25	25	32	2.6	1.7	5.7
Air ft^3 (approx.)	Per 100 Btu	1	1	1	1	1	1	1

1Btu/ft^3 = 3788 J/m^3; 1 ft/sec = 0.305 m/sec; 1 ft^3 = 0.028 m^3; 1 Btu =1056 J.

FIGURE 1301.1
PROPERTIES OF FUEL GAS

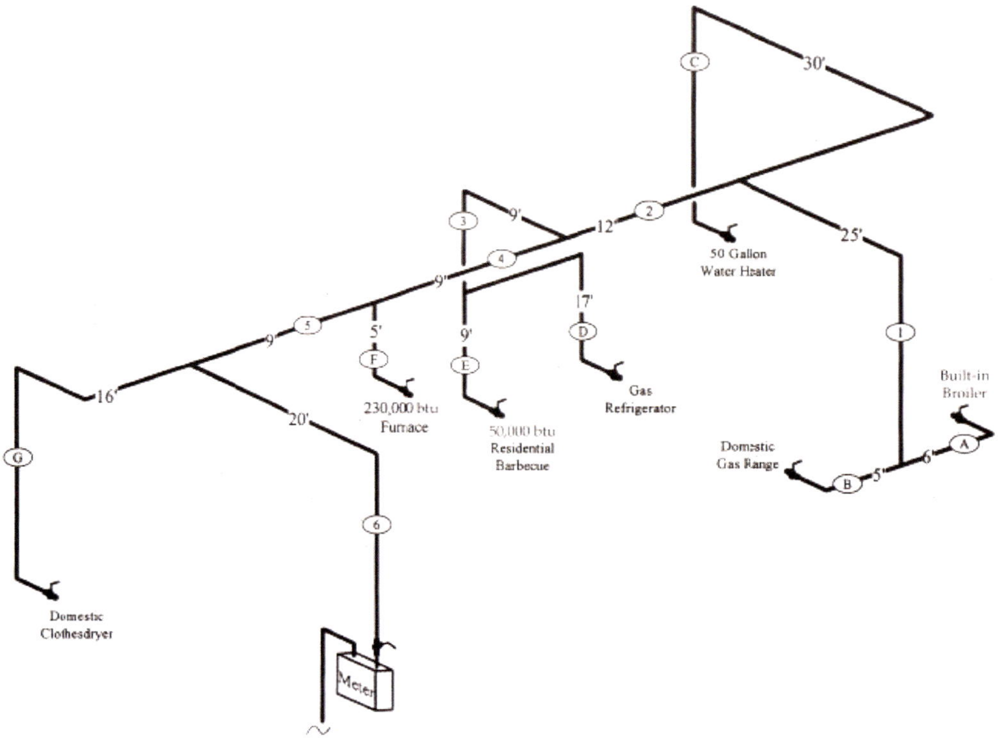

FIGURE 1302.0
FUEL GAS PIPING SYSTEM

1302.1 General. Coverage of piping systems shall extend from the point of delivery to the appliance connections. For other than undiluted liquefied petroleum gas (LP-Gas) systems, the point of delivery shall be the outlet of the service meter assembly or the outlet of the service regulator or service shutoff valve where no meter is provided. For undiluted LP-Gas systems, the point of delivery shall be considered to be the outlet of the final pressure regulator, exclusive of line gas regulators where no meter is installed. Where a meter is installed, the point of delivery shall be the outlet of the meter. [NFPA 54:1.1.1.1(A)]

1302.2 Piping System Requirements. Requirements for piping systems shall include design, materials, components, fabrication, assembly, installation, testing, inspection, operation, and maintenance. [NFPA 54:1.1.1.1(E)]

The scope of this chapter will include all parts of the fuel-gas system, from the design of the piping system, to its installation, inspection, use and maintenance. The scope of the chapter will not apply to the following enumerated systems and their components.

1302.3 Applications. This code shall not apply to the following items:

(1) Portable LP-Gas appliances and equipment of all types that are not connected to a fixed fuel piping system.

(2) Installation of appliances such as brooders, dehydrators, dryers, and irrigation equipment used for agricultural purposes.

(3) Raw material (feedstock) applications except for piping to special atmosphere generators.

(4) Oxygen-fuel gas cutting and welding systems.

(5) Industrial gas applications using such gases as acetylene and acetylenic compounds, hydrogen, ammonia, carbon monoxide, oxygen, and nitrogen.

(6) Petroleum refineries, pipeline compressor or pumping stations, loading terminals, compounding plants, refinery tank farms, and natural gas processing plants.

(7) Large integrated chemical plants or portions of such plants where flammable or combustible liquids or gases are produced by chemical reactions or used in chemical reactions.

(8) LP-Gas installations at utility gas plants.

(9) Liquefied natural gas (LNG) installations.

(10) Fuel gas piping in electric utility power plants.

(11) Proprietary items of equipment, apparatus, or instruments such as gas-generating sets, compressors, and calorimeters.

(12) LP-Gas equipment for vaporization, gas mixing, and gas manufacturing.

(13) LP-Gas piping for buildings under construction or renovations that is not to become part of the permanent building piping system—that is, temporary fixed piping for building heat.

(14) Installation of LP-Gas systems for railroad switch heating.

(15) Installation of LP-Gas and compressed natural gas (CNG) systems on vehicles.

(16) Gas piping, meters, gas pressure regulators, and other appurtenances used by the serving gas supplier in distribution of gas, other than undiluted LP-Gas.

(17) Building design and construction, except as specified herein.

(18) Fuel gas systems on recreational vehicles manufactured in accordance with NFPA 1192.

(19) Fuel gas systems using hydrogen as a fuel.

(20) Construction of appliances. [NFPA 54:1.1.1.2]

1303.0 Inspection.

1303.1 Inspection Notification. Upon completion of the installation, alteration, or repair of gas piping, and prior to the use thereof, the Authority Having Jurisdiction shall be notified that such gas piping is ready for inspection.

All new gas piping and portions of existing systems that may be affected by new work or any changes shall be inspected by the Authority Having Jurisdiction (AHJ) to ensure compliance with all requirements of the code.

It is the responsibility of the permit holder to ensure that the work will stand the tests prescribed and that the installation meets the minimum requirements contained in the code.

No gas-piping system, or part thereof, shall be covered, concealed or put into use until it has been tested, inspected and accepted as prescribed in the code.

1303.2 Excavation. Excavations required for the installation of underground piping shall be kept open until such time as the piping has been inspected and approved. Where such piping is covered or concealed before such approval, it shall be exposed upon the direction of the Authority Having Jurisdiction.

In order for the AHJ to perform its inspection, it is important that they are able to observe the entire gas-piping system to ensure compliance with the code. The inspector should be checking that the material used is approved for underground installations and that it was installed per the manufacturer's installation requirements and per its listing.

1303.3 Type of Inspections. The Authority Having Jurisdiction shall make the following inspections and either shall approve that portion of the work as completed or shall notify the permit holder wherein the same fails to be in accordance with this code.

Inspections for gas-piping systems are conducted in two phases. The AHJ shall inspect the gas-piping installation in the rough stage (open frame) of construction and at the final (all construction complete) to verify compliance with all code requirements and to check for damage from construction. A correction notice should be given to the permit holder for any work that is in violation with the code.

1303.3.1 Rough Piping Inspection. This inspection shall be made after gas piping authorized by the permit has been installed before such piping has been covered or concealed, or before fixture or appliance has been attached thereto. This inspection shall include a determination that the gas piping size, material, and installation meet the requirements of this code.

The AHJ performs a rough inspection to determine that the correct material allowed by the code was used. This inspection also verifies that the installation meets all the requirements of the code. There should be pipe without improper bend or strain, correct spacing of hangers and supports, and correct sizing of the piping system. Although a pressure test is not specifically required in the rough inspection, it is necessary that the system is tested for leaks before any work is covered or concealed. Otherwise, a great amount of time and money will be wasted to repair a leak once walls are covered and painted.

1303.3.2 Final Piping Inspection. This inspection shall be made after piping authorized by the permit has been installed and after portions thereof that are to be covered or concealed are so concealed and before fixture, appliance, or shutoff valve has been attached thereto. This inspection shall comply with Section 1313.1. Test gauges used in conducting tests shall be in accordance with Section 1303.3.3 through Section 1303.3.3.4.

The second inspection (final inspection) shall be made after all the gas piping to be covered or concealed is so concealed and before any fixture, appliances or shutoff valves have been attached.

These tests shall be made using only air, carbon dioxide (CO_2) or nitrogen pressure and shall be made in the presence of the AHJ. All necessary apparatus for conducting tests shall be furnished by the permit holder.

An accurate determination of line tightness should be made within a reasonable period of time. This can only be done if the gauge recording the line test pressure is sensitive, accurate and of such graduations that small leaks can be detected quickly. Be sure to consult Section 1303.3.3 Test Gauges, for the proper gauge requirement.

For gas pressures up to and including 14 inches water column, the piping shall be subjected to an air pressure test of not less than 10 psi (69 kPa) gauge pressure and shall be held for a length of time satisfactory to the AHJ but, in no case, less than 15 minutes.

After the final inspection has been made and it is found that the installation complies with the requirements of the code, the AHJ may issue to the permit holder a certificate of completion. Some jurisdictions will install a gas tag at the gas meter. This tag indicates that the gas-piping system has been inspected, tested and complies with the code.

1303.3.3 Test Gauges. Tests required by this code, which are performed utilizing dial gauges, shall be limited to gauges having the following pressure graduations or increments.

1303.3.3.1 Pressure Tests (10 psi or less). Required pressure tests of 10 psi (69 kPa) or less shall be performed with gauges of 0.10 psi (0.69 kPa) increments or less.

1303.3.3.2 Pressure Tests (greater than 10 psi to 100 psi). Required pressure tests exceeding 10 psi (69 kPa) but less than or equal to 100 psi (689 kPa) shall be performed with gauges of 1 psi (7 kPa) increments or less.

1303.3.3.3 Pressure Tests (exceeding 100 psi). Required pressure tests exceeding 100 psi (689 kPa) shall be performed with gauges of 2 percent increments or less of the required test pressure.

1303.3.3.4 Pressure Range. Test gauges shall have a pressure range not exceeding twice the test pressure applied.

1303.4 Inspection Waived. In cases where the work authorized by the permit consists of a minor installation of additional piping to piping already connected to a gas meter, the foregoing inspections shall be permitted to be waived at the discretion of the Authority Having Jurisdiction. In this event, the Authority Having Jurisdiction shall make such inspection as deemed advisable in order to be assured that the work has been performed in accordance with the intent of this code.

Minor changes to an existing gas piping installation, such as extending an existing line for the relocation of a gas appliance, may be performed without requiring a pressure test of the entire system if approved by the AHJ. The AHJ may require that a soap solution be applied to the exterior of the joint, which will bubble if a leak is present. Whatever method is used for testing the joint, the system must be gastight.

1304.0 Certificate of Inspection.

1304.1 Issuance. Whereupon final piping inspection, the installation is found to be in accordance with the provisions of this code, a certificate of inspection shall be permitted to be issued by the Authority Having Jurisdiction.

1304.2 Gas Supplier. A copy of the certificate of such final piping inspection shall be issued to the serving gas supplier supplying gas to the premises.

The AHJ, in addition to the certificate it issues to the permit holder, will also issue a certificate to the gas supplier that indicates that the gas system has been inspected and tested. This certificate will allow the gas supplier to turn on the gas supply at the location indicated on the certificate.

1304.3 Unlawful. It shall be unlawful for a serving gas supplier or person furnishing gas, to turn on or cause to be turned on, fuel gas or a gas meter or meters until such certificate of final inspection, as herein provided, has been issued.

Because of the hazards associated with fuel gas, it is extremely important to ensure that the gas system has been inspected and tested and that it is safe to turn on the gas supply to the premises. The final inspection ensures that the gas piping has not been damaged during construction and that all openings have been capped or plugged.

1305.0 Authority to Render Gas Service.

1305.1 Authorized Personnel. It shall be unlawful for a person, firm, or corporation, excepting an authorized agent or employee of a person, firm, or corporation engaged in the business of furnishing or supplying gas and whose service pipes supply or connect with the particular premises, to turn on or reconnect gas service in or on a premises where gas service is, at the time, not being rendered.

Only an employee of the gas supplier can turn on or reconnect the gas supply to any premise. The gas supplier will perform a test to ensure that there are no open lines or leaks in the building before connecting or turning on the gas supply. Once the gas is turned on, the gas supplier will light all installed appliances with pilots and check that all appliances are functioning correctly.

1305.2 Outlets. It shall be unlawful to turn on or connect gas in or on the premises unless outlets are securely connected to gas appliances or capped or plugged with screw joint fittings.

Before the gas is turned on or reconnected to the building, all the gas outlets shall be connected to the appliances, and any unused outlets shall be capped or plugged. A gas valve cannot be used as a plug or cap due to the potential of leaking or accidentally being turned on. All venting to the appliances shall also be installed. Following the requirements in this section will provide a safe environment for the appliances to function properly without causing any danger to the occupants or the property.

1306.0 Authority to Disconnect.

1306.1 Disconnection. The Authority Having Jurisdiction or the serving gas supplier is hereby authorized to disconnect gas piping or appliance or both that shall be found not to be in accordance with the requirements of this code or that are found defective and in such condition as to endanger life or property.

When a hazard is discovered, the AHJ or the gas supplier has the authority to disconnect the faulty gas appliance, the gas piping or both. Examples of faulty appliances might be a safety control valve that is leaking or a crack in the heat exchanger of a forced-air furnace that could allow products of combustion (carbon monoxide) to spill into the building.

1306.2 Notice. Where such disconnection has been made, a notice shall be attached to such gas piping or appliance or both that shall state the same has been disconnected, together with the reasons thereof.

Anytime a gas line or gas appliance has been disconnected because of a code violation, a notice shall be attached explaining what the hazardous condition is and what correction is needed in order to restore the gas supply.

1306.3 Capped Outlets. It shall be unlawful to remove or disconnect gas piping or gas appliance without capping or plugging with a screw joint fitting, the outlet from which said pipe or appliance was removed. Outlets to which gas appliances are not connected shall be left capped and gastight on a piping system that has been installed, altered, or repaired.

Exception: Where an approved listed quick-disconnect device is used.

When an appliance is removed or disconnected, the gas line supplying the appliance shall be capped or plugged. This precaution is necessary to ensure that no gas will leak into the building interior and that no appliances will be reconnected until the violation is corrected. The only exception to this requirement is when an approved disconnect device is used. This device stops the flow of gas when the connector is disconnected from the appliance.

1307.0 Temporary Use of Gas.

1307.1 General. Where temporary use of gas is desired, and the Authority Having Jurisdiction deems the use necessary, a permit shall be permitted to be issued for such use for a period of time not to exceed that designated by the Authority Having

Jurisdiction, provided that such gas piping system otherwise is in accordance with to the requirements of this code regarding material, sizing, and safety.

There are times when a temporary gas system is needed, such as to supply heat during a remodel of an existing space or when a business is going to occupy a space on a temporary basis. The AHJ may allow a gas system for a specific amount of time when deemed necessary, provided that all code regarding material, sizing, installation and safety are enforced.

1308.0 Gas Piping System Design, Materials, and Components.

1308.1 Installation of Piping System. Where required by the Authority Having Jurisdiction, a piping sketch or plan shall be prepared before proceeding with the installation. The plan shall show the proposed location of piping, the size of different branches, the various load demands, and the location of the point of delivery. [NFPA 54:5.1.1]

The installation of any fuel-gas system or an addition to a fuel-gas system, large or small, will require a permit. To obtain the permit, a plan or design must be submitted to the building department. The plan must show the system layout, sizing parameters and, of course, the appliances to be installed. The plan check personnel of the building department will then review the plans and verify that the fuel-gas system is code compliant.

See **Figure 1315.1.1** preceding Table 1315.2(1) for a typical gas-piping plan.

1308.1.1 Addition to Existing System. When additional appliances are being connected to a gas piping system, the existing piping shall be checked to determine whether it has adequate capacity. If the capacity of the system is determined to be inadequate for the additional appliances, the existing system shall be enlarged as required, or separate gas piping of adequate capacity shall be provided. [NFPA 54:5.1.2]

Preparing a plan or sketch of the proposed piping system is necessary, whether it is new or an extension of an existing system. By showing the loads and lengths of runs, the sizing can be determined by methods described in this chapter (see Section 1308.4) and can help to prevent added gas utilization equipment from overloading the existing gas-piping system.

1308.2 Provision for Location of Point of Delivery. The location of the point of delivery shall be acceptable to the serving gas supplier. [NFPA 54:5.2]

A natural gas supplier will determine the point of delivery based upon company policies, the regulations of the U.S. Department of Transportation or the state public utility commission. State law may also govern the point of delivery for LP-gas systems. The installer must know the point of delivery so the length of the piping system can be properly determined to provide sufficient gas for proper appliance operation.

1308.3 Interconnections Between Gas Piping Systems. Where two or more meters, or two or more service regulators where meters are not provided, are located on the same premises and supply separate users, the gas piping systems shall not be interconnected on the outlet side of the meters or service regulators. [NFPA 54:5.3.1]

Interconnected gas systems can cause a potential safety problem. Maintenance, emergency and utility personnel have no way of knowing that all services must be turned off to take the system safely out of service.

Interconnected gas systems are not to be confused with a natural gas supplier providing multiple meters or service regulators for the purpose of capacity to the one user. This situation can occur when the supplier has a limitation in the fuel supply, or the meter or regulator capacity would be exceeded, especially for large-volume users. In this case, the outlet piping systems are interconnected at the immediate outlet of the meter or service regulator assembly. Such an interconnection falls under the meter or service regulator assembly, which is governed by the U.S. Department of Transportation.

1308.3.1 Interconnections for Standby Fuels. Where a supplementary gas for standby use is connected downstream from a meter or a service regulator where a meter is not provided, equipment to prevent backflow shall be installed. A three-way valve installed to admit the standby supply and at the same time shut off the regular supply shall be permitted to be used for this purpose. [NFPA 54:5.3.2 – 5.3.2.2]

As with water supply backflow, there should be no situation where a cross connection or a backflow of gasses can occur. Different types of gas supplies must not be mixed. If a building is served by a natural gas pipeline and uses a propane system for standby, it is most important that propane not be fed into the natural gas main and that natural gas not be fed into the propane storage tanks. This would destroy the integrity of the fuel supply and could cause combustion problems. A properly approved three-way valve may be used for this purpose instead of single valves with backflow devices.

1308.4 Sizing of Gas Piping Systems. Gas piping systems shall be of such size and so installed as to provide a supply of gas sufficient to meet the maximum demand and supply gas to each appliance inlet at not less than the minimum supply pressure required by the appliance. [NFPA 54:5.4.1]

Historically, the measurement of the British thermal unit (Btu) for natural gas has varied from 800 Btu to 1200 Btu per cubic feet. Gas meters determine usage in cubic feet, natural gas is sold in therms (100,000 Btu), appliances are rated in Btu/hour and sizing charts are in cubic feet per hour (ft^3/h). Tables 1315.2(1) through 1315.2(23) are used to determine pipe size based on distance from the meter and demand in ft^3/h at a specific outlet. To determine the cubic foot per hour value of the gas, the user will need to know the average Btu heating value per cubic foot of the gas being supplied. The local gas supplier can furnish this information. However, it may be easier to review a local gas bill looking for the term: billing factor, therm multiplier or an equivalent factor or multiplier. If, for example the therm multiplier or factor was 1.038, the Btu/ft^3 would be 1,038. If the factor was .947 the Btu/ft^3 would be 947.

Because propane gas has a consistent heating value of about 2,500 Btu/ft^3 the Undiluted Propane-Gas tables

1315.2(24) through 1315.2(36) in Chapter 13 are calculated with the flow data in thousands of Btu per hour, negating the need for conversion from cubic feet/hr. to Btu/hr.

Table 1308.4.1 has been updated to current appliance designs. Remember that these values are only estimates and should be used only if the appliances are not known.

In some commercial and industrial applications, equipment requires pressure greater than the normal 7-inch water column (1.7 kPa) for natural gas or 11-inch water column (2.7 kPa) for propane. In this case, the serving supplier should be contacted to determine the availability of gas at a higher pressure, and the piping system should be designed according to that supply pressure.

1308.4.1 Maximum Gas Demand. The volumetric flow rate of gas to be provided shall be the sum of the maximum input of the appliances served. The volumetric flow rate of gas to be provided shall be adjusted for altitude where the installation is above 2 000 feet (610 m). [NFPA 54:5.4.2.1 – 5.4.2.2] Where the input rating is not indicated, the gas supplier, appliance manufacturer, or a qualified agency shall be contacted or the rating from Table 1308.4.1 shall be used for estimating the volumetric flow rate of gas to be supplied.

The total connected hourly load shall be used as the basis for piping sizing, assuming all appliances are operating at full capacity simultaneously.

Exception: Sizing shall be permitted to be based upon established load diversity factors. [NFPA 54:5.4.2.3]

Table 1308.4.1 shows estimated values for current appliance designs and should only be used if the appliances are not known.

1308.4.2 Sizing Methods. Gas piping shall be sized in accordance with one of the following:

(1) Pipe sizing tables or sizing equations in this chapter.

(2) Other approved engineering methods.

(3) Sizing tables included in a listed piping system manufacturer's installation instructions. [NFPA 54:5.4.3]

Sizing methods are described in Section 1315.0. The statement "other approved engineering methods" refers to calculations for sizing the system completed by a design engineer using equations other than those contained in Section 1316.3. These calculations must be approved by the AHJ.

The statement "sizing tables from a listed piping manufacturer's installation instructions" refers to piping materials that are not included in sizing Tables 1315.2(1) through 1315.2 (36). This includes corrugated stainless steel tubing (CSST). The tables included in this chapter for CSST refer to the equivalent hydraulic diameter (EHD) rather than the pipe size. The CSST manufacturer's sizing tables will refer to a specific pipe size. This is due to the fact that the different manufacturers of CSST use slightly different diameters and fittings for their pipe.

1308.4.3 Allowable Pressure Drop. The design pressure loss in any piping system under maximum probable flow conditions, from the point of delivery to the inlet connection of the appliance, shall be such that the supply pressure at the appliance is greater than or equal to the minimum pressure required by the appliance. [NFPA 54:5.4.4]

TABLE 1308.4.1
APPROXIMATE GAS INPUT FOR TYPICAL APPLIANCES
[NFPA 54: TABLE A.5.4.2.1]

APPLIANCE	INPUT (Btu/h approx.)
Space Heating Units	
Warm air furnace	
Single family	100 000
Multifamily, per unit	60 000
Hydronic boiler	
Single family	100 000
Multifamily, per unit	60 000
Space and Water Heating Units	
Hydronic boiler	
Single family	120 000
Multifamily, per unit	75 000
Water Heating Appliances	
Water heater, automatic storage	
30 to 40 gallon tank	35 000
Water heater, automatic storage	
50 gallon tank	50 000
Water heater, automatic instantaneous	
Capacity at 2 gallons per minute	142 800
Capacity at 4 gallons per minute	285 000
Capacity at 6 gallons per minute	428 400
Water heater, domestic, circulating or side-arm	35 000
Cooking Appliances	
Range, freestanding, domestic	65 000
Built-in oven or broiler unit, domestic	25 000
Built-in top unit, domestic	40 000
Other Appliances	
Refrigerator	3000
Clothes dryer, Type 1 (domestic)	35 000
Gas fireplace direct vent	40 000
Gas log	80 000
Barbecue	40 000
Gaslight	2500

For SI units: 1000 British thermal units per hour = 0.293 kW

Gas-piping systems must supply the volume of gas required by each appliance at a pressure within the design range established by the appliance manufacturer. The sizing tables take pressure loss into consideration and most systems will fall within the pressure loss values of the tables. If the system does not fall within the range of the tables, then engineered calculations must be used or the use of a higher inlet pressure should be considered.

The gas inlet pressure to the connected equipment must also remain within the manufacturer's design limits when the equipment "cycles off." Gas-piping systems that operate at pressures exceeding the connected gas appliance's rated inlet pressure or that deliver gas at a pressure that varies outside the equipment manufacturer's design inlet pressure limits must incorporate pressure regulators into the system so that gas delivered to the equipment will be within the design pressure range of the appliance.

1308.5 Acceptable Piping Materials and Joining Methods. Materials used for piping systems shall either comply with the requirements of this chapter or be acceptable to the Authority Having Jurisdiction. [NFPA 54:5.6.1.1]

1308.5.1 Used Materials. Pipe, fittings, valves, or other materials shall not be used again unless they are free of foreign materials and have been ascertained to be adequate for the service intended. [NFPA 54:5.6.1.2]

1308.5.1.1 Other Materials. Material not covered by the standards specifications listed herein shall meet the following criteria:

(1) Be investigated and tested to determine that it is safe and suitable for the proposed service.

(2) Be recommended for that service by the manufacturer.

(3) Be acceptable to the Authority Having Jurisdiction. [NFPA 54:5.6.1.3]

Any piping material listed in this chapter is allowed for piping systems and other materials can be approved by the AHJ. Specific criteria for approval of piping materials are not provided in the code. However, the AHJ should require proof that the material can hold the anticipated pressure, is compatible with the gas being used, has mechanical strength needed for the installation, and has resistance to any corrosive environment in which the piping system may be placed.

1308.5.2 Metallic Pipe. Cast-iron pipe shall not be used. [NFPA 54:5.6.2.1]

1308.5.2.1 Steel, Stainless Steel, and Wrought-Iron. Steel, stainless steel, and wrought-iron pipe shall be at least Schedule 40 and shall comply with the dimensional standards of ASME B36.10M and one of the following:

(1) ASTM A53

(2) ASTM A106

(3) ASTM A312 [NFPA 54:5.6.2.2]

Stainless steel pipe and tube for fuel gas distribution is most commonly used in industrial facilities where, because of the environment, it may be the preferred material. Stainless steel pipe and tubing is manufacturered in a wide range of grades and sizes. Stainless streel piping usually has a nominal inside dimension whereas stainless steet tubing has an almost exact outside dimension.

1308.5.2.2 Copper and Copper Alloy Pipe. Copper and copper alloy pipe shall not be used if the gas contains more than an average of 0.3 grains of hydrogen sulfide per 100 standard cubic feet (scf) of gas (0.7 mg/100 L). [NFPA 54:5.6.2.3]

Threaded copper, copper alloy, or aluminum alloy pipe shall not be used with gases corrosive to such material. [NFPA 54:5.6.2.4]

Copper for natural gas systems has been used in North America and throughout the world for many years; however, its use in the western part of the United States was prohibited in the 1950s. This was due to systems failing in this region because of corrosion caused by the "sulfidation" of the copper pipe and copper components of gas meters. Studies were conducted and found that hydrogen sulfide, occurring naturally in the natural gas supplied, was in high concentrations in these areas and was the cause of this corrosion.

Natural gas is a mixture of many hydrocarbon gases and other elements. One of those elements is hydrogen sulfide (H_2S). Hydrogen sulfide will react with copper, creating a coating of black or dark brown "dust," referred to as "sulfidation" (see **Figure 1308.5.2.2**). This sulfidation is perceived to cause two possible corrosion problems in copper natural gas systems:

- Continual flaking of the copper sulfide, thinning the pipe and eventually causing pinholes and, thus, leaks; and
- Continual flaking of the copper sulfide, causing the flakes themselves to fall and be carried into the appliance and possibly block burners or be deposited into gas valves, causing the valves to foul.

The second problem with sulfidation, the possible continual flaking and thus the plugging of burners and the fouling of valves, is the serious concern with these copper systems.

The hydrogen sulfide concentration level is equivalent to a trace and should not have any significant, corrosive effect on copper or brass pipe.

When gas contains more than a trace of hydrogen sulfide, the use of copper and copper alloy is prohibited. All natural gas that is distributed in pipelines and all LP-gas in the United States today are treated to remove hydrogen sulfide. If natural gas is obtained directly from a gas well, this prohibition is important because there is no removal of hydrogen sulfide when gas is used directly from a well.

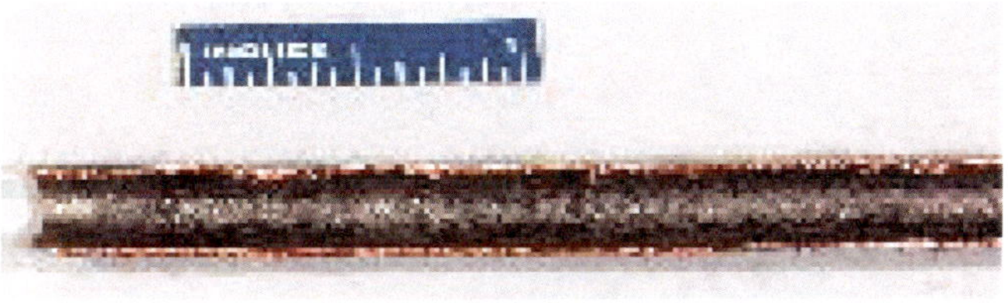

FIGURE 1308.5.2.2
SULFIDATION OF COPPER PIPE

1308.5.2.3 Aluminum Alloy Pipe. Aluminum alloy pipe shall comply with ASTM B241 (except that the use of alloy 5456 is prohibited), and shall be marked at each end of each length indicating compliance. Aluminum alloy pipe shall be coated to protect against external corrosion where it is in contact with masonry, plaster, or insulation or is subject to repeated wettings by such liquids as water, detergents, or sewage. [NFPA 54:5.6.2.5]

Aluminum alloy pipe shall not be used in exterior locations or underground. [NFPA 54:5.6.2.6]

Aluminum tubing is to be marked so that technicians and inspectors can see readily that installed tubing meets the proper specification. The coating requirement and prohibition of underground and exterior applications minimize corrosion.

Strong acids and strong bases readily corrode aluminum and its alloys. The assistance of a metallurgist or qualified corrosion specialist is recommended strongly in selecting an aluminum alloy, because the corrosion behavior of aluminum alloys varies significantly. To our knowledge, aluminum pipe and tubing are not used currently for gas piping in the United States, but they are used in some other countries. Aluminum tubing is used for piping that is part of appliances; however, such piping is not covered by this code.

1308.5.3 Metallic Tubing. Tubing shall not be used with gases corrosive to the tubing material. [NFPA 54:5.6.3.1]

1308.5.3.1 Stainless Steel Tubing. Stainless steel tubing shall comply with one of the following:

(1) ASTM A268

(2) ASTM A269 [NFPA 54:5.6.3.3]

1308.5.3.2 Steel Tubing. Steel tubing shall comply with ASTM A254. [NFPA 54:5.6.3.2]

1308.5.3.3 Copper and Copper Alloy Tubing. Copper and copper alloy tubing shall not be used if the gas contains more than an average of 0.3 grains of hydrogen sulfide per 100 scf of gas (0.7 mg/100 L). Copper tubing shall comply with standard Type K or L of ASTM B88 or ASTM B280.

See commentary for Section 1308.5.2.2.

1308.5.3.4 Aluminum Alloy Tubing. Aluminum alloy tubing shall comply with ASTM B210 or ASTM B241. Aluminum alloy tubing shall be coated to protect against external corrosion where it is in contact with masonry, plaster, or insulation or is subject to repeated wettings by such liquids as water, detergent, or sewage. Aluminum alloy tubing shall not be used in exterior locations or underground. [NFPA 54:5.6.3.5]

1308.5.3.5 Corrugated Stainless Steel Tubing. Corrugated stainless steel tubing shall be listed in accordance with CSA LC-1. [NFPA 54:5.6.3.6]

The code permits the installation of systems using CSST listed in accordance with the requirements of CSA LC1b, Standard for Fuel Gas Piping Systems Using Corrugated Stainless Steel Tubing. Note that the reference to LC1b is only for testing and listing. Installation must be in accordance with the code, and the requirements are the same as for other types of tubing.

CSST consists of a continuous, flexible, stainless steel pipe with an exterior PVC covering. The piping is produced in coils that are air-tested for leaks. It is most often installed in a central manifold configuration (also called "parallel configuration") with "home run" lines that extend to gas appliances. Flexible gas piping is lightweight and requires fewer connections than traditional gas piping because it can be bent easily and routed around obstacles. Never substitute fittings from one manufacturer to another unless specifically allowed by the manufacturer (see **Figure 1308.5.3.4**).

1308.5.4 Plastic Pipe, Tubing, and Fittings. Polyethylene plastic pipe, tubing, and fittings used to supply fuel gas shall conform to ASTM D2513. Pipe to be used shall be marked "gas" and "ASTM D2513." [NFPA 54:5.6.4.1.1] Polyamide pipe, tubing, and fittings shall be identified in and conform to ASTM F2945. Pipe to be used shall be marked "gas" and "ASTM F2945." [NFPA 54:5.6.4.1.2] Polyvinyl chloride (PVC) and chlorinated polyvinyl chloride (CPVC) plastic pipe, tubing, and fittings shall not be used to supply fuel gas. [NFPA 54:5.6.4.1.3]

FIGURE 1308.5.3.4
PLASTIC PIPE AND FITTINGS

Polyethylene is the most common plastic pipe used for gas distribution. There is also a new material called "polyamide" (PA11), a polymer thermoset plastic (Thermoplastic). Note their listings limit use to underground installation. Methods of joining are discussed in Section 1308.5.9.

1308.5.4.1 Regulator Vent Piping. Plastic pipe and fittings used to connect regulator vents to remote vent terminations shall be PVC conforming to UL 651. PVC vent piping shall not be installed indoors. [NFPA 54:5.6.4.2]

1308.5.4.2 Anodeless Risers. Anodeless risers shall comply with Section 1308.5.4.2.1 through Section 1308.5.4.2.3. [NFPA 54:5.6.4.3]

Anodeless risers are used to make the transition between underground PE pipe or tubing and metal pipe aboveground. As PE pipe must be installed below ground, risers are commonly used to connect the underground PE to aboveground piping materials. Anodeless risers are available as factory-assembled units and field-assembled kits. Anodeless risers are made from PE pipe inside a protective metal sheath, usually Schedule 40 steel pipe (see **Figure 1308.5.4.2**). The metal is protected from corrosion by a factory-applied coating and a separate anode is not required; hence the name "anodeless." Factory-assembled risers usually have a 90-degree bend at the PE connection end and come in several lengths depending on the burial depth of the PE pipe.

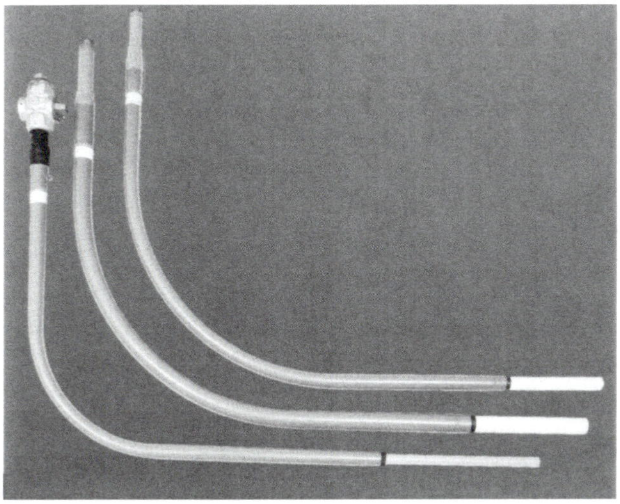

FIGURE 1308.5.4.2
RISERS

1308.5.4.2.1 Factory-Assembled Anodeless Risers. Factory-assembled anodeless risers shall be recommended by the manufacturer for the gas used and shall be leak-tested by the manufacturer in accordance with written procedures. [NFPA 54:5.6.4.3(1)]

1308.5.4.2.2 Service Head Adapters and Field-Assembled Anodeless Risers. Service head adapters and field assembled anodeless risers incorporating service head adapters shall be recommended by the manufacturer for the gas used and shall be design-certified to meet the requirements of Category I of ASTM D2513 and 49 CFR 192.281(e). The manufacturer shall provide the user qualified installation instructions as prescribed by 49 CFR 192.283(b). [NFPA 54:5.6.4.3(2)]

See **Figure 1308.5.4.2.2**.

1308.5.4.2.3 Undiluted Liquefied Petroleum Gas Piping. The use of plastic pipe, tubing, and fittings in undiluted LP-Gas piping systems shall be in accordance with NFPA 58. [NFPA 54:5.6.4.3(3)]

1308.5.5 Workmanship and Defects. Gas pipe, tubing, and fittings shall be clear and free from cutting burrs and defects in structure or threading, and shall be thoroughly brushed and chip and scale blown. Defects in pipe, tubing, and fittings shall not be repaired. Defective pipe, tubing, and fittings shall be replaced. [NFPA 54:5.6.5]

Fabricating gas pipe systems on site involves cutting, threading, and joining with fittings. Whenever a pipe is cut using a pipe cutter, a burr will be formed around the inside diameter of the pipe. A pipe reamer or deburring tool is required to remove the burr. Defective threads resulting from threading the pipe shall not used. Any manufacturing defect found in a fitting, pipe or tubing shall not be repaired, but replaced.

1308.5.6 Metallic Pipe Threads. Metallic pipe and fitting threads shall be taper pipe threads and shall comply with ASME B1.20.1. [NFPA 54:5.6.6.1]

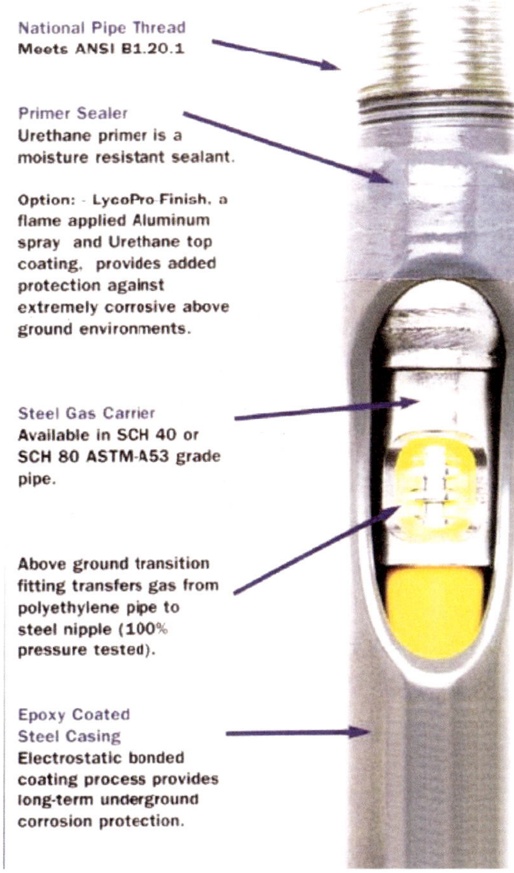

FIGURE 1308.5.4.2.2
ANODELESS RISER

1308.5.6.1 Damaged Threads. Pipe with threads that are stripped, chipped, corroded, or otherwise damaged shall not be used. Where a weld opens during the operation of cutting or threading, that portion of the pipe shall not be used. [NFPA 54:5.6.6.2]

1308.5.6.2 Number of Threads. Field threading of metallic pipe shall be in accordance with Table 1308.5.6.2. [NFPA 54:5.6.6.3]

TABLE 1308.5.6.2
SPECIFICATIONS FOR THREADING METALLIC PIPE
[NFPA 54: TABLE 5.6.6.3]

IRON PIPE SIZE (inches)	APPROXIMATE LENGTH OF THREADED PORTION (inches)	APPROXIMATE NUMBER OF THREADS TO BE CUT
½	¾	10
¾	¾	10
1	⅞	10
1¼	1	11
1½	1	11
2	1	11
2½	1½	12
3	1½	12
4	1⅝	13

For SI units: 1 inch = 25.4 mm

Properly cut pipe threads are tapered (the threads nearest the end of the pipe have a smaller diameter than the threads farthest away). If common threading dies (e.g., fixed cutting heads) are run too far onto the pipe during the threading process, the resulting threads will be straight for a portion of the threads, then will taper. These straight threads are known as running threads, because the thread has been "run" down the pipe. Because pipe fittings depend on a tapered thread on the pipe to seal properly, an improperly threaded pipe can easily result in a leak at the joint. Proper use of a fixed die requires the threading operation to stop as soon as the end of the pipe exits the die.

1308.5.6.3 Thread Joint Compounds. Thread joint compounds shall be resistant to the action of LP-Gas or to any other chemical constituents of the gases to be conducted through the piping. [NFPA 54:5.6.6.4]

When threaded pipe and fittings are assembled, the meshed threads (called an interference fit) form a mechanical seal that holds the pipe together but does not quite provide a gas-tight seal. Gas tightness is provided by a thread compound, which must be used for a gas-tight piping system. Thread compounds should be applied only to the male pipe connection to avoid having the compound enter the piping system during assembly. Listed pipe thread tape is an acceptable thread joint compound. If a listed pipe joint tape is used, it must be applied properly only to the threads and must not overlap the inside diameter of the pipe where it could tear off and enter the gas pipe. The component manufacturer's instructions (for example, for equipment such as gas valves) should be consulted regarding restrictions on the use of pipe thread tape.

1308.5.7 Metallic Piping Joints and Fittings. The type of piping joint used shall be suitable for the pressure and temperature conditions and shall be selected giving consideration to joint tightness and mechanical strength under the service conditions. The joint shall be able to sustain the maximum end force due to the internal pressure and any additional forces due to temperature expansion or contraction, vibration, fatigue, or the weight of the pipe and its contents. [NFPA 54:5.6.7]

1308.5.7.1 Pipe Joints. Schedule 40 and heavier pipe joints shall be threaded, flanged, brazed, welded, or assembled with press-connect fittings listed to CSA LC 4.

(1) Where nonferrous pipe is brazed, the brazing materials shall have a melting point in excess of 1000°F (538°C).

(2) Brazing alloys shall not contain more than 0.05 percent phosphorus. {NFPA 54:5.6.7.1}

1308.5.7.2 Copper Tubing Joints. Copper tubing joints shall be assembled with approved gas tubing fittings, shall be brazed with a material having a melting point in excess of 1000°F (538°C), or shall be assembled with press-connect fittings listed to CSA LC 4, Press-Connect Metallic Fittings for Use in Fuel Gas Distribution Systems. Brazing alloys shall not contain more than 0.05 percent phosphorus. [NFPA 54:5.6.7.2]

1308.5.7.3 Stainless Steel Tubing Joints. Stainless steel joints shall be welded, assembled with approved tubing fittings, brazed with a material having a melting point in excess of 1000°F (538°C), or assembled with press-connect fittings listed to CSA LC 4, Press-Connect Metallic Fittings for Use in Fuel Gas Distribution Systems. Brazing alloys and fluxes shall be recommended by the manufacturer for use on stainless steel alloys. [NFPA 54:5.6.7.3]

1308.5.7.4 Flared Joints. Flared joints shall be used only in systems constructed from nonferrous pipe and tubing where experience or tests have demonstrated that the joint is suitable for the conditions and where provisions are made in the design to prevent separation of the joints. [NFPA 54:5.6.7.4]

1308.5.7.5 Metallic Pipe Fittings. Metallic fittings shall comply with the following:

(1) Threaded fittings in sizes exceeding 4 inches (100 mm) shall not be used.

(2) Fittings used with steel, stainless steel, or wrought-iron pipe shall be steel, stainless steel, copper alloy, malleable iron, or cast-iron.

(3) Fittings used with copper or copper alloy pipe shall be copper or copper alloy.

(4) Fittings used with aluminum alloy pipe shall be aluminum alloy.

(5) Cast-iron fittings shall comply with the following:

  (a) Flanges shall be permitted.

  (b) Bushings shall not be used.

  (c) Fittings shall not be used in systems containing flammable gas-air mixtures.

  (d) Fittings in sizes 4 inches (100 mm) and larger shall not be used indoors unless approved by the Authority Having Jurisdiction.

  (e) Fittings in sizes 6 inches (150 mm) and larger shall not be used unless approved by the Authority Having Jurisdiction.

(6) Aluminum alloy fitting threads shall not form the joint seal.

(7) Zinc-aluminum alloy fittings shall not be used in systems containing flammable gas-air mixtures.

(8) Special fittings such as couplings, proprietary-type joints, saddle tees, gland-type compression fittings, and flared, flareless, or compression-type tubing fittings shall be as follows:

  (a) Used within the fitting manufacturer's pressure-temperature recommendations.

  (b) Used within the service conditions anticipated with respect to vibration, fatigue, thermal expansion, or contraction.

  (c) Acceptable to the Authority Having Jurisdiction.

(9) When pipe fittings are drilled and tapped in the field, the operation shall be in accordance with the following:

  (a) The operation shall be performed on systems having operating pressures of 5 psi (34 kPa) or less.

  (b) The operation shall be performed by the gas supplier or their designated representative.

  (c) The drilling and tapping operation shall be

performed in accordance with written procedures prepared by the gas supplier.

(d) The fittings shall be located outdoors.

(e) The tapped fitting assembly shall be inspected and proven to be free of leaks. [NFPA 54:5.6.7.5]

The use of different materials for fittings than the material for the piping shall not be allowed. For example, steel fittings shall not be used for aluminum alloy pipe and so on. Not only would this intermingling of materials look like shoddy workmanship but, in most cases, it will cause corrosion or at least a possible reduction in pipe size, depending on the materials used. Only properly listed and approved transition fittings may be used to join together different materials.

1308.5.8 Plastic Piping, Joints, and Fittings. Plastic pipe, tubing, and fittings shall be joined in accordance with the manufacturer's instructions. Section 1308.5.8.1 through Section 1308.5.8.4 shall be observed when making such joints. [NFPA 54:5.6.8]

1308.5.8.1 Joint Design. The joint shall be designed and installed so that the longitudinal pullout resistance of the joint will be at least equal to the tensile strength of the plastic piping material. [NFPA 54:5.6.8(1)]

1308.5.8.2 Heat Fusion Joint. Heat fusion joints shall be made in accordance with qualified procedures that have been established and proven by test to produce gastight joints at least as strong as the pipe or tubing being joined. Joints shall be made with the joining method recommended by the pipe manufacturer. Heat fusion fittings shall be marked "ASTM D2513." [NFPA 54:5.6.8(2)]

1308.5.8.3 Compression-Type Mechanical Joints. Where compression-type mechanical joints are used, the gasket material in the fitting shall be compatible with the plastic piping and with the gas distributed by the system. An internal tubular rigid stiffener shall be used in conjunction with the fitting. The stiffener shall be flush with the end of the pipe or tubing and shall extend at least to the outside end of the compression fitting when installed. The stiffener shall be free of rough or sharp edges and shall not be a force fit in the plastic. Split tubular stiffeners shall not be used. [NFPA 54:5.6.8(3)]

1308.5.8.4 Liquefied Petroleum Gas Piping Systems. Plastic piping joints and fittings for use in LP-Gas piping systems shall be in accordance with NFPA 58. [NFPA 54:5.6.8(4)]

1308.5.9 Flange Specifications. Cast iron flanges shall be in accordance with ASME B16.1. [NFPA 54:5.6.9.1.1]

1308.5.9.1 Steel Flanges. Steel flanges shall be in accordance with the following:

(1) ASME B16.5 or

(2) ASME B16.47. [NFPA 54:5.6.10.1.2]

1308.5.9.2 Non-Ferrous Flanges. Non-ferrous flanges shall be in accordance with ASME B16.24. [NFPA 54:5.6.9.1.3]

1308.5.9.3 Ductile Iron Flanges. Ductile iron flanges shall be in accordance with ASME B16.42. [NFPA 54:5.6.9.1.4]

1308.5.9.4 Dissimilar Flange Connections. Raised-face flanges shall not be joined to flat-faced cast iron, ductile iron or nonferrous material flanges. [NFPA 54:5.6.9.2]

1308.5.9.5 Flange Facings. Standard facings shall be permitted for use under this code. Where 150 psi (1034 kPa) steel flanges are bolted to Class 125 cast-iron flanges, the raised face on the steel flange shall be removed. [NFPA 54:5.6.9.3]

1308.5.9.6 Lapped Flanges. Lapped flanges shall be used only aboveground or in exposed locations accessible for inspection. [NFPA 54:5.6.9.4]

1308.5.10 Flange Gaskets. The material for gaskets shall be capable of withstanding the design temperature and pressure of the piping system and the chemical constituents of the gas being conducted without change to its chemical and physical properties. The effects of fire exposure to the joint shall be considered in choosing the material. [NFPA 54:5.6.10]

1308.5.10.1 Flange Gasket Materials. Acceptable materials shall include the following:

(a) Metal (plain or corrugated)

(b) Composition

(c) Aluminum "O" rings

(d) Spiral-wound metal gaskets

(e) Rubber-faced phenolic

(f) Elastomeric [NFPA 54:5.6.10.1]

1308.5.10.2 Metallic Flange Gaskets. Metallic flange gaskets shall be in accordance with ASME B16.20. [NFPA 54:5.6.10.2.1]

1308.5.10.3 Non-Metallic Flange Gaskets. Non-metallic flange gaskets shall be in accordance with ASME B16.21. [NFPA 54:5.6.10.2.2]

1308.5.10.4 Full-Face Flange Gasket. Full-face flange gaskets shall be used with all non-steel flanges. [NFPA 54:5.6.10.3]

1308.5.10.5 Separated Flanges. When a flanged joint is separated, the gasket shall be replaced. [NFPA 54:5.6.10.4]

1308.6 Gas Meters. Gas meters shall be selected for the maximum expected pressure and permissible pressure drop. [NFPA 54:5.7.1]

1308.6.1 Location. Gas meters shall be located in ventilated spaces readily accessible for examination, reading, replacement, or necessary maintenance. [NFPA 54:5.7.2.1]

1308.6.1.1 Subject to Damage. Gas meters shall not be placed where they will be subjected to damage, such as adjacent to a driveway, under a fire escape, in public passages, halls, or where they will be subject to excessive corrosion or vibration. [NFPA 54:5.7.2.2]

See **Figure 1308.6.1.1**.

1308.6.1.2 Extreme Temperatures. Gas meters shall not be located where they will be subjected to extreme temperatures or sudden extreme changes in temperature or in areas where they are subjected to temperatures beyond those recommended by the manufacturer. [NFPA 54:5.7.2.3]

1308.6.2 Supports. Gas meters shall be supported or connected to rigid piping so as not to exert a strain on the meters. Where flexible connectors are used to connect a gas meter to downstream piping at mobile homes in mobile home parks, the meter shall be supported by a post or bracket placed

FIGURE 1308.6.1.1
GAS METER PROTECTION

in a firm footing or by other means providing equivalent support. [NFPA 54:5.7.3]

1308.6.3 Meter Protection. Meters shall be protected against overpressure, backpressure, and vacuum. [NFPA 54:5.7.4]

1308.6.4 Identification. Gas piping at multiple meter installations shall be marked by a metal tag or other permanent means designating the building or the part of the building being supplied and attached by the installing agency. [NFPA 54:5.7.5]

See **Figure 1308.6.4**.

FIGURE 1308.6.4
GAS METER INSTALLATION AND METER TAG

1308.7 Gas Pressure Regulators. A line pressure regulator shall be installed where the gas supply pressure exceeds the maximum allowable inlet pressure of the appliance served. [NFPA 54:5.8.1]

1308.7.1 Listing. Line pressure regulators shall be listed in accordance with CSA Z21.80 where the outlet pressure is set to 2 psi (14 kPa) or less. [NFPA 54:5.8.2]

1308.7.2 Location. The gas pressure regulator shall be accessible for servicing. [NFPA 54:5.8.3]

1308.7.3 Regulator Protection. Pressure regulators shall be protected against physical damage. [NFPA 54:5.8.4]

1308.7.4 Venting of Line Pressure Regulators. Line pressure regulators shall comply with all of the following:

(1) An independent vent to the exterior of the building, sized in accordance with the regulator manufacturer's instructions, shall be provided where the location of a regulator is such that a ruptured diaphragm will cause a hazard.

  (a) Where more than one regulator is at a location, each regulator shall have a separate vent to the outdoors or, if approved by the Authority Having Jurisdiction, the vent lines shall be permitted to be manifolded in accordance with accepted engineering practices to minimize backpressure in the event of diaphragm failure.

  (b) Materials for vent piping shall be in accordance with Section 1308.5 through Section 1308.5.10.5.

Exception: A regulator and vent limiting means combination listed as complying with CSA Z21.80 shall be permitted to be used without a vent to the outdoors.

(2) The vent shall be designed to prevent the entry of water, insects, or other foreign materials that could cause blockage.

(3) The regulator vent shall terminate at least 3 feet (914 mm) from a source of ignition.

(4) At locations where regulators might be submerged during floods, a special antiflood-type breather vent fitting shall be installed, or the vent line shall be extended above the height of the expected flood waters.

(5) A regulator shall not be vented to the appliance flue or exhaust system. [NFPA 54:5.8.5.1]

Gas pressure regulators have a vent that allows the air above the regulator diaphragm to be displaced so the diaphragm can move. This air is vented out of the regulator and may also contain fuel gas products, which in large enough quantities or if the diaphragm fails, could cause an explosion within the building. The regulator, if required, must be vented according to the requirements of this section.

1308.7.5 Venting of Gas Appliance Pressure Regulators. For venting of gas appliance pressure regulators, see Section 902.15. [NFPA 54:5.8.5.2]

1308.7.6 Bypass Piping. Valved and regulated bypasses shall be permitted to be placed around gas line pressure regulators where continuity of service is imperative. [NFPA 54:5.8.6]

1308.7.7 Identification. Line pressure regulators at multiple regulator installations shall be marked by a metal tag or other permanent means designating the building or the part of the building being supplied. [NFPA 54:5.8.7]

Gas pressure regulators are designed to maintain constant downstream gas pressure, regardless of change in the gas flow or in the conditions of the upstream pressure. Ideally, they restrict the flow from the inlet pressure side to balance exactly what the downstream gas pressure should be. There are four basic areas where regulators are used:

1. A service pressure regulator is installed to reduce the main gas service pressure to the design pressure of the piping system.

2. A line gas pressure regulator may be used to reduce the gas piping main pressure to a lower pressure. A typical example would be in a hybrid pressure system – for example, a 2 psi (13.8 kPa) main pressure to an 11-inch water column pressure to the branches serving the appliances
3. An appliance regulator may be needed at the inlet to the appliance, which will reduce the branch pressure to the specific working pressure of the appliance.
4. There may be a pressure regulator in the appliance itself, which will reduce the outlet pressure to the working pressure within the appliance.

1308.8 Overpressure Protection. Where the serving gas supplier delivers gas at a pressure greater than 2 psi for piping systems serving appliances designed to operate at a gas pressure of 14 inches water column or less, overpressure protection devices shall be installed. Piping systems serving equipment designed to operate at inlet pressures greater than 14 inches water column (3.5 kPa) shall be equipped with overpressure protection devices as required by the appliance manufacturer's installation instructions. [NFPA 54:5.9.1]

With increasing uses for gas utilization equipment and with the economy of using higher pressures for smaller-diameter gas-piping systems, a need frequently exists for line gas pressure regulators to deliver proper gas pressures to equipment or groups of equipment. This system parallels using higher electrical voltage (pressure) and smaller wires (pipes) with transformers (regulators) to supply the loads properly (see **Figure 1308.8a**).

Note that this section refers to line pressure regulators and not appliance regulators. Appliance regulators are covered in Section 902.14.

Sections 1308.8 through 1308.9.4 are direct extractions from the latest version of NFPA 54 dealing with overpressure protection devices (OPD) for fuel gas systems exceeding 2 psi pressure. The requirements for these sections are based, in part, on the provisions found in standard ANSI Z21.80/Can 6.22 Line Pressure Regulators, Section 3.8 (see **Figure 1308.8b**).

This section of the standard applies to overpressure protection devices which are integral parts, or provided with Class I or Class II regulators with inlet pressures exceeding 2 psi pressure and outlet pressures limited to less than 1⁄2 psi pressure. Class I regulators can be certified for inlet pressures of 2, 5 and 10 psi with maximum outlet pressure settings of 1⁄2 psi. Class II regulators can be certified for inlet pressures of 5 or 10 psi with maximum outlet pressure of 2 psi. Per Section 1308.9.1, overpressure protection devices must also be provided for Class II regulators that have outlet pressures that exceed 1⁄2 psi. Line pressure regulators with separate overpressure protection devices shall be factory preassembled, tested and supplied to the field as a unit. In lieu of an overpressure protection device, a pressure relief valve may be provided downstream of the regulator. In the event that the line pressure regulator fails wide open the relief valve must provide overpressure protection that does not exceed the regulator's operating inlet pressure.

1308.9 Pressure Limitation Requirements. Where piping systems serving appliances designed to operate with a gas supply pressure of 14 inches water column or less are required to be equipped with overpressure protection by Section 1308.8, each overpressure protection device shall be adjusted to limit the gas pressure to each connected appliance to 2 psi or less upon a failure of the line pressure regulator. [NFPA 54:5.9.2.1]

1308.9.1 Overpressure Protection Required. Where piping systems serving appliances designed to operate with a gas supply pressure greater than 14 inches water column are required to be equipped with overpressure protection by

FIGURE 1308.8A
GAS LINE REGULATOR WITH VENT-LIMITING MEANS
(Courtesy of Maxitrol Company)

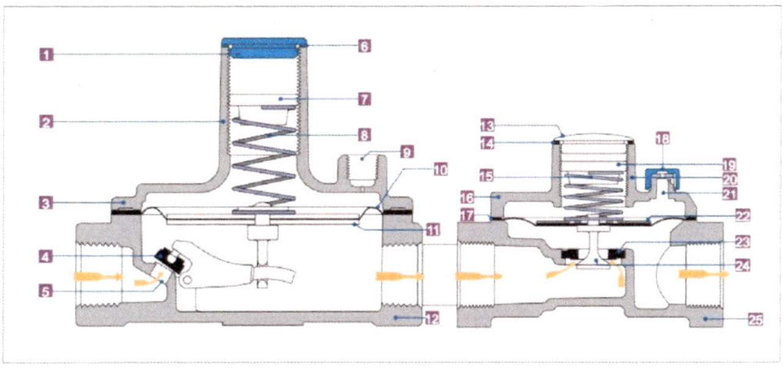

NOTE: Diagrams are graphical representations only and may differ from actual product.

1	Seal Cap	8	Spring	15	Spring	22	Diaphragm Plate
2	Stack	9	Vent Connection	16	Top Housing	23	Rubber Seat
3	Top Housing	10	Diaphragm	17	Diaphragm	24	Stem & Valve
4	Rubber Valve	11	Diaphragm Plates	18	Dust Cap	25	Bottom Housing
5	Valve Seat	12	Bottom Housing	19	Adjusting Screw		
6	Seal Cap Gasket	13	Seal Cap	20	Stack		
7	Adjusting Screw	14	Seal Cap Gasket	21	Vent		

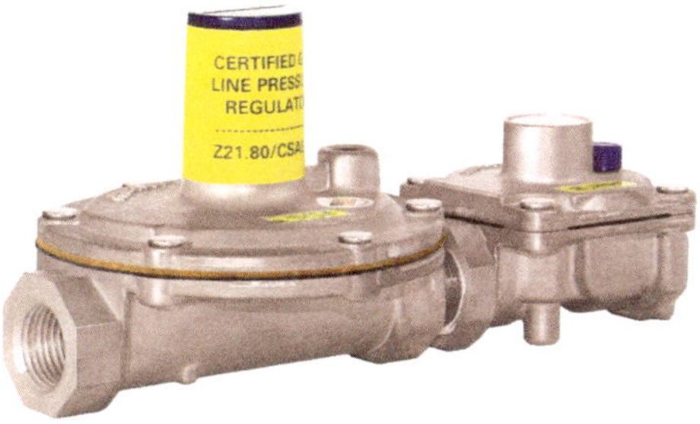

FIGURE 1308.8B
LINE PRESSURE REGULATOR
(Courtesy of Maxitrol Company)

Section 1308.8, each overpressure protection device shall be adjusted to limit the gas pressure to each connected appliance as required by the appliance manufacturer's installation instructions. [NFPA 54:5.9.2.2]

1308.9.2 Overpressure Protection Devices. Each overpressure protection device installed to meet the requirements of this section shall be capable of limiting the pressure to its connected appliance(s) as required by this section independently of any other pressure control equipment in the piping system. [NFPA 54:5.9.2.3]

1308.9.3 Detection of Failure. Each gas piping system for which an overpressure protection device is required by this section shall be designed and installed so that a failure of the primary pressure control device(s) is detectable. [NFPA 54:5.9.2.4]

1308.9.4 Flow Capacity. If a pressure relief valve is used to meet the requirements of this section, it shall have a flow capacity such that the pressure in the protected system is maintained at or below the limits specified in Section 1308.9 under the following conditions:

(1) The line pressure regulator for which the relief valve is providing overpressure protection has failed wide open.

(2) The gas pressure at the inlet of the line pressure regulator for which the relief valve is providing overpressure protection is not less than the regulator's normal operating inlet pressure. [NFPA 54:5.9.2.5]

1308.10 Overpressure Protection Devices. Overpressure protection devices shall be one of the following:

(1) Pressure relief valve.

(2) Monitor regulator.

(3) Series regulator installed upstream from the line regulator and set to continuously limit the pressure on the inlet of the line regulator to the maximum values specified by Section 1308.9 or less.

(4) Automatic shutoff device installed in series with the line pressure regulator and set to shut off when the pressure on the downstream piping system reaches the maximum values specified by Section 1308.9 or less. This device shall be designed so that it will remain closed until manually reset. [NFPA 54:5.9.3.1]

1308.10.1 Separate Devices. The devices in Section 1308.10 shall be installed either as an integral part of the service or line pressure regulator or as separate units. Where separate overpressure protection devices are installed, they shall comply with Section 1308.10.2 through Section 1308.10.7. [NFPA 54:5.9.3.2]

1308.10.2 Construction and Installation. All overpressure protection devices shall meet the following requirements:

(1) Be constructed of materials so that the operation of the device is not impaired by corrosion of external parts by the atmosphere or of internal parts by the gas.

(2) Be designed and installed so they can be operated to determine whether the valve is free. The devices shall also be designed and installed so they can be tested to determine the pressure at which they operate and be examined for leakage when in the closed position. [NFPA 54:5.9.4]

1308.10.3 External Control Piping. External control piping shall be designed and installed so that damage to the control piping of one device does not render both the regulator and the overpressure protective device inoperative. [NFPA 54:5.9.5]

1308.10.4 Setting. Each pressure limiting or pressure relieving device shall be set so that the gas pressure supplied to the connected appliance(s) does not exceed the limits specified in Section 1308.9 and Section 1308.9.1. [NFPA 54:5.9.6]

1308.10.5 Unauthorized Operation. Where unauthorized operation of any shutoff valve could render a pressure relieving valve or pressure limiting device inoperative, one of the following shall be accomplished:

(1) The valve shall be locked in the open position. Instruct authorized personnel in the importance of leaving the shutoff valve open and of being present while the shutoff valve is closed so that it can be locked in the open position before leaving the premises.

(2) Duplicate relief valves shall be installed, each having adequate capacity to protect the system, and arrange the isolating valves or three-way valve so that only one relief valve can be rendered inoperative at a time. [NFPA 54:5.9.7]

1308.10.6 Discharge of Vents. The discharge stacks, vents, or outlet parts of all pressure relieving and pressure limiting devices shall be located so that gas is safely discharged to the outdoors. Discharge stacks or vents shall be designed to prevent the entry of water, insects, or other foreign material that could cause blockage. The discharge stack or vent line shall be at least the same size as the outlet of the pressure relieving device. [NFPA 54:5.9.8.1, 5.9.8.2]

1308.10.7 Size of Fittings, Pipe, and Openings. The fittings, pipe, and openings located between the system to be protected and the pressure relieving device shall be sized to prevent hammering of the valve and to prevent impairment of relief capacity. [NFPA 54:5.9.9]

1308.11 Backpressure Protection. Protective devices shall be installed as close to the equipment as practical where the design of equipment connected is such that air, oxygen, or standby gases could be forced into the gas supply system. Gas and air combustion mixers incorporating double diaphragm "zero" or "atmosphere" governors or regulators shall require no further protection unless connected directly to compressed air or oxygen at pressures of 5 psi (34 kPa) or more. [NFPA 54:5.10.1.1, 5.10.1.2]

1308.11.1 Protective Devices. Protective devices shall include but not be limited to the following:

(1) Check valves.

(2) Three-way valves (of the type that completely closes one side before starting to open the other side).

(3) Reverse flow indicators controlling positive shutoff valves.

(4) Normally closed air-actuated positive shutoff pressure regulators. [NFPA 54:5.10.2]

Gas systems connected to zero governor-equipped mixing equipment require no further back-pressure prevention devices because the standby gas will not flow until the blower is operating to "pull" the gas through the zero governor. This setup is very resistant to failure. If the system uses compressed air, back-pressure protection is required because of the possibility of air being injected into the gas-piping system.

Venturi-type standby fuel systems require that the served gas-piping system be provided with backflow protection because a failure of the venturi gas valves could cause an overpressure of the gas-piping system.

1308.12 Low-Pressure Protection. A protective device shall be installed between the meter and the appliance or equipment if the operation of the appliance or equipment is such that it could produce a vacuum or a dangerous reduction in gas pressure at the meter. Such protective devices include, but are not limited to, mechanical, diaphragm-operated, or electrically operated low-pressure shutoff valves. [NFPA 54:5.11]

If a low-pressure shutoff valve is installed, a manual-reset device is recommended. This device ensures that the gas will not come back on unexpectedly. It is very important to protect the gas supply system from low pressure and vacuum. If the equipment could cause low pressure or vacuum in the supply system, it could adversely affect the supply pressure of other customers on the line. The low-pressure situation may create a dangerous condition for an unsuspecting user, and it would be difficult to determine the source if caused by another customer.

1308.13 Shutoff Valves. Shutoff valves shall be approved and shall be selected giving consideration to pressure drop, service involved, emergency use, and reliability of operation. Shutoff valves of size 1 inch (25 mm) National Pipe Thread and smaller shall be listed. [NFPA 54:5.12]

Shutoff valves are required at the meter, before a pressure regulator and at the appliance itself. Always ensure that the valve is for use with the gas in service and for the temperatures and pressures required for the system.

Shutoff valves that are one-inch National Pipe Thread (NPT) and smaller are required to be listed in recognition that larger valves usually are not listed.

1308.14 Expansion and Flexibility. Piping systems shall be designed to prevent failure from thermal expansion or contraction. [NFPA 54:5.14.1]

1308.14.1 Special Local Conditions. Where local conditions include earthquake, tornado, unstable ground, or flood hazards, special consideration shall be given to increased strength and flexibility of piping supports and connections. [NFPA 54:5.14.2]

1309.0 Excess Flow Valve.

1309.1 General. Where automatic excess flow valves are installed, they shall be listed to CSA Z21.93 and shall be sized and installed in accordance with the manufacturer's instructions. [NFPA 54:5.13]

Excess flow valves automatically shut off the gas when the flow to a residence or commercial facility exceeds design limits. This excess in gas flow can be caused by a break in the service line from ground movement, natural disasters or third-party damage. Areas of the country where earthquakes occur may require an excess flow valve on the main service line or at the appliance outlet (see **Figure 1309.1**).

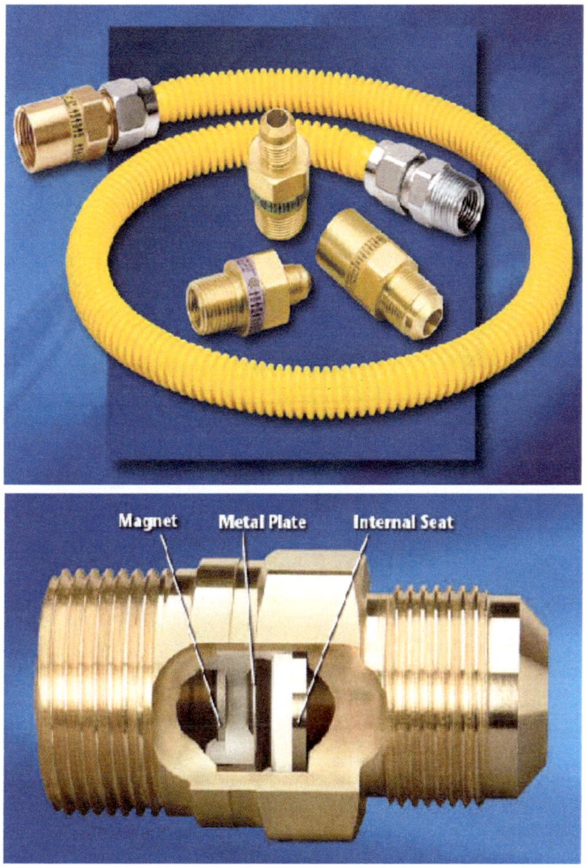

FIGURE 1309.1
EXCESS FLOW VALVE

1310.0 Gas Piping Installation.

1310.1 Piping Underground. Underground gas piping shall be installed with sufficient clearance from any other underground structure to avoid contact therewith, to allow maintenance, and to protect against damage from proximity to other structures. In addition, underground plastic piping shall be installed with sufficient clearance or shall be insulated from any source of heat so as to prevent the heat from impairing the serviceability of the pipe. [NFPA 54:7.1.1]

1310.1.1 Cover Requirements. Underground piping systems shall be installed with a minimum of 12 inches (305 mm) of cover. The minimum cover shall be increased to 18 inches (457 mm) if external damage to the pipe or tubing from external forces is likely to result. Where a minimum of 12 inches (305 mm) of cover cannot be provided, the pipe shall be installed in conduit or bridged (shielded). [NFPA 54:7.1.2.1 – 7.1.2.1(B)]

This section recognizes the possible hazard from the pipe being in contact with, or in close proximity to, other underground structures. Such contact can be a hazard and can cause harm to the pipe during backfilling or, later, when settlement of the structure(s) occurs. The intent of all of the coverage for underground piping is to provide protection for the installed piping from corrosion or physical damage.

Protection from heavy vehicle loads may require deeper burial at a depth of 18 inches or more. An alternative is the use of a casing for the pipe carrying the gas. Deeper burial applies to cultivated areas, with depth-of-cover requirements depending on the anticipated risk, which can vary from hand garden tools to tillage equipment pulled by large tractors. Gas companies with distribution pipes in the area are a good resource with which to consult when questions about burial depth of piping are raised.

Damage to pipes is not limited to plastic pipe. Copper tubing is severed easily with a shovel, and digging tools can damage the coating on steel pipe, leading to corrosion. Digging with powered digging equipment, such as a backhoe, can also damage steel pipe.

There are two ways to protect the pipe from physical damage. One is to protect the pipe from damage caused by digging, so the protection from digging tools must be of sufficient strength to protect the pipe. Such protection can be provided by installing the gas pipe inside a larger pipe, which becomes a shield, or by locating a protective material, such as a steel plate with corrosion protection, above the pipe. In most cases, the pipe is buried at a sufficient depth so that protective devices are not needed.

The second means is to provide excavators with an indicator that a pipe is buried below. Many gas utilities use a "terra tape" (a yellow plastic tape printed with a warning that gas pipe is buried below) placed a few inches above the pipe to warn future diggers that a gas pipe is buried below; the tape is an indicator and not a protective device (see **Figure 1310.1.1**).

If protective devices are needed, they must be of sufficient strength to protect the pipe and be made of a material that will not corrode over time. Plastic pipe or PVC is often used as a protective material. Split along its length, plastic pipe or PVC provides the necessary strength and does not corrode underground.

1310.1.2 Trenches. The trench shall be graded so that the pipe has a firm, substantially continuous bearing on the bottom of the trench. [NFPA 54:7.1.2.2]

1310.1.2.1 Backfilling. Where flooding of the trench is done to consolidate the backfill, care shall be exercised to see that the pipe is not floated from its firm bearing on the trench bottom. [NFPA 54:7.1.2.3]

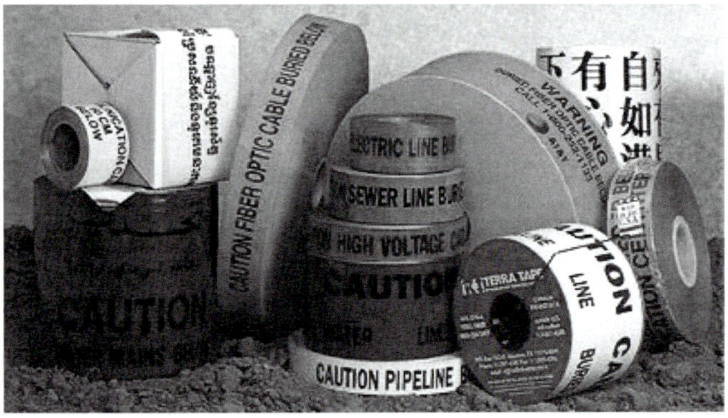

FIGURE 1310.1.1
TERRA TAPE

1310.1.3 Protection Against Corrosion. Steel pipe and steel tubing installed underground shall be installed in accordance with Section 1310.1.3.1 through Section 1310.1.3.9. [NFPA 54:7.1.3]

With the availability of polyethylene and other corrosion-resistant piping systems (e.g., plastic-coated metal pipe and tubing, copper tube in most soils), the need to provide corrosion protection for steel often makes this material the last choice for underground installation. Steel pipe that is installed underground or is in contact with water deteriorates through galvanic corrosion, which is an electrochemical process. With the availability of polyethylene and other corrosion-resistant piping systems, this electrochemical process can be prevented by stopping the flow of electricity between the pipe and the soil. Stopping electricity flow is accomplished by coating the pipe with an electrically insulating material. Factory-applied coatings include polyethylene over a mastic primer and fusion-bonded epoxy. Field-applied coatings usually consist of either hot- or cold-applied tapes over a complementary primer.

Another step in properly protecting a steel pipe underground is to maintain an electrical charge on the pipe relative to the surrounding soil and adequate to ensure that the corrosion-causing transfer of electrons will not occur. Maintaining this electrical charge is called "cathodic protection" and is accomplished by an impressed voltage supplied by either a passive sacrificial anode system or by an active rectifier system. Sizing and spacing of either the anodes or the rectifiers are important. Manufacturers of these components are a good source of guidance.

Partial steel pipe corrosion solutions, such as coating the pipe but not using cathodic protection, usually fare worse than using unprotected pipe. Uncoated pipe will corrode over its entire surface area, whereas corrosion of coated pipe lacking cathodic protection will be concentrated at any pinhole or other imperfection (called a "holiday") in the coating. Because obtaining and maintaining a perfect coating application is virtually impossible, protection of underground steel pipe should include cathodic protection. Any cathodically protected pipe should be electrically isolated from upstream and downstream components to prevent the loss of the required electrical charge. This isolation is accomplished with the use of dielectric unions or flange-insulating kits.

The requirement for protection against corrosion also recognizes that threaded and socket-welded steel pipe joints are especially prone to the effects of corrosion and should not be used where corrosion is anticipated. Similarly, the use of cinders as backfill is prohibited because they accelerate corrosion.

1310.1.3.1 Zinc Coating. Zinc coating (galvanizing) shall not be deemed adequate protection for underground gas piping. [NFPA 54:7.1.3.1]

1310.1.3.2 Underground Piping. Underground piping shall comply with one or more of the following unless approved technical justification is provided to demonstrate that protection is unnecessary:

(1) The piping shall be made of corrosion-resistant material

that is suitable for the environment in which it will be installed.

(2) Pipe shall have a factory-applied, electrically insulating coating. Fittings and joints between sections of coated pipe shall be coated in accordance with the coating manufacturer's instructions.

(3) The piping shall have a cathodic protection system installed, and the system shall be maintained in accordance with Section 1310.1.3.3 or Section 1310.1.3.6. [NFPA 54:7.1.3.2]

1310.1.3.3 Cathodic Protection. Cathodic protection systems shall be monitored by testing and the results shall be documented. The test results shall demonstrate one of the following:

(1) A pipe-to-soil voltage of −0.85 volts or more negative is produced, with reference to a saturated copper-copper sulfate half cell.

(2) A pipe-to-soil voltage of −0.78 volts or more negative is produced, with reference to a saturated KCl calomel half cell.

(3) A pipe-to-soil voltage of −0.80 volts or more negative is produced, with reference to a silver-silver chloride half cell.

(4) Compliance with a method described in Appendix D of Title 49 of the Code of Federal Regulations, Part 192. [NFPA 54:7.1.3.3]

Piping carrying hazardous materials is often protected by a manufactured coating supplemented with cathodic protection. Cathodic protection is a means of preventing metal structures such as pipelines from reacting with the environment and corroding. Carbon steel piping exposed to the elements will break down electrochemically and ultimately fail. Cathodic protection systems prevent the oxidation process from occurring by creating a current flow from the cathodic protection system to the piping. There are two types of cathodic protection systems: galvanic and impressed current.

Galvanic cathodic protection utilizes an electrical-chemical process where one metal is more susceptible to corrosion than another when both metals are linked electrically. Sacrificial anodes, many times made of magnesium, are used to protect steel piping. (see **Figure 1310.1.3.3a**).

In many applications, the difference between the sacrificial anode and the steel piping is not adequate to generate sufficient current for cathodic protection to occur. In these cases, a power supply, or rectifier, is used to generate larger potential differences, enabling more current to flow to the piping being protected. This is referred to as impressed current cathodic protection. In these systems DC current with an output of up to 50 amperes and 50 volts is used to protect the piping system. The voltage depends on several factors, such as size of the pipeline, coating quality and environmental conditions. The typical configuration of the wiring for this system is connecting cables from the positive DC terminal of the rectifier to the anode group and the negative terminal to the pipe. Anodes can be installed in a ground bed consisting of vertical holes backfilled with a material that improves the performance and life of the anodes called coke. They may also be laid in a trench surrounded by coke and backfilled (see **Figure 1310.1.3.3b**).

It is important that these systems be designed by professionals possessing expertise in cathodic protection and knowledge of the piping systems to be protected.

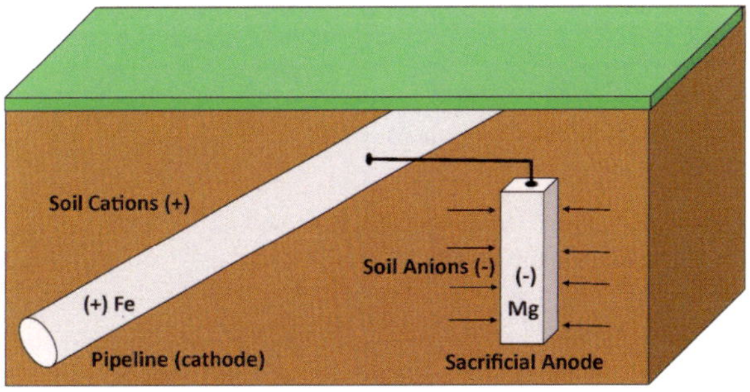

The magnesium rod (sacrificial anode) will protect the steel pipeline from corrosion. Magnesium is more easily oxidized than iron therefore acting as an anode in a galvanic cell. The steel pipeline becomes the cathode and oxygen is reduced protecting the steel pipeline.

FIGURE 1310.1.3.3A
GALVANIC CATHODIC PROTECTION

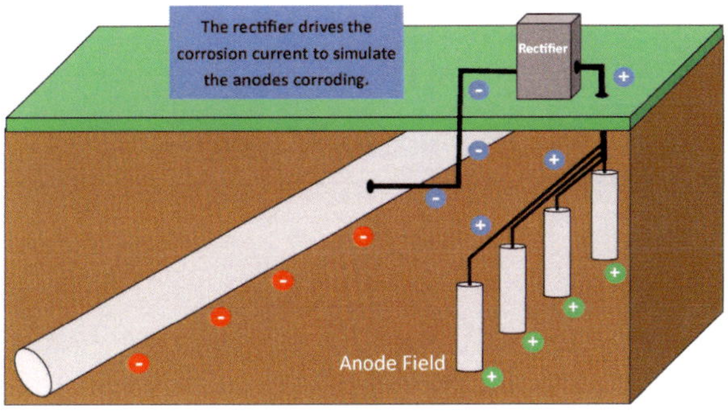

FIGURE 1310.1.3.3B
IMPRESSED CURRENT CATHODIC PROTECTION

1310.1.3.4 Sacrificial Anodes. Sacrificial anodes shall be tested in accordance with the following:

(1) Upon installation of the cathodic protection system, except where prohibited by climatic conditions, in which case the testing shall be performed not later than 180 days after the installation of the system.

(2) 12 to 18 months after the initial test.

(3) Upon successful verification testing in accordance with Section 1310.1.3.4(1) and Section 1310.1.3.4(2), periodic follow-up testing shall be performed at intervals not to exceed 36 months. [NFPA 54:7.1.3.4]

1310.1.3.5 System Failing Tests. Systems failing a test shall be repaired not more than 180 days after the date of the failed testing. The testing schedule shall be restarted as required in Section 1310.1.3.4(1) and Section 1310.1.3.4(2), and the results shall comply with Section 1310.1.3.3. [NFPA 54:7.1.3.5]

1310.1.3.6 Impressed Current Cathodic Protection. Impressed current cathodic protection systems shall be inspected and tested in accordance with the following schedule:

(1) The impressed current rectifier voltage output shall be checked at intervals not exceeding two months.

(2) The pipe-to-soil voltage shall be tested at least annually. [NFPA 54:7.1.3.6]

1310.1.3.7 Documentation. Documentation of the results of the two most recent tests shall be retained. [NFPA 54:7.1.3.7]

1310.1.3.8 Dissimilar Metals. Where dissimilar metals are joined underground, an insulating coupling or fitting shall be used. [NFPA 54:7.1.3.8]

1310.1.3.9 Steel Risers. Steel risers, other than anodeless risers, connected to plastic piping shall be cathodically protected by means of a welded anode. [NFPA 54:7.1.3.9]

1310.1.4 Protection Against Freezing. Where the formation of hydrates or ice is known to occur, piping shall be protected against freezing. [NFPA 54:7.1.4]

Natural gas may contain water vapor that could cause ice build-up if installed in areas with colder temperatures. If the gas used is not dry gas, then precautions must be taken to protect the gas from freezing. Ice problems are all but nonexistent with pipeline natural gas and propane, but they can still exist in gas from alternative sources such as local wells, landfills or waste treatment plants. Consult the gas supplier about the possibility of ice.

1310.1.5 Piping Through Foundation Wall. Underground piping, where installed through the outer foundation or basement wall of a building, shall be encased in a protective sleeve or protected by an approved device or method. The space between the gas piping and the sleeve and between the sleeve and the wall shall be sealed to prevent entry of gas and water. [NFPA 54:7.1.5]

If gas leaks in outdoor, underground piping it can migrate along the trench and into the basement instead of venting to the atmosphere above grade. This condition can be extremely hazardous. The requirement to seal the casing against water entry is in keeping with the desire to exercise good workmanship. See **Figure 1310.1.5** for an illustration of a below-grade foundation wall penetration.

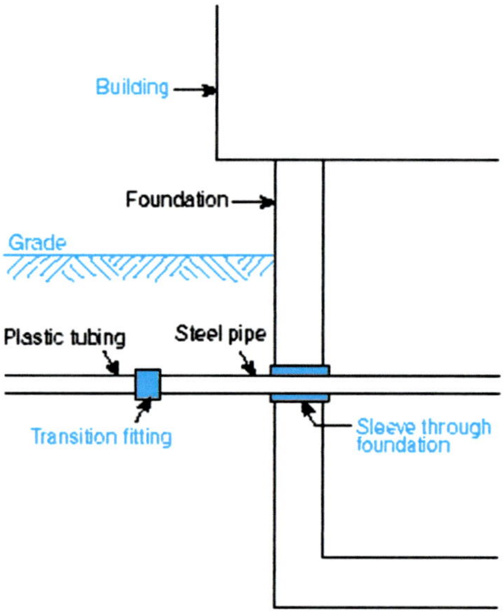

FIGURE 1310.1.5
PIPING THROUGH A FOUNDATION WALL

The use of an inert wrapping material provides an additional method of protection. The material must be of an approved type and be suitable for the application. The material must also be able to seal around the pipe and the wall to prevent the entry of gas or water.

1310.1.6 Piping Underground Beneath Buildings. Where gas piping is installed underground beneath buildings, the piping shall be either of the following:

(1) Encased in an approved conduit designed to withstand the imposed loads and installed in accordance with Section 1310.1.6.1 or Section 1310.1.6.2.

(2) A piping/encasement system listed for installation beneath buildings. [NFPA 54:7.1.6]

Gas leaking from a pipe installed underneath a building can migrate into the building. The code requires that any such pipes be cased and that the annular space between the pipe and the case be vented to the outdoors. Although the resulting installation is perfectly safe, the installation is difficult and the repairs are more so. For these reasons, practicality demands that most pipes run through buildings, rather than under them (see **Figure 1310.1.6**).

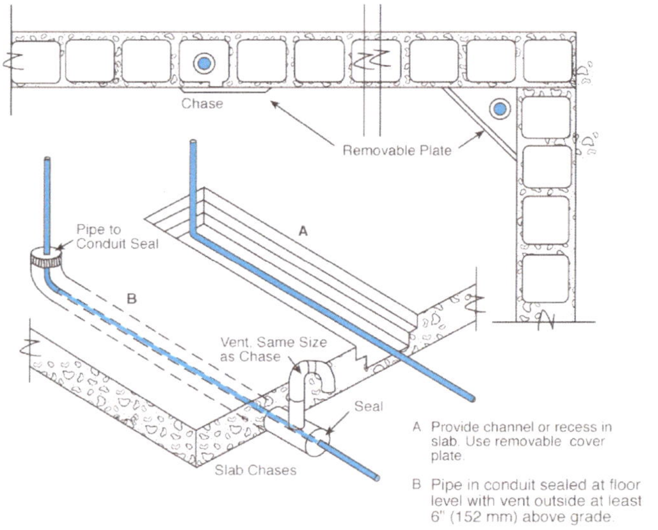

FIGURE 1310.1.6
APPROVED PIPING MAY BE INSTALLED IN OTHER LOCATIONS WITH PERMISSION FROM THE AUTHORITY HAVING JURISDICTION

1310.1.6.1 Conduit with One End Terminating Outdoors. The conduit shall extend into an accessible portion of the building and, at the point where the conduit terminates in the building, the space between the conduit and the gas piping shall be sealed to prevent the possible entrance of any gas leakage. Where the end sealing is of a type that retains the full pressure of the pipe, the conduit shall be designed for the same pressure as the pipe. The conduit shall extend at least 4 inches (102 mm) outside the building, be vented outdoors above finished ground level, and be installed so as to prevent the entrance of water and insects. [NFPA 54:7.1.6.1]

1310.1.6.2 Conduit with Both Ends Terminating Indoors. Where the conduit originates and terminates within the same building, the conduit shall originate and terminate in an accessible portion of the building and shall not be sealed. [NFPA 54:7.1.6.2]

1310.1.7 Plastic Piping. Plastic piping shall be installed outdoors, underground only.

Exceptions:

(1) Plastic piping shall be permitted to terminate aboveground where an anodeless riser is used.

(2) Plastic piping shall be permitted to terminate with a wall head adapter aboveground in buildings, including basements, where the plastic piping is inserted in a piping material permitted for use in buildings. [NFPA 54:7.1.7.1]

Anodeless risers, addressed in Exception (1), are constructed with the plastic pipe running inside steel all the way to the top end, where the pressure seal and the transition from plastic to steel are located (see **Figures 1310.1.7a and 1310.1.7b**).

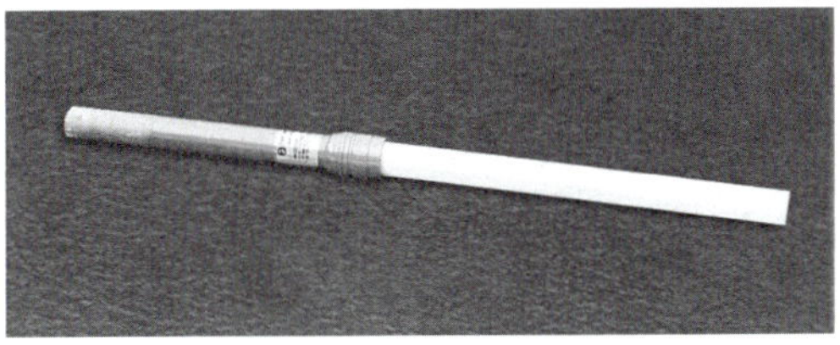

FIGURE 1310.1.7A
TRANSITION FITTING

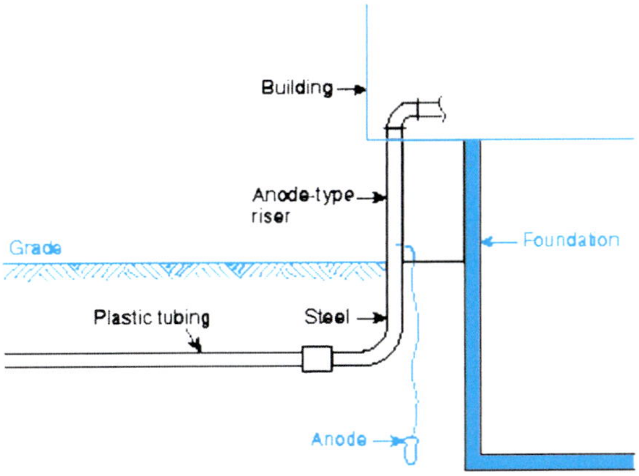

FIGURE 1310.1.7B
ANODE-TYPE RISER INSTALLATION

Exception (2) permits plastic gas piping to be terminated above ground and outside of buildings. This exception permits the use of fittings, known as "wall head adapters," to terminate buried plastic pipe inside of buildings in basement areas. Wall head adapters (see **Figure 1310.1.7c**) often are used by utilities for insertion renewals of steel service lines. Within the scope of this code, wall head adapters can be used for renewals of steel pipes from meters at the property line or in renewals of existing underground steel pipes between buildings, as well as for making a clean-sleeved, below-grade foundation wall penetration.

Wall head adapters (also called "service head adapters") are used to install polyethylene pipe and tubing beyond the gas meter. For more than 30 years, polyethylene has been used successfully in gas distribution service upstream of the gas meter.

Plastic pipe and tubing is more susceptible to inadvertent damage during installation than most metallic pipe. For this reason, special attention should be given to proper compaction below the pipe, to the elimination of shear points on connections during backfilling and to the materials in the backfill, making certain that angular and large materials are not used near the pipe.

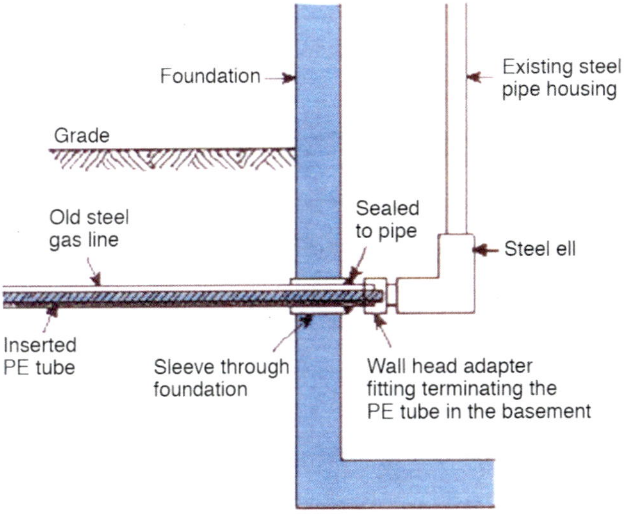

FIGURE 1310.1.7C
WALL HEAD ADAPTER

1310.1.7.1 Connections Between Metallic and Plastic Piping. Connections made between metallic and plastic piping shall be made with fittings conforming to one of the following:

(1) ASTM D2513 Category I transition fittings

(2) ASTM F1973

(3) ASTM F2509 [NFPA 54:7.1.7.2]

1310.1.7.2 Tracer Wire. An electrically continuous corrosion-resistant tracer shall be buried with the plastic pipe to facilitate locating. The tracer shall be one of the following:

(1) A product specifically designed for that purpose.

A product specifically designed for that purpose is not the typical yellow plastic tape printed with a warning that gas pipe is buried below and used as an indicator. The product the UMC refers to is tape that is manufactured with the capacity to identify the specific location of buried plastic gas piping.

(2) Insulated copper conductor not less than 14 AWG.

Where tracer wire is used, access shall be provided from aboveground or one end of the tracer wire or tape shall be brought aboveground at a building wall or riser. [NFPA 54:7.1.7.3 – 7.1.7.3.2]

1310.2 CSST Piping Systems. CSST piping systems shall be installed in accordance with this code and the manufacturer's installation instructions. [NFPA 54:7.1.8]

1310.3 Installation of Aboveground Piping. Piping

installed aboveground shall be securely supported and located where it will be protected from physical damage. Where passing through an exterior wall, the piping shall also be protected from corrosion by coating or wrapping with an inert material approved for such applications. The piping shall be sealed around its circumference at the point of the exterior penetration to prevent the entry of water, insects, and rodents. Where piping is encased in a protective pipe sleeve, the annular spaces between the gas piping and the sleeve and between the sleeve and the wall opening shall be sealed. [NFPA 54:7.2.1]

Piping that passes through an outside wall must be protected from corrosion because this area can collect water. In addition, piping in contact with building materials can corrode due to chemical reactions with those materials or due to galvanic corrosion from contact with metallic siding. The inert wrapping material must be approved for these applications.

The requirement that the pipe be sealed at the wall is often overlooked. Usually the pipe is protected only when the pipe passes through a concrete wall; however, when the pipe passes through a noncorrosive wall, such as wood, it must be sealed to prevent the entry of insects and water.

1310.3.1 Protective Coating. Where piping is in contact with a material or an atmosphere corrosive to the piping system, the piping and fittings shall be coated with a corrosion-resistant material. Any such coating used on piping or components shall not be considered as adding strength to the system. [NFPA 54:7.2.2]

Materials commonly used for protective coating include paint, polyethylene jacketing over a mastic primer and fusion-bonded epoxy. Generally, paint is used on above-grade applications and the others on underground pipe. Using a corrosion-resistant material for these applications is often more satisfactory than providing a corrosion-protection system that inherently has a limited life. When using coated steel pipe underground, consideration generally is given to using cathodic protection as the means of protection from corrosion.

There are two types of cathodic protection systems: impressed current systems, which require a DC power source, and passive systems, which rely on a sacrificial material (zinc and aluminum are commonly used).

1310.3.2 Building Structure. The installation of gas piping shall not cause structural stresses within building components to exceed allowable design limits. Approval shall be obtained before any beams or joists are cut or notched. [NFPA 54:7.2.3.1 – 7.2.3.2]

The installation of gas piping should not cause damage to the building. Before granting approval to cut or notch beams or joists, the AHJ should be satisfied that the installer has consulted with the building architect, structural engineer, building official or builder. Review Section 316.9 and commentary.

1310.3.3 Gas Piping to be Sloped. Piping for other than dry gas conditions shall be sloped not less than ¼ inch in 15 feet (1.4 mm/m) to prevent traps. [NFPA 54:7.2.4]

Piping must be sloped to prevent low spots or traps when delivering gases containing significant water vapor. Pipeline-quality natural gas is considered to be dry. The installer determines the best way to install the pipe to prevent traps. In general, the horizontal piping should be sloped so that any accumulation of liquids will be trapped by the drips or sediment traps.

1310.3.3.1 Ceiling Locations. Gas piping shall be permitted to be installed in accessible spaces between a fixed ceiling and a dropped ceiling, whether or not such spaces are used as a plenum. Valves shall not be located in such spaces.

Exception: Appliance or equipment shutoff valves required by this code shall be permitted to be installed in accessible spaces containing vented appliances.

1310.3.4 Prohibited Locations. Gas piping inside any building shall not be installed in or through a clothes chute, chimney or gas vent, dumbwaiter, elevator shaft, or air duct, other than combustion air ducts. [NFPA 54:7.2.5]

Exception: Ducts used to provide ventilation air in accordance with Section 701.0 or to above-ceiling spaces in accordance with Section 1310.3.3.1.

1310.3.5 Hangers, Supports, and Anchors. Piping shall be supported with metal pipe hooks, metal pipe straps, metal bands, metal brackets, metal hangers, or building structural components, suitable for the size of piping, of adequate strength and quality, and located at intervals so as to prevent or damp out excessive vibration. Piping shall be anchored to prevent undue strains on connected appliances and equipment and shall not be supported by other piping. Pipe hangers and supports shall conform to the requirements of MSS SP-58. [NFPA 54:7.2.6.1]

Table 1310.3.5.1 lists the minimum distances allowable for support of piping, including steel pipe and smooth-wall tubing. Threaded pipe should also be supported at each change of direction to help support the fittings used. The pipe and tubing at vertical drops to appliances should be supported by hangers and not by the appliance it connects to. When installing vertical gas piping, it is important that the piping be supported and secured to a wall.

1310.3.5.1 Spacing. Spacing of supports in gas piping installations shall not be greater than shown in Table 1310.3.5.1. Spacing of supports of CSST shall be in accordance with the CSST manufacturer's instructions. [NFPA 54:7.2.6.2]

TABLE 1310.3.5.1
SUPPORT OF PIPING
[NFPA 54: TABLE 7.2.6.2]

STEEL PIPE, NOMINAL SIZE OF PIPE(inches)	SPACING OF SUPPORTS (feet)	NOMINAL SIZE OF TUBING SMOOTH-WALL (inches O.D.)	SPACING OF SUPPORTS (feet)
½	6	½	4
¾ or 1	8	⅝ or ¾	6
1¼ or larger (horizontal)	10	⅞ or 1 (horizontal)	8
1¼ or larger (vertical)	Every floor level	1 or larger (vertical)	Every floor level

For SI units: 1 inch = 25 mm, 1 foot = 304.8 mm

1310.3.5.2 Expansion and Contraction. Supports, hangers, and anchors shall be installed so as not to interfere with the free expansion and contraction of the piping between anchors. All parts of the supporting system shall be designed and installed so they are not disengaged by movement of the supported piping. [NFPA 54:7.2.6.3]

1310.3.5.3 Piping on Roof Tops. Gas piping installed on the roof surfaces shall be elevated above the roof surface and shall be supported in accordance with Table 1310.3.5.1. [NFPA 54:7.2.6.4]

1310.3.6 Removal of Piping. Where piping containing gas is to be removed, the line shall be first disconnected from sources of gas and then thoroughly purged with air, water, or inert gas before cutting or welding is done.

Residual gas fumes remaining in the fuel-gas piping system must be removed when any type of repair or replacement occurs that will include fire or sparks. Although this section permits the use of air for purging, an inert gas or water may provide an increase in safety for larger volume systems because it will not result in a flammable gas/air mixture. Cutting is not limited to flame cutting because mechanical methods, such as cutting with a chop saw, can produce sparks that can ignite a gas/air mixture.

1310.4 Concealed Piping in Buildings. Gas piping in concealed locations shall be installed in accordance with this section. [NFPA 54:7.3.1]

1310.4.1 Connections. Where gas piping is to be concealed, connections shall be of the following type:

(1) Pipe fittings, such as elbows, tees, couplings, and right/left nipple/couplings.

(2) Joining tubing by brazing (see Section 1308.5.7.1).

(3) Press-connect fittings listed to CSA LC 4.

(4) CSST fittings listed to CSA LC 1.

Fittings manufactured to CSA LC1 (CSST fittings) or CSA LC4 (press connect fittings) should be installed according to the manufacturer's installation requirements. These requirements are separate and distinct for each type of system.

(5) Where necessary to insert fittings in gas pipe that has been installed in a concealed location, the pipe shall be reconnected by welding, flanges, or the use of a right/left nipple/coupling.

The list of approved fittings in this subsection consists of those that are least likely to leak in concealed installations. Left and right nipple and couplings therefore, are acceptable in this type of installation.

Fittings that are specifically listed and approved for use in concealed locations are a part of a listed system designed to meet the concerns that prohibit other tubing fittings in concealed locations.

1310.4.2 Piping in Partitions. Concealed gas piping shall not be located in solid partitions. [NFPA 54:7.3.3]

1310.4.3 Tubing in Partitions. This provision shall not apply to tubing that pierces walls, floors, or partitions. Tubing installed vertically and horizontally inside hollow walls or partitions without protection along its entire concealed length shall meet the following requirements:

(1) A steel striker barrier not less than 0.0508 of an inch (1.3 mm) thick, or equivalent, is installed between the tubing and the finished wall and extends at least 4 inches (102 mm) beyond concealed penetrations of plates, firestops, wall studs, and so on.

(2) The tubing is installed in single runs and is not rigidly secured. [NFPA 54:7.3.4]

This section applies to tubing within a wall or partition (see **Figure 1310.4.3**). Material of 0.0508-inch thickness is approximately 16 gauge. Plates are required near penetrations. At this point, nails and screws are more apt to be able to pin down and penetrate a tube because it is restrained by the penetration. This hazard is also the basis for the requirement that tubing away from penetrations must not be secured. If the tube is not secured, it can move. This protection is not only for initial construction but for the home or building owner who may want to attach something to the wall. Some jurisdictions may require the same protection for steel piping.

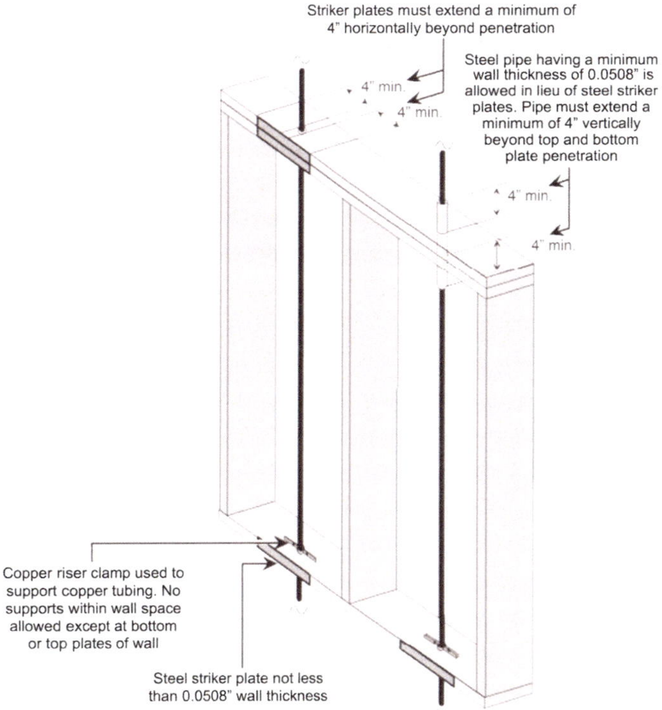

FIGURE 1310.4.3
TUBING INSTALLATION WITHIN PARTITIONS

1310.4.4 Piping in Floors. In industrial occupancies, gas piping in solid floors such as concrete shall be laid in channels in the floor and covered to permit access to the piping with a minimum of damage to the building. Where piping in floor channels could be exposed to excessive moisture or corrosive substances, the piping shall be protected in an approved manner. [NFPA 54:7.3.5.1]

Exception: In other than industrial occupancies and where approved by the Authority Having Jurisdiction, gas piping embedded in concrete floor slabs constructed with Portland cement shall be surrounded with a minimum of 1½ inches (38 mm) of concrete and shall not be in physical contact with other metallic structures such as reinforcing rods or electrically neutral conductors. All piping, fittings, and risers shall be protected against corrosion in accordance with Section 1308.5.6. Piping shall not be embedded in concrete slabs containing quick-set additives or cinder aggregate. [NFPA 54:7.3.5.2]

Concrete containing cinder aggregate or quickset additives, usually calcium chloride, is not permitted because such products attack the metals from which gas piping is made.

It is important to remember that different requirements apply to piping in floors in industrial occupancies than in other occupancies. In industrial occupancies, access to the piping is necessary because corrosive materials that can damage piping are often used in industrial occupancies.

1310.5 Piping in Vertical Chases. Where gas piping exceeding 5 psi (34 kPa) is located within vertical chases in accordance with Section 1310.6, the requirements of Section 1310.5.1 through Section 1310.5.3 shall apply. [NFPA 54:7.4]

Special requirements for fuel-gas piping systems containing gas pressures above 5 psi are needed due to the dangers of what a leak in this system might do. A leak in this high-pressure system would allow far greater quantities of gas into the building. For this reason, the piping must be welded rather than using another joining system. The piping must be located in a ventilated chase rather than in walls and through trusses where normal fuel-gas systems are installed.

1310.5.1 Pressure Reduction. Where pressure reduction is required in branch connections for compliance with Section 1310.6, such reduction shall take place either inside the chase or immediately adjacent to the outside wall of the chase. Regulator venting and downstream overpressure protection shall comply with Section 1308.7.4, Section 1308.7.5 and Section 1308.8 through Section 1308.10.7. The regulator shall be accessible for service and repair, and vented in accordance with one of the following:

(1) Where the fuel gas is lighter than air, regulators equipped with a vent limiting means shall be permitted to be vented into the chase. Regulators not equipped with a vent limiting means shall be permitted to be vented either directly to the outdoors or to a point within the top 1 foot (305 mm) of the chase.

(2) Where the fuel gas is heavier than air, the regulator vent shall be vented only directly to the outdoors. [NFPA 54:7.4.1]

1310.5.2 Chase Construction. Chase construction shall comply with local building codes with respect to fire resistance and protection of horizontal and vertical openings. [NFPA 54:7.4.2]

1310.5.3 Ventilation. A chase shall be ventilated to the outdoors and only at the top. The opening(s) shall have a minimum free area [in square inches (square meters)] equal to the product of one-half of the maximum pressure in the piping [in pounds per square inch (kilopascals)] times the largest nominal diameter of that piping [in inches (millimeters)], or the cross-sectional area of the chase, whichever is smaller. Where more than one fuel gas piping system is present, the free area for each system shall be calculated and the largest area used. [NFPA 54:7.4.3]

1310.6 Maximum Operating Pressure in Buildings. The maximum operating pressure for any piping systems located inside buildings shall not exceed 5 psi (34 kPa) unless one or more of the following conditions are met:

(1) The piping joints are welded or brazed.

(2) The piping joints are flanged and all pipe-to-flange connections are made by welding or brazing.

(3) The piping is located in a ventilated chase or otherwise enclosed for protection against accidental gas accumulation.

(4) The piping is located inside buildings or separate areas of buildings used exclusively for one of the following:

   (a) Industrial processing or heating

   (b) Research

   (c) Warehousing

   (d) Boiler or mechanical rooms

(5) The piping is a temporary installation for buildings under construction.

(6) The piping serves appliances or equipment used for agricultural purposes.

(7) The piping system is an LP-Gas piping system with an operating pressure greater than 20 psi (138 kPa) and complies with NFPA 58. [NFPA 54:5.5.4]

1310.6.1 LP-Gas Systems. LP-Gas systems designed to operate below -5°F (-21°C) or with butane or a propane-butane mix shall be designed to either accommodate liquid LP-Gas or to prevent LP-Gas vapor from condensing back into a liquid. [NFPA 54:5.5.5]

1310.7 Appliance Overpressure Protection. The maximum operating pressure for piping systems serving appliances designed to operate at 14 inches water column (3.5 kPa) inlet pressure or less shall be 2 pounds-force per square inch gauge (psig) (14 kPa) unless an over pressure protection device designed to limit pressure at the appliance to 2 psig (14 kPa) upon failure of the line gas pressure regulator is installed.

1310.8 Gas Pipe Turns. Changes in direction of gas pipe shall be made by the use of fittings, factory bends, or field bends. [NFPA 54:7.5]

1310.8.1 Metallic Pipe. Metallic pipe bends shall comply with the following:

(1) Bends shall be made only with bending tools and procedures intended for that purpose.

(2) All bends shall be smooth and free from buckling, cracks, or other evidence of mechanical damage.

(3) The longitudinal weld of the pipe shall be near the neutral axis of the bend.

(4) Pipe shall not be bent through an arc of more than 90 degrees.

(5) The inside radius of a bend shall be not less than 6 times the outside diameter of the pipe. [NFPA 54:7.5.1]

This section allows a method of bending welded pipe without stressing the longitudinal weld. The neutral axis of the bend is the area where the metal is not stretched or compressed during the bending operation. The axis lies at the sides of the pipe at the approximate midpoint between the inside and the outside of the bend.

1310.8.2 Plastic Pipe. Plastic pipe bends shall comply with the following:

(1) The pipe shall not be damaged, and the internal diameter of the pipe shall not be effectively reduced.

(2) Joints shall not be located in pipe bends.

(3) The radius of the inner curve of such bends shall not be less than 25 times the inside diameter of the pipe.

(4) Where the piping manufacturer specifies the use of special bending tools or procedures, such tools or procedures shall be used. [NFPA 54:7.5.2]

The practice of bending metallic pipe, especially steel pipe, is not used much these days; however, plastic pipe is bent on a regular basis. Gas-piping mains and service pipe are often installed in polyethylene pipe. Turns made by bending the pipe are the norm and the manufacturer's installation instructions must always be followed.

1310.8.3 Elbows. Factory-made welding elbows or transverse segments cut therefrom shall have an arc length measured along the crotch of at least 1 inch (25 mm) for pipe sizes 2 inches (50 mm) and larger. [NFPA 54:7.5.3]

See **Figure 1310.8.3**.

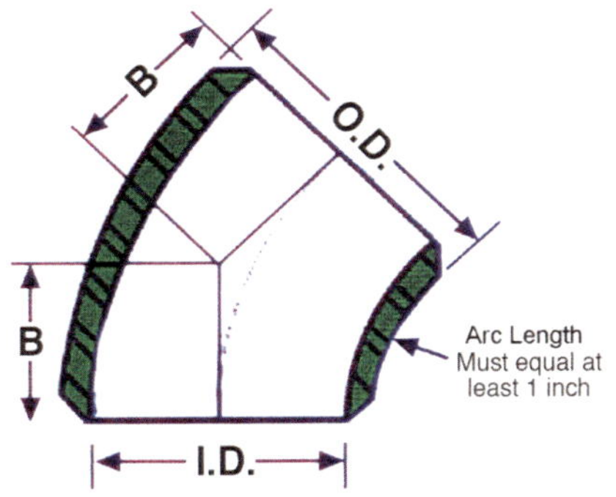

FIGURE 1310.8.3
ARC LENGTH OF CUT FITTING

1310.9 Drips and Sediment Traps. For other than dry gas conditions, a drip shall be provided at any point in the line of pipe where condensate could collect. Where required by the Authority Having Jurisdiction or the serving gas supplier, a drip shall also be provided at the outlet of the meter. This drip shall be installed so as to constitute a trap wherein an accumulation of condensate shuts off the flow of gas before it runs back into the meter. [NFPA 54:7.6.1]

Drips collect condensate from gases that contain condensable products (usually water) or oil added in the gas transmission process and are required by the code and if required by the serving gas supplier or the AHJ (see **Figure 1310.9**).

1310.9.1 Location of Drips. All drips shall be installed only in such locations that they are readily accessible to permit cleaning or emptying. A drip shall not be located where the condensate is likely to freeze. [NFPA 54:7.6.2]

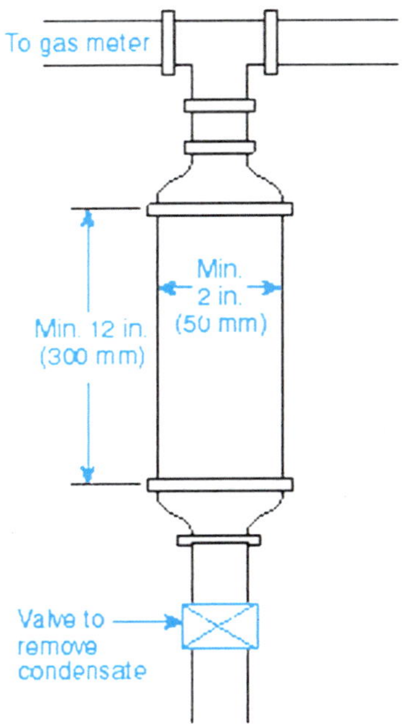

FIGURE 1310.9
DRIP TRAP

Drips are usually located near the outlet of the meter or the service entrance and at other locations where condensate could collect. The drip is installed at the bottom of a downward-flowing line by placing a tee at the bottom of the line, installing a nipple and cap in the run of the tee and continuing the pipe run out the side of the tee. Drips must be located so that they can be emptied to prevent liquid from causing an obstruction to the flow of gas. Drip locations must also be protected from freezing.

1310.9.2 Sediment Traps. The installation of sediment traps shall be in accordance with Section 1312.9. [NFPA 54:7.6.3]

Sediment traps are referenced to Section 1312.9 so that the requirements are not duplicated.

Sediment traps, sometimes confused with drip traps, are installed to collect solid foreign particles to prevent such material from entering close-fitting parts or small passageways (e.g., valves and orifices). Although some gases can contain foreign solids, this situation is not likely to be a problem in either utility gas or LP-gas because of the methods and equipment used in handling these products. Dirt and pipe material cuttings that make their way into the system during construction and are present in limited amounts are the target of sediment traps. Thus, sediment traps seldom need to be opened for service or cleaning.

Many appliance manufacturers provide sediment traps in their appliances, and a number of the ANSI Z21 standards for gas appliances require the installation of a sediment trap.

Section 1312.9 requires the installation of a sediment trap at the time of installation of most appliances if the appliance is not already equipped with one.

1310.10 Outlets. Outlets shall be located and installed in accordance with the following requirements:

(1) The outlet fittings or piping shall be securely fastened in place.

(2) Outlets shall not be located behind doors.

(3) Outlets shall be located far enough from floors, walls, patios, slabs, and ceilings to permit the use of wrenches without straining, bending, or damaging the piping.

(4) The unthreaded portion of gas piping outlets shall extend not less than 1 inch (25.4 mm) through finished ceilings or indoor or outdoor walls.

(5) The unthreaded portion of gas piping outlets shall extend not less than 2 inches (51 mm) above the surface of floors or outdoor patios or slabs.

(6) The provisions of Section 1310.10(4) and Section 1310.10(5) shall not apply to listed quick-disconnect devices of the flush-mounted type or listed gas convenience outlets. Such devices shall be installed in accordance with the manufacturer's installation instructions. [NFPA 54:7.7.1.1 – 7.7.1.6]

1310.10.1 Cap Outlets. Each outlet, including a valve, shall be closed gastight with a threaded plug or cap immediately after installation and shall be left closed until the appliance or equipment is connected thereto. When an appliance or equipment is disconnected from an outlet and the outlet is not to be used again immediately, it shall be capped or plugged gastight.

Exceptions:

(1) Laboratory appliances installed in accordance with Section 1312.3.1 shall be permitted.

(2) The use of a listed quick-disconnect device with integral shutoff or listed gas convenience outlet shall be permitted. [NFPA 54:7.7.2.1]

A plug or cap is required for all gas pipe openings. Closing a valve is not enough to satisfy this requirement because the valve can be opened inadvertently or accidentally. No temporary or makeshift closure is permitted because anything except a properly tightened plug or cap could leak (see **Figure 1310.10.1**). Listed quick-disconnect devices and gas convenience outlets are permitted to go uncapped or unplugged because they are required by their listing to have valves that will automatically shut off the gas either prior to or during the disconnect.

Exception (1) is made for laboratory equipment, such as a Bunsen burner, that is connected and disconnected from hose-end valves as a normal, everyday operation.

1310.10.1.1 Appliance Shutoff Valves. Appliance shutoff valves installed in fireplaces shall be removed and the piping capped gastight where the fireplace is used for solid fuel burning. [NFPA 54:7.7.2.2]

If a fireplace is converted to solid fuel burning, any appliance shutoff valve installed in the fireplace is required to be removed and the gas pipe is to be capped gastight. Experience has shown that these valves will leak when subjected to the heat of a wood fire.

1310.11 Manual Gas Shutoff Valves. An accessible gas shutoff valve shall be provided upstream of each gas pressure regulator. Where two gas pressure regulators are installed in series in a single gas line, a manual valve shall not be required at the second regulator. [NFPA 54:7.8.1]

FIGURE 1310.10.1
MULTIPLE OUTLETS CAPPED AND READY FOR TEST

1310.11.1 Valves Controlling Multiple Systems. Main gas shutoff valves controlling several gas piping systems shall be readily accessible for operation and installed so as to be protected from physical damage. They shall be marked with a metal tag or other permanent means attached by the installing agency so that the gas piping systems supplied through them can be readily identified. [NFPA 54:7.8.2.1]

1310.11.1.1 Shutoff Valves for Multiple House Lines. In multiple-tenant buildings supplied through a master meter, through one service regulator where a meter is not provided, or where meters or service regulators are not readily accessible from the appliance or equipment location, an individual shutoff valve for each apartment or tenant line shall be provided at a convenient point of general accessibility. In a common system serving a number of individual buildings, shutoff valves shall be installed at each building. [NFPA 54:7.8.2.2]

The reason for requiring a shutoff valve for each line is to allow for the system to be separated for repair or maintenance (see **Figure 1310.11.1.1**). In a case where only part of the system has to be shut down for repair, the remaining system can continue to operate. Shutoff valves must be plainly marked and accessible so that, in the future, service persons can determine which lines service which units or buildings.

The requirement of a shutoff valve for each building provides emergency responders with the ability to shut off the flow of gas to a building involved in a fire. In natural gas installations, this shutoff valve is part of the utilities service and meter installation. The natural gas supplier provides a shutoff valve for the supply to its gas meter under the U.S. Department of Transportation regulations (49 CFR 192).

The installations just described are presumed to be

obvious enough to meet the requirement of being "plainly marked." The requirement of posting the location is not treated uniformly by different gas suppliers and local fire departments. Where an installation is not obvious (e.g., an underground propane tank or a meter/regulator concealed by vegetation), a sign or other marking may be needed or required by the AHJ. The intent of this provision is to enable emergency responders to shut off the gas supply to a building in the event of a fire. The local fire department is normally the AHJ for this requirement.

In LP-gas systems, the emergency shutoff valve is normally the propane tank shutoff valve.

FIGURE 1310.11.1.1
MANUAL GAS SHUTOFF VALVE

1310.11.2 Emergency Shutoff Valves. An exterior shutoff valve to permit turning off the gas supply to each building in an emergency shall be provided. The emergency shutoff valves shall be plainly marked as such and their locations posted as required by the Authority Having Jurisdiction. [NFPA 54:7.8.2.3]

The emergency shutoff valve required in this section can be the same valve that is required in Section 1311.11.1.1 for multiple house lines. If the natural gas supplier installs a shutoff valve and gas meter (or shutoff valve if no meter is provided), then the shutoff valve can also serve as the emergency shutoff valve. If the fuel line is run from this location to another building underground, the line must be brought up outside and another shutoff valve installed. This second shutoff will be the emergency shutoff valve and the valve controlling multiple systems. In any piping system, there must be an accessible exterior shutoff valve.

1310.11.3 Shutoff Valve for Laboratories. Each laboratory space containing two or more gas outlets installed on tables, benches, or in hoods in educational, research, commercial, and industrial occupancies shall have a single shutoff valve through which all such gas outlets are supplied. The shutoff valve shall be accessible, located within the laboratory or adjacent to the laboratory's egress door, and identified. [NFPA 54:7.8.2.4]

1310.12 Prohibited Devices. Devices shall not be placed within the interior of gas piping or fittings where such devices reduce the cross-sectional area or otherwise obstruct the free flow of gas, except where allowance in the piping system design has been made for such devices. [NFPA 54:7.9]

1310.13 Systems Containing Gas-Air Mixtures Outside the Flammable Range. Where gas-air mixing machines are employed to produce mixtures above or below the flammable range, they shall be provided with stops to prevent adjustment of the mixture to within or approaching the flammable range. [NFPA 54:7.10]

1310.14 Systems Containing Flammable Gas-Air Mixtures. Systems containing flammable gas-air mixtures shall be in accordance with Section 1310.14.1 through Section 1310.14.6.

1310.14.1 Required Components. A central premix system with a flammable mixture in the blower or compressor shall consist of the following components:

(1) Gas-mixing machine in the form of an automatic gas-air proportioning device combined with a downstream blower or compressor.
(2) Flammable mixture piping, minimum Schedule 40.
(3) Automatic firecheck(s).
(4) Safety blowout(s) or backfire preventers for systems utilizing flammable mixture lines above 2½ inches (64 mm) nominal pipe size or the equivalent. [NFPA 54:7.11.1]

1310.14.2 Optional Components. The following components shall also be permitted to be utilized in any type of central premix system:

(1) Flowmeter(s)
(2) Flame arrester(s) [NFPA 54:7.11.2]

1310.14.3 Additional Requirements. Gas-mixing machines shall have nonsparking blowers and shall be constructed so that a flashback does not rupture machine casings. [NFPA 54:7.11.3]

1310.14.4 Special Requirements for Mixing Blowers. A mixing blower system shall be limited to applications with minimum practical lengths of mixture piping, limited to a maximum mixture pressure of 10 inches water column (2.5 kPa) and limited to gases containing no more than 10 percent hydrogen. The blower shall be equipped with a gas control valve at its air entrance arranged so that gas is admitted to the airstream, entering the blower in proper proportions for correct combustion by the type of burners employed, the said gas control valve being of either the zero governor or mechanical ratio valve type that controls the gas and air adjustment simultaneously. No valves or other obstructions shall be installed between the blower discharge and the burner or burners. [NFPA 54:7.11.4]

1310.14.5 Installation of Gas-Mixing Machines. Installation of gas-mixing machines shall comply with the following:

(1) The gas-mixing machine shall be located in a well-ventilated area or in a detached building or cutoff room provided with room construction and explosion vents in accordance with sound engineering principles. Such rooms or below-grade installations shall have adequate positive ventilation.

(2) Where gas-mixing machines are installed in well-ventilated areas, the type of electrical equipment shall be in accordance with NFPA 70 for general service conditions unless other hazards in the area prevail. Where gas-mixing machines are installed in small detached buildings or cutoff rooms, the electrical equipment and wiring shall be installed in accordance with NFPA 70 for hazardous locations (Articles 500 and 501, Class I, Division 2).

(3) Air intakes for gas-mixing machines using compressors or blowers shall be taken from outdoors whenever practical.

(4) Controls for gas-mixing machines shall include interlocks and a safety shutoff valve of the manual reset type in the gas supply connection to each machine arranged to automatically shut off the gas supply in the event of high or low gas pressure. Except for open burner installations only, the controls shall be interlocked so that the blower or compressor stops operating following a gas supply failure. Where a system employs pressurized air, means shall be provided to shut off the gas supply in the event of air failure.

(5) Centrifugal gas-mixing machines in parallel shall be reviewed by the user and equipment manufacturer before installation, and means or plans for minimizing the effects of downstream pulsation and equipment overload shall be prepared and utilized as needed. [NFPA 54:7.11.5.1 – 7.11.5.5]

1310.14.6 Use of Automatic Firechecks, Safety Blowouts, or Backfire Preventers. Automatic firechecks and safety blowouts or backfire preventers shall be provided in piping systems distributing flammable air-gas mixtures from gas-mixing machines to protect the piping and the machines in the event of flashback, in accordance with the following:

(1) Approved automatic firechecks shall be installed upstream as close as practical to the burner inlets following the firecheck manufacturer's instructions.

(2) A separate manually operated gas valve shall be provided at each automatic firecheck for shutting off the flow of the gas-air mixture through the firecheck after a flashback has occurred. The valve shall be located upstream as close as practical to the inlet of the automatic firecheck.

Caution: These valves shall not be reopened after a flashback has occurred until the firecheck has cooled sufficiently to prevent re-ignition of the flammable mixture and has been reset properly.

(3) A safety blowout or backfiring preventer shall be provided in the mixture line near the outlet of each gas-mixing machine where the size of the piping is larger than 2½ inches (65 mm) NPS, or equivalent, to protect the mixing equipment in the event of an explosion passing through an automatic firecheck. The manufacturers' instructions shall be followed when installing these devices, particularly after a disc has burst. The discharge from the safety blowout or backfire preventer shall be located or shielded so that particles from the ruptured disc cannot be directed toward personnel. Wherever there are interconnected installations of gas-mixing machines with safety blowouts or backfire preventers, provision shall be made to keep the mixture from other machines from reaching any ruptured disc opening. Check valves shall not be used for this purpose.

(4) Large-capacity premix systems provided with explosion heads (rupture discs) to relieve excessive pressure in pipelines shall be located at and vented to a safe outdoor location. Provisions shall be provided for automatically shutting off the supply of the gas-air mixture in the event of rupture. [NFPA 54:7.11.6]

1311.0 Electrical Bonding and Grounding.

1311.1 Pipe and Tubing Other than CSST. Each above-ground portion of a gas piping system, other than CSST, that is likely to become energized shall be electrically continuous and bonded to an effective ground-fault current path. Gas piping, other than CSST, shall be considered to be bonded when it is connected to appliances that are connected to the appliance grounding conductor of the circuit supplying that appliance. [NFPA 54:7.12.1]

This is an electrical requirement designed to provide consistency between the National Fuel Gas Code and NFPA 70, National Electrical Code® (NEC®). This requirement recognizes the text of the NEC. There has been confusion on the meaning of this requirement and of the corresponding requirements of the NEC. The key phrase is "likely to become energized." There is normally one way that gas piping becomes electrically energized: through a failure of appliance wiring or the electrical components that energize the appliance and potentially any gas piping connected to it. If this failure occurs, the grounding electrode provided to the appliance (where a three-wire circuit is provided) will protect the gas piping from becoming electrically energized. If there are no appliances with electrical components connected to the piping system, then the gas-piping system is not likely to become energized.

The net result is that gas piping does not require a separate bonding connection unless the following situations occur:

1. There are gas appliances with electrical connections that are connected to ungrounded wiring systems (two-prong plugs).
2. There are reasons to believe that the gas-piping system can become grounded by a source of electricity outside the piping system. This is highly unlikely.

1311.2 Bonding of CSST Gas Piping. CSST gas piping systems, and gas piping systems containing one or more segments of CSST, shall be electrically continuous and bonded to the electrical service grounding electrode system or, where provided, lightning protection grounding electrode system. [NFPA 54:7.12.2]

CSST systems are required by the manufacturer to be electrically bonded (see **Figure 1311.2a**).

Grounding vs Bonding

Grounding is a type of bonding, in which the conductive objects are connected to the earth using a conductor or metallic rod. The connection between an electrical circuit or instrument and the earth is known as Grounding. Grounding,

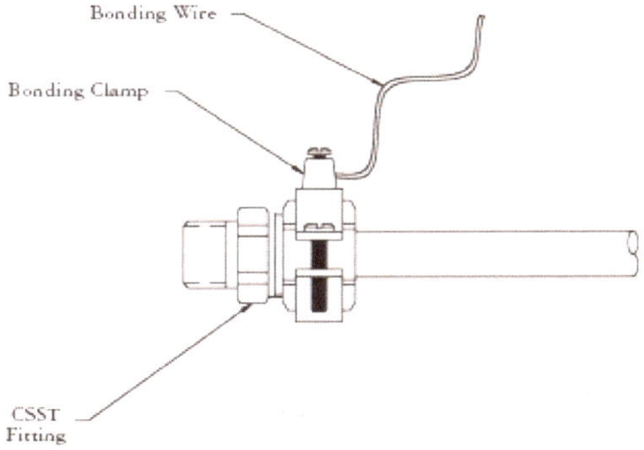

FIGURE 1311.2A
ELECTRICAL BONDING OF CSST SYSTEM

sometimes called Earthing, ensures that all metal parts of an electrical circuit that an individual might contact are connected to the earth, thus ensuring zero voltage.

Bonding, by definition, is the joining of metallic pieces to form a conducting path which ensures safe electrical continuity. Bonding is generally done as protection from electrical shocks. Two or more conductive objects are required for a bonding connection, which is accomplished by connecting the metallic pieces together by means of a conductor (wire). Bonding provides the safety in the case of fault current. In the case of CSST (corrugated stainless steel tubing) used in fuel gas piping, bonding may reduce the risk of damage and fire from lighting strike by reducing the risk of arcing. According to the National Electrical Code (NEC), gas piping systems are considered to be direct-bonded if connected to the electrical service equipment enclosure, the grounded conductor at the electrical service, the grounding electrode conductor or one of more of the grounding electrodes used.

A common bonding installation for single or multi-family structures would be a single connection on the steel nipple downstream of the gas meter and prior to the first CSST connection (see **Figure 1311.2b**). The conductor should be no smaller than a 6 AWG copper wire and should be attached in accordance with the requirements found in the NEC. Bonding clamps shall be installed in accordance with their listings per UL 467 and need to make metal-to-metal contact with a steel pipe component or the first CSST fitting (see **Figure 1311.2c**).

1311.2.1 Bonding Jumper Connection. The bonding jumper shall connect to a metallic pipe, pipe fitting, or CSST fitting. [NFPA 54:7.12.2.1]

1311.2.2 Bonding Jumper Size. The bonding jumper shall not be smaller than 6 AWG copper wire or equivalent. [NFPA 54:7.12.2.2]

1311.2.3 Bonding Jumper Length. The length of the jumper between the connection to the gas piping system and the grounding electrode system shall not exceed 75 feet (22 860 mm). Any additional grounding electrodes installed to meet this requirement shall be bonded to the electrical service grounding electrode system or, where provided, lightning protection grounding electrode system. [NFPA 54:7.12.2.3]

FIGURE 1311.2B
BONDING CONNECTION AT METER

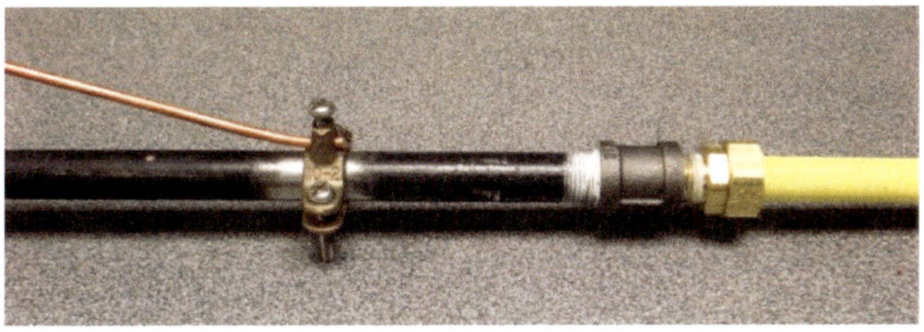

FIGURE 1311.2C
BONDING CLAMP AND CONDUCTOR

1311.2.4 Bonding Connections. Bonding connections shall be in accordance with NFPA 70. [NFPA 54:7.12.2.4]

1311.2.5 Devices Used for Bonding. Devices used for the bonding connection shall be listed for the application in accordance with UL 467. [NFPA 54:7.12.2.5]

1311.3 Arc-Resistant Jacketed CSST. CSST listed with an arc-resistant jacket or coating system in accordance with CSA LC 1 shall be electrically continuous and bonded to an effective ground fault current path. Where any CSST component of a piping system does not have an arc-resistant jacket or coating system, the bonding requirements of Section 1311.2 shall apply. Arc-resistant jacketed CSST shall be considered to be bonded when it is connected to appliances that are connected to the appliance grounding conductor of the circuit supplying that appliance. [NFPA 54:7.12.3]

1311.4 Prohibited Use. Gas piping shall not be used as a grounding conductor or electrode. [NFPA 54:7.12.4]

1311.5 Lightning Protection System. Where a lightning protection system is installed, the bonding of the gas piping shall be in accordance with NFPA 780. [NFPA 54:7.12.5]

1311.6 Electrical Circuits. Electrical circuits shall not utilize gas piping or components as conductors.

Exception: Low-voltage (50V or less) control circuits, ignition circuits, and electronic flame detection device circuits shall be permitted to make use of piping or components as a part of an electric circuit. [NFPA 54:7.13]

1311.7 Electrical Connections. All electrical connections between wiring and electrically operated control devices in a piping system shall conform to the requirements of NFPA 70. [NFPA 54:7.14.1]

1311.7.1 Safety Control. Any essential safety control depending on electric current as the operating medium shall be of a type that shuts off (fail safe) the flow of gas in the event of current failure. [NFPA 54:7.14.2]

The safest way to ensure that the fuel-gas system is free from becoming energized is to follow the manufacturer's installation requirements for the appliance, this code for installation of the piping and the National Electrical Code (NEC) for installation and connection to electrical sources.

1312.0 Appliance and Equipment Connections to Building Piping.

1312.1 Connecting Appliances and Equipment. Appliances and equipment shall be connected to the building piping in compliance with Section 1312.6 through Section 1312.8 by one of the following:

(1) Rigid metallic pipe and fittings.

See **Figure 1312.1a**.

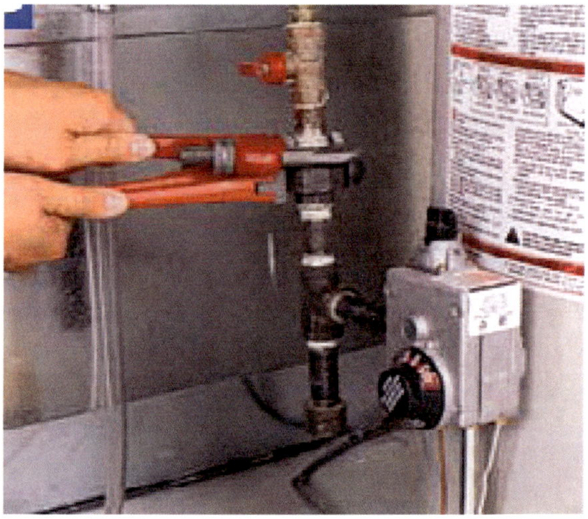

FIGURE 1312.1A
RIGID CONNECTION TO APPLIANCE

(2) Semirigid metallic tubing and metallic fittings. Aluminum alloy tubing shall not be used in exterior locations.

(3) A listed connector in compliance with CSA Z21.24. The connector shall be used in accordance with the manufacturer's installation instructions and shall be in the same room as the appliance. Only one connector shall be used per appliance.

See **Figure 1312.1b**.

(4) A listed connector in compliance with CSA Z21.75. Only one connector shall be used per appliance.

(5) CSST where installed in accordance with the manufacturer's installation instructions. CSST shall connect only to appliances that are fixed in place.

(6) Listed nonmetallic gas hose connectors in accordance with Section 1312.3.

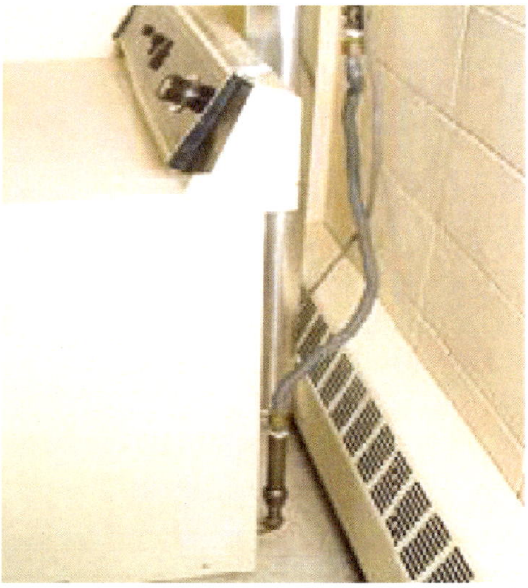

FIGURE 1312.1B
FLEXIBLE APPLIANCE CONNECTION

(7) Unlisted gas hose connectors for use in laboratories and educational facilities in accordance with Section 1312.4. [NFPA 54:9.6.1]

1312.1.1 Commercial Cooking Appliances. Connectors used with commercial cooking appliances that are moved for cleaning and sanitation purposes shall be installed in accordance with the connector manufacturer's installation instructions. Such connectors shall be listed in accordance with CSA Z21.69. [NFPA 54:9.6.1.3]

See **Figure 1312.1.1**.

FIGURE 1312.1.1
KITCHEN EQUIPMENT ON CASTERS WITH QUICK DISCONNECT AND SWIVEL

1312.1.2 Restraining Device. Movement of appliances with casters shall be limited by a restraining device installed in accordance with the connector and appliance manufacturer's installation instructions. [NFPA 54:9.6.1.4]

1312.2 Suspended Low-Intensity Infrared Tube Heaters. Suspended low-intensity infrared tube heaters shall be connected to the building piping system with a connector listed for the application in accordance with CSA Z21.24 as follows:

(1) The connector shall be installed in accordance with the tube heater installation instructions, and shall be in the same room as the appliance.

(2) Only one connector shall be used per appliance. [NFPA 54:9.6.1.5]

1312.3 Use of Nonmetalic Gas Hose Connectors. Listed gas hose connectors shall be used in accordance with the manufacturer's installation instructions and in accordance with Section 1312.3.1 or Section 1312.3.2. [NFPA 54:9.6.2]

Two important requirements for maximum lengths of appliance connectors are contained in Sections 1312.3.1 and 1312.3.2. First, indoor appliance connectors, no matter of what material, shall be a maximum of 6 feet in length. Second, outdoor gas hose connections shall be a maximum of 15 feet in length (see **Figure 1312.3**). Section 1312.5 gives no measurement as to the allowable length for flexible hoses for mobile equipment. The maximums above could be used for this equipment, but to be sure consult the AHJ.

FIGURE 1312.3
CONNECTION TO OUTDOOR EQUIPMENT

1312.3.1 Indoor. Indoor gas hose connectors shall be used only to connect laboratory, shop, and ironing appliances requiring mobility during operation and installed in accordance with the following:

(1) An appliance shutoff valve shall be installed where the connector is attached to the building piping.

(2) The connector shall be of minimum length and shall not exceed 6 feet (1829 mm).

(3) The connector shall not be concealed and shall not extend from one room to another or pass through wall partitions, ceilings, or floors. [NFPA 54:9.6.2(1)]

1312.3.2 Outdoor. Where outdoor gas hose connectors are used to connect portable outdoor appliances, the connector shall be listed in accordance with CSA Z21.54 and installed in accordance with the following:

(1) An appliance shutoff valve, a listed quick-disconnect device, or a listed gas convenience outlet shall be installed where the connector is attached to the supply piping and in such a manner so as to prevent the accumulation of water or foreign matter.

(2) This connection shall be made only in the outdoor area where the appliance is to be used. [NFPA 54:9.6.2(2)]

(3) The connector length shall not exceed 15 feet (4572 mm).

1312.4 Injection (Bunsen) Burners. Injection (Bunsen) burners used in laboratories and educational facilities shall be permitted to be connected to the gas supply by an unlisted hose. [NFPA 54:9.6.3]

1312.5 Connection of Portable and Mobile Industrial Appliances. Where portable industrial appliances or appliances requiring mobility or subject to vibration are connected to the building gas piping system by the use of a flexible hose, the hose shall be suitable and safe for the conditions under which it can be used. [NFPA 54:9.6.4.1]

1312.5.1 Swivel Joints or Couplings. Where industrial appliances requiring mobility are connected to the rigid piping by the use of swivel joints or couplings, the swivel joints or couplings shall be suitable for the service required and only the minimum number required shall be installed. [NFPA 54:9.6.4.2]

1312.5.2 Metal Flexible Connectors. Where industrial appliances subject to vibration are connected to the building piping system by the use of all metal flexible connectors, the connectors shall be suitable for the service required. [NFPA 54:9.6.4.3]

1312.5.3 Flexible Connectors. Where flexible connections are used, they shall be of the minimum practical length and shall not extend from one room to another or pass through any walls, partitions, ceilings, or floors. Flexible connections shall not be used in any concealed location. They shall be protected against physical or thermal damage and shall be provided with gas shutoff valves in readily accessible locations in rigid piping upstream from the flexible connections. [NFPA 54:9.6.4.4]

1312.6 Appliance Shutoff Valves and Connections. Each appliance connected to a piping system shall have an accessible, approved manual shutoff valve with a nondisplaceable valve member, or a listed gas convenience outlet. Appliance shutoff valves and convenience outlets shall serve a single appliance only. [NFPA 54:9.6.5] The shutoff valve shall be located within 6 feet (1829 mm) of the appliance it serves. [NFPA 54:9.6.5.1] Where a connector is used, the valve shall be installed upstream of the connector. A union or flanged connection shall be provided downstream from the valve to permit removal of appliance controls. [NFPA 54:9.6.5.1(A)]

Exceptions:

(1) Shutoff valves serving decorative appliances in a fireplace shall not be located within the fireplace firebox except where the valve is listed for such use. [NFPA 54: 9.6.5.1(B)]

(2) Shutoff valves shall be permitted to be accessibly located

inside wall heaters and wall furnaces listed for recessed installation where necessary maintenance is performed without removal of the shutoff valve.

1312.7 Quick-Disconnect Devices. Quick-disconnect devices used to connect appliances to the building piping shall be listed to CSA Z21.41. Where installed indoors, an approved manual shutoff valve with a nondisplaceable valve member shall be installed upstream of the quick-disconnect device. [NFPA 54:9.6.6 – 9.6.6.2]

1312.8 Gas Convenience Outlets. Appliances shall be permitted to be connected to the building piping by means of a listed gas convenience outlet, in conjunction with a listed appliance connector, installed in accordance with the manufacturer's installation instructions.

Gas convenience outlets shall be listed in accordance with CSA Z21.90 and installed in accordance with the manufacturer's installation instructions. [NFPA 54:9.6.7]

1312.9 Sediment Trap. Where a sediment trap is not incorporated as a part of the appliance, a sediment trap shall be installed downstream of the appliance shutoff valve as close to the inlet of the appliance as practical at the time of appliance installation. The sediment trap shall be either a tee fitting with a capped nipple in the bottom outlet, as illustrated in **Figure 1312.9**, or another device recognized as an effective sediment trap. Illuminating appliances, gas ranges, clothes dryers, decorative appliances for installation in vented fireplaces, gas fireplaces, and outdoor cooking appliances shall not be required to be so equipped. [NFPA 54:9.6.8]

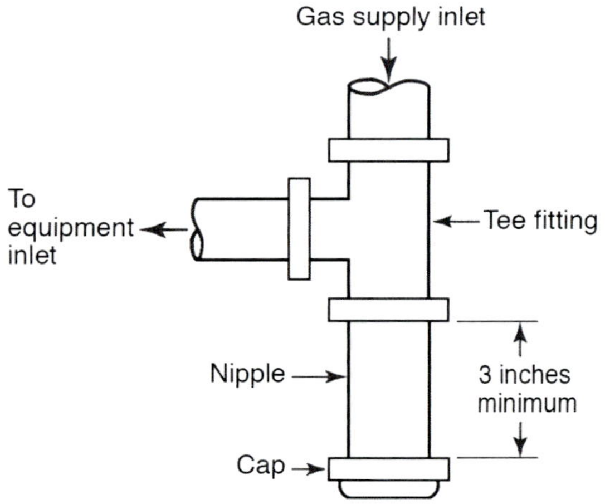

For SI units: 1 inch = 25.4 mm

FIGURE 1312.9
METHOD OF INSTALLING A TEE FITTING SEDIMENT TRAP
[NFPA 54: FIGURE 9.6.8]

1312.10 Installation of Piping. Piping shall be installed in a manner not to interfere with inspection, maintenance, or servicing of the appliances. [NFPA 54:9.6.9]

When installing the fuel-gas system, care must be taken to ensure that the arrangement of the piping and connectors will allow the removal and maintenance of the appliance when necessary.

1312.11 Liquefied Petroleum Gas Facilities and Piping. Liquefied petroleum gas facilities shall comply with NFPA 58.

1313.0 Pressure Testing, Inspection, and Purging.

Before being put into operation, piping installations must be inspected and tested to determine that the total project complies with the code. The piping installation includes all fixed piping from the point of delivery to the equipment. Shutoff valves shall only be tested if they are designed to safely withstand the pressure (Section 1313.1.4), otherwise the piping shall be capped at the appliance (Section 1313.2.3). Appliances and appliance connectors are not part of the fixed piping system and are not included in this test (Section 1313.2.3). The system should be tested at the rough inspection when all piping is exposed to ensure that later concealed piping does not leak (Sections 1303.3.1 and 1313.2.1). The second test is the final inspection prior to initial operation of the piping system (Section 1303.3.2). The system could be tested one more time if the piping system is modified or repaired.

1313.1 Piping Installations. Prior to acceptance and initial operation, all piping installations shall be visually inspected and pressure tested to determine that the materials, design, fabrication, and installation practices comply with the requirements of this code. [NFPA 54:8.1.1.1]

1313.1.1 Inspection Requirements. Inspection shall consist of visual examination, during or after manufacture, fabrication, assembly, or pressure tests. [NFPA 54:8.1.1.2]

1313.1.2 Repairs and Additions. Where repairs or additions are made following the pressure test, the affected piping shall be tested. Minor repairs and additions are not required to be pressure tested, provided that the work is inspected and connections are tested with a noncorrosive leak-detecting fluid or other leak-detecting methods approved by the Authority Having Jurisdiction. [NFPA 54:8.1.1.3]

1313.1.3 New Branches. Where new branches are installed to new appliance(s), only the newly installed branch(es) shall be required to be pressure tested. Connections between the new piping and the existing piping shall be tested with a noncorrosive leak-detecting fluid or approved leak-detecting methods. [NFPA 54:8.1.1.4]

1313.1.4 Piping System. A piping system shall be tested as a complete unit or in sections. Under no circumstances shall a valve in a line be used as a bulkhead between gas in one section of the piping system and test medium in an adjacent section, unless a double block and bleed valve system is installed. A valve shall not be subjected to the test pressure unless it can be determined that the valve, including the valve-closing mechanism, is designed to safely withstand the pressure. [NFPA 54:8.1.1.5]

1313.1.5 Regulators and Valves. Regulator and valve assemblies fabricated independently of the piping system in which they are to be installed shall be permitted to be tested with inert gas or air at the time of fabrication. [NFPA 54:8.1.1.6]

1313.1.6 Test Medium. The test medium shall be air, nitrogen, carbon dioxide, or an inert gas. Oxygen shall not be used as a test medium. [NFPA 54:8.1.2]

1313.2 Test Preparation. Test preparation shall comply with Section 1313.2.1 through Section 1313.2.6.

1313.2.1 Pipe Joints. Pipe joints, including welds, shall be left exposed for examination during the test.

Exception: Covered or concealed pipe end joints that have been previously tested in accordance with this code. [NFPA 54:8.1.3.1]

1313.2.2 Expansion Joints. Expansion joints shall be provided with temporary restraints, if required, for the additional thrust load under test. [NFPA 54:8.1.3.2]

1313.2.3 Appliances and Equipment. Appliances and equipment that are not to be included in the test shall be either disconnected from the piping or isolated by blanks, blind flanges, or caps. Flanged joints at which blinds are inserted to blank off other equipment during the test shall not be required to be tested. [NFPA 54:8.1.3.3]

1313.2.4 Designed for Operating Pressures Less than Test Pressure. Where the piping system is connected to appliances or equipment designed for operating pressures of less than the test pressure, such appliances or equipment shall be isolated from the piping system by disconnecting them and capping the outlet(s). [NFPA 54:8.1.3.4]

1313.2.5 Designed for Operating Pressures Equal to or Greater than Test Pressure. Where the piping system is connected to appliances or equipment designed for operating pressures equal to or greater than the test pressure, such appliances or equipment shall be isolated from the piping system by closing the individual appliance or equipment shutoff valve(s). [NFPA 54:8.1.3.5]

1313.2.6 Safety. All testing of piping systems shall be performed in a manner that protects the safety of employees and the public during the test. [NFPA 54:8.1.3.6]

1313.3 Test Pressure. This inspection shall include an air, CO_2, or nitrogen pressure test, at which time the gas piping shall stand a pressure of not less than 10 psi (69 kPa) gauge pressure. Test pressures shall be held for a length of time satisfactory to the Authority Having Jurisdiction but in no case less than 15 minutes with no perceptible drop in pressure. For welded piping, and for piping carrying gas at pressures in excess of 14 inches water column (3.5 kPa) pressure, the test pressure shall be not less than 60 psi (414 kPa) and shall be continued for a length of time satisfactory to the Authority Having Jurisdiction, but in no case for less than 30 minutes. For CSST carrying gas at pressures in excess of 14 inches water column (3.5 kPa) pressure, the test pressure shall be 30 psi (207 kPa) for 30 minutes. These tests shall be made using air, CO_2, or nitrogen pressure and shall be made in the presence of the Authority Having Jurisdiction. Necessary apparatus for conducting tests shall be furnished by the permit holder. Test gauges used in conducting test shall be in accordance with Section 1303.3.3.1 through Section 1303.3.3.4.

1313.4 Detection of Leaks and Defects. The piping system shall withstand the test pressure specified without showing any evidence of leakage or other defects. Any reduction of test pressures as indicated by pressure gauges shall be deemed to indicate the presence of a leak unless such reduction can be readily attributed to some other cause. [NFPA 54:8.1.5.1]

1313.4.1 Detecting Leaks. The leakage shall be located by means of an approved gas detector, a noncorrosive leak detection fluid, or other approved leak detection methods. [NFPA 54:8.1.5.2]

See **Figure 1313.4.1**.

1313.4.2 Repair or Replace. Where leakage or other defects are located, the affected portion of the piping system shall be repaired or replaced and retested. [NFPA 54:8.1.5.3]

1313.5 Piping System Leak Test. Leak checks using fuel gas shall be permitted in piping systems that have been pressure-tested in accordance with Section 1313.0 through Section 1313.4.2. [NFPA 54:8.2.1]

The pressure test, required in Section 1313.3 and leak test required in Section 1313.5 are often confused with one another. This initial pressure test is required for new installations and additions to existing piping installations. The leak test is in addition and subsequent to the pressure test. This second test is required after the gas system is placed into service or when the gas is turned back on after being turned off.

The medium used for a leak test is the fuel gas itself at its normal supply pressure and applied to the total system (i.e., piping, equipment and equipment connections and valves).

1313.5.1 Turning Gas On. During the process of turning gas on into a system of new gas piping, the entire system shall be inspected to determine that there are no open fittings or

FIGURE 1313.4.1
GAS LEAK DETECTION - LIQUID AND ELECTRONIC

ends and that all valves at unused outlets are closed and plugged or capped. [NFPA 54:8.2.2]

1313.5.2 Leak Check. Immediately after the gas is turned on into a new system or into a system that has been initially restored after an interruption of service, the piping system shall be checked for leakage. Where leakage is indicated, the gas supply shall be shut off until the necessary repairs have been made. [NFPA 54:8.2.3]

1313.5.3 Placing Appliances and Equipment in Operation. Appliances and equipment shall not be placed in operation until after the piping system has been checked for leakage in accordance with Section 1313.5.2, the piping system is purged in accordance with Section 1313.6, and connections to the appliance are checked for leakage. [NFPA 54:8.2.4]

1313.6 Purging Requirements. The purging of piping shall be in accordance with Section 1313.6.1 through Section 1313.6.3. [NFPA 54:8.3]

The purging requirements for purging indoors and outdoors provide increased safety procedures including gas detection, monitoring, and the location of discharge points. Purging requirements are determined based on diameter, length, and operating pressure. The pipe diameter/length and pressure criteria are selected to distinguish between large commercial systems (industrial systems) and small commercial systems (including residential systems). Large systems have pipe volumes or potential for higher flow rates that require procedures to ensure that large volumes of fuel gases are not released indoors and that flammable mixtures do not occur within the piping itself. Installers of these complex systems deal with considerably more variables that may result in a higher potential for discharge of large gas volumes during purging operations. All large systems are required to be purged outdoors and the use of a combustible gas indicator is mandated. For smaller systems, the purging requirements allow five options that have been shown to be effective and are widely used. These include purging directly to the outdoors; purging through the appliance's burner; purging through a standalone burner; purging to the indoors using a combustible gas detector; and purging in accordance with the written purging procedures of a gas supplier. The purging requirements recognize that the gas supplier has been successfully conducting purging operations with their trained personnel. Installers of smaller systems have familiarity with purging these systems and the potential for discharge of large gas volumes during purging operation is low.

1313.6.1 Piping Systems Required to be Purged Outdoors. The purging of piping systems shall be in accordance with Section 1313.6.1.1 through Section 1313.6.1.4 where the piping system meets either of the following:

(1) The design operating gas pressure is greater than 2 psig (14 kPag).

(2) The piping being purged contains one or more sections of pipe or tubing meeting the size and length criteria of Table 1313.6.1. [NFPA 54:8.3.1]

TABLE 1313.6.1
SIZE AND LENGTH OF PIPING
[NFPA 54: TABLE 8.3.1]*

NOMINAL PIPING SIZE (inches)	LENGTH OF PIPING (feet)
≥ 2½ < 3	> 50
≥ 3 < 4	> 30
≥ 4 < 6	> 15
≥ 6 < 8	> 10
≥ 8	Any length

For SI units: 1 inch = 25 mm; 1 foot = 304.8 mm

* CSST EHD size of 62 is equivalent to nominal 2 inches (50 mm) pipe or tubing size.

1313.6.1.1 Removal from Service. Where existing gas piping is opened, the section that is opened shall be isolated from the gas supply and the line pressure vented in accordance with Section 1313.6.1.3. Where gas piping meeting the criteria of Table 1313.6.1 is removed from service, the residual fuel gas in the piping shall be displaced with an inert gas. [NFPA 54:8.3.1.1]

1313.6.1.2 Placing in Operation. Where gas piping containing air and meeting the criteria of Table 1313.6.1 is placed in operation, the air in the piping shall first be displaced with an inert gas. The inert gas shall then be displaced with fuel gas in accordance with Section 1313.6.1.3. [NFPA 54:8.3.1.2]

1313.6.1.3 Outdoor Discharge of Purged Gases. The open end of a piping system being pressure vented or purged shall discharge directly to an outdoor location. Purging operations shall comply with all of the following requirements:

(1) The point of discharge shall be controlled with a shutoff valve.

(2) The point of discharge shall be located at least 10 feet (3048 mm) from sources of ignition, at least 10 feet (3048 mm) from building openings and at least 25 feet (7620 mm) from mechanical air intake openings.

(3) During discharge, the open point of discharge shall be continuously attended and monitored with a combustible gas indicator that complies with Section 1313.6.1.4.

(4) Purging operations introducing fuel gas shall be stopped when 90 percent fuel gas by volume is detected within the pipe.

(5) Persons not involved in the purging operations shall be evacuated from all areas within 10 feet (3048 mm) of the point of discharge. [NFPA 54:8.3.1.3]

1313.6.1.4 Combustible Gas Indicator. Combustible gas indicators shall be listed and calibrated in accordance with the manufacturer's instructions. Combustible gas indicators shall numerically display a volume scale from 0 percent to 100 percent in 1 percent or smaller increments. [NFPA 54:8.3.1.4]

1313.6.2 Piping Systems Allowed to be Purged Indoors or Outdoors. The purging of piping systems shall be in accordance with the provisions of Section 1313.6.2.1 where the piping system meets both of the following:

(1) The design operating pressure is 2 psig (14 kPag) or less.

(2) The piping being purged is constructed entirely from

pipe or tubing not meeting the size and length criteria of Table 1313.6.1. [NFPA 54:8.3.2]

1313.6.2.1 Purging Procedure. The piping system shall be purged in accordance with one or more of the following:

(1) The piping shall be purged with fuel gas and shall discharge to the outdoors.

(2) The piping shall be purged with fuel gas and shall discharge to the indoors or outdoors through an appliance burner not located in a combustion chamber. Such burner shall be provided with a continuous source of ignition.

(3) The piping shall be purged with fuel gas and shall discharge to the indoors or outdoors through a burner that has a continuous source of ignition and that is designed for such purpose.

(4) The piping shall be purged with fuel gas that is discharged to the indoors or outdoors, and the point of discharge shall be monitored with a listed combustible gas detector in accordance with Section 1313.6.2.2. Purging shall be stopped when fuel gas is detected.

(5) The piping shall be purged by the gas supplier in accordance with written procedures. [NFPA 54:8.3.2.1]

1313.6.2.2 Combustible Gas Detector. Combustible gas detectors shall be listed and calibrated or tested in accordance with the manufacturer's instructions. Combustible gas detectors shall be capable of indicating the presence of fuel gas. [NFPA 54:8.3.2.2]

1313.6.3 Purging Appliances and Equipment. After the piping system has been placed in operation, appliances and equipment shall be purged before being placed into operation. [NFPA 54:8.3.3]

1314.0 Required Gas Supply.

1314.1 General. The following regulations, shall comply with this section and Section 1315.0, shall be the standard for the installation of gas piping. Natural gas regulations and tables are based on the use of gas having a specific gravity of 0.60 and for undiluted liquefied petroleum gas having a specific gravity of 1.50.

Where gas of a different specific gravity is to be delivered, the serving gas supplier shall be permitted to be contacted for specific gravity conversion factors to use in sizing piping systems from the pipe sizing tables in this chapter.

This section, along with Section 1315.0, gives specific requirements to accurately size gas piping. The regulations and tables used in this chapter are based on natural gas having a specific gravity of 0.60, supplied at 6 to 8 inches of water column pressure (1/4 to 1/3 psi) at the meter outlet. For undiluted liquefied petroleum, the specific gravity is 1.50 supplied at 11-inch water column (3/8 psi) and sized for 2,500 Btu per cubic foot (25.9 Watt-hours/L).

Always check with the local gas supplier if a different specific gravity is being delivered. The gas supplier should give a conversion factor to use when sizing a gas-piping system with a different gravity.

1314.2 Volume. The hourly volume of gas required at each piping outlet shall be taken as not less than the maximum hourly rating as specified by the manufacturer of the appliance or appliances to be connected to each such outlet.

Every appliance will have a Btu rating located on its information plate. The Btu input rating is to be used when determining the hourly volume of gas required at the piping outlet.

1314.3 Gas Appliances. Where the gas appliances to be installed have not been definitely specified, Table 1308.4.1 shall be permitted to be used as a reference to estimate requirements of typical appliances. To obtain the cubic feet per hour (m^3/h) of gas required, divide the input of the appliances by the average Btu (kW•h) heating value per cubic foot (m^3) of the gas. The average Btu (kW•h) per cubic foot (m^3) of the gas in the area of the installation shall be permitted to be obtained from the serving gas supplier.

If the appliance does not have an input rating on its rating plate, then Table 1308.4.1 may be used as an estimate requirement for typical appliances.

The gas supplier will supply gas from different wells. In order to determine the cubic foot per hour value of the gas, the user will need to know the average Btu heating value per cubic foot of the gas being supplied. The gas supplier will furnish this information. To determine the cubic feet per hour of gas required, divide the input rating of the appliance by the average Btu heating value per cubic foot of the gas.

Example: The gas being supplied has a value of 1,000 Btu per cubic foot and the input rating of an appliance is 125,000 Btus.

$125{,}000 \div 1{,}000 = 125$ cubic feet per hour (cfh)

1314.4 Size of Piping Outlets. The size of the supply piping outlet for a gas appliance shall be not less than ½ of an inch (15 mm).

The size of a piping outlet for a mobile home shall be not less than ¾ of an inch (20 mm).

1315.0 Required Gas Piping Size.

1315.1 Pipe Sizing Methods. Where the pipe size is to be determined using any of the methods in Section 1315.1.1 through Section 1315.1.3, the diameter of each pipe segment shall be obtained from the pipe sizing tables in Section 1315.2 or from the sizing equations in Section 1315.3. [NFPA 54:6.1]

1315.1.1 Longest Length Method. The pipe size of each section of gas piping shall be determined using the longest length of piping from the point of delivery to the most remote outlet and the load of the section. [NFPA 54:6.1.1]

The Longest Length Method requires only one distance measurement – the length from the meter to the most remote outlet. That measurement is then used to size the entire system. In all of the sizing tables the far left column is the "Length" column. For example, if there is a system with 137 feet of pipe from the meter to the most remote outlet and Table 1315.2(1) is being used, you would use the 150 foot row to size all pipe segments in the system.

This is a very simple method for sizing a fuel gas system and, for relatively small systems, it yields very similar sizes of pipe compared to using the Branch Length Method.

However, if the system is a large installation, the use of the Branch Length Method could give smaller sizes in branch segments, and therefore, provide cost savings for the owner.

1315.1.2 Branch Length Method. Pipe shall be sized as follows:

(1) Pipe size of each section of the longest pipe run from the point of delivery to the most remote outlet shall be determined using the longest run of piping and the load of the section.

(2) The pipe size of each section of branch piping not previously sized shall be determined using the length of piping from the point of delivery to the most remote outlet in each branch and the load of the section. [NFPA 54:6.1.2]

The Branch Length Method also requires the distance of the pipe to be from the meter to the most remote outlet but only for the purpose of sizing the system main. The piping from the remote outlet to the meter is also sized using the "Length" column in the sizing tables, but each branch off the main is measured from the meter to the furthest outlet on that branch. The piping from the remote outlet on that branch to the meter is then sized from that "Length" column.

For example, if there is a distance from the meter to the most remote outlet of 137 feet, the 150 foot length row in Table 1315.2(1) would be used again to size the system main. Next, each branch off of the main would have to be measured from the meter to the remote outlet on the branch, and then the piping from the remote outlet on the branch to the main would be sized. If there is a distance of 47 feet from the remote outlet on the branch to the meter, use the 50 foot row would be used to size the branch piping. Each branch would have to be sized in this manner.

1315.1.3 Hybrid Pressure. The pipe size for each section of higher pressure gas piping shall be determined using the longest length of piping from the point of delivery to the most remote line pressure regulator. The pipe size from the line pressure regulator to each outlet shall be determined using the length of piping from the regulator to the most remote outlet served by the regulator. [NFPA 54:6.1.3]

Large systems are sometimes designed to have a higher pressure than the normal fuel-gas system. This inlet pressure is the pressure at the meter outlet, which is also the piping inlet. Using a higher inlet pressure will allow smaller main and branch pipe sizes. However, most appliances require a pressure regulator to bring the pressure down to the working pressure of the appliance. In a system such as this, with regulators at each appliance, it would be termed a "medium" or "high-pressure" system, depending on the pressure of the gas. Sizing would be accomplished using only one sizing table.

The Hybrid Pressure Method is used to size fuel-gas systems utilizing two or more different inlet pressure values. A hybrid pressure system is one that would require a line pressure regulator to step the higher main line pressure to a lower pressure, which is usually the appliance working pressure. In this design, only one line regulator is needed rather than several appliance regulators, as in a high-pressure system. Two sizing tables will have to be used to size the system – one to size the higher pressure portion of the system and another to size the reduced pressure portion of the system. Either sizing method, Branch or Longest Length, could be used in sizing the hybrid system. **Figure 1315.1.3** illustrates a complicated hybrid pressure system.

1315.2 Sizing of Gas Piping Systems. Sizing of piping systems shall be in accordance with Section 1315.2.1 for natural gas piping systems and Section 1315.2.2 for propane piping systems.

Each sizing table is designed for a specific type of gas, a specific piping material, a specific inlet pressure, a specific pressure drop and a specific gravity of fuel gas. For example, Table 1315.2(1) is for natural gas with an inlet pressure less than 2 psi for Schedule 40 Metallic Pipe, at a pressure drop of 0.5 inches of water column and for a gas with a specific gravity of 0.60.

Be sure to use the proper table when sizing the system. Not using the proper table will not only cause a plumber to fail a sizing exam but, if installed, the system could be undersized or oversized and thus might not provide the proper amount of fuel gas for the appliances used.

1315.2.1 Natural Gas Piping Systems. Table 1315.2(1) through Table 1315.2(23) shall be used in conjunction with one of the methods described in Section 1315.1.1 through Section 1315.1.3 for piping materials other than non-corrugated stainless steel tubing. Section 1315.3 shall be used in conjunction with one of the methods described in Section 1315.1.1 through Section 1315.1.3 for non-corrugated stainless steel tubing. [NFPA 54:6.2.1, 6.2.2]

1315.2.2 Propane Piping Systems. Table 1315.2(24) through Table 1315.2(36) shall be used in conjunction with one of the methods described in Section 1315.1.1 through Section 1315.1.3 for piping materials other than non-corrugated stainless steel tubing. Section 1315.3 shall be used in conjunction with one of the methods described in Section 1315.1.1 through Section 1315.1.3 for non-corrugated stainless steel tubing. [NFPA 54:6.3.1, 6.3.2]

1315.3 Sizing Equations. The inside diameter of smooth wall pipe or tubing shall be determined by Equation 1315.3(1), Equation 1315.3(2) and Table 1315.3 using the equivalent pipe length determined by the methods in Section 1315.1.1 through Section 1315.1.3. [NFPA 54:6.4]

LOW-PRESSURE GAS FORMULA (LESS THAN 1.5 psi) (10.3 kPa) [NFPA 54:6.4.1]

[Equation 1315.3(1)]

$$D = \frac{Q^{0.381}}{19.17\left(\frac{\Delta H}{Cr \times L}\right)^{0.206}}$$

Where:

D = inside diameter of pipe, inches

Q = input rate appliance(s), cubic feet per hour at 60°F and 30 inch mercury column

L = equivalent length of pipe, feet

ΔH = pressure drop, inches of water column
Cro = in accordance with Table 1315.3

HIGH-PRESSURE GAS FORMULA (1.5 psi (10.3 kPa) AND ABOVE) [NFPA 54:6.4.2]

[Equation 1315.3(2)]

$$D = \frac{Q^{0.381}}{18.93\left[\frac{(P_1^2 - P_2^2) \cdot Y}{Cr \times L}\right]^{0.206}}$$

Where:

D = inside diameter of pipe, inches
Q = input rate of appliance(s), cubic feet per hour at 60°F and 30 inch mercury column
P_1 = upstream pressure, psia (P_1 + 14.7)
P_2 = downstream pressure, psia (P_2 + 14.7)
L = equivalent length of pipe, feet
Cr = in accordance with Table 1315.3
Y = in accordance with Table 1315.3

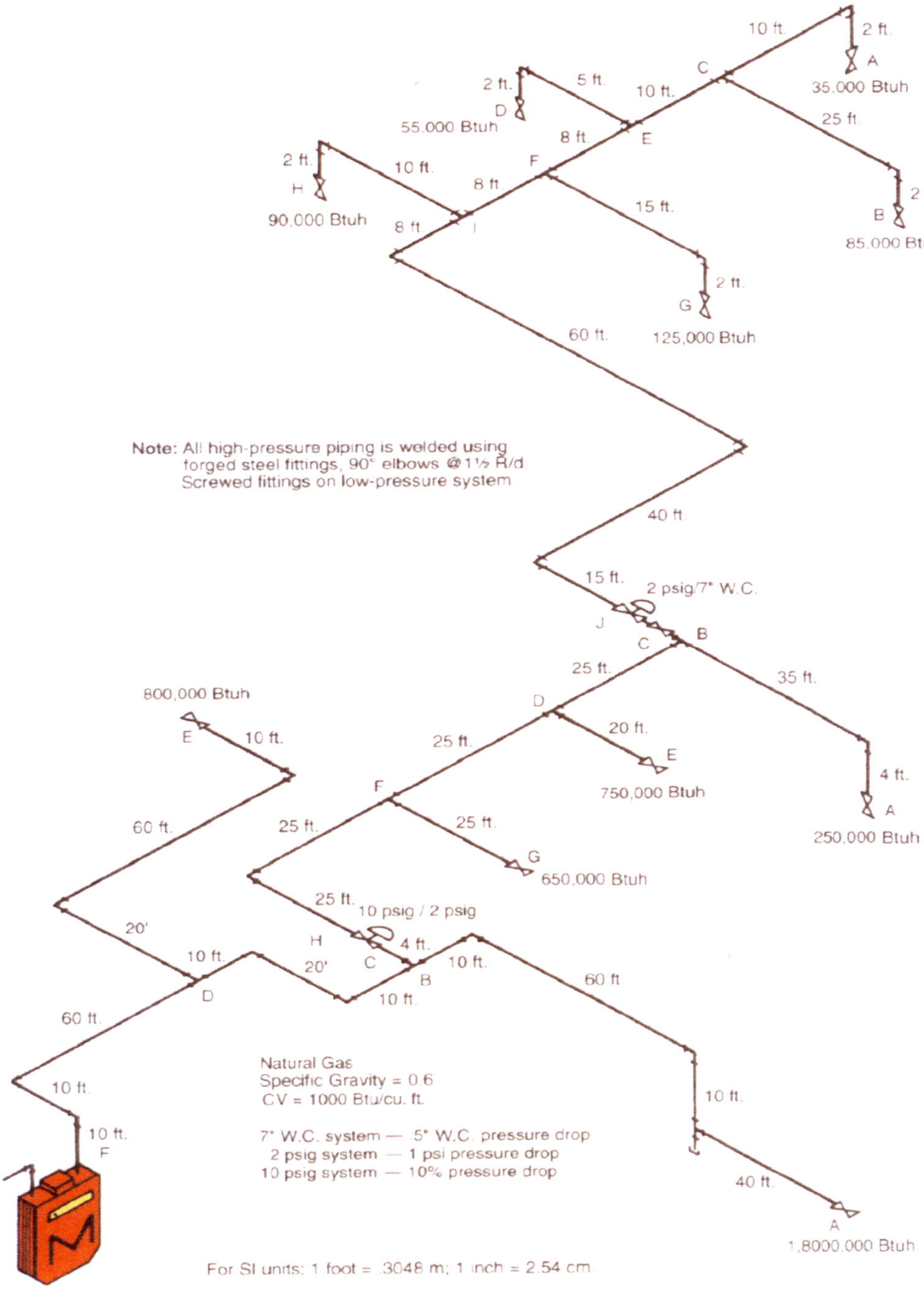

FIGURE 1315.1.3
HYBRID PRESSURE GAS SYSTEM

For SI units: 1 cubic foot = 0.0283 m^3, 1000 British thermal units per hour = 0.293 kW, 1 inch = 25 mm, 1 foot = 304.8 mm, 1 pound-force per square inch = 6.8947 kPa, °C = (°F-32)/1.8, 1 inch mercury column = 3.39 kPa, 1 inch water column = 0.249 kPa

TABLE 1315.3
***Cr* AND *Y* FOR NATURAL GAS AND UNDILUTED PROPANE AT STANDARD CONDITIONS**
[NFPA 54: TABLE 6.4.2]

GAS	FORMULA FACTORS	
	Cr	*Y*
Natural Gas	0.6094	0.9992
Undiluted Propane	1.2462	0.9910

These are the equations that generated the Tables. The equations can be easily used once the givens are known. Suppose the natural gas is low pressure less than 2psi. The input rating of the appliance is 50 cubic feet per hour and the length is 100 feet. The pressure drop is 0.5 inches water column. Using Equation 1315.3(1), find the required diameter. The givens are,

Q = 50

L = 100

ΔH = 0.5

Cr = 0.6094 (Table 1315.3, Natural Gas)

$$D = \frac{50^{0.381}}{19.17\left(\frac{0.5}{0.6094x100}\right)^{0.206}}$$

Therefore, the equation will look like this,

D = 0.6229

Refer to Table 1315.2(1) and find 100 feet in the first column and 50 cubic feet per hour in the second column. The required diameter is ½ inch or 0.622 inches ID, which is the same that the equation yielded. These equations can be utilized when exact lengths and input ratings are known.

1315.4 Sizing of Piping Sections. To determine the size of each section of pipe in a system within the range of Table 1315.2(1) through Table 1315.2(36), proceed as follows:

(1) Measure the length of the pipe from the gas meter location to the most remote outlet on the system.

(2) Select the length in feet column and row showing the distance, or the next longer distance where the table does not give the exact length.

(3) Starting at the most remote outlet, find in the just selected the gas demand for that outlet. Where the exact figure of demand is not shown, choose the next larger figure in the row.

(4) At the top of the column in the table will be found the correct size of pipe.

(5) Using this same row, proceed in a similar manner for each section of pipe serving this outlet. For each section of pipe, determine the total gas demand supplied by that section. Where gas piping sections serve both heating and cooling appliances and the installation prevents both units from operating simultaneously, the larger of the two demand loads needs to be used in sizing these sections.

(6) Size each section of branch piping not previously sized by measuring the distance from the gas meter location to the most remote outlet in that branch and follow the procedures of steps 2, 3, 4, and 5 above. Size branch piping in the order of their distance from the meter location, beginning with the most distant outlet not previously sized.

The steps enumerated here are for sizing by the Branch Length Method. For the Longest Length Method, (6) is eliminated and at (5) each outlet and pipe segment in the entire system is sized from the same length row.

1315.5 Engineering Methods. For conditions other than those covered by Section 1315.1, such as longer runs or greater gas demands, the size of each gas piping system shall be determined by standard engineering methods acceptable to the Authority Having Jurisdiction, and each such system shall be so designed that the total pressure drop between the meter or other point of supply and an outlet where full demand is being supplied to outlets, shall be in accordance with the requirements of Section 1308.4.

1315.6 Variable Gas Pressures. Where the supply gas pressure exceeds 5 psi (34 kPa) for natural gas and 10 psi (69 kPa) for undiluted propane or is less than 6 inches (1.5 kPa) of water column, or where diversity demand factors are used, the design, pipe, sizing, materials, location, and use of such systems first shall be approved by the Authority Having Jurisdiction. Piping systems designed for pressures exceeding the serving gas supplier's standard delivery pressure shall have prior verification from the gas supplier of the availability of the design pressure.

Sizing the Gas System

Sizing the gas system can be easy if the proper steps are followed. The following steps summarize all of the sizing criteria that have been discussed in this chapter.

1. Determine the type of gas piping material to be used, the inlet pressure of the gas and its specific gravity and the allowable pressure drop. Use this information to determine which sizing table will be used to size the system.
2. Identify the appliances on the system and their Btu input rating.
3. Determine the heating value of the gas supplied and convert the appliance Btu/h input rating to cubic feet per hour (cfh).
4. Note the cubic feet per hour at each appliance as in **Figure 1315.1.1** Example Illustrating Use of Tables 1308.4.1 and 1316.2(1).
5. Measure each appliance outlet from the meter to the appliance along the piping system and note the distance for each appliance on the print or paper. The branch with the longest distance will be the most remote outlet. The piping from this outlet back to the meter will be the system main.

Outlet	Appliance	Cubic Feet of Gas/hr	Nominal Pipe Diameter
A	Water Heater, 30gal	32	½"
B	Gas Refrigerator	3	½"
C	Range	59	½"
D	Furnace	136	¾"

6. Note the total cubic feet per hour for each pipe segment of the branches and main in the system starting from the most remote outlet from the meter and working back to the meter, totaling the demand for each segment.

Pipe Section	Cubic Feet of Gas/hr	Nominal Pipe Diameter
1	35	1/2"
2	94	3/4"
3	230	1"

For this example use **Figure 1315.1.1**. The given piping material is Schedule 40 metallic pipe. The gas to be used is natural gas with a specific gravity of 0.60 and a heating value of 1,100 Btu per cubic foot, delivered at less than 2 inches w.c. and with a pressure drop of 0.5 inches water column. Based on these givens, Table 1315.2(1) will be used to size the system. The system is now ready to be sized using either of the two sizing methods: Longest Length Method or the Branch Length Method. The Hybrid Pressure Method would not be used because there is only one pressure value within the system.

Longest Length Method

1. Measure the length of the pipe from the gas meter location to the most remote outlet on the system. In this example, the most remote outlet is A at 60 feet from the meter.
2. In Table 1315.2(1) select the distance in the "Length" column that is equal to or greater than 60 feet. This will determine the row used for pipe sizing (see highlighted row in Table 1315.2(1) extraction below).
3. Convert each appliance's Btu/h rating into cubic feet per hour by dividing the Btu/h rating by 1,100. If the Btu/h rating is not known, use Table 1308.4.1.

   30 gallon water heater: 32 cfh
   Gas Refrigerator: 3 cfh
   Range: 59 cfh
   Furnace: 136 cfh

4. Starting at the most remote outlet, find in this same row the gas demand for that outlet. If the exact number for the demand is not shown, choose the next larger number in the row. At the top of this column, the nominal pipe size will be found.
5. Using the same method in the previous step, proceed in a similar manner using the 60 feet length row for each section.

Branch Length Method

1. Follow the same steps 1-3 as the Longest Length Method.
2. Follow the same step 5 as the Longest Length Method. In the Branch Length Method, the longest pipe run is sized first. Since the longest run is 60 feet, use the 60 feet Length row in Table 1315.2(1) to size each section of the longest run. The results are the same as tabulated in step 5 of the previous method.
3. Size each branch as determined by the length from the point of delivery to the outlet. This is the differing step from the Longest Length Method. Determine the length of the pipe run from the beginning of the meter to each outlet. Then use that Length row in Table 1315.2(1) that is equal to or greater than the length to find the diameter of pipe.

For outlets A, B, and C, the 60 feet Length row is used in Table 1315.2(1). Outlet D uses the 50 feet Length row. The sizing results are the same using either method in this small example. Actual gas systems will be more complicated, and each method may yield differing results allowing the choice for the more economical method.

SCHEDULE 40 METALLIC PIPE [NFPA 54: TABLE 6.2(b)][1, 2]

												GAS:	NATURAL	
												INLET PRESSURE:	LESS THAN 2 psi	
												PRESSURE DROP:	0.5 in. w.c.	
												SPECIFIC GRAVITY:	0.60	
	PIPE SIZE (inch)													
NOMINAL:	½	¾	1	1¼	1½	2	2½	3	4	5	6	8	10	12
ACTUAL ID:	0.622	0.824	1.049	1.380	1.610	2.067	2.469	3.068	4.026	5.047	6.065	7.981	10.020	11.938
LENGTH (feet)	CAPACITY IN CUBIC FEET OF GAS PER HOUR													
10	172	360	678	1390	2090	4020	6400	11 300	23 100	41 800	67 600	139 000	252 000	399 000
20	118	247	466	957	1430	2760	4400	7780	15 900	28 700	46 500	95 500	173 000	275 000
30	95	199	374	768	1150	2220	3530	6250	12 700	23 000	37 300	76 700	139 000	220 000
40	81	170	320	657	985	1900	3020	5350	10 900	19 700	31 900	65 600	119 000	189 000
50	72	151	284	583	873	1680	2680	4740	9660	17 500	28 300	58 200	106 000	167 000
60	65	137	257	528	791	1520	2430	4290	8760	15 800	25 600	52 700	95 700	152 000
70	60	126	237	486	728	1400	2230	3950	8050	14 600	23 600	48 500	88 100	139 000
80	56	117	220	452	677	1300	2080	3670	7490	13 600	22 000	45 100	81 900	130 000
90	52	110	207	424	635	1220	1950	3450	7030	12 700	20 600	42 300	76 900	122 000
100	50	104	195	400	600	1160	1840	3260	6640	12 000	19 500	40 000	72 600	115 000

EXAMPLE - LONGEST LENGTH METHOD - EXTRACTED FROM TABLE 1315.2(1)

FIGURE 1315.1.1
EXAMPLE ILLUSTRATING USE OF TABLE 1308.4.1 AND TABLE 1315.2(1)

Problem: Determine the required pipe size of each section and outlet of the piping system shown in **Figure 1315.1.1**. Gas to be used has a specific gravity of 0.60 and 1100 British thermal units (Btu) per cubic foot (0.0114 kW•h/L), delivered at 8 inch water column (1.9 kPa) pressure.

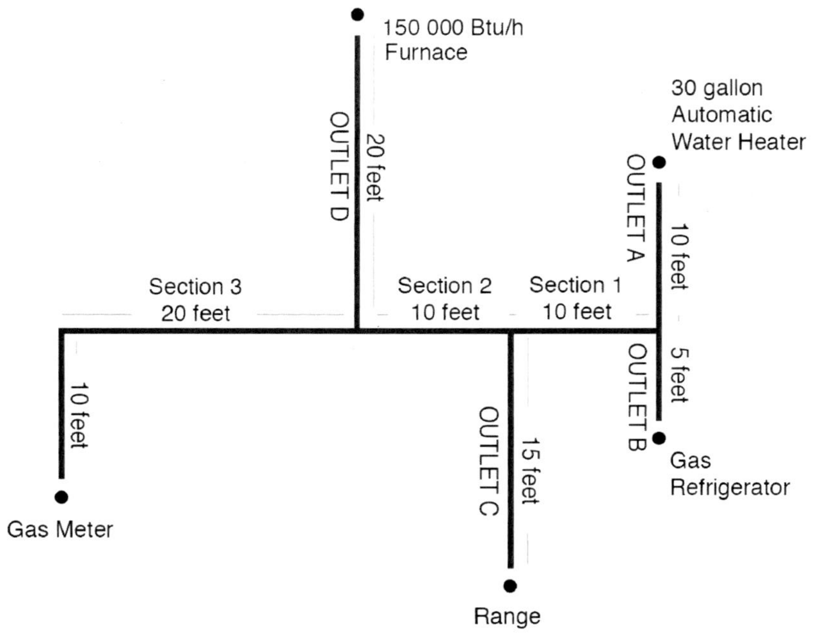

For SI units: 1 foot = 304.8 mm, 1 gallon = 3.785 L, 1000 British thermal units per hour = 0.293 kW, 1 cubic foot per hour = 0.0283 m^3/h

Solution:

(1) Maximum gas demand of Outlet A –
32 cubic feet per hour (0.91 m^3/h) (from Table 1308.4.1).
Maximum gas demand of Outlet B –
3 cubic feet per hour (0.08 m^3/h) (from Table 1308.4.1).
Maximum gas demand of Outlet C –
59 cubic feet per hour (1.67 m^3/h) (from Table 1308.4.1).
Maximum gas demand of Outlet D –
136 cubic feet per hour (3.85 m^3/h) [150 000 Btu/hour (44 kW)] divided by 1100 Btu per cubic foot (0.0114 kW•h/L)

(2) The length of pipe from the gas meter to the most remote outlet (Outlet A) is 60 feet (18 288 mm).

(3) Using the length in feet column row marked 60 feet (18 288 mm) in Table 1315.2(1):
Outlet A, supplying 32 cubic feet per hour (0.91 m^3/h), requires ½ of an inch (15 mm) pipe.
Section 1, supplying Outlets A and B, or 35 cubic feet per hour (0.99 m^3/h) requires ½ of an inch (15 mm) pipe.
Section 2, supplying Outlets A, B, and C, or 94 cubic feet per hour (2.66 m^3/h) requires ¾ of an inch (20 mm) pipe.
Section 3, supplying Outlets A, B, C, and D, or 230 cubic feet per hour (6.51 m^3/h), requires 1 inch (25 mm) pipe.

(4) Using the column marked 60 feet (18 288 mm) in Table 1315.2(1):
Outlet B supplying 3 cubic feet per hour (0.08 m^3/h), requires ½ of an inch (15 mm) pipe.
Outlet C, supplying 59 cubic feet per hour (1.67 m^3/h), requires ½ of an inch (15 mm) pipe.
Using the column marked 60 feet (18 288 mm) in Table 1315.2(1):
Outlet D, supplying 136 cubic feet per hour (3.85 m^3/h), requires ¾ of an inch (20 mm) pipe.

TABLE 1315.2(1)
SCHEDULE 40 METALLIC PIPE [NFPA 54: TABLE 6.2.1(b)][1,2]

									GAS:	NATURAL			
									INLET PRESSURE:	LESS THAN 2 psi			
									PRESSURE DROP:	0.5 in. w.c.			
									SPECIFIC GRAVITY:	0.60			
	PIPE SIZE (inch)												
NOMINAL:	½	¾	1	1¼	1½	2	2½	3	4	5	6	8	10	12
ACTUAL ID:	0.622	0.824	1.049	1.380	1.610	2.067	2.469	3.068	4.026	5.047	6.065	7.981	10.020	11.938
LENGTH (feet)	CAPACITY IN CUBIC FEET OF GAS PER HOUR													
10	172	360	678	1390	2090	4020	6400	11 300	23 100	41 800	67 600	139 000	252 000	399 000
20	118	247	466	957	1430	2760	4400	7780	15 900	28 700	46 500	95 500	173 000	275 000
30	95	199	374	768	1150	2220	3530	6250	12 700	23 000	37 300	76 700	139 000	220 000
40	81	170	320	657	985	1900	3020	5350	10 900	19 700	31 900	65 600	119 000	189 000
50	72	151	284	583	873	1680	2680	4740	9660	17 500	28 300	58 200	106 000	167 000
60	65	137	257	528	791	1520	2430	4290	8760	15 800	25 600	52 700	95 700	152 000
70	60	126	237	486	728	1400	2230	3950	8050	14 600	23 600	48 500	88 100	139 000
80	56	117	220	452	677	1300	2080	3670	7490	13 600	22 000	45 100	81 900	130 000
90	52	110	207	424	635	1220	1950	3450	7030	12 700	20 600	42 300	76 900	122 000
100	50	104	195	400	600	1160	1840	3260	6640	12 000	19 500	40 000	72 600	115 000
125	44	92	173	355	532	1020	1630	2890	5890	10 600	17 200	35 400	64 300	102 000
150	40	83	157	322	482	928	1480	2610	5330	9650	15 600	32 100	58 300	92 300
175	37	77	144	296	443	854	1360	2410	4910	8880	14 400	29 500	53 600	84 900
200	34	71	134	275	412	794	1270	2240	4560	8260	13 400	27 500	49 900	79 000
250	30	63	119	244	366	704	1120	1980	4050	7320	11 900	24 300	44 200	70 000
300	27	57	108	221	331	638	1020	1800	3670	6630	10 700	22 100	40 100	63 400
350	25	53	99	203	305	587	935	1650	3370	6100	9880	20 300	36 900	58 400
400	23	49	92	189	283	546	870	1540	3140	5680	9190	18 900	34 300	54 300
450	22	46	86	177	266	512	816	1440	2940	5330	8620	17 700	32 200	50 900
500	21	43	82	168	251	484	771	1360	2780	5030	8150	16 700	30 400	48 100
550	20	41	78	159	239	459	732	1290	2640	4780	7740	15 900	28 900	45 700
600	19	39	74	152	228	438	699	1240	2520	4560	7380	15 200	27 500	43 600
650	18	38	71	145	218	420	669	1180	2410	4360	7070	14 500	26 400	41 800
700	17	36	68	140	209	403	643	1140	2320	4190	6790	14 000	25 300	40 100
750	17	35	66	135	202	389	619	1090	2230	4040	6540	13 400	24 400	38 600
800	16	34	63	130	195	375	598	1060	2160	3900	6320	13 000	23 600	37 300
850	16	33	61	126	189	363	579	1020	2090	3780	6110	12 600	22 800	36 100
900	15	32	59	122	183	352	561	992	2020	3660	5930	12 200	22 100	35 000
950	15	31	58	118	178	342	545	963	1960	3550	5760	11 800	21 500	34 000
1000	14	30	56	115	173	333	530	937	1910	3460	5600	11 500	20 900	33 100
1100	14	28	53	109	164	316	503	890	1810	3280	5320	10 900	19 800	31 400
1200	13	27	51	104	156	301	480	849	1730	3130	5070	10 400	18 900	30 000
1300	12	26	49	100	150	289	460	813	1660	3000	4860	9980	18 100	28 700
1400	12	25	47	96	144	277	442	781	1590	2880	4670	9590	17 400	27 600
1500	11	24	45	93	139	267	426	752	1530	2780	4500	9240	16 800	26 600
1600	11	23	44	89	134	258	411	727	1480	2680	4340	8920	16 200	25 600
1700	11	22	42	86	130	250	398	703	1430	2590	4200	8630	15 700	24 800
1800	10	22	41	84	126	242	386	682	1390	2520	4070	8370	15 200	24 100
1900	10	21	40	81	122	235	375	662	1350	2440	3960	8130	14 800	23 400
2000	NA	20	39	79	119	229	364	644	1310	2380	3850	7910	14 400	22 700

For SI units: 1 inch = 25 mm, 1 foot = 304.8 mm, 1 cubic foot per hour = 0.0283 m^3/h, 1 pound-force per square inch = 6.8947 kPa, 1 inch water column = 0.249 kPa

Notes:

[1] Table entries are rounded to 3 significant digits.

[2] NA means a flow of less than 10 ft^3/h (0.283 m^3/h).

TABLE 1315.2(2)
SCHEDULE 40 METALLIC PIPE [NFPA 54: TABLE 6.2.1(c)]*

					GAS:	NATURAL		
					INLET PRESSURE:	LESS THAN 2 psi		
					PRESSURE DROP:	3.0 in. w.c.		
					SPECIFIC GRAVITY:	0.60		
INTENDED USE: INITIAL SUPPLY PRESSURE OF 8.0 IN. W.C. OR GREATER									
	PIPE SIZE (inch)								
NOMINAL:	½	¾	1	1¼	1½	2	2½	3	4
ACTUAL ID:	0.622	0.824	1.049	1.380	1.610	2.067	2.469	3.068	4.026
LENGTH (feet)	CAPACITY IN CUBIC FEET OF GAS PER HOUR								
10	454	949	1790	3670	5500	10 600	16 900	29 800	60 800
20	312	652	1230	2520	3780	7280	11 600	20 500	41 800
30	250	524	986	2030	3030	5840	9310	16 500	33 600
40	214	448	844	1730	2600	5000	7970	14 100	28 700
50	190	397	748	1540	2300	4430	7060	12 500	25 500
60	172	360	678	1390	2090	4020	6400	11 300	23 100
70	158	331	624	1280	1920	3690	5890	10 400	21 200
80	147	308	580	1190	1790	3440	5480	9690	19 800
90	138	289	544	1120	1670	3230	5140	9090	18 500
100	131	273	514	1060	1580	3050	4860	8580	17 500
125	116	242	456	936	1400	2700	4300	7610	15 500
150	105	219	413	848	1270	2450	3900	6890	14 100
175	96	202	380	780	1170	2250	3590	6340	12 900
200	90	188	353	726	1090	2090	3340	5900	12 000
250	80	166	313	643	964	1860	2960	5230	10 700
300	72	151	284	583	873	1680	2680	4740	9660
350	66	139	261	536	803	1550	2470	4360	8890
400	62	129	243	499	747	1440	2290	4050	8270
450	58	121	228	468	701	1350	2150	3800	7760
500	55	114	215	442	662	1280	2030	3590	7330
550	52	109	204	420	629	1210	1930	3410	6960
600	50	104	195	400	600	1160	1840	3260	6640
650	47	99	187	384	575	1110	1760	3120	6360
700	46	95	179	368	552	1060	1690	3000	6110
750	44	92	173	355	532	1020	1630	2890	5890
800	42	89	167	343	514	989	1580	2790	5680
850	41	86	162	332	497	957	1530	2700	5500
900	40	83	157	322	482	928	1480	2610	5330
950	39	81	152	312	468	901	1440	2540	5180
1000	38	79	148	304	455	877	1400	2470	5040
1100	36	75	141	289	432	833	1330	2350	4780
1200	34	71	134	275	412	794	1270	2240	4560
1300	33	68	128	264	395	761	1210	2140	4370
1400	31	65	123	253	379	731	1160	2060	4200
1500	30	63	119	244	366	704	1120	1980	4050
1600	29	61	115	236	353	680	1080	1920	3910
1700	28	59	111	228	342	658	1050	1850	3780
1800	27	57	108	221	331	638	1020	1800	3670
1900	27	56	105	215	322	619	987	1750	3560
2000	26	54	102	209	313	602	960	1700	3460

For SI units: 1 inch = 25 mm, 1 foot = 304.8 mm, 1 cubic foot per hour = 0.0283 m^3/h, 1 pound-force per square inch = 6.8947 kPa, 1 inch water column = 0.249 kPa

* Table entries are rounded to 3 significant digits.

TABLE 1315.2(3)
SCHEDULE 40 METALLIC PIPE [NFPA 54: TABLE 6.2.1(d)]*

						GAS:	NATURAL		
						INLET PRESSURE:	LESS THAN 2 psi		
						PRESSURE DROP:	6.0 in. w.c.		
						SPECIFIC GRAVITY:	0.60		
INTENDED USE: INITIAL SUPPLY PRESSURE OF 11.0 IN. W.C. OR GREATER									
	PIPE SIZE (inch)								
NOMINAL:	½	¾	1	1¼	1½	2	2½	3	4
ACTUAL ID:	0.622	0.824	1.049	1.38	1.61	2.067	2.469	3.068	4.026
LENGTH (feet)	CAPACITY IN CUBIC FEET OF GAS PER HOUR								
10	660	1380	2600	5340	8000	15 400	24 600	43 400	88 500
20	454	949	1790	3670	5500	10 600	16 900	29 800	60 800
30	364	762	1440	2950	4410	8500	13 600	24 000	48 900
40	312	652	1230	2520	3780	7280	11 600	20 500	41 800
50	276	578	1090	2240	3350	6450	10 300	18 200	37 100
60	250	524	986	2030	3030	5840	9310	16 500	33 600
70	230	482	907	1860	2790	5380	8570	15 100	30 900
80	214	448	844	1730	2600	5000	7970	14 100	28 700
90	201	420	792	1630	2440	4690	7480	13 200	27 000
100	190	397	748	1540	2300	4430	7060	12 500	25 500
125	168	352	663	1360	2040	3930	6260	11 100	22 600
150	153	319	601	1230	1850	3560	5670	10 000	20 500
175	140	293	553	1140	1700	3270	5220	9230	18 800
200	131	273	514	1056	1580	3050	4860	8580	17 500
250	116	242	456	936	1400	2700	4300	7610	15 500
300	105	219	413	848	1270	2450	3900	6890	14 100
350	96	202	380	780	1170	2250	3590	6340	12 900
400	90	188	353	726	1090	2090	3340	5900	12 000
450	84	176	332	681	1020	1960	3130	5540	11 300
500	80	166	313	643	964	1860	2960	5230	10 700
550	76	158	297	611	915	1760	2810	4970	10 100
600	72	151	284	583	873	1680	2680	4740	9660
650	69	144	272	558	836	1610	2570	4540	9250
700	66	139	261	536	803	1550	2470	4360	8890
750	64	134	252	516	774	1490	2380	4200	8560
800	62	129	243	499	747	1440	2290	4050	8270
850	60	125	235	483	723	1390	2220	3920	8000
900	58	121	228	468	701	1350	2150	3800	7760
950	56	118	221	454	681	1310	2090	3690	7540
1000	55	114	215	442	662	1280	2030	3590	7330
1100	52	109	204	420	629	1210	1930	3410	6960
1200	50	104	195	400	600	1160	1840	3260	6640
1300	47	99	187	384	575	1110	1760	3120	6360
1400	46	95	179	368	552	1060	1690	3000	6110
1500	44	92	173	355	532	1020	1630	2890	5890
1600	42	89	167	343	514	989	1580	2790	5680
1700	41	86	162	332	497	957	1530	2700	5500
1800	40	83	157	322	482	928	1480	2610	5330
1900	39	81	152	312	468	901	1440	2540	5180
2000	38	79	148	304	455	877	1400	2470	5040

For SI units: 1 inch = 25 mm, 1 foot = 304.8 mm, 1 cubic foot per hour = 0.0283 m^3/h, 1 pound-force per square inch = 6.8947 kPa, 1 inch water column = 0.249 kPa

* Table entries are rounded to 3 significant digits.

TABLE 1315.2(4)
SCHEDULE 40 METALLIC PIPE [NFPA 54: TABLE 6.2.1(e)]*

					GAS:			NATURAL	
					INLET PRESSURE:			2.0 psi	
					PRESSURE DROP:			1.0 psi	
					SPECIFIC GRAVITY:			0.60	
	PIPE SIZE (inch)								
NOMINAL:	½	¾	1	1¼	1½	2	2½	3	4
ACTUAL ID:	0.622	0.824	1.049	1.380	1.610	2.067	2.469	3.068	4.026
LENGTH (feet)	CAPACITY IN CUBIC FEET OF GAS PER HOUR								
10	1510	3040	5560	11 400	17 100	32 900	52 500	92 800	189 000
20	1070	2150	3930	8070	12 100	23 300	37 100	65 600	134 000
30	869	1760	3210	6590	9880	19 000	30 300	53 600	109 000
40	753	1520	2780	5710	8550	16 500	26 300	46 400	94 700
50	673	1360	2490	5110	7650	14 700	23 500	41 500	84 700
60	615	1240	2270	4660	6980	13 500	21 400	37 900	77 300
70	569	1150	2100	4320	6470	12 500	19 900	35 100	71 600
80	532	1080	1970	4040	6050	11 700	18 600	32 800	67 000
90	502	1010	1850	3810	5700	11 000	17 500	30 900	63 100
100	462	934	1710	3510	5260	10 100	16 100	28 500	58 200
125	414	836	1530	3140	4700	9060	14 400	25 500	52 100
150	372	751	1370	2820	4220	8130	13 000	22 900	46 700
175	344	695	1270	2601	3910	7530	12 000	21 200	43 300
200	318	642	1170	2410	3610	6960	11 100	19 600	40 000
250	279	583	1040	2140	3210	6180	9850	17 400	35 500
300	253	528	945	1940	2910	5600	8920	15 800	32 200
350	232	486	869	1790	2670	5150	8210	14 500	29 600
400	216	452	809	1660	2490	4790	7640	13 500	27 500
450	203	424	759	1560	2330	4500	7170	12 700	25 800
500	192	401	717	1470	2210	4250	6770	12 000	24 400
550	182	381	681	1400	2090	4030	6430	11 400	23 200
600	174	363	650	1330	2000	3850	6130	10 800	22 100
650	166	348	622	1280	1910	3680	5870	10 400	21 200
700	160	334	598	1230	1840	3540	5640	9970	20 300
750	154	322	576	1180	1770	3410	5440	9610	19 600
800	149	311	556	1140	1710	3290	5250	9280	18 900
850	144	301	538	1100	1650	3190	5080	8980	18 300
900	139	292	522	1070	1600	3090	4930	8710	17 800
950	135	283	507	1040	1560	3000	4780	8460	17 200
1000	132	275	493	1010	1520	2920	4650	8220	16 800
1100	125	262	468	960	1440	2770	4420	7810	15 900
1200	119	250	446	917	1370	2640	4220	7450	15 200
1300	114	239	427	878	1320	2530	4040	7140	14 600
1400	110	230	411	843	1260	2430	3880	6860	14 000
1500	106	221	396	812	1220	2340	3740	6600	13 500
1600	102	214	382	784	1180	2260	3610	6380	13 000
1700	99	207	370	759	1140	2190	3490	6170	12 600
1800	96	200	358	736	1100	2120	3390	5980	12 200
1900	93	195	348	715	1070	2060	3290	5810	11 900
2000	91	189	339	695	1040	2010	3200	5650	11 500

For SI units: 1 inch = 25 mm, 1 foot = 304.8 mm, 1 cubic foot per hour = 0.0283 m^3/h, 1 pound-force per square inch = 6.8947 kPa

* Table entries are rounded to 3 significant digits.

TABLE 1315.2(5)
SCHEDULE 40 METALLIC PIPE [NFPA 54: TABLE 6.2.1(f)]*

						GAS:	NATURAL		
						INLET PRESSURE:	3.0 psi		
						PRESSURE DROP:	2.0 psi		
						SPECIFIC GRAVITY:	0.60		
	PIPE SIZE (inch)								
NOMINAL:	½	¾	1	1¼	1½	2	2½	3	4
ACTUAL ID:	0.622	0.824	1.049	1.380	1.610	2.067	2.469	3.068	4.026
LENGTH (feet)	CAPACITY IN CUBIC FEET OF GAS PER HOUR								
10	2350	4920	9270	19 000	28 500	54 900	87 500	155 000	316 000
20	1620	3380	6370	13 100	19 600	37 700	60 100	106 000	217 000
30	1300	2720	5110	10 500	15 700	30 300	48 300	85 400	174 000
40	1110	2320	4380	8990	13 500	25 900	41 300	73 100	149 000
50	985	2060	3880	7970	11 900	23 000	36 600	64 800	132 000
60	892	1870	3520	7220	10 800	20 800	33 200	58 700	120 000
70	821	1720	3230	6640	9950	19 200	30 500	54 000	110 000
80	764	1600	3010	6180	9260	17 800	28 400	50 200	102 000
90	717	1500	2820	5800	8680	16 700	26 700	47 100	96 100
100	677	1420	2670	5470	8200	15 800	25 200	44 500	90 800
125	600	1250	2360	4850	7270	14 000	22 300	39 500	80 500
150	544	1140	2140	4400	6590	12 700	20 200	35 700	72 900
175	500	1050	1970	4040	6060	11 700	18 600	32 900	67 100
200	465	973	1830	3760	5640	10 900	17 300	30 600	62 400
250	412	862	1620	3330	5000	9620	15 300	27 100	55 300
300	374	781	1470	3020	4530	8720	13 900	24 600	50 100
350	344	719	1350	2780	4170	8020	12 800	22 600	46 100
400	320	669	1260	2590	3870	7460	11 900	21 000	42 900
450	300	627	1180	2430	3640	7000	11 200	19 700	40 200
500	283	593	1120	2290	3430	6610	10 500	18 600	38 000
550	269	563	1060	2180	3260	6280	10 000	17 700	36 100
600	257	537	1010	2080	3110	5990	9550	16 900	34 400
650	246	514	969	1990	2980	5740	9150	16 200	33 000
700	236	494	931	1910	2860	5510	8790	15 500	31 700
750	228	476	897	1840	2760	5310	8470	15 000	30 500
800	220	460	866	1780	2660	5130	8180	14 500	29 500
850	213	445	838	1720	2580	4960	7910	14 000	28 500
900	206	431	812	1670	2500	4810	7670	13 600	27 700
950	200	419	789	1620	2430	4670	7450	13 200	26 900
1000	195	407	767	1580	2360	4550	7240	12 800	26 100
1100	185	387	729	1500	2240	4320	6890	12 200	24 800
1200	177	369	695	1430	2140	4120	6570	11 600	23 700
1300	169	353	666	1370	2050	3940	6290	11 100	22 700
1400	162	340	640	1310	1970	3790	6040	10 700	21 800
1500	156	327	616	1270	1900	3650	5820	10 300	21 000
1600	151	316	595	1220	1830	3530	5620	10 000	20 300
1700	146	306	576	1180	1770	3410	5440	9610	19 600
1800	142	296	558	1150	1720	3310	5270	9320	19 000
1900	138	288	542	1110	1670	3210	5120	9050	18 400
2000	134	280	527	1080	1620	3120	4980	8800	18 000

For SI units: 1 inch = 25 mm, 1 foot = 304.8 mm, 1 cubic foot per hour = 0.0283 m^3/h, 1 pound-force per square inch = 6.8947 kPa

* Table entries are rounded to 3 significant digits.

TABLE 1315.2(6)
SCHEDULE 40 METALLIC PIPE [NFPA 54: TABLE 6.2.1(g)]*

							GAS:	NATURAL	
							INLET PRESSURE:	5.0 psi	
							PRESSURE DROP:	3.5 psi	
							SPECIFIC GRAVITY:	0.60	
	PIPE SIZE (inch)								
NOMINAL:	½	¾	1	1¼	1½	2	2½	3	4
ACTUAL ID:	0.622	0.824	1.049	1.380	1.610	2.067	2.469	3.068	4.026
LENGTH (feet)	CAPACITY IN CUBIC FEET OF GAS PER HOUR								
10	3190	6430	11 800	24 200	36 200	69 700	111 000	196 000	401 000
20	2250	4550	8320	17 100	25 600	49 300	78 600	139 000	283 000
30	1840	3720	6790	14 000	20 900	40 300	64 200	113 000	231 000
40	1590	3220	5880	12 100	18 100	34 900	55 600	98 200	200 000
50	1430	2880	5260	10 800	16 200	31 200	49 700	87 900	179 000
60	1300	2630	4800	9860	14 800	28 500	45 400	80 200	164 000
70	1200	2430	4450	9130	13 700	26 400	42 000	74 300	151 000
80	1150	2330	4260	8540	12 800	24 700	39 300	69 500	142 000
90	1060	2150	3920	8050	12 100	23 200	37 000	65 500	134 000
100	979	1980	3620	7430	11 100	21 400	34 200	60 400	123 000
125	876	1770	3240	6640	9950	19 200	30 600	54 000	110 000
150	786	1590	2910	5960	8940	17 200	27 400	48 500	98 900
175	728	1470	2690	5520	8270	15 900	25 400	44 900	91 600
200	673	1360	2490	5100	7650	14 700	23 500	41 500	84 700
250	558	1170	2200	4510	6760	13 000	20 800	36 700	74 900
300	506	1060	1990	4090	6130	11 800	18 800	33 300	67 800
350	465	973	1830	3760	5640	10 900	17 300	30 600	62 400
400	433	905	1710	3500	5250	10 100	16 100	28 500	58 100
450	406	849	1600	3290	4920	9480	15 100	26 700	54 500
500	384	802	1510	3100	4650	8950	14 300	25 200	51 500
550	364	762	1440	2950	4420	8500	13 600	24 000	48 900
600	348	727	1370	2810	4210	8110	12 900	22 900	46 600
650	333	696	1310	2690	4030	7770	12 400	21 900	44 600
700	320	669	1260	2590	3880	7460	11 900	21 000	42 900
750	308	644	1210	2490	3730	7190	11 500	20 300	41 300
800	298	622	1170	2410	3610	6940	11 100	19 600	39 900
850	288	602	1130	2330	3490	6720	10 700	18 900	38 600
900	279	584	1100	2260	3380	6520	10 400	18 400	37 400
950	271	567	1070	2190	3290	6330	10 100	17 800	36 400
1000	264	551	1040	2130	3200	6150	9810	17 300	35 400
1100	250	524	987	2030	3030	5840	9320	16 500	33 600
1200	239	500	941	1930	2900	5580	8890	15 700	32 000
1300	229	478	901	1850	2770	5340	8510	15 000	30 700
1400	220	460	866	1780	2660	5130	8180	14 500	29 500
1500	212	443	834	1710	2570	4940	7880	13 900	28 400
1600	205	428	806	1650	2480	4770	7610	13 400	27 400
1700	198	414	780	1600	2400	4620	7360	13 000	26 500
1800	192	401	756	1550	2330	4480	7140	12 600	25 700
1900	186	390	734	1510	2260	4350	6930	12 300	25 000
2000	181	379	714	1470	2200	4230	6740	11 900	24 300

For SI units: 1 inch = 25 mm, 1 foot = 304.8 mm, 1 cubic foot per hour = 0.0283 m^3/h, 1 pound-force per square inch = 6.8947 kPa

* Table entries are rounded to 3 significant digits.

TABLE 1315.2(7)
SEMI-RIGID COPPER TUBING [NFPA 54: TABLE 6.2.1(h)][1, 2]

							GAS:	NATURAL	
							INLET PRESSURE:	LESS THAN 2 psi	
							PRESSURE DROP:	0.3 in. w.c.	
							SPECIFIC GRAVITY:	0.60	
	TUBE SIZE (inch)								
NOMINAL: K & L:	¼	⅜	½	⅝	¾	1	1¼	1½	2
NOMINAL: ACR:	⅜	½	⅝	¾	⅞	1⅛	1⅜	–	–
OUTSIDE:	0.375	0.500	0.625	0.750	0.875	1.125	1.375	1.625	2.125
INSIDE:[3]	0.305	0.402	0.527	0.652	0.745	0.995	1.245	1.481	1.959
LENGTH (feet)	CAPACITY IN CUBIC FEET OF GAS PER HOUR								
10	20	42	85	148	210	448	806	1270	2650
20	14	29	58	102	144	308	554	873	1820
30	11	23	47	82	116	247	445	701	1460
40	10	20	40	70	99	211	381	600	1250
50	NA	17	35	62	88	187	337	532	1110
60	NA	16	32	56	79	170	306	482	1000
70	NA	14	29	52	73	156	281	443	924
80	NA	13	27	48	68	145	262	413	859
90	NA	13	26	45	64	136	245	387	806
100	NA	12	24	43	60	129	232	366	761
125	NA	11	22	38	53	114	206	324	675
150	NA	10	20	34	48	103	186	294	612
175	NA	NA	18	31	45	95	171	270	563
200	NA	NA	17	29	41	89	159	251	523
250	NA	NA	15	26	37	78	141	223	464
300	NA	NA	13	23	33	71	128	202	420
350	NA	NA	12	22	31	65	118	186	387
400	NA	NA	11	20	28	61	110	173	360
450	NA	NA	11	19	27	57	103	162	338
500	NA	NA	10	18	25	54	97	153	319
550	NA	NA	NA	17	24	51	92	145	303
600	NA	NA	NA	16	23	49	88	139	289
650	NA	NA	NA	15	22	47	84	133	277
700	NA	NA	NA	15	21	45	81	128	266
750	NA	NA	NA	14	20	43	78	123	256
800	NA	NA	NA	14	20	42	75	119	247
850	NA	NA	NA	13	19	40	73	115	239
900	NA	NA	NA	13	18	39	71	111	232
950	NA	NA	NA	13	18	38	69	108	225
1000	NA	NA	NA	12	17	37	67	105	219
1100	NA	NA	NA	12	16	35	63	100	208
1200	NA	NA	NA	11	16	34	60	95	199
1300	NA	NA	NA	11	15	32	58	91	190
1400	NA	NA	NA	10	14	31	56	88	183
1500	NA	NA	NA	NA	14	30	54	84	176
1600	NA	NA	NA	NA	13	29	52	82	170
1700	NA	NA	NA	NA	13	28	50	79	164
1800	NA	NA	NA	NA	13	27	49	77	159
1900	NA	NA	NA	NA	12	26	47	74	155
2000	NA	NA	NA	NA	12	25	46	72	151

For SI units: 1 inch = 25 mm, 1 foot = 304.8 mm, 1 cubic foot per hour = 0.0283 m^3/h, 1 pound-force per square inch = 6.8947 kPa, 1 inch water column = 0.249 kPa

Notes:

1 Table entries are rounded to 3 significant digits.

2 NA means a flow of less than 10 ft^3/h (0.283 m^3/h).

3 Table capacities are based on Type K copper tubing inside diameter (shown), which has the smallest inside diameter of the copper tubing products.

TABLE 1315.2(8)
SEMI-RIGID COPPER TUBING [NFPA 54: TABLE 6.2.1(i)][1, 2]

						GAS:		NATURAL	
						INLET PRESSURE:		LESS THAN 2 psi	
						PRESSURE DROP:		0.5 in. w.c.	
						SPECIFIC GRAVITY:		0.60	
	TUBE SIZE (inch)								
NOMINAL: K & L:	¼	⅜	½	⅝	¾	1	1¼	1½	2
NOMINAL: ACR:	⅜	½	⅝	¾	⅞	1⅛	1⅜	–	–
OUTSIDE:	0.375	0.500	0.625	0.750	0.875	1.125	1.375	1.625	2.125
INSIDE:[3]	0.305	0.402	0.527	0.652	0.745	0.995	1.245	1.481	1.959
LENGTH (feet)	CAPACITY IN CUBIC FEET OF GAS PER HOUR								
10	27	55	111	195	276	590	1060	1680	3490
20	18	38	77	134	190	406	730	1150	2400
30	15	30	61	107	152	326	586	925	1930
40	13	26	53	92	131	279	502	791	1650
50	11	23	47	82	116	247	445	701	1460
60	10	21	42	74	105	224	403	635	1320
70	NA	19	39	68	96	206	371	585	1220
80	NA	18	36	63	90	192	345	544	1130
90	NA	17	34	59	84	180	324	510	1060
100	NA	16	32	56	79	170	306	482	1000
125	NA	14	28	50	70	151	271	427	890
150	NA	13	26	45	64	136	245	387	806
175	NA	12	24	41	59	125	226	356	742
200	NA	11	22	39	55	117	210	331	690
250	NA	NA	20	34	48	103	186	294	612
300	NA	NA	18	31	44	94	169	266	554
350	NA	NA	16	28	40	86	155	245	510
400	NA	NA	15	26	38	80	144	228	474
450	NA	NA	14	25	35	75	135	214	445
500	NA	NA	13	23	33	71	128	202	420
550	NA	NA	13	22	32	68	122	192	399
600	NA	NA	12	21	30	64	116	183	381
650	NA	NA	12	20	29	62	111	175	365
700	NA	NA	11	20	28	59	107	168	350
750	NA	NA	11	19	27	57	103	162	338
800	NA	NA	10	18	26	55	99	156	326
850	NA	NA	10	18	25	53	96	151	315
900	NA	NA	NA	17	24	52	93	147	306
950	NA	NA	NA	17	24	50	90	143	297
1000	NA	NA	NA	16	23	49	88	139	289
1100	NA	NA	NA	15	22	46	84	132	274
1200	NA	NA	NA	15	21	44	80	126	262
1300	NA	NA	NA	14	20	42	76	120	251
1400	NA	NA	NA	13	19	41	73	116	241
1500	NA	NA	NA	13	18	39	71	111	232
1600	NA	NA	NA	13	18	38	68	108	224
1700	NA	NA	NA	12	17	37	66	104	217
1800	NA	NA	NA	12	17	36	64	101	210
1900	NA	NA	NA	11	16	35	62	98	204
2000	NA	NA	NA	11	16	34	60	95	199

For SI units: 1 inch = 25 mm, 1 foot = 304.8 mm, 1 cubic foot per hour = 0.0283 m^3/h, 1 pound-force per square inch = 6.8947 kPa, 1 inch water column = 0.249 kPa

Notes:

[1] Table entries are rounded to 3 significant digits.

[2] NA means a flow of less than 10 ft^3/h (0.283 m^3/h).

[3] Table capacities are based on Type K copper tubing inside diameter (shown), which has the smallest inside diameter of the copper tubing products.

TABLE 1315.2(9)
SEMI-RIGID COPPER TUBING [NFPA 54: TABLE 6.2.1(j)][1, 2]

					GAS:		**NATURAL**		
					INLET PRESSURE:		**LESS THAN 2 psi**		
					PRESSURE DROP:		**1.0 in. w.c.**		
					SPECIFIC GRAVITY:		**0.60**		
INTENDED USE: TUBE SIZING BETWEEN HOUSE LINE REGULATOR AND THE APPLIANCE									
	TUBE SIZE (inch)								
NOMINAL: K & L:	¼	⅜	½	⅝	¾	1	1¼	1½	2
NOMINAL: ACR:	⅜	½	⅝	¾	⅞	1⅛	1⅜	–	–
OUTSIDE:	0.375	0.500	0.625	0.750	0.875	1.125	1.375	1.625	2.125
INSIDE:[3]	0.305	0.402	0.527	0.652	0.745	0.995	1.245	1.481	1.959
LENGTH (feet)	**CAPACITY IN CUBIC FEET OF GAS PER HOUR**								
10	39	80	162	283	402	859	1550	2440	5080
20	27	55	111	195	276	590	1060	1680	3490
30	21	44	89	156	222	474	853	1350	2800
40	18	38	77	134	190	406	730	1150	2400
50	16	33	68	119	168	359	647	1020	2130
60	15	30	61	107	152	326	586	925	1930
70	13	28	57	99	140	300	539	851	1770
80	13	26	53	92	131	279	502	791	1650
90	12	24	49	86	122	262	471	742	1550
100	11	23	47	82	116	247	445	701	1460
125	NA	20	41	72	103	219	394	622	1290
150	NA	18	37	65	93	198	357	563	1170
175	NA	17	34	60	85	183	329	518	1080
200	NA	16	32	56	79	170	306	482	1000
250	NA	14	28	50	70	151	271	427	890
300	NA	13	26	45	64	136	245	387	806
350	NA	12	24	41	59	125	226	356	742
400	NA	11	22	39	55	117	210	331	690
450	NA	10	21	36	51	110	197	311	647
500	NA	NA	20	34	48	103	186	294	612
550	NA	NA	19	32	46	98	177	279	581
600	NA	NA	18	31	44	94	169	266	554
650	NA	NA	17	30	42	90	162	255	531
700	NA	NA	16	28	40	86	155	245	510
750	NA	NA	16	27	39	83	150	236	491
800	NA	NA	15	26	38	80	144	228	474
850	NA	NA	15	26	36	78	140	220	459
900	NA	NA	14	25	35	75	135	214	445
950	NA	NA	14	24	34	73	132	207	432
1000	NA	NA	13	23	33	71	128	202	420
1100	NA	NA	13	22	32	68	122	192	399
1200	NA	NA	12	21	30	64	116	183	381
1300	NA	NA	12	20	29	62	111	175	365
1400	NA	NA	11	20	28	59	107	168	350
1500	NA	NA	11	19	27	57	103	162	338
1600	NA	NA	10	18	26	55	99	156	326
1700	NA	NA	10	18	25	53	96	151	315
1800	NA	NA	NA	17	24	52	93	147	306
1900	NA	NA	NA	17	24	50	90	143	297
2000	NA	NA	NA	16	23	49	88	139	289

For SI units: 1 inch = 25 mm, 1 foot = 304.8 mm, 1 cubic foot per hour = 0.0283 m^3/h, 1 pound-force per square inch = 6.8947 kPa, 1 inch water column = 0.249 kPa

Notes:

1 Table entries are rounded to 3 significant digits.

2 NA means a flow of less than 10 ft^3/h (0.283 m^3/h).

3 Table capacities are based on Type K copper tubing inside diameter (shown), which has the smallest inside diameter of the copper tubing products.

TABLE 1315.2(10)
SEMI-RIGID COPPER TUBING [NFPA 54: TABLE 6.2.1(k)][2]

							GAS:	NATURAL	
							INLET PRESSURE:	LESS THAN 2 psi	
							PRESSURE DROP:	17.0 in. w.c.	
							SPECIFIC GRAVITY:	0.60	
	TUBE SIZE (inch)								
NOMINAL: K & L:	¼	⅜	½	⅝	¾	1	1¼	1½	2
NOMINAL: ACR:	⅜	½	⅝	¾	⅞	1⅛	1⅜	–	–
OUTSIDE:	0.375	0.500	0.625	0.750	0.875	1.125	1.375	1.625	2.125
INSIDE:[1]	0.305	0.402	0.527	0.652	0.745	0.995	1.245	1.481	1.959
LENGTH (feet)	CAPACITY IN CUBIC FEET OF GAS PER HOUR								
10	190	391	796	1390	1970	4220	7590	12 000	24 900
20	130	269	547	956	1360	2900	5220	8230	17 100
30	105	216	439	768	1090	2330	4190	6610	13 800
40	90	185	376	657	932	1990	3590	5650	11 800
50	79	164	333	582	826	1770	3180	5010	10 400
60	72	148	302	528	749	1600	2880	4540	9460
70	66	137	278	486	689	1470	2650	4180	8700
80	62	127	258	452	641	1370	2460	3890	8090
90	58	119	243	424	601	1280	2310	3650	7590
100	55	113	229	400	568	1210	2180	3440	7170
125	48	100	203	355	503	1080	1940	3050	6360
150	44	90	184	321	456	974	1750	2770	5760
175	40	83	169	296	420	896	1610	2540	5300
200	38	77	157	275	390	834	1500	2370	4930
250	33	69	140	244	346	739	1330	2100	4370
300	30	62	126	221	313	670	1210	1900	3960
350	28	57	116	203	288	616	1110	1750	3640
400	26	53	108	189	268	573	1030	1630	3390
450	24	50	102	177	252	538	968	1530	3180
500	23	47	96	168	238	508	914	1440	3000
550	22	45	91	159	226	482	868	1370	2850
600	21	43	87	152	215	460	829	1310	2720
650	20	41	83	145	206	441	793	1250	2610
700	19	39	80	140	198	423	762	1200	2500
750	18	38	77	135	191	408	734	1160	2410
800	18	37	74	130	184	394	709	1120	2330
850	17	35	72	126	178	381	686	1080	2250
900	17	34	70	122	173	370	665	1050	2180
950	16	33	68	118	168	359	646	1020	2120
1000	16	32	66	115	163	349	628	991	2060
1100	15	31	63	109	155	332	597	941	1960
1200	14	29	60	104	148	316	569	898	1870
1300	14	28	57	100	142	303	545	860	1790
1400	13	27	55	96	136	291	524	826	1720
1500	13	26	53	93	131	280	505	796	1660
1600	12	25	51	89	127	271	487	768	1600
1700	12	24	49	86	123	262	472	744	1550
1800	11	24	48	84	119	254	457	721	1500
1900	11	23	47	81	115	247	444	700	1460
2000	11	22	45	79	112	240	432	681	1420

For SI units: 1 inch = 25 mm, 1 foot = 304.8 mm, 1 cubic foot per hour = 0.0283 m^3/h, 1 pound-force per square inch = 6.8947 kPa, 1 inch water column = 0.249 kPa

Notes:

[1] Table capacities are based on Type K copper tubing inside diameter (shown), which has the smallest inside diameter of the copper tubing products.

[2] Table entries are rounded to 3 significant digits.

TABLE 1315.2(11)
SEMI-RIGID COPPER TUBING [NFPA 54: TABLE 6.2.1(I)][2]

						GAS:	NATURAL		
						INLET PRESSURE:	2.0 psi		
						PRESSURE DROP:	1.0 psi		
						SPECIFIC GRAVITY:	0.60		
	TUBE SIZE (inch)								
NOMINAL: K & L:	¼	⅜	½	⅝	¾	1	1¼	1½	2
NOMINAL: ACR:	⅜	½	⅝	¾	⅞	1⅛	1⅜	–	–
OUTSIDE:	0.375	0.500	0.625	0.750	0.875	1.125	1.375	1.625	2.125
INSIDE:[1]	0.305	0.402	0.527	0.652	0.745	0.995	1.245	1.481	1.959
LENGTH (feet)	CAPACITY IN CUBIC FEET OF GAS PER HOUR								
10	245	506	1030	1800	2550	5450	9820	15 500	32 200
20	169	348	708	1240	1760	3750	6750	10 600	22 200
30	135	279	568	993	1410	3010	5420	8550	17 800
40	116	239	486	850	1210	2580	4640	7310	15 200
50	103	212	431	754	1070	2280	4110	6480	13 500
60	93	192	391	683	969	2070	3730	5870	12 200
70	86	177	359	628	891	1900	3430	5400	11 300
80	80	164	334	584	829	1770	3190	5030	10 500
90	75	154	314	548	778	1660	2990	4720	9820
100	71	146	296	518	735	1570	2830	4450	9280
125	63	129	263	459	651	1390	2500	3950	8220
150	57	117	238	416	590	1260	2270	3580	7450
175	52	108	219	383	543	1160	2090	3290	6850
200	49	100	204	356	505	1080	1940	3060	6380
250	43	89	181	315	448	956	1720	2710	5650
300	39	80	164	286	406	866	1560	2460	5120
350	36	74	150	263	373	797	1430	2260	4710
400	33	69	140	245	347	741	1330	2100	4380
450	31	65	131	230	326	696	1250	1970	4110
500	30	61	124	217	308	657	1180	1870	3880
550	28	58	118	206	292	624	1120	1770	3690
600	27	55	112	196	279	595	1070	1690	3520
650	26	53	108	188	267	570	1030	1620	3370
700	25	51	103	181	256	548	986	1550	3240
750	24	49	100	174	247	528	950	1500	3120
800	23	47	96	168	239	510	917	1450	3010
850	22	46	93	163	231	493	888	1400	2920
900	22	44	90	158	224	478	861	1360	2830
950	21	43	88	153	217	464	836	1320	2740
1000	20	42	85	149	211	452	813	1280	2670
1100	19	40	81	142	201	429	772	1220	2540
1200	18	38	77	135	192	409	737	1160	2420
1300	18	36	74	129	183	392	705	1110	2320
1400	17	35	71	124	176	376	678	1070	2230
1500	16	34	68	120	170	363	653	1030	2140
1600	16	33	66	116	164	350	630	994	2070
1700	15	31	64	112	159	339	610	962	2000
1800	15	30	62	108	154	329	592	933	1940
1900	14	30	60	105	149	319	575	906	1890
2000	14	29	59	102	145	310	559	881	1830

For SI units: 1 inch = 25 mm, 1 foot = 304.8 mm, 1 cubic foot per hour = 0.0283 m^3/h, 1 pound-force per square inch = 6.8947 kPa

Notes:

[1] Table capacities are based on Type K copper tubing inside diameter (shown), which has the smallest inside diameter of the copper tubing products.

[2] Table entries are rounded to 3 significant digits.

TABLE 1315.2(12)
SEMI-RIGID COPPER TUBING [NFPA 54: TABLE 6.2.1(m)][3]

GAS:	NATURAL
INLET PRESSURE:	2.0 psi
PRESSURE DROP:	1.5 psi
SPECIFIC GRAVITY:	0.60

INTENDED USE: PIPE SIZING BETWEEN POINT OF DELIVERY AND THE HOUSE LINE REGULATOR. TOTAL LOAD SUPPLIED BY A SINGLE HOUSE LINE REGULATOR NOT EXCEEDING 150 CUBIC FEET PER HOUR[2].

	TUBE SIZE (inch)								
NOMINAL: K & L:	¼	⅜	½	⅝	¾	1	1¼	1½	2
NOMINAL: ACR:	⅜	½	⅝	¾	⅞	1⅛	1⅜	–	–
OUTSIDE:	0.375	0.500	0.625	0.750	0.875	1.125	1.375	1.625	2.125
INSIDE:[1]	0.305	0.402	0.527	0.652	0.745	0.995	1.245	1.481	1.959
LENGTH (feet)	CAPACITY IN CUBIC FEET OF GAS PER HOUR								
10	303	625	1270	2220	3150	6740	12 100	19 100	39 800
20	208	430	874	1530	2170	4630	8330	13 100	27 400
30	167	345	702	1230	1740	3720	6690	10 600	22 000
40	143	295	601	1050	1490	3180	5730	9030	18 800
50	127	262	532	931	1320	2820	5080	8000	16 700
60	115	237	482	843	1200	2560	4600	7250	15 100
70	106	218	444	776	1100	2350	4230	6670	13 900
80	98	203	413	722	1020	2190	3940	6210	12 900
90	92	190	387	677	961	2050	3690	5820	12 100
100	87	180	366	640	907	1940	3490	5500	11 500
125	77	159	324	567	804	1720	3090	4880	10 200
150	70	144	294	514	729	1560	2800	4420	9200
175	64	133	270	472	670	1430	2580	4060	8460
200	60	124	252	440	624	1330	2400	3780	7870
250	53	110	223	390	553	1180	2130	3350	6980
300	48	99	202	353	501	1070	1930	3040	6320
350	44	91	186	325	461	984	1770	2790	5820
400	41	85	173	302	429	916	1650	2600	5410
450	39	80	162	283	402	859	1550	2440	5080
500	36	75	153	268	380	811	1460	2300	4800
550	35	72	146	254	361	771	1390	2190	4560
600	33	68	139	243	344	735	1320	2090	4350
650	32	65	133	232	330	704	1270	2000	4160
700	30	63	128	223	317	676	1220	1920	4000
750	29	60	123	215	305	652	1170	1850	3850
800	28	58	119	208	295	629	1130	1790	3720
850	27	57	115	201	285	609	1100	1730	3600
900	27	55	111	195	276	590	1060	1680	3490
950	26	53	108	189	268	573	1030	1630	3390
1000	25	52	105	184	261	558	1000	1580	3300
1100	24	49	100	175	248	530	954	1500	3130
1200	23	47	95	167	237	505	910	1430	2990
1300	22	45	91	160	227	484	871	1370	2860
1400	21	43	88	153	218	465	837	1320	2750
1500	20	42	85	148	210	448	806	1270	2650
1600	19	40	82	143	202	432	779	1230	2560
1700	19	39	79	138	196	419	753	1190	2470
1800	18	38	77	134	190	406	731	1150	2400
1900	18	37	74	130	184	394	709	1120	2330
2000	17	36	72	126	179	383	690	1090	2270

For SI units: 1 inch = 25 mm, 1 foot = 304.8 mm, 1 cubic foot per hour = 0.0283 m^3/h, 1 pound-force per square inch = 6.8947 kPa

Notes:

[1] Table capacities are based on Type K copper tubing inside diameter (shown), which has the smallest inside diameter of the copper tubing products.

[2] Where this table is used to size the tubing upstream of a line pressure regulator, the pipe or tubing downstream of the line pressure regulator shall be sized using a pressure drop no greater than 1 inch water column (0.249 kPa).

[3] Table entries are rounded to 3 significant digits.

TABLE 1315.2(13)
SEMI-RIGID COPPER TUBING [NFPA 54: TABLE 6.2.1(n)][2]

					GAS:	NATURAL			
					INLET PRESSURE:	5.0 psi			
					PRESSURE DROP:	3.5 psi			
					SPECIFIC GRAVITY:	0.60			
	TUBE SIZE (inch)								
NOMINAL: K & L:	¼	⅜	½	⅝	¾	1	1¼	1½	2
NOMINAL: ACR:	⅜	½	⅝	¾	⅞	1⅛	1⅜	–	–
OUTSIDE:	0.375	0.500	0.625	0.750	0.875	1.125	1.375	1.625	2.125
INSIDE:[1]	0.305	0.402	0.527	0.652	0.745	0.995	1.245	1.481	1.959
LENGTH (feet)	CAPACITY IN CUBIC FEET OF GAS PER HOUR								
10	511	1050	2140	3750	5320	11 400	20 400	32 200	67 100
20	351	724	1470	2580	3650	7800	14 000	22 200	46 100
30	282	582	1180	2070	2930	6270	11 300	17 800	37 000
40	241	498	1010	1770	2510	5360	9660	15 200	31 700
50	214	441	898	1570	2230	4750	8560	13 500	28 100
60	194	400	813	1420	2020	4310	7750	12 200	25 500
70	178	368	748	1310	1860	3960	7130	11 200	23 400
80	166	342	696	1220	1730	3690	6640	10 500	21 800
90	156	321	653	1140	1620	3460	6230	9820	20 400
100	147	303	617	1080	1530	3270	5880	9270	19 300
125	130	269	547	955	1360	2900	5210	8220	17 100
150	118	243	495	866	1230	2620	4720	7450	15 500
175	109	224	456	796	1130	2410	4350	6850	14 300
200	101	208	424	741	1050	2250	4040	6370	13 300
250	90	185	376	657	932	1990	3580	5650	11 800
300	81	167	340	595	844	1800	3250	5120	10 700
350	75	154	313	547	777	1660	2990	4710	9810
400	69	143	291	509	722	1540	2780	4380	9120
450	65	134	273	478	678	1450	2610	4110	8560
500	62	127	258	451	640	1370	2460	3880	8090
550	58	121	245	429	608	1300	2340	3690	7680
600	56	115	234	409	580	1240	2230	3520	7330
650	53	110	224	392	556	1190	2140	3370	7020
700	51	106	215	376	534	1140	2050	3240	6740
750	49	102	207	362	514	1100	1980	3120	6490
800	48	98	200	350	497	1060	1910	3010	6270
850	46	95	194	339	481	1030	1850	2910	6070
900	45	92	188	328	466	1000	1790	2820	5880
950	43	90	182	319	452	967	1740	2740	5710
1000	42	87	177	310	440	940	1690	2670	5560
1100	40	83	169	295	418	893	1610	2530	5280
1200	38	79	161	281	399	852	1530	2420	5040
1300	37	76	154	269	382	816	1470	2320	4820
1400	35	73	148	259	367	784	1410	2220	4630
1500	34	70	143	249	353	755	1360	2140	4460
1600	33	68	138	241	341	729	1310	2070	4310
1700	32	65	133	233	330	705	1270	2000	4170
1800	31	63	129	226	320	684	1230	1940	4040
1900	30	62	125	219	311	664	1200	1890	3930
2000	29	60	122	213	302	646	1160	1830	3820

For SI units: 1 inch = 25 mm, 1 foot = 304.8 mm, 1 cubic foot per hour = 0.0283 m^3/h, 1 pound-force per square inch = 6.8947 kPa

Notes:

[1] Table capacities are based on Type K copper tubing inside diameter (shown), which has the smallest inside diameter of the copper tubing products.

[2] Table entries are rounded to 3 significant digits.

TABLE 1315.2(14)
CORRUGATED STAINLESS STEEL TUBING (CSST) [NFPA 54: TABLE 6.2.1(o)][1, 2]

										GAS:	**NATURAL**			
										INLET PRESSURE:	**LESS THAN 2 psi**			
										PRESSURE DROP:	**0.5 in. w.c.**			
										SPECIFIC GRAVITY:	**0.60**			
	TUBE SIZE (EHD)[3]													
FLOW DESIGNATION:	**13**	**15**	**18**	**19**	**23**	**25**	**30**	**31**	**37**	**39**	**46**	**48**	**60**	**62**
LENGTH (feet)	**CAPACITY IN CUBIC FEET OF GAS PER HOUR**													
5	46	63	115	134	225	270	471	546	895	1037	1790	2070	3660	4140
10	32	44	82	95	161	192	330	383	639	746	1260	1470	2600	2930
15	25	35	66	77	132	157	267	310	524	615	1030	1200	2140	2400
20	22	31	58	67	116	137	231	269	456	536	888	1050	1850	2080
25	19	27	52	60	104	122	206	240	409	482	793	936	1660	1860
30	18	25	47	55	96	112	188	218	374	442	723	856	1520	1700
40	15	21	41	47	83	97	162	188	325	386	625	742	1320	1470
50	13	19	37	42	75	87	144	168	292	347	559	665	1180	1320
60	12	17	34	38	68	80	131	153	267	318	509	608	1080	1200
70	11	16	31	36	63	74	121	141	248	295	471	563	1000	1110
80	10	15	29	33	60	69	113	132	232	277	440	527	940	1040
90	10	14	28	32	57	65	107	125	219	262	415	498	887	983
100	9	13	26	30	54	62	101	118	208	249	393	472	843	933
150	7	10	20	23	42	48	78	91	171	205	320	387	691	762
200	6	9	18	21	38	44	71	82	148	179	277	336	600	661
250	5	8	16	19	34	39	63	74	133	161	247	301	538	591
300	5	7	15	17	32	36	57	67	95	148	226	275	492	540

For SI units: 1 inch = 25 mm, 1 foot = 304.8 mm, 1 cubic foot per hour = 0.0283 m^3/h, 1 pound-force per square inch = 6.8947 kPa, 1 inch water column = 0.249 kPa

Notes:

[1] Table entries are rounded to 3 significant digits.

[2] Table includes losses for four 90 degree (1.57 rad) bends and two end fittings. Tubing runs with larger numbers of bends, fittings, or both shall be increased by an equivalent length of tubing to the following equation: $L = 1.3\ n$, where L is additional length (ft) of tubing and n is the number of additional fittings, bends, or both.

[3] EHD = Equivalent Hydraulic Diameter, which is a measure of the relative hydraulic efficiency between different tubing sizes. The greater the value of EHD, the greater the gas capacity of the tubing.

TABLE 1315.2(15)
CORRUGATED STAINLESS STEEL TUBING (CSST) [NFPA 54: TABLE 6.2.1(p)][1, 2]

										GAS:	NATURAL			
										INLET PRESSURE:	LESS THAN 2 psi			
										PRESSURE DROP:	3.0 in. w.c.			
										SPECIFIC GRAVITY:	0.60			
INTENDED USE: INITIAL SUPPLY PRESSURE OF 8.0 INCH WATER COLUMN OR GREATER														
	TUBE SIZE (EHD)[3]													
FLOW DESIGNATION:	13	15	18	19	23	25	30	31	37	39	46	48	60	62
LENGTH (feet)	CAPACITY IN CUBIC FEET OF GAS PER HOUR													
5	120	160	277	327	529	649	1180	1370	2140	2423	4430	5010	8800	10 100
10	83	112	197	231	380	462	828	958	1530	1740	3200	3560	6270	7160
15	67	90	161	189	313	379	673	778	1250	1433	2540	2910	5140	5850
20	57	78	140	164	273	329	580	672	1090	1249	2200	2530	4460	5070
25	51	69	125	147	245	295	518	599	978	1123	1960	2270	4000	4540
30	46	63	115	134	225	270	471	546	895	1029	1790	2070	3660	4140
40	39	54	100	116	196	234	407	471	778	897	1550	1800	3180	3590
50	35	48	89	104	176	210	363	421	698	806	1380	1610	2850	3210
60	32	44	82	95	161	192	330	383	639	739	1260	1470	2600	2930
70	29	41	76	88	150	178	306	355	593	686	1170	1360	2420	2720
80	27	38	71	82	141	167	285	331	555	644	1090	1280	2260	2540
90	26	36	67	77	133	157	268	311	524	609	1030	1200	2140	2400
100	24	34	63	73	126	149	254	295	498	579	974	1140	2030	2280
150	19	27	52	60	104	122	206	240	409	477	793	936	1660	1860
200	17	23	45	52	91	106	178	207	355	415	686	812	1440	1610
250	15	21	40	46	82	95	159	184	319	373	613	728	1290	1440
300	13	19	37	42	75	87	144	168	234	342	559	665	1180	1320

For SI units: 1 foot = 304.8 mm, 1 cubic foot per hour = 0.0283 m^3/h, 1 pound-force per square inch = 6.8947 kPa, 1 inch water column = 0.249 kPa

Notes:

1 Table entries are rounded to 3 significant digits.

2 Table includes losses for four 90 degree (1.57 rad) bends and two end fittings. Tubing runs with larger numbers of bends, fittings, or both shall be increased by an equivalent length of tubing to the following equation: $L = 1.3\ n$, where L is additional length (ft) of tubing and n is the number of additional fittings, bends, or both.

3 EHD = Equivalent Hydraulic Diameter, which is a measure of the relative hydraulic efficiency between different tubing sizes. The greater the value of EHD, the greater the gas capacity of the tubing.

TABLE 1315.2(16)
CORRUGATED STAINLESS STEEL TUBING (CSST) [NFPA 54: TABLE 6.2.1(q)][1, 2]

											GAS:	NATURAL		
											INLET PRESSURE:	LESS THAN 2 psi		
											PRESSURE DROP:	6.0 in. w.c.		
											SPECIFIC GRAVITY:	0.60		
INTENDED USE: INITIAL SUPPLY PRESSURE OF 11.0 INCH WATER COLUMN OR GREATER														
	TUBE SIZE (EHD)[3]													
FLOW DESIGNATION:	**13**	**15**	**18**	**19**	**23**	**25**	**30**	**31**	**37**	**39**	**46**	**48**	**60**	**62**
LENGTH (feet)	CAPACITY IN CUBIC FEET OF GAS PER HOUR													
5	173	229	389	461	737	911	1690	1950	3000	3375	6280	7050	12 400	14 260
10	120	160	277	327	529	649	1180	1370	2140	2423	4430	5010	8800	10 100
15	96	130	227	267	436	532	960	1110	1760	1996	3610	4100	7210	8260
20	83	112	197	231	380	462	828	958	1530	1740	3120	3560	6270	7160
25	74	99	176	207	342	414	739	855	1370	1564	2790	3190	5620	6400
30	67	90	161	189	313	379	673	778	1250	1433	2540	2910	5140	5850
40	57	78	140	164	273	329	580	672	1090	1249	2200	2530	4460	5070
50	51	69	125	147	245	295	518	599	978	1123	1960	2270	4000	4540
60	46	63	115	134	225	270	471	546	895	1029	1790	2070	3660	4140
70	42	58	106	124	209	250	435	505	830	956	1660	1920	3390	3840
80	39	54	100	116	196	234	407	471	778	897	1550	1800	3180	3590
90	37	51	94	109	185	221	383	444	735	848	1460	1700	3000	3390
100	35	48	89	104	176	210	363	421	698	806	1380	1610	2850	3210
150	28	39	73	85	145	172	294	342	573	664	1130	1320	2340	2630
200	24	34	63	73	126	149	254	295	498	579	974	1140	2030	2280
250	21	30	57	66	114	134	226	263	447	520	870	1020	1820	2040
300	19	27	52	60	104	122	206	240	409	477	793	936	1660	1860

For SI units: 1 foot = 304.8 mm, 1 cubic foot per hour = 0.0283 m^3/h, 1 pound-force per square inch = 6.8947 kPa, 1 inch water column = 0.249 kPa

Notes:

[1] Table entries are rounded to 3 significant digits.

[2] Table includes losses for four 90 degree (1.57 rad) bends and two end fittings. Tubing runs with larger numbers of bends, fittings, or both shall be increased by an equivalent length of tubing to the following equation: $L = 1.3\ n$, where L is additional length (ft) of tubing and n is the number of additional fittings, bends, or both.

[3] EHD = Equivalent Hydraulic Diameter, which is a measure of the relative hydraulic efficiency between different tubing sizes. The greater the value of EHD, the greater the gas capacity of the tubing.

TABLE 1315.2(17)
CORRUGATED STAINLESS STEEL TUBING (CSST) [NFPA 54: TABLE 6.2.1(r)][1, 2, 3, 4]

									GAS:		NATURAL			
									INLET PRESSURE:		2.0 psi			
									PRESSURE DROP:		1.0 psi			
									SPECIFIC GRAVITY:		0.60			
	TUBE SIZE (EHD)[5]													
FLOW DESIGNATION:	13	15	18	19	23	25	30	31	37	39	46	48	60	62
LENGTH (feet)	CAPACITY IN CUBIC FEET OF GAS PER HOUR													
10	270	353	587	700	1100	1370	2590	2990	4510	5037	9600	10 700	18 600	21 600
25	166	220	374	444	709	876	1620	1870	2890	3258	6040	6780	11 900	13 700
30	151	200	342	405	650	801	1480	1700	2640	2987	5510	6200	10 900	12 500
40	129	172	297	351	567	696	1270	1470	2300	2605	4760	5380	9440	10 900
50	115	154	266	314	510	624	1140	1310	2060	2343	4260	4820	8470	9720
75	93	124	218	257	420	512	922	1070	1690	1932	3470	3950	6940	7940
80	89	120	211	249	407	496	892	1030	1640	1874	3360	3820	6730	7690
100	79	107	189	222	366	445	795	920	1470	1685	3000	3420	6030	6880
150	64	87	155	182	302	364	646	748	1210	1389	2440	2800	4940	5620
200	55	75	135	157	263	317	557	645	1050	1212	2110	2430	4290	4870
250	49	67	121	141	236	284	497	576	941	1090	1890	2180	3850	4360
300	44	61	110	129	217	260	453	525	862	999	1720	1990	3520	3980
400	38	52	96	111	189	225	390	453	749	871	1490	1730	3060	3450
500	34	46	86	100	170	202	348	404	552	783	1330	1550	2740	3090

For SI units: 1 foot = 304.8 mm, 1 cubic foot per hour = 0.0283 m^3/h, 1 pound-force per square inch = 6.8947 kPa

Notes:

1 Table does not include effect of pressure drop across the line regulator. Where regulator loss exceeds 0.75 psi (5.17 kPa), DO NOT USE THIS TABLE. Consult with regulator manufacturer for pressure drops and capacity factors. Pressure drops across a regulator are capable of varying with flow rate.

2 CAUTION: Capacities shown in table are capable of exceeding maximum capacity for a selected regulator. Consult with regulator or tubing manufacturer for guidance.

3 Table includes losses for four 90 degree (1.57 rad) bends and two end fittings. Tubing runs with larger numbers of bends, fittings, or both shall be increased by an equivalent length of tubing according to the following equation: $L = 1.3\ n$, where L is additional length (ft) of tubing and n is the number of additional fittings, bends, or both.

4 Table entries are rounded to 3 significant digits.

5 EHD = Equivalent Hydraulic Diameter, which is a measure of the relative hydraulic efficiency between different tubing sizes. The greater the value of EHD, the greater the gas capacity of the tubing.

TABLE 1315.2(18)
CORRUGATED STAINLESS STEEL TUBING (CSST) [NFPA 54: TABLE 6.2.1(s)][1, 2, 3, 4]

												GAS:	NATURAL	
												INLET PRESSURE:	5.0 psi	
												PRESSURE DROP:	3.5 psi	
												SPECIFIC GRAVITY:	0.60	
	TUBE SIZE (EHD)[5]													
FLOW DESIGNATION:	**13**	**15**	**18**	**19**	**23**	**25**	**30**	**31**	**37**	**39**	**46**	**48**	**60**	**62**
LENGTH (feet)	**CAPACITY IN CUBIC FEET OF GAS PER HOUR**													
10	523	674	1080	1300	2000	2530	4920	5660	8300	9140	18 100	19 800	34 400	40 400
25	322	420	691	827	1290	1620	3080	3540	5310	5911	11 400	12 600	22 000	25 600
30	292	382	632	755	1180	1480	2800	3230	4860	5420	10 400	11 500	20 100	23 400
40	251	329	549	654	1030	1280	2420	2790	4230	4727	8970	10 000	17 400	20 200
50	223	293	492	586	926	1150	2160	2490	3790	4251	8020	8930	15 600	18 100
75	180	238	403	479	763	944	1750	2020	3110	3506	6530	7320	12 800	14 800
80	174	230	391	463	740	915	1690	1960	3020	3400	6320	7090	12 400	14 300
100	154	205	350	415	665	820	1510	1740	2710	3057	5650	6350	11 100	12 800
150	124	166	287	339	548	672	1230	1420	2220	2521	4600	5200	9130	10 500
200	107	143	249	294	478	584	1060	1220	1930	2199	3980	4510	7930	9090
250	95	128	223	263	430	524	945	1090	1730	1977	3550	4040	7110	8140
300	86	116	204	240	394	479	860	995	1590	1813	3240	3690	6500	7430
400	74	100	177	208	343	416	742	858	1380	1581	2800	3210	5650	6440
500	66	89	159	186	309	373	662	766	1040	1422	2500	2870	5060	5760

For SI units: 1 foot = 304.8 mm, 1 cubic foot per hour = 0.0283 m^3/h, 1 pound-force per square inch = 6.8947 kPa

Notes:

1 Table does not include effect of pressure drop across the line regulator. Where regulator loss exceeds 1 psi (7 kPa), DO NOT USE THIS TABLE. Consult with regulator manufacturer for pressure drops and capacity factors. Pressure drops across regulator are capable of varying with the flow rate.

2 CAUTION: Capacities shown in table are capable of exceeding the maximum capacity of selected regulator. Consult tubing manufacturer for guidance.

3 Table includes losses for four 90 degree (1.57 rad) bends and two end fittings. Tubing runs with larger numbers of bends, fittings, or both shall be increased by an equivalent length of tubing to the following equation: $L = 1.3\ n$, where L is additional length (feet) of tubing and n is the number of additional fittings, bends, or both.

4 Table entries are rounded to 3 significant digits.

5 EHD = Equivalent Hydraulic Diameter, which is a measure of the relative hydraulic efficiency between different tubing sizes. The greater the value of EHD, the greater the gas capacity of the tubing.

TABLE 1315.2(19)
POLYETHYLENE PLASTIC PIPE [NFPA 54: TABLE 6.2.1(t)]*

					GAS:	NATURAL	
					INLET PRESSURE:	LESS THAN 2 psi	
					PRESSURE DROP:	0.3 in. w.c.	
					SPECIFIC GRAVITY:	0.60	
	PIPE SIZE (inch)							
NOMINAL OD:	½	¾	1	1 ¼	1 ½	2	3	4
DESIGNATION:	SDR 9.3	SDR 11	SDR 11	SDR 10	SDR 11	SDR 11	SDR 11	SDR 11
ACTUAL ID:	0.660	0.860	1.077	1.328	1.554	1.943	2.864	3.682
LENGTH (feet)	CAPACITY IN CUBIC FEET OF GAS PER HOUR							
10	153	305	551	955	1440	2590	7170	13 900
20	105	210	379	656	991	1780	4920	9520
30	84	169	304	527	796	1430	3950	7640
40	72	144	260	451	681	1220	3380	6540
50	64	128	231	400	604	1080	3000	5800
60	58	116	209	362	547	983	2720	5250
70	53	107	192	333	503	904	2500	4830
80	50	99	179	310	468	841	2330	4500
90	46	93	168	291	439	789	2180	4220
100	44	88	159	275	415	745	2060	3990
125	39	78	141	243	368	661	1830	3530
150	35	71	127	221	333	598	1660	3200
175	32	65	117	203	306	551	1520	2940
200	30	60	109	189	285	512	1420	2740
250	27	54	97	167	253	454	1260	2430
300	24	48	88	152	229	411	1140	2200
350	22	45	81	139	211	378	1050	2020
400	21	42	75	130	196	352	974	1880
450	19	39	70	122	184	330	914	1770
500	18	37	66	115	174	312	863	1670

For SI units: 1 inch = 25 mm, 1 foot = 304.8 mm, 1 cubic foot per hour = 0.0283 m^3/h, 1 pound-force per square inch = 6.8947 kPa, 1 inch water column = 0.249 kPa

* Table entries are rounded to 3 significant digits.

TABLE 1315.2(20)
POLYETHYLENE PLASTIC PIPE [NFPA 54: TABLE 6.2.1(u)]*

					GAS:		NATURAL	
					INLET PRESSURE:		LESS THAN 2 psi	
					PRESSURE DROP:		0.5 in. w.c.	
					SPECIFIC GRAVITY:		0.60	
	PIPE SIZE (inch)							
NOMINAL OD:	½	¾	1	1 ¼	1 ½	2	3	4
DESIGNATION:	SDR 9.3	SDR 11	SDR 11	SDR 10	SDR 11	SDR 11	SDR 11	SDR 11
ACTUAL ID:	0.660	0.860	1.077	1.328	1.554	1.943	2.864	3.682
LENGTH (feet)	CAPACITY IN CUBIC FEET OF GAS PER HOUR							
10	201	403	726	1260	1900	3410	9450	18 260
20	138	277	499	865	1310	2350	6490	12 550
30	111	222	401	695	1050	1880	5210	10 080
40	95	190	343	594	898	1610	4460	8630
50	84	169	304	527	796	1430	3950	7640
60	76	153	276	477	721	1300	3580	6930
70	70	140	254	439	663	1190	3300	6370
80	65	131	236	409	617	1110	3070	5930
90	61	123	221	383	579	1040	2880	5560
100	58	116	209	362	547	983	2720	5250
125	51	103	185	321	485	871	2410	4660
150	46	93	168	291	439	789	2180	4220
175	43	86	154	268	404	726	2010	3880
200	40	80	144	249	376	675	1870	3610
250	35	71	127	221	333	598	1660	3200
300	32	64	115	200	302	542	1500	2900
350	29	59	106	184	278	499	1380	2670
400	27	55	99	171	258	464	1280	2480
450	26	51	93	160	242	435	1200	2330
500	24	48	88	152	229	411	1140	2200

For SI units: 1 inch = 25 mm, 1 foot = 304.8 mm, 1 cubic foot per hour = 0.0283 m^3/h, 1 pound-force per square inch = 6.8947 kPa, 1 inch water column = 0.249 kPa

* Table entries are rounded to 3 significant digits.

TABLE 1315.2(21)
POLYETHYLENE PLASTIC PIPE [NFPA 54: TABLE 6.2.1(v)]*

					GAS:	NATURAL		
					INLET PRESSURE:	2.0 psi		
					PRESSURE DROP:	1.0 psi		
					SPECIFIC GRAVITY:	0.60		
	PIPE SIZE (inch)							
NOMINAL OD:	½	¾	1	1 ¼	1 ½	2	3	4
DESIGNATION:	SDR 9.3	SDR 11	SDR 11	SDR 10	SDR 11	SDR 11	SDR 11	SDR 11
ACTUAL ID:	0.660	0.860	1.077	1.328	1.554	1.943	2.864	3.682
LENGTH (feet)	CAPACITY IN CUBIC FEET OF GAS PER HOUR							
10	1860	3720	6710	11 600	17 600	31 600	87 300	169 000
20	1280	2560	4610	7990	12 100	21 700	60 000	116 000
30	1030	2050	3710	6420	9690	17 400	48 200	93 200
40	878	1760	3170	5490	8300	14 900	41 200	79 700
50	778	1560	2810	4870	7350	13 200	36 600	70 700
60	705	1410	2550	4410	6660	12 000	33 100	64 000
70	649	1300	2340	4060	6130	11 000	30 500	58 900
80	603	1210	2180	3780	5700	10 200	28 300	54 800
90	566	1130	2050	3540	5350	9610	26 600	51 400
100	535	1070	1930	3350	5050	9080	25 100	48 600
125	474	949	1710	2970	4480	8050	22 300	43 000
150	429	860	1550	2690	4060	7290	20 200	39 000
175	395	791	1430	2470	3730	6710	18 600	35 900
200	368	736	1330	2300	3470	6240	17 300	33 400
250	326	652	1180	2040	3080	5530	15 300	29 600
300	295	591	1070	1850	2790	5010	13 900	26 800
350	272	544	981	1700	2570	4610	12 800	24 700
400	253	506	913	1580	2390	4290	11 900	22 900
450	237	475	856	1480	2240	4020	11 100	21 500
500	224	448	809	1400	2120	3800	10 500	20 300
550	213	426	768	1330	2010	3610	9990	19 300
600	203	406	733	1270	1920	3440	9530	18 400
650	194	389	702	1220	1840	3300	9130	17 600
700	187	374	674	1170	1760	3170	8770	16 900
750	180	360	649	1130	1700	3050	8450	16 300
800	174	348	627	1090	1640	2950	8160	15 800
850	168	336	607	1050	1590	2850	7890	15 300
900	163	326	588	1020	1540	2770	7650	14 800
950	158	317	572	990	1500	2690	7430	14 400
1000	154	308	556	963	1450	2610	7230	14 000
1100	146	293	528	915	1380	2480	6870	13 300
1200	139	279	504	873	1320	2370	6550	12 700
1300	134	267	482	836	1260	2270	6270	12 100
1400	128	257	463	803	1210	2180	6030	11 600
1500	124	247	446	773	1170	2100	5810	11 200
1600	119	239	431	747	1130	2030	5610	10 800
1700	115	231	417	723	1090	1960	5430	10 500
1800	112	224	404	701	1060	1900	5260	10 200
1900	109	218	393	680	1030	1850	5110	9900
2000	106	212	382	662	1000	1800	4970	9600

For SI units: 1 inch = 25 mm, 1 foot = 304.8 mm, 1 cubic foot per hour = 0.0283 m^3/h, 1 pound-force per square inch = 6.8947 kPa

* Table entries are rounded to 3 significant digits.

TABLE 1315.2(22)
POLYETHYLENE PLASTIC TUBING [NFPA 54: TABLE 6.2.1(w)][2, 3]

	GAS:	NATURAL
	INLET PRESSURE:	LESS THAN 2.0 psi
	PRESSURE DROP:	0.3 in. w.c.
	SPECIFIC GRAVITY:	0.60
	PLASTIC TUBING SIZE (CTS)[1] (inch)	
NOMINAL OD:	½	1
DESIGNATION:	SDR 7	SDR 11
ACTUAL ID:	0.445	0.927
LENGTH (feet)	CAPACITY IN CUBIC FEET OF GAS PER HOUR	
10	54	372
20	37	256
30	30	205
40	26	176
50	23	156
60	21	141
70	19	130
80	18	121
90	17	113
100	16	107
125	14	95
150	13	86
175	12	79
200	11	74
225	10	69
250	NA	65
275	NA	62
300	NA	59
350	NA	54
400	NA	51
450	NA	47
500	NA	45

For SI units: 1 inch = 25 mm, 1 foot = 304.8 mm, 1 cubic foot per hour = $0.0283 m^3/h$, 1 pound-force per square inch = 6.8947 kPa, 1 inch water column = 0.249 kPa

Notes:

1 CTS = Copper tube size.

2 Table entries are rounded to 3 significant digits.

3 NA means a flow of less than 10 ft^3/h (0.283 m^3/h).

TABLE 1315.2(23)
POLYETHYLENE PLASTIC TUBING [NFPA 54: TABLE 6.2.1(x)][2, 3]

	GAS:	NATURAL
	INLET PRESSURE:	LESS THAN 2.0 psi
	PRESSURE DROP:	0.5 in. w.c.
	SPECIFIC GRAVITY:	0.60
	PLASTIC TUBING SIZE (CTS)[1] (inch)	
NOMINAL OD:	½	1
DESIGNATION:	SDR 7	SDR 11
ACTUAL ID:	0.445	0.927
LENGTH (feet)	CAPACITY IN CUBIC FEET OF GAS PER HOUR	
10	72	490
20	49	337
30	39	271
40	34	232
50	30	205
60	27	186
70	25	171
80	23	159
90	22	149
100	21	141
125	18	125
150	17	113
175	15	104
200	14	97
225	13	91
250	12	86
275	11	82
300	11	78
350	10	72
400	NA	67
450	NA	63
500	NA	59

For SI units: 1 inch = 25 mm, 1 foot = 304.8 mm, 1 cubic foot per hour = $0.0283 m^3/h$, 1 pound-force per square inch = 6.8947 kPa, 1 inch water column = 0.249 kPa

Notes:

1 CTS = Copper tube size.

2 Table entries are rounded to 3 significant digits.

3 NA means a flow of less than 10 ft^3/h (0.283 m^3/h).

TABLE 1315.2(24)
SCHEDULE 40 METALLIC PIPE [NFPA 54: TABLE 6.3.1(a)]*

							GAS:	UNDILUTED PROPANE	
							INLET PRESSURE:	10.0 psi	
							PRESSURE DROP:	1.0 psi	
							SPECIFIC GRAVITY:	1.50	
INTENDED USE: PIPE SIZING BETWEEN FIRST STAGE (HIGH PRESSURE) REGULATOR AND SECOND STAGE (LOW PRESSURE) REGULATOR									
	PIPE SIZE (inch)								
NOMINAL INSIDE:	½	¾	1	1¼	1½	2	2½	3	4
ACTUAL:	0.622	0.824	1.049	1.380	1.610	2.067	2.469	3.068	4.026
LENGTH (feet)	CAPACITY IN THOUSANDS OF BTU PER HOUR								
10	3320	6950	13 100	26 900	40 300	77 600	124 000	219 000	446 000
20	2280	4780	9000	18 500	27 700	53 300	85 000	150 000	306 000
30	1830	3840	7220	14 800	22 200	42 800	68 200	121 000	246 000
40	1570	3280	6180	12 700	19 000	36 600	58 400	103 000	211 000
50	1390	2910	5480	11 300	16 900	32 500	51 700	91 500	187 000
60	1260	2640	4970	10 200	15 300	29 400	46 900	82 900	169 000
70	1160	2430	4570	9380	14 100	27 100	43 100	76 300	156 000
80	1080	2260	4250	8730	13 100	25 200	40 100	70 900	145 000
90	1010	2120	3990	8190	12 300	23 600	37 700	66 600	136 000
100	956	2000	3770	7730	11 600	22 300	35 600	62 900	128 000
125	848	1770	3340	6850	10 300	19 800	31 500	55 700	114 000
150	768	1610	3020	6210	9300	17 900	28 600	50 500	103 000
175	706	1480	2780	5710	8560	16 500	26 300	46 500	94 700
200	657	1370	2590	5320	7960	15 300	24 400	43 200	88 100
250	582	1220	2290	4710	7060	13 600	21 700	38 300	78 100
300	528	1100	2080	4270	6400	12 300	19 600	34 700	70 800
350	486	1020	1910	3930	5880	11 300	18 100	31 900	65 100
400	452	945	1780	3650	5470	10 500	16 800	29 700	60 600
450	424	886	1670	3430	5140	9890	15 800	27 900	56 800
500	400	837	1580	3240	4850	9340	14 900	26 300	53 700
550	380	795	1500	3070	4610	8870	14 100	25 000	51 000
600	363	759	1430	2930	4400	8460	13 500	23 900	48 600
650	347	726	1370	2810	4210	8110	12 900	22 800	46 600
700	334	698	1310	2700	4040	7790	12 400	21 900	44 800
750	321	672	1270	2600	3900	7500	12 000	21 100	43 100
800	310	649	1220	2510	3760	7240	11 500	20 400	41 600
850	300	628	1180	2430	3640	7010	11 200	19 800	40 300
900	291	609	1150	2360	3530	6800	10 800	19 200	39 100
950	283	592	1110	2290	3430	6600	10 500	18 600	37 900
1000	275	575	1080	2230	3330	6420	10 200	18 100	36 900
1100	261	546	1030	2110	3170	6100	9720	17 200	35 000
1200	249	521	982	2020	3020	5820	9270	16 400	33 400
1300	239	499	940	1930	2890	5570	8880	15 700	32 000
1400	229	480	903	1850	2780	5350	8530	15 100	30 800
1500	221	462	870	1790	2680	5160	8220	14 500	29 600
1600	213	446	840	1730	2590	4980	7940	14 000	28 600
1700	206	432	813	1670	2500	4820	7680	13 600	27 700
1800	200	419	789	1620	2430	4670	7450	13 200	26 900
1900	194	407	766	1570	2360	4540	7230	12 800	26 100
2000	189	395	745	1530	2290	4410	7030	12 400	25 400

For SI units: 1 inch = 25 mm, 1 foot = 304.8 mm, 1000 British thermal units per hour = 0.293 kW, 1 pound-force per square inch = 6.8947 kPa

* Table entries are rounded to 3 significant digits.

TABLE 1315.2(25)
SCHEDULE 40 METALLIC PIPE [NFPA 54: TABLE 6.3.1(b)]*

							GAS:	UNDILUTED PROPANE
							INLET PRESSURE:	10.0 psi
							PRESSURE DROP:	3.0 psi
							SPECIFIC GRAVITY:	1.50
INTENDED USE: PIPE SIZING BETWEEN FIRST STAGE (HIGH PRESSURE) REGULATOR AND SECOND STAGE (LOW PRESSURE) REGULATOR									
	PIPE SIZE (inch)								
NOMINAL INSIDE:	½	¾	1	1¼	1½	2	2½	3	4
ACTUAL:	0.622	0.824	1.049	1.380	1.610	2.067	2.469	3.068	4.026
LENGTH (feet)	CAPACITY IN THOUSANDS OF BTU PER HOUR								
10	5890	12 300	23 200	47 600	71 300	137 000	219 000	387 000	789 000
20	4050	8460	15 900	32 700	49 000	94 400	150 000	266 000	543 000
30	3250	6790	12 800	26 300	39 400	75 800	121 000	214 000	436 000
40	2780	5810	11 000	22 500	33 700	64 900	103 000	183 000	373 000
50	2460	5150	9710	19 900	29 900	57 500	91 600	162 000	330 000
60	2230	4670	8790	18 100	27 100	52 100	83 000	147 000	299 000
70	2050	4300	8090	16 600	24 900	47 900	76 400	135 000	275 000
80	1910	4000	7530	15 500	23 200	44 600	71 100	126 000	256 000
90	1790	3750	7060	14 500	21 700	41 800	66 700	118 000	240 000
100	1690	3540	6670	13 700	20 500	39 500	63 000	111 000	227 000
125	1500	3140	5910	12 100	18 200	35 000	55 800	98 700	201 000
150	1360	2840	5360	11 000	16 500	31 700	50 600	89 400	182 000
175	1250	2620	4930	10 100	15 200	29 200	46 500	82 300	167 800
200	1160	2430	4580	9410	14 100	27 200	43 300	76 500	156 100
250	1030	2160	4060	8340	12 500	24 100	38 400	67 800	138 400
300	935	1950	3680	7560	11 300	21 800	34 800	61 500	125 400
350	860	1800	3390	6950	10 400	20 100	32 000	56 500	115 300
400	800	1670	3150	6470	9690	18 700	29 800	52 600	107 300
450	751	1570	2960	6070	9090	17 500	27 900	49 400	100 700
500	709	1480	2790	5730	8590	16 500	26 400	46 600	95 100
550	673	1410	2650	5450	8160	15 700	25 000	44 300	90 300
600	642	1340	2530	5200	7780	15 000	23 900	42 200	86 200
650	615	1290	2420	4980	7450	14 400	22 900	40 500	82 500
700	591	1240	2330	4780	7160	13 800	22 000	38 900	79 300
750	569	1190	2240	4600	6900	13 300	21 200	37 400	76 400
800	550	1150	2170	4450	6660	12 800	20 500	36 200	73 700
850	532	1110	2100	4300	6450	12 400	19 800	35 000	71 400
900	516	1080	2030	4170	6250	12 000	19 200	33 900	69 200
950	501	1050	1970	4050	6070	11 700	18 600	32 900	67 200
1000	487	1020	1920	3940	5900	11 400	18 100	32 000	65 400
1100	463	968	1820	3740	5610	10 800	17 200	30 400	62 100
1200	442	923	1740	3570	5350	10 300	16 400	29 000	59 200
1300	423	884	1670	3420	5120	9870	15 700	27 800	56 700
1400	406	849	1600	3280	4920	9480	15 100	26 700	54 500
1500	391	818	1540	3160	4740	9130	14 600	25 700	52 500
1600	378	790	1490	3060	4580	8820	14 100	24 800	50 700
1700	366	765	1440	2960	4430	8530	13 600	24 000	49 000
1800	355	741	1400	2870	4300	8270	13 200	23 300	47 600
1900	344	720	1360	2780	4170	8040	12 800	22 600	46 200
2000	335	700	1320	2710	4060	7820	12 500	22 000	44 900

For SI units: 1 inch = 25 mm, 1 foot = 304.8 mm, 1000 British thermal units per hour = 0.293 kW, 1 pound-force per square inch = 6.8947 kPa

* Table entries are rounded to 3 significant digits.

TABLE 1315.2(26)
SCHEDULE 40 METALLIC PIPE [NFPA 54: TABLE 6.3.1(c)]*

							GAS:	UNDILUTED PROPANE	
							INLET PRESSURE:	2.0 psi	
							PRESSURE DROP:	1.0 psi	
							SPECIFIC GRAVITY:	1.50	
INTENDED USE: PIPE SIZING BETWEEN 2 PSIG SERVICE AND LINE PRESSURE REGULATOR									
	PIPE SIZE (inch)								
NOMINAL:	½	¾	1	1¼	1½	2	2½	3	4
ACTUAL ID:	0.622	0.824	1.049	1.380	1.610	2.067	2.469	3.068	4.026
LENGTH (feet)	CAPACITY IN THOUSANDS OF BTU PER HOUR								
10	2680	5590	10 500	21 600	32 400	62 400	99 500	176 000	359 000
20	1840	3850	7240	14 900	22 300	42 900	68 400	121 000	247 000
30	1480	3090	5820	11 900	17 900	34 500	54 900	97 100	198 000
40	1260	2640	4980	10 200	15 300	29 500	47 000	83 100	170 000
50	1120	2340	4410	9060	13 600	26 100	41 700	73 700	150 000
60	1010	2120	4000	8210	12 300	23 700	37 700	66 700	136 000
70	934	1950	3680	7550	11 300	21 800	34 700	61 400	125 000
80	869	1820	3420	7020	10 500	20 300	32 300	57 100	116 000
90	815	1700	3210	6590	9880	19 000	30 300	53 600	109 000
100	770	1610	3030	6230	9330	18 000	28 600	50 600	103 000
125	682	1430	2690	5520	8270	15 900	25 400	44 900	91 500
150	618	1290	2440	5000	7490	14 400	23 000	40 700	82 900
175	569	1190	2240	4600	6890	13 300	21 200	37 400	76 300
200	529	1110	2080	4280	6410	12 300	19 700	34 800	71 000
250	469	981	1850	3790	5680	10 900	17 400	30 800	62 900
300	425	889	1670	3440	5150	9920	15 800	27 900	57 000
350	391	817	1540	3160	4740	9120	14 500	25 700	52 400
400	364	760	1430	2940	4410	8490	13 500	23 900	48 800
450	341	714	1340	2760	4130	7960	12 700	22 400	45 800
500	322	674	1270	2610	3910	7520	12 000	21 200	43 200
550	306	640	1210	2480	3710	7140	11 400	20 100	41 100
600	292	611	1150	2360	3540	6820	10 900	19 200	39 200
650	280	585	1100	2260	3390	6530	10 400	18 400	37 500
700	269	562	1060	2170	3260	6270	9990	17 700	36 000
750	259	541	1020	2090	3140	6040	9630	17 000	34 700
800	250	523	985	2020	3030	5830	9300	16 400	33 500
850	242	506	953	1960	2930	5640	9000	15 900	32 400
900	235	490	924	1900	2840	5470	8720	15 400	31 500
950	228	476	897	1840	2760	5310	8470	15 000	30 500
1000	222	463	873	1790	2680	5170	8240	14 600	29 700
1100	210	440	829	1700	2550	4910	7830	13 800	28 200
1200	201	420	791	1620	2430	4680	7470	13 200	26 900
1300	192	402	757	1550	2330	4490	7150	12 600	25 800
1400	185	386	727	1490	2240	4310	6870	12 100	24 800
1500	178	372	701	1440	2160	4150	6620	11 700	23 900
1600	172	359	677	1390	2080	4010	6390	11 300	23 000
1700	166	348	655	1340	2010	3880	6180	10 900	22 300
1800	161	337	635	1300	1950	3760	6000	10 600	21 600
1900	157	327	617	1270	1900	3650	5820	10 300	21 000
2000	152	318	600	1230	1840	3550	5660	10 000	20 400

For SI units: 1 inch = 25 mm, 1 foot = 304.8 mm, 1000 British thermal units per hour = 0.293 kW, 1 pound-force per square inch = 6.8947 kPa

* Table entries are rounded to 3 significant digits.

TABLE 1315.2(27)
SCHEDULE 40 METALLIC PIPE [NFPA 54: TABLE 6.3.1(d)]*

						GAS:	UNDILUTED PROPANE		
						INLET PRESSURE:	11.0 in. w.c.		
						PRESSURE DROP:	0.5 in. w.c.		
						SPECIFIC GRAVITY:	1.50		
INTENDED USE: PIPE SIZING BETWEEN SINGLE OR SECOND STAGE (LOW PRESSURE) REGULATOR AND APPLIANCE									
	PIPE SIZE (inch)								
NOMINAL INSIDE:	½	¾	1	1¼	1½	2	2½	3	4
ACTUAL ID:	0.622	0.824	1.049	1.380	1.610	2.067	2.469	3.068	4.026
LENGTH (feet)	CAPACITY IN THOUSANDS OF BTU PER HOUR								
10	291	608	1150	2350	3520	6790	10 800	19 100	39 000
20	200	418	787	1620	2420	4660	7430	13 100	26 800
30	160	336	632	1300	1940	3750	5970	10 600	21 500
40	137	287	541	1110	1660	3210	5110	9030	18 400
50	122	255	480	985	1480	2840	4530	8000	16 300
60	110	231	434	892	1340	2570	4100	7250	14 800
80	101	212	400	821	1230	2370	3770	6670	13 600
100	94	197	372	763	1140	2200	3510	6210	12 700
125	89	185	349	716	1070	2070	3290	5820	11 900
150	84	175	330	677	1010	1950	3110	5500	11 200
175	74	155	292	600	899	1730	2760	4880	9950
200	67	140	265	543	814	1570	2500	4420	9010
250	62	129	243	500	749	1440	2300	4060	8290
300	58	120	227	465	697	1340	2140	3780	7710
350	51	107	201	412	618	1190	1900	3350	6840
400	46	97	182	373	560	1080	1720	3040	6190
450	42	89	167	344	515	991	1580	2790	5700
500	40	83	156	320	479	922	1470	2600	5300
550	37	78	146	300	449	865	1380	2440	4970
600	35	73	138	283	424	817	1300	2300	4700
650	33	70	131	269	403	776	1240	2190	4460
700	32	66	125	257	385	741	1180	2090	4260
750	30	64	120	246	368	709	1130	2000	4080
800	29	61	115	236	354	681	1090	1920	3920
850	28	59	111	227	341	656	1050	1850	3770
900	27	57	107	220	329	634	1010	1790	3640
950	26	55	104	213	319	613	978	1730	3530
1000	25	53	100	206	309	595	948	1680	3420
1100	25	52	97	200	300	578	921	1630	3320
1200	24	50	95	195	292	562	895	1580	3230
1300	23	48	90	185	277	534	850	1500	3070
1400	22	46	86	176	264	509	811	1430	2930
1500	21	44	82	169	253	487	777	1370	2800
1600	20	42	79	162	243	468	746	1320	2690
1700	19	40	76	156	234	451	719	1270	2590
1800	19	39	74	151	226	436	694	1230	2500
1900	18	38	71	146	219	422	672	1190	2420
2000	18	37	69	142	212	409	652	1150	2350

For SI units: 1 inch = 25 mm, 1 foot = 304.8 mm, 1000 British thermal units per hour = 0.293 kW, 1 inch water column = 0.249 kPa

* Table entries are rounded to 3 significant digits.

TABLE 1315.2(28)
SEMI-RIGID COPPER TUBING [NFPA 54: TABLE 6.3.1(e)][2]

					GAS:	UNDILUTED PROPANE			
					INLET PRESSURE:	10.0 psi			
					PRESSURE DROP:	1.0 psi			
					SPECIFIC GRAVITY:	1.50			
INTENDED USE: TUBE SIZING BETWEEN FIRST STAGE (HIGH PRESSURE) REGULATOR AND SECOND STAGE (LOW PRESSURE) REGULATOR									
	TUBE SIZE (inch)								
NOMINAL: K & L:	¼	⅜	½	⅝	¾	1	1¼	1½	2
NOMINAL: ACR:	⅜	½	⅝	¾	⅞	1⅛	1⅜	–	–
OUTSIDE:	0.375	0.500	0.625	0.750	0.875	1.125	1.375	1.625	2.125
INSIDE:[1]	0.305	0.402	0.527	0.652	0.745	0.995	1.245	1.481	1.959
LENGTH (feet)	**CAPACITY IN THOUSANDS OF BTU PER HOUR**								
10	513	1060	2150	3760	5330	11 400	20 500	32 300	67 400
20	352	727	1480	2580	3670	7830	14 100	22 200	46 300
30	283	584	1190	2080	2940	6290	11 300	17 900	37 200
40	242	500	1020	1780	2520	5380	9690	15 300	31 800
50	215	443	901	1570	2230	4770	8590	13 500	28 200
60	194	401	816	1430	2020	4320	7780	12 300	25 600
70	179	369	751	1310	1860	3980	7160	11 300	23 500
80	166	343	699	1220	1730	3700	6660	10 500	21 900
90	156	322	655	1150	1630	3470	6250	9850	20 500
100	147	304	619	1080	1540	3280	5900	9310	19 400
125	131	270	549	959	1360	2910	5230	8250	17 200
150	118	244	497	869	1230	2630	4740	7470	15 600
175	109	225	457	799	1130	2420	4360	6880	14 300
200	101	209	426	744	1060	2250	4060	6400	13 300
250	90	185	377	659	935	2000	3600	5670	11 800
300	81	168	342	597	847	1810	3260	5140	10 700
350	75	155	314	549	779	1660	3000	4730	9840
400	70	144	292	511	725	1550	2790	4400	9160
450	65	135	274	480	680	1450	2620	4130	8590
500	62	127	259	453	643	1370	2470	3900	8120
550	59	121	246	430	610	1300	2350	3700	7710
600	56	115	235	410	582	1240	2240	3530	7350
650	54	111	225	393	558	1190	2140	3380	7040
700	51	106	216	378	536	1140	2060	3250	6770
750	50	102	208	364	516	1100	1980	3130	6520
800	48	99	201	351	498	1060	1920	3020	6290
850	46	96	195	340	482	1030	1850	2920	6090
900	45	93	189	330	468	1000	1800	2840	5910
950	44	90	183	320	454	970	1750	2750	5730
1000	42	88	178	311	442	944	1700	2680	5580
1100	40	83	169	296	420	896	1610	2540	5300
1200	38	79	161	282	400	855	1540	2430	5050
1300	37	76	155	270	383	819	1470	2320	4840
1400	35	73	148	260	368	787	1420	2230	4650
1500	34	70	143	250	355	758	1360	2150	4480
1600	33	68	138	241	343	732	1320	2080	4330
1700	32	66	134	234	331	708	1270	2010	4190
1800	31	64	130	227	321	687	1240	1950	4060
1900	30	62	126	220	312	667	1200	1890	3940
2000	29	60	122	214	304	648	1170	1840	3830

For SI units: 1 inch = 25 mm, 1 foot = 304.8 mm, 1000 British thermal units per hour = 0.293 kW, 1 pound-force per square inch = 6.8947 kPa

Notes:

[1] Table capacities are based on Type K copper tubing inside diameter (shown), which has the smallest inside diameter of the copper tubing products.

[2] Table entries are rounded to 3 significant digits.

TABLE 1315.2(29)
SEMI-RIGID COPPER TUBING [NFPA 54: TABLE 6.3.1(f)][2, 3]

						GAS:	UNDILUTED PROPANE		
						INLET PRESSURE:	11.0 in. w.c.		
						PRESSURE DROP:	0.5 in. w.c.		
						SPECIFIC GRAVITY:	1.50		
INTENDED USE: TUBE SIZING BETWEEN SINGLE OR SECOND STAGE (LOW PRESSURE) REGULATOR AND APPLIANCE									
	TUBE SIZE (inch)								
NOMINAL: K & L:	¼	⅜	½	⅝	¾	1	1¼	1½	2
NOMINAL: ACR:	⅜	½	⅝	¾	⅞	1⅛	1⅜	–	–
OUTSIDE:	0.375	0.500	0.625	0.750	0.875	1.125	1.375	1.625	2.125
INSIDE:[1]	0.305	0.402	0.527	0.652	0.745	0.995	1.245	1.481	1.959
LENGTH (feet)	CAPACITY IN THOUSANDS OF BTU PER HOUR								
10	45	93	188	329	467	997	1800	2830	5890
20	31	64	129	226	321	685	1230	1950	4050
30	25	51	104	182	258	550	991	1560	3250
40	21	44	89	155	220	471	848	1340	2780
50	19	39	79	138	195	417	752	1180	2470
60	17	35	71	125	177	378	681	1070	2240
70	16	32	66	115	163	348	626	988	2060
80	15	30	61	107	152	324	583	919	1910
90	14	28	57	100	142	304	547	862	1800
100	13	27	54	95	134	287	517	814	1700
125	11	24	48	84	119	254	458	722	1500
150	10	21	44	76	108	230	415	654	1360
175	NA	20	40	70	99	212	382	602	1250
200	NA	18	37	65	92	197	355	560	1170
250	NA	16	33	58	82	175	315	496	1030
300	NA	15	30	52	74	158	285	449	936
350	NA	14	28	48	68	146	262	414	861
400	NA	13	26	45	63	136	244	385	801
450	NA	12	24	42	60	127	229	361	752
500	NA	11	23	40	56	120	216	341	710
550	NA	11	22	38	53	114	205	324	674
600	NA	10	21	36	51	109	196	309	643
650	NA	NA	20	34	49	104	188	296	616
700	NA	NA	19	33	47	100	180	284	592
750	NA	NA	18	32	45	96	174	274	570
800	NA	NA	18	31	44	93	168	264	551
850	NA	NA	17	30	42	90	162	256	533
900	NA	NA	17	29	41	87	157	248	517
950	NA	NA	16	28	40	85	153	241	502
1000	NA	NA	16	27	39	83	149	234	488
1100	NA	NA	15	26	37	78	141	223	464
1200	NA	NA	14	25	35	75	135	212	442
1300	NA	NA	14	24	34	72	129	203	423
1400	NA	NA	13	23	32	69	124	195	407
1500	NA	NA	13	22	31	66	119	188	392
1600	NA	NA	12	21	30	64	115	182	378
1700	NA	NA	12	20	29	62	112	176	366
1800	NA	NA	11	20	28	60	108	170	355
1900	NA	NA	11	19	27	58	105	166	345
2000	NA	NA	11	19	27	57	102	161	335

For SI units: 1 inch = 25 mm, 1 foot = 304.8 mm, 1000 British thermal units per hour = 0.293 kW, 1 inch water column = 0.249 kPa

Notes:

[1] Table capacities are based on Type K copper tubing inside diameter (shown), which has the smallest inside diameter of the copper tubing products.

[2] Table entries are rounded to 3 significant digits.

[3] NA means a flow of less than 10 000 Btu/h (2.93 kW).

TABLE 1315.2(30)
SEMI-RIGID COPPER TUBING [NFPA 54: TABLE 6.3.1(g)][2]

						GAS:	UNDILUTED PROPANE		
						INLET PRESSURE:	2.0 psi		
						PRESSURE DROP:	1.0 psi		
						SPECIFIC GRAVITY:	1.50		
INTENDED USE: TUBE SIZING BETWEEN 2 PSIG SERVICE AND LINE PRESSURE REGULATOR									
	TUBE SIZE (inch)								
NOMINAL: K & L:	1/4	3/8	1/2	5/8	3/4	1	1 1/4	1 1/2	2
NOMINAL: ACR:	3/8	1/2	5/8	3/4	7/8	1 1/8	1 3/8	–	–
OUTSIDE:	0.375	0.500	0.625	0.750	0.875	1.125	1.375	1.625	2.125
INSIDE:[1]	0.305	0.402	0.527	0.652	0.745	0.995	1.245	1.481	1.959
LENGTH (feet)	CAPACITY IN THOUSANDS OF BTU PER HOUR								
10	413	852	1730	3030	4300	9170	16 500	26 000	54 200
20	284	585	1190	2080	2950	6310	11 400	17 900	37 300
30	228	470	956	1670	2370	5060	9120	14 400	29 900
40	195	402	818	1430	2030	4330	7800	12 300	25 600
50	173	356	725	1270	1800	3840	6920	10 900	22 700
60	157	323	657	1150	1630	3480	6270	9880	20 600
70	144	297	605	1060	1500	3200	5760	9090	18 900
80	134	276	562	983	1390	2980	5360	8450	17 600
90	126	259	528	922	1310	2790	5030	7930	16 500
100	119	245	498	871	1240	2640	4750	7490	15 600
125	105	217	442	772	1100	2340	4210	6640	13 800
150	95	197	400	700	992	2120	3820	6020	12 500
175	88	181	368	644	913	1950	3510	5540	11 500
200	82	168	343	599	849	1810	3270	5150	10 700
250	72	149	304	531	753	1610	2900	4560	9510
300	66	135	275	481	682	1460	2620	4140	8610
350	60	124	253	442	628	1340	2410	3800	7920
400	56	116	235	411	584	1250	2250	3540	7370
450	53	109	221	386	548	1170	2110	3320	6920
500	50	103	209	365	517	1110	1990	3140	6530
550	47	97	198	346	491	1050	1890	2980	6210
600	45	93	189	330	469	1000	1800	2840	5920
650	43	89	181	316	449	959	1730	2720	5670
700	41	86	174	304	431	921	1660	2620	5450
750	40	82	168	293	415	888	1600	2520	5250
800	39	80	162	283	401	857	1540	2430	5070
850	37	77	157	274	388	829	1490	2350	4900
900	36	75	152	265	376	804	1450	2280	4750
950	35	72	147	258	366	781	1410	2220	4620
1000	34	71	143	251	356	760	1370	2160	4490
1100	32	67	136	238	338	721	1300	2050	4270
1200	31	64	130	227	322	688	1240	1950	4070
1300	30	61	124	217	309	659	1190	1870	3900
1400	28	59	120	209	296	633	1140	1800	3740
1500	27	57	115	201	286	610	1100	1730	3610
1600	26	55	111	194	276	589	1060	1670	3480
1700	26	53	108	188	267	570	1030	1620	3370
1800	25	51	104	182	259	553	1000	1570	3270
1900	24	50	101	177	251	537	966	1520	3170
2000	23	48	99	172	244	522	940	1480	3090

For SI units: 1 inch = 25 mm, 1 foot = 304.8 mm, 1000 British thermal units per hour = 0.293 kW, 1 pound-force per square inch = 6.8947 kPa

Notes:

[1] Table capacities are based on Type K copper tubing inside diameter (shown), which has the smallest inside diameter of the copper tubing products.

[2] Table entries are rounded to 3 significant digits.

TABLE 1315.2(31)
CORRUGATED STAINLESS STEEL TUBING (CSST) [NFPA 54: TABLE 6.3.1(h)][1, 2]

										GAS:	UNDILUTED PROPANE			
										INLET PRESSURE:	11.0 in. w.c.			
										PRESSURE DROP:	0.5 in. w.c.			
										SPECIFIC GRAVITY:	1.50			
INTENDED USE: CSST SIZING BETWEEN SINGLE OR SECOND STAGE (LOW PRESSURE) REGULATOR AND APPLIANCE SHUTOFF VALVE														
	TUBE SIZE (EHD)[3]													
FLOW DESIGNATION:	**13**	**15**	**18**	**19**	**23**	**25**	**30**	**31**	**37**	**39**	**46**	**48**	**60**	**62**
LENGTH (feet)	**CAPACITY IN THOUSANDS OF BTU PER HOUR**													
5	72	99	181	211	355	426	744	863	1420	1638	2830	3270	5780	6550
10	50	69	129	150	254	303	521	605	971	1179	1990	2320	4110	4640
15	39	55	104	121	208	248	422	490	775	972	1620	1900	3370	3790
20	34	49	91	106	183	216	365	425	661	847	1400	1650	2930	3290
25	30	42	82	94	164	192	325	379	583	762	1250	1480	2630	2940
30	28	39	74	87	151	177	297	344	528	698	1140	1350	2400	2680
40	23	33	64	74	131	153	256	297	449	610	988	1170	2090	2330
50	20	30	58	66	118	137	227	265	397	548	884	1050	1870	2080
60	19	26	53	60	107	126	207	241	359	502	805	961	1710	1900
70	17	25	49	57	99	117	191	222	330	466	745	890	1590	1760
80	15	23	45	52	94	109	178	208	307	438	696	833	1490	1650
90	15	22	44	50	90	102	169	197	286	414	656	787	1400	1550
100	14	20	41	47	85	98	159	186	270	393	621	746	1330	1480
150	11	15	31	36	66	75	123	143	217	324	506	611	1090	1210
200	9	14	28	33	60	69	112	129	183	283	438	531	948	1050
250	8	12	25	30	53	61	99	117	163	254	390	476	850	934
300	8	11	23	26	50	57	90	107	147	234	357	434	777	854

For SI units: 1 foot = 304.8 mm, 1000 British thermal units per hour = 0.293 kW, 1 inch water column = 0.249 kPa

Notes:

1 Table includes losses for four 90 degree (1.57 rad) bends and two end fittings. Tubing runs with larger numbers of bends, fittings, or both shall be increased by an equivalent length of tubing to the following equation: $L = 1.3\ n$, where L is additional length (ft) of tubing and n is the number of additional fittings, bends, or both.

2 Table entries are rounded to 3 significant digits.

3 EHD = Equivalent Hydraulic Diameter, which is a measure of the relative hydraulic efficiency between different tubing sizes. The greater the value of EHD, the greater the gas capacity of the tubing.

TABLE 1315.2(32)
CORRUGATED STAINLESS STEEL TUBING (CSST) [NFPA 54: TABLE 6.3.1(i)][1, 2, 3, 4]

									GAS:	**UNDILUTED PROPANE**				
									INLET PRESSURE:	**2.0 psi**				
									PRESSURE DROP:	**1.0 psi**				
									SPECIFIC GRAVITY:	**1.50**				
INTENDED USE: CSST SIZING BETWEEN 2 PSIG SERVICE AND LINE PRESSURE REGULATOR														
	TUBE SIZE (EHD)[5]													
FLOW DESIGNATION:	**13**	**15**	**18**	**19**	**23**	**25**	**30**	**31**	**37**	**39**	**46**	**48**	**60**	**62**
LENGTH (feet)	**CAPACITY IN THOUSANDS OF BTU PER HOUR**													
10	426	558	927	1110	1740	2170	4100	4720	7130	7958	15 200	16 800	29 400	34 200
25	262	347	591	701	1120	1380	2560	2950	4560	5147	9550	10 700	18 800	21 700
30	238	316	540	640	1030	1270	2330	2690	4180	4719	8710	9790	17 200	19 800
40	203	271	469	554	896	1100	2010	2320	3630	4116	7530	8500	14 900	17 200
50	181	243	420	496	806	986	1790	2070	3260	3702	6730	7610	13 400	15 400
75	147	196	344	406	663	809	1460	1690	2680	3053	5480	6230	11 000	12 600
80	140	189	333	393	643	768	1410	1630	2590	2961	5300	6040	10 600	12 200
100	124	169	298	350	578	703	1260	1450	2330	2662	4740	5410	9530	10 900
150	101	137	245	287	477	575	1020	1180	1910	2195	3860	4430	7810	8890
200	86	118	213	248	415	501	880	1020	1660	1915	3340	3840	6780	7710
250	77	105	191	222	373	448	785	910	1490	1722	2980	3440	6080	6900
300	69	96	173	203	343	411	716	829	1360	1578	2720	3150	5560	6300
400	60	82	151	175	298	355	616	716	1160	1376	2350	2730	4830	5460
500	53	72	135	158	268	319	550	638	1030	1237	2100	2450	4330	4880

For SI units: 1 foot = 304.8 mm, 1000 British thermal units per hour = 0.293 kW, 1 pound-force per square inch = 6.8947 kPa

Notes:

[1] Table does not include effect of pressure drop across the line regulator. Where regulator loss exceeds 0.5 psi (3.4 kPa) [based on 13 inch water column (3.2 kPa) outlet pressure], DO NOT use THIS TABLE. Consult with regulator manufacturer for pressure drops and capacity factors. Pressure drops across a regulator are capable of varying with flow rate.

[2] CAUTION: Capacities shown in table are capable of exceeding the maximum capacity for a selected regulator. Consult with regulator or tubing manufacturer for guidance.

[3] Table includes losses for four 90 degree (1.57 rad) bends and two end fittings. Tubing runs with larger numbers of bends, fittings, or both shall be increased by an equivalent length of tubing to the following equation: $L = 1.3\,n$, where L is additional length (ft) of tubing and n is the number of additional fittings, bends, or both.

[4] Table entries are rounded to 3 significant digits.

[5] EHD = Equivalent Hydraulic Diameter, which is a measure of the relative hydraulic efficiency between different tubing sizes. The greater the value of EHD, the greater the gas capacity of the tubing.

TABLE 1315.2(33)
CORRUGATED STAINLESS STEEL TUBING (CSST) [NFPA 54: TABLE 6.3.1(j)][1, 2, 3, 4]

											GAS:	**UNDILUTED PROPANE**		
											INLET PRESSURE:	**5.0 psi**		
											PRESSURE DROP:	**3.5 psi**		
											SPECIFIC GRAVITY:	**1.50**		
	TUBE SIZE (EHD)[5]													
FLOW DESIGNATION:	**13**	**15**	**18**	**19**	**23**	**25**	**30**	**31**	**37**	**39**	**46**	**48**	**60**	**62**
LENGTH (feet)	**CAPACITY IN THOUSANDS OF BTU PER HOUR**													
10	826	1070	1710	2060	3150	4000	7830	8950	13 100	14 441	28 600	31 200	54 400	63 800
25	509	664	1090	1310	2040	2550	4860	5600	8400	9339	18 000	19 900	34 700	40 400
30	461	603	999	1190	1870	2340	4430	5100	7680	8564	16 400	18 200	31 700	36 900
40	396	520	867	1030	1630	2030	3820	4400	6680	7469	14 200	15 800	27 600	32 000
50	352	463	777	926	1460	1820	3410	3930	5990	6717	12 700	14 100	24 700	28 600
75	284	376	637	757	1210	1490	2770	3190	4920	5539	10 300	11 600	20 300	23 400
80	275	363	618	731	1170	1450	2680	3090	4770	5372	9990	11 200	19 600	22 700
100	243	324	553	656	1050	1300	2390	2760	4280	4830	8930	10 000	17 600	20 300
150	196	262	453	535	866	1060	1940	2240	3510	3983	7270	8210	14 400	16 600
200	169	226	393	464	755	923	1680	1930	3050	3474	6290	7130	12 500	14 400
250	150	202	352	415	679	828	1490	1730	2740	3124	5620	6390	11 200	12 900
300	136	183	322	379	622	757	1360	1570	2510	2865	5120	5840	10 300	11 700
400	117	158	279	328	542	657	1170	1360	2180	2498	4430	5070	8920	10 200
500	104	140	251	294	488	589	1050	1210	1950	2247	3960	4540	8000	9110

For SI units: 1 foot = 304.8 mm, 1000 British thermal units per hour = 0.293 kW, 1 pound-force per square inch = 6.8947 kPa

Notes:

[1] Table does not include effect of pressure drop across the line regulator. Where regulator loss exceeds 0.5 psi (3.4 kPa) [based on 13 inch water column (3.2 kPa) outlet pressure], DO NOT USE THIS TABLE. Consult with regulator manufacturer for pressure drops and capacity factors. Pressure drops across a regulator are capable of varying with flow rate.

[2] CAUTION: Capacities shown in table are capable of exceeding the maximum capacity for a selected regulator. Consult with regulator or tubing manufacturer for guidance.

[3] Table includes losses for four 90 degree (1.57 rad) bends and two end fittings. Tubing runs with larger numbers of bends, fittings, or both shall be increased by an equivalent length of tubing to the following equation: $L = 1.3\,n$, where L is additional length (ft) of tubing and n is the number of additional fittings, bends, or both.

[4] Table entries are rounded to 3 significant digits.

[5] EHD = Equivalent Hydraulic Diameter, which is a measure of the relative hydraulic efficiency between different tubing sizes. The greater the value of EHD, the greater the gas capacity of the tubing.

TABLE 1315.2(34)
POLYETHYLENE PLASTIC PIPE [NFPA 54: TABLE 6.3.1(k)]*

					GAS:	UNDILUTED PROPANE		
					INLET PRESSURE:	11.0 in. w.c.		
					PRESSURE DROP:	0.5 in. w.c.		
					SPECIFIC GRAVITY:	1.50		
INTENDED USE: PE PIPE SIZING BETWEEN INTEGRAL SECOND-STAGE REGULATOR AT TANK OR SECOND-STAGE (LOW PRESSURE) REGULATOR AND BUILDING								
	PIPE SIZE (inch)							
NOMINAL OD:	½	¾	1	1¼	1½	2	3	4
DESIGNATION:	SDR 9.3	SDR 11	SDR 11	SDR 10	SDR 11	SDR 11	SDR 11	SDR 11
ACTUAL ID:	0.660	0.860	1.077	1.328	1.554	1.943	2.864	3.682
LENGTH (feet)	CAPACITY IN THOUSANDS OF BTU PER HOUR							
10	340	680	1230	2130	3210	5770	16 000	30 900
20	233	468	844	1460	2210	3970	11 000	21 200
30	187	375	677	1170	1770	3180	8810	17 000
40	160	321	580	1000	1520	2730	7540	14 600
50	142	285	514	890	1340	2420	6680	12 900
60	129	258	466	807	1220	2190	6050	11 700
70	119	237	428	742	1120	2010	5570	10 800
80	110	221	398	690	1040	1870	5180	10 000
90	103	207	374	648	978	1760	4860	9400
100	98	196	353	612	924	1660	4590	8900
125	87	173	313	542	819	1470	4070	7900
150	78	157	284	491	742	1330	3690	7130
175	72	145	261	452	683	1230	3390	6560
200	67	135	243	420	635	1140	3160	6100
250	60	119	215	373	563	1010	2800	5410
300	54	108	195	338	510	916	2530	4900
350	50	99	179	311	469	843	2330	4510
400	46	92	167	289	436	784	2170	4190
450	43	87	157	271	409	736	2040	3930
500	41	82	148	256	387	695	1920	3720

For SI units: 1 inch = 25 mm, 1 foot = 304.8 mm, 1000 British thermal units per hour = 0.293 kW, 1 inch water column = 0.249 kPa

* Table entries are rounded to 3 significant digits.

TABLE 1315.2(35)
POLYETHYLENE PLASTIC PIPE [NFPA 54: TABLE 6.3.1(l)]*

						GAS:	UNDILUTED PROPANE	
						INLET PRESSURE:	2.0 psi	
						PRESSURE DROP:	1.0 psi	
						SPECIFIC GRAVITY:	1.50	
INTENDED USE: PE PIPE SIZING BETWEEN 2 PSI SERVICE REGULATOR AND LINE PRESSURE REGULATOR								
	PIPE SIZE (inch)							
NOMINAL OD:	½	¾	1	1¼	1½	2	3	4
DESIGNATION:	SDR 9.3	SDR 11	SDR 11	SDR 10	SDR 11	SDR 11	SDR 11	SDR 11
ACTUAL ID:	0.660	0.860	1.077	1.328	1.554	1.943	2.864	3.682
LENGTH (feet)	CAPACITY IN THOUSANDS OF BTU PER HOUR							
10	3130	6260	11 300	19 600	29 500	53 100	147 000	284 000
20	2150	4300	7760	13 400	20 300	36 500	101 000	195 000
30	1730	3450	6230	10 800	16 300	29 300	81 100	157 000
40	1480	2960	5330	9240	14 000	25 100	69 400	134 100
50	1310	2620	4730	8190	12 400	22 200	61 500	119 000
60	1190	2370	4280	7420	11 200	20 100	55 700	108 000
70	1090	2180	3940	6830	10 300	18 500	51 300	99 100
80	1010	2030	3670	6350	9590	17 200	47 700	92 200
90	952	1910	3440	5960	9000	16 200	44 700	86 500
100	899	1800	3250	5630	8500	15 300	42 300	81 700
125	797	1600	2880	4990	7530	13 500	37 500	72 400
150	722	1450	2610	4520	6830	12 300	33 900	65 600
175	664	1330	2400	4160	6280	11 300	31 200	60 300
200	618	1240	2230	3870	5840	10 500	29 000	56 100
250	548	1100	1980	3430	5180	9300	25 700	49 800
300	496	994	1790	3110	4690	8430	23 300	45 100
350	457	914	1650	2860	4320	7760	21 500	41 500
400	425	851	1530	2660	4020	7220	12 000	38 600
450	399	798	1440	2500	3770	6770	18 700	36 200
500	377	754	1360	2360	3560	6390	17 700	34 200
550	358	716	1290	2240	3380	6070	16 800	32 500
600	341	683	1230	2140	3220	5790	16 000	31 000
650	327	654	1180	2040	3090	5550	15 400	29 700
700	314	628	1130	1960	2970	5330	14 700	28 500
750	302	605	1090	1890	2860	5140	14 200	27 500
800	292	585	1050	1830	2760	4960	13 700	26 500
850	283	566	1020	1770	2670	4800	13 300	25 700
900	274	549	990	1710	2590	4650	12 900	24 900
950	266	533	961	1670	2520	4520	12 500	24 200
1000	259	518	935	1620	2450	4400	12 200	23 500
1100	246	492	888	1540	2320	4170	11 500	22 300
1200	234	470	847	1470	2220	3980	11 000	21 300
1300	225	450	811	1410	2120	3810	10 600	20 400
1400	216	432	779	1350	2040	3660	10 100	19 600
1500	208	416	751	1300	1960	3530	9760	18 900
1600	201	402	725	1260	1900	3410	9430	18 200
1700	194	389	702	1220	1840	3300	9130	17 600
1800	188	377	680	1180	1780	3200	8850	17 100
1900	183	366	661	1140	1730	3110	8590	16 600
2000	178	356	643	1110	1680	3020	8360	16 200

For SI units: 1 inch = 25 mm, 1 foot = 304.8 mm, 1000 British thermal units per hour = 0.293 kW, 1 pound-force per square inch = 6.8947 kPa

* Table entries are rounded to 3 significant digits.

TABLE 1315.2(36)
POLYETHYLENE PLASTIC TUBING [NFPA 54: TABLE 6.3.1(m)][2]

	GAS:	UNDILUTED PROPANE
	INLET PRESSURE:	11.0 in. w.c.
	PRESSURE DROP:	0.5 in. w.c.
	SPECIFIC GRAVITY:	1.50
INTENDED USE: SIZING BETWEEN INTEGRAL 2-STAGE REGULATOR AT TANK OR SECOND-STAGE (LOW PRESSURE) REGULATOR AND BUILDING		
	PLASTIC TUBING SIZE (CTS)[1] (inch)	
NOMINAL OD:	½	1
DESIGNATION:	SDR 7	SDR 11
ACTUAL ID:	0.445	0.927
LENGTH (feet)	CAPACITY IN THOUSANDS OF BTU PER HOUR	
10	121	828
20	83	569
30	67	457
40	57	391
50	51	347
60	46	314
70	42	289
80	39	269
90	37	252
100	35	238
125	31	211
150	28	191
175	26	176
200	24	164
225	22	154
250	21	145
275	20	138
300	19	132
350	18	121
400	16	113
450	15	106
500	15	100

For SI units: 1 inch = 25 mm, 1 foot = 304.8 mm, 1000 British thermal units per hour = 0.293 kW, 1 inch water column = 0.249 kPa

Notes:

[1] CTS = Copper tube size.

[2] Table entries are rounded to 3 significant digits.

CHAPTER 14
PROCESS PIPING

1401.0 General.

Process piping is a form of piping that is used to transport liquids or gasses used in industrial processes and manufacturing. These piping systems are designed to ensure that the piping systems will meet the health and safety standards in addition to the needs of the manufacturing process.

1401.1 Applicability. Except as otherwise addressed in this code, this chapter shall govern the installation of process piping in or in conjunction with a building or structure located upon the premises.

1402.0 Permit.

1402.1 General. It shall be unlawful to install, alter, or repair or cause to be installed, altered, or repaired process material piping without first obtaining a permit.

Permits for process piping shall show the total number of outlets to be provided for on each system and such other information as required by the Authority Having Jurisdiction.

Fees for process piping permits are included in Table 104.5.

Section 104.1 of the UMC states that no mechanical system regulated by this code shall be installed, altered, repaired, replaced, or remodeled unless a separate permit for each building or structure has been obtained. Section 104.5 and Table 104.5 requires permit fees for a process piping system to be determined by the Authority Having Jurisdiction (AHJ).

1403.0 Plans Required.

1403.1 General. Plans, engineering calculations, diagrams, and other data shall be submitted in one or more sets with each application for a permit. The Authority Having Jurisdiction shall be permitted to require plans, computations, and specifications to be prepared and designed by a registered design professional.

Where plans or other data are submitted for review, a plan review fee shall be paid, as provided in Section 104.3.2.

1404.0 Workmanship.

1404.1 General. Process piping shall not be strained or bent, nor shall tanks, vessels, vats, appliances, or cabinets be supported by or develop strain or stress on the piping.

See **Figure 1404.1**.

1405.0 Inspections.

1405.1 General. Upon completion of the installation, alteration, or repair of process piping, and prior to the use thereof, the Authority Having Jurisdiction shall be notified that such piping is ready for inspection.

Excavations required for the installation of underground piping shall be kept open until such time as the piping has been inspected and approved. Where such piping is covered or concealed before such approval, it shall be exposed upon the direction of the Authority Having Jurisdiction.

1405.2 Required Inspections. The Authority Having Jurisdiction shall make the following inspections and shall either approve that portion of the work as completed or shall notify the permit holder wherein the same fails to be in accordance with this code.

When the AHJ finds that the system or portion of the system being inspected fails to meet the criteria of this code, he shall notify the person requesting the inspection in person, by mail or by notice posted on the job (see Section 106.2).

1405.2.1 Rough Piping Inspection. This inspection shall be made after process piping authorized by the permit has been installed and before piping has been covered or concealed. This inspection shall include a determination that the piping size, material, and installation are in accordance with the requirements of this code.

1405.2.2 Final Piping Inspection. This inspection shall be made after piping authorized by the permit has been installed and after portions thereof that are to be covered or concealed are so concealed. This inspection shall include a pressure test, at which time the piping shall stand a pressure of not less than one-and-one-half times the maximum designed operating pressure where hydraulic testing is conducted or 110 percent where testing is conducted pneumatically. Test pres-

FIGURE 1404.1
SUPPORTS

sures shall be held for a length of time satisfactory to the Authority Having Jurisdiction, but in no case for less than 30 minutes with no perceptible drop in pressure. HPM drain, waste, and vent piping shall be tested in accordance with the plumbing code. Tests shall be made in the presence of the Authority Having Jurisdiction. Necessary apparatus for conducting tests shall be furnished by the permit holder.

For the final inspection, everything that is to be concealed in chases, behind walls, underground, and above the ceiling should already be covered. Where the system can be tested hydrostatically the testing pressure shall be no less than 1-1/2 times the maximum designed operating pressure. A system that is going to operate at 80 psi shall be tested at no less than 120 psi for a length of time that satisfies the AHJ, but at no time less than 30 minutes. If the system is going to be tested pneumatically it needs to be at 110 percent of the operating pressure. That same system that is designed to operate at 80 psi will need to be tested at a minimum of 88 psi for the same length of time. Drain, waste, and vent piping from rooms that store or dispense HPM (Hazardous Production Material) shall be tested in accordance with the plumbing code.

1405.3 Other Inspections. In addition to the inspections required by this section, the Authority Having Jurisdiction shall be permitted to require a special inspector, as specified in the building code, during installation of piping systems. In cases where the work authorized was installed in accordance with plans and specifications prepared by a registered design professional, the Authority Having Jurisdiction shall be permitted to require a final signed report stating that the work was installed in accordance with approved plans and specifications and the applicable provisions of this chapter.

The AHJ can require that a special inspector be present to witness the installation of piping to ensure that all materials and procedures that are used are within the specifications of the approved design and any other provisions of this code. They may also require the engineer that designed the system provide a signed report stating everything was installed per his engineered design.

1406.0 Pipe, Tubing, and Fittings.

1406.1 General. Process pipe, tubing, and fittings shall comply with ASME B31.3 and shall be installed in accordance with the manufacturer's installation instructions. Materials shall be rated for the operating temperatures and pressures of the system, and shall be compatible with the type of liquid.

Table 1701.1 of the Uniform Mechanical Code lists all of the referenced national standards used in this code. Although none of these standards are referenced in this chapter, they cover the manufacturing and installation practices of the materials that are to be installed.

1406.2 Hazardous Process Piping (HPP). HPP supply piping or tubing in service corridors shall be exposed to view. HPP piping shall be identified in accordance with nationally recognized standards to indicate the material being transported. Liquid HPP piping shall have an approved means for directing spilled materials to an approved containment or drainage system.

Liquid HPP waste or drainage systems shall be installed in accordance with the plumbing code.

All HPP supply piping installed in service corridors shall be exposed to view so that leaks can be easily observed and corrected. All HPP piping shall be labeled in accordance with ASME A13.1-2007, which is ANSI designated as an American National Standard. The standard requires that pipe labels are placed adjacent to all valves and flanges, all changes in pipe direction, on both sides of wall, floor and ceiling penetrations, and every 25 to 50 feet on straight runs.

An approved means for directing spilled materials may be secondary containment piping, containment trenches or drip pans, so long as they are constructed from a material that is compatible with the material that the piping system was designed to service.

1406.2.1 Installation in Exit Corridors and Above Other Occupancies. Hazardous process supply pipe shall not be located within exit corridors, within a portion of a means of egress required to be enclosed in fire-resistive construction, or in concealed spaces in or above areas not classified as Group H Occupancies, except as permitted by this subsection.

Hazardous production material piping and tubing shall be permitted to be installed within the space defined by the walls of exit corridors and the floor or roof above, or in concealed spaces above other occupancies in accordance with Section 1406.2.1.1 through Section 1406.2.1.6.

Refer to the building code for definition of a Group H Occupancy.

1406.2.1.1 Automatic Sprinklers. Automatic sprinklers shall be installed within the space unless the space is less than 6 inches (152 mm) in the least dimension.

1406.2.1.2 Ventilation. Ventilation at not less than 6 air changes per hour (ACH) shall be provided. The space shall not be used to convey air from other areas.

1406.2.1.3 Receptor. Where the piping or tubing is used to transport HPP liquids, a receptor shall be installed below such piping or tubing. The receptor shall be designed to collect discharge or leakage and drain it to an approved location. The 1 hour enclosure shall not be used as part of the receptor.

1406.2.1.4 Separation. HPP supply piping and tubing and HPP nonmetallic waste lines shall be separated from the exit corridor and from an occupancy other than a semi-conductor fabrication facility classified as a Group H Occupancy by construction, as required for walls or partitions that have a fire-protection rating of not less than 1 hour. Where gypsum wallboard is used, joints on the piping side of the enclosure need not be taped, provided the joints occur over framing members. Access openings into the enclosure shall be protected by approved fire assemblies.

1406.2.1.5 Emergency Shutoff Valves. Readily accessible manual or automatic remotely activated fail-safe emergency shutoff valves shall be installed on piping and tubing other than waste lines at the following locations:

(1) At branch connections into the fabrication area.

(2) At entries into exit corridors. Excess flow valves shall be installed as required by the fire code.

Section 203.0 defines readily accessible as "Having a direct access without the necessity of removing a panel, door, or similar obstruction." See **Figure 1406.2.1.5** for an example of installation of readily accessible emergency shutoff valves.

1406.2.1.6 Electrical Wiring. Electrical wiring and equipment located in the piping space shall be approved for Class I, Division 2, Hazardous Locations.

Exception: Occasional transverse crossing of the corridors by supply piping that is enclosed within the corridor need not comply with Section 1406.2.1.1 through Section 1406.2.1.6.

The National Electrical Code (NEC) defines hazardous locations as those areas "where fire or explosion hazards may exist due to flammable gases or vapors, flammable liquids, combustible dust, or ignitable fibers or flyings."A substantial part of the NEC is devoted to the discussion of hazardous locations. That's because electrical equipment can become a source of ignition in these volatile areas. Articles 500 through 504, and 510 through 517 provide classification and installation standards for the use of electrical equipment in these locations. The writers of the NEC developed a short-hand method of describing areas classified as hazardous locations. One of the purposes of this discussion is to explain this classification system. Hazardous locations are classified in three ways by the National Electrical Code: TYPE, CONDITION, and NATURE.

There are three types of hazardous conditions: Class I - gas and vapor, Class II - dust, and Class III - fibers and flyings. There are two kinds of hazardous conditions: Division 1 - normal, and Division 2 - abnormal.

1406.3 Special Requirements for HPP Gases. In addition to other requirements of this section, HPP gases shall comply with this subsection and the fire code.

NFPA standards such as 30, Flammable and Combustible Liquids Code, 50, the Standard for Bulk Oxygen Systems, 50A, the Standard for Gaseous Hydrogen systems, along with other such standards will need to be considered along with this section of the UMC.

1406.3.1 Special Provisions. Where HPP supply gas is carried in pressurized piping, a fail-safe system shall shut off flow due to a rupture in the piping. Where the piping originates from outside the building, the valve shall be located outside the building as close to the bulk source as practical.

1406.3.2 Piping and Tubing Installation. Piping and tubing shall be installed in accordance with approved standards. Supply piping for hazardous production materials having a health hazard ranking of 3 or 4 shall have welded connections throughout, unless an exhausted enclosure is provided.

Exception: Material that is incompatible with ferrous piping shall be permitted to be installed in nonmetallic piping with approved connections.

The NFPA standard 704, Standard System for Identification of the Hazards of Materials for Emergency Response, uses a diamond shaped chart with different symbols, colors and numbers to rate the health risks associated with certain chemicals or materials. This chapter requires all welded joints on systems designed to carry chemicals or materials with a health rating of 3 or 4. The standard states a rating of 3 or 4 is for highly to extremely toxic materials. The only exception to this rule is if the joint or connection is in an exhausted enclosure to ensure no public exposure to that chemical. See **Figure 1406.3.2**.

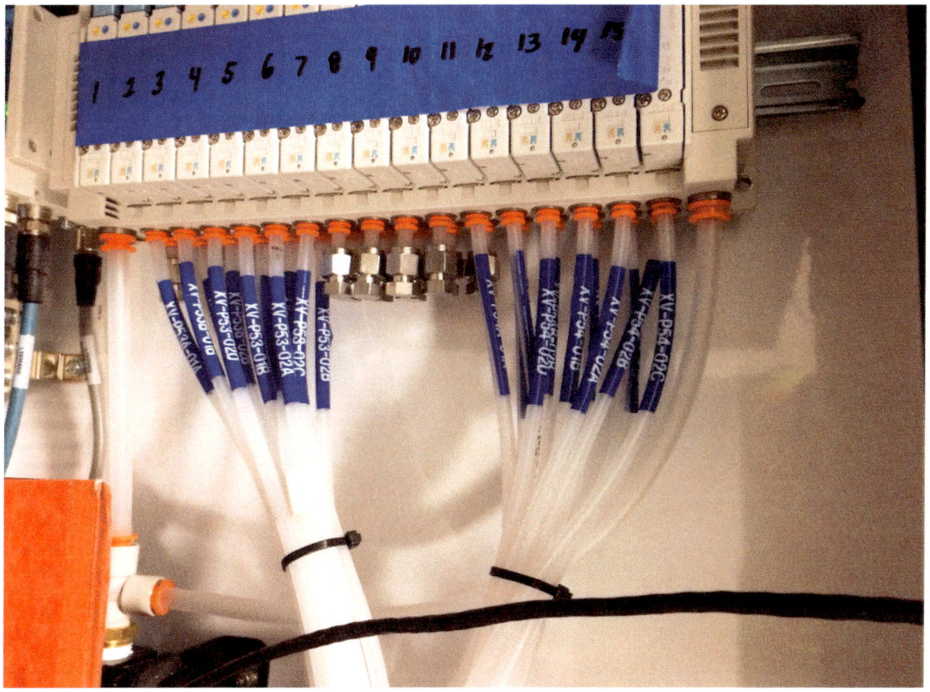

FIGURE 1406.2.1.5
INSTALLATION IN EXIT CORRIDORS

FIGURE 1406.3.2
PIPING AND TUBING INSTALLATION

1406.3.3 Gas-Detection System. Where hazardous production material gas is used or dispensed and the physiological warning properties of the gas are at a higher level than the accepted permissible exposure limit of the gas, a continuous gas-monitoring system shall be provided to detect the presence of the short-term hazard condition. Where dispensing occurs and flammable gases or vapors are capable of being present in quantities in excess of 25 percent of the lower explosive limit, a continuous gas-monitoring system shall be connected to the emergency control station.

Section 218.0 defines PEL (Permissible Exposure Limit) as the time-weighted average concentration [set by the U.S. Occupational Safety and Health Administration (OSHA)] for a normal 8-hour workday and a 40-hour workweek to which nearly all workers can be repeatedly exposed without adverse effect. If the level that a person can detect a gas through physical means is above the PEL, then a continuous monitoring system shall be installed to warn workers of the impeding danger. If the vapors are present and flammable in quantities of 20 percent above the LEL (see Section 214.0 Definitions) a continuous gas-monitoring system shall be provided.

CHAPTER 15
SOLAR ENERGY SYSTEMS

1501.0 General.

1501.1 Applicability. See Section 1203.0 and the Uniform Solar Energy and Hydronics Code (USEHC), published by the International Association of Plumbing and Mechanical Officials. The Uniform Solar Energy and Hydronics Code (USEHC) provides requirements that shall be permitted to be adopted as part of the code by the Authority Having Jurisdiction.

CHAPTER 16
STATIONARY POWER PLANTS

1601.0 Stationary Fuel Cell Power Plants.

1601.1 General. Fuel cell power plants with a power output of less than 50 kW shall be listed and installed in accordance with the manufacturer's instructions. Fuel cell power plants with a power output of greater than 50 kW shall be installed in accordance with NFPA 853. [NFPA 54:10.30] Stationary fuel cell power plants shall be tested in accordance with CSA FC-1.

There are five primary types of fuel cells based on their unique electrolyte use, and each has specific characteristics that make it better in certain applications over others. All can be used in a stationary application and some are suited to be mobile as well.

Polymer Electrolyte Fuel Cell (PEFC) or Proton Exchange Membrane Fuel Cell (PEMFC). The electrolyte in this type of fuel cell is an ion exchange membrane made of some type of polymer that is a good conductor of protons. This type of fuel cell runs at low temperatures with electrical efficiencies of about 45 percent, and is the primary candidate for automotive, small stationary, and portable power applications. PEMFCs require very pure hydrogen as the fuel.

Phosphoric Acid Fuel Cell (PAFC). The electrolyte in this type of fuel cell is phosphoric acid, concentrated to 100. PAFCs have a high operating temperature and achieve an electrical efficiency of about 40 percent. Buses and stationary applications currently use PAFCs.

Molten Carbonate Fuel Cell (MCFC). The electrolyte in this type of fuel cell is usually a combination of alkali carbonates, retained in a ceramic matrix. The MCFC operates at very high temperatures which enables the end user to utilize both the electricity and the thermal energy generated by the fuel cell, resulting in electrical efficiencies of more than 70 percent. MCFCs are well-suited to large-scale stationary applications, and are currently being demonstrated for powering buildings.

Solid Oxide Fuel Cell (SOFC). The electrolyte in the SOFC is a solid, nonporous metal oxide. At temperatures over 650 degrees Celsius, the SOFC can utilize a hydrocarbon fuel directly, without reforming, similar to the MCFC. Also similar to the MCFC, the SOFC generates both electricity and usable thermal energy. High-temperature SOFCs are being demonstrated for stationary power applications, while low-temperature SOFCs are also being looked at for automotive applications.

Alkaline Fuel Cell (AFC). This was one of the first modern fuel cells to be developed and was used to provide on-board electric power for the Apollo space vehicle. The electrolyte in this fuel cell is Alkaline. AFCs require pure hydrogen and pure oxygen as the reactants. The operating temperature for this type of fuel cell is around 200 degrees Celsius.

1602.0 Stationary Gas Engines and Generators.

1602.1 General. The installation of gas engines shall conform to NFPA 37. [NFPA 54:10.23]

1602.2 Connection to the Gas Supply Piping. Stationary gas engines shall not be rigidly connected to the gas supply piping. [NFPA 54:10.23.1]

1602.3 Stationary Engine Generators. Stationary engine generators shall be tested in accordance with UL 2200, and shall be installed in accordance with NFPA 37 and the manufacturer's installation instructions.

See **Figure 1602.0**.

FIGURE 1602.0
RECIPROCATING PISTON STATIONARY ENGINE GENERATOR

CHAPTER 17

REFERENCED STANDARDS

1701.0 General.

Code Requirements and Mandatory Referenced Standards: Section 302.1 provides for the minimum standards for all material related to pipe, fittings, appliances, appurtenances, equipment, materials, and devices used in a mechanical system. Each device must be listed (third-party certified) by a listing agency (accredited conformity assessment body) and must conform to approved applicable recognized standards referenced in this code. In order to conform to minimum standards, all devices must be free from defects and meet the performance requirements contained in the standards. Products are referred to as "listed" indicating that the minimum quality required in the referenced standard is upheld through the rigors of the third-party certification process.

A standard is a set of technical definitions, requirements and guidelines for the manufacture of devices and products that establishes test methods, specifications, classifications, practices and scoping requirements. Standards are documents that provide specific guidelines for manufacturing all sorts of products. Their purpose is to ensure that products are made to be safe and efficient for the consumer. They also provide dimensional requirements that ensure that the product can be installed and interchanged as required.

Although each standard is different, there are common requirements. All standards must:

- Define the scope of the product, system or process that is covered.
- Be able to be used repeatedly and not so specific that it does not apply to many applications.
- Include a test methods and performance criteria, with a clear and concise description for measurement and evaluation of one or more properties, qualities or characteristics. A test method is a kind of standard that produces a test result, and it is recommended practice to reference a test method along with the applicable performance requirements. (e.g. A test method may describe how to measure thickness. A performance standard will specify what the thickness must be when measured in accordance with the test method).
- Specify the requirements in simple, precise understandable language free from ambiguous terminology.
- Contain terms and explanations of symbols, abbreviations or acronyms that are relevant to the standard and its application.
- Be enforceable to the extent of its use and application, which includes being written in mandatory language and does not mandate the use of proprietary materials or agencies.

Standard requirements are typically based on unambiguous specifications, which include physical, mechanical and chemical properties. Additionally, standards identify the applicable test methods and performance criteria to determine that each requirement is met or satisfied.

Codes and standards work together to protect public health and safety. A standard is considered a basis of comparison or an approved model. Simply stating, codes tell the user what to do and when and under what circumstances to do it. Codes are often legal requirements that are enforced by local jurisdictions that enforce their provisions. Standards provide the user with approved materials and tell the user that such products have been tested to performance requirements and are applicable only to the extent the code references the standard. For example, a standard may have performance specifications for materials and use of application or installation requirements. The standard is only applicable to the extent suggested in the text of the code. The code text takes precedence when the requirements of the standard conflict with the requirements of the code.

The terms "Listed," "Listing Agency," "Approved," and "Approved Testing Agency" are defined in Chapter 2 of the Uniform Mechanical Code (UMC). A product must be listed and presented to the Authority Having Jurisdiction (AHJ) for approval. Listing provides independent confirmation that a product conforms to standards and that it is verified to comply with those standards. For a product to be "listed," it must be tested to the applicable recognized standards for that product, be found safe for use in a specific manner, and then be put on a list of such products by a "listing agency." These lists (usually called "directories" of listed plumbing products) are used by juris- dictions when inspecting installations for code compliance. The listing agency performs periodic inspections to ensure the product continues to be manufactured to the correct standards. A listed product is labeled with the mark of the listing agency so inspectors in the field can identify it as listed.

In order to be listed, products are required to comply with standards that are designed to prove the products ability to provide safe operation over long periods. A product typically undergoes many different tests that simulate use and extended wear and tear. Testing criteria are used to develop maximum tolerances for products and materials. All testing procedures and results must be verifiable.

Types of Standards

Nationally Recognized Consensus Standards: The following is a partial list in Table 1701.1 of different standards developing organizations (SDO) that create Nationally Recognized Consensus Standards with the typical designation of the American National Standards Institute (ANSI). ANSI is not the developer of the standards, but is the accreditation body of SDOs that use the consensus process for development of national standards. It promotes developing American National Standards by accrediting the procedures of standards developing organizations. Accreditation by ANSI represents the procedures used by the standards body with developing American National Standards that meet the essential requirements for openness, balance, consensus and due process.

- American Society of Mechanical Engineers (ASME)
- American Society of Sanitary Engineering (ASSE)
- American Welding Society (AWS)
- American Water Works Association (AWWA)
- Canadian Standards Association (CSA)
- ASTM International
- International Association of Plumbing and Mechanical Officials (IAPMO)
- Industry Safety Equipment Association (ISEA)
- National Fire Protection Association (NFPA)
- NSF International (NSF)
- Underwriters Laboratories (UL)

The above are examples of standard developing organizations (SDOs) that provide a consensus by a group or body that is open to representatives from all affected and interested parties. Standards are developed by presenting a rationale for the need of a particular standard. If the proposed rationale for the standard is approved, the next process is development of a draft text by a committee of subject matter experts, which constitute a wide spectrum of stakeholders, including manufacturers, testing agencies, environmental experts and end users. When the developing committee reaches a consensus on the content of the draft, it is sent out to all interested parties (public review period) and the standards committee for review and comments (ballot form). All comments are reviewed and considered during this period and are responded to through ballot form back to the standards committee for review. All changes that are found to be persuasive are incorporated into the draft and further public review and ballots are sent out to the standards committee. A vote is taken and any comments received are resolved. Once a level of consensus is gained among the committee the standard is then presented to the Board of Standards Review (BSR) for approval as an American National Standards from ANSI. Once the standard is published, it will then be reviewed every five years for possible revision, reaffirmation or withdrawal.

Other Standards: Standards written by other organizations are listed in Table 1701.1. IAPMO has developed more than 300 quasi-consensus standards which:

- cover innovative new plumbing products not covered by existing standards;
- are widely accepted by industry and regulators (jurisdictions);
- are jointly developed by IAPMO staff, stakeholders, and public comments;
- are approved by the IAPMO Standards Review Committee (SRC); and
- can evolve into ANSI standards.

The development process for IAPMO quasi-consensus standards parallels the ANSI process for consensus standards, but it is much shorter (usually 1 to 3 months) thus allowing manufacturers timely and efficient access to the marketplace.

IAPMO Guide Criteria

The IAPMO Standards Department works with stakeholders to provide an opportunity for the development of a new performance test standard known as an IAPMO Guide Criteria (IGC) when no applicable standard exists for an innovatively new product. Often, new products or new technologies surge ahead far faster than most standards can keep pace. Through IAPMO Guide Criteria (IGC), IAPMO provides manufacturers and product developers an opportunity to use IGC standards as a vehicle for testing, listing, and introducing new products to the marketplace. Once an IGC is accepted, IAPMO R&T can list products manufactured and tested in compliance with the new requirements.

The IAPMO Standards Review Committee (SRC) meets every month to review proposals for IGCs and various types of standards. SRC meetings are open to the public whereby all proponents and participants are given the opportunity to present their views and supporting information on the proposals discussed at these meetings. IAPMO standards, including IGCs, approved by the SRC are posted on the IAPMO Standards public review webpage for public comment over a 20-day period.

IAPMO Material and Property Standards: If after three years an IGC continued in use in active listings, then the IGC evolved to become an IAPMO material and property standards (PS) to continue acceptable quality and performance level for products in the absence of approved nationally recognized standards. While PS type standards are no longer developed, those in existence continue to be used.

Form and Style for Standards: Each standard includes a title; scope; purpose; referenced documents; terminology or definitions; materials or classification; test methods and performance requirements; mandatory installation practices; product marking and quality assurance. In addition, mandatory and non-mandatory information with recommended installation practices are included.

Scope: The scope defines without ambiguity the subject of the document and the aspects covered, thereby indicating the limits of applicability of the standard. It provides a common understanding of the standard with defining the standard's overall boundaries by suggesting what it will and will not perform (objectives) on the subject product. In addition, scoping sections include physical and performance testing methods with material component references.

Reference Standards: Many standards make reference to other standards or documents as a part of the entire standard for compliance with physical or performance requirements for various material or device components listed in the standard. For example, the standard may reference another standard for a test method for dimensions. Therefore, the test method does not need to be repeated in the standard; only the referenced standard test method is mentioned and collaborated. In the mechanical industry, the code references various product standards, which define the product's performance and design.

Terminology or Definitions: Standards have definitions that are specific to the applicability of the standard and apply only to the requirements of the standard. These definitions aid the user in understanding the application and extent of requirements within the standard. It is important to stress that these definitions are applicable to this specific standard. There may be instances where the standard promulgator uses the same definitions in many referenced standards; however, the user should be aware that only those definitions indicated within the standards are applicable to the performance and testing requirements.

Material or Physical Requirements: Material or physical requirements specifically define the materials used, such as its origin, composition or properties. Within this, such specific requirements may include physical properties, mechanical or chemical properties. In addition, many physical and chemical properties reference various standards or cell classifications and that such materials must be in accordance with the applicable standard. The material or physical characteristics must be precise and specific to satisfy the performance criteria.

Test Methods and Performance Requirements

Test methods and performance requirements identify the methods for deciding whether each of the requirements is satisfied. Many standards include provisions for sampling or test specimens; test conditions and procedures; calculations and acceptance criteria. Each standard must have standardized methods for testing and evaluating each product or device. Test methods are the key to ensuring product performance is measured accurately and consistently. As with all standards, test requirements include mandatory and optional elements that vary according to the test methods.

Installation Practices: Many standards include some installation instructions that give specific application, cleaning, installation, preparation, and training methods. Typically, these are in the form of a set of instructions for performing specific methods to ensure installing the product or device complies with recommended practices. These practices are important for people who install such products or devices, as they contain significant input from many subject experts in the field and ensure the health and safety of the consumer. Codes contain many of these installation practices as specific code requirements based on the specific standard.

Product Marking and Quality Assurance: Standards specify the required product markings and detail, especially if the markings are required to be permanently applied. Methods such as stamping or embossing with specific lettering and coloring are often required. Each standard has specific requirements, which may include the manufacturer's name or trademark, standard designation, model number, sizing and spacing intervals.

Nonmandatory Information: Nonmandatory information is located in the appendix and provides a guide without recommending a specific action. The purpose of this is to provide the user with help based on a consensus of subject matter experts. This information is intended to provide added techniques in supplementing the standard. For example, the installation guidelines and joining methods for a particular piping material may be covered with information about other uses and practices.

In summary, standards are important and work with the plumbing code to promote health and safety. They help to produce consistent performance requirements and lead to better specification of materials so the end user knows the properties of that material. This means designers and installers can specify the material that best suits their needs. Standards lead to harmonization of performance requirements, which provides for the greatest opportunities for consumers, businesses and the industry.

IAPMO Product Markings: The Research and Testing (R&T) division of IAPMO tests and evaluates products to determine whether they comply with select standards and applicable codes. Each tested product is then assessed as meeting or not meeting the standards. Compliant products may qualify for application of one of seven possible IAPMO marking labels (see **Figure 1701.1a**).

By its shape or markings, an IAPMO R&T certification mark may denote that a product is either "listed" or "classified." Each of these labels carries considerably different implications. A product that bears the "listed" label has demonstrated compliance with the specific standard or standards to which it was tested. Additionally, a listed product is defined as compliant with code stipulations, engineering concepts or other fundamental principles found in the UPC or Uniform Mechanical Code (UMC).

Conversely, a "classified" label represents compliance with all standards to which the product was tested but does not ensure compliance with UPC or UMC code stipulations or requirements. Classified products are published in a separate product directory, the objective being to avoid user confusion regarding the propriety of their application. Section 302.1.2 of the UMC allows the use of these products, but only with specific approval from the local AHJ. Listed products require no special approval or acceptance by local authorities when installed within the limits of their listing.

The remaining five IAPMO R&T labels shown in **Figure 1701.1a** have specialty applications, e.g., solar, recreational vehicles, etc. These five labels represent specialty listings that have limited applications but that are code compliant when applied within the limitations of their listings and each applicable code.

There are no approvals issued by IAPMO R&T. In all cases, R&T evaluations are solely for the purpose of verifying that the tested product has demonstrated compliance with a specified test standard. Satisfying a given standard does not ensure compliance with applicable codes. Products labeled as "listed" products will not conflict with either the language or the intent of the UPC or the UMC. Other products labeled as "classified" products meet the standard to which they were tested, but cannot comply with certain requirements found in the UPC or UMC.

The use of specialized labeling is intended to provide the consumer with guidance regarding the limits of product application.

Compliance with an evaluation standard neither suggests that a product is suitable for any specific application nor that a product is code compliant nor approved for use in a particular application. IAPMO R&T tests to standards specified by the person who submits the product for evaluation (see **Figure 1701.1b**). Because all test standards are not necessarily related to code requirements or standard engineering practices, test compliance is essentially unrelated to the appropriate utilization of a particular product.

1701.1 Standards. The standards listed in Table 1701.1 are referenced in various sections of this code and shall be considered part of the requirements of this document. The standards are listed herein by the standard number and effective date, the title, application and the section(s) of this code that reference the standard. The application of the referenced standard(s) shall be as specified in Section 302.1.2.

The promulgating agency acronym referred to in Table 1701.1 are defined in a list found at the end of the table.

Table 1701.1 lists standards referenced within the body of Code and are considered as part of the requirements of the Code. Table 1701.2 consists of installation and testing standards that are not referenced in other sections of this code but may aid in the user's ability to quickly find an applicable standard. The application of both sets of referenced standards are specified in Section 302.1.2 of this Code.

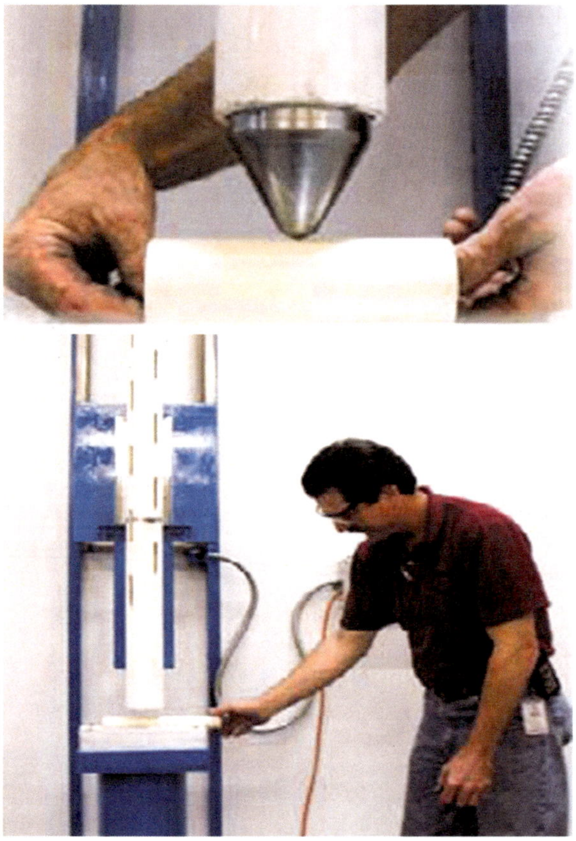

FIGURE 1701.1B
IAPMO R&T TESTING

NSF 61-9

The product complies with Section 9 of the NSF 61 standard.

C IAPMO™ · C IAPMO C™

Classified Marking Certification Marks

The product complies with only the product's performance standard BUT not acceptable or recognized by the UPC.

UMC™ IAPMO - UMC™

Uniform Mechanical Code Certification Marks

The product complies with BOTH the product's performance AND the UMC

UPC® UPC®

Uniform Plumbing Code Certification Marks

The product complies with BOTH the product's performance AND the UPC

USEC™ USEC®

Uniform Solar Energy Code Certification Marks

The product complies with BOTH the product's performance AND the Uniform Solar Energy Code

USPC® USPC®

Uniform Pool, Spa & Hot Tub Code Certification Marks

The product complies with BOTH the product's performance AND the Pool, Spa and Hot Tub Code.

IAPMO T® IAPMO-T®

Manufactured Housing/Recreational Vehicle Certification Marks

The product complies with IAMPO's MHRV product's performance standard.

CM®

Materials/Components Certification Marks

The product material and/or component complies with applicable sections of the product's performance standard.

FIGURE 1701.1A
IAPMO MARKS

TABLE 1701.1
REFERENCED STANDARDS

STANDARD NUMBER	STANDARD TITLE	APPLICATION	REFERENCED SECTION
49 CFR 192.281	Plastic Pipe	Piping, Plastic	1308.5.4.2.2
49 CFR 192.283	Plastic Pipe: Qualifying Joining Procedures	Piping, Plastic	1308.5.4.2.2
AABC-2016	Total System Balance, 7th Edition	Balancing	314.1(1)
ACCA Manual B-2009	Balancing and Testing Air and Hydronic Systems	Balancing	314.1(2)
ACCA Manual D-2016	Residential Duct Systems	Ducts, Balancing	601.2
ACCA Manual J-2016	Residential Load Calculations	Ducts	1105.1
ACCA Manual S-2014	Residential Equipment Selection	Equipment	1105.1
ACCA Manual Zr-2018	Residential Zoning	Ducts	601.2
ACCA 4 QM-2019	Maintenance of Residential HVAC Systems	HVAC Systems	102.3.2
AHRI 700-2017a	Specifications for Refrigerants	Refrigerants	1104.7, 1104.7.3
AMCA 540-2013	Louvers Impacted by Wind Borne Debris	Louvers	315.1.2
AMCA 550-2015	Test Method for High Velocity Wind Driven Rain Resistant Louvers	Louvers	315.1.1
ASHRAE 15-2016	Safety Standard for Refrigeration Systems	Refrigeration Systems	1102.1, 1106.1, Table 1113.5
ASHRAE 34-2016	Designation and Safety Classification of Refrigerants	Refrigeration Classifications	1102.3, 1103.1, Table 1102.3, Table 1106.2.5.2
ASHRAE 62.1-2016	Ventilation for Acceptable Indoor Air Quality	Indoor Air Quality Ventilation	402.4.1
ASHRAE 111-2008	Measurement, Testing, Adjusting, and Balancing of Building HVAC Systems	Balancing	314.1(3)
ASHRAE 129-1997 (R2002)	Measuring Air-Change Effectiveness	Air Change Effectiveness	Table 403.2.2
ASHRAE 154-2016	Ventilation for Commercial Cooking Operations	Commercial Kitchens	510.5.6
ASHRAE 170-2013	Ventilation of Health Care Facilities	Ventilation	402.1.2
ASHRAE/ACCA 180-2018	Inspection and Maintenance of Commercial Building HVAC Systems	Maintenance	102.3.1, 1013.3
ASHRAE Handbook-2017	Fundamentals	Climatic Conditions	Figure 803.1.2(6)

TABLE 1701.1 (continued)
REFERENCED STANDARDS

STANDARD NUMBER	STANDARD TITLE	APPLICATION	REFERENCED SECTION
ASME B1.20.1-2013	Pipe Threads, General Purpose (Inch)	Joints	1211.2(3), 1211.4(7), 1211.12(3), 1211.13(2), 1308.5.6
ASME B16.1-2015	Gray Iron Pipe Flanges and Flanged Fittings: Classes 25, 125, and 250	Fittings	1308.5.9
ASME B16.3-2016	Malleable Iron Threaded Fittings: Classes 150 and 300	Fittings	Table 1210.1
ASME B16.5-2017	Pipe Flanges and Flanged Fittings: NPS ½ through NPS 24 Metric/Inch	Fittings	Table 1210.1, 1308.5.9.1(1)
ASME B16.9-2018	Factory-Made Wrought Buttwelding Fittings	Fittings	Table 1210.1
ASME B16.11-2016	Forged Fittings, Socket-Welding and Threaded	Fittings	Table 1210.1
ASME B16.15-2018	Cast Copper Alloy Threaded Fittings: Classes 125 and 250	Fittings	Table 1210.1
ASME B16.18-2018	Cast Copper Alloy Solder Joint Pressure Fittings	Fittings	Table 1210.1
ASME B16.20-2017	Metallic Gaskets for Pipe Flanges	Joints	1308.5.10.2
ASME B16.21-2016	Nonmetallic Flat Gaskets for Pipe Flanges	Fuel Gas Piping	1308.5.10.3
ASME B16.22-2018	Wrought Copper and Copper Alloy Solder-Joint Pressure Fittings	Fittings	Table 1210.1
ASME B16.23-2016	Cast Copper Alloy Solder Joint Drainage Fittings: DWV	Fittings	Table 1210.1
ASME B16.24-2016	Cast Copper Alloy Pipe Flanges, Flanged Fittings, and Valves: Classes 150, 300, 600, 900, 1500, and 2500	Fittings	Table 1210.1, 1308.5.9.2
ASME B16.26-2018	Cast Copper Alloy Fittings for Flared Copper Tubes	Fittings	Table 1210.1
ASME B16.29-2017	Wrought Copper and Wrought Copper Alloy Solder-Joint Drainage Fittings – DWV	Fittings	Table 1210.1
ASME B16.42-2016	Ductile Iron Pipe Flanges and Flanged Fittings: Classes 150 and 300	Fuel Gas Piping	1308.5.9.3
ASME B16.47-2017	Large Diameter Steel Flanges: NPS 26 Through NPS 60 Metric/Inch	Fuel Gas Piping	1308.5.9.1(2)
ASME B16.51-2018	Copper and Copper Alloy Press-Connect Pressure Fittings	Fittings	Table 1210.1
ASME B31.3-2016	Process Piping	Process Piping	1406.1
ASME B31.5-2016	Refrigeration Piping and Heat Transfer Components	Refrigeration Piping	1109.1.1, 1109.1.3
ASME B36.10M-2018	Welded and Seamless Wrought Steel Pipe	Fuel Gas Piping	1308.5.2.1
ASME BPVC Section I-2017	Rules for Construction of Power Boilers	Boilers	1002.1(1), Table 1003.2.1
ASME BPVC Section IV-2017	Rules for Construction of Heating Boilers	Boilers	1002.1(2)
ASME BPVC Section VIII.1-2017	Rules for Construction of Pressure Vessels Division 1	Pressure Vessels	1002.1, 1004.3, 1112.10.1, 1112.13, 1113.1, 1115.4, 1115.4.1, 1117.1(2), 1117.1(3), 1117.2, 1209.4

**TABLE 1701.1 (continued)
REFERENCED STANDARDS**

STANDARD NUMBER	STANDARD TITLE	APPLICATION	REFERENCED SECTION
ASME CSD-1-2018	Controls and Safety Devices for Automatically Fired Boilers	Boilers	1003.2.1
ASSE 1013-2011	Reduced Pressure Principle Backflow Preventers and Reduced Pressure Principle Fire Protection Backflow Preventers	Backflow Protection	1202.2
ASSE 1017-2009	Temperature Actuated Mixing Valves for Hot Water Distribution Systems	Valves	1207.3.1
ASSE 1061-2015	Push-Fit Fittings	Fittings	1211.2(1), 1211.4(5), Table 1210.1
ASSE 1079-2012	Dielectric Pipe Fittings	Fittings	1211.14.1
ASSE Z359.1-2016	The Fall Protection Code	Miscellaneous	303.8.4.1
ASTM A53/A53M-2018	Pipe, Steel, Black and Hot-Dipped, Zinc-Coated, Welded and Seamless	Piping	1308.5.2.1(1), Table 1210.1
ASTM A106/A106M-2018	Seamless Carbon Steel Pipe for High-Temperature Service	Piping	1308.5.2.1(2), Table 1210.1
ASTM A126-2004 (R2014)	Gray Iron Castings for Valves, Flanges, and Pipe Fittings	Piping	Table 1210.1
ASTM A254/A254M-2012	Copper-Brazed Steel Tubing	Piping	1308.5.3.2, Table 1210.1
ASTM A268/A268M-2010 (R2016)	Seamless and Welded Ferritic and Martensitic Stainless Steel Tubing for General Service	Tubing	1308.5.3.1(1)
ASTM A269/A269M-2015a	Seamless and Welded Austenitic Stainless Steel Tubing for General Service	Piping, Tubing	1308.5.3.1(2), Table 1210.1
ASTM A312/A312M-2018a	Seamless, Welded, and Heavily Cold Worked Austenitic Stainless Steel Pipes	Piping Ferrous	1308.5.2.1(3), Table 1210.1
ASTM A420/A420M-2016	Piping Fittings of Wrought Carbon Steel and Alloy Steel for Low-Temperature Service	Fittings	Table 1210.1
ASTM A554-2016	Welded Stainless Steel Mechanical Tubing	Piping	Table 1210.1
ASTM A778/A778M-2016	Welded, Unannealed Austenitic Stainless Steel Tubular Products	Piping	Table 1210.1
ASTM B32-2008 (R2014)	Solder Metal	Joints	1211.4(6)
ASTM B42-2015a	Seamless Copper Pipe, Standard Sizes	Piping	Table 1210.1
ASTM B43-2015	Seamless Red Brass Pipe, Standard Sizes	Piping	Table 1210.1
ASTM B75/B75M-2011	Seamless Copper Tube	Piping	Table 1210.1
ASTM B88-2016	Seamless Copper Water Tube	Piping	1308.5.3.3, Table 1210.1
ASTM B135/B135M-2017	Seamless Brass Tube	Piping	Table 1210.1
ASTM B210-2012	Aluminum and Aluminum-Alloy Drawn Seamless Tubes	Piping	1308.5.3.4
ASTM B241/B241M-2016	Aluminum and Aluminum-Alloy Seamless Pipe and Seamless Extruded Tube	Piping	1308.5.2.3, 1308.5.3.4
ASTM B251/B251M-2017	General Requirements for Wrought Seamless Copper and Copper-Alloy Tube	Piping	Table 1210.1
ASTM B280-2018	Seamless Copper Tube for Air Conditioning and Refrigeration Field Service	Piping	1109.1.2, 1308.5.3.3

TABLE 1701.1 (continued)
REFERENCED STANDARDS

STANDARD NUMBER	STANDARD TITLE	APPLICATION	REFERENCED SECTION
ASTM B302-2017	Threadless Copper Pipe, Standard Sizes	Piping	Table 1210.1
ASTM B447-2012a	Welded Copper Tube	Piping	Table 1210.1
ASTM B813-2016	Liquid and Paste Fluxes for Soldering of Copper and Copper Alloy Tube	Joints	1211.4(6)
ASTM B828-2016	Making Capillary Joints by Soldering of Copper and Copper Alloy Tube and Fittings	Joints	1211.4(6)
ASTM B1003-2016	Seamless Copper Tube for Linesets	Piping	1109.1.2
ASTM C315-2007 (R2016)	Clay Flue Liners and Chimney Pots	Venting Systems	803.1.11.1
ASTM C411-2017	Hot-Surface Performance of High-Temperature Thermal Insulation	Duct Coverings and Linings	605.1.2
ASTM D1693-2015	Environmental Stress-Cracking of Ethylene Plastics	Piping	Table 1210.1
ASTM D1785-2015^{e1}	Poly (Vinyl Chloride) (PVC) Plastic Pipe, Schedules 40, 80, and 120	Piping	Table 1210.1
ASTM D2241-2015	Poly (Vinyl Chloride) (PVC) Pressure-Rated Pipe (SDR Series)	Piping	Table 1210.1
ASTM D2464-2015	Threaded Poly (Vinyl Chloride) (PVC) Plastic Pipe Fittings, Schedule 80	Fittings	Table 1210.1
ASTM D2466-2017	Poly (Vinyl Chloride) (PVC) Plastic Pipe Fittings, Schedule 40	Fittings	Table 1210.1
ASTM D2467-2015	Poly (Vinyl Chloride) (PVC) Plastic Pipe Fittings, Schedule 80	Fittings	Table 1210.1
ASTM D2513-2018a	Polyethylene (PE) Gas Pressure Pipe, Tubing, and Fittings	Piping	1308.5.4, 1308.5.4.2.2, 1308.5.8.2, 1310.1.7.1(1), Table 1210.1
ASTM D2564-2012 (R2018)	Solvent Cements for Poly (Vinyl Chloride) (PVC) Plastic Piping Systems	Joints	1211.12(2)
ASTM D2609-2015	Plastic Insert Fittings for Polyethylene (PE) Plastic Pipe	Fittings	Table 1210.1
ASTM D2683-2014	Socket-Type Polyethylene Fittings for Outside Diameter-Controlled Polyethylene Pipe and Tubing	Fittings	Table 1210.1
ASTM D2737-2012a	Polyethylene (PE) Plastic Tubing	Piping, Plastic	Table 1210.1
ASTM D2846/D2846M-2017b^{e1}	Chlorinated Poly (Vinyl Chloride) (CPVC) Plastic Hot- and Cold-Water Distribution Systems	Piping	1211.2(2), 1211.3(2), Table 1210.1
ASTM D3035-2015	Polyethylene (PE) Plastic Pipe (DR-PR) Based on Controlled Outside Diameter	Piping	Table 1210.1
ASTM D3139-1998 (R2011)	Joints for Plastic Pressure Pipes Using Flexible Elastomeric Seals	Joints	1211.12(1)
ASTM D3261-2016	Butt Heat Fusion Polyethylene (PE) Plastic Fittings for Polyethylene (PE) Plastic Pipe and Tubing	Fittings	Table 1210.1
ASTM D3350-2014	Polyethylene Plastics Pipe and Fittings Materials	Piping, Fittings	Table 1210.1
ASTM E84-2018b	Surface Burning Characteristics of Building Materials	Miscellaneous	508.3.4, 602.2, 605.1.1, 605.1.2, 1201.2
ASTM E779-2019	Determining Air Leakage Rate by Fan Pressurization	Air Ducts	405.2.1
ASTM E814-2013a (R2017)	Fire Tests of Penetration Firestop Systems	Miscellaneous	507.4.4, 507.4.5

TABLE 1701.1 (continued)
REFERENCED STANDARDS

STANDARD NUMBER	STANDARD TITLE	APPLICATION	REFERENCED SECTION
ASTM E2231-2018	Specimen Preparation and Mounting of Pipe and Duct Insulation Materials to Assess Surface Burning Characteristics	Insulation of Ducts	605.1.2
ASTM E2336-2016	Fire Resistive Grease Duct Enclosure Systems	Grease Ducts	507.4.2.2, 507.4.5
ASTM F437-2015	Threaded Chlorinated Poly (Vinyl Chloride) (CPVC) Plastic Pipe Fittings, Schedule 80	Fittings	Table 1210.1
ASTM F438-2017	Socket-Type Chlorinated Poly (Vinyl Chloride) (CPVC) Plastic Pipe Fittings, Schedule 40	Fittings	Table 1210.1
ASTM F439-2013	Chlorinated Poly (Vinyl Chloride) (CPVC) Plastic Pipe Fittings, Schedule 80	Fittings	Table 1210.1
ASTM F441/F441M-2015	Chlorinated Poly (Vinyl Chloride) (CPVC) Plastic Pipe, Schedules 40 and 80	Piping	Table 1210.1
ASTM F442/F442M-2013^{e1}	Chlorinated Poly (Vinyl Chloride) (CPVC) Plastic Pipe (SDR-PR)	Piping	Table 1210.1, 1211.2(2)
ASTM F493-2014	Solvent Cements for Chlorinated Poly (Vinyl Chloride) (CPVC) Plastic Pipe and Fittings	Joints	1211.2(2), 1211.3(2)
ASTM F656-2015	Primers for Use in Solvent Cement Joints of Poly (Vinyl Chloride) (PVC) Plastic Pipe and Fittings	Joints	1211.2(2), 1211.3(2), 1211.12(2)
ASTM F714-2013	Polyethylene (PE) Plastic Pipe (DR-PR) Based on Outside Diameter	Piping, Plastic	Table 1210.1
ASTM F876-2017	Crosslinked Polyethylene (PEX) Tubing	Piping	1211.5, Table 1210.1
ASTM F877-2018a	Crosslinked Polyethylene (PEX) Hot- and Cold-Water Distribution Systems	Piping	Table 1210.1
ASTM F1055-2016a	Electrofusion Type Polyethylene Fittings for Outside Diameter Controlled Polyethylene and Crosslinked Polyethylene (PEX) Pipe and Tubing	Fittings	Table 1210.1
ASTM F1281-2017	Crosslinked Polyethylene/Aluminum/Crosslinked Polyethylene (PEX-AL-PEX) Pressure Pipe	Piping	Table 1210.1
ASTM F1282-2017	Polyethylene/Aluminum/Polyethylene (PE-AL-PE) Composite Pressure Pipe	Piping	Table 1210.1
ASTM F1476-2007 (R2013)	Performance of Gasketed Mechanical Couplings for Use in Piping Applications	Fittings	Table 1210.1
ASTM F1548-2001 (R2018)	Performance of Fittings for Use with Gasketed Mechanical Couplings Used in Piping Applications	Fittings	Table 1210.1
ASTM F1807-2018a	Metal Insert Fittings Utilizing a Copper Crimp Ring for SDR9 Cross-linked Polyethylene (PEX) Tubing and SDR9 Polyethylene of Raised Temperature (PE-RT) Tubing	Fittings	Table 1210.1
ASTM F1960-2018a	Cold Expansion Fittings with PEX Reinforcing Rings for Use with Cross-linked Polyethylene (PEX) and Polyethylene of Raised Temperature (PE-RT) Tubing	Fittings	Table 1210.1
ASTM F1970-2012^{e1}	Special Engineered Fittings, Appurtenances or Valves for Use in Poly (Vinyl Chloride) (PVC) or Chlorinated Poly (Vinyl Chloride) (COPE) Systems	Piping	Table 1210.1
ASTM F1973-2013 (R2018)	Factory Assembled Anodeless Risers and Transition Fittings in Polyethylene (PE) and Polyamide 11 (PA11) and Polyamide 12 (PA12) Fuel Gas Distribution Systems	Fuel Gas	1310.1.7.1(2)
ASTM F1974-2009 (R2015)	Metal Insert Fittings for Polyethylene/Aluminum/Polyethylene and Crosslinked Polyethylene/Aluminum/ Crosslinked Polyethylene Composite Pressure Pipe	Fittings	1211.6(1), 1211.9(1), Table 1210.1

TABLE 1701.1 (continued)
REFERENCED STANDARDS

STANDARD NUMBER	STANDARD TITLE	APPLICATION	REFERENCED SECTION
ASTM F2080-2018	Cold-Expansion Fittings with Metal Compression-Sleeves for Crosslinked Polyethylene (PEX) Pipe and SDR9 Polyethylene of Raised Temperature (PE-RT) Pipe	Fittings	Table 1210.1
ASTM F2098-2018	Stainless Steel Clamps for Securing SDR9 Cross-linked Polyethylene (PEX) Tubing to Metal Insert and Plastic Insert Fittings	Joints	Table 1210.1
ASTM F2158-2008 (R2016)	Residential Central-Vacuum Tube and Fittings	Exhaust Systems	506.1(3), 506.2(2)
ASTM F2159-2018a	Plastic Insert Fittings Utilizing a Copper Crimp Ring for SDR9 Cross-linked Polyethylene (PEX) Tubing and SDR9 Polyethylene of Raised Temperature (PE-RT) Tubing	Fittings	Table 1210.1
ASTM F2389-2017a	Pressure-Rated Polypropylene (PP) Piping Systems	Piping	1211.11(1), Table 1210.1
ASTM F2434-2018	Metal Insert Fittings Utilizing a Copper Crimp Ring for SDR9 Cross-linked Polyethylene (PEX) Tubing and SDR9 Cross-linked Polyethylene/Aluminum/Cross-linked Polyethylene (PEX-AL-PEX) Tubing	Fittings	1211.6(1), Table 1210.1
ASTM F2509-2015	Field-Assembled Anodeless Riser Kits for Use on Outside Diameter Controlled Polyethylene and Polyamide-11 (PA11) Gas Distribution Pipe and Tubing	Fuel Gas	1310.1.7.1(3)
ASTM F2620-2013	Standard Practice for Heat Fusion Joining of Polyethylene Pipe and Fittings	Joints	1211.8(1), 1211.8(3)
ASTM F2623-2014	Polyethylene of Raised Temperature (PE-RT) SDR9 Tubing	Piping	Table 1210.1
ASTM F2735-2018	Plastic Insert Fittings for SDR9 Cross-linked Polyethylene (PEX) and Polyethylene of Raised Temperature (PE-RT) Tubing	Fittings	Table 1210.1
ASTM F2769-2018	Polyethylene of Raised Temperature (PE-RT) Plastic Hot and Cold-Water Tubing and Distribution Systems	Piping, Fittings	Table 1210.1
ASTM F2855-2012	Specification for Chlorinated Poly (Vinyl Chloride)/Aluminum/Chlorinated Poly (Vinyl Chloride) (CPVC-AL-CPVC) Composite Pressure Tubing	Piping, Plastic	Table 1210.1
ASTM F2945-2018	Polyamide 11 Gas Pressure Pipe, Tubing, and Fittings	Tubing, Fittings	1308.5.4
ASTM F3226/F3226M-2016^{e1}	Metallic Press-Connect Fittings for Piping and Tubing Systems	Fittings	Table 1210.1
ASTM F3253-2018	Crosslinked Polyethylene (PEX) Tubing with Oxygen Barrier for Hot- and Cold-Water Hydronic Distribution Systems	Piping	Table 1210.1
AWS A5.8M/A5.8-2011-AMD 1	Filler Metals for Brazing and Braze Welding	Joints	1211.4(1)
AWWA C110-2012	Ductile-Iron and Gray-Iron Fittings	Fittings	Table 1210.1
AWWA C111-2017	Rubber-Gasket Joints for Ductile-Iron Pressure Pipe and Fittings	Joints	1211.7(1), 1211.7(2)
AWWA C115-2011	Flanged Ductile-Iron Pipe with Ductile-Iron or Gray-Iron Treaded Flanges	Piping	Table 1210.1
AWWA C151-2017	Ductile-Iron Pipe, Centrifugally Cast	Piping, Ferrous	Table 1210.1
AWWA C153-2011	Ductile-Iron Compact Fittings	Fittings	Table 1210.1

TABLE 1701.1 (continued)
REFERENCED STANDARDS

STANDARD NUMBER	STANDARD TITLE	APPLICATION	REFERENCED SECTION
AWWA C901-2017	Polyethylene (PE) Pressure Pipe and Tubing, ¾ in. (19 mm) Through 3 in. (76 mm) for Water Service	Piping, Plastic	Table 1210.1
CSA B137.1-2017	Polyethylene (PE) Pipe, Tubing, and Fittings for Cold-Water Pressure Services	Piping	Table 1210.1
CSA B137.2-2017	Polyvinylchloride (PVC) Injection-Moulded Gasketed Fittings for Pressure Applications	Piping, Plastic	Table 1210.1
CSA B137.3-2017	Rigid Polyvinylchloride (PVC) Pipe and Fittings for Pressure Applications	Piping, Plastic	Table 1210.1
CSA B137.5-2017	Crosslinked Polyethylene (PEX) Tubing Systems for Pressure Applications	Piping	Table 1210.1
CSA B137.6-2017	Chlorinated Polyvinylchloride (CPVC) Pipe, Tubing, and Fittings for Hot- and Cold-Water Distribution Systems	Piping, Plastic	Table 1210.1
CSA B137.9-2017	Polyethylene/Aluminum/Polyethylene (PE-AL-PE) Composite Pressure-Pipe Systems	Piping	Table 1210.1
CSA B137.10-2017	Crosslinked Polyethylene/Aluminum/Crosslinked Polyethylene (PEX-AL-PEX) Composite Pressure-Pipe Systems	Piping	Table 1210.1
CSA B137.11-2017	Polypropylene (PP-R) Pipe and Fittings for Pressure Applications	Piping	1211.11(1), Table 1210.1
CSA B137.18-2017	Polyethylene of Raised Temperature Resistance (PE-RT) Tubing Systems for Pressure Applications	Piping, Plastic	Table 1210.1
CSA FC 1-2014	Fuel Cell Technologies-Part 3-100: Stationary Fuel Cell Power Systems-Safety	Fuel Cell Power Plants	1601.1
CSA LC 1-2018	Fuel Gas Piping Systems Using Corrugated Stainless Steel Tubing (same as CSA 6.26)	Fuel Gas	1308.5.3.5, 1310.4.1(4), 1311.3
CSA LC 4a-2013 (R2017)	Press-Connect Metallic Fittings For Use in Fuel Gas Distribution Systems (same as CSA 6.32a)	Fuel Gas	1308.5.7.1, 1308.5.7.2, 1308.5.7.3, 1310.4.1(3)
CSA NGV 5.1-2016	Residential Fueling Appliances	Appliances	938.1
CSA Z21.8-1994 (R2017)	Installation of Domestic Gas Conversion Burners	Fuel Gas, Appliances	909.1
CSA Z21.10.1-2017	Gas Water Heaters, Volume I, Storage Water Heaters with Input Ratings of 75,000 Btu Per Hour or Less (same as CSA 4.1)	Fuel Gas, Appliances	Table 1203.2
CSA Z21.10.3-2017	Gas-Fired Water Heaters, Volume III, Storage Water Heaters with Input Ratings Above 75,000 Btu Per Hour, Circulating and Instantaneous (same as CSA 4.3)	Fuel Gas, Appliances	Table 1203.2
CSA Z21.24-2015	Connectors for Gas Appliances (same as CSA 6.10)	Fuel Gas	1312.1(3), 1312.2
CSA Z21.41-2014	Quick Disconnect Devices for Use with Gas Fuel Appliances (same as CSA 6.9)	Fuel Gas	1312.7
CSA Z21.54-2014	Gas Hose Connectors for Portable Outdoor Gas-Fired Appliances (same as CSA 8.4)	Fuel Gas	1312.3.2
CSA Z21.69-2015	Connectors for Moveable Gas Appliances (same as CSA 6.16)	Fuel Gas	1312.1.1
CSA Z21.75-2016	Connectors for Outdoor Gas Appliances and Manufactured Homes (same as CSA 6.27)	Appliances	1312.1(4)
CSA Z21.80a-2012	Line Pressure Regulators (same as CSA 6.22a)	Fuel Gas	1308.7.1, 1308.7.4(1)
CSA Z21.90-2015	Gas Convenience Outlets and Optional Enclosures (same as CSA 6.24)	Gas Outlets	1312.8
CSA Z21.93-2017	Excess Flow Valves for Natural and Propane Gas with Pressures up to 5 psig (same as CSA 6.30)	Fuel Gas	1309.1

TABLE 1701.1 (continued)
REFERENCED STANDARDS

STANDARD NUMBER	STANDARD TITLE	APPLICATION	REFERENCED SECTION
CSA Z83.4-2017	Non-Recirculating Direct Gas-Fired Heating and Forced Ventilation Appliances for Commercial and Industrial Application (same as CSA 3.7)	Air Heaters, Non-Recirculating, Non-Recirculating Heaters	914.1
CSA Z83.18-2017	Recirculating Direct Gas-Fired Heating and Forced Ventilation Appliances for Commercial and Industrial Applications	Air Heaters, Industrial Heaters, Recirculating	915.1
IAPMO PS 117-2017	Press and Nail Connections	Fittings	Table 1210.1
IIAR 2-2014	Standard for Safe Design of Closed-Circuit Ammonia Refrigeration Systems	Ammonia Refrigeration Systems	1102.2
IIAR 3-2017	Ammonia Refrigeration Valves	Ammonia Refrigeration Systems	1102.2
IIAR 4-2015	Installation of Closed-Circuit Ammonia Refrigeration Systems	Ammonia Refrigeration	1102.2
IIAR 5-2013	Start-Up and Commissioning of Closed Circuit Ammonia Refrigeration Systems	Ammonia Refrigeration Systems	1102.2
MSS SP-58-2018	Pipe Hangers and Supports-Materials, Design, Manufacture, Selection, Application, and Installation	Miscellaneous	1310.3.5
NEBB-2015	Procedural Standard for Testing, Adjusting and Balancing of Environmental Systems, 8th Edition	Balancing	314.1(4)
NFPA 2-2016	Hydrogen Technologies Code	Gaseous Hydrogen Systems	937.1
NFPA 10-2018	Portable Fire Extinguishers	Fire Extinguishing	513.10, 513.10.1, 513.11, 517.7.4
NFPA 12-2018	Carbon Dioxide Extinguishing Systems	Fire Extinguishing	513.2.3(1)
NFPA 13-2019	Installation of Sprinkler Systems	Miscellaneous	513.2.3(2), 517.7.6
NFPA 17-2017	Dry Chemical Extinguishing Systems	Fire Extinguishing	513.2.3(3), 513.3.5
NFPA 17A-2017	Wet Chemical Extinguishing Systems	Fire Extinguishing	513.2.3(4), 513.2.5.6, 513.3.5
NFPA 30A-2018	Motor Fuel Dispensing Facilities and Repair Garages	Miscellaneous	303.11.1
NFPA 31-2016	Installation of Oil-Burning Equipment	Fuel Gas, Appliances	301.5, 1002.2.2, 1301.1
NFPA 37-2018	Installation and Use of Stationary Combustion Engines and Gas Turbines	Generators	1602.1, 1602.3
NFPA 51-2018	Design and Installation of Oxygen-Fuel Gas Systems for Welding, Cutting, and Allied Processes	Fuel Gas Systems	902.7
NFPA 52-2019	Vehicular Natural Gas Fuel Systems Code	CNG Vehicle Fuel Systems	938.1
NFPA 54/Z223.1-2018	National Fuel Gas Code	Fuel Gas	516.2.1
NFPA 58-2017	Liquefied Petroleum Gas Code	Fuel Gas	303.7, 516.2.1, 1308.5.4.2.3, 1308.5.8.4, 1310.6(7), 1312.11
NFPA 68-2018	Explosion Protection by Deflagration Venting	Product Conveying Ducts	505.3.3
NFPA 69-2019	Explosion Prevention Systems	Explosion Prevention	505.3.1

TABLE 1701.1 (continued)
REFERENCED STANDARDS

STANDARD NUMBER	STANDARD TITLE	APPLICATION	REFERENCED SECTION
NFPA 70-2017	National Electrical Code	Miscellaneous	301.4(1), 301.4(3), 511.1.6, 512.2.5, 516.2.7, 516.2.9(4), 602.2.1, 905.8.2, 1104.4(5), 1107.1.7, 1107.1.8, 1217.8.1, 1310.14.5(2), 1311.2.4, 1311.7
NFPA 80-2019	Fire Doors and Other Opening Protectives	Fire Doors	510.7.7
NFPA 82-2019	Incinerators and Waste and Linen Handling Systems and Equipment	Incinerator Chutes	802.2.8, Table 802.4, 925.1
NFPA 85-2019	Boiler and Combustion Systems Hazards Code	Appliances	1002.1(3), 1011.1, Table 1003.2.1
NFPA 86-2019	Ovens and Furnaces	Product Conveying Ducts	505.3.2
NFPA 88A-2019	Parking Structures	Miscellaneous	303.11
NFPA 90A-2018	Installation of Air-Conditioning and Ventilating Systems	HVAC	604.1
NFPA 90B-2018	Installation of Warm Air Heating and Air-Conditioning Systems	HVAC	604.1
NFPA 211-2019	Chimneys, Fireplaces, Vents, and Solid Fuel-Burning Appliances	Fuel Gas Appliances	517.7, 517.7.1, 801.2, 801.3, 802.5.2, 802.5.3, 802.5.7.1, 802.5.7.3, 902.10
NFPA 262-2019	Standard Method of Test for Flame Travel and Smoke of Wires and Cables for Use in Air-Handling Spaces	Certification	602.2.1
NFPA 409-2016	Aircraft Hangars	Miscellaneous	303.12
NFPA 654-2017	Prevention of Fire and Dust Explosions from the Manufacturing, Processing, and Handling of Combustible Particulate Solids	Dust Explosion Prevention, Gutters	506.6, Table 505.9
NFPA 780-2017	Installation of Lightning Protection Systems	Fuel Gas	1311.5
NFPA 853-2015	Installation of Stationary Fuel Cell Power Systems	Fuel Cell Power Plants, Fuel Cells	1601.1
NFPA 1192-2018	Recreational Vehicles	Fuel Gas Piping	1302.3(18)
NSF 358-1-2017	Polyethylene Pipe and Fittings for Water-Based Ground-Source “Geothermal” Heat Pump Systems	Piping, Plastic	Table 1210.1
NSF 358-2-2017	Polypropylene Pipe and Fittings for Water-Based Ground-Source “Geothermal” Heat Pump Systems	Piping, Plastic	Table 1210.1
NSF 358-3-2016	Cross-linked polyethylene (PEX) pipe and fittings for water-based ground-source (geothermal) heat pump systems	Piping, Plastic	Table 1210.1
SMACNA-2002	HVAC Systems Testing, Adjusting and Balancing, 3rd Edition	Balancing	314.1(5)
SMACNA-2017	HVAC Duct Construction Standards Metal and Flexible, 4th Edition	Ducts, Metal and Flexible	504.4.5, 506.2, 602.3, 603.3, 603.4, 603.7.1, 603.9, 603.11, 605.1
SMACNA-2012	HVAC Air Duct Leakage Test Manual, 2nd Edition	Ducts	603.9.2, 603.10
SMACNA-2015	Phenolic Duct Construction Standards	Duct Systems	602.4.1

TABLE 1701.1 (continued)
REFERENCED STANDARDS

STANDARD NUMBER	STANDARD TITLE	APPLICATION	REFERENCED SECTION
UL 17-2008	Vent or Chimney Connector Dampers for Oil-Fired Appliances (with revisions through September 25, 2013)	Vent Dampers	802.15.1
UL 103-2010	Factory-Built Chimneys for Residential Type and Building Heating Appliances (with revisions through March 15, 2017)	Fuel Gas, Appliances	802.5.1, 802.5.1.1
UL 127-2011	Factory-Built Fireplaces (with revisions through July 27. 2016)	Fireplaces	802.5.1.1, 913.1, 913.1.1
UL 181-2013	Factory-Made Air Ducts and Air Connectors (with revisions through April 18, 2017)	Air Connectors, Air Ducts	602.3, 602.4.1, 602.4.4, 603.1.4, 603.1.5, 603.4, 603.9.1, 605.1.1, Table 603.9.1
UL 181A-2013	Closure Systems for Use with Rigid Air Ducts (with revisions through March 22, 2017)	Air Ducts	603.9.1, Table 603.9.1
UL 181B-2013	Closure Systems for Use with Flexible Air Ducts and Air Connectors (with revisions through March 21, 2017)	Air Connectors, Air Ducts	603.9.1, Table 603.9.1
UL 197-2010	Commercial Electric Cooking Appliances (with revisions through January 26, 2018)	Appliances, Commercial Cooking, Electric Appliances	921.1
UL 207-2009	Refrigerant-Containing Components and Accessories, Nonelectrical (with revisions through June 27, 2014)	Refrigeration Components	1109.2
UL 268A-2008	Smoke Detectors for Duct Application (with revisions through August 12, 2016)	Smoke Detectors	609.1
UL 295-2017	Commercial-Industrial Gas Burners	Gas Burners	910.2
UL 296-2017	Oil Burners (with revisions through November 29, 2017)	Fuel Gas, Appliances	910.1
UL 300-2005	Fire Testing of Fire Extinguishing Systems for Protection of Commercial Cooking Equipment (with revisions through December 16, 2014)	Certification	513.2.2, 513.2.5, 517.3.1.1(6)
UL 378-2006	Draft Equipment (with revisions through September 17, 2013)	Fuel Gas, Appliances	802.3.3, 802.15.1
UL 391-2010	Solid-Fuel and Combination-Fuel Central and Supplementary Furnaces (with revisions through June 12, 2014)	Furnaces, Solid Fuel	904.10
UL 412-2011	Refrigeration Unit Coolers (with revisions through August 28, 2018)	Refrigeration	934.2
UL 427-2011	Refrigerating Units (with revisions through February 10, 2017)	Refrigeration Systems	934.3
UL 441-2016	Gas Vents (with revisions through July 27, 2016)	Fuel Gas	802.1
UL 467-2013	Grounding and Bonding Equipment	Grounding and Bonding	1311.2.5
UL 471-2010	Commercial Refrigerators and Freezers (with revisions through November 8, 2018)	Freezers, Refrigerators	934.1
UL 499-2014	Electric Heating Appliances (with revisions through February 23, 2017)	Kilns	931.4
UL 555-2006	Fire Dampers (with revisions through October 21, 2016)	Dampers	606.2
UL 555C-2014	Ceiling Dampers (with revisions through May 1, 2017)	Dampers	606.3
UL 555S-2014	Smoke Dampers (with revisions through October 27, 2016)	Dampers	606.1
UL 641-2010	Type L Low-Temperature Venting Systems (with revisions through April 23, 2018)	Equipment	802.1
UL 651-2011	Schedule 40, 80, Type EB and A Rigid PVC Conduit and Fittings (with revisions through June 15, 2016)	Piping, Plastic	1308.5.4.1

TABLE 1701.1 (continued)
REFERENCED STANDARDS

STANDARD NUMBER	STANDARD TITLE	APPLICATION	REFERENCED SECTION
UL 705-2017	Power Ventilators (with revisions through October 8, 2018)	Power Ventilators	504.4.2.3
UL 710-2012	Exhaust Hoods for Commercial Cooking Equipment (with revisions through June 25, 2018)	Exhaust Hoods, Hoods	507.3.1, 508.2, 508.2.1
UL 710B-2011	Recirculating Systems (with revisions through August 14, 2014)	Exhaust Hoods	508.1, 513.2.2, 516.2.2, 516.2.3
UL 710C-2006	Ultraviolet Radiation Systems For Use In The Ventilation Control of Commercial Cooking Operations	Exhaust Hoods, Hoods	508.2.1
UL 723-2018	Test for Surface Burning Characteristics of Building Materials	Miscellaneous	508.3.4, 602.2, 605.1.1, 605.1.2, 1201.2
UL 726-1995	Oil-Fired Boiler Assemblies (with revisions through October 9, 2013)	Fuel Gas, Appliances	1002.2.1
UL 727-2018	Oil-Fired Central Furnaces	Fuel Gas, Appliances	904.11
UL 729-2003	Oil-Fired Floor Furnaces (with revisions through November 22, 2016)	Furnaces, Floor, Oil Fired Furnaces	906.13
UL 730-2003	Oil-Fired Wall Furnaces (with revisions through November 22, 2016)	Furnaces, Wall, Oil Fired	907.4
UL 731-2018	Oil-Fired Unit Heaters	Heaters, Oil Fired	917.6
UL 737-2011	Fireplace Stoves (with revisions through August 19, 2015)	Fireplace Stoves	913.2
UL 762-2013	Power Roof Ventilators for Restaurant Exhaust Appliances	Ventilators	511.1
UL 778-2016	Motor Operated Water Pumps (with revisions through October 20, 2017)	Pumps	1208.1
UL 791-2006	Residential Incinerators (with revisions through November 7, 2014)	Incinerators	925.3
UL 795-2016	Commercial-Industrial Gas Heating Equipment	Heating Equipment, Gas Fired	904.12
UL 834-2004	Heating, Water Supply, and Power Boilers - Electric (with revisions through September 24, 2018)	Appliances	1002.3, Table 1203.2
UL 858-2014	Household Electric Ranges (with revisions through June 4, 2018)	Electric Ranges, Ranges	920.1
UL 867-2011	Electrostatic Air Cleaners (with revisions through August 7, 2018)	Filters	936.1
UL 896-1993	Oil-Burning Stoves (with revisions through November 22, 2016)	Stoves, Oil Fired	921.3
UL 900-2015	Air Filter Units	Air Filters, Filters	311.2
UL 907-2016	Fireplace Accessories	Fireplace Accessories	913.3
UL 921-2016	Commercial Dishwashers (with revisions through September 20, 2017)	Appliances	519.1
UL 923-2013	Microwave Cooking Appliances (with revisions through July 19, 2017)	Microwaves	920.3.2(3), 920.4.2(3)
UL 959-2010	Medium Heat Appliance Factory-Built Chimneys (with revisions through June 12, 2014)	Fuel Gas, Appliances	802.5.1
UL 1046-2010	Grease Filters for Exhaust Ducts (with revisions through April 17, 2017)	Filters-Grease	509.1, 509.1.1, 518.1(4)
UL 1240-2005	Electric Commercial Clothes-Drying Equipment (with revisions through March 16, 2018)	Clothes Dryers, Commercial	908.1

**TABLE 1701.1 (continued)
REFERENCED STANDARDS**

STANDARD NUMBER	STANDARD TITLE	APPLICATION	REFERENCED SECTION
UL 1479-2015	Fire Tests of Penetration Firestops	Miscellaneous	507.4.4, 507.4.5
UL 1482-2011	Solid-Fuel Type Room Heaters (with revisions through August 19, 2015)	Room Heaters, Solid Fuel Heaters	802.5.1.1, 916.3
UL 1738-2010	Venting Systems for Gas-Burning Appliances Categories II, III, and IV (with revisions through November 7, 2014)	Fuel Gas, Appliances	802.4.1, 802.4.2, 802.4.3
UL 1777-2015	Chimney Liners	Chimneys, Liners	802.5.3(2), 803.1.11.2
UL 1812-2013	Ducted Heat Recovery Ventilators (with revisions through July 19, 2018)	Heat Recovery Ventilators	504.5
UL 1815-2012	Nonducted Heat Recovery Ventilators (with revisions through July 19, 2018)	Heat Recovery Ventilators	504.5
UL 1820-2004	Fire Test of Pneumatic Tubing for Flame and Smoke Characteristics (with revisions through July 13, 2017)	Surface Burning Test, Pneumatic Tubing	602.2.3
UL 1887-2004	Fire Test of Plastic Sprinkler Pipe for Visible Flame and Smoke Characteristics (with revisions through July 13, 2017)	Surface Burning Test, Fire Sprinkler Pipe	602.2.2
UL 1963-2011	Refrigerant Recovery/Recycling Equipment (with revisions through October 11, 2013)	Refrigerant Equipment	1101.2
UL 1978-2010	Grease Ducts (with revisions through April 28, 2017)	Ducts, Grease	507.3.1, 510.4.1, 510.5.2, 510.5.3
UL 1995-2015	Heating and Cooling Equipment	HVAC, Electric	903.1, 904.13
UL 1996-2009	Electric Duct Heaters (with revisions through July 15, 2016)	Duct Heaters	905.8
UL 2021-2015	Fixed and Location-Dedicated Electric Room Heaters (with revisions through December 14, 2016)	Product, Heating, Electric	916.1
UL 2043-2013	Fire Test for Heat and Visible Smoke Release For Discrete Products and Their Accessories Installed in Air-Handling Spaces (with revisions through July 13, 2018)	Surface Burning Test, Discrete Products	602.2.4
UL 2158-2018	Electric Clothes Dryers	Clothes Dryers, Electric	908.1
UL 2158A-2013	Clothes Dryer Transition Duct (with revisions through April 18, 2017)	Clothes Dryer Exhaust	504.4
UL 2162-2014	Commercial Wood-Fired Baking Ovens-Refractory Type	Baking Ovens	921.2
UL 2200-2012	Stationary Engine Generator Assemblies (with revisions through July 29, 2015)	Assemblies	1602.3
UL 2221-2010	Tests of Fire Resistive Grease Duct Enclosure Assemblies	Duct Wrap, Grease Duct Enclosure	507.4.4.1
UL 2518-2016	Air Dispersion Systems	Duct Systems	602.4.3, 603.12
UL 2523-2009	Solid Fuel-Fired Hydronic Heating Appliances, Water Heaters, and Boilers (with revisions through March 16, 2018)	Appliances	1002.4, Table 1203.2
UL 2561-2016	1400 Degree Fahrenheit Factory-Built Chimneys (with revisions through April 19, 2018)	Chimneys	802.5.1
UL 2790-2010	Commercial Incinerators (with revisions through October 8, 2014)	Incinerators	925.2
UL 60335-2-40-2017	Household and Similar Electrical Appliances – Safety – Part 2-40: Particular Requirements for Electrical Heat Pumps, Air-Conditioners and Dehumidifiers	Appliances	903.1, 904.13

TABLE 1701.1 (continued)
REFERENCED STANDARDS

STANDARD NUMBER	STANDARD TITLE	APPLICATION	REFERENCED SECTION
UL 60335-2-89-2017	Household and Similar Electrical Appliances – Safety – Part 2-89: Particular Requirements for Commercial Refrigerating Appliances with an Incorporated or Remote Refrigerant Unit or Compressor	Appliances	934.1, 934.2, 934.3

1701.2 Standards, Publications, Practices, and Guides. The standards, publications, practices and guides listed in Table 1701.2 are not referenced in other sections of this code. The application of the referenced standards, publications, practices and guides shall be as specified in Section 302.1.2. The promulgating agency acronyms are found at the end of the table.

A list of additional standards, publications, and guides that are not referenced in specific sections of this code appear in Table 1701.2. The standards from Table 1701.2 shall be permitted only after they have been approved by the AHJ. See also Section 301.2.2.

TABLE 1701.2
STANDARDS, PUBLICATIONS, PRACTICES, AND GUIDES

DOCUMENT NUMBER	DOCUMENT TITLE	APPLICATION
ACCA Manual N-2012	Commercial Load Calculations	Ducts
ACCA Manual Q-1990	Low Pressure, Low Velocity Duct System Design	Ducts
AHRI 870-2016	Performance Rating of Direct Geoexchange Heat Pumps	Heat Pumps
AHRI 1200-2013	Performance Rating of Commercial Refrigerated Display Merchandisers and Storage Cabinets	Commercial Refrigerated Display Merchandisers and Storage Cabinets
AHRI 1230-2014a	Performance Rating of Variable Refrigerant Flow (VRF) Multi-Split Air-Conditioning and Heat Pump Equipment	Refrigerants
AMCA 500-D-2018	Laboratory Methods of Testing Dampers for Rating	Dampers
ASCE 25-2016	Earthquake-Actuated Automatic Gas Shutoff Devices	Fuel Gas
ASHRAE 52.2-2012	General Ventilation Air-Cleaning Devices for Removal Efficiency by Particle Size	Cleaning Devices
ASHRAE 55-2017	Thermal Environmental Conditions for Human Occupancy	Miscellaneous
ASHRAE 62.2-2016	Ventilation and Acceptable Indoor Air Quality in Residential Buildings	Ventilation
ASHRAE/IES 90.1-2016	Energy Standard for Buildings Except Low-Rise Residential Buildings	Energy
ASHRAE/IES 90.2-2018	Energy-Efficient Design of Low-Rise Residential Buildings	Energy, Dwellings
ASHRAE 127-2012	Method of Testing for Rating Computer and Data Processing Room Unitary Air Conditioners	Air Conditioners
ASHRAE/ACCA 183-2007 (R2017)	Peak Cooling and Heating Load Calculations in Buildings Except Low-Rise Residential Buildings	Cooling and Heating Load, Miscellaneous
ASHRAE 194-2012	Method of Test for Direct-Expansion Ground-Source Heat Pumps	Ground-Source Heat Pumps
ASHRAE Handbook-2016	HVAC Systems and Equipment	Design
ASME A13.1-2015	Scheme for the Identification of Piping Systems	Piping
ASME A112.18.6/CSA B125.6-2017	Flexible Water Connectors	Piping
ASME B1.20.3-1976 (R2018)	Dryseal Pipe Threads (Inch)	Joints
ASME B16.33-2012 (R2017)	Manually Operated Metallic Gas Valves for Use in Gas Piping Systems up to 175 psi (Sizes NPS ½ through NPS 2)	Valves
ASME B16.50-2018	Wrought Copper and Copper Alloy Braze-Joint Pressure Fittings	Fittings
ASTM A568/A568M-2017a	Steel, Sheet, Carbon, Structural, and High-Strength, Low-Alloy, Hot-Rolled and Cold-Rolled, General Requirements for	Piping
ASTM A653/A653M-2017	Steel Sheet, Zinc-Coated (Galvanized) or Zinc-Iron Alloy-Coated (Galvannealed) by the Hot-Dip Process	Piping, Ferrous
ASTM A733-2016	Welded and Seamless Carbon Steel and Austenitic Stainless Steel Pipe Nipples	Piping, Ferrous

TABLE 1701.2 (continued)
STANDARDS, PUBLICATIONS, PRACTICES, AND GUIDES

DOCUMENT NUMBER	DOCUMENT TITLE	APPLICATION
ASTM B370-2012	Standard Specification for Copper Sheet and Strip for Building Construction	Miscellaneous
ASTM B687-1999 (R2016)	Brass, Copper, and Chromium-Plated Pipe Nipples	Piping, Copper Alloy
ASTM C518-2017	Steady-State Thermal Transmission Properties by Means of the Heat Flow Meter Apparatus	Certification: Calibration, Error Analysis, Heat Flow Meter Apparatus, Heat Flux, Instrument Verification, Thermal Conductivity, Thermal Resistance, Thermal Testing, Tunnel Test
ASTM D93-2018	Flash Point by Pensky-Martens Closed Cup Tester	Certification
ASTM D396-2018a	Fuel Oils	Boiler
ASTM D2517-2018	Reinforced Epoxy Resin Gas Pressure Pipe and Fittings	Piping, Plastic
ASTM E96/E96M-2016	Water Vapor Transmission of Materials	Miscellaneous
ASTM E136-2016a	Behavior of Materials in a Vertical Tube Furnace at 750°C	Furnace
AWS B2.4-2012	Welding Procedure and Performance Qualification for Thermoplastics	Joints, Certification
CSA/IGSHPA C448-2016	Design and Installation of Ground Source Heat Pump Systems for Commercial and Residential Buildings	Miscellaneous
CSA C448.1-2016	Design and Installation of Ground Source Heat Pump Systems for Commercial and Institutional Buildings	Ground-Source Heat Pumps
CSA C448.2-2016	Design and Installation of Ground Source Heat Pump Systems for Residential and Other Small Buildings	Ground-Source Heat Pumps
CSA Z21.1-2016	Household Cooking Gas Appliances	Household Cooking Appliances, Gas Fired
CSA Z21.5.1-2017	Gas Clothes Dryers, Volume I, Type 1 Clothes Dryers (same as CSA 7.1)	Fuel Gas, Appliances
CSA Z21.5.2-2016	Gas Clothes Dryers,Volume II, Type 2 Clothes Dryers (same as CSA 7.2)	Fuel Gas, Appliances
CSA Z21.11.2-2016	Gas-Fired Room Heaters, Volume II, Unvented Room Heaters	Room Heaters, Unvented Heaters
CSA Z21.12b-1994 (R2015)	Draft Hoods	Fuel Gas, Appliances
CSA Z21.13-2017	Gas-Fired Low Pressure Steam and Hot Water Boilers (same as CSA 4.9)	Fuel Gas, Appliances
CSA Z21.15b-2013 (R2014)	Manually Operated Gas Valves for Appliances, Appliance Connector Valves and Hose End Valves (same as CSA 9.1b)	Fuel Gas
CSA Z21.17a-2008 (R2014)	Domestic Gas Conversion Burners (same as CSA 2.7a)	Conversion Burner Installation, Gas Burners
CSA Z21.18b-2012 (R2016)	Gas Appliance Pressure Regulators (same as CSA 6.3b)	Appliance Regulators, Gas Refrigerators, Pressure Regulators
CSA Z21.19-2014	Refrigerators Using Gas Fuel (same as CSA 1.4)	Gas Refrigerators, Refrigerators
CSA Z21.20a-2008 (R2016)	Automatic Gas Ignition Systems and Components	Automatic Ignition, Gas Ignition, Ignition

TABLE 1701.2 (continued)
STANDARDS, PUBLICATIONS, PRACTICES, AND GUIDES

DOCUMENT NUMBER	DOCUMENT TITLE	APPLICATION
CSA Z21.21-2015	Automatic Valves for Gas Appliances (same as CSA 6.5)	Appliance Valves, Automatic Gas Valves, Valves
CSA Z21.22-2015	Relief Valves for Hot Water Supply Systems (same as CSA 4.4)	Valves
CSA Z21.40.1a-1997/CGA 2.91a-M97 (R2017)	Gas-Fired, Heat Activated Air-Conditioning and Heat Pump Appliances	Gas Fired Air Conditioning, Gas Fired Heat Pump
CSA Z21.42-2013	Gas-Fired Illuminating Appliances	Illuminating Appliances
CSA Z21.47-2016	Gas-Fired Central Furnaces (same as CSA 2.3)	Fuel Gas, Appliances
CSA Z21.50-2016	Vented Decorative Gas Appliances (same as CSA 2.22)	Appliances, Decorative Appliances
CSA Z21.56-2017	Gas-Fired Pool Heaters (same as CSA 4.7)	Fuel Gas, Swimming Pools, Spas, and Hot Tubs
CSA Z21.58-2018	Outdoor Cooking Gas Appliances (same as CSA 1.6)	Cooking Appliances
CSA Z21.60-2017	Decorative Gas Appliances for Installation in Solid-Fuel Burning Fireplaces (same as CSA 2.26)	Decorative Appliances, Decorative Fireplace Appliances
CSA Z21.61-1983 (R2013)	Gas-Fired Toilets	Toilets
CSA Z21.66-2015	Automatic Damper Devices for Use with Gas-Fired Appliances (same as CSA 6.14)	Vent Dampers, Automatic Dampers
CSA Z21.71a-2005 (R2016)	Automatic Intermittent Pilot Ignition Systems for Field Installation	Automatic Pilot Ignition, Pilot Ignition
CSA Z21.86-2016	Vented Gas-Fired Space Heating Appliances (same as CSA 2.32)	Fuel Gas Appliances
CSA Z21.88-2017	Vented Gas Fireplace Heaters (same as CSA 2.33)	Fireplace Heaters
CSA Z83.8-2016	Gas Unit Heaters, Gas Packaged Heaters, Gas Utility Heaters, and Gas-Fired Duct Furnaces (same as CSA 2.6)	Fuel Gas Appliances
CSA Z83.11-2016	Gas Food Service Equipment (same as CSA 1.8)	Fuel Gas, Appliances
CSA Z83.19-2017	Gas-Fired High-Intensity Infrared Heaters (same as CSA 2.35)	High Intensity Heaters, Infrared Heaters
CSA Z83.20-2016	Gas-Fired Low-Intensity Infrared Heaters (same as CSA 2.34)	Infrared Heaters, Low Intensity Heaters
IAPMO IGC 293-2012	Tubing and Fittings for Special Hydronic Radiant Drywall Panels	HVAC, Fittings, Tubing
IAPMO IGC 315-2016	Water Manifold Systems	Fittings
IAPMO IGC 327-2016	Flexible Metallic Expansion Joints for Pressure Systems	Joints
IAPMO IGC 332-2017a	Hydronic Radiators	Miscellaneous
IAPMO PS 120-2004	Flashing and Stand Combination for Air Conditioning Units (Residential or Commercial Unit Curb)	Air Conditioning Flashing Stand
ISO 13256-1-1998 (R2012)	Water Source Heat Pumps-Testing and Rating for Performance-Water-to-Air and Brine-to-Air Heat Pumps	Water-Source Heat Pumps
ISO 13256-2-1998	Water Source Heat Pumps-Testing and Rating for Performance-Water-to-Water and Brine-to-Water Heat Pumps	Water-Source Heat Pumps

TABLE 1701.2 (continued)
STANDARDS, PUBLICATIONS, PRACTICES, AND GUIDES

DOCUMENT NUMBER	DOCUMENT TITLE	APPLICATION
MSS SP-67-2017	Butterfly Valves	Valves
MSS SP-80-2013	Bronze Gate, Globe, Angle, and Check Valves	Valves
MSS SP-104-2018	Wrought Copper Solder-Joint Pressure Fittings	Fittings
MSS SP-106-2012	Cast Copper Alloy Flanges and Flanged Fittings: Class 125, 150, and 300	Fittings
MSS SP-109-2018	Weld-Fabricated Copper Solder-Joint Pressure Fittings	Fittings
NFPA 30-2018	Flammable and Combustible Liquids Code	Combustible Liquids, Flammable Liquids
NFPA 91-2015	Exhaust Systems for Air Conveying of Vapors, Gases, Mists, and Particulate Solids	Product Conveying Ducts
NFPA 96-2017	Ventilation Control and Fire Protection of Commercial Cooking Operations	Commercial Cooking
NFPA 221-2018	High Challenge Fire Walls, Fire Walls, and Fire Barrier Walls	Building Fire Walls, Fire Barrier
NFPA 259-2018	Test Method for Potential Heat of Building Materials	Testing, Building Materials
NFPA 274-2018	Test Method to Evaluate Fire Performance Characteristics of Pipe Insulation	Pipe Insulation
NFPA 501A-2017	Fire Safety Criteria for Manufactured Home Installations, Sites, and Communities	Miscellaneous
NFPA 5000-2018	Building Construction and Safety Code	Miscellaneous
NGWA-01-2014	Water Well Construction Standard	Water Wells
NSF 60-2017	Drinking Water Treatment Chemicals - Health Effects	Miscellaneous
NSF 358-4-2018	Polyethylene of Raised Temperature (PE-RT) Tubing and Fittings for Water-Based Ground-Source (Geothermal) Heat Pump Systems	Ground-Source Heat Pumps
SAE J512-1997	Automotive Tube Fittings	Fittings
SMACNA-2003	Fibrous Glass Duct Construction Standard, 7th Edition	Fiberglass Ducts
SMACNA-2008	IAQ Guidelines for Occupied Buildings Under Construction, 2nd Edition	Ventilation
UL 21-2014	LP-Gas Hose (with revisions through July 28, 2017)	Hose, LPG Hose
UL 33-2010	Heat Responsive Links for Fire-Protection Service (with revisions through April 14, 2015)	Fusible Links
UL 51-2013	Power-Operated Pumps and Bypass Valves for Anhydrous Ammonia, LP-Gas, and Propylene (with revisions through October 5, 2018)	Ammonia Pumps, LP-Gas Pumps, Pumps
UL 80-2007	Steel Tanks for Oil-Burner Fuels and Other Combustible Liquids (with revisions through January 16, 2014)	Fuel Gas
UL 125-2014	Flow Control Valves for Anhydrous Ammonia and LP-Gas (with revisions through January 12, 2018)	Fuel Gas
UL 132-2015	Safety Relief Valves for Anhydrous Ammonia and LP-Gas (with revisions through January 12, 2018)	Fuel Gas
UL 144-2012	LP-Gas Regulators (with revisions through November 05, 2014)	Fuel Gas
UL 174-2004	Household Electric Storage Tank Water Heaters (with revisions through December 15, 2016)	Appliances

TABLE 1701.2 (continued)
STANDARDS, PUBLICATIONS, PRACTICES, AND GUIDES

DOCUMENT NUMBER	DOCUMENT TITLE	APPLICATION
UL 180-2012	Liquid-Level Gauges for Oil Burner Fuels and other combustible liquids (with revisions through May 12, 2017)	Gauges, Level Gauges
UL 252-2017	Compressed Gas Regulators (with revisions through August 10, 2018)	Fuel Gas
UL 343-2008	Pumps for Oil-Burning Appliances (with revisions through December 14, 2017)	Fuel Gas, Appliances
UL 353-1994	Limit Controls (with revisions through November 8, 2011)	Controls
UL 404-2010	Gauges, Indicating Pressure, for Compressed Gas Service (with revisions through February 11, 2015)	Fuel Gas
UL 429-2013	Electrically Operated Valves	Valves
UL 443-2006	Steel Auxiliary Tanks for Oil-Burner Fuel (with revisions through March 8, 2013)	Fuel Gas
UL 525-2008	Flame Arresters (with revisions through October 13, 2017)	Flame Arrestors
UL 565-2013	Liquid-Level Gauges for Anhydrous Ammonia and LP-Gas (with revisions through February 23, 2018)	Fuel Gas
UL 569-2013	Pigtails and Flexible Hose Connectors for LP-Gas (with revisions through July 28, 2017)	Fuel Gas
UL 732-2018	Oil-Fired Storage Tank Water Heaters (with revisions through August 9, 2018)	Fuel Gas, Appliances
UL 733-1993	Oil-Fired Air Heaters and Direct-Fired Heaters (with revisions through October 9, 2013)	Water Heaters, Direct Fired, Oil Fired
UL 842-2015	Valves for Flammable Fluids (with revisions through October 27, 2017)	Valves
UL 984-1996	Hermetic Refrigerant Motor-Compressors (with revisions through September 23, 2005)	Compressors, Refrigeration
UL 1453-2016	Electric Booster and Commercial Storage Tank Water Heaters (with revisions through May 18, 2018)	Appliances
UL 1746-2007	External Corrosion Protection Systems for Steel Underground Storage Tanks (with revisions through December 19, 2014)	Tanks

ABBREVIATIONS IN TABLE 1701.1 AND TABLE 1701.2

AABC	Associated Air Balance Council, 1220 19th Street NW, Suite 410, Washington, DC 20036.
ACCA	Air Conditioning Contractors of America Association, Inc., 2800 S Shirlington Road, Suite 300, Arlington, VA 22206.
AHRI	Air-Conditioning, Heating, and Refrigeration Institute, 2111 Wilson Boulevard, Suite 500, Arlington, VA 22201.
AMCA	Air Movement and Control Association, 30 West University Drive, Arlington Heights, IL 60004-1806.
ANSI	American National Standards Institute, Inc., 25 W. 43rd Street, 4th Floor, New York, NY 10036.
ASCE	American Society of Civil Engineers, 1801 Alexander Bell Drive, Reston, VA 20191-4400.
ASHRAE	American Society of Heating, Refrigerating, and Air Conditioning Engineers, Inc., 1791 Tullie Circle, NE, Atlanta, GA 30329-2305.
ASME	American Society of Mechanical Engineering, Two Park Avenue, New York, NY 10016-5990.
ASSE	American Society of Sanitary Engineering, 18927 Hickory Creek Drive, Suite 220, Mokena, IL 60448.
ASTM	ASTM International, 100 Barr Harbor Drive, West Conshohocken, PA 19428-2959.
AWS	American Welding Society, 8669 NW 36 Street, #130, Miami, FL 33166-6672.
AWWA	American Water Works Association, 6666 W. Quincy Avenue, Denver, CO 80235.
CFR	Code of Federal Regulations, U.S Government Publishing Office, 723 North Capitol Street, N.W. Washington, DC 20401-001.
CGA	Canadian Gas Association, 243 Consumers Road, Suite 1200, North York, Ontario, Canada M2J 5E3.
CSA	Canadian Standards Association, 178 Rexdale Boulevard, Toronto, Ontario, Canada M9W 1R3.
e1	An editorial change since the last revision or reapproval.
IAPMO	International Association of Plumbing and Mechanical Officials, 4755 E. Philadelphia Street, Ontario, CA 91761.
IES	Illuminating Engineering Society, 120 Wall St. Fl 17, New York, NY 10005-4026.
IGSHPA	International Ground Source Heat Pump Association, 1723 W. Tyler Ave. Stillwater, OK 74078.
IIAR	International Institute of Ammonia Refrigeration, 1001 N. Fairfax Street, Suite 503, Alexandria, VA 22314.
ISO	International Organization for Standardization, 1 ch. de la Voie-Creuse, Casa Postale 56, CH-1211 Geneva 20, Switzerland.
MSS	Manufacturers Standardization Society of the Valve and Fittings Industry, 127 Park Street NE, Vienna, VA 22180.
NEBB	National Environmental Balancing Bureau, 8575 Grovemont Circle, Gaithersburg, MD 20877.
NFPA	National Fire Protection Association, 1 Batterymarch Park, Quincy, MA 02169-7471.
NGWA	National Ground Water Association, 601 Dempsey Road, Westerville, Ohio 43081-8978.
NSF	NSF International, 789 N. Dixboro Road, Ann Arbor, MI 48105.
SAE	Society of Automotive Engineers, 400 Commonwealth Drive, Warrendale, PA 15096.
SMACNA	Sheet Metal and Air Conditioning Contractors National Association, 4201 Lafayette Center Drive, Chantilly, VA 20151-1219.
UL	Underwriters Laboratories, Inc., 333 Pfingsten Road, Northbrook, IL 60062.

APPENDICES

The appendices are intended to supplement the provisions of the installation requirements of this code. The definitions in Chapter 2 are also applicable to the appendices.

TABLE OF CONTENTS

APPENDIX A

RESIDENTIAL PLANS EXAMINER REVIEW FORM FOR HVAC SYSTEM DESIGN (Loads, Equipment, Ducts) [ACCA]

The following Residential Plans Examiner Review Form for HVAC System Design (Loads, Equipment, Ducts), Form RPER1, is included here for the convenience of the users of the Uniform Mechanical Code.

ACCA Air Conditioning Contractors of America

Residential Plans Examiner Review Form for HVAC System Design (Loads, Equipment, Ducts)

Form RPER 1
15 Mar 09

County, Town, Municipality, Jurisdiction
Header Information

Contractor ______________________

Mechanical License # ______________________

Building Plan # ______________________

REQUIRED ATTACHMENTS	ATTACHED	
Manual J1 Form (and supporting worksheets):	Yes ☐	No ☐
or MJ1AE Form* (and supporting worksheets):	Yes ☐	No ☐
OEM performance data (heating, cooling, blower):	Yes ☐	No ☐
Manual D Friction Rate Worksheet:	Yes ☐	No ☐
Duct distribution system sketch:	Yes ☐	No ☐

Home Address (Street or Lot#, Block, Subdivision) ______________________

HVAC LOAD CALCULATION (See Section 1105.1)

Design Conditions

Winter Design Conditions

Outdoor temperature ________ °F

Indoor temperature ________ °F

Total heat loss ________ Btu

Summer Design Conditions

Outdoor temperature ________ °F

Indoor temperature ________ °F

Grains difference ________ Δ Gr @ ____ % Rh

Sensible heat gain ________ Btu

Latent heat gain ________ Btu

Total heat gain ________ Btu

Building Construction Information

Building

Orientation (Front door faces) ________
North, East, West, South, Northeast, Northwest, Southeast, Southwest

Number of bedrooms ________

Conditioned floor area ________ Sq Ft

Number of occupants ________

Windows

Eave overhang depth ________ Ft

Internal shade ________
Blinds, drapes, etc

Number of skylights ________

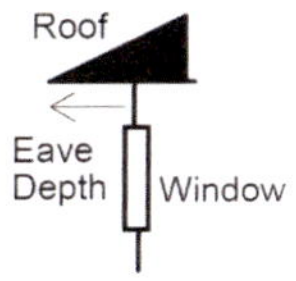

HVAC EQUIPMENT SELECTION

Heating Equipment Data

Equipment type ________
Furnace, Heat pump, Boiler, etc.

Model ________

Heating output capacity ________ Btu
Heat pumps - capacity at winter design outdoor conditions

Auxiliary heat output capacity ________ Btu

Cooling Equipment Data

Equipment type ________
Air Conditioner, Heat pump, etc

Model ________

Sensible cooling capacity ________ Btu

Latent cooling capacity ________ Btu

Total cooling capacity ________ Btu

Blower Data

Heating CFM ________ CFM

Cooling CFM ________ CFM

Static pressure ________ IWC
Fan's rated external static pressure for design airflow

HVAC DUCT DISTRIBUTION SYSTEM DESIGN (See Section 601.2)

Design airflow ________ CFM

External Static Pressure (ESP) ________ IWC

Component Pressure Losses (CPL) ________ IWC

Available Static Pressure (ASP) ________ IWC
ASP = ESP - CPL

Longest supply duct: ________ Ft

Longest return duct: ________ Ft

Total Effective Length (TEL) ________ Ft

Friction Rate: ________ IWC
Friction Rate = (ASP × 100) ÷ TEL

Duct Materials Used (circle)

Trunk Duct: Duct board, Flex, Sheet metal, Lined sheet metal, Other (specify) ________

Branch Duct: Duct board, Flex, Sheet metal, Lined sheet metal, Other (specify) ________

I declare the load calculation, equipment selection, and duct system design were rigorously performed based on the building plan listed above, I understand the claims made on these forms will be subject to review and verification.

Contractor's Printed Name ________________ Date ________

Contractor's Signature ________________

Reserved for County, Town, Municipality, or Authority having jurisdiction use.

* Home qualifies for MJ1AE Form based on Abridged Edition Checklist.

APPENDIX B

PROCEDURES TO BE FOLLOWED TO PLACE GAS EQUIPMENT IN OPERATION

B 101.0 Adjusting the Burner Input.

B 101.1 Adjusting Input. The input rate of the burner shall be adjusted to the proper value in accordance with the appliance manufacturer's instructions. Firing at a rate in excess of the nameplate rating shall be prohibited. The input rate can be adjusted by either changing the size of a fixed orifice, changing the adjustment of an adjustable orifice, or readjusting the appliance's gas pressure regulator outlet pressure (where a regulator is provided in the appliance). [NFPA 54:11.1.1, 11.1.1.1]

B 101.2 High Altitude. Gas input ratings of appliances shall be used for elevations up to 2000 feet (610 m). The input ratings of appliances operating at elevations above 2000 feet (610 m), shall be reduced in accordance with one of the following methods:

(1) At the rate of 4 percent for each 1000 feet (305 m) above sea level before selecting appropriately sized appliances.

(2) As permitted by the Authority Having Jurisdiction.

(3) In accordance with the manufacturer's installation instructions. [NFPA 54:11.1.2]

B 102.0 Primary Air Adjustment.

B 102.1 General. The primary air for injection (Bunsen)-type burners shall be adjusted for proper flame characteristics in accordance with the appliance manufacturer's instructions. After setting the primary air, the adjustment means shall be secured in position. [NFPA 54:11.2]

B 103.0 Safety Shutoff Devices.

B 103.1 General. Where a safety shutoff device is provided, it shall be checked for proper operation and adjustment in accordance with the appliance manufacturer's instructions. Where the device does not function properly to turn off the gas supply in the event of pilot outage or other improper operation, it shall be properly serviced or replaced with a new device. [NFPA 54:11.3]

B 104.0 Automatic Ignition.

B 104.1 General. Appliances supplied with means for automatic ignition shall be checked for operation within the parameters provided by the manufacturer. Any adjustments made shall be in accordance with the manufacturer's installation instructions. [NFPA 54:11.4]

B 105.0 Protective Devices.

B 105.1 General. Where required by the manufacturer's installation instructions, all protective devices furnished with the appliance such as a limit control, fan control to blower, temperature and pressure relief valve, low-water cutoff device, or manual operating features, shall be checked for operation within the parameters provided by the manufacturer. Any adjustments made shall be in accordance with the manufacturer's installation instructions. [NFPA 54:11.5]

B 106.0 Checking the Draft.

B 106.1 General. Draft hood-equipped appliances shall be checked to verify that there is no draft hood spillage after 5 minutes of main burner operation. [NFPA 54:11.6]

B 107.0 Operating Instructions.

B 107.1 General. Operating instructions shall be furnished and shall be left in a prominent position near the appliance for use by the consumer. [NFPA 54:11.7]

APPENDIX C
INSTALLATION AND TESTING OF OIL (LIQUID) FUEL-FIRED EQUIPMENT

C 101.0 General.

C 101.1 Applicability. Appendix C governs the installation, testing, or repair of oil or liquid fuel burners, oil or liquid fuel-burning systems, oil or liquid fuel-burning equipment, and the oil or liquid fuel piping systems used in connection with buildings or structures and equipment within the property lines of the premises.

C 102.0 Definitions.

Anti-Flooding Device. A primary safety control that causes the flow of oil or fuel to be shut off after a rise in oil or fuel level, or after receiving excess oil or fuel, and that operates before the hazardous discharge of oil or fuel can occur.

Burner, Automatically Ignited. A burner equipped so that main burner fuel may be turned on and ignited automatically.

Burner, Manually Ignited. A burner equipped, so that main burner fuel is turned on only by hand and ignited under supervision.

Burner, Mechanical Draft Type. A burner that includes a power-driven fan, blower, or other mechanism as the primary means for supplying the air for combustion.

Burner, Natural Draft Type. A burner that depends primarily on the natural draft created in the chimney or venting system to induce air required for combustion into the burner.

Constant Level Valve. A device for maintaining within a reservoir a constant level of oil or fuel for delivery to a burner.

Control Limit. An automatic safety control that is responsive to changes in fluid flow or level, pressure, or temperature and that is normally set beyond the operating range for limiting the operation of the controlled equipment by shutting off the energy supply.

Control Safety. Automatic interlock controls, including relays, switches, and other auxiliary equipment used in conjunction with them, to form a safety control system that is intended to prevent unsafe operation of the controlled equipment.

Draft Booster. A power-operated fan, blower, or other device installed in the chimney connector to increase the natural draft developed in the connected chimney.

Draft Regulator, Barometric. A device built into a fuel-burning appliance or made part of a chimney connector or vent connector that functions to reduce excessive draft through an appliance to a desired value by admitting ambient air into the appliance chimney, chimney connector, vent, or vent connector.

Fuel. Natural, manufactured, or liquefied petroleum gas, or a mixture of these gases; all grades of fuel oil, wood, or any other combustible or flammable material or any mixture of combustible or flammable materials.

Fuel Burner. A device used to convey the appropriate fuel into the combustion chamber zone in close proximity to its primary and secondary air supply to permit a stable controlled heat release compatible with the burner design, listing, and applicable approvals in a boiler, furnace, device or appliance. It includes but is not limited to burning oil or liquid fuel.

Fuel Burner System. The fuel burner and a conveyance system or piping system for the purpose of introducing the appropriate fuel into the combustion chamber zone.

Fuel-Burning Equipment/Appliance. An oil or fuel burner of any type including all oil or liquid fuel burners, oil or liquid fuel-fired units, dual, or multi-fuel burners and heating and cooking appliances with their fuel burner system and with their tank or fuel storage system, piping system, vent connectors, vent flues, fans, blowers, valves, control devices, combustion air, wiring, controls, and related devices including all accessories and appurtenances for safe and proper operation of the appliance.

Fuel Oil. Hydrocarbon oil as specified by ASTM D396, or the Canadian Government Specification Board, 3-GP-28, and having a flashpoint of not less than 100°F (38°C).

Fuel-Piping System. Method of conveying liquid, vapor, steam, gases, or slurry from one point to another, including accessories, appurtenances, and equipment necessary for its proper operation.

Indirect-Fired Appliance. An oil or fuel-burning appliance in which products of combustion (flue gasses) are not mixed in the appliance with the air or other medium being heated.

Labeled. Having attached a label, symbol, or other identifying mark of an organization acceptable to the Authority Having Jurisdiction and concerned with product evaluation that maintains periodic inspection of production of labeled equipment or materials and by whose labeling the manufacturer indicates compliance with appropriate standards or performance in a specified manner.

Premixing. A power burner in which all or nearly all of the air for combustion is mixed with the gas as primary air.

Pump, Oil or Fuel Transfer. An oil or fuel pump, automatically or manually operated, that transfers oil or fuel through continuous piping from a supply tank to an oil or fuel-burning appliance or to an auxiliary tank, and that is not designed to stop pumping automatically in case of total breakage of the oil or fuel supply line between the pump and the appliance.

Tank, Auxiliary. A tank having a capacity of not over 60 gallons (227 L) listed for installation in the supply piping between a burner and its main fuel supply tank. It shall be permitted to be included as an integral part of an automatic pump or a transfer pump, or it shall be permitted to be a separate tank.

Tank, Gravity. A supply tank from which the oil or fuel is delivered directly to the burner by gravity.

Tank, Integral. A tank that is furnished by the manufacturer as an integral part of an oil or fuel-burning appliance.

Tank Storage. A separate tank that is not connected to the oil or fuel-burning appliance.

Tank Supply. A separate tank connected directly or by a pump to the oil or fuel-burning appliance.

Tank, Vacuum or Barometric. A tank not exceeding 5 gallons (19 L) capacity that maintains a definite level of oil or fuel in a sump or similar receptacle by barometric feed. Fuel is delivered from the sump to the burner by gravity.

Valve, Manual Oil, Gas, or Fuel Shutoff. A manually operated valve in a fuel line for the purpose of turning on or completely shutting off the fuel supply to the burner.

Valve, Oil, Gas, or Fuel Control. An automatically or manually operated device consisting essentially of a fuel valve for controlling the fuel supply to a burner.

C 103.0 Standards and Accepted Practices.

C 103.1 General. The installation, testing, and repair of oil or liquid fuel-burning equipment systems shall be in accordance with Section C 103.0, the standards listed in Chapter 17, and other information outlined in this code such as, but not limited to, combustion air, flue and breeching requirements, room clearances and dimensions, and control requirements.

C 104.0 Approval of Equipment.

C 104.1 General. Oil or liquid fuel-burning equipment shall be approved.

C 105.0 Placing Equipment in Operation.

C 105.1 General. After completion of installations, the installer shall test safety and operating controls and venting before placing the burner in service. The correct input of liquid fuel shall be determined, and the fuel-to-air ratio shall be set. Each oil or liquid fuel burner shall be adjusted to its input according to the manufacturer's instructions. Overrating the burners or the appliance is prohibited. The input range shall be appropriate to the appliance:

(1) For conversion burners installed in hot water (liquid) boilers or warm air furnaces, the rate of flow of the oil or liquid fuel in British thermal units per hour (Btu/h) (kW) shall be adjusted to within plus or minus 5 percent of the design load, not to exceed the design rate of the appliance.

(2) For conversion burners installed in steam boilers, the oil or liquid fuel hourly input demand shall be adjusted to be in accordance with the steam load requirements. The oil or liquid fuel input demand necessitated by an oversized boiler shall be established and added to total input demand.

C 106.0 Pilot Operation.

C 106.1 General. Igniter or pilot flames shall be effective to ignite the oil or liquid fuel at the main burner or burners and shall be adequately protected from drafts. Pilot flames shall not become extinguished during the pilot cycle where the main burner or burners are turned on or off in a normal manner either manually or by automatic controls.

C 107.0 Burner Operation.

C 107.1 General. In making tests to determine compliance with the requirements of Section C 107.1, the following care shall be exercised to prevent the accumulation of unburned liquid fuel in the appliance that will result in an explosion or fire:

(1) The flames from the burner shall ignite freely the liquid fuel where operating at the lowest firing position.

(2) Burner flames shall not flash back where the liquid fuel is turned on or off by an automatic control mechanism.

(3) Main burner flames shall ignite freely from the pilot where the pilot flame is reduced to a minimum point that will actuate the pilot safety device.

(4) Where ignition is made in a normal manner, the flame will not flash outside the appliance.

(5) Burners shall not expel liquid fuel through air openings where operating at prevailing pressure.

(6) Burners shall have a liquid-fuel air mixture to ensure smooth ignition of the main burner.

C 108.0 Method of Test.

C 108.1 General Test Methods.

(1) The flue gas, venting, safety and operating controls of the appliance shall be checked by the installer to ensure their proper and safe operation.

(2) Method of test – atmospheric, induced draft or fan assisted types.

   (a) The appliance shall be allowed to operate until the stack temperature becomes stabilized, after which a sample of the undiluted flue products shall be taken from the appliance flue outlet. The sample taken shall be analyzed for carbon monoxide, carbon dioxide, and oxygen. Stack temperature shall be noted.

   Appliance designs incorporating induced-draft assemblies require a flue gas sample to be taken ahead of the draft regulator or induced draft fan.

(3) Performance standards for atmospheric type shall be provided in accordance with the following:

   (a) Shall be not less than 75 percent efficiency as determined by flue gas analysis method at the appliance flue outlet.

   (b) Carbon monoxide concentration in flue gas shall not exceed 0.04 percent.

   (c) Stack temperature shall not exceed 700°F (371°C) plus ambient.

   (d) Carbon dioxide concentration shall be between 8 percent and 13 percent.

   (e) Oxygen concentration shall be between 4 percent and 10 percent.

   (f) Smoke test shall not exceed number 2 for light oils or number 4 for oils heavier than number 4.

   (g) Draft shall comply with the burner manufacturer's instructions.

(4) Performance standards for induced-draft or fan-assisted types shall comply with the following:

(a) Shall be not less than 75 percent efficiency, as determined by flue gas analysis method at appliance flue outlet.

(b) Carbon monoxide concentration in flue gas not exceeding 0.04 percent.

(c) Stack temperature shall not exceed 700°F (371°C) plus ambient.

(d) Carbon dioxide concentration shall be between 8 percent and 13 percent.

(e) Oxygen concentration shall be between 4 percent and 10 percent.

(f) Smoke test shall not exceed number 2 for light oils or number 4 for oils heavier than number 4.

(g) Draft shall comply with the burner manufacturer's instructions.

Induced-draft and fan-assisted types of appliances require a sample be taken after the induced-draft fan that will cause oxygen figures in excess of the limits stated. In such cases, safe liquid fuel combustion ratios shall be maintained and be consistent with approvals and listings of the appliance.

(5) Method of test – power type.

(a) The appliance shall be allowed to operate until the stack temperature becomes stabilized; after that, a sample of the undiluted flue products shall be taken from the appliance flue outlet. The sample shall be analyzed for carbon monoxide, carbon dioxide, and oxygen. Stack temperature shall be noted.

(6) Performance standards for power type.

(a) Shall be not less than 80 percent efficiency as determined by flue gas analysis method at the appliance flue outlet.

(b) Carbon monoxide concentration in the flue gas shall not exceed 0.04 percent.

(c) Stack temperature shall not exceed 700°F (371°C) plus ambient.

(d) Carbon dioxide concentration shall be between 8 percent and 13 percent.

(e) Oxygen concentration shall be between 4 percent and 10 percent.

(f) Smoke test shall not exceed number 2 for light oils or number 4 for oils heavier than number 4.

(g) Draft shall comply with the burner manufacturer's instructions.

(7) After completion of the test of newly installed oil or liquid fuel burner equipment as provided in this section, the installer shall file with the Authority Having Jurisdiction complete records of the test on a form approved by the Authority Having Jurisdiction. The tag stating the date of the test and the name of the installer shall be attached to the appliance at the main valve.

(8) Listing and approval.

(a) The concentration of oxygen in the undiluted flue products of oil or liquid fuel burners shall in no case be less than 3 percent nor exceed 10 percent and shall be in accordance with performance standards and shall be consistent with the listing and approval of the equipment.

(b) The allowable limit of carbon monoxide shall not exceed 0.04 percent.

(c) The flue gas temperature of an oil appliance, as taken on the appliance side of the draft regulator, shall not exceed applicable performance standards and shall be consistent with the listing and approvals of the equipment.

(9) The oxygen figures shall not apply where there is an approved oxygen trim system on the burner that is designed for that use, including a low oxygen interlock where approved by the Authority Having Jurisdiction.

(10) Supervision shall be as follows:

(a) Supervised startup shall be required to verify the safe operation of an oil or liquid fuel burner and to provide documentation that operation is consistent with this code, listing, and approval. Supervised startup shall be required for liquid fuel burners in Section C 109.1(2), Section C 109.1(3), and Section C 109.1(4). Supervised startup requires that the liquid-fuel burner shall be tested in the presence of the mechanical official in a manner set forth by the Authority Having Jurisdiction before the installation is approved. Testing shall include safety and operating controls, input, flue gas analysis, and venting. Flue gas shall be tested at high, medium, and low fires. Provisions shall be made in the system to allow a firing test in warm weather. After completion of the test of newly installed oil or liquid fuel burner equipment, as provided in this section, the installer shall file with the Authority Having Jurisdiction complete records of the test on a form approved by the Authority Having Jurisdiction. The tag stating the date of the test and the name of the tester shall be attached to the appliance at the main valve.

(b) Oil and liquid fuel burners of 1 000 000 Btu/h (293 kW) input or more require a supervised startup in accordance with Section C 108.1(10)(a).

(c) Installation of oxygen trim systems, modulating dampers, or other draft control or combustion devices require a supervised startup in accordance with Section C 108.1(10)(a).

(d) Direct-fired heaters shall require a supervised startup in accordance with Section C 108.1(10)(a).

(11) The complete control diagram of the installation and operating instructions shall be supplied and posted by the installer of the appliance.

C 109.0 Special Requirements Based on Btu/h Input.

C 109.1 General.

(1) Zero to 400 000 Btu/h (0 kW to 117 kW) per burner.

(a) One approved manual shutoff valve lever handle.

(b) One approved fuel oil filter, installed on the supply piping.

(c) Approved automatic safety shutoff valve to provide 100 percent shutoff of all oil.

(d) A flame safeguard control capable of providing 100 percent shutoff in the event of flame failure. Flame failure response timing shall not exceed the control manufacturer's instructions.

(e) Two controls, one operating and one high limit, activated by temperature or pressure, as appropriate.

(f) Burners relying on mechanical means to provide air for combustion shall have actual proof-of-air interlock device.

(g) Installations with dampered combustion air openings shall prove damper open position before trial for burner ignition.

(h) Vent dampers and flue dampers shall be properly interlocked to prevent burner ignition unless safely open.

(2) Four hundred thousand and one to 999 999 Btu/h (117.2 kW to 292.9 kW) per burner.

(a) One approved manual shutoff valve lever handle.

(b) One approved fuel-oil filter, installed on the supply piping.

(c) Two safety shutoff valves in series, maximum five seconds closing time.

(d) One electronic flame safeguard pilot control providing a separately supervised and proven pilot, 100 percent shutoff manual reset. Flame failure response time shall not exceed the control manufacturer's instructions.

Direct-spark ignition shall be allowed where approved by the Authority Having Jurisdiction and where used on number 2 or lighter oil.

(e) Two controls, one operating and one high limit, activated by temperature or pressure, as appropriate.

(f) Burners relying on mechanical means to provide air for combustion shall have actual proof-of-air interlock device.

(g) Power burners shall include proven prepurge of not less than 60 seconds at high-fire damper settings. This prepurge shall occur before every burner cycle, regardless of reason.

(h) Installations with dampered combustion air openings shall prove damper open position before trial for burner ignition.

(i) Vent dampers and flue dampers shall be interlocked to prevent burner ignition unless safely open.

(j) One high oil or liquid fuel-pressure interlock, reset from flame safeguard or manually.

(k) Where hot water or steam, one low water cutoff.

(l) An atomizing medium proving switch.

(m) A low oil temperature switch for oil or liquid fuel requiring preheating.

(n) A high oil temperature interlock for oil or liquid fuel requiring preheating.

(o) The burner oil pump shall automatically not operate or rotate while the alternate fuel is firing.

(p) A pressure-relief valve shall be provided between safety shutoff valves and between pump and safety valves where an integral valve is used with a pump.

(q) A separate relief device is required on each transfer pump.

(3) One million to 2 499 999 Btu/h (293 kW to 732 kW) input per burner.

(a) One approved manual shutoff valve lever handle.

(b) One approved fuel-oil filter, installed on the supply piping.

(c) Two safety shutoff valves in series with a combined flame failure response and valve closing time not to exceed 5 seconds with strainer directly before the valves.

(d) Programmed electronic flame safeguard including proven low-fire start, manual reset lockout, 100 percent shutoff (both pilot and main burner), and a separately supervised and proven pilot.

Flame-sensing systems utilizing a UV scanner shall prove pilot and interrupt ignition spark prior to main burner valves being energized.

(e) Two controls, one operating and one high limit, activated by temperature or pressure.

(f) Burners relying on mechanical means to provide air for combustion shall have actual proof-of-air interlock device.

(g) Power burners shall include proven prepurge of not less than 60 seconds at high-fire damper settings. This prepurge shall occur before every burner cycle, regardless of reason.

(h) Installations with dampered combustion air openings shall prove damper open position before trial for burner ignition.

(i) Vent dampers and flue dampers shall be interlocked to prevent burner ignition unless safely open.

(j) One high oil or liquid fuel-pressure interlock, reset from flame safeguard or manually.

(k) Where hot water or steam, two low water cutoffs.

(l) An atomizing medium proving switch.

(m) A low oil temperature switch for oil or liquid fuel requiring preheating.

(n) A high oil temperature interlock for oil or liquid fuel requiring preheating.

(o) The burner oil pump shall automatically not operate or rotate while the alternate fuel is firing.

(p) A pressure-relief valve shall be provided between

safety shutoff valves and between pump and safety valves where an integral valve is used with a pump.

(q) A separate relief device is required on each transfer pump.

(r) One low oil or liquid fuel-pressure interlock, reset from flame safeguard or manually.

(s) Burners with automatic controls, prepurge, proof-of-closure, modulation, or postpurge shall not use relays external to the flame safeguard to accomplish these functions.

(4) Two million five hundred thousand to 12 499 999 Btu/h (733 kW to 3663.3 kW) per burner.

(a) One approved manual shutoff valve lever handle.

(b) One approved fuel-oil filter, installed on the supply piping.

(c) Two safety shutoff valves in series, with a combined flame failure response and valve closing time not to exceed 5 seconds with strainer directly before the valves.

(d) Programmed electronic flame safeguard including proven low-fire start, manual reset lockout, 100 percent shutoff (both pilot and main burner), and a separately supervised and proven pilot.

Flame-sensing systems utilizing a UV scanner shall prove pilot and interrupt ignition spark prior to main burner valves being energized.

(e) Two controls, one operating and one high limit, activated by temperature or pressure.

(f) Burners relying on mechanical means to provide air for combustion shall have actual proof-of-air interlock device.

(g) Power burners shall include proven prepurge of not less than 60 seconds at high-fire damper settings. This prepurge shall occur before every burner cycle, regardless of reason.

(h) Installations with dampered combustion air openings shall prove damper open position before trial for burner ignition.

(i) Vent dampers and flue dampers shall be interlocked to prevent burner ignition unless safely open.

(j) One high oil or liquid fuel-pressure interlock, reset from flame safeguard or manually.

(k) Where hot water or steam, two low water cutoffs.

(l) An atomizing medium proving switch.

(m) A low oil temperature switch for oil or liquid fuel requiring preheating.

(n) A high oil temperature interlock for oil or liquid fuel requiring preheating.

(o) A separate firing rate control valve.

(p) The burner oil pump shall automatically not operate or rotate while the alternate fuel is firing.

(q) A pressure-relief valve shall be provided between safety shutoff valves and between pump and safety valves where an integral valve is used with a pump.

(r) A separate relief device is required on each transfer pump.

(s) One low oil or liquid fuel-pressure interlock reset from flame safeguard or manually.

(t) Burners with automatic controls, prepurge, proof-of-closure, modulation, or postpurge shall not use relays external to the flame safeguard to accomplish these functions.

(5) More than 12 500 000 Btu/h (3663.4 kW) per burner inputs. These burners shall comply with the requirements of the appropriate standards listed in Chapter 17 and the following:

(a) One approved manual shutoff valve lever handle.

(b) One approved fuel-oil filter, installed on the supply piping.

(c) Two safety shutoff valves in series, one with proof of closure, with a combined flame failure response and valve closing time not to exceed 2 seconds with strainer directly before the valves.

(d) Programmed electronic flame safeguard including proven low-fire start, manual reset lockout, 100 percent shutoff (both pilot and main burner), and a separately supervised and proven pilot.

Flame-sensing systems utilizing a UV scanner shall prove pilot and interrupt ignition spark prior to main burner valves being energized.

(e) Two controls, one operating and one high limit, activated by temperature or pressure.

(f) Burners relying on mechanical means to provide air for combustion shall have actual proof-of-air interlock device.

(g) Power burners must include proven prepurge of not less than 60 seconds at high-fire damper settings. This prepurge shall occur before every burner cycle, regardless of reason.

(h) Installations with dampered combustion air openings shall prove damper open position before trial for burner ignition.

(i) Vent dampers and flue dampers shall be interlocked to prevent burner ignition unless safely open.

(j) One high oil or liquid fuel-pressure interlock; reset from flame safeguard or manually.

(k) A manual firing cock.

(l) Where hot water or steam, two low water cut-offs.

(m) An atomizing medium proving switch.

(n) A low oil temperature switch for oil or liquid fuel requiring preheating.

(o) A high oil temperature interlock for oil or liquid fuel requiring pre-heating.

(p) A separate firing rate control valve.

(q) The burner oil pump shall automatically not operate or rotate while the alternate fuel is firing.

(r) A pressure-relief valve shall be provided between safety shutoff valves and between pump and safety valves where an integral valve is used with a pump.

(s) A separate relief device is required on each transfer pump.

(t) One low oil or liquid fuel-pressure interlock, reset from flame safeguard or manually.

(u) Burners with automatic controls, prepurge, proof-of-closure, modulation, or postpurge shall not use relays external to the flame safeguard to accomplish these functions.

(6) Shutoff Valve.

(a) Oil or liquid fuel burner installations shall include a non-electric shutoff valve that is held open by a fusible link designed to close at 165°F (74°C), installed near the burner in the same room as the burner. This shall prevent the flow of oil or liquid fuel to the burner through the supply pipe. A check valve is required in the return line if the tank is higher than the burner.

APPENDIX D
FUEL SUPPLY: MANUFACTURED/MOBILE HOME PARKS AND RECREATIONAL VEHICLE PARKS

D 101.0 Fuel Gas Piping Systems.

D 101.1 General. All fuel gas piping systems serving manufactured homes, accessory buildings, or structures and communities shall be designed and constructed in accordance with any applicable provisions of Chapter 13 and NFPA 58. NFPA 31 shall apply to oil fuel-burning systems and shall conform to the criteria of the Authority Having Jurisdiction. {NFPA 501A:4.1.1.1 – 4.1.1.2}

D 101.2 Gas Supply Connections. Gas supply connections at sites, where provided from an underground gas supply piping system, shall be located and arranged to permit attachment to a manufactured home occupying the site. For the installation of liquefied petroleum gas (LP-Gas) storage systems, the provisions of NFPA 58 shall be followed. [NFPA 501A:4.1.2.1 – 4.1.2.2]

D 101.3 Location of Gas Supply Connection. The gas supply to the manufactured home shall be located within 4 feet (1219 mm) of the manufactured home stand.

Exception: The requirement of Section D 101.3 shall not apply to gas supply connections for manufactured homes located on all-weather wood, concrete, or concrete block foundation systems or on foundations constructed in accordance with the local building code or, in the absence of a local code, with a recognized model building code. [NFPA 501A:4.1.3]

D 101.4 Recreational Vehicle Park Fuel-Gas Equipment and Installations. Fuel gas equipment and installations shall comply with this appendix, except as otherwise permitted or required by this code.

D 102.0 Single and Multiple Manufactured Home Site Fuel Supply Systems.

D 102.1 Underground Installation. Underground gas piping system installations shall comply with any applicable building code and Section D 102.1.1 and Section D 102.1.2. [NFPA 501A:4.2.1]

D 102.1.1 Open-Ended Gastight Conduit. Underground gas piping shall not be installed beneath that portion of a manufactured home site reserved for the location of a manufactured home or manufactured home accessory building or structure unless installed in the open-ended gastight conduit of Section D 102.1.2. [NFPA 501A:4.2.1.1]

D 102.1.2 Requirements. The open-ended gastight conduit shall conform to the requirements in the following:

(1) The conduit shall be not less than Schedule 40 pipe that is approved for underground installation beneath buildings.

(2) The interior diameter of the conduit shall be not less than ½ of an inch (15 mm) larger than the outside diameter of the gas piping.

(3) The conduit shall extend to a point not less than 4 inches (102 mm) beyond the outside wall of the manufactured home or accessory building or structure, and the outer ends shall not be sealed.

(4) Where the conduit terminates within a manufactured home or accessory building or structure, it shall be accessible and the space between the conduit and the gas piping shall be sealed to prevent leakage of gas into the building. [NFPA 501A:4.2.1.2– 4.2.1.2.4]

D 103.0 Manufactured Home Site Gas Shutoff Valve.

D 103.1 General. Each manufactured home site shall have a listed gas shutoff valve installed upstream of the manufactured home site gas outlet. The gas shutoff valve shall be located on the outlet riser at a height of not less than 6 inches (152 mm) above grade. A gas shutoff valve shall not be located under any manufactured home. The outlet shall be equipped with a cap or plug to prevent discharge of gas whenever the manufactured home site outlet is not connected to a manufactured home. [NFPA 501A:4.2.2.1 – 4.2.2.4]

Exception: Gas shutoff valves shall conform to Section D 103.1, except for manufactured homes located on foundations constructed in accordance with the local building code or, in the absence of a local code, with a recognized model building code. [NFPA 501A:4.2.2]

D 104.0 Gas Meters.

D 104.1 Support of Meters. Where installed, gas meters shall be supported by a post or bracket placed on a firm footing or other means providing equivalent support and shall not depend on the gas outlet riser for support. [NFPA 501A:4.2.3.1]

D 104.2 Location of Meters. Each gas meter shall be installed in an accessible location and shall be provided with unions or other fittings so that the meter can be removed easily and placed in an upright position. Meters shall not be installed in unventilated or inaccessible locations or closer than 3 feet (914 mm) to sources of ignition. [NFPA 501A:4.2.3.2.1 – 4.2.3.2.2]

D 104.3 Meter Shutoff Valve or Cock. All gas meter installations shall be provided with shutoff valves or cocks located adjacent to and on the inlet side of the meters. In the case of a single meter installation utilizing an LP-Gas container, the container service valve shall be permitted to be used in lieu of the shutoff valve or cock. All gas meter installations shall be provided with test tees located adjacent to and on the outlet side of the meters. [NFPA 501A:4.2.4.1 – 4.2.4.3]

D 105.0 Cathodic Protection Requirements.

D 105.1 General. Cathodic protection shall be installed for corrosion control of buried or submerged metallic gas piping in accordance with the following requirements:

(1) Where amphoteric metals are included in a buried or submerged pipeline containing a metal of different anodic potential the following protection shall be provided:

(a) The buried or submerged pipeline shall be cathodically protected at a negative (cathodic) voltage of 0.85 volt, measured between the structure surface and a saturated copper-copper sulfate half cell contacting the electrolyte.

(b) The amphoteric metals shall be electrically isolated from the remainder of the pipeline with insulating flanges, or equivalent, and cathodically protected.

(2) The amount of cathodic protection shall be such that the protective coating and the pipe are not damaged.

D 106.0 Manufactured Home Community LP-Gas Supply Systems.

D 106.1 General. Where 10 or more customers are served by one LP-Gas supply system, the installation of the gas supply system shall be in accordance with 49 CFR 192. Other types of liquefied petroleum gas supply systems and the storage and handling of LP-Gas shall be in accordance with NFPA 58 (see Section D 111.1). [NFPA 501A:4.3.2.1 – 4.3.2.2]

D 107.0 Required Gas Supply.

D 107.1 General. The minimum hourly volume of gas required at each manufactured home site outlet or any section of the manufactured home community gas piping system shall be calculated as shown in Table D 107.1. [NFPA 501A:4.3.4.1]

In extreme climate areas, additional capacities other than those shown in Table D 107.1 shall be considered.

D 108.0 Gas Pipe Sizing and Pressure.

D 108.1 Size. The size of each section of a gas piping system shall be determined in accordance with NFPA 54, or by other standard engineering methods acceptable to the Authority Having Jurisdiction. [NFPA 501A:4.3.5.1]

TABLE D 107.1
DEMAND FACTORS FOR USE IN CALCULATING GAS PIPING SYSTEMS IN MANUFACTURED HOME COMMUNITIES
[NFPA 501A: TABLE 4.3.4.1]

NUMBER OF MANUFACTURED HOME SITES	BTU/H PER MANUFACTURED HOME SITE
1	125 000
2	117 000
3	104 000
4	96 000
5	92 000
6	87 000
7	83 000
8	81 000
9	79 000
10	77 000
11–20	66 000
21–30	62 000
31–40	58 000
41–60	55 000
Over 60	50 000

For SI units: 1000 British thermal units per hour = 0.293 kW

D 108.2 Pressure. Where all connected appliances are operated at their rated capacity, the gas supply pressure shall be not less than 7 inches water column (1.7 kPa). The gas supply pressure shall not exceed 14 inches water column (3.5 kPa). [NFPA 501A:4.3.5.2]

D 109.0 Gas Piping Materials.

D 109.1 Metal. Metal gas pipe shall be standard-weight wrought iron or steel (galvanized or black), yellow brass containing not more than 75 percent copper, or internally tinned or treated copper of iron pipe size. Galvanizing shall not be considered protection against corrosion.

Seamless copper or steel tubing shall be permitted to be used with gases not corrosive to such material. Steel tubing shall comply with ASTM A254. Copper tubing shall comply with ASTM B88 (Type K or Type L) or ASTM B280. Copper tubing (unless tin-lined) shall not be used if the gas contains more than an average of 0.3 grains of hydrogen sulfide per 100 standard cubic feet (0.7 mg/100 L) of gas. [NFPA 501A:4.3.6.1 – 4.3.6.1.6]

D 109.2 Protection Coatings for Metal Gas Piping. All buried or submerged metallic gas piping shall be protected from corrosion by approved coatings or wrapping materials. All gas pipe protective coatings shall be approved types, shall be machine applied, and shall conform to recognized standards. Field wrapping shall provide equivalent protection and is restricted to those short sections and fittings that are necessarily stripped for threading or welding. Risers shall be coated or wrapped to a point at least 6 inches (152 mm) aboveground. [NFPA 501A:4.3.6.2 – 4.3.6.2.4]

D 109.3 Plastic. Plastic piping shall only be used underground and shall meet the requirements of ASTM D2513 or ASTM D2517, as well as the design pressure and design limitations of 49 CFR (Section 192.123), and shall otherwise conform to the installation requirements thereof. [NFPA 501A:4.3.6.3]

D 110.0 Gas Piping Installations.

D 110.1 Minimum Burial Below Ground Level and Clearances. All gas piping installed belowground level shall have a minimum earth cover of 18 inches (457 mm) and shall be installed with at least 12 inches (305 mm) of clearance in any direction from any other underground utility system. [NFPA 501A:4.3.7.1]

D 110.2 Metallic Gas Piping. All metallic gas piping systems shall be installed in accordance with approved plans and specifications, including provisions for cathodic protection. Each cathodic protection system shall be designed and installed to conform to the provisions of 49 CFR 192. [NFPA 501A:4.3.7.2.1, 4.3.7.2.2]

D 110.2.1 Cathodic Protection. Where the cathodic protection system is designed to protect only the gas piping system, the gas piping system shall be electrically isolated from all other underground metallic systems or installations. Where only the gas piping system is cathodically protected against corrosion, a dielectric fitting shall be used in the manufactured home gas connection to insulate the manufactured home from the underground gas piping system. [NFPA 501A:4.3.7.2.3, 4.3.7.2.4]

D 110.2.2 Underground Metallic Systems. Where a cathodic protection system is designed to provide all underground metallic systems and installations with protection against corrosion, all such systems and installations shall be electrically bonded together and protected as a whole. [NFPA 501A:4.3.7.2.5]

D 110.3 Plastic Gas Piping. Plastic gas piping shall be used underground and shall be installed with an electrically conductive wire for locating the pipe. The wire used to locate the plastic pipe shall be copper, not smaller in size than 18 AWG, with insulation approved for direct burial. Every portion of a plastic gas piping system consisting of metallic pipe shall be cathodically protected against corrosion. [NFPA 501A:4.3.7.3.1 – 4.3.7.3.3]

D 110.4 Gas Piping System Shutoff Valve. An accessible and identifiable shutoff valve controlling the flow of gas to the entire manufactured home community gas piping system shall be installed in a location acceptable to the Authority Having Jurisdiction and near the point of connection to the service piping or to the supply connection of an LP-Gas container. [NFPA 501A:4.3.7.4]

D 111.0 Liquefied Petroleum Gas Appliances.

D 111.1 General. LP-Gas equipment shall be installed in accordance with the applicable provisions of NFPA 58. [NFPA 501A:4.3.8]

D 112.0 Oil Supply.

D 112.1 General. The following three methods of supplying oil to an individual manufactured home site shall be permitted:

(1) Supply from an outside underground tank (see Section D 113.6).

(2) Supply from a centralized oil distribution system designed and installed in accordance with accepted engineering practices and in compliance with NFPA 31.

(3) Supply from an outside aboveground tank (see Section D 113.6). [NFPA 501A:4.3.9]

D 112.2 Minimum Oil Supply Tank Size. Oil supply tanks shall have a minimum capacity equal to 20 percent of the average annual oil consumption. [NFPA 501A:4.3.10]

D 112.3 Oil Supply Connections. Oil supply connections at manufactured home sites, where provided from a centralized oil distribution system, shall be located and arranged to permit attachment to a manufactured home utilizing the stand. [NFPA 501A:4.3.11.1] The installation of such facilities shall comply with the following requirements:

(1) The main distribution pipeline shall be permitted to be connected to a tank or tanks having an aggregate capacity not exceeding 20 000 gallons (75 708 L) at a point below the liquid level.

(2) Where this piping is so connected, a readily accessible internal or external shutoff valve shall be installed in the piping as close as practicable to the tank.

(3) If external and aboveground, the shutoff valve and its tank connections shall be made of steel.

(4) Connections between the tank(s) and the main pipeline shall be made with double swing joints or flexible connectors, or shall otherwise be arranged to permit the tank(s) to settle without damaging the system.

(5) If located aboveground, the connections specified in Section D 112.3(4) shall be located within the diked area.

(6) A readily accessible and identified manual shutoff valve shall be installed either inside or outside of the structure in each branch supply pipeline that enters a building, mobile home, travel trailer, or other structure. If outside, the valve shall be protected from weather and damage. If inside, the valve shall be located directly adjacent to the point at which the supply line enters the structure.

(7) A device shall be provided in the supply line at or ahead of the point where it enters the interior of the structure that will automatically shut off the oil supply, if the supply line between this device and the appliance is broken. This device shall be located on the appliance side of the manual shutoff valve required in Section D 112.3(6) and shall be solidly supported and protected from damage.

(8) Means shall be provided to limit the oil pressure at the appliance inlet to a maximum gauge pressure of 3 pound-force per square inch gauge (psig) (21 kPa). If a pressure-reducing valve is used, it shall be a type approved for the service.

(9) A device shall be provided that will automatically shut off the oil supply to the appliance if the oil pressure at the appliance inlet exceeds a gauge pressure of 8 psig (55 kPa). The device shall not be required under either of the following conditions:
   (a) Where the distribution system is supplied from a gravity tank and the maximum hydrostatic head of oil in the tank is such that the oil pressure at the appliance inlet will not exceed a gauge pressure of 8 psig (55 kPa).
   (b) Where a means is provided to automatically shut off the oil supply if the pressure-regulating device provided in accordance with Section D 112.3(8) fails to regulate the pressure as required.

(10) Only appliances equipped with primary safety controls specifically listed for the appliance shall be connected to a centralized oil distribution system. [NFPA 31:9.2.10–9.2.15]

D 113.0 Fuel Supply Systems Installation.

D 113.1 Flexible Gas Connector. Except for manufactured homes located on an all-weather wood, concrete, or concrete block foundation system or on a foundation constructed in accordance with the local building code or, in the absence of a local code, with a recognized model building code, each gas supply connector shall be listed for outside manufactured home use, shall be not more than 6 feet (1829 mm) in length, and shall have a capacity rating to supply the connected load. [NFPA 501A:4.4.1]

D 113.2 Use of Approved Pipe and Fittings of Extension. Where it is necessary to extend a manufactured home inlet to permit connection of the 6 foot (1829 mm) listed connector to the site gas outlet, the extension shall be of approved materials of the same size as the manufactured home inlet and shall be adequately supported at no more than 4 foot (1219 mm) intervals to the manufactured home. [NFPA 501A:4.4.2]

D 113.3 Mechanical Protection. All gas outlet risers, regulators, meters, valves, and other exposed equipment shall be protected against accidental damage. [NFPA 501A:4.4.3]

D 113.4 Special Rules on Atmospherically Controlled Regulators. Atmospherically controlled regulators shall be installed in such a manner that moisture cannot enter the regulator vent and accumulate above the diaphragm. Where the regulator vent is obstructed due to snow and icing conditions, shields, hoods, or other suitable devices shall be provided to guard against closing of the vent opening. [NFPA 501A:4.4.4.1 – 4.4.4.2]

D 113.5 Fuel Gas Piping Test. The manufactured home fuel gas piping system shall be tested only with air before it is connected to the gas supply. The manufactured home gas piping system shall be subjected to a pressure test with all appliance shutoff valves in their closed positions. [NFPA 501A:4.4.5]

D 113.5.1 Procedures. The fuel gas piping test shall consist of air pressure at not less than 10 inches water column or more than 14 inches water column (2.5 kPa to 3.5 kPa). The fuel gas piping system shall be isolated from the air pressure source and shall maintain this pressure for not less than 10 minutes without perceptible leakage. Upon satisfactory completion of the fuel gas piping test, the appliance valves shall be opened and the gas appliance connectors shall be tested with soapy water or bubble solution while under the pressure remaining in the piping system. Solutions used for testing for leakage shall not contain corrosive chemicals. Pressure shall be measured with a manometer, slope gauge, or gauge that is calibrated in either water inch (mm) or psi (kPa) with increments of either 1⁄10 of an inch (2.5 mm) or 1⁄10 psi (0.7 kPa) gauge, as applicable. Upon satisfactory completion of the fuel gas piping test, the manufactured home gas supply connector shall be installed and the connections shall be tested with soapy water or bubble solution. [NFPA 501A:4.4.5.1.1 – 4.4.5.1.6]

D 113.5.2 Warning. The following warning shall be supplied to the installer:

WARNING: Do not overpressurize the fuel gas piping system. Damage to valves, regulators, and appliances can occur due to pressurization beyond the maximums specified. [NFPA 501A:4.4.5.2]

D 113.5.3 Vents. Gas appliance vents shall be visually inspected to ensure that they have not been dislodged in transit and are connected securely to the appliance. [NFPA 501A:4.4.5.3]

D 113.6 Oil Tanks. Oil tank capacities shall comply with the following:

(1) No more than one 660 gallon (2498 L) tank or two tanks with an aggregate capacity of 660 gallons (2498 L) or less shall be connected to one oil-burning appliance.

(2) Two supply tanks, where used, shall be cross-connected and provided with a single fill and single vent, as described in NFPA 31 and shall be on a common slab and rigidly secured, one to the other.

(3) Tanks having a capacity of 660 gallons (2498 L) or less shall be securely supported by rigid, noncombustible supports to prevent settling, sliding, or lifting. [NFPA 501A:4.4.6]

D 113.6.1 Installation. Oil supply tanks shall be installed in accordance with the applicable provisions of NFPA 31. [NFPA 501A:4.4.6.1]

D 113.6.2 Capacity. A tank with a capacity no larger than 60 gallons (227 L) shall be permitted to be a DOT-5 shipping container (drum), and so marked, or a tank meeting the provisions of UL 80. Tanks other than DOT-5 shipping containers having a capacity of not more than 660 gallons (2498 L) shall meet the provisions of UL 80. Pressure tanks shall be built in accordance with Section VIII, Pressure Vessels, of the ASME Boiler and Pressure Vessel Code. [NFPA 501A:4.4.6.2.1 – 4.4.6.2.2]

D 113.6.3 Location. Tanks, as described in Section D 113.6 and Section D 113.6.2, that are adjacent to buildings shall be located not less than 10 feet (3048 mm) from a property line that is permitted to be built upon. [NFPA 501A:4.4.6.3]

D 113.6.4 Vent. Tanks with a capacity no larger than 660 gallons (2498 L) shall be equipped with an open vent no smaller than 1½ inch (40 mm) iron pipe size; tanks with a 500 gallon (1892 L) or less capacity shall have a vent of 1¼ inch (32 mm) iron pipe size. [NFPA 501A:4.4.6.4]

D 113.6.5 Liquid Level. Tanks shall be provided with a means of determining the liquid level. [NFPA 501A:4.4.6.5]

D 113.6.6 Fill Opening. The fill opening shall be a size and in a location that permits filling without spillage. [NFPA 501A:4.4.6.6]

D 114.0 Manufactured Home Accessory Building Fuel Supply Systems.

D 114.1 General. Fuel gas supply systems installed in a manufactured home accessory building or structure shall comply with the applicable provisions of NFPA 54 and NFPA 58. Fuel oil supply systems shall comply with the applicable provisions of NFPA 31. [NFPA 501A:4.5.1 – 4.5.2]

D 115.0 Community Building Fuel Supply Systems in Manufactured Home Communities.

D 115.1 Fuel Gas Piping and Equipment Installations. Fuel gas piping and equipment installed within a permanent building in a manufactured home community shall comply with nationally recognized appliance and fuel gas piping codes and standards adopted by the Authority Having Jurisdiction. Where the state or other political subdivision does not assume jurisdiction, such fuel gas piping and equipment installations shall be designed and installed in accordance with the applicable provisions of NFPA 54 or NFPA 58. [NFPA 501A:4.6.1.1 – 4.6.1.2]

D 115.2 Oil Supply Systems in Manufactured Home Communities. Oil-burning equipment and installation within a manufactured home community shall be designed and constructed in accordance with the applicable codes and standards adopted by the Authority Having Jurisdiction. Where the state or other political subdivision does not assume jurisdiction, such installations shall be designed and constructed in accordance with the applicable provisions of NFPA 31. [NFPA 501A:4.6.2.1 – 4.6.2.2]

D 115.3 Oil-Burning Equipment and Installation. Oil-burning equipment and installation within a building constructed in a manufactured home community in accordance with the local building code or a nationally recognized building code shall comply with nationally recognized codes and standards adopted by the Authority Having Jurisdiction. Where the state or other political subdivision does not assume jurisdiction, such oil-burning equipment and installations shall be designed and installed in accordance with the applicable provisions of NFPA 31. [NFPA 501A:4.6.3.1 – 4.6.3.2]

D 115.4 Inspection and Tests. Inspections and tests for fuel gas piping shall be made in accordance with Chapter 1 and Chapter 13 of this code.

APPENDIX E
SUSTAINABLE PRACTICES

E 101.0 General.

E 101.1 Applicability. The purpose of this appendix is to provide a comprehensive set of technically sound provisions that encourage sustainable practices and works towards enhancing the design and construction of mechanical systems that result in a positive long-term environmental impact. This appendix is not intended to circumvent the health, safety, and general welfare requirements of this code.

E 101.2 Definition of Terms. For the purposes of this code, the definitions shall apply to this appendix.

No attempt is made to define ordinary words, which are used in accordance with their established dictionary meanings, except where a word has been used loosely, and it is necessary to define its meaning as used in this appendix to avoid misunderstanding.

The definitions of terms are arranged alphabetically according to the first word of the term.

E 201.0 Definitions.

E 201.1 Cycles of Concentration for Cooling Towers. Cycles of concentration equals the specific conductance of the water in the cooling tower basin divided by the combined flow-weighted average specific conductance of the makeup water(s) to the cooling tower.

E 201.2 Duct Wall Penetrations. Includes pipe, tubing, rods, and wire. Screws and other fasteners are not considered to be ductwork penetrations.

E 201.3 Energy Star. A joint program of the U.S. Environmental Protection Agency and the U.S. Department of Energy. Energy Star is a voluntary program designed to identify and promote energy-efficient products and practices.

E 201.4 Geothermal. Renewable energy generated by deep-earth.

E 201.5 Heating Seasonal Performance Factor (HSPF). The total heating output of a heat pump during its normal annual usage period for heating in British thermal units (Btu) (kW•h) divided by the total electric energy input during the same period. [ASHRAE 90.1:3.2]

E 201.6 Integrated Energy Efficiency Ratio (IEER). A single-number figure of merit expressing cooling part-load EER efficiency for commercial unitary air-conditioning and heat pump equipment on the basis of weighted operation at various load capacities for the equipment. [ASHRAE 90.1:3.2]

E 201.7 Integrated Part-Load Value (IPLV). A single-number figure of merit based on part-load EER, COP_C, or kW/kW expressing part-load efficiency for air-conditioning and heat pump equipment on the basis of weighted operation at various load capacities for the equipment. [ASHRAE 90.1:3.2]

E 201.8 Joint, Transverse. Connections of two duct sections oriented perpendicular to airflow.

E 201.9 Maintenance. The upkeep of property or equipment by the owner of the property in accordance with the requirements of this appendix.

E 201.10 Minimum Efficiency Reporting Value (MERV). Filter minimum efficiency reporting value, in accordance with ASHRAE 52.2.

E 201.11 Multi-Occupant Spaces. Indoor spaces used for presentations and training, including classrooms and conference rooms.

E 201.12 Recirculation System. A system of hot water supply and return piping with shutoff valves, balancing valves, circulating pumps, and a method of controlling the circulating system.

E 201.13 Seam, Longitudinal. Joints oriented in the direction of airflow.

E 201.14 Seasonal Energy Efficiency Ratio (SEER). The total cooling output of an air conditioner during its normal annual usage period for cooling in Btu (kW•h) divided by the total electric energy input during the same period in Btu (kW•h). [ASHRAE 90.1:3.2]

E 301.0 General Regulations.

E 301.1 Installation. Mechanical systems covered by this appendix shall be installed in accordance with this code, other applicable codes, and the manufacturer's installation and operating instructions.

E 301.2 Qualifications. Where permits are required, the Authority Having Jurisdiction shall have the authority to require contractors, installers, or service technicians to demonstrate competency. Where determined by the Authority Having Jurisdiction, the contractor, installer or service technician shall be licensed to perform such work.

E 302.0 Disposal of Liquid Waste.

E 302.1 Disposal. It shall be unlawful for a person to cause, suffer, or permit the disposal of liquid wastes, in a place or manner, except through and by means of an approved drainage system, installed and maintained in accordance with the provisions of the plumbing code.

E 302.2 Connections to Plumbing System Required. Equipment and appliances, used to receive or discharge liquid wastes or sewage, shall be connected to the drainage system of the building or premises in accordance with the requirements of the plumbing code and this appendix.

E 303.0 Abandonment.

E 303.1 General. An abandoned system or part thereof covered under the scope of this appendix shall be disconnected from remaining systems, drained, plugged, and capped in an approved manner.

E 401.0 Water Conservation and Efficiency.

E 401.1 General. The provisions of this section establish the means of conserving potable and nonpotable water used in and around a building.

E 402.0 Meters.

E 402.1 Required. A water meter shall be required for buildings connected to a public water system, including municipally supplied reclaimed (recycled) water. In other than single-family houses, multi-family structures not exceeding three stories above grade, and modular houses, a separate meter or submeter shall be installed in the following locations:

(1) The makeup water supply to cooling towers, evaporative condensers, and fluid coolers.

(2) The makeup water supply to one or more boilers collectively exceeding 1 000 000 British thermal units per hour (Btu/h) (293 kW).

(3) The water supply to a water-using process where the consumption exceeds 1000 gallons per day (gal/d) (0.0438 L/s), except for manufacturing processes.

(4) The makeup water supply to an evaporative cooler having an air flow exceeding 30 000 cubic feet per minute (ft^3/min) (14.1584 m^3/s).

E 402.2 Consumption Data. A means of communicating water consumption data from submeters to the water consumer shall be provided.

E 402.3 Access. Meters and submeters shall be accessible.

E 403.0 HVAC Water Use.

E 403.1 Once-Through Cooling. Once-through cooling using potable water is prohibited.

E 403.2 Cooling Towers and Evaporative Coolers. Cooling towers and evaporative coolers shall be equipped with makeup water and blow down meters, conductivity controllers, and overflow alarms. Cooling towers shall be equipped with efficiency drift eliminators that achieve drift reduction to 0.002 percent of the circulated water volume for counterflow towers and 0.005 percent for cross-flow towers.

E 403.3 Cooling Tower Makeup Water. Not less than 5 cycles of concentration is required for air-conditioning cooling tower makeup water having a total hardness of less than 11 grains per gallon (gr/gal) (188 mg/L) expressed as calcium carbonate. Not less than 3.5 cycles of concentration is required for air-conditioning cooling tower makeup water having a total hardness equal to or exceeding 11 gr/gal (188 mg/L) expressed as calcium carbonate.

Exception: Air-conditioning cooling tower makeup water having discharge conductivity range not less than 7 gr/gal (120 mg/L) to 9 gr/gal (154 mg/L) of silica measured as silicon dioxide.

E 403.4 Evaporative Cooler Water Use. Evaporative cooling systems shall use 3.5 gallons (13.2 L) or less of water per ton-hour (kW•h) of cooling where system controls are set to maximum water use. Water use expressed in maximum water use per ton-hour (kW•h) of cooling, shall be marked on the device and included in the product user manual, product information literature, and manufacturer's installation instructions. Water use information shall be readily available at the time of code compliance inspection.

E 403.4.1 Overflow Alarm. Cooling systems shall be equipped with an overflow alarm to alert building owners, tenants, or maintenance personnel where the water refill valve continues to allow water to flow into the reservoir where the reservoir is full. The alarm shall have a sound pressure level rating of not less than 85 dBa measured at a distance of 10 feet (3048 mm).

E 403.4.2 Automatic Pump Shut-Off. Cooling systems shall automatically cease pumping water to the evaporation pads where airflow across evaporation pads ceases.

E 403.4.3 Cooler Reservoir Discharge. A water quality management system (either timer or water quality sensor) shall be provided. Where timers are used, the time interval between discharge of reservoir water shall be set to 6 or more hours of cooler operation. Where water quality sensors are used, the discharge of reservoir water shall be set for 800 ppm or more of total dissolved solids (TDS). Continuous discharge or continuous bleed systems shall not be installed.

E 403.4.4 Discharge Water Reuse. Discharge water shall be reused where applications exist on site. Where a nonpotable water source system exists on site, evaporative cooler discharge water shall be collected and discharged to the collection system.

Exception: Where the reservoir water affects the quality of the nonpotable water supply making the nonpotable water unusable for its intended purposes.

E 403.4.5 Discharge Water to Drain. Where discharge water is not recovered for reuse, the sump overflow line shall not be directly connected to a drain. Where the discharge water is discharged into a sanitary drain, an air gap of not less than 6 inches (152 mm) shall be provided between the termination of the discharge line and the drain opening. The discharge line shall terminate in a location that is visible to the building owner, tenants, or maintenance personnel.

E 403.5 Use of Reclaimed (Recycled) and Onsite Treated Nonpotable Water for Cooling. Where approved for use by the water or wastewater utility and the Authority Having Jurisdiction, reclaimed (recycled), or onsite treated nonpotable water shall be permitted to be used for industrial and commercial cooling or air-conditioning.

E 403.5.1 Drift Eliminator. A drift eliminator shall be utilized in a cooling system, utilizing alternate sources of water, where the aerosolized water is capable of coming in contact with employees or members of the public.

E 403.5.2 Disinfection. A biocide shall be used to treat the cooling system recirculation water where the recycled water is capable of coming in contact with employees or members of the public.

E 501.0 Heating, Ventilation, and Air-Conditioning Systems and Equipment – Energy Efficiency.

E 501.1 Scope. The provisions of this section shall establish the means of enhancing energy efficiency associated with mechanical systems in a building.

E 502.0 Heating, Ventilation, and Air-Conditioning Low-Rise Residential Buildings.

E 502.1 General. The heating, ventilating, air-conditioning, for single-family houses, multi-family structures not exceeding three stories above grade, and modular houses shall be in accordance with Section E 502.2 through Section E 502.12. The heating, ventilation, and air-conditioning system of other buildings shall be in accordance with Section E 503.0.

E 502.2 Heating, Ventilating, and Air-Conditioning Systems and Equipment. This section shall regulate only equipment using single-phase electric power, air conditioners, and heat pumps with rated cooling capacities less than 65 000 British thermal units per hour (Btu/h) (19 kW), warm air furnaces with rated heating capacities less than 225 000 Btu/h (66 kW), boilers less than 300 000 Btu/h (88 kW) input, and heating-only heat pumps with rated heating capacities less than 65 000 Btu/h (19 kW). [ASHRAE 90.2:6.2]

E 502.2.1 Nonresidential Type Systems and Equipment. Heating, ventilating, and air-conditioning systems and equipment that do not fall under the requirements of Section E 502.0 shall be in accordance with the applicable requirements of Section E 503.0.

E 502.3 Balancing. The air distribution system design, including outlet grilles, shall provide a means for balancing the air distribution system unless the design procedure provides a system intended to operate within plus or minus 10 percent of design air quantities. [ASHRAE 90.2:6.3]

E 502.3.1 Balancing Dampers. Balancing dampers shall be installed in branch ducts, and the axis of the damper shall be installed parallel to the direction of airflow in the main duct.

E 502.4 Ducts. Ducts shall be sized, installed, and tested in accordance with Section E 502.4.1 through Section E 502.4.4.

E 502.4.1 Insulation for Ducts. Portions of the air distribution system installed in or on buildings for heating and cooling shall be R-8. Where the mean outdoor dew-point temperature in a month exceeds 60°F (16°C), vapor retarders shall be installed on conditioned-air supply ducts. Vapor retarders shall have a water vapor permeance not exceeding 0.5 perm [2.87 E-11 kg/(Pa•s•m^2)] where tested in accordance with Procedure A in ASTM E96.

Insulation shall not be required where the ducts are within the conditioned space. [ASHRAE 90.2:6.4]

E 502.4.2 Ducts and Register Penetrations. Joints, seams, and penetrations of duct systems shall be made airtight by means of mastics, gasketing, or other means in accordance with this code. Register penetrations shall be sealed to the wall or floor assemblies. Where HVAC duct penetrates a conditioned space, the duct penetration shall be sealed to the wall or floor assembly to prevent leakage into an unconditioned space.

E 502.4.3 Duct Leakage Test. For systems with a duct or air handler outside of the conditioned space, a duct leakage test shall be performed in accordance with Section E 502.4.3.1.

E 502.4.3.1 Duct Leakage Verification Test. Ductwork shall be tested to the maximum permitted leakage in 1 cubic foot per minute (ft^3/min) per 100 square feet [0.0001 (m^3/s)/m^2] of duct surface area in accordance with SMACNA HVAC Air Duct Leakage Test Manual. Register penetrations shall be sealed during the test. The test shall be conducted with a pressure differential of 0.1 inch water gauge (0.02 kPa) across the tested system.

E 502.4.4 Duct Sizing. Duct systems shall be sized in accordance with ACCA Manual D or other methods approved by the Authority Having Jurisdiction with the velocity in the main duct not to exceed 1000 feet per minute (ft/min) (5.08 m/s) and the velocity in the secondary branch duct not to exceed 600 ft/min (3.048 m/s).

E 502.5 Insulation for Piping. HVAC system piping installed to serve buildings and within buildings shall be thermally insulated in accordance with Table E 502.5. [ASHRAE 90.2:6.5]

E 502.6 Ventilation and Combustion Air. The building shall be designed to have the capability to provide the ventilation air specified in Table E 502.6. Mechanical ventilation shall be calculated in accordance with Equation E 502.6. [ASHRAE 90.2:6.6.1]

(Equation E 502.6)

$$\textit{Mechanical Ventilation} = [(0.35 - \textit{Summer}) \times \textit{Volume}] / 60$$

Where:

Mechanical Ventilation = required mechanical ventilation rate to supplement summer infiltration, CFM (m^3/s)

Summer = summer design infiltration rate, ACH

Volume = volume of conditioned space, ft^3 (m^3)

E 502.6.1 Combustion Air. Combustion air for fossil fuel heating equipment shall comply with this code or with one of the following:

(1) Natural gas and propane heating equipment, NFPA 54
(2) Oil heating equipment, NFPA 31
(3) Solid fuel burning equipment, NFPA 211 [ASHRAE 90.2:6.6.2]

E 502.7 Electric Heating Systems. Electric heating systems shall be installed in accordance with the following requirements. [ASHRAE 90.2:6.7]

E 502.7.1 Wall, Floor, or Ceiling Electric-Resistance Heating. Where wall, floor, or ceiling electric-resistance heating units are used, the structure shall be zoned and heaters installed in each zone in accordance with the heat loss of that zone. Where living and sleeping zones are separate, the number of zones shall be not less than two. Where two or more heaters are installed in one room, they shall be controlled by one thermostat. [ASHRAE 90.2:6.7.1]

E 502.7.2 Electric Central Warm Air Heating. Where electric central warm air heating is to be installed, an electric heat pump or an off-peak electric heating system with thermal storage shall be used.

Exceptions:

(1) Electric resistance furnaces where the ducts are located inside the conditioned space, and not less than two zones are provided where the living and sleeping zones are separate.

(2) Packaged air-conditioning units with supplemental electric heat. [ASHRAE 90.2:6.7.2]

E 502.8 Bath Ceiling Units. Bath ceiling units providing a combination of heat, light, or ventilation shall be provided with controls permitting separate operation of the heating function. [ASHRAE 90.2:6.8]

E 502.9 HVAC Equipment, Rated Combinations. HVAC system equipment and system components shall be furnished with the input(s), the output(s), and the value of the appropriate performance descriptor of HVAC products in accordance with federal law or in accordance with Table E 502.9, as applicable. These shall be based on newly produced equipment or components. Manufacturer's instructions shall be furnished with and attached to the equipment. The manufacturer of electric-resistance heating equipment shall furnish full-load energy input over the range of voltages at which the equipment is intended to operate. [ASHRAE 90.2:6.9]

E 502.10 Controls. Each system or each zone within a system shall be provided with not less than one thermostat capable of being set from 55°F (13°C) to 85°F (29°C) and capable of operating the system's heating and cooling. The thermostat or control system, or both, shall have an adjustable deadband, the range of which includes a setting of 10°F (6°C) between heating and cooling where automatic changeover is provided.

TABLE E 502.5
MINIMUM PIPE INSULATION THICKNESS[1, 5]
[ASHRAE 90.2: TABLE 6.5]

INSULATION CONDUCTIVITY			NOMINAL PIPE DIAMETER (inches)				
FLUID DESIGN OPERATING TEMERATURE RANGE (°F)	Btu•inch/(h•ft²•°F)	MEAN RATING TEMPERATURE (°F)	<1	1 TO 1¼	1½ TO 3½	4 TO 6	EQUAL TO OR GREATER THAN 8
HEATING SYSTEMS (STEAM, STEAM CONDENSATE, AND HOT WATER)[2, 3]							
201–250	0.27–0.30	150	1.5	1.5	2.0	2.0	2.0
141–200	0.25–0.29	125	1.0	1.0	1.0	1.5	1.5
105–140	0.22–0.28	100	0.5	0.5	1.0	1.0	1.0
COOLING SYSTEMS (CHILLED WATER, BRINE, AND REFRIGERANT)[4]							
40–55	0.22–0.28	100	0.5	0.5	1.0	1.0	1.0
Below 40	0.22–0.28	100	0.5	1.0	1.0	1.0	1.5

For SI Units: °C= (°F-32)/1.8, 1 British thermal unit inch per hour square foot degree Fahrenheit = [0.1 W/(m•K)], 1 inch = 25 mm

Notes:

1 For insulation outside the stated conductivity range, the minimum thickness (T) shall be determined as follows:

$T = r\{(1 + t/r)^{K/k} - 1\}$

Where:

T = minimum insulation thickness (inches).

r = actual outside radius of pipe (inches) (mm).

t = insulation thickness listed in this table for applicable fluid temperature and pipe size.

K = conductivity of alternate material at mean rating temperature indicated for the applicable fluid temperature [Btu•in/(h•ft²•°F)] [W/(m•K)].

k = the upper value of the conductivity range listed in this table for the applicable fluid temperature.

2 These thicknesses are based on energy efficiency considerations only. Additional insulation is sometimes required relative to safety issues/surface temperature.

3 Piping insulation is not required between the control valve and coil on run-outs where the control valve is located within 4 feet (1219 mm) of the coil and the pipe size is 1 inch (25 mm) or less.

4 These thicknesses are based on energy efficiency considerations only. Issues such as water vapor permeability or surface condensation sometimes require vapor retarders, additional insulation or both.

5 For piping exposed to outdoor air, increase insulation thickness by ½ of an inch (12.7 mm). The outdoor air is defined as any portion of insulation that is exposed to outdoor air. For example, attic spaces and crawl spaces are considered exposed to outdoor air.

TABLE E 502.6
VENTILATION AIR
[ASHRAE 90.2: TABLE 6.6.1]

CATEGORY	MINIMUM REQUIREMENT	CONDITIONS
Mechanical ventilation[1]	50 ft³/min outdoor air	Where summer design infiltration rate calculated in accordance with reference standard (a) or (b) is less than 0.35 ACH[2].
Kitchen exhaust	100 ft³/min intermittent	All conditions
Bath exhaust	intermittent	All conditions

For SI units: 1 cubic foot per minute = 0.00047 m³/s

Notes:

1 Calculate in accordance with Equation E 502.6.

2 Reference standards:

(a) ACCA Manual J

(b) ASHRAE GRP-158

TABLE E 502.9
MINIMUM REQUIREMENTS FOR NON-FEDERALLY COVERED HVAC EQUIPMENT
[ASHRAE 90.2: TABLE 6.9]

EQUIPMENT TYPE	SUBCATEGORY OR RATING CONDITION	MINIMUM EFFICIENCY	TEST PROCEDURE
Groundwater source heat pump*	Cooling Mode	11.0 EER at 70°F Ent. Water	ARI 325
		11.5 EER at 50°F Ent. Water	
	Heating Mode	3.4 COP at 70°F Ent. Water	
		3.0 COP at 50°F Ent. Water	
Unitary A/C	Water cooled split system	9.3 EER at 85°F Ent. Water	ARI 210/240
		8.3 IPLV at 75°F Ent. Water	
	Evaporatively cooled split system	9.3 EER at 95°F Out. Amb.	
		8.5 IPLV at 80°F Out. Amb.	

For SI units: °C = (°F-32)/1.8
* Performance for electrically powered equipment with capacity less than 65 000 Btu/h (19 kW) where rated in accordance with ARI 325.

Wall-mounted temperature controls shall be mounted on an inside wall. [ASHRAE 90.2:6.10.1]

E 502.10.1 Initial Control Setting. The control shall initially be set for a maximum heating temperature of 70°F (21°C) and a cooling temperature of not less than 78°F (26°C).

E 502.10.2 Ventilation Control. Each mechanical ventilation system (supply, exhaust, or both) shall be equipped with a readily accessible switch or other means for shutoff. Manual or automatic dampers installed for the purpose of isolating outside air intakes and exhausts from the air distribution system shall be designed for tight shutoff. [ASHRAE 90.2:6.10.2]

E 502.10.3 Humidity Control. Where additional energy-consuming equipment is provided for adding moisture to maintain specific selected relative humidities in spaces or zones, a humidistat shall be provided. This device shall be capable of being set to prevent energy from being used to produce relative humidity within the space above 30 percent. [ASHRAE 90.2:6.10.3.1]

E 502.10.3.1 Cooling. Where additional energy-consuming equipment is provided for reducing humidity, it shall be equipped with controls capable of being set to prevent energy from being used to produce a relative humidity within the space below 50 percent during periods of human occupancy and below 60 percent during unoccupied periods. [ASHRAE 90.2:6.10.3.2]

E 502.10.4 Freeze Protection Systems. Freeze protection systems, such as heat tracing of outdoor piping and heat exchangers, including self-regulating heat tracing, shall include automatic controls capable of and configured to shut off the systems where outdoor air temperatures are above 40°F (4°C) or where the conditions of the protected fluid will prevent freezing. Snow- and ice-melting systems shall include automatic controls capable of and configured to shut off the systems where the pavement temperature is above 50°F (10°C) and no precipitation is falling and an automatic or manual control that will allow shutoff where the outdoor temperature is above 40°F (4°C) so that the potential for snow or ice accumulation is negligible. [ASHRAE 90.1:6.4.3.7]

E 502.10.5 Other Controls. Where setback, zoned, humidity and cooling controls and equipment are provided, they shall be designed and installed in accordance with Section E 502.10 through Section E 502.10.3.1. [ASHRAE 90.2:6.10.3.3]

E 502.11 Whole House Fans. Whole house exhaust fans shall have insulated louvers or covers which close where the fan is off. Covers or louvers shall have an insulation value of not less than R-4.2, and shall be installed in accordance with the manufacturer's installation instructions. The attic openings shall be sufficient to accommodate the ventilation capacity of the whole house fan. The operation of the whole house fan shall be considered in determining the adequacy of providing combustion air in accordance with this code.

E 502.12 Dampers. Dampers shall be installed to close off outdoor air inlets and exhaust outlets where the ventilation system is not operating.

E 503.0 Heating, Ventilation, and Air-Conditioning – Other than Low-Rise Residential Buildings.

E 503.1 General. The heating, ventilation, and air-conditioning in buildings, other than single-family houses, multi-family structures of not more than three stories above grade, and modular houses, shall be in accordance with Section E 503.0.

E 503.1.1 New Buildings. Mechanical equipment and systems serving the heating, cooling, ventilating, or refrigeration needs of new buildings shall be in accordance with the requirements of this section as described in Section E 503.2. [ASHRAE 90.1:6.1.1.1]

E 503.1.2 Additions to Existing Buildings. Mechanical equipment and systems serving the heating, cooling, ventilating, or refrigeration needs of additions to existing buildings shall be in accordance with the requirements of this section as described in Section E 503.2.

Exception: Where HVACR to an addition is provided by existing HVACR systems and equipment, such existing systems and equipment shall not be required to be in accordance with this appendix. A new system or equipment installed shall be in accordance with specific requirements applicable to those systems and equipment. [ASHRAE 90.1:6.1.1.2]

E 503.1.3 Alterations to Heating, Ventilating, Air-Conditioning, and Refrigeration in Existing Buildings. New HVACR equipment as a direct replacement of existing HVACR equipment shall be in accordance with the following sections as applicable for the equipment being replaced:

(1) Section E 503.3
(2) Section E 503.4
(3) Section E 503.4.6
(4) Section E 503.4.6.2
(5) Section E 503.4.6.3
(6) Section E 503.4.6.4
(7) Section E 503.4.6.8
(8) Section E 503.4.6.9
(9) Section E 503.4.6.11
(10) Section E 503.5.1
(11) Section E 503.5.3
(12) Section E 503.5.4
(13) Section E 503.5.6.1.2
(14) Section E 503.5.6.2
(15) Section E 503.5.6.5
(16) Section E 503.5.7
(17) Section E 503.5.7.2
(18) Section E 503.5.8.1. [ASHRAE 90.1:6.1.1.3.1]

E 503.1.3.1 New Cooling Systems. New cooling systems installed to serve previously uncooled spaces shall be in accordance with this section as described in Section E 503.2. [ASHRAE 90.1:6.1.1.3.2]

E 503.1.3.2 Existing Cooling Systems. Alterations to existing cooling systems shall not decrease economizer capability unless the system is in accordance with Section E 503.5 through Section 503.5.4.1. [ASHRAE 90.1:6.1.1.3.3]

E 503.1.3.3 Ductwork. New and replacement ductwork shall comply with Section E 503.4.7.1 through Section E 503.4.7.2.1. [ASHRAE 90.1:6.1.1.3.4]

E 503.1.3.4 Piping. New and replacement piping shall comply with Section E 503.4.7.1.

Exceptions:

(1) For equipment that is being modified or repaired but not replaced, provided that such modifications or repairs will not result in an increase in the annual energy consumption of the equipment using the same energy type.
(2) Where a replacement or alteration of equipment requires extensive revisions to other systems, equipment, or elements of a building, and such replaced or altered equipment is a like-for-like replacement.
(3) For a refrigerant change of existing equipment.
(4) For the relocation of existing equipment.
(5) For ducts and piping where there is insufficient space or access to comply with these requirements. [ASHRAE 90.1:6.1.1.3.5]

E 503.2 Compliance Path(s). Section E 503.0 shall be achieved in accordance with the requirements of Section E 503.1.1 through Section E 503.1.3.4, Section E 503.6, Section E 503.7, and one of the following:

(1) Section E 503.3 and Section E 503.3.1
(2) Section E 503.4
(3) Section E 503.4 and Section E 503.8 [ASHRAE 90.1:6.2.1]

E 503.2.1 Projects Using Energy Cost Budget Method. Projects using the energy cost budget method in accordance with ASHRAE 90.1 shall comply with Section E 503.4, the mandatory provisions of this section, as a portion of that compliance path. [ASHRAE 90.1:6.2.2]

E 503.3 Simplified Approach Option for HVAC Systems. The simplified approach shall be an optional path for compliance where the following conditions are met:

(1) The building is not more than two stories in height.
(2) Gross floor area is less than 25 000 square feet (2322.6 m^2).
(3) The HVAC system in the building is in accordance with the requirements listed in Section E 503.3.1. [ASHRAE 90.1:6.3.1]

E 503.3.1 Criteria. The HVAC system shall comply with the following criteria:

(1) The system serves a single HVAC zone.
(2) The equipment shall comply with the variable flow requirements of Section E 503.5.6.2.
(3) Cooling (where any) shall be provided by a unitary packaged or split-system air conditioner that is either air-cooled or evaporatively cooled, with efficiency that is in accordance with the requirements shown in Table E 503.7.1(1), Table E 503.7.1(2), or Table E 503.7.1(4) for the applicable equipment category.
(4) The system shall have an air economizer in accordance with Section E 503.5 and Section E 503.4.6.13.
(5) Heating (where any) shall be provided by a unitary packaged or split-system heat pump that is in accordance with the applicable efficiency requirements shown in Table E 503.7.1(2) or Table E 503.7.1(4), a fuel-fired furnace that is in accordance with the applicable efficiency requirements shown in Table E 503.7.1(5), an electric resistance heater, or a baseboard system connected to a boiler that is in accordance with the applicable efficiency requirements shown in Table E 503.7.1(6).
(6) The system shall comply with the exhaust air energy recovery requirements in accordance with Section E 503.5.10.
(7) The system shall be controlled by a manual changeover or dual setpoint thermostat.
(8) Where a heat pump equipped with auxiliary internal electric resistance heaters is installed, controls shall be provided that prevent supplemental heater operation where the heating load is capable of being met by the heat pump alone during both steady-state operation and setback recovery. Supplemental heater operation shall be permitted during outdoor coil defrost cycles. The heat pump shall be controlled in accordance with one of the following:
    (a) A digital or electronic thermostat designed for heat pump use that energizes auxiliary heat where the heat pump has insufficient capacity to maintain setpoint or to warm up the space at a sufficient rate.

(b) A multistage space thermostat and an outdoor air thermostat wired to energize auxiliary heat on the last stage of the space thermostat and where outdoor air temperature is less than 40°F (4°C).

Exceptions: Heat Pumps that comply with the following:

(1) Have a minimum efficiency regulated by NAECA.

(2) In accordance with the requirements shown in Table E 503.7.1(2).

(3) Include all usage of internal electric resistance heating.

(9) The system controls shall not permit reheat or other form of simultaneous heating and cooling for humidity control.

(10) Systems serving spaces other than hotel or motel guest rooms, and other than those requiring continuous operation, which have both a cooling or heating capacity more than 15 000 Btu/h (4.4 kW) and a supply fan motor power more than 0.75 horsepower (hp) (0.56 kW), shall be provided with a time clock that is in accordance with the following:

(a) Can start and stop the system under different schedules for seven different day-types per week.

(b) Is capable of retaining programming and time setting during a loss of power for a period of not less than 10 hours.

(c) Includes an accessible manual override that allows temporary operation of the system for up to 2 hours.

(d) Is capable of and configured with temperature setback down to 55°F (13°C) during off hours.

(e) Is capable of and configured with temperature setup to 90°F (32°C) during off hours.

(11) Systems serving hotel/motel guest rooms shall comply with Section E 503.4.6.3.5.

(12) Except for piping within manufacturer's units, HVAC piping shall be insulated in accordance with Table E 503.7.3(1) and Table E 503.7.3(2). Insulation exposed to weather shall be suitable for outdoor service, e.g., protected by aluminum, sheet metal, painted canvas, or plastic cover. Cellular foam insulation shall be protected as above or painted with a coating that is water retardant and provides shielding from solar radiation.

(13) Ductwork and plenums shall be insulated in accordance with Table E 503.7.2 and shall be sealed in accordance with Section E 503.4.7.2.

(14) Construction documents shall require a ducted system to be air balanced in accordance with industry-accepted procedures.

(15) Outdoor air intake and exhaust systems shall comply with Section E 503.4.6.4 through Section E 503.4.6.5.

(16) Where separate heating and cooling equipment serves the same temperature zone, thermostats shall be interlocked to prevent simultaneous heating and cooling.

(17) Systems with a design supply air capacity more than 10 000 ft^3/min (4.7195 m^3/s) shall have optimum start controls.

(18) The system shall comply with the demand control ventilation requirements of Section E 503.4.6.9 and the ventilation design requirements of Section E 503.5.6.6.

(19) The system shall comply with the door switch requirements of Section E 503.5.14. [ASHRAE 90.1:6.3.2]

E 503.3.2 Climate Zone Determination. Climate zones identified in this appendix shall be determined in accordance with ASHRAE 90.1. For locations in the United States and its territories, the assigned climate zone and, where required, the assigned climate zone letter shall be in accordance with ASHRAE 169.

Exception: Where recorded historical climatic data are available for a construction site, it is permitted to be used to determine compliance where approved by the Authority Having Jurisdiction. [ASHRAE 90.1:5.1.4.1]

See Figure E 503.3.2.

E 503.4 Mandatory Provisions. Equipment shown in Table E 503.7.1(1) through Table E 503.7.1(16) shall have a minimum performance at the specified rating conditions where tested in accordance with the specified test procedure. Where multiple rating conditions or performance requirements are provided, the equipment shall satisfy the stated requirements unless otherwise exempted by footnotes in the table. Equipment covered under the Federal Energy Policy Act of 1992 (EPACT) shall have no minimum efficiency requirements for operation at minimum capacity or other than standard rating conditions. Equipment used to provide service water-heating functions as part of a combination system shall satisfy the stated requirements for the appropriate space heating or cooling category.

Tables are as follows:

(1) Table E 503.7.1(1), "Electrically Operated Unitary Air Conditioners and Condensing Units-Minimum Efficiency Requirements"

(2) Table E 503.7.1 (2), "Electrically Operated Unitary and Applied Heat Pumps-Minimum Efficiency Requirements"

(3) Table E 503.7.1 (3), "Water-Chilling Packages-Efficiency Requirements" (See Section E 503.4.1 for water-cooled centrifugal water-chilling packages that are designed to operate at nonstandard conditions.)

(4) Table E 503.7.1 (4), "Electrically Operated Packaged Terminal Air Conditioners, Packaged Terminal Heat Pumps, Single-Package Vertical Air Conditioners, Single-Package Vertical Heat Pumps, Room Air Conditioners, and Room Air Conditioner Heat Pumps-Minimum Efficiency Requirements"

(5) Table E 503.7.1 (5), "Warm-Air Furnaces and Combination Warm-Air Furnaces/Air-Conditioning Units, Warm-Air Duct Furnaces, and Unit Heaters-Minimum Efficiency Requirements"

(6) Table E 503.7.1 (6), "Gas- and Oil-Fired Boilers-Minimum Efficiency Requirements"

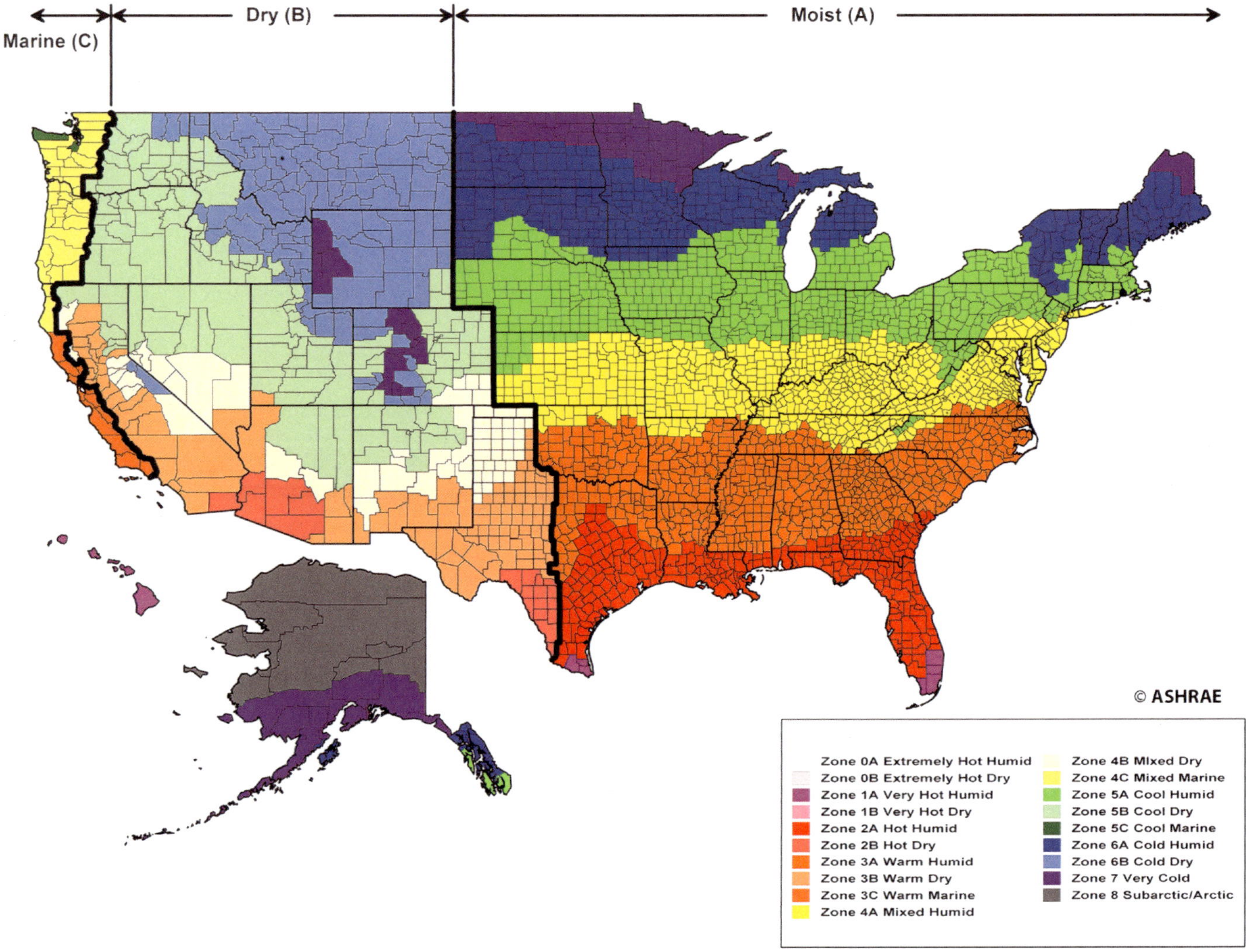

Figure E 503.3.2
ASHRAE CLIMATE ZONE MAP
©ASHRAE, www.ashrae.org. Used with permission from (2013) ASHRAE Standard-(169).

(7) Table E 503.7.1 (7), "Performance Requirements for Heat-Rejection Equipment"

(8) Table E 503.7.1 (8), "Heat Transfer Equipment"

(9) Table E 503.7.1 (9), "Electrically Operated Variable-Refrigerant-Flow Air Conditioners- Minimum Efficiency Requirements"

(10) Table E 503.7.1 (10), "Electrically Operated Variable-Refrigerant-Flow and Applied Heat Pumps-Minimum Efficiency Requirements"

(11) Table E 503.7.1 (11), "Air Conditioners and Condensing Units Serving Computer Rooms"

(12) Table E 503.7.1 (12), "Commercial Refrigerators and Freezers-Minimum Efficiency Requirements"

(13) Table E 503.7.1 (13), "Commercial Refrigeration-Minimum Efficiency Requirements"

(14) Table E 503.7.1 (14), "Vapor-Compression-Based Indoor Pool Dehumidifiers-Minimum Efficiency Requirements"

(15) Table E 503.7.1 (15), "Electrically Operated DX-DOAS Units, Single-Package and Remote Condenser, without Energy Recovery-Minimum Efficiency Requirements"

(16) Table E 503.7.1 (16), "Electrically Operated DX-DOAS Units, Single-Package and Remote Condenser, with Energy Recovery-Minimum Efficiency Requirements"

All furnaces with input ratings of 225 000 Btu/h (66 kW) or more, including electric furnaces, that are not located within the conditioned space shall have jacket losses not exceeding 0.75 percent of the input rating. Air conditioners primarily serving computer rooms and covered by ASHRAE 127 shall comply with the requirements in Table E 503.7.1(11). All other air conditioners shall meet the requirements in Table E 503.7.1(1). [ASHRAE 90.1:6.4.1.1]

E 503.4.1 Water-Cooled Centrifugal Chilling Packages. Equipment not designed for operation in accordance with AHRI 550/590 test conditions of 44.00°F (6.67°C) leaving and 54.00°F (12.22°C) entering chilled-fluid temperatures, and with 85.00°F (29.44°C) entering and 94.30°F

(34.61°C) leaving condenser-fluid temperatures, shall have maximum full-load kW/ton (*FL*) and part-load rating requirements adjusted in accordance with Equation E 503.4.1(1) through Equation E 503.4.1(3):

$FL_{adj} = FL/K_{adj}$ [Equation E 503.4.1(1)]

$PLV_{adj} = IPLV.IP/K_{adj}$ [Equation E 503.4.1(2)]

$K_{adj} = A \times B$ [Equation E 503.4.1(3)]

Where:

FL = full-load kW/ton value from Table E 503.7.1(3)

FL_{adj} = maximum full-load kW/ton rating, adjusted for nonstandard conditions

$IPLV.IP$ = $IPLV.IP$ value from Table E 503.7.1(3)

PLV_{adj} = maximum NPLV rating, adjusted for nonstandard conditions

A = $0.00000014592 \times (LIFT)^4 - 0.0000346496 \times (LIFT)^3 + 0.00314196 \times (LIFT)^2 - 0.147199 \times (LIFT) + 3.93073$

B = $0.0015 \times LvgEvap + 0.934$

$LIFT$ = $LvgCond - LvgEvap$

$LvgCond$ = Full-load condenser leaving fluid temperature (°F)

$LvgEvap$= Full-load evaporator leaving temperature (°F)

The FL_{adj} and PLV_{adj} values shall only be applicable for centrifugal chillers meeting all of the following full-load design ranges:

(1) 36.00°F (2.22°C) ≤ *LvgEvap* ≤ 60.00°F (15.56°C)

(2) *LvgCond* ≤ 115.00°F (46.11°C)

(3) 20.00°F (-6.67°C) ≤ LIFT ≤ 80.00°F (26.67°C)

Manufacturers shall calculate the FL_{adj} and PLV_{adj} before determining whether to label the chiller in accordance with Section E 503.4.4. Chillers that are in accordance with ASHRAE 90.1 shall be labeled on chillers in accordance with the scope of ASHRAE 90.1.

Centrifugal chillers designed to operate outside of these ranges shall not be covered under this appendix.

Example: Path A, 600 ton (600 000 kg) centrifugal chiller Table E 503.7.1(3) efficiencies.

FL = 0.5600 kW/ton

$IPLV.IP$ = 0.5000 kW/ton

$LvgCond$ = 91.16°F

$LvgEvap$ = 42.00°F

$LIFT$ = 91.16°F – 42.00°F = 49.16°F

A = $0.00000014592 \times (49.16)^4 - 0.0000346496 \times (49.16)^3 + 0.00314196 \times (49.16)^2 - 0.147199 \times (49.16) + 3.93073 = 1.02331$

B = $0.0015 \times 42.00 + 0.934 = 0.99700$

K_{adj} = $A \times B$

FL_{ajd} = 0.5600/1.02024 = 0.5489 kW/ton

PLV_{adj} = 0.5000/1.02024 = 0.4901 kW/ton

[ASHRAE 90.1:6.4.1.2.1]

For SI units: 1 metric ton = 1000 kg, 1000 British thermal units per hour = 0.293 kW, 1 gallon per minute = 0.06 L/s, °C = (°F-32)/1.8

E 503.4.1.1 Positive Displacement (air- and water-cooled) Chilling Packages. Equipment with an evaporator leaving fluid temperature higher than 32.00°F (0.00°C) and water-cooled positive displacement chilling packages with a condenser leaving fluid temperature below 115.00°F (46.11°C) shall show compliance with Table E 503.7.1(3) when tested or certified with water at standard rating conditions, in accordance with the referenced test procedure. [ASHRAE 90.1:6.4.1.2.2]

E 503.4.2 Equipment not Listed. Equipment not listed in the tables referenced in Section E 503.4 and Section E 503.4.1 shall be permitted to be used. [ASHRAE 90.1:6.4.1.3]

E 503.4.3 Verification of Equipment Efficiencies. Equipment efficiency information supplied by manufacturers shall be verified in accordance with one of the following:

(1) Equipment covered under EPACT shall be in accordance with U.S. Department of Energy certification requirements.

(2) Where a certification program exists for a covered product, and it includes provisions for verification and challenge of equipment efficiency ratings, then the product shall be listed in the certification program.

(3) Where a certification program exists for a covered product, and it includes provisions for verification and challenge of equipment efficiency ratings, but the product is not listed in the existing certification program, the ratings shall be verified by an independent laboratory test report.

(4) Where no certification program exists for a covered product, the equipment efficiency ratings shall be supported by data furnished by the manufacturer.

(5) Where components such as indoor or outdoor coils from different manufacturers are used, the system designer shall specify component efficiencies whose combined efficiency is in accordance with the minimum equipment efficiency requirements in Section E 503.4 through Section E 503.4.4.1.

(6) Requirements for plate-type liquid-to-liquid heat exchangers are listed in Table E 503.7.1(8). [ASHRAE 90.1:6.4.1.4]

E 503.4.4 Labeling. Mechanical equipment that is not covered by the U.S. National Appliance Energy Conservation Act (NAECA) of 1987 shall carry a permanent label installed by the manufacturer stating that the equipment is in accordance with the requirements of ASHRAE 90.1. [ASHRAE 90.1:6.4.1.5.1]

E 503.4.4.1 Packaged Terminal Air Conditioners. Nonstandard-size packaged terminal air conditioners and heat pumps with existing sleeves having an external wall opening of less than 16 inches (406 mm) high or less than 42 inches (1067 mm) wide and having a cross-sectional area less than 670 square inches (0.432 m^2) shall be factory labeled as follows:

Manufactured for nonstandard-size applications only: Not to be installed in new construction projects. [ASHRAE 90.1:6.4.1.5.2]

E 503.4.5 Load Calculations. Heating and cooling system design loads for the purpose of sizing systems and equipment shall be determined in accordance with ASHRAE/ACCA 183. [ASHRAE 90.1:6.4.2.1]

E 503.4.5.1 Pump Head. Pump differential pressure (head) for the purpose of sizing pumps shall be determined in accordance with generally accepted engineering standards and handbooks acceptable to the Authority Having Jurisdiction. The pressure drop through each device and pipe segment in the critical circuit at design conditions shall be calculated. [ASHRAE 90.1:6.4.2.2]

E 503.4.6 Zone Thermostatic Controls. The supply of heating and cooling energy to each zone shall be individually controlled by thermostatic controls responding to temperature within the zone. For the purposes of Section E 503.4.6, a dwelling unit shall be permitted to be considered a single zone.

Exceptions: Independent perimeter systems that are designed to offset only building envelope loads shall be permitted to serve one or more zones also served by an interior system, provided that:

(1) the perimeter system includes not less than one thermostatic control zone for each building exposure having walls facing only one orientation for 50 contiguous feet (15 240 mm) or more and

(2) the perimeter system heating and cooling supply is controlled by thermostatic controls located within the zones served by the system.

Exterior walls and semiexterior walls are considered to have different orientations where the exposures they face differ by more than 45 degrees (0.79 rad). [ASHRAE 90.1:6.4.3.1.1]

E 503.4.6.1 Dead Band. Where used to control both heating and cooling, zone thermostatic controls shall be capable of and configured to provide a temperature range or dead band of not less than 5°F (3°C) within which the supply of heating and cooling energy to the zone is shut off or reduced to a minimum.

Exceptions:

(1) Thermostats that require manual changeover between heating and cooling modes.

(2) Special occupancy or special applications where wide temperature ranges are not acceptable (such as retirement homes, process applications, museums, some areas of hospitals) and are approved by the Authority Having Jurisdiction. [ASHRAE 90.1:6.4.3.1.2]

E 503.4.6.2 Setpoint Overlap Restriction. Where heating and cooling to a zone are controlled by separate zone thermostatic controls located within the zone, means (such as limit switches, mechanical stops, or, for DDC systems, software programming) shall be provided to prevent the heating setpoint from exceeding the cooling setpoint minus any applicable proportional band. [ASHRAE 90.1:6.4.3.2]

E 503.4.6.3 Off-Hour Controls. HVAC systems shall have the off-hour controls required by Section E 503.4.6.3.1 through Section E 503.4.6.3.4.

Exceptions:

(1) HVAC systems intended to operate continuously.

(2) HVAC systems having a design heating capacity and cooling capacity less than 15 000 Btu/h (4.4 kW) that are equipped with readily accessible manual ON/OFF controls. [ASHRAE 90.1:6.4.3.3]

E 503.4.6.3.1 Automatic Shutdown. HVAC systems shall be equipped with not less than one of the following:

(1) Controls that can start and stop the system under different time schedules for seven different day-types per week, are capable of retaining programming and time setting during loss of power for a period of not less than 10 hours, and include an accessible manual override, or equivalent function, that allows temporary operation of the system for up to 2 hours.

(2) An occupant sensor that is capable of shutting the system off where no occupant is sensed for a period of up to 30 minutes.

(3) A manually operated timer capable of being adjusted to operate the system for up to 2 hours.

(4) An interlock to a security system that shuts the system off where the security system is activated.

Exception: Residential occupancies shall be permitted to use controls that can start and stop the system under two different time schedules per week. [ASHRAE 90.1:6.4.3.3.1]

E 503.4.6.3.2 Setback Controls. Heating systems shall be equipped with controls capable of and configured to automatically restart and temporarily operate the system as required to maintain zone temperatures above an adjustable heating setpoint of not less than 10°F (6°C) below the occupied heating setpoint. Cooling systems shall be equipped with controls capable of and configured to automatically restart and temporarily operate the mechanical cooling system as required to maintain zone temperatures below an adjustable cooling setpoint of not less than 5°F (3°C) above the occupied cooling setpoint or to prevent high space humidity levels.

Exception: Radiant heating systems capable of and configured with a setback heating setpoint at not less than 4°F (2°C) below the occupied heating setpoint. [ASHRAE 90.1:6.4.3.3.2]

E 503.4.6.3.3 Optimum Start Controls. Individual heating and cooling systems with setback controls and DDC shall have optimum start controls. The control algorithm shall, as a minimum, be a function of the difference between space temperature and occupied setpoint, the outdoor temperature, and the amount of time prior to scheduled occupancy. Mass radiant floor slab systems shall incorporate floor temperature into the optimum start algorithm. [ASHRAE 90.1:6.4.3.3.3]

E 503.4.6.3.4 Zone Isolation. HVAC systems serving zones that are intended to operate or be occupied nonsimultaneously shall be divided into isolation areas. Zones shall be permitted to be grouped into a single isolation area provided it does not exceed 25 000 square feet (2322.6 m^2) of conditioned floor area and does not include more than one floor. Each isolation area shall be equipped with isolation devices capable of and configured to automatically shut off the supply of conditioned air and outdoor air to and exhaust air from the area. Each isolation area shall be controlled independently by a device meeting the requirements of Section E 503.4.6.3.1. For central systems and plants, controls and devices shall be provided to allow stable system and equipment operation for any length of time while serving only the smallest isolation area served by the system or plant.

Exceptions: Isolation devices and controls are not required for the following:

(1) Exhaust air and outdoor air connections to isolation zones where the fan system to which they connect is not more than 5000 ft^3/min (2.3597 m^3/s).

(2) Exhaust airflow from a single isolation zone of less than 10 percent of the design airflow of the exhaust system to which it connects.

(3) Zones intended to operate continuously or intended to be inoperative only when all other zones are inoperative. [ASHRAE 90.1:6.4.3.3.4]

E 503.4.6.3.5 Automatic Control of HVAC in Hotel/Motel Guest Rooms. Hotels and motels with more than 50 guest rooms shall be provided with automatic controls for the HVAC equipment serving each guest room capable of and configured according to the requirements in Section E 503.4.6.3.1. [ASHRAE 90.1:6.4.3.3.5]

E 503.4.6.3.5.1 Guest Room HVAC Set-Point Control. Within 30 minutes of all occupants leaving the guest room, HVAC set points shall be automatically raised by not less than 4°F (2°C) from the occupant set point in the cooling mode and automatically lowered by at least 4°F (2°C) from the occupant set point in the heating mode. When the guest room is unrented and unoccupied, HVAC set points shall be automatically reset to 80°F (27°C) or higher in the cooling mode and to 60°F (16°C) or lower in the heating mode. Unrented and unoccupied guest rooms shall be determined by either of the following:

(1) The guest room has been continuously unoccupied for up to 16 hours.

(2) A networked guest room control system indicates the guest room is unrented and the guest room is unoccupied for no more than 30 minutes.

Exceptions:

(1) A networked guest room control system shall be permitted to return the thermostat set points to their default occupied set points 60 minutes prior to the time the room is scheduled to be occupied.

(2) Cooling for humidity control shall be permitted during unoccupied periods.

E 503.4.6.4 Ventilation System Controls. Stair and elevator shaft vents shall be equipped with motorized dampers that are capable of and configured to automatically close during normal building operation and are interlocked to open as required by fire and smoke detection systems. [ASHRAE 90.1:6.4.3.4.1]

E 503.4.6.4.1 Shutoff Damper Controls. Outdoor air intake and exhaust systems shall be equipped with motorized dampers that will automatically shut when the systems or spaces served are not in use. Ventilation outdoor air and exhaust or relief dampers shall be capable of and configured to automatically shut off during preoccupancy building warm-up, cooldown, and setback, except when ventilation reduces energy costs or when ventilation shall be supplied to comply with the code requirements.

Exceptions:

(1) Back-draft gravity (nonmotorized) dampers shall be permitted for exhaust and relief in buildings less than three stories in height and for ventilation air intakes and exhaust and relief dampers in buildings of any height located in Climate Zones 0, 1, 2 and 3. Back-draft dampers for ventilation air intakes shall be protected from direct exposure to wind.

(2) Back-draft gravity (nonmotorized) dampers shall be permitted in systems with a design outdoor air intake or exhaust capacity of 300 ft^3/min (0.142 m^3/s) or less.

(3) Dampers shall not be required in ventilation or exhaust systems serving unconditioned spaces.

(4) Dampers shall not be required in exhaust systems serving Type 1 kitchen exhaust hoods. [ASHRAE 90.1:6.4.3.4.2]

E 503.4.6.4.2 Dampers Leakage. Where outdoor air supply, and exhaust or relief dampers are required in Section E 503.4.6.4, they shall have a maximum leakage rate in accordance with Table E 503.4.6.4.2 where tested in accordance with AMCA 500D. [ASHRAE 90.1:6.4.3.4.3]

E 503.4.6.4.3 Ventilation Fan Controls. Fans with motors more than 0.75 hp (0.56 kW) shall have automatic controls in accordance with Section E 503.4.6.3.1 that are capable of and configured to shut off fans when not required.

Exception: HVAC systems intended to operate continuously. [ASHRAE 90.1:6.4.3.4.4]

E 503.4.6.5 Enclosed Parking Garage Ventilation. Enclosed parking garage ventilation systems shall automatically detect contaminant levels and stage fans or modulate fan airflow rates to 50 percent or less of design capacity, provided acceptable contaminant levels are maintained.

Exceptions:

(1) Garages less than 30 000 square feet (2787.09 m^2) with ventilation systems that do not utilize mechanical cooling or mechanical heating.

(2) Garages that have a garage area to ventilation system motor nameplate horsepower ratio that exceeds 1500 square feet per horsepower (ft^2/hp) (186.8 m^2/kW) and do not utilize mechanical cooling or mechanical heating.

(3) Where not permitted by the Authority Having Jurisdiction. [ASHRAE 90.1:6.4.3.4.5]

E 503.4.6.6 Heat Pump Auxiliary Heat Control. Heat pumps equipped with internal electric resistance heaters shall have controls that prevent supplemental heater operation where the heating load is capable of being met by the heat pump alone during both steady-state operation and setback recovery. Supplemental heater operation shall be permitted during outdoor coil defrost cycles.

Exception: Heat pumps whose minimum efficiency is regulated by U.S. National Appliance Energy Conservation Act

**TABLE E 503.4.6.4.2
MAXIMUM DAMPER LEAKAGE
(cubic foot per minute per square foot) at 1.0 in. w.g
[ASHRAE 90.1: TABLE 6.4.3.4.3]**

	VENTILATION AIR INTAKE		EXHAUST/RELIEF	
CLIMATE ZONE	NONMOTORIZED*	MOTORIZED	NONMOTORIZED*	MOTORIZED
0, 1, 2	–	–	–	–
any height	20	4	20	4
3	–	–	–	–
any height	20	10	20	10
4, 5b, 5c	–	–	–	–
less than 3 stories	not allowed	10	20	10
3 or more stories	not allowed	10	not allowed	10
5a, 6, 7, 8	–	–	–	–
less than 3 stories	not allowed	4	20	4
3 or more stories	not allowed	4	not allowed	4

For SI units: 1 cubic foot per minute = 0.00047 m^3/s, 1 square foot = 0.0929 m^2, 1 inch water gauge = 0.249 kPa

* Dampers smaller than 24 inches (610 mm) in either dimension shall be permitted to have leakage of 40 ft^3/min per square foot [0.203 $(m^3/s)/m^2$].

(NAECA) and whose ratings are in accordance with the requirements shown in Table E 503.7.1(2) and includes the use of an internal electric resistance heating. [ASHRAE 90.1:6.4.3.5]

E 503.4.6.7 Humidification and Dehumidification. Humidity control shall prevent the use of fossil fuel or electricity to produce relative humidity above 30 percent in the warmest zone served by the humidification system and to reduce relative humidity below 60 percent in the coldest zone served by the dehumidification system. Where a zone is served by a system or systems with both humidification and dehumidification capability, means (such as limit switches, mechanical stops, or, for DDC systems, software programming) shall be provided capable of and configured to prevent simultaneous operation of humidification and dehumidification equipment.

Exceptions:

(1) Zones served by desiccant systems, used with direct evaporative cooling in series.

(2) Systems serving zones where specific humidity levels are required, such as museums and hospitals, and approved by the Authority Having Jurisdiction or required by accreditation standards and humidity controls are capable of and configured to maintain a dead band of at least 10 percent relative humidity where no active humidification or dehumidification takes place.

(3) Systems serving zones where humidity levels are required to be maintained with precision of not more than ± 5 percent relative humidity to comply with applicable codes or accreditation standards or as approved by the Authority Having Jurisdiction. [ASHRAE 90.1:6.4.3.6]

E 503.4.6.8 Freeze Protection and Snow or Ice Melting Systems. Freeze protection systems, such as heat tracing of outdoor piping and heat exchangers, including self-regulating heat tracing, shall include automatic controls capable of and configured to shut off the systems when outdoor air temperatures are more than 40°F (4°C) or when the conditions of the protected fluid will prevent freezing. Snow and ice melting systems shall include automatic controls capable of and configured to shut off the systems when the pavement temperature is more than 50°F (10°C) and no precipitation is falling, and an automatic or manual control that will allow shutoff when the outdoor temperature is more than 40°F (4°C) so that the potential for snow or ice accumulation is negligible. [ASHRAE 90.1:6.4.3.7]

E 503.4.6.9 Ventilation Controls for High-Occupancy Areas. Demand control ventilation (DCV) shall be required for spaces that are more than 500 square feet (46.45 m^2) and with a design occupancy for ventilation of not less than 25 people per 1000 square feet (92.9 m^2) of floor area and served by systems with one or more of the following:

(1) Air-economizer.

(2) Automatic modulating control of outdoor air damper.

(3) Design outdoor airflow more than 3000 ft^3/min (1.4158 m^3/s).

Exceptions:

(1) Systems with exhaust air energy recovery in accordance with Section E 503.5.10.

(2) Multiple-zone systems without DDC of individual zones communicating with a central control panel.

(3) Systems with a design outdoor airflow less than 750 ft^3/min (0.3540 m^3/s).

(4) Spaces where more than 75 percent of the space design outdoor airflow is required for makeup air that is exhausted from the space or transfer air that is required for makeup air that is exhausted from other spaces.

(5) Spaces with one of the following occupancy categories in accordance with Chapter 4 or ASHRAE 62.1: correctional cells, daycare sickrooms, science labs, barbers, beauty and nail salons, and bowling alley seating. [ASHRAE 90.1:6.4.3.8]

E 503.4.6.10 Outdoor Heating. Radiant heat systems shall be used to provide heat outdoors. Outdoor radiant heating systems shall be provided with controls that sense the presence of occupants or other device that automatically shuts down the system where no occupants are in the heating area.

E 503.4.6.11 Heated or Cooled Vestibules. Heating for vestibules and for air curtains with integral heating shall include automatic controls capable of and configured to shut off the heating system when outdoor air temperatures are more than 45°F (7.2°C) Vestibule heating and cooling systems shall be controlled by a thermostat in the vestibule capable of and configured to limit heating to a maximum of 60°F (15.5°C) and cooling to a minimum of 85°F (29.4°C).

Exception: Heating or cooling provided by site-recovered energy or by transfer air that would otherwise be exhausted. [ASHRAE 90.1:6.4.3.9]

E 503.4.6.12 Direct Digital Control (DDC) Requirements. Direct digital control shall be required in accordance with Section E 503.4.6.12.1 through Section E 503.4.6.12.3. [ASHRAE 90.1:6.4.3.10]

E 503.4.6.12.1 DDC Applications. DDC shall be provided in the applications and qualifications in accordance with Table E 503.4.6.12.1.

Exception: DDC is not required for systems using the simplified approach to compliance in accordance with Section E 503.3. [ASHRAE 90.1:6.4.3.10.1]

E 503.4.6.12.2 DDC Controls. Where DDC is required by Section E 503.4.6.12.1, the DDC system shall be capable of and configured with all of the following, as required, to provide the control logic required in Section E 503.5:

(1) Monitoring zone and system demand for fan pressure, pump pressure, heating, and cooling.
(2) Transferring zone and system demand information from zones to air distribution system controllers and from air distribution systems to heating and cooling plant controllers.
(3) Automatically detecting those zones and systems that are capable of excessively driving the reset logic and generate an alarm or other indication to the system operator.
(4) Readily allowing operator removal of zone(s) from the reset algorithm. [ASHRAE 90.1:6.4.3.10.2]

E 503.4.6.12.3 DDC Display. Where DDC is required in accordance with Section E 503.4.6.12.1 for new buildings, the DDC system shall be capable of trending and graphically displaying input and output points. [ASHRAE 90.1:6.4.3.10.3]

E 503.4.6.13 Economizer Fault Detection Diagnostics (FDD). Air-cooled direct-expansion cooling units listed in Tables E 503.7.1(1) and E 503.7.1(2), where an air economizer is installed in accordance with Section E 503.5, shall include a fault detection and diagnostics (FDD) system complying with the following:

(1) The following temperature sensors shall be permanently installed to monitor system operation:
    (a) Outdoor air
    (b) Supply air
    (c) Return air, where required for economizer control
(2) The system shall have the capability of displaying the value of each sensor.
(3) The FDD system or unit controls shall be capable of and configured to provide system status by indicating the following:
    (a) Free cooling available
    (b) Economizer enabled

TABLE E 503.4.6.12.1
DDC APPLICATIONS AND QUALIFICATIONS
[ASHRAE 90.1:6.4.3.10.1]

BUILDING STATUS	APPLICATION	QUALIFICATIONS
New building	Air-handling system and all zones served by the system	Individual systems supplying more than three zones and with fan system bhp of 10 hp or more
New building	Chilled-water plant and all coils and terminal units served by the system	Individual plants supplying more than three zones and with design cooling capacity of 300 000 Btu/h or more
New building	Hot-water plant and all coils and terminal units served by the system	Individual plants supplying more than three zones and with design heating capacity of 300 000 Btu/h or more
Alteration or addition	Zone terminal unit such as VAV box	Where existing zones served by the same air-handling, chilled-water, or hot-water system have DDC
Alteration or addition	Air-handling system or fan coil	Where existing air-handling system(s) and fan-coil(s) served by the same chilled- or hot-water plant have DDC
Alteration or addition	New air-handling system and all new zones served by the system	Individual systems with fan system bhp of 10 hp or more and supplying more than three zones and more than 75 percent of zones are new
Alteration or addition	New or upgraded chilled-water plant	Where all chillers are new and plant design cooling capacity is 300 000 Btu/h or more
Alteration or addition	New or upgraded hot-water plant	Where all boilers are new and plant design heating capacity is 300 000 Btu/h or more

For SI units: 1000 British thermal units = 0.293 kW, 1 horsepower = 0.746 kW

(c) Compressor enabled

(d) Heating enabled

(e) Mixed-air low-limit cycle active

(4) The FDD system or unit controls shall have provisions to manually initiate each operating mode so that the operation of compressors, economizers, fans, and the heating system can be independently tested and verified.

(5) The FDD system shall be capable of and configured to detect the following faults:

(a) Air temperature sensor failure/fault

(b) Not economizing when the unit should be economizing

(c) Economizing when the unit should not be economizing

(d) Damper not modulating

(e) Excess outdoor air

(6) The FDD system shall be capable of and configured to report faults to a fault management application or DDC system accessible by operating or service personnel, or annunciated locally on zone thermostats. [ASHRAE 90.1:6.4.3.12]

E 503.4.7 HVAC System Construction and Insulation. HVAC Ducts shall be constructed in accordance with provisions contained in the SMACNA HVAC Duct Construction Standard. HVAC system construction and insulation shall comply with Section E 503.4.7.1 and Section E 503.4.7.2.

E 503.4.7.1 Insulation. Insulation required by this section shall be installed in accordance with industry-accepted standards. These requirements shall not apply to HVAC equipment. Insulation shall be protected from damage, including that due to sunlight, moisture, equipment maintenance, and wind, but not limited to the following:

(1) Insulation exposed to weather shall be suitable for outdoor service (e.g., protected by aluminum, sheet metal, painted canvas, or plastic cover). Cellular foam insulation shall be protected as above or painted with a coating that is water retardant and provides shielding from solar radiation that is capable of causing degradation of the material.

(2) Insulation covering chilled-water piping, refrigerant suction piping, or cooling ducts located outside the conditioned space shall include a vapor retardant located outside the insulation (unless the insulation is inherently vapor retardant), penetrations and joints of which shall be sealed. [ASHRAE 90.1:6.4.4.1.1]

E 503.4.7.1.1 Duct and Plenum Insulation. Supply and return ducts and plenums installed as part of an HVAC air distribution system shall be thermally insulated in accordance with Table E 503.7.2.

Exceptions:

(1) Factory-installed plenums, casings, or ductwork furnished as a part of HVAC equipment tested and rated in accordance with Section E 503.4 through Section E 503.4.4.1.

(2) Ducts or plenums located in heated spaces, semi-heated spaces, or cooled spaces.

(3) For runouts less than 10 feet (3048 mm) in length to air terminals or air outlets, the rated R-value of insulation shall not be required to exceed R-3.5.

(4) Backs of air outlets and outlet plenums exposed to unconditioned or indirectly conditioned spaces with face areas exceeding 5 square feet (0.5 m^2) shall not be required to exceed R-2; those not exceeding 5 square feet (0.5 m^2) shall not be required to be insulated. [ASHRAE 90.1:6.4.4.1.2]

E 503.4.7.1.2 Piping Insulation. Piping shall be thermally insulated in accordance with Table E 503.7.3(1) and Table E 503.7.3(2).

Exceptions:

(1) Factory-installed piping within HVAC equipment tested and rated in accordance with Section E 503.4 through Section E 503.4.4.1.

(2) Piping that conveys fluids having a design operating temperature range between 60°F (16°C) and 105°F (41°C), inclusive.

(3) Piping that conveys fluids that have not been heated or cooled through the use of fossil fuels or electricity (such as roof and condensate drains, domestic cold water supply, and natural gas piping).

(4) Where heat gain or heat loss will not increase energy use (such as liquid refrigerant piping).

(5) In piping 1 inch (25.4 mm) or less, insulation is not required for strainers, control valves, and balancing valves. [ASHRAE 90.1:6.4.4.1.3]

E 503.4.7.1.3 Sensible Heating Panel. Thermally ineffective panel surfaces of sensible heating panels, including U-bends and headers, shall be insulated a minimum of R-3.5. Adjacent building envelope insulation counts toward this requirement. [ASHRAE 90.1:6.4.4.1.4]

E 503.4.7.1.4 Radiant Floor Heating. The bottom surfaces of floor structures incorporating radiant heating shall be insulated with a minimum of R-3.5. Adjacent building envelope insulation counts toward this requirement.

Exception: Heated slab-on-grade floors incorporating radiant heating shall be in accordance with ASHRAE 90.1. [ASHRAE 90.1:6.4.4.1.5]

E 503.4.7.2 Ductwork and Plenum Leakage. Transverse joints, longitudinal seams, and duct wall penetrations shall be sealed. Pressure-sensitive tape shall not be used as the primary sealant, unless it has been certified to comply with UL 181A or UL 181B by an independent testing laboratory and the tape is used in accordance with that certification. All other connections shall be considered transverse joints, including but not limited to spin-ins, taps, other branch connections, access door frames and jambs, and duct connections to equipment.

Exceptions:

(1) Rods that penetrate the duct wall that shall be permitted

to move in order to function properly (control rod for volume damper) shall not be sealed in a fashion that prevents them from working properly.

(2) Spiral lock seams in a round or flat oval duct.

E 503.4.7.2.1 Duct Leakage Tests. Ductwork shall be leak-tested in accordance with the SMACNA HVAC Air Duct Leakage Test Manual. Representative sections totaling not less than 20 percent of the total installed duct area shall be tested. Where the tested 20 percent fail to comply with the requirements of this section, then 40 percent of the total installed duct area shall be tested. Where the tested 40 percent fail to comply with the requirements of this section, then 100 percent of the total installed duct area shall be tested. Sections shall be selected by the building owner or designated representative of the building owner. Positive pressure leakage testing shall be permitted for negative pressure ductwork. The permitted duct leakage shall be not more than the following:

$$L_{max} = C_L P^{0.65} \quad \text{(Equation E 503.4.7.2.1)}$$

Where:

L_{max} = maximum permitted leakage, (ft³/min)/100 square feet [0.0001 (m³/s)/m²] duct surface area.

C_L = Six, duct leakage class, (ft³/min)/100 square feet [0.0001 (m³/s)/m²] duct surface area at 1 inch water column (0.2 kPa).

P = test pressure, which shall be equal to the design duct pressure class rating, inch water column (kPa).

E 503.5 Prescriptive Path, Economizers. Cooling systems shall include either an air economizer or fluid economizer in accordance with Section E 503.5.1 through Section E 503.5.4.1.

Exceptions: Economizers shall not be required for the following systems:

(1) Individual fan-cooling units with a supply capacity less than the minimum listed in Table E 503.5(1).

(2) Chilled-water cooling systems without a fan or that use induced airflow, where the total capacity of these systems is less than 1 000 000 Btu/h (293 kW) in Climate Zones 0, 1B, and 2 through 4; less than 1 400 000 Btu/h (410 kW) in Climate Zones 5 through 8; or any size in Climate Zone 1A.

(3) Systems that include nonparticulate air treatment in accordance with ASHRAE 62.1.

(4) In hospitals and ambulatory surgery centers, where more than 75 percent of the air designed to be supplied by the system is to spaces that are required to be humidified more than 35°F (2°C) dew-point temperature to comply with applicable codes or accreditation standards; in all other buildings, where more than 25 percent of the air designed to be supplied by the system is to spaces that are designed to be humidified more than 35°F (2°C) dew-point temperature to satisfy process needs. This exception shall not apply to computer rooms.

(5) Systems that include a condenser heat recovery system with a minimum capacity in accordance with Section E 503.5.10.1.2.

(6) Systems that serve residential spaces where the system capacity is less than five times the requirement listed in Table E 503.5(1).

(7) Systems that serve spaces whose sensible cooling load at design conditions, excluding transmission and infiltration loads, is less than or equal to transmission and infiltration losses at an outdoor temperature of 60°F (16°C).

(8) Systems expected to operate less than 20 hours per week.

(9) Where the use of outdoor air for cooling will affect supermarket open refrigerated casework systems.

(10) For comfort cooling where the cooling efficiency is not less than the efficiency improvement requirements in accordance with Table E 503.5(2).

(11) Systems primarily serving computer rooms where in accordance with one of the following:

(a) The total design cooling load of all computer rooms in the building is less than 3 000 000 Btu/h (879 kW) and the building in which they are located is not served by a centralized chilled water plant.

TABLE E 503.5(1)
MINIMUM FAN-COOLING UNIT SIZE WHERE AN ECONOMIZER IS REQUIRED
[ASHRAE 90.1: TABLE 6.5.1-1]

CLIMATE ZONES	COOLING CAPACITY WHERE AN ECONOMIZER IS REQUIRED
0A, 0B, 1A, 1B	No economizer requirement
2A, 2B, 3A, 4A, 5A, 6A, 3B, 3C, 4B, 4C, 5B, 5C, 6B, 7, 8	≥54 000 Btu/h

For SI units: 1000 British thermal units per hour = 0.293 kW

TABLE E 503.5(2)
ELIMINATE REQUIRED ECONOMIZER FOR COMFORT COOLING BY INCREASING COOLING EFFICIENCY
[ASHRAE 90.1: TABLE 6.5.1-2]

CLIMATE ZONES	EFFICIENCY IMPROVEMENT*
2A	17%
2B	21%
3A	27%
3B	32%
3C	65%
4A	42%
4B	49%
4C	64%
5A	49%
5B	59%
5C	74%
6A	56%
6B	65%
7	72%
8	77%

* Where a unit is rated with an IPLV, IEER or SEER, to eliminate the required economizer, the minimum cooling efficiency of the HVAC unit shall be increased by the percentage shown. Where the HVAC unit is rated with a full load metric like EER cooling, these shall be increased by the percentage shown.

(b) The room total design cooling load is less than 600 000 Btu/h (176 kW) and the building in which they are located is served by a centralized chilled water plant.

(c) The local water authority does not permit cooling towers.

(d) Less than 600 000 Btu/h (176 kW) of computer room cooling equipment capacity is being added to an existing building.

(12) Dedicated systems for computer rooms where a minimum of 75 percent of the design load serves one of the following:

(a) Spaces classified as an essential facility.

(b) Spaces having a design of Tier IV in accordance with TIA 942.

(c) Spaces classified as Critical Operations Power Systems (COPS) in accordance with NFPA 70.

(d) Spaces where core clearing and settlement services are performed such that their failure to settle pending financial transactions is capable of systemic risk in accordance with "The Interagency Paper on Sound Practices to Strengthen the Resilience of the US Financial System" (April 7, 2003). [ASHRAE 90.1:6.5.1]

E 503.5.1 Air Economizers, Design Capacity. Air economizer systems shall be capable of and configured to modulate outdoor air and return air dampers to provide up to 100 percent of the design supply air quantity as outdoor air for cooling. [ASHRAE 90.1:6.5.1.1.1]

E 503.5.1.1 Control Signal. Economizer controls shall be capable of and configured to sequence the dampers with the mechanical cooling equipment and shall not be controlled by only mixed air temperature.

Exception: The use of mixed air temperature limit control shall be permitted for systems controlled from space temperature (such as single-zone systems). [ASHRAE 90.1:6.5.1.1.2]

E 503.5.1.2 High-Limit Shutoff. Air economizers shall be capable of and configured to automatically reduce outdoor air intake to the design minimum outdoor air quantity where outdoor air intake will no longer reduce cooling energy use. High-limit shutoff control types and associated setpoints for specific climate zones shall be chosen from Table E 503.5.1.2. [ASHRAE 90.1:6.5.1.1.3]

E 503.5.1.3 Dampers. Return, exhaust or relief, and outdoor air dampers shall comply with Section E 503.4.6.4.2. [ASHRAE 90.1:6.5.1.1.4]

E 503.5.1.4 Relief of Excess Outdoor Air. Systems shall provide a means to relieve excess outdoor air during air economizer operation to prevent overpressurizing the building. The relief air outlet shall be located so as to avoid recirculation into the building. [ASHRAE 90.1:6.5.1.1.5]

E 503.5.1.5 Sensor Accuracy. Outdoor air, return air, mixed air, and supply air sensors shall be calibrated within the following accuracies:

(1) Dry-bulb and wet-bulb temperatures shall be accurate to ±2°F (1.1°C) over the range of 40°F (4.4°C) to 80°F (27°C).

(2) Enthalpy and the value of a differential enthalpy sensor shall be accurate to ±3 Btu/lb (7 E+03 J/kg) over the range of 20 Btu/lb (4.6 E+04 J/kg) to 36 Btu/lb (8.4 E+04 J/kg).

(3) Relative humidity shall be accurate to ±5 percent over the range of 20 percent to 80 percent relative humidity. [ASHRAE 90.1:6.5.1.1.6]

E 503.5.2 Fluid Economizers, Design Capacity. Fluid economizer systems shall be capable of providing up to 100 percent of the expected system cooling load at outdoor air temperatures of not more than 50°F (10°C) dry bulb or 45°F (7°C) wet bulb.

TABLE E 503.5.1.2
HIGH-LIMIT SHUTOFF CONTROL SETTINGS FOR AIR ECONOMIZERS[2]
[ASHRAE 90.1: TABLE 6.5.1.1.3]

CONTROL TYPE	ALLOWED ONLY IN CLIMATE ZONE AT LISTED SETPOINT	REQUIRED HIGH LIMIT (ECONOMIZER OFF WHERE):	
		EQUATION	DESCRIPTION
Fixed dry bulb temperature	0B, 1B, 2B, 3B, 3C, 4B, 4C, 5B, 5C, 6B, 7, 8	$T_{oa} > 75°F$	Outdoor air temperature exceeds 75°F
	5A, 6A	$T_{oa} > 70°F$	Outdoor air temperature exceeds 70°F
	0A, 1A, 2A, 3A, 4A	$T_{oa} > 65°F$	Outdoor air temperature exceeds 65°F
Differential dry bulb temperature	0B, 1B, 2B, 3B, 3C, 4B, 4C, 5A, 5B, 5C, 6A, 6B, 7, 8	$T_{oa} > T_{ra}$	Outdoor air temperature exceeds return air temperature
Fixed enthalpy with fixed dry-bulb temperature	All	$h_{oa} > 28$ Btu/lb[1] or $T_{oa} > 75°F$	Outdoor air enthalpy exceeds 28 Btu/lb[1] of dry air[1] or outdoor air temperature exceeds 75°F
Differential enthalpy with fixed dry-bulb temperature	All	$h_{oa} > h_{ra}$ or $T_{oa} > 75°F$	Outdoor air enthalpy exceeds return air enthalpy or outdoor air temperature exceeds 75°F

For SI units: °C = (°F-32)/1.8, 1 British thermal unit per pound = 2326 J/kg

Notes:

[1] At altitudes substantially different than sea level, the fixed enthalpy limit shall be set to the enthalpy value at 75°F (24°C) and 50 percent relative humidity. As an example, at approximately 6000 feet (1829 m) elevation, the fixed enthalpy limit shall be approximately 30.7 Btu/lb (71 408 J/kg).

[2] Devices with selectable rather than adjustable setpoints shall be capable of being set to within 2°F (1°C) and 2 Btu/lb (4649 J/kg) of the setpoint listed.

Exceptions:

(1) Systems primarily serving computer rooms in which 100 percent of the expected system cooling load at the dry bulb and wet bulb temperatures in accordance with Table E 503.5.2 is met with water-cooled fluid economizers.

(2) Systems primarily serving computer rooms in which 100 percent of the expected system cooling load at the dry bulb temperatures listed in Table E 503.5.2 is met with air-cooled fluid economizers.

(3) Systems where dehumidification requirements are not capable of being met using outdoor air temperatures of 50°F (10°C) dry bulb or 45°F (7°C) wet bulb and where 100 percent of the expected system cooling load at 45°F (7°C) dry bulb or 40°F (4°C) wet bulb is met with water-cooled fluid economizers. [ASHRAE 90.1:6.5.1.2.1]

E 503.5.2.1 Maximum Hydronic Pressure Drop. Precooling coils and fluid-to-water heat exchangers used as part of a fluid economizer system shall either have a water-side pressure drop of less than 15 feet of water (45 kPa), or a secondary loop shall be created so that the coil or heat exchanger pressure drop is not seen by the circulating pumps where the system is in the normal cooling (non-economizer) mode. [ASHRAE 90.1:6.5.1.2.2]

E 503.5.3 Integrated Economizer Control. Economizer systems shall be integrated with the mechanical cooling system and be capable of and configured to provide partial cooling even where additional mechanical cooling is required to be in accordance with the remainder of the cooling load. Controls shall not false load the mechanical cooling systems by limiting or disabling the economizer or by other means, such as hot gas bypass, except at the lowest stage of mechanical cooling.

Units that include an air economizer shall comply with the following:

(1) Unit controls shall have the mechanical cooling capacity control interlocked with the air economizer controls such that the outdoor air damper is at the 100 percent open position when mechanical cooling is on, and the outdoor air damper does not begin to close to prevent coil freezing due to minimum compressor run time until the leaving air temperature is less than 45°F (7°C).

(2) DX units with a rated capacity no less than 65 000 Btu/h (19 kW) that control the capacity of the mechanical cooling directly based on occupied space temperature shall have not less than two stages of mechanical cooling capacity.

(3) Other DX units, including those that control space temperature by modulating the airflow to the space, shall comply with the requirements of Table E 503.5.3. [ASHRAE 90.1:6.5.1.3]

E 503.5.4 Economizer Heating System Impact. HVAC system design and economizer controls shall be such that economizer operation does not increase the building heating energy use during normal operation.

Exception: Economizers on variable air valve (VAV) systems that cause zone level heating to increase due to a reduction in supply air temperature. [ASHRAE 90.1:6.5.1.4]

TABLE E 503.5.2
WATER ECONOMIZER SIZING DRY-BULB AND WET-BULB REQUIREMENTS FOR COMPUTER ROOMS*
[ASHRAE 90.1: TABLE 6.5.1.2.1]

CLIMATE ZONE		WATER COOLED		AIR COOLED
		DRY BULB, °F	WET BULB, °F	DRY BULB, °F
0	A	NR	NR	NR
0	B	NR	NR	NR
1	A	NR	NR	NR
1	B	NR	NR	NR
2	A	40.0	35.0	30.0
2	B	35.0	30.0	30.0
3	A	40.0	35.0	25.0
3	B	30.0	25.0	25.0
3	C	30.0	25.0	30.0
4	A	40.0	35.0	25.0
4	B	30.0	25.0	25.0
4	C	30.0	25.0	25.0
5	A	40.0	35.0	20.0
5	B	30.0	25.0	20.0
5	C	30.0	25.0	25.0
6	A	35.0	30.0	20.0
6	B	30.0	25.0	20.0
7	—	30.0	25.0	20.0
8	—	30.0	25.0	20.0

For SI units: °C = (°F-32)/1.8
NR = Not Required

TABLE E 503.5.3
DX COOLING STAGE REQUIREMENTS FOR MODULATING AIRFLOW UNITS
[ASHRAE 90.1:6.5.1.3]

RATING CAPACITY, Btu/h	MINIMUM NUMBER OF MECHANICAL COOLING STAGES	MINIMUM COMPRESSOR DISPLACEMENT*
≥65 000 and <240 000	3	≤35% of full load
≥240 000	4	≤25% full load

For SI units: 1000 British thermal units = 0.293 kW

* For mechanical cooling stage control that does not use variable compressor displacement the percent displacement shall be equivalent to the mechanical cooling capacity reduction evaluated at the full load rating conditions for the compressor.

E 503.5.4.1 Economizer Humidification System Impact. Systems with hydronic cooling and humidification systems designed to maintain inside humidity at a dew-point temperature more than 35°F (2°C) shall use a fluid economizer where an economizer is required in accordance with Section E 503.5 through Section E 503.5.4.1. [ASHRAE 90.1:6.5.1.5]

E 503.5.5 Simultaneous Heating and Cooling Limitation, Zone Controls. Zone thermostatic controls shall prevent the following:

(1) Reheating.

(2) Recooling.

(3) Mixing or simultaneously supplying air that has been previously mechanically heated and air that has been previously cooled, either by mechanical cooling or by economizer systems.

(4) Other simultaneous operation of heating and cooling systems to the same zone.

Exceptions:

(1) Zones for which the volume of air that is reheated, recooled, or mixed is less than the larger of the following:

  (a) Twenty percent of the zone design peak supply for systems with DDC and 30 percent for other systems.

  (b) The outdoor airflow rate required to be in accordance with the ventilation requirements of Chapter 4 or ASHRAE 62.1 for the zone.

  (c) Any higher rate that can be demonstrated, to the satisfaction of the Authority Having Jurisdiction, to reduce overall system annual energy use by offsetting reheat/recool energy losses through a reduction in outdoor air intake for the system.

  (d) The airflow rate required to be in accordance with applicable codes or accreditation standards, such as pressure relationships or minimum air change rates.

(2) Zones with DDC that comply with the following:

  (a) The airflow rate in dead band between heating and cooling does not exceed the larger of the following:

    (1) Twenty percent of the zone design peak supply rate.

    (2) The outdoor airflow rate required to be in accordance with the ventilation requirements of Chapter 4 or ASHRAE 62.1 for the zone.

    (3) A higher rate that can be demonstrated, to the satisfaction of the Authority Having Jurisdiction, to reduce overall system annual energy use by offsetting reheat/recool energy losses through a reduction in outdoor air intake.

    (4) The airflow rate required with applicable codes or accreditation standards, such as pressure relationships or minimum air change rates.

  (b) The airflow rate that is reheated, recooled, or mixed shall be less than 50 percent of the zone design peak supply rate.

  (c) The first stage of heating consists of modulating the zone supply air temperature setpoint up to a maximum setpoint while the airflow is maintained at the dead band flow rate.

  (d) The second stage of heating consists of modulating the airflow rate from the dead band flow rate up to the heating maximum flow rate.

(3) Laboratory exhaust systems that comply with Section E 503.5.11.3.

(4) Zones where at least 75 percent of the energy for reheating or for providing warm air in mixing systems is provided from site-recovered energy (including condenser heat) or site-solar energy. [ASHRAE 90.1:6.5.2.1]

E 503.5.5.1 Supply Air Temperature Reheat Limit. Where reheating is permitted in accordance with this appendix, zones that have both supply and return or exhaust air openings more than 6 feet (1829 mm) above the floor shall not supply heating air more than 20°F (11°C) above the space temperature setpoint.

Exceptions:

(1) Laboratory exhaust systems in accordance with Section E 503.5.11.3.

(2) During preoccupancy building warm-up and setback. [ASHRAE 90.1:6.5.2.1.1]

E 503.5.5.2 Hydronic System Controls. The heating of fluids in hydronic systems that have been previously mechanically cooled and the cooling of fluids that have been previously mechanically heated shall be limited in accordance with Section E 503.5.5.2.1 through Section E 503.5.5.2.3. [ASHRAE 90.1:6.5.2.2]

E 503.5.5.2.1 Three-Pipe System. Hydronic systems that use a common return system for both hot water and chilled water shall not be used. [ASHRAE 90.1:6.5.2.2.1]

E 503.5.5.2.2 Two-Pipe Changeover System. Systems that use a common distribution system to supply both heated and chilled water are acceptable where in accordance with the following:

(1) The system is designed to allow a dead band between

changeover from one mode to the other of not less than 15°F (8°C) outdoor air temperature.

(2) The system is designed to operate and is provided with controls that will allow operation in one mode for not less than 4 hours before changing over to the other mode.

(3) Reset controls are provided that allow heating and cooling supply temperatures at the changeover point to be not more than 30°F (17°C) apart. [ASHRAE 90.1:6.5.2.2.2]

E 503.5.5.2.3 Hydronic (Water Loop) Heat Pump Systems. Hydronic heat pumps connected to a common heat pump water loop with central devices for heat rejection (e.g., cooling tower) and heat addition (e.g., boiler) shall have the following:

(1) Controls that are capable of and configured to provide a heat pump water supply temperature dead band of at least 20°F (11°C) between initiation of heat rejection and heat addition by the central devices (e.g., tower and boiler).

(2) For climate zone 3 through zone 8, where a closed-circuit cooling tower (fluid cooler) is used, either an automatic valve shall be installed to bypass all but a minimal flow of water around the tower (for freeze protection) or low-leakage positive closure dampers shall be provided. Where an open-circuit cooling tower is used directly in the heat pump loop, an automatic valve shall be installed to bypass all heat pump water flow around the tower. Where an open-circuit cooling tower is used in conjunction with a separate heat exchanger to isolate the tower from the heat pump loop then heat loss shall be controlled by shutting down the circulation pump on the cooling tower loop.

Exception: Where a system loop temperature optimization controller is used to determine the most efficient operating temperature based on real-time conditions of demand and capacity, dead bands of less than 20°F (11°C) shall be permitted. [ASHRAE 90.1:6.5.2.2.3]

E 503.5.5.3 Dehumidification. Where humidity controls are provided, such controls shall prevent reheating, mixing of hot and cold airstreams, or other means of simultaneous heating and cooling of the same airstream.

Exceptions:

(1) The system is capable of and configured to reduce supply air volume to 50 percent or less of the design airflow rate or the minimum outdoor air ventilation rate in accordance with Chapter 4 or ASHRAE 62.1 or other applicable federal, state, or local code or recognized standard, whichever is larger before simultaneous heating and cooling takes place.

(2) The individual fan cooling unit has a design cooling capacity of not more than 65 000 Btu/h (19 kW) and is capable of and configured to unload to 50 percent capacity before simultaneous heating and cooling takes place.

(3) The individual mechanical cooling unit has a design cooling capacity of not more than 40 000 Btu/h (11.7 kW). An individual mechanical cooling unit is a single system composed of a fan or fans and a cooling coil capable of providing mechanical cooling.

(4) Systems serving spaces where specific humidity levels are required to satisfy process needs, such as vivariums, museums, surgical suites, pharmacies, and buildings with refrigerating systems, such as supermarkets, refrigerated warehouses, and ice arenas, and where the building includes site-recovered energy or site solar energy that provide energy equal to 75 percent or more of the annual energy for reheating or for providing warm air in mixing systems. This exception shall not apply to computer rooms.

(5) Not less than 90 percent of the annual energy for reheating or for providing warm air in mixing systems is provided from site-recovered energy (including condenser heat) or site-solar energy.

(6) Systems where the heat added to the airstream is the result of the use of a desiccant system and 75 percent of the heat added by the desiccant system is removed by a heat exchanger, either before or after the desiccant system with energy recovery. [ASHRAE 90.1:6.5.2.3]

E 503.5.5.4 Humidifier Preheat. Humidifiers with preheating jackets mounted in the airstream shall be provided with an automatic valve to shut off preheat where humidification is not required. [ASHRAE 90.1:6.5.2.4.1]

E 503.5.5.4.1 Insulation. Humidification system dispersion tube hot surfaces in the airstreams of ducts or air-handling units shall be insulated with a product with an insulating value of not less than R-0.5.

Exception: Systems where mechanical cooling, including economizer operation, does not occur simultaneously with humidification. [ASHRAE 90.1:6.5.2.4.2]

E 503.5.5.5 Preheat Coils. Preheat coils shall have controls that stop their heat output where mechanical cooling, including economizer operation, is occurring. [ASHRAE 90.1:6.5.2.5]

E 503.5.6 Air System Design and Control. HVAC air system design and control shall be in accordance with the provisions of Section E 503.5.6.1 through Section E 503.5.6.6.

E 503.5.6.1 Fan System Power and Efficiency. Each HVAC system having a total fan system motor nameplate horsepower (kW) exceeding 5 hp (3.7 kW) at fan system design conditions shall not exceed the allowable fan system motor nameplate horsepower (kW) (Option 1) or fan system brake horsepower (kW) (Option 2) as shown in Table E 503.5.6.1(1). This shall include supply fans, return or relief fans, exhaust fans, and fan-powered terminal units associated with systems providing heating or cooling capability that operate at fan system design conditions. Single-zone VAV systems shall comply with the constant-volume fan power limitation.

Exceptions:

(1) Hospital, vivarium, and laboratory systems that utilize flow control devices on exhaust, return, or both to maintain space pressure relationships necessary for occupant health and safety, or environmental control shall be permitted to use variable-volume fan power limitation.

(2) Individual exhaust fans with motor nameplate horsepower of 1 hp (0.7 kW) or less. [ASHRAE 90.1:6.5.3.1.1]

E 503.5.6.1.1 Motor Nameplate Horsepower. For each fan, the selected fan motor shall be no larger than the first available motor size greater than the bhp. The fan bhp shall be indicated on the design documents to allow for compliance verification by the Authority Having Jurisdiction.

Exceptions:

(1) For fans less than 6 bhp (4.5 kW), where the first available motor larger than the bhp (kW) has a nameplate rating within 50 percent of the bhp (kW), the next larger nameplate motor size shall be permitted to be selected.

(2) For fans 6 bhp (4.5 kW) and larger, where the first available motor larger than the bhp (kW) has a nameplate rating within 30 percent of the bhp (kW), the next larger nameplate motor size shall be permitted to be selected.

(3) Systems that are in accordance with Section E 503.5.6.1, Option 1.

(4) Fans with motor nameplate horsepower of less than 1 hp (0.7 kW). [ASHRAE 90.1:6.5.3.1.2]

E 503.5.6.1.2 Fan Efficiency. Fans shall have a fan efficiency grade (FEG) of 67 or more, based on manufacturers' certified data in accordance with AMCA 205. The total efficiency of the fan at the design point of operation shall be

TABLE E 503.5.6.1(1)
FAN POWER LIMITATION*
[ASHRAE 90.1: TABLE 6.5.3.1-1]

	LIMIT	CONSTANT VOLUME	VARIABLE VOLUME
Option 1: Fan system motor nameplate (hp)	Allowable nameplate motor (hp)	$hp \leq CFM_S \cdot 0.0011$	$hp \leq CFM_S \cdot 0.0015$
Option 2: Fan system (bhp)	Allowable fan system (bhp)	$bhp \leq CFM_S \cdot 0.00094 + A$	$bhp \leq CFM_S \cdot 0.0013 + A$

For SI units: 1 horsepower = 0.746 kW, 1 cubic foot per minute = 0.00047 m^3/s

* Where:

CFM_S = the maximum design supply airflow rate to conditioned spaces served by the system in cubic feet per minute (m^3/s)
hp = the maximum combined motor nameplate horsepower (kW)
bhp = the maximum combined fan brake horsepower (kW)
A = sum of (PD x $CFM_D/4131$)
PD = each applicable pressure drop adjustment from Table E 503.5.6.1(2) in inch water column (kPa)
CFM_D = the design airflow through each applicable device from Table E 503.5.6.1(2) in cubic feet per minute (m^3/s)

TABLE E 503.5.6.1(2)
FAN POWER LIMITATION PRESSURE DROP ADJUSTMENT
[ASHRAE 90.1: TABLE 6.5.3.1-2]

DEVICE	ADJUSTMENT
CREDITS	
Return or exhaust systems required by code or accreditation standards to be fully ducted, or systems required to maintain air pressure differentials between adjacent rooms	0.5 in. w.c. (2.15 in w.c. for laboratory and vivarium systems)
Return, exhaust, or both airflow control devices	0.5 in. w.c.
Exhaust filters, scrubbers, or other exhaust treatment	The pressure drop of device calculated at fan system design condition
Particulate Filtration Credit: MERV 9 through 12	0.5 in. w.c.
Particulate Filtration Credit: MERV 13 through 15	0.9 in. w.c.
Particulate Filtration Credit: MERV 16 and greater, and electronically enhanced filters	Pressure drop calculated at 2x clean filter pressure drop at fan system design condition
Carbon and other gas-phase air cleaners	Clean filter pressure drop at fan system design condition
Biosafety cabinet	Pressure drop of device at fan system design condition
Energy recovery device, other than coil runaround loop	For each airstream [(2.2 x enthalpy recovery ratio) - 0.5] in w.c.
Coil runaround loop	0.6 in. w.c. for each airstream
Evaporative humidifier or cooler in series with another cooling coil	Pressure drop of device at fan system design condition
Sound attenuation section (fans serving spaces with design background noise goals below NC35)	0.15 in. w.c.
Exhaust system serving fume hoods	0.35 in. w.c.
Laboratory and vivarium exhaust systems in high-rise buildings	0.25 in. w.c. per 100 feet of vertical duct exceeding 75 ft
DEDUCTIONS	
Systems without central cooling device	–0.6 in. wc
Systems without central heating device	–0.3 in. wc
Systems with central electric resistance heat	–0.2 in. wc

For SI units: 1 inch water column = 0.249 kPa, 1 foot = 304.8 mm within 15 percentage points of the maximum total efficiency of the fan.

Exceptions:

(1) Individual fans with a motor nameplate horsepower of 5 hp (3.7 kW) or less that are not part of a group operated as the functional equivalent of a single fan.

(2) Multiple fans in series or parallel (e.g., fan arrays) that have a combined motor nameplate horsepower of 5 hp (3.7 kW) or less and are operated as the functional equivalent of a single fan.

(3) Fans that are part of equipment listed under Section E 503.4.

(4) Fans included in equipment bearing a third party-certified seal for air or energy performance of the equipment package.

(5) Powered wall/roof ventilators (PRV).

(6) Fans outside the scope of AMCA 205.

(7) Fans that are intended to only operate during emergency conditions. [ASHRAE 90.1:6.5.3.1.3]

E 503.5.6.2 Supply Fan Airflow Control. Each cooling system listed in Table E 503.5.6.2 shall be designed to vary the supply fan airflow as a function of load and shall comply with the following requirements:

(1) DX and chilled-water cooling units that control the capacity of the mechanical cooling directly based on space temperature shall have a minimum of two stages of fan control. Low or minimum speed shall not exceed 66 percent of full speed. At low or minimum speed, the fan system shall draw no more than 40 percent of the fan power at full fan speed. Low or minimum speed shall be used during periods of low cooling load and ventilation-only operation.

(2) All other units, including DX cooling units and chilled-water units that control the space temperature by modulating the airflow to the space, shall have modulating fan control. Minimum speed shall not exceed 50 percent of full speed. At minimum speed, the fan system shall draw no more than 30 percent of the power at full fan speed. Low or minimum speed shall be used during periods of low cooling load and ventilation-only operation.

(3) Units that include an air economizer to meet the requirements of Section E 503.5 through Section E 503.5.4.1 shall have a minimum of two speeds of fan control during economizer operation.

Exceptions:

(1) Modulating fan control shall not be required for chilled-water and evaporative cooling units with less than1 hp (0.7 kW) fan motors where the units are not used to provide ventilation air and where the indoor fan cycles with the load.

(2) Where the volume of outdoor air required to meet the ventilation requirements of Chapter 4 or ASHRAE 62.1 at low speed exceeds the air that would be delivered at the speed defined in Section E 503.5.6.2(1), or Section E 503.5.6.2(2), then the minimum speed shall be selected to provide the required ventilation air. [ASHRAE 90.1:6.5.3.2.1]

TABLE E 503.5.6.2
FAN AIRFLOW CONTROL
[ASHRAE 90.1: TABLE 6.5.3.2.1]

COOLING SYSTEM TYPE	FAN MOTOR SIZE, (hp)	MECHANICAL COOLING CAPACITY, (Btu/h)
DX cooling	Any	≥65 000
Chilled-water and evaporative cooling	≥¼	Any

For SI units: 1000 British thermal units per hour = 0.293 kW, 1 horsepower = 0.746 kW, 1 cubic foot per minute = 0.00047 m^3/s

E 503.5.6.2.1 VAV Static Pressure Sensor Location. Static pressure sensors used to control VAV fans shall be located such that the controller setpoint is not more than 1.2 inches water column (0.30 kPa). Where this results in the sensor being located downstream of major duct splits, sensors shall be installed in each major branch to ensure that static pressure is maintained in each.

Exception: Systems that are in accordance with Section E 503.5.6.2.2. [ASHRAE 90.1:6.5.3.2.2]

E 503.5.6.2.2 VAV Setpoint Reset. For multiple-zone VAV systems having a total fan system motor nameplate horsepower exceeding 5 hp (3.7 kW) with DDC of individual zones reporting to the central control panel, static pressure setpoint shall be reset based on the zone requiring the most pressure, such as the setpoint is reset lower until one zone damper is nearly wide open. Controls shall provide the following:

(1) Monitor zone damper positions or other indicator of need for static pressure.

(2) Automatically detect those zones that are capable of excessively driving the reset logic and generate an alarm to the system operator.

(3) Readily allow operator removal of zones from the reset algorithm. [ASHRAE 90.1:6.5.3.2.3]

E 503.5.6.3 Multiple-Zone VAV System Ventilation Optimization Control. Multiple-zone VAV systems with DDC of individual zone boxes reporting to a central control panel shall include means to automatically reduce outdoor air intake flow below design rates in response to changes in system ventilation efficiency in accordance with Section 404.0.

Exceptions:

(1) VAV systems with zonal transfer fans that recirculate air from other zones without directly mixing it with outdoor air, dual-duct dual-fan VAV systems, and VAV systems with fan-powered terminal units.

(2) Systems where total design exhaust airflow is more than 70 percent of total design outdoor air intake flow requirements. [ASHRAE 90.1:6.5.3.3]

E 503.5.6.4 Supply Air Temperature Reset Controls. Multiple zone HVAC systems shall include controls that automatically reset the supply air temperature in response to representative building loads, or to outdoor air temperature. The controls shall reset the supply air temperature to at least 25 percent of the difference between the design supply air temperature and the design room air temperature. Controls that adjust the reset based on zone humidity shall be permitted. Zones that are expected to experience relatively constant loads, such as electronic equipment rooms, shall be designed for the fully reset supply temperature.

Exceptions:

(1) Climate zones 0A, 1A, 2A, and 3A.

(2) Systems that prevent reheating, recooling, or mixing of heated and cooled supply air.

(3) Systems where not less than 75 percent of the energy for reheating (on an annual basis) is from site recovered energy or site-solar energy. [ASHRAE 90.1:6.5.3.5]

E 503.5.6.5 Fractional Horsepower Fan Motors. Motors for fans that are 1/12 hp (62.1 W) or more and less than 1 hp (0.7 kW) shall be electronically-commutated motors or shall have a motor efficiency of not less than 70 percent where rated in accordance with DOE 10 CFR 431. These motors shall also have the means to adjust motor speed for either balancing or remote control. Belt-driven fans shall be permitted to use sheave adjustments for airflow balancing in lieu of a varying motor speed.

Exceptions:

(1) Motors in the airstream within fan coils and terminal units that operate when providing heating to the space served.

(2) Motors installed in space conditioning equipment certified in accordance with Section E 503.4 through Section E 503.4.4.1.

(3) Motors shown in Table E 503.5.6.5(1) or Table E 503.5.6.5(2). [ASHRAE 90.1:6.5.3.6]

E 503.5.6.6 Ventilation Design. The required minimum outdoor air rate is the larger of the minimum outdoor air rate or the minimum exhaust air rate required by Chapter 4, ASHRAE 62.1, ASHRAE 170, or applicable codes or accreditation standards. Outdoor air ventilation systems shall comply with one of the following:

(1) Design minimum system outdoor air provided shall not exceed 135 percent of the required minimum outdoor air rate.

(2) Dampers, ductwork, and controls shall be provided that allow the system to supply no more than the required minimum outdoor air rate with a single setpoint adjustment.

(3) The system includes exhaust air energy recovery complying with Section E 503.5.10. [ASHRAE 90.1:6.5.3.7]

E 503.5.7 Hydronic System Design and Control. Boiler systems with design input of 1 000 000 Btu/h (293 kW) or more shall comply with the turndown ratio in accordance with Table E 503.5.7.

The system turndown requirement shall use multiple single-input boilers, one or more modulating boilers, or a combination of single-input and modulating boilers.

Boilers shall comply with the minimum efficiency requirements in Table E 503.7.1(6). [ASHRAE 90.1:6.5.4.1]

E 503.5.7.1 Hydronic Variable Flow Systems. Chilled- and hot-water distribution systems that include three or more control valves designed to modulate or step open and close as a function of load shall be designed for variable fluid flow and shall be capable of and configured to reduce pump flow rates to not more than the larger of 25 percent of the design flow rate or the minimum flow required by the heating/cooling equipment manufacturer for the proper operation of equipment. Individual or parallel pumps serving variable-flow heating-water or chilled-water systems, where the nameplate horsepower of the motor or combined parallel motors is not less than the power shown in Table E 503.5.7.1, shall have controls or devices that will result in pump motor demand of not more than 30 percent of design wattage at 50 percent of design water flow. The controls or devices shall be controlled as a function of desired flow or to maintain a minimum required differential pressure. Differential pressure shall be measured at or near the most remote heat exchanger or the heat exchanger requiring the greatest differential pressure. The differential pressure setpoint shall not

TABLE E 503.5.6.5(1)
MINIMUM AVERAGE FULL-LOAD EFFICIENCY FOR POLYPHASE SMALL ELECTRIC MOTORS*
[ASHRAE 90.1: TABLE 10.8-3]

	FULL-LOAD EFFICIENCY, %		
NUMBER OF POLES	OPEN MOTORS		
	2	4	6
SYNCHRONOUS SPEED (RPM)	3600	1800	1200
MOTOR HORSEPOWER	EFFICIENCY, %		
0.25	65.6	69.5	67.5
0.33	69.5	73.4	71.4
0.50	73.4	78.2	75.3
0.75	76.8	81.1	81.7
1	77.0	83.5	82.5
1.5	84.0	86.5	83.8
2	85.5	86.5	N/A
3	85.5	86.9	N/A

* Average full-load efficiencies shall be established in accordance with 10 CFR 431.

TABLE E 503.5.6.5(2)
MINIMUM AVERAGE FULL-LOAD EFFICIENCY FOR CAPACITOR-START CAPACITOR-RUN AND CAPACITOR-START INDUCTION-RUN SMALL ELECTRIC MOTORS*
[ASHRAE 90.1: TABLE 10.8-4]

FULL-LOAD EFFICIENCY, %			
NUMBER OF POLES	OPEN MOTORS		
	2	4	6
SYNCHRONOUS SPEED (RPM)	3600	1800	1200
MOTOR HORSEPOWER	EFFICIENCY, %		
0.25	66.6	68.5	62.2
0.33	70.5	72.4	66.6
0.50	72.4	76.2	76.2
0.75	76.2	81.8	80.2
1	80.4	82.6	81.1
1.5	81.5	83.8	N/A
2	82.9	84.5	N/A
3	84.1	N/A	N/A

* Average full-load efficiencies shall be established in accordance with 10 CFR 431.

exceed 110 percent of that required to achieve design flow through the heat exchanger. Where differential pressure control is used to comply with this section, and DDC systems are used, the setpoint shall be reset downward based on valve positions until one valve is nearly wide open.

Exceptions:

(1) Differential pressure set-point reset is not required where valve position is used to comply with Section E 503.5.7.3.

(2) Variable-pump flow control is not required on heating-water pumps where more than 50 percent of annual heat is generated by an electric boiler.

(3) Variable flow is not required for primary pumps in a primary/secondary system.

(4) Variable flow is not required for a coil pump provided for freeze protection.

(5) Variable flow is not required for heat recovery coil runaround loops. [ASHRAE 90.1:6.5.4.2]

E 503.5.7.2 Chiller and Boiler Isolation. Where a chilled-water plant includes more than one chiller, provisions shall be made so that the fluid flow through the chiller is automatically shut off where the chiller is shut down. Chillers piped in series for the purpose of increased temperature differential, shall be considered as one chiller. Where constant-speed chilled-water or condenser water pumps are used to serve multiple chillers, the number of pumps shall be not less than the number of chillers and staged on and off with the chillers. [ASHRAE 90.1:6.5.4.3.1]

E 503.5.7.2.1 Boiler Isolation. Where a boiler plant includes more than one boiler, provisions shall be made so that the flow through the boiler is automatically shut off where the boiler is shut down. Where constant-speed hot-water pumps are used to serve multiple boilers, the number of pumps shall be not less than the number of boilers and staged on and off with the boilers. [ASHRAE 90.1:6.5.4.3.2]

TABLE E 503.5.7
BOILER TURNDOWN
[ASHRAE 90.1: TABLE 6.5.4.1]

BOILER SYSTEM DESIGN INPUT, Btu/h	MINIMUM TURNDOWN RATIO
≥1 000 000 and ≤5 000 000	3 to 1
>5 000 000 and ≤10 000 000	4 to 1
>10 000 000	5 to 1

For SI units: 1000 British thermal units per hour = 0.293 kW

TABLE E 503.5.7.1
PUMP FLOW CONTROL REQUIREMENTS
[ASHRAE 90.1: TABLE 6.5.4.2]

CHILLED WATER PUMPS IN THESE CLIMATE ZONES	HEATING WATER PUMPS IN THESE CLIMATE ZONES	MOTOR NAMEPLATE HORSEPOWER
0A, 0B, 1A, 1B, 2B	NR	≥2 hp
2A, 3B	NR	≥3 hp
3A, 3C, 4A, 4B	7, 8	≥5 hp
4C, 5A, 5B, 5C, 6A, 6B	3C, 5A, 5C, 6A, 6B	≥7.5 hp
—	4A, 4C, 5B	≥10 hp
7, 8	4B	≥15 hp
—	2A, 2B, 3A, 3B	≥25 hp
—	1B	≥100 hp
—	0A, 0B, 1A	≥200 hp

For SI units: 1 horsepower = 0.746 kW

E 503.5.7.3 Chilled- and Hot-Water Temperature Reset Controls. Chilled- and hot-water systems with a design capacity exceeding 300 000 Btu/h (88 kW) supplying chilled or heated water to comfort conditioning systems shall include controls that automatically reset supply water temperatures by representative building loads (including return water temperature) or by outdoor air temperature. Where DDC is used to control valves, the set point shall be reset based on valve positions until one valve is nearly wide open or setpoint limits of the system equipment or application have been reached.

Exceptions:

(1) Where chilled-water supply is already cold, such as chilled water supplied from a district cooling or thermal energy storage system, such that blending would be required to achieve the reset chilled-water supply temperature.

(2) Where a specific temperature is required for a process.

(3) Water temperature reset is not required where valve position is used to comply with Section E 503.5.7. [ASHRAE 90.1:6.5.4.4]

E 503.5.7.4 Hydronic (Water Loop) Heat Pump and Water-Cooled Unitary Air Conditioners. Hydronic heat pumps and water-cooled unitary air-conditioners shall have a two-position automatic valve interlocked to shut off water flow when the compressor is off.

Exception: Units employing water economizers. [ASHRAE 90.1:6.5.4.5.1]

E 503.5.7.4.1 Controls. Hydronic heat pumps and water-cooled unitary air-conditioners having a total pump system power exceeding 5 hp (3.7 kW) shall have controls, devices, or both (such as variable speed control) that will result in pump motor demand of not more than 30 percent of design wattage at 50 percent of design water flow. [ASHRAE 90.1:6.5.4.5.2]

E 503.5.7.5 Pipe Sizing. Chilled-water and condenser-water piping shall be designed such that the design flow rate in each piping segment shall not exceed the values listed in Table E 503.5.7.5 for the appropriate total annual hours of operation. Piping size selections for systems that operate under variable flow conditions (e.g., modulating two-way control valves at coils) and that contain variable-speed pump motors shall be permitted to be made from the "Variable Flow/Variable Speed" columns. All others shall be made from the "Other" columns.

Exceptions:

(1) Design flow rates exceeding the values in Table E 503.5.7.5 shall be permitted in specific sections of piping if the piping in question is not in the critical circuit at design conditions and is not predicted to be in the critical circuit during more than 30 percent of operating hours.

(2) Piping systems that have equivalent or lower total pressure drop than the same system constructed with standard weight steel pipe with piping and fittings sized in accordance with Table E 503.5.7.5. [ASHRAE 90.1:6.5.4.6]

E 503.5.8 Heat Rejection Equipment. Section E 503.5.8 through Section E 503.5.9 applies to heat-rejection equipment used in comfort cooling systems, such as air-cooled condensers, dry coolers, open-circuit cooling towers, closed-circuit cooling towers, and evaporative condensers.

Exception: Heat-rejection devices whose energy use is included in the equipment efficiency ratings listed in Table E 503.7.1(1) through Table E 503.7.1(4). [ASHRAE 90.1:6.5.5.1]

E 503.5.8.1 Fan Speed Control. The fan system on a heat-rejection device powered by an individual motor or an array of motors with a connected power, including the motor service factor, totaling 5 hp (3.7 kW) or more shall have controls and/or devices (such as variable-speed control) that shall result in fan motor demand of no more than 30 percent of design wattage at 50 percent of the design airflow and that shall automatically modulate the fan speed to control the leaving fluid temperature or condensing temperature/pressure of the heat-rejection device.

Exceptions:

(1) Condenser fans serving multiple refrigerant or fluid cooling circuits.

(2) Condenser fans serving flooded condensers.

[ASHRAE 90.1:6.5.5.2.1]

E 503.5.8.2 Variable-Speed Fan Drives. Multicell heat rejection equipment with variable-speed fan drives shall:

(1) Operate the maximum number of fans allowed that comply with the manufacturer's requirements for all system components.

TABLE E 503.5.7.5
PIPING SYSTEM DESIGN MAXIMUM FLOW RATE (gallons per minute)
[ASHRAE 90.1: TABLE 6.5.4.6]

OPERATING HOURS/YEAR	≤2000 HOURS/YEAR		>2000 AND ≤4400 HOURS/YEAR		>4400 HOURS/YEAR	
NOMINAL PIPE SIZE, (inches)	OTHER	VARIABLE FLOW/ VARIABLE SPEED	OTHER	VARIABLE FLOW/ VARIABLE SPEED	OTHER	VARIABLE FLOW/ VARIABLE SPEED
2½	120	180	85	130	68	110
3	180	270	140	210	110	170
4	350	530	260	400	210	320
5	410	620	310	470	250	370
6	740	1100	570	860	440	680
8	1200	1800	900	1400	700	1100
10	1800	2700	1300	2000	1000	1600
12	2500	3800	1900	2900	1500	2300
Maximum velocity for pipes over 14-24 inches in size	8.5 ft/s	13.0 ft/s	6.5 ft/s	9.5 ft/s	5.0 ft/s	7.5 ft/s

For SI units: 1 gallon per minute = 0.06 L/s, 1 foot per second = 0.3048 m/s, 1 inch = 25.4 mm

(2) Control all fans to the same fan speed required for the instantaneous cooling duty, as opposed to staged (on/off) operation. Minimum fan speed shall comply with the minimum allowable speed of the fan drive system per the manufacturer's recommendations. [ASHRAE 90.1:6.5.5.2.2]

E 503.5.9 Limitation on Centrifugal Fan Open-Circuit Cooling Towers. Centrifugal fan open-circuit cooling towers with a combined rated capacity of 1100 gallons per minute (gpm) (69.39 L/s) or greater at 95°F (35°C) condenser water return, 85°F (29°C) condenser water supply, and 75°F (24°C) outdoor air wet-bulb temperature shall comply with the energy efficiency requirement for axial fan open-circuit cooling towers in accordance with Table E 503.7.1(7).

Exception: Centrifugal open-circuit cooling towers that are ducted (inlet or discharge) or require external sound attenuation. [ASHRAE 90.1:6.5.5.3]

E 503.5.9.1 Tower Flow Turndown. Open-circuit cooling towers used on water-cooled chiller systems that are configured with multiple- or variable-speed condenser water pumps shall be designed so that all open-circuit cooling tower cells can be run in parallel with the larger of the following:

(1) The flow that is produced by the smallest pump at its minimum expected flow rate.

(2) Fifty percent of the design flow for the cell. [ASHRAE 90.1:6.5.5.4]

E 503.5.10 Exhaust Air Energy Recovery. Each fan system shall have an energy recovery system where the design supply fan airflow rate exceeds the value listed in Table E 503.5.10(1) and Table E 503.5.10(2), based on the climate zone and percentage of outdoor air at design airflow conditions. Table E 503.5.10(1) shall be used for all ventilation systems that operate less than 8000 hours per year and Table E 503.5.10(2) shall be used for all ventilation systems that operate 8000 or more hours per year.

Energy recovery systems required by this section shall result in an enthalpy recovery ratio of not less than 50 percent. A fifty percent enthalpy recovery ratio shall mean a change in the enthalpy of the outdoor air supply equal to 50 percent of the difference between the outdoor air and entering exhaust air enthalpies at design conditions. Provision shall be provided to bypass or control the energy recovery system to permit air economizer operation in accordance with Section E 503.5.1.

Exceptions:

(1) Laboratory systems that are in accordance with Section E 503.5.11.3.

(2) Systems serving spaces that are not cooled and that are heated to less than 60°F (16°C).

(3) Where more than 60 percent of the outdoor air heating energy is provided from site-recovered energy or site-solar energy.

(4) Heating energy recovery in Climate Zones 0, 1, and 2.

(5) Cooling energy recovery in climate zones 3C, 4C, 5B, 5C, 6B, 7, and 8.

(6) Where the sum of the airflow rates exhausted and relieved within 20 feet (6096 mm) of each other is less than 75 percent of the design outdoor airflow rate, excluding exhaust air that is;

(a) used for another energy recovery system,

(b) not allowed by ASHRAE 170 for use in energy recovery systems with leakage potential, or

(c) of Class 4 as defined in Chapter 2 or ASHRAE 62.1.

(7) Systems requiring dehumidification that employ energy recovery in series with the cooling coil.

(8) Systems expected to operate less than 20 hours per week at the outdoor air percentage in accordance with Table E 503.5.10(1). [ASHRAE 90.1:6.5.6.1]

E 503.5.10.1 Heat Recovery for Service Water Heating. Heat recovery shall comply with Section E 503.5.10.1.1 and Section E 503.5.10.1.2.

E 503.5.10.1.1 Condenser Heat Recovery Systems. Condenser heat recovery systems shall be installed for the heating or preheating of service hot water where the following conditions exist:

(1) The facility operates 24 hours a day.

(2) The total installed heat rejection capacity of the water-cooled system is more than 6 000 000 Btu/h (1757 kW) of heat rejection.

(3) The design service water heating load is more than 1 000 000 Btu/h (293 kW). [ASHRAE 90.1:6.5.6.2.1]

E 503.5.10.1.2 Capacity. The required heat recovery system shall have the capacity to provide the smaller of:

(1) Sixty percent of the peak heat-rejection load at design conditions or

(2) Preheat of the peak service hot-water draw to 85°F (29°C).

Exceptions:

(1) Facilities that employ condenser heat recovery for space heating with a heat recovery design exceeding 30 percent of the peak water-cooled condenser load at design conditions.

(2) Facilities that provide 60 percent of their service water heating from site-solar energy or site-recovered energy or from other sources. [ASHRAE 90.1:6.5.6.2.2]

E 503.5.11 Exhaust Systems. Exhaust systems shall comply with Section E 503.5.11.1 through Section E 503.5.11.3.

E 503.5.11.1 Transfer Air. Conditioned supply air delivered to a space with a mechanical exhaust shall not exceed the greater of the following:

(1) The supply flow required to be in accordance with the space heating or cooling load;

(2) The ventilation rate required by the Authority Having Jurisdiction, the Facility Environmental Health and Safety department, Chapter 4 or ASHRAE 62.1; or

(3) The mechanical exhaust flow minus the available transfer air from conditioned spaces or return air plenums on the same floor, not in different smoke or fire compartments, and that at their closest point are within 15 feet

(4572 mm) of each other. Available transfer air is that portion of outdoor ventilation air that:

(a) is not required to satisfy other exhaust needs,

(b) is not required to maintain pressurization of other spaces, and

(c) is transferable according to applicable codes and standards and to the class of air recirculation limitations in Chapter 4 or ASHRAE 62.1.

Exceptions:

(1) Biosafety level classified laboratories 3 or higher.

(2) Vivarium spaces.

(3) Spaces that are required by applicable codes and standards to be maintained at positive pressure relative to adjacent spaces. For spaces taking this exception, any transferable air that is not directly transferred shall be made available to the associated air-handling unit and shall be used whenever economizer or other options do not save more energy.

(4) Spaces where the demand for transfer air may exceed the available transfer airflow rate and where the spaces have a required negative pressure relationship. For spaces taking this exception, any transferable air that is not directly transferred shall be made available to the associated air-handling unit and shall be used whenever economizer or other options do not save more energy. [ASHRAE 90.1:6.5.7.1]

E 503.5.11.2 Kitchen Exhaust Systems. Replacement air introduced directly into the hood cavity of kitchen exhaust hoods shall not exceed 10 percent of the hood exhaust airflow rate. [ASHRAE 90.1:6.5.7.2.1]

E 503.5.11.2.1 Exhaust Flow Rate. Where a kitchen or dining facility has a total kitchen hood exhaust airflow rate exceeding 5000 ft^3/min (2.3597 m^3/s), each hood shall have an exhaust rate in accordance with Table E 503.5.11.2.1. Where a single hood, or hood section, is installed over appliances with different duty ratings, the maximum allowable flow rate for the hood or hood section shall not exceed the values in Table E 503.5.11.2.1 for the highest appliance duty rating under the hood or hood section. Refer to ASHRAE 154 for definitions of hood type, appliance duty, and net exhaust flow rate.

Exception: Seventy-five percent or more of the total replacement air is transfer air that would otherwise be exhausted. [ASHRAE 90.1:6.5.7.2.2]

E 503.5.11.2.2 Kitchen or Dining Facility. Where a

TABLE E 503.5.10(1)
EXHAUST AIR ENERGY RECOVERY REQUIREMENTS FOR VENTILATION SYSTEMS OPERATING LESS THAN 8000 HOURS PER YEAR*
[ASHRAE 90.1: TABLE 6.5.6.1-1]

CLIMATE ZONE	PERCENT OUTDOOR AIR AT FULL DESIGN AIRFLOW RATE							
	≥10% and <20%	≥20% and <30%	≥30% and <40%	≥40% and <50%	≥50% and <60%	≥60% and <70%	≥70% and <80%	≥80%
	DESIGN SUPPLY FAN AIRFLOW RATE (cubic feet per minute)							
3B, 3C, 4B, 4C, 5B	NR	NR	NR	NR	NR	NR	NR	NR
0B, 1B, 2B, 5C	NR	NR	NR	NR	≥26 000	≥12 000	≥5000	≥4000
6B	≥28 000	≥26 500	≥11 000	≥5500	≥4500	≥3500	≥2500	≥1500
0A, 1A, 2A, 3A, 4A, 5A, 6A	≥26 000	≥16 000	≥5500	≥4500	≥3500	≥2000	≥1000	≥120
7, 8	≥4500	≥4000	≥2500	≥1000	≥140	≥120	≥100	≥80

For SI units: 1 cubic foot per minute = 0.00047 m^3/s
* NR = Not Required

TABLE E 503.5.10(2)
EXHAUST AIR ENERGY RECOVERY REQUIREMENTS FOR VENTILATION SYSTEMS OPERATING NOT LESS THAN 8000 HOURS PER YEAR*
[ASHRAE 90.1: TABLE 6.5.6.1-2]

CLIMATE ZONE	PERCENT OUTDOOR AIR AT FULL DESIGN AIRFLOW RATE							
	≥10% and <20%	≥20% and <30%	≥30% and <40%	≥40% and <50%	≥50% and <60%	≥60% and <70%	≥70% and <80%	≥80%
	DESIGN SUPPLY FAN AIRFLOW RATE (cubic feet per minute)							
3C	NR	NR	NR	NR	NR	NR	NR	NR
0B, 1B, 2B, 3B, 4C, 5C	NR	≥19 500	≥9000	≥5000	≥4000	≥3000	≥1500	≥120
0A, 1A, 2A, 3A, 4B, 5B	≥2500	≥2000	≥1000	≥500	≥140	≥120	≥100	≥80
4A, 5A, 6A, 6B, 7, 8	≥200	≥130	≥100	≥80	≥70	≥60	≥50	≥40

For SI units: 1 cubic foot per minute = 0.00047 m^3/s
* NR = Not Required

TABLE E 503.5.11.2.1
MAXIMUM NET EXHAUST FLOW RATE, CFM PER LINEAR FOOT OF HOOD LENGTH
[ASHRAE 90.1: TABLE 6.5.7.2.2]

TYPE OF HOOD	LIGHT DUTY EQUIPMENT	MEDIUM DUTY EQUIPMENT	HEAVY DUTY EQUIPMENT	EXTRA HEAVY DUTY EQUIPMENT
Wall-mounted canopy	140	210	280	385
Single island	280	350	420	490
Double island (per side)	175	210	280	385
Eyebrow	175	175	Not allowed	Not allowed
Backshelf/ Pass-over	210	210	280	Not allowed

For SI units: 1 foot = 304.8 mm, 1 cubic foot per minute = 0.00047 m^3/s

kitchen or dining facility has a total kitchen hood exhaust airflow rate more than 5000 ft^3/min (2.3597 m^3/s), then one of the following shall be provided:

(1) Fifty percent or more of all replacement air is transfer air that would otherwise be exhausted.

(2) Demand ventilation systems on 75 percent or more of the exhaust air. Such systems shall be capable of and configured to provide 50 percent or more reduction in exhaust and replacement air system airflow rates, including controls necessary to modulate airflow in response to appliance operation and to maintain full capture and containment of smoke, effluent, and combustion products during cooking and idle.

(3) Listed energy recovery devices that result in a sensible energy recovery ratio of 40 percent or more on 50 percent or more of the total exhaust airflow. A 40 percent sensible energy recovery ratio shall mean a change in the dry-bulb temperature of the outdoor air supply equal to 40 percent of the difference between the outdoor air and entering exhaust air dry-bulb temperatures at design conditions. [ASHRAE 90.1:6.5.7.2.3]

E 503.5.11.2.3 Performance Testing. An approved field test method shall be used to evaluate design air flow rates and demonstrate proper capture and containment performance of installed commercial kitchen exhaust systems. Where demand ventilation systems are utilized to be in accordance with Section E 503.5.11.2.2, additional performance testing shall be required to demonstrate proper capture and containment at minimum airflow. [ASHRAE 90.1:6.5.7.2.4]

E 503.5.11.3 Laboratory Exhaust Systems. Buildings with laboratory exhaust systems having a total exhaust rate of more than 5000 ft^3/min (2.3597 m^3/s) shall include not less than one of the following features:

(1) VAV laboratory exhaust and room supply systems capable of and configured to reduce exhaust and makeup airflow rates, incorporate a heat recovery system to precondition makeup air from laboratory exhaust, or both, and shall be in accordance with the following:

$A+B\bullet(E/M) \geq 50\%$ (Equation E 503.5.11.3)

Where:

A = Percentage that the exhaust and makeup airflow rates are capable of being reduced from design conditions.

B = Sensible energy recovery ratio.

E = Exhaust airflow rate through the heat recovery device at design conditions.

M = Makeup airflow rate of the system at design conditions.

(2) VAV laboratory exhaust and room supply systems that are required to have minimum circulation rates to be in accordance with the codes or accreditation standards shall be capable of and configured to reduce zone exhaust and makeup airflow rates to the regulated minimum circulation values, or the minimum required to maintain pressurization relationship requirements. Systems serving nonregulated zones shall be capable of and configured to reduce exhaust and makeup airflow rates to 50 percent of the zone design values, or the minimum required to maintain pressurization relationship requirements.

(3) Direct makeup (auxiliary) air supply of 75 percent or more of the exhaust airflow rate, heated not more than 2°F (1°C) below room setpoint, cooled to not less than 3°F (2°C) above room setpoint, no humidification added, and no simultaneous heating and cooling are used for dehumidification control. [ASHRAE 90.1:6.5.7.3]

E 503.5.12 Radiant Heating Systems. Radiant heating shall be used when heating is required for unenclosed spaces.

Exception: Loading docks equipped with air curtains. [ASHRAE 90.1:6.5.8.1]

E 503.5.12.1 Heating Enclosed Spaces. Radiant heating systems that are used as primary or supplemental heating for enclosed spaces shall be in conformance with the governing provisions of the standard, including, but not limited to the following:

(1) Radiant hydronic ceiling or floor panels (used for heating or cooling).

(2) Combination or hybrid systems incorporating radiant heating (or cooling) panels.

(3) Radiant heating (or cooling) panels used in conjunction with other systems such as VAV or thermal storage systems. [ASHRAE 90.1:6.5.8.2]

E 503.5.13 Hot Gas Bypass Limitation. Cooling systems shall not use hot gas bypass or other evaporator pressure control systems unless the system is designed with multiple steps of unloading or continuous capacity modulation. The capacity of the hot gas bypass shall be limited as indicated in Table E 503.5.13 for VAV units and single-zone

VAV units. Hot-gas bypass shall not be used on constant-volume units. [ASHRAE 90.1:6.5.9]

TABLE E 503.5.13
HOT GAS BYPASS LIMITATION
[ASHRAE 90.1: TABLE 6.5.9]

RATED CAPACITY	MAXIMUM HOT GAS BYPASS (percent of total capacity)
≤240 000 Btu/h	15%
>240 000 Btu/h	10%

For SI units: 1000 British thermal units per hour = 0.293 kW

E 503.5.14 Door Switches. Conditioned spaces with doors, including doors with more than one-half glass, opening to the outdoors shall be provided with controls that when any such door is open, the following shall occur:

(1) Disable mechanical heating or reset the heating setpoint to 55°F (13°C) or lower within five minutes of the door opening.
(2) Disable mechanical cooling or reset the cooling setpoint to 90°F (32°C) or more within five minutes of the door opening. Mechanical cooling shall be permitted to remain enabled where outdoor air temperature is less than the space temperature.

Exceptions:

(1) Building entries with automatic closing devices.
(2) Any space without a thermostat.
(3) Alterations to existing buildings.
(4) Loading docks. [ASHRAE 90.1:6.5.10]

E 503.6 Submittals. The Authority Having Jurisdiction shall require submittal of compliance documentation and supplemental information in accordance with Section E 503.6.1 through Section E 503.6.3.

E 503.6.1 Construction Details. Compliance documents shall show the pertinent data and features of the building, equipment, and systems in sufficient detail to permit a determination of compliance by the building official and to indicate compliance with the requirements of this appendix. [ASHRAE 90.1:4.2.2.1]

E 503.6.2 Supplemental Information. Supplemental information necessary to verify compliance with this appendix, such as calculations, worksheets, compliance forms, vendor literature, or other data, shall be made available where required by the Authority Having Jurisdiction. [ASHRAE 90.1:4.2.2.2]

E 503.6.3 Manuals. Operating and maintenance information shall be provided to the building owner. This information shall include, but not be limited to, the information specified in Section E 503.6.3.1, Section E 503.6.3.2, and Section E 503.6.5.2. [ASHRAE 90.1:4.2.2.3]

E 503.6.3.1 Required Information. Construction documents shall require that an operating manual and maintenance manual be provided to the building owner. The manuals shall include, at a minimum, the following:

(1) Submittal data stating equipment rating and selected options for each piece of equipment requiring maintenance.
(2) Operation manuals and maintenance manuals for each piece of equipment requiring maintenance. Required routine maintenance actions shall be clearly identified.
(3) Names and addresses of not less than one qualified service agency.
(4) A complete narrative of how each system is intended to operate.

The Authority Having Jurisdiction shall only check to ensure that the construction documents require this information to be transmitted to the owner and should not expect copies of any of the materials. [ASHRAE 90.1:8.7.2]

E 503.6.3.2 Lighting Manuals. Construction documents shall require for all lighting equipment and lighting controls that an operating and maintenance manual be provided to the building owner or the designated representative of the building owner within 90 days after the date of system acceptance. These manuals shall include, at a minimum, the following:

(1) Submittal data indicating all selected options for each piece of lighting equipment, including but not limited to lamps, ballasts, drivers, and lighting controls.
(2) Operation and maintenance manuals for each piece of lighting equipment and lighting controls with routine maintenance clearly identified including, as a minimum, a recommended relamping or cleaning program and a schedule for inspecting and recalibrating all lighting controls.
(3) A complete narrative of how each lighting control system is intended to operate including recommended settings. [ASHRAE 90.1:9.7.2.2]

E 503.6.4 Labeling of Material and Equipment. Materials and equipment shall be labeled in a manner that will allow for determination of their compliance with the applicable provisions of this appendix. [ASHRAE 90.1:4.2.3]

E 503.6.5 Completion Requirements. Section E 503.6.5.1 through Section E 503.6.5.4.1 are mandatory provisions and are necessary to comply with this appendix. [ASHRAE 90.1:6.7.2]

E 503.6.5.1 Drawings. Construction documents shall require that, within 90 days after the date of system acceptance, record drawings of the actual installation be provided to the building owner or the designated representative of the building owner. Record drawings shall include, as a minimum, the location and performance data on each piece of equipment, general configuration of duct and pipe distribution system including sizes, and the terminal air or water design flow rates. [ASHRAE 90.1:6.7.2.1]

E 503.6.5.2 Manuals. Construction documents shall require that an operating manual and a maintenance manual be provided to the building owner or the designated representative of the building owner within 90 days after the date of system acceptance. These manuals shall be in accordance with industry-accepted standards and shall include, at a minimum, the following:

(1) Submittal data stating equipment size and selected

options for each piece of equipment requiring maintenance.

(2) Operation manuals and maintenance manuals for each piece of equipment and system requiring maintenance, except equipment not furnished as part of the project. Required routine maintenance actions shall be clearly identified.
(3) Names and addresses of not less than one service agency.
(4) HVAC controls system maintenance and calibration information, including wiring diagrams, schematics, and control sequence descriptions. Desired or field-determined setpoints shall be permanently recorded on control drawings at control devices or, for digital control systems, in programming comments.
(5) A complete narrative of how each system is intended to operate, including suggested setpoints. [ASHRAE 90.1:6.7.2.2]

E 503.6.5.3 System Balancing. Construction documents shall require that HVAC systems be balanced in accordance with generally accepted engineering standards. Construction documents shall require that a written balance report be provided to the building owner or the designated representative of the building owner for HVAC systems serving zones with a total conditioned area exceeding 5000 square feet (464.52 m^2). [ASHRAE 90.1:6.7.2.3.1]

E 503.6.5.3.1 Air System Balancing. Air systems shall be balanced in a manner to first minimize throttling losses. Then, for fans with fan system power greater than 1 hp (0.7 kW), fan speed shall be adjusted to meet design flow conditions. [ASHRAE 90.1:6.7.2.3.2]

E 503.6.5.3.2 Hydronic System Balancing. Hydronic systems shall be proportionately balanced in a manner to first minimize throttling losses; then the pump impeller shall be trimmed or pump speed shall be adjusted to meet design flow conditions.

Exceptions: Impellers need not be trimmed nor pump speed adjusted.

(1) For pumps with pump motors of 10 hp (7.5 kW) or less.
(2) Where throttling results is not greater than 5 percent of the nameplate horsepower draw, or 3 hp (2.2 kW), whichever is greater, above that required where the impeller was trimmed. [ASHRAE 90.1:6.7.2.3.3]

E 503.6.5.4 System Commissioning. HVAC control systems shall be tested to ensure that control elements are calibrated, adjusted, and in proper working condition. For projects larger than 50 000 square feet (4645.15 m^2) conditioned area, except warehouses and semiheated spaces, detailed instructions for commissioning HVAC systems shall be provided by the designer in plans and specifications. [ASHRAE 90.1:6.7.2.4]

E 503.6.5.4.1 Minimum Level of Commission. Commissioning shall be performed for HVAC systems in accordance with Level 1, Basic Commissioning of the SMACNA HVAC Systems Commissioning Manual. (See Section E 801.0 for additional information on HVAC system commissioning)

E 503.7 Minimum Equipment Efficiency Tables. The minimum efficiency requirements for equipment shall comply with Section E 503.7.1; duct insulation shall comply with Section E 503.7.2, and pipe insulation shall comply with Section E 503.7.3.

E 503.7.1 Minimum Efficiency Requirement Listed Equipment – Standard Rating and Operating Conditions. The minimum efficiency requirements for equipment shall comply with Table E 503.7.1(1) through Table E 503.7.1(16).

E 503.7.2 Duct Insulation Tables. Duct insulation shall comply with Table E 503.7.2.

E 503.7.3 Pipe Insulation Tables. Pipe insulation shall comply with Table E 503.7.3(1) through Table E 503.7.3(2).

E 503.8 Alternative Compliance Path. HVAC systems serving heating, cooling, or ventilation needs of a computer room shall be in accordance with Section E 503.1, Section E 503.4, Section E 503.8.1 or Section E 503.8.2, Section E 503.8.3, Section E 502.7 through Section E 502.7.2, and Section E 503.7. [ASHRAE 90.1:6.6.1]

E 503.8.1 Computer Room (PUE_1). The computer room PUE_1 shall be not more than the values listed in Table E 503.8.1. Hourly simulation of the proposed design, for purposes of calculating PUE_1, shall be in accordance with ASHRAE 90.1.

Exception: The compliance path shall not be permitted for a proposed computer room design utilizing a combined heat and power system. [ASHRAE 90.1:6.6.1.1]

E 503.8.2 Computer Room (PUE_0). The computer room PUE_0 is less than or equal to the values listed in Table E 503.8.1, shall be the highest value determined at outdoor cooling design temperatures, and shall be limited to systems only

TABLE E 503.8.1
POWER USAGE EFFECTIVENESS (PUE) MAXIMUM
[ASHRAE 90.1: TABLE 6.6.1]

CLIMATE ZONE	PUE*
0A	1.64
0B	1.62
1A	1.61
1B	1.53
2A	1.49
2B	1.45
3A	1.41
3B	1.42
3C	1.39
4A	1.36
4B	1.38
4C	1.38
5A	1.36
5B	1.33
5C	1.36
6A	1.34
6B	1.33
7	1.32
8	1.30

* PUE_0 and PUE_1 shall not include energy for battery charging.

using electricity for an energy source. PUE_0 shall be calculated for two conditions:

(1) One hundred percent design IT equipment energy and

(2) Fifty percent design IT equipment energy. [ASHRAE 90.1:6.6.1.2]

E 503.8.3 Documentation. Documentation on the following components shall be provided, including a breakdown of energy consumption or demand:

(1) IT equipment

(2) Power distribution losses external to the IT equipment

(3) HVAC systems

(4) Lighting [ASHRAE 90.1:6.6.1.3]

E 504.0 Solar Energy Systems.

E 504.1 General. Solar energy systems shall be installed in accordance with the Uniform Solar, Hydronics and Geothermal Code (USHGC).

E 601.0 Indoor Environment.

E 601.1 Scope. The provisions of this section shall establish the means of reducing the quantity of air contaminants that are odorous, irritating, or harmful to the comfort and well-being of a building's installers, occupants, and neighbors.

E 602.0 Fireplaces.

E 602.1 Requirements. A direct vent sealed-combustion gas or sealed wood-burning fireplace, or a sealed wood stove shall be installed. The fireplace shall comply with Section E 602.1.1 and Section E 602.1.2.

E 602.1.1 Masonry or Factory-Built Fireplace. Masonry and factory-built fireplaces located in conditioned spaces shall be in accordance with Section E 602.1.1.1 through Section E 602.1.1.3.

E 602.1.1.1 Opening Cover. Closeable metal or glass doors covering the entire opening of the firebox shall be installed.

E 602.1.1.2 Combustion Air Intake. A combustion air intake to draw air from the outside of the building directly into the firebox, which is an area of not less than 6 square inches (0.004 m^2) and is equipped with a readily accessible, operable, and tight-fitting damper or combustion air control device.

E 602.1.1.3 Accessible Damper Control. The flue damper shall have a readily accessible control.

Exception: Where a gas log, log lighter, or decorative gas appliance is installed in a fireplace, the flue damper shall be blocked open where required by this code or the manufacturer's installation instructions.

E 602.1.2 Prohibited. Continuous burning pilot lights and the use of indoor air for cooling a firebox jacket, where the indoor air is vented to the outside of the building, are prohibited.

E 603.0 Pollutant Control.

E 603.1 Indoor Air Quality During Construction. Indoor air quality of a building shall be maintained in accordance with Section E 603.1.1 through Section E 603.1.3.

E 603.1.1 Temporary Ventilation During Construction. Temporary ventilation during construction shall be provided in accordance with the following:

(1) Ventilation during construction shall be achieved through openings in the building shell using fans to produce not less than three air changes per hour.

(2) During dust-producing operations, the supply and return HVAC system openings shall be protected from dust in accordance with Section E 603.1.3.

(3) Where the building is occupied during demolition or construction, ventilation shall be provided in accordance with the Control Measures of the SMACNA IAQ Guidelines for Occupied Buildings Under Construction.

(4) The permanent HVAC system shall not be used during construction to condition and ventilate the building within the required temperature range for material and equipment installation. Where required, a supplemental HVAC system shall be used during construction, return air shall be equipped with filters with a minimum efficiency reporting value (MERV) of 8, in accordance with ASHRAE 52.2, or an average efficiency of 30 percent in accordance with ASHRAE 52.2. Before occupancy, filters shall be replaced with filters having a MERV 13 rating in accordance with Section E 603.3.

  Exception: Embedded hydronics system shall be permitted to be used to condition the building during construction.

E 603.1.2 Indoor Air Quality After Construction. After construction ends and interior finishes are installed, flush-out the building to reduce contaminant concentrations by supplying a total outdoor air volume of 14 000 cubic feet per square foot (ft^3/ft^2) (4267.2 m^3/m^2) of occupiable building area. An internal temperature of not less than 60°F (16°C) and relative humidity not higher than 60 percent shall be maintained during the flush-out process. Occupancy shall begin on condition of 3500 ft^3/ft^2 (1066.8 m^3/m^2) of building area, with the remaining 10 500 ft^3/ft^2 (3200.4 m^3/m^2) being accomplished as soon as possible.

Exception: Other means of reducing the contaminant concentration levels shall be permitted where approved by the Authority Having Jurisdiction.

E 603.1.3 Covering of Duct Openings and Protection of Mechanical Equipment During Construction. At the time of rough installation, or during storage on the construction site and until final startup of the heating and cooling equipment, duct and other related air distribution component openings shall be covered with tape, plastic, sheet metal, or other methods acceptable to the enforcing agency to reduce the amount of dust or debris that collects in the system.

**TABLE E 503.7.1(1)
ELECTRICALLY OPERATED UNITARY AIR CONDITIONERS AND CONDENSING UNITS
MINIMUM EFFICIENCY REQUIREMENTS
[ASHRAE 90.1: TABLE 6.8.1-1]**

EQUIPMENT TYPE	SIZE CATEGORY	HEATING SECTION TYPE	SUBCATEGORY OR RATING CONDITION	MINIMUM EFFICIENCY	TEST PROCEDURE[1]
Air conditioners, air cooled	<65 000 Btu/h[2]	All	Split system, three phase	13.0 SEER	AHRI 210/240
			Single package, three phase	14 SEER	
Through the wall, air cooled	≤30 000 Btu/h[2]	All	Split system, three phase	12.0 SEER	AHRI 210/240
			Single package, three phase	12.0 SEER	
Small duct, high velocity, air cooled	<65 000 Btu/h[2]	All	Split system, three phase	11.0 SEER	AHRI 210/240
Air conditioners, air cooled	≥65 000 Btu/h and <135 000 Btu/h	Electric resistance (or none)	Split system and single package	11.2 EER 12.9 IEER	AHRI 340/360
		All other		11.0 EER 12.7 IEER	
	≥135 000 Btu/h and <240 000 Btu/h	Electric resistance (or none)		11.0 EER 12.4 IEER	
		All other		10.8 EER 12.2 IEER	
	≥240 000 Btu/h and <760 000 Btu/h	Electric resistance (or none)		10.0 EER 11.6 IEER	
		All other		9.8 EER 11.4 IEER	
	≥760 000 Btu/h	Electric resistance (or none)		9.7 EER 11.2 IEER	
		All other		9.5 EER 11.0 IEER	
Air conditioners, water cooled	<65 000 Btu/h	All	Split system and single package	12.1 EER 12.3 IEER	AHRI 210/240
	≥65 000 Btu/h and <135 000 Btu/h	Electric resistance (or none)		12.1 EER 13.9 IEER	AHRI 340/360
		All other		11.9 EER 13.7 IEER	
	≥135 000 Btu/h and <240 000 Btu/h	Electric resistance (or none)		12.5 EER 13.9 IEER	
		All other		12.3 EER 13.7 IEER	
	≥240 000 Btu/h and <760 000 Btu/h	Electric resistance (or none)		12.4 EER 13.6 IEER	
		All other		12.2 EER 13.4 IEER	
	≥760 000 Btu/h	Electric resistance (or none)		12.2 EER 13.5 IEER	
		All other		12.0 EER 13.3 IEER	

TABLE E 503.7.1(1) (continued)
ELECTRICALLY OPERATED UNITARY AIR CONDITIONERS AND CONDENSING UNITS
MINIMUM EFFICIENCY REQUIREMENTS
[ASHRAE 90.1: TABLE 6.8.1-1]

EQUIPMENT TYPE	SIZE CATEGORY	HEATING SECTION TYPE	SUBCATEGORY OR RATING CONDITION	MINIMUM EFFICIENCY	TEST PROCEDURE[1]
Air conditioners, evaporatively cooled	<65 000 Btu/h[2]	All	Split system and single package	12.1 EER 12.3 IEER	AHRI 210/240
	≥65 000 Btu/h and <135 000 Btu/h	Electric resistance (or none)		12.1 EER 12.3 IEER	AHRI 340/360
		All other		11.9 EER 12.1 IEER	
	≥135 000 Btu/h and <240 000 Btu/h	Electric resistance (or none)		12.0 EER 12.2 IERR	
		All other		11.8 EER 12.0 IEER	
	≥240 000 Btu/h and <760 000 Btu/h	Electric resistance (or none)		11.9 EER 12.1 IEER	
		All other		11.7 EER 11.9 IEER	
	≥760 000 Btu/h	Electric resistance (or none)		11.7 EER 11.9 IEER	
		All other		11.5 EER 11.7 IEER	
Condensing units, air cooled	≥135 000 Btu/h	–	–	10.5 EER 11.8 IEER	AHRI 365
Condensing units, water cooled	≥135 000 Btu/h	–	–	13.5 EER 14.0 IEER	AHRI 365
Condensing units, evaporatively cooled	≥135 000 Btu/h	–	–	13.5 EER 14.0 IEER	AHRI 365

For SI units: 1000 British thermal units per hour = 0.293 kW

Notes:

[1] ASHRAE 90.1 contains a complete specification of the referenced test procedure, including the referenced year version of the test procedure.

[2] Single-phase, air-cooled air conditioners less than 65 000 Btu/h (19 kW) are regulated by the U.S. Department of Energy Code of Federal Regulations 10 CFR 430. SEER values for single-phase products are set by the U.S. Department of Energy.

TABLE E 503.7.1(2)
ELECTRICALLY OPERATED UNITARY AND APPLIED HEAT PUMPS
MINIMUM EFFICIENCY REQUIREMENTS
[ASHRAE 90.1: TABLE 6.8.1-2]

EQUIPMENT TYPE	SIZE CATEGORY	HEATING SECTION TYPE	SUBCATEGORY OR RATING CONDITION	MINIMUM EFFICIENCY	TEST PROCEDURE[1]
Air cooled (cooling mode)	<65 000 Btu/h[2]	All	Split system, three phase	14 SEER	AHRI 210/240
			Single package, three phase	14 SEER	
Through the wall, air cooled (cooling mode)	≤30 000 Btu/h[2]	All	Split system, three phase	12.0 SEER	AHRI 210/240
			Single package, three phase	12.0 SEER	
Small duct, high velocity, air cooled	<65 000 Btu/h[2]	All	Split System, three phase	11.0 SEER	AHRI 210/240
Air cooled (cooling mode)	≥65 000 Btu/h and <135 000 Btu/h	Electric resistance (or none)	Split system and single package	11.0 EER 12.2 IEER	AHRI 340/360
		All other		10.8 EER 12.0 IEER	
	≥135 000 Btu/h and <240 000 Btu/h	Electric resistance (or none)		10.6 EER 11.6 IEER	
		All other		10.4 EER 11.4 IEER	
	≥240 000 Btu/h	Electric resistance (or none)		9.5 EER 10.6 IEER	
		All other		9.3 EER 10.4 IEER	
Water to air, water loop (cooling mode)	<17 000 Btu/h	All	86°F entering water	12.2 EER	ISO 13256-1
	≥17 000 Btu/h and <65 000 Btu/h			13.0 EER	
	≥65 000 Btu/h and <135 000 Btu/h			13.0 EER	
Water to air, ground-water (cooling mode)	<135 000 Btu/h	All	59°F entering water	18.0 EER	ISO 13256-1
Brine to air, ground loop (cooling mode)	<135 000 Btu/h	All	77°F entering water	14.1 EER	ISO 13256-1
Water to water, water loop (cooling mode)	<135 000 Btu/h	All	86°F entering water	10.6 EER	ISO 13256-2
Water to water, groundwater (cooling mode)	<135 000 Btu/h	All	59°F entering water	16.3 EER	ISO 13256-2
Brine to water, ground loop (cooling mode)	<135 000 Btu/h	All	77°F entering water	12.1 EER	ISO 13256-2
Air cooled (heating mode)	<65 000 Btu/h[2] (cooling capacity)	–	Split system, three phase	8.2 HSPF	AHRI 210/240
			Single package, three phase	8.0 HSPF	
Through the wall, air cooled (heating mode)	≤30 000 Btu/h[2] (cooling capacity)	–	Split system, three phase	7.4 HSPF	AHRI 210/240
			Single package, three phase	7.4 HSPF	
Small duct high velocity, air cooled (heating mode)	<65 000 Btu/h[2]	–	Split system, three phase	6.8 HSPF	AHRI 210/240

TABLE E 503.7.1(2) (continued)
ELECTRICALLY OPERATED UNITARY AND APPLIED HEAT PUMPS
MINIMUM EFFICIENCY REQUIREMENTS
[ASHRAE 90.1: TABLE 6.8.1-2]

EQUIPMENT TYPE	SIZE CATEGORY	HEATING SECTION TYPE	SUBCATEGORY OR RATING CONDITION	MINIMUM EFFICIENCY	TEST PROCEDURE[1]
Air cooled (heating mode)	≥65 000 Btu/hc and <135 000 Btu/h (cooling capacity)	–	47°F db/43°F wb outdoor air	3.3 COP_H	AHRI 340/360
			17°F db/15°F wb outdoor air	2.25 COP_H	
	≥135 000 Btu/hc (cooling capacity)		47°F db/43°F wb outdoor air	3.2 COP_H	
			17°F db/15°F wb outdoor air	2.05 COP_H	
Water to air, water loop (heating mode)	<135 000 Btu/h (cooling capacity)	–	68°F entering water	4.3 COP_H	ISO 13256-1
Water to air, groundwater (heating mode)	<135 000 Btu/h (cooling capacity)	–	50°F entering water	3.7 COP_H	ISO 13256-1
Brine to air, ground loop (heating mode)	<135 000 Btu/h (cooling capacity)	–	32°F entering fluid	3.2 COP_H	ISO 13256-1
Water to water, water loop (heating mode)	<135 000 Btu/h (cooling capacity)	–	68°F entering water	3.7 COP_H	ISO 13256-2
Water to water, groundwater (heating mode)	<135 000 Btu/h (cooling capacity)	–	50°F entering water	3.1 COP_H	ISO 13256-2
Brine to water, ground loop (heating mode)	<135 000 Btu/h (cooling capacity)	–	32°F entering fluid	2.5 COP_H	ISO 13256-2

For SI units: 1000 British thermal units per hour = 0.293 kW, °C = (°F-32)/1.8

Notes:

1 ASHRAE 90.1 contains a complete specification of the referenced test procedure, including the referenced year version of the test procedure.

2 Single-phase, air-cooled heat pumps less than 65 000 Btu/h (19 kW) are regulated by the U.S. Department of Energy Code of Federal Regulations 10 CFR 430. SEER and HSPF values for single-phase products are set by the U.S. Department of Energy.

TABLE E 503.7.1(3)
WATER-CHILLING PACKAGES – MINIMUM EFFICIENCY REQUIREMENTS[1, 2, 5]
[ASHRAE 90.1: TABLE 6.8.1-3]

EQUIPMENT TYPE	SIZE CATEGORY	UNITS	PATH A	PATH B	TEST PROCEDURE[3]
Air-cooled chillers	<150 tons	EER (Btu/Wh)	≥10.100 *FL*	≥9.700 *FL*	AHRI 550/590
			≥13.700 *IPLV.IP*	≥15.800 *IPLV.IP*	
	≥150 tons		≥10.100 *FL*	≥9.700 *FL*	
			≥14.000 *IPLV.IP*	≥16.100 *IPLV.IP*	
Air-cooled without condenser, electrically operated	All capacities	EER (Btu/Wh)	Air-cooled chillers without condenser must be rated with matching condensers and comply with air-cooled chiller efficiency requirements		AHRI 550/590
Water-cooled, electrically operated positive displacement	<75 tons	kW/ton	≤0.750 *FL*	≤0.780 *FL*	AHRI 550/590
			≤0.600 *IPLV.IP*	≤0.500 *IPLV.IP*	
	≥75 tons and <150 tons		≤0.720 *FL*	≤0.750 *FL*	
			≤0.560 *IPLV.IP*	≤0.490 *IPLV.IP*	
	≥150 tons and <300 tons		≤0.660 *FL*	≤0.680 *FL*	
			≤0.540 *IPLV.IP*	≤0.440 *IPLV.IP*	
	≥300 tons and <600 tons		≤0.610 *FL*	≤0.625 *FL*	
			≤0.520 *IPLV.IP*	≤0.410 *IPLV.IP*	
	≥600 tons		≤0.560 *FL*	≤0.585 *FL*	
			≤0.500 *IPLV.IP*	≤0.380 *IPLV.IP*	
Water cooled, electrically operated centrifugal	<150 tons	kW/ton	≤0.610 FL	≤0.695 *FL*	AHRI 550/590
			≤0.550 *IPLV.IP*	≤0.440 *IPLV.IP*	
	≥150 tons and <300 tons		≤0.610 *FL*	≤0.635 *FL*	
			≤0.550 *IPLV.IP*	≤0.400 *IPLV.IP*	
	≥300 tons and <400 tons		≤0.560 *FL*	≤0.595 *FL*	
			≤0.520 *IPLV.IP*	≤0.390 *IPLV.IP*	
	≥400 tons and <600 tons		≤0.560 *FL*	≤0.585 *FL*	
			≤0.500 *IPLV.IP*	≤0.380 *IPLV.IP*	
	≥600 tons		≤0.560 *FL*	≤0.585 *FL*	
			≤0.500 *IPLV.IP*	≤0.380 *IPLV.IP*	
Air-cooled absorption, single effect	All capacities	COP (W/W)	≥0.600 *FL*	NA[4]	AHRI 560
Water-cooled absorption, single effect	All capacities	COP (W/W)	≥0.700 *FL*	NA[4]	AHRI 560
Absorption double effect, indirect fired	All capacities	COP (W/W)	≥1.000 *FL*	NA[4]	AHRI 560
			≥1.050 *IPLV.IP*		
Absorption double effect, direct fired	All capacities	COP (W/W)	≥1.000 *FL*	NA[4]	AHRI 560
			≥1.000 *IPLV*		

For SI units: 1 metric ton = 1000 kg, 1000 British thermal units per hour = 0.293 kW

Notes:

1 The requirements for centrifugal chillers shall be adjusted for nonstandard rating conditions per Section E 503.4.1 and are only applicable for the range of conditions listed there. The requirements for air-cooled, water-cooled positive displacement and absorption chillers are at standard rating conditions defined in the reference test procedure.

2 Both the full-load and *IPLV.IP* requirements must be met or exceeded to comply with this appendix. When there is a Path B, compliance can be with either Path A or Path B for any application.

3 ASHRAE 90.1 contains a complete specification of the referenced test procedure, including the referenced year version of the test procedure.

4 NA means the requirements are not applicable for Path B, and only Path A can be used for compliance.

5 *FL* is the full-load performance requirements, and *IPLV.IP* is for the part-load performance requirements.

TABLE E 503.7.1(4)
ELECTRICALLY OPERATED PACKAGED TERMINAL AIR CONDITIONERS, PACKAGED TERMINAL HEAT PUMPS, SINGLE-PACKAGE VERTICAL AIR CONDITIONERS, SINGLE-PACKAGE VERTICAL HEAT PUMPS, ROOM AIR CONDITIONERS, AND ROOM AIR CONDITIONER HEAT PUMPS – MINIMUM EFFICIENCY REQUIREMENTS
[ASHRAE 90.1: TABLE 6.8.1-4]

EQUIPMENT TYPE	SIZE CATEGORY (INPUT)	SUBCATEGORY OR RATING CONDITION	MINIMUM EFFICIENCY	TEST PROCEDURE[1]
PTAC (cooling mode) standard size	All capacities	95°F db outdoor air	13.8 – (0.300 × Cap/1000)[3] (before 1/1/2015)	AHRI 310/ 380
			14.0 – (0.300 × Cap/1000)[3] (as of 1/1/2015)	
PTAC (cooling mode) nonstandard size[1]	All capacities	95°F db outdoor air	10.9 – (0.213 × Cap/1000)[3] EER	AHRI 310/ 380
PTHP (cooling mode) standard size	All capacities	95°F db outdoor air	14.0 – (0.300 × Cap/1000)[3]	AHRI 310/ 380
PTHP (cooling mode) nonstandard size[2]	All capacities	95°F db outdoor air	10.8 – (0.213 × Cap/1000)[3] EER	AHRI 310/ 380
PTHP (heating mode) standard size	All capacities	–	3.7 – (0.052 × Cap/1000)[3] COP_H	AHRI 310/ 380
PTHP (heating mode) nonstandard size[2]	All capacities	–	2.9 – (0.026 × Cap/1000)[3] COP_H	AHRI 310/ 380
SPVAC (cooling mode)	<65,000 Btu/h	95°F db/75°F wb outdoor air	10.0 EER	AHRI 390
	≥65 000 Btu/h and <135 000 Btu/h		10.0 EER	
	≥135 000 Btu/h and <240 000 Btu/h		10.0 EER	
SPVHP (cooling mode)	<65 000 Btu/h	95°F db/75°F wb outdoor air	10.0 EER	AHRI 390
	≥65 000 Btu/h and <135 000 Btu/h		10.0 EER	
	≥135 000 Btu/h and <240 000 Btu/h		10.0 EER	
SPVHP (heating mode)	<65 000 Btu/h	47°F db/43°F wb outdoor air	3.0 COP_H	AHRI 390
	≥65 000 Btu/h and <135 000 Btu/h		3.0 COP_H	
	≥135 000 Btu/h and <240 000 Btu/h		3.0 COP_H	
Room air conditioners with louvered sides	<6000 Btu/h	–	9.7 SEER	AHAM RAC-1
	≥6000 Btu/h and <8000 Btu/h		9.7 SEER	
	≥8000 Btu/h and <14 000 Btu/h		9.8 EER	
	≥14 000 Btu/h and <20 000 Btu/h		9.7 SEER	
	≥20 000 Btu/h		8.5 EER	
SPVAC (cooling mode), nonweatherized space constrained	≤30 000 Btu/h	95°F db/75°F wb outdoor air	9.2 EER	AHRI 390
	>30 000 Btu/h and ≤36 000 Btu/h		9.0 EER	
SPVHP (cooling mode), nonweatherized space constrained	≤30 000 Btu/h	95°F db/75°F wb outdoor air	9.2 EER	AHRI 390
	>30 000 Btu/h and ≤36 000 Btu/h		9.0 EER	
SPVHP (heating mode), nonweatherized space constrained	≤30 000 Btu/h	47°F db/43°F wb outdoor air	3.0 COP_H	AHRI 390
	>30 000 Btu/h and ≤36 000 Btu/h		3.0 COP_H	

TABLE E 503.7.1(4) (continued)
ELECTRICALLY OPERATED PACKAGED TERMINAL AIR CONDITIONERS, PACKAGED TERMINAL HEAT PUMPS, SINGLE-PACKAGE VERTICAL AIR CONDITIONERS, SINGLE-PACKAGE VERTICAL HEAT PUMPS, ROOM AIR CONDITIONERS, AND ROOM AIR CONDITIONER HEAT PUMPS — MINIMUM EFFICIENCY REQUIREMENTS
[ASHRAE 90.1: TABLE 6.8.1-4]

EQUIPMENT TYPE	SIZE CATEGORY (INPUT)	SUBCATEGORY OR RATING CONDITION	MINIMUM EFFICIENCY	TEST PROCEDURE[1]
Room air conditioners without louvered sides	<8000 Btu/h	–	9.0 EER	AHAM RAC-1
	≥8000 Btu/h and <20 000 Btu/h	–	8.5 EER	
	≥20 000 Btu/h	–	8.5 EER	
Room air conditioner heat pumps with louvered sides	<20 000 Btu/h	–	9.0 EER	AHAM RAC-1
	≥20 000 Btu/h		8.5 EER	
Room air conditioner heat pumps without louvered sides	<14 000 Btu/h	–	8.5 EER	AHAM RAC-1
	≥14 000 Btu/h		8.0 EER	
Room air conditioner, casement only	All capacities	–	8.7 EER	AHAM RAC-1
Room air conditioner, casement slider	All capacities	–	9.5 EER	AHAM RAC-1

For SI units: 1000 British thermal units per hour = 0.293 kW, °C = (°F-32)/1.8

Notes:

1 ASHRAE 90.1 contains a complete specification of the referenced test procedure, including the referenced year version of the test procedure.

2 Nonstandard size units must be factory labeled as follows: "MANUFACTURED FOR NONSTANDARD SIZE APPLICATIONS ONLY; NOT TO BE INSTALLED IN NEW STANDARD PROJECTS." Nonstandard size efficiencies apply only to units being installed in existing sleeves having an external wall opening of less than 16 inch (406 mm) high or less than 42 inch (1067 mm) wide and having a cross-sectional area less than 670 square inches (0.432 m^2).

3 "Cap" means the rated cooling capacity of the product in Btu/h (kW). If the unit's capacity is less than 7000 Btu/h (2.05 kW), use 7000 Btu/h (2.05 kW) in the calculation. Where the unit's capacity is more than 15 000 Btu/h (4.4 kW), use 15 000 Btu/h (4.4 kW) in the calculation.

TABLE E 503.7.1(5)
WARM-AIR FURNACES AND COMBINATION WARM-AIR FURNACES/AIR-CONDITIONING UNITS, WARM-AIR DUCT FURNACES, AND UNIT HEATERS — MINIMUM EFFICIENCY REQUIREMENTS
[ASHRAE 90.1: TABLE 6.8.1-5]

EQUIPMENT TYPE	SIZE CATEGORY (INPUT)	SUBCATEGORY OR RATING CONDITION	MINIMUM EFFICIENCY	TEST PROCEDURE[1]
Warm-air furnace, gas fired	<225 000 Btu/h	Maximum capacity[3]	78% *AFUE* or 80% E_t[2, 4]	DOE 10 CFR Part 430 or Section 2.39, Thermal Efficiency, CSA Z21.47
	≥225 000 Btu/h		80% E_t[4]	Section 2.39, Thermal Efficiency, CSA Z21.47
Warm-air furnace, oil fired	<225 000 Btu/h	Maximum capacity[3]	78% *AFUE* or 80% E_t[2, 4]	DOE 10 CFR Part 430 or Section 42, Combustion, UL 727
	≥225 000 Btu/h		81% E_t[4]	Section 42, Combustion, UL 727
Warm-air duct furnaces, gas fired	All capacities	Maximum capacity[3]	80% E_c[5]	Section 2.10, Efficiency, CSA Z83.8
Warm-air unit heaters, gas fired	All capacities	Maximum capacity[3]	80% E_c[5, 6]	Section 2.10, Efficiency, CSA Z83.8
Warm-air unit heaters, oil fired	All capacities	Maximum capacity[3]	80% E_c[5, 6]	Section 40, Combustion, UL 731

For SI units: 1000 British thermal units per hour = 0.293 kW

Notes:

1 ASHRAE 90.1 contains a complete specification of the referenced test procedure, including the referenced year version of the test procedure.

2 Combination units not covered by the U.S. Department of Energy Code of Federal Regulations 10 CFR 430 [three-phase power or cooling capacity greater than or equal to 65 000 Btu/h (19 kW)] may comply with either rating.

3 Compliance of multiple firing rate units shall be at the maximum firing rate.

4 *Et* = thermal efficiency. Units must also include an interrupted or intermittent ignition device (IID), have jacket losses not exceeding 0.75 percent of the input rating, and have either power venting or a flue damper. A vent damper is an acceptable alternative to a flue damper for those furnaces where combustion air is drawn from the conditioned space.

5 *Ec* = combustion efficiency (100 percent less flue losses). See test procedure for detailed discussion.

6 As of August 8, 2008, according to the Energy Policy Act of 2005, units must also include an interrupted or intermittent ignition device (IID) and have either power venting or an automatic flue damper.

TABLE E 503.7.1(6)
GAS- AND OIL-FIRED BOILERS — MINIMUM EFFICIENCY REQUIREMENTS[2, 3]
[ASHRAE 90.1: TABLE 6.8.1-6]

EQUIPMENT TYPE[1]	SUBCATEGORY OR RATING CONDITION	SIZE CATEGORY (INPUT)	MINIMUM EFFICIENCY	EFFICIENCY AS OF 3/2/2020	TEST PROCEDURE
Boilers, hot water	Gas fired	<300 000 Btu/h[6]	82% *AFUE*	82% *AFUE*	10 CFR Part 430
		≥300 000 Btu/h and ≤2 500 000 Btu/h[4]	80% E_t	80% E_t	10 CFR Part 431
		>2 500 000 Btu/h[1]	82% E_c	82% E_c	
	Oil fired[5]	<300 000 Btu/h[7]	84% *AFUE*	84% *AFUE*	10 CFR Part 430
		≥300 000 Btu/h and ≤2 500 000 Btu/h[4]	82% E_t	82% E_t	10 CFR Part 431
		>2 500 000 Btu/h[1]	84% E_c	84% E_c	
Boilers, steam	Gas fired	<300 000 Btu/h[6]	80% *AFUE*	80% *AFUE*	10 CFR Part 430
	Gas fired— all, except natural draft	≥300 000 Btu/h and ≤2 500 000 Btu/h[4]	79% E_t	79% E_t	10 CFR Part 431
		>2 500 000 Btu/h[1]	79% E_t	79% E_t	
	Gas fired— natural draft	≥300 000 Btu/h and ≤2 500 000 Btu/h[4]	77% E_t	79% E_t	
		>2 500 000 Btu/h[1]	77% E_t	79% E_t	
	Oil fired[5]	<300 000 Btu/h	82% *AFUE*	82% *AFUE*	10 CFR Part 430
		≥300 000 Btu/h and ≤2 500 000 Btu/h[4]	81% E_t	81% E_t	10 CFR Part 431
		>2 500 000 Btu/h[1]	81% E_t	81% E_t	

For SI units: 1000 British thermal units per hour = 0.293 kW

Notes:

1 These requirements apply to boilers with rated input of 8 000 000 Btu/h (2343 kW) or less that are not packaged boilers and to all packaged boilers. Minimum efficiency requirements for boilers cover all capacities of packaged boilers.

2 Ec = combustion efficiency (100 percent less flue losses). See reference document for detailed information.

3 Et = thermal efficiency. See reference document for detailed information.

4 Maximum capacity—minimum and maximum ratings as provided for and allowed by the unit's controls.

5 Includes oil-fired (residual).

6 Boilers shall not be equipped with a constant burning pilot light.

7 A boiler not equipped with a tankless domestic water-heating coil shall be equipped with an automatic means for adjusting the temperature of the water such that an incremental change in inferred heat load produces a corresponding incremental change in the temperature of the water supplied.

TABLE E 503.7.1(7)
PERFORMANCE REQUIREMENTS FOR HEAT REJECTION EQUIPMENT
MINIMUM EFFICIENCY REQUIREMENTS
[ASHRAE 90.1: TABLE 6.8.1-7]

EQUIPMENT TYPE	TOTAL SYSTEM HEAT-REJECTION CAPACITY AT RATED CONDITIONS	SUBCATEGORY OR RATING CONDITION[8]	PERFORMANCE REQUIRED[1,2,3,6,7]	TEST PROCEDURE[4,5]
Propeller or axial fan open-circuit cooling towers	All	95°F entering water 85°F leaving water 75°F entering wb	≥40.2 gpm/hp	CTI ATC-105 and CTI STD-201 RS
Centrifugal fan open-circuit cooling towers	All	95°F entering water 85°F leaving water 75°F entering wb	≥20.0 gpm/hp	CTI ATC-105 and CTI STD-201 RS
Propeller or axial fan closed-circuit cooling towers	All	102°F entering water 90°F leaving water 75°F entering wb	≥16.1 gpm/hp	CTI ATC-105S and CTI STD-201 RS
Centrifugal closed- circuit cooling towers	All	102°F entering water 90°F leaving water 75°F entering wb	≥7.0 gpm/hp	CTI ATC-105S and CTI STD-201 RS
Propeller or axial fan evaporative condensers	All	R-507A test fluid 165°F entering gas temperature 105°F condensing temperature 75°F entering wb	≥157 000 Btu/h·hp	CTI ATC-106
Propeller or axial fan evaporative condensers	All	Ammonia test fluid 140°F entering gas temperature 96.3°F condensing temperature 75°F entering wb	≥134 000 Btu/h·hp	CTI ATC-106
Centrifugal fan evaporative condensers	All	R-507A test fluid 165°F entering gas temperature 105°F condensing temperature 75°F entering wb	≥135 000 Btu/h·hp	CTI ATC-106
Centrifugal fan evaporative condensers	All	Ammonia test fluid 140°F entering gas temperature 96.3°F condensing temperature 75°F entering wb	≥110 000 Btu/h·hp	CTI ATC-106
Air cooled condensers	All	125°F condensing temperature 190°F entering gas temperature 15°F subcooling 95°F entering db	≥176 000 Btu/h·hp	AHRI 460

For SI units: °C = (°F-32)/1.8, 1 gallon per minute per horsepower = 0.085 [(L/s)/kW], 1000 British thermal units per hour = 0.293 kW, 1 horsepower = 0.746 kW

Notes:

1 For purposes of this table, open-circuit cooling tower performance is defined as the water flow rating of the tower at the thermal rating condition listed in Table E 503.7.1(7) divided by the fan motor nameplate power.

2 For purposes of this table, closed-circuit cooling tower performance is defined as the process water flow rating of the tower at the thermal rating condition listed in Table E 503.7.1(7) divided by the sum of the fan motor nameplate power and the integral spray pump motor nameplate power.

3 For purposes of this table, air-cooled condenser performance is defined as the heat rejected from the refrigerant divided by the fan motor nameplate power.

4 ASHRAE 90.1 contains a complete specification of the referenced test procedure, including the referenced year version of the test procedure.

5 The efficiencies and test procedures for both open- and closed-circuit cooling towers are not applicable to hybrid cooling towers that contain a combination of separate wet and dry heat exchange sections. The certification requirements do not apply to field-erected cooling towers.

6 All cooling towers shall comply with the minimum efficiency listed in the table for that specific type of tower with the capacity effect of any project-specific accessories and/or options included in the capacity of the cooling tower.

7 For purposes of this table, evaporative condenser performance is defined as the heat rejected at the specified rating condition in the table, divided by the sum of the fan motor nameplate power and the integral spray pump nameplate power.

8 Requirements for evaporative condensers are listed with ammonia (R-717) and R-507A as test fluids in the table. Evaporative condensers intended for use with halocarbon refrigerants other than R-507A must meet the minimum efficiency requirements listed above with R-507A as the test fluid.

TABLE E 503.7.1(8)
HEAT TRANSFER EQUIPMENT — MINIMUM EFFICIENCY REQUIREMENTS
[ASHRAE 90.1: TABLE 6.8.1-8]

EQUIPMENT TYPE	SUBCATEGORY	MINIMUM EFFICIENCY[1]	TEST PROCEDURE[2]
Liquid-to-liquid heat exchangers	Plate type	NR	AHRI 400

Notes:
1 NR = No Requirement
2 ASHRAE 90.1 contains a complete specification of the referenced test procedure, including the referenced year version of the test procedure.

TABLE E 503.7.1(9)
ELECTRICALLY OPERATED VARIABLE-REFRIGERANT-FLOW AIR CONDITIONERS — MINIMUM EFFICIENCY REQUIREMENTS
[ASHRAE 90.1: TABLE 6.8.1-9]

EQUIPMENT TYPE	SIZE CATEGORY	HEATING SECTION TYPE	SUBCATEGORY OR RATING CONDITION	MINIMUM EFFICIENCY	TEST PROCEDURE
VRF air conditioners, air cooled	<65 000 Btu/h	All	VRF multisplit system	13.0 SEER	AHRI 1230
	≥65 000 Btu/h and <135 000 Btu/h	Electric resistance (or none)	VRF multisplit system	11.2 EER 13.1 IEER (before 1/1/2017) 15.5 IEER (as of 1/1/2017)	
	≥135 000 Btu/h and <240 000 Btu/h	Electric resistance (or none)	VRF multisplit system	11.0 EER 12.9 IEER (before 1/1/2017) 14.9 IEER (as of 1/1/2017)	
	≥240 000 Btu/h	Electric resistance (or none)	VRF multisplit system	10.0 EER 11.6 IEER (before 1/1/2017) 13.9 IEER (as of 1/1/2017)	

For SI units: 1000 British thermal units per hour = 0.293 kW

TABLE E 503.7.1(10)
ELECTRICALLY OPERATED VARIABLE-REFRIGERANT-FLOW AND APPLIED HEAT PUMPS — MINIMUM EFFICIENCY REQUIREMENTS
[ASHRAE 90.1: TABLE 6.8.1-10]

<table>
<tr><th>EQUIPMENT TYPE</th><th>SIZE CATEGORY</th><th>HEATING SECTION TYPE</th><th>SUBCATEGORY OR RATING CONDITION</th><th>MINIMUM EFFICIENCY</th><th>TEST PROCEDURE</th></tr>
<tr><td rowspan="7">VRF air cooled (cooling mode)</td><td><65 000 Btu/h</td><td>All</td><td rowspan="2">VRF multisplit system</td><td>13.0 SEER</td><td rowspan="7">AHRI 1230</td></tr>
<tr><td rowspan="2">≥65 000 Btu/h and <135 000 Btu/h</td><td rowspan="6">Electric resistance (or none)</td><td>11.0 EER
12.9 IEER
(before 1/1/2017)
14.6 IEER
(as of 1/1/2017)</td></tr>
<tr><td>VRF multisplit system with heat recovery</td><td>10.8 EER
12.7 IEER
(before 1/1/2017)
14.4 IEER
(as of 1/1/2017)</td></tr>
<tr><td rowspan="2">≥135 000 Btu/h and <240 000 Btu/h</td><td>VRF multisplit system</td><td>10.6 EER
12.3 IEER
(before 1/1/2017)
13.9 IEER
(as of 1/1/2017)</td></tr>
<tr><td>VRF multisplit system with heat recovery</td><td>10.4 EER
12.1 IEER
(before 1/1/2017)
13.7 IEER
(as of 1/1/2017)</td></tr>
<tr><td rowspan="2">≥240 000 Btu/h</td><td>VRF multisplit system</td><td>9.5 EER
11.0 IEER
(before 1/1/2017)
12.7 IEER
(as of 1/1/2017)</td></tr>
<tr><td>VRF multisplit system with heat recovery</td><td>9.3 EER
10.8 IEER
(before 1/1/2017)
12.5 IEER
(as of 1/1/2017)</td></tr>
<tr><td rowspan="8">VRF water source (cooling mode)</td><td rowspan="2"><65 000 Btu/h</td><td rowspan="8">All</td><td>VRF multisplit systems 86°F entering water</td><td>12.0 EER
16.0 IEER
(as of 1/1/2018)</td><td rowspan="8">AHRI 1230</td></tr>
<tr><td>VRF multisplit systems with heat recovery 86°F entering water</td><td>11.8 EER
15.8 IEER
(as of 1/1/2018)</td></tr>
<tr><td rowspan="2">≥65 000 Btu/h and <135 000 Btu/h</td><td>VRF multisplit system 86°F entering water</td><td>12.0 EER
16.0 IEER
(as of 1/1/2018)</td></tr>
<tr><td>VRF multisplit system with heat recovery 86°F entering water</td><td>11.8 EER
15.8 IEER
(as of 1/1/2018)</td></tr>
<tr><td rowspan="2">≥135 000 Btu/h and <240 000 Btu/h</td><td>VRF multisplit system 86°F entering water</td><td>10.0 EER
14.0 IEER
(as of 1/1/2018)</td></tr>
<tr><td>VRF multisplit system with heat recovery 86°F entering water</td><td>9.8 EER
13.8 IEER
(as of 1/1/2018)</td></tr>
<tr><td rowspan="2">≥240 000 Btu/h</td><td>VRF multisplit system 86°F entering water</td><td>10.0 EER
(before 1/1/2018)
12.0 IEER
(as of 1/1/2018)</td></tr>
<tr><td>VRF multisplit system with heat recovery 86°F entering water</td><td>9.8 EER
(before 1/1/2018)
11.8 IEER
(as of 1/1/2018)</td></tr>
</table>

TABLE E 503.7.1(10) (continued)
ELECTRICALLY OPERATED VARIABLE-REFRIGERANT-FLOW AND APPLIED HEAT PUMPS — MINIMUM EFFICIENCY REQUIREMENTS
[ASHRAE 90.1: TABLE 6.8.1-10]

EQUIPMENT TYPE	SIZE CATEGORY	HEATING SECTION TYPE	SUBCATEGORY OR RATING CONDITION	MINIMUM EFFICIENCY	TEST PROCEDURE
VRF groundwater source (cooling mode)	<135 000 Btu/h	All	VRF multisplit system with heat recovery 59°F entering water	16.2 EER	AHRI 1230
			VRF multisplit system with heat recovery 59°F entering water	16.0 EER	
	≥135 000 Btu/h		VRF multisplit system with heat recovery 59°F entering water	13.8 EER	
			VRF multisplit system with heat recovery 59°F entering water	13.6 EER	
VRF ground source (cooling mode)	<135 000 Btu/h	All	VRF multisplit system 77°F entering water	13.4 EER	AHRI 1230
			VRF multisplit system with heat recovery 77°F entering water	13.2 EER	
	≥135 000 Btu/h		VRF multisplit system 77°F entering water	11.0 EER	
			VRF multisplit system with heat recovery 77°F entering water	10.8 EER	
VRF Air cooled (heating mode)	<65 000 Btu/h (cooling capacity)	—	VRF Multi-split system	7.7 HSPF	AHRI 1230
	≥65 000 Btu/h and <135 000 Btu/h (cooling capacity)	—	VRF Multi-split system 47°F db/43°F wb outdoor air	3.3 COP_H	
			17°F db/15°F wb outdoor air	2.25 COP_H	
	≥135 000 Btu/h (cooling capacity)	—	VRF Multi-split system 47°F db/43°F wb outdoor air	3.2 COP_H	
			17°F db/15°F wb outdoor air	2.05 COP_H	
VRF Water source (heating mode)	<65 000 Btu/h (cooling capacity)	—	*VRF* multisplit system 68°F entering water	4.2 COP_H (before 1/1/2018) 4.3 COP_H (as of 1/1/2018)	AHRI 1230
	≥65 000 Btu/h and <135 000 Btu/h (cooling capacity)	—	*VRF* multisplit system 68°F entering water	4.2 COP_H (before 1/1/2018) 4.3 COP_H (as of 1/1/2018)	
	≥135 000 Btu/h and <240 000 Btu/h (cooling capacity)	—	*VRF* multisplit system 68°F entering water	3.9 COP_H (before 1/1/2018) 4.0 COP_H (as of 1/1/2018)	
	≥240 000 Btu/h (cooling capacity)	—	*VRF* multisplit system 68°F entering water	3.9 COP_H	
VRF Groundwater source (heating mode)	<135 000 Btu/h (cooling capacity)	—	VRF Multi-split system 50°F entering water	3.6 COP_H	AHRI 1230
	≥135 000 Btu/h (cooling capacity)	—	VRF Multi-split system 50°F entering water	3.3 COP_H	
VRF Ground source (heating mode)	<135 000 Btu/h (cooling capacity)	—	VRF Multi-split system 32°F entering water	3.1 COP_H	AHRI 1230
	≥135 000 Btu/h (cooling capacity)	—	VRF Multi-split system 32°F entering water	2.8 COP_H	

For SI units: 1000 British thermal units per hour = 0.293 kW, °C=(°F-32)/1.8

TABLE E 503.7.1(11)
AIR CONDITIONERS AND CONDENSING UNITS SERVING COMPUTER ROOMS MINIMUM EFFICIENCY REQUIREMENTS
[ASHRAE 90.1: TABLE 6.8.1-11]

EQUIPMENT TYPE	NET SENSIBLE COOLING CAPACITY	STANDARD MODEL	MINIMUM NET SENSIBLE COP_C			TEST PROCEDURE
			RETURN AIR DRY-BULB TEMPERATURE/ DEW-POINT TEMPERATURE			
			CLASS 1	CLASS 2	CLASS 3	
			75°F/52°F	85°F/52°F	95°F/52°F	
Air cooled	<65 000 Btu/h	Downflow unit		2.30		AHRI 1360
		Upflow unit—ducted		2.10		
		Upflow unit—nonducted	2.09			
		Horizontal-flow unit			2.45	
	≥65 000 Btu/h and <240 000 Btu/h	Downflow unit		2.20		
		Upflow unit—ducted		2.05		
		Upflow unit—nonducted	1.99			
		Horizontal-flow unit			2.35	
	≥240 000 Btu/h	Downflow unit		2.00		
		Upflow unit—ducted		1.85		
		Upflow unit—nonducted	1.79			
		Horizontal-flow unit			2.15	
Water cooled	<65 000 Btu/h	Downflow unit		2.50		AHRI 1360
		Upflow unit—ducted		2.30		
		Upflow unit—nonducted	2.25			
		Horizontal-flow unit			2.70	
	≥65 000 Btu/h and <240 000 Btu/h	Downflow unit		2.40		
		Upflow unit—ducted		2.20		
		Upflow unit—nonducted	2.15			
		Horizontal-flow unit			2.60	
	≥240 000 Btu/h	Downflow unit		2.25		
		Upflow unit—ducted		2.10		
		Upflow unit—nonducted	2.05			
		Horizontal-flow unit			2.45	
Water cooled with fluid economizer	<65 000 Btu/h	Downflow unit		2.45		AHRI 1360
		Upflow unit—ducted		2.25		
		Upflow unit—nonducted	2.20			
		Horizontal-flow unit			2.60	
	≥65 000 Btu/h and <240 000 Btu/h	Downflow unit		2.35		
		Upflow unit—ducted		2.15		
		Upflow unit—nonducted	2.10			
		Horizontal-flow unit			2.55	
	≥240 000 Btu/h	Downflow unit		2.20		
		Upflow unit—ducted		2.05		
		Upflow unit—nonducted	2.00			
		Horizontal-flow unit			2.40	

TABLE E 503.7.1(11) (continued)
AIR CONDITIONERS AND CONDENSING UNITS SERVING COMPUTER ROOMS
MINIMUM EFFICIENCY REQUIREMENTS
[ASHRAE 90.1: TABLE 6.8.1-11]

EQUIPMENT TYPE	NET SENSIBLE COOLING CAPACITY	STANDARD MODEL	MINIMUM NET SENSIBLE COP_C			TEST PROCEDURE
			RETURN AIR DRY-BULB TEMPERATURE/ DEW-POINT TEMPERATURE			
			CLASS 1	CLASS 2	CLASS 3	
			75°F/52°F	85°F/52°F	95°F/52°F	
Glycol cooled	<65 000 Btu/h	Downflow unit		2.30		AHRI 1360
		Upflow unit—ducted		2.10		
		Upflow unit—nonducted	2.00			
		Horizontal-flow unit			2.40	
	≥65 000 Btu/h and <240 000 Btu/h	Downflow unit		2.05		
		Upflow unit—ducted		1.85		
		Upflow unit—nonducted	1.85			
		Horizontal-flow unit			2.15	
	≥240 000 Btu/h	Downflow unit		1.95		
		Upflow unit—ducted		1.80		
		Upflow unit—nonducted	1.75			
		Horizontal-flow unit			2.10	
Glycol cooled with fluid economizer	<65 000 Btu/h	Downflow unit		2.25		AHRI 1360
		Upflow unit—ducted		2.10		
		Upflow unit—nonducted	2.00			
		Horizontal-flow unit			2.35	
	≥65 000 Btu/h and <240 000 Btu/h	Downflow unit		1.95		
		Upflow unit—ducted		1.80		
		Upflow unit—nonducted	1.75			
		Horizontal-flow unit			2.10	
	≥240 000 Btu/h	Downflow unit		1.90		
		Upflow unit—ducted		1.80		
		Upflow unit—nonducted	1.70			
		Horizontal-flow unit			2.10	

For SI units: 1000 British thermal units per hour = 0.293 kW, °C=(°F-32)/1.8

TABLE E 503.7.1(12)
COMMERCIAL REFRIGERATOR AND FREEZERS — MINIMUM EFFICIENCY REQUIREMENTS
[ASHRAE 90.1: TABLE 6.8.1-12]

EQUIPMENT TYPE	APPLICATION	ENERGY USE LIMITS, KWH/DAY*	TEST PROCEDURE
Refrigerator with solid doors	Holding temperature	$0.10 \times V + 2.04$	AHRI 1200
Refrigerator with transparent doors	Holding temperature	$0.12 \times V + 3.34$	AHRI 1200
Freezers with solid doors	Holding temperature	$0.40 \times V + 1.38$	AHRI 1200
Freezers with transparent doors	Holding temperature	$0.75 \times V + 4.10$	AHRI 1200
Refrigerators/freezers with solid doors	Holding temperature	the greater of $0.12 \times V + 3.34$ or 0.70	AHRI 1200
Commercial refrigerators	Pulldown	$0.126 \times V + 3.51$	AHRI 1200

For SI units: 1000 British thermal units per hour per day = 0.293 kW/day

* V = the chiller or frozen compartment volume (ft^3) as defined in Association of Home Appliance Manufacturers.

TABLE E 503.7.1(13)
COMMERCIAL REFRIGERATION – MINIMUM EFFICIENCY REQUIREMENTS
[ASHRAE 90.1: TABLE 6.8.1-13]

EQUIPMENT TYPE				ENERGY USE LIMITS[2,3] KWH/DAY	TEST PROCEDURE
EQUIPMENT CLASS[1]	FAMILY CODE	OPERATING MODE	RATING TEMPERATURE		
VOP.RC.M	Vertical open	Remote condensing	Medium temperature	$0.82 \times TDA + 4.07$	AHRI 1200
SVO.RC.M	Semivertical open	Remote condensing	Medium temperature	$0.83 \times TDA + 3.18$	AHRI 1200
HZO.RC.M	Horizontal open	Remote condensing	Medium temperature	$0.35 \times TDA + 2.88$	AHRI 1200
VOP.RC.L	Vertical open	Remote condensing	Low temperature	$2.27 \times TDA + 6.85$	AHRI 1200
HZO.RC.L	Horizontal open	Remote condensing	Low temperature	$0.57 \times TDA + 6.88$	AHRI 1200
VCT.RC.M	Vertical transparent door	Remote condensing	Medium temperature	$0.22 \times TDA + 1.95$	AHRI 1200
VCT.RC.L	Vertical transparent door	Remote condensing	Low temperature	$0.56 \times TDA + 2.61$	AHRI 1200
SOC.RC.M	Service over counter	Remote condensing	Medium temperature	$0.51 \times TDA + 0.11$	AHRI 1200
VOP.SC.M	Vertical open	Self contained	Medium temperature	$1.74 \times TDA + 4.71$	AHRI 1200
SVO.SC.M	Semivertical open	Self contained	Medium temperature	$1.73 \times TDA + 4.59$	AHRI 1200
HZO.SC.M	Horizontal open	Self contained	Medium temperature	$0.77 \times TDA + 5.55$	AHRI 1200
HZO.SC.L	Horizontal open	Self contained	Low temperature	$1.92 \times TDA + 7.08$	AHRI 1200
VCT.SC.I	Vertical transparent door	Self contained	Ice cream	$0.67 \times TDA + 3.29$	AHRI 1200
VCS.SC.I	Vertical solid door	Self contained	Ice cream	$0.38 \times V + 0.88$	AHRI 1200
HCT.SC.I	Horizontal transparent door	Self contained	Ice cream	$0.56 \times TDA + 0.43$	AHRI 1200
SVO.RC.L	Semivertical open	Remote condensing	Low temperature	$2.27 \times TDA + 6.85$	AHRI 1200
VOP.RC.I	Vertical open	Remote condensing	Ice cream	$2.89 \times TDA + 8.7$	AHRI 1200
SVO.RC.I	Semivertical open	Remote condensing	Ice cream	$2.89 \times TDA + 8.7$	AHRI 1200
HZO.RC.I	Horizontal open	Remote condensing	Ice cream	$0.72 \times TDA + 8.74$	AHRI 1200
VCT.RC.I	Vertical transparent door	Remote condensing	Ice cream	$0.66 \times TDA + 3.05$	AHRI 1200
HCT.RC.M	Horizontal transparent door	Remote condensing	Medium temperature	$0.16 \times TDA + 0.13$	AHRI 1200
HCT.RC.L	Horizontal transparent door	Remote condensing	Low temperature	$0.34 \times TDA + 0.26$	AHRI 1200
HCT.RC.I	Horizontal transparent door	Remote condensing	Ice cream	$0.4 \times TDA + 0.31$	AHRI 1200
VCS.RC.M	Vertical solid door	Remote condensing	Medium temperature	$0.11 \times V + 0.26$	AHRI 1200
VCS.RC.L	Vertical solid door	Remote condensing	Low temperature	$0.23 \times V + 0.54$	AHRI 1200
VCS.RC.I	Vertical solid door	Remote condensing	Ice cream	$0.27 \times V + 0.63$	AHRI 1200
HCS.RC.M	Horizontal solid door	Remote condensing	Medium temperature	$0.11 \times V + 0.26$	AHRI 1200
HCS.RC.L	Horizontal solid door	Remote condensing	Low temperature	$0.23 \times V + 0.54$	AHRI 1200
HCS.RC.I	Horizontal solid door	Remote condensing	Ice cream	$0.27 \times V + 0.63$	AHRI 1200
HCS.RC.I	Horizontal solid door	Remote condensing	Ice cream	$0.27 \times V + 0.63$	AHRI 1200
SOC.RC.L	Service over counter	Remote condensing	Low temperature	$1.08 \times TDA + 0.22$	AHRI 1200
SOC.RC.I	Service over counter	Remote condensing	Ice cream	$1.26 \times TDA + 0.26$	AHRI 1200
VOP.SC.L	Vertical open	Self contained	Low temperature	$4.37 \times TDA + 11.82$	AHRI 1200
VOP.SC.I	Vertical open	Self contained	Ice cream	$5.55 \times TDA + 15.02$	AHRI 1200
SVO.SC.L	Semivertical open	Self contained	Low temperature	$4.34 \times TDA + 11.51$	AHRI 1200
SVO.SC.I	Semivertical open	Self contained	Ice cream	$5.52 \times TDA + 14.63$	AHRI 1200
HZO.SC.I	Horizontal open	Self contained	Ice cream	$2.44 \times TDA + 9.0$	AHRI 1200
SOC.SC.I	Service over counter	Self contained	Ice cream	$1.76 \times TDA + 0.36$	AHRI 1200
HCS.SC.I	Horizontal solid door	Self contained	Ice cream	$0.38 \times V + 0.88$	AHRI 1200

For SI units: 1000 British thermal units per hour per day = 0.293 kW/day, °C = (°F-32)/1.8

Notes:

1 Equipment class designations consist of a combination [in sequential order separated by periods (AAA).(BB).(C)] of the following:

(a) (AAA)—An equipment family code (VOP = vertical open, SVO = semivertical open, HZO = horizontal open, VCT = vertical transparent doors, VCS = vertical solid doors, HCT = horizontal transparent doors, HCS = horizontal solid doors, and SOC = service over counter).

(b) (BB)—An operating mode code (RC = remote condensing and SC = self contained).

(c) (C)—A rating temperature code (M = medium temperature [38°F], L = low temperature [0°F], or I = ice cream temperature [15°F]). For example, "VOP.RC.M" refers to the "vertical open, remote condensing, medium temperature" equipment class.

2 V is the volume of the case (ft) as measured in accordance with AHRI 1200.

3 TDA is the total display area of the case (ft) as measured in accordance with AHRI 1200.

TABLE E 503.7.1(14)
VAPOR COMPRESSION BASED INDOOR POOL DEHUMIDIFIERS – MINIMUM EFFICIENCY REQUIREMENTS
[ASHRAE 90.1: TABLE 6.8.1-14]

EQUIPMENT TYPE	SUBCATEGORY OR RATING CONDITION	MINIMUM EFFICIENCY	TEST PROCEDURE
Single package indoor* (with or without economizer)	Rating Conditions: A, B, or C	3.5 MRE	AHRI 910
Single package indoor water-cooled (with or without economizer)		3.5 MRE	
Single package indoor air-cooled (with or without economizer)		3.5 MRE	
Split system indoor air-cooled (with or without economizer)		3.5 MRE	

* Units without air-cooled condenser

TABLE E 503.7.1(15)
ELECTRICALLY OPERATED DX-DOAS UNITS, SINGLE-PACKAGE AND REMOTE CONDENSER, WITHOUT ENERGY RECOVERY – MINIMUM EFFICIENCY REQUIREMENTS
[ASHRAE 90.1: TABLE 6.8.1-15]

EQUIPMENT TYPE	SUBCATEGORY OR CONDITION	MINIMUM EFFICIENCY	TEST PROCEDURE
Air cooled (dehumidification mode)	–	4.0 *ISMRE*	AHRI 920
Air source heat pumps (dehumidification mode)	–	4.0 *ISMRE*	AHRI 920
Water cooled (dehumidification mode)	Cooling tower condenser water	4.9 *ISMRE*	AHRI 920
	Chilled Water	6.0 *ISMRE*	
Air source heat pump (heating mode)	–	2.7 *ISCOP*	AHRI 920
Water source heat pump (dehumidification mode)	Ground source, closed loop	4.8 *ISMRE*	AHRI 920
	Ground-water source	5.0 *ISMRE*	
	Water source	4.0 *ISMRE*	
Water source heat pump (heating mode)	Ground source, closed loop	2.0 *ISCOP*	AHRI 920
	Ground-water source	3.2 *ISCOP*	
	Water source	3.5 *ISCOP*	

TABLE E 503.7.1(16)
ELECTRICALLY OPERATED DX-DOAS UNITS, SINGLE PACKAGE AND REMOTE CONDENSER, WITH ENERGY RECOVERY – MINIMUM EFFICIENCY REQUIREMENTS
[ASHRAE 90.1: TABLE 6.8.1-16]

EQUIPMENT TYPE	SUBCATEGORY OR RATING CONDITION	MINIMUM EFFICIENCY	TEST PROCEDURE
Air cooled (dehumidification mode)	–	5.2 *ISMRE*	AHRI 920
Air source heat pumps (dehumidification mode)	–	5.2 *ISMRE*	AHRI 920
Water cooled (dehumidification mode)	Cooling tower condenser water	5.3 *ISMRE*	AHRI 920
	Chilled Water	6.6 *ISMRE*	
Air source heat pump (heating mode)	–	3.3 *ISCOP*	AHRI 920
Water source heat pump (dehumidification mode)	Ground source, closed loop	5.2 *ISMRE*	AHRI 920
	Ground-water source	5.8 *ISMRE*	
	Water source	4.8 *ISMRE*	
Water source heat pump (heating mode)	Ground source, closed loop	3.8 *ISCOP*	AHRI 920
	Ground-water source	4.0 *ISCOP*	
	Water source	4.8 *ISCOP*	

TABLE E 503.7.2
MINIMUM DUCT INSULATION R-VALUE[1]
[ASHRAE 90.1: TABLE 6.8.2]

CLIMATE ZONE	DUCT LOCATION		
	EXTERIOR[2]	UNCONDITIONED SPACE AND BURIED DUCTS	INDIRECTLY CONDITIONED SPACE[3, 4]
SUPPLY AND RETURN DUCTS FOR HEATING AND COOLING			
0 to 4	R-8	R-6	R-1.9
5 to 8	R-12	R-6	R-1.9
SUPPLY AND RETURN DUCTS FOR HEATING ONLY			
0 to 1	none	none	none
2 to 4	R-6	R-6	R-1.9
5 to 8	R-12	R-6	R-1.9
SUPPLY AND RETURN DUCTS FOR COOLING ONLY			
0 to 6	R-8	R-6	R-1.9
7 to 8	R-1.9	R-1.9	R-1.9

Notes:

1. Insulation R-values, measured in [°F•h•ft²/(Btu•in)] [(m•K)/W], are for the insulation as installed and do not include film resistance. The required minimum thicknesses do not consider water vapor transmission and possible surface condensation. Where portions of the building envelope are used as a plenum enclosure, building envelope insulation shall be as required by the most restrictive condition of Section E 503.4.7.1 or ASHRAE 90.1, depending on whether the plenum is located in the roof, wall, or floor. Insulation resistance measured on a horizontal plane in accordance with ASTM C518 at a mean temperature of 75°F (24°C) at the installed thickness.
2. Includes attics above insulated ceilings, parking garages and crawl spaces.
3. Includes return air plenums, with or without exposed roofs above.
4. Return ducts in this duct location do not require insulation.

TABLE E 503.7.3(1)
MINIMUM PIPE INSULATION THICKNESS FOR HEATING AND HOT WATER SYSTEMS[1, 2, 3, 4, 5]
(STEAM, STEAM CONDENSATE, HOT WATER HEATING, AND DOMESTIC WATER SYSTEMS)
[ASHRAE 90.1: TABLE 6.8.3-1]

FLUID OPERATING TEMPERATURE RANGE (F°) AND USAGE	INSULATION CONDUCTIVITY		NOMINAL PIPE SIZE OR TUBE SIZE (inches)				
	CONDUCTIVITY Btu•inch/(h•ft²•°F)	MEAN RATING TEMPERATURE °F	<1	1 to <1½	1½ to <4	4 to <8	≥8
			INSULATION THICKNESS (inches)				
>350	0.32 to 0.34	250	4.5	5.0	5.0	5.0	5.0
251 to 350	0.29 to 0.32	200	3.0	4.0	4.5	4.5	4.5
201 to 250	0.27 to 0.30	150	2.5	2.5	2.5	3.0	3.0
141 to 200	0.25 to 0.29	125	1.5	1.5	2.0	2.0	2.0
105 to 140	0.22 to 0.28	100	1.0	1.0	1.5	1.5	1.5

For SI units: °C=(°F-32)/1.8, 1 inch = 25 mm, 1 British thermal unit inch per hour square foot degree Fahrenheit = [0.1 W/(m•K)]

Notes:

1. For insulation outside the stated conductivity range, the minimum thickness (T) shall be determined as follows:

$$T = r\{(1 + t/r)^{K/k} - 1\}$$

Where:

T = minimum insulation thickness (inches).
r = actual outside radius of pipe (inches).
t = insulation thickness listed in this table for applicable fluid temperature and pipe size.
K = conductivity of alternate material at mean rating temperature indicated for the applicable fluid temperature [Btu•in/(h•ft²•°F)] [W/(m•K)].
k = the upper value of the conductivity range listed in this table for the applicable fluid temperature.

2. These thicknesses are on energy efficiency considerations only. Additional insulation is sometimes required relative to safety issues/surface temperature.
3. For piping smaller than 1½ inches (40 mm) or less and located in partitions within conditioned spaces, reduction of these thicknesses by 1 inch (25.4 mm) shall be permitted (before thickness adjustment required in footnote 1) but not to thicknesses below 1 inch (25.4 mm).
4. For direct-buried heating and hot water system piping, reduction of these thicknesses by 1½ inch (40 mm) shall be permitted (before thickness adjustment required in footnote 1) but not to thicknesses below 1 inch (25.4 mm).
5. Table E 503.7.3(1) is based on steel pipe. Nonmetallic pipes schedule 80 thickness or less shall use the table values. For other nonmetallic pipes having thermal resistance more than that of steel pipe, reduced insulation thicknesses are permitted where documentation is provided showing that the pipe with the proposed insulation has no more heat transfer per foot (mm) than a steel pipe of the same size with the insulation thickness shown in Table E 503.7.3(1).

TABLE E 503.7.3(2)
MINIMUM PIPE INSULATION THICKNESS FOR COOLING SYSTEMS (CHILLED WATER, BRINE, AND REFRIGERANT)[1,2,3,4]
[ASHRAE 90.1: TABLE 6.8.3-2]

INSULATION CONDUCTIVITY			NOMINAL PIPE SIZE OR TUBE SIZE (inches)				
FLUID OPERATING TEMPERATURE RANGE (°F) AND USAGE	CONDUCTIVITY Btu•inch/(h•f²•°F)	MEAN RATING TEMPERATURE °F	<1	1 to <1½	1½ to <4	4 to <8	≥8
			INSULATION THICKNESS (inches)				
40 to 60	021 to 0.27	75	0.5	0.5	1.0	1.0	1.0
<40	0.20 to 0.26	50	0.5	1.0	1.0	1.0	1.5

For SI units: °C = (°F-32)/1.8, 1 inch = 25 mm, 1 British thermal unit inch per hour square foot degree Fahrenheit = [0.1 W/(m•k)]

Notes:

1 For insulation outside the stated conductivity range, the minimum thickness (T) shall be determined as follows:

$$T = r\{(1 + t/r)^{K/k} - 1\}$$

Where:

T = minimum insulation thickness (inches).
r = actual outside radius of pipe (inches).
t = insulation thickness listed in this table for applicable fluid temperature and pipe size.
K = conductivity of alternate material at mean rating temperature indicated for the applicable fluid temperature [Btu•inch/(h•ft²•°F)] [W/(m•K)].
k = the upper value of the conductivity range listed in this table for the applicable fluid temperature.

2 These thicknesses are based on energy efficiency considerations only. Issues such as water, vapor permeability, or surface condensation require vapor retarders or additional insulation.

3 For direct-buried cooling system piping, insulation is not required.

4 Table E 503.7.3(2) is based on steel pipe. Nonmetallic pipes schedule 80 thickness or less shall use the table values. For other nonmetallic pipes having thermal resistance more than that of steel pipe, reduced insulation thicknesses are permitted where documentation is provided showing that the pipe with the proposed insulation has no more heat transfer per foot (mm) than a steel pipe of the same size with the insulation thickness shown in Table E 503.7.3(2).

E 603.2 Isolation of Pollutant Sources. Rooms where activities produce hazardous fumes or chemicals, including commercial kitchens, garages, janitorial or laundry rooms, and copy or printing rooms, shall be exhausted and isolated from adjacent spaces in accordance with this code.

E 603.3 Filters. In mechanically ventilated buildings, particle filters, or air-cleaning devices shall be provided to clean outdoor and return air prior to its delivery to occupied spaces. The particle or air cleaner shall have a MERV of 13.

Exception: A filter or air cleaning device with a lower MERV value shall be permitted provided it is the highest value commercially available for the specific equipment that is installed.

E 603.4 Ozone Depletion and Global Warming Reductions. Installations of HVAC and refrigeration shall not contain CFCs and shall be in accordance with this code.

E 604.0 Indoor Moisture Control.

E 604.1 Rainwater Control. Roof drainage systems shall discharge to a place of disposal in accordance with the plumbing code. Storm water shall be directed away from the building.

E 605.0 Indoor Air Quality for Low-Rise Residential.

E 605.1 General. Rooms or occupied spaces within single-family homes and multifamily structures of three stories or less above grade shall be designed to have ventilation (outdoor) air for occupants in accordance with Section E 605.1.1 through Section E 605.1.3.2, or the applicable local code.

E 605.1.1 Natural Ventilation. Naturally ventilated spaces shall be permanently open to and within 20 feet (6096 mm) of operable wall or roof openings to the outdoors, the openable area of which is not less than 5 percent of the conditioned floor area of the naturally ventilated space. Where openings are covered with louvers or otherwise obstructed, openable area shall be based on the free unobstructed area through the opening.

E 605.1.1.1 Access to Operable Openings. The means to open required operable openings shall be readily accessible to building occupants where the space is occupied.

E 605.1.2 Mechanical Ventilation. Each space that is not naturally ventilated in accordance with Section E 605.1.1 shall be ventilated with a mechanical system capable of providing an outdoor air rate not less than 15 ft³/min (0.007 m³/s) per person times the expected number of occupants. Mechanical ventilation shall comply with this code.

E 605.1.3 Dwelling-Unit Ventilation. A Mechanical exhaust system, supply system, or combination thereof shall be installed to operate for each dwelling unit to provide continuous dwelling-unit ventilation with outdoor air at a rate not less than the rate specified in Section E 605.1.3.1. [ASHRAE 62.2:4.1]

E 605.1.3.1 Total Ventilation Rate. The total required ventilation rate (Q_{tot}) shall be as specified in Table E 605.1.3.1 or, alternatively, calculated using Equation E 605.1.3.1.

(Equation E 605.1.3.1)

$$Q_{tot} = 0.03A_{floor} + 7.5(N_{br} + 1)$$

Where:

Q_{tot} = total required ventilation rate, cubic feet per minute (ft³/min)

A_{floor} = dwelling unit floor area, square foot (ft²)

N_{br} = number of bedrooms (not to be less than one)

For SI units: 1 cubic foot per minute = 0.00047 m³/s, 1 square foot = 0.0929 m²

Exceptions: Dwelling-unit mechanical ventilation systems shall not be required where the Authority Having Jurisdiction determines that window operation is a locally permissible method of providing ventilation and provided one or more of the following conditions is met:

(1) The building has no mechanical cooling and is in zone 1 or 2 of the climate zone map.

(2) The building is thermally conditioned for human occupancy for less than 876 hours per year. [ASHRAE 62.2:4.1.1]

E 605.1.3.2 Effective Annual Average Infiltration Rate (Q_{inf}). Effective Annual Average Infiltration Rate (Q_{inf}) shall be calculated using Equation E 605.1.3.2:

(Equation E 605.1.3.2)

$$Q_{inf}\text{ (CFM)} = (NL \times wsf \times A_{floor}) / (7.3)^*$$

Where:

NL = normalized leakage

wsf = weather and shielding factor from ASHRAE 62.2.

A_{floor} = floor area of residence, ft^2 (m^2)

*Replace 7.3 with 1.44 for metric units. [ASHRAE 62.2:4.1.2(e)]

E 605.1.3.3 Required Mechanical Ventilation Rate (Q_{fan}). Required Mechanical Ventilation Rate (Q_{fan}) shall be calculated using Equation E 605.1.3.3:

(Equation E 605.1.3.3)

$$Q_{fan} = Q_{tot} - (Q_{inf} \times A_{ext})$$

Where:

Q_{fan} = required mechanical ventilation rate, CFM (L/s)

Q_{tot} = total required ventilation rate, CFM (L/s)

Q_{inf} = may be not greater than $\frac{2}{3} \times Q_{tot}$

(see ASHRAE 62.2 for exceptions for existing buildings)

A_{ext} = 1 for single-family detached homes, or the ratio of exterior envelope surface area that is not attached to garages or other dwelling units to total envelope surface area for single-family attached homes. [ASHRAE 62.2:4.1.2(f)]

E 605.1.3.4 Different Occupant Density. Table E 605.1.3.1 and Equation E 605.1.3.1 assume two persons in a studio or one-bedroom dwelling unit and an additional person for each additional bedroom. Where higher occupant densities are known, the rate shall be increased by 7.5 ft^3/min (0.003 m^3/s) for each additional person. Where approved by the Authority Having Jurisdiction, lower occupant densities shall be permitted to be used. [ASHRAE 62.2:4.1.3]

E 605.1.4 System Type. The dwelling-unit mechanical ventilation system shall consist of one or more supply or exhaust fans and associated ducts and controls. Local exhaust fans shall be permitted to be part of a mechanical exhaust system. Where local exhaust fans are used to provide dwelling-unit ventilation, the local exhaust airflow shall be permitted to be credited toward the dwelling-unit ventilation airflow requirement. Outdoor air ducts connected to the return side of an air handler shall be permitted as supply ventilation where manufacturer's requirements for return air temperature are met. See ASHRAE 62.2 for guidance on selection of methods. [ASHRAE 62.2:4.2]

E 605.1.5 Airflow Measurement. The airflow required by this section shall be the quantity of outdoor ventilation air supplied and/or indoor air exhausted by the mechanical ventilation system as installed and shall be measured according to the ventilation equipment manufacturer's instructions, or by using a flow hood, flow grid, or other airflow measuring device at the mechanical ventilation fan's inlet terminals/grilles, outlet terminals/grilles, or in the connected ventilation ducts. Ventilation airflow of systems with multiple operating modes shall be tested in all modes designed to meet this section. [ASHRAE 62.2:4.3]

E 605.1.6 Control and Operation. A readily accessible manual ON-OFF control, including but not limited to a fan switch or a dedicated branch-circuit overcurrent device, shall be provided. Controls shall include text or an icon indicating the system's function.

Exception: For multifamily dwelling units, the manual ON-OFF control shall not be required to be readily accessible. [ASHRAE 62.2:4.4]

TABLE E 605.1.3.1
VENTILATION AIR REQUIREMENTS, (cubic foot per minute)
[ASHRAE 62.2: TABLE 4.1a]

FLOOR AREA (ft^2)	BEDROOMS				
	1	2	3	4	5
<500	30	38	45	53	60
501-1000	45	53	60	68	75
1001-1500	60	68	75	83	90
1501-2000	75	83	90	98	105
2001-2500	90	98	105	113	120
2501-3000	105	113	120	128	135
3001-3500	120	128	135	143	150
3501-4000	135	143	150	158	165
4001-4500	150	158	165	173	180
4501-5000	165	173	180	188	195

For SI units: 1 square foot = 0.0929 m^2, 1 cubic foot per minute = 0.00047 m^3/s

E 605.1.7 Variable Mechanical Ventilation. Dwelling-unit mechanical ventilation systems designed to provide variable ventilation shall comply with Section E 605.1.7.1 or Section E 605.1.7.2 or Section E 605.1.7.3. Section E 605.1.7.2 and Section E 605.1.7.3 also require compliance with ASHRAE 62.2 and require verification with supporting documentation from the manufacturer, designer, or specifier of the ventilation control system that the system meets the requirements of these sections. Where the dwelling-unit ventilation rate varies based on occupancy, occupancy shall be determined by occupancy sensors or by an occupant-programmable schedule. [ASHRAE 62.2:4.5]

E 605.1.7.1 Short-Term Average Ventilation. To comply with this section, a variable ventilation system shall be installed to provide an average dwelling-unit ventilation rate over any three-hour period that is greater than or equal to Q_{fan} as calculated using Section E 605.1.3.3. [ASHRAE 62.2:4.5.1]

E 605.1.7.2 Scheduled Ventilation. This section shall only be allowed to be used where one or more fixed patterns of designed ventilation are known at the time compliance to Section E 605.0 is being determined. Such patterns include those both clock-driven and driven by typical meteorological data. Compliance with this section shall be demonstrated with either Section E 605.1.7.2.1 or Section E 605.1.7.2.2. [ASHRAE 62.2:4.5.2]

E 605.1.7.2.1 Annual Average Schedule. An annual schedule of ventilation complies with this section when the annual average relative exposure during occupied periods is not more than unity as calculated in accordance with ASHRAE 62.2. [ASHRAE 62.2:4.5.2.1]

E 605.1.7.2.2 Block Scheduling. The schedule of ventilation complies with this section when it is broken into blocks of time and each block individually has an average relative exposure during occupied periods that is not more than unity as calculated in ASHRAE 62.2. [ASHRAE 62.2:4.5.2.2]

E 605.1.7.3 Real-Time Control. A real-time ventilation controller complies with this section when it is designed to adjust the ventilation system based on real-time input to the ventilation calculations so that the average relative exposure during occupied periods is not more than unity as calculated in ASHRAE 62.2. The averaging period shall be not less than one day but not more than one year and shall be based on simple, recursive or running average, but not extrapolation. [ASHRAE 62.2:4.5.3]

E 605.1.8 Equivalent Ventilation. A dwelling-unit ventilation system shall be designed and operated in such a way as to provide the same or lower annual exposure as would be provided in accordance with Section E 605.1.3. The calculations shall be based on a single zone with a constant contaminant emission rate. The manufacturer, specifier, or designer of the equivalent ventilation system shall certify that the system is in accordance with this intent and provide supporting documentation. [ASHRAE 62.2:4.6]

E 605.2 Bathroom Exhaust Fans. Except where a whole house energy recovery system is used, a mechanical exhaust fan vented to the outdoors shall be provided in each room containing a bathtub, shower, or tub/shower combination. The ventilation rate shall be not less than 50 ft^3/min (0.02 m^3/s) for intermittent operation and 20 ft^3/min (0.009 m^3/s) for continuous operation. Fans shall comply with the Energy Star Program.

E 605.3 Filters. Heating and air conditioning filters shall have a MERV rating of 6 or higher. The air distribution system shall be designed for the pressure drop across the filter.

E 606.0 Indoor Air Quality for Other than Low-Rise Residential Buildings.

E 606.1 Minimum Indoor Air Quality. The building shall comply with Chapter 4 or ASHRAE 62.1 for ventilation air supply.

E 607.0 Environmental Comfort.

E 607.1 Thermal Comfort Controls. The mechanical systems and controls of building shall be designed to provide and maintain indoor comfort conditions in accordance with ASHRAE 55.

E 607.2 Heating and Air-Conditioning System Design. Heating and air-conditioning systems shall be sized, designed, and have their equipment selected in accordance with the following:

(1) Heat loss and heat gain are established in accordance with ACCA Manual J, ASHRAE handbooks, or other equivalent methods.

(2) Duct systems shall be sized in accordance with ACCA Manual D, ASHRAE handbooks, or other equivalent methods.

(3) Heating and cooling equipment in accordance with ACCA Manual S or other equivalent methods.

E 608.0 Low VOC Solvent Cement and Primer.

E 608.1 General. Primers and solvent cements used to join plastic pipe, and fittings shall be in accordance with Section E 608.1.1 and Section E 608.1.2.

E 608.1.1 Solvent Cement. Solvent cement, including one-step solvent cement, shall have a volatile organic compound (VOC) content of less than or equal to 65 ounces per gallon (oz/gal) (487 g/L) for CPVC cement, 68 oz/gal (509 g/L) for PVC cement, and 43 oz/gal (322 g/L) for ABS cement, as determined by the South Coast Air Quality Management District's Laboratory Methods of Analysis for Enforcement Samples, Method 316A.

E 608.1.2 Primer. Primer shall have a volatile organic compound (VOC) content of less than or equal to 73 oz/gal (546 g/L), as determined by the South Coast Air Quality Management District's Laboratory Methods of Analysis for Enforcement Samples, Method 316A.

E 701.0 Installer Qualifications.

E 701.1 Scope. The provisions of this section address minimum qualifications of installers of mechanical systems covered within the scope of this appendix.

E 702.0 Qualifications.

E 702.1 General. Where permits are required, the Authority Having Jurisdiction shall have the authority to require contractors, installers, or service technicians to demonstrate competency. Where determined by the Authority Having Jurisdiction, the contractor, installer, or service technician shall be licensed to perform such work.

Part I

E 801.0 Heating, Ventilation, and Air Conditioning Systems Commissioning.

E 801.1 Applicability. The provisions of this section apply to the commissioning of commercial and institutional HVAC systems.

E 802.0 Commissioning.

E 802.1 Commissioning Requirements. HVAC commissioning shall be included in the design and construction processes of the project to verify that the HVAC systems and components meet the owner's project requirements and in accordance with this appendix. Commissioning shall be performed in accordance with this appendix by personnel trained and certified in commissioning by a nationally recognized organization. Commissioning requirements shall include the following:

(1) Owner's project requirements
(2) Basis of design
(3) Commissioning measures shown in the construction documents
(4) Commissioning plan
(5) Functional performance
(6) Testing
(7) Post construction documentation and training
(8) Commissioning report

HVAC systems and components covered by this appendix as well as process equipment and controls, and renewable energy systems shall be included in the scope of the commissioning requirements.

E 802.2 Owner's Project Requirements (OPR). The performance goals and requirements of the HVAC system shall be documented before the design phase of the project begins. This documentation shall include not less than the following:

(1) Environmental and sustainability goals
(2) Energy efficiency goals
(3) Indoor environmental quality requirements
(4) Equipment and systems performance goals
(5) Building occupant and O&M personnel expectations

E 802.3 Basis of Design (BOD). A written explanation of how the design of the HVAC system meets the owner's project requirements shall be completed at the design phase of the building project, and updated as necessary during the design and construction phases. The basis of design document shall cover not less than the following systems:

(1) Heating, ventilation, air conditioning (HVAC) systems and controls
(2) Water heating systems
(3) Renewable energy systems

E 802.4 Commissioning Plan. A commissioning plan shall be completed to document the approach to how the project will be commissioned, and shall be started during the design phase of the building project. The commissioning plan shall include not less than the following:

(1) General project information
(2) Commissioning goals
(3) Systems to be commissioned. Plans to test systems and components shall include not less than the following:
    (a) A detailed explanation of the original design intent.
    (b) Equipment and systems to be tested, including the extent of tests.
    (c) Functions to be tested.
    (d) Conditions under which the test shall be performed.
    (e) Measurable criteria for acceptable performance.
(4) Commissioning team information.
(5) Commissioning process activities, schedules, and responsibilities. Plans for the completion of commissioning requirements listed in Section E 802.5 through Section E 802.7 shall be included.

E 802.5 Functional Performance Testing. Functional performance tests shall demonstrate the correct installation and operation of each component, system, and system-to-system interface in accordance with the approved plans and specifications. Functional performance testing reports shall contain information addressing each of the building components tested, the testing methods utilized, and readings and adjustments made.

E 802.6 Post Construction Documentation and Training. A system manual and systems operations training are required.

E 802.6.1 Systems Manual. Documentation of the operational aspects of the HVAC system shall be completed within the systems manual and delivered to the building owner and facilities operator. The systems manual shall include not less than the following:

(1) Site information, including facility description, history, and current requirements.
(2) Site contact information.
(3) Basic O&M, including general site operating procedures, basic troubleshooting, recommended maintenance requirements, and site events log.
(4) Major systems.
(5) Site equipment inventory and maintenance notes.
(6) Equipment/system warranty documentation and information.
(7) "As-Built" design drawings.
(8) Other resources and documentation.

E 802.6.2 Systems Operations Training. The training of the appropriate maintenance staff for each equipment type or system shall include not less than the following:

(1) System/Equipment overview (what it is, what it does, and what other systems or equipment it interfaces with).

(2) Review of the information in the systems manual.

(3) Review of the record drawings on the system/equipment.

E 802.7 Commissioning Report. A complete report of commissioning process activities undertaken through the design, construction, and post-construction phases of the building project shall be completed and provided to the owner.

Part II

E 803.0 Commissioning Acceptance.

E 803.1 General. Part II of this appendix provides a means of verifying the commissioning requirements of Section E 802.1. The activities specified in Part II of this appendix includes three aspects, as described as follows:

(1) Visual inspection of the equipment and installation.

(2) Review of the certification requirements.

(3) Functional tests of the systems and controls.

E 803.2 Construction Documents. Details of commissioning acceptance requirements shall be incorporated into the construction documents, including information that describes the details of the functional tests to be performed. This information shall be permitted to be integrated into the specifications for testing and air balancing, energy management and control system, equipment startup procedures or commissioning. It is possible that the work will be performed by a combination of the test and balance (TAB) contractor, mechanical/electrical contractor, and the energy management control system (EMCS) contractor, so applicable roles and responsibilities shall be clearly called out.

E 803.2.1 Roles and Responsibilities. The roles and responsibilities of the persons involved in commissioning acceptance are included in Section E 803.2.1.1 through Section E 803.2.1.3.

E 803.2.1.1 Field Technician. The field technician shall be responsible for performing and documenting the results of the acceptance procedures on the certificate of acceptance forms. The field technician shall sign the certificate of acceptance to certify that the information he provides on the certificate of acceptance is true and correct.

E 803.2.1.2 Responsible Person. The responsible person shall be the contractor or registered design professional of record. A certificate of acceptance shall be signed by a responsible person to take responsibility for the scope of work specified by the certificate of acceptance document. The responsible person shall perform the field testing and verification work, and where this is the case, the responsible person shall complete and sign both the field technician's signature block and the responsible person's signature block on the certificate of acceptance form. The responsible person assumes responsibility for the acceptance testing work performed by the field technician agent or employee.

E 803.2.1.3 Certificate of Acceptance. The certificate of acceptance shall be submitted to the Authority Having Jurisdiction in order to receive the final certificate of occupancy. The Authority Having Jurisdiction shall not release a final certificate of occupancy unless the submitted certificate of acceptance demonstrates that the specified systems and equipment have been shown to be performing in accordance with the applicable acceptance requirements. The Authority Having Jurisdiction has the authority to require the field technician and responsible person to demonstrate competence, to its satisfaction. Certificate of acceptance forms are located in Section E 806.0.

E 804.0 Commissioning Tests.

E 804.1 General. Functional tests shall be performed on new equipment and systems installed in either new construction or retrofit applications in accordance with this section. The appropriate certificate of acceptance form along with each specific test shall be completed and submitted to the Authority Having Jurisdiction before a final occupancy permit can be granted.

E 804.2 Tests. Functional testing shall be performed on the devices and systems listed in this section. The functional test results are documented using the applicable certificate of acceptance forms shown in parenthesis and located in Section E 806.0. The functional tests shall be performed in accordance with Section E 805.0 using the following forms:

(1) Minimum ventilation controls for constant and variable air volume systems (Form MECH-2A).

(2) Zone temperature and scheduling controls for constant volume, single-zone, unitary air conditioner and heat pump systems (Form MECH-3A).

(3) Duct leakage on a subset of small single-zone systems depending on the ductwork location (Form MECH-4A).

(4) Air economizer controls for economizers that are not factory installed and tested (Form MECH-5A).

(5) Demand-controlled ventilation control systems (Form MECH-6A).

(6) Supply fan variable flow controls (Form MECH-7A).

(7) Valve leakage for hydronic variable flow systems and isolation valves on chillers and boilers in plants with more than one chiller or boiler being served by the same primary pumps through a common header (Form MECH-8A).

(8) Supply water temperature reset control strategies programmed into the building automation system for water systems (e.g., chilled, hot, or condenser water) (Form MECH-9A).

(9) Hydronic variable flow controls on a water system where the pumps are controlled by variable frequency drives (e.g., chilled and hot water systems; water-loop heat pump systems) (Form MECH-10A).

(10) Automatic demand shed control (Form MECH-11A).

(11) Fault detection and diagnostic for DX units (Form MECH-12A).

(12) Automatic fault detection and diagnostic systems (AFDD) (Form MECH-13A).

(13) Distributed energy storage DEC/DX AC systems (Form MECH-14A).

(14) Thermal energy storage (TES) systems (Form MECH-15A).

E 804.3 Acceptance Process. The functional testing process shall comply with Section E 804.3.1 through Section E 804.3.4.

E 804.3.1 Plan Review. The installing contractor, registered design professional of record, owner's agent, or the person responsible for certification of the acceptance testing on the certificate of acceptance (responsible person) shall review the plans and specifications to ensure that they are in accordance with the acceptance requirements. This is typically done prior to signing a certificate of compliance.

E 804.3.2 Construction Inspection. The installing contractor, registered design professional of record, owner's agent, or the person responsible for certification of the acceptance testing on the certificate of acceptance (responsible person) shall perform a construction inspection prior to testing to ensure that the equipment that is installed is capable of complying with the requirements of this appendix and is calibrated. The installation of associated systems and equipment necessary for proper system operation is required to be completed prior to the testing.

E 804.3.3 Acceptance Testing. One or more field technicians shall perform the acceptance testing; identify performance deficiencies; ensure that they are corrected; and where necessary, repeat the acceptance procedures until the specified systems and equipment are performing in accordance with the acceptance requirements. The field technician who performs the testing shall sign the certificate of acceptance to certify the information has been provided to document the results of the acceptance procedures is true and correct.

The responsible person shall review the test results from the acceptance requirement procedures provided by the field technician and sign the certificate of acceptance to certify compliance with the acceptance requirements. The responsible person shall be permitted to perform the field technician's responsibilities, and shall then sign the field technician declaration on the certificate of acceptance to certify that the information on the form is true and correct.

E 804.3.4 Certificate of Occupancy. The Authority Having Jurisdiction shall not issue the final certificate of occupancy until required certificates of acceptance are submitted. Copies of completed, signed certificates of acceptance are required to be posted, or made available with the permit(s), and shall be made available to the Authority Having Jurisdiction.

E 805.0 HVAC System Tests.

E 805.1 Variable Air Volume Systems (Form MECH-2A). This test ensures that adequate outdoor air ventilation is provided through the variable air volume air handling unit at two representative operating conditions. The test consists of measuring outdoor air values at maximum flow and at or near minimum flow. The test verifies that the minimum volume of outdoor air is introduced to the air handling unit where the system is in occupied mode at these two conditions of supply airflow. This test shall be performed in conjunction with supply fan variable flow controls test procedures to reduce the overall system testing time as both tests use the same two conditions of airflow for their measurements.

E 805.1.1 Test Procedure. The procedure for performing a functional test for variable air volume systems shall be in accordance with Section E 805.1.1.1 and Section E 805.1.1.2.

E 805.1.1.1 Construction Inspection. Prior to functional testing, verify and document that the system controlling outside airflow is calibrated either in the field or factory.

E 805.1.1.2 Functional Testing. The functional testing shall be in accordance with the following steps:

Step 1: Where the system has an outdoor air economizer, force the economizer high limit to disable economizer control (e.g., for a fixed drybulb high limit, lower the setpoint below the current outdoor air temperature).

Step 2: Adjust supply airflow to either the sum of the minimum zone airflows or 30 percent of the total design airflow. Verify and document the following:

(1) Measured outside airflow reading is within 10 percent of the total ventilation air called for in the certificate of compliance.

(2) OSA controls stabilize within 5 minutes.

Step 3: Adjust supply airflow to achieve design airflow. Verify and document the following:

(1) Measured outside airflow reading is within 10 percent of the total ventilation air called for in the certificate of compliance.

(2) OSA controls stabilize within 5 minutes.

Step 4: Restore system to "as-found" operating conditions.

E 805.1.2 Acceptance Criteria. System controlling outdoor air flow shall be calibrated in the field or at the factory.

Measured outdoor airflow reading shall be within 10 percent of the total value found on the certificate of compliance under the following conditions:

(1) Minimum system airflow.

(2) Thirty percent of total design flow design supply airflow.

E 805.2 Constant Volume Systems (Form MECH-2A). The purpose of this test is to ensure that adequate outdoor air ventilation is provided through the constant volume air handling unit to the spaces served under operating conditions. The intent of this test is to verify that the minimum volume of outdoor air is introduced to the air handling unit during typical space occupancy.

E 805.2.1 Test Procedure. The procedure for performing a functional test for constant air volume systems shall be in accordance with Section E 805.2.1.1 and Section E 805.2.1.2.

E 805.2.1.1 Construction Inspection. Prior to functional testing, verify and document the following:

(1) Minimum position is marked on the outside air damper.

(2) The system has means of maintaining the minimum outdoor air damper position.

E 805.2.1.2 Functional Testing. Where the system has an outdoor air economizer, force the economizer to the minimum position and stop outside air damper modulation (e.g., for a fixed drybulb high limit, lower the setpoint below the current outdoor air temperature).

E 805.2.2 Acceptance Criteria. The system has a means of maintaining the minimum outdoor air damper position. The minimum damper position is marked on the outdoor air damper. The measured outside airflow reading shall be within 10 percent of the total ventilation air called for in the certificate of compliance.

E 805.3 Constant Volume, Single-Zone, Unitary Air Conditioner and Heat Pumps Systems Acceptance (Form MECH-3A). The purpose of this test is to verify the individual components of a constant volume, single-zone, unitary air conditioner and heat pump system function correctly; including: thermostat installation and programming, supply fan, heating, cooling, and damper operation.

E 805.3.1 Test Procedure. The procedure for performing a functional test for constant volume, single-zone, unitary air conditioner and heat pump systems shall be in accordance with Section E 805.3.1.1 and Section E 805.3.1.2.

E 805.3.1.1 Construction Inspection. Prior to functional testing, verify and document the following:

(1) Thermostat is located within the space-conditioning zone that is served by the HVAC system.

(2) Thermostat shall be in accordance with temperature adjustment and dead band requirements.

(3) Occupied, unoccupied, and holiday schedules shall be programmed per the facility's schedule.

(4) Preoccupancy purge is programmed.

E 805.3.1.2 Functional Testing. The functional testing shall be in accordance with the following steps:

Step 1: Disable economizer and demand control ventilation systems (where applicable).

Step 2: Simulate a heating demand during the occupied condition. Verify and document the following:

(1) Supply fan operates continually.

(2) The unit provides heating.

(3) No cooling is provided by the unit.

(4) Outside air damper is at minimum position.

Step 3: Simulate operation in the dead band during occupied condition. Verify and document the following:

(1) Supply fan operates continually.

(2) Neither heating nor cooling is provided by the unit.

(3) Outside air damper is at minimum position.

Step 4: Simulate cooling demand during occupied condition. Lock out economizer (where applicable). Verify and document the following:

(1) Supply fan operates continually.

(2) The unit provides cooling.

(3) No heating is provided by the unit.

(4) Outside air damper is at minimum position.

Step 5: Simulate operation in the dead band during unoccupied mode. Verify and document the following:

(1) Supply fan is off.

(2) Outside air damper is fully closed.

(3) Neither heating nor cooling is provided by the unit.

Step 6: Simulate heating demand during unoccupied conditions. Verify and document the following:

(1) Supply fan is on (either continuously or cycling).

(2) Heating is provided by the unit.

(3) No cooling is provided by the unit.

(4) Outside air damper is either closed or at minimum position.

Step 7: Simulate cooling demand during unoccupied condition. Lock out economizer (where applicable). Verify and document the following:

(1) Supply fan is on (either continuously or cycling).

(2) Cooling is provided by the unit.

(3) No heating is provided by the unit.

(4) Outside air damper is either closed or at minimum position.

Step 8: Simulate manual override during unoccupied condition. Verify and document the following:

(1) System operates in "occupied" mode.

(2) System reverts to "unoccupied" mode where manual override time period expires.

Step 9: Restore economizer and demand control ventilation systems (where applicable), and remove system overrides initiated during the test.

E 805.3.2 Acceptance Criteria. Thermostat is located within the space-conditioning zone that is served by the respective HVAC system. The thermostat shall comply with temperature adjustment and dead band requirements. Occupied, unoccupied, and holiday schedules shall be programmed per the facility's schedule. Preoccupancy purge is programmed in accordance with the requirements.

E 805.4 Air Distribution Systems (Form MECH-4A). The purpose of this test is to verify duct work associated with non-exempt constant volume, single-zone, HVAC units (e.g., air conditioners, heat pumps, and furnaces) meet the material, installation, and insulation R-values and leakage requirements outlined in this appendix. This test is required for single-zone units serving less than 5000 square feet (464.52 m^2) of floor area where 25 percent or more of the duct surface area is in one of the following spaces:

(1) Outdoors.

(2) In a space directly under a roof where the U-factor of the roof is greater than the U-factor of the ceiling.

(3) In a space directly under a roof with fixed vents or openings to the outside or unconditioned spaces.

(4) In an unconditioned crawl space.

(5) In other unconditioned spaces.

This test applies to both new duct systems and to existing duct systems being extended or the space conditioning system is altered by the installation or replacement of space conditioning equipment, including: replacement of the air handler; outdoor condensing unit of a split system air conditioner or heat pump; cooling or heating coil; or the furnace heat exchanger. Existing duct systems do not have to be tested where they are insulated or sealed with asbestos.

E 805.4.1 Test Procedure. The procedure for performing a functional test for air distribution systems shall be in accordance with Section E 805.4.1.1 and Section E 805.4.1.2.

E 805.4.1.1 Construction Inspection. Prior to functional testing, verify and document the following:

(1) Duct connections shall comply with the requirements of this appendix and this code.

(2) Flexible ducts are not compressed.

(3) Ducts are fully accessible for testing.

(4) Joints and seams are properly sealed in accordance with the requirements of this appendix.

(5) Insulation R-Values shall comply with the minimum requirements of this appendix.

E 805.4.1.2 Functional Testing. Perform duct leakage test in accordance with Section E 503.4.7.2.1.

E 805.4.2 Acceptance Criteria. Flexible ducts are not compressed or constricted. Duct connections shall comply with the requirements of this appendix and this code (new ducts only). Joints and seams are properly sealed in accordance with the requirements of this appendix and this code (new ducts only). Duct R-values shall comply with the minimum requirements of this appendix (new ducts only). Insulation is protected from damage and suitable for outdoor usage where applicable (new ducts only). The leakage shall not exceed the rate in accordance with Section E 503.4.7.2.

E 805.5 Air Economizer Controls Acceptance (Form MECH-5A). The purpose of functionally testing an air economizer cycle is to verify that an HVAC system uses outdoor air to satisfy space cooling loads where outdoor air conditions are acceptable. There are two types of economizer controls; stand-alone packages and DDC controls. The stand-alone packages are commonly associated with small unitary rooftop HVAC equipment, and DDC controls are typically associated with built-up or large packaged air handling systems. Test procedures for both economizer control types are provided.

For units with economizers that are factory installed and certified operational by the manufacturer to economizer quality control requirements, the in-field economizer functional tests do not have to be conducted. A copy of the manufacturer's certificate shall be attached to the Form MECH-5A. However, the construction inspection, including compliance with high-temperature lockout temperature setpoint, shall be completed regardless of whether the economizer is field or factory installed.

E 805.5.1 Test Procedure. The procedure for performing a functional test for air economizer controls shall comply with Section E 805.5.1.1 and Section E 805.5.1.2.

E 805.5.1.1 Construction Inspection. Prior to functional testing, verify and document the following:

(1) Economizer lockout setpoint is in accordance with this appendix.

(2) Economizer lockout control sensor is located to prevent false readings.

(3) System is designed to provide up to 100 percent outside air without over-pressurizing the building.

(4) For systems with DDC controls lockout sensor(s) are either factory calibrated or field calibrated.

(5) For systems with non-DDC controls, manufacturer's startup and testing procedures are applied.

E 805.5.1.2 Functional Testing. The functional testing shall be in accordance with the following steps:

Step 1: Disable demand control ventilation systems (where applicable).

Step 2: Enable the economizer, and simulate a cooling demand large enough to drive the economizer fully open. Verify and document the following:

(1) Economizer damper is 100 percent opened and return air damper is 100 percent closed.

(2) Where applicable, verify that the economizer remains 100 percent open where the cooling demand can no longer be met by the economizer alone.

(3) Applicable fans and dampers operate as intended to maintain building pressure.

(4) The unit heating is disabled.

Step 3: Disable the economizer and simulate a cooling demand. Verify and document the following:

(1) Economizer damper shall close to its minimum position.

(2) Applicable fans and dampers shall operate as intended to maintain building pressure.

(3) The unit heating is disabled.

Step 4: Simulate a heating demand, and set the economizer so that it is capable of operating (e.g., actual outdoor air conditions are below lockout setpoint). Verify the economizer is at minimum position.

Step 5: Restore demand control ventilation systems (where applicable) and remove system overrides initiated during the test.

E 805.5.2 Acceptance Criteria. Air economizer controls acceptance criteria shall be as follows:

(1) Where the economizer is factory installed and certified, a valid factory certificate is required for acceptance. No additional equipment tests are necessary.

(2) Air economizer lockout setpoint is in accordance with this appendix. Outside sensor location accurately reads true outdoor air temperature and is not affected by exhaust air or other heat sources.

(3) Sensors are located to achieve the desired control.

(4) During economizer mode, the outdoor air damper shall modulate open to a maximum position and return air

damper to 100 percent closed.

(5) The outdoor air damper is 100 percent open before mechanical cooling is enabled and for units 75 000 Btu/h (22 kw) and larger remains at 100 percent open while mechanical cooling is enabled (economizer integration where used for compliance).

(6) Where the economizer is disabled, the outdoor air damper closes to a minimum position; the return damper modulates 100 percent open, and mechanical cooling remains enabled.

E 805.6 Demand-Controlled Ventilation Systems Acceptance (Form MECH-6A). The purpose of this test is to verify that systems required to employ demand-controlled ventilation shall be permitted to vary outside ventilation flow rates based on maintaining interior carbon dioxide (CO_2) concentration setpoints. Demand-controlled ventilation refers to an HVAC system's ability to reduce outdoor air ventilation flow below design values where the space served is at less than design occupancy. Carbon dioxide is a good indicator of occupancy load and is the basis used for modulating ventilation flow rates.

E 805.6.1 Test Procedure. The procedure for performing a functional test for demand-control ventilation (DVC) systems shall be in accordance with Section E 805.6.1.1 and Section E 805.6.1.2.

E 805.6.1.1 Construction Inspection. Prior to functional testing, verify and document the following:

(1) Carbon dioxide control sensor is factory calibrated or field-calibrated in accordance with this appendix.

(2) The sensor is located in the high-density space between 3 feet (914 mm) and 6 feet (1829 mm) above the floor or at the anticipated level of the occupants' heads.

(3) DCV control setpoint is at or below the carbon dioxide concentration permitted by this appendix.

E 805.6.1.2 Functional Testing. The functional testing shall be in accordance with the following steps:

Step 1: Disable economizer controls.

Step 2: Simulate a signal at or slightly above the carbon dioxide concentration setpoint required by this appendix. Verify and document the following:

(1) For single zone units, outdoor air damper modulates open to satisfy the total ventilation air called for in the certificate of compliance.

(2) For multiple zone units, either outdoor air damper or zone damper modulate open to satisfy the zone ventilation requirements.

Step 3: Simulate signal well below the carbon dioxide setpoint. Verify and document the following:

(1) For single zone units, outdoor air damper modulates to the design minimum value.

(2) For multiple zone units, either outdoor air damper or zone damper modulate to satisfy the reduced zone ventilation requirements.

Step 4: Restore economizer controls and remove system overrides initiated during the test.

Step 5: With controls restored, apply carbon dioxide calibration gas at a concentration slightly above the setpoint to the sensor. Verify that the outdoor air damper modulates open to satisfy the total ventilation air called for in the certificate of compliance.

E 805.6.2 Acceptance Criteria. Demand-controlled ventilation systems acceptance criteria shall be as follows:

(1) Each carbon dioxide sensor is factory calibrated (with calibration certificate) or field calibrated.

(2) Each carbon dioxide sensor is wired correctly to the controls to ensure proper control of the outdoor air damper.

(3) Each carbon dioxide sensor is located correctly within the space 1 foot (305 mm) to 6 feet (1829 mm) above the floor.

(4) Interior carbon dioxide concentration setpoint is not more than 600 parts per million (ppm) plus outdoor air carbon dioxide value where dynamically measured or not more than 1000 ppm where no OSA sensor is provided.

(5) A minimum OSA setting is provided where the system is in occupied mode in accordance with this appendix regardless of space carbon dioxide readings.

(6) A maximum OSA damper position for DCV control shall be established in accordance with this appendix, regardless of space carbon dioxide readings.

(7) The outdoor air damper shall modulate open where the carbon dioxide concentration within the space exceeds setpoint.

(8) The outdoor air damper modulates closed (toward minimum position) where the carbon dioxide concentration within the space is below setpoint.

E 805.7 Supply Fan Variable Flow Controls (Form MECH-7A). The purpose of this test is to ensure that the supply fan in a variable air volume application modulates to meet system airflow demand. In most applications, the individual VAV boxes serving each space will modulate the amount of air delivered to the space based on heating and cooling requirements. As a result, the total supply airflow provided by the central air handling unit shall vary to maintain sufficient airflow through each VAV box. Airflow shall be controlled using a variable frequency drive (VFD) to modulate supply fan speed and vary system airflow. The most common strategy for controlling the VFD is to measure and maintain static pressure within the duct.

E 805.7.1 Test Procedure. The procedure for performing a functional test for supply fan variable controls shall be in accordance with Section E 805.7.1.1 and Section E 805.7.1.2.

E 805.7.1.1 Construction Inspection. Prior to functional testing, verify and document the following:

(1) Supply fan controls modulate to increase capacity.

(2) Supply fan maintains discharge static pressure within plus or minus 10 percent of the current operating set point.

(3) Supply fan controls stabilize within a 5 minute period.

E 805.7.1.2 Functional Testing. The functional testing shall be in accordance with the following steps:

Step 1: Simulate demand for design airflow. Verify and document the following:

(1) Supply fan controls modulate to increase capacity.

(2) Supply fan maintains discharge static pressure within plus or minus 10 percent of the current operating set point.

(3) Supply fan controls stabilize within a 5 minute period.

Step 2: Simulate demand for minimum airflow. Verify and document the following:

(1) Supply fan controls modulate to decrease capacity.

(2) Current operating setpoint has decreased (for systems with DDC to the zone level).

(3) Supply fan maintains discharge static pressure within plus or minus 10 percent of the current operating setpoint.

(4) Supply fan controls stabilize within a 5 minute period.

Step 3: Restore system to correct operating conditions.

E 805.7.2 Acceptance Criteria. Supply fan variable flow controls acceptance criteria shall be as follows:

(1) Static pressure sensor(s) is factory calibrated (with calibration certificate) or field calibrated.

(2) For systems without DDC controls to the zone level, the pressure sensor setpoint is less than one-third of the supply fan design static pressure.

(3) For systems with DDC controls with VAV boxes reporting to the central control panel, the pressure setpoint is reset by zone demand (box damper position or a trim and respond algorithm).

At full flow:

(1) Supply fan maintains discharge static pressure within plus or minus 10 percent of the current operating control static pressure setpoint.

(2) Supply fan controls stabilizes within a 5 minute period.

(3) At minimum flow (not less than 30 percent of total design flow).

(4) Supply fan controls modulate to decrease capacity.

(5) Current operating setpoint has decreased (for systems with DDC to the zone level).

(6) Supply fan maintains discharge static pressure within plus or minus 10 percent of the current operating setpoint.

E 805.8 Valve Leakage (Form MECH-8A). The purpose of this test is to ensure that control valves serving variable flow systems are designed to withstand the pump pressure over the full range of operation. Valves with insufficient actuators will lift under certain conditions causing water to leak through and loss of control. This test applies to the variable flow systems, chilled and hot-water variable flow systems, chiller isolation valves, boiler isolation valves, and water-cooled air conditioner and hydronic heat pump systems.

E 805.8.1 Test Procedure. The procedure for performing a functional test for valve leakage shall be in accordance with Section E 805.8.1.1 and Section E 805.8.1.2.

E 805.8.1.1 Construction Inspection. Prior to functional testing, verify and document the valve and piping arrangements were installed in accordance with the design drawings.

E 805.8.1.2 Functional Testing. The functional testing shall be in accordance with the following steps:

Step 1: For each pump serving the distribution system, dead head the pumps using the discharge isolation valves at the pumps. Document the following:

(1) Record the differential pressure across the pumps.

(2) Verify that this is within 5 percent of the submittal data for the pump.

Step 2: Reopen the pump discharge isolation valves. Automatically close valves on the systems being tested. Where three-way valves are present, close off the bypass line. Verify and document the following:

(1) The valves automatically close.

(2) Record the pressure differential across the pump.

(3) Verify that the pressure differential is within 5 percent of the reading from Step 1 for the pump that is operating during the valve test.

Step 3: Restore system to correct operating conditions.

E 805.8.2 Acceptance Criteria. System has no flow where coils are closed and the pump is turned on.

E 805.9 Supply Water Temperature Reset Controls (Form MECH-9A). The purpose of this test is to ensure that both the chilled water and hot water supply temperatures are automatically reset based on either building loads or outdoor air temperature, as indicated in the control sequences. Many HVAC systems are served by central chilled and heating hot water plants. The supply water operating temperatures shall meet peak loads where the system is operating at design conditions. As the loads vary, the supply water temperatures shall be permitted to be adjusted to satisfy the new operating conditions. The chilled water supply temperature shall be permitted to be raised as the cooling load decreases, and heating hot water supply temperature shall be permitted to be lowered as the heating load decreases.

This requirement applies to chilled and hot water systems that are not designed for variable flow, and that have a design capacity greater than or equal to 500 000 Btu/h (147 kW).

E 805.9.1 Test Procedure. The procedure for performing a functional test for supply water temperature reset controls shall be in accordance with Section E 805.9.1.1 and Section E 805.9.1.2.

E 805.9.1.1 Construction Inspection. Prior to functional testing, verify and document the supply water temperature sensors shall be either factory or field calibrated.

E 805.9.1.2 Functional Testing. The functional testing shall be in accordance with the following steps:

Step 1: Change reset control variable to its maximum value. Verify and document the following:

(1) Chilled or hot water temperature setpoint is reset to appropriate value.

(2) Actual supply temperature changes to meet setpoint.

(3) Verify that supply temperature is within 2 percent of the control setpoint.

Step 2: Change reset control variable to its minimum value. Verify and document the following:

(1) Chilled or hot water temperature setpoint is reset to appropriate value.

(2) Actual supply temperature changes to meet setpoint.

(3) Verify that supply temperature is within 2 percent of the control setpoint.

Step 3: Restore reset control variable to automatic control. Verify and document the following:

(1) Chilled or hot water temperature setpoint is reset to appropriate value.

(2) Actual supply temperature changes to meet setpoint.

(3) Verify that supply temperature is within 2 percent of the control setpoint.

E 805.9.2 Acceptance Criteria. The supply water temperature sensors are either factory calibrated (with calibration certificates) or field-calibrated. Sensor performance shall comply with the specifications. The supply water reset is operational.

E 805.10 Hydronic System Variable Flow Controls (Form MECH-10A). The purpose of this test is to ensure that hydronic variable flow chilled water and water-loop heat pump systems with circulating pumps larger than 5 hp (3.7 kW) vary system flow rate by modulating pump speed using a variable frequency drive (VFD) or equivalent. As the loads within the building fluctuate, control valves modulate the amount of water passing through each coil and add or remove the desired amount of energy from the air stream to satisfy the load. In the case of water-loop heat pumps, each two-way control valve associated with a heat pump will be closed where that unit is not operating. As each control valve modulates, the pump variable frequency drive (VFD) responds accordingly to meet system water flow requirements. This is not required on heating hot water systems with variable flow designs or for condensing water serving water cooled chillers.

E 805.10.1 Test Procedure. The procedure for performing a functional test for hydronic system variable flow controls shall be in accordance with Section E 805.10.1.1 and Section E 805.10.1.2.

E 805.10.1.1 Construction Inspection. Prior to functional testing, verify and document the pressure sensors are either factory or field calibrated.

E 805.10.1.2 Functional Testing. The functional testing shall comply with the following steps:

Step 1: Open control valves to increase water flow to not less than 90 percent design flow. Verify and document the following:

(1) Pump speed increases.

(2) System pressure is either within plus or minus 5 percent of current operating setpoint, or the pressure is below the setpoint, and the pumps are operating at 100 percent speed.

(3) System operation shall stabilize within 5 minutes after test procedures are initiated.

Step 2: Modulate control valves to reduce water flow to 50 percent of the design flow or less, but not lower than the pump minimum flow. Verify and document the following:

(1) Pump speed decrease.

(2) Current operating setpoint has decreased (for systems with DDC to the zone level).

(3) Current operating setpoint has not increased (for all other systems).

(4) System pressure is within 5 percent of current operating setpoint.

(5) System operation stabilizes within 5 minutes after test procedures are initiated.

E 805.10.2 Acceptance Criteria. The differential pressure sensor is either factory calibrated (with calibration certificates) or field calibrated. The pressure sensor shall be located at or near the most remote HX or control valve. The setpoint system controls shall stabilize.

E 805.11 Automatic Demand Shed Control (Form MECH-11A). The purpose of this test is to ensure that the central demand shed sequences have been properly programmed into the DDC system.

E 805.11.1 Test Procedure. The procedure for performing a functional test for automatic demand shed controls shall be in accordance with Section E 805.11.1.1 and Section E 805.11.1.2.

E 805.11.1.1 Construction Inspection. Prior to functional testing, verify and document that the EMCS interface enables activation of the central demand shed controls.

E 805.11.1.2 Functional Testing. The functional testing shall comply with the following steps:

Step 1: Engage the global demand shed system. Verify and document the following:

(1) That the cooling setpoint in noncritical spaces increases by the proper amount.

(2) That the cooling setpoint in critical spaces do not change.

Step 2: Disengage the global demand shed system. Verify and document the following:

(1) That the cooling setpoint in noncritical spaces return to their original values.

(2) That the cooling setpoint in critical spaces do not change.

E 805.11.2 Acceptance Criteria. The control system changes the setpoints of noncritical zones on activation of a single central hardware or software point then restores the initial setpoints where the point is released.

E 805.12 Fault Detection and Diagnostics (FDD) for Packaged Direct-Expansion (DX) Units (Form MECH-12A). The purpose of this test is to verify proper fault detection and reporting for automated fault detection and diagnostics systems for packaged units. Automated FDD systems ensure proper equipment operation by identifying and diagnosing common equipment problems such as improper refrigerant charge, low airflow, or faulty economizer operation. Qualifying FDD systems receive a compliance credit where using the performance approach. A system that does not meet the eligibility requirements shall be permitted to be installed, but no compliance credit will be given.

E 805.12.1 Test Procedure. The procedure for performing a functional test for fault detection and diagnostics (FDD) for packaged direct-expansion (DX) units shall be in accordance with Section E 805.12.1.1 and Section E 805.12.1.2.

E 805.12.1.1 Construction Inspection. Prior to functional testing, verify and document that the FDD hardware is installed on equipment by the manufacturer, and that equipment make and model include factory-installed FDD hardware that match the information indicated on copies of the manufacturer's cut sheets and on the plans and specifications.

This procedure applies to fault detection and diagnostics (FDD) system for direct-expansion packaged units containing the following features:

(1) The unit shall include a factory-installed economizer and shall limit the economizer dead band to not more than 2°F (-17°C).

(2) The unit shall include direct-drive actuators on outside air and return air dampers.

(3) The unit shall include an integrated economizer with either differential drybulb or differential enthalpy control.

(4) The unit shall include a low temperature lockout on the compressor to prevent coil freeze-up or comfort problems.

(5) Outside air and return air dampers shall have maximum leakage rates in accordance to this appendix.

(6) The unit shall have an adjustable expansion control device such as a thermostatic expansion valve (TXV).

(7) To improve the ability to troubleshoot charge and compressor operation, a high-pressure refrigerant port will be located on the liquid line. A low-pressure refrigerant port will be located on the suction line.

(8) The following sensors shall be permanently installed to monitor system operation, and the controller shall have the capability of displaying the value of each parameter:
   (a) Refrigerant suction pressure
   (b) Refrigerant suction temperature
   (c) Liquid line pressure
   (d) Liquid line temperature
   (e) Outside air temperature
   (f) Outside air relative humidity
   (g) Return air temperature
   (h) Return air relative humidity
   (i) Supply air temperature
   (j) Supply air relative humidity

The controller will provide system status by indicating the following conditions:

(1) Compressor enabled
(2) Economizer enabled
(3) Free cooling available
(4) Mixed air low limit cycle active
(5) Heating enabled

The unit controller shall have the capability to manually initiate each operating mode so that the operation of compressors, economizers, fans, and heating system can be independently tested and verified.

E 805.12.1.2 Functional Testing. The functional testing shall be in accordance with the following steps:

Step 1: Test low airflow condition by replacing the existing filter with a dirty filter or appropriate obstruction.

Step 2: Verify that the fault detection and diagnostics system reports the fault.

Step 3: Verify that the system is able to verify the correct refrigerant charge.

Step 4: Calibrate outside air, return air, and supply air temperature sensors.

E 805.12.2 Acceptance Criteria. The system is able to detect a low airflow condition and report the fault. The system is able to detect where refrigerant charge is low or high and the fault is reported.

E 805.13 Automatic Fault Detection Diagnostics (FDD) for Air Handling Units (AHU) and Zone Terminal Units (Form MECH-13A). The purpose of this test is to verify that the system detects common faults in air handling units and terminal units. FDD systems for air handling units and zone terminal units require DDC controls to the zone level. Successful completion of this test provides a compliance credit where using the performance approach. An FDD system that does not pass this test shall be permitted to be installed, but no compliance credit will be given.

E 805.13.1 Test Procedure. The procedure for performing a functional test for automatic fault detection diagnostics (FDD) for Air Handling Units and Zone Terminal Units shall be in accordance with Section E 805.13.1.1.

E 805.13.1.1 Functional Testing. The functional testing shall be in accordance with Section E 805.13.1.1.1 and Section E 805.13.1.1.2.

E 805.13.1.1.1 Functional Testing for Air Handling Units. The functional testing of AHU with FDD controls shall be in accordance with the following steps:

Step 1: Sensor drift/failure:

(1) Disconnect outside air temperature sensor from unit controller.
(2) Verify that the FDD system reports a fault.
(3) Connect OAT sensor to the unit controller.
(4) Verify that FDD indicates normal system operation.

Step 2: Damper/actuator fault:

(1) From the control system workstation, command the mixing box dampers to full open (100 percent outdoor air).

(2) Disconnect power to the actuator and verify that a fault is reported at the control workstation.

(3) Reconnect power to the actuator and command the mixing box dampers to full open.

(4) Verify that the control system does not report a fault.

(5) From the control system workstation, command the mixing box dampers to a full-closed position (0 percent outdoor air).

(6) Disconnect power to the actuator and verify that a fault is reported at the control workstation.

(7) Reconnect power to the actuator and command the dampers closed.

(8) Verify that the control system does not report a fault during normal operation.

Step 3: Valve/actuator fault:

(1) From the control system workstation, command the heating and cooling coil valves to full open or closed, then disconnect power to the actuator and verify that a fault is reported at the control workstation.

Step 4: Inappropriate simultaneous heating, mechanical cooling, and economizing or all functions:

(1) From the control system workstation, override the heating coil valve and verify that a fault is reported at the control workstation.

(2) From the control system workstation, override the cooling coil valve and verify that a fault is reported at the control workstation.

(3) From the control system workstation, override the mixing box dampers and verify that a fault is reported at the control workstation.

E 805.13.1.1.2 Functional Testing for Zone Terminal Units. The functional testing of one of each type of terminal unit (VAV box) in the project not less than 5 percent of the terminal boxes shall be in accordance with the following steps:

Step 1: Sensor drift/failure:

(1) Disconnect the tubing to the differential pressure sensor of the VAV box.

(2) Verify that control system detects and reports the fault.

(3) Reconnect the sensor and verify proper sensor operation.

(4) Verify that the control system does not report a fault.

Step 2: Damper/actuator fault:

(1) Damper stuck open.

  (a) Command the damper to full open (room temperature above setpoint).

  (b) Disconnect the actuator to the damper.

  (c) Adjust the cooling setpoint so that the room temperature is below the cooling setpoint to command the damper to the minimum position. Verify that the control system reports a fault.

  (d) Reconnect the actuator and restore to normal operation.

(2) Damper stuck closed.

  (a) Set the damper to the minimum position.

  (b) Disconnect the actuator to the damper.

  (c) Set the cooling setpoint below the room temperature to simulate a call for cooling. Verify that the control system reports a fault.

  (d) Reconnect the actuator and restore to normal operation.

Step 3: Valve/actuator fault (for systems with hydronic reheat):

(1) Command the reheat coil valve to full open.

(2) Disconnect power to the actuator. Set the heating setpoint temperature to be lower than the current space temperature, to command the valve closed. Verify that the fault is reported at the control workstation.

(3) Reconnect the actuator and restore normal operation.

Step 4: Feedback loop tuning fault (unstable airflow):

(1) Set the integral coefficient of the box controller to a value 50 times the current value.

(2) The damper cycles continuously and airflow is unstable. Verify that the control system detects and reports the fault.

(3) Reset the integral coefficient of the controller to the original value to restore normal operation.

Step 5: Disconnected inlet duct:

(1) From the control system workstation, command the damper to full closed; then disconnect power to the actuator, and verify that a fault is reported at the control workstation.

E 805.13.2 Acceptance Criteria. The system is able to detect common faults with air-handling units, such as a sensor failure, a failed damper, an actuator, or an improper operating mode.

The system is able to detect and report common faults with zone terminal units, such as a failed damper, an actuator, or a control tuning issue.

E 805.14 Distributed Energy Storage DX AC System (Form MECH-14A). The purpose of this test is to verify the proper operation of distributed energy storage DX systems. Distributed energy systems (DES) reduce peak demand by operating during off-peak hours and storing cooling, usually in the form of ice. During peak cooling hours the ice is melted to avoid compressor operation. The system typically consists of a water tank containing refrigerant coils that cool the water and convert it to ice. As with a standard direction expansion (DX) air conditioner, the refrigerant is compressed in a compressor and then cooled in an air-cooled condenser. The liquid refrigerant then is directed through the coils in the water tank to make ice or to air handler coils to cool the building. This applies to constant or variable volume, direct expansion (DX) systems with distributed energy storage (DES/DXAC).

E 805.14.1 Test Procedure. The procedure for performing a functional test for distributed energy storage DX AC systems shall be in accordance with Section E 805.14.1.1 through Section E 805.14.1.3.

E 805.14.1.1 Construction Inspection. Prior to functional testing, verify and document the following:

(1) The water tank is filled to the proper level.

(2) The water tank is sitting on a foundation with adequate structural strength.

(3) The water tank is insulated and the top cover is in place.

(4) The DES/DXAC is installed correctly (e.g., refrigerant piping, etc.).

(5) Verify that the correct model number is installed and configured.

E 805.14.1.2 Functional Testing. The functional testing shall be in accordance with the following steps:

Step 1: Simulate cooling load during daytime period (e.g., by setting time schedule to include actual time and placing thermostat cooling setpoint below actual temperature). Verify and document the following:

(1) Supply fan operates continually.

(2) Where the DES/DXAC has cooling capacity, DES/DXAC shall run to meet the cooling demand (in ice melt mode).

(3) Where the DES/DXAC has no ice and there is a call for cooling, the DES/DXAC shall run in direct cooling mode.

Step 2: Simulate no cooling load during daytime condition. Verify and document the following:

(1) Supply fan operates in accordance with the facility thermostat or control system.

(2) The DES/DXAC and the condensing unit do not run.

Step 3: Simulate no cooling load during morning shoulder time period. Verify and document the following:

(1) The DES/DXAC is idle.

Step 4: Simulate a cooling load during morning shoulder time period. Verify and document the following:

(1) The DES/DXAC runs in direct cooling mode.

E 805.14.1.3 Calibrating Controls. Set the proper time and date in accordance with the manufacturer's instructions for approved installers.

E 805.14.2 Acceptance Criteria. Distributed energy storage DXAC system acceptance criteria shall be as follows:

(1) Verify night time ice making operation.

(2) Verify that tank discharges during on-peak cooling periods.

(3) Verify that the compressor does not run and the tank does not discharge where there is no cooling demand during on-peak periods.

(4) Verify that the system does not operate during a morning shoulder period where there is no cooling demand.

(5) Verify that the system operates in direct mode (with compressor running) during the morning shoulder time period.

E 805.15 Thermal Energy Storage (TES) System (Form MECH-15A). The purpose of this test is to verify the proper operation of thermal energy storage (TES) systems. TES systems reduce energy consumption during peak demand periods by shifting energy consumption to nighttime. Operation of the thermal energy storage compressor during the night produces cooling energy which is stored in the form of cooled fluid or ice in tanks. During peak cooling hours the thermal storage is used for cooling to prevent the need for chiller operation. This section is limited to the following types of TES systems:

(1) Chilled water storage

(2) Ice-on-coil

(3) Ice harvester

(4) Brine

(5) Ice-slurry

(6) Eutectic salt

(7) Clathrate hydrate slurry (CHS)

E 805.15.1 Test Procedure. The procedure for performing a functional test for thermal energy storage (TES) system shall be in accordance with Section E 805.15.1.1 and Section E 805.15.1.2.

E 805.15.1.1 Construction Inspection. Prior to functional testing, verify and document the following for the chiller and storage tank:

(1) Chiller:

  (a) Brand and Model

  (b) Type (centrifugal, reciprocating, other)

  (c) Capacity (tons) (SIZE)

  (d) Starting efficiency (kW/ton) at beginning of ice production (COMP - kW/TON - START)

  (e) Ending efficiency (kW/ton) at end of ice production (COMP - kW/TON/END)

  (f) Capacity reduction (percent/°F) (PER – COMP - REDUCT/F)

  (g) Verify that the efficiency of the chiller meets or exceeds the requirements of Section E 501.0.

(2) Storage Tank:

  (a) Storage type (TES-TYPE)

  (b) Number of tanks (SIZE)

  (c) Storage capacity per tank (ton-hours) (SIZE)

  (d) Storage rate (tons) (COOL – STORE - RATE)

  (e) Discharge rate (tons) (COOL – SUPPLY - RATE)

  (f) Auxiliary power (watts) (PUMPS + AUX - kW)

  (g) Tank area (CTANK – LOSS - COEFF)

  (h) Tank insulation (R-Value) (CTANK – LOSS – COEFF)

(3) TES System:

  (a) The TES system is one of the above eligible systems.

(b) Initial charge rate of the storage tanks (tons).

(c) Final charge rate of the storage tank (tons).

(d) Initial discharge rate of the storage tanks (tons).

(e) Final discharge rate of the storage tank (tons).

(f) Charge test time (hrs).

(g) Discharge test time (hrs).

(h) Tank storage capacity after charge (ton-hrs).

(i) Tank storage capacity after discharge (ton-hrs).

(j) Tank standby storage losses (UA).

(k) Initial chiller efficiency (kW/ton) during charging.

(l) Final chiller efficiency (kW/ton) during charging.

E 805.15.1.2 Functional Testing. The functional testing shall be in accordance with the following steps:

Step 1: Verify that the TES system and the chilled water plant is controlled and monitored by an energy management system (EMS).

Step 2: Force the time to be between 9:00 p.m. and 9:00 a.m., and simulate a partial or no charge of the tank. Simulate no cooling load by setting the indoor temperature setpoint(s) higher than the ambient temperature.

Where the tank is full or nearly full of ice, it shall be permitted to adjust the control settings for this test. In some cases, the control system will not permit the chiller to start the ice-making process unless a portion of the ice has been melted. The controls designer shall be permitted to use an inventory meter (a 4-20 mA sensor that indicates water level) to determine whether or not ice-making can commence (e.g., not allow ice-making unless the inventory meter signal is less than 17 mA). Where this is the case, this limit can be reset to 20 mA during testing to allow ice making to occur.

Verify that the TES system starts charging (storing energy). This shall be checked by verifying flow and inlet and outlet temperatures of the storage tank, or directly by reading an inventory meter where the system has one.

Step 3: Force the time to be between 6:00 p.m. and 9:00 p.m., and simulate a partial charge on the tank. Simulate a cooling load by setting the indoor temperature setpoint lower than the ambient temperature. Verify that the TES system starts discharging. This shall be checked by observing tank inlet and outlet temperatures and system flow, or directly by reading an inventory meter where the system has one. Where the system has no charge, verify that the system will still attempt to meet the load through storage.

Step 4: Force the time to be between noon and 6:00 p.m., and simulate a cooling load by lowering the indoor air temperature setpoint below the ambient temperature. Verify that the tank starts discharging and the compressor is off.

Step 5: Force the time to be between 9:00 a.m. to noon, and simulate a cooling load by lowering the indoor air temperature setpoint below the ambient temperature. Verify that the tank does not discharge and the cooling load is met by the compressor.

Step 6: Force the time to be between 9:00 p.m. and 9:00 a.m. and simulate a full tank charge. This can be done in a couple of ways:

(1) By changing the inventory sensor limit that indicates tank capacity to the energy management system so that it indicates a full tank.

(2) By resetting the coolant temperature that indicates a full charge to a higher temperature than the current tank leaving temperature. Verify that the tank charging is stopped.

Step 7: Force the time to be between noon and 6:00 p.m. and simulate no cooling load by setting the indoor temperature setpoint above the ambient temperature. Verify that the tank does not discharge and the compressor is off.

E 805.15.2 Acceptance Criteria. Thermal energy storage (TES) system acceptance criteria shall be as follows:

(1) Verify that the system is able to charge the storage tank during off-peak periods where there is no cooling load.

(2) Verify that tank discharges during on-peak cooling periods.

(3) Verify that the compressor does not run and the tank does not discharge where there is no cooling demand during on-peak periods.

(4) Verify that the system does not operate during a morning shoulder period where there is no cooling demand.

(5) Verify that the system operates in direct mode (with compressor running) during the morning shoulder time period.

E 806.0 Certificate of Acceptance Forms.

E 806.1 General. This section includes the certificate of acceptance forms referenced in Section E 804.0 and Section E 805.0.

CERTIFICATE OF ACCEPTANCE	MECH-2A
Outdoor Air Acceptance	**(Page 1 of 3)**
Project Name/Address:	
System Name or Identification/Tag:	System Location or Area Served:

Enforcement Agency:	Permit Number:
Note: Submit one Certificate of Acceptance for each system that must demonstrate compliance.	Enforcement Agency Use: Checked by/Date

FIELD TECHNICIAN'S DECLARATION STATEMENT

- I certify under penalty of perjury the information provided on this form is true and correct.
- I am the person who performed the acceptance requirements verification reported on this Certificate of Acceptance (Field Technician).
- I certify that the construction/installation identified on this form complies with the acceptance requirements indicated in the plans and specifications approved by the enforcement agency, and conforms to the applicable acceptance requirements and procedures specified in Section E 801.0 through Section E 806.0.
- I have confirmed that the Installation Certificate(s) for the construction/installation identified on this form has been completed and is posted or made available with the building permit(s) issued for the building.

Company Name:		
Field Technician's Name:		Field Technician's Signature:
	Date Signed:	Position with Company (Title):

RESPONSIBLE PERSON'S DECLARATION STATEMENT

- I certify under penalty of perjury that I am the Field Technician, or the Field Technician is acting on my behalf as my employee or my agent and I have reviewed the information provided on this form.
- I am a licensed contractor or registered design professional who is eligible per the requirements of the Authority Having Jurisdiction to take responsibility for the scope of work specified on this document and attest to the declarations in this statement (responsible person).
- I certify that the information provided on this form substantiates that the construction/installation identified on this form complies with the acceptance requirements indicated in the plans and specifications approved by the enforcement agency, and conforms to the applicable acceptance requirements and procedures specified in Section E 801.0 through Section E 806.0.
- I have confirmed that the Installation Certificate(s) for the construction/installation identified on this form has been completed and is posted or made available with the permit(s) issued for the building.
- I will ensure that a completed, signed copy of this Certificate of Acceptance shall be posted, or made available with the building permit(s) issued for the building, and made available to the enforcement agency for all applicable inspections. I understand that a signed copy of this Certificate of Acceptance is required to be included with the documentation the builder provides to the building owner at occupancy.

Company Name:		Phone:
Responsible Person's Name:		Responsible Person's Signature:
License:	Date Signed:	Position With Company (Title):

CERTIFICATE OF ACCEPTANCE	MECH-2A
Outdoor Air Acceptance	(Page 2 of 3)
Project Name/Address:	
System Name or Identification/Tag:	System Location or Area Served:

Intent: *Verify measured outside airflow reading is within ± 10% of the total required outside airflow value found in Section E 805.1 through Section E 805.2.2*

Construction Inspection

1. Instrumentation to perform test includes, but not limited to:
   a. Watch.
   b. Calibrated means to measure airflow.
2. Check one of the following:
   ☐ Variable Air Volume (VAV) - Check as appropriate:
      a. Sensor used to control outdoor air flow must have calibration certificate or be field calibrated.
         ☐ Calibration certificate (attach calibration certification).
         ☐ Field calibration (attach results).
   ☐ Constant Air Volume (CAV) - Check as appropriate:
      ☐ System is designed to provide a fixed minimum OSA when the unit is on.

Outdoor Air Acceptance		
A. Functional Testing. (Check appropriate column)	**CAV**	**VAV**
a. Verify unit is not in economizer mode during test - check appropriate column.		
Step 1: CAV and VAV testing at full supply airflow.		
a. Adjust supply to achieve design airflow.	(shaded)	
b. Measured outdoor airflow reading (ft^3/min).		
c. Required outdoor airflow (ft^3/min).		
d. Time for outside air damper to stabilize after VAV boxes open (minutes).	(shaded)	
e. Return to initial conditions (check).	(shaded)	
Step 2: VAV testing at reduced supply airflow.		
a. Adjust supply airflow to either the sum of the minimum zone airflows or 30% of the total design airflow.		
b. Measured outdoor airflow reading (ft^3/min).	(shaded)	
c. Required outdoor airflow (ft^3/min).	(shaded)	
d. Time for outside air damper to stabilize after VAV boxes open and minimum air flow achieved (minutes).	(shaded)	
e. Return to initial conditions (check).	(shaded)	
B. Testing Calculations and Results.	**CAV**	**VAV**
Percent OSA at full supply airflow ($\%OA_{FA}$ for Step 1).		
a. $\%OA_{FA}$ = Measured outside air reading /Required outside air (Step 1b / Step 1c)	%	%
b. $90\% \leq \%OA_{FA} \leq 110\%$	Y / N	Y / N
c. Outside air damper position stabilizes within 15 minutes (Step 1d < 15 minutes)	Y / N	Y / N
Percent OSA at reduced supply airflow ($\%OA_{RA}$ for Step 2).		
a. $\%OA_{RA}$ = Measured outside air reading/required outside air (Step 2b / Step 2c).	%	%
b. $90\% \leq \%OA_{RA} \leq 110\%$.		Y / N
c. Outside air damper position stabilizes within 15 minutes (Step 2d < 15 minutes).	(shaded)	Y / N
Note: Shaded boxes do not apply for CAV systems.	(shaded)	

For SI units: 1 cubic foot per minute = 0.00047 m^3/s

CERTIFICATE OF ACCEPTANCE	MECH-2A
Outdoor Air Acceptance	**(Page 3 of 3)**
Project Name/Address:	
System Name or Identification/Tag:	System Location or Area Served:

C.	**PASS/FAIL Evaluation** (check one):
☐	PASS: All Construction Inspection responses are complete and Testing Calculations & Results responses are positive (Y – yes).
☐	FAIL: Any Construction Inspection responses are incomplete *OR* there is one or more negative (N – no) responses in Testing Calculations & Results section. Provide explanation below. Use and attach additional pages if necessary.

CERTIFICATE OF ACCEPTANCE	MECH-3A
Constant Volume Single Zone Unitary Air Conditioner and Heat Pump Systems	**(Page 1 of 3)**
Project Name/Address:	
System Name or Identification/Tag:	System Location or Area Served:

Enforcement Agency:	Permit Number:
Note: Submit one Certificate of Acceptance for each system that must demonstrate compliance.	Enforcement Agency Use: Checked by/Date

FIELD TECHNICIAN'S DECLARATION STATEMENT

- I certify under penalty of perjury the information provided on this form is true and correct.
- I am the person who performed the acceptance requirements verification reported on this Certificate of Acceptance (Field Technician).
- I certify that the construction/installation identified on this form complies with the acceptance requirements indicated in the plans and specifications approved by the enforcement agency, and conforms to the applicable acceptance requirements and procedures specified in Section E 801.0 through Section E 806.0.
- I have confirmed that the Installation Certificate(s) for the construction/installation identified on this form has been completed and is posted or made available with the building permit(s) issued for the building.

Company Name:		
Field Technician's Name:		Field Technician's Signature:
	Date Signed:	Position with Company (Title):

RESPONSIBLE PERSON'S DECLARATION STATEMENT

- I certify under penalty of perjury that I am the Field Technician, or the Field Technician is acting on my behalf as my employee or my agent and I have reviewed the information provided on this form.
- I am a licensed contractor or registered design professional who is eligible per the requirements of the Authority Having Jurisdiction to take responsibility for the scope of work specified on this document and attest to the declarations in this statement (responsible person).
- I certify that the information provided on this form substantiates that the construction/installation identified on this form complies with the acceptance requirements indicated in the plans and specifications approved by the enforcement agency, and conforms to the applicable acceptance requirements and procedures specified in Section E 801.0 through Section E 806.0.
- I have confirmed that the Installation Certificate(s) for the construction/installation identified on this form has been completed and is posted or made available with the permit(s) issued for the building.
- I will ensure that a completed, signed copy of this Certificate of Acceptance shall be posted, or made available with the building permit(s) issued for the building, and made available to the enforcement agency for all applicable inspections. I understand that a signed copy of this Certificate of Acceptance is required to be included with the documentation the builder provides to the building owner at occupancy.

Company Name:			Phone:
Responsible Person's Name:		Responsible Person's Signature:	
License:	Date Signed:	Position With Company (Title):	

CERTIFICATE OF ACCEPTANCE	MECH-3A
Constant Volume Single Zone Unitary Air Conditioner and Heat Pump Systems	(Page 2 of 3)
Project Name/Address:	
System Name or Identification/Tag:	System Location or Area Served:

Intent: *Verify the individual components of a constant volume, single-zone, unitary air conditioner and heat pump system function correctly, including: thermostat installation and programming, supply fan, heating, cooling, and damper operation.*

Construction Inspection

1. Instrumentation to perform test includes, but not limited to:
   a. None required
2. Installation
   - ☐ Thermostat is located within the space-conditioning zone that is served by the HVAC system.
3. Programming (check all of the following):
   - ☐ Thermostat meets the temperature adjustment and dead band requirements.
   - ☐ Occupied, unoccupied, and holiday schedules have been programmed per the facility's schedule.
   - ☐ Preoccupancy purge has been programmed to meet the requirements of Section E 805.3 through Section E 805.3.2.

A. Functional Testing Requirements. — **Operating Modes**

- A: Heating load during occupied condition
- B: No-load during occupied condition
- C: Heating load during unoccupied condition
- D: No-load during unoccupied condition
- E: Manual override
- F: Cooling load during occupied condition
- G: Cooling load during unoccupied condition

	Step 1: Check and verify the following for each simulation mode required.	A	B	C	D	E	F	G
a.	Supply fan operates continually.							
b.	Supply fan turns off.							
c.	Supply fan cycles on and off.							
d.	System reverts to "occupied" mode to satisfy any condition.							
e.	System turns off when manual override time period expires.							
f.	Gas-fired furnace, heat pump, or electric heater stages on.							
g.	Neither heating or cooling is provided by the unit.							
h.	No heating is provided by the unit.							
i.	No cooling is provided by the unit.							
j.	Compressor stages on.							
k.	Outside air damper is open to minimum position.							
l.	Outside air damper closes completely.							
m.	System returned to initial operating conditions after all tests have been completed:	Y/N						

B. Testing Results	A	B	C	D	E	F	G
Indicate if Passed (P), Failed (F), or N/A (X), fill in appropriate letter.							

CERTIFICATE OF ACCEPTANCE	MECH-3A
Constant Volume Single Zone Unitary Air Conditioner and Heat Pump Systems	**(Page 3 of 3)**
Project Name/Address:	
System Name or Identification/Tag:	System Location or Area Served:

C.	PASS/FAIL Evaluation. (check one):
☐	PASS: All **Construction Inspection** responses are complete and **Testing Results** responses are "Pass" (P).
☐	FAIL: Any **Construction Inspection** responses are incomplete OR there is one or more "Fail" (F) responses in **Testing Results** section. Provide explanation below. Use and attach additional pages if necessary.

CERTIFICATE OF ACCEPTANCE	MECH-4A
Air Distribution Systems Acceptance	(Page 1 of 3)
Project Name/Address:	
System Name or Identification/Tag:	System Location or Area Served:

Enforcement Agency:	Permit Number:
Note: Submit one Certificate of Acceptance for each system that must demonstrate compliance.	Enforcement Agency Use: Checked by/Date

FIELD TECHNICIAN'S DECLARATION STATEMENT

- I certify under penalty of perjury the information provided on this form is true and correct.
- I am the person who performed the acceptance requirements verification reported on this Certificate of Acceptance (Field Technician).
- I certify that the construction/installation identified on this form complies with the acceptance requirements indicated in the plans and specifications approved by the enforcement agency, and conforms to the applicable acceptance requirements and procedures specified in Section E 801.0 through Section E 806.0.
- I have confirmed that the Installation Certificate(s) for the construction/installation identified on this form has been completed and is posted or made available with the building permit(s) issued for the building.

Company Name:		
Field Technician's Name:		Field Technician's Signature:
	Date Signed:	Position with Company (Title):

RESPONSIBLE PERSON'S DECLARATION STATEMENT

- I certify under penalty of perjury that I am the Field Technician, or the Field Technician is acting on my behalf as my employee or my agent and I have reviewed the information provided on this form.
- I am a licensed contractor or registered design professional who is eligible per the requirements of the Authority Having Jurisdiction to take responsibility for the scope of work specified on this document and attest to the declarations in this statement (responsible person).
- I certify that the information provided on this form substantiates that the construction/installation identified on this form complies with the acceptance requirements indicated in the plans and specifications approved by the enforcement agency, and conforms to the applicable acceptance requirements and procedures specified in Section E 801.0 through Section E 806.0.
- I have confirmed that the Installation Certificate(s) for the construction/installation identified on this form has been completed and is posted or made available with the permit(s) issued for the building.
- I will ensure that a completed, signed copy of this Certificate of Acceptance shall be posted, or made available with the building permit(s) issued for the building, and made available to the enforcement agency for all applicable inspections. I understand that a signed copy of this Certificate of Acceptance is required to be included with the documentation the builder provides to the building owner at occupancy.

Company Name:			Phone:
Responsible Person's Name:		Responsible Person's Signature:	
License:	Date Signed:	Position With Company (Title):	

CERTIFICATE OF ACCEPTANCE	MECH-4A
Air Distribution Systems Acceptance	**(Page 2 of 3)**
Project Name/Address:	
System Name or Identification/Tag:	System Location or Area Served:

Intent: *New single zone supply ductwork must be less than 6% leakage rate per Section E 805.4 through Section E 805.4.2, existing single zone ductwork must be less than 15% leakage or other compliance path per Section E 805.4 through Section E 805.4.2.*

Construction Inspection

1. Scope of test – New Buildings – this test required on New Buildings only if all check boxes 1(a) through 1(c) are checked.

   Existing Buildings – this test required if 1(a) through 1(d) are checked.

   Ductwork conforms to the following (note if any of these are not checked, then this test is not required):

   ☐ 1(a) Connected to a constant volume, single zone air conditioners, heat pumps, or furnaces.

   ☐ 1(b) Serves less than 5000 square feet of floor area.

   ☐ 1(c) Has more than 25% duct surface area located in one or more of the following spaces.

   – Outdoors.
   – A space directly under a roof where the U-factor of the roof is greater than U-factor of the ceiling.
   – A space directly under a roof with fixed vents or openings to the outside or unconditioned spaces.
   – An unconditioned crawl space.
   – Other unconditioned spaces.

   ☐ 1(d) A duct is extended or any of the following replaced: air handler, outdoor condensing unit of a split system, cooling or heating coil, or the furnace heat exchanger.

2. Instrumentation to perform test includes:

   a. Duct Pressure Test.

3. Material and Installation. Complying new duct systems shall have a checked box for all of the following categories (a) through (g):

   a. Choice of drawbands. (check one of the following)

☐	Stainless steel worm-drive hose clamps.
☐	UV-resistant nylon duct ties.

☐	b. Flexible ducts are not constricted in any way.
☐	c. Duct leakage tests performed before access to ductwork and connections are blocked.
☐	d. Joints and seams are not sealed with cloth back rubber adhesive tape unless used in combination with mastic and drawbands.
☐	e. Duct R-values are verified R-8 per Section E 805.4 through Section E 805.4.2.
☐	f. Ductwork located outdoors has insulation that is protected from damage and suitable for outdoor service.
☐	g. A sticker has been affixed to the exterior surface of the air handler access door per Section E 805.4 through Section E 805.4.2.

For SI units: 1 square foot = 0.0929 m^2

CERTIFICATE OF ACCEPTANCE	MECH-4A
Air Distribution Systems Acceptance	**(Page 3 of 3)**
Project Name/Address:	
System Name or Identification/Tag:	System Location or Area Served:

Air Distribution System Leakage Diagnostic.
The installing contractor must pressure test every new HVAC systems that meet the requirements of Section E 805.4 through Section E 805.4.2 and every retrofit to existing HVAC systems that meet the requirements of Section E 805.4 through Section E 805.4.2.

RATED FAN FLOW (applies to all systems)		Measured Values	
1.	Cooling capacity or for heating only units heating capacity.		
	(a) Cooling capacity (for all units but heating only units) in tons.		
	(b) Heating capacity (for heating only units) kBtu/h.		
2.	Fan flow calculation		
	(a) Cooling capacity in tons [___________ (Line # 1a) x 400 ft^3/min/ton].		
	(b) Heating only cap. kBtu/h [___________ (Line # 1b) x 21.7 ft^3/min/kBtu/h].		
3.	Total calculated supply fan flow 2(a) or 2(b) ft^3/min.		

NEW CONSTRUCTION OR ENTIRE NEW DUCT SYSTEM ALTERATION:			
	Duct pressurization test results (ft^3/min @ 25 Pa).		
4.	Enter tested leakage flow in ft^3/min:		✓ ✓
5.	Pass if leakage percentage ≤6%: [(Line #4) /________ (Line #3)] x 100	%	☐ Pass ☐ Fail
ALTERATIONS: Pre-existing duct system with duct alteration and/or HVAC equipment change-out.			
6.	Enter tested leakage flow (cubic feet per minute): Pre-test of existing duct system prior to duct system alteration, equipment change-out, or both.		
7.	Enter tested leakage flow (cubic feet per minute): Final test of new duct system or altered duct system for duct system alteration, equipment change-out, or both.		
TEST OR VERIFICATION STANDARDS: For altered duct system and/or HVAC equipment change-out use one of the following three tests or verification standards for compliance:			
8.	Pass if leakage percentage <15% [________ (Line # 7) / ________ (Line # 3)] x 100	%	☐ Pass ☐ Fail
9.	Pass if leakage reduction percentage >60% Leakage reduction = [1 - [________ (Line#7) / ________ (Line#6)] } x 100	%	☐ Pass ☐ Fail
10.	Pass if all accessible leaks are sealed as confirmed by visual inspection and verification by HERS rater (sampling rate 100%).	%	☐ Pass ☐ Fail
	Pass if One of Lines #8 through #10 pass		☐ Pass ☐ Fail

For SI units: 1000 British thermal units per hour = 0.293 kW, 1 cubic foot per minute = 0.00047 m^3/s, 1 metric ton = 1000 kg

CERTIFICATE OF ACCEPTANCE	MECH-5A
Air Economizer Controls Acceptance	**(Page 1 of 3)**
Project Name/Address:	
System Name or Identification/Tag:	System Location or Area Served:

Enforcement Agency:	Permit Number:
Note: Submit one Certificate of Acceptance for each system that must demonstrate compliance.	Enforcement Agency Use: Checked by/Date

FIELD TECHNICIAN'S DECLARATION STATEMENT

- I certify under penalty of perjury the information provided on this form is true and correct.
- I am the person who performed the acceptance requirements verification reported on this Certificate of Acceptance (Field Technician).
- I certify that the construction/installation identified on this form complies with the acceptance requirements indicated in the plans and specifications approved by the enforcement agency, and conforms to the applicable acceptance requirements and procedures specified in Section E 801.0 through Section E 806.0.
- I have confirmed that the Installation Certificate(s) for the construction/installation identified on this form has been completed and is posted or made available with the building permit(s) issued for the building.

Company Name:		
Field Technician's Name:		Field Technician's Signature:
	Date Signed:	Position with Company (Title):

RESPONSIBLE PERSON'S DECLARATION STATEMENT

- I certify under penalty of perjury that I am the Field Technician, or the Field Technician is acting on my behalf as my employee or my agent and I have reviewed the information provided on this form.
- I am a licensed contractor or registered design professional who is eligible per the requirements of the Authority Having Jurisdiction to take responsibility for the scope of work specified on this document and attest to the declarations in this statement (responsible person).
- I certify that the information provided on this form substantiates that the construction/installation identified on this form complies with the acceptance requirements indicated in the plans and specifications approved by the enforcement agency, and conforms to the applicable acceptance requirements and procedures specified in Section E 801.0 through Section E 806.0.
- I have confirmed that the Installation Certificate(s) for the construction/installation identified on this form has been completed and is posted or made available with the permit(s) issued for the building.
- I will ensure that a completed, signed copy of this Certificate of Acceptance shall be posted, or made available with the building permit(s) issued for the building, and made available to the enforcement agency for all applicable inspections. I understand that a signed copy of this Certificate of Acceptance is required to be included with the documentation the builder provides to the building owner at occupancy.

Company Name:		Phone:
Responsible Person's Name:		Responsible Person's Signature:
License:	Date Signed:	Position With Company (Title):

CERTIFICATE OF ACCEPTANCE	MECH-5A
Air Economizer Controls Acceptance	(Page 2 of 3)
Project Name/Address:	
System Name or Identification/Tag:	System Location or Area Served:

Intent: *Verify that airside economizers function properly.*

Construction Inspection

1. Instrumentation to perform test includes, but not limited to:
   a. Handheld temperature probes calibration.
      Date: (must be within last year).
   b. Multimeter capable of measuring ohms and milliamps.
2. Test method (check one of the following):
   - ☐ Economizer comes from HVAC system manufacturer installed by and has been factory calibrated and tested. Attach documentation and complete certification statement. No functional testing required.
   - ☐ Economizer field installed and field tested or factory installed and field tested.
3. Installation (check all of the following first level boxes).
   - ☐ Economizer lockout setpoint complies with Section E 805.5 through Section E 805.5.2.
   - ☐ Economizer lockout control sensor is located to prevent false readings.
   - ☐ System is designed to provide up to 100% outside air without over-pressurizing the building.
   - ☐ For systems with DDC controls lockout sensor(s) are either factory calibrated or field calibrated.
   - ☐ For systems with non-DDC controls, manufacturer's startup and testing procedures have been applied.

A. Functional Testing.	
Step 1: Disable demand control ventilation systems (if applicable).	
Step 2: Enable the economizer and simulate a cooling demand large enough to drive the economizer fully open (check and verify the following).	
☐ Economizer damper modulates 100% open.	
☐ Return air damper modulates 100% closed.	
☐ Where applicable, verify that the economizer remains 100% open when the cooling demand can no longer be met by the economizer alone.	
☐ All applicable fans and dampers operate as intended to maintain building pressure.	
☐ The unit heating is disabled.	
Step 3: Simulate a cooling load and disable the economizer (check and verify the following).	
☐ Economizer damper closes to its minimum position.	
☐ All applicable fans and dampers operate as intended to maintain building pressure.	
☐ The unit heating is disabled.	
Step 4: Simulate a heating demand and enable the economizer (check and verify the following).	
☐ Economizer damper closes to its minimum position.	
Step 5: System returned to initial operating conditions.	**Y/N**

B. Testing Results.	**PASS / FAIL**	
Step 1: Simulate cooling load and enable the economizer (all check boxes are complete).		
Step 2: Simulate cooling load and disable the economizer (all check boxes are complete).		
Step 3: Simulate heating demand and enable the economizer (all check boxes are complete).		

CERTIFICATE OF ACCEPTANCE	MECH-5A
Air Economizer Controls Acceptance	**(Page 3 of 3)**
Project Name/Address:	
System Name or Identification/Tag:	System Location or Area Served:

C.	PASS/FAIL Evaluation (check one):
☐	PASS: All **Construction Inspection** responses are complete and **Testing Results** responses are "Pass."
☐	FAIL: Any **Construction Inspection** responses are incomplete *OR* there is one or more "Fail" responses in **Testing Results** section. Provide explanation below. Use and attach additional pages if necessary.

CERTIFICATE OF ACCEPTANCE	MECH-6A
Demand Control Ventilation Systems Acceptance	**(Page 1 of 3)**
Project Name/Address:	
System Name or Identification/Tag:	System Location or Area Served:

Enforcement Agency:	Permit Number:
Note: Submit one Certificate of Acceptance for each system that must demonstrate compliance.	Enforcement Agency Use: Checked by/Date

FIELD TECHNICIAN'S DECLARATION STATEMENT

- I certify under penalty of perjury the information provided on this form is true and correct.
- I am the person who performed the acceptance requirements verification reported on this Certificate of Acceptance (Field Technician).
- I certify that the construction/installation identified on this form complies with the acceptance requirements indicated in the plans and specifications approved by the enforcement agency, and conforms to the applicable acceptance requirements and procedures specified in Section E 801.0 through Section E 806.0.
- I have confirmed that the Installation Certificate(s) for the construction/installation identified on this form has been completed and is posted or made available with the building permit(s) issued for the building.

Company Name:		
Field Technician's Name:		Field Technician's Signature:
	Date Signed:	Position with Company (Title):

RESPONSIBLE PERSON'S DECLARATION STATEMENT

- I certify under penalty of perjury that I am the Field Technician, or the Field Technician is acting on my behalf as my employee or my agent and I have reviewed the information provided on this form.
- I am a licensed contractor or registered design professional who is eligible per the requirements of the Authority Having Jurisdiction to take responsibility for the scope of work specified on this document and attest to the declarations in this statement (responsible person).
- I certify that the information provided on this form substantiates that the construction/installation identified on this form complies with the acceptance requirements indicated in the plans and specifications approved by the enforcement agency, and conforms to the applicable acceptance requirements and procedures specified in Section E 801.0 through Section E 806.0.
- I have confirmed that the Installation Certificate(s) for the construction/installation identified on this form has been completed and is posted or made available with the permit(s) issued for the building.
- I will ensure that a completed, signed copy of this Certificate of Acceptance shall be posted, or made available with the building permit(s) issued for the building, and made available to the enforcement agency for all applicable inspections. I understand that a signed copy of this Certificate of Acceptance is required to be included with the documentation the builder provides to the building owner at occupancy.

Company Name:		Phone:
Responsible Person's Name:		Responsible Person's Signature:
License:	Date Signed:	Position With Company (Title):

CERTIFICATE OF ACCEPTANCE	MECH-6A
Demand Control Ventilation Systems Acceptance	(Page 2 of 3)
Project Name/Address:	
System Name or Identification/Tag:	System Location or Area Served:

Intent: *Verify that systems required to employ demand controlled ventilation can vary outside ventilation flow rates based on maintaining interior carbon dioxide (CO_2) concentration setpoints.*

Construction Inspection

1. Instrumentation to perform test includes, but not limited to:
   a. Calibrated handheld CO_2 analyzer.
   b. Manufacturer's calibration kit.
   c. Calibrated CO_2/air mixtures.
2. Installation.
   ☐ The sensor is located in the high density space between 3 feet and 6 feet above the floor or at the anticipated level of the occupants heads.
3. Documentation of all carbon dioxide control sensors includes (check one of the following):
   a. Calibration method.
      ☐ Factory-calibration certificate (certificate must be attached).
      ☐ Field calibrated.
   b. Sensor accuracy.
      ☐ Certified by manufacturer to be no more than +/- 75 ppm calibration certificate must be attached.

A. Functional Testing.	Results
a. Disable economizer controls.	
b. Outside air CO_2 concentration (select one of the following).	
☐ Measured dynamically using CO_2 sensor.	______________ ppm
c. Interior CO_2 concentration setpoint (Outside CO_2 concentration + 600 ppm).	______________ ppm
Step 1: Simulate a signal at or slightly above the CO_2 setpoint or follow manufacturers recommended testing procedures.	
☐ For single zone units, outdoor air damper modulates opens to satisfy the total ventilation air called for in the certificate of compliance.	
☐ For multiple zone units, either outdoor air damper or zone damper modulate open to satisfy the zone ventilation requirements.	
Step 2: Simulate signal well below the CO_2 setpoint or follow manufacturers recommended procedures.	
☐ For single zone units, outdoor air damper modulates to the design minimum value.	
☐ For multiple zone units, either outdoor air damper or zone damper modulate to satisfy the reduced zone ventilation requirements.	
Step 3: System returned to initial operating conditions.	**Y/N**

B. Testing Results.	PASS / FAIL	
Step 1: Simulate a high CO_2 load (check box complete).		
Step 2: Simulate a low CO_2 load (check box complete).		

For SI units: 1 inch = 25.4 mm

CERTIFICATE OF ACCEPTANCE	MECH-6A
Demand Control Ventilation Systems Acceptance	**(Page 3 of 3)**
Project Name/Address:	
System Name or Identification/Tag:	System Location or Area Served:

C.	PASS/FAIL Evaluation (check one):
☐	PASS: All **Construction Inspection** responses are complete and **Testing Results** responses are "Pass."
☐	FAIL: Any **Construction Inspection** responses are incomplete *OR* there is one or more "Fail" responses in **Testing Results** section. Provide explanation below. Use and attach additional pages if necessary.

CERTIFICATE OF ACCEPTANCE	MECH-7A
Supply Fan VFD Acceptance	**(Page 1 of 2)**
Project Name/Address:	
System Name or Identification/Tag:	System Location or Area Served:

Enforcement Agency:	Permit Number:
Note: Submit one Certificate of Acceptance for each system that must demonstrate compliance.	Enforcement Agency Use: Checked by/Date

FIELD TECHNICIAN'S DECLARATION STATEMENT

- I certify under penalty of perjury the information provided on this form is true and correct.
- I am the person who performed the acceptance requirements verification reported on this Certificate of Acceptance (Field Technician).
- I certify that the construction/installation identified on this form complies with the acceptance requirements indicated in the plans and specifications approved by the enforcement agency, and conforms to the applicable acceptance requirements and procedures specified in Section E 801.0 through Section E 806.0.
- I have confirmed that the Installation Certificate(s) for the construction/installation identified on this form has been completed and is posted or made available with the building permit(s) issued for the building.

Company Name:		
Field Technician's Name:		Field Technician's Signature:
	Date Signed:	Position with Company (Title):

RESPONSIBLE PERSON'S DECLARATION STATEMENT

- I certify under penalty of perjury that I am the Field Technician, or the Field Technician is acting on my behalf as my employee or my agent and I have reviewed the information provided on this form.
- I am a licensed contractor or registered design professional who is eligible per the requirements of the Authority Having Jurisdiction to take responsibility for the scope of work specified on this document and attest to the declarations in this statement (responsible person).
- I certify that the information provided on this form substantiates that the construction/installation identified on this form complies with the acceptance requirements indicated in the plans and specifications approved by the enforcement agency, and conforms to the applicable acceptance requirements and procedures specified in Section E 801.0 through Section E 806.0.
- I have confirmed that the Installation Certificate(s) for the construction/installation identified on this form has been completed and is posted or made available with the permit(s) issued for the building.
- I will ensure that a completed, signed copy of this Certificate of Acceptance shall be posted, or made available with the building permit(s) issued for the building, and made available to the enforcement agency for all applicable inspections. I understand that a signed copy of this Certificate of Acceptance is required to be included with the documentation the builder provides to the building owner at occupancy.

Company Name:		Phone:
Responsible Person's Name:		Responsible Person's Signature:
License:	Date Signed:	Position With Company (Title):

CERTIFICATE OF ACCEPTANCE	MECH-7A
Supply Fan VFD Acceptance	**(Page 2 of 2)**
Project Name/Address:	
System Name or Identification/Tag:	System Location or Area Served:

Intent: *Verify that the supply fan in a variable air volume application modulates to meet system airflow demand.*

Construction Inspection

1 Instrumentation to perform test includes, but not limited to:

a. Calibrated differential pressure gauge.

2 Installation.

- ☐ Discharge static pressure sensors are either factory calibrated or field-calibrated.
- ☐ The static pressure location, setpoint, and reset control meets the requirements of Section E 805.7 through Section E 805.7.2.

3 Documentation of all discharge static pressure sensors including (check one of the following):

- ☐ Field-calibrated.
- ☐ Calibration complete, all pressure sensors within 10% of calibrated reference sensor.

A. Functional Testing.	**Results**
Step 1: Drive all VAV boxes to achieve design airflow.	
a. Supply fan controls modulate to increase capacity.	**Y / N**
b. Supply fan maintains discharge static pressure within +/-10% of the current operating setpoint.	**Y / N**
c. Supply fan controls stabilize within a 5 minute period.	**Y / N**
Step 2: Drive all VAV boxes to minimum flow.	
a. Supply fan controls modulate to decrease capacity.	**Y / N**
b. Current operating setpoint has decreased (for systems with DDC to the zone level).	**Y / N**
c. Supply fan maintains discharge static pressure within +/-10% of the current operating setpoint.	**Y / N**
d. Supply fan controls stabilize within a 5 minute period.	**Y / N**
Step 3: System returned to initial operating conditions.	**Y / N**

B. Testing Results.	**PASS / FAIL**	
Step 1: Drive all VAV boxes to achieve design airflow.		
Step 2: Drive all VAV boxes to minimum flow.		

C. PASS / FAIL Evaluation (check one):

☐	PASS: All Construction Inspection responses are complete and all Testing Results responses are "Pass."
☐	FAIL: Any Construction Inspection responses are incomplete *OR* there is one or more "Fail" responses in Testing Results section. Provide explanation below. Use and attach additional pages if necessary.

CERTIFICATE OF ACCEPTANCE	MECH-8A
Valve Leakage Test	**(Page 1 of 2)**
Project Name/Address:	
System Name or Identification/Tag:	System Location or Area Served:

Enforcement Agency:	Permit Number:
Note: Submit one Certificate of Acceptance for each system that must demonstrate compliance.	Enforcement Agency Use: Checked by/Date

FIELD TECHNICIAN'S DECLARATION STATEMENT

- I certify under penalty of perjury the information provided on this form is true and correct.
- I am the person who performed the acceptance requirements verification reported on this Certificate of Acceptance (Field Technician).
- I certify that the construction/installation identified on this form complies with the acceptance requirements indicated in the plans and specifications approved by the enforcement agency, and conforms to the applicable acceptance requirements and procedures specified in Section E 801.0 through Section E 806.0.
- I have confirmed that the Installation Certificate(s) for the construction/installation identified on this form has been completed and is posted or made available with the building permit(s) issued for the building.

Company Name:		
Field Technician's Name:		Field Technician's Signature:
	Date Signed:	Position with Company (Title):

RESPONSIBLE PERSON'S DECLARATION STATEMENT

- I certify under penalty of perjury that I am the Field Technician, or the Field Technician is acting on my behalf as my employee or my agent and I have reviewed the information provided on this form.
- I am a licensed contractor or registered design professional who is eligible per the requirements of the Authority Having Jurisdiction to take responsibility for the scope of work specified on this document and attest to the declarations in this statement (responsible person).
- I certify that the information provided on this form substantiates that the construction/installation identified on this form complies with the acceptance requirements indicated in the plans and specifications approved by the enforcement agency, and conforms to the applicable acceptance requirements and procedures specified in Section E 801.0 through Section E 806.0.
- I have confirmed that the Installation Certificate(s) for the construction/installation identified on this form has been completed and is posted or made available with the permit(s) issued for the building.
- I will ensure that a completed, signed copy of this Certificate of Acceptance shall be posted, or made available with the building permit(s) issued for the building, and made available to the enforcement agency for all applicable inspections. I understand that a signed copy of this Certificate of Acceptance is required to be included with the documentation the builder provides to the building owner at occupancy.

Company Name:		Phone:
Responsible Person's Name:		Responsible Person's Signature:
License:	Date Signed:	Position With Company (Title):

CERTIFICATE OF ACCEPTANCE	MECH-8A
Valve Leakage Test	(Page 2 of 2)
Project Name/Address:	
System Name or Identification/Tag:	System Location or Area Served:

Intent: *Ensure that control valves serving variable flow systems are designed to withstand the pump pressure over the full range of operation.*

Construction Inspection

1 Instrumentation to perform test includes, but not limited to:
  a. Calibrated differential pressure gauge.
  b. Pump curve submittals showing the shutoff head.

2 Installation.
  ☐ Valve and piping arrangements were installed per the design drawings.

A. Functional Testing.	**Pump Tag (Id)**		**Results**
Step 1: Determine pump dead head pressure.			
a. Close pump discharge isolation valve.			**Y / N**
b. Measure and record the differential pump pressure.		Feet Water Column =	
c. Record the shutoff head from the submittal.		Feet Water Column =	
d. The measurement across the pump in step 1b is within 5% of the pump submittal in step 1c.			**Y / N**
e. Open pump discharge isolation valve.			**Y / N**
Step 2: Automatically close all valves on the systems being tested. If three-way valves are present, close off the bypass line(s).			
a. The 2-way valves automatically close.			**Y / N**
b. Measure and record the differential pump pressure in feet of water column.		Feet Water Column =	
c. The measurement across the pump in step 2b is within 5% of the measurement in step 1b.			**Y / N**
Step 3: System returned to initial operating conditions.		**Y / N**	

B. Testing Results.	**PASS**	**FAIL**
Step 1: Pressure measurement is within 5% of submittal data for all pumps.	☐	☐
Step 2: Pressure measurements are within 5%.	☐	☐

C. PASS / FAIL Evaluation (check one):	
☐	PASS: All **Construction Inspection** responses are complete and all **Testing Results** responses are "Pass."
☐	FAIL: Any **Construction Inspection** responses are incomplete *OR* there is one or more "Fail" responses in **Testing Results** section. Provide explanation below. Use and attach additional pages if necessary.

For SI units: 1 inch water column = 0.249 kPa

CERTIFICATE OF ACCEPTANCE	MECH-9A
Supply Water Temperature Reset Controls Acceptance	**(Page 1 of 2)**
Project Name/Address:	
System Name or Identification/Tag:	System Location or Area Served:

Enforcement Agency:	Permit Number:
Note: Submit one Certificate of Acceptance for each system that must demonstrate compliance.	Enforcement Agency Use: Checked by/Date

FIELD TECHNICIAN'S DECLARATION STATEMENT

- I certify under penalty of perjury the information provided on this form is true and correct.
- I am the person who performed the acceptance requirements verification reported on this Certificate of Acceptance (Field Technician).
- I certify that the construction/installation identified on this form complies with the acceptance requirements indicated in the plans and specifications approved by the enforcement agency, and conforms to the applicable acceptance requirements and procedures specified in Section E 801.0 through Section E 806.0.
- I have confirmed that the Installation Certificate(s) for the construction/installation identified on this form has been completed and is posted or made available with the building permit(s) issued for the building.

Company Name:		
Field Technician's Name:		Field Technician's Signature:
	Date Signed:	Position with Company (Title):

RESPONSIBLE PERSON'S DECLARATION STATEMENT

- I certify under penalty of perjury that I am the Field Technician, or the Field Technician is acting on my behalf as my employee or my agent and I have reviewed the information provided on this form.
- I am a licensed contractor or registered design professional who is eligible per the requirements of the Authority Having Jurisdiction to take responsibility for the scope of work specified on this document and attest to the declarations in this statement (responsible person).
- I certify that the information provided on this form substantiates that the construction/installation identified on this form complies with the acceptance requirements indicated in the plans and specifications approved by the enforcement agency, and conforms to the applicable acceptance requirements and procedures specified in Section E 801.0 through Section E 806.0.
- I have confirmed that the Installation Certificate(s) for the construction/installation identified on this form has been completed and is posted or made available with the permit(s) issued for the building.
- I will ensure that a completed, signed copy of this Certificate of Acceptance shall be posted, or made available with the building permit(s) issued for the building, and made available to the enforcement agency for all applicable inspections. I understand that a signed copy of this Certificate of Acceptance is required to be included with the documentation the builder provides to the building owner at occupancy.

Company Name:		Phone:
Responsible Person's Name:		Responsible Person's Signature:
License:	Date Signed:	Position With Company (Title):

CERTIFICATE OF ACCEPTANCE	MECH-9A
Supply Water Temperature Reset Controls Acceptance	(Page 2 of 2)
Project Name/Address:	
System Name or Identification/Tag:	System Location or Area Served:

Intent: *Ensure that both the chilled water and hot water supply temperatures are automatically reset based on either building loads or outdoor air temperature, as indicated in the control sequences.*

Construction Inspection

1 Instrumentation to perform test includes, but not limited to:
   a. Calibrated reference temperature sensor or drywell bath.

2 Installation
   ☐ Supply water temperature sensors have been either factory or field calibrated.

3 Documentation of hydronic system supply temperature sensors including (check one of the following):
   ☐ Field-calibrated
   ☐ Calibration complete, hydronic system supply temperature sensors within 1% of calibrated reference sensor or drywell bath.

A. Functional Testing.	
Step 1: Test maximum reset value.	
a. Change reset control variable to its maximum value.	**Y / N**
b. Verify that chilled or hot water temperature setpoint is reset to appropriate value.	**Y / N**
c. Verify that actual system temperature changes to within 2% of the new setpoint.	**Y / N**
Step 2: Test minimum reset value.	
a. Change reset control variable to its minimum value.	**Y / N**
b. Verify that chilled or hot water temperature setpoint is reset to appropriate value.	**Y / N**
c. Verify that actual system temperature changes to within 2% of the new setpoint.	**Y / N**
Step 3: Test maximum reset value.	
a. Restore reset control variable to automatic control.	**Y / N**
b. Verify that chilled or hot water temperature setpoint is reset to appropriate value.	**Y / N**
c. Verify that actual supply temperature changes to meet setpoint.	**Y / N**
d. Verify that actual supply temperature changes to within 2% of the new setpoint.	**Y / N**

B. Testing Results.	**PASS / FAIL**	
System passes criteria in 1c, 2c, and 3d.	☐	☐

C. PASS / FAIL Evaluation (check one):	
☐	PASS: All **Construction Inspection** responses are complete and all **Testing Results** responses are "Pass."
☐	FAIL: Any **Construction Inspection** responses are incomplete *OR* there is one or more "Fail" responses in **Testing Results** section. Provide explanation below. Use and attach additional pages if necessary.

CERTIFICATE OF ACCEPTANCE	MECH-10A
Hydronic System Variable Flow Control Acceptance	**(Page 1 of 3)**
Project Name/Address:	
System Name or Identification/Tag:	System Location or Area Served:

Enforcement Agency:	Permit Number:
Note: Submit one Certificate of Acceptance for each system that must demonstrate compliance.	Enforcement Agency Use: Checked by/Date

FIELD TECHNICIAN'S DECLARATION STATEMENT

- I certify under penalty of perjury the information provided on this form is true and correct.
- I am the person who performed the acceptance requirements verification reported on this Certificate of Acceptance (Field Technician).
- I certify that the construction/installation identified on this form complies with the acceptance requirements indicated in the plans and specifications approved by the enforcement agency, and conforms to the applicable acceptance requirements and procedures specified in Section E 801.0 through Section E 806.0.
- I have confirmed that the Installation Certificate(s) for the construction/installation identified on this form has been completed and is posted or made available with the building permit(s) issued for the building.

Company Name:		
Field Technician's Name:		Field Technician's Signature:
	Date Signed:	Position with Company (Title):

RESPONSIBLE PERSON'S DECLARATION STATEMENT

- I certify under penalty of perjury that I am the Field Technician, or the Field Technician is acting on my behalf as my employee or my agent and I have reviewed the information provided on this form.
- I am a licensed contractor or registered design professional who is eligible per the requirements of the Authority Having Jurisdiction to take responsibility for the scope of work specified on this document and attest to the declarations in this statement (responsible person).
- I certify that the information provided on this form substantiates that the construction/installation identified on this form complies with the acceptance requirements indicated in the plans and specifications approved by the enforcement agency, and conforms to the applicable acceptance requirements and procedures specified in Section E 801.0 through Section E 806.0.
- I have confirmed that the Installation Certificate(s) for the construction/installation identified on this form has been completed and is posted or made available with the permit(s) issued for the building.
- I will ensure that a completed, signed copy of this Certificate of Acceptance shall be posted, or made available with the building permit(s) issued for the building, and made available to the enforcement agency for all applicable inspections. I understand that a signed copy of this Certificate of Acceptance is required to be included with the documentation the builder provides to the building owner at occupancy.

Company Name:		Phone:
Responsible Person's Name:		Responsible Person's Signature:
License:	Date Signed:	Position With Company (Title):

CERTIFICATE OF ACCEPTANCE	MECH-10A
Hydronic System Variable Flow Control Acceptance	(Page 2 of 3)
Project Name/Address:	
System Name or Identification/Tag:	System Location or Area Served:

Intent: *Ensure that when loads within the building fluctuate, control valves modulate the amount of water passing through each coil and add or remove the desired amount of energy from the air stream to satisfy the load.*

Construction Inspection

1. Instrumentation to perform test includes, but not limited to:
   a. Calibrated differential pressure gauge.
2. Installation
   - ☐ Pressure sensors are either factory calibrated or field-calibrated.
   - ☐ Pressure sensor location, setpoint, and reset control meets the requirements of Section E 805.8 through Section E 805.8.2.
3. Documentation of all control pressure sensors including (check one of the following):
   a. Factory-calibrated (proof required).
      - ☐ Factory-calibration certificate.

   b. Field-calibrated.
      - ☐ Calibration complete, all pressure sensors within 10% of calibrated reference sensor.

A. Functional Testing.		**Results**
Step 1: Design flow test.		
a.	Open control valves to achieve a minimum of 90% of design flow.	**Y / N**
b.	Verify that the pump speed increases.	**Y / N**
c.	Are the pumps operating at 100% speed?	**Y / N**
d.	Record the system pressure as measured at the control sensor. (Feet Water Column) =	
e.	Record the system pressure setpoint. (Feet Water Column) =	
f.	Is the pressure reading 1d within 5% of pressure setpoint 1e?	**Y / N**
g.	Did the system operation stabilize within 5 minutes after completion of step 1a?	**Y / N**
Step 2: Low flow test		
a.	Close coil control valves to achieve a maximum of 50% of design flow.	**Y / N**
b.	Verify that the current operating speed decreases (for systems with DDC to the zone level).	**Y / N**
c.	Verify that the current operating speed has not increased (for all other systems that are not DDC).	**Y / N**
d.	Record the system pressure as measured at the control sensor. (Feet Water Column) =	
e.	Record the system pressure setpoint. (Feet Water Column) =	
f.	Is the setpoint in 2e is less than the setpoint in 1d?	**Y / N**
g.	Is the pressure reading 2d within 5% of pressure setpoint 2e?	**Y / N**
h.	Did the system operation stabilize within 5 minutes after completion of step 2a?	**Y / N**
Step 3: System returned to initial operating conditions.		**Y / N**

B. Testing Results	**PASS / FAIL**	
Step 1: Select pass if either 1c or 1f are true.	☐	☐
Step 2: Select pass if 2b, 2e, 2f and 2g are true.	☐	☐

For SI units: 1 inch water column = 0.249 kPa

CERTIFICATE OF ACCEPTANCE	MECH-10A
Hydronic System Variable Flow Control Acceptance	**(Page 3 of 3)**
Project Name/Address:	
System Name or Identification/Tag:	System Location or Area Served:

C. PASS / FAIL Evaluation (check one):	
☐	PASS: All **Construction Inspection** responses are complete and all **Testing Results** responses are "Pass."
☐	FAIL: Any **Construction Inspection** responses are incomplete *OR* there is one or more "Fail" responses in **Testing Results** section. Provide explanation below. Use and attach additional pages if necessary.

CERTIFICATE OF ACCEPTANCE	MECH-11A
Automatic Demand Shed Control Acceptance	**(Page 1 of 2)**
Project Name/Address:	
System Name or Identification/Tag:	System Location or Area Served:

Enforcement Agency:	Permit Number:
Note: Submit one Certificate of Acceptance for each system that must demonstrate compliance.	Enforcement Agency Use: Checked by/Date

FIELD TECHNICIAN'S DECLARATION STATEMENT

- I certify under penalty of perjury the information provided on this form is true and correct.
- I am the person who performed the acceptance requirements verification reported on this Certificate of Acceptance (Field Technician).
- I certify that the construction/installation identified on this form complies with the acceptance requirements indicated in the plans and specifications approved by the enforcement agency, and conforms to the applicable acceptance requirements and procedures specified in Section E 801.0 through Section E 806.0.
- I have confirmed that the Installation Certificate(s) for the construction/installation identified on this form has been completed and is posted or made available with the building permit(s) issued for the building.

Company Name:		
Field Technician's Name:		Field Technician's Signature:
	Date Signed:	Position with Company (Title):

RESPONSIBLE PERSON'S DECLARATION STATEMENT

- I certify under penalty of perjury that I am the Field Technician, or the Field Technician is acting on my behalf as my employee or my agent and I have reviewed the information provided on this form.
- I am a licensed contractor or registered design professional who is eligible per the requirements of the Authority Having Jurisdiction to take responsibility for the scope of work specified on this document and attest to the declarations in this statement (responsible person).
- I certify that the information provided on this form substantiates that the construction/installation identified on this form complies with the acceptance requirements indicated in the plans and specifications approved by the enforcement agency, and conforms to the applicable acceptance requirements and procedures specified in Section E 801.0 through Section E 806.0.
- I have confirmed that the Installation Certificate(s) for the construction/installation identified on this form has been completed and is posted or made available with the permit(s) issued for the building.
- I will ensure that a completed, signed copy of this Certificate of Acceptance shall be posted, or made available with the building permit(s) issued for the building, and made available to the enforcement agency for all applicable inspections. I understand that a signed copy of this Certificate of Acceptance is required to be included with the documentation the builder provides to the building owner at occupancy.

Company Name:		Phone:
Responsible Person's Name:		Responsible Person's Signature:
License:	Date Signed:	Position With Company (Title):

CERTIFICATE OF ACCEPTANCE	MECH-11A
Automatic Demand Shed Control Acceptance	**(Page 2 of 2)**
Project Name/Address:	
System Name or Identification/Tag:	System Location or Area Served:

Intent: *Ensure that the central demand shed sequences have been properly programmed into the DDC system.*

Construction Inspection
1. Instrumentation to perform test includes, but not limited to: a. None. 2. Installation. ☐ The EMCS front end interface enables activation of the central demand shed controls.

A. Functional Testing. **Pump Tag (Id)**	
Step 1: Engage the demand shed controls.	
a. Engage the central demand shed control signal.	**Y / N**
b. Verify that the current operating temperature setpoint in a sample of noncritical spaces increases by the proper amount.	**Y / N**
c. Verify that the current operating temperature setpoint in a sample of critical spaces does not change.	**Y / N**
Step 2: Disengage the demand shed controls.	
a. Disengage the central demand shed control signal.	**Y / N**
b. Verify that the current operating temperature setpoint in the sample of noncritical spaces returns to their original value.	**Y / N**
c. Verify that the current operating temperature setpoint in the sample of critical spaces does not change.	**Y / N**

B. Testing Results.	**PASS / FAIL**	
Test passes if all answers are yes in Step 1 and Step 2.	☐	☐

C. PASS / FAIL Evaluation (check one):	
☐	PASS: All **Construction Inspection** responses are complete and all **Testing Results** responses are "Pass."
☐	FAIL: Any **Construction Inspection** responses are incomplete *OR* there is one or more "Fail" responses in **Testing Results** section. Provide explanation below. Use and attach additional pages if necessary.

CERTIFICATE OF ACCEPTANCE	MECH-12A
Fault Detection and Diagnostics (FDD) for Packaged Direct-Expansion Units	**(Page 1 of 3)**

Project Name/Address:	
System Name or Identification/Tag:	System Location or Area Served:

Enforcement Agency:	Permit Number:
Note: Submit one Certificate of Acceptance for each system that must demonstrate compliance.	Enforcement Agency Use: Checked by/Date

FIELD TECHNICIAN'S DECLARATION STATEMENT

- I certify under penalty of perjury the information provided on this form is true and correct.
- I am the person who performed the acceptance requirements verification reported on this Certificate of Acceptance (Field Technician).
- I certify that the construction/installation identified on this form complies with the acceptance requirements indicated in the plans and specifications approved by the enforcement agency, and conforms to the applicable acceptance requirements and procedures specified in Section E 801.0 through Section E 806.0.
- I have confirmed that the Installation Certificate(s) for the construction/installation identified on this form has been completed and is posted or made available with the building permit(s) issued for the building.

Company Name:		
Field Technician's Name:		Field Technician's Signature:
	Date Signed:	Position with Company (Title):

RESPONSIBLE PERSON'S DECLARATION STATEMENT

- I certify under penalty of perjury that I am the Field Technician, or the Field Technician is acting on my behalf as my employee or my agent and I have reviewed the information provided on this form.
- I am a licensed contractor or registered design professional who is eligible per the requirements of the Authority Having Jurisdiction to take responsibility for the scope of work specified on this document and attest to the declarations in this statement (responsible person).
- I certify that the information provided on this form substantiates that the construction/installation identified on this form complies with the acceptance requirements indicated in the plans and specifications approved by the enforcement agency, and conforms to the applicable acceptance requirements and procedures specified in Section E 801.0 through Section E 806.0.
- I have confirmed that the Installation Certificate(s) for the construction/installation identified on this form has been completed and is posted or made available with the permit(s) issued for the building.
- I will ensure that a completed, signed copy of this Certificate of Acceptance shall be posted, or made available with the building permit(s) issued for the building, and made available to the enforcement agency for all applicable inspections. I understand that a signed copy of this Certificate of Acceptance is required to be included with the documentation the builder provides to the building owner at occupancy.

Company Name:		Phone:
Responsible Person's Name:		Responsible Person's Signature:
License:	Date Signed:	Position With Company (Title):

CERTIFICATE OF ACCEPTANCE	MECH-12A
Fault Detection and Diagnostics (FDD) for Packaged Direct-Expansion Units	**(Page 2 of 3)**
Project Name/Address:	
System Name or Identification/Tag:	System Location or Area Served:

Intent: *The purpose of this test is to verify proper fault detection and reporting for automated fault detection and diagnostics systems for packaged units.*

Construction Inspection

1. Instrumentation to perform test includes, but not limited to:
   a. List of instrumentation may be needed or included.
2. Installation.
   ☐ Verify that FDD hardware is installed on equipment by the manufacturer and that equipment make and model include factory-installed FDD hardware that matches the information indicated on copies of the manufacturer's cut sheets and on the plans and specifications.

A. Eligibility Criteria Results.	**Results**
a. A fault detection and diagnostics (FDD) system for direct-expansion packaged units shall contain the following features to be eligible for credit in the performance calculation method:	
b. The unit shall include a factory-installed economizer and shall limit the economizer dead band to no more than 2°F.	**Y / N**
c. The unit shall include direct-drive actuators on outside air and return air dampers.	**Y / N**
d. The unit shall include an integrated economizer with either differential dry-bulb or differential enthalpy control.	**Y / N**
e. The unit shall include a low temperature lockout on the compressor to prevent coil freeze-up or comfort problems.	**Y / N**
f. Outside air and return air dampers shall have maximum leakage rates conforming to Section E 805.12 through Section E 805.12.2.	**Y / N**
g. The unit shall have an adjustable expansion control device such as a thermostatic expansion valve (TXV).	**Y / N**
h. To improve the ability to troubleshoot charge and compressor operation, a high-pressure refrigerant port will be located on the liquid line. A low-pressure refrigerant port will be located on the suction line.	**Y / N**
i. The following sensors should be permanently installed to monitor system operation and the controller should have the capability of displaying the value of each parameter: ☐ Refrigerant suction pressure ☐ Supply air relative humidity ☐ Return air temp ☐ Supply air relative humidity ☐ Refrigerant suction temp ☐ Outside air relative humidity ☐ Supply air temp ☐ Liquid line pressure ☐ Return air relative humidity ☐ Outside air temp	**Y / N**
j. The controller will provide system status by indicating the following conditions: ☐ Compressor enabled ☐ Economizer enabled ☐ Free cooling available ☐ Heating enabled ☐ Mixed air low limit cycle active	**Y / N**
k. The unit controller shall have the capability to manually initiate each operating mode so that the operation of compressors, economizers, fans, and heating system can be independently tested and verified.	**Y / N**

For SI units: °C = (°F-32)/1.8

CERTIFICATE OF ACCEPTANCE	MECH-12A
Fault Detection and Diagnostics (FDD) for Packaged Direct-Expansion Units	**(Page 3 of 3)**
Project Name/Address:	
System Name or Identification/Tag:	System Location or Area Served:

B. Functional Testing.	**Results**
Step 1: Low airflow test.	
a. Test low airflow condition by replacing the existing filter with a dirty filter or appropriate obstruction.	
b. Verify that the fault detection and diagnostics system reports the fault.	**Y / N**
c. Verify that the system is able to verify the correct refrigerant charge.	**Y / N**
d. Verify that you are able to calibrate the following: ☐ Outside Air Temperature Sensor. ☐ Return Air Temperature Sensors. ☐ Supply Air Temperature Sensors.	**Y / N**

C. Testing Results	**PASS / FAIL**	
Test passes if all answers are yes under **Eligibility Criteria** and **Functional Testing**.	☐	☐

☐	PASS: All **Construction Inspection** responses are complete and all **Testing Results** responses are "Pass."
☐	FAIL: Any **Construction Inspection** responses are incomplete *OR* there is one or more "Fail" responses in **Testing Results** section. Provide explanation below. Use and attach additional pages if necessary.

CERTIFICATE OF ACCEPTANCE	MECH-13A
Automatic Fault Detection and Diagnostics (FDD) for Packaged Direct-Expansion Units and Zone Terminal Units Acceptance	**(Page 1 of 4)**
Project Name/Address:	
System Name or Identification/Tag:	System Location or Area Served:

Enforcement Agency:	Permit Number:
Note: Submit one Certificate of Acceptance for each system that must demonstrate compliance.	Enforcement Agency Use: Checked by/Date

FIELD TECHNICIAN'S DECLARATION STATEMENT

- I certify under penalty of perjury the information provided on this form is true and correct.
- I am the person who performed the acceptance requirements verification reported on this Certificate of Acceptance (Field Technician).
- I certify that the construction/installation identified on this form complies with the acceptance requirements indicated in the plans and specifications approved by the enforcement agency, and conforms to the applicable acceptance requirements and procedures specified in Section E 801.0 through Section E 806.0.
- I have confirmed that the Installation Certificate(s) for the construction/installation identified on this form has been completed and is posted or made available with the building permit(s) issued for the building.

Company Name:		
Field Technician's Name:		Field Technician's Signature:
	Date Signed:	Position with Company (Title):

RESPONSIBLE PERSON'S DECLARATION STATEMENT

- I certify under penalty of perjury that I am the Field Technician, or the Field Technician is acting on my behalf as my employee or my agent and I have reviewed the information provided on this form.
- I am a licensed contractor or registered design professional who is eligible per the requirements of the Authority Having Jurisdiction to take responsibility for the scope of work specified on this document and attest to the declarations in this statement (responsible person).
- I certify that the information provided on this form substantiates that the construction/installation identified on this form complies with the acceptance requirements indicated in the plans and specifications approved by the enforcement agency, and conforms to the applicable acceptance requirements and procedures specified in Section E 801.0 through Section E 806.0.
- I have confirmed that the Installation Certificate(s) for the construction/installation identified on this form has been completed and is posted or made available with the permit(s) issued for the building.
- I will ensure that a completed, signed copy of this Certificate of Acceptance shall be posted, or made available with the building permit(s) issued for the building, and made available to the enforcement agency for all applicable inspections. I understand that a signed copy of this Certificate of Acceptance is required to be included with the documentation the builder provides to the building owner at occupancy.

Company Name:		Phone:
Responsible Person's Name:		Responsible Person's Signature:
License:	Date Signed:	Position With Company (Title):

CERTIFICATE OF ACCEPTANCE	MECH-13A
Automatic Fault Detection and Diagnostics (FDD) for Packaged Direct-Expansion Units and Zone Terminal Units Acceptance	**(Page 2 of 4)**
Project Name/Address:	
System Name or Identification/Tag:	System Location or Area Served:

Intent: *Verify that the system detects common faults in air handling units and zone terminal units.*

Construction Inspection
1. Instrumentation to perform test includes, but not limited to: a. No instrumentation is required – changes are implemented at the building automation system control station. 2. Installation. a. The functional testing verifies proper installation of the controls for FDD for air handling units and zone terminal units. No additional installation checks are required.

A. Eligibility Criteria Results. **Testing of each AHU with FDD controls shall include the following tests:**	**Results**
Step 1: Sensor Drift/Failure:	
a. Disconnect outside air temperature sensor from unit controller.	**Y / N**
b. Verify that the FDD system reports a fault.	**Y / N**
c. Connect OAT sensor to the unit controller.	**Y / N**
d. Verify that FDD indicates normal system operation.	**Y / N**
Step 2: Damper/actuator fault.	
a. From the control system workstation, command the mixing box dampers to full open (100% outdoor air).	**Y / N**
b. Disconnect power to the actuator and verify that a fault is reported at the control workstation.	**Y / N**
c. Reconnect power to the actuator and command the mixing box dampers to full open.	**Y / N**
d. Verify that the control system does not report a fault.	**Y / N**
e. From the control system workstation, command the mixing box dampers to a full-closed position (0% outdoor air).	**Y / N**
f. Disconnect power to the actuator and verify that a fault is reported at the control workstation.	**Y / N**
g. Reconnect power to the actuator and command the dampers closed.	**Y / N**
h. Verify that the control system does not report a fault during normal operation.	**Y / N**
Step 3: Valve/actuator fault.	
a. From the control system workstation, command the heating and cooling coil valves to full open or closed, then disconnect power to the actuator and verify that a fault is reported at the control workstation.	**Y / N**
Step 4: Inappropriate simultaneous heating, mechanical cooling, and/or economizing.	
a. From the control system workstation, override the heating coil valve and verify that a fault is reported at the control workstation.	**Y / N**
b. From the control system workstation, override the cooling coil valve and verify that a fault is reported at the control workstation.	**Y / N**
c. From the control system workstation, override the mixing box dampers and verify that a fault is reported at the control workstation.	**Y / N**

CERTIFICATE OF ACCEPTANCE	MECH-13A
Automatic Fault Detection and Diagnostics (FDD) for Packaged Direct-Expansion Units and Zone Terminal Units Acceptance	(Page 3 of 4)
Project Name/Address:	
System Name or Identification/Tag:	System Location or Area Served:

B. Functional Testing for Zone Terminal Units. **Testing shall be performed on one of each type of terminal unit (VAV box) in the project. A minimum of 5% of results the terminal boxes shall be tested.**	**Results**
Step 1: Sensor Drift/Failure:	
a. Disconnect the tubing to the differential pressure sensor of the VAV box.	**Y / N**
b. Verify that control system detects and reports the fault.	**Y / N**
c. Reconnect the sensor and verify proper sensor operation.	**Y / N**
d. Verify that the control system does not report a fault.	**Y / N**
Step 2: Damper/actuator fault.	
If the damper is stuck open:	
a. Command the damper to be fully open (room temperature above setpoint).	**Y / N**
b. Disconnect the actuator to the damper.	**Y / N**
c. Adjust the cooling setpoint so that the room temperature is below the cooling setpoint to command the damper to the minimum position. Verify that the control system reports a fault.	**Y / N**
d. Reconnect the actuator and restore to normal operation.	**Y / N**
If the damper is stuck closed:	
a. Set the damper to the minimum position.	**Y / N**
b. Disconnect the actuator to the damper.	**Y / N**
c. Set the cooling setpoint below the room temperature to simulate a call for cooling. Verify that the control system reports a fault.	**Y / N**
d. Reconnect the actuator and restore to normal operation.	**Y / N**
Step 3: Valve/actuator fault (for systems with hydronic reheat).	
a. Command the reheat coil valve to full open.	**Y / N**
b. Disconnect power to the actuator. Set the heating setpoint temperature to be lower than the current space temperature, to command the valve closed. Verify that the fault is reported at the control work-station.	**Y / N**
c. Reconnect the actuator and restore normal operation.	**Y / N**
Step 4: Feedback loop tuning fault (unstable airflow).	
a. Set the integral coefficient of the box controller to a value 50 times the current value. Lower the space cooling setpoint to simulate a call for cooling.	**Y / N**
b. The damper cycles continuously and airflow is unstable. Verify that the control system detects and reports the fault.	**Y / N**
c. Reset the integral coefficient of the controller to the original value to restore normal operation.	**Y / N**
Step 5: Disconnected inlet duct.	
a. From the control system workstation, command the damper to full closed, then disconnect power to the actuator and verify that a fault is reported at the control workstation.	**Y / N**

CERTIFICATE OF ACCEPTANCE	MECH-13A
Automatic Fault Detection and Diagnostics (FDD) for Packaged Direct-Expansion Units and Zone Terminal Units Acceptance	**(Page 4 of 4)**
Project Name/Address:	
System Name or Identification/Tag:	System Location or Area Served:

C. Testing Results	**PASS / FAIL**	
Test passes if all answers are yes under **Functional Testing Sections**.	☐	☐

D. PASS / FAIL Evaluation (check one):	
☐	PASS: All **Construction Inspection** responses are complete and all **Testing Results** responses are "Pass."
☐	FAIL: Any **Construction Inspection** responses are incomplete *OR* there is one or more "Fail" responses in **Testing Results** section. Provide explanation below. Use and attach additional pages if necessary.

CERTIFICATE OF ACCEPTANCE	MECH-14A
Distributed Energy Storage DX AC Systems Acceptance	**(Page 1 of 3)**
Project Name/Address:	
System Name or Identification/Tag:	System Location or Area Served:

Enforcement Agency:	Permit Number:
Note: Submit one Certificate of Acceptance for each system that must demonstrate compliance.	Enforcement Agency Use: Checked by/Date

FIELD TECHNICIAN'S DECLARATION STATEMENT

- I certify under penalty of perjury the information provided on this form is true and correct.
- I am the person who performed the acceptance requirements verification reported on this Certificate of Acceptance (Field Technician).
- I certify that the construction/installation identified on this form complies with the acceptance requirements indicated in the plans and specifications approved by the enforcement agency, and conforms to the applicable acceptance requirements and procedures specified in Section E 801.0 through Section E 806.0.
- I have confirmed that the Installation Certificate(s) for the construction/installation identified on this form has been completed and is posted or made available with the building permit(s) issued for the building.

Company Name:		
Field Technician's Name:		Field Technician's Signature:
	Date Signed:	Position with Company (Title):

RESPONSIBLE PERSON'S DECLARATION STATEMENT

- I certify under penalty of perjury that I am the Field Technician, or the Field Technician is acting on my behalf as my employee or my agent and I have reviewed the information provided on this form.
- I am a licensed contractor or registered design professional who is eligible per the requirements of the Authority Having Jurisdiction to take responsibility for the scope of work specified on this document and attest to the declarations in this statement (responsible person).
- I certify that the information provided on this form substantiates that the construction/installation identified on this form complies with the acceptance requirements indicated in the plans and specifications approved by the enforcement agency, and conforms to the applicable acceptance requirements and procedures specified in Section E 801.0 through Section E 806.0.
- I have confirmed that the Installation Certificate(s) for the construction/installation identified on this form has been completed and is posted or made available with the permit(s) issued for the building.
- I will ensure that a completed, signed copy of this Certificate of Acceptance shall be posted, or made available with the building permit(s) issued for the building, and made available to the enforcement agency for all applicable inspections. I understand that a signed copy of this Certificate of Acceptance is required to be included with the documentation the builder provides to the building owner at occupancy.

Company Name:		Phone:
Responsible Person's Name:		Responsible Person's Signature:
License:	Date Signed:	Position With Company (Title):

CERTIFICATE OF ACCEPTANCE	MECH-14A
Distributed Energy Storage DX AC Systems Acceptance	(Page 2 of 3)
Project Name/Address:	
System Name or Identification/Tag:	System Location or Area Served:

Intent: *Verify that the system detects common faults in air handling units and zone terminal units.*

Construction Inspection
1. Instrumentation to perform test includes, but not limited to:
a. No special instrumentation is required to perform these tests.
2. Installation.
Prior to Performance Testing, verify and document the following:
☐ The water tank is filled to the proper level.
☐ The water tank is sitting on a foundation with adequate structural strength.
☐ The water tank is insulated and the top cover is in place.
☐ The DES/DXAC is installed correctly (refrigerant piping, etc.).
☐ Verify that the correct model number is installed and configured.

A. Functional Testing	Results
Step 1: Simulate no cooling load during a nighttime period by setting system time to between 9:00 p.m. and 6:00 a.m. Raise the space temperature setpoint above the current space temperature. Verify and document the following:	
a. The system charges the tank.	**Y / N**
b. The system does not provide cooling to the building.	**Y / N**
Step 2: Simulate cooling load during daytime period (e.g., by setting time schedule to include actual time and placing thermostat cooling set-point below actual temperature). Verify and document the following:	
a. Supply fan operates continually during occupied hours.	**Y / N**
b. If the DES/DXAC has cooling capacity, DES/DXAC runs to meet the cooling demand (in ice melt mode).	**Y / N / N/A**
c. If the DES/DXAC has no ice and there is a call for cooling, the DES/DXAC runs in direct cooling mode.	**Y / N / N/A**
Step 3: Simulate no cooling load during daytime condition. Verify and document the following:	
a. Supply fan operates as per the facility thermostat or control system.	**Y / N**
b. The DES/DXAC and the condensing unit do not run.	
Step 4: Simulate no cooling load during morning shoulder time period. Verify and document the following:	
a. The DES/DXAC is idle (the condensing unit and the refrigerant pumps remain off).	**Y / N**

B. Calibrating Controls.	Results
a. Verify that you are able to set the proper time and date, as per manufacturer's installation manual for approved installers.	**Y / N**

C. Testing Results.	PASS / FAIL	
Test passes if all answers are yes under **Functional Testing** and **Calibrating Controls**.	☐	☐

For SI units: 1 metric ton = 1000 kg, 1000 British thermal units per hour = 0.293 kW

CERTIFICATE OF ACCEPTANCE	MECH-14A
Distributed Energy Storage DX AC Systems Acceptance	**(Page 3 of 3)**
Project Name/Address:	
System Name or Identification/Tag:	System Location or Area Served:

☐	PASS: All **Construction Inspection** responses are complete and all **Testing Results** responses are "Pass."
☐	FAIL: Any **Construction Inspection** responses are incomplete *OR* there is one or more "Fail" responses in **Testing Results** section. Provide explanation below. Use and attach additional pages if necessary.

CERTIFICATE OF ACCEPTANCE	MECH-15A
Thermal Energy Storage (TES) System Acceptance	**(Page 1 of 3)**
Project Name/Address:	
System Name or Identification/Tag:	System Location or Area Served:

Enforcement Agency:	Permit Number:
Note: Submit one Certificate of Acceptance for each system that must demonstrate compliance.	Enforcement Agency Use: Checked by/Date

FIELD TECHNICIAN'S DECLARATION STATEMENT

- I certify under penalty of perjury the information provided on this form is true and correct.
- I am the person who performed the acceptance requirements verification reported on this Certificate of Acceptance (Field Technician).
- I certify that the construction/installation identified on this form complies with the acceptance requirements indicated in the plans and specifications approved by the enforcement agency, and conforms to the applicable acceptance requirements and procedures specified in Section E 801.0 through Section E 806.0.
- I have confirmed that the Installation Certificate(s) for the construction/installation identified on this form has been completed and is posted or made available with the building permit(s) issued for the building.

Company Name:		
Field Technician's Name:		Field Technician's Signature:
	Date Signed:	Position with Company (Title):

RESPONSIBLE PERSON'S DECLARATION STATEMENT

- I certify under penalty of perjury that I am the Field Technician, or the Field Technician is acting on my behalf as my employee or my agent and I have reviewed the information provided on this form.
- I am a licensed contractor or registered design professional who is eligible per the requirements of the Authority Having Jurisdiction to take responsibility for the scope of work specified on this document and attest to the declarations in this statement (responsible person).
- I certify that the information provided on this form substantiates that the construction/installation identified on this form complies with the acceptance requirements indicated in the plans and specifications approved by the enforcement agency, and conforms to the applicable acceptance requirements and procedures specified in Section E 801.0 through Section E 806.0.
- I have confirmed that the Installation Certificate(s) for the construction/installation identified on this form has been completed and is posted or made available with the permit(s) issued for the building.
- I will ensure that a completed, signed copy of this Certificate of Acceptance shall be posted, or made available with the building permit(s) issued for the building, and made available to the enforcement agency for all applicable inspections. I understand that a signed copy of this Certificate of Acceptance is required to be included with the documentation the builder provides to the building owner at occupancy.

Company Name:			Phone:
Responsible Person's Name:		Responsible Person's Signature:	
License:	Date Signed:	Position With Company (Title):	

CERTIFICATE OF ACCEPTANCE	MECH-15A
Thermal Energy Storage (TES) System Acceptance	**(Page 2 of 3)**
Project Name/Address:	
System Name or Identification/Tag:	System Location or Area Served:

Intent: *Verify proper operation of distributed energy storage DX systems.*

Construction Inspection
1. Instrumentation to perform test includes, but not limited to: a. No special instrumentation is required for the acceptance tests.

A. Certificate of Compliance Information

The following Certificate of Compliance information for both the chiller and the storage tank shall be provided on the plans to document the key TES System parameters and allow plan check comparison to the inputs used in the DOE-2 simulation. DOE-2 keywords are shown in ALL CAPITALS in parentheses.

a. Chiller	Brand and Model:			
	Type (centrifugal, reciprocating, etc):			
	Capacity (tons): (Size)			
	Starting Efficiency (kW/ton): (at beginning of ice production) (COMP-kW/TON-START)			
	Ending Efficiency (kW/ton): (at end of ice production) (COMP-kW/TON-END)			
	Capacity Reduction (% / F): (PER-COMP-REDUCT/F)			
b. Storage Tank	Storage Type (Check): (TES-TYPE)	☐ Chilled Water Storage	☐ Ice-on-Coil	☐ CHS
		☐ Ice Harvester	☐ Brine	
		☐ Ice-Slurry	☐ Eutectic Salt	
	Number of tanks (SIZE)			
	Storage Capacity per Tank (ton-hours)			
	Storage Rate (tons): (COOL-STORE-RATE)			
	Discharge Rate (tons): (COOL-SUPPLY-RATE)			
	Auxiliary Power (watts): (PUMP+AUX-kW)			
	Tank Area (square feet): (CTANK-LOSS-COEFF)			
	Tank Insulation (R-Value): (CTANK-LOSS-COEFF)			

For SI units: 1 metric ton = 1000 kg, 1000 British thermal units per hour = 0.293 kW

CERTIFICATE OF ACCEPTANCE	**MECH-15A**
Thermal Energy Storage (TES) System Acceptance	**(Page 3 of 3)**
Project Name/Address:	
System Name or Identification/Tag:	System Location or Area Served:

B. Functional Testing	**Results**
Step 1: TES System Design Verification	
a. In the TES System Design Verification part, the installing contractor shall certify the following information, which verifies proper installation of the TES System consistent with system design expectations: ☐ The TES system is one of the above eligible systems ☐ Initial charge rate of the storage tanks (tons) ☐ Final charge rate of the storage tank (tons) ☐ Tank standby storage losses (UA) ☐ Initial discharge rate of the storage tanks (tons) ☐ Final discharge rate of the storage tank (tons) ☐ Charge test time (hours) ☐ Initial chiller efficiency (kW/ton) during charging ☐ Discharge test time (hours) ☐ Tank storage capacity after charge (ton-hours) ☐ Tank storage capacity after discharge (ton-hours) ☐ Final chiller efficiency (kW/ton) during charging	**Y / N**
Step 2: TES System Controls and Operation Verification	
a. The TES system and the chilled water plant is controlled and monitored by an EMS.	☐ Pass ☐ Fail
b. Force the time between 9:00 p.m. and 9:00 a.m. and simulate a partial or no charge of the tank and simulate no cooling load by setting the indoor temperature setpoint higher than the ambient temperature. Verify that the TES system starts charging (storing energy).	☐ Pass ☐ Fail
c. Force the time to be between 6:00 p.m. and 9:00 p.m. and simulate a partial charge on the tank and simulate a cooling load by setting the indoor temperature set point lower than the ambient temperature. Verify that the TES system starts discharging.	☐ Pass ☐ Fail
d. Force the time to be between noon and 6:00 p.m. and simulate a cooling load by lowering the indoor air temperature set point below the ambient temperature. Verify that the tank starts discharging and the compressor is off. For systems designed to meet partial loads the system should be run until the TES storage is fully depleted. The number of hours of operation must meet or exceed the designed operational hours for the system.	☐ Pass ☐ Fail
e. Force the time to be between 9:00 a.m. to noon, and simulate a cooling load by lowering the indoor air temperature set point below the ambient temperature. Verify that the tank does not discharge and the cooling load is met by the compressor only.	☐ Pass ☐ Fail
f. Force the time to be between 9:00 p.m. and 9:00 a.m. and simulate a full tank charge by changing the output of the sensor to the EMS. Verify that the tank charging is stopped.	☐ Pass ☐ Fail
g. Force the time to be between noon and 6:00 p.m. and simulate no cooling load by setting the indoor temperature set point above the ambient temperature. Verify that the tank does not discharge and the compressor is off.	☐ Pass ☐ Fail

C. PASS / FAIL Evaluation (check one):	
☐	PASS: All **Construction Inspection** responses are complete and all **Testing Results** responses are "Pass."
☐	FAIL: Any **Construction Inspection** responses are incomplete *OR* there is one or more "Fail" responses in **Testing Results** section. Provide explanation below. Use and attach additional pages if necessary.

For SI units: 1 metric ton = 1000 kg, 1000 British thermal units per hour = 0.293 kW

APPENDIX F

GEOTHERMAL ENERGY SYSTEMS

Part I – General.

F 101.0 General.

F 101.1 Applicability. Part I of this appendix shall apply to geothermal energy systems such as, but not limited to, building systems coupled with a ground-heat exchanger, submerged heat exchanger using water-based fluid as a heat transfer medium, or groundwater (well). The regulations of this appendix shall govern the construction, location and installation of geothermal energy systems.

Indoor piping, fittings, and accessories that are part of the groundwater system shall be in accordance with Section F 104.4 and Chapter 12.

F 101.1.1 Prior to Construction. Documents for permits shall be submitted prior to the construction of a building system, ground-heat exchanger, submerged heat exchanger, or water well. Permits shall be issued by the Authority Having Jurisdiction.

F 101.1.2 Equipment, Accessories, Components, and Materials. The mechanical equipment, accessories, components, and materials used shall be of the type and rating approved for the specific use.

F 101.2 Construction Documents. The construction documents for the building system portion of the geothermal energy system shall be submitted to the Authority Having Jurisdiction.

F 101.3 Site Survey. A site survey shall be conducted prior to designing the geothermal system. The requirements for construction documents shall be defined by the Authority Having Jurisdiction. Where no guidance is provided, the following information shall be provided:

(1) Ground-heat exchanger dimensions.

(2) Grout or sealing specifications, as applicable.

(3) Dimensions from building to water well, ground-heat exchanger, or submerged heat exchanger.

(4) Operating temperatures and pressures.

F 101.4 Used Materials. The installation of used pipe, fittings, valves, and other materials shall not be permitted.

F 101.5 Contact with Building Material. A ground source heat pump ground-loop piping system shall not be in direct contact with building materials that cause the piping or fitting material to degrade or corrode, or that interferes with the operation of the system.

F 101.6 Strains and Stresses. Piping shall be installed so as to prevent detrimental strains and stresses in the pipe. Provisions shall be made to protect piping from damage resulting from expansion, contraction, and structural settlement. Piping shall be installed so as to avoid structural stresses or strains within building components.

F 101.7 Flood Hazard. Piping located in a flood hazard area shall be capable of resisting hydrostatic and hydrodynamic loads and stresses, including the effects of buoyancy, during the occurrence of flooding to the design flood elevation.

F 101.8 Pipe Support. Pipe shall be supported in accordance with Section 313.1.

F 101.9 Velocities. Ground source heat pump ground-loop systems shall be designed so that the flow velocities do not exceed the maximum flow velocity recommended by the pipe and fittings manufacturer. Flow velocities shall be controlled to reduce the possibility of water hammer.

F 101.10 Chemical Compatibility. Antifreeze and other materials used in the system shall be chemically compatible with the pipe, tubing, fittings, and mechanical systems.

F 101.11 Transfer Fluid. The transfer fluid shall be compatible with the makeup water supplied to the system.

F 102.0 Definitions.

Geothermal Energy System. A system that uses thermal energy for space heating and cooling, and water heating.

Geothermal Energy System, Closed-Loop. A continuous, sealed, underground, or submerged heat exchanger through which a heat-transfer fluid passes to and returns from a heat pump.

Geothermal Energy System, Open-Loop. A liquid-source heat pump system that uses ground water or surface water to extract or reject heat.

Ground-Heat Exchanger. An underground closed-loop heat exchanger through which a heat-transfer medium passes to and from a heat pump or other rated mechanical equipment. It includes the buried pipe and connecting main(s) up to and terminating with the building.

Groundwater Source. A geothermal energy system that uses the groundwater as a heat source or sink.

Water Well. An excavation that is drilled, cored, bored, washed, driven, dug, jetted, or otherwise constructed for the purposes of extracting groundwater, using the geothermal properties of the earth or injecting water into an aquifer or subsurface reservoir.

F 103.0 Groundwater Systems.

F 103.1 General. The potable water supply connected to a groundwater system shall be protected with an approved backflow prevention device. The connection of a discharge line to the sanitary or storm sewer system, or private sewage disposal system, shall be in accordance with the plumbing code or in accordance with the Authority Having Jurisdiction.

F 104.0 Design of Systems.

F 104.1 Ground-Heat Exchanger Design. The ground-heat exchanger design shall be provided by a licensed professional or a designer with the appropriate certifications or credentials as defined by the Authority Having Jurisdiction.

F 104.2 Piping and Tubing Materials Standards. For water-based systems, ground source heat pump ground-loop pipe and tubing shall comply with the standards listed in Table F 104.2. Piping and tubing used for DX systems shall be of copper in accordance with Section F 401.3.

TABLE F 104.2
PLASTIC GROUND SOURCE LOOP PIPING

MATERIAL	STANDARD
Cross-linked polyethylene (PEX)	ASTM F876, CSA B137.5, CSA C448, NSF 358-3
High Density Polyethylene (HDPE)	ASTM D2737, ASTM D3035, ASTM F714, AWWA C901, CSA B137.1, CSA C448, NSF 358-1
Polypropylene (PP)	ASTM F2389, CSA B137.11, NSF 358-2
Polyethylene Raised Temperature (PE-RT)	ASTM F2623, ASTM F2769, CSA B137.18, CSA C448, NSF 358-4

F 104.3 Fittings. For water-based systems, fittings for ground source heat pump systems shall be approved for installation with the piping materials to be installed, and shall comply with the standards listed in Table F 104.3. Fittings for use in DX systems shall comply with Section F 401.3.

TABLE F 104.3
GROUND SOURCE LOOP PIPE FITTINGS

MATERIAL	STANDARD
Cross-linked polyethylene (PEX)	ASTM F877, ASTM F1055, ASTM F1807, ASTM F1960, ASTMF2080, ASTM F2159, ASTM F2434, CSA B137.5, CSA C448, NSF 358-3
High Density Polyethylene (HDPE)	ASTM D2683, ASTM D3261, ASTM F1055, CSA B137.1, CSA C448, NSF 358-1
Polypropylene (PP)	ASTM F2389, CSA B137.11, NSF 358-2
Polyethylene Raised Temperature(PE-RT)	ASTM D3261, ASTM F1055, ASTM F1807, ASTM F2080, ASTM F2159, ASTM F2769, CSA B137.18, CSA C448; NSF 358-4

F 104.4 Underground Piping and Submerged Materials. Underground and submerged piping for a ground-heat exchanger shall be polyethylene (PE) pipe or tubing in accordance with Section F 104.4.1 and Section F 104.4.1.1, or cross-linked polyethylene (PEX) pipe or tubing in accordance with Section F 104.4.2 and Section F 104.4.2.1.

F 104.4.1 Polyethylene (PE). Polyethylene pipe or tubing shall be manufactured in accordance with the standards listed in Table F 104.2. Pipe or tubing shall have a minimum wall thickness equal to SDR-11 and shall have a minimum pressure rating of not less than 160 psi (1103 kPa) at 73°F (23°C).

Polyethylene pipe and tubing shall be manufactured from a PE compound that has a pipe material designation code of PE 3608, PE 3708, PE 3710, PE 4608, PE 4708, or PE 4710 as defined in the applicable standards referenced in Table F 104.2, with a cell classification in accordance with ASTM D3350 appropriate for the material designation code, and a color and ultraviolet stabilizer code of C or E. Code E compounds shall be stabilized against deterioration from unprotected exposure to ultraviolet rays for not less than 3 years in accordance with the test criteria specified in ASTM D2513.

F 104.4.1.1 Joining Methods for Polyethylene Pipe or Tubing. Joints between high density polyethylene (HDPE) plastic pipe or tubing and fittings shall be installed in accordance with the manufacturer's installation instructions, the appropriate standards listed in accordance with Table F 104.3, and one of the following heat fusion methods:

(1) Butt-fusion joints shall be made in accordance with ASTM F2620.

(2) Socket-fusion joints shall be made in accordance with ASTM F2620.

(3) Electrofusion joints shall be made in accordance with ASTM F1055.

F 104.4.2 Cross-Linked Polyethylene (PEX). Cross-linked polyethylene pipe shall be manufactured in accordance with the standards listed in Table F 104.3. PEX shall have a minimum tubing material designation code of PEX 1206 and shall have a minimum pressure rating of not less than 160 psi (1103 kPa) at 73°F (23°C).

F 104.4.2.1 Joining Methods for Cross-Linked Polyethylene Tubing. Joints between cross-linked polyethylene (PEX) pipe or tubing and fittings shall be installed in accordance with the manufacturer's installation instructions and the appropriate standards in accordance with Table F 104.3.

F 104.5 Indoor Piping. Indoor piping, fittings, and accessories that are part of the groundwater system shall be in accordance with Chapter 12. Such materials shall be rated for the operating temperature and pressures of the system and shall be compatible with the type of transfer medium.

F 105.0 Heat Pumps.

F 105.1 Heat Pump Distribution System. The heat pump distribution system shall be designed as follows:

(1) Individual heat pumps shall have the capacity to handle the peak load for each zone at its peak hour.

(2) Distribution piping and fittings shall be insulated to prevent condensation inside the building.

(3) An isolation valve shall be installed on both supply and return of each unit.

(4) Condensate drains on heat pumps shall be installed in accordance with the manufacturer's installation instructions.

(5) Air filters shall be installed for heat pump units.

(6) Drain valves shall be installed at the base of each supply and return pipe riser for system flushing.

(7) Piping shall be supported in accordance with Section

313.0 and provisions for vibration, expansion or contraction shall be provided.

(8) Specifications for each heat pump, the heating and cooling capacity, the fluid flow rate, the airflow rate, and the external pressure or head shall be provided on the construction documents.

(9) Manually controlled air vents shall be installed at the high points in the system and drains at the low points. Where the heat-transfer fluid is a salt or alcohol, automatic air vents shall not be installed.

(10) Means for flow balancing for the building loop shall be provided.

(11) Supply and return header temperatures and pressures shall be marked.

F 105.2 Circulating Pumps. The circulating pump shall be sized for the operating conditions and the heat transfer fluid properties.

F 106.0 Valves.

F 106.1 Where Required. Shutoff valves shall be installed in ground source-loop piping systems in the locations indicated in Section F 106.2 through Section F 106.8.

F 106.2 Heat Exchangers. Shutoff valves shall be installed on the supply and return side of a heat exchanger, except where the heat exchanger is integral with a boiler or is a component of a manufacturer's boiler and heat exchanger packaged unit, and is capable of being isolated from the hydronic system by the supply and return valves.

F 106.3 Central Systems. Shutoff valves shall be installed on the building supply and return of a central utility system.

F 106.4 Pressure Vessels. Shutoff valves shall be installed on the connection to a pressure vessel.

F 106.5 Pressure-Reducing Valves. Shutoff valves shall be installed on both sides of a pressure-reducing valve.

F 106.6 Equipment and Appliances. Shutoff valves shall be installed on connections to mechanical equipment and appliances. This requirement does not apply to components of a ground source loop system such as pumps, air separators, metering devices, and similar equipment.

F 106.7 Expansion Tanks. Shutoff valves shall be installed at connections to nondiaphragm-type expansion tanks.

F 106.8 Reduced Pressure. A pressure relief valve shall be installed on the low-pressure side of a hydronic piping system that has been reduced in pressure. The relief valve shall be set at the maximum pressure of the system design.

F 107.0 Specific System Components Design.

F 107.1 General. Ground coupled and water source heat pumps shall be certified in accordance with AHRI/ASHRAE/ISO 13256-1 for water-to-air heat pumps and AHRI/ASHRAE/ISO 13256-2 for water-to-water heat pumps. DX heat pumps shall be certified in accordance with ASHRAE 194. All heat pump equipment used in DX systems shall comply with AHRI 870. Heat pumps shall be fitted with a means to indicate that the compressor is locked out.

F 107.2 Heat Exchangers. Heat exchangers used for heat transfer or heat recovery shall protect the potable water system from being contaminated by the heat transfer medium. Single-wall heat exchangers shall comply with Section 1218.1. Double-wall heat exchangers shall separate the potable water from the heat transfer medium by providing a space between the two walls that are vented to the atmosphere.

F 107.3 Heat-Transfer Medium. The heat-transfer medium shall be compatible with components with which it comes into contact. Where antifreeze or corrosion inhibitors are used, such solutions shall be approved by the Authority Having Jurisdiction. The heat-transfer fluid flash point shall be not less than 50°F (27.8°C) above the maximum system operating temperature. For DX systems, the heat transfer medium shall be a refrigerant listed in ASHRAE 34 or this code.

F 107.4 Insulation. The temperature of surfaces within reach of building occupants shall not exceed 140°F (60°C) unless they are protected by insulation. Where sleeves are installed, the sleeve insulation shall retain its full size over the length of the material being protected.

F 108.0 Installation Practices.

F 108.1 Prior to Construction. Documents for permits shall be submitted prior to the construction of a building system, or water well. Permits shall be issued by the Authority Having Jurisdiction.

F 108.2 Equipment, Accessories, Components, and Materials. The mechanical equipment, accessories, components, and materials used shall be of the type and rating approved for the specific use.

F 108.3 Construction Documents. The construction documents for the building system portion of the geothermal energy system shall be submitted to the Authority Having Jurisdiction.

F 108.4 Site Survey Requirements. The site survey shall identify the physical limitations of the land area, including its extent, structures, existing wells of all types, proximity of other existing ground source heat pump systems, pavements, trees, grading, ponds, waterways, easements, overhead and underground services, septic systems, any identified septic repair areas, utility of rights-of-way, and any other elements that could affect an open-loop configuration.

Permission shall be obtained from any adjoining property owner(s), as evidenced by the registration and approval of a formal easement that meets requirements of the Authority Having Jurisdiction. It shall be received prior to the installation of any open-loop system that will extend into, cross, or interfere with the equipment or rights-of-way of utilities, jurisdictions, and other property owners.

The site survey shall include a subsurface investigation that meets the requirements for an open-loop heat exchanger.

F 108.5 Subsurface Investigation. A subsurface investigation shall be performed in accordance with Section F 108.6 as determined by the registered design professional conducting the site survey.

F 108.6 Subsurface Conditions. The water well logs and

other geological records shall be used to anticipate the subsurface conditions of the aquifer and its potential supply of fresh water, multiple aquifers, saltwater intrusions, contaminated soils and groundwater, hazardous gases, and any interference with neighboring water wells and ground source heat exchangers.

Geological issues such as permafrost conditions and building stability shall be considered when reviewing available records.

F 108.7 Ground-Heat Exchanger Installation Practices. A ground-heat exchanger system shall be installed as follows:

(1) Outside piping or tubing located within 5 feet (1524 mm) of any wall or structure shall be continuously insulated with insulation that has a minimum R-5 value. Such pipe or tubing installed under the slab or basement floors shall be insulated within 5 feet (1524 mm) from the structure to the exterior point of exit from the slab.

(2) Freeze protection shall be provided where the design of the ground-heat exchanger system would permit the heat-transfer medium to freeze.

(3) Horizontal piping shall be installed not less than 12 inches (305 mm) below the frost line.

(4) Submerged heat exchangers shall be protected from damage and shall be securely fastened to the bottom of the lake or pond, or other approved submerged structure.

(5) A minimum separation distance shall be maintained between the potable water intake and the submerged heat exchanger system in accordance with the Authority Having Jurisdiction.

(6) Vertical and horizontal ground-heat exchangers shall maintain the following setbacks:

   (a) Ten feet (3048 mm) horizontally from a pressure-tested sewer lateral into a building.

   (b) Twenty feet (6096 mm) horizontally from a non-pressure tested sewer lateral into a building.

   (c) Three feet (914 mm) horizontally from buried utilities such as electrical, gas, or water.

   (d) Fifty feet (15 240 mm) from a water well.

   (e) Fifty feet (15 240 mm) from a septic tank and 100 feet (30 480 mm) from a subsurface sewage leaching field.

   (f) One hundred feet (30 480 mm) from a spring; or at distances specified by the Authority Having Jurisdiction.

(7) Wells and boreholes shall be sealed in accordance with the Authority Having Jurisdiction. Where grout is required, it shall be applied in a single continuous operation from the bottom of the borehole by pumping through a tremie pipe.

F 108.8 Trenching, Excavation, and Backfill. Prior to excavation, trenching, or drilling, buried utilities, drainage, water, and irrigation systems shall be located. Prior to excavation, trenching, or drilling, the contractor, and owner shall agree in writing to site restoration requirements and submit to the Authority Having Jurisdiction for approval. Prior to any excavation, trenching, or drilling, all buried utilities including drainage and irrigation systems shall be located and flagged by the appropriate utility and ground source heat pump system contractor representative.

F 108.9 Trenches, Tunneling, and Driving. Trenches shall comply with Section 317.1. Tunneling and driving shall comply with Section 317.2.

F 108.10 Excavations and Open Trenches. Excavations required to be made for the installation of piping or tubing shall be in accordance with Section 317.3. Piping or tubing shall be supported to maintain its alignment and prevent sagging. Piping in the ground shall be laid on a firm bed for its entire length; where other support is otherwise provided, it shall be approved in accordance with Section 302.0. Piping or tubing shall be backfilled after an inspection in accordance with Section 317.4.

F 108.11 Protection of Piping, Materials, and Structures. Piping and tubing passing under or through walls shall be protected from breakage in accordance with Section 316.1. Piping and tubing shall be installed in accordance with Section 316.2 to provide for expansion, contraction, and structural settlement. An electrically continuous corrosion-resistant tracer wire (not less than AWG 14) or tape shall be buried with the plastic pipe to facilitate locating. One end shall be brought aboveground at a building wall or riser.

F 108.12 Sleeves. In exterior walls, annular space between sleeves and pipes shall be sealed and made watertight and shall not be subject to a load from building construction in accordance with Section 316.7 through Section 316.7.2.

F 108.13 Steel Nail Plates. Steel nail plates shall be installed for plastic and copper piping penetrating framing members to within 1 inch (25.4 mm) of the exposed framing in accordance with Section 316.6.

F 108.14 Exterior Piping Protection. Exterior piping shall be fitted with end caps and protected from freezing, UV radiation, corrosion, and degradation.

F 108.15 Heat Pump and Distribution System Installation. The heat pump and distribution system shall be installed in accordance with the system's design, with this code, and the manufacturer's installation instructions.

F 108.16 Pressurizing During Installation. Ground source heat pump ground loop piping to be embedded in concrete shall be pressure tested prior to pouring concrete. During pouring, the pipe shall be maintained at the proposed operating pressure.

F 108.17 Horizontal Geothermal Piping - Materials and Methods. Horizontal geothermal piping shall be in accordance with Section F 108.17.1 through Section F 108.18.8.

F 108.17.1 Piping Material. Piping materials and joining methods for horizontal piping from the ground-heat exchanger shall be in accordance with Section F 104.2, Section F 104.3, Section F 104.4, Section F 401.3 and Section F 104.5.

F 108.17.2 Dissimilar Materials. Transition fittings between dissimilar materials shall be inside or accessible.

F 108.17.3 Protection of Piping. Pipes passing through walls shall be sleeved and sealed in accordance with Section 316.0.

F 108.18 Trenches, Excavation, and Backfill. Excavation for horizontal piping shall comply with Section F 108.18.1 through Section F 108.18.8, Section F 108.8 through Section F 108.11, and in accordance with the requirements of the Authority Having Jurisdiction. Prior to any excavation, trenching, or drilling, all buried utilities including drainage, and irrigation systems shall be located and flagged by the appropriate utility and ground source heat pump system contractor representative.

F 108.18.1 Trenches. Trenches for underground piping or tubing shall be excavated in accordance with the setback requirements in Section F 301.4.

F 108.18.2 Buried Systems. Buried open-loop system piping, shall be installed not less than 3.3 feet (1006 mm) below the finished grade.

F 108.18.3 Pipe Installation. Piping in horizontal trenches shall be embedded with not less than 6 inches (152 mm) of inert granular material above and below, or in accordance with the Authority Having Jurisdiction and project specifications.

Horizontal piping trenching shall be backfilled with approved material and shall be compacted.

F 108.18.4 Separation. The horizontal piping shall be separated from fluid-based onsite service systems to prevent excessive short-circuiting heat transfer between such systems.

F 108.18.5 Insulation. Insulation shall be provided on the piping where there is close proximity of all site services to prevent thermal interference between fluid-based onsite service systems.

F 108.18.6 Pipe Bends. Sharp bending of pipe shall be prevented or approved elbow fitting shall be used with a bend-radius in accordance with the manufacturer's installation instructions.

F 108.18.7 Closed Cell Insulation. Buried horizontal open-loop system pipes passing parallel within 5 feet (1524 mm) of a wall, structure, or water pipe shall be insulated with R-2 minimum closed cell insulation.

F 108.18.8 Tracer Markings. Means shall be provided for underground detection or utility location of the buried pipe system. This shall include, but is not limited to, metallic detectable tape, with a thickness of not less than 11/64 of an inch (4.4 mm) and a width of 6 inches (152 mm).

This warning marking shall be permanent, conspicuous and resistant to the environmental conditions and shall be placed within 1 foot to 2 feet (305 mm to 610 mm) on top of the horizontal piping of the heat exchanger installation.

F 109.0 System Start-Up.

F 109.1 General. The following requirements shall be verified prior to system start-up:

(1) Piping shall be cleaned, flushed, and purged.

(2) The ground-heat exchanger and building piping shall be cleaned, flushed, and, where required, shall be filled with the heat transfer fluid medium. The ground loop system shall be tested at the design flow rate(s) and differential pressure(s) recorded. Where the actual pressure change at design flow is more than +/- 10 percent of the design flow pressure drop, the cause shall be identified and corrective action taken.

(3) A method for the removal of air and a method for adding heat transfer fluid (where necessary) shall be provided.

(4) The heat pumps shall be operational and adjustments shall be made in accordance with the manufacturer's installation instructions.

(5) All necessary additional flow tests of the ground-heat exchanger shall be completed prior to heat pump start-up.

(6) Ground-heat exchanger and building piping, valves, and operating controls, shall be set, adjusted, and operating as required.

(7) The system shall be labeled at the loop charging valves with a permanent-type label, indicating the type of heat transfer fluid used. Where antifreeze is used, the labels shall indicate the antifreeze type and concentration.

(8) Supply and return lines, as well as associated isolation valves from individual boreholes or water wells, shall be identified and tagged.

(9) Supply and return lines on submerged systems shall be identified in an approved manner, at the point of entry to a surface water resource.

F 109.2 Operation and Maintenance Manual. An operation and maintenance manual for the geothermal system shall be provided to the owner. The manual shall include information on required testing and maintenance of the system. Training shall be provided on the system's operation, maintenance requirements, and on the content of the operation and maintenance manual. The operation and maintenance manual shall contain a layout of the ground-heat exchanger and building loop.

F 109.3 Labeling and Marking. Ground source heat pump ground-loop system piping shall be marked with tape, metal tags, or other methods where it enters a building. The marking shall indicate the following words: "GROUND SOURCE HEAT PUMP-LOOP SYSTEM." The marking shall indicate antifreeze used in the system by name and concentration.

F 109.4 Documentation. The ground source heat pump system as-built installation drawings and instructions shall be provided to the building owner or designated agent.

F 109.5 Maintenance. The periodic maintenance required, in accordance with the design requirements, shall be provided and be made available to the owner or designated agent.

F 109.6 Records. The ground source heat pump system construction documents shall be provided to the owner.

F 109.7 System Start-Up. System start-up shall be in accordance with CSA C448.1, CSA C448.2, and Section F 109.0.

F 109.8 Contaminants. Particulate contaminants shall be removed from the indoor piping system prior to initial start-up.

F 110.0 Decommissioning and Abandonment.

F 110.1 General. Decommissioning of geothermal systems shall comply with CSA C448. Prior to the abandonment or decommissioning of geothermal systems, the owner shall obtain the necessary permits from the Authority Having Jurisdiction.

Part II – Closed-Loop Systems.

F 201.0 General.

F 201.1 Applicability. Part II of this chapter shall apply to geothermal energy systems such as, but not limited to, building systems coupled with a closed-loop system using water-based fluid as a heat transfer medium.

F 201.2 Piping and Tubing. Piping and tubing for closed-loop systems shall be in accordance Section F 104.2 and Table F 104.2.

F 201.3 Borehole Piping and Tubing. Borehole piping or tubing for vertical and horizontally drilled closed-loop systems shall have a minimum wall thickness equal to SDR-11 and shall have a minimum pressure rating of not less than 160 psi (1103 kPa) at 73°F (23°C).

F 201.4 Underground Fittings. Underground fittings for closed-loop systems shall be in accordance with Section F 104.3 and Table F 104.3.

F 201.5 Verification. For closed-loop systems, the system shall be flushed of debris and purged of air after completion of the entire ground-heat exchanger. Flow rates and pressure drops shall be compared to calculated values to assure no blockage or kinking of the pipe. A report shall be submitted to the owner to confirm that the loop flow is in accordance with the construction documents.

F 201.6 Vertical Bores. Vertical bores shall be drilled to a depth to provide complete insertion of the u-bend pipe to its specified depth. The borehole diameter shall be sized for the installation and placement of the heat exchange u-bend and the tremie used to place the grouting material. CSA C448 shall be used for vertical loop depth and borehole diameter sizing guidance. The u-bend joint and pipe shall be visually inspected for integrity in accordance with the manufacturer's installation instructions. The u-bend joint and pipe shall be pressurized to not less than 100 psi (689 kPa), not to exceed the pressure rating of the pipe at the test temperature, for 1 hour to check for leaks before insertion into the borehole.

F 201.6.1 Backfill. Bentonite grout and thermally-enhanced bentonite grout, where used to seal and backfill each borehole, shall comply with NSF 60. Boreholes shall be backfilled in accordance with the Authority Having Jurisdiction.

F 201.6.2 U-Bends and Headers. Headers, u-bends and ground loop pipes shall be pressure-tested in accordance with CSA C448, or as required by the Authority Having Jurisdiction. Before testing, heat fusion joints shall be cooled to ambient temperature. Mechanical joints shall be completely assembled. Flushing and purging to remove air and debris shall be completed before testing. The assembly shall be filled with water (or water/antifreeze solution) and purged at a minimum flow rate of 2 feet per second (0.6 m/s) to remove air, but not more than the maximum flow velocity recommended by the pipe and fittings manufacturer to remove debris.

F 201.6.2.1 Test Pressure. The maximum test pressure shall be 1.5 times the system design pressure, as determined by Section F 201.6.2.3, or Section F 201.6.2.4, not to exceed 100 psi (689 kPa). Components or devices with lower pressure-ratings than the pipe shall be protected from excessive pressure during testing by removing or isolating from the test section.

Exception: Where lower pressure-rated components or devices cannot be removed or isolated from the test section, the maximum test pressure shall not exceed the pressure rating of the component or device.

F 201.6.2.2 Testing Procedure. The test section and the test liquid shall be at the same temperature. The test section shall be filled with liquid and purged of air. The test section shall be brought to the specified test pressure. Test pressure shall be maintained for 4 hours, with makeup fluid added as needed. The test pressure shall be reduced by 10 psi (69 kPa) and monitored for 1 hour with no addition of pressure or makeup fluid. A passing test is indicated where after a period of 1 hour no visual leakage is observed, and pressure remains equal to or greater than 95 percent of the original pressure.

F 201.6.2.3 Calculation of Static Pressure (Water). For water, the static pressure applied shall be equivalent to 0.43 psig (2.96 kPa) per foot (305 mm) of elevation.

F 201.6.2.4 Calculation of Static Pressure (Other Fluids). For fluids of different density, the static pressure shall be calculated using the density of the system fluid.

F 202.0 Ground-Heat Exchanger Testing.

F 202.1 Testing. Pressure-testing of the ground-heat exchanger shall be performed in accordance with the testing method in Section F 201.6.

F 202.2 Individual Loop Pressure Testing. Individual loop testing shall be performed as required by the Authority Having Jurisdiction.

F 202.3 Field Pressure Testing – Final. The ground-heat exchanger and building piping shall be cleaned, flushed, and, where required, shall be filled with the heat transfer fluid medium. The ground loop system shall be tested at the design flow rate(s) and differential pressure(s) recorded. Where the actual pressure change at design flow is more than +/- 10 percent of the design flow pressure drop, the cause shall be identified, and corrective action taken.

F 202.4 Field Flow Testing – Final. Final field flow testing shall be performed as required by the Authority Having Jurisdiction.

Part III – Open-Loop Systems.

F 301.0 General.

F 301.1 Applicability. Part III of this appendix shall apply to geothermal energy systems such as, but not limited to, building systems coupled with a groundwater (well) or surface water open-loop using water-based fluid as a heat transfer medium.

The regulations of this appendix shall govern the construction, location and installation of geothermal energy systems.

Indoor piping, fittings, and accessories that are part of the groundwater system shall be in accordance with Section F 104.4 and Chapter 12.

F 301.2 Test Wells. Test wells drilled to investigate subsurface conditions shall provide details of the groundwater location, chemical and physical characteristics, rock strata, and temperature profiles. The number of test wells shall be determined in accordance with the Authority Having Jurisdiction. Each test well shall be tested for flow rate for a period of not less than 24 hours. Water samples shall be collected in accordance with NGWA-01 from each well to establish existing water quality levels are approved for groundwater system use. Water samples shall be analyzed for standard drinking water, fecal and coliform content, bacterial iron, nitrate, dissolved minerals, pH, hardness, and other compounds in accordance with NGWA-01 or in accordance with the Authority Having Jurisdiction. Wells shall be tested for water production and recovery. Monitoring wells shall be protected, and marked to allow for monitoring of ground temperature, groundwater levels, and groundwater quality.

F 301.3 Installation of Water Wells. Water supply, recharge wells, and pumping equipment shall be hydraulically tested, sealed, and grouted in accordance with approved well construction practices and submitted to the Authority Having Jurisdiction for approval. Wells shall be tested for water production and recovery, water quality before final system design. Wells shall be disinfected upon completion in accordance with NGWA-01 or in accordance with the Authority Having Jurisdiction. A copy of the water quality test results and the log of well construction in accordance with NGWA-01 shall be provided to the owner.

F 301.4 Setbacks. Open-loop ground-heat exchangers shall maintain the following minimum setbacks or at distances specified by the Authority Having Jurisdiction:

(1) Ten feet (3048 mm) horizontally from a pressure-tested sewer lateral into a building.

(2) Twenty feet (6096 mm) horizontally from a non-pressure tested sewer lateral into a building.

(3) Three feet (914 mm) horizontally from buried utilities such as electrical, gas, or water.

(4) Fifty feet (15 240 mm) from a water well.

(5) Fifty feet (15 240 mm) from a septic tank and 100 feet (30 480 mm) from a subsurface sewage leaching field.

(6) One hundred feet (30 480 mm) from a spring.

F 302.0 Open Ground Water Systems.

F 302.1 General. The installation and use of water wells shall be in accordance with the Authority Having Jurisdiction. The water well records shall include well logs, pumping tests, and aquifer information.

F 302.2 Open-Loop Water Well Drilling Logs. The water well drilling logs shall include the following:

(1) The subsurface stratigraphy.

(2) The aquifer type and conditions such as, but not limited to, confined, unconfined, flowing and depth.

(3) The drilling method used and the penetration speed.

(4) The presence of substances known to have a potential risk to health and safety shall be documented in the drill logs and the property owner shall be advised of the potential risk to health and safety.

F 302.3 Design Considerations. A groundwater heat pump system shall be designed by a registered design professional. Due design consideration shall be given to the following:

(1) Where multiple heat pumps or fan coils are connected to a common water loop, a diversified building design load shall be used to design a ground water heat pump.

(2) The water supply well(s) and injection wells, or water discharge system, shall be capable of being operated at sustainable pumping rates that exceed the maximum daily requirements without causing an adverse impact to existing or future offsite uses of groundwater or surface water bodies.

(3) The water temperature and the quality and chemical composition of the water resource are in accordance with the system manufacturer's recommendations.

(4) The groundwater and surface water resources shall be protected by returning water to the source aquifer or an aquifer with the same water quality, or a surface water body.

(5) The return capacity of the injection, or surface water body discharge system, shall be suitable under winter conditions.

(6) The temperature of the return water shall have no adverse thermal impacts on offsite existing or future uses of groundwater, or on surface water bodies, in accordance with the requirements of the Authority Having Jurisdiction.

(7) Pressure gauges shall be provided to aid in start-up and monitoring of the system during operation.

(8) The ability to switch over operation of supply and return wells for 100 percent standby, redevelopment, cleaning of wells, and the thermal balancing of the ground and aquifer shall be provided.

(9) There shall be no adverse effects on the quality and quantity of offsite existing or future users of groundwater, in accordance with the requirements of the Authority Having Jurisdiction.

F 302.4 Water Wells and Injection Wells. Water wells and injection wells for groundwater heat pump systems shall be installed and tested by a registered design professional who is qualified to drill wells that comply with the requirements of the Authority Having Jurisdiction.

Water supply wells and injection wells shall be developed in accordance with NGWA-01.

F 302.5 Testing and Sampling. Pumping tests and water sampling shall be done as required by the registered design professional.

F 302.6 Disinfection. Water wells shall be disinfected upon completion in accordance with the requirements of the Authority Having Jurisdiction and NGWA-01.

F 303.0 Testing and Verification.

F 303.1 Pumping Test. Water supply wells and injection wells shall undergo a stop and start pumping test to demonstrate the sand-free yield.

F 303.2 Retesting. Where sediment is present, the problem shall be corrected and the test shall be repeated until acceptable results are obtained.

F 303.3 Variable Rate Pump Test. The operating conditions of the water supply wells and injection wells shall be evaluated and verified with a variable rate pumping.

F 303.4 Constant Rate Pump Test. The sustainable well yield, aquifer coefficients, and zones of influences on the groundwater flow requirements shall be confirmed with a constant rate-pumping test. The constant rate-pumping test shall be done on the water supply and injection wells at rates and durations as specified by the registered design professional.

F 303.5 Water Level Monitoring. Water levels shall be monitored in the pumping well and observation wells during pumping and recovery periods. The monitoring time intervals shall be as specified by the registered design professional.

F 303.6 Injection Wells. Injection testing shall be performed on water wells that are designated to be used as injection wells at rates specified by the registered design professional. The results of the drilling and pumping tests shall be provided to the owner or the owner's representative and provided in accordance with requirements of the Authority Having Jurisdiction.

F 303.7 Re-Injected Water. The water quality of re-injected water into the earth shall comply with the requirements of the Authority Having Jurisdiction.

Part IV – Direct Exchange (DX) Systems.

F 401.0 Direct Exchange (DX) Systems.

F 401.1 General. The installation and use of Direct Exchange (DX) wells shall be in accordance with the Authority Having Jurisdiction. The DX well records shall include well logs, pressure tests, and aquifer information.

F 401.2 Applicability. Part IV of this chapter shall apply to geothermal energy systems such as, but not limited to, building systems coupled with a DX closed-loop using refrigerant as a heat transfer medium. The regulations of this Chapter shall govern the construction, location and installation of geothermal energy systems.

Indoor piping, fittings, and accessories that are part of the ground source system shall be in accordance with Section F 104.5 and Chapter 12.

F 401.3 DX Systems. Copper pipe and tubing installed for DX systems shall be manufactured in accordance with ASTM B280 and copper fittings in accordance with ASME B16.22. Joints shall be purged with an inert gas and brazed with a brazing alloy having 15 percent silver content in accordance with AWS A5.8. Underground piping and tubing shall have a cathodic protection system installed.

F 401.4 DX System Testing. For direct exchange (DX) systems, each u-bend shall be tested and proved tight with an inert gas at not less than 315 psi (2172 kPa) and maintained for 15 minutes without pressure drop. The pressure reading after tremie grouting of the boreholes shall be maintained in the ground-heat exchanger for not less than 2 hours, in accordance with CSA C448.

F 401.5 Indoor Piping. For DX systems, joints shall be purged with an inert gas and brazed with a brazing alloy having 15 percent silver content in accordance with AWS A5.8.

F 401.6 On Site Storage. For DX systems, copper piping and fittings shall be stored to prevent physical damage, contamination, and each pipe or tubing shall be pressurized with an inert gas and sealed with a cap.

F 401.7 System Start-Up. DX system start-up shall be in accordance with Section F 109.1 and the following:

(1) DX systems shall be pressurized using nitrogen for not less than 1 hour. There shall be no allowable variance to the test pressure after being corrected for ambient temperature changes during the test. The test pressure shall not exceed 150 psig (1034 kPa) when pressure testing the compressor unit and indoor system components.

(2) DX systems shall have permanent type labels installed and affixed on the compressor unit with the refrigerant type and quantity.

(3) For DX systems, refrigerant liquid and vapor lines from the loop system shall be identified and tagged.

F 401.8 DX Piping. DX Piping should be installed in accordance with approved plans and specifications, including provisions for cathodic protection.

APPENDIX G
SIZING OF VENTING SYSTEMS AND OUTDOOR COMBUSTION AND VENTILATION OPENING DESIGN

(The content of this Appendix is based on Annex F and Annex I of NFPA 54)

G 101.0 General.

G 101.1 Applicability. This appendix provides general guidelines for sizing venting systems serving appliances equipped with draft hoods, Category I appliances, and appliances listed for use with Type B vents.

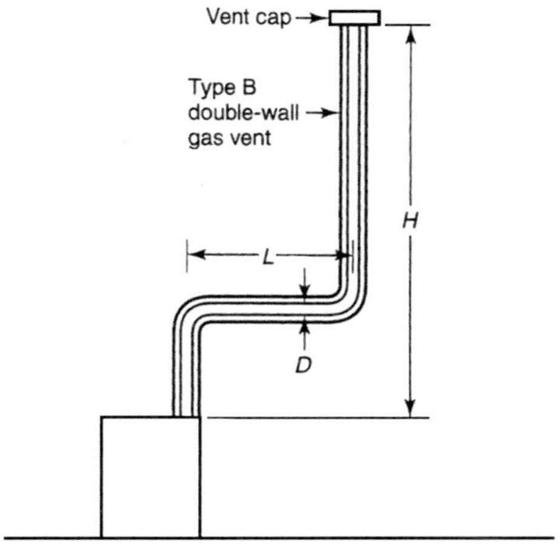

Table 803.1.2(1) is used when sizing a Type B double-wall gas vent connected directly to the appliance.

Note: The appliance can be either Category I draft hood-equipped or fan-assisted type.

FIGURE G 101.2(1)
TYPE B DOUBLE-WALL VENT SYSTEM SERVING A SINGLE APPLIANCE WITH A TYPE B DOUBLE-WALL VENT

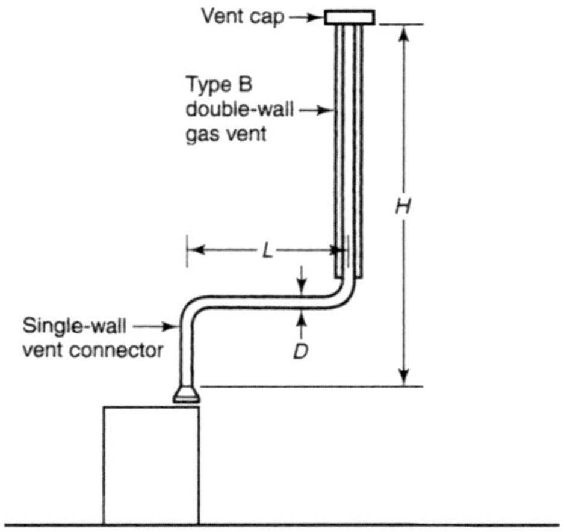

Table 803.1.2(2) is used when sizing a single-wall metal vent connector attached to a Type B double-wall gas vent.

Note: The appliance can be either Category I draft hood-equipped or fan-assisted type.

FIGURE G 101.2(2)
TYPE B DOUBLE-WALL VENT SYSTEM SERVING A SINGLE APPLIANCE WITH A SINGLE-WALL METAL VENT CONNECTOR

G 101.2 Examples Using Single Appliance Venting Tables. See **Figure G 101.2(1) through Figure G 101.2(14)**.

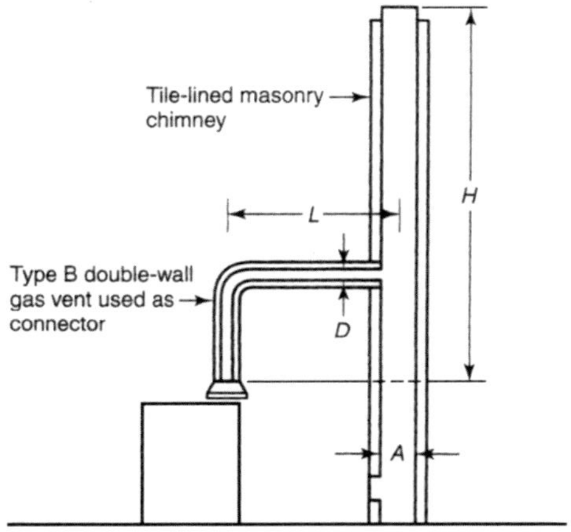

Table 803.1.2(3) is used when sizing a Type B double-wall gas vent connector attached to a tile-lined masonry chimney.

Notes:
1. *A* is the equivalent cross-sectional area of the tile liner.
2. The appliance can be either Category I draft hood-equipped or fan-assisted type.

FIGURE G 101.2(3)
VENT SYSTEM SERVING A SINGLE APPLIANCE WITH A MASONRY CHIMNEY AND A TYPE B DOUBLE-WALL VENT CONNECTOR

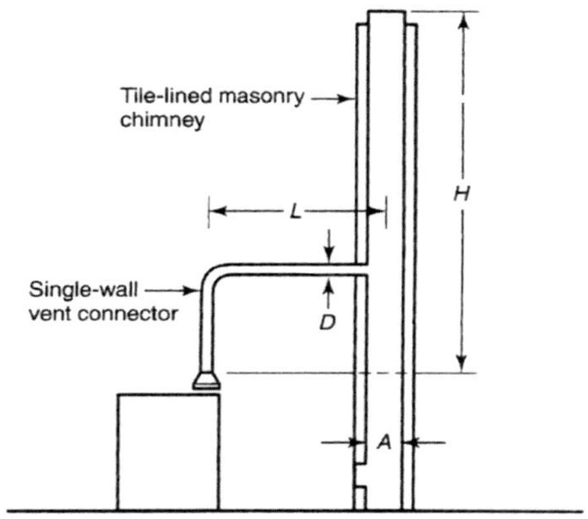

Table 803.1.2(4) is used when sizing a single-wall vent connector attached to a tile-lined masonry chimney.

Notes:
1. *A* is the equivalent cross-sectional area of the tile liner.
2. The appliance can be either Category I draft hood-equipped or fan-assisted type.

FIGURE G 101.2(4)
VENT SYSTEM SERVING A SINGLE APPLIANCE USING A MASONRY CHIMNEY AND A SINGLE-WALL METAL VENT CONNECTOR

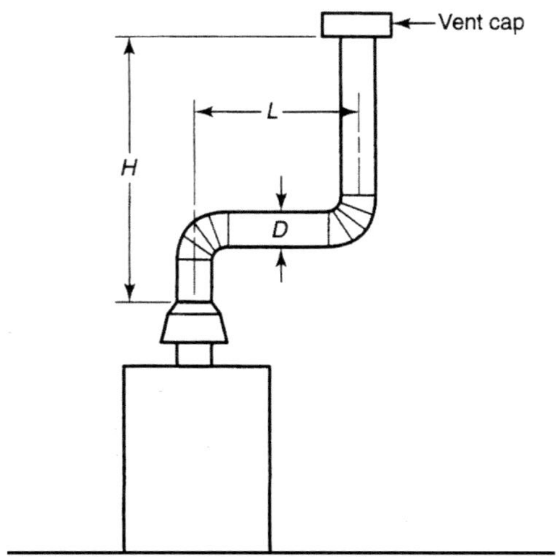

Asbestos cement Type B or single-wall metal vent serving a single draft hood-equipped appliance. [See Table 803.1.2(5)]

FIGURE G 101.2(5)
ASBESTOS CEMENT TYPE B OR SINGLE-WALL METAL VENT SYSTEM SERVING A SINGLE DRAFT HOOD-EQUIPPED APPLIANCE

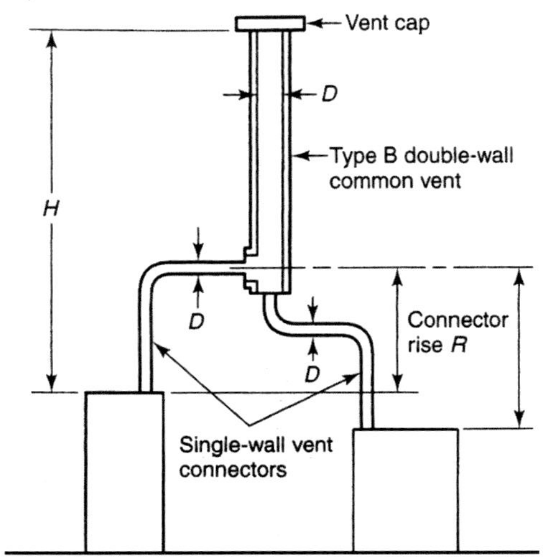

Table 803.2(2) is used when sizing single-wall vent connectors attached to a Type B double-wall common vent.

Note: Each appliance can be either Category I draft hood-equipped or fan-assisted type.

FIGURE G 101.2(7)
VENT SYSTEM SERVING TWO OR MORE APPLIANCES WITH TYPE B DOUBLE-WALL VENT AND SINGLE-WALL METAL VENT CONNECTORS

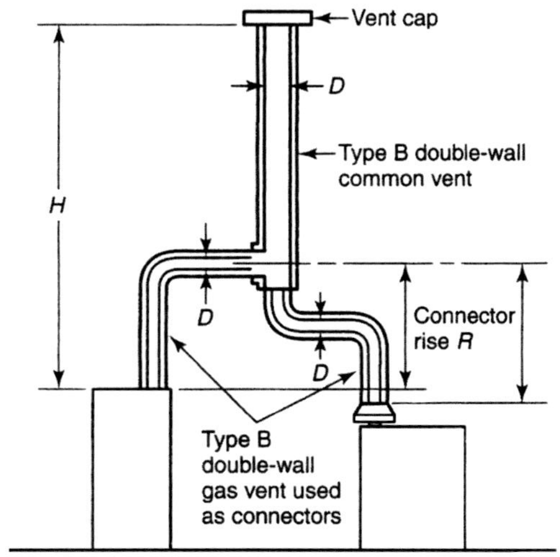

Table 803.2(1) is used when sizing Type B double-wall gas vent connectors attached to a Type B double-wall common vent.

Note: Each appliance can be either Category I draft hood-equipped or fan-assisted type.

FIGURE G 101.2(6)
VENT SYSTEM SERVING TWO OR MORE APPLIANCES WITH TYPE B DOUBLE-WALL VENT AND TYPE B DOUBLE-WALL VENT CONNECTORS

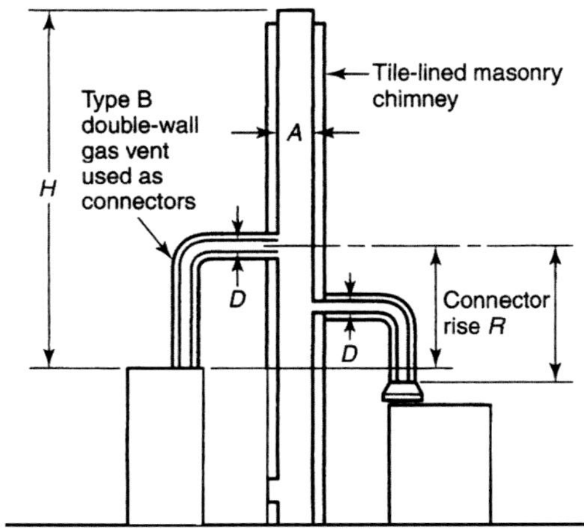

Table 803.2(3) is used when sizing Type B double-wall vent connectors attached to a tile-lined masonry chimney.

Notes:
1. *A* is the equivalent cross-sectional area of the tile liner.
2. Each appliance can be either Category I draft hood-equipped or fan-assisted type.

FIGURE G 101.2(8)
MASONRY CHIMNEY SERVING TWO OR MORE APPLIANCES WITH TYPE B DOUBLE-WALL VENT CONNECTORS

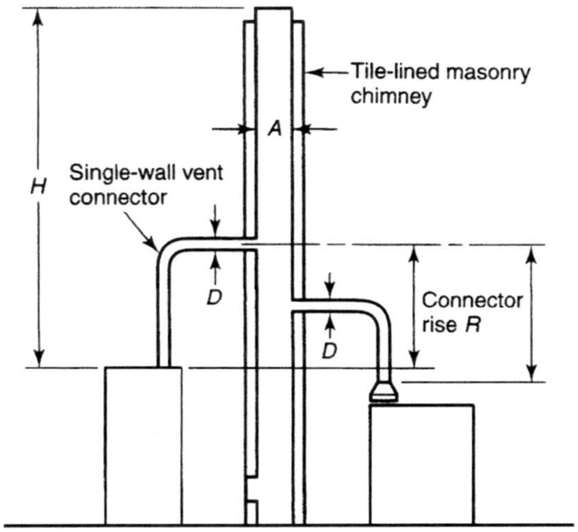

Table 803.2(4) is used when sizing single-wall metal vent connectors attached to a tile-lined masonry chimney.

Notes:

1. *A* is the equivalent cross-sectional area of the tile liner.
2. Each appliance can be either Category I draft hood-equipped or fan-assisted type.

FIGURE G 101.2(9)
MASONRY CHIMNEY SERVING TWO OR MORE APPLIANCES WITH SINGLE-WALL METAL VENT CONNECTORS

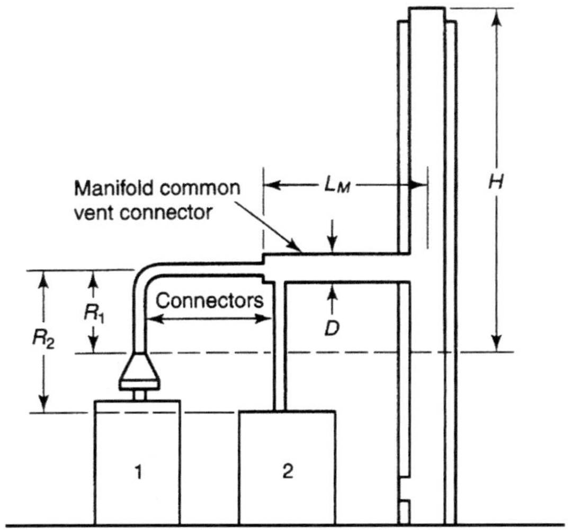

Example: Manifolded common vent connector L_M can be no greater than 18 times the common vent connector manifold inside diameter; that is, a 4 inch (102 mm) inside diameter common vent connector manifold should not exceed 72 inches (1829 mm) in length. [See Section 803.2.3]

Note: This is an illustration of a typical manifolded vent connector. Different appliance, vent connector, or common vent types are possible. [See Section 803.2]

FIGURE G 101.2(11)
USE OF MANIFOLDED COMMON VENT CONNECTOR

See Section 803.2.4

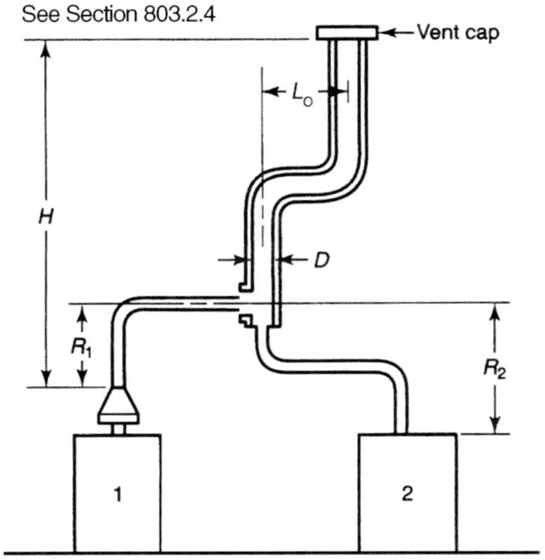

Asbestos cement Type B or single-wall metal pipe vent serving two or more draft hood-equipped appliances. [See Table 803.2(5)]

FIGURE G 101.2(10)
ASBESTOS CEMENT TYPE B OR SINGLE-WALL METAL VENT SYSTEM SERVING TWO OR MORE DRAFT HOOD-EQUIPPED APPLIANCES

See Section 803.2.4

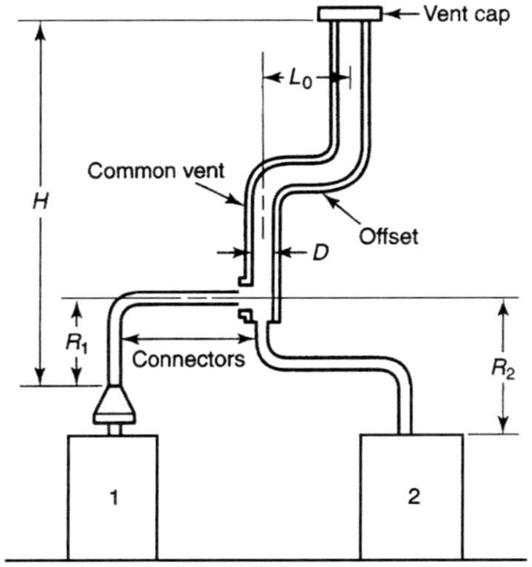

Example: Offset common vent

Note: This is an illustration of a typical offset vent. Different appliance, vent connector, or vent types are possible. [See Section 803.1 and Section 803.2]

FIGURE G 101.2(12)
USE OF OFFSET COMMON VENT

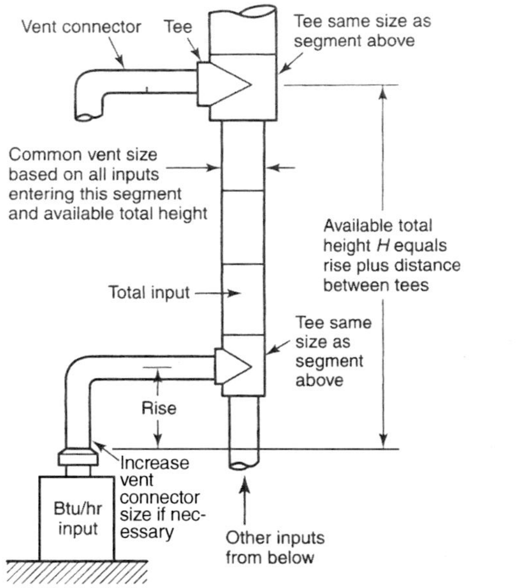

FIGURE G 101.2(13)
MULTISTORY GAS VENT DESIGN PROCEDURE FOR EACH SEGMENT OF SYSTEM

G 101.3 Example 1: Single Draft Hood-Equipped Appliance. An installer has a 120 000 British thermal units per hour (Btu/h) (35 kW) input appliance with a 5 inch (127 mm) diameter draft hood outlet that needs to be vented into a 10 foot (3048 mm) high Type B vent system. What size vent should be used assuming: (1) a 5 foot (1524 mm) lateral single-wall metal vent connector is used with two 90 degree (1.57 rad) elbows or (2) a 5 foot (1524 mm) lateral single-wall metal vent connector is used with three 90 degree (1.57 rad) elbows in the vent system? (See **Figure G 101.3**)

Solution:

Table 803.1.2(2) should be used to solve this problem because single-wall metal vent connectors are being used with a Type B vent, as follows:

(1) Read down the first column in Table 803.1.2(2) until the row associated with a 10 foot (3048 mm) height and 5 foot (1524 mm) lateral is found. Read across this row until a vent capacity greater than 120 000 Btu/h (35 kW) is located in the shaded columns labeled NAT Max for draft hood-equipped appliances. In this case, a 5 inch (127 mm) diameter vent has a capacity of 122 000 Btu/h (35.7 kW) and can be used for this application.

(2) If three 90 degree (1.57 rad) elbows are used in the vent system, the maximum vent capacity listed in the tables must be reduced by 10 percent. This implies that the 5 inch (127 mm) diameter vent has an adjusted capacity of only 110 000 Btu/h (32 kW). In this case, the vent system must be increased to 6 inches (152 mm) in diameter. See the following calculations:

122 000 Btu/h (35.7 kW) x 0.90 = 110 000 Btu/h (32 kW) for 5 inch (127 mm) vent

From Table 803.1.2(2), select 6 inches (152 mm) vent.

186 000 Btu/h (54.5 kW) x 0.90 = 167 000 Btu/h (49 kW)

This figure is greater than the required 120 000 Btu/h (35 kW). Therefore, use a 6 inch (152 mm) vent and connector where three elbows are used.

G 101.4 Example 2: Single Fan-Assisted Appliance. An installer has an 80 000 Btu/h (23.4 kW) input fan-assisted appliance that must be installed using 10 feet (3048 mm) of

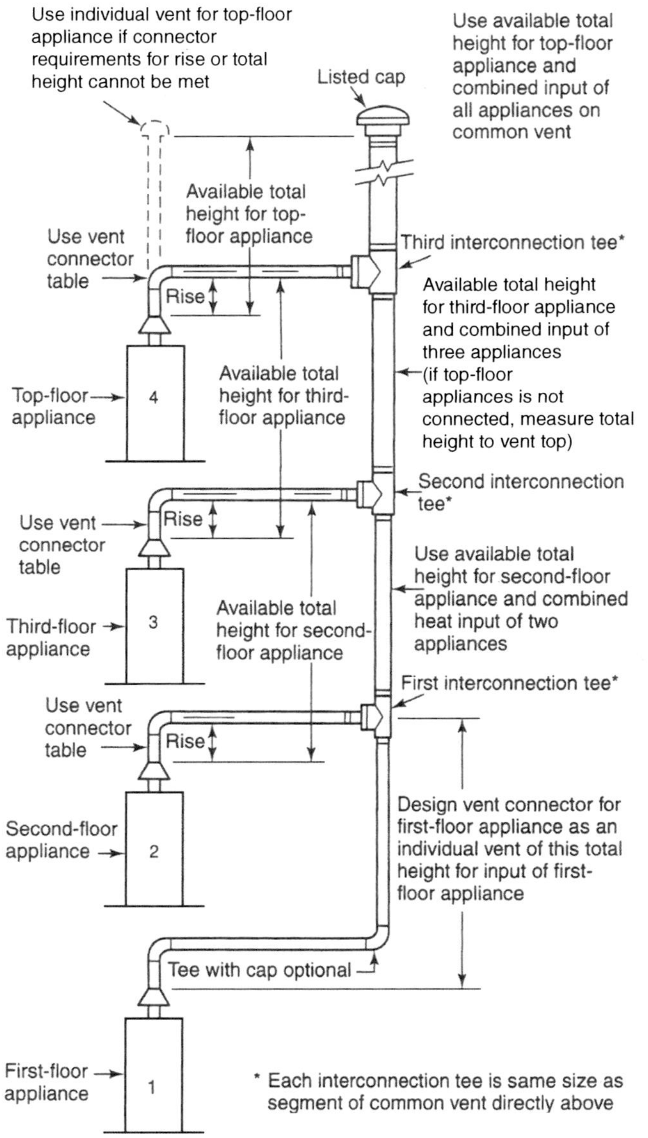

FIGURE G 101.2(14)
PRINCIPLES OF DESIGN OF MULTISTORY VENTS USING VENT CONNECTOR AND COMMON VENT DESIGN TABLES
[See Section 803.2.13 through Section 803.2.16]

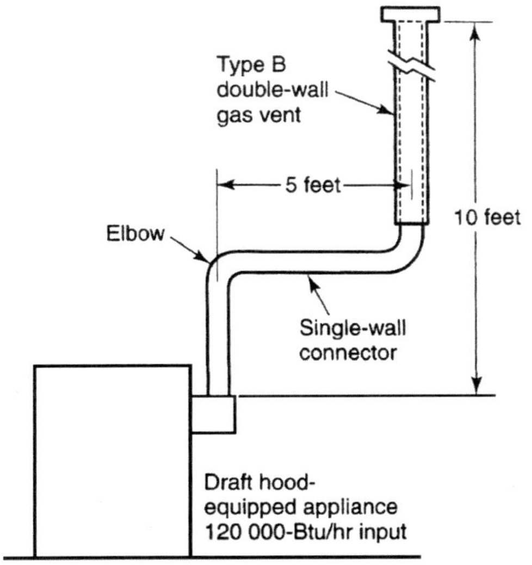

For SI units: 1 foot = 304.8 mm, 1000 British thermal units per hour = 0.293 kW

FIGURE G 101.3
SINGLE DRAFT HOOD-EQUIPPED APPLIANCE – EXAMPLE 1

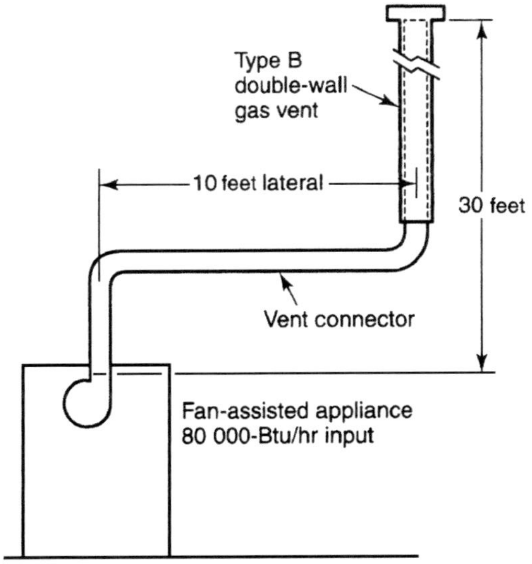

For SI units: 1 foot = 304.8 mm, 1000 British thermal units per hour = 0.293 kW

FIGURE G 101.4
SINGLE FAN-ASSISTED APPLIANCE – EXAMPLE 2

lateral connector attached to a 30 foot (9144 mm) high Type B vent. Two 90-degree (1.57 rad) elbows are needed for the installation. Can a single-wall metal vent connector be used for this application? (See **Figure G 101.4**)

Solution:

Table 803.1.2(2) refers to the use of single-wall metal vent connectors with Type B vent. In the first column find the row associated with a 30 foot (9144 mm) height and a 10 foot (3048 mm) lateral. Read across this row, looking at the FAN Min and FAN Max columns, to find that a 3 inch (76 mm) diameter single-wall metal vent connector is not recommended. Moving to the next larger size single-wall connector [4 inch (102 mm)] we find that a 4 inch (102 mm) diameter single-wall metal connector has a recommended minimum vent capacity of 91 000 Btu/h (26.7 kW) and a recommended maximum vent capacity of 144 000 Btu/h (42 kW). The 80 000 Btu/h (23.4 kW) fan-assisted appliance is outside this range, so the conclusion is that a single-wall metal vent connector cannot be used to vent this appliance using 10 feet (3048 mm) of lateral for the connector. However, if the 80,000 Btu/h (23.4 kW) input appliance could be moved to within 5 feet (1524 mm) of the vertical vent, a 4 inch (102 mm) single-wall metal connector could be used to vent the appliance. Table 803.1.2(2) shows the acceptable range of vent capacities for a 4 inch (102 mm) vent with 5 feet (1524 mm) of lateral to be between 72 000 Btu/h (21.1 kW) and 157 000 Btu/h (46 kW).

If the appliance cannot be moved closer to the vertical vent, a Type B vent could be used as the connector material. In this case, Table 803.1.2(1) shows that, for a 30 foot (9144 mm) high vent with 10 feet (3048 mm) of lateral, the acceptable range of vent capacities for a 4 inch (102 mm) diameter vent attached to a fan-assisted appliance is between 37 000 Btu/h (10.8 kW) and 150 000 Btu/h (44 kW).

G 101.5 Example 3: Interpolating Between Table Values. An installer has an 80 000 Btu/h (23.4 kW) input appliance with a 4 inch (102 mm) diameter draft hood outlet that needs to be vented into a 12 foot (3658 mm) high Type B vent. The vent connector has a 5 foot (1524 mm) lateral length and is also Type B. Can this appliance be vented using a 4 inch (102 mm) diameter vent?

Solution:

Table 803.1.2(1) is used in the case of an all Type B Vent system. However, Table 803.1.2(1) does not have an entry for a height of 12 feet (3658 mm), and interpolation must be used. Read down the 4 inch (102 mm) diameter NAT Max column to the row associated with 10 foot (3048 mm) height and 5 foot (1524 mm) lateral to find the capacity value of 77 000 Btu/h (22.6 kW). Read further down to the 15 foot (4572 mm) height, 5 foot (1524 mm) lateral row to find the capacity value of 87 000 Btu/h (25.5 kW). The difference between the 15 foot (4572 mm) height capacity value and the 10 foot (3048 mm) height capacity value is 10 000 Btu/h (3 kW). The capacity for a vent system with a 12 foot (3658 mm) height is equal to the capacity for a 10 foot (3048 mm) height plus two-fifths of the difference between the 10 foot (3048 mm) and 15 foot (4572 mm) height values, or 77 000 Btu/h (22.6 kW) + $\frac{2}{5}$ x 10 000 Btu/h (3 kW) = 81 000 Btu/h (23.7 kW). Therefore, a 4 inch (102 mm) diameter vent can be used in the installation.

G 102.0 Examples Using Common Venting Tables.

G 102.1 Example 4: Common Venting Two Draft Hood-Equipped Appliances. A 35 000 Btu/h (10.3 kW) water heater is to be common vented with a 150 000 Btu/h (44 kW) furnace, using a common vent with a total height of 30 feet (9144 mm). The connector rise is 2 feet (610 mm) for the water heater with a horizontal length of 4 feet (1219 mm).

The connector rise for the furnace is 3 feet (914 mm) with a horizontal length of 8 feet (2438 mm). Assume single-wall metal connectors will be used with Type B vent. What size connectors and combined vent should be used in this installation? (See **Figure G 102.1**)

Solution:

Table 803.2(2) should be used to size single-wall metal vent connectors attached to Type B vertical vents. In the vent connector capacity portion of Table 803.2(2), find the row associated with a 30 foot (9144 mm) vent height. For a 2 foot (610 mm) rise on the vent connector for the water heater, read the shaded columns for draft hood-equipped appliances to find that a 3 inch (76 mm) diameter vent connector has a capacity of 37 000 Btu/h (10.8 kW). Therefore, a 3 inch (76 mm) single-wall metal vent connector can be used with the water heater. For a draft hood-equipped furnace with a 3 foot (914 mm) rise, read across the appropriate row to find that a 5 inch (127 mm) diameter vent connector has a maximum capacity of 120 000 Btu/h (35 kW) (which is too small for the furnace), and a 6 inch (152 mm) diameter vent connector has a maximum vent capacity of 172 000 Btu/h (50 kW). Therefore, a 6 inch (152 mm) diameter vent connector should be used with the 150 000 Btu/h (44 kW) furnace. Because both vent connector horizontal lengths are less than the maximum lengths listed in Section 803.2.1, the table values can be used without adjustments.

In the common vent capacity portion of Table 803.2(2), find the row associated with a 30 foot (9144 mm) vent height and read over to the NAT + NAT portion of the 6 inch (152 mm) diameter column to find a maximum combined capacity of 257 000 Btu/h (75 kW). Since the two appliances total only 185 000 Btu/h (54 kW), a 6 inch (152 mm) common vent can be used.

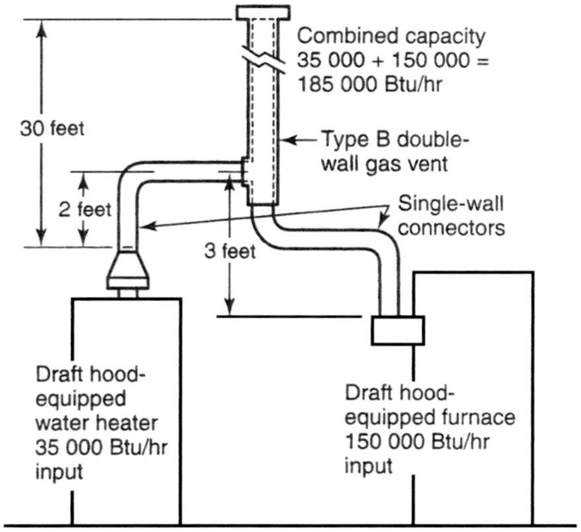

For SI units: 1 foot = 304.8 mm, 1000 British thermal units per hour = 0.293 kW

FIGURE G 102.1
COMMON VENTING TWO DRAFT HOOD-EQUIPPED APPLIANCES – EXAMPLE 4

G 102.2 Example 5(a): Common Venting a Draft Hood-Equipped Water Heater with a Fan-Assisted Furnace into a Type B Vent. In this case, a 35 000 Btu/h (10.3 kW) input draft hood-equipped water heater with a 4 inch (102 mm) diameter draft hood outlet, 2 feet (610 mm) of connector rise, and 4 feet (1219 mm) of horizontal length is to be common vented with a 100 000 Btu/h (29 kW) fan-assisted furnace with a 4 inch (102 mm) diameter flue collar, 3 feet (914 mm) of connector rise, and 6 feet (1829 mm) of horizontal length. The common vent consists of a 30 foot (9144 mm) height of Type B vent. What are the recommended vent diameters for each connector and the common vent? The installer would like to use a single-wall metal vent connector. (See **Figure G 102.2**)

Solution:

Water Heater Vent Connector Diameter. Since the water heater vent connector horizontal length of 4 feet (1219 mm) is less than the maximum value listed in Table 803.2(2), the venting table values can be used without adjustments. Using the Vent Connector Capacity portion of Table 803.2(2), read down the Total Vent Height (*H*) column to 30 feet (9144 mm) and read across the 2 feet (610 mm) Connector Rise (*R*) row to the first Btu/h rating in the NAT Max column that is equal to or greater than the water heater input rating. The table shows that a 3 inch (76 mm) vent connector has a maximum input rating of 37 000 Btu/h (10.8 kW). Although this rating is greater than the water heater input rating, a 3 inch (76 mm) vent connector is prohibited by Section 803.2.18. A 4 inch (102 mm) vent connector has a maximum input rating of 67 000 Btu/h (19.6 kW) and is equal to the draft hood outlet diameter. A 4 inch (102 mm) vent connector is selected. Since the water heater is equipped with a draft hood, there are no minimum input rating restrictions.

Furnace Vent Connector Diameter. Using the Vent Connector Capacity portion of Table 803.2(2), read down the Total Vent Height (*H*) column to 30 feet (9144 mm) and across the 3 feet (914 mm) Connector Rise (*R*) row. Because the furnace has a fan-assisted combustion system, find the first FAN Max column with a Btu/h rating greater than the furnace input rating. The 4 inch (102 mm) vent connector has a maximum input rating of 119 000 Btu/h (34.9 kW) and a minimum input rating of 85 000 Btu/h (24.9 kW).

The 100 000 Btu/h (29 kW) furnace in this example falls within this range, so a 4 inch (102 mm) connector is adequate. Because the furnace vent connector horizontal length of 6 feet (1829 mm) is less than the maximum value listed in Section 803.2.1, the venting table values can be used without adjustment. If the furnace had an input rating of 80 000 Btu/h (23.4 kW), a Type B vent connector would be needed in order to meet the minimum capacity limit.

Common Vent Diameter. The total input to the common vent is 135 000 Btu/h (40 kW). Using the Common Vent Capacity portion of Table 803.2(2), read down the Total Vent Height (*H*) column to 30 feet (9144 mm) and across this row to find the smallest vent diameter in the FAN + NAT column that has a Btu/h rating equal to or greater than 135 000 Btu/h

(40 kW). The 4 inch (102 mm) common vent has a capacity of 132 000 Btu/h (39 kW) and the 5 inch (127 mm) common vent has a capacity of 202 000 Btu/h (59 kW). Therefore, the 5 inch (127 mm) common vent should be used in this example.

Summary: In this example, the installer can use a 4 inch (102 mm) diameter, single-wall metal vent connector for the water heater and a 4 inch (102 mm) diameter, single-wall metal vent connector for the furnace. The common vent should be a 5 inch (127 mm) diameter Type B vent.

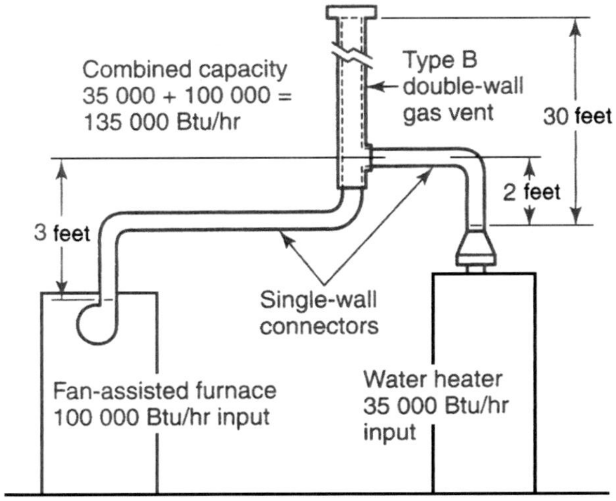

For SI units: 1 foot = 304.8 mm, 1000 British thermal units per hour = 0.293 kW

FIGURE G 102.2
COMMON VENTING A DRAFT HOOD-EQUIPPED WATER HEATER WITH A FAN-ASSISTED FURNACE INTO A TYPE B DOUBLE-WALL COMMON VENT – EXAMPLE 5(a)

G 102.3 Example 5(b): Common Venting into an Interior Masonry Chimney. In this case, the water heater and fan-assisted furnace of Example 5(a) are to be common-vented into a claytile-lined masonry chimney with a 30 foot (9144 mm) height. The chimney is not exposed to the outdoors below the roof line. The internal dimensions of the clay tile liner are nominally 8 inches (203 mm) by 12 inches (305 mm). Assuming the same vent connector heights, laterals, and materials found in Example 5(a), what are the recommended vent connector diameters, and is this an acceptable installation?

Solution:

Table 803.2(4) is used to size common venting installations involving single-wall connectors into masonry chimneys.

Water Heater Vent Connector Diameter. Using Table 803.2(4), Vent Connector Capacity, read down the Total Vent Height *(H)* column to 30 feet (9144 mm), and read across the 2 feet (610 mm) Connector Rise *(R)* row to the first Btu/h rating in the NAT Max column that is equal to or greater than the water heater input rating. The table shows that a 3 inch (76 mm) vent connector has a maximum input of only 31 000 Btu/h (9 kW), while a 4 inch (102 mm) vent connector has a maximum input of 57 000 Btu/h (16.7 kW). A 4 inch (102 mm) vent connector must therefore be used.

Furnace Vent Connector Diameter. Using the Vent Connector Capacity portion of Table 803.2(4), read down the Total Vent Height (*H*) column to 30 feet (9144 mm) and across the 3 feet (914 mm) Connector Rise (*R*) row. Because the furnace has a fan-assisted combustion system, find the first FAN Max column with a Btu/h rating greater than the furnace input rating. The 4 inch (102 mm) vent connector has a maximum input rating of 127 000 Btu/h (37 kW) and a minimum input rating of 95 000 Btu/h (27.8 kW). The 100 000 Btu/h (29 kW) furnace in this example falls within this range, so a 4 inch (102 mm) connector is adequate.

Masonry Chimney. From Table G 102.3, the Equivalent Area for a Nominal Liner size of 8 inches (203 mm) by 12 inches (305 mm) is 63.6 of a square inches (0.041 m^2). Using Table 803.2(4), Common Vent Capacity, read down the FAN + NAT column under the Minimum Internal Area of Chimney value of 63 to the row for 30 foot (9144 mm) height to find a capacity value of 739 000 Btu/h (217 kW). The combined input rating of the furnace and water heater, 135 000 Btu/h (40 kW), is less than the table value so this is an acceptable installation.

Section 803.2.17 requires the common vent area to be no greater than seven times the smallest listed appliance categorized vent area, flue collar area, or draft hood outlet area. Both appliances in this installation have 4 inch (102 mm) diameter outlets. From Table G 102.3, the equivalent area for an inside diameter of 4 inches (102 mm) is 12.2 of a square inches (0.008 m^2). Seven times 12.2 equals 85.4, which is greater than 63.6, so this configuration is acceptable.

G 102.4 Example 5(c): Common Venting into an Exterior Masonry Chimney. In this case, the water heater and fan-assisted furnace of Examples 5(a) and 5(b) are to be common-vented into an exterior masonry chimney. The chimney height, clay-tile-liner dimensions, and vent connector heights and laterals are the same as in Example 5(b). This system is being installed in Charlotte, North Carolina. Does this exterior masonry chimney need to be relined? If so, what corrugated metallic liner size is recommended? What vent connector diameters are recommended? [See Table G 102.3 and **Figure 803.1.2(6)**]

Solution:

According to Section 803.2.20, Type B vent connectors are required to be used with exterior masonry chimneys. Use Table 803.2(8) and Table 803.2(9) to size FAN+NAT common venting installations involving Type-B double-wall connectors into exterior masonry chimneys.

The local 99 percent winter design temperature needed to use Table 803.2(8) and Table 803.2(9) can be found in ASHRAE Handbook – Fundamentals. For Charlotte, North Carolina, this design temperature is 19°F (-7.2°C).

Chimney Liner Requirement. As in Example 5(b), use the 63 square inch (0.04 m^2) internal area columns for this size clay tile liner. Read down the 63 square inches (0.04 m^2) column of Table 803.2(8) to the 30 foot (9144 mm) height row to find that the combined appliance maximum input is 747 000 Btu/h (218.9 kW). The combined input rating of the appliances in this installation, 135 000 Btu/h (40 kW), is less than the maximum value, so this criterion is satisfied. Table 803.2(9), at a 19°F (-7.2°C) design temperature, and at the same vent height and internal area used earlier, shows that the minimum allowable input rating of a space-heating appliance is 470 000 Btu/h (137.7 kW). The furnace input rating of 100 000 Btu/h (29 kW) is less than this minimum value. So this criterion is not satisfied, and an alternative venting design needs to be used, such as a Type B vent shown in Example 5(a) or a listed chimney liner system shown in the remainder of the example.

According to Section 803.2.19, Table 803.2(1) or Table 803.2(2) is used for sizing corrugated metallic liners in masonry chimneys, with the maximum common vent capacities reduced by 20 percent. This example will be continued assuming Type B vent connectors.

Water Heater Vent Connector Diameter. Using Table 803.2(1) Vent Connector Capacity, read down the total Vent Height *(H)* column to 30 feet (9144 mm), and read across the 2 feet (610 mm) Connector Rise *(R)* row to the first Btu/hour rating in the NAT Max column that is equal to or greater than the water heater input rating. The table shows that a 3 inch (76 mm) vent connector has a maximum capacity of 39 000 Btu/h (11.4 kW). Although this rating is greater than the water heater input rating, a 3 inch (76 mm) vent connector is prohibited by Section 803.2.20. A 4 inch (102 mm) vent connector has a maximum input rating of 70 000 Btu/h (20.5 kW) and is equal to the draft hood outlet diameter. A 4 inch (102 mm) vent connector is selected.

Furnace Vent Connector Diameter. Using Table 803.2(1), Vent Connector Capacity, read down the total Vent Height *(H)* column to 30 feet (9144 mm), and read across the

TABLE G 102.3
MASONRY CHIMNEY LINER DIMENSIONS WITH CIRCULAR EQUIVALENTS*

NOMINAL LINER SIZE (inches)	INSIDE DIMENSIONS OF LINER (inches)	INSIDE DIAMETER OR EQUIVALENT DIAMETER (inches)	EQUIVALENT AREA (square inches)
4 x 8	2½ x 6½	4.0	12.2
		5.0	19.6
		6.0	28.3
		7.0	38.3
8 x 8	6¾ x 6¾	7.4	42.7
		8.0	50.3
8 x 12	6½ x 10½	9.0	63.6
		10.0	78.5
12 x 12	9¾ x 9¾	10.4	83.3
		11.0	95.0
12 x 16	9½ x 13½	11.8	107.5
		12.0	113.0
		14.0	153.9
16 x 16	13¼ x 13¼	14.5	162.9
		15.0	176.7
16 x 20	13 x 17	16.2	206.1
		18.0	254.4
20 x 20	16¾ x 16¾	18.2	260.2
		20.0	314.1
20 x 24	16½ x 20½	20.1	314.2
		22.0	380.1
24 x 24	20¼ x 20¼	22.1	380.1
		24.0	452.3
24 x 28	20¼ x 24¼	24.1	456.2
28 x 28	24¼ x 24¼	26.4	543.3
		27.0	572.5
30 x 30	25½ x 25½	27.9	607.0
		30.0	706.8
30 x 36	25½ x 31½	30.9	749.9
		33.0	855.3
36 x 36	31½ x 31½	34.4	929.4
		36.0	1017.9

For SI units, 1 inch. = 25.4 mm, 1 square inch = 0.000645 m^2

* Where liner sizes differ dimensionally from those shown in this table, equivalent diameters can be determined from published tables for square and rectangular ducts of equivalent carrying capacity or by other engineering methods.

3 feet (914 mm) Connector Rise *(R)* row to the first Btu/h rating in the FAN MAX column that is equal to or greater than the furnace input rating. The 100 000 Btu/h (29 kW) furnace in this example falls within this range, so a 4 inch (102 mm) connector is adequate.

Chimney Liner Diameter. The total input to the common vent is 135 000 Btu/h (40 kW). Using the Common Vent Capacity portion of Table 803.2(1), read down the total Vent Height *(H)* column to 30 feet (9144 mm) and across this row to find the smallest vent diameter in the FAN + NAT column that has a Btu/h rating greater than 135 000 Btu/h (40 kW). The 4 inch (102 mm) common vent has a capacity of 138 000 Btu/h (40.4 kW). Reducing the maximum capacity by 20 percent results in a maximum capacity for a 4 inch (102 mm) corrugated liner of 110 000 Btu/h (32 kW), less than the total input of 135 000 Btu/h (40 kW). So a larger liner is needed. The 5 inch (127 mm) common vent capacity listed in Table 803.2(1) is 210 000 Btu/h (62 kW), and after reducing by 20 percent is 168 000 Btu/h (49.2 kW). Therefore, a 5 inch (127 mm) corrugated metal liner should be used in this example.

Single Wall Connectors. Once it has been established that relining the chimney is necessary, Type B double-wall vent connectors are not specifically required. This example could be redone using Table 803.2(2) for single-wall vent connectors. For this case, the vent connector and liner diameters would be the same as found for Type B double-wall connectors.

G 103.0 Example of Combination Indoor and Outdoor Combustion Air Opening Design. Determine the required combination of indoor and outdoor combustion air opening sizes for the following appliance installation example.

Example Installation: A fan-assisted furnace and a draft hood-equipped water heater with the following inputs are located in a 15 foot by 30 foot (4572 mm by 9144 mm) basement with an 8 foot (2438 mm) ceiling. No additional indoor spaces can be used to help meet the appliance combustion air needs.

Fan-Assisted Furnace Input: 100 000 Btu/h (29 kW)

Draft Hood-Equipped Water Heater Input: 40 000 Btu/h (11.7 kW)

Solution:

(1) Determine the total available room volume.

Appliance room volume.

15 feet by 30 feet (4572 mm by 9144 mm) with an 8 foot (2438 mm) ceiling = 3600 cubic feet (101.94 m^3)

(2) Determine the total required volume.

The Standard Method to determine combustion air is used to calculate the required volume.

The combined input for the appliances located in the basement is calculated as follows:

100 000 Btu/h (29 kW) + 40 000 Btu/h (11.7 kW) = 140 000 Btu/h (41 kW)

The Standard Method requires that the required volume be determined based on 50 cubic feet per 1000 Btu/h (4.83 m^3/kW).

Using Table G 103.0 the required volume for a 140 000 Btu/h (41 kW) water heater is 7000 cubic feet (198.22 m^3).

Conclusion:

The indoor volume is insufficient to supply combustion air since the total of 3600 cubic feet (101.94 m^3) does not meet the required volume of 7000 cubic feet (198.22 m^3). Therefore, additional combustion air must be provided from the outdoors.

(3) Determine the ratio of the available volume to the required volume:

$$\frac{3600 \text{ cubic feet}}{7000 \text{ cubic feet}} = 0.51$$

(4) Determine the reduction factor to be used to reduce the full outdoor air opening size to the minimum required based on ratio of indoor spaces:

1.00 – 0.51 (from Step 3) = 0.49

(5) Determine the single outdoor combustion air opening size as though all combustion air is to come from outdoors. In this example, the combustion air opening directly communicates with the outdoors:

$$\frac{140\,000 \text{ Btu/h}}{3000 \text{ British thermal units per square inch (Btu/in}^2\text{)}} = 47 \text{ square inches } (0.03 \text{ m}^2)$$

(6) Determine the minimum outdoor combustion air opening area:

Outdoor opening area = 0.49 (from Step 4) x 47 square inches (0.03 m^2)
= 23 square inches (0.01 m^2)

Section 701.7.3(3) requires the minimum dimension of the air opening should not be less than 3 inches (76 mm).

TABLE G 103.0
STANDARD METHOD: REQUIRED VOLUME, ALL APPLIANCES
[NFPA 54: TABLE A.9.3.2.1]

APPLIANCE INPUT (Btu/h)	REQUIRED VOLUME (cubic feet)
5000	250
10 000	500
15 000	750
20 000	1000
25 000	1250
30 000	1500
35 000	1750
40 000	2000
45 000	2250
50 000	2500
55 000	2750
60 000	3000
65 000	3250
70 000	3500
75 000	3750
80 000	4000
85 000	4250
90 000	4500
95 000	4750
100 000	5000
105 000	5250
110 000	5500
115 000	5750
120 000	6000
125 000	6250
130 000	6500
135 000	6750
140 000	7000
145 000	7250
150 000	7500
160 000	8000
170 000	8500
180 000	9000
190 000	9500
200 000	10 000
210 000	10 500
220 000	11 000
230 000	11 500
240 000	12 000
250 000	12 500
260 000	13 000
270 000	13 500
280 000	14 000
290 000	14 500
300 000	15 000

For SI units: 1000 British thermal units per hour = 0.293 kW, 1 cubic foot = 0.0283 m^3

APPENDIX H
EXAMPLE CALCULATION OF OUTDOOR AIR RATE

H 101.0 Example Calculation of Outdoor Air Rate.

H 101.1 Example Calculation. Determine the outdoor air rate required for a single zone AC unit serving an interior 2000 square feet (185.81 m^2) conference/meeting room with a design occupancy of 100 people. The system supplies and returns air from the ceiling. (See Chapter 4 of this code for guidelines)

Solution:

In accordance with Table 403.2.2, the zone air distribution effectiveness is 1.0 since the system supplies cooling only from the ceiling. Using the rates from Table 402.1 for a conference/meeting room, the minimum system outdoor air rate is calculated to be:

$$V_{ot} = \frac{R_p P_z + R_a A_z}{E_z} \quad \text{(Equation H 101.1)}$$

$$= \frac{5 \text{ x } 100 + 0.06 \text{ x } 2000}{1.0}$$

$$= 620 \text{ cubic feet per minute (ft}^3\text{/min)}$$

Where:

A_z = zone floor area: the net occupiable floor area of the zone in square feet.

P_z = zone population: The largest number of people expected to occupy the zone during typical usage. Where the number of people expected to occupy the zone fluctuates, P_z shall be permitted to be estimated based on averaging approaches described in Section 403.6.1. Where P_z cannot be accurately predicted during design, it shall be estimated based on the zone floor area and the default occupant density in accordance with Table 402.1.

R_p = outdoor airflow rate required per person in accordance with Table 402.1.

R_a = outdoor airflow rate required per unit area in accordance with Table 402.1.

E_z = zone air distribution effectiveness in accordance with Table 403.2.2.

For SI units: 1 square foot = 0.0929 m^2, 1 cubic foot per minute = 0.00047 m^3/s

USEFUL TABLES

CONVERSION TABLES

Note: The information contained in these tables are not part of this American National Standard (ANS) and have not been processed in accordance with ANSI's requirements for an ANS. As such, these tables may contain material that has not been subjected to public review or a consensus process. In addition, they do not contain requirements necessary for conformance to the standard.

UNIT CONVERSIONS

MULTIPLY	BY	TO OBTAIN
Acres	43 560	Square feet (ft^2)
Acre-feet	43 560	Cubic feet (ft^3)
Acre-feet	325 851	Gallons (U. S. liquid)
Atmosphere (atm)	76.0	Centimeters of mercury (0°C)
Atmosphere (standard) (atm)	76.0	Centimeters of mercury (0°C)
Atmosphere (standard) (atm)	33.90	Feet of water (4°C)
Atmosphere (standard) (atm)	29.92	Inches of mercury
Atmosphere (standard) (atm)	101.32501	Kilopascals (kPa)
Atmosphere (standard) (atm)	14.70	Pounds-force/square inch (lbf/in^2)
Barrels	42	Gallons (gal)
Barrels	158.9873	Liters (L)
British thermal units (Btu)	1055.055	Joules (J)
British thermal units/hour (Btu/h)	0.000293	Kilowatts (kW)
British thermal units/hour (Btu/h)	0.293	Watts (W)
British thermal unit hour (Btu/h)	1.055056	Kilojoules (kJ)
British thermal unit hour (Btu/h)	1055.056	Joule (J)
British thermal units/minute (Btu/min)	12.97	Foot pounds-force/second (ft•lbf/s)
British thermal units/minute (Btu/min)	0.02358	Horsepower (hp) (international)
British thermal unit per hour square foot degree Fahrenheit [Btu/(h•ft^2•°F)]	5.678263	Watt per square meter kelvin [W/(m^2•K)]
British thermal unit inch per hour foot degree Fahrenheit [Btu•in/(h•ft•°F)]	0.14442279	Watt per meter kelvin [W/(m•K)]
Candelas per square inch (cd/in^2)	1.550003×10^3	Candela per square meters (cd/m^2)
Celsius (°C)	°C + 273.15	Kelvin (k)
Celsius (°C)	1.8 x °C + 32	Fahrenheit (°F)
Centimeters (cm)	0.3937	Inches (in)
Centimeters of mercury (0°C)	0.01316	Atmosphere (standard)
Centimeters of mercury (0°C)	0.446	Feet of water (4°C)
Centimeters of mercury (0°C)	27.84	Pounds-force/square feet (lbf/ft^2)
Centimeters of mercury (0°C)	0.1934	Pounds-force/square inch (lbf/in^2)
Circumference	6.283	Radians (rad)
Cubic feet (ft^3)	1728	Cubic inches (in^3)
Cubic feet (ft^3)	0.0283	Cubic meters (m^3)
Cubic feet (ft^3)	0.03704	Cubic yards (yd^3)
Cubic feet (ft^3)	7.48052	Gallons (U.S. liquid) (gal)
Cubic feet (ft^3)	28.32	Liters (L)
Cubic feet (ft^3)	29.92	Quarts (U.S. liquid)
Cubic feet/hour (ft^3/h)	0.0283	Cubic meters/hour (m^3/h)

UNIT CONVERSIONS (continued)

MULTIPLY	BY	TO OBTAIN
Cubic feet/minute (ft^3/min)	0.000472	Cubic meters/second (m^3/s)
Cubic feet/minute (ft^3/min)	0.1247	Gallons/second
Cubic feet/minute (ft^3/min)	0.47194	Liters/second (L/s)
Cubic feet/second (ft^3/s)	646 316.89	Gallons/day
Cubic feet/second (ft^3/s)	448.831	Gallons/minute (gpm)
Cubic inches (in^3)	1.64×10^{-5}	Cubic meters (m^3)
Cubic inches (in^3)	0.01639	Liters (L)
Cubic meters (m^3)	264.17	Gallons (U.S. liquid)
Cubic yards (yd^3)	27	Cubic feet (ft^3)
Cubic yards (yd^3)	0.76455	Cubic meters (m^3)
Cubic yards (yd^3)	201.97	Gallons (U.S. liquid)
Degrees (deg)	0.0174	Radians (rad)
Fahrenheit (°F)	(°F-32)/1.8	Celsius (°C)
Feet (ft)	0.3048	Meters (m)
Feet (ft)	304.8	Millimeters (mm)
Feet of water (4°C)	0.0295	Atmosphere (standard) (atm)
Feet of water (4°C)	0.8827	Inches of mercury (0°C)
Feet of water (4°C)	62.43	Pounds-force/square feet
Feet of water (4°C)	0.4335	Pounds-force/square inch
Feet/minute (ft/min)	0.01667	Feet/second (ft/s)
Feet/minute (ft/min)	0.01136	Miles/hour (mi/h)
Feet/second (ft/s)	0.3048	Meters/second (m/s)
Feet/second (ft/s)	0.6818	Miles/hour (mi/h)
Feet/second (ft/s)	0.01136	Miles/minute (mi/min)
Foot lambert (fL)	3.426259	Candela per square meter (cd/m^2)
Foot pounds-force (ft•lbf)	1.355	Joules (J)
Foot pounds-force/minute (ft•lbf/min)	2.260×10^{-5}	Kilowatts (kW)
Foot pounds-force/second (ft•lbf/s)	1.356×10^{-3}	Kilowatts (kW)
Gallons (U.S. liquid) (gal)	231	Cubic inches (in^3)
Gallons (U.S. liquid) (gal)	0.003785	Cubic meters (m^3)
Gallons (U.S. liquid) (gal)	0.1337	Cubic feet (ft^3)
Gallons (U.S. liquid) (gal)	3.785	Liters (L)
Gallons (U.S. liquid) (gal)	4	Quarts (U.S. liquid)
Gallons/day	4.3×10^{-5}	Liters/second
Gallons/minute (gal/min) (gpm)	8.0208	Cubic feet/hour (ft^3/h)
Gallons/minute (gal/min) (gpm)	0.003785412	Cubic meters/minute (m^3/min)
Gallons/minute (gal/min) (gpm)	0.00223	Cubic feet/second (ft^3/s)
Gallons/minute (gal/min) (gpm)	6.309020×10^{-5}	Cubic meters/second (m^3/s)
Gallons/minute (gal/min) (gpm)	0.06309	Liters/second (L/s)
Grains (gr)	0.00006479	Kilograms (kg)
Horsepower (hp)	0.7457	Kilowatts (kW)
Horsepower-hours	$2.684 \times 10^{+6}$	Joules (J)
Horsepower-hours	0.7457	Kilowatt-hours (kW•h)
Inch (water column at 60°F)	0.24884	Kilopascals (kPa)
Inches (in)	2.54	Centimeters (cm)

UNIT CONVERSIONS (continued)

MULTIPLY	BY	TO OBTAIN
Inches (in)	25.4	Millimeters (mm)
Inches/hour	25.4	Millimeters/hour (mm/h)
Inches of mercury (0°C)	0.03342	Atmosphere (standard) (atm)
Inches of mercury (0°C)	1.133	Feet of water (4°C)
Inches mercury (0°C)	3.3863	Kilopascals (kPa)
Inches of mercury (0°C)	0.4912	Pounds-force/square inch
Inches of water (4°C)	0.002458	Atmosphere (standard) (atm)
Inches of water (4°C)	0.07356	Inches of mercury (0°C)
Inches of water (4°C)	5.202	Pounds-force/square feet
Inches of water (4°C)	0.03613	Pounds-force/square inch
Joules (J)	9.480×10^{-4}	British thermal units (Btus)
Joules (J)	0.7376	Foot-pounds
Joules (J)	2.778×10^{-4}	Watt-hours
Kelvin (K)	°K - 273.15	Celsius (°C)
Kilograms (kg)	2.2046	Pounds (lbs)
Kilograms (kg)	1.102×10^{-3}	Tons (short)
Kilopascals (kPa)	0.145038	Pounds-force/square inch
Kilometers (km)	0.6214	Miles (statute)
Kilometers/hour (km/h)	0.6214	Miles/hour (mi/h)
Kilowatts (kW)	3412.14	British thermal units/hour (Btus/hour)
Kilowatts (kW)	1.341	Horsepower (hp)
Kilowatt-hours	3413	British thermal units (Btus)
Kilowatt-hours	$2.655 \times 10^{+6}$	Foot-pounds (ft•lbs)
Kilowatt-hours	$3.6 \times 10^{+6}$	Joules (J)
Kip (1000 lbf)	4.448222	Kilonewtons (kN)
Kip-foot	1.35671	Kilonewton meters (kN•m)
Kips per square inch (kip/in^2)	6.8947	Megapascals (MPa)
Lambert (la)	3.183099×10^{3}	Candela per square meter (cd/m^2)
Liters (L)	0.03531	Cubic feet (ft^3)
Liters (L)	61.02	Cubic inches (in^3)
Liters (L)	0.001	Cubic meters (m^3)
Liters (L)	0.2642	Gallons (U.S. liquid)
Lumens per square foot (lm/ft^2) (footcandle)	10.76391	Lux (lx)
Meters (m)	3.281	Feet (ft)
Meters (m)	39.37	Inches (in)
Meters (m)	1.094	Yards
Meters/second (m/s)	3.281	Feet/second (ft/s)
Meters/second (m/s)	2.237	Miles/hour (mi/h)
Miles (mi)	5280	Feet (ft)
Miles (statute)	1.609	Kilometers (km)
Miles/hour (mi/h)	88	Feet/minute (ft/min)
Miles/hour (mi/h)	1.467	Feet/second (ft/s)
Miles/hour (mi/h)	1.609344	Kilometers/hour (km/h)
Miles/hour (mi/h)	26.82	Meters/minute (m/min)
Miles/hour (mi/h)	0.44704	Meters/second (m/s)

UNIT CONVERSIONS (continued)

MULTIPLY	BY	TO OBTAIN
Millimeters (mm)	0.1	Centimeters (cm)
Millimeters (mm)	0.03937	Inches (in)
Millimeters (mm)	0.001	Meters (m)
Minutes (min)	2.908882×10^{-4}	Radians (rads)
Ounces/square inch (oz/in^2)	43.94185	Kilograms/square meter (kg/m^2)
Ounces/square feet (oz/ft^2)	0.03051517	Kilograms/square meter (kg/m^2)
Ounces (oz)	0.02834	Kilograms (kg)
Ounces (oz)	29.57353	Millimeters (mL)
Pints	0.4731765	Liters (L)
Pound-force feet (lbf•ft)	1.355818	Newton meters (N•m)
Pound-force inch (lbf•in)	0.1129848	Newton meters (N•m)
Pound-force per foot (lbf/ft)	14.5939	Newton meters (N•m)
Pound-force per square foot (lbf/ft^2)	47.88026	Pascals (Pa)
Pound-force per square inch (lbf/in^2)	6.89476	Kilopascals (kPa)
Pound-force per inch (lbf/in)	175.1268	Newton meters (N•m)
Pounds/cubic inch (lb/in^3)	2.767990×10^4	Kilograms/cubic meter (kg/m^3)
Pounds/cubic yard (lb/yd^3)	0.5932764	Kilograms/cubic meter (kg/m^3)
Pounds (lb)	0.45359	Kilograms (kg)
Pounds/cubic foot (lb/ft^3)	16.0184	Kilograms/cubic meters (kg/m^3)
Pounds/foot (lb/ft)	1.4881	Kilograms/meters (kg/m)
Pounds/square inch (lb/in^2)	703.1	Kilograms-force/square meter (kg/m^2)
Pounds/square foot (lb/ft^2)	4.882427	Kilograms-force/square meter (kg/m^2)
Pounds-force (lbf)	4.4482	Newtons (N)
Pounds-force/square inch (psi)	0.06805	Atmosphere (standard) (atm)
Pounds-force/square inch (psi)	2.307	Feet of water (4°C)
Pounds-force/square inch (psi)	2.036	Inches of mercury (0°C)
Pounds-force/square inch (psi)	6.89476	Kilopascals (kPa)
Quarts (U.S. dry) (dry qt)	67.20	Cubic inches (in^3)
Quarts (U.S. liquid) (liq qt)	57.75	Cubic inches (in^3)
Quarts (liquid)	0.9463529	Liters (L)
Radians (rads)	57.30	Degrees (deg)
Seconds (s)	4.848137×10^{-6}	Radians (rads)
Square acre	0.404687	Square kilometers (km^2)
Square feet (ft^2)	144	Square inches (in^2)
Square feet (ft^2)	0.0929	Square meters (m^2)
Square inches (in^2)	645.16	Square millimeters (mm^2)
Square inches (in^2)	0.000645	Square meters (m^2)
Square meters (m^2)	1550	Square inches (in^2)
Square miles (mi^2)	640	Acres
Square miles (mi^2)	2.589988	Square kilometers (km^2)
Square millimeters (mm^2)	1.550×10^{-3}	Square inches (in^2)
Square yards (yd^2)	9	Square feet (ft^2)
Square yards (yd^2)	0.8361274	Square meters (m^3)
Temperature (°C) + 17.28	1.8	Temperature (°F)
Temperature (°F) – 32	$5/9$	Temperature (°C)

UNIT CONVERSIONS (continued)

MULTIPLY	BY	TO OBTAIN
Ton-force (tonf) (2000 lbf)	8.896443	Kilonewtons (kN)
Ton-force foot (tonf•f/ft)	2.71342	Kilonewton meters (kN•m)
Ton-force per square foot (tonf/ft^2)	95.7605	Kilopascals (kPa)
Ton-force per square inch (tonf/in^2)	13.7895	Megapascals (MPa)
Tons (metric)	1000	Kilograms (kg)
Tons (long) (2240 lbs)	1016.047	Kilograms (kg)
Tons (short)	2000	Pounds (lbs)
Watts	3.4121	British thermal units per hour (Btus/hour)
Watts	1.341 x 10^{-3}	Horsepower (hp)
Yards (y)	0.9144	Meters (m)

SI SYMBOLS AND PREFIXES

SI PREFIXES		
MULTIPLICATION FACTOR	**PREFIX**	**SYMBOL**
1 000 000 000 000 000 000 = E+18	exa	E
1 000 000 000 000 000 = E+15	peta	P
1 000 000 000 000 = E+12	tera	T
1 000 000 000 = E+09	giga	G
1 000 000 = E+06	mega	M
1 000 = E+03	kilo	k
100 = E+02	hecto	h
10 = E+01	deka	da
0.1 = E-01	deci	d
0.01 = E-02	centi	c
0.001 = E-03	milli	m
0.000 001 = E-06	micro	µ
0.000 000 001 = E-09	nano	n
0.000 000 000 001 = E-12	pico	p
0.000 000 000 000 001 = E-15	femto	f
0.000 000 000 000 000 001 = E-18	atto	a

AREAS AND CIRCUMFERENCES OF CIRCLES

DIAMETER		CIRCUMFERENCE		AREA	
Inches	mm	Inches	mm	Inches²	mm²
1/8	6	0.40	10	0.01227	8.0
1/4	8	0.79	20	0.04909	31.7
3/8	10	1.18	30	0.11045	71.3
1/2	15	1.57	40	0.19635	126.7
3/4	20	2.36	60	0.44179	285.0
1	25	3.14	80	0.7854	506.7
1 1/4	32	3.93	100	1.2272	791.7
1 1/2	40	4.71	120	1.7671	1140.1
2	50	6.28	160	3.1416	2026.8
2 1/2	65	7.85	200	4.9087	3166.9
3	80	9.43	240	7.0686	4560.4
4	100	12.55	320	12.566	8107.1
5	125	15.71	400	19.635	12 667.7
6	150	18.85	480	28.274	18 241.3
7	175	21.99	560	38.485	24 828.9
8	200	25.13	640	50.265	32 428.9
9	225	28.27	720	63.617	41 043.1
10	250	31.42	800	78.540	50 670.9

EQUAL PERIPHERIES

$S = 0.7854\ D$

$D = 1.2732\ S$

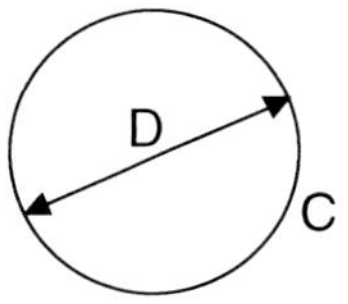

$S = 0.8862\ D$

$D = 1.1284\ S$

$S = 0.2821\ C$

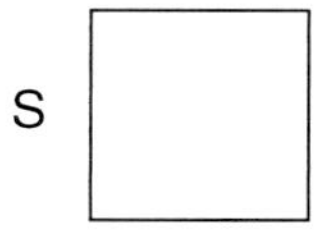

EQUAL AREAS

Area of square (S') = 1.2732 x area of circle

Area of square (S) = 0.6366 x area of circle

$C = \pi D = 2\pi R$

$C = 3.5446\ \sqrt{\text{area}}$

$D = 0.3183\ C = 2R$

$D = 1.1283\ \sqrt{\text{area}}$

$\text{Area} = \pi R^2 = 0.7854\ D^2$

$\text{Area} = 0.07958\ C^2 = \frac{\pi D^2}{4}$

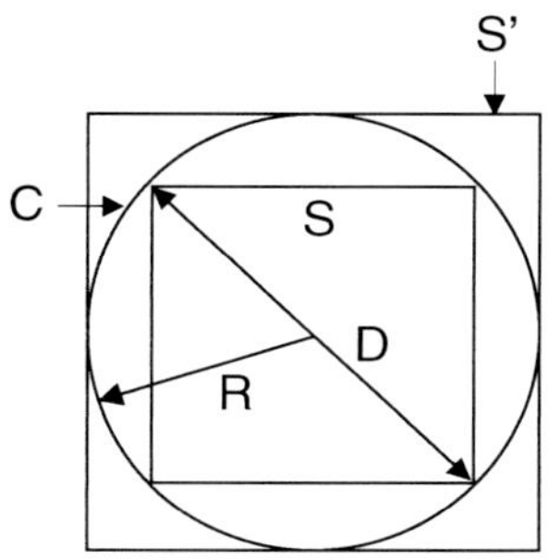

$\pi = 3.1416$

APPROXIMATE MINIMUM THICKNESS FOR CARBON SHEET STEEL CORRESPONDING TO MANUFACTURER'S STANDARD GAUGE AND GALVANIZED SHEET GAUGE NUMBERS

CARBON SHEET STEEL			GALVANIZED SHEET		
MANUFACTURER'S STANDARD GAUGE NO.	DECIMAL AND NOMINAL THICKNESS EQUIVALENT (inch)	RECOMMENDED MINIMUM THICKNESS EQUIVALENT* (inch)	GALVANIZED SHEET GAUGE NO.	DECIMAL AND NOMINAL THICKNESS EQUIVALENT (inch)	RECOMMENDED MINIMUM THICKNESS EQUIVALENT* (inch)
8	0.1644	0.156	8	0.1681	0.159
9	0.1495	0.142	9	0.1532	0.144
10	0.1345	0.127	10	0.1382	0.129
11	0.1196	0.112	11	0.1233	0.114
12	0.1046	0.097	12	0.1084	0.099
13	0.0897	0.083	13	0.0934	0.084
14	0.0747	0.068	14	0.0785	0.070
15	0.0673	0.062	15	0.0710	0.065
16	0.0598	0.055	16	0.0635	0.058
17	0.0538	0.050	17	0.0575	0.053
18	0.0478	0.044	18	0.0516	0.047
19	0.0418	0.038	19	0.0456	0.041
20	0.0359	0.033	20	0.0396	0.036
21	0.0329	0.030	21	0.0366	0.033
22	0.0299	0.027	22	0.0336	0.030
23	0.0269	0.024	23	0.0306	0.027
24	0.0239	0.021	24	0.0276	0.024
25	0.0209	0.018	25	0.0247	0.021
26	0.0179	0.016	26	0.0217	0.019
27	0.0164	0.014	27	0.0202	0.017
28	0.0149	0.013	28	0.0187	0.016
—	—	—	29	0.0172	0.014
—	—	—	30	0.0157	0.013

For SI units: 1 inch = 25.4 mm

* The thickness of the sheets set forth in the code correspond to the thickness shown under these columns. They are the approximate minimum thicknesses and are based on the following references:

Carbon Sheet Steel—Thickness 0.071 inch (1.803 mm) and over:
ASTM A 568, Table 3, Thickness Tolerances of Hot-Rolled Sheet (Carbon Steel).

Carbon Sheet Steel—Thickness less than 0.071 inch (1.803 mm):
ASTM A 568, Table 23, Thickness Tolerances of Cold-Rolled Sheet (Carbon and High-Strength Low Alloy).

Galvanized Sheet Steel—All thicknesses:
ASTM A 653, Table 4, Thickness Tolerances of Hot-Dip Galvanized Sheet.

Minimum thickness is the difference between the thickness equivalent of each gauge and the maximum negative tolerance for the widest rolled width.